TABLE F.5 The SI (Système International) Units

Basic (fundamental) SI units

Physical quantity	SI unit*	Symbol
Length	meter	m
Mass	kilogram	kg
Time	second	s
Electric current	ampere	A
Temperature	kelvin	K
Amount of substance	mole	mol

Derived SI units

Physical quantity	SI unit	Symbol for SI unit	Unit in terms of base units	Unit in terms of other SI units
Velocity (speed)	....	....	m/s	
Acceleration	....	....	m/s^2	
Force	newton	N	$kg \cdot m/s^2$	N/kg
Pressure	pascal	Pa	$kg/(m \cdot s^2)$	J/m
Energy (work; quantity of heat)	joule	J	$kg \cdot m^2/s^2$	N/m^2
Power	watt	W	$kg \cdot m^2/s^3$	$N \cdot m$
Momentum	....	....	$kg \cdot m/s$	J/s
Angle	radian	rad	m/m (dimensionless)	
Frequency	hertz	Hz	s^{-1}	
Electric charge	coulomb	C	$A \cdot s$	
Potential difference (voltage; emf)	volt	V	$kg \cdot m^2/(A \cdot s^3)$	$V \cdot F$
Electric field	....	....	$kg \cdot m/(A \cdot s^3)$	W/A; C/F; J/C
Electric resistance	ohm	Ω	$kg \cdot m^2/(A^2 \cdot s^3)$	V/m; N/C
Capacitance	farad	F	$A^2 \cdot s^4/(kg \cdot m^2)$	V/A
Inductance	henry	H	$kg \cdot m^2/(A^2 \cdot s^2)$	C/V
Magnetic flux	weber	Wb	$kg \cdot m^2/(A \cdot s^2)$	Wb/A; $V \cdot s/A$
Magnetic field (magnetic flux density)	tesla	T	$kg/(A \cdot s^2)$	$V \cdot s$ Wb/m^2; $N/(A \cdot m)$

*Definitions of the basic units are given at appropriate places in the text.

TABLE F.6 The Periodic Table of the Elements

The atomic number is given above the chemical symbol, and the mass number below. The masses are in atomic mass units (u).

1 H 1.0078																	2 He 4.003
3 Li 6.942	4 Be 9.012											5 B 10.81	6 C 12.01	7 N 14.01	8 O 16.00	9 F 19.00	10 Ne 20.18
11 Na 22.99	12 Mg 24.31											13 Al 26.98	14 Si 28.09	15 P 30.97	16 S 32.06	17 Cl 35.45	18 Ar 39.95
19 K 39.10	20 Ca 40.08	21 Sc 44.96	22 Ti 47.90	23 V 50.94	24 Cr 52.00	25 Mn 54.94	26 Fe 55.85	27 Co 58.93	28 Ni 58.71	29 Cu 63.54	30 Zn 65.37	31 Ga 69.72	32 Ge 72.59	33 As 74.92	34 Se 78.96	35 Br 79.91	36 Kr 83.80
37 Rb 85.47	38 Sr 87.62	39 Y 88.91	40 Zr 91.22	41 Nb 92.91	42 Mo 95.94	43 Tc [98]	44 Ru 101.1	45 Rh 102.91	46 Pd 106.4	47 Ag 107.9	48 Cd 112.4	49 In 114.8	50 Sn 118.7	51 Sb 121.8	52 Te 127.6	53 I 126.9	54 Xe 131.3
55 Cs 132.9	56 Ba 137.3	57–71 La Series*	72 Hf 178.5	73 Ta 180.9	74 W 183.8	75 Re 186.2	76 Os 190.2	77 Ir 192.2	78 Pt 195.1	79 Au 197.0	80 Hg 200.6	81 Tl 204.4	82 Pb 207.2	83 Bi 209.0	84 Po [210]	85 At [210]	86 Rn [222]
87 Fr [223]	88 Ra [226]	89–103 Ac Series†	(104) [257]	(105) Ha [260]	(106)	(107)	(108)										

*Lanthanide series

57 La 138.9	58 Ce 140.1	59 Pr 140.9	60 Nd 144.2	61 Pm [147]	62 Sm 150.4	63 Eu 152.0	64 Gd 157.3	65 Tb 158.9	66 Dy 162.5	67 Ho 164.9	68 Er 167.3	69 Tm 168.9	70 Yb 173.0	71 Lu 175.0

†Actinide series

89 Ac [227]	90 Th 232.0	91 Pa [231]	92 U 238.0	93 Np [237]	94 Pu [242]	95 Am [243]	96 Cm [247]	97 Bk [247]	98 Cf [251]	99 Es [254]	100 Fm [253]	101 Md [256]	102 No [254]	103 Lw [257]

Introductory
College Physics

As an additional learning tool, McGraw-Hill publishes a study guide to enhance students' understanding of this textbook. This useful book, *A Study Guide to Accompany Introductory College Physics* (ISBN 0-07-044101-4), prepared by Professor David A. Jerde of St. Cloud State University, uses a question-and-answer approach to lead students through the solution of problems similar to those in this textbook.

Also available at your bookstore are the following McGraw-Hill publications, which students may find helpful in studying the present textbook:

- *College Physics* (8th edition), a volume in Schaum's Outline Series in Science, which includes basic theory, definitions, and hundreds of solved problems and supplementary problems with answers.
- *3000 Solved Problems in Physics,* a volume in Schaum's Solved Problems Series, which contains thousands of problems with completely worked out solutions.

SECOND EDITION

Introductory College Physics

Joseph F. Mulligan
University of Maryland Baltimore County

McGraw-Hill, Inc.

New York   St. Louis   San Francisco   Auckland   Bogotá   Caracas   Hamburg   Lisbon
London   Madrid   Mexico   Milan   Montreal   New Delhi   Paris   San Juan
São Paulo   Singapore   Sydney   Tokyo   Toronto

To my wife, Eleanor,
for her constant encouragement, sage advice,
and devoted assistance.

Introductory College Physics

1 2 3 4 5 6 7 8 9 0   VNH   VNH   9 5 4 3 2 1 0

ISBN 0-07-044057-3

Acknowledgments appear on page 881, and on this page by reference.

This book was set in Times Roman by Waldman Graphics, Inc.
The editors were Denise T. Schanck and David A. Damstra;
the designer was Joan Greenfield;
the production supervisor was Richard A. Ausburn.
Von Hoffmann Press, Inc., was printer and binder.

Cover photo: Fabry-Perot interferometer fringes (see p. 668) from a low-voltage cadmium arc, produced by the PEPSIOS purely interferometric high-resolution scanning spectrometer in the physics department at the University of Wisconsin, Madison. (*Photograph by F. E. Barmore and F. L. Roesler; courtesy of Professor F. Roesler and Design, Inc.*)

Library of Congress Cataloging-in-Publication Data

Mulligan, Joseph F. (Joseph Francis), (date).
    Introductory college physics / Joseph F. Mulligan.—2nd ed.
        p.   cm.
    Includes bibliographical references.
    ISBN 0-07-044057-3
    1. Physics.   I. Title.   II. Title: College physics.
QC21.2.M85   1991   90-34089
530—dc20

Contents

List of Tables xv

Preface xix

CHAPTER 1

Introduction: Important Concepts and Techniques
1

1.1	The Scope of Physics	1
1.2	The Approach of the Physicist	2
1.3	Physical Theories and Laws	4
1.4	Theory and Experiment in Physics	5
1.5	The Fundamental Quantities of Mechanics	6
	Brief Biography: Enrico Fermi	7
1.6	Units and Dimensions	11
1.7	Accuracy of Physical Data	14
1.8	Scalars and Vectors	16
1.9	Graphical Methods for Vectors	19
1.10	Analytical Methods for Vectors	21

CHAPTER 2

Rectilinear Motion
33

2.1	Frames of Reference	34
2.2	Velocity Vectors for Motion in a Plane	34
2.3	Acceleration Vectors for Motion in a Plane	37
2.4	Motion in a Straight Line	39
2.5	Graphical Techniques for Straight-Line Motion	41
2.6	Uniformly Accelerated Motion in a Straight Line	47
2.7	Free Fall and the Acceleration due to Gravity	52
2.8	Kinematics of Motion on an Inclined Plane	56
	Brief Biography: Galileo Galilei	57

CHAPTER 3

Forces and Newton's Laws of Motion
64

3.1	The Nature of Force	64
3.2	Newton's First Law of Motion	65
3.3	Newton's Second Law of Motion	66
	Brief Biography: Isaac Newton	67
3.4	Newton's Third Law of Motion	71

v

3.5 Using Newton's Laws of Motion — 72

3.6 The Force of Friction — 77

3.7 Air Resistance — 80

3.8 Forces for Motion on an Inclined Plane — 81

CHAPTER 4

Motion in a Plane
90

4.1 Newton's Law of Universal Gravitation — 91

4.2 Weight—a Gravitational Force — 92

4.3 Equations of Motion for a Particle in a Plane — 95

4.4 Projectile Motion — 95

4.5 Uniform Circular Motion — 100

4.6 Centripetal Acceleration and Centripetal Force — 101

4.7 Motion in a Vertical Circle under Gravity — 107

4.8 Planetary Motion: Kepler's Laws — 109

4.9 Earth Satellites — 111

CHAPTER 5

Equilibrium
119

5.1 Contact and Reaction Forces — 119

5.2 Tensile and Compressive Forces — 121

5.3 Static Equilibrium: Translation — 122

5.4 How Forces Produce Rotation: Torques — 124

5.5 Static Equilibrium: Rotation — 127

5.6 Center of Gravity — 130

5.7 Dynamic Equilibrium — 134

CHAPTER 6

Work and Energy
143

6.1 The Physical Meaning of Work — 143

6.2 Work Converted into Kinetic Energy — 145

6.3 Work Converted into Gravitational Potential Energy — 147

6.4 Work Converted into Elastic Potential Energy — 151

6.5 Work Converted into Thermal Energy — 153

Brief Biography: Benjamin Thompson, Count Rumford — 154

6.6 The Extended Work-Energy Theorem — 155

6.7 Conservation of Mechanical Energy — 156

6.8 Simple Machines — 160

6.9 Power — 163

CHAPTER 7

Linear Momentum and Collisions
171

7.1 Linear Momentum — 171

7.2 Impulse; the Impulse-Momentum Theorem — 173

7.3 Conservation of Linear Momentum — 174

Brief Biography: Christiaan Huygens — 178

7.4 Collisions in One Dimension: Elastic Collisions — 179

7.5 Collisions in One Dimension: Inelastic Collisions — 181

7.6	Collisions in Two Dimensions	183
7.7	Recoil and Rockets	186

CHAPTER 8

Rotational Motion
194

8.1	Relationship of Rotational Motion to Translational Motion	195
8.2	Rotational Motion with Constant Angular Velocity	196
8.3	Rotational Motion with Constant Angular Acceleration	198
8.4	Work and Rotational Kinetic Energy	200
8.5	Newton's Second Law for Rotational Motion	202
8.6	Rotational Inertias	205
8.7	Angular Momentum and Rotational Impulse	207
8.8	Conservation of Angular Momentum	210

CHAPTER 9

The Structure of Atoms, Molecules, and Gases
218

9.1	Subatomic Particles	218
9.2	The Structure of Atoms	219
	Brief Biography: Ernest Rutherford	220
9.3	The Mole and Avogadro's Number	223
9.4	The Structure of Gases	224
9.5	Temperature	226
9.6	The Kinetic Theory of a Gas	227
9.7	The Ideal Gas Law	230
9.8	Kinetic Theory and the Ideal Gas Law	235
9.9	Distribution of Molecular Velocities in Gases	237

CHAPTER 10

Liquids and Solids
243

10.1	Important Physical Quantities for Liquids	243
10.2	Liquids at Rest	245
10.3	Density Determinations	250
10.4	Liquids in Motion	253
10.5	Viscosity	258
10.6	Surface Tension	261
10.7	Solids: Structure and Density	263
10.8	Elastic Properties of Solids	263
10.9	Hooke's Law	266

CHAPTER 11

Periodic Motion and Waves
272

11.1	Periodic Motion	272
11.2	Simple Harmonic Motion	274
11.3	The Simple Pendulum	277
11.4	Damped and Forced Oscillatory Motion	280
11.5	Vibrating Bodies as the Source of Waves	281
11.6	Transverse and Longitudinal Waves	284
11.7	Mathematical Description of Wave Motion	286

11.8   The Superposition Principle: Phase   287
11.9   Standing Waves on a String   290
11.10   Forced Vibrations of a String   292

CHAPTER 12

Sound Waves
299

12.1   The Nature of Sound Waves   299
12.2   The Speed of Sound Waves   302
12.3   Standing Sound Waves; Resonance   304
12.4   Interference of Sound Waves of Different Frequencies:
        Beats   308
12.5   The Detection of Sound   310
12.6   Intensity and Loudness   311
12.7   Quality of Sound and Pitch   315
        Brief Biography: Hermann von Helmholtz   316
12.8   The Doppler Effect in Sound   317
12.9   Supersonic Speeds and Ultrasonic Waves   320

CHAPTER 13

Temperature and Heat
328

13.1   Temperature   328
13.2   Gas Thermometers   329
13.3   Other Kinds of Thermometers   330
13.4   Temperature Scales   331
13.5   Expansion of Solids and Liquids with Temperature   334
13.6   Heat as a Form of Energy   337
13.7   Specific Heat Capacities   339
13.8   The Mechanical Equivalent of Heat (Joule's Equivalent)   340
        Brief Biography: James Prescott Joule   341
13.9   Change of Phase: Heats of Fusion and Vaporization   342

CHAPTER 14

**The Transfer of Heat
and the First Law of
Thermodynamics**
351

14.1   The Conduction of Heat   351
14.2   The Convection of Heat   354
14.3   Heat Transfer by Radiation   355
14.4   Radiation Balance   357
14.5   Some Applications of Heat Transfer Processes   358
14.6   Heat and Work   362
14.7   The First Law of Thermodynamics   364
14.8   Applications of the First Law of Thermodynamics   368
14.9   Molar Heat Capacities of an Ideal Gas   370

CHAPTER 15

The Second Law of Thermodynamics
378

15.1 Reversible and Irreversible Processes — 379
15.2 Entropy — 383
15.3 Entropy Changes — 385
15.4 The Second Law of Thermodynamics — 388
15.5 Heat Engines; The Carnot Engine — 389
   Brief Biography: Nicolas Léonard Sadi Carnot — 390
15.6 Practical Heat Engines — 393
15.7 The Absolute Thermodynamic Temperature Scale — 396
15.8 Some Applications: Heat Pumps and Refrigerators — 398

CHAPTER 16

Electric Charges and Electric Fields
407

16.1 Elementary Electric Charges and Their Interaction — 407
   Brief Biography: Benjamin Franklin — 408
16.2 Conductors and Insulators — 409
16.3 Electroscopes and Electrometers — 410
16.4 Coulomb's Law — 413
16.5 Electrostatic Problems Involving Point Charges — 415
16.6 The Electric Field — 416
16.7 Gauss' Law — 421

CHAPTER 17

Electric Potential Energy, Potential Difference, and Capacitance
428

17.1 Electric Potential Energy and Electric Potential — 428
17.2 Potential Difference and Voltage — 430
17.3 Electric Potential and the Electric Field — 431
17.4 Electric Potential in the Vicinity of a Point Charge — 433
17.5 The Storage of Electric Charge — 436
17.6 Capacitors and Capacitance — 437
17.7 Dielectrics — 440
17.8 Capacitors in Series and in Parallel — 442
17.9 The Energy Stored in a Capacitor — 444

CHAPTER 18

Direct Current Electricity
450

18.1 EMF's and Direct Currents — 450
18.2 Electric Currents in Electrolytes — 453
18.3 Electric Currents in Metals — 455
18.4 Resistance and Resistivity — 457
18.5 Kirchhoff's Rules — 460
   Brief Biography: Gustav Robert Kirchhoff — 461
18.6 Some Applications of Kirchhoff's Rules — 464
18.7 Energy and Power in DC Circuits — 469
18.8 Joule Heating — 470

CHAPTER 19

Magnetism
479

19.1	Magnets and the Magnetic Field	479
19.2	Force on a Current-Carrying Conductor in a Magnetic Field	481
19.3	Force on a Charged Particle in a Magnetic Field	483
19.4	J. J. Thomson's Determination of e/m for the Electron	489
	Brief Biography: Joseph John Thomson	490
19.5	Magnetic Field of a Straight, Current-Carrying Wire	492
19.6	Ampère's Circuital Law	495
19.7	Circular Coils and Solenoids	496
19.8	Magnetic Properties of Materials	498
19.9	The Earth's Magnetic Field	500
19.10	Galvanometers; Electric Meters	502

CHAPTER 20

Electromagnetic Induction
510

20.1	Motional EMF's	510
20.2	Faraday's Induction Law	512
	Brief Biography: Michael Faraday	513
20.3	Lenz' Law	516
20.4	Mutual Induction and Self-Induction	517
20.5	Energy Stored in an Inductor	519
20.6	Circuit with Inductance and Resistance: The RL Circuit	521
20.7	Circuit with Capacitance and Resistance: The RC Circuit	524
20.8	Electric Generators	527
20.9	Electric Motors	531

CHAPTER 21

Alternating Current Circuits
538

21.1	AC Circuit Containing Only Resistance	538
21.2	AC Circuit Containing Only Inductance	541
21.3	AC Circuit Containing Only Capacitance	543
21.4	The LC Circuit	545
21.5	RLC Series Circuit with Applied AC Voltage	548
21.6	Series Resonance	550
21.7	Transformers	553
21.8	Practical Aspects of Electric Power	556

CHAPTER 22

Electromagnetic Waves
564

22.1	Maxwell's Equations	564
22.2	Maxwell's Displacement Current	565
	Brief Biography: James Clerk Maxwell	567
22.3	The Nature of Electromagnetic Waves	569
22.4	The Predicted Speed of Electromagnetic Waves	569
22.5	Energy in Electromagnetic Waves; Intensity	572
22.6	The Complete Electromagnetic Spectrum	574

22.7	Measurements of the Speed of Electromagnetic Waves	575
22.8	The Generation of Radio Waves	579
22.9	The Detection and Modulation of Radio Waves	581
22.10	Practical Radio and Television	583

CHAPTER 23

Geometrical Optics: Mirrors and Lenses
588

23.1	Ray (Geometrical) Optics	588
23.2	Reflection and Refraction on a Ray Model	589
23.3	Dispersion of Light by a Prism	595
23.4	Spherical Mirrors	596
23.5	Plane Mirrors	603
23.6	Thin Converging Lenses	604
23.7	Thin Diverging Lenses	607
23.8	Combinations of Lenses	609
23.9	Lens Aberrations	611

CHAPTER 24

The Wave Nature of Light: Interference, Diffraction, and Polarization
619

24.1	Huygens' Principle	620
24.2	Interference: Young's Double-Slit Experiment	621
24.3	Multislit Interference: The Diffraction Grating	624
	Brief Biography: Thomas Young	625
24.4	Interference in Thin Films	628
24.5	Some Applications of Interference	631
24.6	The Diffraction of Light	632
24.7	The Polarization of Light	638

CHAPTER 25

Optical Instruments
647

25.1	Cameras	647
25.2	The Human Eye	650
25.3	The Magnifying Glass (Simple Magnifier)	653
25.4	The Compound Microscope	655
25.5	Telescopes	657
25.6	The Resolution of Telescopes and Microscopes	661
25.7	Specialized Kinds of Microscopes	664
25.8	Other Important Optical Instruments	665
	Brief Biography: Albert Abraham Michelson	667

CHAPTER 26

The Theory of Relativity
675

26.1	Galilean Relativity	675
26.2	The Aether and the Speed of Light	679
26.3	The Michelson-Morley Experiment	680
	Brief Biography: Albert Einstein	681
26.4	Einstein's Special Theory of Relativity	684
26.5	The Lorentz Transformation Equations	686

26.6	Consequences of the Lorentz Transformation Equations	687
26.7	Mass, Energy, and Momentum	691
26.8	Experimental Tests of the Special Theory of Relativity	696
26.9	The General Theory of Relativity	698

CHAPTER 27

The Experimental Basis of Quantum Mechanics
705

27.1	Particle Properties of Light; the Photon	706
27.2	Blackbody Radiation	706
	Brief Biography: Max Planck	709
27.3	The Photoelectric Effect	711
27.4	X-Rays	714
27.5	The Compton Effect	717
27.6	The Nature of Light	720
27.7	Wave Properties of Matter: De Broglie Waves	720
27.8	Electron Diffraction and Interference	723
27.9	The Wave-Particle Duality	726
27.10	The Heisenberg Uncertainty Principle	727

CHAPTER 28

The Structure of Atoms
736

28.1	The Spectrum of the Hydrogen Atom	736
28.2	The Bohr-Rutherford Atom	738
	Brief Biography: Niels Bohr	739
28.3	Bohr's Theory of the Hydrogen Atom	741
28.4	Energy Levels in Hydrogen and Other Atoms	745
28.5	The Franck-Hertz Experiment	747
28.6	The Emission and Absorption of Radiation	750
28.7	Quantum Numbers	754
28.8	The Pauli Exclusion Principle and the Periodic Table	756
28.9	Successes and Failures of the Bohr Theory	758
28.10	Wave Mechanics and Quantum Mechanics	761

CHAPTER 29

Nuclear Physics
770

29.1	The Structure of the Atomic Nucleus	770
29.2	Nuclear Forces and Binding Energies	772
29.3	Nuclear Reactions I: Natural Radioactivity	776
29.4	Rules Governing Nuclear Reactions	780
29.5	Nuclear Reactions II: Artificial Radioactivity	781
29.6	Nuclear Fission and Nuclear Fusion	783
	Brief Biography: Lise Meitner	784
29.7	The Nucleus as a Practical Energy Source	788
29.8	Biological Effects of Nuclear Radiations	791
29.9	Use of Radioactive Isotopes in Medicine and Technology	795

CHAPTER 30

Solid-State Physics
804

30.1	The Structure of Solids	804
30.2	Thermal Properties of Solids	807
30.3	The Band Theory of Solids	810
30.4	Semiconductors	812
30.5	Solid-State Devices	814
30.6	Applications of Solid-State Devices in Modern Technology	817
30.7	Superconductivity: Theory and Experiment	820

CHAPTER 31

High-Energy and Elementary-Particle Physics
827

31.1	The Four Forces of Physics	827
31.2	Particle Accelerators	832
31.3	Pions and Muons	836
31.4	Antiparticles and Neutrinos	837
31.5	More New Particles	839
31.6	Quarks	840
31.7	Conservation Principles for Elementary Particles	842
31.8	The Present State of Elementary-Particle Physics	844
	Brief Biography: Richard P. Feynman	846
31.9	The Future of Elementary-Particle Physics	848

APPENDIX 1

Review of Basic Mathematics
854

1.A	Algebra	854
1.B	Geometry	855
1.C	Trigonometry	855
1.D	Series Expansions	856
1.E	Logarithms and Exponential Functions	856

APPENDIX 2

Functional Equations and Graphs
858

2.A	Functional Equations	858
2.B	Graphs	859
2.C	Log-log and Semilog Scales	861

APPENDIX 3

Solving Physics Problems
864

3.A	Some General Hints on Problem Solving	864
3.B	Powers-of-10 Notation	866
3.C	Order-of-Magnitude Calculations	867

APPENDIX 4

**Masses and Abundances
of Some Important
Isotopes**
869

APPENDIX 5

**Answers to Odd-
Numbered Problems**
871

Captions and Credits for Chapter-Opening Photographs   881

Index   883

List of Tables

F.1	Prefixes and Abbreviations Commonly Used in the SI System of Units	**Inside Front Cover**
F.2	Astronomical Data on the Solar System	**Inside Front Cover**
F.3	Numerical Constants	**Inside Front Cover**
F.4	The Greek Alphabet	**Inside Front Cover**
F.5	The SI (Système International) Units	**Inside Front Cover**
F.6	The Periodic Table of the Elements	**Inside Front Cover**
B.1	Useful Conversion Factors	**Inside Back Cover**
B.2	Values of the Fundamental Physical Constants	**Inside Back Cover**
3.1	Coefficients of Static and Sliding Friction	79
8.1	Connection of Linear and Angular Quantities	196
8.2	Rotational Inertia of Various Solid Objects	205
8.3	Comparison Between Translational and Rotational Quantities	212
9.1	Properties of the Particles in Atoms	218
9.2	Constituent Parts of Some Important Isotopes	222
9.3	Useful Pressure Units and Their Conversion Factors	233
9.4	Speeds and Kinetic Energies of Some Important Gases at 0°C	237
10.1	Densities and Specific Gravities for Some Liquids and Gases at STP	244
10.2	Coefficient of Viscosity of Liquids (at 20°C)	259
10.3	Densities of Solids	263
10.4	Elastic Moduli of Metals	265
11.1	Table of Sines and Tangents for Small Angles	279
12.1	Elastic and Inertial Properties and the Speed of Sound	303
12.2	Intensity and Intensity Levels of Some Common Sounds	313
13.1	Important Physical Temperatures on the Fahrenheit, Celsius, and Kelvin Temperature Scales	333
13.2	Thermal Coefficients of Linear Expansion	334
13.3	Thermal Coefficients of Volume Expansion	335

13.4 Specific Heat Capacities — 339

13.5 Heats of Fusion and Vaporization — 343

13.6 Melting and Boiling Temperatures for Some Important Substances — 344

14.1 Thermal Conductivities — 352

14.2 Color Temperature of Stars — 356

14.3 Experimental Values of Molar Heat Capacities and γ for Selected Gases — 371

17.1 Dielectric Constants of Various Materials — 440

18.1 Electric Resistivity and Its Temperature Coefficient at $0°C$ — 458

19.1 Relative Magnetic Permeabilities at Room Temperature — 499

21.1 AC Quantities — 540

21.2 Comparison of AC Circuits Containing R, L, and C — 544

21.3 Comparison of Mechanical and Electrical Oscillations — 547

22.1 Maxwell's Equations — 565

22.2 Results of Measurements of the Speed of Electromagnetic Waves — 578

22.3 Frequencies Used in Radio and Television — 583

23.1 Indices of Refraction — 591

23.2 Important Equations and Sign Conventions for Single Mirrors and Lenses — 614

26.1 Variation of Mass m of a Particle with Its Speed v (Theoretical Values) — 692

26.2 Measured and Predicted Values for Time Differences between Stationary and Moving Clocks — 698

27.1 Experimental Results and Theoretical Predictions for the Photoelectric Effect — 712

27.2 Photoelectric Work Function ϕ for Clean Metal Surfaces — 713

27.3 Wave and Particle Properties of Matter and Radiation — 720

28.1 Important Spectral Series for the Hydrogen Atom — 737

28.2 Quantum Numbers for a Single Electron in Hydrogen — 755

28.3 Quantum Numbers for Electrons in Some Atoms — 757

29.1 Properties of Elementary Particles in Atoms — 771

29.2 Half-Lives of Some Important Radioactive Isotopes — 778

29.3 Energy Released in the Fission of $^{235}_{92}U$ by a Neutron — 786

29.4 Relative Biological Effectiveness (RBE) of Nuclear Radiations — 793

29.5 Sources of Ionizing Radiation in the United States — 793

30.1 Melting Points of Various Crystals — 807

30.2 Thermal Conductivities K (at approximately $20°C$) — 807

30.3 Molar Heat Capacities of Solids ($20°C$) — 808

30.4 Superconducting Temperatures Achieved before 1986   **821**
31.1 The Four Known Forces of Physics   **828**
31.2 Types of Neutrinos   **838**
31.3 Properties of the Hadrons   **839**
31.4 Families of Elementary Particles (as of May, 1990)   **842**
31.5 Types of Interactions and Associated Quanta   **845**
         Color Plate on Optical Spectra   **Facing page 742**

Preface

The scientist does not study nature because it is useful; he studies it because he delights in it, and he delights in it because it is beautiful. If nature were not beautiful, it would not be worth knowing, and if nature were not worth knowing, life would not be worth living.

Henri Poincaré (1854–1912)

This second edition of *Introductory College Physics* is intended for a one-year course in physics for students majoring in biology, health and environmental sciences, and architecture. It is also appropriate for those students in the arts, humanities, history, and the social sciences who desire a thorough introduction to physics to understand better the world in which they live, or simply to fulfill science distribution requirements.

The material covered is that of the standard two-semester, noncalculus, introductory college physics course. No previous physics courses are required, and the only mathematical prerequisites are high school algebra, geometry, and a little trigonometry.

This book presents physics as a unified discipline in which conservation principles and symmetry considerations tie together its various parts. An understanding of the similarities among diverse fields of physics (for example, between a mass vibrating at the end of a spring and an oscillating electric circuit) can make physics both easier and more fascinating.

This edition continues the emphasis on the historical and humanistic dimensions of physics that users of the first edition found appealing. It attempts to show that physics is a vital, unfolding subject marked by the intuition, hard work, and sacrifice of physicists such as Isaac Newton, Albert Einstein, and Marie Curie, and is still being advanced by similarly dedicated scientists today. This historical emphasis is reflected in the quotations at the beginning of each chapter, the 21 one-page biographies, and the captions to the many photographs of famous physicists scattered throughout the text. Historical discussions in the text itself have been used sparingly, however, so as not to hinder the logical development of the material.

Many applications of physics to health, medicine, energy, technology, and everyday life are included. An understanding of the electron microscope, solar cells, and the use of lasers, optical fibers, and radioisotopes in medical diagnosis and treatment help the student to see the relevance of physics to contemporary life. This is, however, a *physics* course, and so health-related applications that include more biology or medicine than physics have been avoided.

Changes in the Second Edition

A number of chapters have undergone major revision in response to comments of users and reviewers. Among the important changes from the first edition are the following.

1   The first five chapters on mechanics have been completely rewritten. Displacement vectors are now introduced in the first chapter and form the basis for the discussion of graphical and analytical techniques useful in handling vectors. Chapter 2 introduces velocity and acceleration vectors in two dimensions, but the subsequent discussion of kinematics in this chapter is restricted to one-dimensional motion.

Forces and Newton's laws are introduced for one-dimensional motion in Chapter 3. Chapter 4 discusses the kinematics and dynamics of motion in two dimensions. Equilibrium is then introduced in Chapter 5 as a special case of Newton's second law in which the net force acting on a system is zero. This major revision and the addition of many new problems strengthens substantially the mechanics in the second edition.

2   The chapter on electrostatics in the first edition has been divided into two—Chapter 16 on Electric Charges and Electric Fields and Chapter 17 on Electric Potential Energy, Potential Difference, and Capacitance—and there are almost twice as many problems as before.

3   Chapter 22 on Electromagnetic Waves has been made more qualitative than in the first edition. The derivation of a numerical value for the speed of light from Maxwell's equations has been merely outlined, since any attempt at a proper derivation is now judged to be beyond the scope of this course.

4   The chapter on geometrical optics now comes—more conventionally—before that on physical optics, unlike the order in the first edition.

5   The chapters on modern physics, one of the strong points of the first edition, have been made current to the date when the book went to press (May, 1990). Included are discussions of high-temperature superconductors, families of elementary particles, the existing colliding-beam accelerators, and the planned superconducting supercollider (SSC).

In addition, much of the text has been rewritten in the interest of clarity and conciseness. Some less important topics (radius of gyration, coefficient of restitution) have been dropped, while more important ones for the students in this course (fluorescence, phosphorescence, new kinds of telescopes) have been added.

The second edition, while slimmer than the first, still provides more than enough material for the standard course in physics, including all topics required for the MCATs and other standard tests. There are no sections marked optional and no physics material relegated to appendixes, because teachers are the best judges of what is important for their students.

For teachers who desire to reduce the amount of material in the course, Chapters 8 (Rotational Motion), 21 (Alternating Current Circuits), 25 (Optical Instruments), 26 (The Theory of Relativity), and the last three chapters (29 to 31) on modern physics can be omitted without interrupting the logical flow. Also, where feasible, applications have been placed in special sections, often at the end of chapters. Depending on class interest, sections such as 15.8, 19.10, 21.8, 22.8, 22.9, 22.10, 23.9, 24.7, 25.7, and 25.8 can be omitted or assigned as outside reading without loss to the continuity of the course. Some more mathematical topics, such as Gauss' law and Ampère's circuital law (needed for the discussion of Maxwell's equations), and the mathematical treatment of wave motion in Sections 11.7 to 11.10 can be omitted in courses in which a more qualitative approach to these topics is preferred.

Features of the Second Edition

Units SI (Système International) units are used almost exclusively. At the beginning, some British engineering units are introduced to convey a feeling for the size of the SI units and to provide practice in converting units. Other non-SI units are occasionally introduced for informational purposes. The kilocalorie is retained in the discussion of heat because of the importance of Joule's equivalent for the historical development of physics.

Worked Examples More than 300 worked examples appear within the chapters to provide models for problem solving. In the second edition, more difficult examples replace simpler ones in many places, there is increased emphasis on analyzing the physical situation in solving problems, and occasional exercises based on the worked examples have been added.

Significant Figures Considerable care has been devoted to the consistent use of significant figures in the worked examples in the text. Students are expected to show a like care in their answers to the end-of-chapter problems.

Questions Over 350 discussion questions are given at the ends of the chapters to test the student's understanding of the material. Additional qualitative multiple-choice exercises have been included for the same reason.

Problems The second edition contains about 1700 problems at the end of chapters, of which over 350 are new. Many old problems have been revised and improved, and all the problems have been rearranged into four categories: multiple-choice exercises, simple exercises, standard problems, and difficult problems. The excessively complex numbering system for problems in the first edition has been dropped, and problems are now numbered consecutively.

No assignment of problems to chapter sections has been made, because an important element in solving physics problems is the ability to recognize what previously acquired knowledge is important for a problem's solution. Assigning problems to individual text sections merely helps students locate an equation in the designated section that "works," whether or not the student understands *why* it works. For the convenience of instructors in assigning homework, however, all problems are listed by chapter section in the *Instructor's Manual*.

Illustrations The second edition contains more than 100 photographs and 900 line drawings, many of which have been considerably improved in this edition. These are intended to be an integral part of the student's learning experience and are referred to at appropriate places in the text.

Summaries   Each chapter ends with a brief summary that includes all-important definitions, principles, and equations in the chapter. The equations in the summary are numbered in this edition to allow the student easy access to the section where they are derived and discussed.

Tables   Sixty-six tables of useful experimental data are provided at pertinent places throughout the book. A list of these tables follows the table of contents. Tables frequently needed, e.g., those containing conversion factors, the physical constants, and the periodic table, are placed inside the front and back covers.

Biographies   The text contains 21 one-page biographies of famous physicists, chosen for the contributions they made to the development of the physics in the chapter containing the biography. Two biographies in the first edition have been dropped, and one (R. Feynman) has been added.

Suggestions for Further Reading   The annotated list of additional readings at the end of each chapter includes books and articles that can be read profitably by students taking their first physics course.

Appendixes   The appendixes include brief reviews of important mathematical techniques and equations, some useful ideas about graphs, and hints on the solution of physics problems, including a discussion of powers-of-10 notation and order-of-magnitude calculations.

Answers to Odd-Numbered Problems   Answers are provided in Appendix 5 for all odd-numbered problems. Every problem in the book has been solved independently by at least three university physics teachers, including the author. Students and instructors can therefore use them with considerable confidence in their correctness.

Index   A good index is an efficient learning tool whose use a student should master. For this reason the index is comprehensive, and students should be able to find a useful page reference to clarify any concept they find difficult. I hope that students will find this book a useful reference tool for future courses in chemistry or biology, or for answers to scientific questions arising in everyday life, and I have prepared the index with this in mind.

Additional Materials for Instructors in the Course

1   An *Instructor's Manual*, prepared by Brian W. Holmes, California State University, San Jose. This includes the answers to all problems (both odd- and even-numbered) in the text, and worked solutions to all difficult problems and to a selection of other important typical problems. The sections of the text for which each problem is intended are also given. The *Instructor's Manual* includes hints on teaching the course, notes on demonstrations, films and other teaching aids, and sample examinations.

2   A selection of 148 overhead projector transparencies. These include some of the most important and complicated line drawings from the text, selected by the author as meriting special attention in lectures based on this textbook.

3 A computerized test-bank of multiple-choice questions for examinations.

4 A *Study Guide* for students, prepared by David A. Jerde, St. Cloud State University, provides additional assistance to students in their use of the text and in the solution of problems. Objectives for the individual chapters are included in the *Study Guide*.

It is hoped that all these teaching aids will make the course more convenient for the instructor and more profitable for the student.

Acknowledgments

I would like to thank the many people who contributed in various ways to improving this second edition. Among these are the following physics professors who read the manuscript and made suggestions on useful changes: Edward Adelson, Ohio State University; John P. Barach, Vanderbilt University; Walter Benenson, Michigan State University; Kenneth C. Clark, University of Washington; Peter G. Debrunner, University of Illinois, Urbana-Champaign; Richard E. Garrett, University of Florida; Arthur S. Hobson, University of Arkansas; James Hurley, University of California, Davis; Quentin C. Kessel, University of Connecticut; Ian Littlewood, University of Central Florida; Christopher Martin, Columbia University; David M. McKinstry, Eastern Washington University; Mary E. Mogge, California State Polytechnic University, Pomona; J. Ronald Mowery, Harrisburg Area Community College; Lewis J. Oakland, University of Minnesota, Duluth; William F. Pelham, Towson State University; Wayne Repko, Michigan State University; Dennis K. Ross, Iowa State University; Edgar B. Singleton, Bowling Green State University; Glenn Sowell, University of Nebraska, Lincoln; John R. Townsend, University of Pittsburgh; James Tressel, Massasoit Community College; L. L. Van Zandt, Purdue University; Thomas A. Weber, Iowa State University; Henry W. White, University of Missouri, Columbia; and Junru Wu, UCLA.

In particular, I would like to thank Brian W. Holmes of California State University, San Jose, and Terrence C. Dymski of the University of Maryland Baltimore County. Brian Holmes read the entire manuscript with great care and made many constructive and beneficial suggestions for its improvement; he has also prepared the *Instructor's Manual* for the course. Terry Dymski read the revised chapters of the second edition and provided many helpful ideas based on his experience in teaching a course using the first edition as text. I would also like to thank Dr. Alfons Weber of the National Institute of Standards and Technology for suggesting the cover photograph, and Professor Frederick L. Roesler of the University of Wisconsin for providing the photograph.

I am especially grateful to the editorial staff of McGraw-Hill, Inc., and in particular to Denise T. Schanck, David A. Damstra, Joan Greenfield, and Alice Goehring, for their assistance in bringing the second edition into final book form.

Any errors that remain are, of course, fully my responsibility, and I welcome corrections and comments to improve future editions.

A Few Words to the Student

Learning physics—like learning any subject—is a highly personal endeavor. As a consequence, all I can do here is offer a few hints I hope will be helpful to you in this course. Your instructor can provide additional suggestions.

1   First, read the Preface to discover this book's special features. These features are intended to help you learn physics. Be sure to use them.

2   Start by reading over the assigned material in the text *before coming to class.* After attending class and taking notes on the lecture, go back and study the text material again. In doing this, remember that the goal is *understanding,* not memorization, and that understanding implies the ability to apply what you have learned.

3   Work through the examples in the text carefully with pencil and paper until you feel that you really understand how to do these sample problems.

4   After you are convinced that you understand the chapter and the worked examples, turn to the questions and problems at the end of the chapter and do as many of these as possible (you are not limited to those assigned!). For helpful hints on solving physics problems, see Appendix 3.

5   If you find a problem you cannot do, even after reading over your class notes and relevant worked examples, discuss it with some of your friends in the class (one of the best ways to learn physics is to argue about it) or consult an instructor. If this is not possible, it often helps to do something else (perhaps read one of the biographies in the text to learn of the frustrations and triumphs of a great physicist) and come back fresh to the problem the next day.

6   Keep up with your assignments, devoting 2 to 3 hours to study and homework for each hour of physics lecture. Physics has a logical structure, with each chapter building on the previous one. If you fall behind, you may find it very difficult to recover by the end of the semester. To the consistent worker goes the prize (perhaps even an A!).

7   Physics is not an easy subject even for physics teachers. But, if you think about it, you will soon realize that nothing really worthwhile comes without effort. The Dutch physicist Paul Ehrenfest (1880–1933) said that he found great joy in struggling to make something clear to himself, and then in making it clear to others so that they could share his joy. As you work through this book, I hope that you may share the joy I found in writing it.

Joseph F. Mulligan

Introductory College Physics

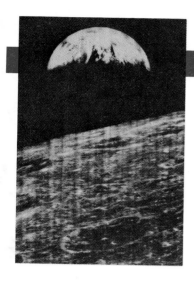

CHAPTER 1

Introduction: Important Concepts and Techniques

Is space infinite, and in what sense? Is the material world infinite in extent, and are all places within that extent equally full of matter? Do atoms exist, or is matter infinitely divisible? The discussion of questions of this kind has been going on ever since men began to reason, and to each of us, as soon as we obtain the use of our faculties, the same old questions arise as fresh as ever.

James Clerk Maxwell (1831–1879)

During college you will study many subjects—literature, history, psychology, mathematics, music, biology, art, computer science, physics, philosophy. Each of these disciplines has its own subject matter and its own approach to reality, and each will make an important contribution to your knowledge of the world. In this chapter we first define the scope and approach of one such discipline—physics. This is but an introduction, however; the nature of physics will gradually become clearer to you as you make progress through this course.

After this brief introduction, we proceed immediately to a discussion of some basic concepts used throughout the remainder of the course—units, dimensions, scalars, and vectors—and to the mathematical techniques needed to handle these quantities.

1.1 The Scope of Physics

Physics, the oldest and most basic of the sciences, is *the science of matter and energy and of the relations between them.* Because it is a science, physics is built on observation and experiment, and its theories and laws must stand the test of continuing comparison with the results of observation and experiment.

The domain of physics includes *matter* in all its forms—solids, liquids, gases, plasmas, molecules, atoms, and the particles out of which atoms are made. It also includes *energy* in all its forms—mechanical, electromagnetic, nuclear, thermal, and radiant energy. Physicists attempt to understand these different kinds of matter and energy that constitute the universe. Some physicists are interested in the particles inside the protons and neutrons that make up the nuclei of all atoms; others do research on the radiation produced in the "big bang" that began our universe, and have succeeded in detecting the remnants of this radiation almost 20 billion years after it was produced. Physicists firmly believe that the principles of physics are the same now as they were 20 billion years ago and that by uncovering these principles we will learn more about the early history of our universe.

Although the forms of matter and energy are many, the basic principles that govern their interactions are few. In this book we will emphasize the basic principles of physics, particularly the principle of conservation of energy. We will first apply these principles to the fields of mechanics, sound, heat, light, electricity, and magnetism. Then we will discuss the revolution produced in this century by Einstein's theory of relativity and by Planck, Bohr, and their followers in the development of the quantum mechanics of atomic systems. Finally, we will discuss briefly new developments in solid-state physics and elementary-particle physics, the fields of greatest research activity in physics today.

The concerns of physics are not remote from everyday life; all branches of engineering are based on the fundamental principles of physics, and our planes fly and our cars run because of the laws of physics. Physicists also continue to make important contributions to fields like biology and medicine by developing sensitive instruments of great utility such as the electron microscope and the CAT scanner. Some eminent biologists like Francis Crick, one of the developers of the Watson-Crick double-helix theory of DNA, and Rosalyn Yalow (Fig. 1.1), winner of the 1977 Nobel Prize for medicine, were trained as physicists and used the techniques of physics in their research.

1.2 The Approach of the Physicist

Physics is one of the so-called hard sciences, not because it is difficult but because it is based on hard, quantitative data and makes predictions in quantitative (i.e., numerical) form. These predictions can then be checked against measurements made in the laboratory. (In the so-called soft sciences such as psychology and sociology, human behavior plays a crucial role, and so the precision that is characteristic of physics is unattainable.) The physicist is able to extract quantitative data from nature only by focusing on the purely physical and quantitative aspects of a problem to the exclusion of other considerations. This fundamental and in some ways oversimplified approach does not take into account the interaction of physical processes with the economy, political events, or the quality of human life but concentrates on matter and energy and their interrelationships.

Once experimental data have been collected in a physics research laboratory of the type shown in Fig. 1.2, theoretical physicists try to develop a theory to correlate and explain these data. In so doing, they rely on the known laws of physics and their own creative imaginations. At times their explanation of the available data may involve nothing more than an application of a simple physical law such as Newton's second law of motion, which relates forces to accelerations; in other cases a break with existing physical theories may be required, as with Einstein's theory of relativity.

The test of the success of a physical theory is twofold: first, it must fit the available data that have been collected; second, it must be fruitful in making predictions about phenomena that have never been observed or measured. If these predictions are confirmed in the real world, the theory is retained; if not, it is modified or replaced by a better one that does agree with the available data. As Albert Einstein (Fig. 1.3) wrote, ''The scientific theorist is not to be envied. For Nature, or more precisely experiment, is an inexorable and not very friendly judge of his work. It never says 'Yes' to a theory. In the most favorable cases it says 'Maybe,' and in the great majority of cases simply 'No.' . . . Probably every theory will some day experience its 'No'—most theories, soon after conception.''

FIGURE 1.1 Dr. Rosalyn Yalow (born 1921), who shared the 1977 Nobel Prize for physiology and medicine for her part in creating the technique of radioimmunoassay, which uses radioactive tracers to locate antibodies present in the human body in quantities so minute that they are detectable in no other way. [*American Institute of Physics (AIP) Niels Bohr Library*]

FIGURE 1.2 Laser research. The two long rectangular boxes from which intense light is being emitted are gas lasers used for research at the Lawrence Livermore Laboratory in California. (*U.S. Department of Energy.*)

FIGURE 1.3 Albert Einstein (1879–1955), with another famous theoretical physicist, J. Robert Oppenheimer (1904–1967). (*International Communication Agency; courtesy of AIP Niels Bohr Library.*)

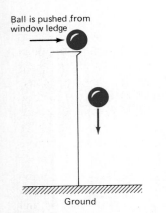

Ball is pushed from
window ledge

Ground

FIGURE 1.4 A ball being pushed from a window ledge and falling freely to the ground.

Physics has been extremely successful over the years in explaining matter and energy at ever-increasing levels of rigor and sophistication. In all cases, however, its secret of success has been the same—the ability to select out of a complicated situation in the real world the crucial physical quantities that determine what is going on. In this process of abstraction, much of the situation may be passed over as irrelevant for physics. For example, a physicist may see a ball falling from a window ledge (Fig. 1.4). Who dropped the ball, its color, and the material out of which it is made are usually of little importance in analyzing the physics of the falling ball. For a physicist, the only important things for the subsequent motion of the ball are the height from which it falls, the force of gravity pulling it down, and air resistance. Since the force of air resistance is often much smaller than the force of gravity, physicists initially ignore even air resistance in calculating a first approximation to the speed with which the ball is moving at any distance below the ledge. Then they can, if necessary, modify the theory later to include the effects of air resistance.

It is this ability to separate the essential points of a physical problem from the nonessentials that is the key to progress in physics. For this reason it can be said that physics has made progress more by what it has learned to ignore than by what it has chosen to take into account. We will see many examples of this important technique in the pages that follow. Solving the problems at the ends of the chapters can help you develop a similar technique—a technique, by the way, that may be of considerable assistance to you in other facets of life.

Physics is still far from complete. There are many things we do not know about the elementary particles, about the behavior of matter at very high and very low temperatures, about plasma physics and astrophysics. All over the world physicists are trying to solve very difficult problems in these and other fields. It will be their ability to penetrate to the heart of a complicated physical situation and then to express the essentials of the situation in the form of an equation that will be the key to their success in solving these problems.

1.3 Physical Theories and Laws

Books have been written on how physicists arrive at definitive descriptions of reality such as those found in Newton's laws of motion or Maxwell's equations for the electromagnetic field. In fact, individual physicists arrive at important conclusions in very different ways, and hence any attempt to systematize their procedures runs the risk of oversimplification. Still it is worthwhile to attempt some kind of broad outline of how physics progresses from experimental data to hypothesis to theory to physical law. The process used by physicists and all scientists for this purpose is called the *scientific method*.

The first step in developing a valid physical understanding of some aspect of reality is to collect as much good, unbiased data as possible, preferably data gathered by different physicists, perhaps in different parts of the world. Physicists then begin to suggest possible explanations of the data before them. Such tentative explanations, assumed for the sake of argument, are called *hypotheses* and usually include intellectual *models* (like the model of a gas made up of molecules flying around in random directions) constructed in an attempt to make sense of the available data. Finally one hypothesis seems to fit the data well and is advanced to the stage of a *physical theory,* i.e., a theoretical explanation, often in mathematical terms, that correlates and makes understandable the data already collected.

Once a theory has been developed, more experiments are needed to test it, perhaps by varying the circumstances of the original experiments or extending their range. Also, if the theory is a good one, it should make predictions about other aspects of reality that have not yet been measured or understood. If these predictions are verified, the theory is well on its way to acceptance by the physics community.

Theories that agree so well with experimental observations that they seem to reflect the consistent behavior of nature under a variety of conditions are called *physical laws*. Newton's law of universal gravitation, which describes the attractive force exerted by every object in the universe on every other object, is an example of such a law. The fact that this law explained not merely the behavior of apples falling from trees but also the motion of the planets around the sun provided the convincing evidence needed to convert Newton's theory into an accepted physical law. Other examples of physical laws are Newton's three laws of motion and Coulomb's law for the attraction or repulsion of electrically charged objects.

Some laws accepted by physicists for centuries have turned out to be inadequate in certain circumstances. Thus Newton's laws of motion need modification to describe the motion of high-speed particles, and in particular particles inside atoms and nuclei. For this purpose the theories of relativity and quantum mechanics have been developed.* For ordinary objects moving at speeds small compared to the speed of light, however, Newton's laws are perfectly adequate, and we apply them to a great variety of physical situations in the following chapters.

There are also a group of "superlaws," usually called *principles*, which physicists believe are so built into the structure of nature that no exceptions are ever possible. The principles of conservation of energy, of linear momentum, and of angular momentum fall in this category and are among the greatest achievements of the human mind. We apply these principles to every branch of physics in the rest of this course.

1.4 Theory and Experiment in Physics

For physics to make progress, both experimental and theoretical physicists are needed. Experimental physicists make observations and perform controlled experiments to collect the physical data that any theory must explain. Theoretical physicists (Fig. 1.5) seek a theory, preferably in the form of an equation, that will fit the observed data. If successful, the theory may make additional predictions, sending the experimentalists back into their laboratories to confirm these also. If the predictions are not completely verified by experiment, the theory will have to be modified somewhat to fit the experimental data. Further predictions of the modified theory are then again checked in the laboratory. Physics thus progresses by the continual interplay of theory and experiment. Without one, the other would be of limited value.

Few physicists have been equally good at theory and experiment. Those who have include, among others, Galileo Galilei, Isaac Newton, and Enrico Fermi. Fermi (see Fig. 1.6 and accompanying biography) had an ability to turn from theory to experiment and then back again to theory in a manner almost unique among twentieth-century physicists.

---

*We delay discussion of relativistic effects until Chap. 26. Our treatment of mechanics will be totally nonrelativistic.

FIGURE 1.5 Three famous theoretical physicists, Victor F. Weisskopf (born 1908), Maria Goeppert-Mayer (1906–1972; Nobel Prize in physics, 1963), and Max Born (1882–1970; Nobel Prize in physics, 1954) in Germany in 1930. (*AIP Niels Bohr Library*.)

1.5 The Fundamental Quantities of Mechanics

To test the validity of equations relating various physical quantities, these quantities must be measured accurately. If physicists in different parts of the world are to agree on such measurements, they must make them in terms of some internationally accepted standard for the quantity being measured. For example, to measure the length of a rod, we must compare its length with an arbitrary but agreed-upon standard length so that we can describe the rod's length as a certain multiple of that standard. If the standard is of length 1.00 meter, and we find that the rod is exactly three times as long as the standard, then its length is 3.00 meters.

In the field of mechanics (which is the subject matter of the first eight chapters of this course) there are three *fundamental physical quantities*—time, length, and mass—in terms of which all other physical quantities (called *derived quantities*) are defined. We now discuss the standards used for these three fundamental quantities of mechanics.

Time

St. Augustine once wrote that we all think we know what time is until we attempt to define it. Fortunately for us, it is much easier to measure the passage of time than it is to define its nature precisely. Time can be measured in terms of the swings of a clock's pendulum or the vibrations of the quartz crystal in a watch. The *period,* or time for one complete vibration, usually designated by the symbol T, is then the fundamental unit in terms of which times can be measured. For ages the time for the earth to make one complete rotation on its axis, which determines the length of a day, was used as the time standard. The fundamental unit of time, the second (s), was then defined as equal to $1/(24 \times 60 \times 60) = 1/86,400$ of a mean solar day.

In 1967 a new definition of the second was adopted by the Thirteenth General Conference on Weights and Measures. *The second is now defined as the duration of 9,192,631,770 periods of a certain microwave radiation emitted by the cesium*

FIGURE 1.6 *(AIP Niels Bohr Library.)*

Enrico Fermi (1901–1954)

Born in Rome in 1901, Enrico Fermi obtained his Ph.D. degree in 1922 from the University of Pisa, the city in which Galileo had been born in 1564. His Ph.D research was on x-rays, and he had already published several research papers before he received his degree. He then did postdoctoral research at the University of Göttingen in Germany and at Leiden in Holland.

After returning to Italy, Fermi taught for 2 years at the University of Florence, where he soon established a reputation by developing what are now known as Fermi-Dirac statistics. In 1926, at the age of only 25, he was appointed to a full professorship in physics at the University of Rome. He gathered a group of outstanding young faculty members and students around him and immediately began to make a name for Rome in the fields of nuclear physics and quantum mechanics. In 1933 his theory of beta decay caused a great stir in world physics circles. He won the 1938 Nobel Prize in physics for his experimental research on neutron absorption by heavy elements, in the course of which he discovered 40 new artificial radioactive isotopes and observed that water and paraffin slow down neutrons. This latter observation was one of the keys to the success of the first nuclear reactor.

Because Fermi's wife, Laura, was Jewish, their life in Italy became increasingly difficult as the power of Mussolini and Hitler grew during the 1930s. Hence, when the Fermis left for Sweden for the Nobel Prize ceremony in 1939, they resolved never to return to Italy. They immigrated to the United States, and Enrico taught at Columbia University in New York City from 1939 to 1942 and at the University of Chicago from 1942 to 1954.

On December 2, 1942, nearly 1 year after the bombing of Pearl Harbor and the entry of the United States into World War II, Fermi directed the production of the first self-sustaining chain reaction in uranium under the football stands at Stagg Field in Chicago. Arthur Compton telephoned the good news of this achievement to James Conant at Harvard by the coded message: "The Italian navigator has landed in the New World." That evening Laura Fermi gave a dinner party for some of her husband's colleagues and their spouses. As the guests arrived, each physicist warmly congratulated Fermi on his great accomplishment. Laura was mystified by this but was unable to find out until years later why congratulations were in order. Such was the secrecy surrounding the atomic weapons project, to which Fermi continued to make significant contributions both during and after World War II.

Fermi had great skill in carrying out what physicists call *order-of-magnitude calculations* to test the approximate correctness of a physical theory or the likely success of a physics experiment.* With a small slide rule and the information stored in his head, he could solve a problem faster and more accurately than lesser physicists could with large calculators and libraries of books.

Fermi was a warm and friendly person who loved physics and the outdoors. He was universally admired and loved by his many students and colleagues. The clarity of his thinking helped make him a great teacher; his probing imagination and his ability to get to the heart of any problem made him a great research physicist. In the words of Emilio Segrè, Fermi's pupil and colleague from Rome, "Fermi gave science his utmost, and with him disappeared the last individual of our times to reach the highest summits in both theory and experiment and to dominate all of physics."

*For order-of-magnitude calculations see Appendix 3.C.

FIGURE 1.7 The Primary Cesium Atomic Frequency Standard—NBS-6 at the U.S. National Institute of Standards and Technology (NIST) laboratory in Boulder, Colorado. This sixth-generation atomic clock is about 6 meters long and is used to define the second to very high accuracy. (*Courtesy of NIST*).

atom. An oscillator is tuned to this frequency* in a device called an *atomic clock* (Fig. 1.7), believed to be accurate to two parts in 10^{14}, that is, to less than 1 s in 1 million years. Other clocks can then be compared with a standard atomic clock and afterward used as secondary time standards for laboratory measurement of the times of physical events.

Because of the extreme accuracy of atomic clocks, times and frequencies are the most accurately measured quantities in physics. Often experimental physicists try to arrange their experiments so that the ratio of two masses, or of two velocities, can be reduced to the ratio of two frequencies, for then the accuracy of the experiment can be enormously improved.

Length

We all have some feeling for what we mean by *length*. It is the distance from one end of an object to the other as we measure it with a ruler or tape measure. The standard unit of length is the *meter* (m).

A new definition of the meter was introduced by the General Conference on Weights and Measures in Paris in November 1983. It is based on the fact that the speed of light has been measured to such high accuracy by laser techniques in the last 20 years that it is now internationally agreed that its value is 299,792,458 m/s, where all nine figures are accurately known. As a result, the speed of light has now been *defined* to have this exact value. Since light travels 299,792,458 m in 1 s and since an extremely accurate value for the second already exists, the meter can be defined as follows (Fig. 1.8):

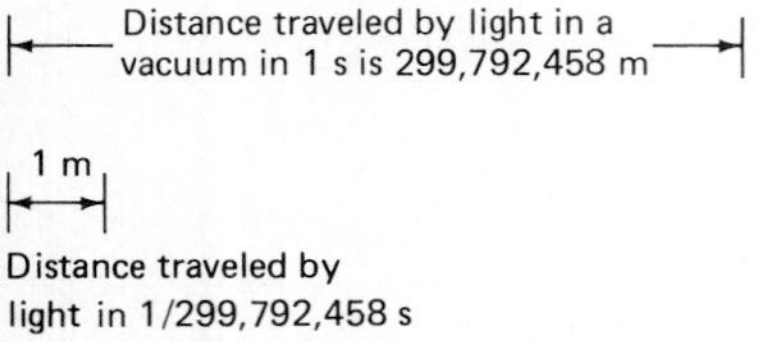

One meter: The distance traveled by light through space in 1/299,792,458 second.

|← Distance traveled by light in a vacuum in 1 s is 299,792,458 m →|

|1 m|
|←→|

Distance traveled by light in 1/299,792,458 s

FIGURE 1.8 The definition of the meter in terms of the speed of light. This schematic diagram shows that since in 1 s light travels 299,792,458 m, the meter can be defined as the distance traveled by light in 1/299,792,458 s.

---

*By *frequency* we mean the number of vibrations per unit time. Frequencies are measured in *hertz* (Hz), which means "cycles per second." The frequency of a vibration is the reciprocal of its period, or $f = 1/T$. We will discuss this more thoroughly in Chap. 11.

Using lasers and optical devices called *interferometers,* secondary-standard meter bars can be constructed whose length agrees exactly with the above definition of the meter. All metric rulers and tape measures can then be checked against such secondary standards maintained at laboratories like the National Institute of Standards and Technology at Gaithersburg, Maryland, or the National Physical Laboratory at Teddington, England.

Mass

FIGURE 1.9 A secondary mass standard maintained at the National Institute of Standards and Technology in Gaithersburg, Maryland. (*U.S. National Institute of Standards and Technology.*)

The third fundamental quantity of mechanics is *mass,* a measure of the amount of matter present in an object. For example, a large truck has more mass than a small automobile. We know from experience that if these two vehicles break down, it is much more difficult to move the truck by pushing it than it is to move the car. Physicists say that the truck has more *inertia* than the car, where by inertia they mean the resistance the vehicle offers to a change in its condition of rest or motion. *Mass is a measure of inertia.* This important attribute of mass is discussed further in Chap. 3.

One reason that mass is such a fundamental quantity is that the mass of a body is the same at all places in the universe. An object has mass whether or not it is near enough to the earth (or some other large object like the moon) to have measurable weight. The *weight* of an object on earth is produced by the attraction of the mass of the earth for the mass of the object and rapidly goes to zero as the distance between the earth and the object increases. This change in distance has no effect, however, on the *mass* of the object, which remains the same no matter where the object is in the universe. (A more careful discussion of mass and weight will be found in Sec. 4.2.)

The standard of mass is a platinum-iridium cylinder, preserved at Sèvres, near Paris, whose mass is defined to be exactly one kilogram (1 kg). Secondary standards of mass that have been compared with the Sèvres primary standard are maintained at standards laboratories throughout the world (see Fig. 1.9). Unknown masses can then be compared with these secondary standards to high accuracy by using sensitive equal-arm balances similar to the one shown in Fig. 1.10.

These three physical quantities—time, length, and mass—are the fundamental quantities we use in our study of mechanics. Since we will also need angles to

FIGURE 1.10 A precision equal-arm two-pan balance, used until 1960 by the U.S. National Bureau of Standards for the comparison of standard masses. It has now been replaced by a more accurate one-pan balance (*Courtesy of NIST*).

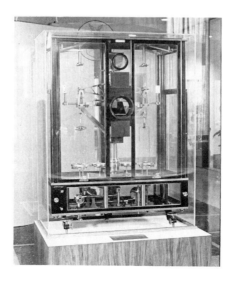

determine directions and to describe circular motion, it may be useful to introduce here some ideas on how angular quantities are measured.

Angle Measurement

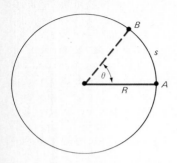

FIGURE 1.11 Definition of a radian: If the arc s (= AB) is equal to the radius R, then the angle θ is equal to one radian.

Consider the circle in Fig. 1.11. If an object moves a distance s along the arc of the circle from A to B, then the angle swept out is θ. Angles are often indicated by Greek letters; here θ is the Greek letter *theta*. We measure θ in radians (rad), where *one radian is the angle subtended by an arc equal to the radius (R) of the circle*. That is, if $s = R$, then θ is equal to 1 rad. The radian is a pure number, without units, since it is the ratio of one length (the arc) to another length (the radius).

The value of θ in radians for any given arc s is then

$$\theta = \frac{s}{R} \tag{1.1}$$

Since for a full circle $s = 2\pi R$, the number of radians in a full circle is

$$\theta = \frac{2\pi R}{R} = 2\pi \text{ rad}$$

The angle subtended by the full circle is usually called *one revolution* (rev). Since there are 360° in a circle, 2π rad = 360°, or

$$\boxed{1 \text{ rad} = \frac{360°}{2\pi} = 57.3°} \tag{1.2}$$

The sizes of these angular quantities are illustrated in Fig. 1.12.

Example 1.1

   If, in Fig. 1.11, the arc length $s = AB = 0.315$ m and the radius $R = 1.20$ m, what is the angle θ in **(a)** radians, **(b)** degrees, and **(c)** revolutions?

SOLUTION

(a)   From our definition of θ,

$$\theta = \frac{s}{R} = \frac{0.315 \text{ m}}{1.20 \text{ m}} = 0.263$$

This result has no units, since the identical units in the numerator and denominator cancel. To make clear, however, what angular unit we are using, it is often useful (but not necessary) to explicitly introduce the term *rad*. So our answer is

$$\theta = \boxed{0.263 \text{ rad}}$$

(b)   Since 1 rad = 57.3°,

$$\theta = 0.263 \text{ rad} \left(\frac{57.3°}{\text{rad}}\right) = \boxed{15.1°}$$

(c)   The angle subtended by a complete circle (or 1 rev) is 360°, and the fraction of a revolution subtended by an angle of 15.1° is

$$\theta = \frac{15.1°}{360°/\text{rev}} = \boxed{0.0419 \text{ rev}}$$

   The same result may be obtained using radian measure for θ. We have $\theta = s/R$, and for one complete revolution, $\theta_c = 2\pi$, and the fraction of a complete revolution represented by θ is

$$\frac{\theta}{\theta_c} = \frac{s/R}{2\pi/\text{rev}} = \frac{0.263 \text{ rev}}{2\pi} = 0.0419 \text{ rev}$$

as found above.

   Notice that if we want to find s by working backward from our answer for θ, we obtain

$$s = R\theta = (1.20 \text{ m})(0.263 \text{ rad}) = 0.315 \text{ m·rad}$$

But s must be in meters, since it is a length. Hence we should, in this case, drop the rad, just as we added it before in the interest of clarity. This yields an answer $s = 0.315$ m, which is consistent with the statement of the problem.

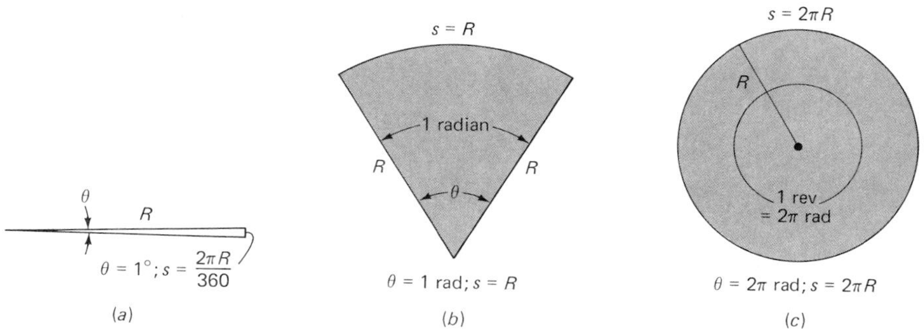

FIGURE 1.12 The angular units: (a) 1 degree; (b) 1 radian; (c) 1 revolution.

1.6 Units and Dimensions

Any physical measurement leads to a numerical result that must be expressed in proper *units*. For example, the length of a house may be expressed in feet and inches or in meters and centimeters. Both the numerical value and the units differ when the length is measured in meters instead of feet, but the actual length of the house is the same in the two cases. Thus lengths of 100 feet (ft) and 30.5 m are equivalent values for the length of a particular house. Similarly a time of 1.50 hours (h) can be expressed as 90.0 minutes (min) or as 5400 s.

The units in which a physical quantity is measured are completely arbitrary. In principle we are free to choose any convenient system of units as long as we use this system consistently.

SI System of Units

The units to be used in this book are the metric SI (for the French *Système International*) units. A list of the fundamental (basic) and derived SI units are given in Table F.5 (inside front cover). All mechanical SI units can be expressed in terms of the three basic units (second, meter, kilogram), as shown in this table.

In the first few chapters we will also use length units like feet and miles that are common in the United States (but not in the rest of the civilized world) to practice converting these units into SI units. We will also, in our discussion of heat, occasionally use the calorie, a non-SI unit, because of its historical and pedagogical importance.

One advantage of the SI system is that since it is a metric system, units used for any one physical quantity are simple powers-of-10* times the SI unit. For example, $1 \text{ m} = 10^2$ cm, and $1 \text{ g} = 10^{-3}$ kg. The abbreviations commonly used for powers-of-10 in the SI system of units are listed in Table F.1 (inside the front cover). For example, 10^{-9} s is called a nanosecond (ns) and 10^{-12} s is a picosecond (ps).

To give us some feeling for these fundamental units, the actual sizes of some length units are shown in Fig. 1.13. A millimeter (10^{-3} m) is about the thickness of the wire in a large paper clip, a centimeter (10^{-2} m) is about the width of the nail on the little finger, and a kilometer (10^3 m) is the length of 11 football fields.

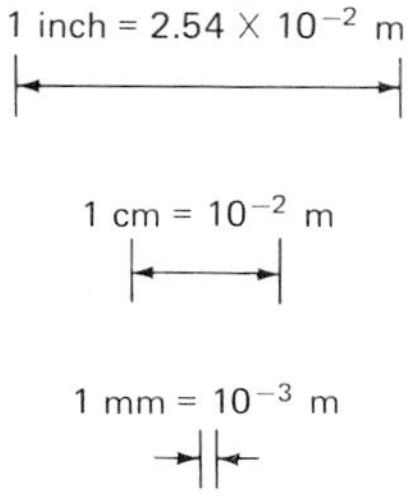

1 inch = 2.54×10^{-2} m

1 cm = 10^{-2} m

1 mm = 10^{-3} m

FIGURE 1.13 Actual size of some common units of length.

---

*Students not familiar with powers-of-10 notation should consult Appendix 3.B. Problems involving powers-of-10 notation are included at the end of this chapter.

Also, a quart of milk has a mass a little less than a kilogram [actually 975 grams (g), where 1 kg $= 10^3$ g], and a jelly bean a mass very close to 1 g.

The British engineering units of length, which are (unfortunately!) still much used for nonscientific work in the United States, are defined in terms of the standard meter. Thus 1 inch (in) is exactly equal to 0.0254 m, and 1 yard (yd) is 0.91440 m. Useful conversion factors are found in Table B.1 inside the back cover.

Sometimes it is necessary to convert nonmetric units, such as feet and miles, into SI units. Take, for example, a distance of 50.0 miles (mi). According to Table B.1, 1.00 mi is equal to 1.61 km, or 1.61×10^3 m. Therefore we have

$$50.0 \text{ mi} = (50.0 \text{ mi}) \left(\frac{1.61 \times 10^3 \text{ m}}{1.00 \text{ mi}} \right) = 8.05 \times 10^4 \text{ m}$$

Note that in converting from miles to meters we have multiplied by a factor $(1.61 \times 10^3 \text{ m})/(1.00 \text{ mi})$, which is equal to 1, since numerator and denominator are equal lengths. Units are treated like any other algebraic factors in multiplication and division. We can cancel similar units in the numerator and denominator and obtain our answer in the desired units (meters), as above. In this section cancellation slashes will be drawn through the units that cancel, in the same way that equal numerical factors in the numerator and denominator of fractions can be canceled. (In the rest of the book the units will be written at every step of a problem for clarity, but it is assumed that you can cancel the units for yourself without having the cancellation explicitly indicated.)

When this technique and a table of conversion factors such as Table B.1 are used, converting units becomes a straightforward process. Sometimes confusion arises about whether a conversion factor is being used correctly. If the units are explicitly written out and do not cancel as they should, then the conversion factor is not being used properly.

Example 1.2

(a) Convert the legal highway speed limit of 55.0 mi/h to meters per second. **(b)** Use your result to find the SI equivalent of a speed of 1.00 mi/h and the equivalent of 1.00 m/s in miles per hour.

SOLUTION

(a) Using the conversion factors in Table B.1, we have

$$55.0 \text{ mi/h} = \left(\frac{55.0 \text{ mi}}{h} \right) \left(\frac{1.61 \times 10^3 \text{ m}}{1.00 \text{ mi}} \right) \left(\frac{1.00 \text{ h}}{3600 \text{ s}} \right)$$

$$= \boxed{24.6 \text{ m/s}}$$

so that 55 mi/h is close to 25 m/s. Notice that the units cancel properly here to yield the answer in the desired SI unit m/s.

(b) Dividing the above equation by the number 55.0, we obtain

$$1.00 \text{ mi/h} = \frac{55.0 \text{ mi/h}}{55.0} = \frac{24.6 \text{ m/s}}{55.0} = \boxed{0.447 \text{ m/s}}$$

Also $\quad 1.00 \text{ m/s} = \frac{24.6 \text{ m/s}}{24.6} = \frac{55.0 \text{ mi/h}}{24.6} = \boxed{2.24 \text{ mi/h}}$

Dimensions

The length of a fence remains the same no matter what units are used to express this length. The property of the fence described by the word *length* we call its *dimension*. It is conventional to write [L] for the dimension of length.

Similarly, the dimensions of the other two fundamental physical quantities in mechanics, time and mass, are expressed as [T] and [M].

Since the width (and height) of an object is the same kind of physical quantity as its length, any area has the dimension $[L] \times [L] = [L^2]$ and any volume the dimension $[L^3]$.

Other quantities of mechanics—velocity, acceleration, force, energy, power—are all combinations of the three fundamental quantities and are therefore called *derived* quantities. For example, velocity is dimensionally the ratio of a distance to a time and has the dimension $[L]/[T]$, while acceleration has the dimension $[L]/[T^2]$, since it is the rate of change of velocity with time. In general, the dimension of any physical quantity is specified by expressing it in terms of the dimensions of the three fundamental physical quantities—length, time, and mass—as was done above for area and velocity.

Note that whether a speed is expressed in units of m/s, mi/h, or mm/year, its dimensions are always the same, that is, $[L]/[T]$.

Numerical constants such as π, which is the ratio of one length (the circumference of a circle) to another length (the diameter), have no dimensions and are referred to as *dimensionless constants* (or pure numbers). An angle expressed in radians is also dimensionless for the same reason.

Dimensional Analysis

The usefulness of introducing dimensions becomes immediately clear when we consider how they can be used to check the validity of calculations or equations, and even to predict the functional dependence of complicated physical properties on simpler properties of a system. The technique used to accomplish these goals is called *dimensional analysis*.

Suppose you are given a very involved problem to solve in which you are asked to find the velocity of an object at the end of a certain period of time. You set up the problem, solve it, and conclude that the velocity is 64 m/s². Now this is certainly wrong, since the answer has the dimension $[L]/[T^2]$, which, as we have seen, is not proper for a velocity. The answer is therefore dimensionally incorrect. There is no point in checking your arithmetic to see if the numerical answer is right, since the dimension indicates a more serious error in solving the problem. *In doing all problems in this book, check your answers to see if they are at least dimensionally correct.* If they are not, you should redo the problem from the beginning.

A second use of dimensional analysis is to check that an equation being used is dimensionally correct. In this case all additive terms in the equation must have the same dimensions. Otherwise they cannot be added. For example, suppose that during a physics examination, you think you remember an equation $a = g + 2vt$ and want to test it for correctness. Here a and g are accelerations with dimension $[L]/[T^2]$, and v is a velocity with dimension $[L]/[T]$. You write a dimensional equation corresponding to this equation by replacing each physical quantity by its proper dimension, ignoring all constants, since they have no dimensions. You obtain

$$\frac{[L]}{[T^2]} = \frac{[L]}{[T^2]} + \frac{[L]}{[T]}[T] \quad \text{or} \quad \frac{[L]}{[T^2]} = \frac{[L]}{[T^2]} + [L]$$

This equation is dimensionally incorrect and cannot be used, since the dimension of the third term is different from those of the first two. A dimensional check such as this will tell you only when an equation is wrong; it will never tell you that it is right, since it can never yield the numerical factors in the equation.

Example 1.3

Prove that the following equation for a simple pendulum is dimensionally correct:

$$T = 2\pi \sqrt{\frac{l}{g}}$$

where l is the length of the pendulum, g the acceleration due to gravity, and T the time for one complete vibration, the *period* of the pendulum.

SOLUTION

The left side of this equation has the dimension of time $[T]$. The right side has the dimension

$$\left[\frac{[L]}{[L]/[T^2]} \right]^{1/2} = \frac{[L^{1/2}]}{[L^{1/2}]} [T] = [T]$$

Both sides have the same dimension, and the equation is dimensionally correct.

1.7 Accuracy of Physical Data

When physicists collect data in the laboratory, they express their results in a form that includes three essential elements: a number giving the magnitude of the quantity measured, a unit in terms of which the quantity is being measured, and an estimated uncertainty in the measured value, since no physical measurement is ever perfectly accurate. For example, a measurement of the diameter d of a coin with an accurate steel ruler might lead to a value $d = 2.45$ centimeters (cm), where the last figure is an estimate, but the measurement of the same diameter with a precision measuring instrument called a *micrometer caliper* (Fig. 1.14) might yield a value $d = 2.446$ cm. Here the same quantity is measured in the two cases, but the larger number of digits given in the second measurement indicates that it was judged to be more accurate than the first by the person making the measurement.

Significant Figures

The digits given in reporting the results of an experiment, or in stating the data for a problem to be solved, are called *significant figures* (or significant digits). For example, 2.45 cm has three significant figures, whereas 2.446 cm has four. The greater the number of significant figures, the more accurate the data are presumed to be. It is misleading for a physicist or physics student to use more significant figures than are warranted by the apparatus used or the data given in the problem.

Zeros written at the right end of numbers are assumed to be significant figures, for by convention they are included to indicate that the data are certain to this last decimal place. For example, the charge on an electron is 1.60×10^{-19} coulombs (C), where the last zero is a significant figure. Zeros to the left of the first nonzero digit are not significant, since they merely locate the decimal point and say nothing about the accuracy of the number given. Thus 0.00164 is the same as 1.64×10^{-3} in powers-of-10 notation (Appendix 3.B), and in both cases the number has only three significant figures. Because power-of-10 notation makes clear the number of significant figures, it is always used in examples and problems in this book if the number of significant figures would otherwise be unclear.

The following rules are useful in handling significant figures:

1 *The final result of an addition or subtraction of two or more quantities should not contain more decimal places than the quantity with the smallest number of decimal places.*

2 *The final result of a multiplication or division should have only as many significant digits as the quantity in the calculation with the smallest number of significant digits.*

FIGURE 1.14 A micrometer caliper. The instrument is designed so that the movable rod moves 1 millimeter (mm) for each complete turn of the screw. A dial reads fractions of a turn to the nearest 1/100 of a turn. The micrometer can therefore measure distances to 0.01 mm, or 0.001 cm. (*Sargent-Welch Scientific Co.*)

The basic rationale behind these rules is that if we add an uncertain digit to a certain one, the resulting digit is uncertain and should be omitted from the final sum. Similarly, if a number with two significant figures is multiplied by one with three significant figures, the multiplication of the completely uncertain (actually unknown) third figure of the two-digit number destroys the certainty of the third significant figure in the product. The result can therefore only be given to two significant figures.

The examples and problems in this book have data given in most cases only to two or three significant figures, depending on the nature of the problem. All answers are therefore expected to include two or three significant figures, and no more, for that is all the data can yield. This should be kept in mind when you are tempted to write down the eight digits indicated on your calculator as the answer to a problem or laboratory experiment. Computer programmers have a saying: "Garbage in, garbage out." For our purposes insignificant figures are "garbage." The result of any calculation can be no more certain (contain no more significant figures) than the least certain piece of data used in that calculation.

Example 1.4

How many significant figures are there in the following results for quantities measured in the laboratory?
(a) 2.997924×10^8 m/s **(b)** 3.0120 s **(c)** 0.00124 m **(d)** 100 s

---

SOLUTION

(a) Seven significant figures.
(b) Five significant figures. The final zero is significant, because it indicates that the last decimal place was measured to be zero.
(c) Three significant figures. The zeros are not significant and are needed only to position the decimal point. For clarity this result would be better written as 1.24×10^{-3} m.
(d) Three significant figures. The final zeros are significant. To make sure that this is what the experimenter intended, it would be clearer if the result were given as 1.00×10^2 s.

Example 1.5

(a) Use Table B.2 (inside back cover) to find the mass difference in kilograms between the rest masses of a proton and a neutron.
(b) If seven significant figures are used for the proton and neutron rest masses, how many significant figures does the mass difference have?

SOLUTION

(a) This is a case in which many more significant figures than three are required to yield a meaningful result. From Table B.2 we have

$$m_p = 1.672623 \times 10^{-27} \text{ kg} \qquad m_n = 1.674929 \times 10^{-27} \text{ kg}$$

The mass difference is then

$$\Delta m = (1.674929 - 1.672623) \times 10^{-27} \text{ kg}$$

$$= \boxed{0.002306 \times 10^{-27} \text{ kg}}$$

(b) The mass difference has only four significant figures and is better written as 2.306×10^{-30} kg. This example illustrates how, after subtraction, the result may have a much smaller number of significant figures than did the original numbers because of cancellations.

Example 1.6

(a) Write 0.00000027 in powers-of-10 notation. (b) Calculate $(2.56 \times 10^3) + (7.92 \times 10^4)$. (c) Calculate $(2.56 \times 10^3) \times (7.92 \times 10^4)$. (d) Calculate $\dfrac{2.56 \times 10^3}{7.92 \times 10^4}$.

SOLUTION

(a) $\boxed{2.7 \times 10^{-7}}$

(b) $(2.56 \times 10^3) + (7.92 \times 10^4) = (2.56 + 79.2) \times 10^3$

$$= \boxed{81.8 \times 10^3}$$

Notice that on adding 2.56 and 79.2, the second decimal place no longer is significant, since 79.2 has only one significant figure to the right of the decimal point. Hence we round off 2.56 to 2.6 and add it to 79.2 to obtain 81.8.

(c) $(2.56 \times 10^3) \times (7.92 \times 10^4) = \boxed{20.3 \times 10^7}$

Here we multiply 2.56 by 7.92, rounding off the product to three significant figures, and add the powers of 10 to obtain $10^{3+4} = 10^7$.

(d) $\dfrac{2.56 \times 10^3}{7.92 \times 10^4} = \boxed{0.323 \times 10^{-1}}$

Here we divide 2.56 by 7.92, round off the quotient to three significant figures, and subtract the powers of 10 to obtain $10^{3-4} = 10^{-1}$. The result can also be written as 3.23×10^{-2}.

1.8 Scalars and Vectors

In this book we will be dealing with two very different kinds of physical quantities: scalars and vectors. If a sack of flour has a mass of 10 kg, that mass is not dependent on where the flour is, whether it is at rest in a storeroom on land or in motion on a ship at sea. The mass is what we call a *scalar* quantity; i.e., it is a physical property of the sack of flour that can be completely described by its magnitude only (10 kg), since the mass does not depend on the position or direction of motion of the flour. Other examples of scalar quantities are energy, temperature, and volume. All can be completely specified by a single number—the magnitude of the scalar. Scalars combine by simple algebraic addition: two 10-kg masses of flour add to give 20 kg of flour, no matter what their position or direction of motion.

A quantity such as velocity is quite different. To a passenger in Denver desiring to go to New York City on a train moving at 25 m/s, it obviously makes a big difference whether the train is moving in the direction of New York City or of Los Angeles. Here both *magnitude* and *direction* are vitally important. Quantities such as velocity are called *vectors*. In adding vectors it is not enough simply to add their magnitudes; their directions must also be taken into account.

Let us define carefully these two important kinds of physical quantities:

Scalar: A physical quantity (like mass or energy) that has no direction and is completely specified by its magnitude alone.

Vector: A physical quantity (like velocity or force) that is completely specified only when both its magnitude and its direction are given.

A vector is often represented by a directed line segment (arrow) whose length represents the vector's magnitude and whose direction shows the vector's direction. Many physical quantities are vectors and must be combined by using directional rules that we now discuss.

Position Vectors

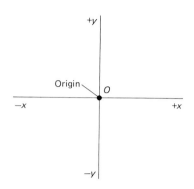

FIGURE 1.15   A two-dimensional rectangular coordinate system used to specify the position of an object. This consists of two mutually perpendicular axes labeled the *x* and *y* axes, relative to which the location of a point in the *xy* plane can be specified. These two axes intersect at the origin (*O*) of the coordinate system, as shown in the figure.

One simple use of a vector is to specify the position of an object with respect to some fixed point in space. A *coordinate system* like that in Fig. 1.15 is used for this purpose. For example, consider a billiard ball resting on a billiard table, as in Fig. 1.16. We set up a coordinate system with the positive *x* axis along the bottom edge of the table and the positive *y* axis along the left edge. These two perpendicular axes intersect at the *origin O*. If the ball is 30 cm from the origin along the *x* axis and 40 cm from the origin along the *y* axis, then its position can be shown on the coordinate system by choosing a convenient scale (say, 1.0 cm on the paper corresponding to 10 cm on the table) and marking off 3.0 cm along the *x* axis and 4.0 cm along the *y* axis. The ball's position is then at point *P* on the diagram. We say that the *x* coordinate of the ball is 30 cm and its *y* coordinate is 40 cm.

There is a better way to represent the position (*P*) of the ball with respect to the origin *O*. This is to draw an arrow with its tail at *O* and its tip at *P*, that is, to locate *P* by means of a *position vector* **r**, as in Fig. 1.17. The length of this vector is then the distance of *P* from *O*, and its direction indicates the precise position of *P* with respect to *O*. If all we knew were the distance *r*, *P* could be any place on a full circle of radius *r* drawn around the origin *O*, and we would have no knowledge of the ball's actual position. We therefore need both the *magnitude r* and the *direction* of the arrow to locate the position *P* with respect to the origin.

We use boldface type, as in **r** or **v**, to indicate vector quantities and ordinary type, as in *r* or *v*, for their magnitudes.

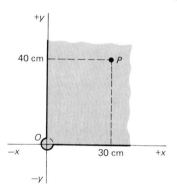

FIGURE 1.16   A billiard ball located on a billiard table by its *x* and *y* coordinates with respect to the origin *O*.

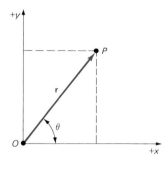

FIGURE 1.17   Position vector of the billiard ball in Fig. 1.16.

Example 1.7

A billiard ball has an *x* coordinate of 30 cm and a *y* coordinate of 40 cm with respect to the corner *O* of a billiard table.

What is the length of the position vector representing the position of the ball with respect to *O*?

SOLUTION

From Fig. 1.16 we have, using the Pythagorean theorem, $r^2 = x^2 + y^2$, and so

$$r^2 = (30 \text{ cm})^2 + (40 \text{ cm})^2 = 2500 \text{ cm}^2$$

and $r = \boxed{50 \text{ cm}}$

This is the length of the position vector **r** in Fig. 1.17. We can also find the angle θ between the position vector and the *x* axis, either graphically or by using some simple trigonometry, as we will see in Secs. 1.9 and 1.10.

Displacement Vectors

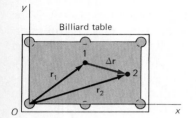

FIGURE 1.18 Positon and displacement vectors for a billiard ball on a billiard table: $\Delta \mathbf{r} = \mathbf{r}_2 - \mathbf{r}_1$.

Suppose our billiard ball is at some instant of time at position 1, specified by the position vector $\mathbf{r}_1$ in Fig. 1.18. We then strike the ball with the billiard cue and it finally comes to rest at some new position 2, where the ball's position vector is $\mathbf{r}_2$. We say that the ball has undergone a *displacement* from position 1 to position 2, and we represent that displacement by a *displacement vector* $\Delta \mathbf{r} = \mathbf{r}_2 - \mathbf{r}_1$.

Displacement vector: A directed line segment (arrow) $\Delta \mathbf{r}$ whose length indicates the magnitude of the displacement and whose direction is the direction of the displacement.

In this case the vector $\Delta \mathbf{r}$ has its tail at position 1 and its tip at position 2. The symbol Δ, the Greek capital *delta*, is used throughout this book to mean "the difference in" or "the change in." Thus we read $\Delta \mathbf{r}$ as "the change in $\mathbf{r}$."

Notice that the *displacement* in not necessarily the same as the *distance* traveled. If the ball moves from 1 to 2 by the shortest possible path, a straight line from 1 to 2, then the distance traveled is equal to the magnitude of the displacement vector. But suppose the ball bounces around the table before coming to rest at position 2. Then the distance the ball travels from 1 to 2 is clearly greater than the length of the displacement vector $\Delta \mathbf{r}$. The *distance traveled is not merely the magnitude of the displacement.*

The *distance* traveled is not a vector, since it can be the same for motion in a variety of different directions; the *displacement,* however, is a vector, since both its magnitude and its direction are unambiguously fixed by the initial and final positions of the ball.

Two vectors are equal if they have the same magnitude and direction, no matter from what position they start. For example, the displacement ($\Delta \mathbf{r}$) of a billiard ball from a position with coordinates $x_1 = 10$ cm, $y_1 = 20$ cm to $x_2 = 30$ cm, $y_2 = 40$ cm is exactly the same as a displacement ($\Delta \mathbf{r}$) from $x_1 = 10$ cm, $y_1 = 0$ cm to $x_2 = 30$ cm, $y_2 = 20$ cm, as shown in Fig. 1.19. These two vectors have the same magnitude and the same direction and therefore correspond to the same displacement, even though they appear at different positions in the diagram. It is for this reason that displacement vectors can always be moved from one position to another as long as their magnitude and direction remain unchanged.

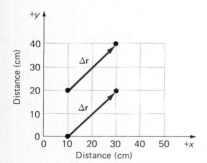

FIGURE 1.19 Two identical displacement vectors corresponding to different initial and final positions of a billiard ball.

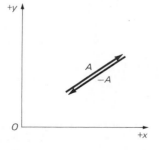

FIGURE 1.20 Adding two displacement vectors in opposite directions results in zero displacement.

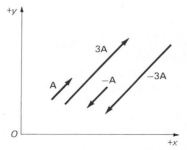

FIGURE 1.21 Graphs of some vectors.

A negative sign before a vector merely reverses its direction; i.e., it interchanges tip and tail without changing the length of the vector. A vector **A** and its negative −**A** are shown in Fig. 1.20. Note that if a billiard ball undergoes two successive displacements **A** and −**A**, the ball arrives back where it started. Its net *displacement* is therefore zero, but it has traveled a *distance* of 2A, if it takes the shortest possible path for the trip.

A numerical factor before a vector changes the magnitude of the vector without changing its direction. Thus 3**A** means a vector with three times the length of **A** but in the same direction; −3**A** means a vector three times as long as **A** but in the opposite direction. These vectors are shown in Fig. 1.21. The multiplication of a vector by a scalar always produces a new *vector*.

1.9   **Graphical Methods for Vectors**

Scalar quantities may be added using simple arithmetic, but special techniques must be used to add vector quantities and obtain their vector sum, or *resultant*. The resultant vector obtained by the addition of two or more vectors is equivalent to the combination of the original vectors and replaces them in any given physical situation. A simple technique for obtaining the resultant of two vectors is a graphical one. It requires only a ruler, a protractor, and a sharp pencil.

Suppose we are told that a cyclist on flat land pedaled 40 km to the east and then 40 km to the northeast with respect to the origin of a coordinate system. What is the cyclist's final position? These two displacement vectors are shown in Fig. 1.22, using a convenient scale of 1 cm = 20 km. Then each vector is 2.0 cm in length, one along the *x* axis, the other at an angle of 45° with respect to the *x* axis as laid off with a protractor. The resultant of these two vectors can be obtained by drawing them with the tail of **B** at the tip of **A**, as in Fig. 1.22*a*.

Then all we need do is complete the triangle by joining the tail of **A** to the tip of **B** to find the magnitude and direction of the resultant **R**, which can be measured with a ruler and protractor. In this case, **R** is 74 km at an angle of 23° above the *x* axis.

Such an approach, usually called the *triangle method,* conforms well with the idea of one displacement following the other in succession, as is the case in this example. The order in which the vectors are added makes no difference: **A** + **B** = **B** + **A**, as Fig. 1.22*b* shows.

This method may be easily extended to more than two vectors, using the *tail-to-tip method* of vector addition.

The resultant, or sum of several vectors, is the single vector leading from the tail of the first vector to the tip of the last vector when the vectors are drawn tail to tip in any order.

Figure 1.23*a* shows this for a situation where five displacement vectors are added. The resultant of the five vectors **A**, **B**, **C**, **D**, and **E** is the vector **R**, constructed by connecting the tail of **A** to the tip of **E**. Here **R** is the vector sum, or resultant, of the five displacement vectors.

Note that, in obtaining the resultant of displacement vectors in this way, the order in which the vectors are added is of no importance, and any vector may be displaced parallel to itself to a new position, since this changes neither its magnitude nor its direction. This is shown in Fig. 1.23*b*, where the resultant is the same as in Fig. 1.23*a*.

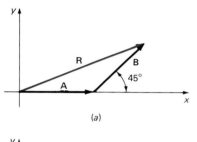

(*a*)

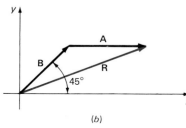

(*b*)

FIGURE 1.22   (*a*) Triangle method for adding two displacement vectors **A** and **B**; (*b*) demonstration that **B** + **A** = **A** + **B**.

FIGURE 1.23 (*a*) Graphical method for adding five vectors **A, B, C, D**, and **E** to obtain the resultant vector **R**, using the tail-to-tip method; (*b*) the same resultant **R** obtained by adding the vectors in a different order; therefore we have **A** + **B** + **C** + **D** + **E** = **R** = **C** + **B** + **A** + **E** + **D**.

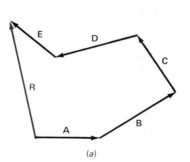

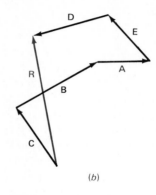

(*a*) (*b*)

Subtraction of Vectors

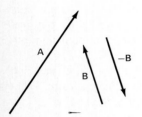

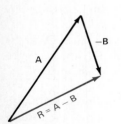

FIGURE 1.24 The subtraction of two vectors. **R** is the difference vector **A** − **B**, obtained by reversing the direction of **B** and then adding the resulting vector to **A**.

The above discussion showed us how to add two vectors **A** and **B** to obtain the resultant **R**, namely, **A** + **B** = **R**, where the boldface type indicates that we are adding *vectors*. What about the subtraction of two vectors, say, **A** − **B**? Since, from our definition of the negative of a vector,

$$\mathbf{A} - \mathbf{B} = \mathbf{A} + (-\mathbf{B}) \tag{1.3}$$

all we need do to subtract **B** from **A** is to reverse the direction of **B** and then add the resulting vector graphically to **A**. An example is given in Fig. 1.24, where two vectors **A** and **B** are shown and we are asked to find their difference **A** − **B**. The diagram shows **B** and −**B**. By graphically adding **A** and −**B** together, we get the difference vector **R** = **A** − **B**.

In a parallelogram constructed with vectors **A** and **B** as its sides (Fig. 1.25), the long diagonal of the parallelogram is the sum **A** + **B**. The short diagonal is **A** − **B**, since it is clear from the figure that **B** + (**A** − **B**) = **A**. It can be seen from the figure that this result for **A** − **B** is the same in both magnitude and direction as was obtained in Fig. 1.24 using the triangle method.

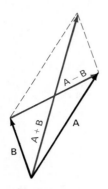

FIGURE 1.25 Parallelogram method: The longer diagonal is the sum **A** + **B**; the shorter diagonal is the difference **A** − **B**. Note that in this diagram the side opposite **A** is also **A** and the side opposite **B** is also **B**.

Example 1.8

A marine recruit in boot camp is commanded to walk due east for 5.00 km, then due north for 3.00 km, and finally 4.00 km northwest. When he finishes, what is his resultant dis-placement **R** (both magnitude and direction) with respect to his starting point?

SOLUTION

We set up a coordinate system with its origin O at the starting position in the training camp. Then we draw the three displacements to scale, one after the other, with the first displacement starting from the origin of the coordinate system. The resultant displacement is then the vector sum $\mathbf{R} = \mathbf{A} + \mathbf{B} + \mathbf{C}$, as shown in Fig. 1.26. From the diagram, the measured magnitude of the resultant R is 6.2 km, and its direction is 21° east of north. The results are given only to two significant figures, since it is difficult to measure quantities on an ordinary graph to more than this accuracy.

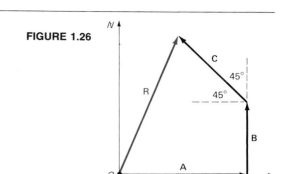

FIGURE 1.26

Example 1.9

Vector $\mathbf{A}$ has a magnitude of 10 m and is directed along the positive x axis. Vector $\mathbf{B}$ has a magnitude of 20 m and is directed at an angle of 60° above the x axis, as shown in Fig. 1.27. Find the vector $\mathbf{A} - \mathbf{B}$.

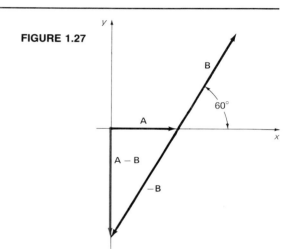

FIGURE 1.27

SOLUTION

We draw the two vectors $\mathbf{A}$ and $\mathbf{B}$ to scale, as in Fig. 1.27. To subtract $\mathbf{B}$ from $\mathbf{A}$, we first find $-\mathbf{B}$ and add this to $\mathbf{A}$. On the graph we find $-\mathbf{B}$ by reversing the direction of $\mathbf{B}$ while keeping its magnitude the same. The resultant is found from measurements on the diagram to have a magnitude of 17 m and to be directed at an angle of $-90°$ with respect to the positive x axis; that is, $\mathbf{A} - \mathbf{B}$ is along the negative y axis.

1.10 Analytical Methods for Vectors

In the previous section, we used graphical methods to find the vector sum and difference of several vectors. In this section we develop another approach, the addition of vectors by using their rectangular components.

Rectangular Components of Vectors

First let us consider two displacement vectors, one along the x axis, the other along the y axis. We call them $\mathbf{A}_x$ and $\mathbf{A}_y$. The resultant of these two vectors is the vector $\mathbf{A}$ shown in Fig. 1.28. This vector displacement is completely equivalent to the combination of the two vector displacements $\mathbf{A}_x$ and $\mathbf{A}_y$ and can replace them in any physical problem.

We now reverse this process. Suppose we have a vector $\mathbf{A}$ in the xy plane, as in Fig. 1.29. It is always possible to replace this vector by two other vectors

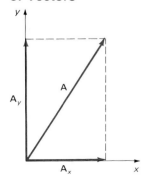

FIGURE 1.28 Two vectors $\mathbf{A}_x$ and $\mathbf{A}_y$ replaced by their vector resultant $\mathbf{A}$.

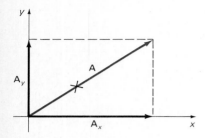

FIGURE 1.29 Replacement of a vector **A** by two rectangular vector components A_x and A_y. The cancellation mark over **A** means that these two vectors take the place of **A**, which is no longer needed.

whose vector sum is equal to **A**. We choose the direction of these two vectors as along the x and y axes. Since these axes are at right angles, we call these vectors $\mathbf{A}_x$ and $\mathbf{A}_y$ the rectangular components of the vector **A**. They are found by dropping perpendiculars from the tip of the vector **A** to the x axis to find $\mathbf{A}_x$ and to the y axis to find $\mathbf{A}_y$. A vector **A** can always be replaced by its two vector components $\mathbf{A}_x$ and $\mathbf{A}_y$. Once **A** has been resolved into $\mathbf{A}_x$ and $\mathbf{A}_y$, these components take the place of the original vector **A**, since $\mathbf{A}_x + \mathbf{A}_y = \mathbf{A}$. To make this clear, we cross out the original vector in Fig. 1.29.

We have chosen $\mathbf{A}_x$ and $\mathbf{A}_y$, the vector components of **A**, as along the x and y axes. Their direction is therefore specified, and only a single number (the length) is needed to describe each one, except for a possible negative sign. We further specify that A_x is positive when $\mathbf{A}_x$ points in the direction of the positive x axis, and A_x is negative when in the opposite direction. The two numbers A_x and A_y are then the scalar magnitudes of the vectors $\mathbf{A}_x$ and $\mathbf{A}_y$ and are called the scalar components of vector **A**. In what follows we will refer to these scalar components A_x and A_y simply as the *rectangular components* of **A**.

Trigonometric Methods

We can find rectangular components most easily by trigonometry (see Appendix 1.C). For example, in Fig. 1.30 the vector **A**, which has magnitude A and makes an angle θ with the positive x axis, has rectangular components

$$A_x = A \cos \theta \qquad A_y = A \sin \theta \tag{1.4}$$

This can be justified by recalling that the sine of an angle is the ratio of the side *opposite the angle* θ to the *hypotenuse* of the right triangle, as in Fig. 1.31. Hence $\sin \theta = A_y/A$, or $A_y = A \sin \theta$. Similarly, the cosine of θ is the ratio of the side adjacent to the angle θ to the hypotenuse so that $\cos \theta = A_x/A$, or $A_x = A \cos \theta$. Also

$$\text{tangent } \theta = \frac{\sin \theta}{\cos \theta} = \frac{A_y/A}{A_x/A} = \frac{A_y}{A_x}$$

The magnitude of the vector **A** is, from the Pythagorean theorem,

$$A = [A_x^2 + A_y^2]^{1/2} \tag{1.5}$$

and its direction is given by the angle θ between the vector **A** and the x axis in Fig. 1.30, where

$$\tan \theta = \frac{A_y}{A_x} \tag{1.6}$$

FIGURE 1.30 Use of trigonometry to obtain the rectangular components A_x and A_y of a vector **A**.

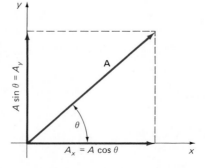

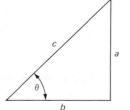

FIGURE 1.31 Definition of trigonometric functions in terms of the sides of a right triangle. Here $\sin \theta = a/c$; $\cos \theta = b/c$; $\tan \theta = a/b$.

Example 1.10

A vector displacement $\mathbf{A}$ has a magnitude of 25.0 cm and a direction specified by an angle of 70.0° above the positive x axis. **(a)** Find the rectangular components of $\mathbf{A}$. **(b)** Use these rectangular components to calculate the magnitude and direction of $\mathbf{A}$ as a check on your answers to part (a).

SOLUTION

(a) From Eqs. (1.4) for the rectangular components of a vector, we have

$$A_x = A \cos \theta = (25.0 \text{ cm})(\cos 70.0°)$$

$$= (25.0 \text{ cm})(0.342) = \boxed{8.55 \text{ cm}}$$

$$A_y = A \sin \theta = (25.0 \text{ cm})(\sin 70.0°)$$

$$= (25.0 \text{ cm})(0.940) = \boxed{23.5 \text{ cm}}$$

(b) The magnitude of $\mathbf{A}$ is, from Eq. (1.5),

$$A = [A_x^2 + A_y^2]^{1/2} = [73.1 + 552]^{1/2} \text{ cm} = \boxed{25.0 \text{ cm}}$$

The direction of $\mathbf{A}$ is, from Eq. (1.6),

$$\tan \theta = \frac{A_y}{A_x} = \frac{23.5}{8.55} = 2.75 \qquad \theta = \boxed{70.0°}$$

These results agree perfectly with the data given in the statement of the problem and therefore provide a convincing proof that our answers to part (a) are correct. Working backward in this fashion from the answer obtained can provide assurance that our answer is right, and gradually build up confidence in our ability to do physics problems correctly.

Rectangular resolution provides a convenient method for adding vectors in two dimensions. We resolve each vector into its rectangular components along the x and y axes and then combine all the x components (as scalars, but with the proper signs) to obtain the x component of the resultant and all the y components to obtain the y component of the resultant. These two components can then be combined using Eqs. (1.5) and (1.6) to find the magnitude and direction of the resultant vector $\mathbf{R}$.

The same method may be used for vectors in a three-dimensional space by including a third component along the z axis.

This general method for adding any number of vectors, using their rectangular components, can be summarized as follows: The resultant $\mathbf{R}$ of the vectors $\mathbf{A}_1$, $\mathbf{A}_2$, $\mathbf{A}_3$, etc., can be found by first obtaining the rectangular components of $\mathbf{R}$:

$$R_x = A_{1x} + A_{2x} + A_{3x} + \cdots \qquad \text{and} \qquad R_y = A_{1y} + A_{2y} + A_{3y} + \cdots$$

or $\qquad R_x = A_1 \cos \theta_1 + A_2 \cos \theta_2 + A_3 \cos \theta_3 + \cdots$

and $\qquad R_y = A_1 \sin \theta_1 + A_2 \sin \theta_2 + A_3 \sin \theta_3 + \cdots$

or $\qquad R_x = \sum_{i=1}^{N} (A_i)_x = \sum_{i=1}^{N} A_i \cos \theta_i$

and $\qquad R_y = \sum_{i=1}^{N} (A_i)_y = \sum_{i=1}^{N} A_i \sin \theta_i$

(1.7)

This is called *summation notation*. It means that we sum over the index i, which runs through the values 1, 2, . . . , N, where N is the number of vectors being added. Once we have found R_x and R_y, we can find the magnitude R of the vector $\mathbf{R}$ from

$$R = [R_x^2 + R_y^2]^{1/2}$$

(1.8)

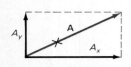

and the direction of **R** can then be found from

$$\tan \theta_R = \frac{R_y}{R_x} \qquad (1.9)$$

where θ_R is the angle between the vector **R** and the positive x axis as measured in a counterclockwise direction.

Once the magnitude and direction of **R** have been found, **R** is completely determined. This is shown for the addition of two vectors in Figs. 1.32 and 1.33. Note that even though R_x and R_y are along the x and y axes, the sides opposite R_x and R_y in the figure also have magnitudes R_x and R_y.

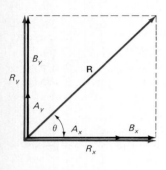

FIGURE 1.32 Addition of two vectors **A** and **B** using rectangular resolution.

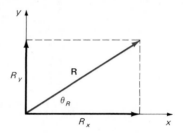

FIGURE 1.33 Determining **R** and θ_R from R_x and R_y.

Example 1.11

Redo Example 1.8, using rectangular resolution of vectors and trigonometry.

SOLUTION

From Fig. 1.26 and the statement of Example 1.8, the three displacement vectors are the following, where all angles are measured counterclockwise *with respect to the positive x axis:*

A: 5.00 km at $\theta_A = 0°$.

B: 3.00 km at $\theta_B = 90°$.

C: 4.00 km at $\theta_C = 135°$.

The components of these vectors are

$A_x = A \cos \theta_A \qquad\qquad A_y = A \sin \theta_A$

$\quad = A \cos 0° = 5.00 \text{ km} \qquad = A \sin 0° = 0$

$B_x = B \cos 90° = 0 \qquad\quad B_y = B \sin 90° = 3.00 \text{ km}$

$C_x = C \cos 135° \qquad\qquad C_y = C \sin 135°$

$\quad = (4.00 \text{ km})(-0.707) \qquad = (4.0 \text{ km})(0.707)$

$\quad = -2.83 \text{ km} \qquad\qquad = 2.83 \text{ km}$

and so

$R_x = A_x + B_x + C_x \qquad\quad R_y = A_y + B_y + C_y$

$\quad = (5.00 + 0 - 2.83) \text{ km} \qquad = (0 + 3.00 + 2.83) \text{ km}$

$\quad = 2.17 \text{ km} \qquad\qquad\quad = 5.83 \text{ km}$

We finally have

$$R = [R_x^2 + R_y^2]^{1/2} = \boxed{6.22 \text{ km}}$$

$$\tan \theta_R = \frac{R_y}{R_x} = \frac{5.83}{2.17} = 2.69 \qquad \theta_R = \boxed{69.6°}$$

where the 69.6° angle is again with respect to the positive x axis. This answer agrees with the less-accurate result found in Example 1.8 by graphical means, since 21° east of north means an angle of $90° - 21° = 69°$, measured counterclockwise with respect to the positive x axis.

In practice it is often simpler, especially with a hand calculator containing trigonometric functions, to find θ_R first from $\tan \theta_R = R_y/R_x$. Then

$$R_x = R \cos \theta_R \qquad \text{or} \qquad R = \frac{R_x}{\cos \theta_R}$$

$$= \frac{2.17 \text{ km}}{0.349} = 6.22 \text{ km}$$

and $\quad R_y = R \sin \theta_R \qquad$ or $\qquad R = \dfrac{R_y}{\sin \theta_R}$

$$= \frac{5.83 \text{ km}}{0.937} = 6.22 \text{ km}$$

so that our results are consistent.

It should be pointed out that the same resultant will be obtained no matter what rectangular coordinate system we choose for the resolution into components. The resultant has a magnitude and direction determined completely by the magnitudes and directions of the individual vectors being added. The use of a set of rectangular axes to combine these vectors is a mathematical device that plays no role in the final result. The individual vectors have the same magnitude no matter what axes are chosen for the resolution, since

$$\sqrt{A_x^2 + A_y^2} = \sqrt{A^2 \sin^2 \theta + A^2 \cos^2 \theta} = A \sqrt{\sin^2 \theta + \cos^2 \theta} = A$$

This is true for all angles, since $\sin^2 \theta + \cos^2 \theta = 1$ for all values of θ (see Appendix 1.C). The direction of each vector will be different, of course, in different coordinate systems. But, no matter what coordinate system is chosen, the direction of the vectors *with respect to one another* does not change, and so the direction of the resultant relative to the individual vectors also does not change.

For this reason the x and y axes used for the rectangular resolution may be chosen to make the calculation as simple as possible. For example, the x axis can be chosen to lie along one vector so that this vector has no y component. For motion on an inclined plane, the direction along the inclined plane can be taken as the x axis, making the y axis perpendicular to the plane.

In what follows we will use the above trigonometric method almost exclusively. It is also possible to use trigonometry to calculate the magnitude and direction of the resultant of two vectors directly, without using rectangular components. This procedure involves using the law of sines and the law of cosines given in Appendix 1.C.

Example 1.12

Given two displacements $\mathbf{A} = 52.6$ m at an angle of $20.0°$ with the positive x axis and $\mathbf{B} = 34.8$ m at an angle of $150°$ with the positive x axis. **(a)** Find the x and y components of the two vectors. **(b)** Find the magnitude and direction of the resultant $\mathbf{R}$. **(c)** Take the direction of vector $\mathbf{A}$ as along the x' axis of a new set of rectangular coordinates x', y'. Find the new rectangular components of vectors $\mathbf{A}$ and $\mathbf{B}$. **(d)** Find the magnitude and direction of the resultant $\mathbf{R}$. Show that it is physically identical with the resultant obtained in part (b).

SOLUTION

The two displacements are shown drawn from a common origin in Fig. 1.34, and their x and y components are found from Eq. (1.4).

(a) $A_x = A \cos \theta_A = (52.6 \text{ m})(\cos 20.0°)$

$= (52.6 \text{ m})(0.940) = \boxed{49.4 \text{ m}}$

$A_y = A \sin \theta_A = (52.6 \text{ m})(\sin 20.0°)$

$= (52.6 \text{ m})(0.342) = \boxed{18.0 \text{ m}}$

$B_x = B \cos \theta_B = (34.8 \text{ m})(\cos 150°)$

$= (34.8 \text{ m})(-0.866) = \boxed{-30.1 \text{ m}}$

$B_y = B \sin \theta_B = (34.8 \text{ m})(\sin 150°)$

$= (34.8 \text{ m})(0.500) = \boxed{17.4 \text{ m}}$

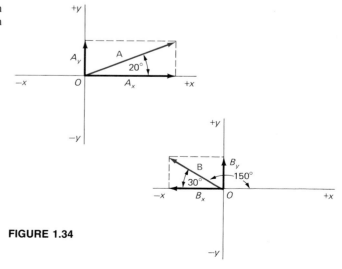

FIGURE 1.34

(b) $R_x = A_x + B_x = (49.4 - 30.1)\ \text{m} = 19.3\ \text{m}$

$R_y = A_y + B_y = (18.0 + 17.4)\ \text{m} = 35.4\ \text{m}$

$R = [R_x^2 + R_y^2]^{1/2}$

$\quad = [(19.3\ \text{m})^2 + (35.4\ \text{m})^2]^{1/2} = \boxed{40.3\ \text{m}}$

$\tan \theta_R = \dfrac{R_y}{R_x} = \dfrac{35.4}{19.3} = 1.83 \qquad \theta_R = \boxed{61.4°}$

This is the angle made by **R** with the positive x axis.

(c) The new rectangular axes are shown in Fig. 1.35, with the vector **A** along the positive x' axis. Then **B** is at an angle of

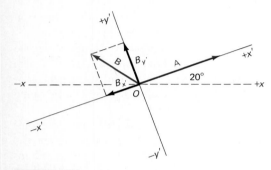

FIGURE 1.35

$150° - 20° = 130°$ with respect to the positive x' axis. The $x'y'$ coordinate system yields simpler results than the xy system, since $A_{y'}$ is zero in the new system. We have

$A_{x'} = 52.6\ \text{m} \cos 0° = 52.6\ \text{m}$

$B_{x'} = 34.8 \cos 130° = -22.3\ \text{m}$

$A_{y'} = 52.6\ \text{m} \sin 0° = 0$

$B_{y'} = 34.8\ \text{m} \sin 130° = 26.7\ \text{m}$

(d) $R_{x'} = A_{x'} + B_{x'} = 30.3\ \text{m}$
$\quad\;\; R_{y'} = A_{y'} + B_{y'} = 26.7\ \text{m}$

$$R = [R_{x'}^2 + B_{y'}^2]^{1/2} = \boxed{40.3\ \text{m}}$$

as found in (b).

$\tan \theta_R = \dfrac{R_{y'}}{R_{x'}} = 0.881 \qquad \theta_R = \boxed{41.4°}$

This is $41.4°$ with respect to the x' axis, which was in the direction of vector **A**. The direction of **R** with respect to the original x axis is then $41.4° + 20° = 61.4°$, as found in part (b). The same result is therefore found for the direction of the resultant with respect to the original vectors whether we use the unprimed or the primed system of coordinates. This must be the case, since the resultant vector cannot depend on the coordinate system chosen but only on the direction of the vectors **A** and **B**.

A Note about Angles and Hand Calculators

In Example 1.12 vector **B** has $\theta_B = 150°$, which puts it in the second quadrant in Fig. 1.34. In this quadrant B_x is negative (in the direction of the negative x axis), and B_y is positive, as shown in the figure. These components can be found in two different ways. We can merely use the definitions of B_x and B_y, as we did in Example 1.12a, and obtain

$B_x = B \cos 150° = 34.8\ \text{m}\ (-0.866) = -30.1\ \text{m}$

$B_y = B \sin 150° = 34.8\ \text{m}\ (0.500) = 17.4\ \text{m}$

Another approach is to look at the diagram and see that since B_x is negative and B_y positive for angles in the second quadrant, we have

$B_x = -B \cos 30° = -34.8\ \text{m}\ (0.866) = -30.1\ \text{m}$

$B_y = +B \sin 30° = 34.8\ \text{m}\ (0.500) = 17.4\ \text{m}$

The same result is found by either method, and you should use the method with which you feel more comfortable.

Hand calculators accept angles of any size and sign and faithfully produce the correct values for sines, cosines, and tangents of these angles. *The same is not true of the inverse process.* A hand calculator will only give the correct angle corresponding to a particular value of the sine, cosine, or tangent *if that angle is in the first or fourth quadrant.* The signs of the x and y components of four vectors in each of the four quadrants are shown in Fig. 1.36.

As an example, suppose that we find as the answer to a problem that a resultant displacement vector has rectangular components $R_x = -2.0\ \text{cm}$ and $R_y = 4.0\ \text{cm}$

FIGURE 1.36 The signs of the x and y components of four vectors, one in each of the four quadrants.

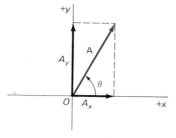

(a) First quadrant:
$0 < \theta < 90°$

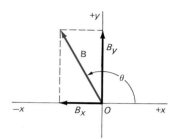

(b) Second quadrant:
$90° < \theta < 180°$

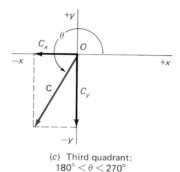

(c) Third quadrant:
$180° < \theta < 270°$

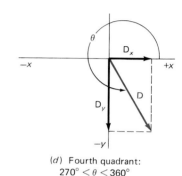

(d) Fourth quadrant:
$270° < \theta < 360°$

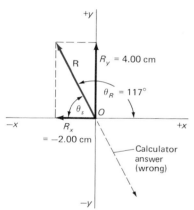

FIGURE 1.37 Finding the correct angle for a resultant in the second quadrant. The dashed vector is drawn in the wrong direction obtained from a hand calculator.

so that $\tan \theta_R = -2.0$. If we blindly find the angle whose tangent is -2.0 on a hand calculator, we obtain $\theta_R = -63°$. But this is clearly wrong: this is an angle in the fourth quadrant for which R_x is plus and R_y is minus, as in Fig. 1.36d, just the opposite of our case where R_x is minus and R_y is plus, as in Fig. 1.36b. The correct result is actually 117°, since θ_s (the supplementary angle to θ_R) in Fig. 1.37 has a tangent given by $\tan \theta_s = 4.0/2.0 = 2.0$, and so $\theta_s = 63°$. Then $\theta_R = 180° - 63° = 117°$.

The calculator therefore gave us a resultant displacement in a direction opposite that of the actual displacement. This suggests that we should resist placing excessive reliance on a hand calculator. If we understand the physical situation, we will be better prepared to reject a calculator's answers in the few cases in which it yields erroneous results.

Summary: Important Definitions and Equations

Physics: The science of matter and energy, and of the relations between them.

Hypothesis: A tentative assumption for the sake of argument to explain data obtained from observation and experiment.

Physical theory: An explanation, perhaps in mathematical terms and based on a physical model, that makes understandable the data obtained from observation and experiment.

Physical law: A physical theory that agrees so well with experimental observations that it seems to reflect the constant behavior of nature.

Principles of physics: "Superlaws," so built into the structure of nature that no exceptions are ever possible. These include the principles of conservation of energy, of linear momentum, and of angular momentum.

Fundamental quantities of mechanics: Those basic quantities in terms of which all other physical quantities in mechanics can be expressed.

Quantity	SI unit
Time	second (s)
Length	meter (m)
Mass	kilogram (kg)

Measurement of angles: The value of the angle θ in *radians* subtended by an arc s of a circle of radius R is

$$\theta = \frac{s}{R}, \tag{1.1}$$

One radian: The angle subtended by an arc equal to the radius of the circle. One radian (rad) = 57.3°. A radian is a dimensionless number.

Dimension: The physical property described by the words time, length, or mass is called a dimension. This property is the same no matter in what units it is expressed.

Dimensional analysis: A technique for establishing the validity of a solution to a problem, a unit, or an equation by checking for dimensional consistency.

Accuracy of physical data

Significant figures (or significant digits): Digits given in reporting the results of an experiment or in stating the data for a problem to be solved. A small number of significant figures indicates low accuracy; a large number of significant figures, high accuracy.
   Rules for handling significant figures:
   1   The final result of an addition or subtraction of two or more quantities should not contain more decimal places than the quantity with the smallest number of decimal places.
   2   The final result of a multiplication or division should have only as many significant digits as the quantity in the calculation with the smallest number of significant digits.

Scalars and vectors:

Scalar: A physical quantity that has no direction and is completely specified by its magnitude alone.

Vector: A physical quantity that is completely specified only when both its magnitude and its direction are given. A vector is often represented by a directed line segment (arrow) whose length represents the vector's magnitude and whose direction shows the vector's direction.
   Position vector: A vector drawn from the origin of some coordinate system to a point in space to indicate the position of an object with respect to the origin.

Displacement vector: A directed line segment (arrow) whose length indicates the magnitude of the displacement and whose direction is the direction of the displacement. A displacement vector is the vector difference between two position vectors. *Displacement* is a *vector: distance* is a *scalar*.

Graphical methods for vectors:
   Addition of vectors: $\mathbf{R} = \mathbf{A} + \mathbf{B}$, where the boldface type means that we are dealing with vector addition, and so both the magnitude and the direction of $\mathbf{A}$ and $\mathbf{B}$ must be used to obtain $\mathbf{R}$.
   Triangle method: If $\mathbf{A}$ and $\mathbf{B}$ are drawn to scale with the tail of $\mathbf{B}$ at the tip of $\mathbf{A}$, then $\mathbf{R}$ is the vector from the tail of $\mathbf{A}$ to the tip of $\mathbf{B}$.
   Tail-to-tip method (or polygon method): An extension of the triangle method from two to more than two vectors.
   Subtraction of vectors:

$$\mathbf{A} - \mathbf{B} = \mathbf{A} + (-\mathbf{B}) \tag{1.3}$$

Analytical methods for vectors
   Rectangular components of a vector $\mathbf{A}$: The magnitudes A_x and A_y (with the proper signs) of the projections of the vector $\mathbf{A}$ on the x and y axes of a chosen coordinate system. If θ is the angle between $\mathbf{A}$ and the positive x axis, then

$$A_x = A \cos \theta \qquad A_y = A \sin \theta \tag{1.4}$$

$$\text{and} \qquad A = [A_x^2 + A_y^2]^{1/2} \tag{1.5}$$

$$\tan \theta = \frac{A_y}{A_x} \tag{1.6}$$

Adding vectors by rectangular resolution:

$$R_x = \sum_{i=1}^{N} (A_i)_x = \sum_{i=1}^{N} A_i \cos \theta_i \tag{1.7}$$

$$R_y = \sum_{i=1}^{N} (A_i)_y = \sum_{i=1}^{N} A_i \sin \theta_i \tag{1.7}$$

$$R = [R_x^2 + R_y^2]^{1/2} \tag{1.8}$$

$$\tan \theta_R = \frac{R_y}{R_x} \tag{1.9}$$

Questions

   1   How does physics differ from philosophy? Which would you say leads to more certain results? Why?
   2   (*a*)   How does physics differ from a science like astronomy? Is there much overlap between these two?
   (*b*)   How does physics differ from astrology? Does astrology have any experimental evidence against which to check its predictions?

   3   Which would you expect to be greater, the contributions made by biology to physics or the contributions made by physics to biology? Why?
   4   Physics has led to the development of a large number of precision instruments of great usefulness to chemists, biologists, and medical scientists, such as electron microscopes, nuclear magnetic resonance (NMR) spectrometers, mass spectrom-

eters, infrared spectrometers, and x-ray diffraction units.

(a)   Is it necessary for a biologist or chemist to understand all the physics behind the working of such instruments in order to use the instruments effectively in research?

(b)   Is it helpful if the biologist or chemist understands at least the basic physical principles behind the working of such instruments?

5   Why is biology in at least one respect a much more difficult science than physics?

6   François de La Rochefoucauld (1613–1680) once wrote: "One of the tragedies of life is the murder of a beautiful theory by a brutal gang of facts." Comment on this statement as it applies to the progress of the science of physics.

7   In a speech delivered on receiving the 1923 Nobel Prize in physics, the American physicist Robert A. Millikan made the following comment:

The fact that Science walks forward on two feet, namely, theory and experiment, is nowhere better illustrated than in the two fields for slight contributions to which you have done me the great honor of awarding me the Nobel Prize in Physics for the year 1923. Sometimes it is one foot which is put forward first, sometimes the other, but continuous progress is only made by the use of both—by theorizing and then testing, or by finding new relations in the process of experimenting and then bringing the theoretical foot up and pushing it on beyond, and so on in unending alternations.

Discuss how well Millikan's statement describes the progress of physics.

8   Do you think that physics will ever be "complete" in the sense that there will no longer be any need for theoretical and experimental research in physics?

9   One of Enrico Fermi's favorite questions on Ph.D. oral examinations in physics was "How many piano tuners are there in New York City?" How would you go about providing an order-of-magnitude answer to this question? (Looking up the number listed in the New York City phone book is against the rules of the game!)

10   Ernest Rutherford, the great nuclear physicist, once wrote: "The value of any working theory depends upon the number of experimental facts it serves to correlate, and upon its power of suggesting new lines of work." Discuss the validity of this statement with respect to the development of a science like physics.

11   (a)   Why is it possible in solving problems involving displacement vectors to move a vector **B** parallel to itself to a new position so as to add it to a vector **A**?

(b)   Can you suggest a situation (not involving displacement vectors) in which a change in the position of a vector would produce a different physical result from that produced by the vector in its original position, even though the magnitude and direction of the vector remained the same?

12   Name a few scalar quantities. Is a scalar quantity dependent on the coordinate system used?

13   Name a few vector quantities. Is a vector quantity dependent on the coordinate system used?

14   Is the sum of two vector displacements dependent on the coordinate system used?

15   Two vectors **A** and **B** have different magnitudes, with A greater than B. Can these two vectors be combined vectorially so that:

(a)   The resultant has a magnitude $2A$?

(b)   The resultant has a magnitude $2B$?

(c)   The resultant is zero?

16   Why is it always possible to replace a vector **A** by rectangular components A_x and A_y along any two mutually perpendicular axes?

Multiple-Choice and Simple Exercises

Note: The numerical problems in this book are broken down into two categories:

1 *Multiple-choice and simple exercises* are fairly simple problems involving only one equation or idea. Some of the multiple-choice questions are qualitative and are included to make sure that the student really understands important points in the text.

2 *Problems* are somewhat more difficult questions requiring the ability to reason, combine equations and ideas, and apply important ideas developed in the chapter. Some of these are more difficult than the others and are marked with asterisks (∗). These require more involved reasoning or deeper understanding of the physics involved. Some are derivations that require considerable use of algebra; others require relating the material in the chapter under study to earlier chapters in the book.

1.1   The number of seconds in a year is:
(a)   3600   (b) 8.64×10^4 (c) 3.15×10^4
(d) 3.15×10^7 (e) 8.64×10^7

1.2   A speed of 40 mi/h is equivalent to:
(a)   34.0 m/s   (b)   10.0 m/s   (c)   340 m/s
(d)   17.9 m/s   (e)   89.6 m/s

1.3   The number 0.03210 has the following number of significant figures:
(a)   6   (b)   5   (c)   4   (d)   3
(e)   None of the above

1.4   The number of significant figures in pi (π), the ratio of the circumference to the diameter of a circle, is:
(a)   3   (b)   5   (c)   2   (d)   4
(e)   As many as you want to calculate

1.5   A metal block of height 3.75 cm is resting on a metal shim whose thickness is known accurately to be 0.1250 cm. The height of the combination of shim and metal block is:
(a)   3.875 cm   (b)   3.8750 cm   (c)   3.9 cm
(d)   3.88 cm   (e)   None of the above

1.6   The result of the calculation

$$\left(\frac{3.98 \times 10^5}{6.31 \times 10^{-3}} \right) (5.6 \times 10^{-2}) \text{ is:}$$

(a) 3.5322×10^6 (b)   3.53   (c) 3.53×10^4
(d) 3.5×10^6 (e) 3.5×10^{-6}

1.7 The sum of 3.89×10^3 and 2.56×10^5 is:
(a) 2.60×10^5 (b) 2.5989×10^5
(c) 2.60×10^3 (d) 2.5989×10^3
(e) 6.45×10^8

1.8 The product of 6.915×10^2 and 2.16×10^{-3} is:
(a) 1.4936 (b) 1.49×10^{-1}
(c) 1.4936×10^{-1} (d) 14.9 (e) 1.49

1.9 Two displacement vectors represent displacements of 7.00 and 5.00 m. One resultant that cannot be obtained from the vector addition of these two displacements is:
(a) 2.00 m (b) 12.0 m (c) 7.00 m
(d) 1.00 m (e) 4.00 m

1.10 A displacement vector **A** has a magnitude of 6.00 m and is directed along the positive x axis. A second displacement vector **B** has a magnitude of -5.00 m along the same axis; i.e., it is directed along the negative x axis. The vector **B** − **A** is then:
(a) 11.0 m (b) 1.00 m (c) -1.00 m
(d) -11.0 m (e) None of the above

1.11 An airplane flies 400 km due east, makes a right-angle turn, and then flies 300 km due north. The magnitude of the plane's displacement from its starting point is:
(a) 700 km (b) 100 km (c) 350 km
(d) 500 km (e) 200 km

1.12 How many degrees correspond to 2.55 rad?

1.13 An angle of 245° is equivalent to how many radians?

1.14 A circle has a radius of 1.20 m. The length of an arc on the circumference of this circle is 0.320 m. What is the angle subtended by this arc in (a) radians, (b) degrees, and (c) revolutions?

1.15 The Washington Monument in the District of Columbia has a height of 555 ft, 5 in. Express this height in meters to four significant figures.

1.16 Express the following distances in meters:
(a) The length of a 100-yd dash.
(b) The length of a marathon run (26 mi, 385 yd).
(c) The distance from Seattle, Washington, to Miami, Florida (3273 mi).

1.17 The elevators in the John Hancock building in Chicago move 900 ft in 30 s. What is this speed in (a) meters per second and (b) miles per hour?

1.18 Measure the length of the line below in meters, centimeters, millimeters, kilometers, feet, and inches. Express all answers in powers-of-10 notation.

1.19 A displacement vector of magnitude 50.0 cm is directed at an angle of 83.0° above the positive x axis. What are the rectangular components of this displacement along the x and y axes?

1.20 A truck travels 1000 m uphill along a road that makes a constant angle of 5.00° with the horizontal. Find the magnitude of the truck's horizontal and vertical components of displacement.

Problems

1.21 Determine the apparent angular diameters of the moon and the sun, as seen from the earth, using the average moon-earth and sun-earth distances given in Table F.2. Show that these two angular diameters are approximately the same for the sun and the moon, as seen from the earth, in agreement with our visual observations. Express your answers in (a) radians and (b) degrees.

1.22 The proton synchrotron at Fermilab in Batavia, Illinois, has a diameter of 1.24 mi.
(a) What is its circumference in meters?
(b) What is the area enclosed by the accelerator ring in square meters?

1.23 If s is a distance and t is a time, what must be the dimensions of the quantities $c_1, c_2, c_3, c_4,$ and c_5 in each of the following equations:
(a) $s = c_1 t$
(b) $s = c_2 t^2$
(c) $s = c_3 \cos (c_4 t)$
(d) $\theta = s/c_5$
(*Hint:* The argument of any trigonometric function, like $c_4 t$ above, must be dimensionless.)

1.24 A 400-m race is run on a circular track of length 800 m, with the runners starting at the north end of the circle and traveling counterclockwise around the circle.
(a) What is the displacement of the finishing line of the race from the starting line?
(b) Is the magnitude of the displacement larger or smaller than the actual distance the runners travel?

1.25 A man walks 1.8 km toward the south, then turns and walks 3.0 km toward the east.
(a) How far did he walk?
(b) What is his displacement from his original position (both magnitude and direction)?

1.26 Carry out the following vector additions and subtractions graphically, where **A** has a magnitude of 5.00 cm and is directed along the positive x axis, **B** has a magnitude of 4.00 cm along the positive y axis, and **C** has a magnitude of 6.00 cm at an angle of 235° above the positive x axis:
(a) **A** + **B** (d) **A** + **B** − **C**
(b) **A** − **B** (e) **A** − **B** − **C**
(c) **A** + **B** + **C**

1.27 Perform the following vector additions and subtractions graphically, where **A**, **B**, and **C** are the displacement vectors shown in Fig. 1.38 and the magnitudes of the vectors are $A = 6.00$ cm, $B = 8.00$ cm, $C = 8.00$ cm.
(a) **A** + **B**
(b) **A** + **B** + **C**
(c) **A** − **B**
(d) **A** + **B** − **C**

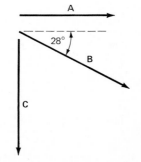

FIGURE 1.38
Diagram for Problem 1.27.

1.28 A girl walks 3.00 km east and then 6.00 km northeast.

(a) What is the total distance she traveled on the trip?

(b) What is her displacement?

(c) If she wants to return to her starting point, what is the required displacement for her return trip?

1.29 A boy bikes 200 m due west. He then changes direction and bikes along a straight road in a different direction. At the end of the second leg he finds he is exactly 150 m northeast of his starting point.

(a) What was the distance traveled on the bike trip along the straight road?

(b) In what direction did the bike move along the straight road?

1.30 In Fig. 1.39, we are given two vectors **A** and **B**. Express the vectors **C**, **D**, **R**, and **P** in terms of **A** and **B**.

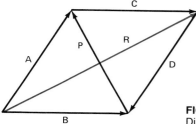

FIGURE 1.39
Diagram for Problem 1.30.

1.31 A box has a length of 5.00 cm, a width of 4.00 cm, and a height of 6.00 cm. A bee starts at one corner of the box and flies in a straight line to the corner diagonally opposite.

(a) What is the magnitude of the displacement vector for the bee's flight?

(b) Could the bee fly along another path that would be shorter than the displacement vector between the two opposite corners of the box?

(c) Could the bee fly along another path that would be longer than the displacement vector between the two corners?

1.32 A helicopter carrying food to a starving tribe in Ethiopia flies 10.0 km due north from its base, lands, and delivers food to the people on the landing field. It then flies 25.0 km in a direction 30.0° N of E.

(a) How far due north of its original position is the final position of the helicopter?

(b) What is the magnitude and direction of its displacement from its original position?

(c) What distance has the plane covered during the whole trip?

1.33 What are the x and y components of a vector displacement of magnitude 25.0 m at an angle of 65.0° above the positive x axis?

1.34 Find the x and y components of a 50.0-m displacement at an angle of 250° with the positive x axis.

1.35 Solve Prob. 1.26 using rectangular components.

1.36 (a) Compute algebraically the resultant of the following displacements: 20.0 m at 30°; 40.0 m at 60°; 30.0 m at 180°, all with respect to the positive x axis.

(b) Check your answer by solving the problem graphically.

1.37 A rectangular building lot has a length of 30.0 m fronting on the street. A surveyor finds, in sighting diagonally across the property, that the line of sight makes an angle of 55.0° with the front edge of the property. What is the area of the building lot?

1.38 A surveyor is attempting to determine the width of the Hudson River at a point north of New York City. She sets up her transit at a point A on the edge of the east bank, and sights directly across the river to a flag placed on the west bank at a point B. She then moves parallel to the bank up the river 500 m to a point C and again sights on the flag at B. She now finds that her line of sight makes an angle of 70.0° with respect to the line from A to C. What is the width AB of the river?

1.39 A vector displacement **A** has a magnitude of 15.0 m and is in a direction 55.0° counterclockwise with respect to the positive x axis.

(a) Find its x and y components.

(b) Repeat for a vector of the same magnitude at an angle of 130°.

(c) Repeat for an angle of 245°.

(d) Repeat for an angle of 285°.

1.40 A displacement of magnitude 2.50 m makes an angle θ with the x axis. Its y component is 1.35 m.

(a) Find the x component of the displacement.

(b) Find the angle θ.

1.41 Consider three vectors **A**, **B**, and **C** with rectangular components $A_x = 15.0$ m, $A_y = 10.0$ m, $B_x = -10.0$ m, $B_y = 20.0$ m, $C_x = -5.00$ m, $C_y = 25.0$ m. Find the magnitude and direction of the vector $\mathbf{A} + \mathbf{B} + \mathbf{C}$, using rectangular resolution. Check your answer graphically.

1.42 Use the vectors in Prob. 1.41 to find $\mathbf{A} + \mathbf{B} - \mathbf{C}$. Again, check your result graphically.

1.43 For the vectors in Prob. 1.41, find the vector **D** such that $\mathbf{D} + \mathbf{A} + \mathbf{B} = 0$, where **0** is the *null vector* (a vector of zero length).

1.44 An ice skater skates 10.0 m southwest, then 20.0 m east, and finally 30.0 m in a direction 50.0° north of east.

(a) Find the rectangular components of each displacement in a suitably chosen coordinate system.

(b) Find the rectangular components of the resultant displacement.

(c) Find the magnitude and direction of the resultant displacement.

(d) Find the displacement required to bring the skater back to her starting point.

*1.45** Find the rectangular components of the vectors **A**, **B**, and **C** given in Prob. 1.41 with respect to a new set of rectangular axes found by rotating the original axes through 65.0° in a counterclockwise direction.

*1.46** Use the new components found for **A**, **B**, and **C** in Prob. 1.45 to find the vector $\mathbf{A} + \mathbf{B} + \mathbf{C}$. Does the resultant vector agree as to magnitude and direction with Prob. 1.41?

1.47 A displacement **R** of 10.0 m from the origin at an angle of 37.0° above the $+x$ axis is the result of three successive displacements: **A**, which is 10.0 m along the negative x axis; **B**, which is 20.0 m at an angle of 150° above the $+x$ axis; and **C**. Find the displacement **C**.

*1.48** What is the displacement of the point on a wheel initially in contact with the ground when the wheel, of radius $R = 30.0$ cm, rolls forward half a revolution? Assume that the wheel is moving along the positive x axis.

1.49 The length of the minute hand on a clock is 10.2 cm.

(*a*) What is the magnitude of the displacement of a point at the very end of the hand between 3:00 P.M. and 3:20 P.M.?

(*b*) What distance was actually traveled by the end of the hand during this time interval?

(*c*) Through what angle did the hand move in this 20-min period?

*1.50 By means of the rules for vector addition, show that for any triangle with sides *a*, *b*, and *c* opposite the angles α, β, and γ, as in Fig. 1.40, the following relations are valid:

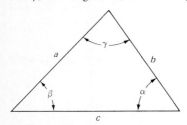

FIGURE 1.40
Diagram for Problem 1.50.

Law of cosines: $c^2 = a^2 + b^2 - 2ab \cos \gamma$

Law of sines: $\dfrac{a}{\sin \alpha} = \dfrac{b}{\sin \beta} = \dfrac{c}{\sin \gamma}$

*1.51 Solve the problem discussed at the beginning of Sec. 1.9 (page 19) and illustrated in Fig. 1.22*a* analytically, using the law of cosines and the law of sines.

*1.52 You are given two displacement vectors and asked to find their resultant. Vector **A** is of magnitude 7.00 cm along the positive *x* axis; vector **B** is of magnitude 5.00 cm at an angle of 55.0° with respect to the positive *x* axis.

(*a*) Find the resultant of these two vectors by using the law of sines and the law of cosines.

(*b*) Check your result, using resolution into rectangular components.

Additional Readings

Astin, Allen V.: "Standards of Measurement," *Scientific American*, vol. 218, no. 6, June 1968, pp. 50–62. A very informative discussion of the standards used for length, mass, time, and temperature, but now a bit dated.

Conant, James B.: *On Understanding Science: An Historical Approach*, Yale University Press, New Haven, Conn., 1951. This book emphasizes the importance of the history of science to our understanding of culture and civilization.

Cottrell, Alan: *Portrait of Nature*, Scribner, New York, 1975. Probably the best single reference for someone wanting to know what science is all about.

Fermi, Laura: *Atoms in the Family*, University of Chicago Press, Chicago, 1954. A delightful, perceptive book by Enrico Fermi's wife, Laura.

Macaulay, David: *The Way Things Work*, Houghton Mifflin, Boston, 1988. An excellent visual guide to the gadgets and machines we use every day, with beautiful, hand-drawn illustrations by the author. It provides fascinating insights into the physical principles behind everything from levers to lasers.

McMahon, Thomas A., and John Tyler Bonner: *On Size and Life*, Scientific American Library, Freeman, New York, 1983. This fascinating book (especially for premedical students) contains a good discussion of dimensional analysis in chapter 3.

Morrison, Philip, Phylis Morrison, and the Office of Charles and Ray Eames: *Powers of Ten*, Scientific American Library, Freeman, San Francisco, 1982. A fascinating trip through the universe in steps of powers of 10.

Ramsey, Norman F.: "Precise Measurement of Time," *American Scientist*, vol. 76, no. 1, January-February 1988, pp. 42–49. An excellent, up-to-date account of how time is measured to high accuracy. The author received the 1989 Nobel Prize in physics for his contributions to the development of the atomic clock.

Schwartz, Clifford: *Prelude to Physics*, Wiley, New York, 1983. A more extended treatment of many of the topics treated in this chapter and in the appendices to the present text, with emphasis on getting the student past the most common stumbling blocks in the introductory physics course. Particularly good on graphical techniques.

Segrè, Emilio: *Enrico Fermi, Physicist*, University of Chicago Press, Chicago, 1970. An excellent biography by a friend and scientific colleague of Enrico Fermi.

Weinberg, Steven: *The First Three Minutes* (updated version), Basic Books, New York, 1988. This contains the modern scientific view of the origin of the universe as presented in a remarkably clear fashion by a Nobel Prize physicist. It is highly recommended for reading at this time.

CHAPTER 2

Rectilinear Motion

Philosophy is written in that great book—I mean the universe—that forever stands open before our eyes, but you cannot read it until you have first learned to understand the language and recognize the symbols in which it is written. It is written in the language of mathematics and its symbols are triangles, circles, and other geometrical figures without which one does not understand a word, without which one wanders through a dark labyrinth in vain.

Galileo Galilei (1564–1642)

The great British historian Herbert Butterfield, in his *Origins of Modern Science* (1965), aptly indicates the importance of our study of motion: "Of all the intellectual hurdles which the human mind has been faced with and has overcome in the last fifteen hundred years, the one which seems to me to have been the most amazing in character, and the most stupendous in the scope of its consequences, is the one relating to the problem of motion."

From the time of Aristotle (384–322 B.C.) to that of Galileo, learned people believed and taught some very wrong ideas about the motion of inanimate objects. Such objects were thought to be intrinsically heavy or light and to fall or rise with a speed proportional to their heaviness or lightness, since they always sought their "natural places" in the universe. It was only when Galileo and Newton combined careful observation and experiment with mathematical analysis that a correct view of the motion of objects emerged.

Mechanics is the name given to the study of the motion of objects and the causes of that motion. In this chapter we discuss the motion of objects without considering its causes. This is *kinematics,* one branch of mechanics. The other branch, called *dynamics,* takes up the causes of motion and is discussed in the next chapter.

In this chapter, after an introductory discussion of displacements, velocities, and accelerations in two dimensions, we limit our discussion to rectilinear motion, i.e., motion in a straight line. Examples include the motion of an automobile along a straight superhighway and the motion of an electron down the straight 2-mile-long evacuated tube of the Stanford linear accelerator at Palo Alto, California. The ideas to be developed here are as important for our understanding of the world of automobiles and airplanes as they are for comprehending the world of the atom. They form a foundation for the imposing structure we call physics.

2.1 Frames of Reference

In this chapter we discuss the position and velocity of various moving objects. To do this scientifically we must first answer the questions "Position with respect to what?" and "Velocity with respect to what?" For example, the elevation of a book on a shelf is quite different if measured with respect to the upper floor of a two-story house than if measured with respect to the ground outside the house.

Similarly, if a person walks forward inside a train at 2 m/s with respect to the train while the train is moving at 25 m/s with respect to the ground, the person's velocity with respect to the ground is 27 m/s. To state the person's velocity without specifying in what reference frame that velocity is being measured has no scientific meaning.

When we discuss motion on the surface of the earth, it is convenient to take the earth's surface as our frame of reference. For the motion of the planets, the sun is a good choice, whereas for the motion of the electrons in an atom, the nucleus of the atom is preferred. The choice of a frame of reference is an arbitrary one, but in all cases it is necessary to specify the reference frame being used, since all motion is measured relative to that frame.

Whatever reference frame is chosen, a rectangular coordinate system can be set up in that reference frame and all motion measured with respect to the axes of this coordinate system. If the motion is in a straight line, we can set up a coordinate system in such a way that the motion is along the x axis of our coordinate system, whereas in the case of falling objects we can choose a vertical y axis as more apt.

2.2 Velocity Vectors for Motion in a Plane

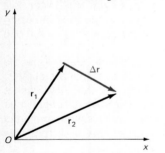

FIGURE 2.1 A displacement vector $\Delta\mathbf{r}$: the vector difference between the position vectors $\mathbf{r}_1$ and $\mathbf{r}_2$.

Average Velocity

Velocity and acceleration are important physical quantities to be discussed in this chapter. We choose to introduce them as vectors, because their vector character will be important in later chapters, such as Chap. 4, which treats motion in more than one dimension. In this chapter, however, we consider rectilinear motion, that is, motion in a straight line; for this kind of motion, the vector character of velocity and acceleration is less important.

By *velocity* we mean the displacement of an object in a given time interval divided by that time interval. In Sec. 1.8 we introduced the *displacement vector* $\Delta\mathbf{r}$ to represent the displacement of an object from one point to another in a plane, where these points are located by position vectors $\mathbf{r}_1$ and $\mathbf{r}_2$ drawn from the origin of the coordinate system, as in Fig. 2.1, so that

$$\Delta\mathbf{r} = \mathbf{r}_2 - \mathbf{r}_1 \tag{2.1}$$

If the object undergoes a displacement $\Delta\mathbf{r}$ in time Δt, then its average velocity during that time interval is defined to be:

Average velocity (a vector): The vector displacement of an object during a time interval divided by that time interval. In the form of an equation,

$$\overline{\mathbf{v}} = \frac{\Delta\mathbf{r}}{\Delta t} = \frac{\mathbf{r}_2 - \mathbf{r}_1}{t_2 - t_1} \tag{2.2}$$

where the bar above the symbol means "average value."

Since displacement is a vector and time is a scalar, the average velocity is a

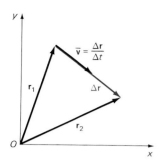

FIGURE 2.2 Average velocity (a vector). Its magnitude is $\Delta r/\Delta t$, and its direction is that of the displacement vector $\Delta\mathbf{r}$.

vector. Its direction is that of the displacement vector $\Delta\mathbf{r}$, and its magnitude depends on how rapidly the object changes position, as in Fig. 2.2.

The dimensions of velocity are $[L]/[T]$, and in the SI system the units are meters per second (m/s).

Constant (uniform) *velocity* is just a special case of average velocity, in which there is no change in either the magnitude or the direction of the velocity over the entire path of the motion.

By *average speed* we mean the total distance traveled during a time interval divided by that time interval. Speed is a scalar quantity, since it is the ratio of one scalar, the distance traveled (which can be in any direction), to another scalar, the time. Since speed is a scalar, we write the average speed as $\bar{v}$.

$$\text{Average speed } (\bar{v}) = \frac{\text{total distance traveled}}{\text{total time elapsed}}$$

Notice that the average speed is not equal to the magnitude of the average velocity, except in very special cases.

In everyday usage, people make no distinction between *velocity* and *speed*. Physicists, however, use these words in a very precise way. The basic difference is that velocity is a *vector*, whereas speed is a *scalar*. Keep this in mind in what follows.

Example 2.1

A car travels north along a straight road for a distance of 200 m, then turns right at an intersection and travels 100 m east along another straight road. The total time elapsed for the trip is $\frac{1}{2}$ min. **(a)** What is the car's average velocity for the trip? **(b)** What is its average speed?

2699032

SOLUTION

The car's path is diagramed in Fig. 2.3.
(a) The average velocity is a vector. We must therefore find both its magnitude and its direction. From the figure, the magnitude of the car's displacement is

$$\Delta r = [(200\text{ m})^2 + (100\text{ m})^2]^{1/2} = 224\text{ m}$$

and its direction is given by

$$\tan\theta = \frac{100}{200} = 0.500 \qquad \theta = 26.6° \text{ east of north}$$

The average velocity is

$$\bar{\mathbf{v}} = \frac{\Delta\mathbf{r}}{\Delta t} = \frac{224\text{ m}}{30.0\text{ s}} = \boxed{7.46\text{ m/s}}$$

at an angle $\boxed{26.6° \text{ east of north}}$.

(b) The average speed is the total distance traveled (irrespective of the direction) divided by the time elapsed:

$$\bar{v} = \frac{\text{total distance}}{\text{time elapsed}} = \frac{200\text{ m} + 100\text{ m}}{30.0\text{ s}} = \boxed{10.0\text{ m/s}}$$

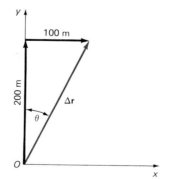

FIGURE 2.3

Notice that unless the direction of the *average velocity* is included in the answer to part (*a*), the answer is *wrong*. Velocity is a vector, and both its *magnitude* and *direction* must always be specified.

Instantaneous Velocity

Suppose a car starts from rest and the driver pushes down on the gas pedal so that the car changes its velocity along a straight road from 0 to 25 m/s over a 10-s period. In this case the velocity at any instant (the *instantaneous velocity* **v**) is continually increasing in magnitude. Suppose we want to know the velocity at some particular instant of time. We can obtain it by evaluating the average velocity over a very, very, short time interval Δt that includes the particular time t at which the velocity is desired. We define the instantaneous velocity as follows:

Instantaneous velocity (the velocity at a particular instant of time): The limit of the average velocity over a time interval that approaches zero but always includes the desired instant of time.

In the form of an equation,

$$\mathbf{v} = \lim_{\Delta t \to 0} \frac{\Delta \mathbf{r}}{\Delta t} \tag{2.3}$$

where we read the right side of this equation as "the limit as Δt approaches zero of $\Delta \mathbf{r}/\Delta t$."

The *instantaneous speed* (a scalar) is the magnitude of the instantaneous velocity at some particular instant of time. The speedometer on a car measures the instantaneous speed. The direction of the instantaneous velocity is the direction of the car's motion at that instant of time. If the car moves at a constant velocity, both its speed and its direction do not change.

In what follows the term *velocity* will be used to mean "instantaneous velocity" unless some other velocity is specifically mentioned. We will clarify the meaning of instantaneous velocity further when we discuss rectilinear motion.

Addition of Velocities

Velocities are vectors and must be added as vectors. As we have seen, velocities are specified with respect to a reference frame. For example, consider a passenger walking along the aisle of a moving train. If we take the train as the frame of reference, then the passenger's velocity may be represented by the vector $\mathbf{v}_{pt}$, where the subscript means "passenger with respect to train." But if the train has a velocity $\mathbf{v}_{tg}$ with respect to the ground as reference frame, then the passenger's velocity with respect to the ground is

$$\mathbf{v}_{pg} = \mathbf{v}_{pt} + \mathbf{v}_{tg} \tag{2.4}$$

This is shown in Fig. 2.4. Since these are instantaneous velocities, Eq. (2.4) is therefore true at any instant of time, no matter how the passenger or the train moves.

The equation above represents a *vector addition*, and the same methods are used to add velocities as were introduced for displacements in Chap. 1.

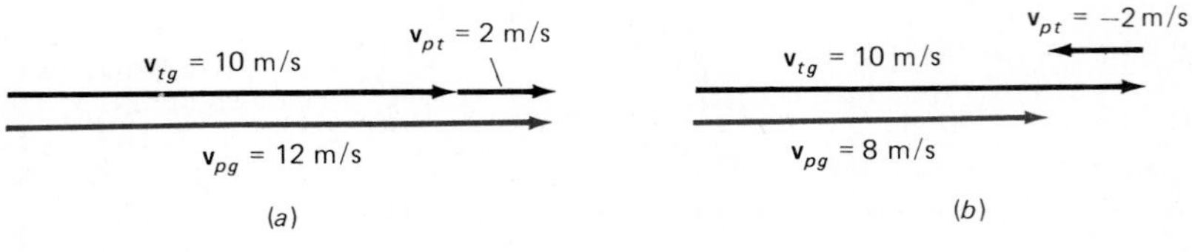

FIGURE 2.4 Addition of two velocities in one dimension. (a) Passenger on train is walking in same direction in which train is moving. (b) Passenger is walking in opposite direction to that of train's motion. Note the difference in the resultant velocity of the passenger with respect to the ground.

Example 2.2

A rower starts out from a dock at the edge of a river. He tries to row directly across the river perpendicular to the riverbank. There is a constant current in the river directed downstream and equal to 1.10 m/s with respect to the riverbank. The man rows at a constant speed of 2.25 m/s with respect to the water. **(a)** What is the magnitude of the man's velocity with respect to the riverbank? **(b)** What is his direction with respect to the riverbank?

SOLUTION

The physical situation is diagramed in Fig. 2.5. We let $\mathbf{v}_{mw}$, $\mathbf{v}_{wb}$, and $\mathbf{v}_{mb}$ represent the instantaneous velocities of the man with respect to the water, the water with respect to the riverbank, and the man with respect to the riverbank.

Since both $\mathbf{v}_{mw}$ and $\mathbf{v}_{wb}$ are constant, all three velocities are constant for the trip, and we have

$$\mathbf{v}_{mb} = \mathbf{v}_{mw} + \mathbf{v}_{wb},$$

(a) The magnitude of the man's velocity is, by vector addition,

$$v_{mb} = [(v_{mw})^2 + (v_{wb})^2]^{1/2}$$
$$= [(2.25 \text{ m/s})^2 + (1.10 \text{ m/s})^2]^{1/2} = \boxed{2.50 \text{ m/s}}$$

(b) The direction of the man's velocity is given by

$$\tan \theta = \frac{v_{wb}}{v_{mw}} = \frac{1.10}{2.25} = 0.488 \qquad \theta = \boxed{26.1°}$$

The actual velocity of the rower with respect to the riverbank at any instant is therefore 2.50 m/s at an angle of 26.1° with respect to a line perpendicular to the riverbank.

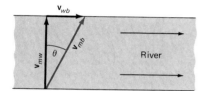

FIGURE 2.5

2.3 Acceleration Vectors for Motion in a Plane

If a car starts from rest and is gradually picking up speed, we say that the car is *accelerating*. By this we mean that its velocity is changing in time. *Acceleration is the time rate of change of velocity.*

Average Acceleration

Suppose that at time t_1 the velocity is specified by the velocity vector $\mathbf{v}_1$, whereas at some later time t_2 the velocity vector has a new value $\mathbf{v}_2$. Then we define the average acceleration as follows:

Average acceleration (a vector quantity): The change in the velocity of an object during a given time interval divided by that time interval. In the form of an equation,

$$\bar{\mathbf{a}} = \frac{\mathbf{v}_2 - \mathbf{v}_1}{t_2 - t_1} = \frac{\Delta \mathbf{v}}{\Delta t} \tag{2.5}$$

where $\Delta \mathbf{v}$ (the change in v) is equal to $\mathbf{v}_2 - \mathbf{v}_1$. Since $\Delta \mathbf{v}$ is a vector and Δt a scalar, $\bar{\mathbf{a}}$ is a vector whose direction is the same as that of $\Delta \mathbf{v}$. If the velocity retains the same direction but decreases in value, then $\bar{\mathbf{a}}$ is negative, meaning that it is in the opposite direction to the original $\mathbf{v}$.

Since the SI units for $\Delta \mathbf{v}/\Delta t$ are $(\text{m/s})/\text{s} = \text{m/s}^2$, these are the units of average acceleration. We read this either as "meters per second per second" or as "meters per second squared." The first reading makes clear the nature of an acceleration: it is the change in a velocity (in meters per second) in an interval of time (in seconds).

Example 2.3

A car is heading east at 20.0 m/s. It accelerates for 20.0 s to pass a truck, and at the end of that time interval it has a velocity of 30.0 m/s in its initial direction. What is its average acceleration for the 20.0-s interval?

SOLUTION

In this case the direction of the velocity does not change, only its magnitude. We do not know how the acceleration may have varied during the 20-s time interval, but we do not need to know this to obtain the *average acceleration*, which is

$$\bar{\mathbf{a}} = \frac{\Delta \mathbf{v}}{\Delta t} = \frac{\mathbf{v}_2 - \mathbf{v}_1}{t_2 - t_1} = \frac{30.0 \text{ m/s} - 20.0 \text{ m/s}}{20.0 \text{ s}}$$

$$= \boxed{0.500 \text{ m/s}^2}$$

This is the magnitude of the acceleration; its direction is *east*. The reason for this is that $\Delta \mathbf{v}$ is positive, since $\mathbf{v}_2$ is greater than $\mathbf{v}_1$. Therefore $\bar{\mathbf{a}}$ is also positive and in the same direction as $\mathbf{v}_1$, which was *east*. In this case Eq. (2.5) leads to $\mathbf{v}_2 = \mathbf{v}_1 + \bar{\mathbf{a}} \, \Delta t$, as shown in Fig. 2.6a.

If, however, the car were slowing down, its acceleration would be negative, since $\mathbf{v}_2$ would be less than $\mathbf{v}_1$ in magnitude. In this case $\bar{\mathbf{a}}$ is in the opposite direction to both $\mathbf{v}_1$ and $\mathbf{v}_2$, as in Fig. 2.6b, and the direction of the acceleration is *west*.

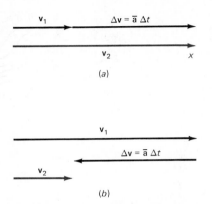

(a)

(b)

FIGURE 2.6 Change in the *magnitude* of the velocity from v_1 to v_2, with no change in direction, resulting in accelerated motion. (a) The acceleration is in the same direction as $\mathbf{v}_1$; (b) the acceleration is in the opposite direction.

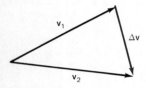

FIGURE 2.7 Change in the *direction* of the velocity from that of $\mathbf{v}_1$ to $\mathbf{v}_2$, with no change in magnitude, resulting in accelerated motion.

There is another way the velocity may change. The magnitude of the velocity may remain the same but its direction may change, as in Fig. 2.7. Here $\mathbf{v}_1$ and $\mathbf{v}_2$ are vectors of the same length so that the magnitude of the velocity does not change in the time Δt. Despite this fact, if the direction changes, we have from the vector diagram,

$$\mathbf{v}_2 = \mathbf{v}_1 + \Delta \mathbf{v} \qquad \text{or} \qquad \Delta \mathbf{v} = \mathbf{v}_2 - \mathbf{v}_1$$

Therefore the average acceleration $\bar{\mathbf{a}} = \Delta \mathbf{v}/\Delta t$ is not zero, and we have accelerated motion. In this case the direction of $\bar{\mathbf{a}}$ is the direction of $\Delta \mathbf{v}$, which is not the direction of either $\mathbf{v}_1$ or $\mathbf{v}_2$.

This is the situation, for example, with a car going around a curve at constant speed. Its direction is changing and so is its velocity. The motion is therefore *accelerated*, even though the car's speed remains constant.

If a car slows down to go around a sharp curve, then both the magnitude and the direction of the velocity change. Accelerated motion therefore corresponds to changes in either the magnitude of the velocity, its direction, or both.

If an acceleration is *constant* over some time interval, then it is equal to the average acceleration for the same time interval.

Instantaneous Acceleration

If the acceleration is not constant, it changes from one instant of time to another (for example, a driver constantly adjusts the gas pedal to feed the motor more or less gasoline). The acceleration at any instant of time we call the *instantaneous acceleration*. Its definition is similar to our definition of instantaneous velocity:

Instantaneous acceleration (the acceleration at a particular instant of time): The limit of the average acceleration over a time interval that approaches zero but always includes the desired instant of time.

$$\mathbf{a} = \lim_{\Delta t \to 0} \frac{\Delta \mathbf{v}}{\Delta t} \tag{2.6}$$

Here $\Delta \mathbf{v}/\Delta t$ is the rate at which the *instantaneous velocity* changes in time.

The units of instantaneous acceleration are the same as for average acceleration, m/s^2 in the SI system.

Instantaneous acceleration is a more useful concept than average acceleration, so when the term *acceleration* is used in this book, it usually means *instantaneous acceleration*. Of course, if the acceleration is constant, the instantaneous acceleration is equal to the constant acceleration (and also to the average acceleration).

In what follows we confine our discussion to several important types of motion in which the acceleration is either constant or varies in some relatively simple way. In Sec. 2.4 we apply the above ideas to motion in a straight line, and in Sec. 2.6 we consider rectilinear motion with constant acceleration.

Adding Accelerations

FIGURE 2.8
The instantaneous acceleration when both the magnitude and the direction of the velocity at point P are changing.

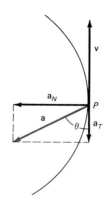

Since accelerations are vectors, they also must be added as vectors. For example, consider a car reducing its speed as it goes around a curve, as in Fig. 2.8. What is its instantaneous acceleration? It consists of two components. The first is parallel to the path of the car at every point and measures the change in the *magnitude* of the velocity with time. Since the car is slowing down, this component of the acceleration is negative and therefore in a direction opposite to that of the velocity. We call this the *tangential component* a_T, since it is tangent to the path of the car at the point of interest. The other component measures the change in the *direction* of the velocity and is perpendicular (or normal*) to the velocity at the point of interest. (We will see more of this in Chap. 4.) The resulting instantaneous velocity can be obtained from these two rectangular components, as shown in the diagram. Its magnitude is given by

$$a = [a_T^2 + a_N^2]^{1/2}$$

and its direction by $\tan \theta = a_N/a_T$.

2.4   Motion in a Straight Line

In the remainder of this chapter we limit ourselves to motion in a straight line (rectilinear motion), like that of a runner in a 50-m dash. This will enable us to apply the somewhat abstract ideas of the last sections to some simple, but important, problems. It will also enable us to introduce some useful graphical techniques.

In Sec. 1.10 we showed that a displacement vector could be replaced by two vector displacements along the x and y axes of a rectangular coordinate system. If the displacement vector is $\Delta \mathbf{r} = \mathbf{r}_2 - \mathbf{r}_1$, these components are $\Delta \mathbf{x}$ and $\Delta \mathbf{y}$.

Because the direction of these two vectors is along the x and y axes, only a single number (the magnitude) is needed to describe each one, since the direction may now be specified simply by a plus or minus sign. The two scalar magnitudes Δx and Δy are then called the *rectangular components* of $\Delta \mathbf{r}$.

---

*By *normal* we mean "perpendicular" or "at right angles." The term derives from the construction trade. In the building of houses, beams must be at right angles to each other, in which case they are referred to as "squared" or "normal." Hence, normal came to mean "at right angles," and this is its meaning in physics.

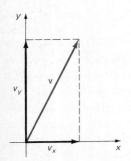

FIGURE 2.9 Rectangular components v_x and v_y of the velocity **v**.

Exactly the same approach may be taken with velocities and accelerations. A velocity **v** in the xy plane can be replaced by two components, one along the x axis, of magnitude v_x, and the other along the y axis, of magnitude v_y, as in Fig. 2.9. Since each of these velocity components is restricted to a straight line, we take care of the direction of v_x by assigning a positive value to v_x if it is in the direction of the positive x axis, and a negative value to v_x if it is in the direction of the negative x axis. Exactly the same approach can be used for the y component of the velocity and for the x and y components of the acceleration. This allows us to drop the vector notation for x and y components.

The vector equations introduced for displacement, velocity, and acceleration in the previous sections can therefore be reduced to component form as follows:

Physical quantity	Vector equation	Equations for rectangular components		
Displacement	$\Delta \mathbf{r} = \mathbf{r}_2 - \mathbf{r}_1$	$\Delta x = x_2 - x_1$	$\Delta y = y_2 - y_1$	(2.7)
Average velocity	$\bar{\mathbf{v}} = \dfrac{\Delta \mathbf{r}}{\Delta t}$	$\bar{v}_x = \dfrac{\Delta x}{\Delta t}$	$\bar{v}_y = \dfrac{\Delta y}{\Delta t}$	
	$= \dfrac{\mathbf{r}_2 - \mathbf{r}_1}{t_2 - t_1}$	$= \dfrac{x_2 - x_1}{t_2 - t_1}$	$= \dfrac{y_2 - y_1}{t_2 - t_1}$	(2.8)
Instantaneous velocity	$\mathbf{v} = \lim\limits_{\Delta t \to 0} \dfrac{\Delta \mathbf{r}}{\Delta t}$	$v_x = \lim\limits_{\Delta t \to 0} \dfrac{\Delta x}{\Delta t}$	$v_y = \lim\limits_{\Delta t \to 0} \dfrac{\Delta y}{\Delta t}$	(2.9)
Average acceleration	$\bar{\mathbf{a}} = \dfrac{\Delta \mathbf{v}}{\Delta t}$	$\bar{a}_x = \dfrac{\Delta v_x}{\Delta t}$	$\bar{a}_y = \dfrac{\Delta v_y}{\Delta t}$	
	$= \dfrac{\mathbf{v}_2 - \mathbf{v}_1}{t_2 - t_1}$	$= \dfrac{v_{2x} - v_{1x}}{t_2 - t_1}$	$= \dfrac{v_{2y} - v_{1y}}{t_2 - t_1}$	(2.10)
Instantaneous acceleration.	$\mathbf{a} = \lim\limits_{\Delta t \to 0} \dfrac{\Delta \mathbf{v}}{\Delta t}$	$a_x = \lim\limits_{\Delta t \to 0} \dfrac{\Delta v_x}{\Delta t}$	$a_y = \lim\limits_{\Delta t \to 0} \dfrac{\Delta v_y}{\Delta t}$	(2.11)

In the following sections we restrict our discussion to motion along the x axis. (Since the x axis can be chosen in any desired direction to facilitate the solution of a problem, this does not really limit us at all.)

Example 2.4

A car is traveling east at a velocity of 25.0 m/s on a straight road. The driver takes his foot off the gas pedal, and the car comes to a complete stop in 12.5 s. What was the average acceleration of the car during that 12.5-s interval?

SOLUTION

Since plus and minus signs determine the direction of the velocity and acceleration along the x axis, it is very important that we be specific about the coordinate system used.

In this case, it seems reasonable to take motion toward the east as in the direction of the positive x axis. Then $v_{1x} = +25.0$ m/s, since the velocity is in the positive x direction, and $v_{2x} = 0$. Then we have

$$\bar{a}_x = \frac{\Delta v_x}{\Delta t} = \frac{v_{2x} - v_{1X}}{t_2 - t_1} = \frac{0 - (+25.0 \text{ m/s})}{12.5 \text{ s}} = \boxed{-2.00 \text{ m/s}^2}$$

Since the average acceleration is negative, its direction is along the negative x axis. The average acceleration is therefore west. In this way it has the *opposite* direction to the initial velocity, and so the car loses velocity and finally stops.

EXERCISE 1 Suppose that the driver, instead of taking his foot off the gas pedal, gives the car more gas, and it acquires a final velocity of 35.0 m/s in a 5.00-s interval. What is the average acceleration in this case? (Note that since acceleration is a vector, both its *magnitude* and its *direction* must always be specified.)

2.5 Graphical Techniques for Straight-Line Motion

In this section we introduce graphical techniques both to clarify the concepts of displacement, velocity, and acceleration and to help us solve problems in rectilinear motion using these concepts. For clarity, we first repeat our definitions of these important physical quantities from above but make them specific with respect to motion in a straight line. Then we indicate how these definitions apply to graphs of the rectilinear motion of an object.

Displacement: A change in the position of an object along the x axis from x_1 to x_2, where $\Delta x = x_2 - x_1$.

Consider a small, pointlike object, which we call a "particle," at a position $x_1 = 2.0$ m from the origin O, as in Fig. 2.10a. If this particle moves to a new position $x_2 = 5.0$ m from O, then the particle's displacement is

$$\Delta x = x_2 - x_1 = (5.0 - 2.0) \text{ m} = + 3.0 \text{ m}$$

since x_2 is greater than x_1. This is a displacement in the direction of the positive x axis, as shown in Fig. 2.10a.

On the other hand, if the particle moves from $x_1 = 6.0$ m to $x_2 = 4.0$ m, then its displacement is $\Delta x = x_2 - x_1 = (4.0 - 6.0) \text{ m} = -2.0$ m, with the minus sign meaning that the displacement is in the direction of the negative x axis, as in Fig. 2.10b.

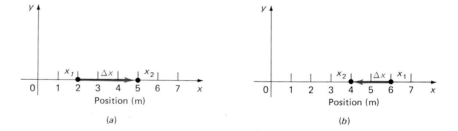

FIGURE 2.10 (a) Positive and (b) negative displacements for rectilinear motion.

(a)

(b)

Example 2.5

A particle's initial position is at $x_1 = -6.00$ cm with respect to the origin of the coordinate system. Its final position is at $x_2 = +4.00$ cm. **(a)** What is its displacement? **(b)** What is the direction of this displacement?

SOLUTION

(a) The displacement is

$$\Delta x = x_2 - x_1 = +4.00 \text{ cm} - (-6.00 \text{ cm}) = \boxed{+10.0 \text{ cm}}$$

(b) Since Δx is positive, the displacement is in the direction of the positive x axis.

Velocity

In Fig. 2.11 we plot the motion of a particular particle along the x axis as a function of time. Time is the abscissa of the graph, and the position of the particle along the x axis is the ordinate. We see that in this case no matter what pairs of values of x and t we choose, the ratio of the displacement Δx to the time interval Δt during which the displacement occurs is constant. In other words, the graph is a straight line with constant slope. But the slope $\Delta x/\Delta t$ is dimensionally a velocity, and in this case it is constant. We therefore have

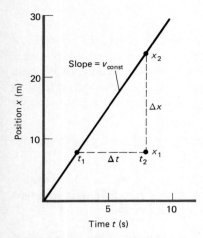

FIGURE 2.11 One-dimensional motion with constant velocity (i.e., direction is fixed and speed is unchanged).

FIGURE 2.12 Rectilinear motion with changing velocity. The instantaneous velocity is equal to the average velocity only at point P.

$$(v_x)_{\text{const}} = \frac{\Delta x}{\Delta t} = \frac{x_2 - x_1}{t_2 - t_1} \tag{2.12}$$

The velocity for rectilinear motion is the slope of the graph of position x plotted against time t. If the graph is a straight line, then the velocity is constant.

The position of a second particle is graphed as a function of time in Fig. 2.12. Since the graph is not a straight line, we see that the velocity is changing with time. In this case the velocity is increasing in magnitude, since the slope of the graph constantly increases. We can calculate the *average velocity* from the graph of the motion.

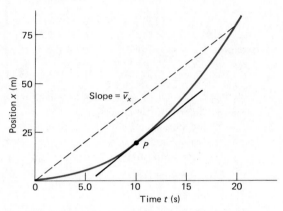

Average velocity: The displacement (Δx) of a particle along the x axis during a time interval (Δt) divided by that time interval.

$$\bar{v}_x = \frac{\Delta x}{\Delta t} = \frac{x_2 - x_1}{t_2 - t_1} \tag{2.13}$$

The average velocity of the particle during the 20-s interval shown is, from the graph,

$$\bar{v}_x = \frac{\Delta x}{\Delta t} = \frac{80 \text{ m} - 0}{20 \text{ s} - 0} = 4.0 \text{ m/s}$$

This is also the slope of the dashed line shown on the graph. The direction of the motion is in the positive x direction, since $\Delta x = x_2 - x_1$ is positive, and so is $\bar{v}_x$.

In this situation the particle's velocity gradually increases from zero to some maximum value much larger than 4.0 m/s during the first 20 s of its motion. The velocity at any instant (the instantaneous velocity v_x) is continually increasing in magnitude and is equal to the average velocity of 4.0 m/s for only a single instant of the 20-s interval. This is at point P, where the straight black line is parallel to the dashed line representing $\bar{v}_x$.

To find the instantaneous velocity for any time during the motion, we use our definition of instantaneous velocity:

Instantaneous velocity: The limit of the average velocity $\bar{v}_x$ over a time interval that approaches zero but always includes the desired instant of time.

$$v_x = \lim_{\Delta t \to 0} \frac{\Delta x}{\Delta t} \tag{2.9}$$

FIGURE 2.13 The instantaneous velocity is the limiting value of the average velocity as the time interval over which the average velocity is evaluated gets smaller and smaller.

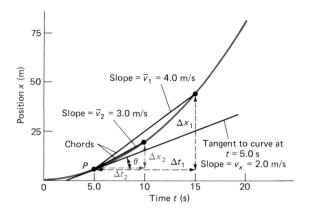

TABLE 2.1 Average Values of the Velocity for Time Intervals near $t = 5.0$ s in Fig. 2.13

Time interval Δt (s)	Displacement Δx (m)	Average velocity $\bar{v}_x = \Delta x / \Delta t$ (m/s)
15.0–5.0	45–5	$\bar{v}_1 = 4.0$
10.0–5.0	20–5	$\bar{v}_2 = 3.0$
7.0–5.0	9.8–5	$\bar{v}_3 = 2.4$
6.0–5.0	7.2–5	$\bar{v}_4 = 2.2$
Limit as $\Delta t \to 0$:		$v_x = 2.0$ m/s

We now apply this definition to Fig. 2.13, which shows the graph of the same motion as in Fig. 2.12.

Suppose we want to know the instantaneous velocity at time $t = 5.0$ s. We obtain the average velocities by calculating the slope of the graph between the time 5.0 s and a series of later times, shortening the time interval Δt involved in each case. We find the results given in Table 2.1 (check the accuracy of the data by your own calculations). We see from the table that as the time interval becomes smaller (as Δt approaches zero), the average velocities get smaller and smaller and also closer and closer together in value. In the limit in which Δt becomes infinitesimally small, the chords drawn to show the slopes approach the tangent to the curve at the point $t = 5.0$ s. This limiting value of the slope is the instantaneous velocity at the time $t = 5.0$ s.

The tangent at any point on a position-against-time graph gives the slope of the graph at that point and therefore the instantaneous velocity v_x at the time corresponding to that point.

On the graph of Fig. 2.13, the tangent to the curve at $t = 5.0$ s is the line of slope v_x. This slope can be obtained from the ratio $\Delta x / \Delta t$ calculated for any two points on this line. For example, taking the values $x = 0$ and $x = 25$ m, we have

$$v_x = \frac{25 \text{ m} - 0}{15 \text{ s} - 2.5 \text{ s}} = 2.0 \text{ m/s}$$

The limit therefore of the series of values obtained for the average velocity for different time intervals $\bar{v}_1 = 4.0$ m/s, $\bar{v}_2 = 3.0$ m/s, $\bar{v}_3 = 2.4$ m/s, $\bar{v}_4 = 2.2$ m/s, and so on, is $v_x = 2.0$ m/s, and this is the instantaneous velocity of the particle at time $t = 5.0$ s. (To avoid crowding the graph, only a few of these average velocities are shown in Fig. 2.13, but you can check for yourself that they do indeed converge to the value 2.0 m/s.)

Example 2.6

Figure 2.14 is a plot of the position of a railway locomotive moving along a straight track. For what time interval is: **(a)** The velocity constant (and not zero)? **(b)** The velocity increasing? **(c)** The velocity decreasing? **(d)** The velocity zero? **(e)** What is the average velocity in the time interval between 15 and 20 s? **(f)** What is the instantaneous velocity at $t = 5.0$ s?

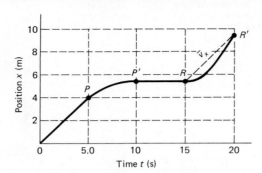

FIGURE 2.14

SOLUTION

(a) The slope of the position-versus-time graph is constant for the interval 0 to 5 s, and so the velocity is constant during this time interval.

(b) In the interval 15 to 20 s, the slope of the curve is increasing from zero to some large value, and so the velocity is also increasing. (If the curve is concave upward, the velocity is increasing.)

(c) In the interval 5 to 10 s, the slope is decreasing from its value at point P to its value at point P', and so the velocity is decreasing. (The curve is concave downward.)

(d) In the interval 10 to 15 s, Δx is zero, and so the velocity is also zero. The train therefore moves at constant speed in the positive x direction between O and P, slows down from P to P', stops for 5 s, and then speeds up again between R and R'.

(e) The average velocity in the time interval between 15 and 20 s is the slope of the chord connecting points R and R'. This is

$$\bar{v}_x = \frac{\Delta x}{\Delta t} = \frac{(9.5 - 5.5) \text{ m}}{(20 - 15) \text{ s}} = \boxed{0.80 \text{ m/s}}$$

(f) At point P the curve is continuous with the straight line between O and P. Hence the instantaneous velocity at P is the same as the constant velocity for the first 5 s. This is

$$v_x = \frac{\Delta x}{\Delta t} = \frac{4.0 \text{ m} - 0}{5.0 \text{ s} - 0} = \boxed{0.80 \text{ m/s}}$$

Acceleration

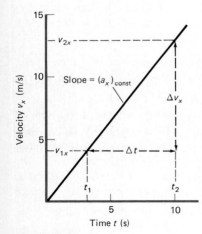

FIGURE 2.15
Graph of the velocity of a particle against time for the first 10 s of its motion when the acceleration is constant.

For rectilinear motion the direction of the velocity cannot change (except to go from the $+x$ to the $-x$ direction), and so the acceleration is always the rate at which the *magnitude* of the velocity changes with time.

In this case, just as the slope of a position-against-time graph yields the velocity v_x, so too the slope of a velocity-against-time graph yields the acceleration a_x. For example, Fig. 2.15 shows a plot of the velocity of a particle against time for the first 10 s of its motion. The graph in this case is a straight line; its slope is therefore constant, and so the acceleration a_x has the same value at all instants of time. The magnitude of this *constant acceleration* is

$$(a_x)_{\text{const}} = \frac{\Delta v_x}{\Delta t} = \frac{v_{2x} - v_{1x}}{t_2 - t_1} \tag{2.14}$$

The velocity of another particle as a function of time is shown in Fig. 2.16. The velocity does not increase linearly with time, and so the acceleration is not constant. We can, however, calculate the *average acceleration* for the motion from the graph.

Average acceleration: The change in the velocity (Δv_x) of an object during a given time interval (Δt) divided by that time interval.

$$\bar{a}_x = \frac{\Delta v_x}{\Delta t} = \frac{v_{2x} - v_{1x}}{t_2 - t_1} \tag{2.10}$$

FIGURE 2.16 Average and instantaneous accelerations. The average acceleration $\bar{a}_x$ in the interval between 5 and 10 s is the slope of the solid straight line a_x. The instantaneous acceleration a_x is the tangent to the curve at the point of interest P, shown here by the dashed straight line.

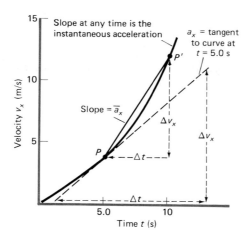

For example, in Fig. 2.16 the velocity increases from 3.8 m/s at point P to 12.0 m/s at point P' during the time interval between 5.0 and 10.0 s. The average acceleration is therefore

$$\bar{a}_x = \frac{(12.0 - 3.8) \text{ m/s}}{(10.0 - 5.0) \text{ s}} = +1.6 \text{ m/s}^2$$

The acceleration is positive, meaning that the velocity is increasing in time. The acceleration is therefore directed along the $+x$ axis, the same direction as the displacement and the velocity. If the velocity were decreasing in time, the acceleration would be negative, directed along the negative x axis.

To find the acceleration at any instant of time during the motion, we use our definition of instantaneous acceleration to calculate its value at that time.

Instantaneous acceleration: The limit of the average acceleration over a time interval that approaches zero but always includes the desired instant of time.

$$a_x = \lim_{\Delta t \to 0} \frac{\Delta v_x}{\Delta t} \tag{2.11}$$

Suppose we want to find the instantaneous acceleration at point P ($t = 5.0$ s) in Fig. 2.16. We can start with points P' and P and calculate the average acceleration for the 5.0-s time interval involved. This leads to the value of 1.6 m/s^2 found above. If we reduce the time interval by bringing P' closer and closer to P, we find that the instantaneous acceleration a_x gradually gets smaller and smaller. Finally when the time interval is a very, very small one in the vicinity of P, the chord connecting P with P' becomes the tangent to the curve at P. This tangent therefore yields the slope of the velocity-against-time curve in the limit as P' approaches P. The tangent at P is equal to a_x and, as found from the graph, is

$$a_x = \frac{(11 - 0) \text{ m/s}}{(13 - 1) \text{ s}} = 0.92 \text{ m/s}^2$$

The acceleration therefore decreases as we come down the curve from P' to P to a limiting value of 0.92 m/s^2. This is the instantaneous acceleration at the time $t = 5.0$ s that corresponds to point P.

Example 2.7

A car is moving along on a straight road at a constant velocity of 20 m/s. The driver presses down on the accelerator and maintains a constant acceleration of 2.0 m/s² for 5.0 s. At the end of this time the driver hears a police siren and is worried that the police are after her for speeding. **(a)** What was the car's final velocity? **(b)** Was the driver exceeding the speed limit of 55 mi/h? **(c)** Draw a velocity-against-time graph for the motion.

SOLUTION

(a) The increase in velocity may be obtained from the expression for constant acceleration:

$$(a_x)_{const} = \frac{\Delta v_x}{\Delta t} = \frac{v_{2x} - v_{1x}}{t_2 - t_1}$$

$$\Delta v_x = (a_x)_{const} \, \Delta t = (2.0 \text{ m/s}^2)(5.0 \text{ s}) = 10 \text{ m/s}$$

The initial velocity of the car was $v_{1x} = 20$ m/s. Its final velocity is, therefore,

$$v_{2x} = v_{1x} + \Delta v_x = 20 \text{ m/s} + 10 \text{ m/s} = \boxed{30 \text{ m/s}}$$

(b) Since the speed limit of 55 mi/h corresponds to 25 m/s, the driver was indeed exceeding the speed limit.

(c) In Fig. 2.17, let us take our zero of time as 2.0 s before the car begins to accelerate so that we can show the initial motion at a constant velocity of 20 m/s. The car accelerates, and the velocity increases at a constant rate to 30 m/s over a 5.0-s interval. At the end of that interval the car is moving at a velocity $v_{2x} = 30$ m/s, as shown on the graph.

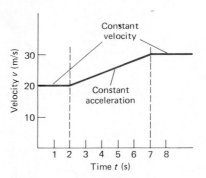

FIGURE 2.17 Diagram for Example 2.7: velocity-against-time graph.

Example 2.8

Figure 2.18 shows the position of a locomotive which moves along a perfectly straight track at a speed that varies with time. Discuss the velocity and acceleration of the locomotive at points A to G.

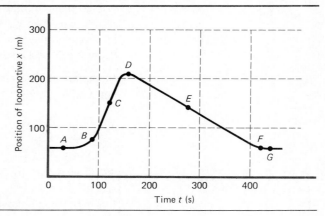

FIGURE 2.18

SOLUTION

We take the track as along the x axis and measure all distances from an origin 0. The essential idea here is that the slope of this graph at any point is the *velocity* of the locomotive at that point.

Point A: Since its position is unchanged as time changes, the locomotive has zero velocity and remains stopped at the position $x_0 = 60$ m with respect to 0.

Point B: Since the slope of the curve is positive, the velocity is increasing from 0 to its constant value at point C as it moves in the positive x direction. Since the slope increases in time, the velocity increases in time, and there is positive acceleration, that is, acceleration in the direction of the motion.

Point C: The locomotive is moving at a constant velocity in the forward direction, since the slope of the curve is constant near C. The magnitude of this velocity is

$$\frac{\Delta x}{\Delta t} = \frac{(200 - 100) \text{ m}}{(140 - 100) \text{ s}} = 2.5 \text{ m/s}.$$

Since the locomotive went from rest to 2.5 m/s in about 40 s, as read from the graph, the average acceleration of the locomotive in the vicinity of B is about

$$\frac{(2.5 \text{ m/s})}{40 \text{ s}} = +6.3 \times 10^{-2} \text{ m/s}^2$$

Point D: Here the velocity of the locomotive becomes zero at a position about 210 m from the origin, for there is no change in x with time at point D. Zero velocity means that the locomotive *stops* and begins to move backward towards its original position. The acceleration is negative, since the velocity continually decreases between points C and E.

Point E: The velocity is now negative (since the slope is negative) and approximately constant. Its value is

$$\frac{\Delta x}{\Delta t} = \frac{(68 - 186)\ \text{m}}{(400 - 200)\ \text{s}} = -0.59\ \text{m/s}$$

The locomotive moves at a slower speed in reverse than it moved forward. Since the velocity is constant, the acceleration is zero.

Point F: The locomotive is slowing down and therefore has a negative acceleration (or "deceleration").

Point G: The locomotive stops when it has returned to its original position, that is, 60 m from the origin. It then has zero velocity and also zero acceleration.

The overall motion of the locomotive described by the graph is that it starts from rest, reaches a constant forward speed of about 2.5 m/s, then slows down, stops, reverses its direction, reaches a constant reverse speed of about 0.59 m/s, slows down again, and finally stops at the same place it started.

Taking the time to understand this graph fully is a very good investment, for it provides physical insights into rectilinear motion not provided by juggling equations.

Example 2.9

Figure 2.19 is a graph of the velocity of a particle against time for a situation in which the acceleration is *not* constant but is increasing with time. From the graph obtain approximate values of the following quantities: **(a)** The instantaneous acceleration at point P, when t = 5.0 s. **(b)** The instantaneous acceleration at point P', when t = 10.0 s. **(c)** The average acceleration in the time interval between points P and P'.

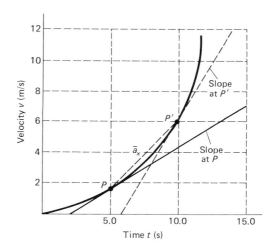

FIGURE 2.19

SOLUTION

(a) The slope of the tangent to the velocity curve at point P is, from the graph,

$$a_x = \frac{v_{2x} - v_{1x}}{t_2 - t_1} = \frac{(7.0 - 0)\ \text{m/s}}{(15 - 2)\ \text{s}} = \boxed{0.54\ \text{m/s}^2}$$

This is the instantaneous acceleration at point P, which corresponds to t = 5.0 s.

(b) The slope of the tangent at point P' is, similarly,

$$a_x = \frac{(13.0 - 0)\ \text{m/s}}{(15.0 - 6.0)\ \text{s}} = \boxed{1.4\ \text{m/s}^2}$$

This is the instantaneous acceleration at time t = 10.0 s.

(c) The average acceleration in the time interval from 5.0 to 10.0 s is the slope of the dashed line marked $\bar{a}_x$ on the graph. This is

$$\bar{a}_x = \frac{(6.0 - 1.8)\ \text{m/s}}{(10.0 - 5.0)\ \text{s}} = \boxed{0.84\ \text{m/s}^2}$$

The average acceleration in this time interval falls between the values of the instantaneous acceleration at P and at P', as expected.

2.6 Uniformly Accelerated Motion in a Straight Line

We now want to apply these ideas about velocity and acceleration to a special kind of motion, *uniformly accelerated motion*, which is *motion in a straight line with an acceleration of constant magnitude*. Such uniformly accelerated motion is simpler to handle than motion with nonconstant acceleration and occurs in a large number of important physical problems.

Position and Average Velocity

We first want to modify slightly the notation we have been using thus far. We take the initial time t_1 in equations such as (2.7) to (2.11) as zero and designate the final time as simply t. The initial position and velocity with respect to a suitable reference frame at time zero are called x_0 and v_0, and the final position and velocity are designated as x and v.* Then, if a car moves with average velocity v from position x_0 to x in the time interval from 0 to t, we have, from Eq. (2.8),

$$\bar{v} = \frac{\Delta x}{\Delta t} = \frac{x - x_0}{t - 0}$$

On solving this equation for x, we obtain

$$x = x_0 + \bar{v}t \tag{2.15}$$

This is the equation of a straight line with slope v and intercept x_0 on the vertical axis when x is plotted as a function of t, as in Fig. 2.20. In many cases, the initial position x_0 may be made zero by the proper choice of the coordinate system in which the object moves. For example, we could choose the origin of our coordinate system so that x is zero when t is zero. This then makes x_0 in Eq. (2.15) equal to zero. Equation (2.15) is one of the fundamental equations of kinematics, and we will use it a great deal in what follows.

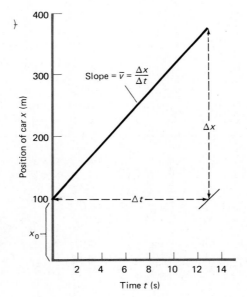

FIGURE 2.20 Position of a car moving with average velocity $\bar{v}$ as a function of time. The car starts from a point $x_0 = 100$ m. Here $\bar{v} = (380 - 100)$ m/13s = 22 m/s.

Constant Acceleration

Consider a train moving along a straight stretch of track with constant acceleration. Since the acceleration is constant, the velocity of the train changes linearly with time (i.e., in direct proportion to the time) from some initial value v_0 (which might be zero) to a final value v. Its average magnitude during the interval in which the acceleration is constant is, according to the definition of an arithmetic average,

---

*Notice that to simplify our notation we drop the subscript x on all velocities and accelerations in what follows. Since the motion is assumed to be along the x axis, all displacements, velocities, and accelerations are along the x axis.

$$\bar{v} = \frac{v_0 + v}{2} \qquad \text{(2.16)}$$

The train's final position x can be found from Eq. (2.15), where the average velocity is given by Eq. (2.16), and so we have

$$x = x_0 + \left(\frac{v_0 + v}{2}\right) t \qquad \text{(2.17)}$$

The constant acceleration is given by Eq. (2.14), which in this case becomes

$$a = \frac{\Delta v}{\Delta t} = \frac{v - v_0}{t - 0}$$

and so

$$v = v_0 + at \qquad \text{(2.18)}$$

On substituting this value of v in Eq. (2.17), we have

$$x = x_0 + \left(\frac{v_0 + v_0 + at}{2}\right) t$$

or

$$x = x_0 + v_0 t + \frac{at^2}{2} \qquad \text{(2.19)}$$

This is our perfectly general equation for the position of the train (or any other object) moving with uniform acceleration a and with initial position x_0 and initial velocity v_0 at time $t = 0$.

We have now developed most of the equations governing uniformly accelerated motion, but it is sometimes important to have an equation connecting v and x in which the time does not explicitly appear. To find this, we replace t in Eq. (2.15) by its value from Eq. (2.18), $t = (v - v_0)/a$, and obtain

$$x = x_0 + \bar{v}\left(\frac{v - v_0}{a}\right)$$

But, from Eq. (2.16), $\bar{v} = (v_0 + v)/2$, and so

$$x - x_0 = \left(\frac{v + v_0}{2}\right)\left(\frac{v - v_0}{a}\right) = \frac{v^2 - v_0^2}{2a}$$

or

$$v^2 = v_0^2 + 2a(x - x_0) \qquad \text{(2.20)}$$

We have now found five equations governing *uniformly accelerated motion* in one dimension:

$x = x_0 + \bar{v}t$	(2.15)	$v = v_0 + at$	(2.18)
$x = x_0 + v_0 t + \dfrac{at^2}{2}$	(2.19)	$\bar{v} = \dfrac{v_0 + v}{2}$	(2.16)
$v^2 = v_0^2 + 2a(x - x_0)$	(2.20)		

These five equations enable us to find the values of the position x and the velocity v of an object moving in one dimension as a function of time t, if we know the constant acceleration a and the initial position and initial velocity of the object.

Notice that Eq. (2.18) does not contain x, (2.15) does not contain a, (2.19) does not contain v, and (2.20) does not contain t. This enables us always to select an equation that relates a desired unknown quantity to quantities given in the statement of the problem.

Figure 2.21 shows how, in uniformly accelerated motion, the graph of position versus time departs from the straight line that would result if the speed were constant. It also shows the graph of velocity and acceleration versus time for uniformly accelerated motion. Notice that the acceleration does not change in time, the velocity increases in time with slope a, and the position increases faster than linearly with time because of the $\frac{1}{2}at^2$ term.

Figure 2.22 shows how velocity depends on time for three different situations: no acceleration, uniform acceleration, and an acceleration increasing with time.

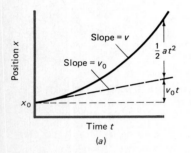

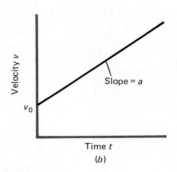

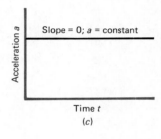

FIGURE 2.21 Uniformly accelerated motion. (a) Position as a function of time. (b) Velocity as a function of time. (c) Acceleration as a function of time.

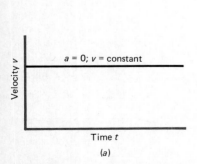

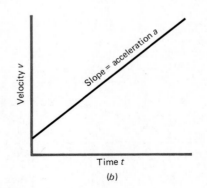

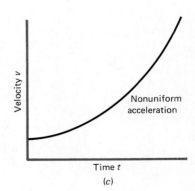

FIGURE 2.22 Velocity versus time for three different accelerations: (a) $a = 0$; (b) $a =$ constant (uniform acceleration); (c) nonuniform acceleration; a increases with time.

Example 2.10

A car is moving along at a constant speed of 20.0 m/s. The driver presses down on the accelerator and maintains a constant acceleration of 2.00 m/s² for 3.00 s in the direction of the motion. What is the displacement of the car in this 3.00-s interval? Do this problem in as many different ways as possible to check your results.

SOLUTION

Here we are given $v_0 = 20.0$ m/s, $t = 3.00$ s, $a = +2.00$ m/s², and we are asked to find the displacement $x - x_0$. Since the acceleration is constant, and $v = v_0 + at$, we find immediately that

$$v = 20.0 \text{ m/s} + (2.00 \text{ m/s}^2)(3.00 \text{ s}) = 26.0 \text{ m/s}$$

This is the final velocity at the end of the 3.0-s period, and it is in the same direction as the initial velocity.

1   From Eq. (2.19) we have

$$x - x_0 = v_0 t + \tfrac{1}{2}at^2$$

$$= (20.0 \text{ m/s})(3.00 \text{ s}) + \frac{(2.00 \text{ m/s}^2)(3.00 \text{ s})^2}{2}$$

$$= \boxed{69.0 \text{ m}}$$

2   A second approach is to use Eq. (2.20). Then

$$x - x_0 = \frac{v^2 - v_0^2}{2a} = \frac{(26.0 \text{ m/s})^2 - (20.0 \text{ m/s})^2}{2(2.00 \text{ m/s}^2)} = \boxed{69.0 \text{ m}}$$

3   A third way is to first calculate the average speed for the 3.0-s interval. We can use Eq. (2.16) to do this, since the motion is uniformly accelerated.

$$\bar{v} = \frac{v_0 + v}{2} = \frac{20.0 \text{ m/s} + 26.0 \text{ m/s}}{2} = 23.0 \text{ m/s}$$

The distance traveled is then, from Eq. (2.15),

$$x - x_0 = \bar{v}t = (23.0 \text{ m/s})(3.00 \text{ s}) = \boxed{69.0 \text{ m}}$$

These different approaches to the same problem are introduced to indicate that there is more than one correct way to do most physics problems. One of the best ways to master the solution of physics problems is, after arriving at an answer by one method, to try an entirely different method and see if the same answer is obtained. This approach will gradually build a mastery of problem-solving techniques and instill confidence in your ability to solve complicated problems not only in physics but in other fields. Trying different methods will also develop insight into which method is faster.

Example 2.11

A car is being driven at a speed of 25.0 m/s (the speed limit) when a bus enters the intersection ahead and stalls directly in the path of the car. **(a)** If the car is 60.0 m from the intersection when the driver applies the brakes, and if the brakes can reduce the car's velocity at a constant rate of 7.50 m/s², will the car escape hitting the bus? **(b)** If the driver of the car has a slow reaction time of 1.00 s so that the car moves along for 1.00 s before the brakes are applied, will an accident be avoided? **(c)** If a collision results (part b), how fast is the car traveling when it hits the bus?

SOLUTION

(a)   This is a problem where the acceleration is constant. We know the initial speed (25.0 m/s), the final speed (zero, if a collision is to be avoided), and the acceleration (−7.50 m/s²), and need to find the car's displacement ($x - x_0$) before it stops. Since Eq. (2.16) contains the three known quantities along with the one unknown, it seems like a good starting point for the problem. Other equations for uniformly accelerated motion include the time it takes for the car to stop, which is not known (although it can be found). Since $v^2 = v_0^2 + 2a(x - x_0)$, we have

$$x - x_0 = \frac{v^2 - v_0^2}{2a} = \frac{0 - (25.0 \text{ m/s})^2}{2(-7.50 \text{ m/s}^2)} = \boxed{41.7 \text{ m}}$$

Since car B is initially 60.0 m from car A, it will stop 18.3 m from the truck, and no accident will occur.

(b)   The solution to part (a) is unrealistic, since it does not include the reaction time of the driver, which is given as 1.00 s. During this time, the car continues to travel at a speed of 25.0 m/s, and the distance traveled in 1.0 s is 25 m. The total distance traveled before the car comes to a stop is then 41.7 m + 25.0 m = 66.7 m, and a collision would result, since the car is initially only 60.0 m from the truck.

(c)   Because of the driver's 1.00-s reaction time, the car starts to decelerate only at a distance of (60.0 − 25.0) m = 35.0 m from the truck, and so $x - x_0$ in Eq. (2.16) is in this case 35.0 m. The final speed of the car at the end of this 35.0 m is then

$$v^2 = v_0^2 + 2a(x - x_0)$$

$$= (25.0 \text{ m/s})^2 + 2(-7.50 \text{ m/s}^2)(35.0 \text{ m}) = 100 \text{ m}^2/\text{s}^2$$

and so $\boxed{v = 10.0 \text{ m/s}}$. (We take the plus sign for the square root, since the car must still be moving in its original direction.) The car is therefore still moving at a speed of 10.0 m/s when the collision occurs.

EXERCISE 2   Solve part (a) of this problem in a different way; for example, first find the time it takes the car to stop, and then calculate the distance it travels in that time.

2.7   Free Fall and the Acceleration Due to Gravity

The constant acceleration with which all objects, whether it be a small coin or a huge wrecking ball, fall near the surface of the earth* is called the *acceleration due to gravity* and is designated by the symbol g. Near the earth's surface the magnitude of g is approximately 9.80 m/s². Since the acceleration due to gravity is a vector, its direction must also be specified; this is toward the center of the earth.

The fall of an object under the influence of gravity (a ball dropped from your hand, for example) is usually referred to as *free fall* and can be completely described by the equations of the preceding section, since the acceleration is constant. Suppose a camera falls out of a stationary blimp anchored at some height h above the ground so that the initial downward velocity is $v_0 = 0$. Here a good choice of a coordinate system is one with an origin at the bottom of the blimp and all downward displacements, velocities, and accelerations taken as positive.

In such a coordinate system the position of the camera as a function of time is, from Eq. (2.19),

$$y = y_0 + v_0 t + \frac{gt^2}{2}$$

where y_0 is zero because it is at the origin of the coordinate system and the initial velocity v_0 is also zero. We have

$$y = \frac{gt^2}{2} \tag{2.21}$$

or, solving this equation for t,

$$t = \sqrt{\frac{2y}{g}} \tag{2.22}$$

where y is the position of the camera at time t.

Just before the camera hits the earth, its speed is given by Eq. (2.18) as

$$v = v_0 + at = 0 + gt = gt$$

On substituting the value of t from Eq. (2.22), we obtain

$$v = gt = g \sqrt{\frac{2y}{g}} = +\sqrt{2gy}$$

The positive root has been taken because the motion is directed downward and we have defined this direction as positive. Since at the earth's surface y is numerically equal to the initial height h, we have

$$v = \sqrt{2gh} \tag{2.23}$$

Figure 2.23 shows the acceleration, velocity, and position of the camera as a function of time. At the moment of release, its acceleration changes from 0 to 9.80 m/s² and remains at that value until it strikes the earth, since g is constant. Its velocity therefore increases by 9.80 m/s in a downward direction for each second of fall, leading to the velocities shown in the diagram. The positions of the camera

---

*Throughout the rest of this chapter we assume that air resistance is negligible. In Sec. 3.7 we will consider the effects of air resistance.

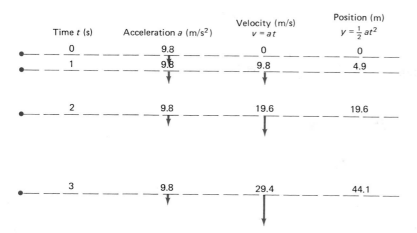

Time t (s)	Acceleration a (m/s²)	Velocity (m/s) $v = at$	Position (m) $y = \frac{1}{2}at^2$
0	9.8	0	0
1	9.8	9.8	4.9
2	9.8	19.6	19.6
3	9.8	29.4	44.1

FIGURE 2.23   Acceleration, velocity, and position of a falling object as a function of time.

FIGURE 2.24   Flash photograph of a falling billiard ball. The ball is illuminated by rapid light flashes at equal time intervals of $\frac{1}{30}$ s as it falls freely. The position scale is in centimeters. The increased distances traveled in equal time intervals show that the motion is accelerated. Measurements on the photograph show that the acceleration is constant and equal to the acceleration due to gravity g. (*Educational Development Center, Newton, Mass.*)

at 1-s intervals after it is dropped as calculated from Eq. (2.21) are shown in Fig. 2.23 and agree well with the positions measured experimentally. Figure 2.24 shows an actual flash photograph of a falling ball.

Since Eq. (2.19) does not include any property peculiar to the camera or any other falling object, all objects will fall with the same acceleration at any point near the earth's surface and reach the same velocity after falling for the same time.

Objects with an Initial Velocity

Suppose a baseball catcher throws a baseball high into the air and catches it when it returns to earth. The vertical motion of the ball is another example of uniformly accelerated motion under the influence of gravity. In this situation, even though the velocity of the ball is upward at first and downward later, *its acceleration is always downward*. When the ball is on its way up, its velocity decreases constantly because of this acceleration until finally the ball stops. Since it is accelerated downward, its velocity then increases at a constant rate as it falls. At the top of its path, its velocity is zero, but *the acceleration is still numerically equal to g and directed downward*.

Since the vector nature of velocities and accelerations is particularly important in this situation where the ball moves both up and down, it is essential that we be clear about the coordinate system used and the sign convention adopted.

The best choice of a coordinate system in this case is one with the origin at the height of the catcher's hand at the moment the ball is released so that the initial position of the baseball can be assumed to be $y_0 = 0$. We choose a sign convention in which displacements, velocities, and accelerations are positive when directed upward and negative when directed downward. This sign convention is convenient because the initial velocity of the ball is upward.

Then, using the equations for uniformly accelerated motion, we have for the position of the ball at any instant,

$$y = y_0 + v_0 t + \frac{at^2}{2} = v_0 t - \frac{gt^2}{2} \tag{2.24}$$

where the acceleration is negative because it is directed downward. Similarly,

$$v = v_0 + at = v_0 - gt \tag{2.25}$$

These two equations enable us to find y and v at any instant of time. Negative values of y mean that the ball has fallen below the level of the catcher's hand, negative values of v that the velocity is directed downward.

For example, to find the time when the ball stops at the top of its flight, we can make use of the fact that $v = 0$ at this point. We have $v = v_0 - gt = 0$, which leads to $t = v_0/g$. So the maximum height to which the ball rises is, from Eq. (2.24),

$$y = v_0 t - \frac{gt^2}{2} = v_0 \left(\frac{v_0}{g}\right) - \frac{g}{2}\left(\frac{v_0}{g}\right)^2 = \frac{v_0^2}{2g}$$

The ball then falls back to the earth from this height, as shown in Fig. 2.25.

The velocity with which the ball returns to the catcher's mitt can be found from Eq. (2.20):

$$v^2 = v_0^2 + 2a(y - y_0) = v_0^2 - 2g(y - y_0)$$

When the ball returns to the catcher, y is equal to its initial value y_0. Therefore $y - y_0 = 0$, and so $v^2 = v_0^2$ and $v = \pm v_0$. The negative sign must be chosen here because the velocity is downward and, according to our choice of coordinate

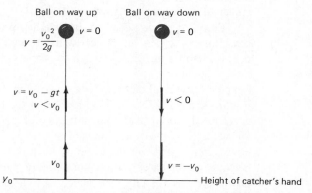

FIGURE 2.25 The motion of a ball thrown upward by a baseball catcher.

system, this means a negative velocity. The ball therefore reaches the catcher's mitt with the same speed it had when it left his hand.

The time for the ball to fall back to the catcher's mitt can be obtained from Eq. (2.25), $v = v_0 - gt$, with $v = -v_0$ when the ball hits the mitt, as just found above. Therefore

$$-v_0 = v_0 - gt \qquad \text{from which} \qquad t = \frac{2v_0}{g}$$

This is just twice the time for the ball to reach its highest point.

In problems of this sort, the sign convention used is extremely important. We are free to call downward displacements, velocities, and accelerations either positive or negative. The choice can be made in the way most convenient for solving a particular problem. Once made, however, this choice must be maintained consistently throughout the problem.

Example 2.12

A very good baseball pitcher can throw a baseball at speeds over 90 mi/h. If a pitcher throws a ball at 90.0 mi/h straight up in the air, how high does it go if we assume no air resistance?

SOLUTION

The speed of 90.0 mi/h must be converted into SI units if we want an answer in SI units. Since 1 mi/h is equal to 0.447 m/s, 90.0 mi/h is equal to 40.2 m/s.

Since the ball is thrown upward, we take displacements, velocities, and accelerations upward as positive. We therefore have $v_0 = 40.2$ m/s, $a = -g = -9.80$ m/s². Since the ball stops at its highest position, we seek $y - y_0$ for $v = 0$. The most useful equation for this uniformly accelerated motion is Eq. (2.20). (Why is this so?)

$$v^2 = v_0^2 + 2a(y - y_0)$$

$$h = y - y_0 = \frac{v^2 - v_0^2}{2a} = \frac{0 - (40.2 \text{ m/s})^2}{2(-9.80 \text{ m/s}^2)} = \boxed{82.5 \text{ m}}$$

Example 2.13

With what speed must a rock be thrown upward if it is to return to the thrower in 5.00 s?

SOLUTION

Since the rock is thrown upward, we take displacements, velocities, and accelerations upward as positive. The acceleration due to gravity is therefore negative. We have $t = 5.00$ s, $a = -g = -9.80$ m/s², $y - y_0 = 0$, since the rock returns to its original position, for which $y = y_0$. We are asked to find v_0.

The equation containing all the known quantities and the one unknown v_0 is Eq. (2.19):

$$y = y_0 + v_0 t + \tfrac{1}{2}at^2 \qquad \text{or} \qquad y - y_0 = v_0 t + \tfrac{1}{2}at^2 = 0$$

from which $\qquad v_0 = -\tfrac{1}{2}at$

$$= -\frac{(-9.80 \text{ m/s}^2)(5.00 \text{ s})}{2} = \boxed{24.5 \text{ m/s}}$$

EXERCISE 3 Use the result just obtained to calculate (a) the height the rock reaches above the ground; (b) the time it takes to reach that height. (c) Is your result consistent with the fact that the total time is 5.00 s?

Example 2.14

A boy throws a stone straight up with an initial speed of 15.0 m/s. **(a)** How long does it take the stone to reach a point 10.0 m above the ground on its way down? **(b)** How fast is the stone moving at this point?

SOLUTION

We have $v_0 = +15.0$ m/s, $y - y_0 = 10.0$ m, $a = -9.80$ m/s^2, with both t and v to be determined. Here we have taken the upward direction as positive.

If we try to solve part (a) first, we obtain a quadratic equation that is a bit complicated. Since we are always free to solve the parts of a problem in any order we desire, let us try to solve part (b) first, using Eq. (2.20).

(b) $v^2 = v_0^2 + 2a(y - y_0)$

$= (15.0 \text{ m/s})^2 + 2(-9.80 \text{ m/s}^2)(10.0 \text{ m}) = 29.0 \text{ m}^2/\text{s}^2$

and so $v = \pm 5.39$ m/s

The plus sign applies to the motion in the upward direction when the stone is at the height of 10 m. Since we are interested in the downward motion at the same point, the correct answer is

$$v = \boxed{-5.39 \text{ m/s}}$$

(a)   Now it is simple to solve the first part of this problem. We have, from Eq. 2.14),

$$t = \frac{v - v_0}{a} = \frac{-5.39 \text{ m/s} - 15.0 \text{ m/s}}{-9.80 \text{ m/s}^2} = \boxed{2.08 \text{ s}}$$

EXERCISE 4   Solve part (a) of this problem by finding t directly, using the formula for the solution of a quadratic equation (Appendix 1.A).

2.8   Kinematics of Motion on an Inclined Plane

A correct understanding of uniformly accelerated motion originated with the great Italian scientist Galileo Galilei (see Fig. 2.26 and accompanying biography). Galileo discussed the mathematics of a possible type of motion that he named *uniformly accelerated motion*. He defined such motion as follows: "A body is said to be uniformly accelerated when, starting from rest, it acquires equal increments of velocity during equal time intervals." In other words, it is motion with *constant acceleration*. Galileo then proposed the hypothesis that the free fall of an object near the earth's surface is an example of uniformly accelerated motion, and that as long as air resistance is small, all objects fall with the *same acceleration*. The velocities and the displacements should therefore increase in the same way for all objects falling freely at any particular place near the earth's surface.

Galileo could not test this hypothesis experimentally with the scientific instruments he had available, since they could not accurately measure the very short times involved in free fall. He therefore argued that he could confirm his hypothesis by studying the rolling of a ball down an inclined track. If this motion were found to be uniformly accelerated and if this acceleration were independent of the size and mass of the rolling ball, then he felt confident that the same would be the case for freely falling objects.

Galileo's argument was that free fall is just the limiting case of motion down an inclined track in which the track becomes vertical. As the track's direction deviates from the vertical, the acceleration due to gravity still determines how the velocity changes, but it is only the component of the acceleration along the direction of the track that is effective. If the angle of the track with the horizontal is made very small, the ball rolls down the track slowly enough to make it possible to measure its position and velocity as a function of time with reasonable accuracy. Then the equation $s = \frac{1}{2}at^2$, derived above for uniformly accelerated motion, can be applied to the motion along the track. If the acceleration is found to be constant for different values of s and t, it can be concluded that this motion is uniformly accelerated and that the same is true of free fall, according to Galileo's hypothesis.

Galileo then carried out a series of experiments on balls rolling down inclined planes and convinced himself and his contemporaries that his hypothesis was indeed

Galileo Galilei (1564–1642)

FIGURE 2.26 *(AIP Niels Bohr Library.)*

Galileo was born in Pisa, Italy, on February 15, 1564. His father, Vincenzio Galilei, was an able mathematician and musician. Galileo studied Latin, Greek, and logic as a boy and in 1581 entered the University of Pisa to study medicine. He found medicine little to his liking, however, and turned to mathematics. In 1583 he was forced to leave the university without a degree because his parents could no longer afford his tuition. This ended his formal education.

In 1589 a wealthy patron, the Marchese del Monte, recognized Galileo's mathematical abilities and had him appointed as lecturer in mathematics at the University of Pisa. Here he carried out experiments on the motion of bodies that led to results so contradictory to those of Aristotle that they aroused the strong antagonism of his Aristotelian colleagues at the uni-

versity. He defended his ideas with such determination and argued with such vehemence that he earned the nickname "the Wrangler." He remained at Pisa until 1592, when he was appointed to teach mathematics at the University of Padua. There he remained for 18 years in what was to prove the happiest and most productive period of his life.

In 1609 Galileo heard of a Dutch lens grinder who had developed an optical instrument that made distant objects seem close at hand. He immediately set to work to construct one himself, using his knowledge of the refraction of light, and in a few days succeeded in combining two glass lenses in a tube to reveal sights never before seen by human eyes. He turned his "telescope" to the skies and discovered all sorts of wonderful things: the rough surface of the moon, the satellites of Jupiter, the phases of Venus, sunspots, the huge number of stars in the Milky Way. The publication of these results in his *Sidereus Nuncius* (1610) made him famous all over Europe.

Galileo's observations made him more and more certain that the Copernican theory, which hypothesized that the earth moves around the sun, was correct. In 1616 he was admonished by church authorities not to uphold or teach Copernicanism as a "true" picture of the universe, although it was permissible as a mathematical hypothesis. In 1623 Galileo's old friend Cardinal Maffeo Barberini was elected Pope Urban VIII. This encouraged Galileo to believe that his ideas would eventually prevail, and in 1632 he published his major treatise on the Copernican system, the *Dialogue*

Concerning the Two Chief World Systems.

After the *Dialogue* appeared, Galileo's world collapsed around him. Pope Urban VIII was furious, perhaps because advisers told him that Simplicio ("the simpleton") in the dialogue was intended to voice the views of the Pope. Galileo was summoned to Rome and in April 1633 was made to stand trial before the Holy Office. In June it was decided that Galileo had rendered himself "vehemently suspect of heresy" because of his defense of Copernicanism. He was forced to abjure publicly all beliefs and writings upholding the Copernican system and was sentenced to imprisonment for the rest of his life. This sentence was later commuted to permanent house detention at the home of his friend Ascanio Piccolomini, the Archbishop of Siena.

Later Galileo received permission to move to his villa at Arcetri near Florence. There he resumed his scientific work on mechanics and in 1636 published what is probably his best work, *Two New Sciences*, which summarized most of his work on the physics of motion, his single most important contribution to physics. After he had completed this book his eyesight began to fail; in 1638 he went completely blind, and he died at Arcetri at age 78 on January 8, 1642, the same year in which Isaac Newton was born.

More important than any of Galileo's discoveries in physics was his approach to the study of nature, an approach which combined mathematics with experiment in a manner so in tune with what physicists do today that he is rightly considered the first modern physicist.

correct. This inspired combination of theory and experiment was the first major breakthrough in our understanding of motion. In principle Galileo's experiments can lead to a value for g, but Galileo reported no such value because at that time he was unable to include in a consistent theory both the displacement of the ball down the plane and its rotational motion about its center.

Description of Motion on an Inclined Plane

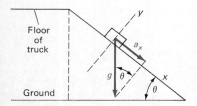

FIGURE 2.27 Coordinate axes and acceleration a_x along the plane for a block on a frictionless inclined loading ramp.

Since Galileo's researches on motion on an inclined plane, such motion has been important in physics. A good example of an inclined plane is a loading ramp used to slide crates to the ground from the body of a truck, as in Fig. 2.27. The angle of the ramp with the ground is fixed at a value θ. In this case it is convenient to choose a coordinate system with the x axis along the plane of the ramp and the y axis at right angles to this plane, as in the figure.

In this chapter we are only interested in analyzing the motion of objects (kinematics), not in the causes of that motion. The vector representing the acceleration due to gravity is directed downward toward the center of the earth and is of magnitude g. It is this acceleration that, in the absence of friction and air resistance, completely determines the motion of objects either in free fall or on inclined planes.

Like any other vector in two dimensions, we may replace the acceleration due to gravity by its two rectangular components along our chosen x and y axes. From Fig. 2.27 the angle between the vertical direction and the y axis is the same as the angle of the plane θ. We therefore have

$$a_x = g \sin \theta$$

Notice that θ is *not* the angle between g and a_x but between g and the normal to the plane.

The y component of g can have no effect on the motion because the solid plane prevents all motion in the y direction. Thus the only acceleration is along x and is $a_x = g \sin \theta$. We can use this value of the acceleration to find the displacement and velocity of the block along the plane at any instant of time, using the equations of uniformly accelerated motion. If the block starts down the plane from rest at a position x_0 (measured along the plane), then its velocity down the plane at any time t is, from Eq. (2.18),

$$v_x = v_0 + a_x t = a_x t = (g \sin \theta)t \tag{2.26}$$

and its displacement at time t is, from Eq. (2.19),

$$x - x_0 = v_0 t + \tfrac{1}{2}a_x t^2 = \frac{(g \sin \theta)t^2}{2} \tag{2.27}$$

These equations provide us with any information we need about the block's motion.

Notice that these results are the same for all objects, no matter their size, shape, or mass, since these properties do not occur in the above equations of motion (Again, we neglect friction and air resistance).

Since Eq. (2.27) is of the form $x - x_0 = \tfrac{1}{2}a_x t^2$, with $a_x = g \sin \theta$ constant, it clearly predicts that motion on an inclined plane is uniformly accelerated. Galileo's experimental proof that this is actually the case led him to the conclusion that both motion on an inclined plane and free fall were examples of uniformly accelerated motion.

We will return to a more complete treatment of motion on an inclined plane in the next chapter.

Example 2.15

A skier starts from rest at the top of a 30.0° hill. The friction between her skis and the snow, and air resistance, are both small enough to be neglected. **(a)** How far has the skier traveled at the end of 5.00 s? **(b)** How fast is she traveling at this time?

SOLUTION

Here we know that $\theta = 30.0°$, $g = 9.80 \text{ m/s}^2$, $v_0 = 0$, and $t = 5.00$ s. We are asked to find the skier's displacement down the slope, $x - x_0$, and her velocity along the slope when $t = 5.00$ s.

(a) The displacement of the skier along the slope is

$$x - x_0 = \tfrac{1}{2}a_x t^2 = \tfrac{1}{2}(g \sin \theta)t^2$$

$$= \tfrac{1}{2}(9.80 \text{ m/s}^2)(\sin 30.0°)(5.00 \text{ s})^2$$

$$= \boxed{61.3 \text{ m}}$$

(b) Her speed v_x along the slope after 5.00 s is

$$v_x = a_x t = (g \sin \theta)t$$

$$= (9.80 \text{ m/s}^2)(\sin 30.0°)(5.00 \text{ s}) = \boxed{24.5 \text{ m/s}}$$

Summary: Important Definitions and Equations

Mechanics: The name given to the study of the motion of objects and the causes of that motion.

Kinematics: The study of the motion of objects without considering the causes of that motion.

Dynamics: The study of the causes of the motion of objects.

Frame of reference: A set of coordinate axes in terms of which the position of objects in space may be specified.

Motion in a plane

Position vector: A vector **r** drawn from the origin of the coordinate system to the point in space at which the object is located.

Displacement vector: The vector difference between two position vectors:

$$\Delta \mathbf{r} = \mathbf{r}_2 - \mathbf{r}_1 \tag{2.1}$$

Velocity: The vector change in an object's position per unit time.

$$\text{Average velocity } \bar{\mathbf{v}} = \frac{\Delta \mathbf{r}}{\Delta t} = \frac{\mathbf{r}_2 - \mathbf{r}_1}{t_2 - t_1} \tag{2.2}$$

$$\text{Average speed} = \frac{\text{total distance traveled}}{\text{total time elapsed}}$$

$$\text{Instantaneous velocity } \mathbf{v} = \lim_{\Delta t \to 0} \frac{\Delta \mathbf{r}}{\Delta t} \tag{2.3}$$

Instantaneous speed: The magnitude of the *instantaneous velocity*.

Acceleration: The rate of change of the velocity with time.

$$\text{Average acceleration } \bar{\mathbf{a}} = \frac{\Delta \mathbf{v}}{\Delta t} = \frac{\mathbf{v}_2 - \mathbf{v}_1}{t_2 - t_1} \tag{2.5}$$

$$\text{Instantaneous acceleration } \mathbf{a} = \lim_{\Delta t \to 0} \frac{\Delta \mathbf{v}}{\Delta t} \tag{2.6}$$

Rectilinear motion (motion along the x axis)

$$\text{Displacement } \Delta x = x_2 - x_1 \tag{2.7}$$

$$\text{Average velocity } \bar{v}_x = \frac{\Delta x}{\Delta t} = \frac{x_2 - x_1}{t_2 - t_1} \tag{2.8}$$

$$\text{Instantaneous velocity } v_x = \lim_{\Delta t \to 0} \frac{\Delta x}{\Delta t} \tag{2.9}$$

$$\text{Average acceleration } \bar{a}_x = \frac{\Delta v_x}{\Delta t} = \frac{v_{2x} - v_{1x}}{t_2 - t_1} \tag{2.10}$$

$$\text{Instantaneous acceleration } a_x = \lim_{\Delta t \to 0} \frac{\Delta v_x}{\Delta t} \tag{2.11}$$

Uniformly accelerated motion: Motion in a straight line with an acceleration whose magnitude remains constant.
Equations for uniformly accelerated motion

$$x = x_0 + \bar{v}t \tag{2.15}$$

$$\bar{v} = \frac{v_0 + v}{2} \tag{2.16}$$

$$v = v_0 + at \tag{2.18}$$

$$x = x_0 + v_0 t + \frac{at^2}{2} \tag{2.19}$$

$$v^2 = v_0^2 + 2a(x - x_0) \tag{2.20}$$

Questions

1 One automobile increases its speed from 5.0 to 10 m/s. A second automobile increases its speed from 20 to 25 m/s in the same time. Is the acceleration the same in these two cases? (Assume that the two cars are moving in the same direction on the same road.)

2 If the automobiles in Question 1 were moving in opposite directions on the same road, is the acceleration the same in the two cases? Why?

3 Must the velocity of a moving object always be in the same direction as its acceleration? Give some examples that support your answer.

4 Can the velocity of an object be zero when its acceleration is not zero? If so, give an example to illustrate your point.

5 A car moving along a straight highway at constant speed slows down as it enters a small town. It stops for the one traffic light in the town, then gradually picks up speed, and finally resumes its original highway speed when it leaves the town. Draw graphs, one above the other, of the car's position, its velocity, and its acceleration as a function of time.

6 (a) Draw a velocity-against-time graph for the flight of a ball thrown straight up in the air and caught on its return to the ground.

(b) Draw an acceleration-against-time graph for the same motion of the ball.

7 If you have a position-against-time graph for a moving particle, is it possible to use this graph to obtain a velocity-against-time graph and then to use this velocity-against-time graph to obtain a graph of the acceleration as a function of time? How would you do this?

8 Is it possible for an object to have a downward acceleration greater than 9.8 m/s²? If so, how could this be accomplished?

9 A person throws a tennis ball straight up in the air and catches it again when it comes down. What is the average velocity of the ball during the up-and-down trip? What is its average speed for the trip? Neglect air resistance, and assume that the initial velocity of the ball is 10 m/s.

10 A flowerpot falls off the ledge of a fourth-floor apartment house window. As it is passing the second-story window of the same house, a boy drops a ball from this window.

(a) Which object hits the ground first?

(b) Which object hits the ground at the higher speed?

Multiple-Choice and Simple Exercises

Before tackling these exercises and problems, you are urged to read over the Note at the end of Chap. 1 (p. 29) and the Hints on Problem Solving in Appendix 3.A.

In this book the numerical data in the statement of the problems are usually given to two or three significant figures, depending on the nature of the problem. Therefore answers to problems are expected to have the same number of significant figures as the data. If graphical methods are used to solve a problem, usually only two significant figures are expected in the answer.

2.1 An acceleration of 3.00 (mi/h)/s is equal to:
(a) 0.447 m/s² (b) 10.3 m/s² (c) 1.34 m/s²
(d) 4.83 m/s² (e) 1.34 m/s

2.2 A baseball is thrown vertically upward, stops at its highest point, and falls back to the ground. The following statement is true about this motion:
(a) The acceleration is always upward.
(b) The acceleration is always downward.
(c) The acceleration is always in the direction of the displacement of the ball.
(d) The acceleration is always in the direction of the velocity of the ball.
(e) The acceleration is always opposed in direction to the velocity of the ball.

2.3 The time it takes for light, traveling at a speed of 3.0×10^8 m/s, to cover 300 m is:
(a) 10^6 s (b) 10^{-6} s (c) 10^{11} s
(d) 10^{-11} s (e) None of the above

2.4 A speed of 40 mi/h is equivalent to:
(a) 34 m/s (b) 10 m/s (c) 90 m/s
(d) 18 m/s (e) 25 m/s

2.5 A car accelerates from rest to 60 mi/h in 6.0 s. Its acceleration in SI units is:
(a) 4.5 m/s (b) 4.5 m·s² (c) 27 m/s²
(d) 9.0 m/s² (e) 4.5 m/s²

2.6 A car traveling with an initial speed v_0 stops when the driver applies the brakes. The negative acceleration during the time t is constant. The distance covered by the car during the time t is:
(a) $v_0 t$ (b) $v_0 t^2/2$ (c) $3v_0 t/2$ (d) $v_0 t/2$
(e) $v_0/2t$

2.7 A car starts from rest and accelerates uniformly for 10 s. The distance traveled in the 10 s is greater than the distance traveled in the first second by a factor of:
(a) 5 (b) 50 (c) 25 (d) 10 (e) 100

2.8 A sailboat is moving at a speed of 3.0 m/s. A strong wind blows up and accelerates the boat forward with a constant acceleration of 0.20 m/s² for 10 s. The final speed of the sailboat at the end of the 10-s period is:
(a) 30 m/s (b) 5.0 m/s (c) 10 m/s
(d) 5.0 m/s² (e) 10 m/s²

2.9 A stone is dropped from a cliff 200 m high. When the stone is 100 m above the ground, its downward acceleration is:
(a) 0 (b) 9.80 m/s² (c) 19.6 m/s²
(d) 100 m/s² (e) 4.90 m/s²

2.10 A flowerpot falls from a third-story windowsill that is 10 m above the ground. When it passes a first-story window 2.0 m above the ground, its speed is:

(a) 9.8 m/s   (b) 8.0 m/s   (c) 13 m/s
(d) 13 m/s²   (e) 1.6 × 10² m/s

2.11 In 1984 Carlos Lopes of Portugal won the 26.2-mi Olympic Marathon in a time of 2 h, 9 min, and 21 s.

(a) What was his average speed in meters per second?
(b) What was his average time for 1.00 km?

2.12 A quarter-miler runs one lap around a 400-m track in 65.0 s.

(a) What is her average speed?
(b) What is her average velocity?

2.13 The stoplights on the north-south avenues in New York City are designed to keep traffic moving at 30 mi/h. The average length of a block between traffic lights is about 80 m. What must the time delay between green lights on successive blocks be to keep the traffic moving continuously?

2.14 It is believed that the universe began with a big bang about 20 billion years ago. Since light travels faster than anything else in the universe, the farthest that the cosmic debris produced in the big bang could have spread by now must be less than the distance light could have traveled in 2.0 × 10¹⁰ years. If the speed of light is 3.0 × 10⁸ m/s, what is the maximum possible radius of the universe at the present?

2.15 A window washer drops a brush from a scaffold on a tall office building.

(a) What is the speed of the falling brush at the end of 2.60 s?
(b) How far has it fallen in that time?

2.16 A car starts from rest and accelerates for 5.00 s with an acceleration of 3.00 m/s². How far does it travel?

2.17 A car is moving at a constant speed of 14.0 m/s when the driver presses down on the gas pedal and accelerates for 10.0 s with an acceleration of 1.00 m/s². What is the average speed of the car during the 10-s interval?

2.18 A wrench falls out of the gondola of a balloon that is 500 m above the ocean. What is its speed just before it hits the water, if we neglect air resistance?

Problems

2.19 An auto travels at 25.0 m/s for 4.00 min, then at 15.0 m/s for 8.00 min, and finally at 20.0 m/s for 2.00 min, all along the same straight highway.

(a) What is the total distance covered?
(b) What is the average speed for the complete trip?

2.20 For the falling ball shown in Fig. 2.24, measure for the $\frac{1}{30}$-s time intervals shown the displacement of the ball during each successive $\frac{1}{30}$ s; then calculate the average velocity during the time interval and the change in the average velocity from one time interval to the next. Use these data to obtain the average value for the acceleration due to gravity g.

(a) What is your value for g?
(b) How does your value compare with the accepted value for g?

2.21 The Pepsi Challenge 10,000-m national championship road race was held in New York City in July 1982. The winner in the men's division was Rod Dixon of New Zealand with a time of 28 min, 12.3 s. The women's division was won by Ellen Hart of Boulder, Colorado, with a time of 32 min, 59.7 s. What is the difference in meters per second between the speeds of Dixon and Hart?

2.22 A train starts from rest with a uniform acceleration of 0.200 m/s².

(a) How fast is it moving after 1.0 min?
(b) At what time during the first minute is its instantaneous speed equal to its average speed?

2.23 A racing car is moving along a straight track at a constant speed of 50.0 m/s. The driver presses down on the gas pedal and maintains a constant acceleration of 6.00 m/s² for 5.00 s.

(a) At the end of the 5.00 s, what is the car's speed?
(b) During the 5.00 s, how much ground does the car cover?

2.24 A ball has an initial upward speed of 8.00 m/s. At the end of 3.00 s, what is the velocity (both magnitude and direction) if we neglect air resistance?

2.25 A girl throws a Frisbee by accelerating it from rest, while it is held in her hand, through a distance of 1.30 m before releasing it. If the Frisbee leaves the girl's hand at a speed of 3.00 m/s, what average acceleration was imparted to it during the throwing motion?

2.26 A reasonable value for the deceleration (negative acceleration) of a skidding car (with brakes locked) on a dry pavement is about 7.0 m/s². How long does it take a car going 25 m/s to come to rest once the brakes are engaged?

2.27 A racing car is one lap behind the lead car, which has 50 laps to go in a race. If the speed of the lead car is 60 m/s, what must the average speed of the second car be to catch the lead car before the end of the race, if 1 lap = 1.50 km?

2.28 A skier starts from rest, accelerates uniformly, and slides 10.0 m down a slope in 5.00 s.

(a) What is the skier's acceleration?
(b) At what time after starting does the skier acquire a speed of 20.0 m/s?

2.29 A train is moving at 80.0 km/h when the engineer applies the brakes and the train decelerates at a rate of 1.50 m/s². How long does it take for the train to travel the next 100 m after the brakes are applied?

2.30 A subway train leaves a station with a constant acceleration of 2.50 m/s². Find (a) its speed at the end of 6.00 s; (b) its average speed during the 6.00-s interval; (c) its displacement in the 6.00 s.

2.31 An express train traveling at 15.0 m/s passes through a subway station without stopping. It constantly accelerates between this station and the next station, and its average velocity for the trip between the two stations is 20.0 m/s. How fast is the train going when it passes the second station?

*2.32 The electrons that produce the picture in a TV set are accelerated by a very large electric force as they pass through a small region in the neck of the picture tube. This region is 1.50 cm in length, and the electrons enter with a speed of 1.00×10^5 m/s and leave with a speed of 1.50×10^8 m/s.

    (*a*)   What is their acceleration over this 1.50-cm length?

    (*b*)   How long does the acceleration take?

2.33   Figure 2.28 is the graph of the position of a train on a straight track as a function of time. Calculate approximate values of the speed of the train at points A, B, C, and D.

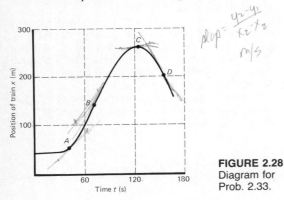

FIGURE 2.28
Diagram for
Prob. 2.33.

2.34   Figure 2.29 is the graph of the velocity of a train on a straight track as a function of time. Calculate approximate values of the acceleration of the train at A, B, C, and D.

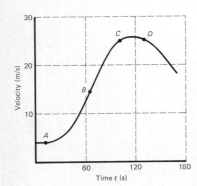

FIGURE 2.29
Diagram for
Prob. 2.34.

2.35   A train is moving along a straight track between Washington, D.C., and Baltimore. When it is 1.00 km from the Washington station, it is moving away from Washington with a speed of 10.0 m/s. It then accelerates uniformly at a rate of 0.100 m/s² for 4.00 min.

    (*a*)   What is the train's speed at the end of the 4.00-min interval?

    (*b*)   How far is it from the Washington station at the end of the 4.00-min time interval?

2.36   A car initially traveling at a constant speed accelerates at the rate of 1.00 m/s² for 15.0 s. If the car travels 200 m during this 15-s interval, what was its speed when it started to accelerate?

2.37   A racing car can accelerate from rest to 45.0 m/s in 6.00 s.

    (*a*)   If we assume that the acceleration is constant, what is its magnitude?

    (*b*)   How far will the racing car travel from rest before reaching a speed of 55.0 m/s?

2.38   An airport has runways only 150 m long. A small plane must reach a ground speed of 40.0 m/s before it can become airborne. What average acceleration must the plane's engines provide if it is to take off safely from this airport?

2.39   Boris Becker of West Germany, the 1989 winner of the Wimbledon tennis tournament, hit serves with measured speeds of 156 mi/h. How long would it take a tennis ball hit at this speed to cover the length of the tennis court, which is officially 78 ft long?

2.40   An engineer moving a locomotive sees a car stuck on the track at a railroad crossing in front of the train. When the engineer first sees the car, the locomotive is 200 m from the crossing and moving at 30.0 m/s. If the engineer's reaction time is 0.600 s, how rapidly does the locomotive have to decelerate to avoid an accident?

2.41   Check by your own calculations the values of the acceleration, velocity, and position of a falling object shown in Fig. 2.23.

2.42   A person drops a stone from the top of a building, and it takes 5.60 s for the stone to reach the ground. How tall is the building?

2.43   A baseball player throws a ball straight up in the air with an initial speed of 35.0 m/s.

    (*a*)   How high does the ball go?

    (*b*)   What is its speed just before it hits the ground?

    (*c*)   How long is it in the air?

2.44   Find, by dimensional analysis, which of the following equations are dimensionally correct:

    (*a*) $x = x_0 + v_0 t + \dfrac{at^2}{2}$

    (*b*) $v^2 = v_0^2 + 2a(x - x_0)$

    (*c*) $v - v_0 = at^2$

2.45   A baseball is thrown at a large gong 50.0 m away. The sound of the gong is heard by the thrower 2.00 s later. The speed of sound in air is about 330 m/s. With what speed was the ball thrown? (Assume that the ball travels in a straight line to the gong.)

∗**2.46**   The Russian juggler Sergei Ignatov can juggle 11 hoops at one time. To do this, he must throw the hoops high into the air to allow himself time to catch and throw aloft each of the 11 hoops in turn. Show that doubling the height reached by a hoop from 5.0 to 10 m increases by only about 40 percent the time required for a hoop to leave the juggler's hand and return again.

2.47   A rocket-driven sled running on a straight, level track has been used to study the physiological effects of large accelerations on astronauts. One such sled can attain a speed of 444 m/s in 1.80 s starting from rest.

    (*a*)   What is the acceleration of the sled, if we assume it is constant?

    (*b*)   How many times the acceleration due to gravity, $g = 9.80$ m/s², is this?

    (*c*)   How far does the sled travel in 1.80 s, starting from rest?

2.48   An electron is shot with a speed of 4.00×10^6 m/s into a region where it is electrically slowed down with a negative acceleration of 2.15×10^{13} m/s². How far does the electron travel before coming to rest?

2.49 A skier starts from rest down a 40.0° slope. Neglect friction and air resistance.

(*a*) How long does it take for the skier to reach a speed of 20.0 m/s?

(*b*) What distance has the skier traveled along the slope in that time?

(*c*) What horizontal distance has the skier traveled? What vertical distance?

2.50 A block is pushed from the top of a 35.0° friction-less inclined plane so that its initial speed at the top of the plane is 50.0 cm/s. The length of the inclined surface is 1.20 m. What is the speed of the block at the bottom of the plane?

2.51 Mary and Jane are racing each other on foot. Jane starts 10.0 m behind Mary but runs at an average speed of 8.00 m/s, whereas Mary runs at a speed of only 7.00 m/s.

(*a*) How far does Jane have to run to catch up with Mary?

(*b*) How long will it take Jane to catch Mary?

***2.52** A police car is at rest beside a road, looking for speeders. A car passes, traveling at a constant speed of 29.0 m/s. The police car starts up 4.00 s later and accelerates from rest with an acceleration of 1.80 m/s² until it reaches a speed of 40.0 m/s. If the police car maintains this speed until it over-takes the speeding car, how far does the police car travel before catching the speeding car?

2.53 A train moving at 20.0 m/s slows down uniformly to 12.0 m/s in 4.00 s.

(*a*) Determine the acceleration of the train.

(*b*) Determine the distance it moves during the fourth second of the 4.00-s interval.

***2.54** A rock is dropped from a seaside cliff and is heard by the thrower hitting the water 2.20 s later. If the speed of sound in air is 330 m/s, how high is the cliff?

***2.55** A person leans out a window 20.0 m above the ground in an apartment house and throws a stone straight up in the air at a speed of 4.00 m/s. The stone rises, stops, and falls to the courtyard below the window.

(*a*) What is the speed of the stone when it reaches the ground?

(*b*) How much time elapses from the moment the person throws the stone until it hits the ground?

***2.56** A camera falls from a blimp that is 500 m above the ground and rising at a speed of 10.0 m/s. Neglecting air resistance, find (*a*) the maximum height reached by the camera with respect to the ground; (*b*) its position and velocity 5.00 s after it leaves the blimp; (*c*) the time it takes to reach the ground; (*d*) the downward speed with which it strikes the ground.

***2.57** A stone is dropped from a bridge at a point 50.0 m above a river, and 1.10 s later another stone is thrown downward after the first one. Both stones hit the water at exactly the same time. What was the initial speed of the second stone?

***2.58** The acceleration due to gravity on the moon is about one-sixth its value on earth. If a baseball reaches a height of 50 m when thrown upward by someone on the earth, what height would it reach when thrown in the same way on the surface of the moon?

***2.59** A research balloon is rising straight up with a con-stant speed of 2.00 m/s. When it is 30.0 m above the ground, a crew member on board throws a ball straight up in the air outside the balloon at a speed of 10.0 m/s relative to the balloon.

(*a*) What is the maximum height above the ground reached by the ball?

(*b*) How far above the ground is the balloon when the ball reaches its maximum height?

(*c*) How long does it take for the ball to return to the level of the balloon?

2.60 A bolt is dislodged from a railroad bridge by a passing train. The bridge is 50.0 m above the level of the road underneath. The bolt falls directly on top of a car that was 20.0 m from the point of impact when the bolt began to fall. How fast was the car traveling?

***2.61** A car and a truck are traveling in opposite directions along a very narrow road at night, each at a speed of 25.0 m/s. When they are 80.0 m apart, they see each other's lights and immediately apply their brakes. The car decelerates at 7.20 m/s² and the truck at 6.30 m/s².

(*a*) If we neglect the drivers' reaction times, will the two collide if they do not leave the road to avoid a collision?

(*b*) If they collide, at what distance from the original position of the car will the collision occur?

(*c*) If they collide, at what speeds will the car and the truck be traveling when they collide?

$V = V_0^x$

Additional Readings

Asimov, Isaac: *Asimov's Biographical Encyclopedia of Science and Technology*, 2d rev. ed., Doubleday, Garden City, N.Y., 1982. This contains the lives and achievements of 1510 outstanding scientists. It is a gold mine of information and is written in Asimov's lively, interesting style.

Butterfield, Herbert: *The Origins of Modern Science*, rev. ed., Free Press, New York, 1965. This stimulating book is the source of the quotation used at the beginning of this chapter.

Cohen, I. Bernard: *The Birth of a New Physics*, Doubleday Anchor, Garden City, N.Y., 1960. An account of the de-velopment of physics from Copernicus to Newton by a well-known historian of science.

De Santillana, Georgio: *The Crime of Galileo*, University of Chicago Press, Chicago, 1955. An interesting discussion of Galileo's problems with the Roman Inquisition. Santil-lana draws some intriguing parallels between the Galileo case and the case of J. Robert Oppenheimer.

Geymonat, Ludovico: *Galileo Galilei*, McGraw-Hill, New York, 1965. A very good brief biography of Galileo.

Gingerich, Owen: "The Galileo Affair," *Scientific American*, vol. 247, no. 2, August 1982, pp. 132–143. A balanced account of Galileo's confrontation with the church.

McCloskey, Michael: "Intuitive Physics," *Scientific American*, vol. 248, no. 4, April 1983, pp. 122–130. Tests show that the misconceptions of people, including physics students, about motion neatly parallel the beliefs about motion in the three centuries *before* Newton.

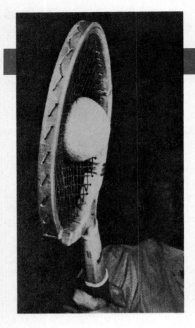

CHAPTER 3

Forces and Newton's Laws of Motion

Nature to him [Newton] was an open book, whose letters he could read without effort. In one person he combined the experimenter, the theorist, the mechanic and, not least, the artist in expression.

Albert Einstein (1879–1955)

We have all wondered at one time or another why certain physical events occur. Why does the earth continue to move in its orbit about the sun instead of going off into the darkness of space? Why do compass needles always point north? Why do electrons in our TV sets strike the fluorescent screen with sufficient speeds to make the tube face glow? In every case the answer is that *forces* control the motion of the objects in question. As far as we now know, there are four kinds of basic forces in the universe: a *gravitational force* that keeps the earth in its orbit around the sun, *electromagnetic forces* that align the compass needle in a north-south direction and accelerate the electrons in a TV tube, and the *strong* and *weak nuclear forces* that explain the stability (and the occasional instability) of atomic nuclei.

In this chapter we begin a discussion of forces that continues throughout the rest of the book. We consider how these forces produce accelerations, in accordance with Newton's laws, and how these accelerations determine the motion of objects. In so doing, we again limit ourselves to *rectilinear motion*. Once the basic ideas of forces and motion are clear, it will be easier to apply them to two-dimensional motion.

3.1 The Nature of Force

Suppose a man wants to move a shopping cart, initially at rest, down an aisle in a supermarket at a speed of 1 m/s. What does he do? He pushes or pulls the cart and thus *changes its velocity* from 0 to 1 m/s in the direction he wants to go. In more technical terms we say that he exerts a *force* on the cart, and this force produces an *acceleration*.

Force: A push or pull that, when applied to an object, tends to accelerate that object.*

---

*A more complete definition of force is given after the idea of *momentum* is introduced in Chap. 7.

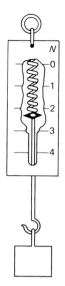

FIGURE 3.1 A spring balance used to measure forces.

Force is a *vector*, since the direction of the applied force determines the direction of the acceleration (another vector) that results. For instance, the force a golf club exerts on a golf ball must be in the general direction of the proper green on the course. We designate a force by the symbol $\mathbf{F}$, and its magnitude by F.

Forces may be measured with spring balances of the kind shown in Fig. 3.1. Pulling on the end of the spring causes an elongation that is directly proportional to the applied force. If the force doubles, the spring stretches twice as much. The numerical scale on the spring balance indicates the magnitude of the force directly. In the SI system, the unit of force is the newton (N), which is defined in Sec. 3.3. In the British engineering system, the unit of force is the pound, where 1 lb = 4.45 N.

Since forces are vectors, they must be added as vectors. All the techniques for adding vectors discussed in Chap. 1 therefore apply to forces. For example, if two men engage in a test of strength by pushing in opposite directions on a heavy crate, the direction each man pushes is of crucial importance. If one man exerts a force of magnitude 100 N along the positive x axis and the other a force of 90 N along the negative x axis, as in Fig. 3.2, the net external force on the crate along the x axis is then

$$\mathbf{F}_{net} = \mathbf{F}_1 + \mathbf{F}_2 = 100 \text{ N} - 90 \text{ N} = +10 \text{ N}$$

This resultant force is in the same direction as $\mathbf{F}_1$. In what follows we use the symbol $\mathbf{F}_{net}$ for the resultant of all the external forces acting on any object of interest. This resultant we call the *net external force*. It is this net external force that is important in determining the acceleration of objects, since all the external forces on an object are equivalent to this one resultant force and can be replaced by it.

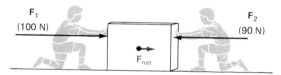

FIGURE 3.2 Two men attempting to push a heavy crate in opposite directions. The net force acting is 10 N along the positive *x* axis.

3.2 Newton's First Law of Motion

The contributions to the study of motion made by many physicists, especially Galileo and Sir Isaac Newton (see Fig. 3.3 and accompanying biography), were combined by Newton in his *Principia* of 1687 into three generalizations now called Newton's laws of motion.* The first of these is:

Newton's First Law of Motion: An object at rest remains at rest and an object in motion continues to move in a straight line at constant speed unless some net external force changes this state of rest or motion.

---

*Although the name *Newton's laws of motion* is commonly used, a better title might be *Newton's generalizations about motion*. They describe how nature behaves under certain conditions. Although these conditions are most often met, there are situations (very high velocities) where Newton's laws are inadequate and a more exact description of motion requires Einstein's theory of relativity.

(Rest and motion are here measured with respect to some frame of reference such as the earth's surface.)

The second part of this statement may seem to contradict our experience that a shopping cart soon comes to a stop unless we keep pushing it. This is because frictional forces (e.g., the rubbing together of rough spots on the hub of the wheel and the wheel supports) act to slow the cart down. If we lived in a world without friction or air resistance, however, the cart would keep moving with constant velocity until it collided with another object (friction and air resistance are treated in more detail in Secs. 3.6 and 3.7).

The general validity of Newton's first law can be seen if we examine situations in which we gradually reduce the forces opposing the motion. For example, if we push a metal puck over a smooth metal surface, it soon comes to a stop after being released, because the rubbing of the surfaces produces a retarding, frictional force. If we pour water over the surface, the friction is somewhat reduced and the puck will travel farther after being released. If we freeze the water on the surface, the puck will travel even farther. Finally, if we float the puck on a layer of air between it and the table, as is done in air hockey games at arcades, a puck set in motion will travel a very long distance on the table before coming to rest. In the limit, i.e., in the ideal case of no friction and no air resistance, we can therefore expect that Newton's first law will be satisfied; the puck will indeed continue to move in a straight line at a constant speed forever.

Before Galileo's time it was believed that a force was necessary to keep an object moving with constant velocity. The great contribution of Galileo and Newton was to show that no net force is necessary to do this: forces are necessary only to produce *changes* in velocities, i.e., to produce *accelerations*, not to keep an object moving with constant speed in a straight line. For example, a baseball hit by a batter would continue to move in a straight line forever if there were no air resistance to slow it down and no force of gravity to pull it toward the earth. No external force is required to keep it moving in a straight line once the batter has imparted an acceleration to the ball with the bat. What keeps the baseball moving, once it has been hit, is not any force applied to it, but its *inertia*.

Newton gave the name *inertia* to that property of an object which keeps its velocity unchanged unless some net external force acts on it. Newton's first law is therefore sometimes called the *law of inertia*. Expressed loosely, it says that material objects have an innate sluggishness, a reluctance to change, which we call *inertia*. As a consequence of its inertia, a ball thrown from a rocket ship in free space far from any stellar object would move along a straight line with the same speed that it had when it left the thrower's hand. The ball would change its velocity only when it got close enough to other objects that forces could be exerted on it.

3.3 Newton's Second Law of Motion

If a net external force acts on an object, it changes the velocity of the object. But a change in velocity with time is an *acceleration*. Hence *the effect of a net external force acting on an object is to accelerate that object.*

Suppose we have a metal puck of mass m on a frictionless air table and apply known forces to the puck by pulling the puck with an accurately calibrated spring balance, as in Fig. 3.4. We first apply a force of magnitude F and find that the acceleration is a in the direction of the force. We then double the force to $2F$ and find that the resulting acceleration is $2a$. If we do this for a large number of forces,

Sir Isaac Newton (1642–1727)

FIGURE 3.3 *(AIP Niels Bohr Library.)*

Born in 1642 at Woolsthorpe, Lincolnshire, England, Isaac Newton is considered by many the greatest scientist who ever lived. He was a delicate child who lived a somewhat solitary life, preferring to construct a variety of mechanical models and devices rather than play games with his schoolmates.

In 1661 he entered Trinity College of Cambridge University but was forced to interrupt his studies when the plague broke out in London. This outbreak was severe enough to close down the university. Newton therefore spent the years 1665 and 1666 at his home in Woolsthorpe, devoting all his energies to private study in mathematics and physics. This period saw some of his greatest discoveries in mechanics, optics, and the calculus, although he did not complete and publish his findings in these fields until much later. As an old man Newton reflected on the roughly 18 months in which he accomplished so much: "For in those days I was in the prime of my age for invention, and minded mathematics and philosophy more than at any time since."

Two years after Newton returned to Cambridge, his professor, Isaac Barrow, became convinced that Newton already knew more mathematics than he did and resigned the Lucasian Professorship of Mathematics in Newton's favor. Newton assumed this prestigious position in 1669 at the age of 26.

During the following years he lectured at the university, but he was a poor lecturer who often seemed to be talking to himself rather than to a class. In 1687 Newton deposited his Lucasian lectures in the Cambridge University library. It is likely that he never delivered these lectures because no students came. His amanuensis recalled that "oftentime he did . . . for want of Hearers, read to ye Walls."

He made his own lens-grinding equipment, ground lenses and mirrors, and built the first successful reflecting telescope. For this work he was elected a Fellow of the Royal Society at the age of 30.

In 1687 Newton published his *Principia* (or *Mathematical Principles of Natural Philosophy*). This was probably the greatest single event in the history of science. The *Principia* combined Newton's three laws of motion and his law of universal gravitation into a system capable of explaining a great variety of phenomena both on earth and in the heavens. Here was the great synthesis of mechanics that scientists had hoped for.

Newton's election in 1689 as a member of Parliament from Cambridge University began the more public side of his life. He moved to London, served inconspicuously in Parliament (an associate claimed that the only speech he ever made was to ask the usher to close the window), suffered a nervous breakdown that may have been due to mercury poisoning, and in 1695 was appointed warden and later master of the Mint. In these posts he discharged his duties as custodian of Britain's currency in an efficient and dedicated fashion.

In 1703 he was elected president of the Royal Society of London, and held that post until his death in 1727. In 1704 he published the *Opticks*, an account of some of his ingenious work on the physics of the spectrum, interference, color vision, and the rainbow. Increasingly, however, his thoughts were directed to biblical and theological studies and to historical chronology rather than to natural science.

Newton never married and at one time protested bitterly to John Locke, the famous philosopher, about Locke's efforts to "embroil him with women." Newton was absent-minded and inconsistent in his behavior, sometimes open and generous, at other times irritable and spiteful. He was extremely diffident about publishing his important ideas but vicious in attacking those who claimed priority in discoveries he believed to be rightly his. The most famous incident along these lines was his controversy with Leibnitz over the invention of the calculus.

Newton died in 1727 at the age of 84 and was buried with great pomp and ceremony in Westminster Abbey, London. Alexander Pope produced an apt epitaph for Newton:

"Nature and Nature's laws lay hid in night:
God said, Let Newton be! and all was light."

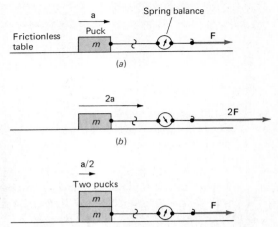

FIGURE 3.4 Measuring the relationship between the force applied and the acceleration produced in a mass on a frictionless table. (*a*) A force *F* produces an acceleration *a*. (*b*) If *F* is doubled, *a* is doubled. (*c*) If *m* is doubled, *a* is cut in half for the same applied force.

we always find that the acceleration is directly proportional to the net force applied and is in the same direction. We can therefore write $a \propto F$, where the proportionality sign ($\propto$) means that a plot of a against F is a straight line.

If we next put two identical pucks on top of each other so as to double the mass, we find that a force of magnitude F produces an acceleration that is exactly half of what it was for one puck. If we use three pucks and the same force, the acceleration is found to be one-third of what it was originally. We conclude that the magnitude of the acceleration a is inversely proportional to the mass accelerated if the applied force is the same, or that $a \propto 1/m$. This means that a plot of a against $1/m$ is a straight line.

If we combine these two results and take into account the vector nature of force and acceleration, we have

$$\mathbf{a} \propto \frac{\mathbf{F}}{m} \quad \text{or} \quad \mathbf{F} = km\mathbf{a}$$

with k a constant of proportionality that depends on the units.* We now define the SI unit of force, the *newton* (N), as follows:

Newton (N): That force which gives a mass of one kilogram an acceleration of one meter per second squared.

$$1 \text{ N} = 1 \text{ (kg)}(1 \text{ m/s}^2)$$

Because of this definition, the constant k must be unity in Eq. (3.1), leading to the result $\mathbf{F} = m\mathbf{a}$, or, for more than one force applied to the object at the same time,

$$\boxed{\mathbf{F}_{\text{net}} = m\mathbf{a}} \tag{3.1}$$

Since this is equivalent to $\mathbf{a} = \mathbf{F}_{\text{net}}/m$, it is customary to express the second law as follows:

Newton's second law of motion: The acceleration of an object, produced by a net

---

*On this change from a direct proportion to an equality, see Appendix 2.A.

external force acting on the object, is directly proportional to the force and in the same direction and is inversely proportional to the mass of the object.

Newton's second law of motion is the most important equation in mechanics and one of the most important equations in physics. It applies to an entire system or to any part of a system. In applying it in particular situations, it is important to isolate one particular object and then to consider all the external forces acting on this object. The vector sum of these external forces we call $\mathbf{F}_{net}$.

Newton's second law, $\mathbf{F}_{net} = m\mathbf{a}$, for motion in a plane, can be resolved into two component equations along perpendicular axes in the plane:

$$(F_{net})_x = ma_x \qquad (F_{net})_y = ma_y \tag{3.2}$$

In this way many practical problems can be reduced to two (or three) one-dimensional problems that can be solved using the methods developed in the previous chapter. In this chapter we limit ourselves to motion in one dimension.

If we graph F_{net} as a function of a for constant m, we obtain a straight line of slope m, as in Fig. 3.5a, since $F_{net} = ma$ is the equation of a straight line of slope m. The larger m is, the steeper the slope and the larger the force needed to produce a given acceleration, as shown in Fig. 3.5b. In other words, the greater an object's *inertia*, the more difficult it is to accelerate it.

There are many forces in physics in addition to the pushes and pulls produced by human or animal muscles. Some, such as frictional forces, act by making immediate contact with the object to be accelerated. Others, such as gravitational and electromagnetic forces, act at a distance without contact.

The many different forces in physics all have the same dimensions and are therefore the same kind of physical quantity. They produce results qualitatively the same (i.e., accelerations), although quantitatively of different orders of magnitude. For example, the gravitational force between two electrons is about 42 orders of magnitude smaller (i.e., smaller by a factor of 10^{42}) than the electric force between the same two electrons separated by the same distance. The acceleration of the electrons produced by these two very different forces would therefore be 10^{42} times less in the gravitational case than in the electrical case. Newton's second law is still applicable in the two cases, however, despite this huge difference in the strength of the two forces and in the resulting accelerations.

FIGURE 3.5 (a) Graph of force applied to an object of mass m against the acceleration produced by that force. In this case the slope of the straight line shows that $m = 8.0$ kg. (b) Graphs of force against acceleration for different masses m.

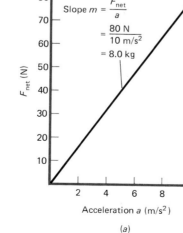

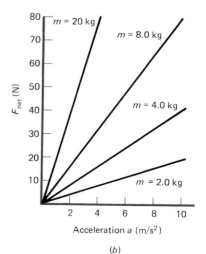

Relationship between First and Second Laws

There is a close relationship between Newton's first two laws. The second law, $F_{net} = ma$, implies that if the net force F_{net} on an object is zero, then the acceleration a is also zero for any value of m. However, if a is zero, then the velocity of the object changes neither in direction nor in magnitude. In other words, if the net external force on an object is zero, an object at rest remains at rest and an object in motion continues to move in a straight line at constant speed. We see, therefore, that the first law is a special case of the second law.

Even though Newton's first law is not independent of his second law, the first law is still important because it leads to a qualitative understanding of the important physical quantities *force* and *inertia*. It also leads to the concept of an *inertial reference system* (to be discussed in Chap. 26), which is defined to be a reference system in which Newton's first law is valid.

Example 3.1

(a) What is the magnitude of the net force required to give a 2000-kg automobile an acceleration of 2.00 m/s? **(b)** What is the displacement of the car after 10.0 s if it starts from rest and the acceleration remains constant over the 10.0-s period?

SOLUTION

(a) We take the direction of the acceleration, and of the motion, as that of the positive x axis. From Newton's second law, the magnitude of the force required is

$$F_{net} = ma = (2000 \text{ kg})(2.00 \text{ m/s}^2)$$

$$= 4.00 \times 10^3 \text{ kg·m/s}^2 = \boxed{4.00 \times 10^3 \text{ N}}$$

since 1 N is equal to 1 kg·m/s².

(b) From Eq. (2.19) for uniformly accelerated motion, we have

$$x = x_0 + v_0 t + \tfrac{1}{2}at^2$$

and so, since $v_0 = 0$, we have for the displacement $x - x_0$:

$$x - x_0 = \tfrac{1}{2}at^2 = \tfrac{1}{2}(2.00 \text{ m/s}^2)(10.0 \text{ s})^2 = \boxed{100 \text{ m}}$$

This displacement is in the same direction as a and F_{net}, along the positive x axis.

EXERCISE 1 How would the required force change if the mass of the auto were reduced to 1000 kg?

Example 3.2

A 15.0-g bullet is accelerated from rest to a speed of 600 m/s as it travels 30.0 cm in a rifle barrel. If we assume that the acceleration is constant, what is the magnitude of the accelerating force?

SOLUTION

We must first find the acceleration of the bullet, since this will lead us, by way of Newton's second law, to the net force required to produce this acceleration. We first use the equations for uniformly accelerated motion to find the needed acceleration. We have, from Eq. (2.20), $v^2 = v_0^2 + 2ax$, with $v_0 = 0$, and $v = 600$ m/s for $x = 0.300$ m, and

$$a = \frac{v^2}{2x} = \frac{(600 \text{ m/s})^2}{2(0.300 \text{ m})} = 6.00 \times 10^5 \text{ m/s}^2$$

Then, from Newton's second law, we have

$$F_{net} = ma = (15.0 \times 10^{-3} \text{ kg})(6.00 \times 10^5 \text{ m/s}^2)$$

$$= \boxed{9.00 \times 10^3 \text{ N}}$$

Example 3.3

Two blocks with masses 10.0 and 20.0 kg are connected by a massless string, as in Fig. 3.6, and rest on a frictionless surface. A constant force of 15.0 N is applied to the heavier block, and both blocks slide along the surface. **(a)** What is the tension in the string? **(b)** What is the acceleration of the two masses?

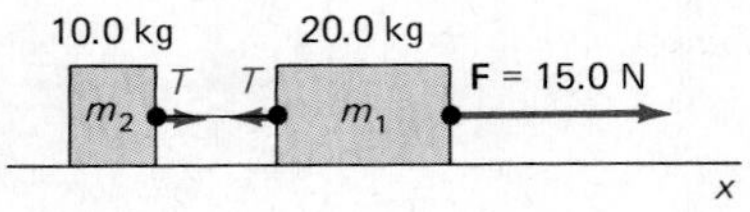

FIGURE 3.6

SOLUTION

This is a good example of Newton's second law applied to a problem that includes more than one object. One approach in this case is to apply Newton's second law to each block separately and then to combine the results to find the acceleration of the two blocks and the tension in the string, which is the force acting to stretch the string.

We choose the positive x axis to be in the direction of the motion. The forces on the heavier block are the applied force **F** along the positive x axis and **T**, the tension in the string, along the negative x axis. The only force on the lighter block is the tension in the string pulling on the block in the positive x direction. Then, from the figure, Newton's second law becomes for the two blocks:

$$(F_{\text{net}})_{1x} = m_1 a_1 \qquad (F_{\text{net}})_{2x} = m_2 a_2$$

or $\qquad F - T = m_1 a_1 \qquad T = m_2 a_2$

But the acceleration of the two blocks is the same, since they are tied together by the string, and so $a_1 = a_2 = a$. Then, on substituting $a = T/m_2$ for a_1 in the first equation, we obtain

(a) $\quad F - T = (m_1)\dfrac{T}{m_2}$

from which $\qquad F = \left(1 + \dfrac{m_1}{m_2}\right)T$

and $\quad T = \dfrac{F}{1 + m_1/m_2} = \dfrac{15.0 \text{ N}}{1 + 20.0/10.0} = \boxed{5.00 \text{ N}}$

(b) Since $T = m_2 a_2 = m_2 a$, we have

$$a = \frac{T}{m_2} = \frac{5.00 \text{ N}}{10.0 \text{ kg}} = \boxed{0.500 \text{ m/s}^2}$$

Notice that the tension is along the length of the string and, for a massless string, has the same magnitude at every point along the string. At one end, the string pulls on the mass m_1 in the negative x direction and, at the other end, on mass m_2 in the positive x direction. The tension in any part of the string may be determined experimentally by replacing that part of the string by a spring balance that will indicate the tension directly.

EXERCISE 2 Solve part (b) of this problem first by writing Newton's second law for the system of the two blocks together to find their acceleration. Once the acceleration is known, the tension is easily found by applying Newton's second law to either block.

3.4 Newton's Third Law of Motion

Another important generalization arrived at by Newton was the following:

Newton's third law of motion: Whenever an object A exerts a force on a second object B, object B exerts a force back on A; these two forces are equal in magnitude and opposite in direction.

In this law the force A exerts on B is called the *action*, and the force B exerts on A is called the *reaction*. For example, the sun exerts a force on the earth according to Newton's law of universal gravitation (Sec. 4.1). This can be called the action force. The earth, in turn, exerts an equal and opposite force on the sun. This is the reaction force.

Newton's third law is often referred to as the *action-reaction law*, which states that to every action there is an equal and opposite reaction. This is a less satisfactory statement of the third law, since it gives the impression that there is one preferred force which must be called the action, with the other force then called the reaction. Actually either force can be chosen to be the action, and the other is then the reaction.

Note that these two forces, the action and the reaction, *are not exerted on the same object.* For example, a physicist pushing a box across a rough floor must exert a force on the box; according to Newton's third law, the box exerts an equal and opposite force on the physicist (Fig. 3.7). This is true whether the box moves or remains at rest. What is important for the motion of the box is the *net force acting on the box*, in this case the force the physicist exerts to move the box forward minus any forces acting on the box in the opposite direction. If the force the physicist exerts is larger than the resisting forces, the box will accelerate. If

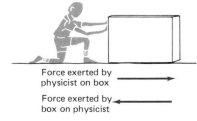

Force exerted by physicist on box ⟶

Force exerted by box on physicist ⟵

FIGURE 3.7 Newton's third law applied to a physicist pushing a box over a rough floor.

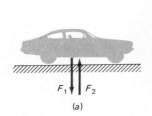

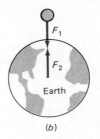

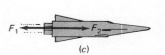

FIGURE 3.8 Some other examples of Newton's third law of motion: Action equals reaction. (a) Car pushes down on road; road pushes up on car. (b) Earth pulls ball down; ball pulls earth up. (c) Rocket pushes gas back; gas pushes rocket forward.

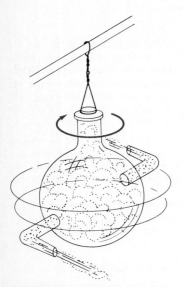

the box is at rest and the force exerted is not sufficient to overcome the opposing forces, then the box will remain at rest. In both cases, however, the force exerted *on the physicist's hands* by the box is *always* equal and opposite to the force the physicist exerts *on the box*, according to Newton's third law. Figure 3.8 gives some other important examples of Newton's third law.

An interesting example of Newton's third law is an early form of the steam engine, called Hero's engine, shown in Fig. 3.9. This engine was developed in Greece around the third century A.D. When the water boils in the spherical container, the steam jets emerging from the two sides drive the container in rotational motion at high speed. This is because the force on the steam driving it out of the container is accompanied by an equal and opposite reaction force exerted back on the container by the emerging steam, according to Newton's third law. It is this reaction force that drives Hero's engine. A similar reaction force drives a rocket or jet plane; gases emerging from the rear of the engine produce reaction forces in the opposite direction that push the rocket or plane forward.

A note of caution: Action-reaction forces *are* equal and opposite, but all equal and opposite forces *are not* action-reaction pairs.

FIGURE 3.9 Hero's engine. The reaction forces from the steam jets on the container set it into rapid rotation.

3.5 Using Newton's Laws of Motion

Because of the importance of Newton's laws of motion in so many fields of physics, we now focus our attention on how these laws can be used to solve practical problems. (For some general ideas about solving physics problems, see Appendix 3.)

Free-Body Diagrams

When we use Newton's second law to solve problems, usually the first thing we must find is $\mathbf{F}_{net}$, the net external force exerted on a particular object in the problem. The best way to obtain $\mathbf{F}_{net}$ is to draw a diagram that eliminates all extraneous detail and shows only the individual forces acting on the object of interest. Such a diagram is called a *free-body diagram* (FBD). Once we have drawn the FBD for the forces acting on the object, we can use the methods of vector arithmetic to obtain $\mathbf{F}_{net}$ and its components. We can then use these components and Newton's second law to calculate the components of the acceleration, from which the complete motion of the object may be obtained.

FIGURE 3.10 Free-body diagram for the horizontal forces in Example 3.3.

For example, the free-body diagrams for the horizontal forces acting on each of the two blocks in Fig. 3.6 are shown in Fig. 3.10. For the forces in the positive x direction, we see from the FBD that $(F_{net})_{1x} = F - T$ and $(F_{net})_{2x} = T$.

Vertical Forces

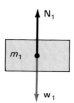

FIGURE 3.11 Free-body diagram for the vertical forces in Example 3.3.

Steps in Applying Newton's Laws

We have seen that all objects near the surface of the earth fall freely with an acceleration $g = 9.80 \text{ m/s}^2$. According to Newton's second law, there must be a force producing this acceleration. This is the attractive gravitational force that the earth exerts on the object, a force that we call the *weight* of the object. Then Newton's second law, $\mathbf{F}_{net} = m\mathbf{a}$, when applied to the force of gravity, becomes $\mathbf{w} = m\mathbf{g}$, where $\mathbf{w}$ is the weight of the object.

The weight is a force in the same direction as $\mathbf{g}$ at any place on the earth's surface. The magnitude of the weight (in newtons) is just the mass of the object (in kilograms) times the acceleration due to gravity (9.80 m/s^2). We will treat this topic more carefully in the next chapter, but it is necessary here to understand the vertical forces acting on a block like m_1 in Fig. 3.10.

These vertical forces are shown in Fig. 3.11. The force of gravity pulls the block m_1 down with a force of magnitude $w_1 = m_1 g$. But the block is in contact with a horizontal surface that supports it by exerting an upward force $\mathbf{N}_1$, a normal force because it is perpendicular to the plane of contact of the block with the horizontal surface. This force $\mathbf{N}_1$ must be equal and opposite to $\mathbf{w}_1$, since there is no acceleration in the vertical direction and therefore no net force in that direction. We have

$$(F_{net})_{1y} = N_1 - w_1 = 0 \qquad \text{or} \qquad N_1 = w_1 = m_1 g$$

so that the normal force is numerically equal to the weight of the block.

Newton's second law of motion is the most important equation in all mechanics. In applying it and Newton's two other laws to problems, a systematic approach is needed. The following is suggested:

1 Draw a diagram of the physical problem indicating all masses, lengths, and angles on the diagram, but no forces.

2 Select one object that you wish to consider first (say, block m_1 in Example 3.3).

3 Draw a *free-body diagram* (FBD) showing the forces acting on this object. Be sure to include all forces acting *on* the object of interest but no forces exerted by the object on other parts of the system. (Forces exerted *by* the object, according to Newton's third law, are exerted *on* other objects. Such forces must therefore be included in the FBDs of the other objects.) In general, all objects in contact with the object of interest exert forces on it, and the force of gravity acts on objects without contact, since gravity acts at a distance.

4 *Choose* a convenient coordinate system, and find the components of $\mathbf{F}_{net}$ along the chosen rectangular axes by combining the components of the external forces acting on the object.

5 Apply Newton's second law to the forces acting on the object under consideration, using the component equations:

$$(F_{net})_x = ma_x \qquad (F_{net})_y = ma_y$$

and calculate the accelerations produced along the chosen axes. These accelerations may then be used to find the velocity and displacement of the object as a function of time, if this is desired.

6 Select another object in the original diagram of the physical situation and repeat the above steps, including a new FBD. This must be done for all important objects in your original diagram. Often you will have to combine results from this step with those from step 5 to solve a problem completely.

These ideas are illustrated in the problems that follow.

Example 3.4

A block of mass $m_2 = 2.00$ kg moves on a level, frictionless surface. It is connected by a light, flexible, nonstretching cord passing over a light, frictionless pulley to a falling block of mass $m_1 = 1.00$ kg, as in Fig. 3.12a. **(a)** What is the acceleration of the system? **(b)** What is the tension in the cord between the two blocks?

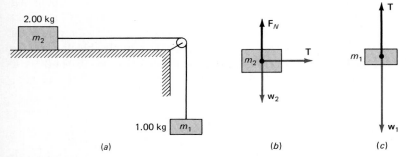

(a)   (b)   (c)

FIGURE 3.12

SOLUTION

This is another example of a problem where it is helpful to consider each of the two masses separately and then combine the results.

(a) The FBDs are given in Fig. 3.12b and c. Applying Newton's second law to the forces on m_1, we have, choosing downward as the positive y direction,

$$(F_{net})_{1y} = w_1 - T = m_1 g - T = m_1 a$$

The vertical forces on m_2 cancel and, for the horizontal forces, we have

$$(F_{net})_{2x} = T = m_2 a$$

choosing motion to the right as in the positive x direction. In these equations m_1 and m_2 are known and T and a are to be found. The tension in the cord must be the same throughout, since both the cord and the pulley are assumed to be massless. Also, the acceleration a must be the same in magnitude for m_1 and m_2, since they are tied together by a massless, unstretchable cord. We can therefore eliminate T between the two equations above by adding them. We obtain

$$m_1 g = (m_1 + m_2)a$$

or $$a = \frac{m_1 g}{m_1 + m_2}$$

$$= \left[\frac{1.00 \text{ kg}}{(1.00 + 2.00) \text{ kg}}\right](9.80 \text{ m/s}^2) = \boxed{3.27 \text{ m/s}^2}$$

(b) Since $T = m_2 a$, we have

$$T = (2.00 \text{ kg})(3.27 \text{ m/s}^2) = \boxed{6.54 \text{ N}}$$

This checks out well, since 6.54 N produces the same acceleration in the 2.00-kg mass as does $(9.80 - 6.54)$ N = 3.26 N in the 1.00-kg mass. Notice that the 1-kg mass acted on by gravity is able to move the 2-kg mass along the surface of the table, even though the 1-kg mass is smaller. This should not surprise us, since there is assumed to be no frictional or other opposing force in the horizontal direction. For this reason even the smallest tension in the string would still be enough to accelerate the 2-kg block along the table.

Example 3.5

An Atwood's machine (Fig. 3.13) is sometimes used in physics laboratories to explore the motion of falling objects and to measure the acceleration due to gravity. It consists of two masses tied together by a light cord that is hung over a pulley.

Both pulley and cord are assumed to be massless, and the pulley is frictionless. If the masses are $m_1 = 22.5$ kg and $m_2 = 35.6$ kg, what is their acceleration?

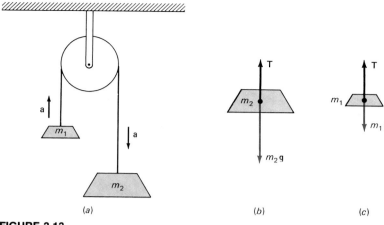

FIGURE 3.13

SOLUTION

In this case the best approach is again to apply Newton's second law to the two separate parts of the system and then to combine the results. Since the larger mass will move down and the smaller mass up, we take motion downward as positive for m_2 and motion upward as positive for m_1. The application of Newton's second law then yields (see the FBDs in Fig. 3.13b and c)

For mass 1: $(F_{net})_1 = T - m_1 g = m_1 a$ [1]

For mass 2: $(F_{net})_2 = m_2 g - T = m_2 a$ [2]

The acceleration a is the same for the two masses since they are tied together by the cord. The tension T must also be the same throughout the cord, since both the cord and the pulley are assumed to have no mass.

We can add Eqs. [1] and [2] above and obtain

$$m_2 g - m_1 g = (m_1 + m_2)a$$ [3]

or $a = \dfrac{(m_2 - m_1)}{m_1 + m_2} g$

$$= \left[\frac{(35.6 - 22.5)\ \text{kg}}{(35.6 + 22.5)\ \text{kg}} \right] (9.80\ \text{m/s}^2) = \boxed{2.21\ \text{m/s}^2}$$

The tension is the same throughout the cord only if the cord can be considered to have zero mass and if the pulley is massless and its bearing frictionless. If the cord has a measurable mass, a net unbalanced force must be applied to accelerate it and the tension at the two ends of the cord must be different. Also, the pulley must be considered massless, for the presence of any mass would require a net difference in the forces on the two sides of the pulley to rotate the pulley.

The secret of physics' success in describing reality is precisely its ability to omit complications like the masses of the cord and the pulley in order to solve highly idealized problems such as this one. Once the physical principles are grasped, we can reintroduce the omitted complications and solve the more realistic problem.

EXERCISE 3 Show that Eq. [3] can also be obtained by applying Newton's second law to the system as a whole rather than to the two masses separately.

Example 3.6

A passenger of mass 100 kg is standing on a spring scale on the floor of an elevator. What apparent weight does the scale read for the passenger when **(a)** the elevator is moving with constant speed; **(b)** the elevator is accelerating upward at 2.00 m/s²; **(c)** the elevator is accelerating downward at 2.00 m/s²; **(d)** the elevator cable breaks and the elevator falls freely under gravity?

SOLUTION

This is a different kind of problem in which the acceleration of the passenger is known, since it is the same as that of the elevator. We desire to find the upward force exerted by the scale on the passenger, since, by Newton's third law, this is equal and opposite to the downward force exerted by the passenger on the scale. This latter force is the *apparent weight* of the passenger as read off the scale.

As usual, the solution of the problem depends on Newton's

second law. We choose a coordinate system fixed with respect to the elevator shaft and take upward forces and accelerations as positive. Then Newton's second law, when applied to all the forces acting on the passenger, yields

$$(F_{net})_y = F - w = ma$$

In this equation F is the upward force exerted on the passenger by the scale, w is the passenger's true weight, and m is the passenger's mass. To solve for F all we need do is substitute the correct value for the passenger's acceleration in the above equation. The forces acting on the passenger in each case are shown in the FBDs in Fig. 3.14a to d.

(a) In this case $a = 0$, and so $F = w$. The scale reads the passenger's true weight, which is

$$mg = (100 \text{ kg})(9.80 \text{ m/s}^2) = \boxed{980 \text{ N}}$$

(b) In this case the acceleration is positive, since the elevator accelerates upward.

$$F_{net} = F - mg = ma \qquad \text{or} \qquad F = m(g + a)$$

$$F = (100 \text{ kg})(9.80 + 2.00) \text{ m/s}^2 = \boxed{1.18 \times 10^3 \text{ N}}$$

The passenger's apparent weight is therefore larger than the real weight by 200 N.

(c) In this case the acceleration is negative, since it is downward, and so

$$F = mg + ma = (100 \text{ kg})(9.80 - 2.00) \text{ m/s}^2 = \boxed{780 \text{ N}}$$

It is seen that the passenger's apparent weight is smaller than the true weight.

(d) If the cable breaks, a is negative and numerically equal to g. Then

$$F = mg + ma = mg - mg = 0$$

The passenger thus appears to be *weightless*, since there is no upward push of the scale on the passenger's feet. This is true of all objects in any freely falling system like this elevator. In this case the weightlessness is only *apparent* and is produced by the accelerating reference frame. *True* weightlessness can only occur for an object in space so far removed from other objects that it experiences no measurable gravitational force.

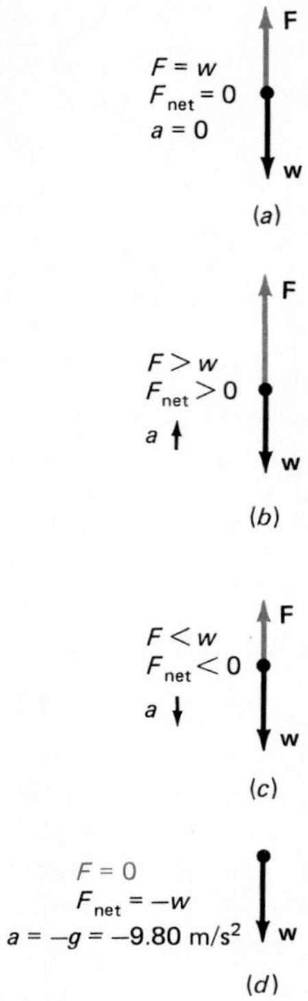

$F = w$
$F_{net} = 0$
$a = 0$

(a)

$F > w$
$F_{net} > 0$
$a \uparrow$

(b)

$F < w$
$F_{net} < 0$
$a \downarrow$

(c)

$F = 0$
$F_{net} = -w$
$a = -g = -9.80 \text{ m/s}^2$

(d)

FIGURE 3.14

Example 3.7

A man is pulling his daughter on a sled over packed snow at a constant speed. He applies a pulling force of 50.5 N with a light rope at an angle 37.0° above the horizontal. The mass of the sled and girl together is 31.5 kg. **(a)** What is the force acting to retard the sled's motion? **(b)** What is the upward force exerted by the packed snow on the sled?

SOLUTION

The problem is illustrated in Fig. 3.15a, and an FBD is given in Fig. 3.15b.

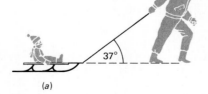

FIGURE 3.15

(a)

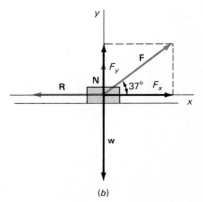

(b)

(a) We must apply Newton's second law to the two components of the motion separately. We are told that the motion along the x axis is at constant speed, and so the acceleration a_x is zero. We have

$$(F_{net})_x = ma_x = 0 \quad \text{or} \quad F \cos 37.0° - R = 0$$

where R is the force retarding the sled's motion and preventing it from picking up speed. Therefore

$$R = F \cos 37.0° = (50.5 \text{ N})(0.800) = \boxed{40.4 \text{ N}}$$

(b) The upward force exerted by the snow on the sled is the reaction to the force exerted by the sled on the snow. We call this reaction force N because it is normal to the snow surface.

If we apply Newton's second law to the y components of the forces, we have

$$(F_{net})_y = N + F_y - w = ma_y = 0$$

since there is no acceleration in the vertical direction. Therefore

$$N = w - F_y = mg - F \sin 37°$$

$$= (31.5 \text{ kg})(9.80 \text{ m/s}^2) - (50.5 \text{ N})(0.600) = \boxed{278 \text{ N}}$$

The upward force is therefore less than the weight of the sled and girl (309 N) because the force along the rope has a component that lifts the sled and reduces the force it exerts on the snow.

3.6 The Force of Friction

We now want to consider some of the forces that oppose motion, like the force in Example 3.7. These we call *resistive* or *dissipative forces*. Examples include friction, air resistance, and the viscosity of fluids. The effect of these forces is to resist the motion of an object and, if motion occurs, to dissipate energy, i.e., to convert some mechanical energy from the moving object into another kind of energy such as heat.

Friction is a good example of a resistive force.

Force of friction: A force, produced by the contact of two surfaces, that opposes the relative motion of one object with respect to the other.

The force on each object is opposite to the direction of its motion with respect to the other object. The forces on the two objects therefore make up an action-reaction pair. Figure 3.16 shows a crate being pushed from left to right over the horizontal surface of a driveway by an applied force **F**. The frictional force **f** of the driveway on the crate acts to the left to oppose the motion of the crate to the right. (Similarly, the frictional force the crate exerts on the driveway acts to pull the driveway to the right.) The net external force acting on the crate in the positive x direction is in this case $F_{net} = F - f$.

Frictional forces can also act when there is no relative motion. A horizontal push given a stalled automobile with its brakes locked may not be enough to start it moving because the frictional force exerted by the road on the tires just balances the push exerted on the car and keeps the car from sliding. The sum of the external forces acting on the car is zero and the car remains at rest, according to Newton's first law. There is, of course, some upper limit to the frictional force possible. This depends on the surfaces of the tires and the road and on the weight of the car. Once this maximum frictional force is exceeded, the car starts to slide and, according to Newton's second law, picks up speed because there is now a net external force acting.

Experimental investigations reveal that the force of friction depends on the nature of the two surfaces, e.g., whether they are rough or smooth, and also on how intimately they are in contact. The great artist and engineer Leonardo da Vinci (1452–1519) first found that the frictional force depends on the normal force between the two surfaces, i.e., the force exerted by one on the other perpendicular to the surface of contact, since this increases the closeness of contact. In the

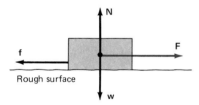

FIGURE 3.16 Forces on a crate being pushed over a driveway by a force **F**. The frictional force **f** between the driveway and the crate opposes the applied force.

example of the crate in Fig. 3.16, the normal force N is equal and opposite to the weight of the crate. We can write in general that the force of friction f is proportional to N:

$$f \propto N \qquad \text{or} \qquad f = \mu N \tag{3.3}$$

where N is the *normal force* and μ is the *coefficient of friction*, which depends on the nature of the two surfaces in contact. Notice that μ is a dimensionless number, since it is a ratio of two forces.

It is important to distinguish between static and sliding friction:

Static friction: The frictional force exerted by one surface on another when there is no relative motion of the two surfaces.

Sliding (kinetic) friction: The frictional force exerted by one surface on another when one surface slides over the other.

Experimentally it is found that sliding friction is less than the maximum value of static friction, since it takes a smaller force to keep an object moving with constant velocity against a frictional force than it does to start the object moving from rest. Also, as noted above, the static frictional force exerted on an object always just balances out the applied force until the point is reached at which the object starts to move. The static frictional force can therefore assume any value from zero up to the maximum value at which sliding begins. Once sliding starts, the sliding frictional force does not usually depend on the speed of the sliding object. We can summarize these facts in the form of two equations for the forces of sliding and static friction:

$$f_{\text{sliding}} = f_k = \mu_k N \tag{3.4}$$

$$f_{\text{static}} = f_s \leq \mu_s N \tag{3.5}$$

where μ_k is the coefficient of sliding (kinetic) friction, μ_s is the coefficient of static friction, and $\mu_s N$ is the maximum value of the force of static friction. These equations are crude approximations used to describe a very complicated problem and are far from fundamental laws of nature.

The frictional forces can be graphed approximately, as in Fig. 3.17. Table 3.1 contains coefficients of friction for various substances in contact. The low value

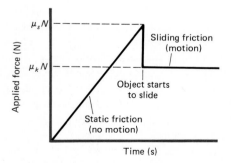

FIGURE 3.17 Schematic diagram for frictional forces. The applied force is gradually increased until the object starts to slide, at which point the force of static friction is $\mu_s N$. The force required to keep the object moving at a constant speed then falls to some constant value $\mu_k N$.

TABLE 3.1 Coefficients of Static and Sliding Friction

Material	Coefficient of static friction μ_s	Coefficient of sliding friction μ_k
Dry, unlubricated surfaces:		
Hard steel on hard steel	0.78	0.42
Mild steel on lead	0.95	0.95
Aluminum on steel	0.61	0.47
Glass on glass	0.94	0.4
Oak on oak (parallel to grain)	0.62	0.48
Oak on oak (perpendicular to grain)	0.54	0.32
Teflon on Teflon	0.04	0.04
Teflon on steel	0.04	0.04
Rubber tire on concrete (low speed)	0.9	0.7
Rubber tire on concrete (high speed)	0.6	0.35
Wet surfaces:		
Rubber tire on wet concrete (low speed)	0.7	0.5
Steel on ice (ice skates)	0.02	0.01
Waxed skis on wet snow	0.14	0.10

of μ for Teflon indicates why Teflon powder is often used as a lubricant in place of oil or grease, and why Teflon pots and pans have nonstick surfaces.

Although friction has many disadvantages and produces heat and wear, it also has many important uses. It lets the wheels of a locomotive grip the track and move a train forward. It makes walking easier by keeping our shoes from slipping on the sidewalk. Walking on ice, however, is difficult because the coefficient of friction is so low.

Example 3.8

A steel puck slides over a steel surface. **(a)** If the mass of the puck is 2.00 kg, what is the force of friction opposing the motion? **(b)** If the initial speed of the puck is 3.00 m/s, how long will it take for the puck to come to rest?

SOLUTION

(a) From Eq. (3.4), the force of sliding friction is $f_k = \mu_k N$. In this case the normal force N is equal to the weight of the puck w, since no other vertical force acts. We have for N, therefore,

$$N = w = mg = (2.00 \text{ kg})(9.80 \text{ m/s}^2) = 19.6 \text{ N}$$

Then, using Table 3.1 to find μ_k, we have

$$f_k = \mu_k N = (0.42)(19.6 \text{ N}) = \boxed{8.2 \text{ N}}$$

Note that we give this result to only two significant figures, since the value given for μ_k was only good to two significant figures.
(b) We choose the positive x axis as the direction of the initial motion. Because the force of friction opposes the motion of the puck, the puck's acceleration is opposite to its velocity and is therefore in the negative x direction. This is the only force acting in the horizontal direction, and so Newton's second law yields

$$a = \frac{f_k}{m} = \frac{-8.2 \text{ N}}{2.00 \text{ kg}} = -4.1 \text{ m/s}^2$$

We can then use Eq. (2.18) to obtain the time required for the puck to come to rest.

$$t = \frac{v - v_0}{a} = \frac{0 - 3.00 \text{ m/s}}{-4.1 \text{ m/s}^2} = \boxed{0.73 \text{ s}}$$

So the force of friction causes the puck to come to rest in less than 1 s. If the surfaces were lubricated, the force of friction and thus the coefficient of friction would be reduced, and the puck would continue moving for a longer time before stopping.

EXERCISE 4 In Example 3.7 what is the coefficient of sliding friction μ_k between the sled and the snow if sliding friction is the only retarding force acting? Note that in this case the normal force is *not* equal to the weight of the sled and girl.

3.7 Air Resistance

When a golf ball soars through the air, it has to push the air out of the way. This sets up a reaction force that resists the motion of the ball and slows it down. This force of air resistance depends in a complicated way on the size, shape, and speed of the ball and on the properties of the air.

For small objects moving at low velocities, this resistive force is found, to a good approximation, to be directly proportional to the velocity of the moving object but in the opposite direction. We can therefore write for the force of air resistance

$$\mathbf{f} = -K\mathbf{v} \tag{3.6}$$

where K is a constant that depends on the properties of the moving object and of the air.

Terminal Velocity

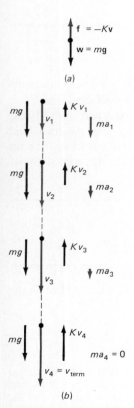

(a)

(b)

FIGURE 3.18 Fall of a ball under the force of gravity in air. (a) FBD showing the force of air resistance $-Kv$. (b) The acceleration is gradually reduced because the air resistance increases with the speed of the falling object. Finally, when $v_4 = v_{term}$, the resistive force is exactly equal to the force of gravity on the object, and the resulting acceleration is zero.

When air resistance acts, we find that the velocity of a freely falling object does not increase without limit. Rather it approaches a *terminal velocity* that depends on the nature of the falling object and the resisting medium. Once that velocity is reached, no further acceleration occurs and the object continues to move at its terminal velocity.

To clarify this idea of terminal velocity, let us consider an object falling under the influence of gravity in air that produces a resistive force $-Kv$ on the object. The FBD for the forces on this falling object is shown in Fig. 3.18a, where the downward direction is taken as the direction of the positive y axis. Then Newton's second law becomes

$$F_{net} = mg - Kv = ma \tag{3.7}$$

The net force acting is the sum of the downward force of the object's weight (mg) and the upward force due to the resistance of the air ($-Kv$). Because of our choice of a coordinate system, the downward forces, velocities, and accelerations are all positive. As the object falls, v increases continuously, and so Kv also increases and approaches the value mg, as shown in Fig. 3.18b. When the magnitude of the resistive force Kv equals that of the gravitational force mg, we have

$$Kv = mg \qquad \text{and} \qquad F_{net} = mg - Kv = 0$$

and so $\qquad ma = 0 \qquad$ and $\qquad a = 0$

There is no further acceleration, and the subsequent motion of the object is at a terminal velocity obtained from the condition that

$$mg - Kv = 0 \qquad \text{and so} \qquad v_{term} = \frac{mg}{K} \tag{3.8}$$

We see therefore that the terminal velocity is proportional to the weight of the object and is inversely proportional to the air resistance. In this sense Aristotle's belief that the velocity of a falling object was proportional to its weight was partially valid, although he overlooked the role that air resistance plays.

By experiment we find that, for objects of the same shape and density falling in the same resistive material, the heavier the object, the greater its terminal velocity. For example, a raindrop 2 mm in diameter falls with a terminal velocity in air of about 4.5 m/s, whereas a hailstone of the same size has a terminal velocity of 4.0 m/s because ice has a lower density than water. Similarly, large raindrops have larger terminal velocities than small raindrops.

Example 3.9

A raindrop of mass 4.21×10^{-6} kg falls freely in air, with a terminal velocity of 4.50 m/s. **(a)** What is the value of the constant K that determines the resistive force? **(b)** If the raindrop is replaced by an iron pellet of exactly the same size and shape but of mass 3.30×10^{-5} kg, what is its terminal velocity? **(c)** For how long would the iron pellet have to fall in a vacuum to acquire a velocity equal to its terminal velocity in air?

SOLUTION

(a) We have, from Eq. (3.8),

$$K = \frac{mg}{v}$$

$$= \frac{(4.21 \times 10^{-6} \text{ kg})(9.80 \text{ m/s}^2)}{4.50 \text{ m/s}} = \boxed{9.17 \times 10^{-6} \text{ kg/s}}$$

(b) For the iron pellet, K remains approximately the same, since the pellet has the same size and shape as the raindrop. Therefore the terminal velocities are directly proportional to the weights, and so

$$v'_{\text{term}} = \frac{3.30 \times 10^{-5} \text{ g}}{4.21 \times 10^{-6} \text{ g}} (4.50 \text{ m/s}) = \boxed{35.3 \text{ m/s}}$$

(c) In the absence of air resistance, the acceleration is 9.80 m/s^2 for the whole trip for both raindrop and iron pellet. We have motion with uniform acceleration under the force of gravity, and so, for the iron pellet,

$$v = v_0 + at = v_0 + gt$$

and $$t = \frac{v - v_0}{g} = \frac{35.3 \text{ m/s} - 0}{9.80 \text{ m/s}^2} = \boxed{3.60 \text{ s}}$$

3.8 Forces for Motion on an Inclined Plane

A good illustration of many of the ideas developed in this chapter is provided by a block sliding down an inclined plane. In Sec. 2.8 we considered the kinematics of this problem; we now consider the dynamics of the same problem, starting with the idealized case of no friction, and then consider how the motion changes when friction is taken into account.

Motion without Friction

Consider a block of mass m sliding down a frictionless plane, as in Fig. 3.19a. The angle of inclination of the plane is θ. We choose a coordinate system with the x axis along the plane and the y axis normal to the plane. Since we assume there is no friction, the only forces acting on the block are the force of gravity, $\mathbf{w} = m\mathbf{g}$, and the normal force of the plane. The weight of the block can be resolved into two components, one along the x axis and the other along the y axis. The component normal to the plane is balanced by the normal force $\mathbf{N}$ of the plane on the block, as shown in the FBD of Fig. 3.19b. The component parallel to the plane is $w \sin \theta$, as shown in Fig. 3.19a and c. Since there is no friction, $w \sin \theta$ is the only force along the plane. So it is the net force on the block. We have, therefore,

FIGURE 3.19 (a) A block sliding down a frictionless plane of inclination angle θ. (b) FBD for the balanced forces normal to the plane. (c) FBD for the unbalanced force accelerating the block down the plane.

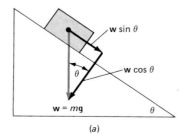

(a)

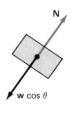

(b)

(c)

$$F_{\text{net}} = w \sin \theta = mg \sin \theta \qquad (3.9)$$

$$\text{and} \quad a = \frac{F_{\text{net}}}{m} = \frac{mg \sin \theta}{m} = g \sin \theta$$

This is the same result for the acceleration already obtained in Sec. 2.8. Note that the acceleration does not depend on the mass of the block.

Example 3.10

A 10.0-kg block (m_2) is on a frictionless plane inclined at 30.0° the horizontal, as shown in Fig. 3.20a. A light cord runs from m_2 to a pulley and then to a 5.00-kg block (m_1) that hangs at the end of the cord. Assume that the cord and pulley are both massless. **(a)** Describe the motion of this system, including the acceleration of the 10.0-kg block. **(b)** Repeat for a 7.50-kg block hanging at the end of the cord.

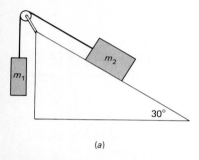

(a)

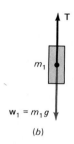

(b)

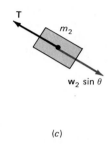

(c)

FIGURE 3.20 (a) Diagram for Example 3.10. (b) FBD for block m_1. (c) FBD for block m_2.

SOLUTION

The physical situation is shown in Fig. 3.20a. The forces on the 5.00-kg block are $w_1 = m_1 g$ directed downward and the tension T in the cord pulling upward in the opposite direction, as in the FBD of Fig. 3.20b. For the 10.0-kg block, the force along the plane away from the pulley is

$$F_2 = w_2 \sin \theta = m_2 g \sin \theta$$

as shown above [Eq. (3.9)], and the force toward the pulley is T, the tension in the cord, as in Fig. 3.20c.

(a) For the motion of the block of mass m_1, we have, assuming motion down as positive,

$$F_{\text{net}} = m_1 g - T = m_1 a$$

For the block on the incline, we assume motion toward the pulley as positive, since the system of two blocks can only move as a unit. We have

$$F_{\text{net}} = T - m_2 g \sin \theta = m_2 a$$

The tensions and accelerations in these two equations must be the same, from the statement of the problem. Adding these two equations, we have

$$m_1 g - m_2 g \sin \theta = m_1 a + m_2 a$$

$$\text{or} \quad a = \frac{m_1 - m_2 \sin \theta}{m_1 + m_2} g.$$

For $m_1 = 5.00$ kg and $m_2 = 10.0$ kg, we have

$$a = \frac{5.00 - 10.0 \sin 30.0°}{5.00 + 10.0} g = \boxed{0}$$

In this case there is no acceleration, since the net force on each of the blocks is zero. The system therefore either remains at rest or, if in motion with a constant speed v, continues in motion with the same speed v.

(b) Here $m_1 = 7.50$ kg, and

$$a = \frac{m_1 - m_2 \sin 30.0°}{m_1 + m_2} = \frac{7.50 - 5.00}{7.50 + 10.0} = \boxed{0.143 \text{ m/s}^2}$$

Since a is positive, the block m_1 moves down, and the block m_2 is pulled up the plane with this same acceleration.

Motion with Friction

If the plane is not frictionless, the motion of a block will be opposed by kinetic friction f_k, as shown in the FBD of Fig. 3.21. The frictional force f_k depends on the normal force N according to $f_k = \mu_k N$. However, the normal force N is equal

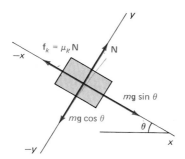

FIGURE 3.21 The forces on a block moving down an inclined plane, with friction present.

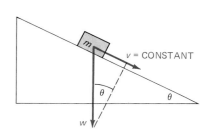

FIGURE 3.22 A block moving down an inclined plane at constant speed so that the acceleration along the plane is zero.

to the component of the weight normal to the plane, and so $N = mg \cos \theta$ and $f_k = \mu_k N = \mu_k mg \cos \theta$. Then for the motion of the block down the plane, Newton's second law yields

$$F_{\text{net}} = mg \sin \theta - \mu_k mg \cos \theta = ma$$

and $\quad a = g \sin \theta - \mu_k g \cos \theta$ (3.10)

In this case also the acceleration a is independent of the mass.

Suppose we observe a block sliding down a plane of angle θ *at a constant speed*, as in Fig. 3.22. Then the sum of the forces both along the plane (x direction) and normal to the plane (y direction) must cancel, since there is no acceleration and so no net force. We have from Eq. (3.10), for the motion in the x direction with the forces the same as in Fig. 3.21 but the acceleration $a = 0$,

$$g \sin \theta = \mu_k g \cos \theta$$

and so $\quad \mu_k = \dfrac{g \sin \theta}{g \cos \theta} = \tan \theta$ (3.11)

In a simple laboratory experiment the angle of an inclined plane can be varied until the block moves down the plane at constant speed. Then the coefficient of sliding friction can be obtained from $\mu_k = \tan \theta$. This is a convenient, though not highly accurate, way of measuring μ_k.

Example 3.11

A 5.0-kg block (m_2) on a 40° inclined plane is attached to a very light cord passing over a massless pulley. A 6.2-kg block (m_1) hangs freely at the other end of the cord. Find the acceleration of the 5.0-kg block along the plane. The coefficient of sliding friction between the block and the surface of the plane is 0.35.

SOLUTION

The physical situation is shown in Fig. 3.23a, and FBDs are given in Fig. 3.23b and c.

The problem is identical to that in Example 3.10 except for the presence of friction. This adds an extra retarding force along the plane of magnitude $f_k = \mu_k N = \mu_k m_2 g \cos \theta$ (see the FBD). In the case of no friction, we found for the acceleration along the plane,

$$a = \frac{m_1 - m_2 \sin \theta}{m_1 + m_2} g$$

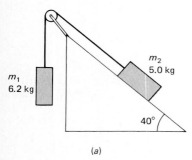

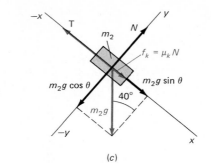

(a)   (b)   (c)

FIGURE 3.23 (a) Diagram for Example 3.11. (b) FBD for block m_1. (c) FBD for block m_2.

In the frictional case this is modified by the introduction of the frictional force, which acts to reduce the acceleration along the plane. This modifies the above equation by the introduction of the additional opposing force $\mu_k m_2 g \cos\theta$:

$$a = \frac{m_1 - m_2 \sin\theta - \mu_k m_2 \cos\theta}{m_1 + m_2} g$$

In this case we have

$$a = \left[\frac{6.2 - (5.0)(0.642) - 0.35(5.0)(0.766)}{6.2 + 5.0}\right](9.80 \text{ m/s}^2)$$

$$= \boxed{1.4 \text{ m/s}^2}$$

EXERCISE 5 Work through this problem from the beginning without assuming the results for the nonfrictional case and show that the same result is obtained as above.

Summary: Important Definitions and Equations

Force: A push or pull that, when applied to an object, tends to accelerate that object. Force is a *vector*.

Newton (N): That force which gives a one-kilogram mass an acceleration of one meter per second squared.

$$1 \text{ N} = (1 \text{ kg})(1 \text{ m/s}^2)$$

Newton's laws of motion

1 An object at rest remains at rest and an object in motion continues to move in a straight line at constant speed, unless some net external force changes this state of rest or motion (law of inertia).

2 The acceleration of an object produced by a net external force acting on the object is directly proportional to the force and in the same direction and is inversely proportional to the mass of the object.

$$\mathbf{F}_{net} = m\mathbf{a} \qquad (3.1)$$

3 Whenever an object A exerts a force on a second object B, object B exerts a force back on A; these two forces are equal in magnitude and opposite in direction.

Free-body diagram (FBD): A vector diagram showing all the individual forces acting on an object.

Special forces

Weight: The attractive gravitational force the earth exerts on an object. The magnitude of this force is $w = mg$. Weight is a force (a gravitational force); mass is a measure of inertia.

Force of friction: A force, produced by the contact of two surfaces, that opposes the relative motion of one object with respect to the other.

Static friction: The frictional force exerted by one surface on another when there is no relative motion of the two surfaces.

$$f_s \leq \mu_s N \qquad (3.5)$$

where μ_s is the coefficient of static friction and N is the normal force between the two surfaces.

Sliding (kinetic) friction: The frictional force exerted by one surface on another when one surface slides over the other.

$$f_k = \mu_k N \qquad (3.4)$$

where μ_k is the coefficient of sliding friction.

Normal force: A force exerted by one object on another perpendicular to the surface of contact between the two.

Air resistance: The reaction force exerted by the air on any object moving through it.

$$\mathbf{f} = -K\mathbf{v} \qquad (3.6)$$

(an approximation valid for small objects moving at slow speeds).

Terminal velocity: The limiting downward speed reached by objects falling in a resistive medium when the force opposing the motion becomes equal to the force of gravity on the object.

$$v_{term} = \frac{mg}{K} \qquad (3.8)$$

Questions

1 Newton's third law tells us that if a horse pulls on a cart with a force of 100 N, then the cart pulls back on the horse with a force of 100 N. If this is so, why do the cart and horse move forward?

2 Why is it necessary to feed more gasoline to a car's motor when accelerating to pass another car than when the car is moving at a constant speed?

3 A man holding a bouquet of flowers is riding in an elevator when the cable breaks. In his panic he drops the flowers. Will the flowers stay near his hand? Hit the floor of the elevator? Hit the roof of the elevator? Give reasons for your answer.

4 Automobile riders often suffer whiplash when their car is suddenly struck from the rear by another car. In whiplash the head is thrown violently backward. Explain why whiplash occurs, using Newton's laws of motion.

5 A brick hangs by a thin cord from a ceiling, and another thin cord dangles from the bottom of the brick.

(*a*) If the bottom cord is suddenly jerked downward, what will happen?

(*b*) If a slow, steady pull is applied to the bottom cord, what will happen? Why?

6 A frictionless puck is given a push along a very long, flat plane that makes contact with the earth's surface at one point. Describe qualitatively the motion of the puck along the plane.

7 A trick that courageous party-goers sometimes perform is to pull the tablecloth out from under the china and glassware. If the trickster is lucky enough to pull this off, which of Newton's laws provides the secret to his or her success? Explain your answer.

8 It has been suggested that, in line with the United States' conversion from British engineering to metric units, a "quarter-pounder" at McDonald's should be renamed a Newton-burger.

(*a*) Why would this be an apt name?

(*b*) If the quarterpounder and the Newtonburger had the same price, which would be the better buy?

9 A man stands on the platform of a large, accurate, platform-type spring scale. Discuss how the scale reading changes when the man tosses a bowling ball into the air and then catches the ball on its way down.

10 Assume that the retarding force of friction is the same regardless of the velocity of a car. Show that, under this assumption, the minimum distance in which the car can be brought to a stop by braking is proportional to the square of the speed of the car at the instant the brakes are applied.

11 A ball is thrown vertically into the air. If air resistance is taken into account, will the time during which the ball rises be equal to the time during which the ball falls? If not, will the time of rise be longer or shorter than the time of fall? Why? (*Hint:* Draw an FBD)

12 Is the terminal velocity the greatest speed a projectile can have? Explain.

13 According to Newton's third law, all action-reaction pairs are equal and opposite. Give an example showing that a pair of equal and opposite forces need not be an action-reaction pair.

14 Consider Atwood's machine (Example 3.5), and show that a massless cord must have the same tension throughout. (*Hint:* Assume that the tension is different in the two vertical segments of the cord. Draw an FBD for the cord. Then apply Newton's second law to the cord and show what happens if the mass of the cord is negligible.)

Multiple-Choice and Simple Exercises

3.1 The net external force required to keep a hockey puck moving with constant speed in a straight line is:

(*a*) Constant (and not zero)

(*b*) Directly proportional to the mass of the puck

(*c*) Directly proportional to the weight of the puck

(*d*) Zero

(*e*) Inversely proportional to the puck's velocity

3.2 A catapult for launching fighter planes from an aircraft carrier produces an acceleration of 30 m/s^2 on a 1.0×10^4 kg plane. The net unbalanced force required to produce this acceleration is:

(*a*) 9.8×10^4 N   (*b*) 1.0×10^4 N

(*c*) 3.0×10^5 N   (*d*) 3.1×10^4 N

(*e*) 2.9×10^6 N

3.3 A book rests on the shelf of a bookcase. The reaction force to the force of gravity acting on the book is:

(*a*) The force of the shelf holding the book up

(*b*) The weight of the book

(*c*) The frictional force between book and shelf

(*d*) The force exerted by the book on the earth

(*e*) None of the above

3.4 A rock of mass 50 kg is resting on a concrete driveway. The magnitude of the force exerted by the driveway on the rock is:

(*a*) 0   (*b*) 50 N

(*c*) 4.9×10^2 N   (*d*) 5.1 N

(*e*) None of the above

3.5 A 2000-kg car is moving along a road with a constant speed of 20 m/s. The net force exerted on the car in the direction of its motion is:

(*a*) 4.0×10^4 N   (*b*) 100 N

(*c*) 2.0×10^4 N   (*d*) 1.96×10^4 N

(*e*) None of the above

3.6 The average net force required to accelerate a 2000-kg car from rest to a speed of 30 m/s in 10 s is:

(*a*) 6.0×10^2 N   (*b*) 6.0×10^3 N

(*c*) 6.0×10^5 N   (*d*) 6.0×10^4 N

(*e*) None of the above

3.7 The weight of a woman on the surface of the earth is 550 N. Her mass is:

(*a*) 550 kg   (*b*) 5.39×10^3 kg

(*c*) 5.39×10^2 kg   (*d*) 56.1 lb   (*e*) 56.1 kg

3.8 A man has a mass of 80 kg. His weight in newtons on the surface of the earth is:

    (a) 80 N   (b) 2.5×10^2 N   (c) 7.8×10^2 N
    (d) 32 N   (e) 9.8 N

3.9 A sky diver has a mass of 100 kg. She jumps from a plane, and at a certain point in her descent the air resistance is 250 N. Her downward acceleration at that point in the descent is:

    (a) 9.80 m/s²   (b) 7.30 m/s²   (c) 0
    (d) 9.80 m/s   (e) 4.90 m/s²

3.10 A sailboat of mass 550 kg is moving over the water at constant velocity.

    (a) Draw a free-body diagram showing the forces acting on the sailboat in both the x and y directions.
    (b) Is the net external force F_x zero?
    (c) Is the net external force F_y zero?

3.11 A trunk of mass 50.0 kg rests on a floor. What constant horizontal force is required to give it a velocity of 20.0 m/s in 10.0 s if we neglect air resistance and friction?

3.12 A force is applied to a 2.0-kg mass and produces an acceleration of 4.0 m/s². The same force applied to a 10-kg mass will produce what acceleration?

3.13 A tow truck of mass 5000 kg is towing a car with a short rope. If the rope has negligible mass, the tow truck is accelerating at 1.00 m/s², and the mass of the car is 1500 kg, what is the tension in the rope? (Neglect friction.)

3.14 (a) An elevator is accelerating upward with an acceleration of 1.20 m/s². What is the upward force exerted by the floor of the elevator on a 90.0-kg passenger?

    (b) If the same elevator accelerates downward with the same acceleration as in part (a), what is the upward force exerted by the elevator floor on the passenger?

3.15 A block of mass 3.0 kg is resting on a frictionless metal table. The block is connected by a light, flexible cord passing over a light, frictionless pulley to a 4.0-kg mass that can fall freely. If the 4.0-kg mass is falling, what is the acceleration of the system?

3.16 A boy is pushing horizontally on a heavy table with a force of 100 N. The table, which weighs 500 N, refuses to budge.

    (a) What is the force of friction acting on the table?
    (b) What is the coefficient of static friction at the instant that the applied force is 100 N?

3.17 A box of 5.0-kg mass is being pulled by a horizontal rope along a floor. If the coefficient of sliding friction between box and floor is 0.62, what is the frictional retarding force on the box?

3.18 The force of air resistance on a raindrop is 5.0×10^{-5} N when it falls with a terminal velocity of 5.5 m/s. What is the mass of the raindrop?

3.19 A skier of mass 100 kg comes down a slope of constant angle 25.0° with the horizontal.

    (a) What is the force on the skier parallel to the slope?
    (b) What force normal to the slope is exerted by his skis?

3.20 A man pulling a 15.0-kg sled with a rope exerts a force of 80.0 N at an angle of 40.0° above the horizontal.

    (a) What is the net force accelerating the sled along the ground, if we neglect friction?
    (b) What is the downward force of the sled on the snow?

Problems .

3.21 A 10.0-kg block is acted on by a single force with components $F_x = 25.0$ N and $F_y = 50.0$ N. Find the magnitude and direction of the block's acceleration.

3.22 Two men decide to pull a truck stuck in mud with their cars. They attach ropes and one pulls horizontally with a force of 1000 N at an angle of 35° with respect to the direction in which the truck is pointing, while the other car pulls with a force of 1200 N at an angle of 29° with respect to the same direction. What is the net forward force pulling the truck out of the mud?

3.23 Two square blocks are in contact on a horizontal frictionless surface. The larger block has a mass $m_1 = 3.00$ kg; the smaller block a mass $m_2 = 2.00$ kg.

    (a) If a horizontal force of 5.00 N is applied to block m_1 to push the two blocks along the surface, what force does m_1 exert on m_2?
    (b) If the force of 5.00 N is applied to block m_2 to move the blocks in the opposite direction, what is the force exerted by block m_2 on block m_1?
    (c) Why are not these two forces the same in magnitude but in opposite directions?

3.24 An empty 200-kg elevator accelerates upward at 1.50 m/s². What is the tension in the cable that lifts the elevator cab?

3.25 A net horizontal force of 2.00×10^3 N is applied to a 1500-kg car being towed by a tow truck. If this force is applied for 5.00 s and the car starts from rest:

    (a) What is the acceleration of the car?
    (b) What is its average speed during these 5.00 s?

3.26 In a test of the safety of automobiles a car is deliberately crashed into a brick wall. If the 2000-kg car is traveling at 40.0 m/s when it hits the wall and its speed is reduced to zero in 0.0100 s:

    (a) What force does the wall exert on the car?
    (b) What force does the car exert on the wall?

3.27 A 90.0-kg astronaut lies on a couch in a rocket accelerating upward with an acceleration of 9.80 m/s². What force does the astronaut exert on the couch?

3.28 A golfer exerts an average force of 50.0 N for 0.035 s on a golf ball weighing 0.451 N.

    (a) Find the average acceleration of the golf ball.
    (b) Find the speed of the ball at the end of the 0.035-s interval.

3.29 An average person jumping straight up in the air by pushing off from the floor with his or her legs can exert a force on the floor roughly equal to twice the person's weight. If this force is applied over a distance of about 0.50 m (from a crouching to a stretching position):

    (a) With what speed does the person leave the ground?
    (b) How high will the person be able to jump?

3.30 A light cord passes over a light, frictionless pulley. It has a 5.00-kg mass at one end and a 10.0-kg mass at the other, both hanging freely.

(a) Find the acceleration of the system.

(b) Find the tension in the cord.

3.31 A tow truck is pulling a disabled car by using a rope that will only stand a tension of 1200 N. If the car has a mass of 1000 kg, what is the largest acceleration possible for the car?

3.32 A 1500-kg car is traveling at 25.0 m/s when the driver takes his foot off the gas pedal. If it takes 5.00 s for the car to slow down to 20.0 m/s, how large is the net force slowing down the car?

3.33 Two masses are suspended from the ends of a massless string over a frictionless pulley. One mass (m_1) is equal to 7.50 kg, and the other mass (m_2) is unknown. If the 7.50-kg mass falls 0.523 m in 1.50 s, starting from rest, what is the mass m_2?

3.34 A 85.0-kg man stands on a spring platform scale on the floor of a moving elevator. Under what conditions of motion for the elevator will the scale read a 30 percent higher weight for this man than when the elevator is at rest?

3.35 A 2500-kg truck starts from rest and reaches a velocity of 20.0 m/s in 35.0 s.

(a) What was the average acceleration of the truck during this 35.0-s interval?

(b) What was the net force acting on the truck to accelerate it?

3.36 A man pushes a 5.0-kg crate along the floor by exerting a force of 80 N on the crate at an angle of 30° downward from the horizontal. The coefficient of sliding friction between crate and floor is 0.45.

(a) What is the normal force exerted by the floor on the crate?

(b) What is the acceleration of the crate?

(c) What distance does the crate move in 5.0 s?

*3.37 A 6.5-kg block is placed on top of a 10-kg block that rests on a table, as in Fig. 3.24. The 6.5-kg block is tied to a wall by a string, and a force of 155 N is applied to the 10-kg block in a horizontal direction. The coefficient of sliding friction between the two blocks, and between the bottom block and the table is 0.35.

(a) Draw FBDs for each block.

(b) Determine the acceleration of the 10-kg block.

(c) Determine the tension in the string connecting block m_1 to the wall.

*3.38 Two blocks are arranged at the ends of a massless string, one (m_2) resting on a table, the other (m_1) hanging freely, as in Fig. 3.12; m_1 is 1.5 kg and m_2 is 2.5 kg. The system starts from rest and, when m_1 has fallen through 0.40 m, its downward speed is 1.25 m/s.

(a) What is the frictional force between the block and the table?

(b) What is the coefficient of sliding friction between the block and the table?

3.39 A steel shuffleboard puck slides over an outdoor concrete surface, where the coefficient of sliding friction between the two surfaces is 0.40. If the mass of the puck is 0.30 kg, how fast must the puck be traveling when it leaves the player's "pusher" if it is to come to rest in the scoring area, which is 15 m away?

3.40 A 1000-kg car is coasting along a level road at 25.0 m/s.

(a) How large a retarding force is required to stop it in a distance of 60.0 m, if the force is constant over the whole 60 m?

(b) If the brakes stop the wheels completely and the car skids the entire 60 m, what is the minimum coefficient of sliding friction between tires and road that will make this possible?

3.41 A 50-kg crate is at rest on a level floor. The coefficient of static friction is 0.45; the coefficient of kinetic friction is 0.40.

(a) What horizontal force will start the crate in motion from rest?

(b) If the crate is pushed horizontally with a force of 400 N, how far does it move in 3.0 s?

3.42 A bookcase of mass 10 kg is to be moved from one room to another by sliding the bookcase along the floor on a piece of cardboard which acts to reduce the friction that would otherwise result between wood and wood. If the coefficient of sliding friction between cardboard and floor is 0.35, what constant horizontal force is required to give the bookcase a speed of 0.50 m/s in 1.5 s, starting from rest?

*3.43 Two blocks A and C are attached to a third block B resting on a level surface, as in Fig. 3.25. There is friction between block B and the surface with a coefficient of sliding friction $\mu_k = 0.20$. If block A has a mass of 2.0 kg and block C a mass of 10.0 kg, and block B moves to the right with an acceleration of 3.0 m/s²:

(a) What is the magnitude of the mass B?

(b) What is the tension in each cord?

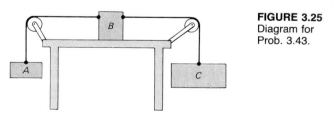

FIGURE 3.25
Diagram for Prob. 3.43.

3.44 A trunk of mass 20 kg rests on a floor. What constant horizontal force pushing the trunk is required to give it a velocity of 10 m/s in 20 s if the coefficient of sliding friction between trunk and floor is 0.65?

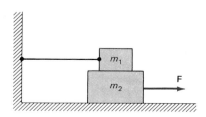

FIGURE 3.24   Diagram for Prob. 3.37.

3.45 The coefficient of kinetic friction between automobile tires and wet concrete is 0.50 at low speeds. What is the minimum distance in which a car traveling at 15 m/s can be brought to a full stop in rain if no skidding occurs?

3.46 A delivery man is loading his truck by pushing a heavy crate of vegetables up a wooden plank into the back of the truck. If the crate has a mass of 30 kg, the plank is at an angle of 25° with respect to the ground, and the coefficients of static and sliding friction are $\mu_s = 0.54$ and $\mu_k = 0.32$:

  (a) How much force must be applied along the ramp to start the crate moving?

  (b) Once the crate starts moving, how much force is required to keep it moving at constant speed?

  (c) If the applied force continues to be that found in part (a), what will be the crate's acceleration?

3.47 A 5.0-kg mass slides on a horizontal surface. The coefficient of sliding friction between the block and the surface is 0.25; the coefficient of static friction is 0.38. The block has an initial speed of 8.0 m/s.

  (a) What is the acceleration of the block?

  (b) What is the speed of the block after 2.0 s?

3.48 Sky divers with open parachutes are often said to hit the ground with terminal velocities about equal to the velocity reached by a person who jumps out of a second-story window (say, 15 ft above the ground).

  (a) Approximately what is this velocity?

  (b) What is the value of the drag coefficient K in this case for a 100-mg sky diver?

3.49 A raindrop of mass 6.0×10^{-6} kg falls in air and reaches a terminal speed of 5.5 m/s.

  (a) What is the value of the constant K that determines the resistive force?

  (b) If there were no air resistance, how long would it take the raindrop to reach the same speed?

3.50 A block of mass 10 kg starts from rest and slides a distance of 8.0 m down an inclined plane making an angle of 40° with the horizontal. The coefficient of sliding friction between block and plane is 0.40.

  (a) What is the net force on the block along the plane?

  (b) What is the speed of the block after sliding 8.0 m?

  (c) What would be its speed if friction were negligible?

3.51 A 10-kg block on an inclined plane of inclination angle 37° is attached to a light cord passing over a pulley to a 20-kg block hanging freely at the end of the cord. Find the acceleration of the 10-kg block (a) if there is no friction; (b) if the coefficient of sliding friction between the block and the surface of the plane is 0.35.

3.52 An elevator starts from rest with a constant upward acceleration and moves 1.00 m in the first 4.00 s. A passenger in the elevator is holding a 10.0-kg bundle at the end of a vertical cord. What is the tension in the cord as the elevator accelerates?

3.53 A 100-kg mass hangs by a cord from the ceiling of an elevator. If we neglect the mass of the cord, what is the tension in the cord when:

  (a) The elevator is accelerating upward at 1.00 m/s²?

  (b) The elevator is accelerating downward at 1.00 m/s²?

  (c) The elevator is accelerating downward at 9.80 m/s²?

3.54 A person is standing on a spring bathroom scale on the floor of an elevator which is being accelerated. If the person's mass is 90.0 kg and the elevator accelerates at 2.50 m/s², what does the scale read for the person's weight (a) when the elevator is accelerating upward; (b) when the elevator is accelerating downward?

3.55 A spring balance is hung from the roof of an elevator cab, and a 10.0-kg mass is hung from the bottom of the spring balance. If the balance is calibrated in newtons, what does it read when:

  (a) The elevator is stopped at a floor?

  (b) The elevator is accelerating upward at 1.50 m/s²?

  (c) The elevator is accelerating downward at 1.50 m/s²?

*3.56 Two blocks A and B are connected by a heavy uniform rope of mass 2.0 kg, as shown in Fig. 3.26. The masses of A and B are 10.0 and 5.00 kg, respectively. If an upward force of 200 N is applied to the top block:

  (a) What is the acceleration of the system?

  (b) What is the tension at point P_1 on the heavy rope?

  (c) What is the tension at point P_2 on the heavy rope?

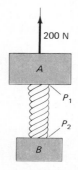

FIGURE 3.26 Diagram for Prob. 3.56.

*3.57 A U-Haul trailer is being pulled by a heavy chain attached to a car. The mass of the trailer is 100 kg and the mass of the chain is 5.00 kg. If the car pulls on the chain with a force of 50.0 N:

  (a) What is the acceleration of the trailer?

  (b) What is the force applied to the trailer by the chain?

  (c) Why is the force different from the force applied by the car?

3.58 A thin string will break under a tension exceeding 30.0 N. This string is attached to the roof of an elevator, and a mass m is hung from the end of the string. If the elevator's acceleration is 2.00 m/s² upward, what is the largest mass that the string can support without breaking?

3.59 An electron in a vacuum tube is accelerated from rest at the cathode to the anode, travels 2.00 cm in a straight line, and reaches the anode with a speed of 5.00×10^6 m/s. We assume that the electric force applied and the resulting acceleration are both constant.

  (a) If the electron's mass is 9.11×10^{-31} kg, what is the force on the electron?

  (b) Are we justified in neglecting the gravitational force on the electron? Compare the gravitational force on the electron with the electric force in part (a).

*3.60 A mass of 2.00 kg hangs from the end of a cord attached to a support on a small cart that rolls on horizontal trolley tracks, as in Fig. 3.27. Find the angle θ made with the

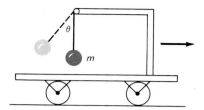

FIGURE 3.27   Diagram for Prob. 3.60.

vertical support, and the tension in the cord, if (*a*) the cart moves with constant speed of 3.00 m/s in a straight line along the tracks; (*b*) the cart moves with a constant acceleration of 3.00 m/s² along the same tracks. (*c*) Draw a FBD to show the forces acting on the mass in each of these cases.

*3.61   A 2.0-kg mass has an initial velocity of 4.0 m/s up an inclined plane. The angle of the inclined plane is 37° above the horizontal. The coefficient of sliding friction is 0.25, and the coefficient of static friction is 0.50.

(*a*)   How much time elapses before the mass returns to its initial position?

(*b*)   How far up the incline does the mass travel before changing directions?

*3.62   A 10-kg block sits at rest on a horizontal surface whose coefficient of sliding friction is 0.25 and whose coefficient of static friction is 0.50. A rope pulls on the block at an angle of 60° above the horizontal.

(*a*)   What is the minimum tension in the rope that will cause the mass to accelerate?

(*b*)   Suppose that the tension in the rope were twice the value obtained in part (*a*) above. Find the normal force and the force of friction acting on the block.

Additional Readings

Andrade, E. N. da C.: *Sir Isaac Newton*, Doubleday Anchor, Garden City, N.Y., 1958. A good brief biography of Newton.

Cohen, I. Bernard: ''Newton's Discovery of Gravity,'' *Scientific American*, vol. 244, no. 3, March 1981, pp. 166–173. A careful presentation of the historical facts about Newton's discovery of the law of universal gravitation.

Gamow, George: *Gravity*, Doubleday Anchor, Garden City, N.Y., 1962. A popular account of both classical and modern theories of the gravitational field.

Heiskanen, Weikko A.: ''The Earth's Gravity,'' *Scientific American*, vol. 193, no. 3, September 1955, pp. 164–174. An interesting account of why *g* varies from place to place on the earth's surface.

Palmer, Frederic: ''Friction,'' *Scientific American*, vol. 184, no. 2, February 1951, pp. 54–58. An article pointing out how difficult it is to explain one of our most common physical phenomena.

Westfall, Richard A.: *Never at Rest*, Cambridge University Press, New York, 1981. A recent and perhaps the most scholarly biography of Newton in English.

$$\frac{kg/m/s^2}{kg} = m/s^2$$

CHAPTER 4

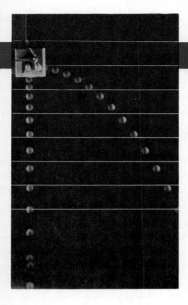

Motion in a Plane

The inner logic of Galilean mechanics was so strong that Newton was able to take the great step of applying it to the motion of the stars.

Max Born (1882–1970)

Thus far we have confined our discussion of mechanics to motion in a straight line. Although this has enabled us to introduce the basic ideas and laws of mechanics, very few objects in the universe move in only one dimension. Motion in a plane, i.e., motion on a flat, two-dimensional surface, is, however, quite common in our experience. If we neglect the slight curvature of the earth, ice skaters, ballroom dancers, and restaurant workers all execute motion in a plane. So too do basketballs, golf balls (if not hooked or sliced!), and various other kinds of projectiles whose initial straight-line motion is bent into a planar curve by the force of gravity.

In this chapter we discuss a number of interesting kinds of planar motion, including the motion of projectiles, the planets, and earth satellites. Because of the importance of the force of gravity for such motions, we begin with a discussion of the nature of this force.

4.1 Newton's Law of Universal Gravitation

The most familiar force on our planet is gravity, which attracts all objects to the earth. Newton was the first to explain the nature of this gravitational force. The story goes that he was sitting under an apple tree, saw an apple fall to the ground, and was struck not by the apple but by the insight that the force pulling the apple to the ground must be the same kind of force which keeps the moon moving in its orbit about the earth. Whether or not this story is factual, there is no doubt it is accurate in spirit. Newton certainly was the first physicist to write an equation that explained both the motion of the apple and the motion of the moon, and he was able to use his insight into the universal nature of the force of gravity to calculate the period of the moon's motion around the earth.

> *Newton's law of universal gravitation:* Every mass in the universe attracts every other mass by a force that is directly proportional to the product of the two masses and inversely proportional to the square of the distance between them.

The magnitude of the gravitational force F_G is thus given by

$$F_G \propto \frac{m_1 m_2}{r^2}$$

and so

$$F_G = G \frac{m_1 m_2}{r^2}$$ **(4.1)**

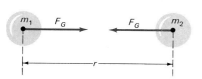

FIGURE 4.1 The gravitational attractive force **F**$_G$ between two masses m_1 and m_2 separated by a distance r.

where G is a constant called the *universal gravitational constant*, equal to 6.672×10^{-11} $m^3/(kg \cdot s^2)$. The direction of **F**$_G$ is along the line joining the masses m_1 and m_2, and r is the distance between m_1 and m_2, as shown in Fig. 4.1.

Two objects of arbitrary shape and size may be regarded as made up of point masses. The sum of the gravitational forces that all the point masses of one object exert on all the point masses of the other is the gravitational force between the two objects. If the objects have spherical symmetry, the mass of each object may be considered as concentrated at the center of the object. The distance between the two centers is then used in Newton's law of gravitation to find the attractive force.

For example, the gravitational force between the earth and the sun is directed along the line connecting the center of the earth with the center of the sun, and its magnitude is inversely proportional to the square of the distance between their two centers. The force exerted by the sun on the earth is exactly the same in magnitude as the force exerted by the earth on the sun but is in the opposite direction. These two forces are a good example of an action-reaction pair.

Example 4.1

Calculate the magnitude of the gravitational force exerted by the sun on the earth, using the data given in Table F.2.

SOLUTION

From Newton's law of universal gravitation, the magnitude of **F**$_G$ is

$$F_G = G \frac{m_S m_E}{r^2}$$

where m_S and m_E are the masses of the sun and the earth, respectively, and r is the distance between their centers. Therefore

$$F_G = \left(6.67 \times 10^{-11} \ \frac{m^3}{kg \cdot s^2}\right) \frac{(1.99 \times 10^{30} \ kg)(5.98 \times 10^{24} \ kg)}{(1.49 \times 10^{11} \ m)^2}$$

$$= 3.58 \times 10^{22} \ kg \cdot m/s^2 = \boxed{3.58 \times 10^{22} \ N}$$

This enormous force keeps the earth in its elliptical orbit about the sun. If this force ceased to exist, the earth would go flying off in a straight line tangent to its original orbit.

EXERCISE 1 Calculate the gravitational force exerted by the earth on the moon, and compare the result with the sun-earth force.

The Gravitational Constant

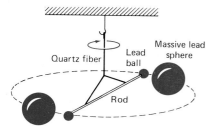

FIGURE 4.2 A gravitational torsion balance.

The constant G in Newton's law of universal gravitation must be measured experimentally. This was first done in 1798 in a famous experiment performed by Henry Cavendish (1731–1810), after whom the Cavendish Physics Laboratory at Cambridge University in England is named. In this experiment two small lead balls are placed at the ends of a thin rod that is suspended at its center by a very thin quartz fiber hung from a support. This arrangement allows the system to rotate about the fiber as an axis (Fig. 4.2). Two large lead balls are then brought up at opposite sides of the small balls. It is found that the large balls attract the small balls and twist the suspended system about its axis. By attaching a small mirror to the fiber and bouncing a light beam off the mirror, the motion of the system can be studied. The twist of the quartz fiber can be obtained from the deflection of the light beam, and the force needed to produce this twist can be found by producing the same twist with a known force. In this way the gravitational force F_G can be determined.

Since this experiment yields a value for F_G, and since the masses m_1 and m_2 and the distance r can be easily measured, we can solve Eq. (4.1) for G. The experimental value obtained is $G = 6.672 \times 10^{-11} \text{ m}^3/(\text{kg}\cdot\text{s}^2)$. Note that these must be the units of G if Eq. (4.1) is to be dimensionally correct. The units of F_G are newtons, or $\text{kg}\cdot\text{m}/\text{s}^2$. From Eq. (4.1) the units of G must be

$$\frac{\text{kg}\cdot\text{m}}{\text{s}^2} \cdot \frac{\text{m}^2}{\text{kg}^2} = \frac{\text{m}^3}{\text{kg}\cdot\text{s}^2} = \frac{\text{N}\cdot\text{m}^2}{\text{kg}^2}, \quad \text{since} \quad 1 \text{ N} = \frac{1 \text{ kg}\cdot\text{m}}{\text{s}^2}$$

The *universal gravitational constant G* is one of the least accurately known constants of nature. The gravitational force between most objects on earth is so small that it is difficult to measure with much accuracy. In astronomical systems the gravitational forces are much greater, but the masses and separations are not precisely known.

This gravitational force holds our complex universe together: it keeps the moon in orbit around the earth; it holds the planets in orbit around the sun; it anchors all of us to the earth so that we don't float off into space; and it explains the presence and structure of globular star clusters and galaxies.

Newton's great contribution was the recognition of the truly *universal* nature of the gravitational force. The constant G has the same value at every place in the universe. Newton's discovery of the law of universal gravitation is sometimes called the greatest single discovery in the history of science.

4.2 Weight—A Gravitational Force

To us the most familiar example of Newton's law of gravitation is the attractive force exerted by the earth on objects near its surface. If m_E is the mass of the earth and m the mass of a book on a shelf in the library, then the earth exerts a gravitational force on the book of magnitude

$$F_G = \frac{Gm_E m}{r^2} \tag{4.2}$$

where r is the distance from the center of the earth to the center of the book.

As long as the shelf supports the book, the downward force F_G is balanced by the upward force of the shelf on the book. If, however, the book is pushed off the shelf, the only force acting on the book is F_G. Thus F_G is the net force on the book, and the resulting acceleration, according to Newton's second law, is given by

$$a = \frac{F_G}{m} \tag{4.3}$$

The book then falls with this acceleration.

Careful experiments have shown that *this acceleration is the same for all objects, no matter what their mass, at any given point near the earth, as long as air resistance is neglected.* Even two objects as different as a feather and a coin fall with the same acceleration in a glass tube from which the air has been evacuated.

How do we explain this identical acceleration for all masses? Suppose we replace the one book by two books of different masses m_1 and m_2. They are attracted by the earth with forces of magnitudes given by Eq. (4.2):

$$F_{G1} = \frac{Gm_E m_1}{r^2} \qquad F_{G2} = \frac{Gm_E m_2}{r^2} \tag{4.4}$$

If the two books are pushed off the same shelf, they fall with accelerations a_1 and a_2, where, from Eq. (4.3), the magnitudes of the accelerations are

$$a_1 = \frac{F_{G1}}{m_1} \qquad a_2 = \frac{F_{G2}}{m_2}$$

On substituting the values of F_{G1} and F_{G2} from Eq. (4.4), we obtain

$$a_1 = \frac{Gm_E m_1}{m_1 r^2} \qquad a_2 = \frac{Gm_E m_2}{m_2 r^2}$$

Then, on canceling the masses m_1 and m_2,*

$$a_1 = \frac{Gm_E}{r^2} \qquad a_2 = \frac{Gm_E}{r^2}$$

and so $a_1 = a_2$. The acceleration is therefore *the same* for all objects at the same place near the earth's surface.

We can now relate the acceleration g due to gravity to more fundamental quantities like G and m_E. Near the earth's surface, the magnitude of g is

$$g = \frac{Gm_E}{r_E^2} = \frac{(6.672 \times 10^{-11} \text{ m}^3/\text{kg·s}^2)(5.98 \times 10^{24} \text{ kg})}{(6.37 \times 10^6 \text{ m})^2} = 9.83 \text{ m/s}^2 \tag{4.5}$$

The value of g varies slightly with location on the earth's surface and with altitude, since these factors change the value of the distance r_E in Eq. (4.5). The value of g is a maximum at sea level and decreases with increasing altitude. It varies from about 9.832 m/s² at the north and south poles to about 9.780 m/s² at the equator, indicating that the earth is not a perfect sphere but bulges out at the equator. In this book g will always be taken as equal to 9.80 m/s².

Weight and Mass

As we just saw, all objects at the same place near the earth's surface fall with the same acceleration g. According to Newton's second law, this acceleration must be produced by a force, which is the *weight* of the object.

Weight: The gravitational force exerted on an object.

In this case of a gravitational force, Newton's second law, $\mathbf{F}_{net} = m\mathbf{a}$, becomes

$$\boxed{\mathbf{w} = m\mathbf{g}} \tag{4.6}$$

where $\mathbf{w}$ and $\mathbf{g}$ are both directed toward the center of the earth. Equation (4.6) clarifies the relationship between the mass of an object and its weight. On earth, to obtain the weight of an object in newtons from its mass in kilograms, we must multiply by $g = 9.80$ m/s².

If we calculate a value for the acceleration due to gravity at the surface of the moon (g_M) from Eq. (4.5), we find that it is only about one-sixth the acceleration

---

*There is a subtle assumption here that the mass m_1 appearing in Newton's second law is the same mass m_1 appearing in the gravitational force equation. This may seem like a reasonable assumption, but it was only after Einstein's theory of general relativity was developed that it was possible to justify it.

due to gravity at the surface of the earth. Near the surface of the moon an astronaut would fall with an acceleration of $9.80/6$ m/s^2 $= 1.6$ m/s^2 and would have a weight one-sixth her weight on earth. The astronaut's *mass* would, however, be *exactly the same* on the moon as on the earth.

This shows us the essential difference between mass and weight:

Weight is a force (a gravitational force), whereas mass is a measure of inertia.

Mass is therefore an *invariant* quantity. By this we mean that it does not vary from place to place, whereas weight changes as g changes from one place to another.

Spring balances of the type shown in Fig. 3.1 measure weights directly, since they measure the force of gravity acting down on an object hung from the end of the balance. They give correct values for the weight of an object at any place on or off the earth, but they cannot yield absolute values for its mass unless the value of g at the place is known, for then $m = w/g$.

We can, however, compare two masses by using a spring balance. Since, for two masses m_1 and m_2, we have $w_1 = m_1 g$ and $w_2 = m_2 g$, we have $w_2/w_1 = m_2/m_1$. At any place, therefore, the ratio of the masses of two objects is the same as the ratio of their weights as found with a spring balance. If one mass is known, the other can therefore be obtained.

Example 4.2

(a) What is the magnitude of the force exerted on a 2.00-kg ball at the surface of the earth by the gravitational attraction of the earth? **(b)** What reaction force is exerted on the earth by the ball? **(c)** What acceleration of the earth would this force produce?

SOLUTION

(a) The weight of the ball is simply

$$w = mg = (2.00 \text{ kg})(9.80 \text{ m/s}^2)$$

$$= 1.96 \times 10^1 \text{ kg·m/s}^2 = \boxed{19.6 \text{ N}}$$

(b) The ball, by Newton's third law, exerts an equal and opposite force of 19.6 N on the earth, pulling the earth toward the ball!

(c) The earth's acceleration is given by Newton's second law:

$$a = \frac{F_{\text{net}}}{m_E} = \frac{19.6 \text{ N}}{5.98 \times 10^{24} \text{ kg}} = \boxed{3.28 \times 10^{-24} \text{ m/s}^2}$$

This infinitesimal acceleration is much too small to be observed.

Example 4.3

An astronaut has a mass of 57.0 kg on the surface of the earth. If she were on the surface of the moon, where the acceleration due to gravity is only one-sixth that on earth: **(a)** What would her mass be? **(b)** What would her weight be?

SOLUTION

(a) Since mass is an invariant property, it does not change on the surface of the moon or anywhere else. Hence the astronaut's mass on the surface of the moon is still $\boxed{57.0 \text{ kg}}$.

(b) Her weight, however, is different than it was on earth. Since the acceleration due to gravity near the surface of the moon is one-sixth the gravitational acceleration near the surface of the earth, we have

$$w_M = (57.0 \text{ kg}) \left(\frac{9.80 \text{ m/s}^2}{6} \right) = \boxed{93 \text{ N}}$$

Her weight has been reduced by a factor of 6 from its value on earth:

$$w = mg = (57.0 \text{ kg})(9.80 \text{ m/s}^2) = 5.59 \times 10^2 \text{ N}$$

4.3   Equations of Motion for a Particle in a Plane

The motion of a particle in a plane may be described by resolving the displacement, velocity, and acceleration vectors for the motion into components along the mutually perpendicular x and y axes. In this way motion in a plane can be reduced to a combination of two rectilinear motions. For purposes of convenience and clarity, it may be worthwhile to repeat here (from Sec. 2.4) the equations of motion that we will be using throughout this chapter.

x components   *y* components

Uniform motion (constant *v*)

$$x = x_0 + v_x t \qquad\qquad y = y_0 + v_y t \qquad\qquad (4.7)$$

Uniformly accelerated motion (constant *a*)

$$x = x_0 + \bar{v}_x t \qquad\qquad y = y_0 + \bar{v}_y t \qquad\qquad (4.8)$$

$$x = x_0 + v_{0x} t + \tfrac{1}{2} a_x t^2 \qquad\qquad y = y_0 + v_{0y} t + \tfrac{1}{2} a_y t^2 \qquad\qquad (4.9)$$

$$v_x = v_{0x} + a_x t \qquad\qquad v_y = v_{0y} + a_y t \qquad\qquad (4.10)$$

$$\bar{v}_x = \frac{v_{0x} + v_x}{2} \qquad\qquad \bar{v}_y = \frac{v_{0y} + v_y}{2} \qquad\qquad (4.11)$$

$$v_x^2 = v_{0x}^2 + 2a_x(x - x_0) \qquad\qquad v_y^2 = v_{0y}^2 + 2a_y(y - y_0) \qquad\qquad (4.12)$$

Newton's second law

$$(F_{\text{net}})_x = ma_x \qquad\qquad (F_{\text{net}})_y = ma_y \qquad\qquad (4.13)$$

(It would be worthwhile to spend a few minutes recalling the meaning of these equations before going further with this chapter.)

In many problems it is possible to treat the x and y components of the motion separately, using the above equations. Since the choice of a coordinate system is arbitrary, we can always choose the x axis as the direction that makes the analysis of the problem as simple as possible. Then we can combine the resulting equations for the component motions to find the actual motion of the object in the xy plane. The path of an object moving in a plane is, in general, curved, with the x and y components of the motion being tied together by their common dependence on the time t. By eliminating t between the component equations, we can obtain an equation involving only x and y to describe the path of the object in the xy plane.

4.4   Projectile Motion

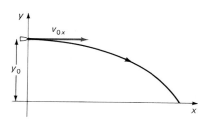

FIGURE 4.3 Path of a bullet fired from a gun parallel to the ground.

We live at a time when the motion of projectiles is of great interest and practical importance. The task of NASA engineers and scientists at launching sites like Cape Canaveral is to predict the path that will be followed by the projectiles they launch. This path we call the *trajectory* of the projectile. We have already learned enough physics to calculate the trajectories of some simple projectiles.

As a first example, consider a bullet fired from a gun parallel to the ground, as in Fig. 4.3. It is convenient to choose a coordinate system in which the muzzle velocity of the gun is v_{0x} and the bullet is fired at a height y_0 above the ground. (We assume that the effect of air resistance is negligible.) For simplicity we choose the origin of the coordinate system at $x_0 = 0$ so that the displacement in the x

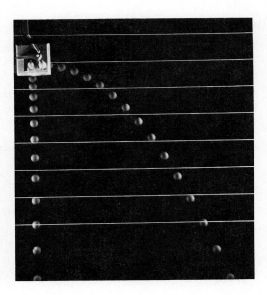

FIGURE 4.4 A flash photograph of two golf balls, one falling from rest and the other projected horizontally. The interval between light flashes is $\frac{1}{30}$ s, and the horizontal lines are 15 cm apart. It can be seen that the horizontal motion of the right-hand ball has no effect on its vertical motion, since its vertical position is everywhere identical with that of the ball falling from rest. (*Educational Development Center, Newton, Mass.*)

direction is simply $x - x_0 = x$. We can break up the motion into two components, one in the x direction, where the velocity is constant ($a_x = 0$), and another in the y direction, where the velocity increases downward ($a_y = -g$). Projectile motion is therefore a combination of *horizontal motion with constant velocity and vertical motion with constant acceleration.* The two components of motion are independent of each other, as shown in Fig. 4.4. The actual motion is then the superposition, or combination, of its x and y components.

We can therefore use the fact that $v_x = $ constant and $a_y = -g$ to describe the motion. The distance traveled in the x direction in time t is, from Eq. (4.9),

$$x = v_{0x}t$$

where v_{0x} is the muzzle velocity. For the y motion, the time it takes for the bullet to fall a distance y_0 from rest to the earth can be obtained from Eq. (4.9):

$$y = y_0 - \tfrac{1}{2}gt^2 = 0$$

where $y = 0$ at the surface of the earth. And so we have

$$t = \left(\frac{2y_0}{g}\right)^{1/2} \tag{4.14}$$

Notice that this is the same result found in Sec. 2.7 for an object falling from a height y_0. The bullet therefore travels a distance in the x direction of

$$x = v_{0x}t = v_{0x}\left(\frac{2y_0}{g}\right)^{1/2}$$

before it hits the ground. Since $y = y_0 - \tfrac{1}{2}gt^2$, we have

$$y = y_0 - \tfrac{1}{2}g\left(\frac{x}{v_{0x}}\right)^2 = y_0 - \frac{gx^2}{2v_{0x}^2} \tag{4.15}$$

An equation like this, in which one variable y depends on the square of another variable x, defines a parabola, shown in Fig. 4.3. The trajectory of a bullet fired parallel to the surface of the earth is therefore a *parabola*.

Example 4.4

A bullet is fired from a rifle at a speed of 1000 m/s. The gun is held parallel to the surface of the earth at a height of 1.50 m. If we neglect air resistance: **(a)** How long will it take for the bullet to fall to the ground? **(b)** How far will the bullet travel before it hits the ground if we neglect the curvature of the earth?

SOLUTION

(a) The time it takes for the bullet to reach the ground is, from Eq. (4.14), simply $t = (2y_0/g)^{1/2}$, the time it would take for the bullet to fall this vertical distance from rest. In this case,

$$t = \left[\frac{2(1.50 \text{ m})}{9.80 \text{ m/s}^2} \right]^{1/2} = \boxed{0.553 \text{ s}}$$

(b) In this time the horizontal distance traveled is

$$x = v_{0x}t = \left(1000 \frac{\text{m}}{\text{s}} \right)(0.553 \text{ s}) = \boxed{553 \text{ m}}$$

Projectile Fired at an Angle

Another interesting case is that of a projectile fired at an angle θ_0 with respect to the horizontal, as in Fig. 4.5. How far does it travel, and how does its maximum range depend on the initial angle θ_0?

We set up a coordinate system in the xy plane with the origin at the point where the projectile begins its flight, i.e., just outside the muzzle of the gun. Then $a_x = 0$ and $a_y = -g$, and, at any instant of time,

$$v_x = v_{0x} = v_0 \cos \theta_0 = \text{constant}$$

$$v_y = v_{0y} - gt = v_0 \sin \theta_0 - gt$$

The magnitude of the velocity at time t is therefore

$$v = \sqrt{v_x^2 + v_y^2}$$

where v_x is a constant, but v_y obviously depends on t. The direction of the shell's velocity at time t is determined by

$$\tan \theta = \frac{v_y}{v_x}$$

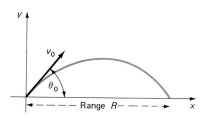

where θ is shown at point 1 in Fig. 4.6 and is the angle v makes with the horizontal at any instant. Also, since v_y is constantly decreasing, θ goes from its original value θ_0 to zero at the top of the trajectory and then becomes progressively more negative as time goes on.

The position of the projectile at time t can be found from Eqs. (4.7) and (4.9), where x_0 and y_0 are both zero in this case.

$$x = v_{0x}t = v_0 \cos \theta_0 \, t \tag{4.16}$$

$$y = v_{0y}t + \tfrac{1}{2}a_y t^2 = v_0 \sin \theta_0 \, t - \tfrac{1}{2}gt^2 \tag{4.17}$$

FIGURE 4.5 Path of a projectile fired at an angle θ_0 with respect to the horizontal.

To find the trajectory of the projectile, we must eliminate t between these two equations. From Eq. (4.16) we have $t = x/(v_0 \cos \theta_0)$. Substituting this value into Eq. (4.17), we obtain

$$y = (v_0 \sin \theta_0) \left(\frac{x}{v_0 \cos \theta_0} \right) - \frac{1}{2} g \left(\frac{x}{v_0 \cos \theta_0} \right)^2$$

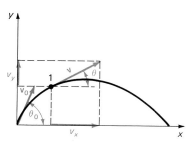

FIGURE 4.6 The velocity components and direction of the velocity at point 1 on the trajectory of a projectile.

or $\quad y = (\tan \theta_0)x - \dfrac{1}{2} \dfrac{g}{v_0^2 \cos^2 \theta_0} x^2 \tag{4.18}$

Since v_0 and θ_0 are constants, this takes the form

$$y = Ax - Bx^2$$

where A and B are constants. This is again the equation of a parabola. The path of a projectile fired at an angle with respect to the ground and acted on only by the force of gravity is therefore a parabola, as shown in Fig. 4.7.

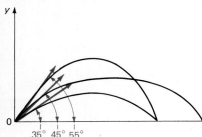

FIGURE 4.7 Trajectories of projectiles fired at different angles with respect to the horizontal with the same initial speed v_0.

Range of the Projectile

The projectile rises to some maximum height and then returns to earth. *The range of a projectile is the horizontal distance R from the point of firing to the point where it returns to earth.* At the point where the projectile returns to earth, $y = 0$, and so, from Eq. (4.18),

$$y = 0 = \tan \theta_0 x - \frac{1}{2} \frac{g}{v_0^2 \cos^2 \theta_0} x^2$$

$$x = \frac{2v_0^2 \cos^2 \theta_0}{g} \tan \theta_0 = \frac{2v_0^2 \cos^2 \theta_0}{g} \frac{\sin \theta_0}{\cos \theta_0} = \frac{2v_0^2}{g} \sin \theta_0 \cos \theta_0$$

Now the value of x at which $y = 0$ is precisely the range R. Therefore we have

$$R = \frac{2v_0^2}{g} \sin \theta_0 \cos \theta_0$$

but, from trigonometry (Appendix 1.C), $2 \sin \theta_0 \cos \theta_0 = \sin 2\theta_0$, and so

$$R = \frac{v_0^2}{g} \sin 2\theta_0 \tag{4.19}$$

This is the general expression for the range of the projectile, which, in the absence of other forces like air resistance and wind, depends only on the initial velocity v_0 and the angle of projection θ_0.

What is the value of θ_0 for which the range R is a maximum for a given v_0? It is the value of R when $\sin 2\theta_0$ takes on its maximum value of 1, that is, when $\theta_0 = 45°$; in which case

$$R_{\max} = \frac{v_0^2}{g} \tag{4.20}$$

For angles above and below 45°, the range diminishes, as shown in Fig. 4.7. Since $\sin 70° = \sin 110°$, the range is the same for 35° and 55°, in accordance with Eq. (4.19). Note that the range in Eqs. (4.19) and (4.20) refers to a projectile fired from one elevation and returning to the *same* elevation.

The vertical motion of the projectile in this case is identical with that of a ball thrown up in the air with a velocity $v_0 \sin \theta_0$. This must be the case, since if we looked at the projectile from a frame of reference moving to the right with a velocity $v_0 \cos \theta_0$, it would appear that the shell was shot straight up in the air, stopped at its highest point, and then fell back to earth again.

Example 4.5

A golfer hits a ball at an angle of 45.0° with respect to the ground at an initial speed of 30.0 m/s. The ball does not slice or hook, so its trajectory is planar. We assume that air resistance can be neglected. **(a)** Find the time at which the golf ball reaches its highest point. What time is required for the ball to return to the ground? **(b)** How high is the highest point of the trajectory above the ground? **(c)** What is the horizontal range of the ball? **(d)** What is the ball's velocity as it strikes the ground?

SOLUTION

(a) At its highest point, $v_y = 0$, and so

$$v_0 \sin \theta_0 - gt = 0$$

$$t = \frac{v_0 \sin \theta_0}{g} = \frac{(30.0 \text{ m/s})(0.707)}{(9.80 \text{ m/s}^2)} = \boxed{2.16 \text{ s}}$$

The time for the total trajectory is just twice this time, since the ball takes the same time to rise as to fall, just as with a ball thrown straight up in the air. The same result can be obtained from Eq. (4.9) with $y = 0$.

(b) The maximum height can be obtained from Eq. (4.9) by substituting t from part (a):

$$y = v_0 \sin \theta_0\, t - \tfrac{1}{2}gt^2$$

$$= (30.0 \text{ m/s})(0.707)(2.16 \text{ s}) - \tfrac{1}{2}(9.80 \text{ m/s}^2)(2.16 \text{ s})^2$$

$$= 45.8 \text{ m} - 22.9 \text{ m} = \boxed{22.9 \text{ m}}$$

(c) For 45°, the range is a maximum and equal to $R_{\max} = v_0^2/g$ from Eq. (4.20).

$$R_{\max} = \frac{(30.0 \text{ m/s})^2}{9.80 \text{ m/s}^2} = \boxed{91.8 \text{ m}}$$

(d) $v_x = v_0 \cos \theta_0 = 21.2 \text{ m/s}$

$$v_y = v_0 \sin \theta_0 - gt$$

$$= 21.2 \text{ m/s} - (9.80 \text{ m/s}^2)(4.32 \text{ s}) = -21.2 \text{ m/s}$$

Then

$$v = \sqrt{v_x^2 + v_y^2}$$

$$= \sqrt{(21.2)^2 + (-21.2)^2} = \boxed{30.0 \text{ m/s}}$$

and

$$\tan \theta = \frac{v_y}{v_x} = \frac{-21.2}{21.2} = -1$$

and so

$$\theta = \boxed{-45.0°}$$

This is an angle of 45.0° below the x axis. The golf ball thus returns to the ground with the same speed it had originally and with $\theta = -\theta_0$. This is what we would expect from the symmetry of the trajectories shown in Fig. 4.7.

Example 4.6

Expert sharpshooters often use a technique which makes it almost certain that they will hit their targets in shooting competitions. If a target is thrown straight up in the air, the contestant waits until it reaches the top of its trajectory before firing directly at the target. The target therefore starts to fall at the instant the contestant pulls the trigger.

Will the sharpshooter always hit the target? Does this depend on the speed with which the bullet is shot out of the gun?

SOLUTION

The situation is as diagramed in Fig. 4.8. The target is a height y_0 above the ground and a horizontal distance x_0 along the ground from the gun. The bullet is shot with an initial velocity v_0 at an angle θ_0 with respect to the ground and is aimed directly at the target when it is at the top of its trajectory.

Here $y_0 = x_0 \tan \theta_0$. Since the bullet is fired at time $t = 0$, the height of the target above the ground at time t, as it falls under gravity, is

$$y_t = y_0 - \tfrac{1}{2}gt^2 = x_0 \tan \theta_0 - \tfrac{1}{2}gt^2$$

The height of the bullet at a horizontal distance x from the gun is, from our discussion in the previous section, with $y_0 = 0$ for the bullet,

$$y_b = v_0 \sin \theta_0\, t - \tfrac{1}{2}gt^2 \quad \text{or, since} \quad t = \frac{x}{v_0 \cos \theta_0}$$

$$y_b = v_0 \sin \theta_0 \frac{x}{v_0 \cos \theta_0} - \tfrac{1}{2}gt^2$$

When $x = x_0$, $y_b = x_0 \tan \theta_0 - \tfrac{1}{2}gt^2$, and so $y_b = y_t$. The bullet will therefore *always* strike the target, *no matter what the value of* v_0.

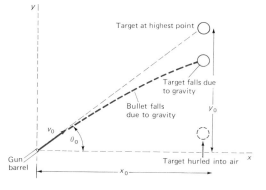

FIGURE 4.8

4.5 Uniform Circular Motion

A very important kind of motion in a plane is circular motion at constant speed. (We cannot say at constant *velocity*, since the direction of the velocity continually changes.) Such motion we call *uniform circular motion*. This case is quite different from projectile motion, since the force acting to keep the moving object in a circular path is always directed toward the center of the circle. Consider, for example, a ball moving in a horizontal circle at the end of a string. The only possible direction for the force is along the string. This force is therefore constantly changing direction as the object moves.

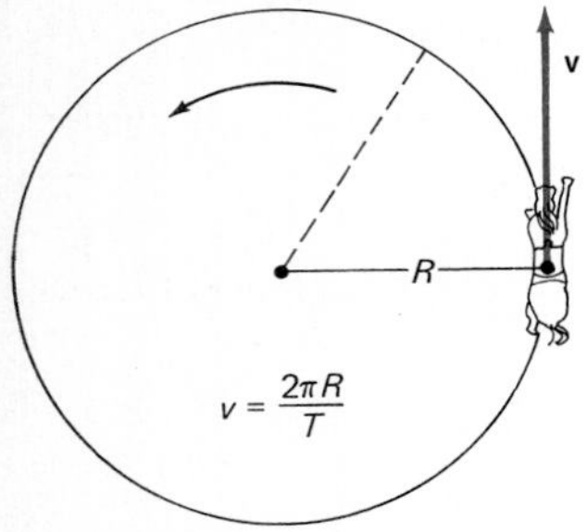

As an example of circular motion, consider one of the stationary horses on a carousel moving at constant speed. As the carousel rotates through one complete turn, the horse travels around the circumference of a circle through a linear distance $2\pi R$, where R is the radius of the circle in which the horse moves, as in Fig. 4.9. If it takes T s for the carousel, moving at constant speed, to make one complete revolution, then the linear speed of the horse along the circumference of the circle is

$$v = \frac{\Delta s}{\Delta t} = \frac{2\pi R}{T} \qquad \text{(4.21)}$$

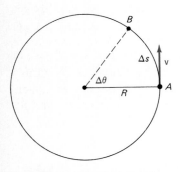

Circumference
= $2\pi R$

$v = \dfrac{2\pi R}{T}$

FIGURE 4.9 Motion of a horse on a carousel.

where Δs is the linear distance traveled by the horse around the circle in time Δt.

The period T is the time for one complete revolution. Related to the concept of the period in circular motion is the idea of *frequency*. For example, if the carousel takes 10 s for one complete revolution when it is up to constant speed, then its frequency is $\frac{1}{10}$ of a revolution per second ($\frac{1}{10}$ rev/s). *The frequency is the number of revolutions per second.* Frequency is related to the period by the equation

$$f = \frac{1}{T} \qquad \text{(4.22)}$$

If T is in seconds, the frequency is in cycles per second, or hertz (Hz), where 1 Hz = 1 cycle/s.

As the horse moves in its circular path through an arc Δs (from A to B), a line connecting the horse to the center of the carousel sweeps out an angle $\Delta\theta$, as in Fig. 4.10. If this angle is swept out in time Δt, then we define the constant angular velocity ω as follows:

$$\omega = \frac{\Delta\theta}{\Delta t} \qquad \text{(4.23)}$$

FIGURE 4.10 The angular velocity $\omega = \Delta\theta/\Delta t$, where $\Delta\theta$ is the angle swept out in time Δt.

(The symbol for angular velocity is ω, the Greek letter *omega*, the last letter in the Greek alphabet.) The units of angular velocity are radians per second, since we measure angles in radians (see Sec. 1.5).*

It should be noted that all the horses on a carousel have exactly the same angular velocity, since they all sweep out the same angle in time Δt. Their linear velocities differ, however, depending on how far they are from the center of the carousel.

For one complete revolution at constant angular velocity, we have

---

*A fuller treatment of angular velocity and angular acceleration will be found in Chap. 8.

$$\omega = \frac{\Delta\theta}{\Delta t} = \frac{2\pi}{T} = 2\pi f \qquad \text{(4.24)}$$

since the angle swept out in one complete revolution is 2π. The angular velocity is therefore 2π times the frequency of revolution.

We can also obtain an important relationship between the linear and angular velocities. From Eq. (4.21),

$$v = \frac{2\pi R}{T} = R\frac{2\pi}{T} = R\omega \qquad \text{(4.25)}$$

The magnitude of the linear velocity of a point a distance R from the center of rotation is equal to R times the angular velocity of the system. This equation is only valid when ω is expressed in radians per second.

Example 4.7

A car in one of the rides at an amusement park is rotating in a circle of radius 5.00 m and makes one complete revolution every 3.00 s. **(a)** What is the angular velocity of the car in radians per second? **(b)** What is the angular velocity of the car in degrees per second? **(c)** What is its linear speed?

SOLUTION

(a) Since the car makes one complete revolution every 3.00 s and each revolution corresponds to an angle of 2π rad, the angular velocity is

$$\omega = \left(\frac{1\ \text{rev}}{3.00\ \text{s}}\right)\left(2\pi\ \frac{\text{rad}}{\text{rev}}\right) = \boxed{2.09\ \text{rad/s}}$$

(b)
$$\omega = \left(2.09\ \frac{\text{rad}}{\text{s}}\right)\left(\frac{57.3°}{1\ \text{rad}}\right) = \boxed{120°/\text{s}}$$

which is just the $\frac{1}{3}$ rev/s stated in the problem.

(c) The car's linear speed can be found from Eq. (4.25):

$$v = R\omega = (5.00\ \text{m})(2.09\ \text{rad/s}) = \boxed{10.5\ \text{m/s}}$$

4.6 Centripetal Acceleration and Centripetal Force

In the preceding section we discussed angular displacements and angular velocities and the corresponding linear quantities that relate to motion along the arc of a circle. But what produces motion in a circle in the first place? Consider, for example, a model airplane flying in a horizontal circle at the end of a string, as in Fig. 4.11. If no forces act on the plane, it would simply fly off in a straight line tangent to the circle, as Newton's first law requires. If the airplane moves in a circle, some force must be provided by the string to produce this circular motion. We now consider the nature of this force, its magnitude and direction, and the acceleration it produces.

FIGURE 4.11 (a) A model airplane moving at constant speed in a horizontal circle. (b) The vector $\Delta\mathbf{v} = \mathbf{v}_B - \mathbf{v}_A$. (c) The triangle formed by the chord AB and the radii OA and OB.

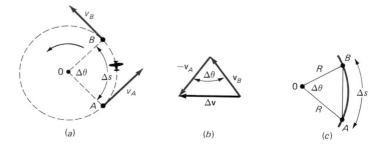

(a)   (b)   (c)

In the example illustrated in this figure, the magnitude of the linear velocity is constant, but its direction is continually changing. Therefore *the motion is accelerated*. The acceleration of the airplane is equal to the change in velocity divided by the change in time Δt in the limit as $\Delta t \rightarrow 0$. It is given approximately by the average acceleration

$$\bar{\mathbf{a}} = \frac{\Delta \mathbf{v}}{\Delta t} = \frac{\mathbf{v}_B - \mathbf{v}_A}{\Delta t} \tag{4.26}$$

where $\mathbf{v}_A$ and $\mathbf{v}_B$ are the plane's velocity at points A and B in Fig. 4.11a; these velocities are vectors, of course, directed tangent to the circle.

The vector $\Delta \mathbf{v} = \mathbf{v}_B - \mathbf{v}_A$ is shown in Fig. 4.11b, where $\mathbf{v}_A$ and $\mathbf{v}_B$ have the same length, since the magnitude v of the velocity is constant. Notice that $\Delta \mathbf{v}$ points inward and is approximately directed toward the center of the circle.

The triangle in Fig. 4.11b is similar to the triangle OAB formed by the chord AB and the radii OA and OB (Fig. 4.11c), since they are both isosceles triangles with the same vertex angle. These triangles are similar because the two radii are perpendicular to the two velocity vectors. Because the corresponding sides of similar triangles are proportional, and because $\mathbf{v}_A$ and $\mathbf{v}_B$ have magnitude v, we may write approximately

$$\frac{\Delta v}{v} = \frac{AB}{R} = \frac{\Delta s}{R} = \frac{v\,\Delta t}{R} \qquad \text{or} \qquad \frac{\Delta v}{\Delta t} = \frac{v^2}{R}$$

This is not exact because we have taken the chord AB as equal to the arc length Δs, which is $v\,\Delta t$, and the arc and chord are only approximately equal. However, as Δt becomes smaller and smaller, this approximate relationship becomes more exact, since the arc and the chord approach the same length. In addition, as Δt becomes smaller, $\Delta \mathbf{v}$ becomes normal to $\mathbf{v}$ and points to the exact center of the circle. In this limit the acceleration is

$$a_C = \lim_{\Delta t \rightarrow 0} \frac{\Delta v}{\Delta t} = \frac{v^2}{R} \qquad \text{or} \qquad \boxed{a_C = \frac{v^2}{R}} \tag{4.27}$$

Our approximate expression became exact in the limit as $\Delta t \rightarrow 0$. The acceleration $\mathbf{a}_C$, directed toward the center of the circle at right angles to the velocity, is called the *centripetal acceleration*, where *centripetal* means *center-seeking*. Although Eq. (4.27) was derived for uniform circular motion, it turns out to be true in general for all circular motion. The direction of the centripetal acceleration $\mathbf{a}_C$ is shown in Fig. 4.12.

The centripetal acceleration for circular motion is directed toward the center of the circle and is of magnitude v^2/R.

The dimensions of v^2/R are $[L/T]^2/[L] = [L]/[T^2]$, which are the correct dimensions for an acceleration. The units of a_C are therefore meters per second squared. Since $v = R\omega = R(2\pi f)$ from Eq. (4.25),

$$a_C = \frac{R^2(2\pi f)^2}{R} = 4\pi^2 f^2 R = \omega^2 R \tag{4.28}$$

where f is the frequency, or number of revolutions per second.

In addition to the centripetal acceleration, which is normal to the velocity, it is also possible that the speed of our model airplane may change as it moves along. (For example, the plane might be moving in a vertical circle so that the force of

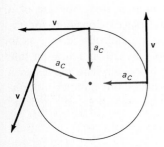

FIGURE 4.12 The centripetal acceleration $\mathbf{a}_C$ is a vector perpendicular to the linear velocity; it is directed toward the center of the circle of motion.

gravity acts to speed the plane up as it falls and to slow it down as it rises.) This change in the *magnitude* of the velocity gives rise to an acceleration:

$$a_T = \lim_{\Delta t \to 0} \frac{\Delta v}{\Delta t}$$

where Δv is the change in the *magnitude* of the velocity **v**. The acceleration $\mathbf{a}_T$ is called the *tangential acceleration*, because any change in the magnitude of the velocity must be along the direction of the motion and therefore tangent to the direction of the velocity. The *tangential* acceleration $\mathbf{a}_T$ therefore indicates the change in the *magnitude* of the velocity **v**, whereas the *centripetal acceleration* $\mathbf{a}_C$ arises from a change in the *direction* of **v**.

The resultant acceleration is the vector sum of the centripetal and tangential accelerations. Since $\mathbf{a}_C$ is perpendicular to $\mathbf{a}_T$, we have for the magnitude of the total acceleration:

$$a = [a_T^2 + a_C^2]^{1/2} \qquad (4.29)$$

The direction of the resultant acceleration can be seen from Fig. 4.13 to be given by:

$$\tan \phi = \frac{a_C}{a_T} \qquad (4.30)$$

ϕ is thus the angle between the resultant acceleration **a** and $\mathbf{a}_T$; since $\mathbf{a}_T$ is parallel to the velocity, ϕ is also the angle between the acceleration and the velocity. Note that since a centripetal acceleration exists, the direction of the resultant acceleration is *not* that of the velocity.

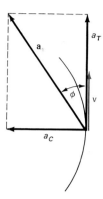

FIGURE 4.13 Acceleration for circular motion: The resultant acceleration is the vector sum of the tangential acceleration $\mathbf{a}_T$ (due to a change in speed) and the centripetal acceleration $\mathbf{a}_C$ (due to a change in direction).

Centripetal Force

Since every acceleration must be produced by a nonzero net force, any object executing circular motion with a speed v must be acted on by a net force of magnitude $F_C = ma_C$ so that

$$\text{Centripetal force } F_C = \frac{mv^2}{R} \qquad (4.31)$$

This centripetal force is directed toward the center of the circle.

The equation for the centripetal force tells us nothing about how this force is provided physically. It merely says that, for an object moving in a circle, a net force of magnitude mv^2/R must be acting to keep the object moving in a circle. For a toy airplane moving in a horizontal circle at the end of a string, this force is the pull of the string on the airplane; for a planet it is the gravitational attraction of the sun.

We can also use Eq. (4.28) and write the centripetal force as

$$F_C = 4\pi^2 f^2 mR = mR\omega^2 \qquad (4.32)$$

Example 4.8

An automobile of mass 1500 kg is moving at a constant speed of 15.0 m/s around a curve of radius of curvature 100 m on an unbanked road. **(a)** What minimum coefficient of friction (static, since static friction opposes slippage between the tires and the road) is needed to keep the car on the road? **(b)** How does this result change if the car has a mass of 3000 kg?

SOLUTION

(a) The weight of the car is balanced out by the normal force exerted upward on the car by the road, and so $N = mg$.

To keep the car moving in a circle of radius $R = 100$ m, a net (centripetal) force of magnitude $F_C = mv^2/R$ is required. This force can only be provided, for an unbanked curve, by the friction between the road and the tires, so that, from Eqs. (3.5) and (4.31),

$$f_s = \mu_s N = \mu_s mg = \frac{mv^2}{R}$$

or $\quad \mu_s = \dfrac{mv^2}{mgR} = \dfrac{v^2}{gR} = \dfrac{(15.0 \text{ m/s})^2}{(9.80 \text{ m/s}^2)(100 \text{ m})} = \boxed{0.230}$

The coefficient of static friction must be at least this much or the car will leave the road. It can be seen from Table 3.1 that such coefficients of friction are easily obtainable between rubber tires and concrete.

(b) The mass of the car cancels out of the final equation for the coefficient of friction. Our answer is therefore just as valid for a Cadillac as it is for a Hyundai.

Banking of Curves

A car going around a curve requires a centripetal force to keep it moving in a circle. On a flat road this force is provided by the friction between the road and the car's tires, as in Example 4.8. To reduce the chance of skidding, however, it is better to bank the curve. Let us see if we can find the angle θ that the banked curve should make with the flat road surface to reduce to a minimum the chances of skidding.

Figure 4.14a shows a top view of the situation. Figure 4.14b provides an end view showing the banking angle θ.

To simplify the problem, we at first assume that there is no friction and that the banking alone provides the centripetal force. If the curve is banked at an angle θ, the road exerts a normal force of magnitude N perpendicular to its surface. This force has two components: $N \cos \theta$ is vertical, opposite in direction to w, the weight of the car; $N \sin \theta$ is horizontal and directed inward along a radius of the circle of which the curve is a part. A free-body diagram of the forces on the car is shown in Fig. 4.14c.

Since the car does not accelerate vertically, the net vertical force is zero, and $N \cos \theta$ just balances the weight of the car. The net (centripetal) force thus is $N \sin \theta$, and we have

$$N \cos \theta = mg \qquad \text{and} \qquad N \sin \theta = \frac{mv^2}{R}$$

On dividing the second equation by the first, we obtain

$$\tan \theta = \frac{v^2}{Rg} \tag{4.33}$$

The banking angle θ is therefore determined by the sharpness of the curve (the smaller R is, the larger the amount of banking needed) and by the speed of the car. The mass of the car plays no role in the result. For a given radius R, the tangent of the banking angle should vary as v^2. For this reason it is impossible to bank a curve for all speeds, so curves are banked for the average speed of cars at that point on the road. For speeds in excess of this, there is a greater chance of skidding.

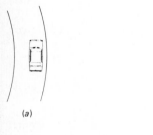

(a)

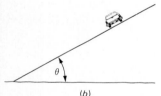

(b)

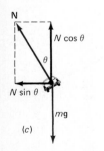

(c)

FIGURE 4.14 (a) Top view of a car going around a curve. (b) End view of a car going around a curve. (c) Free-body diagram for the car on a banked curve.

Example 4.9

It is desired to make the same curve as in Example 4.8 safe for cars moving at speeds of 15.0 m/s by banking the road and neglecting the effect of friction. What must the banking angle be?

SOLUTION

In this case the centripetal force required to keep the car on the road as it goes around the curve is $mv^2/R = N \sin \theta$ (Fig. 4.14c). This leads directly to Eq. (4.33) for the banking angle:

$$\tan \theta = \frac{v^2}{Rg} = \frac{(15.0 \text{ m/s})^2}{(100 \text{ m})(9.80 \text{ m/s}^2)} = 0.230$$

and so the banking angle is

$$\theta = \boxed{12.9°}$$

Example 4.10

A ball at the end of a string is rotating at constant speed in a horizontal circle of radius 1.50 m. The mass of the ball is 0.250 kg, and the string makes an angle of 10.0° with the hor- izontal. **(a)** Draw a free-body diagram showing *all the forces* acting on the ball. **(b)** What is the rotational velocity of the ball? **(c)** How many revolutions does the ball make per second?

SOLUTION

At first glance, this problem may seem different from the previous one; in fact, it is quite similar. The physical situation is shown in Fig. 4.15a.
(a) The FBD diagram is shown in Fig. 4.15b. The string can- not be perfectly horizontal, since there must be a vertical com- ponent of the tension in the string sufficient to balance out the weight of the ball.
(b) We write Newton's second law of motion for the vertical and horizontal components, where the tension in the string pro- vides the centripetal force needed to keep the ball moving in a circle. We have

$$(F_{net})_y = 0 \qquad (F_{net})_x = \frac{mv^2}{R}$$

$$T \sin \theta - mg = 0$$

$$T \sin \theta = mg \qquad T \cos \theta = \frac{mv^2}{R} = mR\omega^2$$

On dividing the left-hand equation by the right-hand one, we obtain

$$\tan \theta = \frac{g}{R\omega^2}$$

Therefore

$$\omega^2 = \frac{9.80 \text{ m/s}^2}{(1.50 \text{ m})(\tan 10.0°)} = \frac{37.1}{\text{s}^2}$$

and

$$\omega = \boxed{6.09 \text{ rad/s}}$$

(c) $\quad f = \dfrac{\omega}{2\pi} = \dfrac{6.09 \text{ rad/s}}{2\pi \text{ rad/rev}} = \boxed{0.969 \text{ rev/s}}$

The ball makes about 1 rev/s. The faster the ball goes, the greater the tension in the string, and hence the smaller the angle made by the string with the horizontal, since $\sin \theta = mg/T$.
It is worthwhile to emphasize the similarities of this prob- lem to that of the banking of curves. In both cases the mass (of the car or ball) is of no importance; the optimum banking angle and the angle made by the string with the horizontal both depend on the velocity of the motion. This is an example of something that frequently occurs in physics: a technique learned in con- nection with one branch of physics often turns out to be of great value in some very different branch.

FIGURE 4.15 (a) Diagram for Example 4.10. (b) Free- body diagram.

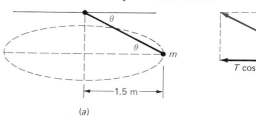

(a)

(b)

Centrifuges and Ultracentrifuges

The centrifuge is a device frequently used in biological, chemical, and medical laboratories to separate substances of different densities. It provides an interesting example of centripetal acceleration.

One type of centrifuge, the so-called swinging bucket rotor, consists of a rotating cylinder (the rotor) to which holders for test tubes are attached. The test- tube holders hang vertically when the rotor is stopped, but swing out to horizontal positions when the rotor gets up to speed, as shown in Fig. 4.16a.

Axis of
rotation

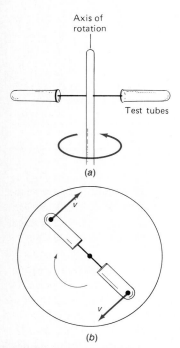

Test tubes

(a)

(b)

FIGURE 4.16 (a) Side view of a rotating centrifuge. (b) Top view of a rotating centrifuge.

It is often important in medical laboratories to take a sample of whole blood and determine the ratio of the volume of red blood cells to total blood volume. The blood cells are slightly denser than the rest of the liquid and hence would slowly fall through the liquid, under the force of gravity, and collect at the bottom of the test tube. This process can take days or even weeks. With a centrifuge the same process can be carried out in minutes or even seconds, since the centrifuge can increase the force acting to separate the blood cells by a factor of thousands over the force of gravity.

When a blood sample in the test tube is whirling around in the centrifuge at a very high speed, as in Fig. 4.16b, a red blood cell behaves somewhat like a passenger going around a sharp turn in a car. The passenger feels pushed against the side door, even though actually the passenger's body merely tries to maintain motion in a straight line, according to Newton's first law. The car turns so that the door gets in the passenger's way and pulls the body around the corner with the car. Similarly, the red blood cells in the centrifuge test tube want to keep moving in a straight line, but the whirling test tube gets in the way. The denser blood cells therefore fall quickly through the liquid to the bottom of the tube, where the glass wall of the tube provides the centripetal force to keep the blood cells moving in a circular path. The force that must be provided by the test tube wall is mv^2/R, and this provides a centripetal acceleration of $v^2/R = R\omega^2$. This is very much larger than the acceleration due to gravity for the blood cells, which would normally determine the rate at which the cells would fall through the liquid. The effective acceleration in this case is $g_{eff} = R\omega^2$. For a blood cell 10 cm from the center of the rotor, rotating at 100 rev/s, this is

$$g_{eff} = (0.10 \text{ m}) \left[(100) \left(2\pi \frac{\text{rad}}{\text{s}} \right) \right]^2 \cong 40,000 \text{ m/s}^2 \cong 4000g$$

It is because of this greatly increased "effective value of g" that the red blood cells can be separated from the solution so quickly in a centrifuge.

Ordinary centrifuges rotate at speeds up to 100 rev/s and thus produce effective g values in the thousands. To separate out the lipoproteins from the heavier proteins and albumin in a blood sample, an ordinary centrifuge is inadequate and *ultracentrifuges* become necessary.

Ultracentrifuges have rotating parts supported by air jets or magnetic fields to

FIGURE 4.17 Large centrifuge used by NASA to study the effect of large g values on the human body at the Manned Spacecraft Center, Houston, Texas. The centrifuge has a 50-ft-long arm at the end of which a gondola can hold three astronauts. The rapidly rotating arm creates g values similar to those that astronauts experience during lift-off and reentry into the earth's atmosphere. (*Courtesy of NASA.*)

reduce frictional heating and are constructed to withstand the forces produced by speeds of as much as 3000 rev/s. They can produce effective g values above 10^6 m/s^2.

Large centrifuges are used to study the effect of large values of g on the human body. In these devices men and women are rotated rapidly in a circle in carriages at the end of a rigid arm. Such research is particularly important in the space program (Fig. 4.17), since astronauts are subjected to large effective g's when they are blasting off from the earth or making their return landing.

4.7 Motion in a Vertical Circle under Gravity

The acceleration of an object moving in a vertical circle on the earth's surface includes changes in both the magnitude and the direction of the velocity. In Fig. 4.18a we show a small car rotating around the inside of a circular track in a vertical plane. In this case the tangential component of the acceleration is produced by the component of the force of gravity tangent to the circle, which makes the car speed up as it falls and slow down as it rises. The component of the acceleration normal to the track is produced by a combination of the push of the track (P) and the component of the force of gravity normal to the track. This normal component of the acceleration is the centripetal acceleration that keeps the car moving in a circle.

The forces on the car are its weight $w = mg$, always directed down, and the push of the track P, which is everywhere normal to the track (assumed to be frictionless). The weight of the car at point A can be resolved into a component normal to the path, $w \cos \theta$, and a component tangent to the path, $w \sin \theta$, as in Fig. 4.18a. The net forces acting are therefore [we label the normal component $(F_{net})_C$, since it is a centripetal force]

$$(F_{net})_C = P - w \cos \theta \qquad (F_{net})_T = w \sin \theta$$

The tangential acceleration is therefore

FIGURE 4.18 (a) Forces acting on a small car at point A rotating around the inside of a circular, vertical track. (b) Components of the acceleration for the car's motion at point A. (c) Forces on the car at point B at bottom of track; at this point P, the normal push of the track on the car, has its largest value. (d) Forces on the car at point D at the top of the track; at this point P has its smallest value.

$$a_T = \frac{(F_{net})_T}{m} = \frac{mg \sin \theta}{m} = g \sin \theta$$

while the normal acceleration is

$$a_C = \frac{(F_{net})_C}{m} = \frac{P - w \cos \theta}{m} = \frac{v^2}{R}$$

since the normal component of the acceleration keeps the car moving in a circle

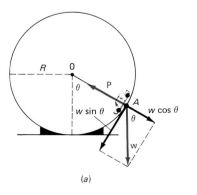

(a)

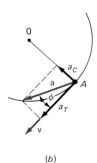

(b)

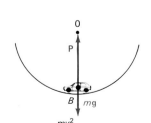

$F_{net} = P - mg = \dfrac{mv^2}{R}$

(c)

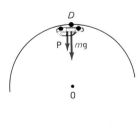

$F_{net} = P + mg = \dfrac{mv^2}{R}$

(d)

and therefore provides the required centripetal force mv^2/R. The push of the track on the car at any point along the track is therefore

$$P = m\left(\frac{v^2}{R} + g \cos \theta\right) \tag{4.34}$$

where v and θ both vary with position along the track.

The resultant acceleration of the car at point A is then $a = (a_T^2 + a_C^2)^{1/2}$, and its direction is specified by the angle ϕ, where $\tan \phi$ is a_C/a_T, as shown in Fig. 4.18b.

As the car falls from point A to point B at the bottom of the track, its speed increases. At point B (Fig. 4.18c), $\theta = 0$, $\sin \theta = 0$, $\cos \theta = 1$, and

$$a_T = g \sin \theta = 0 \qquad P = m\left(\frac{v^2}{R} + g\right)$$

so that the push of the track both opposes the weight of the car and provides the centripetal force to keep the car moving in a circle, as in Fig. 4.18c.

At the top of the track (point D in Fig. 4.18d), $\theta = 180°$, $\sin \theta = 0$, $\cos \theta = -1$, and

$$a_T = 0 \qquad P = m\left(\frac{v^2}{R} - g\right) \qquad \text{and} \qquad P + mg = \frac{mv^2}{R}$$

In this case the push of the track is greatly reduced, since the weight of the car contributes to the centripetal force needed to keep the car moving in a circular path.

Motion in a vertical circle is therefore a good example of how the acceleration of an object moving in a plane is, in general, the vector combination of tangential and normal components; the tangential component corresponds to a change in the speed of the object, and the normal component to a change in its direction.

Example 4.11

A small car of mass 25.0 g is rotating in a vertical circle of radius 30.0 cm. What is the minimum speed the car must have at the top of the track if it is not to fall off?

SOLUTION

At the top of the track, θ in Fig. 4.18a is 180°, $\sin \theta = 0$, and $\cos \theta = -1$. The push of the track on the car at point D in Fig. 4.18d is, as shown in the diagram,

$$P = m\left(\frac{v^2}{R} - g\right)$$

If the speed of the car is too low, $P \to 0$ and the car begins to fall off the track. (Note that $P < 0$ is impossible, since surfaces can only push, never pull.) This happens when $v^2/R - g = 0$ or $v = \sqrt{gR}$.

In the present case

$$v = \sqrt{(9.80 \text{ m/s}^2)(0.300 \text{ m})} = \boxed{1.71 \text{ m/s}}$$

Notice that the mass of the car, although given, is not needed to solve this problem.

Example 4.12

A ball of mass 200 g swings in a vertical circle at the end of a cord of length $R = 0.800$ m. When the cord makes an angle of 25° with the downward direction, the ball's speed is 2.00 m/s. **(a)** Find the radial and tangential components of the acceleration. **(b)** Find the resultant acceleration. **(c)** Find the tension T in the cord at this point. **(d)** How does the tension at the top and bottom of the circle compare with the value found in part (c) if we assume that the ball has the same speed at all points on its circular path?

SOLUTION

(a) This problem is identical physically with that of the motion of a car in a vertical circle, with the push of the track replaced by the tension in the cord. If line OA in Fig. 4.18*b* makes an angle of 25° with the vertical, the tangential acceleration at A, from our discussion above, is

$$a_T = g \sin \theta = (9.80 \text{ m/s}^2)(\sin 25°) = \boxed{4.14 \text{ m/s}^2}$$

The radial or centripetal acceleration at A is

$$a_C = \frac{v^2}{R} = \frac{(2.00 \text{ m/s})^2}{0.800 \text{ m}} = \boxed{5.00 \text{ m/s}^2}$$

(b) The resultant acceleration therefore has a magnitude

$$a = \sqrt{a_T^2 + a_C^2} = \boxed{6.49 \text{ m/s}^2}$$

and a direction given by $\tan \phi = a_C/a_T = 5.00/4.14 = 1.20$. The angle ϕ between the resultant acceleration and a_T, which is tangent to the curve, as shown in Fig. 4.18*b*, is then

$$\phi = \boxed{50°}.$$

(c) From the above discussion, the tension in the cord is, from Eq. (4.34),

$$T = m \left(\frac{v^2}{R} + g \cos \theta \right)$$

$$= (0.200 \text{ kg})(5.00 + 8.88) \text{ m/s}^2 = \boxed{2.78 \text{ N}}$$

(d) At the bottom of the circle (point B in Fig. 4.18*c*) the tension is greater than at A, since v^2 is the same, and $\cos \theta$ is larger. At the top (point D) the tension is less than at A, since v^2 is the same, and $\cos \theta$ has its maximum negative value.

4.8 Planetary Motion: Kepler's Laws

Although Galileo believed that the earth and the other planets rotated about the sun, he was never able to provide convincing proof for this Copernican model of planetary motion. It was left for Newton to provide this proof at a later date.

The observational evidence on which Newton's theory was based was in great part due to the Danish astronomer Tycho Brahe (1546–1601), who had collected astronomical data on the motion of the planets good to about 1 minute of arc (1′), or $\frac{1}{60}°$. This precision was all the more remarkable since Brahe's observations were made with the naked eye because telescopes had not yet been invented. Brahe's secret was the use of very large sighting devices mounted on strong, rigid support systems, and of highly precise scales for his position and angle measurements.

The German astronomer Johannes Kepler (1571–1630) took Brahe's data and came up with a detailed description of the motion of the planets about the sun. The word *description* is purposely used here because Kepler's three laws merely *described* the behavior of the planets. We will here merely state Kepler's three laws, as a preliminary to presenting Newton's explanation of them.

1   *Law of elliptical orbits:* The orbit of each planet about the sun is an ellipse with the sun at one focus (Fig. 4.19).

2   *Law of equal areas:* Each planet moves so that a line drawn from the sun to

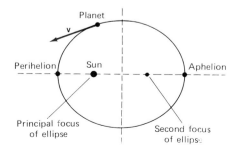

FIGURE 4.19 Elliptical path of a planet around the sun. The perihelion is the position of closest approach of the planet to the sun; the aphelion is the position at which the planet is farthest from the sun.

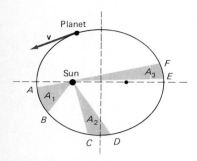

FIGURE 4.20 Kepler's second law, the law of equal areas. The planet sweeps out equal areas in equal times. If area A_1 is equal to A_2 and to A_3, then the times for the planet to go from A to B, from C to D, and from E to F are all equal. Hence the planet travels faster in going from A to B than from E to F.

the planet sweeps out equal areas in equal times (Fig. 4.20). The planets thus move faster when they are closer to the sun.

3 *Law of periods (harmonic law):* The ratio of the squares of the periods of any two planets revolving about the sun is equal to the ratio of the cubes of their average distances from the sun. If A and B are two planets, then

$$\frac{T_B^2}{T_A^2} = \frac{R_B^3}{R_A^3} \qquad (4.35)$$

In these three laws Kepler made sense of the motion of the planets for the first time and thus achieved one of the basic goals of physics—the understanding of the universe on the basis of a small number of physical principles. Kepler was not able to derive these three laws from the more basic laws of mechanics; it was Sir Isaac Newton who first accomplished this goal.

Newton's Explanation of Kepler's Laws

Newton had, in his three laws of motion and the law of universal gravitation, all that was needed to describe the motion of the planets and to derive Kepler's laws. Here we present only Newton's proof of Kepler's third law.

In deriving Kepler's third law, we introduce the following approximation: Since the elongation of the ellipses is quite small,* we use a simplified model, which assumes that the planets move in circles of radius R about the sun, where R is each planet's average distance from the sun in its elliptical orbit. Some form of physical force is needed to provide the centripetal force required for circular motion. Newton's great discovery was that this force was the same kind which causes an apple to fall from a tree to the ground, i.e., a gravitational force.

Let us assume that the gravitational force of the sun on a planet of mass m_P keeps the planet moving in a circle. Since $\mathbf{F}_{net} = m\mathbf{a}$, we have from Eqs. (4.1) and (4.31),

$$F_G = \frac{Gm_S m_P}{R^2} = \frac{m_P v^2}{R}$$

On canceling like terms, we have

$$v^2 = \frac{Gm_S}{R} \qquad \text{or} \qquad v^2 R = Gm_S \qquad (4.36)$$

Then, since $v = 2\pi R/T$ from Eq. (4.21), $v^2 R = 4\pi^2 R^3/T^2 = Gm_S$ and

$$\frac{R^3}{T^2} = \frac{Gm_S}{4\pi^2} \qquad (4.37)$$

---

*The earth is about 4.8×10^9 m closer to the sun in January than in July, but the average earth-sun distance is about 1.5×10^{11} m. The change in the earth-sun distance is therefore only about 3 percent.

Now the right side of this equation is independent of the properties of any particular planet and is thus a constant for all the planets. For two different planets A and B we have

$$\frac{R_B^3}{T_B^2} = \frac{Gm_S}{4\pi^2} = \frac{R_A^3}{T_A^2} \quad \text{and so} \quad \frac{T_B^2}{T_A^2} = \frac{R_B^3}{R_A^3}$$

which is Eq. (4.35), Kepler's third law.

Before Newton's time many people believed that the laws governing the motion of the heavens were different from those describing motion on earth. Newton's demonstration that the laws of physics worked equally well on earth and in the heavens was a momentous contribution to our understanding of the universe.

4.9 Earth Satellites

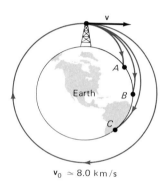

$v_0 \simeq 8.0 \text{ km/s}$

FIGURE 4.21 Newton's suggestion of the possibility of an earth satellite. He pointed out that if an object were fired parallel to the surface of the earth, as its initial speed increased it would travel greater and greater distances before returning to earth. In the diagram, as the speeds increase, the object travels first to point A, then to B, then to C, and so on. If a high enough speed v_0 could be reached, the object would circle the earth forever. This speed v_0 is about 8.0 km/s.

Isaac Newton predicted that satellites could orbit the earth as the moon does (Fig. 4.21), but seventeenth-century technology was not up to putting satellites into permanent orbits about the earth.

Why does the moon or a satellite in orbit not fall back toward the earth? The fact is that it does! The moon is constantly accelerated toward the earth by the gravitational force of the earth, with an acceleration that depends on its distance from the center of the earth. Despite this, the moon does not get any closer to the surface of the earth. In the absence of the force of gravity, the moon would move in a straight line, tangent to its former orbit, as in Fig. 4.22. Such a straight path would cause the moon to get increasingly far away from the earth. In fact, however, the force of gravity pulls the moon toward the earth, as shown in the figure. If the moon is moving at the right speed and is at the correct distance from the earth, the gravitational force provides the exact centripetal force (mv^2/R) necessary to keep the satellite moving in a closed orbit around the earth.

Consider a case in which a satellite is in orbit very close to the surface of the earth. The Russian satellite *Vostok 6*, the first satellite to carry a woman into space, had an average height above the earth of about 193 km. Since the radius of the earth is 6360 km, this makes the radius of *Vostok*'s orbit 6.55×10^6 m. Then, equating the centripetal force to the gravitational attraction of the earth for the satellite, we have

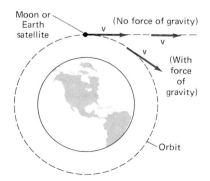

FIGURE 4.22 An earth satellite in orbit around the earth. The gravitational attraction of the earth causes the satellite to fall from the straight-line path it would follow in the absence of gravity. It falls just enough to keep it moving in a circular orbit around the earth.

$$\frac{mv^2}{R} = \frac{Gmm_E}{R^2} = w_S = mg_S$$

since the gravitational force on the satellite is precisely what we mean by its weight. We have, therefore,

$$\frac{v^2}{R} = g_S = \frac{Gm_E}{R^2} \tag{4.38}$$

where g_S is the acceleration due to gravity of the satellite at a height 193 km above the earth's surface and is, with $R = 6.55 \times 10^6$ m,

$$g_S = \left(\frac{6.67 \times 10^{-11} \text{ m}^3}{\text{kg·s}^2}\right)\left(\frac{5.98 \times 10^{24} \text{ kg}}{(6.55 \times 10^6 \text{ m})^2}\right) = 9.30 \text{ m/s}^2$$

Then

$$v = \sqrt{g_S R} = \sqrt{(9.30 \text{ m/s}^2)(6.55 \times 10^6 \text{ m})} = 7.80 \text{ km/s}$$

A speed of 7.80×10^3 m/s was therefore needed to keep *Vostok 6* in orbit at a height of 193 km above the surface of the earth.

The period of the satellite in such an orbit is

$$T = \frac{s}{v} = \frac{2\pi R}{v} = \frac{2\pi(6.55 \times 10^6 \text{ m})}{7.80 \times 10^3 \text{ m/s}} = 5.27 \times 10^3 \text{ s} = 87.8 \text{ min}$$

This result agrees well with the observed fact that *Vostok 6* completed 1 rev around the earth every 88.34 min.

Weightlessness

We have probably all seen television pictures of astronauts floating around inside their spaceship, in which the force of gravity does not seem to be acting, as in Fig. 4.23. A spaceship in orbit about the earth is being accelerated toward the earth with an acceleration g_S, which differs from g on the earth's surface because of the different value of R. For an astronaut, it is like being in an elevator falling freely through space. We found, in Example 3.6, that a passenger standing on a spring scale in a freely falling elevator appears to be weightless, since there is no upward push from the scale on the passenger's feet. This is true of all objects in any freely falling system like the elevator.

An earth satellite is merely a freely falling spaceship; it falls toward the earth with an acceleration g_S [see Eq. (4.38)]. The astronauts in such a spaceship also fall toward the earth with the same acceleration g_S and therefore experience no upward pushes from the floors or rest pads in the satellite. As a result, they do not feel the compression of their vertebrae that is normally produced by a gravitational force and conclude that they are in a weightless state. This is true whether they are falling freely toward the earth, the moon, or some other planet or star. As long as the spaceship is falling freely in space, i.e., as long as its rocket engines are not operating, everything within the spaceship appears to be weightless. Objects therefore float freely in the cabin, as seen in Fig. 4.23. This is an *apparent weightlessness*, produced by the accelerated reference system in which the astronauts find themselves.

Weightlessness gives astronauts a lightsome feeling, since no effort is needed to hold up their arms when they are in free-fall. Prolonged exposure to weightlessness can, however, have bad effects. Red blood cell counts decrease, bones become more brittle, and muscles lose their tone through lack of use. These effects

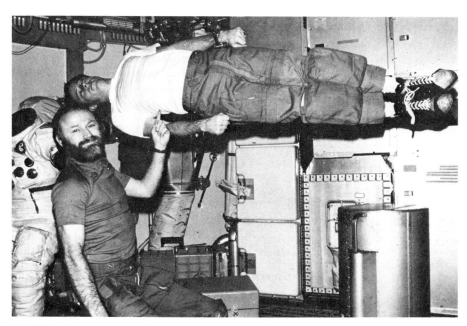

FIGURE 4.23 Astronauts Gerald P. Carr (left) and Edward G. Gibson (floating) demonstrate the effects of weightlessness during the *Skylab 4* space mission. Carr is pretending to support Gibson on one finger, something made possible by the apparent weightlessness of the occupants of the orbiting spacecraft. (*Courtesy of NASA.*)

arise because muscles and bones are no longer needed to support the body's weight. Astronauts can counteract some of the undesirable side effects of weightlessness through suitable physical exercise. Much research is being done on the physical effects of weightlessness by NASA in connection with the U.S. space program.

Example 4.13

A communications satellite is used to relay radio and TV signals around the earth. Such satellites keep a fixed position above one spot on the earth's surface and move eastward with the earth as it rotates on its axis. These are sometimes referred to as *stationary satellites*, or *synchronous satellites*. **(a)** What must be the distance of such a satellite from the center of the earth? **(b)** How fast must it travel in its orbit? **(c)** How far above the earth's surface must the satellite be? **(d)** Is Kepler's third law valid for this synchronous satellite and for *Vostok 6* (discussed at the beginning of this section).

SOLUTION

(a) The satellite, like the earth, makes one complete revolution each day, or every 24 h. Since the force needed to keep it in orbit is provided by the gravitational attraction of the earth, the equation governing its motion is of the same form as that governing the motion of the earth around the sun. We have, from Eq. (4.37), with the mass of the sun replaced by the mass of the earth,

$$R^3 = \frac{Gm_E T^2}{4\pi^2}$$

where $T = 24$ h $= 8.64 \times 10^4$ s. Therefore

$$R^3 = \frac{\left(6.67 \times 10^{-11} \ \dfrac{\text{m}^3}{\text{kg} \cdot \text{s}^2}\right)(5.98 \times 10^{24} \ \text{kg})(8.64 \times 10^4 \ \text{s})^2}{4(3.14)^2}$$

$$= 7.55 \times 10^{22} \ \text{m}^3$$

$$R = \boxed{4.23 \times 10^7 \ \text{m}}$$

(b) The speed of the satellite must be

$$v = \frac{2\pi R}{T} = \frac{2\pi(4.23 \times 10^7 \text{ m})}{8.64 \times 10^4 \text{ s}} = \boxed{3.08 \times 10^3 \text{ m/s}}$$

(c) Since the radius of the earth is 6.37×10^6 m, the height of the satellite above the earth's surface must be

$$4.23 \times 10^7 \text{ m} - 0.64 \times 10^7 \text{ m} = \boxed{3.59 \times 10^7 \text{ m}}$$

Synchronous communications satellites therefore circle the earth at a height about 3.59×10^7 m, or 2.23×10^4 mi, above the earth's surface.

(d) To test Kepler's third law, we make use of the information above. We have, for the synchronous communication satellite,

$$R_S = 4.23 \times 10^7 \text{ m} \qquad T_S = 24 \text{ h} = 8.64 \times 10^4 \text{ s}$$

and for *Vostok 6*,

$$R_V = 6.55 \times 10^6 \text{ m} \qquad T_V = 88.3 \text{ min} = 5.30 \times 10^3 \text{ s}$$

Let us try to predict the period of the *Vostok* satellite from the other data given. Using Kepler's third law [Eq. (4.35)], we obtain (carry out the calculation yourself) $T_V = \boxed{5.26 \times 10^3 \text{ s}}$.

This is good agreement with the observed 5.30×10^3 s, especially given the approximation used in assuming a constant radius for the elliptical orbit.

Summary: Important Definitions and Equations

Newton's law of universal gravitation:

$$F_G = \frac{Gm_1m_2}{r^2} \tag{4.1}$$

where $G = 6.672 \times 10^{-11}$ m³/kg·s² (or N·m²/kg²). For earth's gravitation (near surface of earth),

$$g = \frac{F_G}{m} = \frac{Gm_E}{r_E^2} \tag{4.5}$$

At any point near the earth's surface, the acceleration due to gravity, **g**, has the same magnitude for all objects.

Weight: The gravitational force exerted on an object:

$$\mathbf{w} = m\mathbf{g} \tag{4.6}$$

Weight is a *force* (a gravitational force), whereas *mass* is a measure of *inertia*.

Motion in a plane: For motion in a plane the acceleration a has a component a_T parallel to the velocity that specifies the change in the magnitude of the velocity and a component a_C perpendicular to the velocity that specifies the change in the direction of the velocity.

$$a = [a_T^2 + a_C^2]^{1/2} \tag{4.29}$$

$$\tan \phi = \frac{a_C}{a_T} \tag{4.30}$$

Projectile motion: A combination of horizontal motion with constant velocity and vertical motion with constant acceleration.

Uniform circular motion (motion in a circle with constant speed):
Period (T): The time for one complete revolution.
Frequency: The number of revolutions per unit time:

$$f = \frac{1}{T} \tag{4.22}$$

Linear velocity:

$$v = \frac{2\pi R}{T} \tag{4.21}$$

Constant angular velocity:

$$\omega = \frac{\Delta\theta}{\Delta t} = 2\pi f = \frac{2\pi}{T} = \frac{v}{R} \tag{4.23–4.25}$$

Centripetal acceleration and centripetal force:

$$a_C = \frac{v^2}{R} \tag{4.27}$$

$$F_C = \frac{mv^2}{R} = mR\omega^2 \tag{4.31–4.32}$$

Kepler's three laws of planetary motion:
1 *Law of elliptical orbits:* The orbit of each planet about the sun is an ellipse, with the sun at one focus.
2 *Law of equal areas:* Each planet moves so that a line drawn from the sun to the planet sweeps out equal areas in equal times.
3 *Law of periods* (harmonic law): The ratio of the squares of the periods of any two planets revolving about the sun is equal to the ratio of the cubes of their average distances from the sun.

$$\frac{T_B^2}{T_A^2} = \frac{R_B^3}{R_A^3} \tag{4.35}$$

Earth satellites: Objects that orbit the earth in a closed orbit similar to the way planets orbit the sun.

Weightlessness: The apparent absence of weight that exists for persons and things inside freely falling objects, including freely falling elevators and earth satellites.

Questions

1 A sailor climbs to the top of a ship's mast to repair a sail while the ship is moving at a fast speed. At the top of the mast a pocketknife falls out of the sailor's pocket. Where will the knife hit the deck of the ship: behind the bottom of the mast, in front of the bottom of the mast, or exactly at the foot of the mast? Why?

2 What is the gravitational force produced by a thin uniform spherical shell on a small ball placed exactly at the center of the sphere?

3 (*a*) If a rock is buried deep beneath the surface of the earth, is the force of gravity on the rock greater or less than it would be at the surface of the earth?

(*b*) What would be the force of gravity on the rock at the exact center of the earth? (*Hint:* Consider the forces on the rock in all directions.)

4 A 300-lb football player is hard to lift on the surface of the earth and equally hard to knock down with a football block. If you and he were on the surface of the moon, which of these would you rather do? Why?

5 At what point in its path does a projectile have its greatest speed? Its smallest speed? (Assume that it is fired above a horizontal surface.)

6 Show from the formula for the magnitude of the centripetal acceleration ($a_C = v^2/R$) that for motion in a circle at constant speed, the centripetal acceleration must be normal to the direction of motion.

7 Give a precise definition of an ellipse. Can you in-dicate a practical method of constructing an ellipse with a string, two pins, and a pencil?

8 A coin rests on a record that is on a turntable. The turntable starts to move and gradually picks up speed. Eventually the coin slides off the record.

(*a*) Explain why this happens.

(*b*) In what direction does the coin slide?

9 If the earth is constantly falling toward the sun, why does it not gradually move closer to the sun? (If it did, the increased heating of the earth's surface by solar radiation would soon destroy all life on earth.)

10 The distinction between mass and weight can be a matter of life and death in space. Suppose a space station is being built in orbit around the earth. Astronauts building the station are moving huge girders into position in the space station.

(*a*) Are these girders weightless?

(*b*) If they are weightless, are they any easier to move around in space than they would be on earth? Why?

(*c*) Could an astronaut be crushed between two huge girders moving toward each other in space, even if they were weightless?

11 The earth actually moves faster in its orbit around the sun in the winter than it does in the summer. Is the earth therefore closer to the sun in the winter or in the summer?

12 The force of the sun on the moon is twice as large as the force of the earth on the moon. Why doesn't the sun pull the moon away from the earth?

Multiple-Choice and Simple Exercises

4.1 A batter hits a baseball at an angle of exactly 45° with respect to the horizontal. The ball first hits the ground 110 m from the bat. The ball must have left the bat with an initial speed of about:

(*a*) 1.1×10^3 m/s (*b*) 3.4 m/s (*c*) 9.8 m/s
(*d*) 33 m/s (*e*) 11 m/s

4.2 A horse fixed on a carousel is moving with a constant horizontal speed of 5.0 m/s. The horse is 7.0 m from the center of the carousel. The period of the horse's motion is:

(*a*) 1.4 s (*b*) 0.11 s (*c*) 8.8 s
(*d*) 0.71 s (*e*) 4.4 s

4.3 The angular velocity of the carousel horse in the previous exercise is, in degrees per second:

(*a*) 0.71°/s (*b*) 41°/s (*c*) 57°/s
(*d*) 35°/s (*e*) 80°/s

4.4 The centripetal acceleration of a fly sitting on a phonograph record at a distance of 20 cm from the center of the record, while the record is rotating at 33.3 rev/min, is:

(*a*) 2.4 m/s^2 (*b*) 2.4 m/s (*c*) 0.60 m/s^2
(*d*) 1.5 m/s^2 (*e*) 1.5 m/s

4.5 An ultracentrifuge is rotating at 1.0×10^5 rev/min. The effective g value experienced by a piece of sand in a test tube at a distance of 12 cm from the center of the rotating system is:

(*a*) 1.6×10^4 g (*b*) 3.3×10^4 g
(*c*) 2.1×10^5 g (*d*) 1.3×10^7 g
(*e*) 1.3×10^6 g

4.6 A small car is moving in a vertical circle. The acceleration of the car at the bottom of the circle is:

(*a*) In the same direction as its velocity
(*b*) In the opposite direction to its velocity
(*c*) Directed toward the center of the circle
(*d*) Zero (*e*) Directed downward

4.7 A curve of radius 175 m is to be banked to minimize skidding at 55 mi/h. The banking angle for a 1500-kg car should be:

(*a*) 20° (*b*) 20 rad (*c*) 0.83 rad
(*d*) 0.83° (*e*) 0.36°

4.8 For an earth satellite to remain in an orbit at a distance of 10,000 km from the center of the earth, it would have to travel at a speed of:

(*a*) 6.3 km/s (*b*) 6.3×10^4 km/s
(*c*) 6.3 m/s (*d*) 6.3×10^3 km/s
(*e*) None of the above

4.9   For the earth satellite in Prob. 4.8, the period of the satellite would be:
(a) 1.0×10^4 s   (b) 6.4×10^3 s
(c) 2.0×10^3 s   (d) 3.2×10^3 s
(e) 1.6×10^{-4} s

4.10   At a certain instant of time an automobile is going around a curve. Its acceleration in its direction of motion is 3.0 m/s^2 and its centripetal acceleration is 4.0 m/s^2.
(a)   What is the magnitude of the resultant acceleration at that instant?
(b)   What is the direction of this acceleration with respect to the direction of motion of the car at that instant?

4.11   A bullet is fired from a rifle at a speed of 400 m/s. The rifle is held level to the ground and 1.50 m off the surface of the ground. How long will it take for the bullet to hit the earth at the end of its trajectory?

4.12   What is the range of a tennis ball hit at an angle of 5.0° above the horizontal at a speed of 30 m/s? (Neglect spin and air resistance.)

4.13   A ball at the end of a string 0.75 m in length rotates at constant speed in a horizontal circle. It makes 3 complete revolutions per second.
(a)   What is the period of the ball's motion?
(b)   What is the frequency of the motion?
(c)   What is the angular velocity of the ball?

4.14   How far above the earth's surface would an earth satellite have to be for the force of gravity on it (due to the earth) to be reduced to one-tenth its value on the earth?

4.15   A car is going around a curve of radius 100 m at a speed of 20 m/s. At what angle should this curve be banked to reduce to a minimum the possibility of skidding?

4.16   A grinding wheel is rotating at 1000 rev/min. If the radius of the wheel is 10 cm, find the centripetal force on a 2.0-g particle at the rim of the wheel.

4.17   The moon is rotating in its orbit around the earth at a distance of 3.84×10^8 m from the earth. What is its linear velocity? (*Hint:* Use Table F.2 inside front cover.)

4.18   A synchronous communications satellite is in a circular orbit at a distance of about 22,300 mi above the earth's surface.
(a)   What is the speed of the satellite?
(b)   What is the component of its acceleration parallel to its direction of motion?
(c)   What is the component of its acceleration normal to its direction of motion?
(d)   If its mass is 100 kg, to what net force is it subject?

4.19   A bicycle has wheels 65.0 cm in diameter. If the bicycle covers ground at a speed of 10.0 m/s, what is the angular velocity of the wheels?

Problems

4.20   A ball of mass 5.00 kg is at a height of 200 km above the earth's surface.
(a)   What is the mass of the ball at this height?
(b)   What is the weight of the ball at this height?
(c)   How does this weight compare with the ball's weight at the earth's surface?

4.21   Use the data in Table F.2 to calculate how much an astronaut with a mass of 100 kg would weigh (a) on the earth's surface, (b) on the surface of Mars, (c) on the surface of Jupiter.

*__4.22__   The distance between the earth and the moon is 3.84×10^8 m. At what point between the earth and the moon is the net gravitational force on any object equal to zero? (See Table F.2)

*__4.23__   The earth has a bulge at the equator which makes its radius at the equator larger than its radius at the poles by about 21 km. What is the difference in the value of g at the north pole and at the equator? (*Hint:* Use the power series expansion discussed in Appendix 1.D.)

*__4.24__   In the state of California, San Francisco and Mount Hamilton have about the same latitude and longitude, but Mount Hamilton is 1282 m above sea level whereas San Francisco is only 114 m above sea level. If the acceleration due to gravity in San Francisco is 9.7997 m/s^2, what would you expect its value to be at the top of Mount Hamilton?

4.25   An archer is trying to hit a bull's-eye 1.50 m off the ground and 20.0 m from the archer. If the arrow is fired from a point also 1.50 m above the ground and at a speed of 40.0 m/s, in what direction must the arrow be pointed to hit the bull's-eye?

4.26   A plane drops a hamper of medical supplies from a height of 3.20×10^3 m during a practice run over the ocean.
(a)   If we neglect the effects of air resistance, how long does it take the hamper to reach the water?
(b)   How fast is the hamper moving downward at the instant it strikes the water?
(c)   If the plane's horizontal velocity was 100 m/s at the instant the hamper was dropped, what is the overall velocity of the hamper at the instant it strikes the ground (both speed and direction)?

4.27   A gun fires a shell at a speed of 85 m/s at an angle of 53° above the horizontal.
(a)   Find the position and velocity of the shell when $t = 2.0$ s.
(b)   Find the time at which the shell reaches its highest point and the elevation at that point.
(c)   Find the range of the shell.
(d)   Find the velocity when it returns to earth, neglecting air resistance.

4.28   A bullet is fired from a gun at a speed of 400 m/s in a direction parallel to the ground and from an original height of 1.50 m.
(a)   Find the x and y coordinates of its position at a time 0.100 s after it was fired.
(b)   Repeat this calculation for 0.2 s, 0.3 s, 0.4 s, . . . , until the bullet strikes the ground.
(c)   Graph the position of the bullet as a function of time.

4.29   A baseball is thrown with an initial speed of 40 m/s at an angle of 40° above the horizontal. How far does the baseball travel horizontally before reaching its maximum height?

4.30 A boy throws a ball from the roof of a building 25.0 m high toward a building 75.0 m tall, which is a horizontal distance of 15.0 m from the first building. The ball is thrown at a speed of 10.0 m/s at an angle of 35° above the horizontal.

(*a*) How far above or below the level from which the ball is thrown will it strike the wall of the second building?

(*b*) What is the velocity of the ball when it strikes the wall of the second building?

4.31 Two golf balls are hit at a speed of 21.0 m/s, one at an angle of 35° with the horizontal, the other at an angle of 55°.

(*a*) What are the ranges of the two balls?

(*b*) What are the times of flight of the two balls? (Neglect air resistance.)

*4.32 In a circus act a performer is being shot out of a cannon whose muzzle is at an angle of 30° with the horizontal, as in Fig. 4.24. The circus performer is supposed to land in a net 20 m away and at an elevation 1.5 m higher than the gun muzzle. With what minimum speed *v* must the performer leave the gun in order to land safely in the net?

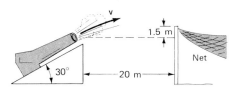

FIGURE 4.24 Diagram for Prob. 4.32.

*4.33 A 1500-kg car travels around a curve of radius 60 m banked at an angle of 10°. The car is traveling at a speed of 25 m/s.

(*a*) Will the banking be sufficient to keep the car on the road?

(*b*) If not, what additional frictional force will be required?

(*c*) What must be the coefficient of friction between the car and the road to keep the car from skidding?

4.34 (*a*) Calculate the centripetal acceleration of a racing car traveling on a circular race course of radius 600 m at a speed such that it makes 1 complete revolution of the track each minute.

(*b*) If the mass of the car is 1500 kg, what is the centripetal force on the car?

(*c*) At what angle would the track have to be banked to prevent skidding at this speed?

4.35 A block of mass 100 g rests at a point 1.2 m from the center of a rotating wooden platform (horizontal). The coefficient of static friction between the metal block and the platform is 0.30.

(*a*) Draw a free-body diagram for this situation.

(*b*) If the rate of rotation of the platform is gradually increased from rest, at what angular speed will the block first begin to slide?

*4.36 A curve of 150 m radius on a level road is banked at the correct angle for a speed of 20 m/s. If a car rounds this curve at 30 m/s, what coefficient of friction between tires and road is needed to keep the car from skidding?

4.37 A highway designer is planning a curve with a radius of 250 m. How much must the roadway be banked if cars traveling at 25 m/s are to negotiate the curve without any help from friction?

4.38 (*a*) Calculate the centripetal force on your body due to the earth's rotation on its axis.

(*b*) Compare this force with the gravitational attraction of the earth for your body.

(*c*) Why are these two forces not equal?

4.39 (*a*) Calculate the centripetal force needed to keep your body moving with the earth in orbit about the sun.

(*b*) Show that this force is exactly equal to the gravitational attraction of the sun for your body.

4.40 A miniature ultracentrifuge is driven by air jets at 9.0×10^4 rev/min and produces an effective *g* value of 1.5×10^5 *g*. What is the radius of the rotor on the centrifuge?

4.41 (*a*) A person is standing on a spring scale at the equator when the earth suddenly stops rotating! If the person survived the resulting catastrophe, by what percentage would his or her apparent weight increase?

(*b*) Answer the same question for a person at the south pole.

4.42 A metal block of mass 2.00 kg is attached to one end of a string 1.00 m long, with breaking strength 600 N. The block is then whirled in a horizontal circle on a frictionless tabletop about a fixed point at the center of the table. What is the maximum velocity the block can attain before the string breaks?

4.43 A ball of mass 10.0 g is attached to a cord of length $R = 50.0$ cm and rotates in a vertical circle.

(*a*) What is the minimum speed the ball must have at the top of the circle if the cord is not to become slack and the ball begin to fall?

(*b*) If the mass is rotating at this same speed at the bottom of the same circular path, what is the tension in the cord?

4.44 A ball is fastened to one end of a string 0.20 m long and the other end is held fixed at point *O* so that the string makes an angle of 30° with the vertical, as in Fig. 4.25. The ball rotates in a horizontal circle with the string kept at this 30° angle. Find the speed of the ball in its circular path required to preserve this 30° angle.

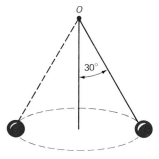

FIGURE 4.25 Diagram for Prob. 4.44.

4.45 A car in a ride at an amusement park does a loop-the-loop in a vertical circle of radius 5.0 m. What is the minimum speed that will keep the car moving in a vertical circle at the top of the loop without the car falling off the tracks?

4.46 A bucket of water is to be swung in a vertical circle without a drop of water being spilled. If the circle has a radius of 1.0 m, what must be the speed of the upside-down bucket at the top of the arc in order for the water to stay in the bucket?

∗4.47 A boy stands at the edge of a carousel 5.0 m from its center. He holds in his hand a string of length 1.0 m with a ball on its end. The carousel starts from rest and when it is up to its full speed of 1 rev every 5.0 s, the boy notices that the string now makes an angle θ with the vertical.

(*a*) Explain the cause of this change in position of the pendulum bob.

(*b*) What is the angle θ in this case? (*Hint:* The solution may lead to a complicated equation that must be solved either graphically or by successive approximations.)

∗4.48 A 1.0-kg metal sphere hangs at the end of a string 5.0 m long to form a pendulum. The sphere is pulled to one side so that it makes an angle of 40° with the vertical and is then released.

(*a*) Find the tension in the cord just after the ball is released.

(*b*) Find the tension in the cord when the ball passes through the bottom of its swing.

∗4.49 A binary star system consists of two identical stars, with centers separated by 1.20×10^{11} m, rotating about a point midway between them. The stars make 1 complete revolution every 13.1 years.

(*a*) What keeps the two stars from colliding with each other?

(*b*) What is the gravitational force between the two stars?

(*c*) What are the masses of the two stars?

4.50 How small would the earth's period of rotation on its axis have to be to make objects weightless on the equator?

4.51 Use the average distance of the earth from the sun and the period of the earth's motion around the sun to find the mass of the sun.

4.52 Use the distance of Mars from the sun, the distance of the Earth from the sun, and the period of the Earth's motion around the sun to obtain the period of Mars in its orbit around the sun.

4.53 (*a*) What is the shortest possible period for an earth satellite consistent with the values of the earth's mass and radius?

(*b*) What is the shortest possible period for a satellite orbiting around Mars?

4.54 An artificial satellite of mass 2000 kg orbits the earth in a circular orbit of radius 6.60×10^6 m. What is its period of revolution?

4.55 It is desired to launch a satellite into a circular orbit 1000 km above the earth. What initial velocity tangent to this orbit must be imparted to the satellite if this orbit is to be a stable one for the satellite?

4.56 (*a*) What is the effect of the rotation of the earth on the value of g at the equator?

(*b*) What percentage change in g does this represent?

4.57 During the *Apollo II* lunar mission the command module circled the moon with a period of 2 h, 20 min, while two of the astronauts landed on the moon's surface in the lunar module. How high above the moon's surface did the command module circle?

4.58 An apple on the surface of the earth falls 4.9 m in its first second of fall from rest. The moon is about 60 times farther away from the earth's center than is the apple. How far does the moon fall toward the earth each second?

4.59 With what speed would a moon satellite have to rotate in a circular orbit 2.0×10^6 m from the center of the moon?

Additional Readings

Beams, Jesse W.: "Ultrahigh-Speed Rotation," *Scientific American*, vol. 204, no. 4, April 1961, pp. 134–147. A fascinating account of work by Beams on high-speed rotation and the ultracentrifuge.

Casper, B. M., and R. J. Noer: *Revolutions in Physics*, Norton, New York, 1972. Contains a good discussion of the Copernican revolution.

Cohen, I. Bernard: "Isaac Newton," in *Lives in Science*, a *Scientific American* book, Simon and Schuster, New York, 1957, pp. 21–30. A brief account of Newton's life and contributions to science.

Gray, G. W.: "The Ultracentrifuge," *Scientific American*, vol. 184, no. 6, June 1951, pp. 42–52. An account of the development of the ultracentrifuge and its use in accelerating sedimentation.

Koestler, Arthur: *The Sleepwalkers: A History of Man's Changing View of the Universe*, Macmillan, New York, 1959. A fascinating account of Copernicus, Kepler, and Galileo by a gifted writer.

Wood, Elizabeth A.: *Science for the Airplane Passenger*, Ballantine, New York, 1968. The chapters on frames of reference and quantitative measurements present some interesting material relevant to this chapter.

Zee, A.: *An Old Man's Toy: Gravity at Work and Play in Einstein's Universe*, Macmillan, New York, 1989. An illuminating and often entertaining account of the force of gravity and of new attempts to penetrate its mysteries.

CHAPTER 5

Equilibrium

From being the preoccupation of a few curious spirits science has grown to be a universal study, on the fruits of which peace among people and the prosperity of nations depend, but the great principles enunciated by Newton and their orderly development by him remain as the foundations of the discipline and as a shining example of the exalted power of the human mind.

E. N. da C. Andrade (1887–1971)

Chapter 3 might lead us to think that forces always lead to motion. Nothing could be further from the truth. Any nonzero net force on an object accelerates that object, but a large number of forces may act on an object and it may still remain at rest, as long as the vector sum of the forces is zero. If the vector sum is zero, there is no net force and hence no acceleration. This condition we call *equilibrium.*

When the net force on an object is zero, then, according to Newton's first law of motion, an object at rest remains at rest and an object in motion continues to move in a straight line at constant speed. The first kind of equilibrium, in which the object under consideration remains at rest, is called *static equilibrium.* The second type, in which the object moves at constant speed in a straight line, is referred to as *dynamic equilibrium.* We consider both in this chapter.

Many equilibrium situations are of great practical importance. A gigantic ocean-going oil tanker at anchor in a harbor is at rest, but it experiences both the very large gravitational pull of the earth and the equally large buoyant force of the water. Since these two forces are equal and opposite, the oil tanker remains at rest. Again, the arches in a cathedral are subjected to forces in a variety of directions. But the architects have so planned the cathedral that all the forces on the arches cancel out. If this were not the case, the arches would soon shift position and the cathedral would collapse.

5.1 Contact and Reaction Forces

In discussing equilibrium, it is crucial that we first isolate the object in equilibrium and consider all the forces *acting on this object.* Some of these forces may be produced by contact of the object with its surroundings.

For example, consider a book resting on a table, as in Fig. 5.1a. The book is pulled toward the center of the earth by the force of gravity, which gives the book a weight $\mathbf{w} = m\mathbf{g}$. Since the book is not accelerated, the table must be pushing up on the book with a force $\mathbf{F}_N$ that just balances the weight of the book pushing down on the table. The force exerted by the table on the book is really

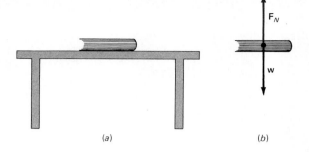

the reaction to the force exerted by the book on the table, which equals the weight of the book. Since the two forces acting *on the book* are its weight acting down and the push of the table acting up, and since these two forces are equal and opposite, the book is in translational equilibrium and remains at rest. This is shown in the FBD of Fig. 5.1*b*.

Another example of static equilibrium is provided by a ladder resting against a smooth wall. For the ladder to remain at rest, all the forces acting on it must cancel.

In this case, however, there are forces in both the horizontal (*x*) direction and the vertical (*y*) direction (Fig. 5.2). First we have the weight of the ladder, which is a force directed downward and which we assume acts at the geometrical center of the ladder. This force must be balanced by the *y* component of the force exerted by the ground on the base of the ladder so that this *y* component (F_y) and *w* are equal in magnitude but in opposite directions. This assumption that F_y and *w* are equal is valid only if we assume that the wall is completely smooth so that there is no frictional force parallel to the wall keeping the ladder from moving up or down the wall. Then the push F_{wall} of the wall is horizontal and there is no vertical force produced by the wall.

The ground cannot be smooth, though, for if it were, the ladder would merely slide until it assumed a horizontal position parallel to the ground. The friction between the ground and the ladder keeps the ladder from slipping. The base of the ladder "tries" to move along the ground, and therefore sets up a horizontal force on the ground directed away from the wall. The reaction force to this is the *x* component (F_x) of the force exerted by the ground on the ladder, a force that is directed toward the wall. At equilibrium the sum of the horizontal forces acting on the ladder must be zero, irrespective of where these forces are applied. Hence F_x is equal to F_{wall}, the outward push exerted on the ladder by the wall. This force is a reaction force to that exerted on the wall by the ladder. Since no net force acts on the ladder, it is not accelerated but is in static equilibrium.

In discussing static equilibrium it is therefore very important that all the forces acting *on the object* in equilibrium be considered. Some of these may at first glance appear to be forces exerted by the object on its surroundings. Remember that to every action force there is a reaction force. Associated with every force exerted by an object on its surroundings must be a reaction force exerted by the surroundings on the object. All these forces acting *on the object* must be taken into account in describing the equilibrium of the object.

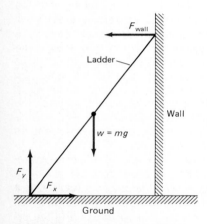

FIGURE 5.2 Forces on a ladder in an equilibrium position against a wall.

5.2 Tensile and Compressive Forces

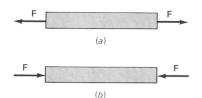

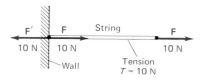

FIGURE 5.3 (*a*) Tensile forces acting on a metal rod. (*b*) Compressive forces acting on a metal rod.

FIGURE 5.4 Forces on a string attached to a wall at the left and pulled by a force **F** to the right. The forces acting on the string are the tensile forces **F** and **F**′ at the two ends, where **F**′ is a reaction force.

FIGURE 5.5 FBD for the right half of the string. For equilibrium the tension **T** must be equal in magnitude to the applied force **F**. This is true at any point along the length of the string.

We have already seen in Chap. 3 examples of two important kinds of forces that occur frequently in equilibrium problems: tensile and compressive forces.

Tensile forces: Forces that act to increase the length of an object by pulling on its two ends (Fig. 5.3*a*).

Compressive forces: Forces that act to decrease the length of an object by pushing its two ends together (Fig. 5.3*b*).

Let us consider the tensile forces acting on an ideal string (or cord, or rope), one which is perfectly flexible, unstretchable, and without mass. No such string exists, but in many real physical situations the strings involved approach this ideal. We assume that this is the case in what follows, unless the opposite is specified.

Suppose the string is attached to a wall, and a force **F** equal to 10 N is exerted on it, as in Fig. 5.4. Since the string does not stretch, it exerts the same force **F** on the wall. The wall, by Newton's third law, exerts an equal and opposite force **F**′ on the string at its left end. The forces acting *on the string* are therefore the two tensile forces **F** and **F**′, which are equal in magnitude and act on its two ends in opposite directions. In such a situation we say that the string is under *tension*, by which we mean that the molecules of one segment of the string exert forces on the molecules of the adjacent segment to keep them from pulling apart under the tensile forces. This tension **T** is equal in magnitude to the applied force **F**.

If we draw an FBD for the forces acting on the right half of the string, as in Fig. 5.5, we see that the force exerted by the molecules in the left half of the string on the right half must be equal and opposite to **F**, since the right half of the string is in equilibrium. The tension **T** in the string must therefore be equal in magnitude to the applied tensile force **F**. No matter at what point in the string this calculation is carried out, the result is the same. *The tension T is the same at every point in a stretched, massless string.*

Compressive forces are the opposite of tensile forces. We can attempt to compress a rigid rod attached to a wall by pushing it toward the wall at its free end with a force **F**. In this case the rod exerts an equal force **F** on the wall, directed toward the wall. This sets up a reaction force **F**′ whose magnitude is equal to that of the applied force **F** and whose direction is the opposite of **F**. The rod is thus acted on by compressive forces at its two ends and is said to be under *compression*.

Clearly a metal rod can sustain compressive forces; strings and ropes cannot: they simply buckle under such forces. They can sustain tensile forces and are therefore used, together with systems of pulleys or wheels, to change the effective direction of applied forces without changing their magnitude. We will see examples of this throughout the rest of this chapter.

An interesting example of these ideas is provided by *concrete*, the popular building material. Concrete is quite strong under compressive forces but is easily broken when subjected to tensile forces. For this reason it is necessary to reinforce the concrete used for beams in buildings. The reinforcement is provided by iron rods embedded in the concrete before it hardens. An even better solution is to prestress the concrete, that is, to keep the iron rods under tension when the concrete is being poured. When the concrete dries, the external tensile forces are removed from the rods, which tend to revert to their original length and thus subject the concrete to compressive forces. The amount of compression built into the prestressed concrete is made greater than the tensile forces to which a beam may be subjected in a building. In this way the concrete is never subjected to net tensile forces and therefore does not crack.

5.3 Static Equilibrium: Translation

In this section we consider one kind of static equilibrium—*translational equilibrium*. If an object does not move from one place to another in space, it is clearly not being accelerated and, from Newton's second law, it must therefore have no net force acting on it. The ladder resting against the wall in the preceding section is a good illustration of translational equilibrium.

For translational equilibrium, the vector sum of the external forces acting on an object must be zero; that is,

$$\mathbf{F}_{\text{net}} = \sum_{i=1}^{N} \mathbf{F}_i = 0 \qquad (5.1)$$

or, equivalently,

$$\sum_{i=1}^{N} (F_i)_x = 0 \qquad \sum_{i=1}^{N} (F_i)_y = 0 \qquad \sum_{i=1}^{N} (F_i)_z = 0 \qquad (5.2)$$

In accordance with Newton's third law, the internal forces cancel in pairs and only the external forces need be considered. In this chapter we will consider only problems in which the external forces all lie in the same plane (taken as the xy plane) so that only the x and y equations above need be considered.

Equations (5.2) state the *first condition of equilibrium*:

For translational equilibrium the components in any given direction of all the forces acting on an object must add to zero.

In solving problems on objects in translational equilibrium, it is useful to adopt the following straightforward procedure:

1 Draw a diagram of the physical problem, indicating all masses, lengths, and angles on the diagrams.

2 Select one object. Draw a free-body diagram for *all* the forces, including all reaction forces acting *on the object*. Label each force with its own letter or symbol.

3 Select a set of convenient rectangular coordinates for the problem, and resolve all the forces acting on the object into x and y components.

4 Apply Eqs. (5.2) to the x and y components of the forces acting. These equations will provide two independent relationships that can then be solved for any two unknown quantities in the problem.

5 Repeat this procedure, as needed, for all objects of interest in the problem. In so doing, you may have to use results obtained from earlier applications of the equilibrium conditions.

Examples 5.1 and 5.2 illustrate this procedure.

Example 5.1

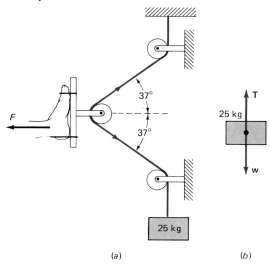

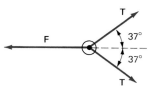

The pulley arrangement shown in Fig. 5.6a applies traction to a hapless skier's leg. A 25.0-kg mass acted on by gravity produces the tension in the rope that applies the traction. If the angles are as shown, what is the horizontal force applied to the leg? Assume that the pulleys are frictionless.

FIGURE 5.6 (a) Diagram for Example 5.1. (b) and (c) Free-body diagrams for same situation.

(a)   (b)   (c)

SOLUTION

Let us first isolate the 25.0-kg mass and consider the forces on it, as in Fig. 5.6b, since this is an easy way to obtain the tension in the rope that we need to solve the problem. If T is the tension in the rope, for equilibrium we must have

$$\sum F_y = 0 \quad \text{and so} \quad T - w = T - mg = 0$$

where we have written ΣF_y for $\Sigma(F_i)_y$ in Eq. (5.2) for convenience. Therefore

$$T = mg = (25.0 \text{ kg})(9.80 \text{ m/s}^2) = 2.45 \times 10^2 \text{ N}$$

Since the pulleys are frictionless, this tension remains the same throughout the rope, and the rope pulls down on the ceiling with a force of 2.45×10^2 N. The reaction force of the ceiling on the rope is what ultimately supports the 25-kg mass.

Now let us isolate the pulley attached to the skier's foot consider the forces acting on it. These are the pull of the to the left (F) and the pull of the two segments of the rope and below the central pulley to the right, as shown in Fig. lence we have

$$\sum F_x = 0 \quad -F + T \cos 37.0° + T \cos 37.0° = 0$$

$$\text{or} \quad F = 2T \cos 37.0°$$

$$= 2(2.45 \times 10^2 \text{ N})(0.799) = 3.92 \times 10^2 \text{ N}$$

This is the force exerted on the pulley by the skier's leg. The reaction to this is the equal and opposite force of 3.92×10^2 N exerted on the skier's leg by the pulley. This is the traction force, which is therefore $\boxed{3.92 \times 10^2 \text{ N}}$.

Example 5.2

A 10.0-kg painting hangs on a wall, as illustrated in Fig. 5.7a. The supporting wire, attached to the frame of the painting by eyelets, passes over a hook at A in the wall. The wire makes an angle of 60.0° with the vertical at each side. **(a)** What is the tension in the supporting wire? **(b)** What is the vertical force exerted by the tension in the wire on each eyelet? **(c)** What is the horizontal force exerted by the tension in the wire on each eyelet?

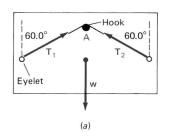

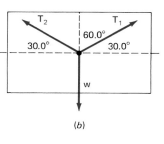

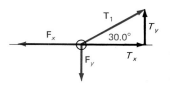

FIGURE 5.7 (a) Diagram for Example 5.2 of a painting hanging on a wall. (b) Free-body diagram of the forces on the painting. (c) Free-body diagram of the forces exerted on the left eyelet.

(a)   (b)   (c)

SOLUTION

(a) Let us isolate the painting and consider it as the object in equilibrium. Then the horizontal forces are the horizontal components of the tensions in the wire, or $T \cos 30.0°$. For equilibrium,

$$\sum F_x = 0 \quad \text{and so} \quad T_1 \cos 30.0° - T_2 \cos 30.0° = 0$$

and $\quad T_1 = T_2$

as we would expect.

The vertical forces on the painting are the vertical components of the tension in the wire directed upward and the weight of the painting directed downward, as in the free-body diagram of Fig. 5.7b. For equilibrium

$$\sum F_y = 0 \quad T_1 \sin 30.0° + T_2 \sin 30.0° - w = 0$$

Setting $T_1 = T_2 = T$, we have $2T \sin 30.0° - mg = 0$ and

$$T = \frac{mg}{2 \sin 30.0°} = \frac{(10.0 \text{ kg})(9.80 \text{ m/s}^2)}{2(0.500)} = \boxed{98.0 \text{ N}}$$

(b) The vertical force exerted by the tension in the wire on each eyelet is, from Fig. 5.7c,

$$T_y = T \sin 30.0° = \boxed{49.0 \text{ N}}$$

Since the eyelet is in equilibrium, the sum of the vertical forces acting on it must be zero. The force F_y shown in the figure must therefore be half the weight of the painting pulling down on the eyelet.

(c) The magnitude of the horizontal force produced by the tension of the wire on the eyelet is $T_x = T \cos 30.0° = \boxed{84.9 \text{ N}}$. This force is opposed by the pull of the frame in which the eyelet is embedded, so the eyelet too is in equilibrium.

Note that in this case the horizontal forces on the eyelets are greater than the vertical forces (half the weight of the painting) each eyelet must support. The smaller the angle the wire makes with the horizontal, the larger the horizontal forces on the eyelets. Thus, if the angle with the horizontal is only 10.0°, we find that $T = 2.82 \times 10^2$ N, and $T_x = 2.78 \times 10^2$ N.

In this case, the eyelets must be able to sustain forces *three times as great* as the weight of the painting, because the near-horizontal direction of the supporting wire requires that the tension be very great to produce vertical components sufficient to support the weight of the painting. This large tension and small angle make the horizontal forces very large. This must always be considered when hanging paintings or mirrors, since the horizontal forces set up can become great enough to pull eyelets out of the frame or even to break the frame, if the supporting wire is too close to being horizontal.

5.4 How Forces Produce Rotation: Torques

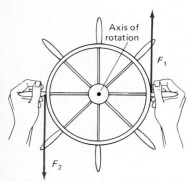

FIGURE 5.8 The forces exerted by the captain's hands on a ship's wheel. The two forces add to zero, guaranteeing translational equilibrium. However, the forces produce a rotation of the wheel about its axis.

For translational equilibrium it is of no importance at what point on the object the forces are applied. If we consider the possibility that the object may rotate, however, then the point at which the forces are applied becomes of great importance. Consider, for example, the wheel with which a ship's captain steers a vessel. This wheel is mounted securely in such a way that it cannot be translated from one position in space to another without moving the whole ship along with it. The captain, however, can easily move the steering wheel circularly about the shaft on which it is mounted by pushing, for example, the right side of the wheel up and the left side down, as in Fig. 5.8. This causes the wheel to move, but the motion is a *rotation* about the axis of the wheel, not a translation in space.

The reason for this is that the two forces are applied *at different points* on the wheel. If the two forces F_1 and F_2 are equal in magnitude and opposite in direction, the first condition of equilibrium tells us that no translational motion can result. But the effect of these two equal and opposite forces in this case is to produce a counterclockwise rotation about the axis of the wheel, as Fig. 5.8 shows. We now want to consider what additional requirement must be satisfied by two forces to make *both* translation and rotation impossible. To do this we first need some way to measure the effectiveness of a given force in producing a rotation about an axis.

Torque

The rotational effect produced by a force depends on two things: the force's *magnitude* and the force's *line of action* with respect to the axis of rotation. By the *line of action* of a force we mean the extended straight line of which the force vector forms a part.

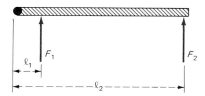

FIGURE 5.9 Two forces F_1 and F_2 applied to open a door (seen from above) have different lines of action. The force F_2 is much more effective than F_1 in opening the door because its lever arm (ℓ_2) is greater.

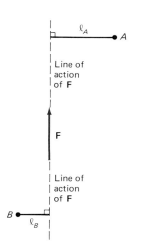

FIGURE 5.10 Lever arms of the force **F** about two different axes of rotation, one at *A*, the other at *B*. The corresponding lever arms are ℓ_A and ℓ_B. About point *A* the torque is clockwise; about *B* it is counterclockwise.

Thus a force of 20 N may be enough to open a door easily if applied at the edge of the door away from the hinges, but will not be very effective in opening the door if applied near the hinges (Fig. 5.9). The effectiveness of any given force in producing rotation depends, therefore, both on its magnitude and on the distance of its line of action from the axis of rotation, a distance we call the *lever arm* of the force.

We make these ideas quantitative by introducing the idea of *torque* (sometimes called *moment of a force*).

Torque: The product of the magnitude of a force (F) and its lever arm (ℓ).

$$\tau = F\ell \qquad\qquad (5.3)$$

where τ (Greek tau) is the torque and ℓ is the perpendicular distance from the line of action of the force to the axis of rotation.

By convention we call torques producing counterclockwise rotations positive and those producing clockwise rotations negative, although the opposite convention would work equally well. The torques exerted on the door in Fig. 5.9 are therefore positive.

The lever arm, being the perpendicular distance from the line of action of the force to the axis of rotation, is the *shortest distance* from the axis of rotation to the line of action of the force. Notice in Fig. 5.10 that the line of action of the force extends indefinitely in both directions along the force vector. In this figure the lever arm for the force **F** about an axis at right angles to the paper at *A* is shown, as is that for a different axis of rotation *B*. It can be seen that each lever arm makes a right angle with the line of action of the corresponding force vector at the place where they meet. In this case the torque about *A* is negative, while the torque about *B* is positive. Be sure to notice that the lever arm is *not* the distance from the axis of rotation to the point where the force acts on the object.

The units for torque are the units of force times distance, and hence in the SI system they are newton-meters (N·m).

Vector Nature of Torque

Our discussion of torque has thus far been confined to cases in which all forces lie in the same (*xy*) plane. In such cases the vector nature of torque is not important. But torque is more properly defined as a *vector*, with its direction along the axis

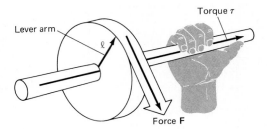

FIGURE 5.11 A force applied to a wheel to rotate it about an axis. The torque vector τ is along the axis of rotation, in the direction in which the thumb of the right hand points.

of rotation (*not* in the direction of the applied *force*).* The direction of a torque can be found from the following right-hand rule (see Fig. 5.11):

If the fingers of the right hand curl in the direction of the rotation produced by the applied force, then the extended thumb points in the direction of the torque.

For forces restricted to the xy plane, a clockwise rotation therefore corresponds to a torque pointing into the plane of the paper (as we look down on the xy plane) and a counterclockwise rotation to a torque out of the plane of the paper. All torques are therefore along the axis of rotation. Their magnitudes are given by $+F\ell$ if the torque is counterclockwise and by $-F\ell$ if the torque is clockwise. We can then add these torques as we add vectors in one dimension.

Example 5.3

A force of 200 N acts on a bicycle rim 30.0 cm in radius. **(a)** The force acts tangent to the rim at the point of application. Find the torque exerted about the axle of the bicycle wheel.

(b) Find the torque exerted if the force is directed at an angle of 40.0° to a spoke of the wheel.

SOLUTION

(a) The torque is given by $\tau = F\ell$, where ℓ is the perpendicular distance from the line of action of the force to the axis of rotation. In this case $\ell = 30.0$ cm, since F acts perpendicular to ℓ. We have

$$\tau = (200 \text{ N})(0.300 \text{ m}) = \boxed{60.0 \text{ N·m}}$$

(b) In this case the situation is similar to that shown in Fig. 5.12, with $\theta = 40.0°$. Thus $\sin 40.0° = \ell/r$, and so $\ell = r \sin 40°$ and

$$\tau = (200 \text{ N})(0.300 \text{ m})(0.642) = \boxed{38.6 \text{ N·m}}$$

EXERCISE 1 What is the torque if the force is directed along a spoke of the wheel toward the axis of rotation?

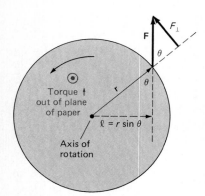

FIGURE 5.12 Vector definition of torque as $\tau = \mathbf{r} \times \mathbf{F} = rF \sin \theta$.

*A more general definition of the torque vector follows from the idea of the *cross product* of two vectors. A torque can be defined in terms of the cross product $\mathbf{r} \times \mathbf{F}$:

$$\tau = \mathbf{r} \times \mathbf{F} = rF \sin \theta$$

where τ is the torque, $\mathbf{F}$ is the applied force, $\mathbf{r}$ is the position vector from the axis of rotation to the point of application of $\mathbf{F}$, and θ is the angle between the lines of action of vectors $\mathbf{r}$ and $\mathbf{F}$, as in Fig. 5.12. The cross product of two vectors is a vector perpendicular to both. The torque vector is therefore perpendicular to both $\mathbf{r}$ and $\mathbf{F}$; if they are both in the xy plane, then τ is along the z axis. The direction of τ is given by the right-hand rule above. In this case the rotation is counterclockwise, and the torque vector is directed out of the plane of the paper. (See arrow marked $\odot$ in Fig. 5.12; here we are introducing the convention that a dot indicates the tip of an arrow pointing out of the plane of the paper and a cross indicates the tail of an arrow pointing into the plane of the paper.)

The magnitude of the torque is given (from the definition of the cross product) as $Fr \sin \theta$. Since, in Fig. 5.12, $r \sin \theta$ is equal to the lever arm ℓ, this is equivalent to $\tau = F\ell$, as in Eq. (5.3), so that our results are consistent. Since $F \sin \theta$ is the component $(F_\perp)$ of $\mathbf{F}$ perpendicular to the position vector $\mathbf{r}$, we can also write the torque as $\tau = F_\perp r$. Either approach will, of course, lead to the same result in any problem.

Example 5.4

A flat piece of tile in the shape of an equilateral triangle with sides 20 cm long is subjected to the three forces in the xy plane shown in Fig. 5.13, where $F_1 = 17.3$ N and $F_2 = F_3 = 10.0$ N. **(a)** What is the resultant force on the tile? **(b)** What is the resultant torque on the tile with respect to an axis through its center at right angles to its surface?

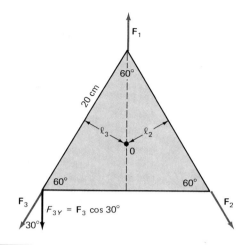

FIGURE 5.13

SOLUTION

(a) By symmetry the horizontal components of forces $\mathbf{F}_2$ and $\mathbf{F}_3$ are equal and opposite, and hence $\Sigma F_x = 0$. The vertical component of the resultant force is

$$\sum F_y = 17.3 \text{ N} - 0.866(10.0 \text{ N}) - 0.866\,(10.0 \text{ N}) = 0$$

and the tile is in *translational equilibrium*.
(b) For the torques about the axis 0, we have

$$\sum \tau_0 = F_1\ell_1 - F_2\ell_2 + F_3\ell_3$$

$$= (17.3 \text{ N})(0) - (10.0 \text{ N})(\ell_2) + (10.0 \text{ N})(\ell_3)$$

Since $\ell_2 = \ell_3$, $\Sigma \tau_0 = 0$, and the torques all cancel. In such a case we say that the tile is also in *rotational equilibrium*. Note that the lever arm of $\mathbf{F}_1$ is zero, since its line of action passes through the axis of rotation.

EXERCISE 2 As a check on this result, try taking the point of application of $\mathbf{F}_1$ as the axis of rotation, and take torques about this axis. Comment on the significance of this new choice of axis.

5.5 Static Equilibrium: Rotation

The requirement that a body be in rotational equilibrium under the action of a number of forces is the xy plane means that the net effect of the forces is to produce no rotation (as in Example 5.4), no matter what axis is chosen about which to calculate torques. *For rotational equilibrium the net torque must be zero,** or

$$\boxed{\sum \tau = 0 \qquad \text{(about any axis in } z \text{ direction)}} \tag{5.4}$$

This is the *second condition of equilibrium*, the condition that must be satisfied if no rotational motion is to result. This condition is completely general and applies to all axes of rotation. We must, of course, use the *same* axis for calculating each torque, and we must be sure to include *all* the forces that act on the body under consideration, including all reaction forces acting on it.

Since we are free to choose the axis about which moments are taken in problems of rotational equilibrium, we can often eliminate the torques of unknown

---

*Since we are only considering forces acting in the xy plane, all torques are directed along the z axis and their directions can be indicated simply by plus and minus signs. For this reason we omit the vector notation in Eq. (5.4).

forces from the torque equation by choosing an axis of rotation about which these forces have zero lever arms. Sometimes this can greatly simplify a problem.

In solving problems involving torques, the following procedures are useful:

1 Draw a diagram of the physical problem, indicating all masses, lengths, and angles on the diagram.

2 Draw a free-body diagram of all the forces acting on one object.

3 Use the procedure described in Sec. 5.3 to find relationships among the forces that act in the problem.

4 Select a convenient axis of rotation for calculational purposes. (This need *not* be an axis about which the system could really rotate.) Choose an axis through which pass the lines of action of as many unknown forces as possible, and thus eliminate the torques of these forces from the problem.

5 Apply Eq. (5.4) to all torques acting about the chosen axis of rotation; be careful to assign the correct signs to clockwise and counterclockwise torques. This should provide enough equations, when used in conjunction with Eq. (5.1), to solve for all unknown quantities.

6 Choose a different axis of rotation and solve the problem of the torques about this new axis as a check on your answer.

Examples 5.5 to 5.7 illustrate these procedures.

Example 5.5

The seesaw shown in Fig. 5.14a (for which the weight of the seesaw plank can be neglected) has a child of weight 200 N on the right side and the child's mother of weight 600 N on the left side. **(a)** If the child is 3.00 m from the fulcrum, i.e., the support of the seesaw, how far from the fulcrum must the mother sit to produce rotational equilibrium? **(b)** Solve the same problem by calculating torques about point B, which is 1.00 m to the right of A.

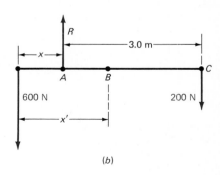

FIGURE 5.14 (a) Diagram of physical situation for Example 5.5 (b) FBD for problem.

(a)   (b)

SOLUTION

(a) The easiest way to solve this problem is to take torques about the fulcrum at A because this is the actual axis of rotation of the system, and so the lever arm of the reaction force R at A is zero. We have, from Fig. 5.14b,

$$\sum \tau_A = (600 \text{ N})(x) - (200 \text{ N})(3.00 \text{ m}) + (R)(0 \text{ m}) = 0$$

or $x = \left(\dfrac{200 \text{ N}}{600 \text{ N}}\right)(3.00 \text{ m}) = \boxed{1.00 \text{ m}}$

The mother must sit 1.00 m from the fulcrum.

(b) If we calculate torques about B, the torque due to the reaction force R at the fulcrum is now not zero. Applying the first condition of equilibrium to forces in the vertical direction, we have

$$\sum F_y = 600 \text{ N} + 200 \text{ N} - R = 0 \quad \text{or} \quad R = 800 \text{ N}$$

Taking torques about B, with x' the mother's distance from B, we have

$$\sum \tau_B = (600 \text{ N})x' - (800 \text{ N})(1.00 \text{ m}) - (200 \text{ N})(2.00 \text{ m}) = 0$$

or

$$x' = \left(\frac{800 \text{ N} + 400 \text{ N}}{600 \text{ N}}\right) \text{m} = \boxed{2.00 \text{ m}}$$

and the mother should sit 2.0 m from B, which is the same position as found in part (a).

EXERCISE 3 Solve the same problem by taking torques about point C.

Example 5.6

A girl of mass 80.0 kg stands on the free end of a diving board of 4.00 m length. The diving board is supported by two pedestals, one 3.00 m from the free end and the other 4.00 m from the same end, as in Fig. 5.15a.

Calculate the magnitude and direction of the forces exerted on the board by the two pedestals. Assume that the weight of the diving board is negligible compared to the other forces acting.

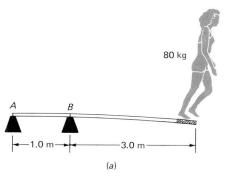

FIGURE 5.15 (a) Diagram for Example 5.6. (b) Free-body diagram.

SOLUTION

A free-body diagram is shown in Fig. 5.15b. From the first condition of equilibrium applied to the forces acting on the diving board,

$$\sum F_y = 0 \quad \text{or} \quad F_A + F_B - w = 0$$

$$F_A + F_B = w$$

$$= (80.0 \text{ kg})(9.80 \text{ m/s}^2) = 7.84 \times 10^2 \text{ N}$$

Here we have assumed that F_A and F_B are both directed upward, which may turn out not to be true.

Since the board is also in rotational equilibrium, we must apply our second condition of equilibrium. In this case we take torques about B to eliminate one unknown force from the problem. Assuming that F_A is directed upward, and taking clockwise torques as negative, we have

$$\sum \tau_B = 0 \quad -F_A(1.00 \text{ m}) - w(3.00 \text{ m}) = 0$$

$$F_A = -3w = -3(7.84 \times 10^2 \text{ N}) = \boxed{-2.35 \times 10^3 \text{ N}}$$

F_A has a magnitude of 2.35×10^3 N and is directed *downward*, contrary to our original assumption.

Then, since

$$F_A + F_B = 7.84 \times 10^2 \text{ N}$$

$$F_B = 7.84 \times 10^2 \text{ N} + 2.35 \times 10^3 \text{ N} = \boxed{3.13 \times 10^3 \text{ N}}$$

We see that the upward thrust of pedestal B is just sufficient to balance the downward forces exerted on the board by pedestal A (2.35×10^3 N) and by the girl's weight (7.84×10^2 N). If a heavier diver uses the board, the upward thrust at B increases to satisfy the condition of translational equilibrium, and the downward pull at A increases to satisfy the condition of rotational equilibrium.

Note that it is possible to make an arbitrary assumption about the direction of unknown forces in solving problems, as in the case of F_A above. As long as the problem is solved correctly, an incorrect assumption merely results in the unknown force having a negative sign, indicating that its direction is opposite that originally assumed. We must, of course, stick to our original assumption consistently throughout the problem.

Example 5.7

Consider a ladder 10.0 m long and of mass 50.0 kg resting against a smooth vertical wall and making an angle of 53.0° with the ground, as in Fig. 5.16a. Find the magnitude of the horizontal force exerted by the wall (f_W) and the magnitude and direction of the force exerted by the ground on the ladder. Assume that the weight of the ladder acts at the midpoint of the ladder.

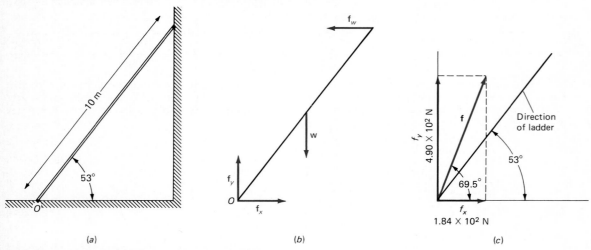

FIGURE 5.16 (a) Diagram for Example 5.7. (b) FBD for forces on ladder. (c) Resultant force exerted on ladder by ground.

SOLUTION

The forces acting on the ladder are shown in the free-body diagram of Fig. 5.16b. For equilibrium,

$$\sum F_y = 0 \qquad f_y - w = 0$$

$$f_y = w = mg = (50.0 \text{ kg})(9.80 \text{ m/s}^2) = 4.90 \times 10^2 \text{ N}$$

$$\sum F_x = 0 \qquad f_x - f_W = 0 \qquad \text{or} \qquad f_x = f_W$$

This is all the information we have from the first condition of equilibrium, but it is not enough, since we do not know either f_x or f_W. We can, however, solve the problem using the second condition of equilibrium. Suppose we take torques about the point 0 where the ladder rests on the ground. (At this point the lever arms of the forces f_x and f_y are zero.) From trigonometry we find that this point is 6.00 m from the wall and that the top of the ladder is 8.0 m from the ground, since cos 53.0° = 0.600 and sin 53.0° = 0.800. There is a clockwise torque produced by the weight of the ladder acting at its center, and a counter-clockwise torque produced by the push of the wall (f_W) on the ladder.

From the second condition of equilibrium,

$$\sum \tau_0 = 0 \qquad \text{and so} \qquad f_W(8.00 \text{ m}) - w(3.00 \text{ m}) = 0$$

or $\quad f_W = \frac{3}{8}w = \frac{3}{8}(4.90 \times 10^2) = \boxed{1.84 \times 10^2 \text{ N}}$

and $\quad f_x = f_W = 1.84 \times 10^2 \text{ N}$

The force applied by the ground to the ladder therefore has components of 4.90×10^2 N in the vertical direction and 1.84×10^2 N in the horizontal direction. This leads to a resultant of $\boxed{5.23 \times 10^2 \text{ N}}$ at an angle of $\boxed{69.5°}$ with respect to the ground. Note that this force is *not parallel to the ladder*, as shown in Fig. 5.16c.

5.6 Center of Gravity

Every particle of an extended object has some weight, and so the gravitational pull of the earth on the object consists of a large number of forces all directed toward the center of the earth. For an object of ordinary size these forces are essentially parallel to one another.

It is found experimentally that any object can be supported in equilibrium by a single upward force, provided that force has a line of action passing through a particular point that we call the object's *center of gravity*. (For a uniform, flat, circular disk, for example, this point would be the center of the disk. In other words, a single upward force equal to the sum of the gravitational forces on all the particles in the object, if applied at the center of gravity (CG) of the object, will balance out all the gravitational forces and torques on the object. The result is both translational and rotational equilibrium.

We are therefore led to define the center of gravity as follows:

Center of gravity (CG): The point at which the entire weight of an object is considered to be concentrated for the purpose of calculating gravitational torques on the object.

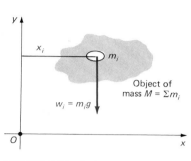

FIGURE 5.17 Defining the center of gravity of an irregularly shaped object. Each bit of mass m_i in the object exerts a torque $\tau_i = w_i x_i = m_i g x_i$ about the y axis. Note that the vector w_i is here directed perpendicular to the plane of the paper.

To locate the CG of an irregularly shaped object resting on a flat table, we consider the torque about the y axis exerted on the object by the force of gravity acting perpendicular to the plane of the paper in Fig. 5.17. A particle of mass m_1 in the object has a weight $w_1 = m_1 g$, and the weight produces a torque about the y axis of magnitude $\tau_1 = w_1 x_1$, where x_1 is the lever arm. The same is true for all the other particles in the object. The total torque on the object about the y axis is then

$$\tau = \sum_i \tau_i = w_1 x_1 + w_2 x_2 + \cdots = \sum_i w_i x_i = \sum_i m_i g x_i \qquad (5.5)$$

while the total weight of the object is

$$W = \sum w_i = \sum_i m_i g = g \sum_i m_i = Mg \qquad (5.6)$$

where M is the mass of the object.

We now ask: if all the weight Mg were concentrated at some mean distance $\bar{x}$ from the y axis, for what value of $\bar{x}$ would the resulting torque $Mg\bar{x}$ be the same as that given by Eq. (5.5)? On equating these two values of τ, we obtain

$$\tau = Mg\bar{x} = \sum_i m_i g x_i$$

$$\boxed{\bar{x} = \frac{\sum m_i x_i}{M}} \qquad (5.7)$$

This is the value of x for which the correct gravitational torque about the y axis would be obtained if all the mass were concentrated there. Hence it is also the x coordinate of the center of gravity.

To find the y coordinate of the center of gravity, we consider the gravitational torque on the object about the x axis, and repeat the above process. We obtain

$$\boxed{\bar{y} = \frac{\sum\limits_i m_i y_i}{M}} \qquad (5.8)$$

by analogy with Eq. (5.7) for the x coordinate.

The coordinates $\bar{x}$ and $\bar{y}$ (and $\bar{z}$, if needed) determine the location of the center of gravity of the object. The gravitational pull of the earth on any object acts as if the weight of the object were all concentrated at the CG, that is, at a point whose coordinates are $\bar{x}$, $\bar{y}$ in the coordinate system of Fig. 5.17.

Center of Mass

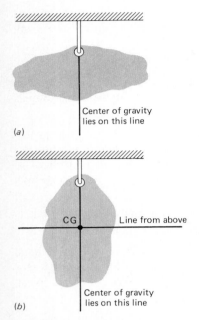

(a)

CG   Line from above

Center of gravity
lies on this line

(b)

FIGURE 5.18   Finding by experiment
the center of gravity of an irregularly
shaped object.

A related concept is that of *center of mass (CM)*, that point at which the whole *mass* of the object may be considered to be concentrated when Newton's laws are applied to the motion of the object. For an object of uniform density, the center of mass is at the center of symmetry of the object. For example, the center of mass of a sphere, cube, circular disk, or rectangular plate is at the geometric center of each, while the center of mass for a cylinder is on the axis of the cylinder halfway between the two ends. For odd-shaped objects such as a doughnut or boomerang, however, the center of mass need not even be within the object.

If the object is situated in a uniform gravitational field that is the same for all parts of the object, then the center of gravity and center of mass of the object are at the same point. This would not be true, however, for a large space platform whose most distant point might be some kilometers farther away from the center of the earth than its nearest point. In this case the platform's center of gravity would be closer to the earth than its center of mass. In most cases, however, the CM coincides with the CG, and its position can be obtained from Eqs. (5.7) and (5.8).

The above analysis leads to a simple technique for finding the center of gravity of an irregularly shaped flat object. If we suspend the object from a hole near one edge, as in Fig. 5.18a, the force of gravity will pull down on the CG, and, when the object is in equilibrium, the CG will assume a position directly below the hole. A straight line drawn from the hole and directed straight down toward the center of the earth must then pass through the CG. If the object is then rotated through some angle (say, 90°) and the process repeated, as in Fig. 5.18b, the two lines will intersect at the CG of the object. As a check, the object can be hung from any other point, and it will be found that a vertical line through that point will also pass through the CG.

Example 5.8

A small metal machine shaft has a constant density throughout but has two sections of different radii. Its dimensions are shown in Fig. 5.19. Where is the center of mass of this machine shaft?

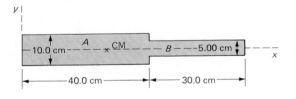

FIGURE 5.19

SOLUTION

We first take the axis of rotation of the shaft as the x axis. Then the y coordinate of the center of mass is zero, since by symmetry the center of mass is clearly on the axis of rotation.

To find the x coordinate of the center of mass, we consider the two parts of the shaft, A and B, separately. For each part the mass can be considered as acting at its center of mass (CM). For mass A this is a distance of 20.0 cm from the y axis, which we take as our axis of rotation to determine the x coordinate of the center of mass of the system. For mass B its CM is a distance of 40.0 cm + $\frac{1}{2}$(30.0 cm) = 55.0 cm from the y axis.

In this way we can use Eq. (5.7) to find the CM of the complete shaft. We have

$$\bar{x} = \frac{\sum_i m_i x_i}{M} = \frac{m_A(20.0) + m_B(55.0)}{m_A + m_B} = \frac{20.0 m_A/m_B + 55.0}{m_A/m_B + 1.00}$$

where we have divided the numerator and denominator by m_B. But the material is homogeneous, and so, since the mass of any substance is its volume times its density ρ, and the volume of a cylinder is $L\pi R^2$,

$$\frac{m_A}{m_B} = \frac{\rho V_A}{\rho V_B} = \frac{L_A \pi R_A^2}{L_B \pi R_B^2} = \frac{40.0(5.00)^2}{30.0(2.50)^2} = 5.33$$

Then

$$\bar{x} = \frac{20.0(5.33) + 55.0}{5.33 + 1.00} = \boxed{25.5 \text{ cm}}$$

The center of mass of the shaft is therefore on the axis of rotation of the shaft and 25.5 cm from the left end in Fig. 5.19.

Example 5.9

A tapered telegraph pole, 5.00 m long and weighing 400 N, can be balanced at a point 2.00 m from its thicker end. **(a)** Where is the CG of the pole? **(b)** If the pole were lifted by two men at the ends of the pole, what force would each man have to exert?

SOLUTION

(a) Since the pole balances at a point 2.00 m from its thicker end, its CG must be at this distance from the thicker end, since the upward push of the fulcrum must have the same line of action as the weight of the pole acting down. Otherwise, a net unbalanced torque would act to rotate the pole about the fulcrum.

(b) If forces F_1 and F_2 are applied upward by the two men to support the pole, then, from the first condition of equilibrium, we have

$$F_1 + F_2 = w = 400 \text{ N}$$

To find F_2, we can take moments about an axis at the thin end (where F_1 is applied), which is the length of the pole away from the point of application of F_2. Then, from the second condition of equilibrium, $\Sigma\tau = 0$, and so

$$-w(3.00 \text{ m}) + F_2(5.00 \text{ m}) = 0$$

$$F_2 = \tfrac{3}{5}w = \boxed{240 \text{ N}}$$

and

$$F_1 = w - F_2 = (400 - 240) \text{ N} = \boxed{160 \text{ N}}$$

The forces exerted by the two men are therefore 240 N at the thicker end and 160 N at the thinner end.

Kinds of Equilibrium

The equilibrium of an object acted on by the force of gravity depends on the effect of a small external impulse (a nudge) applied to the object for a brief instant. The equilibrium is called stable, unstable, or neutral, depending on how the object responds when displaced slightly from its equilibrium position. These three kinds of equilibrium are illustrated for a cone in Fig. 5.20.

In Fig. 5.20*a* the cone is standing on its base and, if tilted slightly and released, will return to its original position. In this case the cone is said to be in *stable equilibrium*.

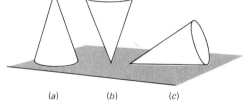

FIGURE 5.20 Three kinds of equilibrium for a solid cone: (*a*) stable; (*b*) unstable; (*c*) neutral.

(a) (b) (c)

The cone resting on its vertex in Fig. 5.20*b*, however, is not in stable equilibrium. It can be in equilibrium only when its center of gravity lies directly above its vertex. Any slight push will displace the CG from its position directly above the vertex, and the torque exerted by the weight of the cone will cause it to fall over. This situation, in which any slight perturbation destroys the cone's equilibrium, is called *unstable equilibrium*.

A third possibility, *neutral equilibrium*, is shown in Fig. 5.20*c*. In this case the cone rests on its side. When displaced slightly, it remains in any position to which it is moved.

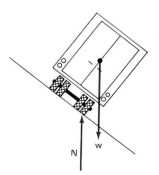

FIGURE 5.21 When a vertical line through the CG of a tractor-trailer falls outside the wheel base of the truck, a torque acts to overturn the truck.

The position of the center of gravity is crucial in determining the stability of an object under the force of gravity. The lower the CG, the greater the stability of the object and the more difficult it is to overturn it. The reason so many tractor-trailers overturn (usually on beltways in rush hours!) is that when fully loaded, they have very high CGs and become unstable when a vertical line through the CG falls outside the wheel base of the truck, as in Fig. 5.21.

When an object is pushed in an effort to overturn it, the CG is raised if the object is in stable equilibrium, is lowered for an object in unstable equilibrium, and remains at the same height for an object in neutral equilibrium. Because of the gravitational pull of the earth, objects will always seek a position where their CG is as low as possible. For this reason, an object in stable equilibrium, when pushed, will merely return to its original position, whereas an object in unstable equilibrium will fall over.

5.7 Dynamic Equilibrium

Thus far we have discussed the conditions under which objects at rest remain at rest and do not undergo translational or rotational motion as a result of forces acting on them. Static equilibrium occurs because the applied forces cancel out, so no translational acceleration results, and similarly the torques cancel, so no rotational acceleration results.

Translational Equilibrium

It is important to realize, however, that equilibrium can and does exist even *when an object is moving*, provided that motion is of a special kind. Suppose you are in a car moving along a straight road at a constant 25 m/s. Since the car moves with constant speed in a straight line, one of two things must be happening: either (1) there are no forces acting on the car or (2) the forces acting on the car have a vector sum of zero and there is no net force acting. The function of the gasoline fed to the motor is to produce just enough force to balance out the retarding forces of friction and air resistance and keep the car moving at a constant speed. This forward force is produced by the reaction of the road to the backward push of the wheels as the car's motor turns them. In this situation the accelerating and retarding forces are in equilibrium, and the car keeps moving at constant speed, according to Newton's first law of motion.

Here is a very important point to remember when solving problems on motion in subsequent chapters: *The first condition of equilibrium does not mean that an object is at rest; it simply means that the object is not accelerated.* The object can be either at rest or in motion with constant speed in a straight line. In both cases, either no forces are acting or the forces that do act cancel each other out and equilibrium results.

Example 5.10

A sailboat is tacking into the wind at an angle of 45° with respect to the wind direction. **(a)** If the resultant force exerted by the wind perpendicular to the sail is 1000 N and the boom that contains the sail makes an angle of 25° with respect to the keel of the boat, find the force that pushes the boat forward. **(b)** If the boat moves along at a constant speed in its original direction, what other forces must be acting on it and how are they provided?

SOLUTION

The situation is diagramed in Fig. 5.22a, with the wind coming from the east and the sailboat moving in a northeasterly direction. When the sail is set properly, the wind exerts a force on it normal to its surface, as shown in the free-body diagram of Fig. 5.22b. The two components of this force are F_x along the direction of motion and F_y at right angles to it.

(a) If the force is 1000 N, then its components are

$$F_x = F \sin 25° = (1000 \text{ N})(0.42) = 4.2 \times 10^2 \text{ N}$$

$$F_y = F \cos 25° = (1000 \text{ N})(0.91) = 9.1 \times 10^2 \text{ N}$$

The force of the wind driving the boat forward is F_x, or

$$\boxed{4.2 \times 10^2 \text{ N}}.$$

(b) If the sailboat moves through the water at a constant speed, then the resistance of the water (and air) to the motion of the

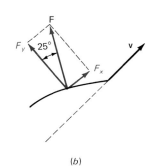

FIGURE 5.22 (a) Diagram for Example 5.10.
(b) FBD for sailboat.

(a)   *(b)*

boat must equal 4.2×10^2 N in the opposite direction to the boat's progress. Otherwise the sailboat would be accelerating.

Similarly, the keel and side of the boat must experience a resistive force of 9.1×10^2 N to balance out F_y and keep the boat from moving sideways. Since the force of the wind acts near the middle of the sail, whereas the water resistance acts on the hull of the boat, there is also a torque set up that tends to overturn the sailboat. This is the reason we see yachtspeople shifting their weight on the decks of sailboats to produce a torque opposed to that produced by the wind.

Example 5.11

In Fig. 5.23a, $m_1 = 10$ kg, $m_2 = 12$ kg, and the pulley and cord connecting them are weightless. When mass m_1 is given an initial push downward, it is observed that it continues to move with constant speed v. What is the coefficient of sliding friction between block m_2 and the inclined plane?

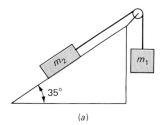

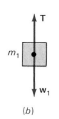

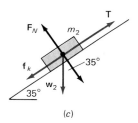

(a)   *(b)*   *(c)*

FIGURE 5.23 (a) Diagram for Example 5.11. (b) and (c) FBDs.

SOLUTION

This is an example of dynamic equilibrium, since both blocks move in a straight line at constant speed, and so the forces on both m_1 and m_2 must cancel. These forces are shown in the free-body diagrams of Fig. 5.23b and c.

For mass m_1,

$$T = w_1 = m_1 g = (10 \text{ kg})(9.8 \text{ m/s}^2) = 98 \text{ N}$$

For the forces acting on m_2, along the plane, with f_k equal to the force of sliding friction, we have

$$\Sigma F = T - f_k - w_2 \sin 35° = 0$$

where the tension T is the same throughout the cord, and so

$$T - \mu_k w_2 \cos 35° - w_2 \sin 35° = 0$$

from which

$$\mu_k = \frac{T - w_2 \sin 35°}{w_2 \cos 35°} = \boxed{0.32}$$

Rotational Equilibrium

In dealing with the second condition of equilibrium, the fact that all torques cancel out means either that there is no rotational motion or that there is no change in the existing rotational motion of the object. An example of the latter would be the spinning front wheel of a bicycle (off the ground) that is driven at a high rotational speed by a man's hand exerting a torque on the rim. In this case, if the wheel rotates at a constant speed, then the retarding torques due to friction and air resistance just balance out the torque applied by the man's hand. A state of dynamic equilibrium exists, and the wheel continues to rotate at constant speed unless a net torque is applied to accelerate it.

If the man ceases to exert a torque on the wheel with his hand, then there is no force to balance the forces of friction and air resistance acting on the wheel, and it gradually slows down and finally comes to rest. In this case external forces provide a net torque to slow down and finally stop the rotation of the wheel.

Summary: Important Definitions and Equations

Tensile and compressive forces: Forces that tend to stretch or to compress an object when applied at the two ends of the object.

Reaction forces: Forces that arise as a consequence of Newton's third law of motion.

Torque:

$$\tau = F\ell \tag{5.3}$$

where ℓ is the lever arm, the perpendicular distance from the line of action of the force **F** to the axis of rotation. Counterclockwise rotations are commonly taken as positive, clockwise rotations as negative.

Equilibrium: A physical situation in which forces act but no acceleration is produced.

Static equilibrium: The object on which the forces act remains at rest.

Dynamic equilibrium: The object on which the forces act continues to move with constant speed in a straight line or to rotate at a constant speed.

Translational equilibrium: No translational acceleration occurs.

Rotational equilibrium: No rotational acceleration occurs.

First condition of equilibrium (translational equilibrium):

$$\sum F_x = 0 \qquad \sum F_y = 0 \tag{5.2}$$

Second condition of equilibrium (rotational equilibrium):

$$\sum \tau = 0 \tag{5.4}$$

for any axis in z direction, where rotation occurs in the xy plane.

Center of gravity (CG): That point in an object at which its entire weight may be considered concentrated for purposes of calculating gravitational torques. Coordinates of center of gravity:

$$\bar{x} = \frac{\sum_i m_i x_i}{M} \qquad \bar{y} = \frac{\sum_i m_i y_i}{M} \tag{5.7, 5.8}$$

Center of mass (CM): That point in a rigid body at which the whole *mass* of the object may be considered to be concentrated when Newton's laws are applied to the object's motion.

For objects over whose extension the gravitational field of the earth is constant, the CG and CM are at the same point.

Kinds of equilibrium:

Stable: After a small external impulse, the object returns to its original position.

Unstable: After a small external impulse, the object falls over.

Neutral: After a small external impulse, the object remains in any position to which it is moved.

Questions

1 A baseball is thrown straight up in the air by a baseball player and momentarily comes to rest at its highest point. Is the baseball in equilibrium at this point?

2 Give an example of a body that is in motion but also in equilibrium.

3 If a spring is attached to a wall and a person pulls on the end of the spring with a force of 50 N, this force is transmitted undiminished all the way to the wall, and the wall pulls back on the spring with a force of 50 N. Since the spring is being subjected to a force of 50 N at both ends, why is the tension in the spring not 100 N instead of 50 N?

4 Suppose that the net force acting on a car is zero.

(*a*) Must the car be at rest in this case? Why?

(*b*) What can you say definitely about the motion (or lack thereof) of the car?

5 (*a*) If two equal and opposite forces are applied to an object, under what conditions is the object in equilibrium?

(*b*) Under what conditions is the object not in equilibrium? What is the nature of the resulting motion?

6 Give an example in which the neglect of a reaction force would invalidate the solution of an equilibrium problem.

7 (*a*) Is it possible for an object to be in translational equilibrium and not in rotational equilibrium? Give examples to support your answer.

(*b*) Is it possible for an object to be in rotational equilibrium and not in translational equilibrium? Give examples to support your answer.

8 (*a*) Distinguish clearly between the center of mass of an object and its center of gravity.

(*b*) Under what conditions are these two the same?

(*c*) Give an example of a situation in which an object would have its center of gravity at a different place from its center of mass.

9 (*a*) Where is the center of mass of a doughnut?

(*b*) Where is its center of gravity?

(*c*) How can the force of gravity act at a point where there is no mass?

10 Does the experimental method suggested in Sec. 5.6 for finding the center of gravity of an object require that the object be homogeneous throughout? Why?

11 Why do you tend to lean backward when carrying a heavy load of groceries in your arms, and forward when you are wearing a backpack?

Multiple-Choice and Simple Exercises

5.1 A 1000-kg car is parked on a 30° incline. The force of friction keeping the car from sliding down the incline is, in units of 10^3 N:

(*a*) 1.0   (*b*) 0.50   (*c*) 4.9   (*d*) 9.8

(*e*) 8.5

5.2 A block of weight 400 N resting on a table is pulled at constant speed by a force of 200 N directed at an angle of 30.0° above the horizontal. The normal force exerted by the table on the block is:

(*a*) 200 N   (*b*) 400 N   (*c*) 500 N

(*d*) 100 N   (*e*) 300 N

5.3 In Prob. 5.2 the coefficient of sliding friction between the table and the block is:

(*a*) 0   (*b*) 0.35   (*c*) 0.43   (*d*) 0.58

(*e*) 0.26

5.4 A uniform horizontal beam weighs 200 N and supports a 600-N weight three-fourths of the way down its length. The beam is supported at its two ends by pillars on which it rests. The force exerted by the support nearest to the added weight is:

(*a*) 800 N   (*b*) 550 N   (*c*) 400 N

(*d*) 250 N   (*e*) 600 N

5.5 The two sides of a seesaw are of equal length. A person of weight 800 N sits halfway from the fulcrum (the support of the seesaw) on one side. To balance the person's weight, sandbags piled at the end of the seesaw on the other side of the fulcrum should weigh:

(*a*) 800 N   (*b*) 1600 N   (*c*) 400 N

(*d*) 2400 N   (*e*) 200 N

5.6 The 0.80-m horizontal rod *AB* in Fig. 5.24 is pivoted at *A* and supported by a rope that makes an angle of 40° with the line *AB*. The lever arm of the force exerted by the rope on the rod is:

(*a*) 0 m   (*b*) 0.80 m   (*c*) 1.3 m

(*d*) 0.61 m   (*e*) 0.51 m

5.7 If the horizontal rod *AB* in Fig. 5.24 is attached to a rigid wall at point *A* and the tension in the rope is $T = 100$ N, the compressional force exerted on the rod is, if we neglect the weight of the rod:

(*a*) 100 N   (*b*) 0 N   (*c*) 64 N   (*d*) 77 N

(*e*) 84 N

5.8 Two horizontal forces are acting on a small puck on a frictionless table. One is a force of 1.0 N along the positive *x* axis. The second is a force of 1.5 N at an angle of 45° above the negative *x* axis.

(*a*) What are the magnitude and direction of the resultant force?

(*b*) If the puck's mass is 0.25 kg, what are the magnitude and direction of the acceleration?

5.9 A force of 50 N is directed at an angle of 63° above the positive *x* axis. What are the *x* and *y* components of this force?

5.10 A 4.0-m-long ladder whose weight is uniformly distributed rests up against the wall of a building, making an angle of 60° with the ground. If the ladder has a weight of 500 N, what torque does the ladder's weight produce about an axis at the point of contact of the ladder with the ground?

5.11 A farmer is trying to lift a 200-kg rock by using a wooden beam as a lever. He supports the beam on a fulcrum placed so that the distance from the fulcrum to the rock is one-fifth the distance from the fulcrum to the point at which he pushes down on the beam. What force must he exert to lift the rock?

5.12 Calculate the torque about the base support of a diving board exerted by a 100-kg person 2.5 m from this support.

5.13 A lamp of mass 100 kg is hung in the corner of a room by suspending it from the ceiling and the adjacent wall by two ropes, as shown in Fig. 5.25.

(*a*) Find the tension T_1 in the rope attached to the wall.

(*b*) Find the tension T_2 in the rope attached to the ceiling.

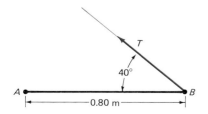

FIGURE 5.24   Diagram for Exercises 5.6 and 5.7.

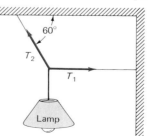

FIGURE 5.25   Diagram for Exercise 5.13.

5.14 A 50-kg block slides down a 40° inclined plane at a constant speed.
(a) How large is the frictional force opposing its motion?
(b) What is the coefficient of kinetic friction between block and plane?

5.15 Two blocks of mass m_1 and m_2 are connected to a light string that passes over a pulley. Mass m_1 sits on a frictionless inclined plane of inclination 35°. Mass m_2 hangs freely at the end of the string. If $m_1 = 100$ kg, what must m_2 be to keep the two blocks at rest?

5.16 A mounting bolt on an automobile engine is supposed to be tightened with an applied torque of 60 N·m.
(a) If a wrench 25 cm long is used to tighten the bolt, what force must be applied perpendicular to the wrench handle to produce the desired torque?
(b) If the force is applied at an angle of 75° with respect to the wrench handle, what force is required to produce the desired torque?

Problems

5.17 Three hollow metal rods, each 1.25 m in length, are connected by pins to form a triangular frame, as in Fig. 5.26. The frame stands upright, and a weight w of 400 N is hung from the vertex V of the triangular frame.
(a) What is the tension in rod UW?
(b) What is the compressional force on rod UV?

5.19 A traffic light of mass 25.0 kg is hanging from two cables, as shown in Fig. 5.28.
(a) Draw a free-body diagram for this situation.
(b) Compute the tension in the two cables A and B.

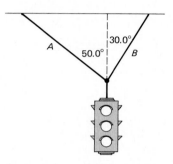

FIGURE 5.28 Diagram for Prob. 5.19.

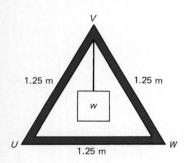

FIGURE 5.26 Diagram for Prob. 5.17.

5.18 An assembly of five identical springs is mounted in a frame, as shown in Fig. 5.27. The tension in spring A is 30.0 N. What are the tensions in springs B, C, D, and E?

5.20 A woman pushes a lawnmower at a steady speed. The lawnmower has a mass of 25.0 kg and the woman exerts a 100-N force along a line making an angle of 45.0° below the horizontal. Find the magnitude of the vertical force the ground exerts on the lawnmower.

5.21 Calculate the vertical traction force exerted on the neck of the woman in Fig. 5.29. The hanging weight has a mass of 3.00 kg. (Assume that the free pulley cannot move horizontally.)

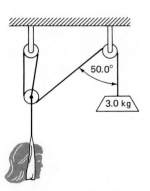

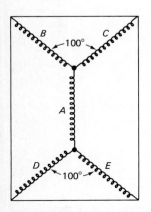

FIGURE 5.27 Diagram for Prob. 5.18.

FIGURE 5.29 Diagram for Prob. 5.21.

5.22 Two ropes support a load of 200 kg. The two ropes are perpendicular to each other, and one rope has twice the tension of the other.

(*a*) Find the tension in each rope.

(*b*) Find the angle each rope makes with the vertical.

5.23 A block-and-tackle system for lifting heavy weights is shown in Fig. 5.30. If the mass to be lifted is 195 kg and the movable pulley has a mass of 5.0 kg, what force F must be exerted on the rope to lift the weight? (Assume the same tension throughout the rope.)

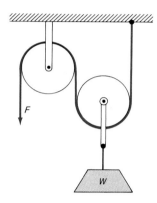

FIGURE 5.30   Diagram for Prob. 5.23.

5.24 Find the tension T_1, T_2, T_3, T_4, and T_5 in the ropes shown in Fig. 5.31, if these ropes support a weight of 500 N at the bottom.

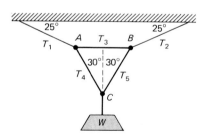

FIGURE 5.31   Diagram for Prob. 5.24.

5.25 Figure 5.32 shows a system of pulleys and weights hung from a ceiling. If $W_3 = 600$ N, what are the values of W_1, W_2, T_1, and T_2?

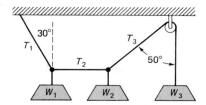

FIGURE 5.32   Diagram for Prob. 5.25.

5.26 An acrobat hangs by his hands from the middle of a tightly stretched horizontal wire so that the angle between the wire and the horizontal is 10.0°.

(*a*) Draw a free-body diagram for this physical situation.

(*b*) If the acrobat's mass is 100 kg, what is the tension in the wire?

5.27 A bowling ball weighs 75 N and rests against the sides of a frictionless angle iron, as shown in Fig. 5.33. What are the magnitudes of the forces exerted by the sides of the angle iron on the bowling ball?

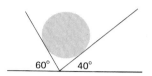

FIGURE 5.33   Diagram for Prob. 5.27.

5.28 A pair of equal parallel forces in opposite directions and not acting through a single point is called a *couple*. The net force produced on any object to which a couple is applied is zero, but the torque is not zero.

(*a*) For the couple shown in Fig. 5.34, show that the torque produced about the axis of rotation is Fd, where d is the distance between the lines of action of the two forces.

(*b*) Prove that the torque produced by a couple is independent of the axis of rotation about which the torque is calculated.

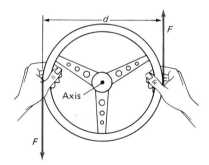

FIGURE 5.34   Diagram for Prob. 5.28: A couple acting to produce rotation about the axle of the wheel.

5.29 Show that in Example 5.4 rotational equilibrium also exists with respect to an axis of rotation at the point of application of force F_3 in Fig. 5.13.

5.30 A wheel of 0.500 m diameter has an axle of radius 4.00 cm. If a force of 100 N is exerted along the rim of the wheel, what force applied to the outside of the axle will result in zero net torque on the wheel?

5.31 Two vehicles are crossing a flat, symmetrical bridge 30.0 m in length that weighs 6.00×10^5 N. A truck weighing 5.00×10^4 N is 9.00 m from one end of the bridge, and a car weighing 1.10×10^4 N is 6.00 m from the other end of the bridge. What forces must be exerted by the supporting piers at the two ends of the bridge to maintain it in equilibrium?

5.32 A uniform horizontal 400-N beam of length L has two weights hanging from it, 200 N at $L/4$ from one end and 300 N at $3L/4$ from the same end.

(*a*) What must be the magnitude of the one additional force on the beam that will produce equilibrium?

(*b*) At what point along the length of the beam must this force be applied for the beam to be in equilibrium?

5.33 A 50.0-kg plank, 2.00 m long and of uniform thickness, rests on two knife-edges attached to two spring balances that measure weights, as in Fig. 5.35. What are the readings of the two balances on which the knife-edges rest?

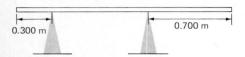

FIGURE 5.35 Diagram for Prob. 5.33.

5.34 A ladder 8.00 m long leans against a vertical frictionless wall and makes an angle of 50.0° with the ground, which is assumed to be rough. The mass of the ladder is 40.0 kg, with its center of gravity at its center.

(*a*) Find the magnitude of the horizontal force exerted by the wall on the ladder.

(*b*) Find the magnitude and direction of the force exerted by the ground on the ladder.

5.35 Repeat Prob. 5.34 for a situation in which a 100-kg person is standing on the ladder three-fourths of the way from the bottom of the ladder.

∗5.36 Two flat wooden boards, each of uniform construction and 2.00 m long, and each with a mass of 2.00 kg, are stood on end 1.00 m apart and tilted until they touch at the top.

(*a*) Find the horizontal force exerted by one board on the other at the top of the assembly, assuming that at that point there are no vertical forces acting.

(*b*) Find the magnitude of the coefficient of static friction needed to keep the bottom ends of the boards from sliding.

5.37 Three girls who weigh 500, 600, and 700 N, respectively, are to be seated on a plank of negligible weight in such a way that the 500-N girl is between the other two girls, the heavier girls are 4.00 m apart, and the center of mass of the system is located at the position of the 500-N girl. What are the distances of the two girls from the first girl?

∗5.38 A uniform door of weight 500 N is held up by two hinges A and B in Fig. 5.36. The weight of the door is shared

equally between the two hinges. The width of the door is 75.0 cm, and the distance between the two hinges is 120 cm. Find the magnitude and direction of the forces exerted on the door at the hinges A and B.

5.39 A uniform ladder of mass 20 kg and length L leans against a smooth wall so that it makes an angle of 53° with respect to the ground. A person of mass 80 kg climbs up the ladder to a point two-thirds of the way to the top. How large must the coefficient of static friction between floor and ladder be if the ladder is not to slip?

5.40 A 50.0 kg sign that is 1.00 m long and 50.0 cm high is held by a wire at a 35.0° angle and by a wall clamp at point A, as in Fig. 5.37. The wire is attached 80.0 cm from point A.

(*a*) What is the tension in the wire?

(*b*) What is the magnitude and direction of the force exerted on the sign at A?

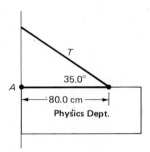

FIGURE 5.37 Diagram for Prob. 5.40.

∗5.41 The boom shown in Fig. 5.38 has a mass of 2.00 kg and a length of 3.00 m and supports a mass of 10.0 kg. A rope is attached two-thirds of the way up the boom. Find (*a*) the tension T in the rope and (*b*) the magnitude and direction of the force exerted on the boom by the wall at A.

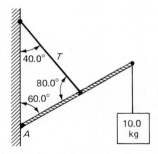

FIGURE 5.38 Diagram for Prob. 5.41.

5.42 A uniform steel meterstick, weighing 10.0 N, rests on two spring scales at its two ends. A 5.60-kg mass is placed on the meterstick at a distance of 30.0 cm from its left end. Find the readings on the two scales, which measure weights in newtons.

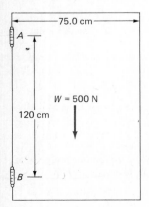

FIGURE 5.36 Diagram for Prob. 5.38.

5.43 The Calder mobile in Fig. 5.39 is designed to have no vertical motion, but the pieces have small enough masses to rotate easily in an outdoor breeze. The masses are $m_1 = 0.500$ kg, $m_2 = 0.750$ kg, and $m_3 = 0.250$ kg. Each horizontal rod is 1.00 m long and has a mass of 0.100 kg. The vertical wires are of negligible weight. What must the distances ℓ_1 and ℓ_2 be for the system to be in static equilibrium under the forces acting vertically?

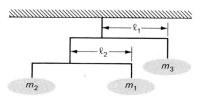

FIGURE 5.39 Diagram for Prob. 5.43.

5.44 A uniform meterstick 1.0 m long has a weight of 4.0 N. It is loaded with two pieces of heavy metal, with a 4.0-N piece being attached at the 25-cm line and a 3.0-N piece at the 85-cm line. Where is the center of gravity of the loaded meterstick?

5.45 A uniform boom weighing 800 N is supported by a horizontal chain-rope connected to it at a point three-fourths of its length from the bottom of the boom, as in Fig. 5.40. A 4.00×10^3 N girder is being lifted into place by the boom.

(a) Find the tension in the supporting chain-rope.

(b) Find the magnitude and direction of the force exerted on the boom by the support (P) at its base.

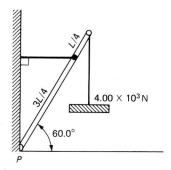

FIGURE 5.40 Diagram for Prob. 5.45.

5.46 A uniform wooden beam is 2.00 m long and has a mass of 6.25 kg. Attached to the beam are masses of 5.00 kg at one end, 4.00 kg at the opposite end, and 3.00 kg in the middle. Where is the center of gravity of this system?

5.47 Three particles of masses 2.00, 3.00, and 7.00 kg are located at the vertices of an equilateral triangle of sides 0.800 m. Find the coordinates of the CG of this system with respect to a coordinate system with origin at the 2.00-kg particle and with the 3.00-kg particle located along the positive y axis.

5.48 Four particles of masses 2.00, 4.00, 6.00, and 8.00 kg are located at the four corners of a rectangle specified by (x, y) coordinates (0, 0), (1.0 m, 0), (1.0 m, 2.0 m), and (0, 2.0 m), respectively.

(a) Find the location of the CM of this system.

(b) In a uniform gravitational field, is the CG at the same point as the CM for this system?

5.49 The front wheels of a truck support 1.80×10^4 N and the rear wheels 3.00×10^4 N. The distance between front and rear axles is 6.25 m.

(a) Find the weight of the truck.

(b) Find the location of the truck's CG.

5.50 A tapered telegraph pole 15.0 m long and weighing 9.00×10^3 N rests on two spring balances. One at the thinner end records a weight of 1.50×10^3 N. The other spring balance is 4.00 m from the thicker end. What distance from the thinner end is the CG of the pole?

*5.51 Two circular disks, one of radius 18.0 cm, the other of radius 6.00 cm, are stamped out of the same sheet of thin brass. The smaller disk is placed on top of the larger disk with its center 10.0 cm from the center of the larger disk. Where is the CG of the combination?

*5.52 A circle of radius 25.0 cm is cut out of a uniform circular disk of radius 55.0 cm. If the circumference of the hole touches the center of the disk, where is the CG of the system?

*5.53 A uniform rectangular metal block, three times as tall as it is wide, rests on an inclined plane, as in Fig. 5.41. The coefficient of static friction is so great that the block cannot slide down the plane. If the angle θ is slowly increased from zero, at what angle θ will the block topple over?

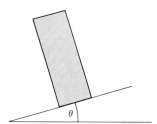

FIGURE 5.41 Diagram for Prob. 5.53.

*5.54 The system shown in Fig. 5.42 is in equilibrium. Mass m_1 is 10 kg and m_2 is 2.0 kg. The coefficient of static friction between m_1 and the surface on which it rests is 0.35.

(a) Find the frictional force exerted on block m_1.

(b) Find the maximum total mass m_2 that can be attached to the hanging string and still preserve the system in equilibrium.

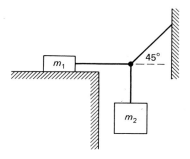

FIGURE 5.42 Diagram for Prob. 5.54.

*5.55 A block of mass m_1 slides down the inclined plane in Fig. 5.43 at constant speed even when another block of mass

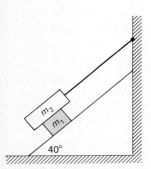

FIGURE 5.43 Diagram for Prob. 5.55.

m_2 rests on top of it and is attached by a cord to a rigid support, as shown.

(a) Draw a free-body diagram of all forces acting on m_1.

(b) If $m_1 = 2.0$ kg and $m_2 = 2.0$ kg and the coefficient of kinetic friction between the two masses is the same as between m_1 and the plane, what is the value of this coefficient of friction?

5.56 A sailboat is tacking into the wind at an angle of 60° with respect to the wind direction (see Fig. 5.22).

(a) If the resultant force exerted by the wind normal to the sail is 800 N and the boom that contains the sail makes an angle of 25° with respect to the keel, find the force exerted on the boat in a forward direction.

(b) What force must the water exert on the keel to keep the boat from moving sideways?

(c) If the sailboat moves along at a constant speed, what force is being produced in a direction opposite to its motion by the resistance of the water and air?

Additional Readings

Andrade, E. N. da C.: "Isaac Newton," in *Newton Tercentenary Celebrations*, Cambridge University Press, New York, 1947. A brief life of the most famous of all physicists, and the source of the quote at the beginning of this chapter (with permission of the Cambridge University Press). This essay is also reprinted in James R. Newman (ed.): *The World of Mathematics*, vol. 1, Simon and Schuster, New York, 1956.

Hobbie, Russell K.: *Intermediate Physics for Medicine and Biology*, Wiley, New York, 1978. The first chapter of this more advanced text contains some interesting examples of the forces acting on various parts of the human body.

Mark, Robert: "Structural Analysis of Gothic Cathedrals," *Scientific American*, vol. 227, no. 5, November 1972, pp. 90–99. An interesting comparison of the cathedrals at Bourges and Chartres, using optical stress analysis of model structures.

Metcalf, Harold J.: *Topics in Classical Biophysics*, Prentice-Hall, Englewood Cliffs, N.J., 1980. The first chapter of this brief book discusses biomechanics, including equilibrium problems.

Steinman, David B.: "Bridges," *Scientific American*, vol. 191, no. 5, November 1954, pp. 60–71. An account of some famous bridge structures, including the collapse of the Tacoma Narrows Bridge.

CHAPTER 6

Work and Energy

Energy will remain in some sense the lord and giver of life, a reality transcending our mathematical descriptions. Its nature lies at the heart of the mystery of our existence as animate beings in an inanimate universe.

Freeman Dyson (b. 1923)

Physics is based on a small number of universally valid principles. Of these none has been more fruitful than the principle of conservation of energy. In nontechnical terms, this principle means that the total energy of an isolated system remains constant over time, even though the individual kinds of energy in the system may change.

This principle has led to important discoveries in physics such as that of the *neutrino*. The existence of this very light, uncharged particle—involved in the emission of electrons by atomic nuclei—was predicted in 1930 on the basis of energy (and momentum) conservation. Only 26 years later was this elusive particle first observed experimentally. Here is a confirmation of the great fruitfulness and enduring importance of this principle in helping us understand our universe.

In this chapter we consider many different kinds of mechanical energy, as well as the transformation of one kind of energy into another. We also study how work can be converted into energy and how, in turn, energy can be converted into useful work. This discussion culminates in the conservation-of-energy principle, which applies not only to abstract theories of modern physics but also to important problems in everyday life.

6.1 The Physical Meaning of Work

Personal experience indicates some kind of relationship between work and energy. Someone who can do large amounts of work is called "energetic." In physics, however, *work* has a more restricted meaning than in everyday usage.

Work: The product of the applied force and the displacement of an object in the direction of the force.

For example, if we apply a force of magnitude F to push a book along the top of a desk, and the book's displacement is of magnitude d in the direction of this force, as in Fig. 6.1, then the work done by the force is

$$\mathcal{W} = Fd$$

The displacement here is that produced while the force is acting. If the force is

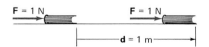

FIGURE 6.1 The work done by force **F** pushing the book 1 m across the desk is $\mathcal{W} = Fd = (1\ N)\,(1\ m) = 1\ J$.

measured in newtons and the displacement in meters, the work is in joules (J), where

$$1 \text{ J} = 1 \text{ N} \times 1 \text{ m} = 1 \text{ N·m}$$

In the British engineering system the unit of work is the foot-pound (ft·lb), which is the work done by a force of one pound moving an object through a distance of one foot in the direction of the force. One joule is equal to 0.737 ft·lb.

When a toy truck is being pulled along the ground with a cord, as in Fig. 6.2, the force exerted may be at an angle θ with respect to the direction of motion. Then the work done is the magnitude of the force in the direction in which the truck moves times the truck's displacement. If we take the direction of motion as along the x axis, and the truck moves from an initial position x_0 to a final position x, the displacement vector from x_0 to x is $\mathbf{d}$. Since F_x is the component of $\mathbf{F}$ in the direction of $\mathbf{d}$, and $F_x = F \cos \theta$, the work is $\mathcal{W} = (F \cos \theta)\, d$. Because $\mathbf{F}$ and $\mathbf{d}$ are both vectors, the *work done by a force* is

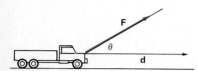

FIGURE 6.2 Pulling a toy truck with a cord that makes an angle θ with respect to the direction of the truck's motion: $\mathcal{W} = \mathbf{F} \cdot \mathbf{d} = Fd \cos \theta$.

$$\mathcal{W} = \mathbf{F} \cdot \mathbf{d} = Fd \cos \theta \qquad (6.1)$$

where the dot product $\mathbf{F} \cdot \mathbf{d}$ is defined as the magnitude of $\mathbf{F}$ times the magnitude of $\mathbf{d}$ times the cosine of the angle between them. The dot product of two vectors is a scalar quantity, and so the work $\mathcal{W}$ is a scalar.

Example 6.1

A 100-kg man pulls his 30.0-kg daughter on a sled with a rope that makes an angle of 40.0° with respect to the ground. **(a)** If the man exerts a force of 50.0 N and the sled moves 100 m, how much work does he do? **(b)** If the man pulls horizontally on the sled, what force would he have to exert to do the same amount of work?

SOLUTION

(a) Here

$$\mathcal{W} = \mathbf{F} \cdot \mathbf{d} = Fd \cos 40.0°$$

$$= (50.0 \text{ N})(100 \text{ m})(0.766) = 3.83 \times 10^3 \text{ N·m}$$

$$= \boxed{3.83 \times 10^3 \text{ J}}$$

Note that the masses of the man and his daughter are irrelevant here. They are given merely to give you practice in separating the "sheep" (the needed data) from the "goats" (the useless data) in solving physics problems.

(b) The horizontal force required in this case is the x component of $\mathbf{F}$ and is therefore

$$F_x = F \cos 40° = (50.0 \text{ N})(0.766) = \boxed{38.3 \text{ N}}$$

Note that this force is smaller than in part (a) because the force here is used *completely* to move the sled forward. In part (a) some of the applied force is used to reduce the force the sled exerts on the ground.

Equation (6.1) applies only to the work done on the moving object by the force $\mathbf{F}$. There may be forces other than $\mathbf{F}$ acting on the object, and the work done by these forces must be calculated separately. If the net force acting is $\mathbf{F}_{\text{net}} = \Sigma_{i=1}^{n} \mathbf{F}_i$, then the net work done is

$$\mathcal{W}_{\text{net}} = \mathbf{F}_{\text{net}} \cdot \mathbf{d} = \sum_{i=1}^{n} (\mathbf{F}_i \cdot \mathbf{d}) \qquad (6.2)$$

As a consequence the net work done may be zero, even though the work done by individual forces is not zero.

Work is, of course, a scalar but can have both positive and negative values. Equation (6.1) shows us that the work done by a force may be negative. For example, if I take a book from a bookshelf and slowly lower it to the floor, the force **F** I must exert to keep the book from falling is upward, whereas the displacement **d** is downward, and so

$$\mathcal{W} = \mathbf{F} \cdot \mathbf{d} = Fd \cos 180° = -Fd$$

In this case, the work I do is *negative*, since my force is in the opposite direction to the book's displacement. The work done by the force of gravity, however, is *positive*, since the gravitational force is in the same direction as the displacement.

Our definition of work leads to the surprising conclusion that, in a technical sense, a person need do no work *on a book* to hold it up for a long period of time, or even to move it horizontally at a fixed height above the ground. In the first case $d = 0$, and so no work is done; in the second case, the force that must be exerted to hold the book up is at right angles to the direction of motion, $\cos \theta = \cos 90° = 0$, and so $\mathcal{W} = 0$. Clearly this physical definition of work is at variance with our biological experience. We all feel that we are using up energy if we hold up a sack of potatoes for a few minutes. Just as a hovering helicopter must burn fuel to stay at a stationary position above the ground, so too our muscles use lots of energy in microscopic, internal motions to hold up the sack of potatoes. But these motions never result in any measurable displacement and therefore never do any work in the sense in which the word is used in physics.

Example 6.2

A boy pushes a crate a distance of 5.00 m along a rough horizontal surface at a constant speed by exerting a force of 12.0 N in the direction of motion. **(a)** What work does the boy do? **(b)** What work is done by the force of friction that opposes the motion? **(c)** What work is done by any other forces acting on the crate? **(d)** What is the net work done on the crate by all the forces acting?

SOLUTION

(a) The work done by the boy is

$$\mathcal{W} = \mathbf{F} \cdot \mathbf{d} = (12.0 \text{ N})(5.00 \text{ m})(\cos 0°) = \boxed{60.0 \text{ J}}$$

(b) Since the crate moves at a constant speed, the net force in the direction of **d** must be zero. Hence the frictional force **f** must be equal to 12.0 N in a direction opposite to that of the displacement **d**. Therefore

$$\mathcal{W}_f = \mathbf{f} \cdot \mathbf{d} = (12.0 \text{ N})(5.00 \text{ m})(\cos 180°) = \boxed{-60.0 \text{ J}}$$

Negative work is therefore done on the crate by the force of friction because the force of friction is in the opposite direction to the displacement.

(c) The only other forces acting on the crate are its weight acting down and the push of the surface upward perpendicular to the surface. The work done by each of these is zero, since these forces are perpendicular to the crate's displacement. Thus $\cos \theta = \cos 90° = 0$.

(d) The net work done by all four forces is then

$$\mathcal{W}_{net} = \mathcal{W} + \mathcal{W}_f = 60.0 \text{ J} + (-60.0 \text{ J}) = \boxed{0}$$

This result is expected, since, from Eq. (6.2), $\mathcal{W}_{net} = \mathbf{F}_{net} \cdot \mathbf{d}$, and in this case $\mathbf{F}_{net} = 0$.

6.2 Work Converted into Kinetic Energy

Energy: That property of an object which enables it to do work.

This definition indicates that energy can be converted into useful work, as, for example, when the electric energy in a car's battery is used to start the engine. The reverse is also true: work can be converted into stored energy. We can, for example, do work to pump the water from a lake to a higher reservoir, as is done at the Ludington pumped-water storage plant on Lake Michigan. That stored water then has added energy because of its height and is used to drive electric generators and produce electric energy needed in peak periods.

Work and energy are therefore directly convertible. They are the same kind of physical quantity and so have the same dimensions. In the SI system of units they also have the same unit, the joule (J), where one joule is equal to one newton-meter.

Before discussing the conversion of work into energy, we introduce two important definitions:

Mechanical energy: The energy possessed by an object by virtue of its motion or position.

Kinetic energy: The energy possessed by an object by virtue of its motion.

We now want to investigate the quantitative relationship between the work done and the kinetic energy produced. In this section we confine our discussion to *translational motion*, where the object actually moves from one position in space to another; rotational kinetic energy will be discussed in Chap. 8.

Let us consider a hockey puck moving slowly with a velocity v_0 on a frictionless surface like those in the air-hockey games found at penny arcades, as in Fig. 6.3. We exert a constant force **F** on the puck in the direction of v_0 for a short period of time and give it a displacement **d** during that time. The work done is then $\mathcal{W} = Fd$, for **F** and **d** are in the same direction. Since there is no resisting force, the force **F** accelerates the puck according to Newton's second law. In this case, therefore, the applied force **F** is equal to the net force F_{net}, because it is the only force acting on the object.

The force **F** produces a constant acceleration **a**, and the velocity increases from its initial value v_0 to some final value v. The work done is thus converted into energy of motion.

FIGURE 6.3 Work converted into kinetic energy. All the work done by the net force F_{net} is converted into kinetic energy.

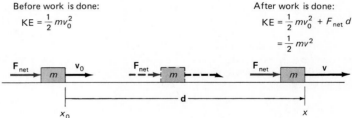

Before work is done:

$$KE = \frac{1}{2} mv_0^2$$

After work is done:

$$KE = \frac{1}{2} mv_0^2 + F_{net} d$$

$$= \frac{1}{2} mv^2$$

From Eq. (2.20) for uniformly accelerated motion in a straight line, the puck's final velocity is related to the initial velocity by the equation

$$v^2 = v_0^2 + 2a(x - x_0)$$

In our case $x - x_0$ is equal to the displacement d, and so we have

$$a = \frac{v^2 - v_0^2}{2d}$$

Now the work done in accelerating the puck is

$$\mathcal{W} = Fd = mad = m\frac{(v^2 - v_0^2)\, d}{2d}$$

or $\quad \mathcal{W} = \frac{1}{2}mv^2 - \frac{1}{2}mv_0^2$ (6.3)

This result [Eq. (6.3)] leads us to define the translational *kinetic energy* of the moving puck as $\frac{1}{2}mv^2$.

Translational kinetic energy: The energy of motion of an object, equal to one-half the product of its mass and the square of its velocity.

$$KE = \frac{1}{2}mv^2$$ (6.4)

Note that, even though **v** is a vector, v^2 is not, since it is the dot product $\mathbf{v} \cdot \mathbf{v}$, which is a scalar. Hence, like work, kinetic energy is a scalar quantity and can be specified by a single number that gives its magnitude.

As Eq. (6.3) indicates, the effect of the work done in this case has been to increase the kinetic energy of the puck from $\frac{1}{2}mv_0^2$ to $\frac{1}{2}mv^2$. For Eq. (6.3) to be correct, the dimensions must be the same on both sides. This can easily be shown using dimensional analysis. Work has the dimensions

$$[\mathcal{W}] = [F][L] = [M][A][L] = [M]\frac{[L]}{[T^2]}[L] = [M]\frac{[L^2]}{[T^2]}$$ (6.5a)

whereas kinetic energy has the dimensions

$$[KE] = [M][V^2] = [M]\frac{[L^2]}{[T^2]}$$ (6.5b)

Since the right sides of Eqs. (6.5a) and (6.5b) are the same dimensionally, Eq. (6.3) is dimensionally correct. The units for kinetic energy are therefore the same as the units for work, that is, joules (J).

Example 6.3

A 60-kg woman pushes a 20.0-kg suitcase on wheels a distance of 10.0 m by exerting a constant force of 2.00 N in the direction of motion, starting from rest. **(a)** How much work does she do? **(b)** If there were no friction acting on the suitcase, what would its final velocity be at the end of the 10-m distance?

SOLUTION

(a) From Eq. (6.1),

$$\mathcal{W} = \mathbf{F} \cdot \mathbf{d} = Fd = (2.00 \text{ N})(10.0 \text{ m}) = \boxed{20.0 \text{ J}}$$

(b) If there is no friction, all the work goes into the kinetic energy of the suitcase, and from Eq. (6.3),

$$\mathcal{W} = \frac{1}{2}mv^2 - \frac{1}{2}mv_0^2$$

$$20.0 \text{ J} = \frac{1}{2}mv^2 - 0 = \frac{1}{2}(20.0 \text{ kg})v^2$$

and so $v^2 = 2.00 \text{ J/kg} = 2.00 \text{ N·m/kg} = 2.00 \text{ m}^2/\text{s}^2$ and

$$v = \boxed{1.4 \text{ m/s}}$$

6.3 Work Converted into Gravitational Potential Energy

Just as the work done on the puck in the previous section was converted into kinetic energy of the puck, so it is possible under different circumstances for work to be converted to another kind of energy, one that depends on the *position* of the puck rather than its motion. We call this kind of energy *potential energy*, because an elevated puck has the *potential* for doing work.

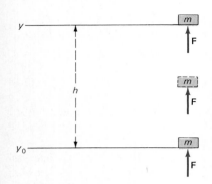

FIGURE 6.4 Work converted into gravitational potential energy. Here the work done by the force **F** increases the gravitational potential energy by increasing the puck's height above the table. Work = mgh = PE$_{grav}$.

Potential energy: The energy possessed by an object by virtue of its position or state.

We first consider the potential energy possessed by an object because of its *position* in a gravitational field. Instead of pushing the puck along the table, suppose we lift it slowly from rest and place it on a shelf directly above the table, as in Fig. 6.4. In this case we presume that no acceleration is imparted to the puck, since it is moved at a constant and very slow velocity. Hence the puck is in equilibrium throughout its motion, and there is no net force acting on it. This means that the upward force **F** exerted on the puck must differ only infinitesimally from the weight **w** = $m\mathbf{g}$ pushing downward. The work done, therefore, by the external force **F** in lifting the puck a distance h above the table against the force of gravity is

$$\mathcal{W} = \mathbf{F} \cdot \mathbf{d} = F(y - y_0) = mgh \tag{6.6}$$

This is a very different situation from the first one we considered. Now, although the puck has no velocity and therefore no kinetic energy, it possesses stored energy by virtue of its raised position. (We could, for example, nudge the puck off the shelf and use it, in conjunction with the force of gravity, to hammer a thumbtack into the table.) The stored energy in this case we call *gravitational potential energy*.

Gravitational potential energy: The energy possessed by an object because of its position in a gravitational field.

$$\boxed{\text{PE}_{grav} = mgh} \tag{6.7}$$

where h is the height above some position chosen as the zero of potential energy and g is the acceleration due to gravity.

There is a certain arbitrariness about Eq. (6.7). It says that the potential energy is zero when h is zero. But from what place do we measure the height h—from the ground outside, the floor of the room, or the top of the table? The answer is that it makes no difference so long as we consistently use the same height as the zero for potential energy throughout any problem. The change in potential energy depends on the displacement d, which is equal to $y - y_0$. This is independent of where we place the origin of our coordinate system.

A good example of the use of gravitational potential energy is a pendulum clock, whose hands are turned by falling weights. Once a week the owner of the clock lifts the weights by doing work on them, thus increasing their potential energy. During the week the weights fall slowly, do work, and, in so doing, give up their stored potential energy to turn the clock's hands.

Hydrologic Cycle

One of the most important uses of gravitational potential energy is in the production of electricity by hydroelectric power plants. Water is raised by the action of the sun to higher elevations through what is called the *hydrologic cycle* (see Fig. 6.5). The sun shines down on the oceans, rivers, and lakes of the world and raises the water temperature. This gives some water molecules sufficient energy to evaporate as water vapor and form clouds in the atmosphere. When these clouds are blown by the winds (which also obtain their energy from solar energy) over high mountains, they are driven upward, cool off, and release their moisture in the form of snow or rain. Some of this water can then be trapped in mountain reservoirs and

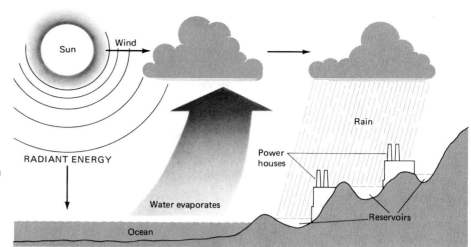

FIGURE 6.5 The hydrologic cycle. In this cycle energy from the sun goes through a series of transformations and is finally converted into gravitational potential energy, which can then be transformed into electric energy by hydroelectric turbines.

used to generate electricity by means of hydroelectric turbines. The water has a potential energy mgh, or 9.80 J/kg per meter of height above the point at which the stored energy in the water is to be used (we call this height the *head* of the water). By opening the gates of the reservoir, this stored energy can be converted into kinetic energy of the water and used to drive turbines and generate electricity.

The great advantage of the use of the hydrologic cycle to produce electricity is, of course, that the process is *renewable* and no money need be spent on fuel. For this reason hydroelectric power is the cheapest form of electric power available at the present time.

The Conservative Nature of the Gravitational Force

The gravitational force is the prime example of what physicists call a *conservative force*, which differs greatly from the resistive (frictional) forces considered in Secs. 3.6 and 3.7. A force is *conservative* if the following is true:

The work done against a conservative force depends only on the initial and final positions of the system, not on the path taken to go from the initial to the final position.

To illustrate this, consider a ball in the gravitational field of the earth, as in Fig. 6.6. We call the ball and earth together the *system* under consideration so that we can be clear about what is internal and what is external to the system. Here the

FIGURE 6.6 The gravitational force is a conservative force: the change in energy of the ball depends only on the difference in height between y_1 and y_2 and is independent of the path taken from A to C.

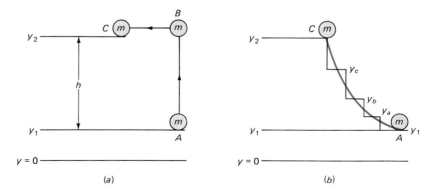

force of gravity is *internal* to the system. Any external agent moving the ball around in the gravitational field is *external* to the system.

If we lift the ball of mass m from position A to position B by exerting a force vertically upward, as in Fig. 6.6a, the work done is $\mathcal{W} = \mathbf{F} \cdot \mathbf{d} = mg(y_2 - y_1) = mgh$, where $h = y_2 - y_1$. No work is done in moving the object from B to C, since they are at the same height. On the other hand, if we take this same object from A to C by the curved path shown in color in Fig. 6.6b, the work is still exactly the same for the following reason.

We can approximate the curved path in the figure by a series of small steps, as shown. Then along the horizontal portions of these steps we do no work, since our lifting force (which opposes the force of gravity) has no component in the horizontal direction. The work along the vertical steps will be the sum of a series of terms like

$$\mathcal{W} = mg(y_a - y_1) + mg(y_b - y_a) + mg(y_c - y_b) + mg(y_2 - y_c)$$

When we add these quantities, all the intermediate distances cancel out and we obtain

$$\mathcal{W} = mg(y_2 - y_1) = mgh$$

as before. Hence the work done is, to a good approximation, independent of the path taken. In the limit, as the number of steps increases and approaches infinity, we find that the work done is *exactly* the same in the two cases. This is true, no matter the path taken.

If a person (who is external to the system) lifts the ball from level y_1 to level y_2 by any conceivable path, then the energy in the system increases by an amount $mg(y_2 - y_1) = mgh$. This work is *completely recoverable* and is converted into an equal amount of kinetic energy if the ball is allowed to fall back to height y_1. This is another indication that the force of gravity is a conservative force. A physical system in which all forces acting are conservative is called a *conservative system*.

Example 6.4

A workman pushes a 200-kg crate up a frictionless metal ramp of length 4.00 m into the back of a truck (Fig. 6.7a). **(a)** If the angle made by the ramp with the ground is 25.0° and the workman pushes the block at a very slow, constant speed, how much work does he do? **(b)** what is the increased gravitational potential energy of the crate?

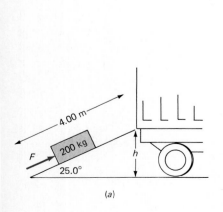

(a)

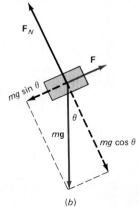

(b)

FIGURE 6.7 (*a*) Diagram for Example 6.4. (*b*) Free-body diagram for Example 6.4.

SOLUTION

(a) The forces acting on the crate are shown in the free-body diagram of Fig. 6.7*b*. The forces perpendicular to the ramp are not important, since there is no friction. The force exerted by the workman up the plane must be equal to the component of the weight of the crate down the plane, for the crate is at every point of its journey in equilibrium (it is not accelerated). Therefore

$$F - mg \sin \theta = 0$$

or $F = mg \sin 25.0° = (200 \text{ kg})(9.80 \text{ m/s}^2)(0.423)$

$$= 829 \text{ N}$$

and so

$$\mathcal{W} = \mathbf{F} \cdot \mathbf{d} = (829 \text{ N})(4.00 \text{ m}) = \boxed{3.31 \times 10^3 \text{ J}}$$

(b) The work done by the workman is, in the absence of friction, converted into gravitational potential energy, as we have seen in the last section. Therefore

$$PE_{grav} = \boxed{3.31 \times 10^3 \text{ J}}$$

The same result can be obtained from Eq. (6.7). Since $h = 4.00 \text{ m} \times \sin 25.0° = 1.69 \text{ m}$,

$$PE_{grav} = mgh = (200 \text{ kg})(9.80 \text{ m/s}^2)(1.69 \text{ m})$$

$$= \boxed{3.31 \times 10^3 \text{ J}}$$

Nonconservative Forces

Not all forces are conservative. A *nonconservative force* is any force for which the work done is dependent on the path, not merely on the initial and final positions of the system. Thus friction and air resistance are nonconservative forces, since work done against these forces depends on the path taken: the longer the path, the more work that must be done. This is because, when work is done against a nonconservative force, mechanical energy is lost in the form of heat, and the longer the path, the more mechanical energy is converted into heat.

6.4 Work Converted into Elastic Potential Energy

There are other kinds of mechanical potential energy in addition to gravitational. For example, consider a horizontal metal spring attached to a wall, as in Fig. 6.8. At its normal length L the spring is in equilibrium, and no net force acts on it. Suppose someone seizes the end of the spring and stretches it through a distance x so that its new length is $L + x$. It is found by experiment that, so long as x is not too great, the spring behaves *elastically*; i.e., its increase in length x is proportional to the stretching force F, or

$$F \propto x$$

and so $F = kx$ **(6.8)**

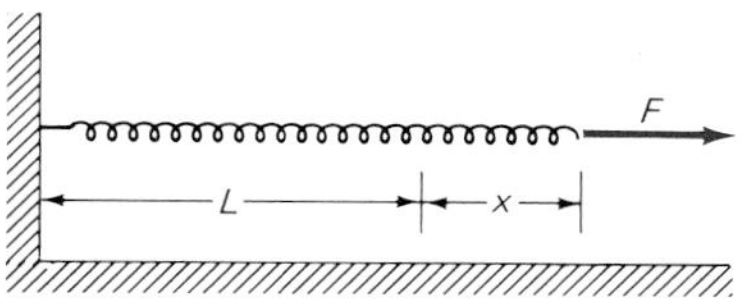

FIGURE 6.8 Work converted into elastic potential energy. When a spring is stretched a distance x, the elastic potential energy stored in the spring is $PE_{elast} = \frac{1}{2}kx^2$.

where k is called the *force constant* of the spring. (Springs that behave in this manner are said to obey *Hooke's law*.) The units of k are newtons per meter. If $F = kx$ is the force the person applies, by Newton's third law of motion the spring exerts an equal and opposite force $(-kx)$ back on the person's hand. If the applied force F is reduced to zero, the spring returns to its original length; that is, $x = 0$ follows from $F = 0$ according to Eq. (6.8). This is what we mean when we say that the spring is *elastic*.

When the spring is stretched to a length $L + x$, it has some stored energy by virtue of its increased length. We could, for example, attach the spring to a door and use this stored energy to open the door. This energy we call *elastic potential energy*, because it is stored energy in the spring.

Elastic potential energy: The energy stored in a spring or other elastic body by virtue of its distortion, or change in size or shape.

Since the force applied to stretch the spring varies from 0 when the length is L to kx when the length is $L + x$, the average force $\overline{F}$ applied is

$$\overline{F} = \frac{0 + kx}{2} = \tfrac{1}{2}kx$$

The work done in stretching the spring is, then,

$$\mathcal{W} = \overline{F} \cdot \mathbf{d} = \tfrac{1}{2}kx \cdot x = \tfrac{1}{2}kx^2$$

This work has been converted into stored potential energy in the spring, and so

$$\boxed{PE_{elast} = \tfrac{1}{2}kx^2} \qquad \text{(6.9)}$$

Here x is the increase (or decrease) in length of the spring. Since $k = F/x$, the units of PE_{elast} are clearly those of $(F/x)x^2$ or Fx. These are newton-meters, or joules, as they must be for any energy.

Note that, in Eq. (6.9), PE_{elast} is positive no matter what the sign of x is; that is, a spring stores energy both when it is compressed and when it is extended.

An example of the use of elastic potential energy is provided by a trampoline, in which the kinetic energy of the jumper is converted into stored elastic potential energy in the springs of the trampoline and then immediately changed back again to kinetic energy. This enables the jumper to reverse direction without losing the mechanical energy acquired initially through muscular exertion.

Elastic Forces as Conservative Forces

Since we have calculated an elastic potential energy ($PE_{elast} = \tfrac{1}{2}kx^2$), we have already assumed that the elastic force is a conservative force. This means that the energy stored in the system depends only on the displacement x of the spring from its equilibrium position. In this case the work done in stretching the spring depends only on its initial and final positions and is completely recoverable, since the system is conservative.

In the field of mechanics, the two main conservative forces are the gravitational and elastic forces. Other examples of conservative forces, to be discussed later, are the electric force between charged particles and the nuclear force that holds protons and neutrons together in atomic nuclei.

Example 6.5

(a) A spring is hung from a ceiling and a 500-g mass is hung from it. This increases the length of the spring by 20.0 cm. What is the force constant k for the spring? **(b)** If the same spring is then moved from the ceiling to a wall and stretched horizontally so that the 500-g mass is displaced 40.0 cm from its equilibrium position, what is the stored potential energy in the spring?

SOLUTION

(a) Since the stretching force $F = kx$, and here $F = mg = (0.500 \text{ kg})(9.80 \text{ m/s}^2)$, we have

$$k = \frac{F}{x} = \frac{(0.500 \text{ kg})(9.80 \text{ m/s}^2)}{0.200 \text{ m}} = \boxed{24.5 \text{ N/m}}$$

(b) From Eq. (6.9),

$$PE_{elast} = \frac{1}{2}kx^2 = \frac{1}{2}\left(24.5 \frac{\text{N}}{\text{m}}\right)(0.400 \text{ m})^2 = \boxed{1.96 \text{ J}}$$

6.5   Work Converted into Thermal Energy

Our next example of work conversion into energy uses the example discussed in Sec. 6.2, that of a small air-hockey puck moving on a table with an initial velocity v_0. In this case, however, we turn off the air jets, allowing a frictional force to oppose the motion of the puck as it slides over the table, as in Fig. 6.9. If we exert on the puck a constant force **F** that is just sufficient to overcome the force of sliding friction, we are able to give the puck a displacement **d** without imparting to it any additional velocity. Because the speed of the puck is the same at the end as it was at the beginning (v_0), the puck's acceleration is zero, and so, from Newton's second law, the sum of all the forces acting on the puck must also be zero. But we are exerting a force **F** on the puck, and so, for the total force to be zero, there must be an equal and opposite force **f** exerted on the puck by some other source such that $\mathbf{F}_{\text{net}} = \mathbf{F} + \mathbf{f} = 0$. Here **f** must be the frictional force of the table rubbing on the bottom surface of the puck.

FIGURE 6.9 Work converted into thermal energy. If the frictional force **f** always remains equal in magnitude to the applied force **F**, no acceleration occurs and all the work done is converted into thermal energy (heat) in the puck and the table. The KE is the same at the end as at the beginning, and there is also no change in the PE.

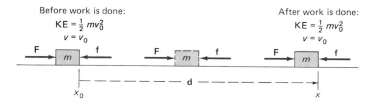

In this case we did an amount of work on the puck equal to $\mathcal{W} = Fd$. Where did this work go? Certainly not into kinetic energy, since the puck's kinetic energy did not change; not into gravitational potential energy, for the height of the puck was the same at the end as at the beginning; and not into elastic potential energy, for the puck and table were in no way deformed or changed in shape. If we observe this simple experiment more closely, the only significant difference we find between it and the other cases we have discussed is that the puck and the table have warmed up a little; i.e., thermal energy (or heat) has been generated while the work was being done. A reasonable explanation would then seem to be that the work done went into thermal energy in the puck and the table. The frictional rubbing of the two surfaces raised their temperatures, in the same way that the rubbing together of two sticks can heat them to a point where a flame results.

Nonconservative forces, such as the force of friction, are sometimes called *dissipative forces*, because mechanical energy is not conserved when such forces act but rather is dissipated or lost. For this reason problems involving dissipative forces are often difficult, since these forces cause the total mechanical energy of the system to decrease over time.

Although the explanation given above for the conversion of work into thermal energy may seem obvious to us, it was not obvious to physicists much before the end of the eighteenth century. Until then mechanics and heat had been treated as two very different disciplines that no one had been able to tie together. Heat was supposed to be material fluid called "caloric," which was added to a body to make it warmer and subtracted from the body to make it colder. In 1778 Benjamin

FIGURE 6.10 (*Courtesy of the Granger Collection.*)

Benjamin Thompson, Count Rumford (1753–1814)

Although born into a simple American family of farmers in Woburn, Massachusetts, Benjamin Thompson spent most of his life in Great Britain, France, and Germany.

Thompson's education as a boy was largely self-acquired, and he soon gained a reputation as a bright lad interested in books and scientific instruments. In 1772 he was employed as a teacher in Concord, New Hampshire, and while there had the good fortune to marry, at age 19, a wealthy widow 11 years his senior. His newly acquired wealth enabled him to live the life of a country gentleman, managing his wife's estate and helping John Wentworth, the British governor of New Hampshire, with some agricultural experiments. To reward Thomp-

son, the governor commissioned him a major in the New Hampshire militia in 1773. There Thompson acted as an informant for the governor, arousing the rancor of his fellow colonists to such an extent that he was almost tarred and feathered. He prudently left for Boston, leaving his wife and baby daughter behind—and never returned.

In Boston Thompson worked for a while for General Gage, the highest-ranking British officer in the Massachusetts Bay Colony. When the American Revolution began in 1776, his loyalty was naturally suspect, and he was forced to flee to England.

In London Thompson parlayed his knowledge of the American scene into a comfortable job with the British Foreign Office. Here he returned to his first love, scientific research, and did some important work on the chemical and physical properties of gunpowder.

After the peace of 1783, Thompson left for the continent and became aide-de-camp and military adviser to Karl Theodor, the ruler of Bavaria. While in Bavaria, Thompson made his most significant contribution to science by his experiments on the boring of cannon and his subsequent realization that the caloric theory of heat was completely inadequate to explain his data. As a consequence of this work, he became one of the most respected men in Bavaria, and in 1792 was raised to the rank of Count of the Holy

Roman Empire, ever afterward to be known as Count Rumford.

In 1795 Rumford returned to Great Britain and devoted his energies to starting the Royal Institution of Great Britain, dedicated to the application of science to the common purposes of life. After getting the Royal Institution started, he went to Paris, where he met and in 1805 married Madame Lavoisier, the widow of the great chemist Antoine Lavoisier. These two strong personalities had a very stormy marriage, however, that ended in divorce after only 2 years.

The last years of Count Rumford's life in Paris were devoted to the scientific development of chimneys, stoves, lamps, heating systems, and even the first drip coffeepot. His interest in the basic scientific principles behind devices such as these made Rumford the world's first great applied physicist.

He died suddenly in August 1814, leaving his entire estate to Harvard College to endow a professorship in his name.

Count Rumford's scientific accomplishments merit him a far greater reputation than he has today. His contemporaries were unwilling to give him full credit for his scientific accomplishments, however, perhaps because of his limitations as a human being. This neglect affects his reputation even today.

Thompson (Count Rumford), while doing some experiments on the boring of cannon in Bavaria (see Fig. 6.10 and accompanying biography) was led to the idea that the heat produced in the process could not be a material substance, since the rubbing seemed to produce an inexhaustible supply of heat.* Rumford concluded that the mechanical motion of the borer was being converted to heat and that heat

*See quotation from Rumford at the beginning of Chap. 13.

154

was therefore a form of motion. This idea was later extended by other physicists to show that thermal energy is the randomized kinetic energy of the individual molecules of a system (see Chaps. 9 and 13). This is a much less useful form of energy than the kinetic or potential energy of a large-scale object like a hammer or of water at the top of a dam.

The examples discussed above show us that work (in the technical sense in which it is used in physics) is always converted into some form of energy. Sometimes this energy is useful kinetic or potential energy. Sometimes it is thermal energy stored in a less useful fashion in the molecules of the system. In all cases, however, the work done is not lost; it is converted into one of the many different forms of energy.

6.6 The Extended Work-Energy Theorem

We first state the following *work-energy theorem*, which is based on our previously derived Eq. (6.3):

The work done by a net force acting on an object is equal to the change in the object's kinetic energy.

In the form of an equation, this is

$$\mathcal{W} = KE_f - KE_i \tag{6.10}$$

where KE_i is the initial kinetic energy of the object before the work is done on the object and KE_f is the final kinetic energy after the force doing the work has ceased to act.

Let us next consider a mechanical system that is not necessarily a conservative system. We denote by $\mathcal{W}$ any work done on the system by an external force (say, a person who gives a shove to a hockey puck); by PE the total potential energy of the system, where $PE = PE_{grav} + PE_{elast}$; by KE the kinetic energy; and by ΔQ the amount of mechanical energy converted into thermal energy. Our *extended work-energy theorem* can then be written

$$\mathcal{W} = KE_f - KE_i + PE_f - PE_i + \Delta Q$$

or $$\mathcal{W} = \Delta KE + \Delta PE + \Delta Q \tag{6.11}$$

This equation states that the work done by an external force is equal to the total increase in mechanical energy of the system plus any thermal energy produced. From a practical point of view, this still may not be useful, since ΔQ is frequently not easy to determine experimentally.

A simpler case is one in which $\Delta Q = 0$; that is, no mechanical energy is converted into thermal energy, and hence all the work done on the system goes into potential and kinetic energy. In that case Eq. (6.11) becomes

$$\mathcal{W} = \Delta KE + \Delta PE \tag{6.12}$$

If the work done by the external force is known and the change in potential energy is given, the final kinetic energy of the system can then be obtained from its initial value.

Example 6.6

A 1.56-kg ball hangs from the end of a string attached to the ceiling, as in Fig. 6.11. The string is 2.00 m long from the ceiling to the CG of the ball. The ball is slowly pulled to the side until the string makes an angle of 25.0° with the vertical. **(a)** How much work is done in moving the ball to this new position? **(b)** If the ball is then released from this position, what is its speed as it passes through its equilibrium position?

FIGURE 6.11

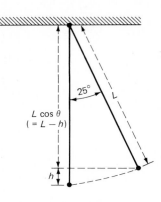

SOLUTION

(a) The direction of the pulling force is not specified, and, even if it were, the pulling force must increase as the ball is drawn to the side. At first sight, therefore, it may appear difficult to find the work done, since the pulling force is unknown. However, since there are no dissipative forces in this system, we know that all the work of the external force goes to change the mechanical energy of the system. Thus Eq. (6.12) applies.

At both the beginning and end of the movement of the ball, its speed is zero, and so Eq. (6.12) becomes $\mathcal{W} = \Delta PE$, and, if we can find the change in gravitational PE of the ball, we can obtain the work done.

If we take the initial gravitational PE of the ball as zero, then, from Fig. 6.11, the height through which the ball is raised is

$$h = L - L \cos \theta = (2.00 \text{ m})(1 - \cos 25.0°)$$

$$= (2.00 \text{ m})(1 - 0.906) = 0.188 \text{ m}$$

Then the final PE of the ball is mgh, and so

$$\mathcal{W} = \Delta PE = mgh - 0$$

$$= (1.56 \text{ kg})(9.80 \text{ m/s}^2)(0.188 \text{ m}) = \boxed{2.87 \text{ J}}$$

(b) If the ball is simply released from its new position, no *external* work is done on it as it falls back toward its rest position. Therefore mechanical energy is conserved, and we have, from Eq. (6.12),

$$0 = \Delta KE + \Delta PE$$

In this case the PE falls by 2.87 J as the ball returns to its rest position. Therefore when it passes through its rest position, we have from above,

$$\Delta KE - 2.87 \text{ J} = 0$$

and so $\quad \Delta KE = \dfrac{1}{2} mv^2 - 0 = 2.87 \text{ J}$

and thus $\quad v^2 = \dfrac{2 (2.87 \text{ J})}{1.56 \text{ kg}} = 3.68 \text{ m}^2/\text{s}^2$

and $\quad v = \boxed{1.92 \text{ m/s}}$

EXERCISE 1   A woman rides a bicycle up a hill that is, at its highest point, 10.0 m above the level road from which she starts (Fig. 6.12). The woman and the bicycle together have a mass of 100 kg. If she does 1.00×10^4 J of work in climbing the hill and her initial speed is 4.00 m/s, what is her speed at the top of the hill (neglecting all dissipative forces)?

FIGURE 6.12

6.7   Conservation of Mechanical Energy

Suppose we have a conservative system on which no work is done by an external force. Then

$$\mathcal{W} = \Delta KE + \Delta PE = 0$$

or $\quad\quad PE_f - PE_i = -(KE_f - KE_i)$

and so $PE_f + KE_f = PE_i + KE_i$

or

Total mechanical energy = PE + KE = constant

(6.13)

This is the *principle of conservation of mechanical energy*:

For any isolated (no external forces acting), conservative ($\Delta Q = 0$) system, the sum of the potential and kinetic energies of the system remains constant.

Falling Objects

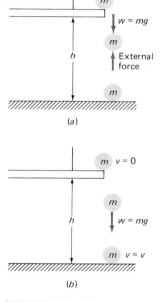

FIGURE 6.13 Energy conservation for the motion of a ball. (*a*) Ball acquires $PE_{grav} = mgh$ by being lifted to a height *h*. (*b*) Ball falls off ledge, and PE_{grav} is converted into KE. Since $mgh = \frac{1}{2}mv^2$, $v = \sqrt{2gh}$.

As an illustration of the usefulness of the principle of conservation of mechanical energy, consider the case of a falling ball, as in Fig. 6.13. Suppose that initially the ball is resting on the ground and that a person does work on it to lift it slowly to a ledge at a height *h* above the ground. We presuppose that there is no friction and no air resistance, and so $\Delta Q = 0$. If the work done on the ball in lifting it from a height $y_i = 0$ to a height $y_f = h$ is $\mathcal{W}$, we have, from Eq. (6.12),

$$\mathcal{W} = \Delta KE + \Delta PE$$

But there has been no change in the kinetic energy of the ball, since it is at rest at both the beginning and end of the lifting process. Hence $\Delta KE = 0$, and so

$$\mathcal{W} = \Delta PE = mg(y_f - y_i) = mgh$$

The work done is therefore converted into an increase in the potential energy by an amount *mgh*.

If we push the ball off the ledge, how rapidly will it be falling just before it hits the ground? Of course, we could find this result by using the laws of uniformly accelerated motion discussed in Secs. 2.6 and 2.7. It turns out to be easier, however, to apply the principle of conservation of mechanical energy, which yields

$$PE_i + KE_i = PE_f + KE_f$$

or $mgh + 0 = 0 + \frac{1}{2}mv^2$ from which $v = \sqrt{2gh}$

This is the speed of the ball just before it hits the ground and is the same result as found in Sec. 2.7.

In this case no *external* force acts on the system, which consists of the earth and the ball, and so no *external* work is done on the system. The force of gravity (which is an internal force for the ball-earth system) acts, however, to convert PE into KE of the ball, in accordance with Eq. (6.13). An *internal* force can act to change KE to PE, and vice versa, but cannot change the total energy of a mechanical system.

Of course, when the ball collides with the ground, some of its kinetic energy is converted into thermal energy (and sound), and mechanical energy is no longer conserved. As a result, the ball bounces up to a height lower than the original ledge. If we consider all forms of energy in the process, however, it still remains true that the total energy of the system is conserved.

Roller Coaster

Another good example of the conservation of mechanical energy is provided by roller coasters of the kind shown in Fig. 6.14. In discussing its motion we simplify the situation by neglecting friction. The roller coaster car is hauled to the top of the first hill, which is the highest point on the track, a distance *H* above the ground. This gives the car and its riders a potential energy equal to *mgH*, as in Fig. 6.15, where *m* is the total mass of car and riders. The car is then released, and its potential energy is gradually converted to kinetic energy as the car descends. At the bottom of the first hill, its kinetic energy is large enough to carry the car to the top of the next hill, which is lower than the first, while in climbing this hill the car's kinetic energy gradually decreases and its potential energy simultaneously increases. The motion of the car therefore becomes a constant conversion from PE to KE and back again, as shown in the figure.

The kinetic energy (and hence the speed) of the car at any point can be obtained from the conservation-of-mechanical-energy principle:

$$\text{Total energy} = \text{PE} + \text{KE} = \text{constant}$$

$$mgh + \tfrac{1}{2}mv^2 = mgH \qquad\qquad (6.14)$$

where mgH is the initial mechanical energy put into the system when the car is hauled to the top of the first hill and v is the speed of the car at a height h. For any height h, the velocity can then be found from this equation, since all the other quantities are known. Note that a plot of the height of the roller coaster track as a function of distance along the ground resembles a graph of the potential energy of the system (similar to Fig. 6.15) because the potential energy of the car is equal to mgh. Since there is friction in a real roller coaster, the system gradually loses mechanical energy. Eventually the car will not have sufficient mechanical energy to climb one of the hills and will be trapped in one of the valleys unless additional energy is provided. This energy can be supplied by using an electric motor to lift the car to the top of another hill and start the process over again.

FIGURE 6.14   A modern roller coaster: the Loch Ness Monster at The Old Country, Busch Gardens, Williamsburg, Virginia. (*Courtesy of The Old Country, Busch Gardens.*)

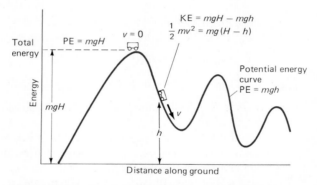

FIGURE 6.15   Energy conservation in a roller coaster.

Example 6.7

Approximately what running speed would an 80-kg pole-vaulter have to acquire to clear a bar 5.0 m off the ground? Assume that the pole-vaulter's center of gravity is initially 1.0 m off the ground and that the vaulter clears the bar if his center of gravity just reaches the height of the bar.

SOLUTION

We can simplify this problem by neglecting any additional energy supplied by the vaulter's arms, any resistance of the air to the jumper's body, the slight forward velocity of the vaulter at the top of the vault, and other secondary effects (we only use two significant figures in our calculations because of the approximate nature of our results). We concentrate solely on the conservation of mechanical energy involved in the process. When the vaulter shoves the pole into the slot provided for it at the end of the runway, he has a certain kinetic energy. At the top of the vault, all this kinetic energy has been converted into gravitational potential energy. We have

$$\text{KE}_i + \text{PE}_i = \text{KE}_f + \text{PE}_f$$

or $KE_i + 0 = 0 + mgh$

where $h = 5.0$ m $- 1.0$ m $= 4.0$ m. Then

$KE_i = (80$ kg$)(9.8$ m/s$^2)(4.0$ m$) = 3.1 \times 10^3$ J

and so

$$v = \left(\frac{2KE_i}{m}\right)^{1/2} = \left[\frac{2(3.1 \times 10^3 \text{ J})}{80 \text{ kg}}\right]^{1/2} = \boxed{8.8 \text{ m/s}}$$

World-class sprinters can run 100 m in about 10 s for an average speed around 10 m/s. Considering that a pole-vaulter must carry a pole while running, 8.8 m/s is a reasonable answer.

Example 6.8

A water slide at an amusement park has a parabolic shape. A girl of mass 80.0 kg starts at the top of the slide, 10.0 m above the ground. **(a)** What would be her speed at a point on the slide 5.00 m above the ground, if we neglect all frictional losses?

(b) If her speed at a height 5.00 m above the ground was found actually to be 8.00 m/s, how much energy was dissipated as heat in the sliding process?

SOLUTION

In this case we do not even know the direction of the slide at height 5.00 m above the ground. But we can still solve the problem by using the conservation-of-mechanical-energy principle to find the KE at this height, and from this the girl's speed.
(a) Since mechanical energy is conserved,

$$PE_i + KE_i = PE_f + KE_f$$

or $mgy_1 + 0 = mgy_2 + KE_f$

from which

$KE_f = mg(y_1 - y_2) = (80.0$ kg$)(9.80$ m/s$^2)(5.00$ m$)$

$$= 3.92 \times 10^3 \text{ J}$$

Then $v^2 = \dfrac{2KE_f}{m} = \dfrac{2(3.92 \times 10^3 \text{ J})}{80.0 \text{ kg}} = 98.0$ m^2/s^2

$$v = \boxed{9.90 \text{ m/s}}$$

This problem shows clearly the usefulness of energy considerations in solving mechanical problems.
(b) In this case we need the generalized work-energy theorem, which includes ΔQ, the energy lost as heat. We have, from Eq. (6.11), with $W = 0$,

$$0 = \Delta KE + \Delta PE + \Delta Q$$

Since $KE_i = 0$, we have

$\Delta Q = PE_i - PE_f - KE_f = mg(y_1 - y_2) - \frac{1}{2}mv_2^2$

$$= 3.92 \times 10^3 \text{ J} - \frac{1}{2}(80.0 \text{ kg})(8.00 \text{ m/s})^2$$

$$= \boxed{1.34 \times 10^3 \text{ J}}$$

Example 6.9

A 2.00-kg mass is attached to the end of a light horizontal spring with force constant $k = 20.0$ N/m. The mass rests on a frictionless surface and is initially displaced until the spring is stretched 0.500 m. **(a)** What is the initial elastic potential energy

stored in the spring? **(b)** What is the kinetic energy of the mass when it passes through its equilibrium position? **(c)** What is the speed of the mass at this position?

SOLUTION

(a) $PE_{elast} = \frac{1}{2}kx^2 = \frac{1}{2}(20.0$ N/m$)(0.500$ m$)^2 = \boxed{2.50 \text{ J}}$

(b) Since there is no friction, the total mechanical energy of the system is conserved, and so $PE_{elast} + KE = $ constant at any position of the spring in the course of its motion. The constant energy of the system must be equal to the initial energy, which is all elastic PE and equal to 2.50 J. Therefore

$PE_{elast} + KE = 2.50$ J   or $\frac{1}{2}kx^2 + \frac{1}{2}mv^2 = 2.50$ J

At the mass' equilibrium position, $x = 0$, and thus $\frac{1}{2}kx^2 = 0$ and

$KE = \frac{1}{2}mv^2 = \boxed{2.50 \text{ J}}$

As the mass vibrates back and forth at the end of the spring, there is a constant conversion of PE_{elast} to KE and then back to PE_{elast}, with the total energy remaining constant.

(c) $v^2 = \dfrac{2KE}{m} = \dfrac{2(2.50 \text{ J})}{2.00 \text{ kg}} = 2.50$ J/kg $= 2.50$ m^2/s^2

$$v = \boxed{1.58 \text{ m/s}}$$

Note two things here: (1) This problem is conceptually identical with the roller coaster problem considered earlier in this section, except that here the PE is elastic rather than gravitational. (2) The mass is not required to find the energies asked for in parts (*a*) and (*b*); it is, however, essential for finding the speed of the mass once the KE is known.

**A More General
Conservation-of-Energy
Principle**

The principle of conservation of mechanical energy can be expanded to include nonmechanical forms of energy such as thermal energy and electric energy. This leads us to a much more general principle of present-day physics, the *principle of conservation of energy,* which is stated as follows:

> In an isolated physical system, energy is neither created nor destroyed; it merely changes from one form to another.

This is one of the most important principles in all of physics. We have already seen many examples of its usefulness in mechanics and will see many more throughout the rest of this course, especially in our discussion of heat. The validity of this principle is based on the fact that no experiment has ever been performed that contradicts it.

If this principle is correct, you may wonder why we are constantly urged to "conserve energy," since we have just said that energy is always conserved and never lost. The drive to conserve energy is more accurately a push to conserve *useful* forms of energy like the fossil fuels—coal, oil, natural gas—rather than to waste them by converting them needlessly into less useful forms of energy like heat. It is still true that energy is never destroyed or lost; but it may be converted from useful forms to relatively useless forms. We will see more of this in Chap. 15.

6.8 Simple Machines

One practical application of the principle of conservation of energy is to simple machines. A *machine* is a device that does useful work when it is provided with energy. If energy is to be conserved, it is clear that such a machine cannot do more work than the energy it receives. In other words, the work output of the machine cannot exceed the energy or work input. At best, under ideal conditions the work output may be equal to the work or energy input; if friction or other dissipative forces are present, the work output will, of necessity, be even smaller than the input.

A good example of a simple machine is the *inclined plane,* already discussed in Example 6.4. We saw there that a workman only had to exert a force of 829 N to move a 200-kg crate up a 25° inclined plane. If he had tried to lift it directly into the truck, the force required would have been $mg = (200 \text{ kg})(9.80 \text{ m/s}^2) = 1960 \text{ N}$.

Of course no energy is created in this process. The amount of work required to lift the crate directly up from the street to the truck is 3.31×10^3 J, exactly the same work required to slide the crate up the plank. The work done is the same in the two cases, but the *force* required with the inclined plane is far less than the force required without it. This is the way all simple machines work: *they allow smaller forces to be used without changing the amount of work done.* In physics we never get something for nothing; in this case we are able to exert *smaller forces* to do a certain amount of work, but we must exert them over *greater distances.*

Levers

A lever is a device such as a wooden plank resting on a sawhorse (which serves as a *fulcrum*) that can be used to reduce the force necessary to do a desired amount of work.

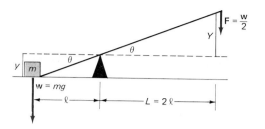

FIGURE 6.16 A lever used to lift a mass *m*. If the lever arm *L* is twice *ℓ*, then the force **F** required to lift the weight **w** is only **w**/2.

Consider, for example, the lever in Fig. 6.16. If the lever arm *L* is twice as long as *l*, and we apply a force **F** in an attempt to lift the weight **w**, then, from the figure

$$\tan \theta = \frac{Y}{L} = \frac{y}{l} \quad \text{or} \quad \frac{Y}{y} = \frac{L}{l} = 2$$

But if the lever is 100 percent efficient, the work input must be equal to the work output that lifts the weight, and so

$$\mathcal{W} = FY = wy \quad \text{or} \quad F = \frac{wy}{Y} = \frac{w}{2}$$

In this example a weight *w* can be lifted with a lever by exerting a force *F* that is only half as large as the weight. If the ratio of the lever arms is increased even more, the force that must be applied to lift the weight can be decreased even further. If the ratio of the lever arms becomes very large, it is possible to lift a very large weight by exerting a very small force. This is the meaning of Archimedes' celebrated claim that, if he had a lever long and rigid enough (and a fulcrum to support it), he could have moved the earth.

It cannot be overemphasized that in this example the work output is *equal* to the work input, but the input can be in the form of a small force working through a large distance. This is the advantage of a lever.

Pulleys

A simple pulley can make it more convenient for a person to do some useful work like lifting a weight. For example, the simple pulley in Fig. 6.17 enables a person to pull down on the rope at a convenient angle and thus lift the weight *w*. In this case the force exerted by the person must be equal to *w*, and there is no advantage other than convenience in using the pulley.

Compound pulleys enable a person to lift weights by the exertion of forces smaller than the weights lifted. In Fig. 6.18 two pulleys *A* and *B* are at fixed positions in space and one pulley *C* is free to move with the weight to be lifted. When we pull down on the rope with a force *F*, the weight *w* is lifted only one-third of the distance *L* through which the end of the rope is moved. This is because three lengths of rope are being shortened, and each one shortens only by *l* = *L*/3. This is also the distance the free pulley and the weight move up. Here we have

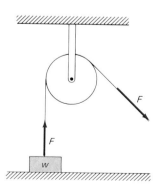

FIGURE 6.17 A simple pulley used to change the direction of a force. The applied force is equal to the lifting force.

$$\mathcal{W} = FL = wl = w\frac{L}{3} \quad \text{or} \quad F = \frac{w}{3}$$

We can lift a weight *w* by exerting a force only one-third as large as *w* if we use a pulley system of the kind shown here, where three segments of rope are attached to the movable pulley. In the absence of friction, the force needed to lift

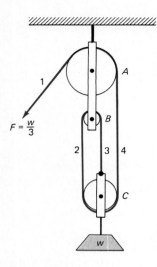

FIGURE 6.18 A compound pulley system. The force required to lift the weight w is only $F = w/3$, since there are three ropes pulling up on the free pulley attached to the weight w.

$F = \dfrac{w}{3}$

a weight w is reduced to w/n, when n rope segments are attached to the free pulley lifting the weight. If we consider in Fig. 6.18 the static situation in which the weight is at rest, the weight w is being supported by three segments of rope so that in equilibrium the tension in each rope is only $w/3$. Since the tension at point 4 is the same as the tension at point 1 and hence the same as the force exerted by the person holding the rope, the force that must be exerted to lift the weight is also only $w/3$ in this case, or w/n for n rope segments.

Mechanical Advantage

In the above examples the ratio of the distance through which the force is exerted to the distance the weight moves is, in the ideal order, the same as the ratio of the weight that is moved to the force needed to move it. We call the ratio of distances the *ideal mechanical advantage* (IMA):

$$\text{IMA} = \frac{L}{l} \tag{6.15}$$

In actual fact, there are losses in the system due to friction and other dissipative forces, and the *actual mechanical advantage* (AMA) is less than the IMA and is given by the ratio of the output force to the input force:

$$\text{AMA} = \frac{w}{F} \tag{6.16}$$

The *efficiency* of any simple machine is defined as the ratio of work output to work input:

$$\text{Efficiency} = \frac{\text{work output}}{\text{work input}} \times 100\% = \frac{wl}{FL} \times 100\%$$

$$= \frac{w/F}{L/l} \times 100\%$$

or $$\boxed{\text{Efficiency} = \frac{\text{AMA}}{\text{IMA}} \times 100\%} \tag{6.17}$$

Example 6.10

The pulley system in Fig. 6.18 is lifting a 10-N weight. To lift the weight it is found that it is necessary to exert a force of 4.0 N. What is **(a)** the ideal mechanical advantage, **(b)** the actual mechanical advantage, and **(c)** the efficiency of this pulley system?

SOLUTION

(a) The ideal mechanical advantage of the pulley system is the ratio of the distance through which the force is exerted to the distance the 10-N weight is moved. For a pulley system with three rope segments doing the lifting, this is

$$\text{IMA} = \frac{L}{l} = \boxed{3}$$

(b) The actual mechanical advantage is the ratio of the weight moved to the force exerted to move it:

$$\text{AMA} = \frac{w}{F} = \frac{10 \text{ N}}{4.0 \text{ N}} = \boxed{2.5}$$

(c) The efficiency of the pulley system is then

$$\text{Efficiency} = \frac{\text{AMA}}{\text{IMA}} = \frac{2.5}{3} = 0.83 = \boxed{83\%}$$

This particular compound pulley system is 83 percent as efficient in lifting the weight as a completely frictionless, ideal pulley system would be. In this case the other 17 percent of the energy goes into thermal energy because of friction in the pulley system.

6.9 Power

The same amount of work is required to lift a 10-kg mass through 2.0 m whether it takes 2.0 s or 2.0 h to do the job. The work required is $\mathcal{W} = mgh$, and time appears nowhere in this equation. The *power* input would be very different, however, in these two cases.

Power: The rate at which work is done or energy is transformed.

$$P = \frac{\Delta \mathcal{W}}{\Delta t}$$

(6.18)

where $\Delta \mathcal{W}$ is the amount of work done in the time Δt.

The unit of power, the joule per second (J/s), we call a watt (W), where 1 W = 1 J/s. Thus, in the case just mentioned, if it takes 2.0 s to move the 10-kg mass, we have

$$P = \frac{mgh}{\Delta t} = \frac{(10 \text{ kg})(9.8 \text{ m/s}^2)(2.0 \text{ m})}{2.0 \text{ s}} = 98 \text{ J/s} = 98 \text{ W}$$

If, on the other hand, it takes 2 h,

$$P = \frac{mgh}{\Delta t} = \frac{(10 \text{ kg})(9.8 \text{ m/s}^2)(2.0 \text{ m})}{2.0 \text{ h}(3600 \text{ s/h})} = 0.027 \text{ J/s} = 0.027 \text{ W}$$

The watt is the same power unit used in electricity. A 100-W electric bulb, when lighted, consumes electric energy at the rate of 100 J/s, that is, 100 W, and converts this energy into heat and light.

The watt is named for Sir James Watt (1736–1819), whose steam engine was the predecessor of today's more powerful engines. Watt himself suggested the rate at which a horse works as the unit of power. This he called a *horsepower,* but horses would have short lives indeed if forced to work at this rate for very long. One horsepower (hp) is equal to 746 W, or about 0.75 kW. Thus if a car's motor has a power output of 100 hp, it is doing work at the rate of about 75 kW, or 75 kJ/s.

While the watt and kilowatt are units of *power,* a kilowatthour is a unit of *energy.* The actual amount of energy used, for example, by a 100-W light bulb depends on how long the bulb is on. To find the actual energy consumed, we must multiply the power (100 W) by the time the bulb is lighted. This is merely to restate Eq. (6.18) in the equivalent form

$$\Delta W = P \, \Delta t \quad \text{or} \quad 1 \text{ J} = \left(\frac{1 \text{ J}}{\text{s}} \right) (1 \text{ s}) = 1 \text{ Ws} \tag{6.19}$$

or one joule is equal to one watt-second. Therefore one kilowatthour is

$$1 \text{ kWh} = 10^3 \text{ Wh} = \left(10^3 \frac{\text{J}}{\text{s}} \right) \left(\frac{3600 \text{ s}}{1 \text{ h}} \right) (1 \text{ h}) = 3.60 \times 10^6 \text{ J}$$

$$\boxed{1 \text{ kWh} = 3.60 \times 10^6 \text{ J}}$$

The kilowatthour is therefore a unit of *energy,* not of power, since the joule is an energy unit. For this reason electric power companies charge for the number of kilowatthours used, not for the number of kilowatts, since both the rate at which the energy is used (kilowatts) and the time (hours) are important.

Power and Velocity

There is a very useful relationship between power and velocity that we can easily derive. Suppose a force **F** does work on an object to move it through a small displacement $\Delta\mathbf{d}$ in time Δt. Then, from Eq. (6.18), the power is

$$P = \frac{\Delta W}{\Delta t} = \frac{\mathbf{F} \cdot \Delta \mathbf{d}}{\Delta t} = \mathbf{F} \cdot \overline{\mathbf{v}}$$

since $\Delta\mathbf{d}/\Delta t$ is the average velocity $\overline{\mathbf{v}}$ of the object during time t. We therefore have, on evaluating the dot product,

$$P = F\overline{v} \cos \theta \tag{6.20}$$

where θ is the angle between the force **F** and the displacement $\Delta\mathbf{d}$. Power can therefore be defined as the product of the applied force and the average velocity of the object in the direction of that force.

To show the usefulness of this expression for power, let us consider its application to an automobile. The work done by a car's engine goes into overcoming friction and air resistance (as when the car is cruising at constant speed on a flat road); into lifting the weight of the car up an inclined plane (when the road is rising), thus increasing its PE; and into accelerating the car (e.g., to pass other cars), thus increasing its KE. The following example shows how useful Eq. (6.20) is in such cases.

Example 6.11

A 1500-kg car is traveling at a constant speed of 25.0 m/s on a level road. The forces of friction and air resistance on the car add up to 800 N. **(a)** What power must be generated by the engine to keep the car moving at this speed on level ground? **(b)** If the car has to climb a 15.0° hill at 25.0 m/s, what additional power is needed? **(c)** If the driver accelerates to 30.0 m/s over a 10.0-s interval on a level road, what additional power must the engine provide?

SOLUTION

(a) To keep the car moving on level ground requires a power, from Eq. (6.20),

$$P_1 = F\overline{v} = (800 \text{ N})(25.0 \text{ m/s}) = \boxed{2.00 \times 10^4 \text{ W}}$$

Since automobile engines are usually rated in horsepower, this is

$$P_1 = \frac{2.00 \times 10^4 \text{ W}}{746 \text{ W/hp}} = \boxed{26.8 \text{ hp}}$$

(b) For the car to climb the 15.0° hill, the car's engine must exert a force

$$F = mg \sin \theta = (1500 \text{ kg})(9.80 \text{ m/s}^2)(\sin 15.0°)$$
$$= (1500 \text{ kg})(9.80 \text{ m/s}^2)(0.259) = 3.80 \times 10^3 \text{ N}$$

$$P_2 = F\overline{v} = (3.80 \times 10^3 \text{ N})(25.0 \text{ m/s})$$

$$= 9.50 \times 10^4 \text{ W} = \boxed{1.27 \times 10^2 \text{ hp}}$$

(c) The acceleration in this case is

$$a = \frac{v_f - v_i}{t} = \frac{30.0 \text{ m/s} - 25.0 \text{ m/s}}{10.0 \text{ s}} = 0.500 \text{ m/s}^2$$

The force required is then

$$F = ma = (1500 \text{ kg})(0.500 \text{ m/s}^2) = 750 \text{ N}$$

The average speed is

$$\bar{v} = \frac{25.0 \text{ m/s} + 30.0 \text{ m/s}}{2} = 27.5 \text{ m/s}$$

and so the power is

$$P_3 = F\bar{v} = (750 \text{ N})(27.5 \text{ m/s})$$
$$= 2.06 \times 10^4 \text{ W} = \boxed{27.6 \text{ hp}}$$

We see that the greatest strain is put on the car's engine when it is climbing hills, and that about 150 hp is adequate even for climbing a 15.0° hill without reducing speed, allowing also for the effects of air resistance and friction. Similarly the greatest strain is put on an airplane's engine when it is taking off and rapidly gaining altitude.

Summary: Important Definitions and Equations

Work: The product of the applied force and the displacement of an object in the direction of the force.

$$\mathcal{W} = \mathbf{F} \cdot \mathbf{d} = Fd \cos \theta \qquad \text{(6.1)}$$

The joule is the unit for both work and energy, where 1 joule (J) = 1 N·m.

Energy: That property of an object which enables it to do work.

Mechanical energy: The energy possessed by an object by virtue of its motion or position.
  Kinetic energy: The energy possessed by an object by virtue of its motion: KE = $\frac{1}{2}mv^2$ (6.4)
  Potential energy: The energy possessed by an object by virtue of its position or state.
  Gravitational potential energy: The energy possessed by an object by virtue of its position in a gravitational field: $PE_{grav} = mgh$ (6.7)
  Elastic potential energy: The energy possessed by an object by virtue of its distortion or change in size or shape: $PE_{elast} = \frac{1}{2}kx^2$ (for a spring) (6.9)
  Conservative forces: Forces that conserve the total mechanical energy of a system. Examples are gravitational and elastic forces.
  Nonconservative forces: Forces that dissipate mechanical energy in the form of heat. Examples are friction and air resistance.

Work-energy theorem: The work done by a net force acting on an object is:

$$\mathcal{W} = KE_f - KE_i \qquad \text{(6.10)}$$

Extended work-energy theorem

$$\mathcal{W} = \Delta KE + \Delta PE + \Delta Q \qquad \text{(6.11)}$$

Principle of conservation of mechanical energy

Total mechanical energy = PE + KE = constant (6.13)
  (if $\Delta Q = 0$ and $\mathcal{W} = 0$)

General principle of conservation of energy: In an isolated physical system energy is neither created nor destroyed; it merely changes from one form to another.

Simple machine: A simple device such as a lever, a jack, or a pulley system that does useful work when energy is supplied to it.

Mechanical advantage

Actual: AMA = w/F (6.16)

Ideal: IMA = L/l (6.15)

$$\text{Efficiency} = \frac{\text{work out}}{\text{work in}} \times 100\%$$
$$= \frac{\text{AMA}}{\text{IMA}} \times 100\% \qquad \text{(6.17)}$$

Power: The rate at which work is done or energy transformed.

$$P = \frac{\Delta \mathcal{W}}{\Delta t} = \mathbf{F} \cdot \mathbf{\bar{v}} \qquad \text{(6.18)}$$

1 watt (W) = 1 J/s

Questions

1 Does the sun lose any energy by doing work on the earth as it moves in its orbit around the sun?

2 The dimensions of torque and of work are the same: $[M][L^2]/[T^2]$. Does this mean that torque and work are the same kind of physical quantity? Point out clearly the distinctions between them.

3 Experienced hikers, rather than step up on a log in their way on a hiking trail and then step down off it on the other side, instead step over the log to the ground on the other side (if the size of the log permits). Does this make sense from an energy point of view?

4 Figure 6.4 represents a highly idealized case, since it is impossible in practice to lift the puck to the shelf by a purely vertical motion. Rather the puck must be moved horizontally to the right at floor level, lifted to the height of the shelf, and then moved back to the left to its final position on the shelf. In the absence of friction and air resistance, will this change the analysis in Sec. 6.3 in any way? Why?

5 Describe the energy transformations taking place as a skier starts from the top of a slope, picks up speed, tries to slow down, and finally plows into a snowbank at the bottom of the slope.

6 Discuss similarities and differences between the hydrologic cycle as a source of electric energy and the use, for the same purpose, of pumped-water storage at high elevations.

7 Discuss the energy transformations taking place from the time a pole-vaulter starts his run to the time he lands in the pit after leaping over (or hitting) the bar. Discuss also the energy transformations undergone by the pole.

8 A person lifts a stone of mass m from the floor a vertical distance h and puts it on a shelf. She does this very slowly so that the stone is not accelerated. What is the work done by the person? What is the work done by the downward gravitational force exerted by the earth on the stone?

9 Why cannot the kinetic energy given to the molecules of two objects when they are rubbed together be converted back into useful work at some later time?

10 We know that in any practical energy conversion scheme such as a hydroelectric plant or a steam-electric generator some mechanical energy is lost in the form of heat. Why then does it make sense to talk about a principle of conservation of energy?

11 Can you suggest a simple machine whose ideal mechanical advantage is exactly equal to 1? What is the usefulness of such a machine?

12 Can you suggest a simple machine whose ideal mechanical advantage is deliberately designed to be less than 1? Give an example of the use of such a machine (one such machine should be very familiar to most students).

13 A new electric power company places an ad in the newspaper saying that it will charge only 7 cents per kilowatt of electricity. Is this reasonable? Why?

Multiple-Choice and Simple Exercises

6.1 A shopper pushes down on a 12-kg shopping cart at an angle of 20° with respect to the horizontal. She exerts a force of 6.0 N and moves the cart 20 m. The work done is:
(a) 1.1×10^2 J   (b) 2.3×10^3 J   (c) 41 J
(d) 2.4×10^2 J   (e) 0

6.2 A man pushes his daughter in a toy truck down the level driveway of their home, starting from rest. He exerts a constant force of 3.0 N parallel to the ground and pushes the truck 10 m in 14.0 s. The mass of child and truck together is 30 kg. If we neglect friction, the kinetic energy of the truck and child together at the end of 14.0 s is:
(a) 30 J   (b) 30 W   (c) 300 J   (d) 38 J
(e) 2.9×10^3 J

6.3 The height of Victoria Falls is 111 m. A drop of water starting from rest at the top of the falls would have, at the bottom, a speed of:
(a) 85 m/s   (b) 47 m/s   (c) 60 m/s
(d) 33 m/s   (e) None of the above

6.4 A spring of mass 4.0 kg has a force constant of 20 N/m. If the spring is compressed 15 cm, its stored potential energy is:
(a) 5.9 J   (b) 0.23 J   (c) 0.45 J
(d) 0.60 J   (e) 2.3 J

6.5 A 1500-kg car is moving along a level road with a kinetic energy of 7.5×10^4 J. The driver accelerates the car, and the engine provides an additional 5.0×10^4 J of energy during the next 10 s. If 1.0×10^4 J of this energy must be used to overcome friction, the final kinetic energy of the car is:
(a) 7.9×10^4 J   (b) 13.5×10^4 J
(c) 11.5×10^4 J   (d) 7.5×10^4 J
(e) None of the above

6.6 A 100-kg roller coaster car has a speed of 20.0 m/s at the bottom of one of the hills. The highest hill it will be able to climb is, in the absence of friction,
(a) 20.0 m   (b) 1.00 m   (c) 200 m
(d) 6.30 m   (e) 160 m

6.7 A lever is being used to lift a 100-kg rock that is 1.0 m from the fulcrum. The person using the lever exerts a force downward at a distance 2.5 m from the fulcrum and finds that lifting the rock requires a force of 500 N. The efficiency of the lever is:
(a) 100%   (b) 128%   (c) 40%   (d) 78%
(e) 22%

6.8 If it takes a crane 100 s to lift a 1000-kg girder to a height of 100 m, the rate at which work is done by the crane is:
(a) 9.80×10^5 J   (b) 9.80×10^5 W
(c) 9.80×10^3 W   (d) 9.80×10^3 J
(e) 1.00×10^3 W

6.9 One joule is equal to:
(a) 3.6×10^6 kWh   (b) 2.8×10^{-7} kWh
(c) 2.8×10^{-7} kW   (d) 2.8×10^{-7} kWs
(e) 3.6×10^6 kW

6.10 As to order of magnitude, 1 hp is equal to:
(a) 1 W   (b) 1 J/s   (c) 1 kJ/s
(d) 10 kW   (e) 10 W

6.11 Use the precise physics definition of work to answer the following questions.
(a) How much work is done by a 70-kg man who carries a 30-kg sack 100 m along a level corridor?
(b) How much work does the same man do in carrying his load up 100 steps, each with a riser 25 cm high and a tread 50 cm wide?
(c) How much work does he do carrying his load up a ramp 100 m long, whose angle with the horizontal has a tangent of 0.10?

6.12 A 70-kg jogger moving at a speed of 2.0 m/s has a kinetic energy of 140 J. If she accelerates to 3.5 m/s, how much additional work must her body do to produce this acceleration?

6.13 What is the kinetic energy of an electron of mass 9.1×10^{-31} kg moving at 3.0×10^7 m/s?

6.14 An 85-kg ski jumper starts at a height of 60 m above the bottom of the takeoff slope. At what speed will he be traveling at the bottom of the slope, in the absence of friction and air resistance?

6.15   What is the gravitational potential energy of 2.0×10^6 kg of water stored in a small mountain lake 1.2×10^3 m above sea level?

6.16   A 100-g ball is dropped out a window at a height of 30 m. Just before the ball reaches the ground, its speed is 22 m/s. How much energy has been lost in the fall because of air resistance?

6.17   A spring of force constant 20 N/m rests on a frictionless table. A 5.0-kg mass is attached to the spring, and it is displaced 25 cm from its equilibrium position and then released. What is the maximum kinetic energy of the system?

6.18   A lever is used to lift a 100-kg mass. If this mass is 1.00 m from the fulcrum of the lever and if the person trying to lift the mass can exert a force of at most 300 N, how long must the lever arm on the lifter's side of the fulcrum be?

6.19   A motor is being used to drive a winch to lift girders into position at a construction site. If each of the girders has a mass of 120 kg and is to be lifted 25 m in 10 s, what must the horsepower of the motor be?

6.20   A jet engine develops a thrust (i.e., a forward-directed force) of 1.90×10^5 N.

(a)   If the plane is flying at a constant speed of 300 m/s, what is the power produced by this engine?

(b)   How much is this power in units of horsepower?

Problems

6.21   A person applies a force of 200 N to a car that is rolling slowly along a level track. Find the work done on the car if it moves 10.0 m while the person pushes (a) in the direction of the car's displacement; (b) in a direction 45° with respect to the car's displacement; (c) in a direction opposite to the car's displacement; (d) in a direction at right angles to the displacement.

6.22   A force **F** is applied at an angle of 30° above the horizontal to move a 100-kg block at constant velocity along a flat surface. The block moves 20 m, and the coefficient of sliding friction between block and surface is 0.30.

(a)   Find the magnitude of the force **F**.

(b)   Find the work done by the force **F**.

(c)   Where did this work go?

6.23   A 10-kg box is pushed 10 m along a rough floor ($\mu_k = 0.50$) by a force of 100 N in the direction of motion.

(a)   What is the work done by this applied force?

(b)   What is the work done by the force of friction on the box?

(c)   What is the total work done by all forces acting on the box?

(d)   What change in KE is produced by the forces acting on the box?

(e)   If the initial speed of the box is 0, what is its final speed?

6.24   A 1500-kg car is moving along a level road at 15.0 m/s. The driver accelerates, and the engine provides 2.00×10^4 J of additional energy during the next 10 s. If 0.30×10^4 J of this energy must be used to overcome friction, what is the final speed of the car?

6.25   A boy pulls a 30-kg sled, initially at rest, along a level surface of ice. The coefficient of friction between sled and ice is 0.10. The boy exerts a force of 50 N parallel to the ice over a 10-m stretch. Find (a) the work done by the boy; (b) the work done against friction; (c) the final kinetic energy of the sled.

6.26   A roller coaster car has a speed of 25.0 m/s at the bottom of the first hill in Fig. 6.15. Will it be able to make it to the top of the second hill, which is 40.0 m high, if we neglect all energy losses in the system?

6.27   A water slide at an amusement park has the shape shown in Fig. 6.19. A girl of mass 70.0 kg on the slide has a speed of 2.00 m/s at height $y_2 = 12.0$ m. What is her speed at height $y_1 = 6.00$ m, if we neglect friction?

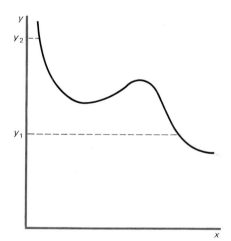

FIGURE 6.19   Diagram for Prob. 6.27.

6.28   A 150-g arrow is shot straight up into the air with a speed of 20.0 m/s. It reaches a maximum height of 19.0 m. Find (a) its initial kinetic energy; (b) its potential energy at its topmost position; (c) the energy lost due to air resistance. (d) If the same frictional force retards the arrow as it drops back to earth, find the speed with which the arrow returns to earth.

6.29   What is the average force needed to stop a bullet of mass 20.0 g and speed 500 m/s as it penetrates a wooden block to a distance of 10.0 cm?

6.30   (a)   A vertical spring is attached to a support and a 100-g mass hung from the bottom of the spring. The length of the spring increases by 25.0 cm. What is the force constant of the spring?

(b)   If the spring is placed on a frictionless table and attached to a wall and the mass is then displaced through 50.0 cm from its equilibrium position, what is the stored potential energy in the spring?

6.31 A spring of force constant 20.0 N/m is attached to a wall and rests on a frictionless table. A mass of 5.00 kg is attached to the spring and displaced 25.0 cm from its equilibrium position and then released. Find (*a*) the maximum PE in the system; (*b*) the maximum KE in the system; (*c*) the total energy of the system; (*d*) the maximum speed of the mass.

6.32 A 4.00-kg block slides on a horizontal frictionless surface with a speed of 1.40 m/s. It is brought to rest when it hits a bumper that compresses a spring. How much is the spring compressed if its force constant is 2.00×10^3 N/m?

6.33 A spring of force constant 24.5 N/m rests on a frictionless table and is attached at one end to a wall. If a 2.00-kg mass is attached to the spring and it is displaced 40.0 cm from its equilibrium position and then released, find (*a*) the maximum KE of the system; (*b*) the total energy of the system; (*c*) the maximum speed of the mass; (*d*) the maximum acceleration of the mass; (*e*) the KE; and (*f*) the PE when the displacement is 10.0 cm.

6.34 A child on a bicycle (combined mass 75.0 kg) wants to climb a 10° hill that is 200 m high.

(*a*) Calculate how much work must be done by the child in climbing the hill at constant speed.

(*b*) If each complete revolution of the pedals moves the bike 5.00 m along the road, and if the pedals move in a circle of radius 18.0 cm, what average force must the child apply to the pedals? (Neglect friction and air resistance.)

6.35 An 80-kg trapeze artist swings on a rope 4.00 m long that is initially horizontal. He starts from rest 6.00 m above the ground, swings along an arc of a circle, and then lets go when the rope is exactly vertical. Find his velocity (*a*) as he lets go and (*b*) as he lands.

*6.36 Use energy considerations to solve the following problem.

(*a*) From what height *h* above the bottom of the loop in Fig. 6.20 must a car start in order to just make it around the loop without falling off?

(*b*) If the car starts from this height *h*, what is the speed of the car at points *A*, *B*, and *C*?

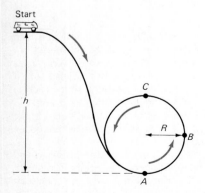

FIGURE 6.20 Diagram for Prob. 6.36.

6.37 A pendulum ball of weight *w* hangs at the end of a string of length *L*. A variable horizontal force F_x that starts at zero and gradually increases pulls the ball very slowly outward until finally the string makes an angle θ with the vertical. Find the work done by the force F_x in terms of *L*, θ, and *w*.

6.38 A 300-g wooden block is attached to a horizontal spring. The block rests on a table for which the coefficient of friction between table and block is 0.38. If the spring is first compressed 20 cm by a force of 15 N and then released, what will be its maximum displacement to the other side of its original position?

*6.39 A block of mass $m = 1.00$ kg, initially at rest, is dropped from a height *h* of 0.500 m onto a vertical spring whose force constant is 2.00×10^3 N/m (see Fig. 6.21). Find the maximum distance that the spring is compressed.

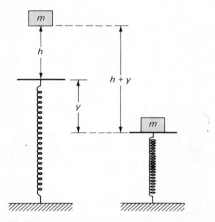

FIGURE 6.21 Diagram for Prob. 6.39.

6.40 From what height would a 1500-kg automobile have to fall to acquire the kinetic energy it has when moving in a straight line at 25.0 m/s?

*6.41 A winch is used to lift a 75.0-kg climber up a mountain from a ledge to which he has fallen. He is lifted 20.0 m vertically with an acceleration that remains constant at $g/20$.

(*a*) How much work does the winch do?

(*b*) What is the kinetic energy of the climber just before he reaches the level of the winch?

*6.42 A 2.0-kg block collides with a horizontal spring of force constant 5.0 N/m. The spring compresses 20 cm. If the coefficient of sliding friction between the block and the surface on which it rests is 0.30, what was the speed of the block at the instant it collided with the spring?

6.43 The 20.0-g bob of a simple pendulum of length 1.00 m is released from a height that is 10.0 cm higher than its equilibrium position, as in Fig. 6.22.

(*a*) What is its speed at the bottom of its swing as it passes through its equilibrium position?

(*b*) What is its kinetic energy at the bottom of its swing?

(*c*) What are its KE and PE when it is a vertical distance of 5.00 cm above its equilibrium position?

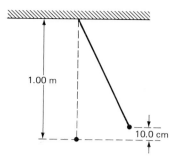

FIGURE 6.22 Diagram for Prob. 6.43.

6.44 A rifle fires a 5.0-g bullet with a muzzle velocity of 500 m/s into a block of wood. The wood brings the bullet to rest after it has traveled 10.0 cm in the wood. Determine the average force of friction between the bullet and the wood.

∗6.45 A skier starts at the top of a 32° slope that is 100 m long. Her initial speed is 2.0 m/s, while at the bottom of the slope her speed is 20 m/s. If we neglect air resistance, what is the coefficient of friction between her skies and the snow on the slope?

6.46 A 100-kg projectile is fired at an angle of 35° with respect to the horizontal. The initial speed of the projectile is 20.0 m/s, air resistance is negligible, and the end point is at the same elevation as the firing point.

(a) What is its kinetic energy at the beginning, the midpoint (the highest elevation of the projectile), and the end of its flight?

(b) What is its potential energy at the same three points on its path?

(c) What is its total energy at the same three points on its path?

6.47 Answer parts (a), (b), and (c) of Prob. 6.46 for an angle of 55° and an initial speed of 20.0 m/s. (d) Is the total energy the same for the two different trajectories at angles of 35° and 55°? Why?

6.48 The automobile jack in Fig. 6.23 has a lever arm of 0.40 m and a pitch of 4.0 mm; i.e., the top of the jack rises 4.0 mm for each full turn of the lever arm. If the efficiency of the jack is 35 percent, what force is required to lift a 1000-kg car with the jack?

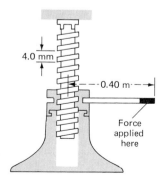

FIGURE 6.23 Diagram for Prob. 6.48.

6.49 By using the wheel and axle shown in Fig. 6.24, a 500-N load can be lifted by the application of a 60-N force to the rim of the wheel. The radii are R = 90 cm and r = 8.0 cm. Determine (a) the IMA, (b) the AMA, and (c) the efficiency of this wheel and axle.

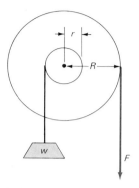

FIGURE 6.24 Diagram for Prob. 6.49.

6.50 Determine the force F needed to lift a weight of 400 N, using each of the three pulley systems shown in Fig. 6.25.

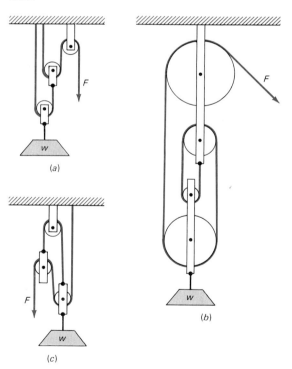

FIGURE 6.25 Diagram for Prob. 6.50.

6.51 A 1000-kg Volkswagen makes the climb from the foot of Pikes Peak, Colorado, to the top, 4.30×10^3 m above the base. If we neglect friction and air resistance, and the climb is made in 0.50 h, what is the average power generated?

*6.52 A 4.0-kg rock falls under the force of gravity through a distance of 30 m. Plot as a function of time (a) the work done by the force of gravity on the rock; (b) the power, which is the rate at which the force of gravity does work on the rock.

6.53 Montmorency Falls east of Quebec City in Canada is 90 m high and has a water flow rate of 35 m^3/s. If the density of water is 1.0×10^3 kg/m^3, what is the maximum electric power that can be generated by these falls at 100 percent efficiency?

6.54 An advertisement claims that a 1500-kg car can accelerate from rest to 25.0 m/s in 8.00 s. What average power must the motor produce to cause this acceleration, if we ignore friction and air resistance?

6.55 A 90.0-kg jogger is moving along at 2.00 m/s and accelerates to a speed of 3.50 m/s over a 4.00-s interval.

(a) How much energy must the jogger's body provide to produce this acceleration?

(b) What is the power provided during the 4.0 s?

6.56 A 2000-kg car accelerates uniformly from 15.0 to 24.0 m/s in 5.00 s. Find (a) the work done by the engine; (b) the average power delivered by the engine.

6.57 If a horse could really work at the exact rate of 1 hp, how much work could it do in one 8-h day? Express your answer in both joules and kilowatthours.

6.58 A rattlesnake can strike with an acceleration equivalent to an increase in speed from 0 to 27 m/s in 0.50 s. If the rattlesnake's head has a mass of 0.15 kg, what average power does the rattlesnake generate during its strike?

6.59 A car dealer is trying to find the actual power output of a car. He accelerates the 1000-kg car on level ground and finds that it will go uniformly from rest to 20 m/s in 10 s. He then finds that the car will coast to a stop from 20 m/s in 700 m, once the motor is turned off. If we assume that the frictional and air resistance forces are the same for both starting and stopping, what is the average power output of the car?

*6.60 The initial velocity a projectile shot vertically upward from the earth must have to escape from the gravitational field of the earth into space is called its *escape velocity*. If we neglect air resistance, it can be shown (we will see the derivation for the similar electrical case in Sec. 17.4) that the gravitational potential energy of a mass m at a distance r from the center of the earth is $PE_G = -Gmm_e/r$, where the negative sign means that the force between m and m_e is attractive.

(a) Use these facts to find the required escape velocity for a projectile fired directly upward from the earth.

(b) Does the escape velocity depend on the mass of the projectile?

Additional Readings

Brown, Sanford C.: *Count Rumford: Physicist Extraordinary,* Doubleday Anchor, Garden City, N.Y., 1962. A lively account of a most unusual physicist.

——: *Benjamin Thompson, Count Rumford,* MIT Press, Cambridge, Mass., 1981. A longer and more scholarly biography than the previous reference.

Chalmers, Bruce: *Energy,* Academic Press, New York, 1963. A very practical approach that stresses applications.

Dyson, Freeman J.: "Energy in the Universe," *Scientific American,* vol. 224, no. 3, September 1971, pp. 50–59. This is one of the best of a group of outstanding articles in an issue of *Scientific American* devoted entirely to energy. The quotation at the beginning of this chapter is taken from page 51 of this article (with permission).

Helmholtz, Hermann von: "On the Conservation of Force" (introduction to a series of lectures delivered in Karlsruhe in the winter of 1862–1863), *Harvard Classics,* vol. 30, P.F. Collier and Son Corporation, New York, 1938, pp. 173–210. An excellent, pedagogical presentation of the principle of conservation of energy based on Helmholtz's famous 1847 mathematical paper on this topic.

Mulligan, Joseph F.: *Practical Physics: The Production and Conservation of Energy,* McGraw-Hill, New York, 1980. This text, at a lower level than the present one, contains many applications of the ideas of this chapter to our present energy problems.

Romer, Robert H.: *Energy: An Introduction to Physics,* Freeman, San Francisco, Calif., 1976. This book, at about the same level as the present text, contains both a great deal of excellent physics and many good applications of physics to the nation's energy problems.

CHAPTER 7

Linear Momentum and Collisions

Experiment and observation provide the only way of arriving at the knowledge of the causes of all that one sees in Nature.

Christiaan Huygens (1629–1695)

In the preceding chapter we introduced the first of the important conservation principles of physics: conservation of energy. In this chapter and the next we consider two additional conservation laws for physical systems: conservation of linear momentum and conservation of angular momentum. These fundamental principles, valid in all fields of physics, contribute to the unity and coherence that make physics so intellectually satisfying. The principle of conservation of linear momentum explains phenomena as disparate as the recoil of a gun, the interaction of two nuclear particles, and the flight of a rocket.

7.1 Linear Momentum

We begin by defining the *linear momentum* of a small object that we call a *particle*.

Linear momentum: The product of the mass of a particle and its velocity.

$$\mathbf{p} = m\mathbf{v} \tag{7.1}$$

Since the velocity $\mathbf{v}$ is a vector, so too is the momentum $\mathbf{p}$, and its direction is the same as that of the velocity. The SI unit of momentum is kilogram-meters per second (kg·m/s), or newton-seconds (N·s). Since a reference frame is required for measuring a velocity, a reference frame is also required for momentum. For example, in Fig. 7.1 the reference frame for both the velocity and the momentum of the football player is the football field.

For a large object, or for a system of particles considered as a single entity, the linear momentum is the total mass of the object or system times the velocity of its center of mass, where the center of mass is defined as in Sec. 5.6.

The word *momentum* has become a misused buzzword in media descriptions of sporting events, but when the word is used correctly, it can contribute significantly to our understanding of some sports. For example, in football a heavy, fast running back is hard to stop because the player has a great deal of momentum, since both m and $\mathbf{v}$ are very large. It therefore requires a very large force to stop the runner in a short time. Similarly, a truck going 30 m/s can do much more serious damage than a motorcycle moving at 15 m/s because the truck's momentum is so much greater.

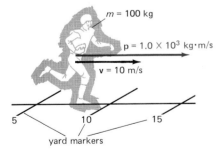

FIGURE 7.1 The linear momentum of a football player is the product of the player's mass and velocity, $\mathbf{p} = m\mathbf{v}$. If $m = 100$ kg and $v = 10$ m/s, the magnitude of the football player's momentum is 1.0×10^3 kg·m/s, and its direction is the same as that of the velocity v.

Newton's Second Law and Momentum

Newton originally expressed his second law in terms of momentum rather than acceleration, although he called $m\mathbf{v}$ the "quantity of motion" rather than the momentum. In modern terminology we may restate *Newton's second law of motion* as follows:

> The time rate of change of the momentum of an object is equal to the net external force applied to the object.

In equation form, this is

$$\mathbf{F}_{net} = \frac{\Delta \mathbf{p}}{\Delta t} \tag{7.2}$$

where $\mathbf{F}_{net}$ is the net external force acting and $\Delta \mathbf{p}$ is the change in momentum in time Δt. We now show that this equation contains Newton's second law in the more usual form discussed in Chap. 3.

Consider an object of fixed mass m undergoing constant acceleration $\mathbf{a}$ from an initial velocity $\mathbf{v}_0$ to a final velocity $\mathbf{v}$ in a time interval Δt:

$$\frac{\Delta \mathbf{p}}{\Delta t} = \frac{m\mathbf{v} - m\mathbf{v}_0}{\Delta t} = \frac{m(\mathbf{v} - \mathbf{v}_0)}{\Delta t} = m\frac{\Delta \mathbf{v}}{\Delta t} = m\mathbf{a}$$

and so $\qquad \mathbf{F}_{net} = \dfrac{\Delta \mathbf{p}}{\Delta t} = m\mathbf{a}$

which is Newton's second law in the form we saw earlier. This equation remains valid in the limit as $\Delta t \to 0$ and therefore is also valid for instantaneous momenta and instantaneous accelerations.

The expression of Newton's second law in the form of Eq. (7.2) is actually more general than $\mathbf{F}_{net} = m\mathbf{a}$. The equation $\mathbf{F}_{net} = \Delta \mathbf{p}/\Delta t$ is always true; the equation $\mathbf{F}_{net} = m\mathbf{a}$ is valid only if m is constant in time.

The change in the momentum $\Delta \mathbf{p}$ is equal to $\Delta(m\mathbf{v})$, and so a change in momentum can be caused by a change in velocity, a change in mass, or changes in both velocity and mass. The possibility of a change in mass is important in rockets, which eject mass backward in order to accelerate forward.

Equation (7.2) is valid both for a single particle or object and for a *system* made up of many particles or objects. The change in the linear momentum of the system is in every case equal to the net external force applied to the system. Internal forces exerted by one particle on another do not produce any change in the linear momentum of the system and are therefore not included in $\mathbf{F}_{net}$.

That linear momentum is a more significant concept than velocity will become clear when we discuss one of the great integrating principles of physics—the conservation of linear momentum. There is no similar principle for velocity.

Example 7.1

A truck weighing 4.00×10^4 N is traveling at 30.0 m/s, and a Volkswagen weighing 6.0×10^3 N is moving along at 15.0 m/s. Compare the net force required to stop each vehicle, assuming that each is brought to rest with constant deceleration in the same time.

SOLUTION

The changes in the magnitude of the momenta in the two cases are

$$\Delta p_1 = 0 - m_1 v_1 = -\frac{4.00 \times 10^4}{g}(30.0 \text{ m/s})$$

$$= -1.22 \times 10^5 \text{ kg·m/s}$$

$$\Delta p_2 = 0 - m_2 v_2 = -\frac{6.00 \times 10^3}{g}(15.0 \text{ m/s})$$

$$= -9.18 \times 10^3 \text{ kg·m/s}$$

Then, since $\mathbf{F}_{net} = \Delta \mathbf{p}/\Delta t$, $\Delta \mathbf{p} = \mathbf{F}_{net}\,\Delta t$, and we have

$$\frac{F_1\,\Delta t}{F_2\,\Delta t} = \frac{\Delta p_1}{\Delta p_2} = \frac{-1.22 \times 10^5}{-9.18 \times 10^3} = \boxed{13.3}$$

or $F_1/F_2 = 13.3$, and a force 13.3 times greater is required to stop the truck in the same time.

7.2 Impulse; the Impulse-Momentum Theorem

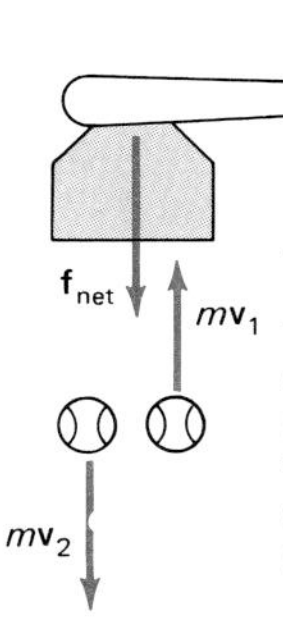

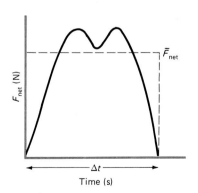

FIGURE 7.2 Application of the impulse-momentum theorem to a bat striking a baseball. The impulse provided by the bat is equal to the change in the ball's momentum.

If we multiply both sides of Eq. (7.2) by Δt, we obtain the *impulse-momentum theorem*:

$$\boxed{\mathbf{F}_{net}\,\Delta t = \Delta \mathbf{p}} \qquad (7.3)$$

The quantity on the left, the product of the applied force and the time interval Δt during which the force acts, is called the *impulse*. Notice that impulse, like momentum, is a *vector* quantity, and has the same units as momentum, kilogram-meters per second (kg·m/s). Equation (7.3) states that *the impulse of the net force acting on an object is equal to the change in linear momentum of the object*. The same change in momentum can be achieved by a large force acting for a short time or by a small force acting for a longer time. Thus a linebacker can stop the forward progress of a tailback by a fierce tackle that lasts for a brief time interval or by applying a lesser force for a longer period of time. However, the fierce, brisk tackle is to be preferred, since it prevents the tailback from gaining much additional yardage during the time he is being stopped.

Equation (7.3) is easiest to apply when the net force is constant. However, if the net force is not constant, then we must use the average value of the net force. Since the average value of the net force may not be known, this may present difficulties. If the net force varies over a long time interval Δt, these difficulties might require the use of calculus for their resolution. Luckily, we can avoid these difficulties (and calculus, too!) in situations in which the time interval Δt is short, as is the case for a baseball bat striking a ball (Fig. 7.2). The magnitude of $\mathbf{F}_{net}$ will look something like Fig. 7.3, and the average value of the force $\overline{\mathbf{F}}_{net}$ can be chosen so that the impulse $\overline{\mathbf{F}}_{net}\,\Delta t$ is a good approximation to the true impulse to be used in the impulse-momentum equation. Figure 7.4 is a photograph taken while a baseball bat is applying an impulse to a baseball. The ball is also applying an impulse to the bat.

The dimensions of linear momentum are $[M][L]/[T]$, and those of impulse are

$$[F][T] = [M]\frac{[L]}{[T^2]}[T] = \frac{[M][L]}{[T]}$$

FIGURE 7.3 Approximate variation of the force over time for a bat striking a baseball. The change in momentum is approximately equal to $\overline{F}_{net}\,\Delta t$, where $\overline{F}_{net}$ is the average value of F_{net} over the time interval Δt, as shown in the figure.

FIGURE 7.4 A high-speed flash photograph of a baseball bat hitting a baseball. Notice how the ball is deformed, indicating the enormous size of the impulsive force at the instant the picture was taken (*Photo by Dr. Harold E. Edgerton, courtesy of Palm Press, Inc.*)

Hence impulse and momentum have the same dimensions, as is required by Eq. (7.3). They also have the same units, kilogram-meters per second, or newton-seconds, since

$$1 \text{ N} \cdot \text{s} = 1 \frac{\text{kg} \cdot \text{m} \cdot \text{s}}{\text{s}^2} = 1 \text{ kg} \cdot \text{m/s}$$

Example 7.2

A baseball pitcher throws a fastball that is moving at 40 m/s when Eddie Murray's bat collides with it. The baseball, with a mass of 0.50 kg, leaves the bat at a speed of 50 m/s headed directly for the pitcher (who ducks). What is the impulse imparted to the ball by the bat?

SOLUTION

Momentum is a vector quantity, and so the momentum of the pitched ball is in the opposite direction to that of the batted ball. We take the direction of the *impulse* as positive, and thus the velocity of the ball leaving the bat is positive, whereas the velocity of the ball thrown by the pitcher is negative. We obtain from the impulse-momentum theorem:

$$\text{Impulse} = \mathbf{F}_{\text{net}} \, \Delta t = \Delta \mathbf{p} = \mathbf{p}_2 - \mathbf{p}_1$$
$$= (0.50 \text{ kg})(50 \text{ m/s}) - (0.50 \text{ g})(-40 \text{ m/s})$$
$$= 45 \text{ kg} \cdot \text{m/s} = \boxed{45 \text{ N} \cdot \text{s}}$$

Note that in cases like this we can find the value of the impulse from the change in momentum, even though we know neither the force $\mathbf{F}_{\text{net}}$ nor the interval Δt during which it acts.

7.3   Conservation of Linear Momentum

Since $\mathbf{F}_{\text{net}} = \Delta \mathbf{p}/\Delta t$, it is clear that if $\mathbf{F}_{\text{net}} = 0$, that is, *if no net external force acts*, then $\Delta \mathbf{p} = 0$ and

$$\boxed{\mathbf{p} = \sum_i m_i \mathbf{v}_i = \text{constant}} \qquad (7.4)$$

This is true for a single particle or for a system of particles. We therefore express the *principle of conservation of linear momentum* in words as follows:

> If no net external force acts on a system, the total linear momentum of the system remains constant.

We now consider some of the implications of this very important principle.

Center-of-Mass Motion

We may conclude from our discussion in Sec. 5.6 that, for an object or group of particles in space, the coordinates of the center of mass (CM) are

$$\bar{x} = \frac{\sum_{i=1}^{N} m_i x_i}{M} \qquad \bar{y} = \frac{\sum_{i=1}^{N} m_i y_i}{M} \qquad \bar{z} = \frac{\sum_{i=1}^{N} m_i z_i}{M}$$

where x_i, y_i, z_i, $\bar{x}$, $\bar{y}$, and $\bar{z}$ are all measured with respect to the origin of our coordinate system and M is the total mass of the system. We now combine these three scalar equations into one vector equation,

$$\mathbf{r}_{CM} = \frac{\sum_{i=1}^{N} m_i \mathbf{r}_i}{M}$$

where $\mathbf{r}_{CM}$ and $\mathbf{r}_i$ are position vectors of the CM and of the ith particle with respect to the origin of the coordinate system.

If the particles in the system change position in time, then the change in the position vectors in time Δt will be

$$M \,\Delta \mathbf{r}_{CM} = \sum_{i=1}^{N} m_i \,\Delta \mathbf{r}_i$$

and on dividing by the time Δt and replacing $\Delta \mathbf{r}/\Delta t$ by $\mathbf{v}$, we have

$$M\mathbf{v}_{CM} = \sum_{i=1}^{N} m_i \mathbf{v}_i$$

(7.5)

or $$\mathbf{p}_{CM} = M\mathbf{v}_{CM} = \sum_{i=1}^{N} m_i \mathbf{v}_i = \sum_{i=1}^{N} \mathbf{p}_i$$

so that *the linear momentum of the system is the product of the total mass of the system and the velocity of the CM*. This is equal to the vector sum of the individual momenta of all the parts of the system and therefore is a true measure of the system's linear momentum.

From our conservation-of-linear-momentum principle we can therefore conclude that the center of mass of the system will continue to move with its initial velocity as long as no net external force acts on it. If an external force does act, the CM will move in accordance with Newton's second law. Internal forces may break the object into pieces, as when a shell disintegrates in the air (Fig. 7.5), but this has no effect on the CM motion of the system or on the total momentum of the shell fragments.

FIGURE 7.5 An exploding shell; an example of the conservation of linear momentum. (*a*) The linear momentum of the shell before it explodes, **p** = M**v**. (*b*) The sum of the linear momenta of all the fragments of the shell after it explodes, $\sum_i m_i \mathbf{v}_i$, is equal to the original linear momentum **p** = M**v**.

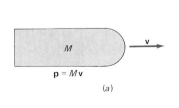

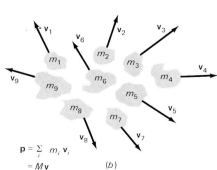

Effect of Internal Forces

Internal forces do not change the CM motion, but they can radically change the motions of the individual particles or pieces of the system.

For simplicity, let us consider two colliding pucks of masses m_1 and m_2 (Fig. 7.6), where we assume that no external forces act on the two particles and so linear momentum is conserved. During the collision m_1 exerts a force $\mathbf{F}_{12}$ on m_2, and m_2 exerts an equal and opposite force $\mathbf{F}_{21}$ back on m_1. We therefore apply the impulse-momentum theorem and obtain

$$\mathbf{F}_{12}\,\Delta t = \Delta\mathbf{p}_2 \qquad \mathbf{F}_{21}\,\Delta t = \Delta\mathbf{p}_1$$

But, from Newton's third law, $\mathbf{F}_{12} = -\mathbf{F}_{21}$, and so

$$\mathbf{F}_{12}\,\Delta t + \mathbf{F}_{21}\,\Delta t = 0$$

or $\qquad \Delta\mathbf{p} = \Delta\mathbf{p}_2 + \Delta\mathbf{p}_1 = 0$

Therefore there is no change in the total momentum of the system (even though the momentum of the individual parts may change radically), and so

$$\mathbf{p}_1 + \mathbf{p}_2 = \text{constant}$$

This is another statement of the conservation-of-momentum principle for a system subject only to *internal* forces.

If, for example, m_1 and m_2 are constant, $\mathbf{v}_1$ and $\mathbf{v}_2$ are the velocities before the collision, and $\mathbf{v}_1'$ and $\mathbf{v}_2'$ are the velocities after the collision, we have

$$m_1\mathbf{v}_1 + m_2\mathbf{v}_2 = m_1\mathbf{v}_1' + m_2\mathbf{v}_2' \tag{7.6}$$

that is, the sum of the particles' momenta is the same before and after the collision.

If we know m_1, m_2, $\mathbf{v}_1$, and $\mathbf{v}_2$ in Eq. (7.6), we can obtain a relationship between $\mathbf{v}_1'$ and $\mathbf{v}_2'$. We cannot, however, obtain definite values for $\mathbf{v}_1'$ and $\mathbf{v}_2'$ without further information about the collision, in particular, about what happens to the *kinetic energy* in the collision. In general, kinetic energy is not conserved in collisions, since some of this energy may be converted into other forms of energy like heat. This is true even when no net external force acts on the system. No matter what happens to the kinetic energy in an isolated collision, however, the linear momentum is *always* conserved.

It is this universal aspect of the conservation-of-linear-momentum principle that makes it so important in physics and so useful in practical situations as different as a collision between two tiny nuclear particles or the crash of two giant diesel locomotives.

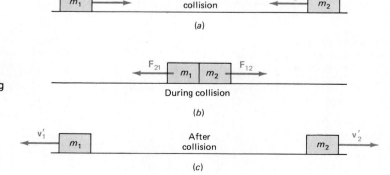

FIGURE 7.6 Head-on collision of two metal pucks: (*a*) before collision; (*b*) during collision; (*c*) after collision. Because action equals reaction during the collision, linear momentum is conserved.

**Both External and
Internal Forces**

When both external and internal forces act on a system of particles, it is often very helpful to consider the motion of the system in two steps. First we calculate the effect of the external forces on the CM of the system and use Newton's laws to find the motion of the CM. Then we consider the internal forces to find the motion of the particles *with respect to the CM*. The total motion of the system is then described by combining the results of these two calculations.

We can often use the conservation-of-linear-momentum principle to solve problems when we do not know the instantaneous forces acting; applying Newton's second law is often difficult in such problems. (Example 7.3 illustrates this fact.) Also, in dealing with many problems in atomic and nuclear physics, it is impossible to specify instantaneous positions, velocities, and accelerations for the particles involved, but it is still possible to apply the principle of conservation of linear momentum.

One of the first physicists to understand the mechanics of collisions was Christiaan Huygens (1629–1695). Huygens (see the accompanying biography on the next page), together with Galileo and Newton, turned mechanics into the exact science it now is.

Example 7.3

An explosive shell is fired from a mortar with an initial speed $v_0 = 200$ m/s at an angle of 45.0° with the ground, as in Fig. 7.7. At the top of its trajectory (P), the shell breaks into two equal fragments, the first of which falls straight down with zero initial vertical velocity (that is, v_{1x} and v_{1y} at the top are both zero). If we neglect the effects of air resistance, **(a)** what can be said about the motion of the two fragments after the shell breaks up? **(b)** at which distance from the origin will the center of mass of the two fragments reach the ground? **(c)** at what distance from the origin will the second fragment land?

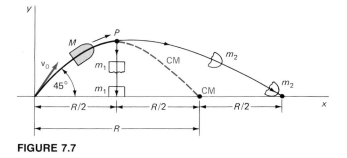

FIGURE 7.7

SOLUTION

(a) If we neglect air resistance, the only *external* force acting after the shell leaves the mortar is the force of gravity. The shell therefore follows a parabolic trajectory, as discussed in Sec. 4.4. After the shell breaks into two, the center of mass of the fragments continues along the same parabolic path shown in the diagram; i.e., if at any instant of time after the separation the center of mass of the two fragments is calculated, it will always be found along the dashed line in the figure. This is the trajectory that would be followed by the shell, if no breakup occurred. The force causing the separation of the two fragments is an *internal force* and therefore has no effect on the motion of the center of mass of the system.

(b) The center of mass of the two fragments reaches the ground at the same position the original shell would have reached the ground. For a firing angle of 45° this is given by Eq. (4.20):

$$\text{Range } R = \frac{v_0^2}{g} \sin 2\theta_0 = \frac{v_0^2}{g} = \frac{(200 \text{ m/s})^2}{9.80 \text{ m/s}^2}$$

$$= \boxed{4.08 \times 10^3 \text{ m}}$$

(c) Let us apply the principle of conservation of linear mo-

mentum to the breakup of the shell at point P. Both the x and y components of the linear momentum must be conserved. Taking the y momentum first, since it is the simpler, we have

$$\sum_i P_{yi} = \sum_i P'_{yi} \quad \text{or} \quad Mv_y = m_1 v_{1y} + m_2 v_{2y} \quad (1)$$

But at the top of the trajectory v_y is zero, since the motion is in the x direction. Also v_{1y} is zero, from the statement of the problem. From Eq. (1) we therefore conclude that $v_{2y} = 0$, and the second fragment is given no momentum in the y direction when the shell breaks up. It therefore continues to move in the x direction but is bent into a parabolic path by the force of gravity.

From the conservation-of-momentum principle applied to the horizontal motion at the top of the trajectory, we have

$$\sum_i P_{xi} = \sum_i P'_{xi} \quad \text{or} \quad Mv_x = m_1 v_{1x} + m_2 v_{2x} \quad (2)$$

But v_{1x} is zero and $m_1 = m_2 = M/2$ from the statement of the problem. Also v_x is everywhere equal to the initial x component of the velocity, that is, v_{0x}, so that

$$Mv_{0x} = 0 + \frac{M}{2}v_{2x} \quad \text{or} \quad v_{2x} = 2v_{0x}$$

Christiaan Huygens (1629–1695)

FIGURE 7.8 Christiaan Huygens: A portrait by G. Netscher. (*Courtesy of Haags Gemeentemuseum.*)

One of the greatest of scientific geniuses, Christiaan Huygens is one of the least known of the world's important physicists. He made significant contributions to astronomy by discovering a satellite circling Saturn; to mechanics by his understanding of centripetal force, of collision theory, and of work and energy; to optics, where he is justly regarded as the founder of the wave theory of light; and to experimental physics, where he developed the first accurate pendulum clock for timing physical events.

Huygens was born on April 14, 1629, in The Hague, the son of the fa-mous Constantine Huygens, who was a poet, philosopher, classical scholar, and diplomat. Christiaan was taught by a private tutor until he was 16, his education including Latin, music, and drawing. He grew to love mathematics more than all other subjects, however, and this interest dominated his later life. In 1645, Huygens entered the University of Leyden to study mathematics and law. After further studies at the University of Breda, he completed his education by traveling widely throughout Europe. Meanwhile he became involved in the scientific ferment of the day that was stimulated by the ideas of René Descartes (1596–1650), a friend of the Huygens family.

Huygens lived at home from 1650 to 1666, except for journeys to Paris and London, and these 16 years were the most productive of his scientific career. His first published work came out in 1651; it was followed by a constant stream of important books on a great variety of scientific subjects.

In 1665, Colbert, minister to Louis XIV, King of France, decided to set up the Académie Royale des Sciences in Paris and to include Huygens as one of its charter members. The king provided financial support to enable Huygens to live in a well-appointed apartment in the Bibliothèque Royale.

He remained in Paris from 1666 until 1681, except for occasional visits to his native Holland for reasons of health. In 1672, when the French armies were attacking the Dutch Republic, he stayed on in Paris. He even dedicated a celebrated book on his invention of the pendulum clock to Louis XIV, who had vowed to destroy the Dutch Republic!

In 1681 Huygens was forced to leave Paris because of ill health, and never returned. His convalescence was slow, and in 1683 Colbert, his patron, died, and opposition mounted to Huygens' return as head of the Académie Royale.

The years after 1683 were hard ones for Huygens. He lived in The Hague, cut off from the scientific centers of the world. He remained single, living alone at his family's estate most of the time, and the income from his family's property was barely enough to support him. In 1689 he visited England, met Newton and Boyle, and returned home even more unhappy about his intellectual isolation in the Netherlands.

He died in The Hague in 1695, survived by no disciples and relatively forgotten by the swiftly moving world of science, which owed so much to his ideas.

Because the second fragment falls back to the ground from the same height as the original shell would have, it takes the same time to fall. Since it is traveling at twice the horizontal speed of the original shell, it travels a horizontal distance from the top of the trajectory not of $R/2$ (see Fig. 7.7) but $2(R/2) = R$. The range of the second fragment is therefore

$$\frac{R}{2} + R = \frac{3}{2}R = \frac{3}{2}(4.08 \times 10^3 \text{ m}) = \boxed{6.12 \times 10^3 \text{ m}}$$

Since the first fragment has a range of only $R/2$ (it falls straight down), the center of mass of the two fragments together has a range $(\frac{1}{2}R + \frac{3}{2}R)/2 = R$, the range of the original shell. This agrees with our conclusions in parts (*a*) and (*b*). Note that nothing actually strikes the ground a distance R from the origin, where the original shell would have landed. But the two fragments that do land have their CM at this place.

7.4 Collisions in One Dimension: Elastic Collisions

We define an elastic collision as follows:

Elastic collision: A collision in which both the KE and the linear momentum of the system are conserved.

Good examples of elastic collisions are the collision of two hard steel ball bearings or of two molecules in a gas. In this section we consider only elastic, head-on collisions between two particles of masses m_1 and m_2. In any collision, linear momentum is conserved. If the collision is elastic, KE is also conserved. Hence we have

$$m_1 \mathbf{v}_1 + m_2 \mathbf{v}_2 = m_1 \mathbf{v}_1' + m_2 \mathbf{v}_2' \qquad \text{(conservation of linear momentum)}$$

$$\tfrac{1}{2} m_1 v_1^2 + \tfrac{1}{2} m_2 v_2^2 = \tfrac{1}{2} m_1 v_1'^2 + \tfrac{1}{2} m_2 v_2'^2 \qquad \text{(conservation of kinetic energy)}$$

If, in addition, the collision is head-on, then the motion is confined to a straight line, and we can drop the vector notation and distinguish the direction of the velocities by $+$ and $-$ signs.

If the masses and initial velocities are known, these two equations can be solved for the two unknowns v_1' and v_2'. Rearranging the two equations, we obtain

$$m_1(v_1 - v_1') = m_2(v_2' - v_2) \tag{7.7}$$

$$m_1(v_1^2 - v_1'^2) = m_2(v_2'^2 - v_2^2) \tag{7.8}$$

Dividing the second equation by the first, we obtain

$$v_1 + v_1' = v_2' + v_2$$

and so

$$v_2' - v_1' = -(v_2 - v_1) \tag{7.9}$$

Here $v_2' - v_1'$ is the velocity of m_2 relative to m_1 after the collision, while $v_2 - v_1$ is the velocity of m_2 relative to m_1 before the collision. Therefore, *the relative velocity of the two particles in an elastic collision is reversed in direction but has the same magnitude as it had before the collision.* This means that the two balls approaching each other before the collision are moving away from each other after the collision, but there is no change in the magnitude of their relative velocity.

Suppose m_2 is initially at rest, so that $v_2 = 0$. Then, from Eq. (7.7),

$$v_2' = \frac{m_1}{m_2} (v_1 - v_1')$$

and, substituting for v_1' from Eq. (7.9),

$$v_2' = \frac{m_1}{m_2} [v_1 - (v_2' - v_1)] = \frac{m_1}{m_2} (2v_1 - v_2')$$

or

$$v_2' \left(1 + \frac{m_1}{m_2}\right) = 2 \left(\frac{m_1}{m_2}\right) v_1$$

and so

$$v_2' = \frac{2m_1}{m_1 + m_2} v_1 \tag{7.10}$$

Also, from Eq. (7.9), with $v_2 = 0$,

$$v_1' = v_2' - v_1 = \frac{2m_1}{m_1 + m_2} v_1 - v_1 = \frac{m_1 - m_2}{m_1 + m_2} v_1 \tag{7.11}$$

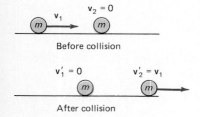

FIGURE 7.9   Head-on collision of a moving billiard ball with a ball at rest that has the same mass. The first ball stops and the second ball moves away with the velocity the first ball had originally.

These two equations allow us to find v_1' and v_2' if we know the initial velocity v_1 and the masses in an elastic head-on collision in which one mass is at rest.

For example, if a Ping-Pong ball strikes a steel ball at rest ($m_1 \ll m_2$), we find from Eqs. (7.10) and (7.11) that $v_1' = -v_1$ (the Ping-Pong ball reverses its direction) and $v_2' = 0$ (the steel ball remains at rest, as it was initially). On the other hand, if a steel ball strikes a Ping-Pong ball at rest ($m_2 \ll m_1$), we find that $v_1' = v_1$ (the steel ball has its velocity unchanged), whereas $v_2' = 2v_1$ (the Ping-Pong ball moves away with a speed twice the original speed of the steel ball).

An interesting example is the case in which the two masses are equal. Then, from Eqs. (7.10) and (7.11), we have $v_2' = v_1$ and $v_1' = 0$. This means that the mass m_1 stops dead and gives all its momentum and energy to the second mass, which moves away with the same velocity that m_1 had originally, as in Fig. 7.9. Anyone who plays pool or billiards will recognize this as an example of what happens on a billiard table, as long as the collision is head-on.

The complete transfer of momentum and energy in an elastic collision from a moving particle to a second particle initially at rest and with the same mass explains why water is used as the "moderator" in boiling-water nuclear reactors. The moderator is used to slow down neutrons and increase the chance that they will cause uranium nuclei to fission, i.e., break into two pieces and release useful energy. The most effective way to accomplish this is to have the neutrons collide with particles of mass equal to the neutron mass, as we have just seen. The nuclei of the hydrogen atoms in ordinary water serve this function very well, since they consist of a single proton of mass almost identical to that of the colliding neutrons. As a consequence, the elastic collisions between the neutrons and protons are very effective in slowing down the neutrons, as required for the efficient running of the reactor.

Example 7.4

Collisions involving nuclear particles are often good examples of elastic collisions. Compare the change in speed experienced by a neutron that emerges from the fission of a uranium-235 nucleus and collides head-on **(a)** with the nucleus of a hydrogen atom (a proton) at rest; **(b)** with a uranium-235 nucleus which has a mass 235 times that of the neutron and is also at rest.

SOLUTION

(a)   For the head-on neutron-proton collision, the proton and neutron masses are almost identical (see Table B.2). Using Eq. (7.11), we have for the velocity of the neutron after the collision,

$$v_1' = \frac{m_1 - m_2}{m_1 + m_2} v_1 = \frac{m_1 - m_1}{m_1 + m_1} v_1 = \boxed{0}$$

(b)   For the head-on collision with the ^{235}U nucleus,

$$v_1' = \frac{m_1 - 235m_1}{m_1 + 235m_1} v_1 = \frac{-234m_1}{236m_1} v_1 = \boxed{-0.99v_1}$$

In the head-on collision with the proton, therefore, the neutron loses all its energy and comes to rest, because the two masses are almost equal. On the other hand, the collision with the uranium nucleus has changed the direction of the neutron's velocity but reduced its speed hardly at all. Hence there is almost no change in its kinetic energy.

Since in practice few collisions are perfect head-on collisions, it usually takes many collisions of a neutron with hydrogen atoms before its speed is reduced almost to zero.

Example 7.5

Consider the head-on collision of two balls on a frictionless surface, one of mass 2.0 kg moving to the right along the x axis with speed 5.0 m/s, and the other of mass 1.0 kg moving to the left along the x axis with speed 10 m/s. If kinetic energy is conserved in the head-on collision, what are the final velocities of the two balls?

SOLUTION

As can be seen from Fig. 7.10, the total momentum of the balls before the collision is zero, since $\mathbf{p} = m_1\mathbf{v}_1 + m_2\mathbf{v}_2 = (2.0 \text{ kg})(5.0 \text{ m/s}) - (1.0 \text{ kg})(10 \text{ m/s}) = 0$, where we have used $+$ and $-$ signs to distinguish the two directions. Since linear momentum is always conserved in any collision, the momentum after the collision is also zero, and so

$$p' = (2.0 \text{ kg})v_1' + (1.0 \text{ kg})v_2' = 0 \quad \text{or} \quad v_2' = -2v_1'$$

Since, as indicated in the statement of the problem, the collision is elastic, we know from Eq. (7.9) that the relative velocity of the two particles is reversed in the collision; that is,

$$v_2' - v_1' = -(v_2 - v_1)$$
$$= -(-10 \text{ m/s} - 5.0 \text{ m/s}) = 15 \text{ m/s}$$

or, from above,

$$-2v_1' - v_1' = 15 \text{ m/s} \quad \text{and so} \quad v_1' = \boxed{-5.0 \text{ m/s}}$$

Then $\quad v_2' = -2v_1' = \boxed{+10 \text{ m/s}}$

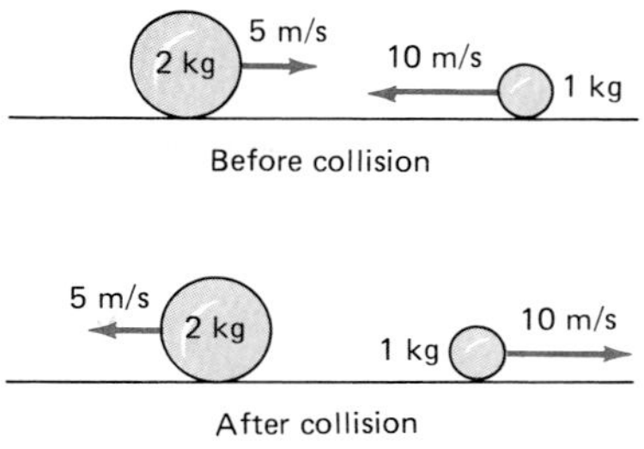

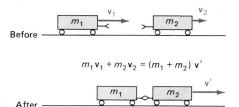

Before collision

After collision

FIGURE 7.10

As a result of this elastic collision the two balls merely reverse their original directions and leave the collision with the same speeds they had before the collision. This keeps the total linear momentum of the system at zero, its original value. If the initial linear momentum had not been zero, this simple reversal of velocities would not have resulted.

7.5   Collisions in One Dimension: Inelastic Collisions

At the opposite extreme from an elastic collision is one in which the colliding objects stick together and move as a unit with zero relative velocity after the collision. Such a collision, in which some of the original kinetic energy of the two objects is lost, is called *completely inelastic*. Two freight cars that collide and are then held together by a coupler serve as a good example of a completely inelastic collision in one dimension (Fig. 7.11). In such a case the principle of conservation of linear momentum, which is *always* valid, leads to the equation

$$m_1v_1 + m_2v_2 = (m_1 + m_2)v' \tag{7.12}$$

where v' is the final velocity of the coupled freight cars. This final velocity (v') can be obtained immediately if m_1, m_2, v_1, and v_2 are known.

The kinetic energy before the collision is

$$\text{KE}_i = \tfrac{1}{2}m_1v_1^2 + \tfrac{1}{2}m_2v_2^2$$

and after the collision it is

$$\text{KE}_f = \tfrac{1}{2}(m_1 + m_2)v'^2$$

Suppose we consider the case where the mass m_2 is originally at rest, so that $v_2 = 0$. Then the ratio of the kinetic energy after the collision to that before is

$$\frac{\text{KE}_f}{\text{KE}_i} = \frac{\tfrac{1}{2}(m_1 + m_2)v'^2}{\tfrac{1}{2}m_1v_1^2}$$

But, from Eq. (7.12), with $v_2 = 0$,

$$v' = \frac{m_1}{m_1 + m_2}v_1$$

and so

FIGURE 7.11   A completely inelastic collision: the collision of two freight cars that are coupled together after the collision. Linear momentum is, as always, conserved, but KE is not.

Before

$$m_1v_1 + m_2v_2 = (m_1 + m_2)v'$$

After

$$\frac{KE_f}{KE_i} = \frac{\frac{1}{2}(m_1 + m_2)}{\frac{1}{2}m_1v_1^2} \frac{m_1^2v_1^2}{(m_1 + m_2)^2} = \frac{m_1}{m_1 + m_2} \qquad (7.13)$$

The total kinetic energy therefore *decreases* in the collision, since m_1 is less than $m_1 + m_2$. If the two masses are equal, $KE_f/KE_i = \frac{1}{2}$. Then half the kinetic energy of the system is lost when the moving mass collides with an equal mass at rest, and they move away stuck together after the collision. In this case, of course, $v' = v_1/2$, as is clear from Eq. (7.12). This means that each mass after the collision has one-fourth the kinetic energy the moving mass had before the collision, and so the total kinetic energy after the collision is one-half what it was before, as we have just seen. The other half of the original kinetic energy is lost in the form of heat (and perhaps sound).

Example 7.6

Two 0.100-kg clay balls are attached to the end of two cords, each 1.00 m long, as in Fig. 7.12, and are then displaced from their equilibrium position in opposite directions. The first ball is raised a vertical distance of 0.400 m above its rest position, and the second a vertical distance of 0.300 m. The higher ball is allowed to drop first, followed by the lower ball, so that they collide head-on at their equilibrium positions. They then stick together and move as a unit. What is the magnitude and direction of their velocity immediately after the collision?

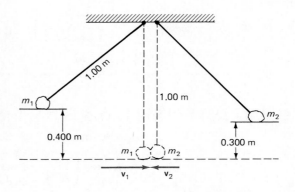

FIGURE 7.12

SOLUTION

The linear momentum of the system is conserved *during* the collision. However, we must first apply the conservation-of-mechanical-energy principle to find the velocity of each ball just before the collision. From $m_1gh = \frac{1}{2}m_1v_1^2$, we find that $v_1 = 2.80$ m/s; similarly, $v_2 = 2.42$ m/s.

Since linear momentum is conserved,

$$\mathbf{p}' = \mathbf{p}$$

and so $\quad (m_1 + m_2)\mathbf{v}' = m_1\mathbf{v}_1 + m_2\mathbf{v}_2$

The total momentum immediately before the collision is in the positive x direction and has a magnitude

$$p = m_1v_1 - m_2v_2 = (0.100 \text{ kg})(2.80 - 2.42) \text{ m/s}$$

$$= 0.038 \text{ kg·m/s}$$

After the collision, we have

$$p' = (m_1 + m_2)v' = 0.038 \text{ kg·m/s}$$

and so $\quad v' = \dfrac{0.038 \text{ kg·m/s}}{0.200 \text{ kg}} = \boxed{0.19 \text{ m/s}}$

in the positive x direction. Notice that there are only two significant figures in the answer since one significant figure is lost in calculating $m_1v_1 - m_2v_2$.

Example 7.7

A bullet of mass $m = 10.0$ g is fired into a soft wooden block of mass $M = 10.0$ kg mounted as in Fig. 7.13 so that it can swing freely to a height h above its equilibrium position. **(a)** Show how this "ballistic pendulum" can be used to find the speed v of the bullet as it leaves the gun. **(b)** How much energy is lost when the bullet collides with the block of wood?

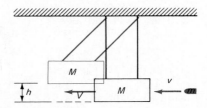

FIGURE 7.13

SOLUTION

(a) Since there is no net external force acting during the collision, linear momentum is conserved and we can write $mv = (m + M)V$, where v is the muzzle velocity of the bullet and V is the velocity of the block (and embedded bullet) as it starts its horizontal motion. Mechanical energy is *not* conserved during this completely inelastic collision. Just after the collision, however, when the block starts to swing, mechanical energy *is* conserved and KE is converted to gravitational PE at a height h. We have

$$\tfrac{1}{2}(m + M)V^2 = (m + M)gh$$

or $V = \sqrt{2gh}$

Then, from our first equation above,

$$v = \frac{m + M}{m} V = \frac{m + M}{m}\sqrt{2gh}$$

If we know m and M and can measure h, the velocity v of the bullet can be determined. The nice feature of this setup is

that a very large velocity, which is difficult to measure directly with any accuracy, can be obtained from a simple height measurement.

Today much more sophisticated techniques than this are used to measure bullet velocities, but the principle illustrated in this example is still useful.

(b) We know that this collision is completely inelastic since the bullet sticks in the block. The ratio of the kinetic energy after the collision to the initial KE is, from Eq. (7.13),

$$\frac{\text{KE}_f}{\text{KE}_i} = \frac{m}{m + M} = \frac{0.0100 \text{ kg}}{10.01 \text{ kg}} = \boxed{10^{-3}}$$

and the final kinetic energy is only 10^{-3}, or 0.1 percent of the initial kinetic energy. The other 99.9 percent is lost as heat, sound, and energy required to deform the block when the bullet bores into the wood. It is only this 0.1 percent of the initial KE that is conserved and converted into PE_{grav} after the collision.

7.6   Collisions in Two Dimensions

It is, of course, likely that two colliding particles will not hit head-on but rather at some glancing angle that will cause them to move off at angles with respect to the original direction of motion, as in Fig. 7.14. We then have to consider the collision in two (or perhaps three) dimensions. For motion in a plane, we choose some suitable coordinate system and break Eq. (7.6) into x and y components:

$$m_1 v_{1x} + m_2 v_{2x} = m_1 v'_{1x} + m_2 v'_{2x}$$
$$m_1 v_{1y} + m_2 v_{2y} = m_1 v'_{1y} + m_2 v'_{2y}$$

(7.14)

where the unprimed and primed symbols refer to velocities before and after the collision, respectively.

Since the choice of x and y coordinate axes is arbitrary, we may conclude that:

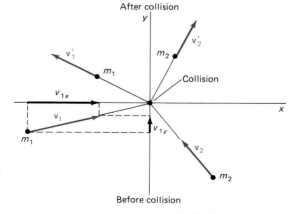

FIGURE 7.14 A collision in two dimensions. The x and y components of the linear momentum are conserved.

If no net external force acts on a system of particles, the components of the total linear momentum of the system along any chosen axis are the same after a collision as before the collision.

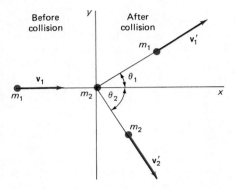

FIGURE 7.15 Collision of a particle of mass m_1 and velocity $\mathbf{v}_1$ with a particle of known mass m_2 that is at rest.

As an example, suppose we have a particle of mass m_1 and velocity $\mathbf{v}_1$ colliding with a particle known to have a mass m_2 and to be at rest. Then the situation is as shown in Fig. 7.15, where the initial velocity is taken in the x direction. After the collision, particle 1 moves off at an angle θ_1 with respect to its original direction and particle 2 moves off at an angle θ_2 with respect to the same direction. Then Eq. (7.14) becomes

x components: $\qquad m_1 v_1 + 0 = m_1 v_1' \cos \theta_1 + m_2 v_2' \cos \theta_2$ **(7.15)**

y components: $\qquad 0 + 0 = m_1 v_1' \sin \theta_1 - m_2 v_2' \sin \theta_2$

or $\qquad\qquad\qquad m_1 v_1' \sin \theta_1 = m_2 v_2' \sin \theta_2$ **(7.16)**

These two scalar equations are, in general, inadequate to solve for the four unknowns (v_1', v_2', θ_1, and θ_2) in the problem. If, however, any two of the unknowns are known (θ_1 and θ_2, for example), then the two remaining unknowns v_1' and v_2' can be obtained by solving the two above equations simultaneously.

In certain cases, additional physical considerations may simplify the solution of two-dimensional problems on momentum conservation. For example, if we know that the collision is *elastic*, then KE is conserved and we have a third equation:

$$\tfrac{1}{2}m_1 v_1^2 + \tfrac{1}{2}m_2 v_2^2 = \tfrac{1}{2}m_1 v_1'^2 + \tfrac{1}{2}m_2 v_2'^2$$ **(7.17)**

In this case, if m_1, m_2, v_1, and v_2 are given, then the measurement of any one of the four unknowns v_1', v_2', θ_1, and θ_2 enables the other three to be obtained from the simultaneous solution of Eqs. (7.15) to (7.17).

Another important case is that of a *completely inelastic collision*, in which the two particles stick together after colliding. This reduces the number of unknowns by two, since $v_1' = v_2' (=v')$ and $\theta_1 = \theta_2 (=\theta)$. If, therefore, we know m_1, m_2, v_1, and v_2, we need only the first two equations above to solve for the two unknowns v' and the angle θ. This is illustrated in Example 7.8.

Much experimental work in nuclear physics involves observing and analyzing collisions between nuclear particles. For example, suppose a target is known to contain a nuclear particle of mass m_2 that is at rest. An unknown nuclear particle is observed to collide with this rest particle in an *elastic collision*. Is it possible to determine the mass m_1 and the velocity $\mathbf{v}_1$ of the unknown particle from conservation-of-momentum considerations?

The motion will be confined to a plane (why?), and we can apply our three equations governing momentum and KE conservation to the collision. In this case we have two knowns (m_2, v_2) and six unknowns (m_1, v_1, v_1', v_2', θ_1, and θ_2). The angles θ_1 and θ_2 can usually be determined experimentally from cloud-chamber or bubble-chamber photographs of the reaction (see Fig. 29.20). Then, if the speed v_2' of the scattering particle after the collision can be determined by experiment, our three equations can be solved for the unknowns m_1, v_1, and v_1'. In this way the nature of the nuclear particle of mass m_1 can be determined.

Example 7.8

Two metal pucks are wrapped with Velcro tape so that they stick together after colliding. The collision is therefore completely inelastic. If one puck of mass $m_1 = 0.800$ kg is initially moving in the positive x direction at a speed $v_1 = 1.65$ m/s, and the second puck of mass $m_2 = 0.500$ kg is moving in the positive y direction at a speed of 2.24 m/s, what is the velocity of the two pucks immediately after the collision?

SOLUTION

Figure 7.16 shows the situation before and after the collision. Since no net external force acts, momentum is conserved, and we have, with v', the common velocity of the two pucks after the collision,

x components: $m_1 v_1 = (m_1 + m_2)v' \cos \theta$

y components: $m_2 v_2 = (m_1 + m_2)v' \sin \theta$

To solve these two equations simultaneously for the two unknowns θ and v', we divide the second equation by the first.

$$\frac{\sin \theta}{\cos \theta} = \tan \theta = \frac{m_2 v_2}{m_1 v_1} = \frac{(0.500 \text{ kg})(2.24 \text{ m/s})}{(0.800 \text{ kg})(1.65 \text{ m/s})} = 0.848$$

and so $\theta = \boxed{40.3°}$

To find the magnitude of the final velocity, we use the y component equation above, and obtain

$$v' = \frac{m_2 v_2}{(m_1 + m_2) \sin \theta} = \frac{(0.500)(2.24 \text{ m/s})}{(1.30)(0.647)} = \boxed{1.33 \text{ m/s}}$$

The final velocity of the two pucks is therefore 1.33 m/s at an angle of 40.3° above the position x axis.

EXERCISE 1   What fraction of the initial KE of the pucks is lost in the collision?

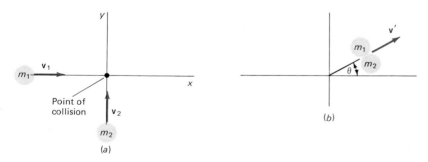

(a)

(b)

FIGURE 7.16   A completely inelastic collision: (a) two pucks before the collision; (b) the two pucks after the collision.

Example 7.9

A metal puck on a steel table has a mass $m_1 = 0.250$ kg and an initial speed of 2.00 m/s along the positive x axis. This puck makes a glancing collision with a stationary puck of mass $m_2 = 0.500$ kg, and immediately after the collision the first puck moves off at a speed of 1.00 m/s at 30° above the x axis. What are **(a)** the direction and **(b)** the magnitude of the final velocity of the second puck? **(c)** Is this an elastic collision?

SOLUTION

The event described here is similar to the one in Fig. 7.15. In this problem we cannot assume that the collision is either elastic or completely inelastic. Linear momentum is still conserved, however, and since this collision is in two dimensions, we may apply Eqs. (7.15) and (7.16). These two equations are adequate in this case, since six quantities are given in the statement of the problem, leaving only v_2' and θ_2 as unknowns. Our two-component conservation-of-momentum equations therefore suffice to find these two unknowns.

(a) We have, from Eq. (7.15),

$$(0.250 \text{ kg})(2.00 \text{ m/s})$$
$$= (0.250 \text{ kg})(1.00 \text{ m/s}) \cos 30° + (0.500 \text{ kg})(v_2' \cos \theta_2)$$

from which $v_2' \cos \theta_2 = 0.567 \text{ m/s}$

From Eq. (7.16), we obtain

$$(0.250 \text{ kg})(1.00 \text{ m/s})(\sin 30°) = (0.500 \text{ kg})(v_2' \sin \theta_2)$$

from which $v_2' \sin \theta_2 = 0.250 \text{ m/s}$

We then combine these two results to obtain

$$\frac{v_2' \sin \theta_2}{v_2' \cos \theta_2} = \frac{0.250}{0.567} = 0.441$$

$$\tan \theta_2 = 0.441 \qquad \theta_2 = \boxed{23.8°}$$

(b) Since $v_2' \sin \theta_2 = 0.250 \text{ m/s}$,

$$v_2' = \frac{0.250 \text{ m/s}}{\sin 23.8°} = \boxed{0.620 \text{ m/s}}$$

The velocity of m_2 is therefore 0.620 m/s at an angle of 23.8° *below* the x axis (see Fig. 7.15).

EXERCISE 2 Check this result by substituting it back in the conservation-of-momentum equations (7.15) and (7.16) to show that it indeed satisfies these equations.

(c) The KE before the collision is

$$KE_i = \tfrac{1}{2} m_1 v_1^2 = \tfrac{1}{2}(0.250 \text{ kg})(2.00 \text{ m/s})^2 = 0.500 \text{ J}$$

After the collision the KE is

$$KE_f = \tfrac{1}{2} m_1 v_1'^2 + \tfrac{1}{2} m_2 v_2'^2$$
$$= \tfrac{1}{2}(0.250 \text{ kg})(1.00 \text{ m/s})^2 + \tfrac{1}{2}(0.500 \text{ kg})(0.620 \text{ m/s})^2$$
$$= 0.125 \text{ J} + 0.096 \text{ J} = 0.221 \text{ J}$$

Since $KE_f < KE_i$, the kinetic energy is not conserved; some KE is converted into thermal energy in the collision and so this is an *inelastic collision*.

7.7 Recoil and Rockets

The principle of conservation of linear momentum explains the recoil of rifles, cannon, and even atomic nuclei that eject high-speed atomic particles.

Suppose a rifle of mass M fires a bullet of mass m, as in Fig. 7.17. The linear momentum of the system is zero *before* the firing. Since the only force that acts in the x direction is internal to the system, a result of the explosion of the gunpowder, the total linear momentum of the system of gun and bullet must also be zero *after* the firing. Therefore

FIGURE 7.17 Recoil of a rifle when it fires a bullet.

$$m\mathbf{v} + M\mathbf{V} = 0 \qquad \text{or} \qquad \mathbf{V} = \frac{-m}{M}\mathbf{v} \tag{7.18}$$

where $\mathbf{v}$ is the velocity of the bullet and $\mathbf{V}$ is the velocity of the rifle. The velocity of the rifle is in the opposite direction to the bullet's motion and is reduced in size by the ratio of the two masses. This backward velocity of a gun when it fires a bullet is referred to as *recoil*. The more massive the gun (compared to the bullet), the smaller the recoil velocity.

We can also obtain the ratio of the kinetic energies of rifle and bullet:

$$\frac{\tfrac{1}{2}mv^2}{\tfrac{1}{2}MV^2} = \frac{m}{M}\left(\frac{v}{V}\right)^2 = \frac{m}{M}\left(\frac{-M}{m}\right)^2 = \frac{M}{m} \tag{7.19}$$

The kinetic energies are therefore inversely proportional to the masses, just as is true for the velocities from Eq. (7.18). The bullet therefore receives more kinetic energy than the rifle by a factor of M/m, which may be 500 or more.

Rockets

The operation of a rocket also depends on the principle of conservation of linear momentum as applied to the system consisting of the rocket and its ejected fuel.

To conserve linear momentum, recoil must occur when the rocket ejects hot gases from its rear. It is this recoil that propels the rocket forward, in the same way that a deflating balloon moves in the opposite direction from the emerging air.

The thrust that powers a rocket is not produced (as is sometimes erroneously supposed) by the ejected gases pushing against the air in the atmosphere but is due to a recoil of the kind that occurs when a rifle fires a bullet. Rocket propulsion is the only propulsive system that can function perfectly in a vacuum, since a rocket carries both its fuel and oxidizer along with it. Although aircraft jet engines also use reaction forces to propel jet planes, they require oxygen from the air to burn the fuel and can operate only in the earth's atmosphere.

Thrust of a Rocket

A rocket expels hot gases from its combustion chamber and does this at a constant rate as long as fuel is being burned by the rocket. Since the rocket exerts a backward force on the gases, these gases exert an equal and opposite forward force on the rocket. We designate by $\mathbf{u}$ the velocity of the ejected gases *relative to the rocket*. We also assume that the mass of the fuel is reduced at a constant rate, so $\Delta m/\Delta t$ is constant as long as any fuel remains.

By Newton's second law in the form of Eq. (7.2), the force exerted *on the gases* by the rocket is

$$\mathbf{F}_G = \frac{\Delta \mathbf{p}_G}{\Delta t} = \frac{\Delta(m\mathbf{u})}{\Delta t} = \frac{\Delta m}{\Delta t}\,\mathbf{u}$$

since $\mathbf{u}$ is constant in time. By Newton's third law this force on the gases is equal and opposite to the reaction force exerted by the gases *on the rocket*. This force *on the rocket* we designate as $\mathbf{F}_R$ and call the *thrust* of the rocket. If the rocket is fired directly upward from the earth, therefore, its thrust is directed away from the earth and is of magnitude

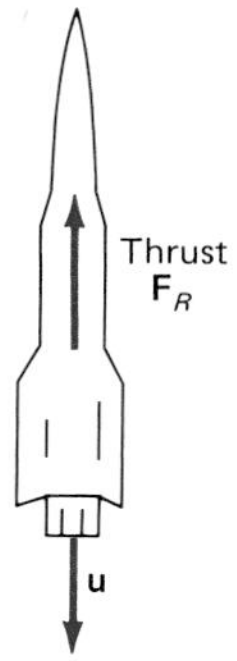

$$F_R = u\,\frac{\Delta m}{\Delta t} \tag{7.20}$$

where u is the exhaust speed of the gases with respect to the rocket, as in Fig. 7.18.

The force of gravity is, of course, also pulling the rocket back toward the earth, so the total force acting *on the rocket* is

$$F_{\text{net}} = F_R - mg = u\,\frac{\Delta m}{\Delta t} - mg$$

and the rocket's acceleration is

$$a = \frac{u\,\Delta m/\Delta t}{m} - g \tag{7.21}$$

FIGURE 7.18 Rocket propulsion. The thrust of the rocket is $F_R = u\,(\Delta m/\Delta t)$ and therefore depends on the exhaust speed of the gas with respect to the rocket (u) and on the rate at which the gas is ejected from the rocket ($\Delta m/\Delta t$).

This acceleration of the rocket with respect to the earth depends therefore on the thrust of the rocket, on the mass of the rocket, and on the value of g. Since the motion is accelerated, the velocity v of the rocket with respect to the earth increases with time.

For large thrusts u must be large, as Eq. (7.20) shows. It is usually about

2 km/s for most chemical propellants. The time rate of change of mass $\Delta m/\Delta t$ must also be large. In most cases the fuel makes up 90 to 95 percent of the original mass of the rocket plus contents, so $\Delta m/m$ is about 0.95 for the whole acceleration process.

As the rocket rises, its acceleration relative to the earth increases for a number of reasons. First of all, as the fuel is expended, the rocket's mass decreases, and since the thrust is constant, the acceleration increases [see Eq. (7.21)]. Second, as the rocket rises, the value of g decreases. This also increases the acceleration of the rocket relative to the earth, since g is in the opposite direction to the acceleration a.

Since most of its fuel is consumed in powering the rocket through the earth's atmosphere, the energy a rocket carries in its fuel supply can be utilized more effectively if the rocket is initially given a boost by another rocket. This is the function of the booster rockets that often serve as the early stages of multistage rockets.

Example 7.10

A rocket of initial total mass 4.00×10^4 kg ejects hot gases from its combustion chamber at a rate of 1000 kg/s. The speed of the gas is 2.00 km/s relative to the rocket. The fuel, which makes up 90 percent of the total mass of the rocket before blast-off, is consumed in 36 s. **(a)** Find the thrust of the rocket engine. **(b)** Find the upward acceleration at time $t = 0$, $t = 20$ s, $t = 36$ s, and $t = 40$ s (neglect any change in g with altitude).

SOLUTION

(a)   Rocket thrust $= u\,\dfrac{\Delta m}{\Delta t} = (2.00 \text{ km/s})(1000 \text{ kg/s})$

$$= 2.00 \times 10^6 \text{ kg·m/s}^2 = \boxed{2.00 \times 10^6 \text{ N}}$$

(b)   The acceleration is given by Eq. (7.21):

$$a = \frac{\text{thrust}}{m} - g$$

If fuel is used up at the rate of 1000 kg/s, then the remaining mass of rocket plus fuel at the four times given is

$$m_0 = 4.00 \times 10^4 \text{ kg} \qquad m_{20} = 2.00 \times 10^4 \text{ kg}$$

$$m_{36} = 0.400 \times 10^4 \text{ kg} \qquad m_{40} = 0.400 \times 10^4 \text{ kg}$$

The corresponding accelerations are

$$a_0 = \frac{2.00 \times 10^6 \text{ N}}{4.00 \times 10^4 \text{ kg}} - 9.80 \text{ m/s}^2 = \boxed{40.2 \text{ m/s}^2}$$

$$a_{20} = \frac{2.00 \times 10^6 \text{ N}}{2.00 \times 10^4 \text{ kg}} - 9.80 \text{ m/s}^2 = \boxed{90.2 \text{ m/s}^2}$$

$$a_{36} = \frac{2.00 \times 10^6 \text{ N}}{0.400 \times 10^4 \text{ kg}} - 9.80 \text{ m/s}^2 = \boxed{490 \text{ m/s}^2}$$

Notice how rapidly the acceleration increases as the mass of the fuel is reduced to zero.

Just after $t = 36$ s the fuel runs out, and at $t = 40$ s there is no fuel left and therefore no thrust, since u in Eq. (7.21) is zero. The acceleration of the rocket is just -9.80 m/s^2, as it would be for any object acted on only by the force of gravity. This force tends to slow the rocket down as it rises. If, however, the rocket reaches a sufficiently high KE, it can escape from the earth's gravitational field and travel freely through space. This *escape velocity* is about 11 km/s (see Prob. 6.60).

Summary: Important Definitions and Equations

Linear momentum

$$\mathbf{p} = m\mathbf{v} \tag{7.1}$$

Units of momentum and impulse are kg·m/s or N·s.

Newton's second law: $\mathbf{F}_{net} = \dfrac{\Delta \mathbf{p}}{\Delta t}$ (7.2)

Impulse-momentum theorem

$$\mathbf{F}_{net} \, \Delta t = \Delta \mathbf{p} \tag{7.3}$$

where $\mathbf{F}_{net} \, \Delta t$ is the impulse of the force $\mathbf{F}_{net}$.

Principle of conservation of linear momentum: If no net external force acts on a physical system, the total linear momentum of the system remains constant.

$$\mathbf{p} = \sum_i m_i \mathbf{v}_i = \text{constant} \tag{7.4}$$

Kinds of collisions

Elastic: Both linear momentum and kinetic energy of the system are conserved.

Inelastic: Linear momentum is conserved, but kinetic energy is not.

Completely inelastic collision: A collision in which the colliding objects stick together and move as a unit with zero relative velocity after the collision.

Collisions in two dimensions: Both the x and y components of the linear momentum are conserved, if no net external force acts in the xy plane.

Rockets as examples of conservation of linear momentum:

$$Rocket\ thrust = u\,\frac{\Delta m}{\Delta t} \tag{7.20}$$

$$Rocket\ acceleration\ a = \frac{thrust}{m} - g \tag{7.21}$$

Recoil:

$$\mathbf{V} = -\frac{m}{M}\,\mathbf{v} \tag{7.18}$$

Questions

1 A ball is dropped from a Goodyear blimp. As it falls its linear momentum continually increases, whereas the momentum of the blimp is unchanged. How can we talk about conservation of linear momentum in this case?

2 A meteor is traveling through the earth's atmosphere at a constant velocity. As it moves, it loses mass because of the intense heating caused by contact with the gases in the atmosphere. Is the momentum of the meteor changing? Why?

3 A billiard ball strikes the cushion of a billiard table at a right angle. Before it hits the cushion the ball's momentum is mv; after hitting the cushion the ball's momentum is approximately $-mv$. Hence the change in momentum is approximately $-2mv$. How can you say that linear momentum is conserved in this collision?

4 Two pucks are sliding toward each other on a wooden table. The coefficient of friction between the pucks and the wooden surface of the table is 0.35.

(*a*) Is linear momentum conserved when the pucks collide?

(*b*) Is linear momentum conserved before and after this collision?

5 Two billiard balls collide head-on and elastically on a billiard table.

(*a*) Draw a graph of linear momentum for the two balls as a function of time, from a few seconds before the collision to a few seconds after.

(*b*) Draw a graph of kinetic energy of the two balls as a function of time for the time interval indicated in part (*a*).

6 Two billiard balls collide head-on and inelastically. Draw the same graphs asked for in Question 5.

7 In Example 7.3 will the fragments of the shell reach the ground at the same average time they would if the shell did not separate at all?

8 Two hockey pucks of equal mass slide toward each other on a metal table. Each puck has the same speed with respect to the table. A piece of soft wax is dropped between the two pucks just before they collide, and they stop dead and stick together.

(*a*) What is the final linear momentum of the system? Is linear momentum conserved in this collision?

(*b*) What is the final kinetic energy of the system? Is kinetic energy conserved?

9 Which results in more damage when two cars collide, high momenta or high kinetic energies?

10 A ball of wax is shot from a toy gun at a wall. Will the wax ball exert a larger impulse on the wall if it sticks to the wall or if it rebounds?

11 A baseball player is trapped in a boxcar on a deserted side track and decides to move the boxcar to a more populated region by constantly throwing a baseball at full speed at the front wall of the boxcar and thus giving the car a forward velocity. Discuss the likelihood that such a technique will succeed. Will the boxcar move at all? If so, in what direction?

12 (*a*) Is it possible to have a system that has kinetic energy but no linear momentum?

(*b*) Is it possible to have a system that has linear momentum but no kinetic energy?

13 A railroad coal car is coasting along at constant speed on a frictionless track. It begins to rain and water accumulates in the bottom of the open car. Does the speed of the car change? Does it increase or decrease? Why?

Multiple-Choice and Simple Exercises

7.1 A Volkswagen weighing 6.0×10^3 N is traveling at 15 m/s. A Mercedes weighing 1.0×10^4 N is traveling at 30 m/s. The ratio of the momentum of the Volkswagen to that of the Mercedes is:
(*a*) 6/1   (*b*) 2/1   (*c*) 1/2   (*d*) 10/3
(*e*) 3/10

7.2 A sponge rubber ball of mass 0.25 kg is thrown with a speed of 30 m/s perpendicularly against a wall. It collides elastically with the wall and rebounds. The momentum of the ball after hitting the wall, with respect to its original direction, is:
(*a*) -7.5 kg·m/s   (*b*) $+7.5$ kg·m/s
(*c*) 225 kg·m/s   (*d*) -225 kg·m/s
(*e*) -30 kg·m/s

7.3 Two balls collide head-on in an elastic collision. The first ball has a mass of 2.0 kg and a speed of 10 m/s in the negative x direction. The second ball has a mass of 1.5 kg and a speed of 20 m/s in the positive x direction. The relative velocity $(v_2 - v_1)$ of the two balls after the collision is:
(*a*) -30 m/s   (*b*) 30 m/s   (*c*) 40 m/s
(*d*) -10 m/s   (*e*) 10 m/s

7.4 A pitcher throws a baseball with a momentum of 20 kg·m/s and the ball leaves the bat with a momentum of 25 kg·m/s. The impulse exerted by the hitter's bat on the ball is:
(*a*) 5.0 kg·m/s   (*b*) 5.0 N·s   (*c*) 45 N·s
(*d*) 45 J   (*e*) 45 N

7.5 A 4000-kg freight car is moving with a speed of 10 m/s when it hits an identical freight car (at rest) and couples to it. The final speed of the two freight cars together is:
(*a*) 10 m/s   (*b*) 20 m/s   (*c*) 5.0 m/s
(*d*) 0   (*e*) None of the above

7.6 A 0.50-kg. ball is thrown at 40 m/s. What is the linear momentum of the ball?
(*a*) 20 kg·m/s   (*b*) 20 m/s
(*c*) 4.0×10^2 kg·m/s   (*d*) 4.0×10^2 J
(*e*) 20 J

7.7 A 3.0-kg rifle fires a 25-g bullet at 500 m/s. The recoil momentum of the rifle is:
(*a*) 1.5 kg·m/s   (*b*) 4.2 m/s   (*c*) 4.2 kg·m/s
(*d*) 12.5 kg·m/s   (*e*) 60 kg·m/s

7.8 An 80-kg man and a 50-kg woman stand at the center of a frozen pond on ice skates. They extend their hands and push off against each other. The man acquires an initial speed of 0.25 m/s. The speed of the woman is:
(*a*) 0.25 m/s   (*b*) 0.16 m/s   (*c*) 0.40 m/s
(*d*) 0.33 m/s   (*e*) None of the above

7.9 In Prob. 7.8 the ratio of the kinetic energy of the woman to that of the man is:
(*a*) 8/5   (*b*) 5/8   (*c*) 64/25   (*d*) 25/64
(*e*) $(8/5)^{1/2}$

7.10 A rocket near the earth's surface ejects exhaust gases at the rate of 500 kg/s and the speed of the gases with respect to the rocket is 1.7 km/s. At the instant when the rocket mass is 5.0×10^4 kg, its acceleration is (here $+$ means up and $-$ down):
(*a*) -9.8 m/s^2 (*b*) 9.8 m/s^2 (*c*) -7.2 m/s^2
(*d*) 7.2 m/s^2 (*e*) 17 m/s^2

7.11 A pitcher throws a 0.50-kg baseball at 40 m/s. What is the ball's linear momentum?

7.12 A billiard ball of mass 0.60 kg and speed 0.70 m/s makes a head-on elastic collision with an identical billiard ball at rest.
(*a*) What is the kinetic energy of the second ball after the collision?
(*b*) What is its linear momentum?

7.13 What average force is needed to stop a hammer with a momentum of 25 kg·m/s in 0.050 s?

7.14 Two identical freight cars, each of mass 4000 kg and moving at 20 m/s in opposite directions, collide and stick together. How much kinetic energy is lost in the collision?

7.15 A particle of mass 0.50 kg has a velocity of 4.0 m/s at an angle of 40° above the positive x axis. What are the x and y components of the particle's momentum?

7.16 Two ice pucks initially at rest are driven apart by an explosion with velocities of 10 and 3.3 cm/s. What is the ratio of the masses of the two pucks?

7.17 A rocket ejects gas at the rate of 500 kg/s and the speed of the gases with respect to the rocket is 2.0 km/s. What is the thrust of the rocket?

7.18 A rocket is moving with a speed of 2000 m/s with respect to the earth (vertically upward). If the rocket is ejecting gas that has a speed of 1.8 km/s relative to the rocket, what is the speed and direction of the gas with respect to the earth?

7.19 A small metal glider of mass 100 g is moving along a frictionless airtrack at a constant speed of 0.50 m/s. A second glider of mass 50 g is then carefully dropped from rest on top of the first one and sticks there. What is the final speed of the two gliders?

7.20 Two stars of masses $m_1 = 2.0 \times 10^{30}$ kg and $m_2 = 4.0 \times 10^{30}$ kg have velocities $v_1 = 3.0 \times 10^4$ m/s in the $+x$ direction and $v_2 = 5.0 \times 10^4$ m/s in the $-y$ direction in a suitably chosen coordinate system. The two stars collide and stick together. What is the final velocity after the collision?

Problems

7.21 A 1.0-kg steel ball bearing moving at 2.50 m/s strikes a steel plate at an angle of 30.0° with respect to the surface of the plate and rebounds elastically at the same angle and speed. What is the change in (*a*) the magnitude and (*b*) the direction of its linear momentum?

7.22 A 10.0-kg shell is fired into the air at 150 m/s at an angle of 55.0° with respect to the horizontal. What are (*a*) the KE and (*b*) the momentum of the shell at the beginning and end of its flight? (Neglect air resistance.)

7.23 In Prob. 7.22 show that the net force acting on the shell during its flight is exactly equal to the time rate of change of the shell's momentum.

7.24   A machine gun fires 20 bullets of mass 50.0 g in quick succession into a 20.0-kg wooden block resting on a frictionless table. If the bullets are fired at 750 m/s parallel to the tabletop, what is the final speed of the block?

7.25   A 0.450-kg softball is pitched at a speed of 15.0 m/s. The batter hits it back directly at the pitcher at a speed of 20.0 m/s. The bat acts on the ball for 0.0150 s.

  (a)   What is the impulse imparted by the bat to the ball?
  (b)   What is the average force exerted by the bat on the ball?

7.26   A freight car of mass 6.40×10^4 kg is rolling slowly along a level track at 0.350 m/s. A rope is attached to the back of the car and trails along the ground. A bystander picks up the rope and tries to stop the freight car by pulling on it.

  (a)   If the maximum force the person can exert is 400 N, how long will it take to bring the freight car to rest?
  (b)   If an automobile is on the track 12.0 m in front of the freight car when the person begins to pull on the rope, will the freight car hit the automobile?

7.27   A horizontal stream of water from a hose strikes a window on the outside wall of a building and then rolls down the side of the window. If 0.100 kg of water strikes the window each second with a speed of 2.50 m/s, find (a) the constant force exerted on the window by the water; (b) the impulse of this force over a 0.100-s interval.

7.28   A 100-g marble is shot along the floor and collides head-on with a 65.0-g marble at rest. The speed of the first marble is reduced from 200 to 100 cm/s in the collision. What is the speed of the second marble after the collision?

7.29   A 70-kg astronaut becomes separated from her earth satellite, which is moving in orbit around the earth. She finds herself 25 m away from the satellite and moving at exactly the same velocity as it is. She has a 0.60-kg camera in her hand and decides to get back to the satellite by throwing the camera at a speed of 10 m/s in a direction away from the satellite.

  (a)   Will this maneuver get her back to the satellite? Why?
  (b)   How long will it take for her to reach the satellite?
  (c)   How fast will she be moving when she reaches it?

7.30   To remove a large 10.0-kg rock from a building site, an explosive charge is placed inside the rock. When the charge is detonated, the rock breaks up into three pieces. Two have a mass of 2.00 kg each and are shot out at right angles to each other at 10.0 m/s.

  (a)   Draw a vector diagram of the momenta in this physical situation, and solve for the magnitude and direction of the velocity of the third piece of rock.
  (b)   Solve for the same quantity mathematically instead of graphically.

*7.31   Tarzan holds on to a vine of negligible weight and swings down from a tree limb 9.0 m above the ground. At the lowest point of his swing he picks up his mate, Jane, who is standing on the ground, and then continues his swing. He and Jane land on another tree limb located at the maximum height reached on their swing. If Tarzan has a mass of 100 kg, and Jane 70 kg, how high above the ground is the limb on which they land?

7.32   A uranium-238 nucleus, which has a mass of 3.95×10^{-25} kg and is at rest, breaks up into an alpha particle ($m = 6.64 \times 10^{-27}$ kg) and a thorium-234 nucleus ($m = 3.88 \times 10^{-25}$ kg). If the alpha particle is emitted at 1.40×10^7 m/s, what is the recoil speed of the thorium nucleus?

7.33   Two blocks, one of mass 3.00 kg and the other of mass 2.00 kg, are used to compress the two ends of a spring. The two blocks, originally at rest, are then released, and the 3.00-kg block moves away at a speed of 1.00 m/s.

  (a)   What is the speed of the 2.00-kg block?
  (b)   What are the kinetic energies of each of the two blocks?

7.34   Two blocks, one of mass 5.00 kg, the other of mass 10.0 kg, are used to compress a spring whose force constant is 100 N/m. The spring is compressed 20.0 cm and the blocks are then released. What are (a) the final speeds and (b) the KEs of each of the blocks?

7.35   An explosion blows a rock into three parts. Two pieces go off at right angles to each other, a 1.00-kg piece at 20.0 m/s and a 2.00-kg piece at 30.0 m/s. The third piece flies off at 50.0 m/s.

  (a)   Draw a vector diagram showing the direction of the third piece (*Hint:* Take the direction of one rock as along the x axis).
  (b)   What is the momentum of the third piece?
  (c)   What was the mass of the rock before the explosion?

*7.36   Derive the generalized form of Eqs. (7.10) and (7.11) when it is not assumed that m_2 is initially at rest.

*7.37   A 2.00-kg ball moving at 10.0 m/s collides head-on with a 50.0-kg ball moving in the opposite direction at 6.00 m/s. Find the speed of each ball after impact if (a) the collision is perfectly elastic; (b) the collision is perfectly inelastic (i.e., the balls stick together).

*7.38   (a)   A head-on elastic collision occurs between two particles, one of which is initially at rest and the second of which is in motion directly toward the first. Find the ratio of the two masses that will lead to the transfer of maximum kinetic energy to the particle initially at rest.

  (b)   Show that if the masses of the two particles do not meet the requirement arrived at in part (a), then better energy transfer can be obtained by inserting a third particle between the two, with the third particle having a mass that is the geometric mean of the masses of the other two particles; that is, $m_3 = \sqrt{m_1 m_2}$.

*7.39   Cloud chambers are often used to provide photographs of nuclear reactions. A proton is seen in such a photograph to have undergone an elastic collision with another particle; in the collision the proton's velocity was deviated by an angle of 60° from its original direction. The struck particle, which is presumed to have been at rest, goes off at an angle of 30° to the other side of the proton's original direction. What was the mass of the struck particle in terms of the proton mass? (*Hint:* Apply principles of conservation of momentum and energy.)

7.40   A 0.30-kg hockey puck is moving in the positive x direction with a speed of 0.20 m/s. A second, 0.15-kg puck, with a piece of wax attached to its front edge, is moving at 0.50 m/s in the negative x direction. The two pucks collide head-on and stick together. What is the velocity of the two pucks after the collision?

7.41   A 10.0-g bullet is fired horizontally into a 10.0-kg block of wood that is initially at rest and that has a speed of 150 cm/s after the impact. The bullet comes to rest in the block. Find the initial speed of the bullet.

7.42 In a ballistic pendulum experiment (compare Example 7.7) a 12-g bullet is fired horizontally into a 5.00-kg block of wood suspended by a long cord. The bullet sticks in the wood, and the two rise together until the center of gravity of the block has been raised 15.0 cm from its original position. Find the initial speed of the bullet.

7.43 In the first nuclear reactor constructed during World War II at Stagg Field in Chicago, graphite blocks were used as the moderator to slow down the neutrons. Graphite is made entirely of carbon atoms ($^{12}_{6}$C) with masses 12 times the neutron mass.

(*a*) In a head-on elastic collision of a neutron with a carbon atom at rest, what percentage of the kinetic energy of the neutron is lost?

(*b*) How many head-on collisions are needed to reduce the kinetic energy of the neutron to 1 percent of its original value?

*7.44 A gun fires a 5.0-g bullet at a speed of 250 m/s horizontally into a 200-g wooden block resting on a table. The bullet stops in the wooden block, and the block and bullet then move together 5.8 m along the table before being stopped by friction.

(a) What is the speed of the block immediately after the bullet hits it?

(b) What is the coefficient of friction between block and table?

7.45 A 10.0-g bullet traveling at 400 m/s passes completely through a 0.500-kg block of wood resting on a table. After the bullet passes through, the block has a speed in the direction of the bullet's original motion of 3.0 m/s. What is the speed of the bullet after it passes through the block?

7.46 A 50-kg iceboat is moving along a frozen lake without friction at a constant speed of 30 m/s. A second iceboat travels in the same direction and catches up with the first one. A person carefully lowers a 10-kg hamper of provisions into the first boat in a direction at right angles to the iceboat's velocity at an instant when the boats are moving side-by-side with the same velocity. What is the change in the first boat's velocity?

*7.47 A bag of jelly beans is emptied at a constant rate into the pan of a spring scale. Each jelly bean has a mass of 1.0 g and is dropped from a height of 0.70 m onto the pan. Ten jelly beans hit the scale pan each second. What is the scale reading at the end of the first 5.0 s, if the jelly beans all stick to the scale pan without bouncing?

7.48 Pieces of coal drop from the bottom of a hopper at a rate of 7.5 kg/s onto a conveyer belt moving horizontally at a speed of 2.0 m/s. What force is needed to drive the conveyer belt? (Neglect friction.)

7.49 A 70.0-kg boy and a 50.0-kg girl are ice-skating. They coast along together at 1.20 m/s in a straight line. The boy then pushes off from the girl at right angles to the original direction of motion and acquires a speed of 2.50 m/s in that direction. If we ignore friction and air resistance, what is (*a*) the girl's final momentum and (*b*) her kinetic energy?

7.50 Two cars, one of mass m_1 = 1000 kg, the other of mass m_2 = 2000 kg, are moving at right angles to each other when they collide and stick together. The initial velocity of

car 1 is 10.0 m/s in the positive x direction and that of car 2 is 18.0 m/s in the positive y direction. What is the velocity of the wreckage of the two cars immediately after the collision?

7.51 A 2000-kg car traveling at 20 m/s crashes into a 4000-kg truck traveling at 25 m/s along a street at right angles to the car's path. The two vehicles lock together, and the wreckage moves 25.2 m before coming to rest.

(*a*) Find the direction in which the wreckage moves after the collision.

(*b*) Find the speed of the wreckage just after the collision.

(*c*) Find the force of friction that brings the cars to rest.

7.52 A rocket stands on its launchpad pointing straight up into the air. Its jet engines are activated and eject exhaust gases at a rate of 1000 kg/s. These gases have a speed of 2.00 km/s with respect to the rocket. What is the largest mass the rocket can have and still be slowly lifted off its launchpad by the thrust of its engines? (By "slowly" we mean almost equal to zero velocity.)

7.53 A polonium nucleus ($^{210}_{84}$Po) of mass 3.48 × 10^{-25} kg emits an alpha particle of mass 6.64 × 10^{-27} kg and speed 1.30 × 10^{7} m/s. Find the speed with which the polonium nucleus recoils.

7.54 A cannon is mounted in such a way that a constant force resists the recoil motion of the cannon when it fires. The cannon has a mass of 500 kg and fires a 1.00-kg shell with a muzzle velocity of 500 m/s.

(*a*) Find the recoil velocity of the cannon.

(*b*) If the resisting force is 4000 N, find the time it takes to stop the cannon.

(*c*) How far does the cannon move in that time?

7.55 A gun mounted on a railway car can recoil only in the horizontal direction.

(*a*) If the gun has a mass of 5.00 × 10^{4} kg and fires a 250-kg artillery shell at an angle of 40.0° and a speed of 200 m/s, what is the recoil velocity of the gun in the horizontal direction?

(*b*) What happens to the vertical component of the recoil momentum?

*7.56 A 500-kg cannon, mounted on wheels, fires a 5.00-kg shell horizontally at 1000 m/s.

(*a*) If the wheels turn freely, what is the recoil speed of the cannon?

(*b*) If the gun is firmly anchored in the ground, what is the recoil speed of cannon and earth together?

(*c*) If the shell accelerated uniformly along the 2.00-m length of the cannon, what is the average force on the shell inside the barrel?

(*d*) What is the average force exerted back on the cannon by the expanding combustion gases in the barrel?

7.57 (*a*) What minimum thrust must the engines of a rocket of initial total mass 5.00 × 10^{4} kg have to lift it from the earth if the rocket is aimed straight upward?

(*b*) If the engines eject gases at the rate of 100 kg/s, how fast must the gases be moving when they leave the engines, if the rocket is to lift off?

(*c*) What is the acceleration of the rocket after 2.00 min?

7.58 A 5000-kg rocket traveling at 200 m/s is moving freely through space on a journey to the moon. The ground controllers find that the rocket has drifted off course and that it must change direction by 12° if it is to hit the moon. By radio control the rocket's engines are fired for a brief instant in a direction perpendicular to the direction of the rocket's motion. If the gases are expelled at a speed of 2000 m/s, what mass of gas must be expelled to make the needed correction in direction?

7.59 A device similar to that in Fig. 7.19 is often seen in toy and novelty shops. All five balls have the same mass. When one ball is pulled back and released, the ball on the opposite end flies out with the same speed that the original ball had just before it collided.

(*a*) Prove that this is to be expected if the collision is elastic.

(*b*) Predict what will happen when two balls are pulled out instead of one. Justify your predictions on the basis of physical principles.

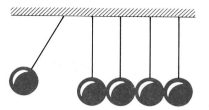

FIGURE 7.19   Diagram for Prob. 7.59.

Additional Readings

Adair, Robert K.: *The Physics of Baseball,* Harper and Row, New York, 1990. This book, by an elementary particle physicist, contains a good discussion of the impulse given by a bat to a baseball and the ball's subsequent trajectory.

Asimov, Isaac: *Asimov's Biographical Encyclopedia of Science and Technology,* 2d rev. ed., Doubleday, Garden City, N.Y., 1982. Contains a good brief account of Huygens' accomplishments.

Bos, H. J. M.: "Christiaan Huygens," in C. C. Gillispie (ed.), *Dictionary of Scientific Biography,* Scribner, New York, 1970–1980, vol. 6. The best available treatment of Huygens' life and work in English. The volumes of this dictionary are a gold mine of information for anyone interested in the history of science.

Feynman, Richard: *The Character of Physical Law,* MIT Press, Cambridge, Mass., 1965. Chapter 3, "The Great Conservation Principles," is particularly relevant.

Haber-Schaim, Uri, John H. Dodge, and James A. Walter: *PSSC Physics,* 5th ed., Heath, Lexington, Mass., 1981. Chapter 6 of this text includes a careful study of momentum conservation accompanied by some excellent photographs of collision processes.

Lewis, H. W.: "Ballistocardiography," *Scientific American,* vol. 198, no. 2, February 1958, pp. 89–95. An interesting attempt to study conservation of linear momentum in the pumping of blood by the heart.

CHAPTER 8

Rotational Motion

From time to time in the kingdom of mankind a man arises who is of universal significance, whose work changes the current of human thought or of human experience, so that all that comes after him bears evidence of his spirit. Such a man was Shakespeare, such a man was Beethoven, such a man was Newton, and, of the three, his kingdom is the most widespread.

E. N. da C. Andrade (1887–1971)

The spoked wheel was introduced about 2700 B.C. during the bronze age. Since that time the wheel has played a fundamental role in agriculture, industry, transportation, and recreation. The rotational motion of rigid bodies such as the wheel is of great importance in modern science and technology.

When a bicycle wheel turns, its motion consists of a rotation about the axle of the wheel plus the translational motion of the axle along with the bicycle (see Fig. 8.1). The most general motion of a moving body consists of a translational motion of its center of mass with respect to some chosen frame of reference plus the rotation of the body about its center of mass. The motion of a Frisbee is a good example: it rotates about its center as it glides through the air, as in Fig. 8.2, and so its motion is a combination of translation and rotation. Other good examples are tumbling footballs, pirouetting ballet dancers, and the earth itself.

We have already discussed translational motion in detail. We have also discussed uniform rotational motion in Sec. 4.5. In this chapter we will expand our treatment of rotational motion and, in particular, discuss the rotational motion of rigid bodies.

FIGURE 8.1 The motion of a bicycle wheel. The wheel rotates about its axle with an angular velocity ω while at the same time the axle executes translational motion in the direction in which the bicycle is moving with velocity **v**.

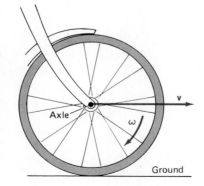

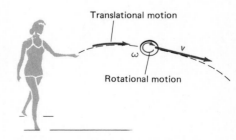

FIGURE 8.2 The motion of a Frisbee is a combination of translational motion and rotational motion.

8.1 Relationship of Rotational Motion to Translational Motion

In what follows, we will show how the basic equations for rotational motion can be obtained for rigid bodies by analogy with those already derived for translational motion. By a *rigid body* we mean a solid object that rotates as a unit when subjected to a torque. Even though rotational motion is in many respects analogous to translational motion, there is one very important difference. For translational motion we assume that all portions of a solid object move with exactly the same translational velocity. For example, when we consider the translational motion of a car, we assume that the front and rear bumpers have the same velocity.

For the *rotational motion of a rigid body,* however, the situation is quite different. In a given time, a point on the rim of a wheel covers a greater linear distance than does a point halfway out from the center, and so these points have different linear speeds. Despite this, the wheel rotates in one piece, and we therefore want to describe the speed of this rotation in some convenient way. As we saw in Sec. 4.5, the most useful description of rotational motion is in terms of the number of *radians* swept out by the wheel per second. Since the radian is an angular measure, rotational velocities are referred to as *angular velocities.*

In Fig. 8.3 point A sweeps through an arc Δs twice as large as the arc $\Delta s'$ swept out by point B in the same time Δt, for $\Delta s = R' \Delta \theta$ [Eq. (1.1)], whereas $\Delta s' = (R'/2) \Delta \theta$. The linear velocity of A is therefore twice that of B, but the angular velocity $\omega = \Delta \theta / \Delta t$ is the same for both and, more generally, for all points on the wheel. This is consistent with the relationship between angular velocity and linear velocity, $v = R\omega$, where R is the distance from the axis of rotation to the point where v is being calculated, as discussed in Sec. 4.5.

By analogy with the linear case we distinguish between the average angular velocity ($\overline{\omega}$) and the instantaneous angular velocity (ω) as follows:

$$\overline{\omega} = \frac{\Delta \theta}{\Delta t} \qquad \omega = \lim_{\Delta t \to 0} \frac{\Delta \theta}{\Delta t} \tag{8.1}$$

If a wheel is started spinning from rest and its angular velocity increases in time, then the motion is accelerated. We again define angular acceleration by analogy with the linear case and distinguish average angular acceleration ($\overline{\alpha}$) from instantaneous angular acceleration (α), where these quantities are defined as follows:

$$\overline{\alpha} = \frac{\Delta \omega}{\Delta t} \qquad \alpha = \lim_{\Delta t \to 0} \frac{\Delta \omega}{\Delta t} \tag{8.2}$$

In most cases considered in this book, the angular acceleration will be constant in time, and there will be no numerical difference between the two quantities above. Angular acceleration is measured in radians per second squared (rad/s^2), since it is an angular velocity (rad/s) divided by a time (s).

Just as linear displacements and velocities are connected to the corresponding angular quantities by $\Delta s = R \Delta \theta$ and $v = R\omega$, so too the tangential acceleration of a point on a wheel is connected to the angular acceleration:

$$a_T = \lim_{\Delta t \to 0} \frac{\Delta v}{\Delta t} = \lim_{\Delta t \to 0} \frac{R \Delta \omega}{\Delta t} = R \lim_{\Delta t \to 0} \frac{\Delta \omega}{\Delta t} = R\alpha \tag{8.3}$$

where a_T measures only the change in the magnitude of the velocity, not in its direction, since a_T is in the direction of v.

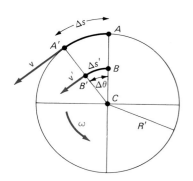

FIGURE 8.3 A rotating wheel; all points on it move with the same *angular velocity* ω but with different *linear* velocities **v**, if they are different distances from the axis of rotation.

TABLE 8.1   Connection of Linear and Angular Quantities

Quantity	Linear motion (translation)		Angular motion (rotation)		Connection	
Displacement	Δs		$\Delta\theta$		$\Delta s = R\,\Delta\theta$	(1.1)
Velocity	$v = \lim\limits_{\Delta t \to 0}\dfrac{\Delta s}{\Delta t}$	(2.3)	$\omega = \lim\limits_{\Delta t \to 0}\dfrac{\Delta\theta}{\Delta t}$	(8.1)	$v = R\omega$	(4.25)
Acceleration	$a_T = \lim\limits_{\Delta t \to 0}\dfrac{\Delta v}{\Delta t}$	(2.6)	$\alpha = \lim\limits_{\Delta t \to 0}\dfrac{\Delta\omega}{\Delta t}$	(8.2)	$a_T = R\alpha$	(8.3)

These important relations are collected in Table 8.1. In the table the relationships are given for instantaneous values of these quantities, although we will be dealing mainly with situations in which either the angular velocity or the angular acceleration is constant. Here the linear quantities are the linear displacement, linear velocity, and linear (tangential) acceleration of a point on a rotating object a distance R from the axis of rotation; $\Delta\theta$, ω, and α are the corresponding angular quantities for rotation about the fixed axis of the object.

Example 8.1

A heavy flywheel of radius 1.60 m starts from rest and rotates with constant angular acceleration until at the end of 15.0 s it is moving with an angular velocity of 20.0 rad/s. Find **(a)** its angular acceleration; **(b)** its average angular velocity during the 15.0-s interval; **(c)** the linear velocity of a point on the rim at the end of 15.0 s; **(d)** the linear (tangential) acceleration of a point on the rim; and **(e)** the linear velocity of a point 0.800 m from its center at the end of 15.0 s.

SOLUTION

(a)   Since the angular acceleration is constant, it is

$$\alpha = \frac{\Delta\omega}{\Delta t} = \frac{\omega - \omega_0}{\Delta t} = \frac{20.0 \text{ rad/s} - 0}{15.0 \text{ s}} = \boxed{1.33 \text{ rad/s}^2}$$

(b)   the average angular velocity is

$$\bar{\omega} = \frac{\omega_0 + \omega}{2} = \frac{0 + 20.0 \text{ rad/s}}{2} = \boxed{10.0 \text{ rad/s}}$$

(c)   From Eq. (4.25),

$$v = R\omega = (1.60 \text{ m})(20.0 \text{ rad/s}) = \boxed{32.0 \text{ m/s}}$$

This velocity is tangent to the rim of the wheel.
Note that the radian is dimensionless and therefore does not appear in the answer for v.

(d)   From Eq. (8.3),

$$a_T = R\alpha = (1.60 \text{ m})(1.33 \text{ rad/s}^2) = \boxed{2.13 \text{ m/s}^2}$$

(e) $v = R\omega = (0.80 \text{ m})(20.0 \text{ rad/s}) = \boxed{16.0 \text{ m/s}}$

The direction of the velocity is at right angles to a line drawn from the center of the wheel to the point at which v is calculated.

8.2   Rotational Motion with Constant Angular Velocity

If a wheel is rotating about an axis with a constant angular velocity ω, then $\omega = \Delta\theta/\Delta t =$ constant and $\Delta\theta = \omega\,\Delta t$. In this case equal angles are swept out in equal time intervals, in the same way that a car moving at a constant speed covers equal distances in equal times. For an example, see Fig. 8.4. If it takes the same time interval Δt for a point on the rim to go from A to B as it takes to go from B to C, then $AB = BC$ and so $\Delta\theta_1 = \Delta\theta_2$. Hence $\Delta\theta_1/\Delta t = \Delta\theta_2/\Delta t = \omega =$

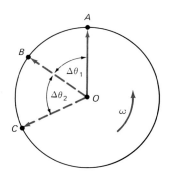

constant. Thus, if a flywheel is rotating at a constant angular velocity of 10.0 rad/s, in 1 min the angular distance covered will be

$$\Delta\theta = \left(10.0 \ \frac{rad}{s}\right)(60.0 \ s) = 600 \ rad$$

Since there is 2π rad in one complete revolution, the number of revolutions made by the wheel in the 1 min is

$$N = 600 \ rad \times \frac{1 \ rev}{2\pi \ rad} = 95.5 \ rev$$

FIGURE 8.4   Rotation with constant angular velocity. Equal angles are swept out in equal times, so that $\Delta\theta_1/\Delta t = \Delta\theta_2/\Delta t = \omega = $ constant.

Example 8.2

A gear wheel on a machine is being rotated by a steel chain at a rate of 3.00 rev/s. **(a)** Through what angle does the wheel turn in 1 min? **(b)** If the wheel's radius is 10.0 cm, how far does a point on the rim move each second? **(c)** Through what distance does the chain move in 1 min?

SOLUTION

(a)   The angular velocity of the wheel is

$$\omega = 3.0 \ rev/s = 3.0 \ (2\pi) \ rad/s = 6\pi \ rad/s$$

The angle swept out in 1 min at this angular velocity is

$$\theta = \omega t = \left(\frac{6\pi \ rad}{s}\right)(60.0 \ s) = \boxed{360\pi \ rad}$$

(b)   When the wheel rotates through an angle θ, a point on the rim moves through a distance

$$s = R\theta = R\omega t = (0.100 \ m)\left(\frac{6\pi}{s}\right)(1.00 \ s) = \boxed{1.88 \ m}$$

Another way to obtain the same result is to note that a point on the rim goes around the circumference of the wheel 3 times per second. The circumference has a length $2\pi(0.100 \ m) = 0.638 \ m$, and so $s = 3(0.638 \ m) = 1.88 \ m$, as before.

(c)   The chain must move along with a point on the rim. Therefore in 1 min the chain moves a distance

$$s = vt = R\omega t = (0.100 \ m)\left(\frac{6\pi}{s}\right)(60.0 \ s) = \boxed{113 \ m}$$

Example 8.3

A bicycle wheel rolls without slipping on the ground, as in Fig. 8.5. The wheel has a radius of 0.330 m and a constant angular velocity of 4.00 rev/s. **(a)** How fast does the bicycle move forward? What is the linear velocity with respect to the ground of a point on the rim that is **(b)** at the top of the wheel, **(c)** at the bottom of the wheel where it makes contact with the ground?

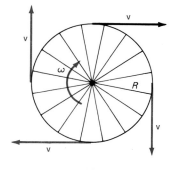

FIGURE 8.5   A bicycle wheel rotating with constant angular velocity ω. (a) The linear speed of a point on the rim is $v = R\omega$. (See next page for parts (b) and (c).)

(a)

SOLUTION

(a) We know that 4 rev/s corresponds to $4(2\pi) = 8\pi$ rad/s, since there is 2π rad in a complete circle, and so

$$\omega = \frac{\Delta\theta}{\Delta t} = \frac{8\pi \text{ rad}}{1.00 \text{ s}} = 25.1 \text{ rad/s}$$

The linear velocity of a point on the rim of the wheel relative to the axle of the wheel is, then, as shown in Fig. 8.5a,

$$v = R\omega = (0.330 \text{ m})(25.1 \text{ rad/s}) = 8.28 \text{ m/s}$$

Now, as a length Δs of the wheel's rim peels off along the ground, the CM of the wheel moves forward a distance Δs with respect to the ground, as shown in Fig. 8.5b, as long as friction keeps the wheel from slipping. The velocity of the CM of the bicycle *with respect to the ground* therefore has a magnitude

$$V = \frac{\Delta s}{\Delta t} = R\omega = \boxed{8.28 \text{ m/s}}$$

(b) For the linear velocities of points at the top and bottom of the rim with respect to the ground, we make use of an equation similar to Eq. (2.4), where all velocities are along the direction of translational motion of the wheel,

$$\mathbf{v}_{RG} = \mathbf{v}_{RA} + \mathbf{v}_{AG}$$

where $\mathbf{v}_{RG}$ is the velocity of a point on the rim with respect to the ground, $\mathbf{v}_{RA}$ is the velocity $\mathbf{v}$ of a point on the rim with respect to the axle (the CM of the wheel), and $\mathbf{v}_{AG}$ is the velocity of the axle with respect to the ground (V). Then, at the top of the wheel, this becomes, from Fig. 8.5c,

$$v_{RG} = v + V = 8.28 \text{ m/s} + 8.28 \text{ m/s} = \boxed{16.6 \text{ m/s}}$$

(c) At the bottom of the wheel,

$$v_{RG} = V - v = 8.28 \text{ m/s} - 8.28 \text{ m/s} = \boxed{0}$$

We see that the point at the bottom of the wheel has zero velocity with respect to the ground. This should be no surprise, since we have assumed that the wheel rolls without slipping.

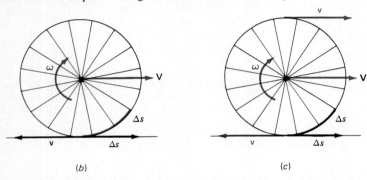

(b)   (c)

FIGURE 8.5 (b) As the rim of the wheel moves through an arc Δs, the wheel moves forward a distance Δs on the ground. The forward speed V of the bicycle is thus numerically equal to the linear speed v of a point on the rim of the wheel with respect to the axle of the wheel. (c) The resultant velocity of the point on the top of the wheel is $2v$ with respect to the ground; the velocity of the bottom of the wheel is zero with respect to the ground.

8.3 Rotational Motion with Constant Angular Acceleration

Let us now consider a slightly more difficult case in which the angular acceleration is constant so that the angular velocity changes with time. The angular motion in this case is completely analogous to the linear motion of a ball falling under the constant acceleration of gravity. To describe the angular motion, first we need the equations governing uniformly accelerated translational motion, already derived in Chap. 2. Then we replace the linear quantities in these equations with the corresponding angular quantities. This gives us the equations governing rotational motion with constant acceleration. These equations are given below, where *it is understood that all angular quantities are expressed in radian measure* and that we replace x in the equations of Chap. 2 by s, since the linear motion is along the arc of a circle.

Linear motion with constant a:		**Angular motion with constant α:**	
$s = s_0 + \bar{v}t$	**(2.15)**	$\theta = \theta_0 + \bar{\omega}t$	**(8.4)**
$\bar{v} = \dfrac{v_0 + v}{2}$	**(2.16)**	$\bar{\omega} = \dfrac{\omega_0 + \omega}{2}$	**(8.5)**
$v = v_0 + at$	**(2.18)**	$\omega = \omega_0 + \alpha t$	**(8.6)**
$s = s_0 + v_0 t + \frac{1}{2}at^2$	**(2.19)**	$\theta = \theta_0 + \omega_0 t + \frac{1}{2}\alpha t^2$	**(8.7)**
$v^2 = v_0^2 + 2a(s - s_0)$	**(2.20)**	$\omega^2 = \omega_0^2 + 2\alpha(\theta - \theta_0)$	**(8.8)**

In the rotational case, the angular velocity changes from an initial value ω_0 (which may be zero) to a final value ω during the time t. Since the angular acceleration is constant, the average value of ω during the time interval t is $\bar{\omega} = (\omega_0 + \omega)/2$, as in Eq. (8.5). The total angular displacement in time t is then $\bar{\omega}t$, and if θ_0 is the initial value of θ, then at time t we have $\theta = \theta_0 + \bar{\omega}t$, which is Eq. (8.4).

The same reasoning can be used to obtain the other equations for angular motion above. But it is easier to apply the equations for linear motion to a point on a rotating wheel of radius R so that s, v, and a_T are the linear displacement, velocity, and tangential acceleration, respectively, of that point as it moves in a circle. Then from Table 8.1, $s = R\theta$, $v = R\omega$, and $a_T = R\alpha$, where R is the distance from the point to the axis of rotation of the wheel. If we divide Eq. (2.18) by R, we obtain

$$\frac{v}{R} = \frac{v_0}{R} + \left(\frac{a_T}{R}\right)t \qquad \text{or} \qquad \omega = \omega_0 + \alpha t$$

which is Eq. (8.6). Similarly, if we divide Eq. (2.19) by R, we obtain

$$\frac{s}{R} = \frac{s_0}{R} + \left(\frac{v_0}{R}\right)t + \frac{1}{2}\left(\frac{a_T}{R}\right)t^2$$

or $\quad \theta = \theta_0 + \omega_0 t + \frac{1}{2}\alpha t^2$

which is Eq. (8.7). Also, we have, on dividing Eq. (2.20) by R^2,

$$\frac{v^2}{R^2} = \frac{v_0^2}{R^2} + \frac{2a_T}{R^2}(s - s_0)$$

or $\quad \left(\dfrac{v}{R}\right)^2 = \left(\dfrac{v_0}{R}\right)^2 + \dfrac{2a_T}{R}\left(\dfrac{s}{R} - \dfrac{s_0}{R}\right)$

and so $\quad \omega^2 = \omega_0^2 + 2\alpha(\theta - \theta_0)$

which is Eq. (8.8).

Solving problems for rotation with constant angular acceleration involves the use of Eqs. (8.4) to (8.8) in a manner very similar to the way we solved problems for linear motion in Chap. 2 by using Eqs. (2.15) to (2.20). Equations (8.4) to (8.8) allow us to solve any problem involving rotational motion with constant acceleration. Figure 8.6 illustrates such motion for a wheel. As time goes on, the angular velocity of the wheel increases so that the line OA sweeps out greater angles in equal times. From Eq. (8.7) the angle swept out varies as t^2, as indicated in the figure.

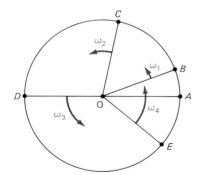

FIGURE 8.6 Rotation of a wheel with constant angular acceleration. Here $\omega = \omega_0 + \alpha t$. If $\omega_0 = 0$, $\omega = \alpha t$, and the following numerical results are obtained: after 1 s, $\omega_1 = 1\alpha$; after 2 s, $\omega_2 = 2\alpha = 2\omega_1$; similarly, $\omega_3 = 3\omega_1$; $\omega_4 = 4\omega_1$; and so forth. The angles swept out in equal time intervals also increase, since, from Eq. (8.7), $\theta = \frac{1}{2}\alpha t^2$ and θ increases as the square of the time. The angles in this diagram correspond to this dependence on t^2.

Example 8.4

The rotor of a medical centrifuge is accelerated from rest with a constant angular acceleration of 7.00 rad/s². **(a)** What is its angular velocity after 5.00 min in radians per second? In revolutions per minute? **(b)** How many revolutions has the centrifuge made in 5.00 min?

SOLUTION

(a) From Eq. (8.6), $\omega = \omega_0 + \alpha t$, and so

$$\omega = 0 + \left(7.00\ \frac{\text{rad}}{\text{s}^2}\right)(5.00\ \text{min})\left(60\ \frac{\text{s}}{\text{min}}\right)$$

$$= \boxed{2.10 \times 10^3\ \text{rad/s}}$$

Then, since 1 rev = 2π rad,

$$\omega = \left(2.10 \times 10^3\ \frac{\text{rad}}{\text{s}}\right)\left(\frac{1\ \text{rev}}{2\pi\ \text{rad}}\right)\left(\frac{60\ \text{s}}{\text{min}}\right)$$

$$= \boxed{2.01 \times 10^4\ \text{rev/min}}$$

(b) From Eq. (8.8), we have $\omega^2 = \omega_0^2 + 2\alpha(\theta - \theta_0)$. Here $\omega_0 = 0$ and $\theta_0 = 0$, and so $\omega^2 = 2\alpha\theta$, or

$$\theta = \frac{\omega^2}{2\alpha} = \frac{(2.10 \times 10^3\ \text{rad/s})^2}{2(7.00\ \text{rad/s}^2)} = 3.15 \times 10^5\ \text{rad}$$

$$= (3.15 \times 10^5\ \text{rad})\left(\frac{1\ \text{rev}}{2\pi\ \text{rad}}\right) = \boxed{5.01 \times 10^4\ \text{rev}}$$

The same result can be obtained by noting that the average angular velocity is, from Eq. (8.5),

$$\bar{\omega} = \frac{\omega_0 + \omega}{2} = \frac{2.01 \times 10^4\ \text{rev/min}}{2} = 1.00 \times 10^4\ \text{rev/min}$$

In 5 min the number of revolutions is then

$$(5\ \text{min})\left(1.00 \times 10^4\ \frac{\text{rev}}{\text{min}}\right) = 5.00 \times 10^4\ \text{rev}$$

which agrees, within a slight rounding error, with the result obtained above.

EXERCISE 1   If a blood sample in the centrifuge is 10.0 cm from the axis of rotation, what is its **(a)** angular acceleration; **(b)** tangential acceleration; **(c)** linear velocity after 5.00 min; **(d)** centripetal acceleration after 5.00 min?

8.4   Work and Rotational Kinetic Energy

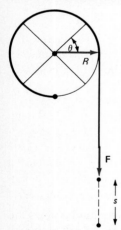

Suppose we set a wheel into rotation about a fixed axis by pulling on a rope wrapped around its circumference, as in Fig. 8.7. If we exert a force of magnitude F and the rope moves a distance s in the direction of the force, then the work done is

$$\mathcal{W} = Fs$$

When we pull on the rope, we exert a torque setting the wheel into rotation. This torque is, as we saw in Sec. 5.4, $\tau = FR$, since the force (as seen in Fig. 8.7) is tangent to the rim so that the lever arm is R, the radius of the wheel. Thus

$$\mathcal{W} = Fs = \frac{\tau s}{R} = \tau\theta$$

since $s/R = \theta$ from Eq. (1.1), and so

$$\boxed{\mathcal{W} = \tau\theta} \tag{8.9}$$

FIGURE 8.7 Torque exerted by a force applied to a rope. The force does an amount of work $\mathcal{W} = \tau\theta$, where the torque τ is equal to FR. The wheel rotates through an angle θ as the rope is pulled down a distance s.

The work done on a wheel in setting it into rotation is the product of the torque acting and the angular displacement of the wheel.

Here the linear displacement s in the usual expression for work is replaced by the angular displacement θ, and the force F by the torque τ.

Similarly, the rate at which work is done, or, in other words, the *power*, is

$$P = \frac{\Delta W}{\Delta t} = \frac{\tau \Delta \theta}{\Delta t} = \tau \omega \qquad (8.10)$$

Note the similarity between this and $P = Fv$ obtained in Sec. 6.9 for linear motion.

Rotational Kinetic Energy

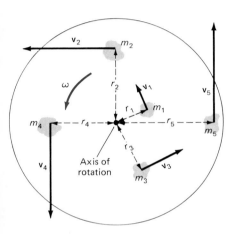

In the absence of friction the effect of the work done by the torque is to increase the wheel's rotational kinetic energy. This energy is the sum of the individual kinetic energies of all the bits of mass m_i making up the wheel, each rotating at its own linear velocity v_i, as in Fig. 8.8. Each bit of kinetic energy is then

$$KE_i = \tfrac{1}{2}m_i v_i^2$$

where the index i refers to a particular bit of mass m_i in the wheel. We also know that $v_i = \omega r_i$ (Eq. 4.25), where ω *is the same for all the elements of mass in the wheel since the wheel rotates as a rigid body*, and r_i is the radial distance of each element of mass from the axis of rotation. Therefore

$$KE_{rot} = \sum_{i=1}^{N} KE_i = \sum_{i=1}^{N} \tfrac{1}{2}m_i v_i^2 = \sum_{i=1}^{N} \tfrac{1}{2}m_i(\omega^2 r_i^2) = \tfrac{1}{2}\left(\sum_{i=1}^{N} m_i r_i^2\right)\omega^2$$

where there are N bits of mass making up the total mass of the wheel.

We call the sum $\sum_{i=1}^{N} m_i r_i^2$ the *rotational inertia I* (or *moment of inertia I*) of the wheel about the axis of rotation. Thus the *rotational kinetic energy* is

$$\boxed{KE_{rot} = \tfrac{1}{2}I\omega^2} \qquad (8.11)$$

FIGURE 8.8 Rotational kinetic energy. KE_{rot} is equal to the sum of the individual kinetic energies of all the bits of mass m_i making up the wheel. As shown in the text, $KE_{rot} = \tfrac{1}{2}I\omega^2$.

where

$$\boxed{I = \sum_{i=1}^{N} m_i r_i^2} \qquad (8.12)$$

is the *rotational inertia* of the rotating object; the units of rotational inertia are kg·m². Note that the rotational inertia of a body is in no way as intrinsic and unique a property of a body as mass is. Rotational inertia *depends on the axis* about which it is calculated, whereas mass does not.

If we compare Eq. (8.11) for rotational kinetic energy with the corresponding expression for translational kinetic energy $KE_{trans} = \tfrac{1}{2}mv^2$, we see that the linear velocity v is replaced by the angular velocity ω, and the mass m (or translational inertia) is replaced by the rotational inertia I. *The rotational inertia therefore plays the same role in rotational motion as does mass in translational motion.* As a result, in extending the parallelism that exists between linear and angular motion to more complicated situations, we must always replace m by its rotational analogue I.

The expression for rotational kinetic energy, $\tfrac{1}{2}I\omega^2$, is correct dimensionally. Since

$$I = \sum_{i=1}^{N} m_i r_i^2 \qquad KE_{rot} \text{ has dimensions } [M][L^2]\frac{1}{[T^2]}$$

which are the same as for translational kinetic energy ($\tfrac{1}{2}mv^2$).

Significance of Rotational Inertia

For translational motion mass is a measure of inertia, that is, of how difficult it is to accelerate an object having that mass. In rotational motion it is not mass alone that counts, but mass times the square of the distance of that mass from the axis of rotation. Consider a rotating disk like a phonograph record. Some bits of mass of the record are close to the axis and some are far away from that axis. The bits close to the axis have a small r_i and therefore a very small $m_i r_i^2$, and so contribute very little to the rotational inertia. Bits of mass m_i far from the axis possess large values of $m_i r_i^2$ and contribute a great deal to I. When we multiply all these bits of mass by their proper r_i^2 and add the results, we obtain the rotational inertia I for the whole record. It is this rotational inertia that determines, for example, the torque needed to produce a given angular acceleration in the record. This will be discussed in Sec. 8.5.

Two objects might have different rotational inertias even though they have the same mass. For example, a disk and a hoop might have the same mass and the same radius, but the hoop, with most of its mass on the rim, has a greater rotational inertia.

In devices like flywheels and gyroscopes, it is desirable to maintain as constant an angular velocity as possible, despite torques due to friction and air resistance that may act to slow down the rotating object. For this reason, the bulk of the mass in a flywheel is placed near the rim. The angular acceleration produced by outside torques is reduced to a minimum because the rotational inertia has been made a maximum. Figure 8.9 shows a rim-loaded flywheel, i.e., a flywheel with almost all its mass at a distance R from the axis of rotation. Since all the bits of mass are at approximately the same distance R, its rotational inertia is

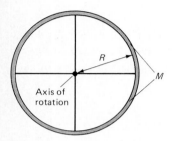

FIGURE 8.9 A rim-loaded flywheel with almost all its mass concentrated at the same distance from the axis of rotation. In this case the rotational inertia is $I = MR^2$.

$$I = \sum_{i=1}^{N} m_i r_i^2 = \left(\sum_{i=1}^{N} m_i \right) R^2 = MR^2$$

Example 8.5

A light rope is wrapped 6 times around the rim of a wheel of radius 0.750 m. The rope is then pulled off the rim by a person exerting a constant force of 20.0 N on the rope tangent to the rim. **(a)** How much work is done in rotating the wheel? **(b)** If we neglect friction and the KE of the rope, what is the final rotational kinetic energy of the wheel? **(c)** If the rotational inertia of the wheel is 10.0 kg·m², what is its angular velocity when it is no longer being accelerated?

SOLUTION

(a) From Eq. (8.9), $\mathcal{W} = \tau\theta$. Here the torque is

$$\tau = FR = (20.0 \text{ N})(0.750 \text{ m}) = 15.0 \text{ N·m}$$

and the angle $\theta = (6 \text{ rev})(2\pi \text{ rad/rev}) = 12\pi \text{ rad}$

and so $\quad \mathcal{W} = (15.0 \text{ N·m})(12\pi \text{ rad}) = \boxed{5.65 \times 10^2 \text{ J}}$

(b) The final kinetic energy of the wheel is the same as the work done, that is, $\boxed{5.65 \times 10^2 \text{ J}}$.

(c) From Eq. (8.11), $\text{KE}_{\text{rot}} = \frac{1}{2}I\omega^2$, and so

$$\omega = \sqrt{\frac{2\text{KE}_{\text{rot}}}{I}} = \sqrt{\frac{2(5.65 \times 10^2 \text{ J})}{10.0 \text{ kg·m}^2}} = \boxed{10.6 \text{ rad/s}}$$

8.5 Newton's Second Law for Rotational Motion

As said before, the most important law in classical mechanics is Newton's second law of motion. Applied to translational motion, this takes the form $\mathbf{F} = m\mathbf{a}$. We now show what form Newton's second law takes when applied to rotational motion.

FIGURE 8.10 Looking down on a small puck forced to rotate about point O on a frictionless table. The net torque needed to give the puck an angular acceleration α is $\tau_{\text{net}} = I\alpha$.

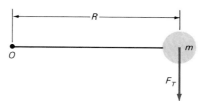

Consider a small puck of mass m attached to a very light rod on a frictionless air table, as in Fig. 8.10. The puck rotates about the end of the rod at point O and therefore in a circle of radius R. If we apply a tangential force F_T on the puck at right angles to the radius R, then from Newton's second law, $F_T = ma_T$, where a_T is the tangential acceleration of the puck. Then, on multiplying by the radius R, we have

$$RF_T = mRa_T = mR(R\alpha)$$

since $a_T = R\alpha$, where α is the angular acceleration of the puck. Now, since RF_T is the product of the force and its lever arm, it is just the torque acting, and so

$$\tau = RF_T = (mR^2)\alpha = I\alpha$$

for in this case mR^2 is equal to the rotational inertia I of the puck about the axis of rotation O. This result for the torque turns out to be true for all objects, no matter what their size or shape. It is also true for any number of forces exerting torques, as long as we replace the vector sum of these forces by $\mathbf{F}_{\text{net}}$. If τ_{net} is the net torque acting, we can write *Newton's second law for rotational motion* as

$$\boxed{\tau_{\text{net}} = I\alpha}$$

(8.13)

This law states that *the angular acceleration of an object, produced by a net external torque acting on the object, is directly proportional to the net torque and inversely proportional to the rotational inertia of the object*. Note that in Eq. (8.13) $\mathbf{F}_{\text{net}}$ in Newton's second law is replaced by its rotational analogue τ_{net}, the acceleration is replaced by its rotational counterpart α, and the mass m by the rotational inertia I, as expected.

The torque needed to stop a 5-ton flywheel of radius 5 m rotating at 10 rev/s is much larger than would be required to stop a small solid disk spinning at 2 rev/s. The mass of the flywheel is larger than that of the disk, and the average distance of this mass from the axis of rotation is far greater for the flywheel than for the disk. A much larger torque is therefore needed to produce the same angular acceleration.

Newton's laws also help explain why a tightrope walker in the circus often carries a long pole, held in such a way as to give it a large rotational inertia with respect to the rope as an axis of rotation. If the circus performer starts to lose balance and fall, he or she exerts a torque on the pole. By the angular analogue of Newton's third law, the pole then exerts a reaction torque back on the tightrope walker. Because the pole has a large rotational inertia, its initial angular acceleration about the wire is very small, but the reaction torque on the performer is much more effective in restoring balance because of the performer's smaller rotational inertia.

Example 8.6

A frictionless wheel of rotational inertia $I = 2.50 \times 10^{-3}$ kg·m² and radius 10.0 cm is mounted as in Fig. 8.11a. A cord wound around it a number of times is attached to a mass $m = 1.00$ kg that is then allowed to fall under gravity. Find **(a)** the linear acceleration of the mass m; **(b)** the tension in the cord (assumed massless); **(c)** the downward speed of the mass after it has fallen a distance of 1.00 m. **(d)** Prove that mechanical energy is conserved in the system by comparing the initial mechanical energy with the mechanical energy after the mass has fallen 1.00 m.

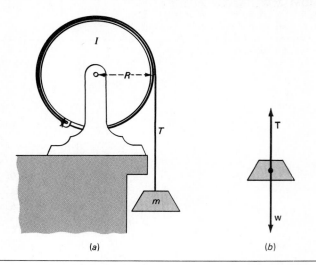

FIGURE 8.11

(a)   (b)

SOLUTION

(a) Since the acceleration of the mass is obviously downward, we take the downward direction as positive. Then the torque τ on the wheel exerted by the tension (T) in the rope produces an angular acceleration α so that

$$\tau_{net} = TR = I\alpha \quad \text{and} \quad T = \frac{I\alpha}{R}$$

or, since $\alpha = a/R$, where a is the linear acceleration,

$$T = \frac{Ia}{R^2}$$

(Note that T means *tension*, not *torque*; τ is the torque.)

Let us now apply Newton's second law to the linear motion of the mass m, as shown in the free-body diagram of Fig. 8.11b.

$$F_{net} = mg - T = ma$$

or, using the result above for T,

$$mg - \frac{Ia}{R^2} = ma$$

$$a\left(m + \frac{I}{R^2}\right) = mg$$

and so

$$a = \frac{m}{m + I/R^2} g$$

The acceleration is therefore less than the acceleration due to gravity, as might be expected.

Putting in numbers, we find that

$$a = \left[\frac{1.00 \text{ kg}}{1.00 \text{ kg} + (2.50 \times 10^{-3} \text{ kg·m}^2)/(0.100 \text{ m})^2}\right]\left(9.80 \frac{\text{m}}{\text{s}^2}\right)$$

$$= \left(\frac{1.00}{1.25}\right)(9.80 \text{ m/s}^2) = \boxed{7.84 \text{ m/s}^2}$$

(b) The tension in the cord is

$$T = \frac{Ia}{R^2} = \frac{(2.50 \times 10^{-3} \text{ kg·m}^2)(7.84 \text{ m/s}^2)}{(0.100 \text{ m})^2} = \boxed{1.96 \text{ N}}$$

(c) The downward speed of the mass after it has fallen a distance of 1.00 m can be found from Eq. (2.20) $v^2 = v_0^2 + 2a(y - y_0)$. Here y_0 and v_0 are both zero, and so we have

$$v^2 = 2ay = 2(7.84 \text{ m/s}^2)(1.00 \text{ m}) = 15.7 \text{ m}^2/\text{s}^2$$

$$v = \boxed{3.96 \text{ m/s}}$$

(d) The system here includes the mass m, the wheel, and the gravitational force provided by the earth. We choose the initial position of the mass as the zero position for calculating PE_{grav}. Then the initial mechanical energy of the system is zero, and at any instant of time, we must have

$$PE_{grav} + KE_{trans} + KE_{rot} = 0$$

After the mass has *fallen* 1.00 m, we have

$$PE_{grav} = -mgy = -(1.00 \text{ kg})(9.80 \text{ m/s}^2)(1.00 \text{ m})$$

$$= -9.80 \text{ J}$$

$$KE_{trans} = \tfrac{1}{2}mv^2 = \tfrac{1}{2}(1.00 \text{ kg})(3.96 \text{ m/s})^2 = 7.84 \text{ J}$$

$$KE_{rot} = \tfrac{1}{2}I\omega^2 = \tfrac{1}{2}I(v/R)^2$$

$$= \tfrac{1}{2}(2.50 \times 10^{-3} \text{ kg·m}^2)\left(\frac{3.96 \text{ m/s}}{0.100 \text{ m}}\right)^2 = 1.96 \text{ J}$$

Then the total mechanical energy is

$$-9.80 \text{ J} + 7.84 \text{ J} + 1.96 \text{ J} = \boxed{0}$$

and so the total energy is conserved. Note that the force of gravity does no work directly on the wheel. It does work on the hanging mass, and the hanging mass, in turn, does work on the wheel to give it KE_{rot}.

8.6   Rotational Inertias

In some simple cases it is possible to calculate rotational inertias. For example, in a bicycle wheel most of the mass is located at the same distance R from the axis of rotation, and so we have

$$I = \sum_{i=1}^{N} m_i r_i^2 = \sum_{i=1}^{N} m_i R^2 = MR^2$$

Next consider a dumbbell of length L rotating about an axis O at its center and perpendicular to the length of the dumbbell, as in Fig. 8.12. If the mass of each ball is m, and the rod has negligible mass, we have

$$I = \sum_{i=1}^{N} m_i r_i^2 = m \left(\frac{L}{2}\right)^2 + m \left(\frac{L}{2}\right)^2 = \frac{mL^2}{2}$$

For most objects, where the mass is irregularly distributed over a considerable region of space, it is necessary to find the rotational inertia by experiment. For some symmetrical objects, calculus can be used to find the average value of r_i^2. In Table 8.2 some values of I obtained in this way are given for a few important symmetrical objects. Table 8.2 shows that the rotational inertia depends on the axis about which it is calculated: the rotational inertia of a thin rod about an axis perpendicular to one end is $\frac{1}{3}ML^2$, but about an axis perpendicular to the rod's center, it is only $\frac{1}{12}ML^2$.

The mass of a body *cannot* be considered as concentrated at the center of mass for the purpose of computing its rotational inertia. For a bicycle wheel the center of mass of the wheel is at its center, which is on the axis of rotation. This means that, if all the mass were considered concentrated at that point, R would be zero, and so would the rotational inertia I—which is clearly wrong.

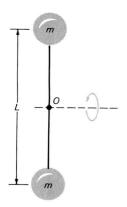

FIGURE 8.12   Rotational inertia of a dumbbell: $I = mL^2/2$.

TABLE 8.2   Rotational Inertia of Various Solid Objects

Solid object	Axis of rotation	Diagram	Rotational inertia I	Solid object	Axis of rotation	Diagram	Rotational inertia I
Cylindrical shell of radius R	Axis of cylinder		MR^2	Solid cylinder of radius R	Axis of cylinder		$\frac{1}{2}MR^2$
Solid sphere of radius R	Any diameter		$\frac{2}{5}MR^2$	Spherical shell of radius R	Any diameter		$\frac{2}{3}MR^2$
Uniform thin rod of length L	Perpendicular to center		$\frac{1}{12}ML^2$	Uniform thin rod of length L	Perpendicular to one end		$\frac{1}{3}ML^2$

Example 8.7

Find the rotational inertia of a uniform thin rod of length 0.215 m and mass 0.423 kg about an axis perpendicular to one end.

SOLUTION

We have, from Table 8.2,

$$I = \tfrac{1}{3}ML^2 = \tfrac{1}{3}(0.423 \text{ kg})(0.215 \text{ m})^2 = \boxed{6.52 \times 10^{-3} \text{ kg·m}^2}$$

Example 8.8

A solid metal sphere of radius 5.00 cm and mass 500 g starts from rest at the top of an incline of height $H = 20.0$ cm and rolls down the incline, as in Fig. 8.13. How fast is it moving along the ground at the bottom of the incline? Assume that it rolls without slipping and that any rolling friction is negligible.

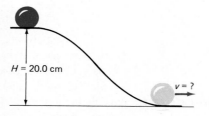

FIGURE 8.13

SOLUTION

This is a good example of the conservation of mechanical energy, since the initial gravitational potential energy is converted to *both* translational kinetic energy and rotational kinetic energy.

The initial gravitational PE with respect to the ground is $\text{PE}_{\text{grav}} = MgH$. The final translational KE is $\tfrac{1}{2}Mv^2$, and the final rotational KE is $\tfrac{1}{2}I\omega^2$. From the conservation-of-energy principle,

$$MgH = \tfrac{1}{2}Mv^2 + \tfrac{1}{2}I\omega^2$$

But $v = R\omega$, and for a solid sphere we have, from Table 8.2, $I = \tfrac{2}{5}MR^2$. Hence

$$MgH = \tfrac{1}{2}Mv^2 + \tfrac{1}{2}(\tfrac{2}{5}MR^2)\omega^2 = \tfrac{1}{2}Mv^2 + \tfrac{1}{2}(\tfrac{2}{5}M)v^2$$

On canceling M throughout, we have

$$v^2 = \frac{gH}{\tfrac{1}{2} + \tfrac{1}{5}} = \tfrac{10}{7}gH \qquad \text{and}$$

$$v = \sqrt{\tfrac{10}{7}gH} = \boxed{1.67 \text{ m/s}}$$

In this case of a solid sphere, therefore, the rotational kinetic energy is two-fifths the translational kinetic energy. Note that v is independent of both M and R. Can you explain why?

Parallel-Axis Theorem

Table 8.2 gives the rotational inertias for various solids about certain special axes within the solid. Sometimes, however, we desire the rotational inertia of a solid object about a different axis inside or outside the object. In this case a very useful relationship is the *parallel-axis theorem*:

If I_{CM} is the rotational inertia of an object of total mass M about any axis through its center of mass (CM), then the rotational inertia about a parallel axis a distance d from the center of mass is

$$\boxed{I = I_{CM} + Md^2}$$

(8.14)

We will not prove this theorem here, but its use is illustrated in Example 8.9. The parallel-axis theorem is particularly useful in problems that combine translational motion with rotational motion.

Example 8.9

If the rotational inertia of a uniform thin rod of length L about an axis through its center and perpendicular to its length is $\frac{1}{12}ML^2$, find its rotational inertia about an axis at one end and perpendicular to its length.

SOLUTION

About an axis at one end, which is a distance $d = L/2$ from the center of mass, we have from the parallel-axis theorem,

$$I = I_{CM} + Md^2 = \tfrac{1}{12}ML^2 + M\left(\frac{L}{2}\right)^2 = \boxed{\tfrac{1}{3}ML^2}$$

This agrees with the result found in Table 8.2.

8.7 Angular Momentum and Rotational Impulse

In the preceding chapter we saw that the linear momentum of a particle or system of particles is defined as $\mathbf{p} = m\mathbf{v}$, and that Newton's second law can then be written in the more general form

$$\mathbf{F}_{net} = \frac{\Delta \mathbf{p}}{\Delta t} \qquad \text{or} \qquad \mathbf{F}_{net}\,\Delta t = \Delta \mathbf{p}$$

where $\mathbf{F}_{net}\,\Delta t$ is called the *impulse* of the force $\mathbf{F}_{net}$.

A perfectly parallel treatment of rotational motion is possible. We define an *angular momentum*, by analogy with the case of linear momentum ($\mathbf{p} = m\mathbf{v}$), as

$$\boxed{L = I\omega} \tag{8.15}$$

Here rotational inertia replaces the ordinary inertia (mass), and angular velocity replaces linear velocity. The impulse-momentum theorem then takes the form

$$\tau_{net} = \frac{\Delta L}{\Delta t}$$

or $\qquad \tau_{net}\Delta t = \Delta L = \Delta(I\omega) \tag{8.16}$

This states that the effect of a constant net torque exerted over a time interval Δt is to change the angular momentum of the system by an amount ΔL. Here $\tau_{net}\Delta t$ is the *rotational impulse* of the torque τ_{net}.

Clearly, if the rotational inertia does not change while the torque acts, we have

$$\tau_{net}\Delta t = I\Delta\omega \qquad \text{or} \qquad \tau_{net} = I\frac{\Delta\omega}{\Delta t} = I\alpha$$

and so $\qquad \boxed{\tau_{net} = I\alpha}$ $\tag{8.13}$

which is Newton's second law for rotational motion, as discussed in Sec. 8.5. Here again we see the significance of I as an inertial term. The larger the rotational inertia I, the greater the torque required to produce the same angular acceleration.

The units of angular momentum are the units of $I\omega$, that is, $\text{kg}\cdot\text{m}^2/\text{s}$. Notice how this differs from the units for linear momentum, which are $\text{kg}\cdot\text{m/s}$.

Example 8.10

For the situation of Example 8.6, shown in Fig. 8.11, find **(a)** the net torque; **(b)** the angular acceleration of the wheel; **(c)** the tangential acceleration of a point on the rim; **(d)** the angular velocity of the wheel after the mass has fallen 1.00 m. **(e)** Show that the torque acting is equal to the time rate of change of the angular momentum.

SOLUTION

(a) Here $\tau_{net} = RT$, where T is the tension in the cord, found in Example 8.6b to be 1.96 N. Therefore

$$\tau_{net} = (0.100 \text{ m})(1.96 \text{ N}) = \boxed{0.196 \text{ N·m}}$$

(b) $\alpha = \dfrac{\tau_{net}}{I} = \dfrac{0.196 \text{ N·m}}{2.50 \times 10^{-3} \text{ kg·m}^2} = \boxed{78.4 \text{ rad/s}^2}$

(c) $a_T = R\alpha = (0.100 \text{ m})(78.4 \text{ rad/s}^2) = \boxed{7.84 \text{ m/s}^2}$

This result agrees with the acceleration of the mass m found in Example 8.6a, as it must, since the rim of the wheel must have the same linear speed as the nonslipping cord attached to it.
(d) When the mass has fallen 1.00 m, the linear displacement of a point on the rim is 1.00 m and the angular displacement of the wheel is $\theta = s/R = 1.00 \text{ m}/0.100 \text{ m} = 10.0 \text{ rad}$. Then

$$\omega^2 = \omega_0^2 + 2\alpha(\theta - \theta_0)$$

$$= 2(78.4 \text{ rad/s}^2)(10.0 \text{ rad})$$

$$= 1.57 \times 10^3 \text{ rad}^2/\text{s}^2$$

$$\omega = \boxed{39.6 \text{ rad/s}}$$

This yields for the linear velocity of a point on the rim:

$$v = R\omega = (0.100 \text{ m})(39.6 \text{ rad/s}) = 3.96 \text{ m/s}$$

in agreement with the speed of the mass found in Example 8.6c.
(e) Let us consider the change in angular momentum between $t = 0$ and the time when the mass has fallen 1.00 m. Since the wheel starts from rest, this is

$$\Delta L = I\omega = (2.50 \times 10^{-3} \text{ kg·m}^2)(39.6 \text{ rad/s})$$

$$= \boxed{9.90 \times 10^{-2} \text{ kg·m}^2/\text{s}}$$

Since the acceleration is constant, the time required to produce this angular momentum is the same time required to produce a linear velocity of 3.96 m/s. This is

$$\frac{v}{a} = \frac{3.96 \text{ m/s}}{7.84 \text{ m/s}^2} = 0.505 \text{ s}$$

Then $\tau_{net}\Delta t = (0.196 \text{ N·m})(0.505 \text{ s})$

$$= \boxed{9.90 \times 10^{-2} \text{ kg·m}^2/\text{s}}$$

and the net torque acting is indeed equal to the time rate of change in angular momentum.

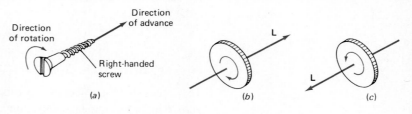

FIGURE 8.14 Vector nature of angular momentum. (a) A right-handed screw (an ordinary carpenter's screw) moves away from the carpenter when it is turned in a clockwise direction and toward the carpenter when it is turned in a counterclockwise direction. The angular momentum is defined to be a vector **L** along the axis of rotation. For a rotating wheel, the direction of **L** is determined by imagining how a right-handed screw would move when turned in the direction of the wheel's rotation. (b) A right-handed screw turned clockwise would move in the direction of **L** shown. (c) A right-handed screw turned counterclockwise would move in the opposite direction, as shown.

Vector Nature of Angular Momentum

Just as torque is a vector, so too is angular momentum. Its direction is along the axis of rotation in the direction given by the right-hand rule discussed in Sec. 5.4* (see Fig. 8.14). Two angular momenta must therefore be added as vectors, and only if the rotations are about the same axis can we simply add or subtract the magnitudes of two angular momentum vectors (The same is true of angular velocities, but we have neglected this complication in the early parts of this chapter).

---

*The angular momentum (about some axis) of a particle of mass m moving with linear momentum $\mathbf{p} = m\mathbf{v}$ can also be written as the cross product $\mathbf{L} = \mathbf{r} \times \mathbf{p}$, where $\mathbf{r}$ is the position vector of the particle with respect to the axis of rotation. The magnitude of $\mathbf{L}$ is then $L = rmv \sin \theta$. If a rigid body is rotating about an axis through its CM, $\mathbf{v}$ is perpendicular to $\mathbf{r}$ and so $\sin \theta = 1$. Therefore $L = rmv = mr^2\omega$, which is the same as $I\omega$ in Eq. (8.15).

Gyroscopes

FIGURE 8.15 A small gyroscope (*Courtesy of Central Scientific Co.*).

The importance of the vector nature of angular momentum is illustrated by a gyroscope of the type shown in Fig. 8.15. This gyro is shown mounted horizontally in Fig. 8.16, with its weight supported by the pivot at O. The wheel of the gyroscope is rotating with an angular velocity $\boldsymbol{\omega}$, shown as a vector along the axis of the gyroscope. The gyro therefore also has an angular momentum $\mathbf{L} = I\boldsymbol{\omega}$ along the rotation axis.

The forces acting on the gyroscope are its weight $\mathbf{w} = m\mathbf{g}$ directed downward and the upward force $\mathbf{N}$ exerted by the pivot at O. These forces are equal in magnitude and opposite in direction, and so no net translational motion results.

There is, however, a net torque about O produced by the force of gravity acting on the CG of the gyro. This torque $\boldsymbol{\tau}$ is about the axis marked by $\boldsymbol{\tau}$ in the figure. It does not, however, cause the spinning gyro to fall: the gyroscope, according to the previous paragraph, is in translational equilibrium. What this net torque actually does is produce a change $\Delta\mathbf{L}$ in the angular momentum $\mathbf{L}$ of the gyroscope.

The change in the angular momentum is of amount $\Delta\mathbf{L} = \boldsymbol{\tau}\Delta t$ in time Δt, where $\Delta\mathbf{L}$ is in the direction of $\boldsymbol{\tau}$ and so at right angles to $\mathbf{L}$. When we add $\Delta\mathbf{L}$ to $\mathbf{L}$, as in Fig. 8.16b, we obtain a new vector $\mathbf{L} + \Delta\mathbf{L}$ in the direction shown. But this must be *the direction of the new angular momentum of the system*, and so the rotational axis moves from the direction of $\mathbf{L}$ to that of the vector $\mathbf{L} + \Delta\mathbf{L}$. This rotation of the axis of the spinning wheel in a horizontal plane is called *precession* and is one of the most impressive features of gyroscope motion.

It is an interesting and paradoxical result of the vector nature of angular momentum that the *downward* force of gravity results in this case in a *horizontal* precession. If the gyro is first set spinning on the pivot with its spin axis almost vertical, as in Fig. 8.15, the precession is very slow. As friction slows the top down, however, and the spin axis assumes a more horizontal position, the torque increases as the lever arm of the force of gravity increases. This means that $\Delta L / \Delta t$ increases, and the gyroscope precesses more and more rapidly until it finally falls to the ground.

FIGURE 8.16 (a) A gyroscope mounted with its axis horizontal. (b) The change $\Delta\mathbf{L}$ in the angular momentum of the gyroscope that causes its axis to precess in a horizontal plane.

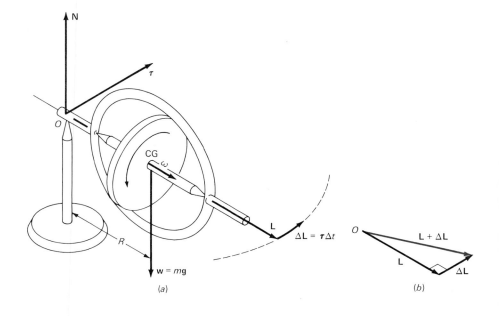

$\Delta L = \tau \Delta t$

$L + \Delta L$

(a)

(b)

8.8 Conservation of Angular Momentum

Since from Eq. (8.16) $\tau_{\text{net}} = \Delta\mathbf{L}/\Delta t$, if no net external torque acts in the time Δt, we have $\Delta\mathbf{L} = 0$ or $\mathbf{L} = I\boldsymbol{\omega} = $ constant, and so

$$I_1\boldsymbol{\omega}_1 = I_2\boldsymbol{\omega}_2 \qquad\qquad\qquad\text{(8.17)}$$

This is a statement of the *principle of conservation of angular momentum*:

> If no net external torque acts on a system, the total angular momentum of the system remains constant.

Since $\mathbf{L} = I\boldsymbol{\omega}$, if no net external torque acts we can conclude:

1 If I is constant, then the angular speed of the system remains constant; in other words, the angles swept out in equal time intervals do not change with time.

2 Since the direction of $\mathbf{L}$ is the direction of the axis of rotation, the direction of the axis of rotation remains unchanged.

Examples of Angular Momentum Conservation

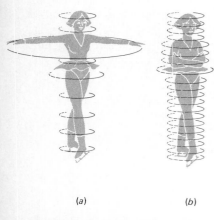

(a) (b)

FIGURE 8.17 Conservation of angular momentum: a figure skater doing a fast pirouette. Her rotational speed increases when her arms are brought close to her body.

Because of its large rotational inertia, a flywheel tends to respond to external torques with only small angular accelerations. As a result, the flywheel's angular velocity does not change much in response to external torques. Such a flywheel is useful for storing energy because you can put a lot of energy into it and because, if no external torques act, its angular momentum remains unchanged as it rotates. Since its rotational inertia I is a constant, its angular velocity also remains unchanged and the flywheel continues to rotate for a very long time at its original angular speed. In the absence of friction, therefore, the initial rotational kinetic energy given to it remains unchanged and stored in the flywheel for future use.

A navigational gyroscope is a heavy rotational wheel like a flywheel, mounted in frictionless gimbal bearings in such a way that even if the mount is moved, no torque is exerted on the wheel of the gyro. The rotating wheel therefore tends to keep its axis pointing in the same fixed direction. The gyro can therefore be used to determine the direction of an airplane or boat with respect to the gyro's fixed orientation.

We have all watched a figure skater doing a fast pirouette. She starts her spin with arms extended, as in Fig. 8.17a. Then she pulls her arms in close to her body, as in Fig. 8.17b, and her rotational speed increases greatly. Since the net external torque (due to air resistance and friction between her skates and the ice) is negligible, her angular momentum $L = I\omega$ remains constant. However, her moment of inertia I decreases as she pulls in her hands. Since $I_1\omega_1 = I_2\omega_2$, her final rotation speed $\omega_2 = \omega_1(I_1/I_2)$ is larger than her initial rotation speed, because her final rotational inertia I_2 is smaller than her initial rotational inertia I_1.

Kepler's Second Law

This law (see Sec. 4.8) states that a planet, in its elliptical path around the sun, sweeps out equal areas in equal times. Consider a planet at two different positions in its orbit around the sun, first at aphelion, the planet's farthest distance from the

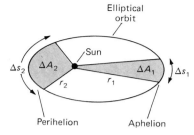

FIGURE 8.18 Kepler's second law. For a planet moving in orbit around the sun, equal areas are swept out in equal times because angular momentum must be conserved.

sun, and then at perihelion, its nearest approach to the sun, as in Fig. 8.18. The rate at which the area is swept out in the vicinity of aphelion is $\Delta A_1/\Delta t$, where ΔA_1 is marked in the figure, and the time rate at which area is swept out near perihelion is $\Delta A_2/\Delta t$. We now show that Kepler's second law follows immediately from the principle of conservation of angular momentum.

Since no net external torque acts on the sun-planet system, we have, from Eq. (8.17), for the motion of the planet around the sun,

$$I_1\omega_1 = I_2\omega_2 \qquad \text{or} \qquad mr_1^2\omega_1 = mr_2^2\omega_2$$

and so $r_1v_1 = r_2v_2$, since $v = r\omega$. But the rate at which area is being swept out is, to a good approximation, equal to one-half the base times the altitude of each approximately triangular segment shown in Fig. 8.18, so

$$\frac{\Delta A_1}{\Delta t} = \tfrac{1}{2}r_1\frac{\Delta s_1}{\Delta t} = \tfrac{1}{2}r_1v_1$$

since $v_1 = \Delta s_1/\Delta t$ is the length of arc swept out in unit time. But $r_1v_1 = r_2v_2$ and so

$$\frac{\Delta A_1}{\Delta t} = \tfrac{1}{2}r_1v_1 = \tfrac{1}{2}r_2v_2 = \frac{\Delta A_2}{\Delta t}$$

or equal areas are swept out in equal times—which is Kepler's second law.

This calculation is only approximate, but it becomes exact when we go to the limit of $\Delta t \to 0$.

Example 8.11

A man stands at the center of a turntable, holding his arms extended horizontally with a 5.0-kg dumbbell in each hand. He is set rotating slowly about a vertical axis, with an angular velocity of 1 rev every 1.5 s. The rotational inertia of the turntable and the man without the dumbbells is 5.0 kg·m², and the change in this value when he brings his empty hands to his sides would be small compared to the effect of the dumbbells. The man holds the dumbbells originally at a distance of 1.0 m from the axis of rotation, while rotating, and then drops them quickly to his sides at a distance of 20 cm from the axis of rotation. What is his angular velocity after he drops the dumbbells to his sides?

SOLUTION

Since there are no external torques acting, angular momentum is conserved, and so $I_1\omega_1 = I_2\omega_2$, where I_1 and I_2 are the initial and final rotational inertias and ω_1 and ω_2 are the initial and final angular velocities. The rotational inertia of a rotating system is just the sum of the rotational inertias of all the parts of the system rotating about the same axis. In this case the two parts are the man and the dumbbells.

We have for the initial and final values of I,

$$I_1 = 5.0 \text{ kg·m}^2 + 2(5.0 \text{ kg})(1.0 \text{ m})^2 = 15 \text{ kg·m}^2$$

$$I_2 = 5.0 \text{ kg·m}^2 + 2(5.0 \text{ kg})(0.20 \text{ m})^2 = 5.4 \text{ kg·m}^2$$

Then, since $\omega_1 = 1 \text{ rev}/1.5 \text{ s} = 2\pi \text{ rad}/1.5 \text{ s} = 4.2 \text{ rad/s}$,

$$\omega_2 = \left(\frac{I_1}{I_2}\right)\omega_1 = \left(\frac{15 \text{ kg·m}^2}{5.4 \text{ kg·m}^2}\right)(4.2 \text{ rad/s})$$

$$= 12 \text{ rad/s} = \boxed{1.9 \text{ rev/s}}$$

The angular velocity has increased from 4.2 to 12 rad/s, or by about a factor of 3. Note that, because of the approximations used in solving this problem, it would be unwarranted to carry more than two significant figures in the answer to the problem.

Example 8.12

When the gravitational collapse of a star occurs, its radius shrinks dramatically. Suppose that the sun undergoes such a collapse some time in the future and that its radius shrinks from its present 6.96×10^8 m to about 20 km, which is typical of some neutron stars. (A neutron star is an extremely small, extremely dense celestial object comparable to the sun in mass but only a few kilometers in radius.) What would happen to the present spin rate of the sun, which is about $\tfrac{1}{30}$ rev/day?

SOLUTION

From the principle of conservation of angular momentum, we have $I_1\omega_1 = I_2\omega_2$, where for a solid sphere $I = \frac{2}{5}MR^2$ (from Table 8.2), and so

$$\omega_2 = \frac{R_1^2}{R_2^2}\omega_1 \quad \text{or} \quad f_2 = \frac{R_1^2}{R_2^2}f_1$$

since $f = \omega/2\pi$ is the frequency of rotation. Therefore

$$f_2 = \frac{(6.96 \times 10^8 \text{ m})^2}{(20 \times 10^3 \text{ m})^2}\left(\frac{1}{30}\frac{\text{rev}}{\text{day}}\right) = \boxed{4.0 \times 10^7 \text{ rev/day}}$$

This is 470 revolutions a second! Such large rotational frequencies have actually been observed in pulsars, which are believed to be rotating neutron stars of enormous densities.

Summary: Important Definitions and Equations

Most of the important ideas and equations of this chapter are summarized in the accompanying Table 8.3.

TABLE 8.3   Comparison between Translational and Rotational Quantities

Linear motion (translation)	Physical quantity	Angular motion (rotation)
Δs(m)	Displacement $\Delta s = R\,\Delta\theta$	$\Delta\theta$(rad)
$v = \lim\limits_{\Delta t \to 0}\dfrac{\Delta s}{\Delta t}$	Velocity $v = R\omega$	$\omega = \lim\limits_{\Delta t \to 0}\dfrac{\Delta\theta}{\Delta t}$
$a_T = \lim\limits_{\Delta t \to 0}\dfrac{\Delta v}{\Delta t}$	Acceleration $a_T = R\alpha$	$\alpha = \lim\limits_{\Delta t \to 0}\dfrac{\Delta\omega}{\Delta t}$
m	Inertia	$I = \sum\limits_i m_i r_i^2$
F	Force; torque	$\tau = F\ell$
$\mathcal{W} = \mathbf{F}\cdot\mathbf{s}$	Work	$\mathcal{W} = \tau\theta$
$P = \mathbf{F}\cdot\mathbf{v}$	Power	$P = \tau\omega$
$\mathbf{F}_{\text{net}} = m\mathbf{a} = \dfrac{\Delta\mathbf{p}}{\Delta t}$	Newton's second law	$\tau_{\text{net}} = I\alpha = \dfrac{\Delta L}{\Delta t}$
$\mathbf{p} = m\mathbf{v}$	Momentum	$L = I\omega$
$\mathbf{F}_{\text{net}}\,\Delta t = \Delta\mathbf{p}$	Impulse	$\tau_{\text{net}}\,\Delta t = \Delta L$
$\frac{1}{2}mv^2$	Kinetic energy	$\frac{1}{2}I\omega^2$
If $F_{\text{net}} = 0$, $p_2 = p_1$ $m_2 v_2 = m_1 v_1$	Conservation of momentum	If $\tau_{\text{net}} = 0$, $L_2 = L_1$ $I_2\omega_2 = I_1\omega_1$

Equations for rotational motion with constant angular acceleration

$$\theta = \theta_0 + \overline{\omega}t \qquad \theta = \theta_0 + \omega_0 t + \tfrac{1}{2}\alpha t^2$$

$$\overline{\omega} = \frac{\omega_0 + \omega}{2} \qquad \omega = \omega_0 + \alpha t \qquad \text{(8.4–8.8)}$$

$$\omega^2 = \omega_0^2 + 2\alpha(\theta - \theta_0)$$

Parallel-axis theorem: If I_{CM} is the rotational inertia of an object of total mass M about any axis through its center of mass (CM), then the rotational inertia about a parallel axis a distance d from the center of mass is

$$I = I_{\text{CM}} + Md^2 \qquad \text{(8.14)}$$

Conservation of angular momentum: If no net external torque acts on a system, the total angular momentum of the system remains constant.

$$I_1\omega_1 = I_2\omega_2 \qquad \text{(8.17)}$$

Questions

1 A fly sits at the outer edge of a phonograph record rotating at a constant speed of 33.3 rev/min.

(a) Is the linear velocity of the fly constant? Why?

(b) Is its angular velocity constant? Why?

2 Is there such a thing as *rotational potential energy?*

Can you give an example of a situation in which this concept might have some importance?

3 Estimate your rotational inertia about a vertical axis through your center of mass when you are standing upright. Repeat for a horizontal axis through your center of mass.

4 A solid sphere and a solid cylinder are rolled down the same incline. If they have exactly the same mass and the same radius, which has the larger linear speed at the bottom of the incline?

5 Which bicycle wheel would be harder to stop rotating about its axle, one with an aluminum rim or one with a lead rim if the rims were the same size?

6 Can you suggest a subway car braking system that employs a flywheel and that conserves energy?

7 There have been proposals to use flywheels (instead of gasoline) to store energy in cars and buses. Explain why such a scheme might lead to problems for a vehicle going around a curve or over a hilltop. Can you suggest a simple means to correct this problem?

8 A fly is sitting on the edge of a Frisbee that is rotating with a constant angular speed about its center. The fly begins to walk toward the center of the Frisbee. What happens to the motion of the Frisbee?

9 Why do helicopters always have two rotors, either one large horizontal rotor to provide the lift and a smaller vertical one at the tail of the plane, or two equal-size horizontal rotors that rotate in opposite directions?

10 Why do football players try to throw or kick "spirals," in which the football rotates about its long axis, which is in line with the direction of the ball's motion?

11 If hard-boiled eggs become mixed with uncooked eggs in the refrigerator, one way to separate them is to spin each egg on a table top, stop it, and then quickly release it. Explain.

Multiple-Choice and Simple Exercises

8.1 A point on the circumference of a bicycle wheel moves 3.0 m in 1.0 s. If the wheel has a radius of 0.33 m, its angular velocity is:
   (a) 1.0°/s   (b) 9.1 rev/s   (c) 1.0 rad/s
   (d) 9.1 s^{-1} (e) 9.1 m/s

8.2 A metal chain is attached to a gear rotating on an axle. The gear has a radius of 5.0 cm and turns at a constant speed of 2.5 rev/s. The linear distance moved by the chain in 1.0 min is:
   (a) 47 rad   (b) 47 m   (c) 4.7 × 10^3 m
   (d) 3π × 10^2 m   (e) 0.78 m

8.3 The work done by a torque of 5.0 N·m in spinning a flywheel through 10 rev is:
   (a) π × 10^2 W   (b) 20π J   (c) 50 W
   (d) π × 10^2 J   (e) 50 J

8.4 The rotational kinetic energy of a 200-g steel ball of radius 1.0 cm rotated at the end of a string 0.80 m in length, if the ball makes 2 complete revolutions per second, is:
   (a) 10 J   (b) 20 J   (c) 0.26 J   (d) 1.6 J
   (e) 5.1 J

8.5 A force of 20 N is exerted on a rope wrapped around a wheel initially at rest. The force is tangent to the rim of the wheel, which has a radius of 0.60 m. This force pulls an amount of rope off the wheel just sufficient to go around the wheel once. The final rotational kinetic energy of the wheel is:
   (a) 75 J   (b) 12 J   (c) 150 J   (d) 38 J
   (e) None of the above

8.6 The torque that must be applied to a flywheel of rotational inertia 1.2 × 10^{-3} kg·m^2 to give it an angular acceleration of 5.0 rad/s^2 is:
   (a) 2.4 × 10^{-4} N·m   (b) 6.0 × 10^{-3} N·m
   (c) 6.0 × 10^{-3} N   (d) 6.0 × 10^{-3} kg·m/s^2
   (e) 4.2 × 10^3 N·m

8.7 The angular momentum of a bicycle wheel of mass 1.6 kg and radius 0.33 m rotating at an angular velocity of 2.0 rev/s is:
   (a) 2.2 kg·m/s   (b) 2.2 kg·m^2/s
   (c) 2.2 kg·m^2/s^2 (d) 6.7 kg·m/s
   (e) 6.7 kg·m^2/s

8.8 A ball is rotating in a horizontal circle at the end of a string of length 1.0 m at an angular velocity of 10 rad/s. The string is gradually shortened to 0.50 m without any force being exerted in the direction of the ball's motion. The new angular velocity of the ball is:
   (a) 40 rad/s   (b) 5.0 rad/s   (c) 20 rad/s
   (d) 10 rad/s   (e) 2.5 rad/s

8.9 In Prob. 8.8 the new linear speed of the ball is:
   (a) 2.5 m/s   (b) 5.0 m/s   (c) 40 m/s
   (d) 10 m/s   (e) 20 m/s

8.10 A turntable is designed to acquire an angular velocity of 33.3 rev/min in 0.50 s, starting from rest. Find the average angular acceleration of the turntable during the 0.50 s.

8.11 A wheel is rolling without slipping so that its axle has a translational speed of 10 m/s along the ground. If the radius of the wheel is 0.40 m, what is its angular velocity?

8.12 The rear wheel of a bicycle is rotating at a rate of 2.0 rev/s. If the radius of the wheel is 0.33 m, how fast is the bicycle moving along the ground?

8.13 The rotor in a centrifuge is accelerated from rest with a constant angular acceleration of 5.0 rad/s^2. What is the average angular velocity of the rotor during the first minute?

8.14 A bicycle wheel of mass 1.0 kg and radius 0.33 m is rotating at a rate of 4.0 rev/s. What is its kinetic energy?

8.15 An automobile engine develops a torque of 475 N·m and is rotating at a speed of 2000 rev/min. What horsepower does the engine generate?

8.16 What is the rotational inertia of a system of four masses, each 100 g, located at the corners of a square with sides 50.0 cm? Take the axis of rotation at the center of the square and perpendicular to the plane of the square.

8.17 A net torque of magnitude 5.0 × 10^4 N·m is exerted on the rotor of an electric generator for 30 s. What is the angular momentum of the rotor at the end of the 30 s, if it was initially at rest?

8.18 What would the period of the earth in its motion around the sun be if the earth were moved to a position half its present average distance from the sun while preserving the an-

2π rad/rev
        10

2000 rev   1 min
————— · ————
  min   40s   kg·m^2/s

kg·m/s   · m

gular momentum it now has and moving in an orbit that is approximately circular?

8.19 A diver comes off the high board with his body straight and with an angular velocity that will enable him to make one complete turn in 2.0 s. With his body straight, his rotational inertia is 19.8 kg·m². He then tucks his body in tightly by bending at both the waist and the knees. This reduces his rotational inertia to 3.8 kg·m². How many revolutions per second can he make in this tucked position?

Problems

8.20 A flywheel starts from rest and accelerates at a constant rate of 2.00 rad/s². When it has rotated through 60π rad, what is its angular velocity?

8.21 A potter's wheel of radius 15.0 cm starts from rest and rotates with constant angular acceleration until at the end of 30.0 s it is moving with an angular velocity of 15.0 rad/s.

(a) What is the angular acceleration?

(b) What is the linear velocity of a point on the rim at the end of the 30.0 s?

(c) What is the average angular velocity of the wheel during the 30.0 s?

(d) Through what angle did the wheel rotate in the 30.0 s?

8.22 A bicycle wheel of radius 0.300 m rolls down a hill without slipping. Its linear velocity increases constantly from 0 to 6.00 m/s in 3.00 s.

(a) What is its angular acceleration?

(b) Through what angle does the wheel turn in the 3 s?

(c) How many revolutions does it make?

8.23 A 200-N force is applied tangentially to the rim of a wheel 50.0 cm in radius and of rotational inertia 2.00 kg·m². The wheel rotates with negligible friction.

(a) Find the work done on the wheel during the first 5.00 s after the force is applied.

(b) If the wheel starts from rest, what is its kinetic energy of rotation after 5.00 s?

(c) Are the answers to parts (a) and (b) the same? Why or why not?

8.24 (a) What is the rotational kinetic energy of the earth, if we assume that it is a solid sphere of uniform density making one complete revolution on its axis every 24 h?

(b) The energy used in the United States each year is of the order of magnitude of 10^{20} J. If it were possible to use the rotational energy of the earth to satisfy this energy need, for how many years would this energy source suffice?

*8.25 Nitrogen molecules, which make up 80% of the air in our atmosphere, are of a dumbbell shape with two nitrogen atoms, each of mass 2.32×10^{-26} kg, separated by a distance of 1.10×10^{-10} m as in Fig. 8.19. Such molecules have an average translational speed of about 490 m/s at 0°C. They also rotate like dumbbells about an axis at right angles to the line connecting the two N atoms, with a rotational KE that is two-thirds of their translational KE. What is the average angular velocity of a nitrogen molecule at 0° C?

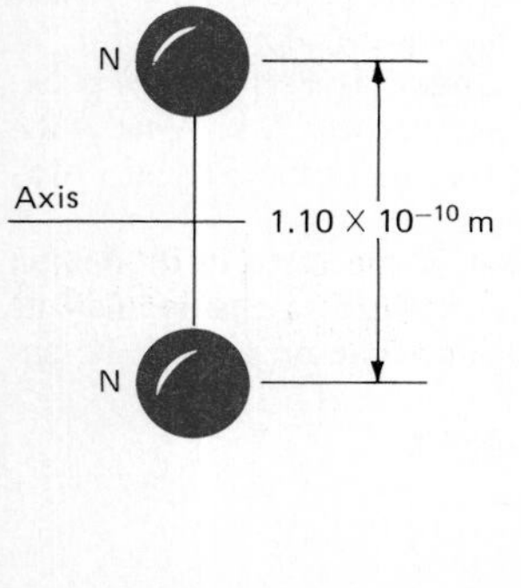

8.26 (a) A phonograph turntable of radius 15.0 cm and mass 0.700 kg acquires an angular velocity of 33.3 rev/min in 1.50 s, starting from rest. What torque must have been exerted on the turntable during this interval?

(b) What was the rotational kinetic energy of the turntable at the end of the 1.50 s?

8.27 The motor driving a grinding wheel with a rotational inertia of 0.150 kg·m² is switched off when the wheel has a rotational speed of 300 rev/min. After 10.0 s the wheel has slowed down to 240 rev/min.

(a) What is the torque exerted by friction to slow the wheel down?

(b) If this torque remains constant, when will the wheel come to rest?

8.28 A 2000-kg car carries a 25.0-kg rim-loaded flywheel of radius 0.300 m to store energy. When the car stops, the flywheel is engaged and the car's kinetic energy is transferred to it. What will be the angular velocity of the flywheel if it absorbs all the kinetic energy as the car decelerates from 25.0 m/s to rest?

8.29 A 100-g mass is suspended from the rim of a wheel, which has a radius of 60.0 cm and a horizontal axis of rotation, by a light string wound around the wheel, as in Fig. 8.20. The wheel is frictionless and has a rotational inertia of 0.100 kg·m². The mass is allowed to fall freely from rest.

(a) What is the linear acceleration of the mass?

(b) What is the tension in the string?

(c) What is the velocity of the mass after 10.0 s?

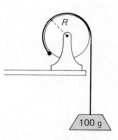

FIGURE 8.20 Diagram for Prob. 8.29.

FIGURE 8.19 Diagram for Prob. 8.25.

8.30 A wheel with rotational inertia 0.0500 kg·m² is spinning freely at 4.00 rev/s. It is then connected to a cord that lifts an 8.00-kg mass from rest on the ground. How high can the mass be lifted before the wheel stops rotating? (Neglect any energy loss when the connection is made.)

8.31 (a) Find the rotational inertia of a solid sphere of mass 1.00 kg and radius 5.00 cm around an axis 50.0 cm from its center. (*Hint:* Use Table 8.2 and the parallel-axis theorem.)

(b) How large is the error in assuming, for purposes of calculating the rotational inertia, that all the mass is concentrated at the center of the sphere?

8.32 (a) Calculate the rotational inertia of a solid sphere of radius 10.0 cm and mass 4.50 kg about an axis through its center.

(b) Repeat this calculation for a spherical shell with the same radius and mass.

(c) Calculate the rotational inertias of the solid sphere and the spherical shell about an axis 1.25 m from the center of each, and compare the results.

8.33 A dumbbell of length 80.0 cm has a small 5.00-kg mass at each end.

(a) What is the rotational inertia of the dumbbell about an axis through its center and perpendicular to the line connecting the two masses?

(b) What is the angular momentum of the dumbbell if it rotates about this same axis with an angular velocity of 2.00 rad/s?

8.34 The rotational inertia about a particular axis through its center of mass of an irregularly shaped 1.00-kg object is found by experiment to be 3.00×10^{-2} kg·m^2. What is the rotational inertia of the same object about an axis parallel to the first and 0.140 m from it?

∗8.35 Find the rotational inertia of a hoop of radius R and mass M about a point on its rim, when the motion is confined to the plane of the hoop.

8.36 A solid cylinder of mass M rolls down a ramp of height H to a level floor below. Compare the rotational and translational kinetic energy of the cylinder as it rolls along the floor.

8.37 A 0.200-kg solid spherical ball of radius 14.0 cm starts from rest and rolls without slipping down a 30.0° incline. The center of mass of the ball falls a vertical distance of 5.00 m in descending the incline. What is the ball's final linear speed?

8.38 A solid metal disk of radius 5.00 cm and mass 400 g starts from rest at the top of an incline of height 25.0 cm and rolls down to the ground.

(a) What is its speed at the bottom, if we neglect rolling friction?

(b) How would this result change if the radius were changed to 10.0 cm and the mass to 800 g?

(c) How does the result obtained in (a) differ if the disk slides down the plane instead of rolling down and we neglect friction?

∗8.39 A thin cylindrical shell and a solid cylinder have the same mass and radius. The two are released side by side and roll down, without slipping, from the top of an inclined plane that is 1.20 m above the ground.

(a) What are the final linear velocities of the two objects?

(b) Which reaches the bottom first?

(c) When the first object reaches the bottom, what is the height above ground of the other object?

∗8.40 Two circular cylinders have the same radius R and mass M, but one is solid and the other is a hollow shell, with all its mass concentrated at its rim, a distance R from the axis

of rotation. If the two cylinders are released together at the top of a hill that makes an angle of 20° with level ground, how far apart will they be after 5.0 s? (Ignore rolling friction.)

∗8.41 A uniform solid cylinder is rolling without slipping on a horizontal surface at 2.50 m/s. It reaches a ramp inclined at an angle of 40.0° with respect to the flat surface. If we neglect rolling friction, (a) how high is the cylinder above the horizontal surface when it stops? (b) How would this answer change if the angle were 30.0°?

8.42 Three objects start from rest together and roll down an inclined plane without slipping. One is a 0.400-kg hollow cylinder, the second a 4.00-kg solid cylinder, and the third a 1.00-kg hollow sphere. All three objects have the same radius R.

(a) Do the three arrive at the bottom of the plane at the same instant?

(b) If not, in what order do they arrive?

(c) Does this result depend on the values of the masses? On R?

8.43 A 500-g mass hangs from the rim of a wheel of radius 20.0 cm, as in Fig. 8.21. When released from rest, the mass falls 0.500 m in 7.50 s. Find the rotational inertia of the wheel. (Neglect friction.)

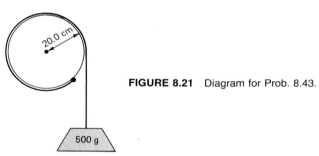

FIGURE 8.21 Diagram for Prob. 8.43.

8.44 In the Bohr model of the hydrogen atom an electron of mass 9.11×10^{-31} kg revolves in a circular orbit of radius 0.528×10^{-10} m about the proton at a linear speed of 2.19×10^6 m/s. What is (a) the angular momentum of the electron in this orbit; (b) its angular velocity; (c) its centripetal acceleration; (d) the centripetal force acting?

8.45 An Atwood's machine has a solid pulley wheel with rotational inertia 0.180 kg·m^2 and radius 50.0 cm. Two masses, one 3.00 kg and the other 5.00 kg, are attached to the two ends of a cord that passes over the pulley wheel, as in Fig. 8.22. The cord does not slip on the wheel but moves with it.

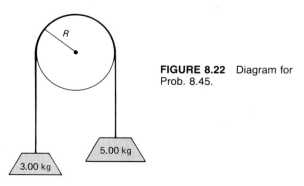

FIGURE 8.22 Diagram for Prob. 8.45.

(*a*) What is the resulting linear acceleration of the two masses?

(*b*) What are the tensions in the two sections of the cord?

(*c*) What is the angular acceleration of the wheel?

(*d*) Prove that after 1 s the work done on the system by the force of gravity is exactly equal to the kinetic energy of the whole system, so energy is conserved.

*8.46 Two pulley wheels, one of radius $R_1 = 10.0$ cm, the other of radius $R_2 = 30.0$ cm, are mounted rigidly on a common axle, as in Fig. 8.23. The rotational inertia of the two pulleys, which are clamped together, is 1.40 kg·m². A 2.00-kg mass and a 1.50-kg mass are connected to cords attached to each of the pulleys, as shown.

(*a*) Find the angular acceleration of the system.

(*b*) Find the tensions T_1 and T_2 in the two cords.

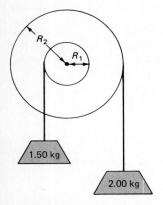

FIGURE 8.23 Diagram for Prob. 8.46.

*8.47 A solid circular disk of mass *m* and radius *R* has a rope wrapped around its outside edge and is allowed to fall, the rope being attached to the ceiling, as in Fig. 8.24. How large is the downward acceleration of the disk compared with the acceleration due to gravity?

FIGURE 8.24 Diagram for Prob. 8.47.

*8.48 Two blocks are connected to a pulley of radius 25.0 cm and rotational inertia 0.100 kg·m², as in Fig. 8.25. The 1.00-kg block hangs freely and the 2.00-kg block slides on a frictionless plane. The cord does not slip.

(*a*) What is the acceleration of the blocks?

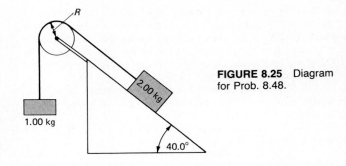

FIGURE 8.25 Diagram for Prob. 8.48.

(*b*) What is the tension in the two sections of the cord?

(*c*) What is the linear acceleration of a point on the rim of the pulley?

*8.49 A thin meterstick of mass 0.20 kg is hinged at the floor so that it can rotate only in a vertical circle. Initially the meterstick is held upright. It is then allowed to fall to the ground, rotating about the hinge at its bottom end, as in Fig. 8.26. What is its angular speed just before it hits the floor?

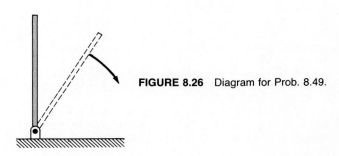

FIGURE 8.26 Diagram for Prob. 8.49.

8.50 A disk of rotational inertia 2.50×10^{-2} kg·m² is rotating with an angular velocity of 10.0 rad/s, when a non-rotating disk of rotational inertia 1.30×10^{-2} kg·m² is carefully dropped onto it. If the two stick together and rotate as one about their symmetry axes, what is the final angular velocity?

8.51 An ice skater has a rotational inertia of 1.5 kg·m² about a vertical axis through his body. If he extends his arms he can increase this to 1.8 kg·m². He first spins at a rotational speed of 2.0 rev/s with arms extended and then quickly pulls his arms tightly to his sides.

(*a*) What is his new rotational speed?

(*b*) How much kinetic energy has he gained?

(*c*) Where did this energy come from?

8.52 A girl of mass 50.0 kg jumps radially onto a moving, frictionless carousel rotating at 0.300 rev/s and lands 2.00 m from the center. The carousel has a rotational inertia of 200 kg·m².

(*a*) Find the angular velocity of the carousel after the girl jumps aboard.

(*b*) Find the kinetic energy of the system both before and after she jumps on.

(*c*) If kinetic energy is lost, where does it go?

8.53 A 40.0-kg boy stands on the edge of a frictionless turntable of radius 4.00 m that is initially at rest. The rotational inertia of the turntable is 750 kg·m². The boy starts to run around the edge of the platform and reaches a speed of 2.00 m/s relative to the ground. What is the angular velocity of the turntable relative to the ground when the boy has reached this speed?

8.54 A solid round platform of mass 100 kg and radius 3.00 m is mounted so that it can rotate freely around a vertical axis, but is initially at rest. A 30.0-kg child comes running at 1.00 m/s in a direction tangential to the platform and jumps on. If she lands at the very edge of the platform, what is the angular velocity of the platform just after she jumps on?

∗8.55 A mass attached to the end of a string that passes through a vertical tube is rotated in a horizontal plane, as in Fig. 8.27. The mass is initially swung in a circle of radius 40 cm at a tangential speed of 60 cm/s.

(a) What will be the tangential speed of the mass if the radius is slowly reduced to 20 cm by pulling the string down through the tube?

(b) Is this changed speed caused by a torque exerted by the pulling force? If not, what explains the changed speed?

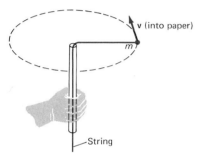

FIGURE 8.27   Diagram for Prob. 8.55.

∗8.56 A 2.00-kg metal ball is supported by a horizontal frictionless surface, as in Fig. 8.28. A cord passes through a hole on the axis about which the ball is rotating. The ball initially moves in a circle of 0.400-m radius at 10.0 rad/s. The radius of rotation is then slowly decreased as the cord is pulled from below the table. If the breaking strength of the cord is 200 N, what will the radius of the circle be when the cord breaks?

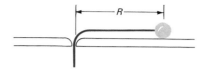

FIGURE 8.28   Diagram for Prob. 8.56.

∗8.57 Two identical, flat, solid disks having moments of inertia I_C are rotating clockwise with angular velocity ω_1, as in Fig. 8.29. They collide at point O on their rims and immediately stick firmly together.

(a) With what angular velocity ω_2 do they rotate about O? (*Hint:* Use the parallel-axis theorem to find I_O.)

(b) Has rotational kinetic energy been conserved? If not, what fraction of the original energy has been lost?

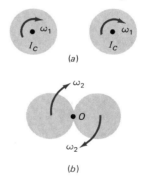

FIGURE 8.29   Diagram for Prob. 8.57.

Additional Readings

Andrade, E. N. da C.: "Isaac Newton," in *Newton Tercentenary Celebration*, Cambridge University Press, New York, 1947. The quote at the beginning of the present chapter is taken from this essay (with permission of Cambridge University Press).

Frohlich, Cliff: "The Physics of Somersaulting and Tumbling," *Scientific American*, vol. 242, no. 3, March 1980, pp. 154–165. A very interesting account of angular momentum conservation in the motion of divers, gymnasts, astronauts, and even cats.

Kreifeldt, John G., and Ming-Chuen Chuang: "Moment of Inertia: Psychophysical Study of an Overlooked Sensation," *Science*, vol. 206, 1979, pp. 588–590. Points out the importance of rotational inertia in determining the "feel" of an object.

Laws, Kenneth: "Physics and Dance," *American Scientist*, vol. 73, September–October 1985, pp. 426–431. A discussion of how familiar physics principles explain the striking movements of ballet dancers.

Post, R. F., and S. F. Post: "Flywheels," *Scientific American*, vol. 229, no. 6, December 1973, pp. 17–23. An interesting account of research on the use of flywheels to store energy.

Walker, Jearl: *Roundabout: The Physics of Rotation in the Everyday World* (readings from *Scientific American*), Freeman, New York, 1985. A fascinating account of amusement-park physics, racquetball, pool, judo, ballet, tops, and boomerangs.

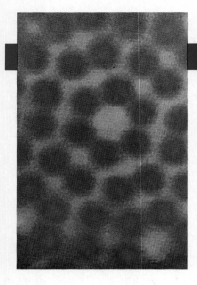

CHAPTER 9

The Structure of Atoms, Molecules, and Gases

It is not in the nature of things for any one man to make a sudden violent discovery; science goes step by step, and every man depends on the work of his predecessors. . . . Scientists are not dependent on the ideas of a single man, but on the combined wisdom of thousands of men, all thinking of the same problem, and each doing his little bit to add to the great structure of knowledge which is gradually being erected.

Ernest Rutherford (1871–1937)

Thus far we have discussed the motion of highly idealized particles whose relevance to the real world may seem remote. The world around us is made up of matter in a variety of forms—gases, liquids, solids, plasmas—but none of these closely resembles the idealized particles of mechanics. To understand the structure of matter, we must first understand the building blocks out of which all matter is constructed, i.e., the particles that make up atoms and molecules. In this chapter we consider first the particles that combine to form atoms and molecules and then how such atoms and molecules behave to produce the observed properties of some very common gases, such as the air we breathe.

9.1 Subatomic Particles

Only three of the many subatomic particles existing in nature are needed to understand the basic structure and behavior of atoms and molecules: the electron, the proton, and the neutron (Table 9.1).

The electron is an elementary particle with a very small mass, 9.110×10^{-31} kg, and a negative electric charge, -1.602×10^{-19} coulombs (C). It is this

TABLE 9.1 Properties of the Particles in Atoms

Particle	Symbol	Charge (C)	Mass (kg)	Mass (u*)
Electron	e	-1.60218×10^{-19}	9.10939×10^{-31}	5.48580×10^{-4}
Nucleons:				
Proton	p	$+1.60218 \times 10^{-19}$	1.67262×10^{-27}	1.00728
Neutron	n	0	1.67493×10^{-27}	1.00866

*The symbol u indicates *unified mass units*, whose significance is clarified in Sec. 9.3.

charge that gives the electron its unique importance in practical electric and electronic circuits. (The nature of electric charge and the use of the coulomb as the basic unit of charge are discussed in Chap. 16.)

The proton has a mass 1836 times that of the electron, or 1.673×10^{-27} kg, still a small mass on any ordinary scale. The proton's charge is positive and exactly equal in magnitude to the charge on the electron.

The neutron has no charge; in other words, it is electrically neutral. Its mass is 1.675×10^{-27} kg, only about 0.14 percent larger than that of the proton. Protons and neutrons are often referred to by the generic name *nucleon*, since they are found in the *nucleus* of atoms.

9.2   The Structure of Atoms

The positively charged proton, the negatively charged electron, and the uncharged neutron are the basic building blocks out of which all atoms are constructed. As a result of the experimental work of Ernest Rutherford in England in 1911 (see Fig. 9.1 and accompanying biography), we know that every atom has a nucleus, or central core, which contains all the protons and neutrons. Since the electrons are so light, the mass of an atom is essentially the sum of the masses of its neutrons and protons. The nuclear charge is positive and numerically equal to the sum of the charges on the protons. The nucleus has a diameter less than 10^{-14} m and contains almost all the mass in the atom. It is therefore extremely dense (see Example 9.1). Since it was well known even in Rutherford's time that the diameter of atoms is about 10^{-10} m, it is clear that atoms contain a great deal of empty space. Thus, if we should attempt to draw to scale an atom in which the nucleus is shown as a round dot 1 mm in diameter, the atom would have to extend over a diameter of 10 m. (For this reason, in diagrams of atoms, molecules, and nuclei in this book, the drawings are not to scale.)

Example 9.1

The nucleus of an oxygen atom contains 8 protons and 8 neutrons and has a radius of about 3.50×10^{-15} m. What is its density?

SOLUTION

From Table 9.1, we know that the mass of each proton and neutron is approximately 1.67×10^{-27} kg. The mass M of the oxygen nucleus is, therefore,

$$M = 16(1.67 \times 10^{-27} \text{ kg}) = 2.67 \times 10^{-26} \text{ kg}$$

The volume V of a sphere of radius 3.50×10^{-15} m is

$$V = (\tfrac{4}{3})\pi R^3 = (\tfrac{4}{3})\pi(3.50 \times 10^{-15} \text{ m})^3 = 1.79 \times 10^{-43} \text{ m}^3$$

Because the density of any object is its mass divided by its volume, the density of the oxygen nucleus is

$$\rho = \frac{M}{V} = \frac{2.67 \times 10^{-26} \text{ kg}}{1.79 \times 10^{-43} \text{ m}^3} = \boxed{1.49 \times 10^{17} \text{ kg/m}^3}$$

Since the density of water, one of the denser liquids, is only 10^3 kg/m³, it is clear that this oxygen nucleus has a density about 10^{14} times that of water!

The region of a neutral atom outside the nucleus contains as many electrons as there are protons in the nucleus. This makes the overall atom electrically neutral, since the positive charge in the nucleus just cancels the negative charge on the electrons. The electrons are in rapid motion in the space surrounding the nucleus. It is impossible to say where any electron is at any precise instant, but the physical

Ernest Rutherford (1871–1937)

FIGURE 9.1 Ernest Rutherford in 1926, when he visited his native New Zealand as Cawthron Lecturer. (*Courtesy of AIP Niels Bohr Library.*)

The great astronomer Sir Arthur Eddington once said that in 1911 Ernest Rutherford introduced the greatest change in our understanding of matter since the time of Democritus (400 B.C.). Eddington was referring to Rutherford's idea of the nuclear atom. It was this idea, and the experimental work leading up to it, that made him the acknowledged founder of modern nuclear physics.

That such a great scientist should have been born in New Zealand, at that time a remote country with few people and almost no scientific traditions, is as remarkable as was Rutherford's innate skill with scientific apparatus. He was born into a poor family, his father being a farmer who also did odd jobs to support his 12 children. Rutherford's simple surroundings in his youth no doubt influenced the development of his character, which remained open and unassuming throughout his life.

Rutherford attended Nelson College in New Zealand, where he excelled in mathematics, and later Canterbury College in the city of Christchurch. After receiving his degree in 1893, he began some physics research in a small, drafty cellar, working on the magnetic properties of iron under the influence of electric discharges. He published his results in two scientific papers, and as a result was awarded in 1895 a scholarship to study in England at Cambridge University. When his mother came to tell him the good news, he was digging potatoes. He flung away his spade with a shout of glee, exclaiming: "That's the last potato I'll dig!"

At Cambridge, Rutherford worked under the great J. J. Thomson (see biography, Chap. 19) on the effect of x-rays on the conduction of electricity in gases, and later in the newly emerging field of radioactivity. From 1898 to 1907 he was professor of physics at McGill University in Montreal, where he did some of his most important work on the nature of the emissions from radioactive substances. For this he received the Nobel Prize (in chemistry*) in 1908. He returned to England in 1907 and spent the years until 1919 as professor and director of the physical laboratory at Manchester. It was here that he did his renowned work on the nuclear atom. On the basis of his work on alpha particles (helium nuclei), he was led in 1911 to a description of the atom as made up of a small, very dense nucleus surrounded by orbital electrons. This description was taken up by Niels Bohr in 1913 and, combined with Bohr's ideas about quantum theory, provided the basic theory of the atom still accepted today.

In 1919 Rutherford succeeded Thomson as professor and director of the Cavendish Laboratory in Cambridge, and he spent his last years there directing work on artificially induced radioactivity, which eventually led to the splitting of the nucleus.

Rutherford loved laboratory work, perhaps because he was so successful at it. He once remarked to a colleague: "I am sorry for the poor fellows who haven't got labs to work in."

Despite his many accomplishments, Rutherford was never a slave to work. He delighted in reading novels and detective stories, avidly played bridge and golf, and loved the outdoors. He would often advise his colleagues to leave their offices and laboratories and "go home and think." He was happily married to his childhood sweetheart from New Zealand.

In 1931 Great Britain named Rutherford a baron, and he took the name Baron Rutherford of Nelson, after the town in New Zealand where he had been born. He immediately sent a cable to his mother in New Zealand, stating simply "Now Lord Rutherford, more your honor than mine, Ernest."

Rutherford lived an exciting and productive life until he died in 1937 at the age of 66. Although he took delight in his many accomplishments as a physicist, he would probably have taken even greater delight in the eulogy of a friend of 30 years: "Rutherford never made an enemy and never lost a friend." He was truly both a great physicist and a great man.

*This is our justification for including Rutherford in this chapter on atoms, molecules, and gases, even though his major contributions to science were in the field of nuclear physics.

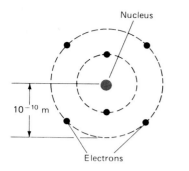

FIGURE 9.2 A schematic model of a carbon atom.

theory called *quantum mechanics* shows that it is possible to specify the regions of the atom in which the electron is most likely to be found. Despite this inherent ambiguity as to the position of the electrons, diagrams of atoms often show the electrons moving in precise circular or elliptical orbits about the nucleus. This is the way electrons were originally pictured in the atom described by Rutherford and Bohr, and for some purposes this idealized picture is still useful today (Fig. 9.2).

All neutral atoms, i.e., those with zero net electric charge, contain the same number of protons as electrons. This number—1 for hydrogen, 2 for helium, 92 for uranium—distinguishes the atoms of one chemical element from those of all other elements and accounts for most of the significant physical and chemical properties of the atom. We call this the *atomic number*, or *charge number*, of the atom and designate it by Z.

Atomic (or charge) number Z: The number of protons in the nucleus of an atom.

Elements with different atomic numbers are assigned different chemical symbols such as H for hydrogen and He for helium. Subscripts preceding the chemical symbol indicate the atomic number, for example, $_1$H, $_2$He, $_{92}$U, or, in general, $_Z X$, where X is the chemical symbol for the particular atom. In a table (see Table F.6) conceived by the Russian chemist Dmitri Ivanovich Mendeléev (1834–1907), the elements are listed in the order of their atomic numbers. Because of the periodic repetitions of the physical and chemical properties of the elements as Z increases, Mendeléev's table is called the *periodic table of the elements*.

Ions

If an electron is removed from a neutral atom, the atom ceases to be electrically neutral and takes on a net positive charge. Such an atom is called a *positive ion*. Thus a positive lithium ion, designated as Li^+, is formed when a neutral Li atom loses one electron. There is also the possibility that an atom may pick up an added electron and become negatively charged. Such an atom is called a *negative ion*, such as F^-. Multiple charged ions, such as Be^{2+}, are also possible.

Ion: An atom that has either lost or gained electrons and therefore has acquired a net positive or negative charge.

Since the chemical and physical properties of atoms are in great part determined by the number of electrons in the atom, positive ions often behave like atoms with an atomic number one less than the ion, and negative ions like atoms with an atomic number one greater.

Isotopes

Two atoms may have the same number of protons and electrons but different numbers of neutrons in the nucleus. Such atoms are distinguished by their mass number A, written as a superscript before the chemical symbol, and are called *isotopes*.

Mass number A: The number of protons plus neutrons in an atomic nucleus.

Isotopes: Atoms having the same charge number Z but different mass numbers A.

The word *isotope* comes from two Greek words meaning the "same place," which appropriately applies to position in the periodic table. Isotopes of the same atom have the same chemical properties, since these depend on the electrons, but their nuclear properties may differ greatly.

TABLE 9.2 Constituent Parts of Some Important Isotopes

Element	Isotope (symbol $^A_Z X$)	Number of protons	Number of electrons	Number of neutrons	Atomic (charge) number Z	Mass number A
Hydrogen	$^1_1 H$	1	1	0	1	1
	$^2_1 H$	1	1	1	1	2
	$^3_1 H$	1	1	2	1	3
Carbon	$^{12}_6 C$	6	6	6	6	12
Oxygen	$^{16}_8 O$	8	8	8	8	16
Uranium	$^{235}_{92} U$	92	92	143	92	235
	$^{238}_{92} U$	92	92	146	92	238

The two uranium isotopes, $^{235}_{92}U$ and $^{238}_{92}U$, for example, both have 92 protons and 92 electrons. The first has $235 - 92 = 143$ neutrons in the nucleus, while the second has 146. The chemical and physical properties of these two isotopes, as found from chemical reactions, spectra, etc., are almost identical, but their nuclear properties are vastly different.

The numbers of protons, neutrons, and electrons in some important isotopes are given in Table 9.2.

Example 9.2

Two very important elements for the semiconductor industry are silicon and germanium. How many electrons, protons, and neutrons do the isotopes **(a)** $^{28}_{14}Si$ and **(b)** $^{72}_{32}Ge$ contain?

SOLUTION

(a) For $^{28}_{14}Si$ the atomic number is 14, and so silicon contains 14 protons in the nucleus and 14 extranuclear electrons. The number of neutrons is

$$A - Z = 28 - 14 = 14$$

(b) By the same sort of analysis for $^{72}_{32}Ge$, this isotope consists of 32 protons and 40 neutrons in the nucleus and 32 extranuclear electrons.

EXERCISE 1 How many protons, neutrons, and electrons are there in the fluorine ion $^{19}_9 F^-$?

Example 9.3

Copper ($_{29}Cu$) has two predominant isotopes, $^{63}_{29}Cu$, which is 69.1 percent of the copper in the earth's crust, and $^{65}_{29}Cu$, which is 30.9 percent. What is the approximate mass number A of copper as found in the earth?

SOLUTION

Mass contributed by $^{63}_{29}Cu = 0.691 \times 63 = \quad 43.5$

Mass contributed by $^{65}_{29}Cu = 0.309 \times 65 = \quad 20.1$

$$\text{Sum} = \boxed{63.6}$$

This value agrees reasonably well with the value given in the periodic table of the elements.

9.3 The Mole and Avogadro's Number

The *mole* is the basic SI unit for an amount of substance. One gram mole, or simply mole (abbreviated mol), is defined as follows:

Mole: One mole of an atomic gas (such as helium) contains the same number of atoms as in 12.00 g of $^{12}_{6}C$. This number is called Avogadro's number and is equal to 6.022×10^{23} per mole. For a molecular gas (such as nitrogen, that is, N_2) one mole of gas contains Avogadro's number of molecules.

One mole* of any substance contains as many atoms or molecules as one mole of any other substance. A mole of any substance may be obtained by taking a mass (in grams) of the substance numerically equal to the atomic mass number (A) or molecular mass number (M) of the substance, where the molecular mass number is the sum of the atomic mass numbers of the atoms making up the molecule.

For example, to a good approximation (since the true mass numbers are not actually integers), 1 mol of $^{4}_{2}He$ consists of 4 g of $^{4}_{2}He$ atoms, and 1 mol of H_2O consists of $2(1) + 16 = 18$ g of H_2O molecules. Both samples contain Avogadro's number of atoms (for $^{4}_{2}He$) or molecules (for H_2O).

Avogadro's number is named for Amedeo Avogadro (1776–1856), the Italian physicist who first proposed the hypothesis that all gases at the same pressure and temperature contain the same number of atoms or molecules per unit volume. Avogadro's number approximately equals the number of grains of sand in a cube 1 mi on each side.

Unified Mass Unit

This introduction of Avogadro's number leads us to another unit of great importance in atomic and nuclear physics, the *unified mass unit,* usually abbreviated u and defined in terms of the mass of $^{12}_{6}C$. From the above discussion, 1 mol of $^{12}_{6}C$ atoms has a mass of 12 g and contains 6.022×10^{23} atoms. The unified mass unit (u) is then defined to be one-twelfth the mass of the $^{12}_{6}C$ atom (including electrons), where the mass of 1 mol of $^{12}_{6}C$ is taken to be exactly 12.00 g. Then

$$1 \text{ u} = \frac{1}{12}\left(\frac{12.00 \text{ g}}{6.022 \times 10^{23}}\right) = 1.661 \times 10^{-24} \text{ g} = 1.661 \times 10^{-27} \text{ kg} \qquad \textbf{(9.1)}$$

This value of the unified mass unit is roughly equal to the mass of the proton and so also to the mass of the hydrogen atom.

The atomic mass numbers found in the periodic table of the elements (inside the front cover) represent the masses, in *unified mass units,* of the elements as found on earth and therefore are averages over the mass numbers of the isotopes of the element existing in nature. Thus the atomic mass of chlorine is 35.45 u, since natural chlorine is made up of the isotopes $^{35}_{17}Cl$ and $^{37}_{17}Cl$ in abundances leading to a weighted average of 35.45 u. Since $1 \text{ u} = 1.661 \times 10^{-24}$ g, and 1 mol of atomic Cl contains N_A atoms, the mass of 1 mol of atomic Cl is

$$m(_{17}Cl) = (35.45 \text{ u})(1.661 \times 10^{-24} \text{ g/u})(6.022 \times 10^{23}) = 35.45 \text{ g}$$

---

*By "mole" we always mean *gram mole* in this book. Some physics books use the kilogram mole, which contains 10^3 times as many atoms (or molecules) as the gram mole. If the kilogram mole is used, Avogadro's number has the value $N_A = 6.022 \times 10^{26}$ per kilogram mole.

We can therefore conclude that *the mass in grams of 1 mol of any substance having just one isotope is equal to the mass in unified mass units of one atom (or molecule) of the substance*. If the substance contains more than one isotope, the mass of 1 mol (in grams) is equal to the average mass of an atom (in unified mass units). This average mass is determined by weighting the isotopes in proportion to their abundance on earth.

Example 9.4

The atomic mass of the boron isotope $^{11}_{5}B$ is 11.0 u. Find the mass in kilograms of one $^{11}_{5}B$ atom.

SOLUTION

Since 1 u = 1.66×10^{-27} kg, we have

$m(^{11}_{5}B) = (11.0 \text{ u})(1.66 \times 10^{-27} \text{ kg/u})$

$= \boxed{1.83 \times 10^{-26} \text{ kg}}$

Another way to arrive at the same result is to note that the mass in grams of 1 mol of any substance is numerically equal to the mass in unified mass units of one atom of the same substance. Hence 1 mol of $^{11}_{5}B$ has a mass of 11.0 g. Then, since

1 mol contains Avogadro's number of $^{11}_{5}B$ atoms, we have

$m(^{11}_{5}B) = \dfrac{11.0 \text{ g/mol}}{6.02 \times 10^{23}/\text{mol}}$

$= 1.83 \times 10^{-23} \text{ g} = 1.83 \times 10^{-26} \text{ kg}$

Similar methods can be used to find the mass of any atom or molecule.

EXERCISE 2   What is the mass in kilograms of one molecule of carbon dioxide (CO_2)?

9.4 The Structure of Gases

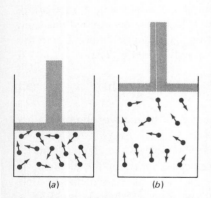

FIGURE 9.3 A gas in a container of variable size. (*a*) Gas atoms confined to a small volume. (*b*) The piston is raised and the gas atoms move to fill uniformly the additional space in the container. The arrows indicate the velocities of the molecules.

Atoms are made up of protons, neutrons, and electrons. Molecules in turn are made up of atoms, in some cases identical atoms like the two oxygen atoms in O_2, in other cases dissimilar atoms, as in H_2O. Molecules, while preserving their separate identities, then combine to form the three states in which matter normally exists: solid, liquid, and gas.* In this and in the following chapter, we see how atomic structure explains the behavior of matter in these three states.

Gas: A *collection of atoms or molecules* that expands to fill uniformly the volume of any container in which it is placed. (See Fig. 9.3.)

In gases the molecular speeds are very great, and the attractive forces among the molecules small. The effect of the attractive forces can be increased either by lowering the temperature or by squeezing the gas down into a very small volume by an increase in pressure. In this way it is possible to liquefy gases.

Some gases are homogeneous throughout and are made up of either single atoms like He, diatomic molecules like nitrogen (N_2), or polyatomic molecules like carbon dioxide (CO_2). In this homogeneous case, all the molecules (or atoms) in the gas are identical, as is shown in Fig. 9.4.

Other gases are mixtures of different molecules, and at the microscopic level they are therefore nonhomogeneous. For example, clean, dry air is 78% nitrogen and 21% oxygen, with the remaining 1% consisting mostly of argon (Ar), carbon

---

Plasmas are sometimes called "the fourth state of matter." These are highly ionized gases that exist only at very high temperatures. We postpone any discussion of plasmas until we take up nuclear fusion in Chap. 29.

FIGURE 9.4 Three types of homogeneous gases: (a) a gas made up of atoms (He); (b) a gas made up of diatomic molecules (N_2); (c) a gas made up of polyatomic molecules (CO_2).

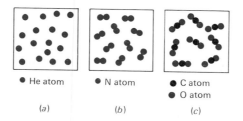

• He atom   • N atom   • C atom
                                  ● O atom

(a)   (b)   (c)

dioxide (CO_2), neon (Ne), and helium (He). Thus if we could extract individual molecules from a small sample of air, there is a high probability that in every five molecules extracted we would find four molecules of N_2 and one of O_2. But there is *zero* probability that we would find a "molecule" of air, no matter how hard we tried, since a molecule of air does not exist. All that exists at the molecular level are molecules of a number of gases that are mixed together in the proper proportions to make the substance we colloquially call "air."

Important Physical Quantities for Gases

There are a number of large-scale (macroscopic) quantities that can be easily measured for gases. These include:

Volume V: The space occupied by the gas.

Since a gas expands to fill any container in which it is placed, the volume of a gas is equal to the interior volume of its container. Volumes are measured in cubic meters (m^3).

Mass $\mathcal{M}$: The total mass of all the atoms or molecules making up the gas.

The total mass of gas is merely the number N of atoms (or molecules) present in the gas times the mass m of each atom (or molecule). Another way of expressing this same mass is by giving the number n of moles of gas present times the mass M of 1 mol of gas:

$$\mathcal{M} = Nm = nM \qquad \textbf{(9.2)}$$

Thus the mass of 5.0 mol of hydrogen gas (H_2) is

$$\mathcal{M} = nM = (5.0 \text{ mol})(2.0 \text{ g/mol}) = 10 \text{ g} = 0.010 \text{ kg}$$

Density ρ: The mass of gas per unit volume.

$$\boxed{\rho = \frac{\mathcal{M}}{V}} \qquad \textbf{(9.3)}$$

In the SI system densities are expressed in kilograms per cubic meter (kg/m^3). For example, the density of H_2 gas is 0.0899 kg/m^3 at *standard temperature and pressure* (STP)* and for radon gas it is 9.73 kg/m^3 at STP. The Greek letter ρ (*rho*) is commonly used for densities.

Pressure P: The force per unit area acting perpendicular to any surface of the container holding the gas or to any imaginary surface drawn inside the gas.

---

*STP means a temperature of 0°C (273 K) and a pressure of 1 atmosphere (1.01 × 10^5 N/m²).

$$\boxed{P = \frac{F}{A}}$$ (9.4)

Pressures are measured in newtons per square meter (N/m^2), where $1\ N/m^2$ is called 1 pascal (Pa). In the British engineering system, pressures are given in pounds per square inch (the units in which tire pressures are still commonly measured in the United States), where $1\ lb/in^2 = 6.90 \times 10^3\ N/m^2$.

The pressure of a gas is caused by the atoms or molecules of the gas moving around in all directions inside the container and colliding with its walls. The force they exert perpendicular to any given surface of the container, divided by the area of that surface, is the pressure of the gas. This pressure is the same throughout the container. Thus the pressure of the air in an automobile tire is the same at any point on the wall of the tire and at any point inside the tire. Although force is a vector, pressure is a scalar, since it is the same in all directions inside the gas.

Example 9.5

How many atoms are contained in a tiny speck of gold foil, which is in the shape of a cube 10^{-5} cm on a side? The atomic mass of naturally occurring gold is 197 u and its density is $19.3 \times 10^3\ kg/m^3$.

SOLUTION

Since the density of any substance is given by $\rho = M/V$, we have for the gold speck

$$M = \rho V = \left(19.3 \times 10^3\ \frac{kg}{m^3}\right)(10^{-7}\ m)^3$$

$$= 1.93 \times 10^{-17}\ kg = 1.93 \times 10^{-14}\ g$$

Now, 1 mol of gold contains Avogadro's number N_A of atoms and has a mass of 197 g. We set up a proportion:

$$\frac{N \text{ atoms of gold}}{1.93 \times 10^{-14}\ g} = \frac{6.02 \times 10^{23}\ atoms}{197\ g}$$

$$N = \left(\frac{1.93 \times 10^{-14}}{197}\right)(6.02 \times 10^{23}\ atoms)$$

$$= \boxed{5.90 \times 10^7\ atoms}$$

This tiny speck of gold foil therefore contains about 59 million atoms!

9.5 Temperature

The last important physical quantity needed to specify the state of a gas is in some ways the most important and yet the most elusive. We all have some idea of what temperature means and could define it loosely as follows:

Temperature: A measure of the hotness or coldness of an object.

We derive our ideas about temperatures from our sense of touch, but, if we are interested in *measuring* temperatures with any scientific accuracy, our sense of touch is of little use. Although we can tell by touching an object whether it is hot or cold, this is *a very subjective* judgment that can easily lead us astray. For example, if you are at a swimming pool and the water is at 70°F (20°C), and you jump in after being in the hot sun for an hour or more, the water seems deliciously cool to you. But, if you take an ice-cold shower before diving into the pool, the same pool water seems warm. A more objective and quantitative method than subjective feeling is needed to determine an object's temperature.

The scientific measurement of temperature is based on the experimental fact that two objects in contact with each other always come to the same temperature if we wait long enough. We know from experience that if we put a frozen dinner in a hot oven and wait a half hour or so, the frozen dinner will become as hot as the inside of the oven. Heat constantly flows from the oven into the dinner, which heats up until it comes into thermal equilibrium with the oven, that is, until there is no net flow of heat into or out of it. When this occurs, we say that the oven and the dinner are in *thermal equilibrium,* or "at the same temperature."

To measure temperature we use a *thermometer.* When put into close contact with the object whose temperature is to be determined, a thermometer gradually comes into thermal equilibrium with the object. For example, when a doctor puts a thermometer into a patient's mouth to measure body temperature, the doctor expects that after a few minutes the thermometer will come into thermal equilibrium with the patient's body. Only then can the doctor be sure that the thermometer temperature corresponds to the patient's body temperature.

The physical basis for the use of thermometers to measure temperature is an experimental fact sometimes referred to as the *zeroth law of thermodynamics* (see Fig. 9.5):

Two systems in thermal equilibrium with a third system are in thermal equilibrium with each other.

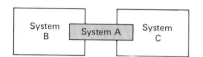

FIGURE 9.5 The zeroth law of thermodynamics: If system *A* is in thermal equilibrium simultaneously with system *B* and system *C*, then *B* and *C* must be in thermal equilibrium with each other; i.e., they must be at exactly the same temperature.

For example, if we immerse a thermometer in a pot of hot water and then in a pot of coffee, in each case allowing thermal equilibrium to be established, and if we find that in both cases the thermometer reads exactly the same, then we say that the water and the coffee are *at the same temperature.*

Temperature is a new kind of physical quantity that cannot be defined in terms of the three fundamental quantities of mechanics—mass, length, and time. It is an *intensive* property of a substance—that is, it measures the degree of hotness or coldness—and *is independent of how much of the substance is present,* as long as the number of atoms or molecules present is statistically large: the temperature of one drop of water can be the same as that of a million cubic meters of water.

We discuss practical thermometers in Sec. 13.3. For the present we consider a thermometer as simply a device having some measurable property that changes with temperature, for example, the length of a column of mercury in a mercury thermometer or the electric resistance of a metal in a resistance thermometer. Such devices can be provided with temperature scales based on assigned values for important temperatures such as the boiling point of water or the melting point of ice. When such a thermometer comes into thermal equilibrium with an object whose temperature is being measured, the change in this measurable property enables us to read the temperature of the object directly from the scale on the thermometer.

9.6 The Kinetic Theory of a Gas

We now seek to understand the nature of temperature at the molecular level. Is it possible to relate temperature directly to quantities like the velocity, momentum, and kinetic energy of the molecules in a gas by applying ideas and equations already developed in our treatment of mechanics? Is classical mechanics applicable to the motion of atoms and molecules, and can it help us understand the properties of a very real and practical substance, such as the air we breathe?

To answer these questions let us assume a simplified model of a gas and apply Newton's laws to the motion of the gas molecules. Then we will see whether such a model yields quantitative results in agreement with laboratory measurements of the volume, pressure, and temperature of a gas.

We assume we are dealing with a gas that has the following characteristics:

1   The gas consists of a large number (N) of identical molecules,* each of mass m. These molecules move at random in all directions with a great variety of speeds. This is why a gas fills any container in which it is placed.

2   The distances between the molecules in the gas are large compared with the size of the molecules, which have diameters of about 10^{-10} m. Each molecule may therefore be considered to be a point particle.

3   The forces on the moving molecules are negligible except when they collide with one another or with the container walls. The effect of the force of gravity on these very light particles is assumed to be negligible.

4   Collisions between two molecules or between a molecule and a wall are assumed to be perfectly elastic (no kinetic energy is converted into heat), and so *both momentum and kinetic energy are conserved* (Sec. 7.4).

Molecular Basis of Gas Pressure

We first use this simplified model of a gas to show how the pressure of a gas depends on the motion of its molecules.

Let our gas, in this case consisting of N identical molecules, be confined to a rectangular box of length L and cross-sectional area A, as in Fig. 9.6. The molecules striking the right end of the box exert an average force F on the wall.

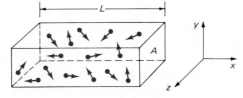

FIGURE 9.6   A box containing a gas. The length of the box is L, which is directed along the x axis, and its cross-sectional area (in the yz plane) is A. The arrows indicate the random direction in which the molecules move.

Even though this force is produced by large numbers of individual molecular collisions with the wall, the collisions are so frequent that, on the average, F is constant over any finite time interval. The pressure on the area A is then, from Eq. (9.4), $P = F/A$. To calculate F, we must find the normal force f each molecule exerts on the wall and then sum over all the molecules striking the wall. We can resolve the velocity **v** of a particular molecule into components v_x, v_y, and v_z along the three edges of the box. If a molecule with velocity component v_x collides elastically with the right end of the box, the x component of its velocity is reversed, as in Fig. 9.7, so that the change in its momentum is

$$\Delta p = p_{\text{final}} - p_{\text{init}} = -mv_x - mv_x = -2mv_x$$

For our purpose, because the pressure is defined in terms of the force *perpendicular* to the wall (which is in the yz plane), only the x component of momentum is important.

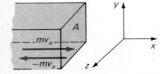

FIGURE 9.7   Transfer of momentum between a gas molecule and the end wall of a container. The momentum of the molecule is reversed on its hitting the rigid wall.

---

*Some gases, such as helium (He) or neon (Ne), are made up of atoms, not molecules. In what follows we will for simplicity use the term *molecule* for the basic constituents of the gas, whether they be atoms or molecules.

If the molecule now moves to the left end of the box and back again without colliding with another molecule, it does so in the time it takes to travel a distance twice the length of the box at a speed v_x, so the time between collisions with the wall is

$$t = \frac{2L}{v_x}$$

(9.5)

The number of collisions per second of a molecule with the wall is then $1/t = v_x/2L$, and the change in momentum of the molecule per second is

$$\frac{\Delta p}{\Delta t} = (-2mv_x)\left(\frac{v_x}{2L}\right) = \frac{-mv_x^2}{L}$$

This rate of change in momentum is, by Newton's second law, equal to the force exerted by the wall on the molecule. But, by Newton's third law, this force exerted by the wall on the molecule is equal in magnitude and opposite in direction to the force exerted by the molecule on the wall. The force exerted by each molecule on the wall is then $f = +mv_x^2/L$.

The total force F on the wall perpendicular to the x direction is obtained by summing over all the molecules, where each molecule is assumed to have a different x component of velocity v_{xi}. Then

$$F = \sum_i f_i = \sum_{i=1}^{N} \frac{mv_{xi}^2}{L} = \frac{m}{L} \sum_{i=1}^{N} v_{xi}^2$$

and the pressure on the wall is

$$P = \frac{F}{A} = \frac{m}{LA} \sum_{i=1}^{N} v_{xi}^2 = \frac{m}{V} \sum_{i=1}^{N} v_{xi}^2$$

where V is the volume of the gas, equal to LA.

On multiplying both sides by V and also multiplying and dividing the right side by N (the number of molecules), we have

$$PV = Nm \sum_{i=1}^{N} \frac{v_{xi}^2}{N}$$

But the quantity $\sum_{i=1}^{N} v_{xi}^2/N$ is simply the average value of v_x^2 for all the molecules. We write this quantity as $\overline{v_x^2}$, which is defined to be

$$\overline{v_x^2} = \sum_{i=1}^{N} \frac{v_{xi}^2}{N}$$

(9.6)

and obtain

$$PV = Nm\overline{v_x^2}$$

(9.7)

We show in Fig. 9.8 that the magnitude of a gas molecule's velocity is related to its three rectangular components by the equation, $v^2 = v_x^2 + v_y^2 + v_z^2$. If we take averages over all the molecules in the gas, we obtain

$$\overline{v^2} = \overline{v_x^2} + \overline{v_y^2} + \overline{v_z^2}$$

In the present case, there is nothing about motion in the x direction to distin-

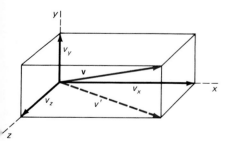

FIGURE 9.8 Application of the Pythagorean theorem to the velocity of gas molecules. **v** is the vector velocity of a gas molecule with components v_x, v_y, and v_z, which are not necessarily equal. In the figure v' is the projection of **v** on the xz plane, and so $v'^2 = v_x^2 + v_z^2$. But, from the diagram, $v^2 = v'^2 + v_y^2$, and so $v^2 = v_x^2 + v_y^2 + v_z^2$.

guish it from motion in the y or z direction, for the pressure is the same on all sides of the box. We would therefore expect that

$$\overline{v_x^2} = \overline{v_y^2} = \overline{v_z^2}$$

and so

$$\overline{v^2} = \overline{v_x^2} + \overline{v_y^2} + \overline{v_z^2} = 3\overline{v_x^2} \tag{9.8}$$

and so $\overline{v_x^2} = \frac{1}{3}\overline{v^2}$, where $\overline{v^2}$ is the square of the speed averaged over all the molecules, without respect to direction. Then we have, from Eqs. (9.7) and (9.8), $PV = \frac{1}{3}Nm\overline{v^2}$, or

$$PV = \frac{2N}{3}(\tfrac{1}{2}m\overline{v^2}) \tag{9.9}$$

Now $\frac{1}{2}m\overline{v^2}$ is the *average translational kinetic energy* of one molecule, and $N(\frac{1}{2}m\overline{v^2})$ is the *total translational kinetic energy of the gas*. We therefore obtain the extremely important result that the product of the pressure and the volume of a gas is equal to two-thirds the total translational kinetic energy of the gas:

$$\boxed{PV = \tfrac{2}{3}\text{KE}} \tag{9.10}$$

This calculation, while involved, is worth the effort required to understand it, since it illustrates how the classical mechanics developed in Chaps. 2 to 8, combined with an atomic model for gases, can lead to a basic understanding of the behavior of these gases.

In this calculation we have neglected collisions between molecules, but the same result is obtained when collisions are taken into account. By neglecting collisions we simplify the problem without introducing any serious error in the result. One of the secrets of theoretical physics is finding such simple models that can be used in calculations without radically distorting the physical situation.

The model we have just used to obtain Eq. (9.10) is called the *kinetic-theory-of-gases* model, from the Greek word *kinesis*, meaning "motion." This model enables us to relate the large-scale or bulk quantities describing a gas to the atomic structure and motion of the molecules of the gas. This is reflected in Eq. (9.9), which connects macroscopic variables (P and V) to the average kinetic energy of the individual molecules making up the gas ($\frac{1}{2}m\overline{v^2}$).

9.7 The Ideal Gas Law

Suppose we have a mass $\mathcal{M}$ of He gas in a container of volume V closed off by a movable piston and sitting on a hot plate, as in Fig. 9.9. The mass of the gas is $\mathcal{M} = nM$, where n is the number of moles of the gas and M is the mass in grams of 1 mol. We perform a series of experiments to see how the volume, the pressure, and the temperature are related for this gas when the number of moles remains unchanged.

Boyle's Law

First we keep T fixed and vary P and V by compressing the gas slowly with the piston. If the pressure and volume change from P_1 and V_1 before the compression to P_2 and V_2 after the compression, we find experimentally that

$$P_1 V_1 = P_2 V_2$$

FIGURE 9.9 A glass container, filled with He gas, on a hot plate.

FIGURE 9.10 Boyle's law for He (an ideal gas). At constant temperature, the volume varies inversely with pressure.

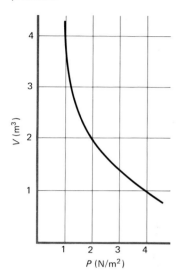

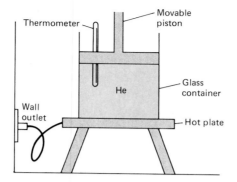

If we repeat this for a great variety of different values of P and V, always keeping the temperature unchanged, we find that in every case

$$PV = \text{constant} \quad \text{(if } T \text{ is constant)} \quad \text{or} \quad V = \frac{\text{constant}}{P} \quad \text{(9.11)}$$

This is called *Boyle's law,* which simply states that, at constant temperature, the volume of a gas is inversely proportional to its pressure, since PV is equal to a constant. This is shown in Fig. 9.10. Boyle's law takes its name from the British chemist and physicist Robert Boyle (1627–1691).

A gas that obeys Boyle's law is called an *ideal gas.* Most gases at reasonably low densities and not-too-low temperatures satisfy Boyle's law. The assumptions made in Sec. 9.6 ensured that we were dealing with an ideal gas there.

Charles' Law

We can perform a second experiment by keeping a fixed pressure on the gas, putting some weights on the piston and allowing it to move up and down freely, so that both the volume and temperature change while the pressure remains constant. The Frenchman Jacques Charles (1746–1823) first performed this experiment and found that, at fixed pressure, the volume increased linearly with temperature. This is shown in Fig. 9.11. We see that we get a straight-line graph for the volume-versus-temperature curve for helium (an ideal gas) so long as the temperature does not come too close to $-269°C$, which is the temperature at which helium liquefies. Even though we cannot reduce the volume to zero, we find that, if we extrapolate* the straight line in Fig. 9.11, it crosses the x axis at $-273.2°C$. As a matter of fact, if we repeat this experiment for a group of different ideal gases, their graphs will have different slopes but all will cross the x axis at the same value of the temperature, $-273.2°C$, as shown in Fig. 9.12. This temperature at which the volume of all ideal gases would be reduced to zero (if they did not liquefy) is called the *absolute zero of temperature.* It is the basis for the introduction of the *absolute,* or *Kelvin, temperature scale.*

The Kelvin temperature scale is named after William Thomson (Lord Kelvin), who first suggested its use in 1848.

We introduce the absolute temperature scale by setting its zero at $-273.2°C$ and choosing the size of a kelvin[†] to be exactly the same as a Celsius (or centigrade) degree. Then the freezing point of water is $0°C$, or 273.2 K, and the boiling point

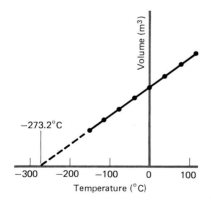

FIGURE 9.11 Charles' law: variation of the volume of helium with the temperature (°C) at constant pressure.

---

*On *extrapolation* see Appendix 2.B.

[†]On the absolute scale an international commission has decided to use *kelvins* (symbol K) to specify temperatures instead of "degrees Kelvin," or °K.

FIGURE 9.12 Variation of the volume of a group of different ideal gases with temperature, at constant pressure. Note how the extrapolations of the straight lines all cross the temperature axis at $-273.2°C$.

FIGURE 9.13 The same graph of volume against temperature for a group of gases as in Fig. 9.12, but with the new zero of the temperature scale taken at $-273.2°C$, and with temperature intervals the same as on the Celsius scale. The result is the absolute (Kelvin) temperature scale.

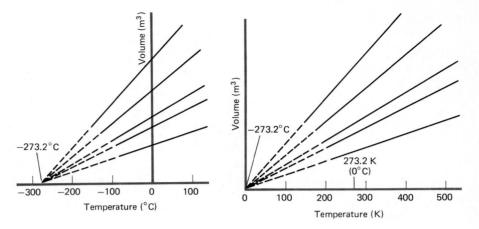

of water is 100°C, or 373.2 K. There are therefore 100 degrees between the freezing and boiling points of water on each of these scales. Any Celsius temperature can be converted to the corresponding Kelvin temperature by simply adding 273.2 to the Celsius value.

When the volume of an ideal gas at constant pressure is plotted against the *absolute temperature*, we obtain a straight line that passes through the origin, i.e., through the zero of temperature, as shown in Fig. 9.13. The volume of an ideal gas at constant pressure is directly proportional to the absolute temperature:

$$V \propto T \qquad \text{(at constant } P\text{)} \tag{9.12}$$

This is *Charles' law,* sometimes also called Gay-Lussac's law, after Joseph Louis Gay-Lussac (1778–1850), who discovered it independently of Jacques Charles.

There is a third related gas law which states that, if the volume is held constant, the pressure of a gas is directly proportional to its *absolute* temperature:

$$P \propto T \qquad \text{(at constant } V\text{)} \tag{9.13}$$

This is the law that governs the pressure in our automobile tires. For a fixed amount of air in a tire, the pressure increases in the summer and is reduced in the winter by amounts that depend on the change in the *absolute temperature*.

Absolute and Gauge Pressures

In equations like (9.13), the pressure is always the absolute or total pressure, not the pressure read by an automobile-tire pressure gauge. A pressure gauge reads the difference between the pressure inside the tire and the atmospheric pressure outside the tire. Atmospheric pressure is the force per unit area exerted by the air near the surface of the earth; this pressure is caused by the weight of all the air in the atmosphere above the point in question, as in Fig. 9.14. Absolute pressure is the gauge pressure plus atmospheric pressure. An atmospheric pressure of 1 atmosphere (1 atm) is equal to 1.013×10^5 N/m², or 14.7 lb/in². A gauge pressure of 1.73×10^5 N/m² (or 25.1 lb/in²) therefore corresponds to an absolute tire pressure

$$P = P_{\text{gauge}} + P_{\text{atm}}$$

$$= 1.73 \times 10^5 \text{ N/m}^2 + 1.013 \times 10^5 \text{ N/m}^2 = 2.74 \times 10^5 \text{ N/m}^2$$

or, in British units,

$$P = 25.1 \text{ lb/in}^2 + 14.7 \text{ lb/in}^2 = 39.8 \text{ lb/in}^2$$

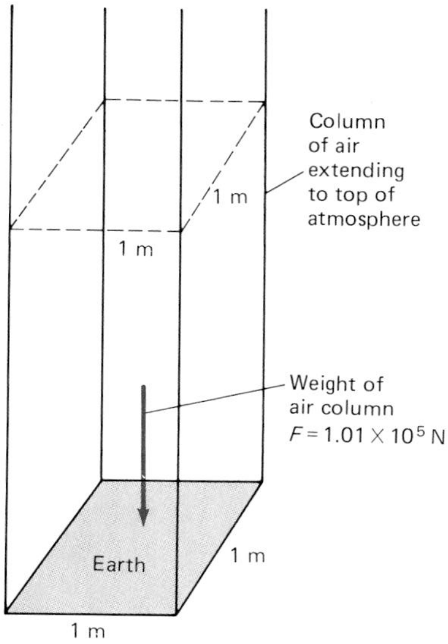

FIGURE 9.14   Atmospheric pressure: the force exerted on a unit area of the ground by the weight of a column of air of the same unit cross section and extending from the earth to the top of the earth's atmosphere.

Column of air extending to top of atmosphere

1 m

1 m

Weight of air column
$F = 1.01 \times 10^5$ N

Earth

1 m

1 m

TABLE 9.3   Useful Pressure Units and Their Conversion Factors

Pressure unit		atm	N/m^2 (Pa)	lb/in^2	mmHg (torr)
			Conversion factor		
1 atm	=	1	1.01×10^5	14.7	760
1 N/m^2 (Pa)	=	9.87×10^{-6}	1	1.45×10^{-4}	7.50×10^{-3}
1 lb/in^2	=	6.80×10^{-2}	6.90×10^3	1	51.7
1 mmHg (torr)	=	1.32×10^{-3}	133	1.93×10^{-2}	1

British units are introduced here because tire pressures are still given (unfortunately!) in these units in the United States. Some useful factors to be employed in converting pressures are given in Table 9.3. The significance of these different pressure units, and how they are to be used, will become clearer in the chapters that follow.

Example 9.6

The pressure in an automobile tire is P_1 on a hot summer day when the temperature is 38°C. What will the pressure be when the temperature is -18°C? What is the percentage change in the pressure?

SOLUTION

According to Eq. (9.13), $P \propto T$, or

$$\frac{P_2}{P_1} = \frac{T_2}{T_1} \quad \text{and} \quad P_2 = \frac{T_2}{T_1} P_1$$

Now 38°C = 38 + 273 = 311 K and -18°C = 273 − 18 = 255 K, and so

$$P_2 = \frac{255}{311} P_1 = \boxed{0.82 P_1}$$

Because of the drop in temperature, the tire pressure has fallen by 18 percent. This shows why it is a good idea to add more air to automobile tires in the winter. Note that this is an 18 percent drop in the *absolute pressure*, not the gauge pressure.

Exercise 3   If the gauge pressure corresponding to the absolute pressure P_1 is 35 lb/in^2, what is the percentage change in the gauge pressure for the same automobile tire?

Ideal Gas Law

If we put together the three gas laws in Eqs. (9.11) to (9.13), we obtain the following general equation:

$$\boxed{PV = nRT} \tag{9.14}$$

where n is the number of moles of gas present and R is a constant. Experimentally R turns out to be the same constant for all gases and is therefore called the *universal gas constant*. Its value in SI units is

$$R = 8.31 \text{ J/(mol·K)} = 8.31 \times 10^3 \text{ J/(kg mol·K)}$$

Equation (9.14) is called the *ideal gas law*, or sometimes the *equation of state of an ideal gas*. It is obeyed by most gases at pressures near or below atmospheric pressure, so long as the temperature is not too low.

For Eq. (9.14) to be applied correctly to problems using SI units, P should be in N/m^2, V in m^3, and T in K. If n is in moles, then R must be in J/(mol·K);

whereas if n is in kg mol, then R must be in J/(kg mol·K). In chemistry it is usual practice to measure P in atmospheres and V in liters. In this case R must have the value 0.0821 liter·atm/(mol·K).

If we are dealing with a *fixed amount of gas*, so that n (the number of moles) remains constant, Eq. (9.14) leads to

$$\frac{P_1 V_1}{T_1} = nR = \frac{P_2 V_2}{T_2} \tag{9.15}$$

so that we can find the final values of pressure, volume, and temperature from the initial values of the same quantities. It is crucial to remember, however, that in Eq. (9.15) the temperature must *always* be expressed on the absolute scale, i.e., in *kelvins*, and the pressure must be the *absolute* or *total pressure, not* the gauge pressure.

Example 9.7

Find the volume of 1 mol of any gas at STP.

SOLUTION

Since $PV = nRT$, we have $V = nRT/P$, or

$$V = \frac{(1.00 \text{ mol}) [8.31 \text{ J/(mol·K)}](273 \text{ K})}{1.01 \times 10^5 \text{ N/m}^2}$$

$$= \boxed{22.4 \times 10^{-3} \text{ m}^3}$$

Since 1 liter is 10^3 cm^3 (see Table B.1) or 10^{-3} m^3, 1 mol of any gas at STP therefore occupies 22.4 liters.

Notice how in this problem, in which the ideal gas equation must be solved explicitly, SI units must be used consistently throughout if a meaningful answer is to be obtained. Example 9.8 illustrates this same point.

Example 9.8

A tank of nitrogen gas at 0°C and an absolute pressure of 50.0 atm has a volume of 0.100 m^3. What is the mass of nitrogen in the tank?

SOLUTION

We have

$P = 50.0 \text{ atm} = 50.0(1.01 \times 10^5 \text{ N/m}^2) = 50.5 \times 10^5 \text{ N/m}^2$

$V = 0.100 \text{ m}^3 \qquad T = 0°C = 273 \text{ K}$

Since $PV = nRT$, the number of moles of N_2 present is

$$n = \frac{PV}{RT} = \frac{(50.5 \times 10^5 \text{ N/m}^2)(0.100 \text{ m}^3)}{[8.31 \text{ J/(mol·K)}](273 \text{ K})} = 2.23 \times 10^2 \text{ mol}$$

But 1 mol of N_2 has a mass of 28.0 g, and so

$$\text{Mass of N}_2 = (2.23 \times 10^2 \text{ mol})\left(\frac{28.0 \text{ g}}{1 \text{ mol}}\right) = \boxed{6.24 \text{ kg}}$$

Example 9.9

The glass cylinder in Fig. 9.9 has a volume of 0.0500 m^3 and contains He gas at 27.0°C and atmospheric pressure. The movable piston at the top has a cross-sectional area of 0.100 m^2. The container is then heated to 127°C with the hot plate. How far will the piston move when the temperature of the He gas goes from 27.0 to 127°C?

SOLUTION

Since no gas enters or leaves the container, the number of moles n is constant. Therefore the ideal gas law becomes [See Eq. (9.15)]

$$\frac{P_1 V_1}{T_1} = \frac{P_2 V_2}{T_2}$$

where the subscripts 1 and 2 refer to the gas properties before and after the heating occurs. Because the movable piston is acted on by the pressure of the atmosphere, the gas pressure must always remain equal to its initial value, and we have $P_1 = P_2$, and so

$$V_2 = \frac{T_2}{T_1} V_1 = \left(\frac{400 \text{ K}}{300 \text{ K}}\right) V_1 = \tfrac{4}{3}V_1$$

The increase in volume is therefore

$$\Delta V = V_2 - V_1 = \tfrac{4}{3}V_1 - V_1 = \tfrac{1}{3}V_1$$

$$= \frac{0.0500 \text{ m}^3}{3} = 0.0167 \text{ m}^3$$

Since the cross-sectional area is fixed at 0.100 m², the distance moved by the piston is

$$\Delta L = \frac{\Delta V}{A} = \frac{0.0167 \text{ m}^3}{0.100 \text{ m}^2} = 0.167 \text{ m} = \boxed{16.7 \text{ cm}}$$

9.8 Kinetic Theory and the Ideal Gas Law

We now want to take our two equations for ideal gases, one based on the molecular structure of the gas and the other on its bulk properties, and merge these into one coherent theory of great importance for our understanding of gases. The result will be one of the most important insights physics can provide.

We start with two equations, one from kinetic theory, the other from experiments on ideal gases,

$$PV = \tfrac{2}{3}N(\tfrac{1}{2}m\overline{v^2}) \qquad\qquad \textbf{(9.9)}$$

and

$$PV = nRT \qquad\qquad \textbf{(9.14)}$$

Putting these two equations together, we have

$$\tfrac{2}{3}N(\tfrac{1}{2}m\overline{v^2}) = nRT$$

Now Nm is the total mass of the gas, since N is the number of molecules and m is the mass of each molecule. This can also be expressed as nM, from Eq. (9.2), where n is the number of moles of gas and M is the mass of 1 mol, and so $\tfrac{2}{3}n\,(\tfrac{1}{2}M\overline{v^2}) = nRT$, or

$$\boxed{\tfrac{1}{2}M\overline{v^2} = \tfrac{3}{2}RT} \qquad\qquad \textbf{(9.16)}$$

The left side is the total translational kinetic energy per mole for the random motion of the gas molecules, since M is the mass of 1 mol.

If we divide Eq. (9.16) by Avogadro's number, we obtain

$$\frac{1}{2}\frac{M\overline{v^2}}{N_A} = \frac{3}{2}\frac{RT}{N_A}$$

or

$$\boxed{\tfrac{1}{2}m\overline{v^2} = \tfrac{3}{2}kT} \qquad\qquad \textbf{(9.17)}$$

Here k is a constant called *Boltzmann's constant* after the Austrian physicist Ludwig Boltzmann (1844–1906). Its value is

$$k = \frac{R}{N_A} = \frac{8.31 \text{ J/(mol·K)}}{6.02 \times 10^{23} \text{ molecules/mol}}$$

$$= 1.38 \times 10^{-23} \text{ J/K per molecule} \qquad\qquad \textbf{(9.18)}$$

The constant k is sometimes called the *gas constant per molecule*; it plays a very important role in many branches of classical and modern physics.

Equation (9.17) shows that *the Kelvin (absolute) temperature of a gas is directly proportional to the average random translational kinetic energy of the gas molecules*. For any set of molecules of mass m, the higher the temperature, the more rapidly the gas molecules move. This is the basic microscopic significance of temperature: it is a measure of the kinetic energy of the gas molecules. Equation (9.17) indicates that all gases at the same temperature (no matter what their volumes, pressures, masses, or other properties) have *exactly the same* average kinetic energy per molecule. Their speeds, of course, depend on their masses, since it is their KE that is directly proportional to the Kelvin temperature. It is worth emphasizing again that the temperature in Eq. (9.17) must be the absolute temperature, if the proportionality between temperature and KE is to be valid.

Example 9.10

Calculate **(a)** the average kinetic energy and **(b)** the approximate average speed of oxygen (O_2) molecules in the air at room temperature (about 20°C).

SOLUTION

(a) From Eq. (9.17) we have for the average kinetic energy of a gas molecule,

$$\tfrac{1}{2}m\overline{v^2} = \tfrac{3}{2}kT = 1.5(1.38 \times 10^{-23} \text{ J/K})(293 \text{ K})$$

$$= \boxed{6.07 \times 10^{-21} \text{ J}}$$

(b) Since $\tfrac{1}{2}m\overline{v^2} = 6.07 \times 10^{-21}$ J

$$\overline{v^2} = \frac{2(6.07 \times 10^{-21} \text{ J})}{m}$$

Here m is the mass of one molecule of oxygen and must be expressed in kilograms if the speed is to be obtained in meters per second. Now

$$m = \frac{M}{N_A} = \frac{32.0 \text{ g/mol}}{(6.022 \times 10^{23})/\text{mol}}$$

$$= 5.31 \times 10^{-23} \text{ g} = 5.31 \times 10^{-26} \text{ kg}$$

The same result could have been obtained by noting that the mass of one oxygen molecule is 32 u, or 32 (1.66×10^{-27} kg) $= 5.31 \times 10^{-26}$ kg, as above.

Then

$$\overline{v^2} = \frac{2(6.07 \times 10^{-21} \text{ J})}{5.31 \times 10^{-26} \text{ kg}} = 2.29 \times 10^5 \text{ m}^2/\text{s}^2$$

and $\sqrt{\overline{v^2}} = \boxed{4.78 \times 10^2 \text{ m/s}}$

This result is not the average speed in the usual sense, but rather the square root of the average of the squared speeds, a quantity called the *root-mean-square speed*, v_{rms}. (Other important rms quantities will be seen in Chap. 21 on ac electric circuits.)

Root-Mean-Square Speed

Table 9.4 gives some useful data on molecular speeds at 0°C. In this table v_{rms}, the *root-mean-square speed*, is $\sqrt{\overline{v^2}}$. Since $\tfrac{1}{2}m\overline{v^2}$ is constant for all gases at any particular temperature T, the larger the mass of a molecule, the smaller v_{rms}. This is clear from the table. On the other hand, from Eq. (9.16) the translational kinetic energy per mole, $\tfrac{1}{2}M\overline{v^2} = \tfrac{1}{2}Mv_{rms}^2$, is expected to be constant for all gases at the same T, since it is equal to $\tfrac{3}{2}RT$. The table shows that this is verified by experiment. Note that in a container 1 m long a He atom would go back and forth the length of the container about 650 times a second because of its high value for v_{rms}.

Gaseous Diffusion

If we have two gases of molecular masses m_1 and m_2 at the same temperature, from Eq. (9.17) the ratio of their molecular speeds is

$$\frac{v_{rms,1}}{v_{rms,2}} = \frac{\sqrt{\overline{v_1^2}}}{\sqrt{\overline{v_2^2}}} = \frac{\sqrt{m_2}}{\sqrt{m_1}} \qquad \text{(9.19)}$$

Thus if we have two isotopes—$^{235}_{92}U$ and $^{238}_{92}U$, for example—at the same temper-

TABLE 9.4   Speeds and Kinetic Energies of Some Important Gases at 0°C

Gas	Atomic or molecular mass M [kg/(kg mol)]	Root-mean-square speed $v_{rms} = \sqrt{\overline{v^2}}$ (10^2 m/s)	Translational kinetic energy $\frac{1}{2}Mv_{rms}^2 = \frac{1}{2}M\overline{v^2}$ [10^6 J/(kg mol)]
H_2	2	18.4	3.39
He	4	13.1	3.43
Ne	20	5.9	3.48
N_2	28	4.9	3.36
O_2	32	4.6	3.39
CO_2	44	3.9	3.35

ature, the lighter isotope will have the greater average speed. Since $^{235}_{92}U$ is the isotope needed for nuclear power plants and weapons and it makes up only 0.7 percent of uranium as found in nature, its concentration needs to be increased to be useful. One way to do this is to use a process called *gaseous diffusion*, in which the gas uranium hexafluoride (UF_6) containing both these isotopes of uranium is allowed to diffuse through porous barriers over long distances. The gas emerging from the barrier first has an increased concentration of the lighter isotope since the lighter atoms have higher average speeds. This emerging gas can then be passed through a second porous barrier to increase even more the concentration of the lighter isotope. Four thousand diffusion stages are required to produce almost pure $^{235}_{92}U$. The gaseous diffusion plants at Oak Ridge, Tennessee, and Hanford, Washington, are among the largest engineering projects ever constructed.

Example 9.11

What is the percentage difference, at room temperature, between the rms speeds of UF_6 molecules containing $^{235}_{92}U$ and those containing $^{238}_{92}U$?

SOLUTION

From the periodic table (Table F.6) we find that a fluorine atom has a mass of 19.00 unified mass units (u). The masses of the UF_6 molecules containing the two different uranium isotopes are thus 235 + 6(19) = 349 u and 238 + 6(19) = 352 u. From Eq. (9.19),

$$\frac{v_{rms}(^{235}_{92}U)}{v_{rms}(^{238}_{92}U)} = \frac{(352)^{1/2}}{(349)^{1/2}} = \frac{18.76}{18.68}$$

and the percentage difference is

$$\frac{18.76 - 18.68}{18.68} \times 100\% = \boxed{0.43\%}$$

It is because this difference in the speeds is so small that so many stages are needed to produce almost pure $^{235}_{92}U$.

9.9   Distribution of Molecular Velocities in Gases

Table 9.4 shows that the rms speed of the nitrogen (N_2) molecules, which make up 80 percent of ordinary air, is 490 m/s at 0°C. This is, of course, only a special kind of average of the actual speeds of all the N_2 molecules in the gas. These speeds actually range from near zero to above 1500 m/s. Figure 9.15 shows a graph of the distribution of speeds in N_2 at three temperatures, 273, 1273, and 2273 K. Notice how the peak of the curve shifts to higher speeds as the temperature rises, since $\overline{v^2} \propto T$, and so $v_{rms} \propto \sqrt{T}$.

The distribution of molecular speeds shown in Fig. 9.15 was first derived theoretically by the British physicist James Clerk Maxwell (1831–1879) in 1859 and is referred to as *Maxwell's distribution law*. This distribution is plotted more carefully for a sample of 10^6 oxygen molecules in Fig. 9.16. Here we plot along the y axis the number of molecules out of the 10^6 that possess speeds in a particular range, say, from 250 to 300 m/s. When this is done for all speeds from zero to very high values, we obtain the given distribution curve, which shows how the speeds are distributed in the gas. The total area under the curve corresponds to the total number of molecules (10^6), and the area of the vertical strip shown in color on the graph corresponds to the number possessing speeds between 250 and 300 m/s.

The most probable speed v_p is the speed at which the curve has its maximum. The average speed $\bar{v}$ is obtained by adding up all the speeds and dividing by the number of molecules. This is somewhat larger than v_p since the speeds must cut off at zero on the low side, while on the high side any speed can exist, in principle, up to the speed of light, 3×10^8 m/s (very high speeds are, however, very improbable). The root-mean-square velocity v_{rms}, since it is obtained by averaging the squares of the speeds, emphasizes the greater speeds even more, and so $v_{rms} > \bar{v} > v_p$, as shown in the figure.

FIGURE 9.15 Maxwell distribution law for the molecular speeds of N_2 molecules at three different temperatures. Note how the peak moves to higher speeds as the temperature increases.

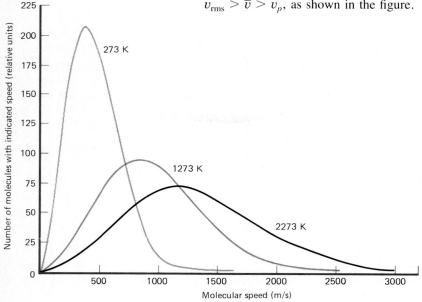

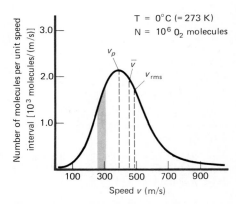

FIGURE 9.16 Distribution of molecular speeds in a sample of 10^6 oxygen (O_2) molecules at 0°C, showing the values of the most probable speed (v_p), the average speed ($\bar{v}$), and the root-mean-square speed (v_{rms}).

Evaporation

Since the distribution of molecular speeds in a liquid is similar to the Maxwell distribution of Fig. 9.16, there will always be a small number of molecules possessing very large speeds, even at room temperature. At high temperatures, of course, this number increases very rapidly, as it does in gases. Some of these high-speed molecules may be near the free surface of the liquid and may therefore be able to escape from the liquid as vapor molecules. This is the process known as *evaporation,* depicted in Fig. 9.17.

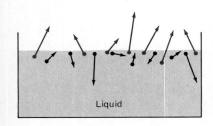

FIGURE 9.17 The evaporation of molecules from a liquid. The colored dots indicate molecules escaping from the liquid and entering the vapor phase; the black dots indicate molecules remaining in the liquid. The arrows indicate their velocities.

Summary: Important Definitions and Equations

Subatomic particles (see Table 9.1)

Notation for atoms: $_{Z}^{A}X$

Z = charge number

= number of protons in nucleus

= number of electrons in neutral atom

A = mass number = $Z + N$

= number of protons plus neutrons in nucleus

Ion: An atom that has either lost or gained electrons and therefore has acquired a net positive or negative charge.

Isotopes: Atoms having the same charge number Z but different mass numbers A.

Mole (or **gram mole**): The amount of a substance in grams that contains Avogadro's number of atoms or molecules.

Avogadro's number (N_A): The number of atoms or molecules in one mole of any substance. $N_A = 6.022 \times 10^{23}$ atoms (or molecules) per mole.

Unified mass unit (u):

$1\ u = \frac{1}{12}$ the mass of the ^{12}C atom = 1.661×10^{-27} kg

Gas: A collection of atoms or molecules that expands to fill uniformly the volume of any container in which it is placed.

Important physical quantities related to gases

Volume V: The space occupied by the gas.

Mass $\mathcal{M}$: The total mass of all the atoms or molecules making up the gas.

Density ρ: The mass per unit volume.

$$\rho = \frac{\mathcal{M}}{V} \tag{9.3}$$

Pressure P: The force per unit area acting perpendicular to any surface of the container holding the gas or to any imaginary surface drawn inside the gas.

$$P = \frac{F}{A} \tag{9.4}$$

Temperature: A measure of the hotness or coldness of an object.

Absolute (Kelvin) temperature: A temperature scale in which the zero is set at $-273.2°C$ and the size of a degree is the same as a Celsius degree. *The absolute temperature of a gas is directly proportional to the average random translational KE of the gas molecules.*

Ideal gas: A gas that obeys the gas laws at all pressures and temperatures.

Ideal gas law

$$PV = nRT \tag{9.14}$$
P = pressure (N/m^2)
V = volume (m^3)
n = number of moles of gas
R = the universal gas constant
= 8.314 J/(mol·K)
T = temperature (K)

Relationship between kinetic energy and absolute temperature of a gas

$$\tfrac{1}{2}M\overline{v^2} = \tfrac{3}{2}RT \quad \text{where } R = 8.314 \text{ J/(mol·K)} \tag{9.16}$$

$$\tfrac{1}{2}m\overline{v^2} = \tfrac{3}{2}kT \quad \text{where } k \text{ (Boltzmann's constant)} = \tag{9.17}$$
$$1.380 \times 10^{-23} \text{J/K}$$

Different speeds of gas molecules

v_p (most probable speed): The speed at which the velocity distribution curve has its maximum.
$\overline{v}$ (average speed): The sum of all speeds divided by the number of molecules
v_{rms} (root-mean-square speed): $v_{\text{rms}} = \sqrt{\overline{v^2}} \propto \sqrt{T}$ (9.17)
$$v_{\text{rms}} > \overline{v} > v_p$$

Questions

1 (a) Would you expect $_{4}Be^{+}$ and $_{3}Li$ to have similar chemical and physical properties? Why?

(b) Would you expect $_{3}Li^{+}$ and $_{4}Be$ to have similar properties? Why?

2 Hydrogen exists in three different isotopic forms, $_{1}^{1}H$ (ordinary hydrogen), $_{1}^{2}H$ (heavy hydrogen, or deuterium), and $_{1}^{3}H$ (tritium). Compare neutral atoms made of these three isotopes, and point out in what respect their properties differ and in what respect they are the same.

3 (a) Why does 1 mol of any gas always contain exactly the same number of atoms or molecules?

(b) Why would you expect equal volumes of any two gases at the same temperature and pressure to contain the same number of moles?

4 Which has the greater number of atoms, a gram of nitrogen (N_2) or a gram of oxygen (O_2)?

5 Two liters of hydrogen and one liter of oxygen gas are pumped into a strong steel drum and ignited by an electric spark so that the following chemical reaction takes place: $2H_2 + O_2 \rightarrow 2H_2O$. At the end only water vapor remains. After the temperature of the drum has returned to its original value, is the pressure greater or less than before the reaction? By how much?

6 Two gas cylinders stand on a laboratory floor. One

contains helium (He), and the other nitrogen (N_2). Both are at exactly the same temperature and pressure. If the first cylinder contains 2 kg of $^4_2$He and the second 8 kg of molecular $^{14}_7$N, which cylinder has the greater volume? Explain.

7 Two bottles of gas sit beside each other on a laboratory shelf. One bottle contains hydrogen gas (H_2) with a total translational kinetic energy of 3.3 J. The other bottle contains nitrogen gas (N_2) with a total translational kinetic energy of 4.5 J. If the nitrogen bottle contains 1.5 times as many molecules as does the hydrogen bottle, which bottle is at the higher temperature?

8 Why doesn't the size of gas molecules enter into either the equation for the kinetic energy of a gas or the ideal gas equation? Can you imagine any circumstances in which the size of the gas molecules would be important?

9 On the moon the acceleration due to gravity is only about one-sixth its value on earth. Can you use this fact and your knowledge of the kinetic theory of gases to explain why the moon has no atmosphere?

10 As one ascends directly up from the earth in a balloon, the percentage of the earth's atmosphere made up of hydrogen increases and the percentage of oxygen decreases. Explain.

11 When we compress a gas by rapidly pushing down on a piston to reduce its volume, the temperature rises. Can you explain this experimental fact on the basis of the kinetic theory of gases? (*Hint:* Consider what happens to the momentum of the gas molecules when they strike the moving piston.)

12 Why does the earth have so little helium in its atmosphere?

Multiple-Choice and Simple Exercises

In all the following problems, STP, or standard temperature and pressure, means a temperature of 0°C (273 K) and a pressure of 1 atmosphere (1.01×10^5 N/m^2). Under these conditions 1 mol of any gas occupies 22.4 liters and 1 kg mol occupies 22.4×10^3 liters, where 1 liter $= 10^3$ cm^3 = 10^{-3} m^3.

9.1 The nucleus of a plutonium atom ($^{239}_{94}$Pu) contains the following number of electrons:
(a) 94   (b) 239   (c) 145   (d) 333
(e) None of the above

9.2 The nucleus of a radium atom ($^{226}_{88}$Ra) contains the following number of neutrons:
(a) 88   (b) 226   (c) 138   (d) 314
(e) None of the above

9.3 The mass of an atom of gold ($^{197}_{79}$Au) is about:
(a) 3.3×10^{-25} kg   (b) 3.3×10^{-25} g
(c) 1.3×10^{-25} kg   (d) 1.3×10^{-25} g
(e) 3.3×10^{-22} kg

9.4 Oxygen gas at 0°C is in a rectangular container of length $L = 2.3$ m and cross-sectional area $A = 0.020$ m. In 1 s an oxygen atom moving along the x axis perpendicular to the area A makes approximately the following number of collisions with one end of the container:
(a) 1   (b) 10   (c) 100   (d) 1000
(e) None of the above

9.5 Helium gas is in a 2.0-liter container at atmospheric pressure. A piston at the end of the container is moved until the volume is reduced to 0.50 liter. If this is done slowly enough that the temperature of the gas does not change, the final pressure of the He is:
(a) 1 atm   (b) 2 atm   (c) $\frac{1}{2}$ atm
(d) 4 atm   (e) None of the above

9.6 The average translational kinetic energy of a He atom in a sample of gas at 20°C is about:
(a) 3.4×10^6 J   (b) 3.4×10^3 J
(c) 4.0×10^{-22} J   (d) 6.0×10^{-21} J
(e) None of the above

9.7 The rms speed of an argon atom ($^{40}_{18}$Ar) at room temperature is about:

(a) 1.8×10^5 m/s   (b) 4.2×10^2 m/s
(c) 0.9×10^5 m/s   (d) 3.0×10^2 m/s
(e) None of the above

9.8 A cylinder contains a mixture of hydrogen and oxygen gases. If the rms speed of the oxygen gas molecules is 5.0×10^2 m/s, the rms speed of the hydrogen molecules is about:
(a) 5.0×10^2 m/s   (b) 20×10^2 m/s
(c) 1.3×10^2 m/s   (d) 14×10^2 m/s
(e) 1.8×10^2 m/s

9.9 The temperature of a gas is increased from 300 to 600 K. The rms speeds of the molecules in the gas increases by a factor of:
(a) 2   (b) 4   (c) $\frac{1}{2}$ (d) 0.71   (e) 1.4

9.10 An ideal gas undergoes a process that doubles its temperature and doubles its pressure at the same time. If V_i is the initial volume of the gas and V_f its final volume, the following is true about the gas:
(a) $V_f = V_i/4$ (b) $V_f = V_i/2$ (c) $V_f = 4V_i$
(d) $V_f = V_i$ (e) $V_f = 2V_i$

9.11 Two containers are at the same temperature. One contains a gas of molecular mass M_1 and the other a gas of molecular mass M_2. The average momenta of the gas molecules in the two containers are related by the equation:
(a) $p_2 = p_1(m_1/m_2)^{1/2}$ (b) $p_2 = p_1$
(c) $p_2 = p_1(m_2/m_1)$ (d) $p_2 = p_1(m_1/m_2)$
(e) $p_2 = p_1(m_2/m_1)^{1/2}$

9.12 A neon atom ($^{20}_{10}$Ne) has a radius of about 1.1×10^{-10} m, and its nucleus has a radius of about 3.3×10^{-15} m.
(a) What is the density of a neon atom?
(b) What is the density of its nucleus?

9.13 Lithium (Li) as found in nature consists of 7.4% $^6_3$Li and 92.6% $^7_3$Li. What is its average mass number?

9.14 The element neon has the following natural abundances: $^{20}_{10}$Ne, 90.92%; $^{22}_{10}$Ne, 8.82%; $^{21}_{10}$Ne, 0.26%. What is the mass of a mole of neon as found on earth?

9.15 How many moles are contained in:
(a) 5.0 kg of $^4_2$He?   (b) 200 g of $^{235}_{92}$U?
(c) 600 g of molecular nitrogen (N_2) consisting of the isotope $^{14}_7$N?

9.16 A sample of unknown gas is contained in a tank of volume 5.0×10^{-2} m^3 at a pressure of 1.0×10^5 N/m^2.

(*a*) Is it possible to obtain the kinetic energy of the gas from these data, even if we do not know the kind of gas present?

(*b*) If so, what is the sum of the kinetic energies of all the gas molecules?

(*c*) Is it possible to obtain the speed of the gas molecules from the data given?

9.17 A cylindrical container of nitrogen gas contains a piston at the top that is free to move to preserve the same atmospheric pressure throughout an experiment. The gas is slowly heated from 0 to 200°C. If the initial volume of the gas was 0.010 m³, what is its final volume?

9.18 A tank of oxygen gas at 20°C and a pressure of 20 atm has a volume of 0.015 m³. How many moles are in the tank?

9.19 What is the percentage difference between the rms speeds of nitrogen molecules in a gas at 900 K and at room temperature (300 K)?

9.20 What is the percentage difference between the rms speeds of $^4_2$He and $^3_2$He atoms at the same temperature?

9.21 What is the rms speed of a He atom ($^4_2$He) in a sample of gas at 600 K?

9.22 What is the volume occupied by 8.0 mol of H_2 gas at STP?

9.23 Boron has two predominant isotopes, $^{10}_5$B, which is 19.8 percent of the boron found in the earth's crust, and $^{11}_5$B which is 80.2 percent. What is the approximate mass number of boron as found in the earth? (Compare your calculated result with that given in the periodic table of the elements.)

9.24 A block of lead has a mass of 2.2 kg. How many lead atoms are in this block?

9.25 What is the mass of 6 mol of carbon dioxide gas (CO_2) in (*a*) unified mass units and (*b*) kilograms?

9.26 How many atoms are there in 0.010 kg of argon (Ar) gas? The atomic mass of argon is 40.0 g/mol.

Problems

9.27 The average molecular mass of air is about 29 g/mol, since air is 80% nitrogen. If the air in a room 3.0 m high, 5.0 m wide, and 8.0 m long is at STP, what is the mass of the air in the room?

9.28 (*a*) How many molecules are there in 1.0 cm³ of air at STP?

(*b*) On the average, how far apart are two air molecules?

(*c*) How does this compare with the approximate size of the atoms and molecules making up the air?

9.29 (*a*) The volume occupied by a neon atom is about 6.0×10^{-30} m³. In a sample of neon gas at STP, what fraction of the volume of the gas is occupied by neon atoms?

(*b*) How does this result fit in with our assumptions about ideal gases in Sec. 9.6?

9.30 In a diesel engine the cylinder compresses air from STP to one-twentieth of its original volume and a pressure of 60 atm. What is the final temperature of the air after it has been compressed, if no heat escapes during the compression?

9.31 A medical syringe contains air at standard atmospheric pressure. The plunger is pushed in and the volume of air reduced to one-sixth its original volume. Find the final pressure of the air (*a*) when the compression is performed so slowly that there is no change in temperature; (*b*) when the compression is performed rapidly and the temperature rises from 293 to 315 K.

9.32 In an "ultra-high" vacuum system the gas pressure can be reduced to 10^{-15} atm. How many molecules of air exist in 1.0 cm³ of this very good vacuum, if the laboratory temperature is about 20°C?

9.33 A balloon has its gas bag only partially filled with 4.0×10^3 m³ of He on the ground, where the pressure is 1.0 atm and the temperature is 27°C. Find the volume of the gas bag when it has risen to such a height that the pressure has fallen to 0.25 atm and the temperature to −18°C.

9.34 A glass bottle filled with nitrogen is sealed off at STP. If the gas is then heated to 550°C, what is the new pressure of the gas? Express your answer in both atmospheres and newtons per meter squared.

9.35 An empty oil can contains 0.30 m³ of air when the cap is screwed tightly on inside a garage at 20°C. It is then set out in the sun on an asphalt driveway on a hot day, where the temperature rises to 40°C. If the original pressure was 1.0 atm, what is the final pressure?

9.36 (*a*) A tank of He gas at 27°C has a volume of 0.010 m³ at 20 atm. How much He is in the tank?

(*b*) What is the density of the He gas?

9.37 An automobile tire (assumed to have constant volume) is filled to a gauge pressure of 29.4 lb/in² on a day when the temperature is 0°C. What is the gauge pressure after the outside temperature has risen to 30°C? (*Hint:* Note the difference between absolute pressure and gauge pressure.)

9.38 A mass of 100 g of nitrogen (N_2) gas is contained in a glass cylinder at 8.0×10^5 N/m² and 27°C. What is the volume of the cylinder?

9.39 What is the mass of the carbon dioxide (CO_2) contained in a 0.10-m³ tank at an absolute pressure of 5.0 atm and at 25°C?

9.40 In a hospital, oxygen gas for patients is stored in large cylinders in a central room that is at 15°C, and the oxygen is then pumped from there to the patients' rooms. If a cylinder is at 50 atm and the pressure must be reduced to 1.0 atm before the oxygen is administered to a patient in a hospital room at 22°C, what volume does 1.0 cm³ of oxygen from the tank occupy when it is administered?

9.41 A cylinder contains 0.0500 m³ of the gas at 100 atm and 300 K. If this gas is liquefied, what will the volume of the liquid helium be? Assume that the density of liquid helium is 125 kg/m³.

*__9.42__ At 0°C and 1.00×10^3 N/m², the density of a gas is 1.24×10^{-2} kg/m³.

(*a*) Find the atomic mass of the gas and identify it.

(*b*) What is v_{rms} for the gas under these conditions?

9.43 A light bulb is sealed off at a pressure of 2.0×10^{-7} N/m² at 20°C. The tube has a volume of 75 cm³.

(*a*) How many gas molecules remain in the tube?

(*b*) If the heating of the filament in the tube raises the tube temperature to 50°C, what is the new pressure in the bulb?

9.44 An automobile tire whose volume is 0.025 m³ at 293 K is found to have a pressure of 20 lb/in² as read from a tire gauge. If it is desired to bring this pressure up to 28 lb/in² without changing the temperature, how much air must be pumped into the tire?

*9.45 A partition separates a container into two parts, one of which is four times the size of the other. The large-volume section holds He gas at 3.0 atm, while the smaller volume contains O_2 gas at a pressure of 1.0 atm. The temperature is the same for both gases and does not change during the experiment. If a small door in the partition is opened by remote control so that the gases are free to move from one side of the partition to the other, what is the final pressure of the system?

9.46 Suggest an experiment that would lead to a value for Avogadro's number if Boltzmann's constant is known.

9.47 Calculate (*a*) the average kinetic energy and (*b*) the rms speed of nitrogen (N_2) molecules in the air at 400 K.

9.48 The temperature of the interior of the sun is of the order of 10^7 K.

(*a*) Find the average translational KE of a hydrogen atom at this temperature.

(*b*) Find the rms speed of a hydrogen atom at this temperature.

(*c*) How does this speed compare with that of a hydrogen atom at 0°C?

9.49 A molecule of oxygen (O_2) at the surface of the earth and at 20°C has a velocity directed upward. If it could continue to rise without colliding with any other molecules, how far from the surface of the earth would it be when it stopped? (Neglect any change in *g* with height.)

9.50 Air contains 78% N_2 molecules and 20% O_2 molecules.

(*a*) What is the ratio of the average translational KE per molecule of these two gases at 0°C?

(*b*) What is the ratio of the rms speeds of these two gases at 0°C?

9.51 How does the mean translational kinetic energy of a hydrogen (H_2) molecule at 27.0°C compare with its change in gravitational potential energy when it is lifted through a distance of 100 m? Assume that the acceleration due to gravity remains fixed at 9.80 m/s².

*9.52 A law of considerable importance in biological situations, for example, for the exchange of oxygen with the blood, is called *Dalton's law of partial pressures*. It states that the total pressure of a gas mixture is the sum of the partial pressures of the component gases, i.e., the pressure that each gas would exert if it alone occupied the whole container. Show that Dalton's law follows directly from our model of an ideal gas.

9.53 The velocity required for a projectile shot from the surface of the earth to escape from the earth is 11.2 km/s.

(*a*) At what temperature would hydrogen molecules reach an rms speed equal to this escape velocity?

(*b*) At what temperature would helium (He) atoms reach this escape velocity?

9.54 Find the temperature at which the rms speed of O_2 molecules is equal to the rms speed of He atoms at 27°C.

9.55 A container of hydrogen gas (H_2) contains a mixture of the three isotopes of hydrogen, ^1_1H, ^2_1H, and ^3_1H.

(*a*) At 0°C what is the translational kinetic energy of molecules of these three isotopes?

(*b*) At 0°C what are the rms speeds of these three isotopes?

9.56 (*a*) Why is it easier to separate two isotopes of hydrogen by gaseous diffusion than it is to separate two isotopes of uranium?

(*b*) What is the percentage difference between the rms speeds of ^1_1H and ^2_1H at room temperature, compared with the same percentage difference for $^{234}_{92}U$ and $^{235}_{92}U$?

9.57 Prove, using a simple numerical example, that $\overline{v^2}$ does not equal $(\overline{v})^2$ for the speeds of the molecules in a gas.

9.58 Suppose it were possible to measure directly the speeds of 10 molecules in a tank of nitrogen (N_2) gas at STP and that the following speeds are obtained, with all speeds in 10^2 m/s:

9   1   3   5   6   5   8   2   5   7

For these 10 molecules, calculate (*a*) the most probable speed, (*b*) the average speed, and (*c*) the rms speed.

9.59 (*a*) Calculate the kinetic energy of a mole of krypton gas ($^{84}_{36}Kr$) at 27°C.

(*b*) What is the rms speed of the krypton atoms at this temperature?

(*c*) What is the total momentum relative to the laboratory of all the krypton atoms in a cylinder sitting on the floor of the laboratory?

Additional Readings

Andrade, E. N. da C.: *Rutherford and the Nature of the Atom*, Doubleday Anchor, Garden City, N.Y., 1964. A simple, brief, and interesting biography by a physicist.

Baker, Jeffrey J. W., and Garland E. Allen: *Matter, Energy, and Life*, 4th ed., Addison-Wesley, Reading, Mass., 1981. A good introduction to the physical and chemical concepts important in the life sciences.

Conant, James B. (ed.): *Robert Boyle's Experiments in Pneumatics*, Harvard University Press, Cambridge, Mass., 1950. An historical account of the pioneering work done by Robert Boyle on the gas laws.

Eve, A. S.: *Rutherford*, Macmillan, New York, 1939. This is considered the definitive biography of Lord Rutherford.

Nash, L. K.: *Atomic-Molecular Theory*, Harvard University Press, Cambridge, Mass., 1950. An account of the historical development of the theories of the structure of atoms.

Wilson, David: *Rutherford: Simple Genius*, MIT Press, Cambridge, Mass., 1983. The most recent biography of the famous physicist.

CHAPTER 10

Liquids and Solids

It is the intimate relation between these activities of hand and mind, which gives to the craft of the experimenter its peculiar charm. It is difficult to find in other professions such a happy mixture of both activities. . . . The experimenter must be enough of a theorist to know what experiments are worth doing and enough of a craftsman to be able to do them. He is only pre-eminent in being able to do both.

P. M. S. Blackett (1897–1974)

Of the three states of matter, gases are the easiest to handle quantitatively. Gas molecules are far enough apart that they can be considered to act independently of one another. We can, to a good approximation, ignore the forces between gas molecules except when they collide, and derive the bulk properties of the gas from the behavior of the individual molecules. This is not true of either liquids or solids, where the intermolecular forces are much greater. Crystalline solids, however, possess an orderly molecular arrangement that repeats itself over and over again as we move through the solid. This periodic structure simplifies the calculation of many solid-state properties. But liquids possess neither the total disorder that characterizes gases nor the order that characterizes crystalline solids, and so liquids are less well understood than either gases or solids. Some noncrystalline solids, such as glass, actually have properties closer to those of liquids than to crystalline solids.

10.1 Important Physical Quantities for Liquids

The intermolecular forces between the molecules in a liquid, unlike those between molecules in a gas, are quite large and bind together the molecules in the liquid. As a consequence, liquids to not expand to fill a container as gases do, but settle down, under the influence of gravity, to the bottom of any container in which they are placed. The intermolecular forces are not as strong in liquids as in solids and do not restrict the molecules to fixed positions in space. If external forces are applied to a liquid, it can therefore easily change shape.

Density and Specific Gravity

The formula for the density ρ of a liquid is the same as that given by Eq. (9.3) for a gas:

$$\rho = \frac{\mathcal{M}}{V}$$

where $\mathcal{M}$ is the mass of the liquid and V is the volume. For liquids (unlike gases) density varies very little with the pressure and temperature, and we usually treat it

TABLE 10.1   Densities and Specific Gravities for Some Liquids and Gases at STP

Substance	Specific gravity	Density (g/cm³)	Density (kg/m³)
Liquids:			
Water (H_2O) (at 4°C)	1.00	1.00	1.00×10^3
Mercury (Hg)	13.6	13.6	13.6×10^3
Alcohol, ethyl	0.79	0.79	0.79×10^3
Gasoline	0.68	0.68	0.68×10^3
Blood, whole	1.05	1.05	1.05×10^3
Gases:			
Hydrogen (H_2)	0.90×10^{-4}	0.90×10^{-4}	0.090
Helium (He)	1.79×10^{-4}	1.79×10^{-4}	0.179
Nitrogen (N_2)	1.26×10^{-3}	1.26×10^{-3}	1.26
Oxygen (O_2)	1.43×10^{-3}	1.43×10^{-3}	1.43
Carbon dioxide (CO_2)	1.98×10^{-3}	1.98×10^{-3}	1.98
Air	1.29×10^{-3}	1.29×10^{-3}	1.29

as constant. The most common liquid, water, is often taken as a standard for density measurements, and the density of all other liquids is expressed in terms of the density of water. This requires the concept of *specific gravity*:

Specific gravity: The ratio of the density of any material to the density of water.

$$Sp\ gr = \frac{\rho_x}{\rho_{water}}$$

(10.1)

Specific gravity is a dimensionless number, since it is the ratio of two quantities that are expressed in the same units. In Eq. (10.1) the density of water is taken at atmospheric pressure and 4°C, the temperature at which water is densest.

Since, under these conditions, water has a density of 1.00×10^3 kg/m³, or 1.00 g/cm³, the density of any substance in grams per cubic centimeter is numerically equal to its specific gravity; its density in kilograms per cubic meter is 10^3 times its specific gravity. Table 10.1 gives the specific gravity and density for a number of important liquids and gases.

Example 10.1

An aqueous solution of 50% albumin at 15.5°C has a specific gravity of 1.135. **(a)** What volume (in cubic meters) would be occupied by 100 g of this solution? **(b)** What would the volume be in liters?

SOLUTION

Since the solution has a specific gravity of 1.135, its density is 1.135×10^3 kg/m³, or 1.135 g/cm³.

(a) Hence the volume occupied by 100 g, or 0.100 kg, is

$$V = \frac{M}{\rho} = \frac{0.100\ kg}{1.135 \times 10^3\ kg/m^3} = \boxed{8.81 \times 10^{-5}\ m^3}$$

(b) Since 1 liter $= 10^{-3}$ m³,

$$V = (8.81 \times 10^{-5}\ m^3) \left(\frac{1\ liter}{10^{-3}\ m^3} \right)$$

$$= 8.81 \times 10^{-2}\ liter = \boxed{0.0881\ liter}$$

Pressure

We have already discussed the pressure of a gas. For a liquid at rest we call the pressure it produces normal to the walls of its container, to the bottom of the container, or to any imaginary surface inside the container the *hydrostatic pressure*. We use the same definition for this pressure as in Eq. (9.4), $P = F/A$, where F is the force exerted by the liquid perpendicular to the area A. Both gases and liquids possess the property that any force not perpendicular to the walls of the container will simply cause one layer of fluid to slide over another. It is this inability to resist such *tangential* forces that enables liquids and gases to *flow* so easily. For this reason both gases and liquids are called *fluids*. The only force that a fluid at rest can sustain is one at right angles to a boundary surface, and for this reason pressure is defined in terms of the force perpendicular to a surface. Pressure is a scalar, since it is the same in all directions inside the liquid, as we will see.

Some properties and behavior are similar for all fluids, but in many respects liquids and gases behave quite differently. For example, gases have low viscosities and are easily compressed; liquids have higher viscosities and are virtually incompressible. Because of these differences we limit our discussion of fluids in this chapter to *liquids*, although some of what follows is applicable also to gases.

10.2 Liquids at Rest

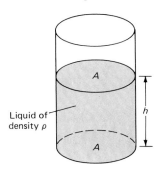

Liquid of density ρ

FIGURE 10.1 A container filled to a height h with liquid of density ρ. The container is assumed to have a constant cross-sectional area A.

One of the most important principles governing the behavior of liquids was first proposed by Blaise Pascal (1623–1662), after whom the pressure unit is named. Pascal's principle states:

The pressure applied to a confined liquid is transmitted undiminished to every point in the liquid.

Thus, if a container of liquid has a surface at the top where liquid meets air, as in Fig. 10.1, the pressure of the atmosphere on the liquid surface must be added to the hydrostatic pressure at each point inside the liquid to obtain the total pressure acting at that point.

Let us find the total pressure P' produced at the bottom of a container of liquid, if the liquid column is of cross-sectional area A and height h. An FBD of the forces acting on the liquid, which is in equilibrium, is given in Fig. 10.2. The force acting on the top surface of the liquid is $P_{atm}A$, where P_{atm} is the atmospheric pressure and A is the cross-sectional area of the liquid. The downward force produced by the weight of the liquid is $w = Mg = \rho Vg$, where ρ is the density of the liquid of volume $V = Ah$, so that $w = \rho Ahg$.

For equilibrium, the force exerted on the liquid by the bottom of the container to hold the liquid up must equal the sum of the weight of the liquid and the force exerted by the atmosphere. Therefore

$$F_{bottom} = P_{atm}A + \rho Ahg$$

But, by Newton's third law, F_{bottom} must be equal in magnitude to the force the liquid exerts on the bottom of the container, which we call F', so that

$$F' = P_{atm}A + \rho Ahg = P'A$$

and so

$$\boxed{P' = P_{atm} + \rho gh} \tag{10.2}$$

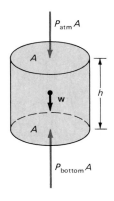

FIGURE 10.2 FBD of the forces acting on the liquid of Fig. 10.1, which is assumed to be in equilibrium.

This equation shows that the pressure in excess of atmospheric at a depth h below the surface of a liquid is

$$P = \rho g h \qquad\qquad (10.3)$$

The difference in pressure between two points in a liquid at different depths is, therefore, $\Delta P = \rho g\, \Delta h$.

The pressure exerted by the earth's atmosphere cannot be found from Eq. (10.3), since the density of the air varies with height above the earth's surface (and g also varies with altitude). The air tends to settle toward the earth so that its density near the earth is greater than in the upper atmosphere. For this reason calculus is needed to determine the pressure produced by the earth's atmosphere, if Eq. (10.3) is used. Of course, atmospheric pressure is easily measured with a barometer, as we will see.

Example 10.2

Find the height of a column of water that would produce the same pressure on the earth as is produced by the earth's atmosphere.

SOLUTION

We have seen that atmospheric pressure is 1.01×10^5 Pa (or N/m^2), and so we have

$$P = \rho g h = 1.01 \times 10^5 \text{ N/m}^2$$

Now, for water $\rho = 1.00 \times 10^3 \text{ kg/m}^3$, and so

$$h = \frac{P}{\rho g} = \frac{1.01 \times 10^5 \text{ N/m}^2}{(1.00 \times 10^3 \text{ kg/m}^3)(9.80 \text{ m/s}^2)}$$

$$= 10.3 \text{ N·s}^2/\text{kg} = \boxed{10.3 \text{ m}}$$

since $1 \text{ N} = 1 \text{ kg·m/s}^2$. The height of the water column must therefore be 10.3 m.

EXERCISE 1   Repeat this problem for a column of liquid mercury rather than of water.

The hydrostatic pressure increases as we go from the top to the bottom of a liquid column because the amount of liquid above different heights increases. What about the pressure at some fixed depth below the surface of the liquid? Is it the same in all directions? Clearly it must be, for if we immerse into the liquid a very light piece of plastic wrap at the end of a string, it is not pushed in one direction or the other by the liquid. The pressure on its front face must therefore be the same as that on its back face, and this is true regardless of its orientation. The hydrostatic pressure at any given depth in the liquid is therefore the same at all places and in all directions and depends only on depth.

This is true even for different amounts of liquid above a fixed point, so long as the height of the column remains the same. In Fig. 10.3 the pressures at A, B, and C, which are at the same distance below the liquid surface, must be the same; otherwise the liquid would flow from one place to the other. Even though there is more liquid above B and less above C than there is above A, the pressures remain the same because in B the slanted glass walls exert upward forces, and in C downward forces, to balance out the different weights of liquid.

Another clear indication that pressures are the same everywhere at the same depth comes from the human body, which is continually subjected to atmospheric pressure. The body normally does not feel this pressure because breathing fills the lungs and tissues with air, which produces a pressure inside the body that results in a force equal in magnitude and opposite in direction to the force produced by the outside pressure. Even the delicate human ear does not detect any pressure difference. If, however, we are in an airplane that is losing altitude rapidly, and the pressurizing system in the cabin is not perfect, our ears feel pressure differences

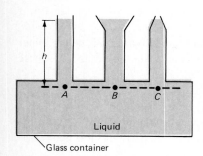

FIGURE 10.3   The pressure beneath three liquid columns of the same height but of different shapes and sizes. This pressure is the same at the same depth for all three columns ($P_A = P_B = P_C$) and is equal to $\rho g h$.

because the pressure in the middle ear on the inside of the eardrum is not able to adjust rapidly enough to the new pressure on the outside of the eardrum (see Fig. 12.13). In such circumstances, swallowing and yawning help to adjust the internal ear pressure to the external pressure.

Example 10.3

A watch manufacturer advertises a woman's wristwatch as waterproof to a water depth of 150 m. **(a)** What is the pressure at this depth? **(b)** What is the net normal force on the watch crystal which has a diameter of 2.00 cm?

SOLUTION

(a) The total pressure at a depth h is, from Eq. (10.2),

$$P' = P_{atm} + \rho g h$$

$$= 1.01 \times 10^5 \text{ N/m}^2$$

$$+ (1.00 \times 10^3 \text{ kg/m}^3)(9.80 \text{ m/s}^2)(150 \text{ m})$$

$$= \boxed{1.57 \times 10^6 \text{ N/m}^2}$$

(b) This pressure is the same in all directions in the water at a depth h. We presuppose that the pressure inside the watch crystal is atmospheric, since the watch was presumably put together at atmospheric pressure.

The pressure difference across the crystal face is then just

$$P' - P_{atm} = 1.57 \times 10^6 \text{ N/m}^2 - 1.01 \times 10^5 \text{ N/m}^2$$

$$= 1.47 \times 10^6 \text{ N/m}^2$$

and the net normal force on the crystal face is

$$F_{net} = PA = (1.47 \times 10^6 \text{ N/m}^2)[\pi(1.00 \times 10^{-2} \text{ m})^2]$$

$$= \boxed{4.62 \times 10^2 \text{ N}}$$

This result could, of course, have been obtained directly from $F_{net} = \rho g h A$.

Hydraulic Devices

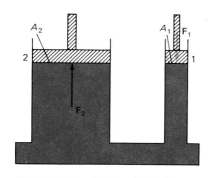

FIGURE 10.4 A hydraulic lift. Here F_2 must be equal to $(A_2/A_1)F_1$ to maintain equilibrium, since the pressure applied by the piston at 1 is transmitted undiminished to every point in the liquid. The liquid therefore exerts a large upward force on the piston at 2.

As an interesting application of Pascal's principle, let us consider a container of liquid with two entry tubes at the top, one of large cross-sectional area A_2 and the other of small cross-sectional area A_1, as in Fig. 10.4. Two pistons close off the entry ports. The liquid assumes the same height in the two tubes for the reasons just discussed. Suppose now that with piston 1 we apply an added force F_1 to A_1 so that an added pressure is exerted on the liquid equal to $P = F_1/A_1$. This pressure is added to the pressure produced by the weight of the liquid, according to Pascal's principle, and the resulting pressure is the same at all points at the same depth below the surface. The pressure on piston 2 is therefore equal to F_2/A_2, so

$$P = \frac{F_1}{A_1} = \frac{F_2}{A_2} \quad \text{or} \quad F_2 = \frac{A_2}{A_1} F_1 \tag{10.4}$$

Since A_2 can be made much larger than A_1, a small downward force F_1 can produce a larger upward force F_2. This is the principle of the hydraulic lift, a device that enables us to move heavy objects by applying relatively small forces to the lift.

Other examples of the usefulness of Pascal's principle are hydraulic presses and hydraulic brakes on cars.

Example 10.4

When a hydraulic lift is used to elevate an automobile, prove that the work output is equal to the work input, and hence that mechanical energy is conserved.

SOLUTION

Suppose we move the small piston in Fig. 10.4 down through a distance y_1 and thus cause the large piston to move upward through a distance y_2. Then the work done by F_1 is $F_1 y_1$, and the volume of the liquid in tube 1 is reduced by an amount $V_1 = y_1 A_1$. The volume of liquid in tube 2 must be increased by the same amount, since the liquid is incompressible. Thus

$$V_2 = y_2 A_2 = V_1 = y_1 A_1 \qquad \text{or} \qquad y_1 = \frac{A_2}{A_1} y_2$$

and so the work done by the applied force F_1 is, using Eq. (10.4),

$$W_1 = F_1 y_1 = \frac{F_1 (A_2 y_2)}{A_1} = F_2 y_2 = W_2$$

The work output is exactly the same as the work input, and no energy is created in the process.

A hydraulic device of this sort is equivalent to a mechanical lever that allows us to exert a small force over a large distance, and with the help of the lever produce a large force working over a short distance. In both the hydraulic lift and the lever no energy is created, but mechanical energy is conserved.

Barometers

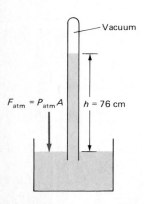

FIGURE 10.5 A mercury barometer. Since a vacuum exists in the tube above the mercury column, the column stands to a height of approximately 76 cm.

A *barometer* is an instrument used to measure the pressure exerted by the earth's atmosphere. The simplest way to make a barometer is to take a piece of glass tubing about a meter long, close it off at one end, fill it with mercury, invert it, and dip it carefully into a container of mercury without allowing any air to enter, as in Fig. 10.5. The mercury level in the tube drops to an equilibrium level of about 76 cm above the pool of mercury. Since the pressure in the pool of mercury is the same at all points at the same depth below the surface, the pressure due to 76 cmHg must be the same as the pressure produced by the air above the mercury pool. A pressure of 1 atm must therefore correspond to the pressure produced by a 76-cm column of Hg. From Eq. (10.2) this is

$$P = \rho g h = (13.6 \times 10^3 \text{ kg/m}^2)(9.80 \text{ m/s}^2)(0.76 \text{ m}) = 1.01 \times 10^5 \text{ N/m}^2$$

Pressures are often given in units of centimeters or millimeters of Hg (cmHg or mmHg) for this reason. As we see from Table 9.3, a pressure of 1 atm = 76.0 cmHg = 760 mmHg.

Note that the pressure at the bottom of the column of mercury is due to the mercury alone, because the top of the tube is closed off, and there is not enough air present in the space above the mercury column to add notably to the pressure at the bottom of the column.

A pressure of 1 mmHg is often referred to as a *torr* after Evangelista Torricelli (1608–1647), the Italian physicist who invented the mercury barometer. The best vacuum on earth, obtainable by using powerful vacuum pumps, is about 10^{-12} torr. In intergalactic space, the pressure is about 10^{-17} torr. This corresponds to less than one molecule per cubic centimeter of space.

Manometers

A *manometer* is another device used to measure pressure, usually the difference between the pressure on a confined gas and atmospheric pressure. It consists of a U-shaped tube partially filled with a liquid such as mercury, as in Fig. 10.6. If a force is exerted on the top of the liquid in tube 2, say, by a piston pushing down on the liquid, the heights of the two columns will change, as shown in the figure. For equilibrium we must have $F_2 = F_1 + \rho A g \, \Delta h$, where Δh is the difference in the liquid levels in the two sides of the tube. Therefore we have

$$P_2 A = P_1 A + \rho A g \, \Delta h$$

from which $\qquad P_2 = P_1 + \rho g \, \Delta h$

If P_1 is the pressure of the atmosphere,

$$P_2 = P_{\text{atm}} + \rho g \, \Delta h \tag{10.5}$$

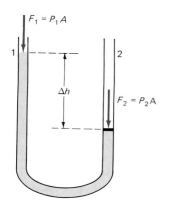

$F_1 = P_1 A$

1

2

Δh

$F_2 = P_2 A$

FIGURE 10.6   An open-ended mercury manometer. The pressure at the surface of the liquid in tube 2 is $P_2 = P_1 + \rho g\,\Delta h$. If P_1 is atmospheric pressure, then $P_2 = P_{atm} + \rho g\,\Delta h$.

This is the pressure on the liquid produced by the piston. Clearly tube 2 could be connected to a container of gas and used in the same way to determine gas pressures both above and below atmospheric pressure.

A manometer can be used to measure the pressure of automobile tires. In this case $\rho g\,\Delta h$ is the gauge pressure, the difference in pressure inside and outside the tires, as explained in Sec. 9.7. The absolute pressure inside the tire is the gauge pressure plus the pressure of the atmosphere.

Example 10.5

A mercury manometer is connected to a gas cylinder. The mercury column connected to the cylinder falls to a point 18 cm below the top of the other mercury column. **(a)** What is the gauge pressure of the gas in the cylinder? **(b)** What is its absolute pressure?

SOLUTION

(a)   The manometer reads the gauge pressure $\Delta P = \rho g\,\Delta h$, where Δh is the difference in height of the two mercury columns. Hence

$$\Delta P = (13.6 \times 10^3 \text{ kg/m}^3)(9.8 \text{ m/s}^2)(0.18 \text{ m})$$

$$= 24 \times 10^3 \text{ N/m}^2 = \boxed{0.24 \times 10^5 \text{ N/m}^2}$$

(b)   The absolute pressure is then

$$P = P_{atm} + \Delta P$$

$$= 1.01 \times 10^5 \text{ N/m}^2 + 0.24 \times 10^5 \text{ N/m}^2$$

$$= \boxed{1.25 \times 10^5 \text{ N/m}^2}$$

Blood Pressure

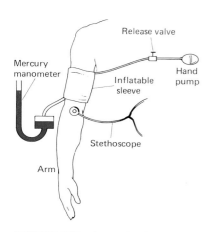

FIGURE 10.7   Apparatus for measuring blood pressure in the human body.

Release valve

Mercury manometer

Hand pump

Inflatable sleeve

Stethoscope

Arm

Mercury manometers can be used in conjunction with an inflatable sleeve to measure blood pressure in the human body. During a single cardiac cycle or heart beat, the blood pressure varies from a maximum during cardiac systole (heart contraction) to a minimum during cardiac diastole (heart relaxation). These two pressures are called *systolic* and *diastolic*, respectively.

The doctor or nurse places an inflatable sleeve on the patient's arm at about the level of the heart, as in Fig. 10.7. The sleeve is inflated with air until the pressure squeezes the arm sufficiently to cut off the blood flow. The person making the measurement then gradually lowers the pressure in the sleeve until it is just below this cutoff value, and so a brief pulse of blood flows with each heart beat. The stethoscope, placed under the sleeve in contact with the artery, can detect these pulsations of blood. At this point the blood pressure inside the artery is equal to the pressure of the inflated sleeve outside the artery. The pressure of the sleeve can then be measured with a manometer or other type of pressure gauge. In this way the systolic pressure can be obtained.

As the pressure in the sleeve is reduced even further, pulses are still heard in the stethoscope, since the blood flow is still cut off for part of each heart cycle. As soon as the sleeve pressure is reduced below that of the diastolic blood pressure, however, the sound disappears since the blood now flows smoothly through the artery. Again the sleeve pressure can be measured with a manometer and the diastolic blood pressure obtained.

Since blood pressure is measured with a manometer of some sort, what is measured is the *gauge pressure*, i.e., the difference between the pressure of the blood and atmospheric pressure. This is the only quantity of physiological interest, since it is the pressure maintained by the human circulatory system.

Normal blood pressure for healthy young adults is about 120/80, i.e., a systolic pressure of 120 mmHg and a diastolic pressure of 80 mmHg. High blood pressure is considered present when the systolic pressure is above 160 to 170 mmHg and the diastolic pressure above 95 to 105 mmHg. High blood pressure indicates that the heart is exerting itself too much to push the blood through the arteries, and the person may suffer a heart attack or stroke.

10.3 Density Determinations

You may have heard the story of how Archimedes (287–212 B.C.), the famous Greek mathematician and scientist, ran through the streets of Athens (naked, according to Vitruvius, since Archimedes had been taking a bath), shouting "*Eureka*," which in Greek means "I have found it." What he had found was a way to determine the density of the material in the king's crown.

King Hiero II of Syracuse was suspicious that a goldsmith had substituted silver for gold in his crown and asked Archimedes to determine whether this was indeed the case. The insight Archimedes had during his bath was that, if the tub were filled to the brim, then the overflow when he stepped into the tub would be exactly equal to the volume of his body. This gave him a way to determine accurately the volume of an irregularly shaped object like Hiero's crown. He placed the crown and an equal weight of pure gold into containers filled with water and measured the overflow in each case as accurately as possible. King Hiero's crown displaced more water than an equal mass of gold. Its density was therefore lower than that of pure gold, and Archimedes concluded that the goldsmith had stolen some gold from the crown and replaced it with some other metal.

This is still the easiest, if not the most accurate, way to obtain the density of an unknown solid object. We simply find the mass of the object on a balance, determine its volume by measuring the overflow when it is immersed in a beaker of water, and then calculate its density from $\rho = M/V$.

Archimedes' Principle

Archimedes is also given credit for a related principle of hydrostatics:

An object partially or totally submerged in a liquid is buoyed up by a force equal to the weight of the liquid the object displaces.

The proof of Archimedes' principle is straightforward for a regularly shaped object like a cylinder. Thus, if a cylinder is immersed in water, as in Fig. 10.8, the force acting downward on the top surface of the cylinder due to the water above the surface, is $F_1 = P_1A = \rho g h_1 A$. We neglect the effect of atmospheric pressure, which is the same on the top and the bottom of the cylinder and cancels out. The upward force on the bottom of the cylinder is

$$F_2 = P_2A = \rho g h_2 A$$

and the net upward or buoyant force acting on the cylinder is, therefore,

$$F_B = F_2 - F_1 = \rho g A(h_2 - h_1)$$

But $A(h_2 - h_1)$ is just the volume V of the cylinder, and so the buoyant force is

$$F_B = \rho g V = Mg$$

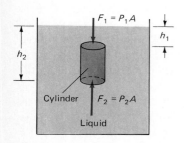

FIGURE 10.8 Proof of Archimedes' principle. The cylinder of cross-sectional area A is buoyed up by a force $w = Mg$, where M is the mass of the displaced liquid.

where $\mathcal{M}$ is the mass of the displaced liquid. The cylinder is therefore buoyed up by a force equal to the weight $\mathcal{M}g$ of the displaced liquid.

Although derived here for a simple case, Archimedes' principle can be shown to be true in general for any object, no matter what its shape.

Example 10.6

A log floats on water with four-fifths of its volume submerged, as in Fig. 10.9. What is the density of the log?

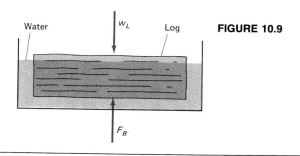

FIGURE 10.9

SOLUTION

The weight of the log is $w_L = \mathcal{M}g = \rho_L V_L g$. The volume of water displaced is given as

$$V_W = \tfrac{4}{5}V_L$$

By Archimedes' principle the buoyant force is

$$F_B = \rho_W V_W g = \rho_W \tfrac{4}{5} V_L g$$

Since the log is floating in equilibrium, its weight is equal to the buoyant force exerted by the water, and so

$$w_L = F_B$$

or $$\rho_L V_L g = \tfrac{4}{5}\rho_W V_L g$$

$$\rho_L = \tfrac{4}{5}\rho_W = \boxed{0.800 \times 10^3 \text{ kg/m}^3}$$

Notice that the fraction of the log submerged is just the ratio of the log's density to the density of water; i.e., it is the *specific gravity* of the log.

EXERCISE 2   A piece of heavy plastic floats beneath the surface of a pool filled with water and neither rises to the top nor sinks to the bottom. What is the specific gravity of the plastic? What is its density?

Use of Archimedes' Principle to Determine Densities of Solids Denser than Water

Suppose we first weigh an object on a spring scale in air. (A spring scale measures weights directly, not masses.) The object's weight is $w = \mathcal{M}g$. We then submerge the object in water at the end of a supporting string, as in Fig. 10.10a, and measure its apparent weight in water, since the force of gravity on the object has been reduced by the buoyant force F_B of the water. From the FBD of Fig. 10.10b, we obtain

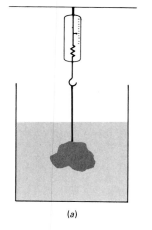
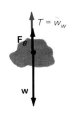

FIGURE 10.10   Determination of the density of a solid object, using Archimedes' principle: (a) an object suspended from a spring balance by a string and submerged in water; (b) FBD for the situation in (a). Here $F_B + T = w$, which leads to $F_B = w - w_W$, since $T = w_W$.

(a)

(b)

$$w_W = w - F_B \quad \text{or} \quad F_B = w - w_W$$

where w is the object's weight in air and w_W its apparent weight in water. By Archimedes' principle,

$$F_B = \rho_W V g = w - w_W \quad \text{or} \quad V = \frac{w - w_W}{\rho_W g}$$

But
$$V = \frac{M}{\rho} = \frac{w - w_W}{\rho_W g}$$

and so
$$\frac{\rho}{\rho_W} = \frac{M g}{w - w_W} = \frac{w}{w - w_W} \tag{10.6}$$

Hence the specific gravity of a solid object (ρ/ρ_W) and therefore its density can be found from two weight measurements of the object—one in air, the other in water. If a liquid other than water is used, the density of the object can still be found from Eq. (10.6), so long as the density of the liquid is known.

Example 10.7

Suppose a crown like the one Archimedes tested has a weight of 30 N in air and only 28 N when submerged in water. **(a)** Is it made out of pure gold or some other material? **(b)** If the crown is made partially of lead and partially of gold, what is the percentage of lead in it?

SOLUTION

(a) The specific gravity of the crown's material can be obtained from Eq. (10.6):

$$\frac{\rho}{\rho_W} = \frac{w}{w - w_W} = \frac{30\ N}{2\ N} = 15$$

The material has a specific gravity of 15, or a density of 15×10^3 kg/m³. From Table 10.3 we see that it certainly is not pure gold, which has a specific gravity of 19.3. Since silver has a density of 10.5×10^3 kg/m³ and lead a density of 11.3×10^3 kg/m³, it is likely that the crown is a mixture of gold with silver or lead.

(b) Let x be the fraction of lead in the crown. Then $1 - x$ is the fraction of gold. These two fractions, each multiplied by the proper specific gravity, must yield the measured specific gravity of 15, so

$$x(11.3) + (1 - x)(19.3) = 15$$

Solving for x, we obtain $x = 0.54$. The crown is therefore

$$\boxed{54\%\ \text{lead}}$$

and only 46% gold.

Example 10.8

A tank of water rests on a spring scale that reads 200 N. Determine the new scale reading when a steel object of volume 0.0100 m³ is submerged in the water and supported by a cord connecting it to a support above the tank, as in Fig. 10.11a.

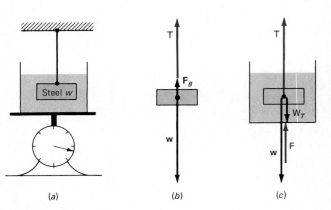

(a) (b) (c)

FIGURE 10.11

SOLUTION

The weight of the tank and water together is $W = 200$ N. If we isolate the steel object and consider all the forces acting on it, as in Fig. 10.11b, we have

$$T + F_B = w \qquad (1)$$

where T is the tension in the string, F_B is the buoyant force of the water on the object, and w is the weight of the steel object in air.

If we now treat the tank and the water (of combined weight W_T) and the steel object as a system as in Fig. 10.11c, the net external force on this system is zero so that

$$T + F = w + W_T \qquad (2)$$

where F is the magnitude of the force exerted by the spring scale on the tank. Note that the buoyant force F_B does not appear in this equation, since it is a force internal to the tank and its contents.

Combining Eqs. (1) and (2), we obtain

$$F = F_B + W_T$$

But F is equal and opposite to the force exerted on the scale, which is the weight the scale registers. The scale reading is then

$$F = F_B + W_T = \rho V g + W_T$$

$$= (1.00 \times 10^3 \text{ kg/m}^3)(0.0100 \text{ m}^3)(9.80 \text{ m/s}^2) + 200 \text{ N}$$

$$= 98.0 \text{ N} + 200 \text{ N} = \boxed{298 \text{ N}}$$

We see that this result does not depend on the object's being steel. All that is important is the *volume* of the liquid displaced by the object, since this determines the buoyant force F_B.

Center of Buoyancy

The point through which a buoyant force acts is called the *center of buoyancy* (CB). The CB is at the center of gravity (CG) of the displaced liquid. If the mass of a ship and its cargo is assumed to be evenly distributed on each side of the ship, and the ship is upright, the CB and the CG of the ship lie on the same vertical line, as shown in Fig. 10.12a. Since the buoyant force $\mathbf{F}_B$ and the weight $\mathbf{w}$ are equal in magnitude and have the same line of action, equilibrium results.

Whereas the CG of the ship and its contents does not change position as the ship rolls, the CB can shift greatly. Rolling therefore displaces the new center of buoyancy (CB') with respect to the CG of the ship. The buoyant force is still equal to the ship's weight, but it produces a torque about the CG of the ship. In the case shown in Fig. 10.12b, this torque acts to restore the ship to its original position, and so we have stable equilibrium. Under different conditions, however, the torque produced by the buoyant force F_B' may capsize the ship.

FIGURE 10.12 (a) For an upright ship, the center of buoyancy (CB) lies on the same vertical line as the CG. (b) If the ship rolls slightly, the new center of buoyancy CB' is displaced with respect to the CG, resulting in a torque about the CG of the ship. In the case shown here this torque acts to right the ship.

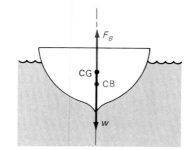

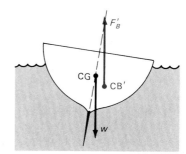

10.4 Liquids in Motion

The flow of liquids, such as the cooling water in the huge pipes of a nuclear power plant or the blood in the arteries of the human body, can be roughly divided into two classes, laminar (streamline) flow and turbulent flow.

By *laminar* (streamline) flow we mean the flow of a liquid along streamlines, similar to those in Fig. 10.13. Any particle of the liquid that is at point D is carried

FIGURE 10.13 Laminar (streamline) flow. Any liquid particle arriving at point D will move along a streamline from D to B to C. Similarly any particle at D' will move from D' to B' to C'. This behavior remains the same over time.

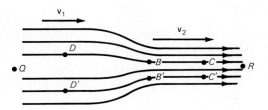

along a particular streamline and arrives at point B and later at point C. The flow is smooth and completely predictable.

In turbulent flow, eddies and vortices appear in the liquid, and much of the energy of the liquid is converted into heat by the turbulent motion. The velocity at any point along the path of the liquid is not constant but varies rapidly and unpredictably in magnitude and direction. In turbulent flow it is impossible to predict where a particle of the liquid at D will be at any later time. A good example of turbulent flow is that of the water in white-water rapids.

A useful dimensionless parameter to characterize the type of liquid flow is the *Reynolds number*: $\text{Re} = \rho v \ell / \eta$, where ρ is the density and v the average velocity of the liquid, η is the viscosity of the liquid (Sec. 10.5), and ℓ is a length characteristic of the geometry of the flow. For example, for a liquid flowing in a pipe, ℓ would be the internal diameter of the pipe. At Re values of about 2000, the flow usually changes from laminar to turbulent.

Use of the Reynolds number allows us to make scale models of fluid flow. The size of a pipe may be reduced by a factor of 5, for example, provided that the values of ρ and η for the fluid used for the modeling are chosen to make Re and v the same. The behavior of the fluid under study can then be found using apparatus of greatly reduced scale.

In this section we consider only the simplest kind of liquid flow—laminar flow (i.e., flow along streamlines without turbulence) and without viscosity.

Equation of Continuity

If we consider a small tube of liquid whose boundaries are set by streamlines, as in Fig. 10.13, the velocity of the liquid may change as it flows along the tube. Suppose that at Q the velocity of the liquid is v_1 and the cross-sectional area A_1, and that at R the velocity is v_2 and the area A_2. Then for smooth, continuous flow, the mass of liquid passing the point Q must be equal to the mass passing the point R in the same time interval Δt. In time Δt the mass passing Q is

$$\Delta \mathcal{M}_1 = \rho_1 \, \Delta V_1 = \rho_1 A_1 v_1 \, \Delta t$$

where $v_1 \, \Delta t$ is the length of the liquid element passing Q in time Δt and ρ_1 is the density of the liquid at Q. Similarly, at R

$$\Delta \mathcal{M}_2 = \rho_2 A_2 v_2 \, \Delta t$$

Then, if we assume that the liquid is incompressible so that $\rho_1 = \rho_2$,

$$\frac{\Delta \mathcal{M}_1}{\Delta t} = \frac{\Delta \mathcal{M}_2}{\Delta t} = \rho_1 A_1 v_1 = \rho_2 A_2 v_2$$

and so
$$\boxed{A_1 v_1 = A_2 v_2}$$
(10.7)

This is the *equation of continuity*, which says that, for streamline flow, the speed of flow of a liquid varies inversely with the cross-sectional area of the tube. For example, if there are constrictions in blood vessels in the human body, the blood must move more rapidly to get through these constrictions.

Example 10.9

In a normal adult at rest the average speed with which the blood flows through the aorta, the main artery leading from the heart, is $\bar{v}_1 = 0.33$ m/s. **(a)** If the aorta has a cross-sectional radius of about 1.0 cm, how much blood flows per second through it? **(b)** If the total cross-sectional area of all the arteries fed by the aorta is 20×10^{-4} m², at what speed does blood flow in these arteries?

SOLUTION

(a) The amount of blood flowing per second through the aorta is that occupying a volume of cross-sectional area $A_1 = \pi(1.0 \times 10^{-2} \text{ m})^2 = 3.1 \times 10^{-4}$ m² and of length 0.33 m, since all the blood in such a length passes through the aorta at a speed of 0.33 m/s. The volume of blood flowing per second through the aorta is therefore

$$A_1\bar{v}_1 = (3.1 \times 10^{-4} \text{ m}^2)(0.33 \text{ m/s}) = \boxed{1.0 \times 10^{-4} \text{ m}^3/\text{s}}$$

(b) From the equation of continuity, we have

$A_1 v_1 = A_2 v_2$ or, for average values of the velocities,

$$\bar{v}_2 = \frac{A_1 \bar{v}_1}{A_2} = \frac{1.0 \times 10^{-4} \text{ m}^3/\text{s}}{20 \times 10^{-4} \text{ m}^2} = \boxed{0.050 \text{ m/s}}$$

The blood therefore moves about six times more slowly through the other arteries than it does through the aorta.

Bernoulli's Equation

The most important equation in fluid dynamics is named for the Swiss mathematician Daniel Bernoulli (1700–1782). It describes the streamline flow of a nonviscous, incompressible liquid, as shown in Fig. 10.14. In the case illustrated in the figure, we assume that the liquid (say, water) is pumped uphill, and the pipe is also assumed to widen as it rises. We concentrate on the liquid between points Q and R in Fig. 10.14a. In time Δt the left end of this liquid has moved from Q to Q' and the right end from R to R', as shown in Fig. 10.14b.

We now apply our extended work-energy theorem to this system. We have, from Eq. (6.12),

$$\text{Work done} = \Delta \text{KE} + \Delta \text{PE}$$

Here the work is done by the net force produced by the unequal pressures applied at the two ends to move the liquid through the pipe. The work done by the force F_1 at Q is $\mathcal{W}_1 = F_1 \Delta L_1 = P_1 A_1 \Delta L_1$. Similarly, the work done by the force F_2 at R is $\mathcal{W}_2 = -P_2 A_2 \Delta L_2$, since F_2 and ΔL_2 are in opposite directions. The total work done is then

$$\mathcal{W} = \mathcal{W}_1 + \mathcal{W}_2 = P_1 A_1 \Delta L_1 - P_2 A_2 \Delta L_2$$

FIGURE 10.14 Bernoulli's equation for streamline flow in a nonviscous, incompressible liquid: (a) initial position of a chosen volume of liquid V_1; (b) final position of the same volume of liquid, now marked as V_2.

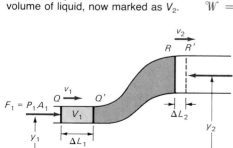

(a)

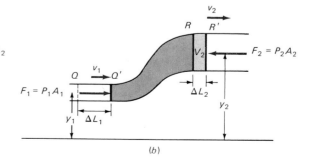

(b)

But, since from the equation of continuity $A_1v_1 = A_2v_2$, we have

$$A_1 \frac{\Delta L_1}{\Delta t} = A_2 \frac{\Delta L_2}{\Delta t} \quad \text{or} \quad A_1\,\Delta L_1 = A_2\,\Delta L_2$$

Now $A_1\,\Delta L_1$ and $A_2\,\Delta L_2$ are the volumes V_1 and V_2, and so

$$V_1 = V_2 = \frac{m}{\rho}$$

where m is the mass of the colored portion of liquid in the diagram. The liquid between points Q' and R is unchanged in the process. If we set $V_1 = V_2 = V$, we have

$$\mathcal{W} = P_1A_1\,\Delta L_1 - P_2A_2\,\Delta L_2 = (P_1 - P_2)V = (P_1 - P_2)\frac{m}{\rho}$$

and our work-energy theorem becomes

$$(P_1 - P_2)\frac{m}{\rho} = \Delta KE + \Delta PE = \tfrac{1}{2}mv_2^2 - \tfrac{1}{2}mv_1^2 + mgy_2 - mgy_1$$

Dividing through by m and multiplying by ρ, we have

$$P_1 - P_2 = \tfrac{1}{2}\rho v_2^2 - \tfrac{1}{2}\rho v_1^2 + \rho gy_2 - \rho gy_1$$

and so $\quad P_1 + \tfrac{1}{2}\rho v_1^2 + \rho gy_1 = P_2 + \tfrac{1}{2}\rho v_2^2 + \rho gy_2$

Since this is true for any two points in the flow, we have in general

$$\boxed{P + \tfrac{1}{2}\rho v^2 + \rho gy = \text{constant}} \tag{10.8}$$

This is *Bernoulli's equation* for liquid flow. All three terms in it have the dimensions of energy per unit volume (you should be able to prove this). It is therefore a statement of the principle of conservation of energy applied to liquid flow. It states that in the streamline flow of an incompressible, nonviscous liquid the pressure, speed of flow, and height of the liquid always adjust themselves so as to keep the total energy of the liquid constant.

Bernoulli's equation contains most other equations of hydrostatics and hydrodynamics as special cases. Thus it tells us that for a liquid at rest, since $v_1 = v_2 = 0$,

$$P_1 + \rho gy_1 = P_2 + \rho gy_2$$

If P_1 is the pressure at the surface of a liquid and P_2 is the pressure a depth h below the surface, then we obtain Eq. (10.5), since

$$P_2 = P_1 + \rho g(y_1 - y_2) = P_1 + \rho g\,\Delta h$$

where y_1 and y_2 are measured from a point at the bottom of the fluid, as in Fig. 10.15.

Also, for steady horizontal flow, where y remains constant, we have

$$P + \tfrac{1}{2}\rho v^2 = \text{constant} \tag{10.9}$$

The pressure of a liquid is therefore reduced at those points where the velocity increases, and vice versa.

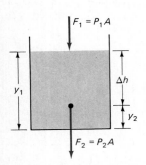

FIGURE 10.15 Bernoulli's equation applied to the pressure exerted by a liquid at rest: $P_2 = P_1 + \rho g\,\Delta h$.

Applications of Bernoulli's Equation

Bernoulli's equation holds strictly only when all the assumptions made in its derivation are satisfied, in particular when we are dealing with an incompressible, nonviscous fluid. But even if these conditions are not perfectly satisfied, e.g., if we are dealing with a slightly viscous gas of variable density, it turns out that many qualitative predictions of Bernoulli's equation remain valid, as long as the density does not change too much. This is true in particular of the relationship between particle speed and pressure found in Eq. (10.9).

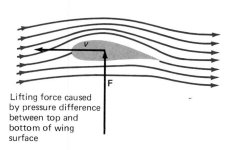

FIGURE 10.16 Lifting force on an airplane as predicted by Bernoulli's equation. On the top wing surface the air speed is higher and the pressure lower. Along the bottom of the wing the air speed is lower and the pressure higher. This generates the lifting force.

Lifting force caused by pressure difference between top and bottom of wing surface

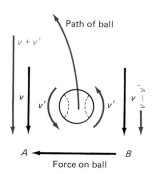

FIGURE 10.17 Why a baseball curves: an application of Bernoulli's equation. Imagine yourself looking down on the baseball from a frame of reference that moves along with the ball through the air. In this reference frame **v** is the velocity of the air with respect to the ball.

The wings of an airplane have a curved shape that forces the air to flow more rapidly above the wing and less rapidly below it when the plane is moved forward by its motors, as in Fig. 10.16. As a result, the pressure above the wing is reduced, while below the wing it is increased, according to the above equation. This provides the "lift" to keep the airplane in the air. If a plane stalls in flight so that this pattern of streamline airflow is interrupted, the lifting force is not sufficient to support the plane's weight and the pilot must put the plane into a shallow dive to restore the airflow and increase the lift.

The curves thrown by baseball pitchers and the slices and hooks of golfers can also be understood, at least in part, in terms of Eq. (10.9). Figure 10.17 shows what happens when a pitcher throws a curve. As the ball moves forward, the air rushes past it at a velocity v. Since the ball is spinning, however, it drags some air with it with a velocity v' with respect to the ball. The resultant speed of the air at A with respect to the ball is then $v + v'$, and at B it is $v - v'$. The pressure is, therefore, by Bernoulli's principle, lower at A than at B, since the velocity is greater at A. Hence a force is set up that continuously deflects the ball in the direction of A.* The rougher the surface of the ball, the greater the effect. This is the reason major league pitchers resort to all kinds of tricks to make the surface of the ball as rough as possible.

Example 10.10

Cold water circulates through a house from the main water pipe in the basement. The water enters the house through an 8.0-cm-diameter pipe at a flow speed of 0.60 m/s under a pressure of 4.00×10^5 N/m² (i.e., four times atmospheric pressure). **(a)** What is the flow speed of the water in a 5.0-cm-diameter pipe 9.0 m above the main water pipe in the basement? **(b)** What is the water pressure in the 5.0-cm pipe?

---

*This is far from the whole story. As a matter of fact, the Reynolds number for a spinning baseball moving through air at about 45 m/s indicates that the flow of air around the ball is not laminar but turbulent. This turbulence must be taken into account in any complete explanation of the curving of a baseball.

SOLUTION

(a) We know that the flow speed and the cross-sectional area of the pipe at any two points along the pipe are related by the equation of continuity $A_1v_1 = A_2v_2$. Therefore

$$v_2 = \frac{A_1}{A_2}v_1 = \frac{(0.040 \text{ m})^2}{(0.025 \text{ m})^2}(0.60 \text{ m/s}) = \boxed{1.5 \text{ m/s}}$$

(b) From Bernoulli's equation we have

$$P_1 + \tfrac{1}{2}\rho v_1^2 + \rho g y_1 = P_2 + \tfrac{1}{2}\rho v_2^2 + \rho g y_2$$

$$P_2 = P_1 + \tfrac{1}{2}\rho(v_1^2 - v_2^2) + \rho g(y_1 - y_2)$$

$$= 4.00 \times 10^5 \text{ N/m}^2 + \tfrac{1}{2}(10^3 \text{ kg/m}^3)[(0.60 \text{ m/s})^2 - (1.5 \text{ m/s})^2]$$

$$+ (10^3 \text{ kg/m}^3)(9.8 \text{ m/s}^2)(0 - 9.0 \text{ m})$$

Since $1 \text{ kg·m/s}^2 = 1 \text{ N}$,

$$P_2 = (4.00 \times 10^5 - 9.5 \times 10^2 - 8.8 \times 10^4) \text{ N/m}^2$$

$$= \boxed{3.12 \times 10^5 \text{ N/m}^2}$$

The pressure on the third floor is less than in the basement because work has been done both to lift the water to the third floor and to increase the kinetic energy of the water.

Example 10.11

A water cooler open to the atmosphere at the top contains water to a height h above the exit hole, as in Fig. 10.18. With what speed will the water flow out through the hole?

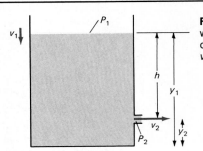

FIGURE 10.18 Speed of water flow from a water cooler. Here $P_1 = P_2$ and $v_2 = \sqrt{2gh}$.

SOLUTION

This is a good illustration of Bernoulli's principle. We have

$$P_1 + \tfrac{1}{2}\rho v_1^2 + \rho g y_1 = P_2 + \tfrac{1}{2}\rho v_2^2 + \rho g y_2$$

Both the top of the cooler and the exit hole are assumed to be exposed to the atmosphere, so $P_1 = P_2$. Also v_1, which is the speed with which the water level falls, is so small that we can take it to be zero to a good approximation. Hence we have

$$\tfrac{1}{2}\rho v_2^2 = \rho g(y_1 - y_2) \quad \text{or} \quad v_2 = \sqrt{2g(y_1 - y_2)}$$

Since $y_1 - y_2$ is here equal to h, the distance of the hole below the level of the water in the cooler, this becomes

$$v_2 = \sqrt{2gh}$$

a result sometimes called *Torricelli's theorem*.

It is interesting to note that the water speed in this case is the same as would be given by mechanics for a mass m of water falling freely under the force of gravity through a distance h.

Figure 10.19 shows how the velocity differs for water flow from holes at different heights.

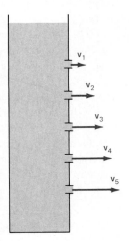

FIGURE 10.19 Velocity of water flow from a cooler as a function of height of water level above exit hole.

10.5 Viscosity

In our discussion of Bernoulli's principle we omitted one very important property of liquids—viscosity. *Viscosity* is a measure of how difficult it is to cause a liquid to flow, or of how resistive the liquid is to motion through it. Thus water has a

FIGURE 10.20 Distribution of speeds for the viscous flow of liquid in a pipe. The length of the arrows indicates the magnitude of the velocity at various distances from the center of the pipe.

FIGURE 10.21 A way to measure the viscosity of a liquid. The greater the viscosity, the larger the size of the force F required to give the upper plate a speed v.

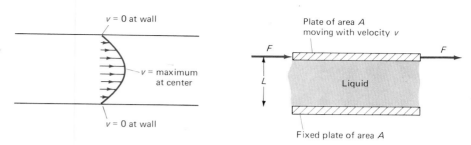

higher viscosity than air; we can easily walk through the air, but it is much more difficult to walk through water. If water did not resist the motion of our bodies, however, swimming would be impossible, because it is the reactive force of the water on our bodies that pushes us forward when we swim, just as it is the reactive force of the ground on our feet that pushes us forward when we walk.

If a liquid is flowing through a pipe, the walls of the pipe rub against it and tend to slow it down. The liquid therefore flows more slowly near the walls of the pipe than at the center, and a distribution of speeds is set up across the width of the pipe, similar to that in Fig. 10.20.

We can approximate this speed distribution by a model in which adjacent thin cylindrical shells of liquid move with a slightly different speed through the pipe. One shell rubs against the next shell in the same way that a book rubs against a desktop when dragged over it. We have seen that friction between the book and the desktop generates heat and wastes useful mechanical energy. The same thing happens with liquids. Mechanical energy is lost and heat is generated by the internal sliding of one layer of liquid over another. Viscosity for liquids is thus analogous to the frictional force between solids and depends on the force required to slide one layer of the liquid over another at a given speed.

One way to measure viscosity is to confine a liquid between two flat plates each of area A and to measure the magnitude of the force **F** required to move the top plate at a constant speed v with respect to the fixed bottom plate. If the distance between the plates is L, as in Fig. 10.21, then we define the *coefficient of viscosity* η (Greek lowercase letter *eta*) as follows:

$$\eta = \frac{F/A}{v/L} \tag{10.10}$$

For fixed A and L, the larger the force needed to produce a speed v, the larger η, and the greater the viscosity of the fluid.

Equation (10.10) shows that the SI units for the coefficient of viscosity η must be

$$\frac{N/m^2}{(m/s)/m} = \frac{N \cdot s}{m^2} \text{ or Pa} \cdot s \qquad \text{since} \qquad 1 \text{ Pa} = 1 \frac{N}{m^2}$$

One pascal-second is equal to 1000 centipoise (cP), a unit frequently used in tabulating viscosities. The viscosity of water at 20.20°C is taken as a reference and is 1.0000 cP.

The viscosities of some important liquids are shown in Table 10.2. Note the large variations. The coefficient of viscosity of liquids depends on temperature, falling as the temperature rises. This is what we would expect, since liquids approach the gaseous state as their temperature rises, and gases have much lower viscosities than liquids.

TABLE 10.2 Coefficient of Viscosity of Liquids (at 20°C)

Liquid	Coefficient of viscosity η (Pa·s, or N·s/m²)
Water	1.00×10^{-3}
Benzene	0.65×10^{-3}
Carbon tetrachloride	0.97×10^{-3}
Ethyl alcohol	1.20×10^{-3}
Glycerin	0.83
Mercury	1.55×10^{-3}
Methyl alcohol	0.60×10^{-3}
Sulfuric acid	22×10^{-3}
Turpentine	1.49×10^{-3}
Whole blood (37°C)	4.0×10^{-3}

Poiseuille's Law

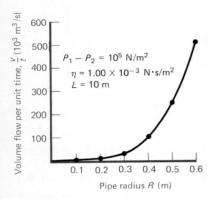

$P_1 - P_2 = 10^5$ N/m^2

$\eta = 1.00 \times 10^{-3}$ N·s/m^2

$L = 10$ m

FIGURE 10.22 The volume flow of water (V/t) through a round pipe as a function of pipe radius (R) for constant pressure difference between the ends of the pipe.

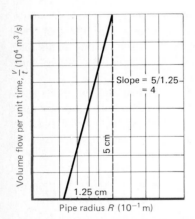

Slope = 5/1.25 = 4

5 cm

1.25 cm

FIGURE 10.23 Data of Fig. 10.22 plotted on log-log graph paper (see Appendix 2.C). The slope is 4, indicating that V/t is proportional to R^4.

It is often important to know the rate at which a liquid flows—e.g., blood through the human body or oil through the Alaskan pipeline. We would expect that the greater the pressure difference between the two ends of a pipe ($\Delta P = P_1 - P_2$) and the shorter the pipeline, the greater the flow rate. We would therefore expect that the volume of liquid flowing out of the pipe per second would vary as the *pressure gradient*, $(P_1 - P_2)/L$. Also, since increased viscosity tends to slow down the flow, we would expect an inverse dependence on the coefficient of viscosity η. The dependence on the radius of the tube R is more difficult to predict. To find this dependence we can resort to an experiment on the flow of water through tubes of radius 0.01, 0.02, 0.04. and 1.0 m for the same values of $P_1 - P_2$, L, and η. The results are shown in Figs. 10.22 and 10.23. In Fig. 10.22 the volume flowing through in unit time, V/t, is plotted as a function of R. The dependence is clearly not linear, but it is hard to tell the exact functional dependence. In Fig. 10.23 we plot log (V/t) versus log R. We find a straight line with a slope of 4. Hence, as shown in Appendix 2.C, V/t varies as R^4.

The Frenchman Jean L. M. Poiseuille (1797–1896) combined this R^4 dependence with the other factors described above to form a complete equation for fluid flow. This is now called *Poiseuille's law*:

$$\frac{V}{t} = \frac{\pi R^4 (P_1 - P_2)}{8 \eta L}$$

(10.11)

The remarkable quantity in this equation is the R^4 term. An oversimplified explanation for this fourth-power dependence on R is that, since both the cross-sectional area of the pipe and the velocity of flow at the center are proportional to R^2, the flow rate is proportional to the product of the two, or to R^4 (a rigorous proof requires the use of calculus).

Poiseuille's major research interest was the physiology of blood circulation. This work led him to his equation, which has a wide applicability in physics and engineering. The blood distribution system in the human body is very complicated, and it is amazing that Poiseuille's law applies to it as well as it does.

Poiseuille's law also points up problems that can result from the filling in of the arteries with deposits of cholesterol or other debris. According to Eq. (10.11), if in some way R is reduced by a factor of 2, with η and L remaining unchanged, then the pumping pressure of the heart must increase by a factor of 16 to keep the blood flowing at its proper rate. Usually the heart cannot completely provide this increase in pressure, and, as the artery fills up, the blood circulation drops. The clogging of arteries leads to two quite undesirable consequences, high blood pressure and poor blood circulation. This *stenosis*, or narrowing of the arteries, is often a contributing factor in chronic heart disease. Hardening of the arteries (*arteriosclerosis*) can have the same results, since it reduces the flexibility of the arteries and keeps them from widening when the pressure increases.

Example 10.12

Blood flows through a thin capillary tube in a person's leg. The tube is 2.0 mm long and 3.0 μm in radius. The speed of the blood through the center of this capillary is 1.0 mm/s. What is the drop in the blood pressure as the blood passes through this capillary?

SOLUTION

From Table 10.2 the coefficient of viscosity for blood at 37°C (body temperature) is $\eta = 4.0 \times 10^{-3}$ Pa·s $= 4.0 \times 10^{-3}$ N·s/m².

From Poiseuille's equation for the viscous flow of a liquid (the case here), we have [Eq. (10.11)]

$$P_1 - P_2 = \frac{8\eta L V}{\pi R^4 t}$$

where V/t is the amount of blood flowing through the capillary per second. As we saw in Example 10.9, this flow rate $V/t = A\bar{v}$, where $\bar{v}$ is the average speed of flow across the diameter of the tube, if v is not constant. The speed of flow varies from 1.0 mm/s at the center of the capillary to 0 at the walls of the capillary, where the viscosity prevents the fluid from moving at all. An approximation to the average flow speed is therefore

$$\bar{v} = \frac{1.0 \text{ mm/s} + 0}{2} = 0.50 \text{ mm/s}$$

Then

$$P_1 - P_2 = \frac{8\eta L}{\pi R^4}(\pi R^2 \bar{v}) = \frac{8\eta L}{R^2}\bar{v}$$

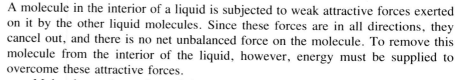

$$= \frac{8(4.0 \times 10^{-3} \text{ N·s/m}^2)(2.0 \times 10^{-3} \text{ m})(0.50 \times 10^{-3} \text{ m/s})}{(3.0 \times 10^{-6} \text{ m})^2}$$

$$= 3.6 \times 10^3 \text{ N/m}^2 = \boxed{3.6 \times 10^3 \text{ Pa}}$$

The pressure drop in this one capillary is about 27 mmHg compared to an average blood pressure in the human body of about 100 mmHg.

Note that only two significant figures are carried through this calculation because of the approximation used in the calculation and of the uncertainty in the value of the viscosity η for whole blood.

10.6 Surface Tension

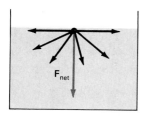

FIGURE 10.24 Forces on a molecule at the surface of a liquid. The resultant of all these forces, **F**net, holds a surface molecule in the liquid.

A molecule in the interior of a liquid is subjected to weak attractive forces exerted on it by the other liquid molecules. Since these forces are in all directions, they cancel out, and there is no net unbalanced force on the molecule. To remove this molecule from the interior of the liquid, however, energy must be supplied to overcome these attractive forces.

Molecules at the surface of the liquid are pulled back toward the interior of the liquid, as in Fig. 10.24, since the molecules attracting them are in the interior of the liquid. A surface molecule is surrounded by only about half as many molecules as one in the interior, and so only about half as much energy is needed to remove a molecule from the surface as from the interior. As a consequence, a molecule in the surface layer has a higher potential energy than a molecule in the interior of the liquid, and so increasing the surface layer increases the potential energy of the liquid. Since the liquid tends to minimize its potential energy, as does a roller-coaster car gliding down a hill, the liquid tries to minimize its surface area for any given volume of liquid. As a result, the surface of a liquid behaves like a stretched membrane and produces a *surface tension* in the liquid.

Surface tension explains the spherical shape of liquid droplets, as in Fig. 10.25, the "hanging" of drops of water at the end of a faucet, the behavior of liquids in capillary tubes, and the floating of objects denser than water, such as needles and small insects, on the tight "skin" of the water.

In this latter case we find that the needle rests in a slight depression in the liquid surface, as in Fig. 10.26. Since this depression increases the surface area of the liquid, forces are set up parallel to the liquid surface trying to reduce the surface area and thus lower the surface energy. These forces have components in the vertical direction that are sufficient to balance out the weight of the needle.

The surface tension γ (Greek letter *gamma*) in the surface film of a liquid is defined as the ratio of the molecular force F parallel to the surface of the liquid

FIGURE 10.25 A high-speed photograph, taken by Dr. Harold Edgerton of MIT, illustrating how spherical droplets are formed when a drop of liquid falls on a liquid surface. The droplets break off in the form of tiny spheres because a sphere has a smaller surface area for a given volume than any other geometric shape. (*Photo courtesy of Omikron/ Photo Researchers.*)

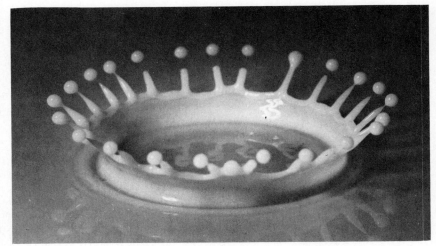

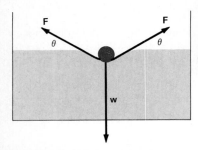

FIGURE 10.26 A needle resting in equilibrium on the surface of a liquid. The vertical components of the surface tension forces balance out the weight of the needle.

(which acts to reduce the surface area of the liquid) to the length L of the surface film across which this force acts:

$$\gamma = \frac{F}{L} \tag{10.12}$$

The SI units for surface tension are newtons per meter (N/m).

The surface tension of a liquid decreases with increasing temperature, since the forces acting in the surface layer are reduced by the increased thermal motion of the molecules. Also soaps and detergents decrease the surface tension of water, thus enabling the water to make closer contact with the dirt molecules and remove them more efficiently.

Example 10.13

A circular wire ring used to measure surface tension is shown in Fig. 10.27. The ring is attached to a spring balance that measures the added force **F** required to pull the ring through the surface of the liquid. If the ring has an inner diameter of 5.0 cm and an outer diameter of 5.2 cm and the balance reads 0.035 N just before the wire ring breaks loose from the liquid surface, what is the surface tension of the liquid?

SOLUTION

The forces due to surface tension are shown (in an exaggerated form) in Fig. 10.28. There are forces on both the outer and inner sides of the ring because of the surface tension of the liquid pulling down on them. The total length over which surface tension forces act in this case is therefore

$$L = \pi(5.0 \times 10^{-2} \text{ m}) + \pi(5.2 \times 10^{-2} \text{ m}) = 0.32 \text{ m}$$

Then $\quad \gamma = \dfrac{F}{L} = \dfrac{0.035 \text{ N}}{0.32 \text{ m}} = \boxed{0.11 \text{ N/m}}$

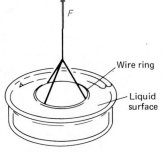

FIGURE 10.27 A circular wire ring used to measure surface tension.

FIGURE 10.28 Surface tension forces on the ring of Fig. 10.27 (exaggerated for clarity).

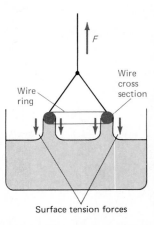

10.7 Solids: Structure and Density

TABLE 10.3 Densities of Solids

Substance	Density (10³ kg/m³)
Ice	0.92
Aluminum	2.70
Cast iron	7.20
Copper	8.89
Gold	19.3
Lead	11.3
Nickel	8.85
Silver	10.5
Steel	7.8
Tungsten	19.0
Platinum	21.4
Uranium	18.7

In atomic solids the atoms vibrate back and forth about fixed positions in space in the way they would if springs connected each atom to its neighbors. If the temperature of a solid is raised, the atoms vibrate much more violently. This increases the average distance of each atom from its neighbors, just as a crowd of people swinging their arms wildly in all directions wind up having larger average separations than if they stood meekly with their hands at their sides. As a result for most solids the density, defined in our usual way as $\rho = \mathcal{M}/V$, decreases as the temperature increases, since the average volume per molecule increases with temperature. This density change is, however, quite small, and for this reason the density of solids is usually stated without specifying the temperature at which the density is measured. Table 10.3 gives the densities of some important solids.

If the temperature of a solid is increased a large amount (the actual amount depending on its structure), the atoms may move so far apart while vibrating that the "springs" holding the solid together break, and the solid begins to flow; i.e., one layer of atoms begins to move with respect to another. When this happens, the substance makes what is called a *phase transition* from the solid phase to the liquid phase, as when ice melts. The temperature at which this transition occurs is very precise and is called the *melting point* of the solid.

10.8 Elastic Properties of Solids

If we consider a solid to be made up of atoms at fixed positions in space, bound together by forces that behave somewhat as if the atoms were connected by tiny springs, we have a microscopic picture that explains much of the macroscopic behavior of solids subjected to external forces.

Three basic changes can occur when a solid is subjected to forces: (1) it can change its length in the direction in which the force is applied, this change being accompanied by smaller changes in the other two directions; (2) it can change its volume when subjected to a constant pressure from all sides; (3) it can change its shape when subjected to a shearing force, i.e., a force that tends to slide one layer of molecules over another.

Stress and Strain

To explain these three kinds of changes in solids, we introduce the concepts of *stress* and *strain*. If a force is applied to a solid, the *stress* is defined as the force per unit area applied to the solid; that is,

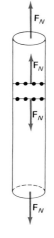

FIGURE 10.29 The stretching of an aluminum wire under a force **F**, which, in this case, is the normal force **F**$_N$.

$$\text{Stress} = \frac{\text{force}}{\text{area}} = \frac{F}{A} \tag{10.13}$$

For example, if we take a piece of aluminum wire and pull it at both ends, as in Fig. 10.29, the effect of this stress is to separate one plane of atoms in the aluminum from the next plane. In other words, the effect is to increase the lengths of the imaginary springs holding the solid together. It is possible, however, that a particular plane of atoms in the aluminum crystal is not perpendicular to **F** but is at some oblique angle with respect to **F**, as in Fig. 10.30. Then the stress **F** can be resolved into two components—F_N, a force normal to the atomic plane, and F_T, a force tangential to the atomic plane. In this case F_N attempts to pull planes of aluminum atoms apart, whereas F_T tends to slide one plane of atoms over another and in this way to *shear* the body apart.

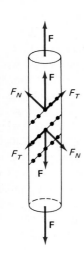

FIGURE 10.30 A crystal atomic plane that is not normal to the applied force **F**. **F** can be broken up into two rectangular components: F_N, which tries to pull the atomic planes apart, and F_T, which tries to shear (or slide) one atomic plane over the other.

Stresses produce changes in either the length, volume, or shape of a solid. Such a change is called a *strain* and is expressed as the ratio of a change in a particular physical quantity to a quantity with the same physical dimensions (usually the original value of the quantity). Thus we have the following types of change:

1 *Change in length:* If an applied force stretches a thin rod or wire, we have

$$\text{Tensile strain} = \frac{L - L_0}{L_0} = \frac{\Delta L}{L_0} \tag{10.14}$$

where ΔL is the change in the original length L_0 under the applied stress. If the rod or wire is not stretched, but compressed instead, we have

$$\text{Compressive strain} = \frac{L_0 - L}{L_0} = \frac{\Delta L}{L_0} \tag{10.15}$$

2 *Change in volume:* Here we have

$$\text{Volume strain} = \frac{\Delta V}{V_0} \tag{10.16}$$

where ΔV is the change in the original volume V_0. In this case it is assumed that the solid is submerged in a liquid or in some other way subjected to equal pressures on all sides.

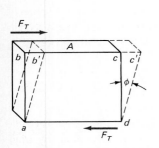

FIGURE 10.31 Change of shape of a solid under a shearing stress.

3 *Change in shape:* The shearing strain is defined in terms of the angle through which the solid is distorted. For example, consider a block with a face *abcd* and top area A, as shown in Fig. 10.31, to which we apply a shearing force F_T parallel to the surface area A to move it to the right. If the new position edge *bc* assumes is $b'c'$, then we define the shearing strain as

$$\text{Shearing strain} = \tan \phi = \frac{cc'}{cd} \tag{10.17}$$

Since for small angles (and ϕ is very small) $\tan \phi = \phi$ (see Table 11.1), the shearing strain is just equal to the angle ϕ in radians. An example of a shear is the deformation of a book, as shown in Fig. 10.32.

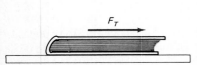

FIGURE 10.32 The deformation of a book by a shearing stress.

As can easily be seen, all strains defined in this way are pure numbers without dimensions and without units.

Elastic Moduli

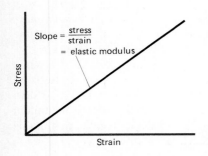

Any solid is changed somewhat in size and shape when a stress is applied to it. *An elastic body is one that returns to its original size and shape when the stress is removed.* For elastic bodies the strain is always directly proportional to the stress as long as the stress is not too great, and so the plot of stress versus strain is a straight line, as in Fig. 10.33. The slope of this straight line is called the *elastic modulus* of the solid. Three separate elastic moduli correspond to the three possible types of change:

FIGURE 10.33 Plot of stress versus strain. The elastic modulus is the slope of the straight line that results.

1 *Length modulus, or Young's modulus*

$$Y = \frac{\text{tensile stress}}{\text{tensile strain}} = \frac{\text{compressive stress}}{\text{compressive strain}} = \frac{F_N/A}{\Delta L/L_0} \qquad (10.18)$$

2 *Torsion, or shear, modulus, or modulus of rigidity*

$$S = \frac{\text{shearing stress}}{\text{shearing strain}} = \frac{F_T/A}{\phi} \qquad (10.19)$$

3 *Volume, or bulk, modulus*

$$B = \frac{-F_N/A}{\Delta V/V_0} = \frac{-P}{\Delta V/V_0} \qquad (10.20)$$

Here P is the pressure, equal to F_N/A, which is assumed to be the same over the whole external surface of the solid. The negative sign arises because an increase in pressure produces a decrease in volume.

In all three cases the dimensions of the elastic modulus are those of the term F/A, since strain is always a dimensionless quantity.

Compressibility

The bulk modulus is a measure of how difficult it is to compress a solid. Its inverse, the *compressibility*, is a measure of how easy it is to compress a solid. We define the compressibility as $K = 1/B$ so that we have

$$K = \frac{1}{B} = \frac{-\Delta V/V_0}{P} \qquad (10.21)$$

or the *compressibility is the fractional decrease in volume per unit increase in pressure*. We can therefore write

$$\Delta V = -KPV_0 \qquad (10.22)$$

so that the decrease in volume is directly proportional both to the original volume and to the applied pressure. These equations for compressibility turn out to be as valid for fluids as they are for solids. On the other hand, tensile or shearing forces cannot be applied to liquids, since liquids merely yield to the applied force.

Table 10.4 gives values for the elastic moduli of some metals. Notice that tungsten is the hardest of all the materials listed, since it requires the largest stress to produce a given strain. Lead is the softest. Notice also that the order of magnitude of all three elastic moduli is approximately the same for any one substance, with the shear modulus being roughly one-third Young's modulus.

TABLE 10.4   Elastic Moduli of Metals

Metal	Young's modulus Y (10^{10} N/m²)	Shear modulus S (10^{10} N/m²)	Bulk modulus B (10^{10} N/m²)
Aluminum	7.0	3.0	7.0
Brass	9.1	3.6	6.1
Copper	11.0	4.2	14.0
Iron	19.0	7.0	10.0
Lead	1.6	0.56	0.77
Steel	20.0	8.4	16.0
Tungsten	36.0	15.0	20.0

Example 10.14

A 2.0-m piece of no. 12 gauge copper wire (diameter = 2.1×10^{-3} m) is used to hang a large pot of flowers from the ceiling of a store. If the pot has a mass of 5.0 kg, how much does the wire stretch?

SOLUTION

Here we want an expression relating the applied stress to the strain produced. Since

$$Y = \frac{F_N/A}{\Delta L/L_0} = \frac{Mg/\pi R^2}{\Delta L/L_0}$$

we have

$$\Delta L = \frac{Mg/\pi R^2}{Y} L_0 = \frac{MgL_0}{\pi R^2 Y}$$

Here $M = 5.0$ kg, $L_0 = 2.0$ m, and

$$R = \frac{D}{2} = 1.05 \times 10^{-3} \text{ m}$$

Also, from Table 10.4, $Y = 11.0 \times 10^{10}$ N/m² for copper, and so

$$\Delta L = \frac{(5.0 \text{ kg})(9.8 \text{ m/s}^2)(2.0 \text{ m})}{\pi(1.05 \times 10^{-3} \text{ m})^2 (11.0 \times 10^{10} \text{ N/m}^2)}$$

$$= \boxed{0.26 \text{ mm}}$$

It is clear from this that a wire stretches very little even under sizable forces.

10.9 Hooke's Law

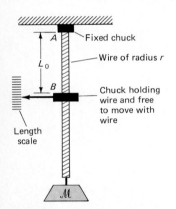

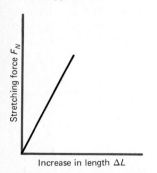

FIGURE 10.34 Young's modulus experiment for measuring the increase in the length of a wire between A and B when a stretching force is applied to it.

FIGURE 10.35 Hooke's law for the stretching of a wire. The increase in length is directly proportional to the force applied.

A standard laboratory experiment in college physics is the measurement of the stretching of a long, thin, metal wire when heavy weights are applied to it. This method for measuring Young's modulus is worth discussing here because of what it can tell us about the properties of solids.

In Fig. 10.34 we consider a wire of radius r and length L_0 between points A and B, which is stretched by weights attached to the bottom of the wire. We wish to obtain Young's modulus for the wire. By definition this is

$$Y = \frac{F_N/A}{\Delta L/L_0} = \frac{Mg/\pi R^2}{\Delta L/L_0} \qquad (10.23)$$

where $F_N = Mg$ and M is the sum of the masses applied to lengthen the wire. All the quantities here can be easily measured except for ΔL, the very small increase in length. ΔL can be measured with a micrometer caliper or, as is usually done, with a device called an *optical lever* that essentially magnifies the small increase in the wire's length into a larger distance which can be more accurately measured.

Equation (10.23) can be turned around to read

$$F_N = \frac{YA \, \Delta L}{L_0} \qquad (10.24)$$

or $\qquad Mg = F_N = k \, \Delta L$

where $k = YA/L_0$. Since Y, A, and L_0 are all constants, we would expect that ΔL would increase linearly with the applied force F_N. We can verify this by adding weights of increasing size to the wire and measuring the corresponding increases in length ΔL. The result is shown in Fig. 10.35. As long as the weights are not too large, the strain and stress are indeed linearly related, and k is a true constant in the equation $F_N = k \, \Delta L$. This is an expression of *Hooke's law*, which says that *the strain produced in an elastic body is directly proportional to the applied stress*. This was first discovered by Robert Hooke (1635–1703), a contemporary, and in some cases a rival, of Sir Isaac Newton. Although called a law, Hooke's discovery is more a generalization from experiment that is only valid under certain conditions.

Hooke's law is also valid for springs, even though in this case the stress is almost a perfect shear rather than a tensile stress. In this case k is called the *force constant* of the spring and has the same dimensions of $[F]/[L]$ as it does in Eq. (10.24). In the SI system the units are newtons per meter.

When we remove the weights in Fig. 10.34, the wire reverts to its original length, as long as Hooke's law is applicable. There is a point, however, beyond which Hooke's law ceases to be applicable, because the spring becomes permanently distorted.

Elastic Limit

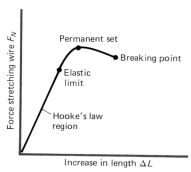

FIGURE 10.36 The elastic limit and breaking point of a metal wire.

If we try to extend further the straight line in Fig. 10.35 by applying more weights, we eventually come to a point where k is no longer a constant. Instead we observe a larger-than-expected increase in length accompanying a normal increase in F_N, as in Fig. 10.36. This point, at which the validity of Hooke's law ceases, is called the *elastic limit*. Beyond this elastic limit the wire assumes a permanent set; i.e., it does not return to its original length when the weights are removed. The wire is no longer hard and elastic, since some of the bonds holding the atoms in the metal together have been broken. If we go even further, the breaking point of the wire is reached and the wire separates into two pieces. The maximum force that can be applied before the wire breaks is called the *ultimate strength* of the wire. The breaking point is indicated on the graph in Fig. 10.36.

Such experimental behavior is consistent with our previously proposed model of a crystal made up of atoms held together by the equivalent of tiny springs. It is the stretching of these "springs" that we observe in experiments such as the Young's modulus experiment just discussed.

Summary: Important Definitions and Equations

Liquids

Density: $\rho = \mathcal{M}/V$
Density of water $= 1.00 \text{ g/cm}^3 = 1.00 \times 10^3 \text{ kg/m}^3$
Specific gravity: The ratio of the density of a substance to the density of water.

$$\text{Sp gr} = \frac{\rho_x}{\rho_{\text{water}}} \qquad (10.1)$$

Liquids at rest: Pressure at a depth h below the surface of a liquid:

$$P' = P_{\text{atm}} + \rho g h \qquad (10.2)$$

Pascal's principle: The pressure applied to a confined liquid is transmitted undiminished to every point in the liquid.

Pascal (unit of pressure): 1 Pa = 1 N/m^2

Barometer: An instrument used to measure the pressure of the earth's atmosphere.
Manometer: A device to measure the difference between the pressure of a substance and that of the earth's atmosphere.
Archimedes' principle: An object partially or totally submerged in a liquid is buoyed up by a force equal to the weight of the liquid the object displaces.
Center of buoyancy (CB): The point at which a buoyant force acts on an object.

Liquids in motion:

Laminar flow: Flow along streamlines that is smooth and predictable.
Turbulent flow: Flow in which the particles experience rapid, unpredictable changes in velocity that lead to eddies and vortices.
Equation of continuity (laminar flow):

$$A_1 v_1 = A_2 v_2 \qquad (10.7)$$

Bernoulli's equation (laminar flow):

$$P + \tfrac{1}{2}\rho v^2 + \rho g y = \text{constant} \qquad (10.8)$$

Viscosity η: A measure of how difficult it is to cause a liquid to flow, or of how resistive the liquid is to motion through it.

$$\eta = \frac{F/A}{v/L} \qquad (10.10)$$

Poiseuille's law: $\dfrac{V}{t} = \dfrac{\pi R^4 (P_1 - P_2)}{8 \eta L} \qquad (10.11)$

Surface tension γ: The property of a liquid that makes it behave as if it had a stretched membrane for a surface.

$$\gamma = \frac{F}{L} \qquad (10.12)$$

Solids

Stress: The force per unit area (F/A) applied to a solid to change its size or shape.

Strain: The change in the size or shape of a solid produced by an applied stress.

Elastic modulus: The ratio of stress to strain.

Length, or Young's, modulus: $Y = \dfrac{F_N/A}{\Delta L/L_0}$ **(10.18)**

Torsion, or shear, modulus: $S = \dfrac{F_T/A}{\phi}$ **(10.19)**

Volume, or bulk, modulus: $B = \dfrac{-P}{\Delta V/V_0}$ **(10.20)**

Compressibility: $K = \dfrac{1}{B} = \dfrac{-\Delta V/V_0}{P}$ **(10.21)**

Hooke's law: The strain produced in a solid is directly proportional to the applied stress, so long as the stress does not exceed the elastic limit.

Questions

1 The density of steel is approximately eight times that of water. How can a steel ship float?

2 Why are bed sores less likely to develop on the body of a hospital patient if the hospital provides the patient with a water bed rather than a regular bed? (*Hint:* What does Pascal's principle indicate about the water in the water bed?)

3 (*a*) When we say that a healthy young adult's blood pressure is 120/80, what precisely do we mean?

(*b*) Does this mean gauge pressures or absolute pressures?

4 Discuss how you would determine the density of an irregularly shaped object in water, using Archimedes' principle, (*a*) when the object is denser than water; (*b*) when the object is less dense than water.

5 Does a submarine on the ocean floor experience exactly the same buoyant force at a depth of only 10 m? Why?

6 A siphon is a tube used to transfer a liquid from a higher to a lower container, as in Fig. 10.37. Explain how this device works.

7 Explain why "water always seeks its own level," using principles discussed in this chapter.

8 When measuring blood pressure, why does a doctor always put the inflatable sleeve on the patient's arm at the level of the heart?

9 Why are the roofs of houses frequently ripped off during hurricanes? (*Hint:* Use Bernoulli's principle.)

10 Prove that the three terms in Bernoulli's equation all have the same dimensions, that is, energy per unit volume.

11 (*a*) Is it possible to pitch a baseball so that it will break upward as it approaches the batter? How?

(*b*) Is it possible to throw a pitch that will break down (a "sinker") when it approaches the batter? How?

(*c*) Is is possible to throw a pitch that will first rise and then almost immediately sink? Why?

12 Why does the speed of the wind increase with height above the earth's surface? What does this tell us about the nature of the fluid flow of the wind?

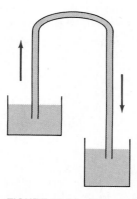

FIGURE 10.37 Diagram for Question 6: a siphon.

Multiple-Choice and Simple Exercises

10.1 The density of chlorobenzene is 1.523×10^3 kg/m^3. Its specific gravity is:

(*a*) 1.523×10^3 (*b*) 1.523 g/cm^3
(*c*) 1.523 kg/m^3 (*d*) 1.523
(*e*) 1.523×10^{-3}

10.2 According to Pascal's principle, the pressure at every point in a liquid:

(*a*) Depends only on the density of the liquid.
(*b*) Is equal to the weight of the liquid.
(*c*) Is the same.

(*d*) Is changed the same amount by an externally applied pressure.

(*e*) Is equal to the externally applied pressure.

10.3 A cylindrical jar open to the atmosphere contains mercury to a depth of 15 cm. The pressure at the bottom of the jar is:

(*a*) 2.0×10^4 Pa (*b*) 2.0×10^4 N/m^2
(*c*) 1.2×10^5 N (*d*) 1.0 atm (*e*) 1.2 atm

10.4 The deepest point in the Pacific Ocean is the Marianas Trench, where the depth is about 11,000 m. The pressure

on a sunken ship at the bottom of the Marianas Trench is:
(a) 1.1×10^8 atm   (b) 1.1×10^3 atm
(c) 1.1×10^3 N/m^2 (d) 1.1×10^5 atm
(e) 1.1×10^5 N/m^2

10.5 A hydraulic lift has one piston of radius 50 cm and another piston of radius 10 cm. If we neglect friction, the ratio of the work input of the lift to the work output is:
(a) 1   (b) 5/1
(c) 25/1   (d) 1/5
(e) 1/25

10.6 A life preserver floats with one-fifth of its volume submerged in water. What is the density of the material out of which the life preserver is made?
(a) 0.80 g/cm^3 (b) 0.20×10^3 g/cm^3
(c) 0.20   (d) 0.20×10^3 kg/m^3
(e) 0.80×10^3 kg/m^3

10.7 Water flows from a pipe of diameter 6.0 cm into one of diameter 8.0 cm. If the speed in the 6.0-cm pipe is 16 m/s, the speed in the 8.0-cm pipe is:
(a) 9.0 m/s   (b) 12 m/s   (c) 21 m/s
(d) 28 m/s   (e) 36 m/s

10.8 A water cooler open to the air at the top contains water to a height of 50 cm above the exit spout. The water emerges from the spout at a speed of:
(a) 9.8 m/s   (b) 9.8 m/s^2 (c) 3.1 m/s
(d) 4.9 m/s   (e) 2.2 m/s

10.9 A baseball is pitched northward in such a manner that when observed from above its rotation is clockwise. The ball will:
(a) Curve because of the force of gravity alone.
(b) Curve upward.
(c) Curve downward.
(d) Curve toward the west.
(e) Curve toward the east.

10.10 A 1.5-N weight stretches a certain spring through a distance of 0.30 m. The work done in stretching the same spring from its unstretched position through 0.70 m is:
(a) 1.2 J   (b) 75 J   (c) -12 J   (d) 8.0 J
(e) Zero; the force exerted by the spring is conservative.

10.11 Prove that a density of 10 g/cm^3 is equal to a density of 10^4 kg/m^3.

10.12 A glass jar has a mass of 0.20 kg when it is empty, and a volume of 1.4×10^{-3} m^3. A liquid is poured into the jar so that it is full. The total mass is 1.80 kg. What is the density of the liquid?

10.13 What is the pressure at the bottom of a column of gasoline 1.0 m high, if the top of the column is open to the atmosphere? (*Hint:* See Table 10.1.)

10.14 What is the force exerted by the earth's atmosphere on an area of 1.0 km^2 on the surface of the earth?

10.15 A hydraulic lift is being used to move a skid filled with mailbags. The small piston has a radius of 5.0 cm and the large piston a radius of 75 cm. If someone applies a force of 200 N to the small piston, what weight can the large piston lift, if we assume 100 percent efficiency?

10.16 A mercury manometer like the one in Fig. 10.6 is connected to a flask containing N$_2$ gas. The mercury column connected to the flask falls to a point 35 cm below the top of the other mercury column, which is exposed to the air. What is the absolute pressure of the N$_2$ gas?

10.17 A piece of metal weighs 16 N in air but only 12 N in water. What is its density?

10.18 A steel wire of length 2.5 m and diameter 3.6 mm is stretched by clamping one end in a vise and exerting a force of 200 N on the other end. How much does the wire stretch?

10.19 A solid brass sphere of radius 5.0 cm is on a ship that sinks and comes to rest at the bottom of the ocean, where the pressure is 500 atm. How much does the volume of the sphere decrease?

10.20 The compressibility of lead is 1.3×10^{-10} m^2/N. What change in volume of a lead cube with sides 10 cm long is produced by a pressure change of 5.0 atm?

Problems

10.21 A dam is constructed to form an artificial lake of surface area 12 km^2. At the dam the lake is 20 m deep. What is the pressure exerted by the water on the dam (a) at its base, i.e., at a depth of 20 m; (b) in the middle of the lake at a depth of 20 m; (c) on the dam at a depth of 10 m?

10.22 Hospital patients often have to be fed intravenously. A needle is inserted in a vein and gravity forces liquid nutrient into the patient's vein. If the liquid has the same density as water, at what height must the liquid container be placed so that the excess pressure at the vein is 50 mmHg? Assume that the blood pressure in the vein is 20 mmHg above atmospheric pressure and that the blood's flow rate is very small.

*_**10.23** Prove that the total force exerted by the water on the wall of a dam of length L, behind which water stands to a height H, is $\frac{1}{2}\rho g H^2 L$, where ρ is the density of the water behind the dam.

*_**10.24** An engineer is designing life preservers for an ocean liner. They must be able to keep one-fifth of a 90-kg body above water. If the average density of a person is about $1.05 \times$ 10^3 kg/m^3 and the life preservers are to be made of a material with a density of 0.55×10^3 kg/m^3, what must be the volume of the life preservers?

10.25 A mercury hydraulic system has a large cylinder with a snug-fitting piston at the top of cross-sectional area 500 cm^2 and a mass of 200 kg (see Fig. 10.4). The smaller cylinder has a cross-sectional area of 100 cm^2 and is open to the atmosphere.
(a) What is the difference in height of the two mercury columns when equilibrium is established?
(b) Which column is higher?

10.26 When a mercury manometer is connected to a gas cylinder, the mercury column connected to the cylinder rises to a point 25 cm above that of the mercury column open to the air. What is (a) the gauge pressure and (b) the absolute pressure of the gas?

10.27 A patient's systolic pressure is 130 mmHg, and the diastolic pressure is 90 mmHg.
(a) What are these gauge pressures in SI units?

(b)   What are the corresponding absolute pressures?

(c)   What is the average pressure of the patient's blood as it leaves the heart, if atmospheric pressure is taken as a reference level?

10.28   A 36-kg child is floating so that its body is just submerged beneath the surface water of a swimming pool. What is the volume of the child's body?

10.29   A glass stirring rod weighs 0.100 N in air, 0.075 N in water, and 0.050 N in an unknown liquid. What is (a) the density and (b) the specific gravity of the unknown liquid?

10.30   (a)   What is the volume of a block of copper of mass 100 g?

(b)   If the block is suspended at the end of a string completely submerged in a water tank, what is the tension in the string?

10.31   A piece of metal weighs 2.0 N in air and 1.7 N in water. What is its (a) specific gravity; (b) density; (c) volume?

10.32   A piece of nickel is suspected to be be hollow. It weighs 3.65 N in air and 3.14 N in water. How large is the hollow space, if any, in the nickel?

10.33   What fraction of the volume of an iceberg is above the water? Assume that the density of ice is 0.92×10^3 kg/m^3 and that of seawater is 1.03×10^3 kg/m^3.

10.34   A hydraulic jack of the kind shown in Fig. 10.38 is used to lift an automobile in a repair shop. The small piston of cross-sectional area A_1 has a radius of 2.0 cm and the large piston of cross-sectional area A_2 a radius of 25 cm. The mechanic lifts the car by exerting force on the small piston with a lever, which has the lever arms shown in the diagram. If no energy is lost in the process, what force must be exerted by the mechanic to lift a 1000-kg car?

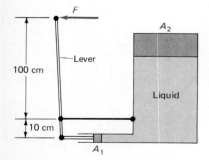

FIGURE 10.38
Diagram for Prob. 10.34: a hydraulic jack.

10.35   A pontoon bridge is laid down to carry provisions to survivors of a flood. The bridge, which rests on the water, is 3.3 m wide and 60 m long. When eight identical trucks are on the bridge, it sinks 20 cm into the water. What is the weight of one truck?

10.36   A solid block of foam plastic of density 500 kg/m^3 and mass 1.0 kg is to be weighted down with lead foil so that it will just sink in water.

(a)   What mass of lead is required?

(b)   What is the volume of the lead?

10.37   Bubbles of N$_2$ gas escape from a rusted gas cylinder at the bottom of a lake 20 m deep. What is the percentage increase in their volume when they reach the surface of the lake (assume no temperature change)?

10.38   At the point where the main water line enters the basement of a five-story building, the water gauge reads a pressure of 1.2×10^5 N/m^2.

(a)   Will this water be able to reach the fifth floor, where the water pipes are 15 m above the basement?

(b)   What is the greatest height the water can reach?

10.39   An 8.0-cm-diameter pipe carries water into the basement of a house at a pressure of 4.0×10^5 N/m^2. The water in the pipe has a speed of 0.50 m/s. The pipe splits into two pipes at the same level to carry cold water to the two sides of the house. If each of these pipes has a diameter of 4.0 cm, what is the speed of the water in the two pipes?

10.40   In an office building a vertical water pipe 6.0 cm in diameter contains water that is not moving. The pressure in the basement of the office building is 5.6×10^5 N/m^2. Fourteen floors up the pressure is 1.4×10^5 N/m^2. What is the length of pipe between the basement and the fourteenth floor?

*__10.41__   A large water tank has a small hole 1.0 cm^2 in area punched in its bottom. Water flows in at the top of the tank at a rate of 2.0×10^{-3} m^3/s. At equilibrium (when the amount of water in the tank remains constant) what will the water level be?

10.42   Air flows past the upper surface of a horizontal airplane wing in streamline flow at 200 m/s and past the lower surface at 150 m/s. The density of the air is 0.90 kg/m^3 at the altitude of the plane and the wing area is 20 m^2.

(a)   What is the pressure difference between the top and bottom surfaces of the wing?

(b)   What is the lifting force on the plane?

10.43   The amount of water flowing out of a pipe in 1.0 min is 0.54 m^3. If the pipe has a radius of 6.0 cm, what is the average speed of the water?

10.44   A horizontal pipe has a constricted area where the diameter is reduced from 8.0 to 2.0 cm. In the 8.0-cm portion the flow rate is 2.0 m/s and the pressure is 6.5×10^5 Pa.

(a)   What is the flow rate in the 2.0-cm portion of the pipe?

(b)   What is the pressure in the 2.0-cm portion?

10.45   What pressure is required to accelerate a column of blood 50 cm long in a rigid tube 1.0 cm in radius from zero velocity to 100 cm/s in 0.10 s (neglect viscosity and gravity)? (Although the aorta is not rigid, this is roughly what the left ventricle must do with every heartbeat.)

10.46   At a certain point along a pipeline the speed of water flow is 1.2 m/s and the gauge pressure 3.0×10^5 N/m^2. What is the gauge pressure in the line at a point 10 m lower than the first point, where the cross-sectional area of the pipe is one-half that at the first point?

*__10.47__   A water cooler tank is resting on the floor. In addition to the regular outlet, which is 4.0 cm from the floor, the tank has another hole directly above the regular outlet and 30.0 cm above the floor. How high must the water be in the tank for the streams of water from the two holes to hit the floor at the same place?

10.48   Water enters the basement of an apartment house through an 8.0-cm-diameter pipe at a flow rate of 0.45 m/s and a pressure of 3.0×10^5 N/m^2. The water flows through the same pipe to the sixth floor, which is 20 m above the basement. What is the water pressure on the sixth floor?

10.49   How much methyl alcohol will flow through a

1.0-m length of tubing of radius 1.0 mm in 1.0 min, if the pressure difference across the two ends of the tubing is 0.20 atm?

10.50 What must the pressure difference across the two ends of a 1.0-km section of the Alaskan pipeline be if oil must move through the pipeline at 0.0010 m^3/s? Assume that the diameter of the pipe is 50 cm, the density of the oil 0.90×10^3 kg/m^3, and the viscosity of the oil 0.50 Pa·s.

*10.51 Suppose a sphere of radius R falls at a constant terminal speed v through a viscous material of viscosity η. Show, using dimensional analysis, that the resistive (drag) force acting on the sphere must be

$$F = k\eta vR \qquad \text{where } k = \text{constant}$$

(*Hint:* Assume that $F \propto \eta^x v^y R^z$ and evaluate x, y, and z.)

*10.52 A blood transfusion is being administered to a patient by using gravity flow to supply blood at a rate of 5.0 cm^3/min. Assume that the pressure in the vein into which the blood is being injected is 20 mmHg and that the density and viscosity of the blood have the values given in Tables 10.1 and 10.2. The needle used has a length of 5.0 mm and a diameter of 0.40 mm. How high must the bottle containing the blood be placed above the patient's arm to provide the desired flow rate?

10.53 An upward force of 1.50×10^{-2} N is required to lift a wire ring of average radius 1.64 cm through the surface of the blood plasma contained in an open beaker. What is the surface tension of blood plasma?

*10.54 A small block of wood of density 0.460×10^3 kg/m^3 is submerged in water and held in place at a depth of 3.68 m. Find (*a*) the acceleration of the block toward the surface when the block is released and (*b*) the time for the block to reach the surface. (Neglect the effects of viscosity.)

10.55 A tugboat is using a steel cable to pull a ship at constant speed. The drag exerted on the ship by the water is 1.5×10^7 N. If any strain above $\Delta L/L_0 = 0.050$ will break the cable, what is the smallest-radius steel cable that can be used?

10.56 A hydraulic press contains 0.15 m^3 of oil. If the compressibility of the oil is 6.3×10^{-10} m^2/N, what is its decrease in volume when subjected to a pressure of 100 atm?

10.57 A wooden platform is suspended from a ceiling by four steel wires at its four corners. The wires are 4.0 m long, and each has a diameter of 2.5 mm. Two people, each weighing 900 N, stand at the center of the platform. How much do the wires stretch?

*10.58 A technician is setting up a laboratory experiment and suspends a 20-kg horizontal bar from the ceiling by using three equally spaced steel wires each of radius 0.60 mm. The two wires at the ends of the rod are exactly 3.000 m long, but the wire at the center has a length of 3.001 m (when all three are not under tension).

(*a*) By how much is each wire stretched?

(*b*) How much of the weight of the bar is supported by each wire?

*10.59 A 10.0-kg metal ball of radius 5.00 cm is suspended from a point 2.000 m above the ground by a steel wire of unstretched length 1.890 m and of diameter 0.100 cm. The ball is then set swinging so that its center is moving at a speed of 4.00 m/s when it passes through its lowest point. How close to the floor is the bottom of the ball at its lowest point? (*Hint:* The increase in length is so small compared to the length that, as a first approximation, it may be neglected in calculating the centripetal force.)

Additional Readings

Blackett, P. M. S.: "The Craft of Experimental Physics," in *University Studies* (edited by Harold Wright), Nicholson and Watson, London, 1933. This is an outstanding paper on the work of experimental physicists by a Nobel Prize–winning physicist. It is the source of the quote at the beginning of this chapter.

Holden, Alan, and Phylis Singer: *Crystals and Crystal Growing.* Doubleday Anchor, Garden City, N.Y., 1960. A fascinating account of the growing and use of crystals.

McDonald, J. E.: "The Shape of Raindrops," *Scientific American,* vol. 190, no. 2, February 1954, pp. 64–68. An interesting article showing how the shape of liquid drops can be explained by the forces acting on them.

Schrier, Eric W., and William F. Allman (eds.): *Newton at the Bat: The Science in Sports,* Charles Scribner's Sons. New York, 1984. The first section of this book is devoted to "Balls and Other Flying Objects."

Shortley, George, and Dudley Williams: *Principles of College Physics,* vol. 1, Prentice-Hall, Englewood Cliffs, N.J., 1959, chaps. 10, 11. This textbook contains a more detailed treatment of the structure of liquid and solids than most books at an introductory level.

Wannier, Gregory H.: "The Nature of Solids," *Scientific American,* vol. 187, no. 6, December 1952, pp. 39–48. A somewhat dated, but still very useful, account of the structure of solids by an eminent solid-state physicist.

CHAPTER 11

Periodic Motion and Waves

Motion appears in many aspects—but there are two obvious kinds, one which appears in astronomy and another which is the echo of that. As the eyes are made for astronomy so are the ears made for the motion which produces harmony.

Plato (427–347 B.C.)

In Chap. 6 we discussed the motion of a mass at the end of a stretched horizontal spring. If there is no loss of mechanical energy in this system, the mass will repeat its back-and-forth motion (usually called a *vibration*, or *oscillation*) indefinitely. This motion is an example of a larger class of motions that we call *periodic*.

Periodic motions include a host of important physical phenomena. For example, violin strings and air columns in organ pipes vibrate periodically, and so the joy of music is really the delight in a special kind of periodic motion. Sound waves and the various kinds of electromagnetic waves—radio, TV, radar, light—are all periodic. A properly functioning heart beats in periodic motion. All kinds of machinery from jackhammers to internal-combustion engines execute periodic motion. For this reason an understanding of periodic motion is basic to an understanding of our universe.

11.1 Periodic Motion

Periodic motion: Motion that repeats itself regularly after a fixed time interval.

Period T: The time during which a physical system completes one full cycle of its motion.

The word *period* comes from the Greek words meaning "round path," or circle. It was coined to describe the first observed periodic motion—the motion of the planets—and it referred to the time a planet took to complete one orbit and return to its original position. The motion of a child on a swing or of the pendulum in a grandfather's clock is periodic.

If we stretch a mass at the end of a horizontal spring a distance X_0 from its equilibrium position, as in Fig. 11.1, and then release it, the mass will speed up, pass through its original rest position at maximum speed, and then slow down as the spring begins to be compressed. The spring will be compressed until the mass reaches a position $-X_0$; there the mass stops instantaneously, and then speeds up until it passes its equilibrium position with maximum speed in the opposite direction; it then slows down and stops instantaneously at its original position $+X_0$. If we assume that there is no friction or other resistive forces acting, the mass will

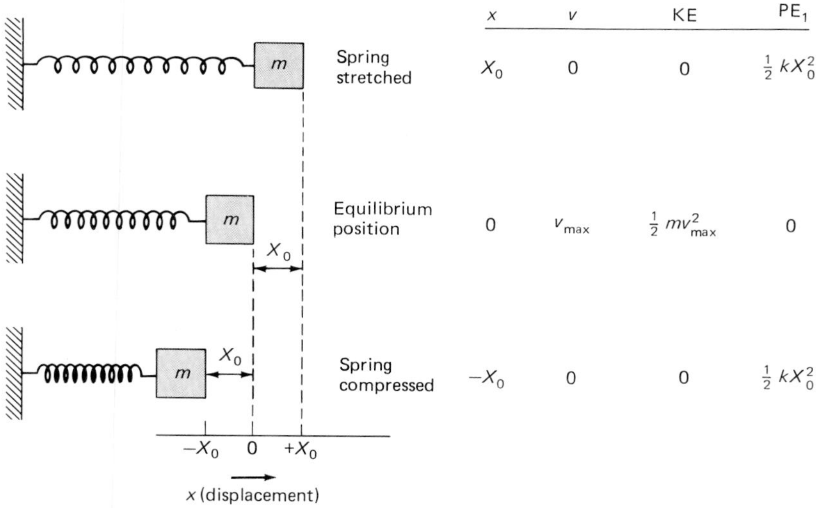

	x	v	KE	PE$_1$
Spring stretched	X_0	0	0	$\frac{1}{2}kX_0^2$
Equilibrium position	0	v_{max}	$\frac{1}{2}mv_{max}^2$	0
Spring compressed	$-X_0$	0	0	$\frac{1}{2}kX_0^2$

$-X_0 \quad 0 \quad +X_0$

x (displacement)

FIGURE 11.1 Motion of a mass at the end of a horizontal spring on a frictionless table. The amplitude of the motion is X_0. The energy oscillates between kinetic and elastic potential energy, as shown here for half a cycle.

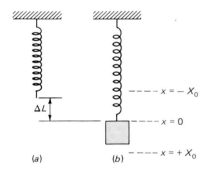

FIGURE 11.2 Mass at the end of a vertical spring. The motion of the mass is the same as that of a mass at the end of a horizontal spring, except that here the end of the spring is displaced a distance ΔL to a new equilibrium position when the mass is attached.

continue to move back and forth between $-X_0$ and $+X_0$ indefinitely. Such a motion is periodic: it repeats itself over and over in successive intervals of equal time.

In this case the *period T* is the time it takes for the mass to go from $+X_0$ to $-X_0$ and then back again to $+X_0$, that is, the time for a complete *cycle*, or round-trip. We can also define some other important quantities for this periodic motion:

Frequency f: The number of complete cycles (or periods) per second: $f = 1/T$.

Displacement x: The distance of the mass from its equilibrium position at any instant of time.

Amplitude X$_0$: The maximum value of the displacement x.

These technical terms, together with the familiar ideas of velocity and acceleration, provide us with all we need to discuss periodic motion quantitatively.

The motion of a mass at the end of a horizontal spring repeats itself over and over again in time. There is also a continuous conversion of energy from the elastic potential energy in the spring to the kinetic energy of the mass, and then back to potential energy in the spring, as the mass goes from maximum displacement in one direction to maximum displacement in the opposite direction. These energy transformations are shown in Fig. 11.1. This energy conversion repeats itself as the mass moves in the opposite direction and finally returns to its original position.

The motion of a mass at the end of a vertical spring is the same as that at the end of a horizontal spring except that in the former the pull of gravity displaces the mass initially to a new equilibrium position, as shown in Fig. 11.2. The vertical motion of the mass about this new equilibrium position is then identical to the motion of a mass at the end of a horizontal spring.

Example 11.1

A mass at the end of a horizontal spring is stretched 0.100 m from its equilibrium position and then released. It takes 0.750 s to move through its equilibrium position to its extreme position in the opposite direction. What is **(a)** the amplitude; **(b)** the period; **(c)** the frequency of the motion? **(d)** If the force constant of the spring is 4.00 N/m, what is the initial energy of the system?

SOLUTION

(a)   Since the amplitude is the maximum displacement,

$X_0 = \boxed{0.100 \text{ m}}$

(b)   Since it takes 0.750 s for the mass to make a one-way trip from one side to the other, the time for a round-trip is 2 × 0.750 s = 1.50 s. Hence

$T = \boxed{1.50 \text{ s}}$

(c) $f = \dfrac{1}{T} = \dfrac{1}{1.50 \text{ s}} = \boxed{0.667 \text{ Hz}}$

(d)   As we have seen in Chap. 6, the energy stored in the spring is, from Eq. (6.9),

$\tfrac{1}{2}kX_0^2 = \tfrac{1}{2}(4.00 \text{ N/m})(0.100 \text{ m})^2 = 0.0200 \text{ N·m} = \boxed{0.0200 \text{ J}}$

11.2   Simple Harmonic Motion

The first scientists to study periodic motion were the Greeks, who were interested in periodic motions in music and astronomy. They called periodic motion *harmonic motion* (see the quotation from the Greek philosopher Plato at the beginning of this chapter). The motion of a violin string is a very complicated kind of harmonic motion and difficult to analyze quantitatively. But there are vibrations of a single frequency that we can analyze more easily. These less complicated vibrations constitute *simple harmonic motion* (SHM).

A particle or system executes simple harmonic motion when its position as a function of time can be described by a single sine or cosine curve, as in Fig. 11.3, which shows the displacement of the mass at the end of a horizontal spring as a function of time. The amplitude of the motion is X_0, as shown in the diagram. The motion repeats itself over and over with a period T, which remains fixed throughout the motion. Such a motion is both periodic and simple harmonic.

Simple harmonic motion of the kind shown in Fig. 11.3 is directly related to the motion of a point in a circle, as we will now show.

Consider a small ball rotating at constant speed in a vertical circle (called the *reference circle*) of radius R. We are interested in the x component of this circular motion. To find this, we might imagine a large spotlight shining down on the rotating ball from the top of the page and projecting on the x axis a shadow of the ball. As the ball rotates through 360°, or 2π rad, its shadow goes from $+X_0$ to $-X_0$ and back again to $+X_0$, as in Fig. 11.4. If we plot the position of the ball's shadow as a function of the angle θ swept out by the rotating ball, we find that x executes a perfect cosine curve similar to the one in Fig. 11.3. This has to be the case because the cosine of θ is the ratio of the adjacent side (x) to the hypotenuse R in the triangle shown in Fig. 11.4. Since R is constant and numerically equal to X_0, as the ball rotates its shadow along the x axis (which is the adjacent side in the triangle) is always proportional to $\cos \theta$. Hence we have

$$\cos \theta = \frac{x}{R} \qquad \text{or} \qquad x = R \cos \theta = X_0 \cos \theta$$

But, from Eqs. (4.23) and (4.24),

$$\theta = \omega t = 2\pi f t$$

where ω is the angular velocity and f is the frequency, i.e., the number of complete rotations per second. We have, therefore, for the motion of the shadow along the x axis

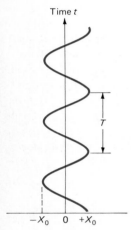

Time t

T

$-X_0$ 0 $+X_0$

Displacement x

FIGURE 11.3   A cosine curve for the displacement of a mass at the end of a horizontal spring as a function of time. The initial displacement of the mass is $+X_0$, and the mass vibrates back and forth between the extreme positions $-X_0$ and $+X_0$ as time goes on. (In this diagram time is plotted along the vertical axis.) The motion can be described as a function of time by an equation of the form $x = X_0 \cos \omega t$.

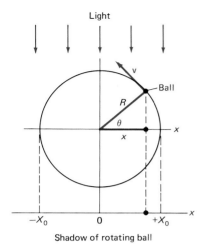

Shadow of rotating ball

FIGURE 11.4 The projection on the x axis of the circular motion in the xy plane of a small ball (i.e., the x component of the ball's motion). The projected shadow executes SHM along the x axis in a way described by the equation $x = X_0 \cos \omega t$.

$$x = X_0 \cos \theta = X_0 \cos \omega t = X_0 \cos 2\pi f t \qquad (11.1)$$

The period of the projected motion along the x axis is the same as the period of the circular motion, since they both complete one cycle in the same time, and so

$$T = \frac{1}{f} = \frac{2\pi}{\omega} \qquad (11.2)$$

Thus the x component of the circular motion in Fig. 11.4 is identical with the motion of a mass m vibrating along the x axis at the end of a horizontal spring of force constant k.

We conclude that the projection of circular motion on a straight line (like the x axis) is simple harmonic motion. This will enable us to reach some important conclusions about the period and frequency in SHM.

Acceleration in SHM and Hooke's Law

A frequently used definition of SHM is that an object executes SHM if the force controlling its motion satisfies Hooke's law. For example, we have seen in Sec. 10.9 that for a spring the increase in length x of the spring is proportional to the applied force F', or

$$F' = kx \qquad (11.3)$$

Now, according to Newton's third law, if a person applies the force F' by pulling on the mass at the end of the spring and thus stretching the spring, then the spring must exert an equal and opposite force $F = -F'$ back on the mass and, through it, on the person's hand. The force exerted by the spring on the mass, the so-called restoring force, is therefore

$$F = -F' = -kx \qquad (11.4)$$

Any system for which the restoring force is proportional to the displacement and in the opposite direction is said to obey Hooke's law. Any such system, when released, will execute SHM. The acceleration of the mass m is given by Newton's second law:

$$a = \frac{F}{m} = \frac{-kx}{m} \qquad \text{or} \qquad \boxed{a = -\frac{k}{m}x} \qquad (11.5)$$

This leads us to define SHM as follows:

Simple harmonic motion: A periodic motion in which the restoring force is proportional to the displacement and in the opposite direction.

We now want to see if the motion of the shadow along the x axis of a ball rotating in a circle satisfies Eq. (11.5). For motion at constant speed in a circle of radius R, the only acceleration that exists is the centripetal acceleration, given by Eq. (4.27) as $a_C = v^2/R$. The projection of this acceleration on the x axis is the acceleration of the shadow of the rotating ball along the axis. From Fig. 11.5 this projection is

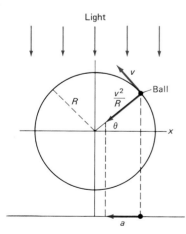

Shadow of rotating ball

FIGURE 11.5 The projection onto the x axis of the acceleration vector for uniform circular motion. When x is positive, the acceleration a is negative, and vice versa.

$$a = \frac{v^2}{R} \cos \theta = \frac{v^2}{R} \frac{x}{R} = \frac{v^2}{R^2} x = \omega^2 x \tag{11.6}$$

where x is the displacement of the shadow along the x axis and ω is the angular velocity, equal to v/R. When x is positive, a is in the negative x direction, and vice versa, as is clear from the figure. We should therefore more properly write

$$a = -\omega^2 x \tag{11.7}$$

Period of Simple Harmonic Motion

We see, therefore, that the projection of circular motion on the x axis satisfies the defining equation for SHM, that is, Eq. (11.5), with

$$\omega^2 = \frac{k}{m} \quad \text{or} \quad (2\pi f)^2 = \frac{k}{m}$$

and so
$$\boxed{f = \frac{1}{2\pi} \sqrt{\frac{k}{m}}} \tag{11.8}$$

The period of the motion is, then,

$$\boxed{T = \frac{1}{f} = 2\pi \sqrt{\frac{m}{k}}} \tag{11.9}$$

This is a remarkable result, telling us that, to obtain the period T for an object vibrating at the end of a spring, all we need know is the mass of the object and the force constant k of the spring. The amplitude X_0 does not affect the period: motions with large amplitudes take just as long as motions with small amplitudes, as long as Hooke's law remains valid.

Speed of a Mass Executing Simple Harmonic Motion

The total energy of a mass moving at the end of a spring of force constant k is the initial energy stored in the spring, which is, from Eq. (6.9),

$$\mathcal{W} = \tfrac{1}{2}kX_0^2 \tag{11.10}$$

where X_0 is the initial extension of the spring and therefore the amplitude of the vibration. When the mass is in motion, its energy is just the sum of a kinetic-energy term and a potential-energy term:

$$\mathcal{W} = \text{KE} + \text{PE} = \tfrac{1}{2}mv^2 + \tfrac{1}{2}kx^2 = \tfrac{1}{2}kX_0^2$$

and so
$$mv^2 = k(X_0^2 - x^2)$$

or
$$v = \pm \sqrt{\frac{k}{m}} \sqrt{X_0^2 - x^2} \tag{11.11}$$

This equation correctly predicts that $v = 0$ when $x = \pm X_0$ so that the mass stops instantaneously at the two ends of its vibration. Also, when $x = 0$, v has its maximum values

$$v_{\max} = \pm \sqrt{\frac{k}{m}} X_0 = \pm \omega X_0$$

where the signs refer to motion in the positive and negative x directions.

Figure 11.6 shows graphs of the displacement, velocity, and acceleration in

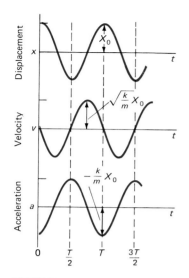

FIGURE 11.6 Graphs of the displacement x, the velocity v, and the acceleration a of a mass moving in SHM as a function of time. Here T is the period of the motion.

SHM as a function of time. The acceleration curve is the negative of the displacement curve except for a change in amplitude, in accordance with Eq. (11.5). The velocity takes the form of a sine curve if x and a are represented by cosine curves. This can be shown analytically since the velocity of the mass is given by

$$v = \sqrt{\frac{k}{m}} \sqrt{X_0^2 - x^2} = \sqrt{\frac{k}{m}} \sqrt{X_0^2 - X_0^2 \cos^2 \omega t}$$

$$= X_0 \sqrt{\frac{k}{m}} \sqrt{1 - \cos^2 \omega t} = X_0 \sqrt{\frac{k}{m}} \sqrt{\sin^2 \omega t} = \pm X_0 \sqrt{\frac{k}{m}} \sin \omega t$$

The square root yields both $+$ and $-$ signs, but the sign here must be negative, since it is clear that, as x decreases from X_0 to 0, the velocity of the mass is in the *negative* x direction. Therefore

$$v = -\sqrt{\frac{k}{m}} X_0 \sin \omega t \qquad \text{or} \qquad v = -\omega X_0 \sin \frac{2\pi t}{T} \qquad \textbf{(11.12)}$$

To summarize, then, the displacement, velocity, and acceleration of a mass moving in SHM at the end of a horizontal spring are given by

$$x = X_0 \cos \omega t \qquad \textbf{(11.1)} \qquad v = -\omega X_0 \sin \omega t \qquad \textbf{(11.12)}$$

$$a = -\omega^2 X_0 \cos \omega t \qquad \textbf{(11.7)} \qquad \text{where } \omega = 2\pi f = \sqrt{\frac{k}{m}} = \frac{2\pi}{T} \qquad \textbf{(11.8)}$$

These values are consistent with the graphs in Fig. 11.6.

Example 11.2

A 0.50-kg mass moves in simple harmonic motion at the end of a horizontal spring of force constant 2.0 N/m. **(a)** What is the period of the motion? **(b)** If the amplitude is 0.10 m, what is the equation for the displacement at any instant of time t? **(c)** What is the equation for the acceleration of the mass as a function of time?

SOLUTION

(a) From Eq. (11.9),

$$T = 2\pi \sqrt{\frac{m}{k}} = 2\pi \sqrt{\frac{0.50 \text{ kg}}{2.0 \text{ N/m}}} = \pi s = \boxed{3.14 \text{ s}}$$

(b) From Eq. (11.1),

$$x = X_0 \cos \omega t = 0.10 \cos \sqrt{\frac{k}{m}} t = \boxed{0.10 \cos 2t}$$

where the amplitude is in meters.

(c) From Eq. (11.7),

$$a = -\omega^2 x = -\left(\frac{k}{m}\right) x$$

$$= \frac{-2.0 \text{ N/m}}{0.50 \text{ kg}} (0.10 \cos 2t) = \boxed{-0.40 \cos 2t}$$

where the acceleration is in meters per second squared.

EXERCISE 1 What is the total energy of this system?

11.3 The Simple Pendulum

A simple pendulum consists of a heavy spherical bob at the end of a light supporting string or rod, as in Fig. 11.7. The length L of such a pendulum is the distance from the point of suspension to the bob's center of mass. The period T of the pendulum is the time it takes for the pendulum bob to execute a complete vibration, from one side to the other and back again.

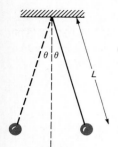

FIGURE 11.7 A simple pendulum.

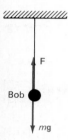

FIGURE 11.8 Forces on a pendulum bob at rest.

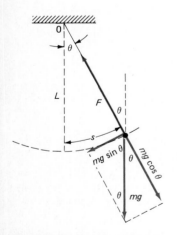

FIGURE 11.9 Forces on a pendulum bob when it is displaced from its equilibrium position.

Finding the period of a simple pendulum would be relatively simple if we were sure the pendulum executed simple harmonic motion. We must first see if this is actually the case. The situation here is quite different from that of a mass at the end of a spring, where the force producing the motion is internal to the system; i.e., the force depends on the elastic properties of the spring itself. A pendulum, however, has no elastic properties and its motion is caused by the external force of gravity pulling down on the bob.

When the pendulum is at rest, the force of gravity is perfectly balanced by the upward force F produced by the tension in the string, as shown in Fig. 11.8. For any other position, however, the force of gravity can be broken up into two components, one parallel to the string and one at right angles to it, as in Fig. 11.9. The component parallel to the string, $mg \cos \theta$, is smaller than the tension F in the string, since a net unbalanced force (a centripetal force) directed along the string toward 0 is required to keep the bob moving in its circular arc. The component normal to the string, $mg \sin \theta$, is the force tending to restore the bob to its equilibrium position. We now want to see whether this restoring force is proportional to the displacement. If it is, pendulum motion is SHM and it will be easy to find the period of the motion.

From Fig. 11.9 the linear displacement of the bob from its rest position at any time is

$$s = L\theta \tag{11.13}$$

where θ is the angular displacement (in radians) of the string and s is the arc of the circle in which the bob moves. The force acting to restore the bob to its equilibrium position is

$$F = mg \sin \theta \tag{11.14}$$

Of course, as the pendulum swings, θ is constantly changing and therefore so is F. When $\theta = 0$, $F = 0$ and there is no restoring force at the equilibrium position of the bob.

If the motion is SHM, the restoring force must be proportional to the displacement, or, in other words, $F = -ks$, where k is the equivalent of the force constant of a spring. We have, from Eqs. (11.13) and (11.14),

$$\frac{F}{s} = \frac{-mg \sin \theta}{L\theta}$$

where we have introduced the minus sign to indicate that the restoring force is always in a direction opposite to the displacement. We can write this equation as

$$F = \frac{-mgs}{L} \left(\frac{\sin \theta}{\theta} \right) = -ks \left(\frac{\sin \theta}{\theta} \right)$$

where $\quad k = \dfrac{mg}{L} \tag{11.15}$

is a constant, sometimes called the *effective force constant*. We have SHM *only if* F is proportional to s and in the opposite direction. This is true only if $(\sin \theta)/\theta$ is a constant, but this is not true in general. The sine of 30° is $\frac{1}{2}$ and the sine of 90° is 1, so an increase in the angle by a factor 3 increases the value of the sine only by a factor of 2. Hence, *in general pendulum motion is not SHM*, for F is not equal to $-ks$ and the restoring force is not proportional to the displacement.

TABLE 11.1   **Table of Sines and Tangents for Small Angles**

Angle θ (degrees)	(rad)	$\sin \theta$	$\dfrac{\sin \theta}{\theta}$	$\tan \theta$
2.0	0.0349	0.0349	1.000	0.0349
4.0	0.0698	0.0698	1.000	0.0699
6.0	0.1047	0.1045	0.998	0.1051
8.0	0.1396	0.1392	0.997	0.1405
10.0	0.1745	0.1736	0.995	0.1763
12.0	0.2094	0.2079	0.993	0.2126
15.0	0.2618	0.2588	0.989	0.2679
20.0	0.3490	0.3420	0.980	0.3640
30.0	0.5236	0.500	0.955	0.5774
45.0	0.7853	0.707	0.900	1.000
60.0	1.047	0.866	0.827	1.732

It is well known, however, that, for small values of θ, $\sin \theta$ and θ (measured in radians) are approximately equal. In such cases where the amplitude of vibration is small, $(\sin \theta)/\theta \simeq 1$, $F \simeq -ks$, and we have a very good approximation to SHM. To see how good the approximation is, we list in Table 11.1 the values of θ, $\sin \theta$, and $(\sin \theta)/\theta$ for angles from 0 to 60°. Note that for angles less than 15° the error introduced by assuming that $\sin \theta$ and θ (in radians) are equal is only about 1 percent. We can therefore conclude:

For pendulum vibrations in which the amplitude of the motion does not exceed 15°, pendulum motion is a very good approximation (but still only an approximation) to SHM.

Since the period for SHM is $T = 2\pi\sqrt{m/k}$ from Eq. (11.9), and since from Eq. (11.15) $k = mg/L$, we have for the period of the pendulum

$$T \simeq 2\pi\sqrt{\frac{m}{mg/L}} \quad \text{or} \quad \boxed{T \simeq 2\pi\sqrt{\frac{L}{g}}} \tag{11.16}$$

To an approximation that is good for small angles, the period of a simple pendulum depends, therefore, only on the length of the pendulum and on the acceleration due to gravity. It does not depend on the mass of the bob, nor on the material out of which it is made, nor on the amplitude of the oscillation (as long as the small angle approximation remains valid).

Practical Applications

Equation (11.16) has a number of very interesting consequences. First of all, if we have a pendulum of fixed length L and we measure its period T, we can find the acceleration due to gravity g, since $g = 4\pi^2 L/T^2$ from Eq. (11.16). This gives us a more accurate way to determine g than by timing the fall of a swiftly moving object. A very accurate pendulum can be used in this way to determine variations in the earth's gravity field with altitude and latitude. Such variations can help locate mineral or fossil-fuel deposits. Pendulum clocks can also be used to tell time, since their period provides a convenient time interval that, for small amplitudes, is independent of the amplitude of the vibration. Pendulum clocks often have a length close to 1.0 m and periods close to 2.0 s.

Example 11.3

An astronaut sets up a pendulum of length 1.00 m on the surface of the moon and finds that its period is 4.90 s. **(a)** What is the acceleration due to gravity on the surface of the moon?

(b) Show that this result is consistent with what you would expect from the lunar data in Table F.2.

SOLUTION

(a) We have, from Eq. (11.16), $T = 2\pi\sqrt{L/g_M}$, where g_M is the acceleration due to gravity on the moon, and thus

$$g_M = \frac{4\pi^2 L}{T^2} = \frac{4\pi^2(1.00 \text{ m})}{(4.9 \text{ s})^2} = \boxed{1.64 \text{ m/s}^2}$$

The acceleration due to gravity on the surface of the moon is therefore about one-sixth what it is on earth.

(b) From Eq. (3.8) the ratio of g_M to g for the earth is

$$\frac{g_M}{g} = \frac{Gm_M/r_M^2}{Gm_E/r_E^2} = \frac{m_M r_E^2}{m_E r_M^2}$$

$$= \left(\frac{7.35 \times 10^{22} \text{ kg}}{5.98 \times 10^{24} \text{ kg}}\right)\left[\frac{(6.37 \times 10^6 \text{ m})^2}{(1.72 \times 10^6 \text{ m})^2}\right]$$

$$= 0.169 \simeq 1/6$$

which agrees with the result in (*a*).

11.4   Damped and Forced Oscillatory Motion

Thus far we have assumed that mechanical energy is conserved when an object oscillates in SHM. In the real world, however, this is seldom the case. A plucked guitar string will vibrate and emit sound only for a few seconds before the sound becomes too weak to be heard. A swinging pendulum, if left to itself, will swing in arcs of smaller and smaller amplitude and gradually come to a full stop. These are examples of *damped* oscillatory motion. By *damping* we mean the conversion of mechanical energy to thermal energy by friction, air resistance, and other processes that dissipate mechanical energy.

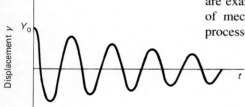

FIGURE 11.10 Damped motion of a mass at the end of a vertical spring. The amplitude of the vibration decreases as time goes on.

Figure 11.10 shows what happens in a damped mechanical system consisting of a mass vibrating at the end of a vertical spring. The amplitude of the motion is gradually reduced, and since the total mechanical energy of the system is proportional to the square of the amplitude [Eq. (11.10)], the mechanical energy of the system grows smaller with time.

Forced Vibrations and Resonance

We have been considering oscillations in mechanical systems given an initial amount of energy and then left alone to vibrate at their own natural frequency. Such an oscillation is called a *free oscillation*. Some of the most interesting physical effects occur, however, when we attempt to drive a vibrating system like a pendulum at a frequency that may or may not be equal to that of the freely vibrating system. Oscillations produced by external forces applied periodically are called *forced oscillations*.

The pushing of a child on a swing is a good example of a forced oscillation. The swing is a good approximation to a simple pendulum and has a natural frequency that depends on the length of the swing's ropes. Anyone who has ever pushed a swing knows that, unless the period of the pushes is timed to the motion of the swing, the swing will never build up a large amplitude of oscillation. It is possible to build up very large amplitudes in the swing by pushing it gently at precisely the right time in each cycle. In other words, the frequency of the applied force must match the natural frequency of the swing.

Resonance: The large-amplitude oscillations of an object or system that result from its being driven at a frequency equal to its natural frequency of oscillation.

Resonance can produce bad effects. Soldiers marching across a bridge have to break step because resonance between their marching frequency and a natural frequency of the bridge (similar to the natural frequency of a swing) could cause the bridge to oscillate with such large amplitudes that it would collapse. Resonance between wind gusts and a natural frequency of the Tacoma Narrows Bridge at Puget Sound in Washington totally destroyed that bridge in 1940, just 4 months after it was completed.

Resonance can also have deleterious effects on living cells. There is evidence that, at certain frequencies related to the size and shape of the cell, the rate at which cells are destroyed by very high frequency sound waves and by microwaves is greatly increased.

Resonance is also very important in radio and TV reception, in the absorption of sound and light, and in the interaction between atoms and radiation that leads to the intense light produced by a laser—as we will see.

11.5 Vibrating Bodies as the Source of Waves

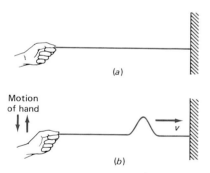

FIGURE 11.11 A pulse produced on a stretched string. (*a*) String is stretched tight by person's hand. (*b*) Hand moves quickly up and down once and produces a pulse along the string.

Suppose we attach one end of a string to a wall and then hold the other end, as shown in Fig. 11.11. If we move the free end rapidly upward and then down to its original position, what we call a "pulse" travels along the string toward the wall. This pulse carries energy, since the particles of the string are set in motion by it, but it carries no mass, since the particles all along the string merely vibrate back and forth at right angles to the direction of the pulse. Once the pulse has gone by, each particle is back at rest at its original position in the string. Some phenomena loosely referred to as "waves" actually involve pulses similar to this one, as, for example, the ripples produced when we drop a stone in a tank of water.

The name *wave,* however, is more properly restricted to changes produced in a medium (in this case the string) by a *periodic* force applied to it. Suppose, for example, we mount a thin strip of spring steel in a vise, as in Fig. 11.12. If the free end of the steel strip is displaced from its equilibrium position and released, it will vibrate in SHM about this equilibrium position, because the restoring force is proportional to the displacement and in the opposite direction. The equation for this vibration is $y = Y_0 \cos \omega t$; this variation of y with t is shown in Fig. 11.13.

If now we attach the free end of this steel strip to the string discussed above and set the strip into SHM, it will send down the string a true wave, i.e., a periodic

FIGURE 11.12 Vibration of a strip of spring steel held rigidly in a vise.

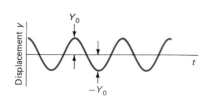

FIGURE 11.13 Displacement of the tip of the steel strip in Fig. 11.12 as a function of time.

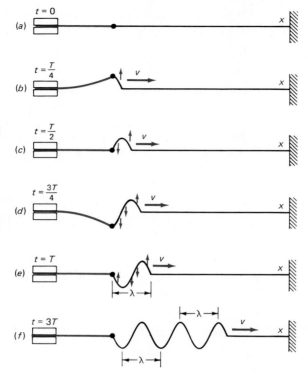

FIGURE 11.14 A wave on a string generated by the vibration of a strip of spring steel. The displacement of the string as a function of distance x along the string is shown at intervals of $T/4$, where T is the period of the vibrating strip. In (e) the time is equal to one full period, and one complete wavelength is moving down the string with a speed v. In (f) the time is equal to $3T$, and the wave fills almost the whole string.

change in the displacement of the particles of the string. Suppose the strip is struck by a blow from a rubber hammer and set into SHM. As it executes SHM, the particles at the end of the string near the strip are also forced into SHM. They, in turn, interact with their neighboring particles and set them into SHM. As a result, a wave travels down the string toward the right. The shape of the string at the end of quarter-period time intervals is shown in Fig. 11.14a to e, where the period is that of the steel strip. As long as the strip continues to vibrate, the particles of the string vibrate in SHM with the same frequency and pass on their motion to their neighbors. Finally, after a longer period of time (3 T in the figure) almost the entire string will be in motion, as shown in Fig. 11.14f. This motion is SHM and takes the form of a sine or cosine function, depending on the initial conditions.

The distance marked in Fig. 11.14e is called the *wavelength* of the wave and is designated by the symbol λ (Greek *lambda*, the equivalent of an English ell). The wavelength is the distance between two consecutive crests or between any two other equivalent points on the wave pattern, as in Fig. 11.14f.

Wavelength λ*:* The distance between any two consecutive equivalent points on a wave pattern.

Let us now find the relationship between the period T of the vibrating steel strip, the wavelength λ of the wave on the string, and the speed of the wave along the string. We see from Fig. 11.14a to e that in the time T required for the strip to execute one full period of its motion in the y direction, the wave has moved a distance λ in the x direction; for when the strip goes through one complete cycle, the wave pattern also traces out one complete cycle and starts to repeat itself again. The speed of the wave is then the distance covered divided by the time:

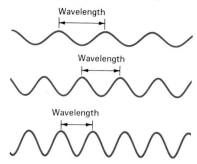

Longer wavelength, lower frequency

Wavelength

Wavelength

Wavelength

Shorter wavelength, higher frequency

FIGURE 11.15 Reciprocal relationship between wavelength and frequency. For waves moving with a constant velocity v, the longer the wavelength, the lower the frequency, and vice versa, since $v = \lambda f$.

$$v = \frac{\lambda}{T}$$

Since the frequency f is equal to $1/T$, we have

$$\boxed{v = \lambda f}$$ **(11.17)**

The speed of a wave is the product of its frequency and its wavelength.

This is one of the most important equations governing wave motion. We have derived it for a particular kind of wave, a mechanical wave on a string, but it applies as well to water waves, sound waves, and electromagnetic waves. Frequently physicists determine the speed of a wave indirectly by measuring its frequency and wavelength and using Eq. (11.17). In this equation, if λ is in meters and f in hertz (i.e., cycles per second), then v is in meters per second, since a cycle is a pure number without dimensions or units. The reciprocal relationship between frequency and wavelength predicted by Eq. (11.17) when the speed of propagation is constant, is illustrated in Fig. 11.15.

Notice that—for a wave on a string—the speed of the wave is determined not by either λ or f but by the properties of the medium (the string) through which the wave travels. This is true of all waves. Once this speed and the frequency are known, the wavelength can be obtained from the relationship $\lambda = v/f$.

Waves

We can summarize the preceding discussion in the following definition of a wave:

Wave: A mechanism for transmitting energy through space that involves periodic variations in the medium through which the energy is transported.

For example, if a wave is moving down a stretched string, the displacement of particles along the string varies periodically in both time and space. If the string is vibrating back and forth in the y direction while the wave moves down the string in the x direction, the displacement of one string particle is periodic in time and can be written

$$y = Y_0 \cos \omega t = Y_0 \cos 2\pi f t = Y_0 \cos \frac{2\pi t}{T}$$ **(11.18)**

Similarly, the displacement of the string particles along the x axis at one instant of time is periodic in the distance x and can be expressed as

$$y = Y_0 \cos \frac{2\pi x}{\lambda}$$ **(11.19)**

This equation states that the value of y for $x = \lambda, 2\lambda, 3\lambda, \ldots, n\lambda$ is the same as its value for $x = 0$; that is, $y = Y_0$. Hence the wave is also periodic in x; the *distance* after which the motion repeats itself is the wavelength λ, just as the *time* after which the motion repeats itself is the period T.

A wave transports energy from one place to another. This is very different from the way a bowling ball knocks over bowling pins. In bowling, mass is moved from one place to another, and kinetic energy goes with the moving mass. In a wave there is *no transport of mass* from one place to another, only a transport of *energy.*

As an example consider the situation in Fig. 11.16 for a wave moving along a string. In (*a*) the wave is moving to the right at a speed v but has not reached point P as yet. A short time later (*b*), the wave has passed by point P and the

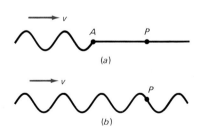

(*a*)

(*b*)

FIGURE 11.16 The transport of energy from A to P by a wave on a string. (*a*) The particles in the string at P have no vibrational kinetic energy. (*b*) The particles at P have vibrational kinetic energy delivered to them by the wave.

string particles at P are vibrating back and forth in SHM along a line perpendicular to the direction of wave propagation. While in (a) the particles at P had no vibrational kinetic energy, once the wave reaches them they have some kinetic energy of vibration. Energy has been transported from A to P, even though there has been no transport of mass.

Example 11.4

(a) A wave travels down a stretched string at a speed of 2.0 m/s. The frequency of the wave is 60 Hz and its amplitude is 0.020 m. Write an expression for the displacement of a particle on the string as a function of time at a fixed distance x from the end of the string. (b) Write a similar equation for the variation of the displacement as a function of distance along the string.

SOLUTION

(a) Because we desire the displacement (y) of the particle as a function of time (t), we use Eq. (11.18), $y = Y_0 \cos (2\pi t/T)$. The amplitude Y_0 is 0.020 m, the frequency is 60 Hz, and so

$$T = \frac{1}{f} = \frac{1}{60} \text{ s,}$$

and our equation is

$$y = 0.020 \cos \frac{2\pi t}{\frac{1}{60}} = \boxed{0.020 \cos 120\pi t}$$

where the amplitude is in meters.

(b) The equation giving the variation of y with x is Eq. (11.19). The wavelength is

$$\lambda = \frac{v}{f} = \frac{2.0 \text{ m/s}}{60 \text{ Hz}} = \frac{1}{30} \text{ m}$$

and our equation is

$$y = Y_0 \cos \frac{2\pi x}{\lambda} = 0.020 \cos \frac{2\pi x}{\frac{1}{30}} = \boxed{0.020 \cos 60\pi x}$$

EXERCISE 2 Write an expression for the velocity of a particle as a function of time at a fixed distance x from the end of the string.

11.6 Transverse and Longitudinal Waves

Nature provides us with a great variety of waves—sound waves, water waves, earthquake (seismic) waves, light waves, and many more. These are all examples of two basic kinds of waves—*transverse* and *longitudinal*.

Longitudinal waves: Waves in which the vibrations in the medium are parallel to the direction of energy transport.

Transverse waves: Waves in which the vibrations in the medium are perpendicular to the direction of energy transport.

The simplest example of a transverse wave is a wave traveling down a stretched string. A water wave is a reasonable approximation to a transverse wave, since the water molecules move up and down at right angles to the direction of the wave. A water wave is not a perfect example of a transverse wave, however, since the molecules also have some motion back and forth parallel to the surface of the liquid as the wave passes along the surface.

Speed of Transverse Waves on a String

We would expect that this speed would depend in some way on the inertia of the string (i.e., its mass) and on the tension F in the string, since the tension determines how tightly the motion of one particle in the string is coupled to that of its neighbors. The tension is the same at any point along a string attached to a wall at one end and stretched by a weight at the other end, as in Fig. 11.17. The total mass of a uniform string is clearly proportional to the length L, but the length of the string should not change the speed of the waves, as long as the tension remains

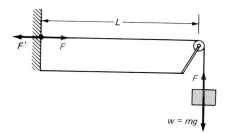

FIGURE 11.17 A string stretched between a wall and a weight pulling at the other end. The tension in the string is constant and equal to $F = mg$.

the same. The relevant factor must therefore not be the total mass m but rather the mass per unit length, which we call the linear density $\mu = m/L$ (μ is the Greek lowercase letter *mu*). It seems reasonable, then, to expect the speed of a transverse wave on a string to depend on F and on μ, and it is difficult to imagine any other property of the string that could affect the speed. (Can you suggest one?)

Using dimensional analysis (Prob. 11.62), it can be shown that the speed of the wave does indeed depend only on F and μ in such a way that

$$v \propto \sqrt{\frac{F}{\mu}} \tag{11.20}$$

We use a proportionality sign instead of an equals sign here because dimensional analysis cannot tell us what the constant of proportionality is. By considering the forces acting and applying Newton's second law, we can prove that the constant of proportionality is exactly 1 in this case, if SI units are used. Therefore

$$v = \sqrt{\frac{F}{\mu}} \tag{11.21}$$

For any given frequency f, the wavelength of a wave on the string is therefore

$$\lambda = \frac{v}{f} = \frac{1}{f}\sqrt{\frac{F}{\mu}} \tag{11.22}$$

Example 11.5

(a) The string of a violin has a mass of 0.600 g and a length of 40.0 cm. What must the tension be if the speed of a wave on the string is to be 200 m/s? **(b)** What is the frequency of the wave for a wavelength of 40.0 cm?

SOLUTION

(a) Here

$$\mu = \frac{m}{L} = \frac{0.600 \times 10^{-3} \text{ kg}}{0.400 \text{ m}} = 1.50 \times 10^{-3} \text{ kg/m}$$

and, from Eq. (11.21),

$$F = \mu v^2 = (1.50 \times 10^{-3} \text{ kg/m})(200 \text{ m/s})^2 = \boxed{60.0 \text{ N}}$$

This is quite a large force, since it is the tension that would be produced in the string if we attached the string to a hook on the ceiling and hung about 6 kg from its free end.

(b) $f = \dfrac{v}{\lambda} = \dfrac{200 \text{ m/s}}{0.40 \text{ m}} = 500/\text{s} = \boxed{500 \text{ Hz}}$

Longitudinal Waves

If we attach a tuning fork to a long spring in such a way that the prongs of the fork move parallel to the length of the spring, we can send a longitudinal wave along the spring, as shown in Fig. 11.18. Here the particles of the spring move back and forth parallel to the length of the spring. Such waves are sometimes called

FIGURE 11.18 Longitudinal wave on a spring. The sine curve shows the periodic nature of the compressional wave moving along the spring.

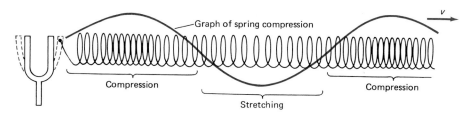

Graph of spring compression

Compression   Compression

Stretching

compressional waves because they consist of a series of compressions and elongations (stretchings) moving along the spring in a periodic fashion.

Sound waves are an especially good example of longitudinal waves. In a sound coming to our ears from a bandstand, for example, the air molecules vibrate back and forth about fixed positions in space along a line drawn from the bandstand to our ears. This causes periodic changes in the density and pressure of the air along the path of the sound. The next chapter is devoted to sound waves.

11.7 Mathematical Description of Wave Motion

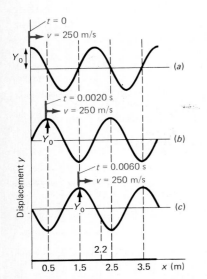

FIGURE 11.19 A transverse wave represented by a cosine function moving along the positive x axis. As time goes on, the original peak of the cosine function is located at larger and larger values of x.

We now want to find a mathematical expression for a transverse wave moving along a string in the positive x direction, since this will simplify many derivations and problems in this and the next chapter. Let us start by returning to Eq. (11.19), which describes, at one instant of time, the displacement of the particles of a string in the y direction as a function of position along the string in the x direction: $y = Y_0 \cos (2\pi x/\lambda)$. In this expression, when $x = 0$, $y = Y_0$, and the cosine function describing the configuration of the string at time $t = 0$ takes the form shown in Fig. 11.19a.

We now show that the equation describing the resulting wave at any point along the string *and* at any instant of time is

$$y = Y_0 \cos \left[2\pi \left(\frac{x}{\lambda} - \frac{t}{T} \right) \right] = Y_0 \cos \left[\frac{2\pi}{\lambda} (x - vt) \right] \qquad (11.23)$$

since $v/\lambda = f = 1/T$, where T is the period of the transverse vibration.

To see that this equation correctly represents the wave, notice that for y to remain at its maximum value Y_0, the argument of the cosine function $(x/\lambda - t/T)$ must remain zero so that the cosine is equal to 1. Since λ and T are constants, this means that, as t increases, x must increase with t in such a way as to keep x/λ equal to t/T in Eq. (11.23), so $x/t = \lambda/T = \lambda f = v$. The peak of the wave, which was originally at $x = 0$ when $t = 0$, therefore moves along the positive x axis at a speed v. The same is true for every other point on the wave pattern.

Equation (11.23) therefore describes the motion of a transverse wave, periodic in both x and t, along the positive x axis with a speed v.

If the wave should be moving in the negative x direction, the corresponding equation is

$$y = Y_0 \cos \left[2\pi \left(\frac{x}{\lambda} + \frac{t}{T} \right) \right] \qquad (11.24)$$

In this case, as t increases, the peak of the wave determined by $y = Y_0$ moves along the negative x axis, since $(x/\lambda + t/T)$ is equal to zero only for negative values of x, because t is necessarily positive. As time goes on, the wave moves to the left along the negative x axis.

Example 11.6

Suppose that, for a transverse wave on a string, the amplitude Y_0 is 0.050 m, the wavelength 2.00 m, and the frequency 125 Hz. At time $t = 0$, the wave has a displacement peak at the point $x = 0$ along the string. **(a)** What is the position of this peak at time $t = 0.0020$ s? **(b)** At time $t = 0.0060$ s?

(c) What is the velocity of the wave? **(d)** Draw graphs showing the motion of the wave along the x axis. **(e)** What is the displacement of a particle on the string at position $x = 2.2$ m and time $t = 0.0060$ s?

SOLUTION

Let us concentrate on the peak originally at the origin, as shown in Fig. 11.19a. This peak moves along the x axis in time in such a way that the displacement y must remain equal to the amplitude Y_0. We must therefore have

$$\cos\left[2\pi\left(\frac{x}{\lambda} - \frac{t}{T}\right)\right] = 1$$

from which

$$\frac{x}{\lambda} - \frac{t}{T} = 0$$

(a) For $t = 0.0020$ s and $T = 1/f = 1/125 = 0.0080$, we have

$$x = \frac{t\lambda}{T} = \frac{0.0020 \text{ s}}{0.0080 \text{ s}} (2.00 \text{ m}) = \boxed{0.50 \text{ m}}$$

(b) For $t = 0.0060$ s,

$$x = \frac{0.0060 \text{ s}}{0.0080 \text{ s}} (2.00 \text{ m}) = \boxed{1.50 \text{ m}}$$

(c) The velocity of the wave is simply the distance the original peak travels in some given time. We have

$$v = \frac{x}{t} = \frac{0.50 \text{ m}}{0.0020 \text{ s}} = 250 \text{ m/s}$$

and also

$$v = \frac{1.50 \text{ m}}{0.0060 \text{ s}} = 250 \text{ m/s}$$

Both these results are consistent with the result obtained from the basic equation for the wave:

$$v = \lambda f = (2.00 \text{ m})(125 \text{ Hz}) = \boxed{250 \text{ m/s}}$$

(d) The graphs in Fig. 11.19 b and c show the motion of the wave in time. It can be seen how the wave peak moves to the right as time goes on.

(e) We use Eq. (11.23) to find the displacement y for any x and t.

$$y = Y_0 \cos\left[2\pi\left(\frac{x}{\lambda} - \frac{t}{T}\right)\right]$$

$$= (0.050 \text{ m}) \cos\left[2\pi\left(\frac{2.2 \text{ m}}{2.0 \text{ m}} - \frac{0.0060 \text{ s}}{0.0080 \text{ s}}\right)\right]$$

$$= (0.050 \text{ m}) \cos[2\pi(1.10 - 0.75)]$$

$$= (0.050 \text{ m}) \cos(0.70\pi) = \boxed{-0.030 \text{ m}}$$

The displacement is negative, meaning that the string is below the x axis at this time and place. The result is consistent with the portrait of the wave shown in Fig. 11.19c. Notice that in this and similar problems the angle whose cosine is needed is given *in radians, not in degrees.*

Example 11.7

The equation of a transverse wave traveling on a string is

$$y = 2.0 \cos[\pi(0.50x - 200t)]$$

where all distances are in centimeters. Find the amplitude, wavelength, frequency, period, and speed of the wave.

SOLUTION

The general expression for a transverse wave of this kind is Eq. (11.23):

$$y = Y_0 \cos\left[2\pi\left(\frac{x}{\lambda} - \frac{t}{T}\right)\right]$$

To find the designated quantities, we must put our equation in a similar form and compare the quantities in the two equations. We have

$$y = 2.0 \cos[2\pi(0.25x - 100t)]$$

On comparing these two equations we see that we have

Amplitude $Y_0 = \boxed{2.0 \text{ cm}}$

Wavelength $\lambda = \dfrac{1}{0.25}$ cm $= \boxed{4.0 \text{ cm}}$

Period $T = \dfrac{1}{100}$ s $= \boxed{0.010 \text{ s}}$ $f = \dfrac{1}{T} = \boxed{100 \text{ Hz}}$

Speed $v = \lambda f = (4.0 \text{ cm})(100 \text{ Hz}) = \boxed{4.0 \text{ m/s}}$

11.8 The Superposition Principle; Phase

Two or more waves can pass through the same region of space independently of one another. Thus in a concert hall the sounds of a violin and an oboe can be clearly distinguished by the audience even though the sound waves from these two instruments are traveling through the same region of space at the same time. The

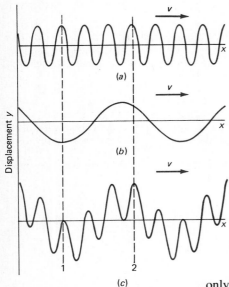

Displacement *y*

(a)

(b)

(c)

1   2

FIGURE 11.20 Superposition of two waves traveling in the same direction along the *x* axis with the same speed but different frequencies and amplitudes. The displacement of the resultant wave at any position *x* (shown in *c*) is the algebraic sum of the displacements produced by the individual waves in (*a*) and (*b*).

only way such behavior is possible is if the two waves act independently on the particles of the air and on the eardrums of the audience, producing a motion that is the *superposition* of the motions due to the two separate waves.

Another example is shown in Fig. 11.20 for two waves on a string. At any instant of time the displacement of the string is simply the algebraic sum of the displacements caused by the individual waves, as in Fig. 11.20. For example, at position 1 in the diagram, the displacements of waves *a* and *b* are in opposite directions and cancel out, whereas at position 2 they are both positive and produce a large positive displacement of the particle at that position on the string. This *superposition principle* is true for situations in which Hooke's law, in the form of Eq. (11.4), is valid, i.e., where the restoring force is proportional to the displacement. This is the case for most small-amplitude waves but is not true for large-amplitude waves, such as shock waves, where the superposition principle is not applicable.

Interference of Waves and Phase Differences

The result of superimposing two or more waves on the same medium is known as *interference*. Thus if we have two transverse waves traveling along a string in the same direction, with the same frequency and the same amplitude, the resulting motion of the string will depend on the algebraic addition (i.e., with the proper sign) of the displacements due to the two waves at every point along the string. How the displacements add depends on the relative *phases* of the two waves.

Phase: The specific stage of a cycle that a wave (or other periodic system) has reached at a particular time or at a particular place.

The phase of a single wave is somewhat arbitrary, since it depends on the initial conditions at the chosen zero of time. Phase becomes really important when we are considering the superposition of two or more waves of different phases, and so we focus our attention on the *relative phase* of one wave with respect to a second wave. For example, in Fig. 11.21 waves *a* and *b* are *in phase* because they reach their maximum values in the positive direction for exactly the same values of *x*. In this case the resulting motion of the string is exactly the same as that due to the motion of either wave alone, but with an amplitude twice as large. This is called

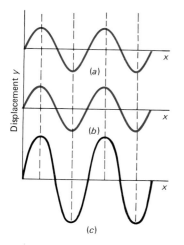

FIGURE 11.21 Constructive interference of two waves of the same frequency and amplitude, and in phase. The wave shown in (c) is the sum of the waves in (a) and (b).

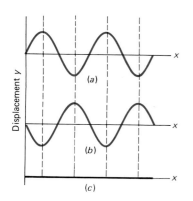

FIGURE 11.22 Destructive interference of two waves of the same frequency and amplitude, but 180°, or π rad, out of phase. This leads to complete cancellation of the two waves for all values of x, as shown in (c).

FIGURE 11.23 The phase difference between (a) $y_1 = \sin x$ and (b) $y_2 = \cos x$. If the cosine function is shifted by $\lambda/4$ along the $+x$ axis, it becomes identical with the sine function.

constructive interference, since at every point the displacements of the string produced by the two waves add constructively.

On the other hand, Fig. 11.22 shows *destructive interference*, where the two waves are 180°, or π rad, *out of phase*. Since their amplitudes are the same, the negative displacements of one wave cancel the positive displacements of the other, and the result is no motion at all of the string. This is the basis of a noise suppression technology being proposed for car exhausts and helicopter transmissions. An electronic sensor would detect the noise being produced, and then generate an equal-intensity wave 180° out of phase with the undesirable sound. The two sounds would cancel out and—at least in a perfect world—the result would be silence.

Another example of a phase difference is that between sine and cosine functions of the form

$$y_1 = Y_0 \cos\left[2\pi\left(\frac{x}{\lambda} - \frac{t}{T}\right)\right] \quad \text{and} \quad y_2 = Y_0 \sin\left[2\pi\left(\frac{x}{\lambda} - \frac{t}{T}\right)\right]$$

Since, from Appendix 1.C, $\sin\theta = \cos(90° - \theta) = \cos(\theta - 90°) = \cos(\theta - \pi/2)$, we can write

$$y_2 = Y_0 \cos\left[2\pi\left(\frac{x}{\lambda} - \frac{t}{T}\right) - \frac{\pi}{2}\right]$$

and we see that y_1 and y_2 are 90°, or $\pi/2$ rad, out of phase. This is clear from Fig. 11.23, where wave b has to be shifted through $\pi/2$ rad, which corresponds to a distance of $\lambda/4$ along the x axis, to be in phase with wave a.

In the general expression for a wave, the phase is usually written as an angle ϕ in the equation

$$y = Y_0 \cos\left[2\pi\left(\frac{x}{\lambda} - \frac{t}{T}\right) + \phi\right]$$ (11.25)

A phase difference of ϕ between two waves means that one wave must be shifted through a distance of $(\phi/2\pi)\lambda$ to be perfectly superimposed on the other wave. This is the case for the sine and cosine functions above, where the phase difference is $\pi/2$. This means that the cosine wave had to be shifted by an amount $(\phi/2\pi)\lambda = [(\pi/2)/2\pi]\lambda = \lambda/4$ to be coincident with the sine wave.

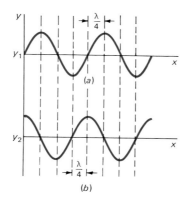

Example 11.8

Two waves of identical frequency and amplitude are moving down the same string in the positive x direction. If the wave produced by their superposition is to have an amplitude 1.50 Y_0, where Y_0 is the amplitude of either wave, what must the phase angle ϕ between the two waves be?

SOLUTION

Since the phase difference between the two waves is given as ϕ, we choose to represent the two waves by the expressions

$$y_1 = Y_0 \cos\left[2\pi\left(\frac{x}{\lambda} - \frac{t}{T}\right) + \frac{\phi}{2}\right] \quad \text{or} \quad y_1 = Y_0 \cos\left(\theta + \frac{\phi}{2}\right)$$

$$y_2 = Y_0 \cos\left[2\pi\left(\frac{x}{\lambda} - \frac{t}{T}\right) - \frac{\phi}{2}\right] \quad \text{or} \quad y_2 = Y_0 \cos\left(\theta - \frac{\phi}{2}\right)$$

where $\theta = 2\pi(x/\lambda - t/T)$. These two functions differ in phase by the angle $\phi/2 - (-\phi/2) = \phi$, as desired. We choose these functions in this way to make it easier to use standard trigonometric identities to solve the problem.

From Appendix 1.C, we have

$$\cos(A + B) + \cos(A - B) = 2\cos A \cos B$$

so that the result of the superposition is, in our example,

$$y = y_1 + y_2 = Y_0\left(2\cos\theta\cos\frac{\phi}{2}\right)$$

$$= \left(2Y_0\cos\frac{\phi}{2}\right)\cos\theta = y'\cos\theta$$

The amplitude of this resulting wave is $y' = 2Y_0 \cos\phi/2$, since $\cos\theta$ contains the dependence on x and t, while the amplitude must be independent of x and t.

To satisfy the condition of the problem, we must have

$$y' = 1.50\,Y_0 = 2Y_0\cos\frac{\phi}{2} \quad \text{or} \quad \cos\frac{\phi}{2} = 0.75$$

and so $\quad \dfrac{\phi}{2} = 41.4° \quad$ and $\quad \phi = \boxed{82.8°}$

11.9 Standing Waves on a String

FIGURE 11.24 Transverse wave on a string. The wave is moving toward the wall at the right with speed v.

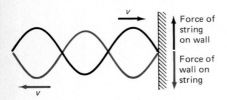

FIGURE 11.25 Transverse wave reflected from a rigid wall at one end of a string. A phase change of π rad occurs on reflection.

Suppose we consider a transverse wave produced by the periodic up-and-down motion of a string. The wave moves along the string in the positive x direction, as in Fig. 11.24. If the right end of the string is attached to a wall, the wave is reflected back from the wall and returns in the negative x direction. The reflected wave will be 180° out of phase with the original wave, since the two must add to give zero displacement at the wall, where the string is fixed. Hence we say that there is a phase change of 180°, or π rad, when the wave is reflected by the wall.

This phase change can also be explained by Newton's third law. When the wave reaches the wall and the string pulls up on the wall at the point of contact, the wall exerts an equal and opposite force down on the string, as in Fig. 11.25. It is this force that starts the reflected wave back down the string and inverts the wave. The reflected wave has had its phase changed from up to down, i.e., by π rad, on reflection by the wall. It is possible to show that, if the end of the string is attached not to an unmovable wall but to a bolt free to slide in a slot on the wall, the bolt will move in the same direction as the string and will start a reflected wave that is *in phase* with the incident wave at the wall. We therefore conclude:

Reflection from a fixed end causes a phase change of 180°, or π rad; reflection from a free end causes no change of phase.

We will see many applications of these ideas when we discuss sound and light.

Standing Waves

Suppose we have two waves of the same amplitude, frequency, and velocity traveling in opposite directions along a string. We want to find the result of the superposition of these two waves.

We can represent the two waves by the equations

$$y_1 = Y_0 \cos\left[2\pi\left(\frac{x}{\lambda} - \frac{t}{T}\right)\right] \quad \text{and} \quad y_2 = Y_0 \cos\left[2\pi\left(\frac{x}{\lambda} + \frac{t}{T}\right)\right]$$

where the minus sign in the first equation means that the wave is traveling along the positive x axis and the plus sign in the second equation means that the wave travels in the negative x direction. The superposition of these two waves leads to

$$y = y_1 + y_2 = Y_0\left[\cos 2\pi\left(\frac{x}{\lambda} - \frac{t}{T}\right) + \cos 2\pi\left(\frac{x}{\lambda} + \frac{t}{T}\right)\right]$$

But, from Appendix 1.C, we have for two angles A and B,

$$\cos(A + B) + \cos(A - B) = 2\cos A \cos B$$

and, on setting $A = 2\pi x/\lambda$ and $B = 2\pi t/T$, our expression above for y becomes

$$y = 2Y_0 \cos\frac{2\pi x}{\lambda} \cos\frac{2\pi t}{T} = Y' \cos\frac{2\pi t}{T} \qquad \text{(11.26)}$$

The resultant wave is therefore periodic in time t [because $\cos(2\pi t/T)$ is periodic in time] but with an amplitude $Y' = 2Y_0 \cos(2\pi x/\lambda)$, which is a periodic function of the distance x along the string from the source.

Thus the amplitude Y' has a minimum value (namely, zero) when $x = \lambda/4$, $3\lambda/4$, $5\lambda/4$, or, in general $(2n + 1)\lambda/4$, with n an integer. These points of no motion we call *nodes*. The amplitude has a maximum value of $2Y_0$ when $x = 0$, $\lambda/2$, λ, $3\lambda/2$, or, in general, $n\lambda/2$, with n an integer. These points are called *antinodes* (or *loops*) since the string has its maximum displacement at these points. The nodes and antinodes are shown in Fig. 11.26. Adjacent pairs of nodes are one-half wavelength apart, as are adjacent pairs of antinodes.

Because of the appearance of the string described by Eq. (11.26), with some points at rest and others in violent motion, and with no appearance of energy flow from one part of the string to the other, the resultant wave is called a *standing wave*. Hence *two superimposed traveling waves of the same frequency, amplitude, and speed, moving in opposite directions in the same medium, produce a standing wave.*

How two waves moving in opposite directions can be combined graphically to produce such a standing wave is shown in Fig. 11.27. A phase difference between two waves of the same frequency, amplitude, and speed does not destroy the standing wave pattern or change the distance between nodes; it merely shifts the position of the nodes along the string.

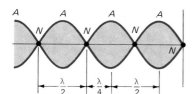

FIGURE 11.26 Nodes (N) and antinodes (A) for a standing wave on a string. Adjacent pairs of nodes are $\lambda/2$ apart, as are adjacent pairs of antinodes.

FIGURE 11.27 Graphical combination of two waves moving in opposite directions to form a standing wave. The wave in (a) moving to the left and the wave in (b) moving to the right are shown in (c), superimposed at intervals of $T/4$. The nodes remain fixed at positions separated by $\lambda/2$, and the antinodes also remain fixed at separations of $\lambda/2$. Hence the name *standing wave.*

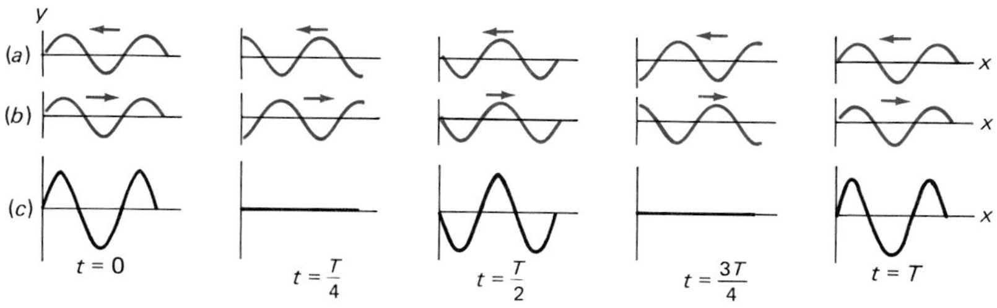

Example 11.9

A standing wave on a string can be described by Eq. (11.26), with $Y_0 = 0.050$ m and $\lambda = 0.10$ m. What is the value of the amplitude of the standing wave at the following positions along the string? **(a)** 0; **(b)** 0.010 m; **(c)** 0.025 m; **(d)** 0.050 m; **(e)** 0.10 m; **(f)** 0.35 m.

SOLUTION

The amplitude of the standing wave is, from Eq. (11.26),

$$Y' = 2Y_0 \cos \frac{2\pi x}{\lambda} = 0.10 \cos \frac{2\pi x}{0.10} = 0.10 \cos 20\pi x$$

(a) When $x = 0$, $\cos 20\pi x = 1$, and so $Y' = \boxed{0.10 \text{ m}}$.

(b) When $x = 0.010$ m, we have

$$Y' = 0.10 \cos [20\pi(0.01)] = 0.10 \cos 0.2\pi$$

But $\cos 0.2\pi = \cos 36° = 0.81$, and so

$$Y' = 0.10(0.81) = \boxed{0.081 \text{ m}}$$

(c) $Y' = 0.10 \cos [20\pi(0.025)] = 0.10 \cos 0.5\pi = \boxed{0}$

Similarly,

(d) $Y' = \boxed{-0.10 \text{ m}}$

(e) $Y' = \boxed{0.10 \text{ m}}$

(f) $Y' = \boxed{-0.10 \text{ m}}$

The minus signs in the amplitudes make no difference here, since as time goes on, the displacement y varies rapidly between $+Y'$ and $-Y'$. Hence we have maximum displacements for the standing wave at intervals of $\lambda/2$, or 0.05 m, that is, for $x = 0, 0.05, 0.10, 0.15, 0.20, 0.25, 0.30, 0.35$ m. Between two consecutive loops we have nodes where $Y' = 0$, for example, at $x = 0.025$ m.

11.10 Forced Vibrations of a String

If the length of a string fixed at both ends is an integral multiple of half wavelengths of the wave generated by a mechanical vibrator attached to the string, then large-amplitude standing waves may be set up in the string. These standing waves are produced by the superposition of waves traveling back and forth along the string.

Resonance for Waves on a String

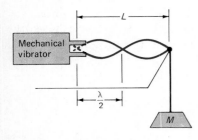

FIGURE 11.28 String being driven by a mechanical vibrator. Resonance can be produced either by keeping the mass M fixed and varying the frequency of the mechanical vibrator or by keeping the frequency fixed and varying M. In the case shown $f_2 = (1/L)\sqrt{F/\mu}$.

Suppose we have a string driven by a mechanical vibrator at one end and passing over a pulley to a hanging weight at the other end, as in Fig. 11.28. Even though the end tied to the vibrator must be in motion, its amplitude is so small that we can, to a good approximation, consider both ends of the string to be nodes. When such a string is driven at some arbitrary frequency, the waves running back and forth along the string tend to cancel each other out, and nothing is observed except a slight blur to indicate that the string is moving. If, however, the tension in the string is kept fixed by the hanging weight and the frequency of the vibrator is varied, large standing waves will be observed in the string for certain discrete frequencies. This is an example of *resonance*, in which the reflected waves combine constructively to produce large-amplitude oscillations of the string at the antinodes, because the applied frequency is equal to one of the natural frequencies of vibration of the string. We will now show what these natural frequencies are.

The speed of the transverse waves is $v = (F/\mu)^{1/2}$, where μ is the linear density of the string and F the tension [see Eq. (11.21)]. To have nodes at the two ends of the string, since the string is prevented from moving at these points, the length L of the string must be an integral multiple (n) of half wavelengths,

$$L = \frac{n\lambda}{2} \quad \text{or} \quad \lambda_n = \frac{2L}{n} \tag{11.27}$$

where $n = 1, 2, 3, 4, \ldots$, and λ_n signifies the wavelength corresponding to a particular value of n. The natural frequencies of the string are, therefore, from Eqs. (11.21) and (11.27),

$$f_n = \frac{v}{\lambda_n} = \frac{\sqrt{F/\mu}}{2L/n}$$

or $$f_n = \frac{n}{2L} \sqrt{\frac{F}{\mu}} \qquad \text{with } n = 1, 2, 3, \ldots \qquad \textbf{(11.28)}$$

These are the natural frequencies of a string of length L, with tension F and linear density μ. Whenever the string is driven at one of these frequencies, large-amplitude standing waves result, because the frequency of the driver is the same as one of the natural frequencies of the string. The vibration in Fig. 11.28, for example, shows resonance for $n = 2$. The wavelength is equal to the length of the string, consistent with setting $n = 2$ in Eq. (11.27).

Resonance can also be achieved by keeping the vibrator's frequency fixed and varying the tension in the string by adding weights at its end until one of the natural frequencies f_n of the string becomes equal to the applied frequency. This is accomplished by changing the speed of the wave (by changing F) to a value such that the wavelength (v/f) satisfies Eq. (11.27) for some integral value of n.

In Eq. (11.28) the frequency with $n = 1$ is called either the *fundamental frequency* or the *first harmonic* (these two terms are synonymous), that with $n = 2$ is called the *second harmonic*, and so forth. The second harmonic therefore has twice the frequency of the fundamental. Sometimes the frequency with $n = 2$ is called the *first overtone*, with $n = 3$ the second overtone, etc. Thus the first overtone is the same as the second harmonic, the second overtone corresponds to the third harmonic, and so on. In what follows, we will always use harmonics instead of overtones because the numerical relationship of a harmonic to the fundamental frequency is clear once we know what harmonic it is. The nth harmonic always corresponds to a frequency nf, where f is the fundamental frequency.

Example 11.10

In Fig. 11.28 the vibrator has a frequency of 30 Hz, while the string is 3.0 m long and has a linear density of 0.40 g/cm. The tension is varied by adding weights to the string passing over the pulley. **(a)** What tension is needed to cause the string to resonate with the vibrator in four segments? (Assume that the points where the string meets the vibrator and the pulley are approximately nodes.) **(b)** What is the speed of the waves in this case?

SOLUTION

(a) From Eq. (11.28) we have

$$f_n = \frac{n}{2L} \sqrt{\frac{F}{\mu}} \qquad \text{from which} \qquad F_n = \frac{4L^2 f_n^2 \mu}{n^2}$$

In this case the frequency does not change but is kept fixed at $f = 30$ Hz, while F_n must vary with n, the number of antinodes in the standing wave. Also $\mu = 0.40$ g/cm $= 0.040$ kg/m.

For four segments, each of which is $\lambda/2$ in length, n is equal to 4 because n gives the numbers of half wavelengths in the length of the string [Eq. (11.27)], and so

$$F_4 = \frac{4L^2 f^2 \mu}{4^2} = \frac{4(3.0 \text{ m})^2 (30/\text{s})^2 (0.040 \text{ kg/m})}{16} = \boxed{81 \text{ N}}$$

(b) For four antinodes we have, from Eq. (11.21),

$$v_4 = \left(\frac{F_4}{\mu}\right)^{1/2} = \left(\frac{81 \text{ N}}{0.040 \text{ kg/m}}\right)^{1/2} = \boxed{45 \text{ m/s}}$$

Summary: Important Definitions and Equations

Periodic motion: Motion that repeats itself after a fixed time interval.
Period (T): The time during which a physical system completes one full cycle of motion.
Frequency (f): The number of complete cycles per second; $f = 1/T$.
Displacement (x): The distance of a moving object from its equilibrium position at any instant of time.
Amplitude (X_0): The maximum value of the displacement.

Simple harmonic motion (SHM):
  1 A periodic motion in which the restoring force is proportional to the displacement and in the opposite direction: $F = -kx$. **(11.4)**
  2 The projection on the x axis of the motion of a particle moving with constant speed in a vertical circle.

Equations for SHM of a mass at end of a spring:

$$x = X_0 \cos \frac{2\pi t}{T} \quad \textbf{(11.1)} \qquad v = -\omega X_0 \sin \omega t \quad \textbf{(11.12)}$$

$$a = -\frac{k}{m} x \quad \textbf{(11.5)} \qquad T = 2\pi \sqrt{\frac{m}{k}} \quad \textbf{(11.9)}$$

Simple pendulum: A heavy bob at the end of a light string. Motion is SHM only for small amplitudes for which $(\sin \theta)/\theta \approx 1$.
For small amplitudes, $T = 2\pi\sqrt{L/g}$. **(11.16)**
Damping: The decrease in amplitude of a periodic mechanical system caused by the conversion of mechanical energy into heat.

Wave: A mechanism for transmitting energy through space that involves periodic variations in the medium through which the energy is transported.
Wavelength (λ): The distance between any two consecutive equivalent points on a wave pattern.
Speed of a wave: $v = \lambda f$

Longitudinal wave: One in which the particle motion is parallel to the direction of energy transport.

Transverse wave: One in which the particle motion is perpendicular to the direction of energy transport.
Speed of transverse wave on a string: $v = \sqrt{F/\mu}$, where F is the tension in the string and $\mu = m/L$ is the linear density.

General equation for a wave:

$$y = Y_0 \cos \left[2\pi \left(\frac{x}{\lambda} - \frac{t}{T} \right) + \phi \right] \quad \textbf{(11.25)}$$

where ϕ is the *phase angle*, which determines the initial value of y when x and t are both zero.
Phase: The specific stage of a cycle that a wave (or other periodic system) has reached at a particular time or at a particular place.

Superposition principle: Two or more waves can pass through the same region of space independently of one another. The resulting displacement of the medium is the sum of the displacements caused by the individual waves.

Standing wave: A wave pattern produced by two superimposed traveling waves of the same frequency, amplitude, and speed, moving in opposite directions in the same medium. At some positions, the amplitude is zero (nodes); at other positions, the amplitude is a maximum (antinodes).

$$y = 2Y_0 \cos \frac{2\pi x}{\lambda} \cos \frac{2\pi t}{T} \quad \textbf{(11.26)}$$

Resonance: Large oscillations of a system that result when the applied frequency is the same as one of the natural frequencies of the system.

Resonant frequencies of a string of length L with nodes at both ends:

$$f_n = \frac{n}{2L} \sqrt{\frac{F}{\mu}} \quad (n = 1, 2, 3, \ldots) \quad \textbf{(11.28)}$$

Questions

  1 Give examples of motions that are periodic but not simple harmonic. Is it possible for a motion to be simple harmonic without being periodic?
  2 The restoring force for a particular spring is proportional to the square of the displacement and in the opposite direction. Is the motion of a mass at the end of this spring SHM?
  3 (a) Draw a graph of the kinetic energy of the mass at the end of the vibrating horizontal spring in Fig. 11.1 as a function of time, starting at $t = 0$ when the displacement is $x = X_0$, and ending at $t = T$ when the displacement is again $x = X_0$.
  (b) Do the same for the system's potential energy.
  4 A mass is vibrating up and down at the end of a vertical spring. When the mass is at its lowest position, ($x = +X_0$ in Fig. 11.2), what are the displacement, speed, and acceleration of the mass?
  5 Explain why the period of a simple pendulum does not depend on the mass of the bob.
  6 At what point in the motion of a mass at the end of a spring is the acceleration zero? Compare this situation with that in which a ball is thrown vertically upward, stops, and falls back to earth again. When is the acceleration zero in this case?
  7 Why would you expect a more accurate value for g by timing the vibrations of a simple pendulum than you could obtain by timing the free-fall of an object?
  8 Will a pendulum clock adjusted to keep perfect time

at the foot of Mount Whitney in California also keep perfect time at the top of Mount Whitney?

9 Two perfectly tuned violins are resting on a table. A musician bows a string on one violin very strongly and then stops the string with a finger, but the sound of the same note continues to come from the second violin. Explain what has happened. Does the table play any role in this phenomenon?

10 When a car is stuck in a rut, a recommended method of getting it moving again is to rock it back and forth, each time changing from forward speed to reverse and then back again to forward. Why do you think such a method might work? Indicate what happens to the kinetic energy and the potential energy of the car with time for the case in which this method succeeds in extricating the car.

11 Suggest two ways in which resonance can be achieved between the natural frequency of a swing holding a child and the frequency of the pushes.

12 (*a*) Is the wave described by the equation

$$y = 0.20 \cos (60\pi t + 120\pi x)$$

moving in the positive or negative *x* direction? Explain.
(*b*) What is the speed of such a wave?

13 A standing wave is set up in a string, and a screwdriver touched to the exact center of the string does not damp the vibration. What are some possible wavelengths for the wave in terms of the string's length?

14 Suppose a tunnel could be drilled completely through the earth along a diameter. What would happen to a rock dropped into the tunnel? (Neglect air resistance.)

15 We have seen that a pendulum has a period which does not depend on the amplitude of the motion, provided that the motion is small. Explain whether the period increases or decreases if the pendulum swings are large instead of small.

16 Why do cars have shock absorbers? How would a car behave without them?

Multiple-Choice and Simple Exercises

11.1 When a body executes simple harmonic motion, its period is:
(*a*) The reciprocal of its speed.
(*b*) Proportional to its acceleration.
(*c*) Inversely proportional to its acceleration.
(*d*) Proportional to its amplitude.
(*e*) Independent of its amplitude.

11.2 It is impossible for two particles executing simple harmonic motion to remain in phase with each other if they have different:
(*a*) Masses   (*b*) Periods   (*c*) Amplitudes
(*d*) Restoring forces   (*e*) Kinetic energies

11.3 The displacement in simple harmonic motion is a maximum when the:
(*a*) Acceleration is zero.
(*b*) Velocity is a maximum.
(*c*) Velocity is zero.
(*d*) Kinetic energy is a maximum.
(*e*) Linear momentum is a maximum.

11.4 The period of a simple pendulum swinging through a small angle is independent of:
(*a*) The acceleration due to gravity.
(*b*) The height above sea level.
(*c*) The mass of the pendulum bob.
(*d*) The length of the pendulum.
(*e*) The location on the earth's surface.

11.5 When a pendulum swings in a small arc, the acceleration tangent to the arc at one end of the path is:
(*a*) Zero   (*b*) A maximum   (*c*) A minimum
(*d*) Changing direction
(*e*) In a direction perpendicular to the path.

11.6 A 0.40-kg mass at the end of a horizontal spring is displaced 0.20 m and released, and moves in SHM at the end of the spring of force constant 2.0 N/m. The PE of the system when *x* = 0.10 m is:
(*a*) 0.040 J   (*b*) 0.020 J   (*c*) 0.010 J
(*d*) 0.045 J   (*e*) 0.030 J

11.7 For a simple pendulum of amplitude 10°, the error involved in assuming that the motion is SHM is about:
(*a*) 5%   (*b*) 0.5%   (*c*) 0.17%   (*d*) 1%
(*e*) Exactly 0; the motion is SHM.
(*Hint:* See Table 11.1.)

11.8 A stretched string is 2.0 m long and fixed at both ends. If standing waves are produced on it by a person varying the applied frequency, the following *cannot* be a value for the wavelength obtained:
(*a*) 4.0 m   (*b*) 2.0 m   (*c*) 1.33 m
(*d*) 3.0 m   (*e*) 1.0 m

11.9 An applied frequency that produces resonance in a stretched string of length 2.0 m, with a tension of 1.0×10^3 N and linear density 0.025 kg/m, is:
(*a*) 25 Hz   (*b*) 125 Hz   (*c*) 150 Hz
(*d*) 175 Hz   (*e*) 225 Hz

11.10 A string has a length 2.0 m and a linear density $\mu = 0.045$ kg/m. A 60-Hz oscillator is applied to the string, and the tension is varied to achieve resonance. The tension in the string required to produce a large-amplitude vibration of the string in three segments is:
(*a*) 288 N   (*b*) 4.5 N
(*c*) 40.5 N   (*d*) 864 N
(*e*) 2.6×10^3 N

11.11 A horizontal spring has a force constant of 5.0 N/m. It is stretched through a distance of 0.060 m and then released. What is the total mechanical energy of the system?

11.12 A vertical spring of force constant 3.0 N/m has a 2.0-kg mass attached to it. It is stretched 8.0 cm from its new equilibrium position and released. What is the numerical relationship between the acceleration of the mass and its displacement?

11.13 A mass at the end of a spring moves in SHM with an amplitude of 2.0 cm and a frequency of 2.5 Hz. Write a general equation for the displacement of the mass as a function of time, if the displacement is 2.0 cm when *t* = 0.

11.14 A mass at the end of a spring moves in SHM of

amplitude 0.15 m and period 2.0 s. Write a general equation for the velocity of this mass as a function of time.

11.15 What is the frequency of a simple pendulum of length 2.0 m at a place where the acceleration due to gravity is 9.79 m/s²?

11.16 A transverse wave moving along a stretched string has a period of 0.010 s and a wavelength of 30 cm. What is its speed?

11.17 A 2.0-g cello string is 0.68 m long. What must the tension in the string be if the speed of a wave traveling along it is to be 150 m/s?

11.18 A traveling wave on a string is described by the equation $y = 0.020 \cos (50\pi x - 100\pi t)$, where all distances are in meters. What is the speed of such a wave?

11.19 A standing wave is set up on a string by two traveling waves, each of amplitude 0.020 m and wavelength 0.80 m, moving in opposite directions. The resultant standing wave may be described by the equation

$$y = \left(2Y_0 \cos \frac{2\pi x}{\lambda}\right)\left(\cos \frac{2\pi t}{T}\right)$$

What is the amplitude of this standing wave at $x = 0.10$ m?

11.20 Hanging weights are tied to one end of a 10-m horizontal string to produce a tension of 500 N. What must the linear density of the string be if it is to resonate to a frequency of 60 Hz and vibrate in three segments?

Problems

11.21 A 0.50-kg object is hung from a spring, and the spring stretches 0.020 m. The object is now pulled down an additional 0.010 m from this equilibrium position and released. Find (a) the amplitude, (b) the period, (c) the frequency of the resulting motion. (d) Write an equation describing the position of the object as a function of time.

11.22 A 0.50-kg mass moves in SHM at the end of a horizontal spring of force constant 2.0 N/m. The amplitude of the motion is 0.20 m.

(a) What is the velocity when the mass passes through its equilibrium position?

(b) Prove that the total energy of the system is the same at $x = 0$ as at $x = X_0$.

11.23 The SHM of a mass at the end of a vertical spring is described by the equation $y = Y_0 \cos (2\pi t/T)$, where $Y_0 = 0.050$ m and $T = 2.0$ s. Draw graphs of the displacement of the mass, its velocity, and its acceleration as a function of time. Use the correct amplitudes and relative phases for the three graphs.

11.24 A 200-g mass vibrates in SHM at the end of a spring. The amplitude of the motion is 0.20 m, and the period is 2.8 s. Find (a) the frequency of the motion, (b) the force constant of the spring, (c) the maximum speed of the mass, (d) the speed when the displacement is 0.10 m, (e) the acceleration when the displacement is 0.10 m.

11.25 An unknown mass m is hung from a spring, and the increase in length of the spring is 0.10 m. The mass is then pulled down farther and released. What is the frequency of the resulting SHM?

11.26 Four students jump into a car and notice that the springs sink 6.0 cm under their combined 350-kg mass. If the total load supported by the springs (car plus students) is 1000 kg, what is the period of vibration of the loaded automobile?

11.27 A 1.0-kg metal block rests on a frictionless surface, as in Fig. 11.29, and is attached to two walls in a room by two springs ($k_1 = 40$ N/m; $k_2 = 25$ N/m). The block is

pushed a small distance to one side of its equilibrium position and released. At what frequency does it vibrate?

11.28. The same two springs as in Prob. 11.27 are connected to a mass of 1.0 kg, as shown in Fig. 11.30. If the mass is displaced a small distance x and released, what is the frequency of the resulting SHM? (*Hint:* The tension must be the same in the two springs at any instant.)

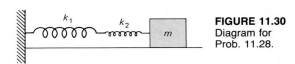

FIGURE 11.30
Diagram for Prob. 11.28.

11.29 A 1.0-kg mass resting on a frictionless surface is attached to the end of a horizontal spring of force constant 4.0 N/m. It is displaced 0.10 m and released.

(a) Graph the kinetic energy of the mass for one cycle.

(b) Graph the elastic potential energy of the spring for one cycle.

(c) Add graphically the results obtained in parts (a) and (b) and prove that the total energy of the system remains constant.

(d) What is the total energy of the system?

11.30 Repeat Prob. 11.29 for a vertical spring instead of a horizontal one. (In this case the gravitational potential energy of the mass changes as the spring oscillates. Take the zero of gravitational PE halfway between the unstretched and equilibrium positions of the spring.)

11.31 Two 0.50-kg masses are suspended from a spring, as in Fig. 11.31, and the spring stretches 0.25 m beyond its

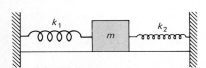

FIGURE 11.29 Diagram for Prob. 11.27.

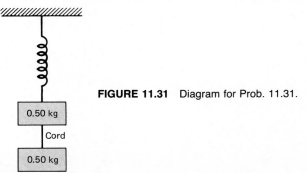

FIGURE 11.31 Diagram for Prob. 11.31.

length with no masses attached. At $t = 0$, when the system is in equilibrium, the cord between the top and bottom masses is cut, and the remaining mass executes SHM.

(a)   What is the frequency of the resultant motion?

(b)   What is the amplitude of vibration of the mass?

(c)   Write an equation to describe the SHM of the mass.

11.32   A 2.0-kg mass oscillates in SHM with an amplitude of 0.50 m at the end of a spring with a spring constant of 500 N/m.

(a)   What is the maximum kinetic energy of this system?

(b)   How does this maximum KE change when the mass is changed to 4.0 kg?

*****11.33**   A 1.0-kg mass at rest on a frictionless surface is attached to a horizontal spring of force constant 600 N/m. A second mass makes an elastic, head-on collision with the first mass and sets it into SHM with an amplitude of 0.20 m. If the incident speed of the colliding mass is 5.0 m/s, what is the numerical value of this mass?

11.34   A 1.0-kg block is attached to a horizontal spring with force constant $k = 400$ N/m. The spring is extended 10 cm, and the block is then released.

(a)   Calculate the velocity of the block as it passes through its equilibrium position (assume the surface is frictionless).

(b)   Repeat the calculation for the velocity at the equilibrium position, if a constant frictional force of 5.0 N retards its motion.

11.35   What is the acceleration due to gravity on a planet where an astronaut sets up a simple pendulum of length 1.00 m and finds that it makes 100 vibrations of small amplitude in 380 s?

11.36   A simple pendulum has a period of 1.98 s at a place where $g = 9.80$ m/s^2. What is the value of g at a place where this same pendulum has a period of 1.97 s?

11.37   A pendulum that has a period of exactly 2.000 s at the Greenwich Observatory in England, where $g = 9.812$ m/s^2, is taken to Paris, where it loses 20 s a day. What is the acceleration due to gravity in Paris?

11.38   A 2.0-kg pendulum bob hangs at the end of a string of length 2.0 m. The bob is displaced until the string makes an angle of 15° with the vertical and is then released. What is the KE of the bob when it passes through its equilibrium position?

*****11.39**   Prove that the natural frequency of a mass oscillating at the end of a spring which obeys Hooke's law is the same as that of a simple pendulum whose length is equal to the increase in length the mass produces when hung on the spring.

*****11.40**   A pendulum vibrates in SHM with an amplitude of 12°. Compare the amount of time spent by the bob between -6 and $+6°$ with the sum of the times spent between -12 and $-6°$ and between $+6$ and $+12°$.

*****11.41**   A ballistic pendulum consists of a 2.5-kg block of wood at the end of a massless string 1.5 m long. A 5.0-g bullet is fired into the block and comes to rest in it, and the system moves like a simple pendulum with an amplitude of 10°. Determine the momentum of the bullet.

11.42   Show that the first equality in Eq. (11.23) can be written

$$y = Y_0 \cos \left[\frac{2\pi}{\lambda} (x - vt) \right]$$

where v is the wave velocity.

11.43   A horizontal string 4.0 m long has a mass of 1.80 g.

(a)   What must the tension in the string be so that a 100-Hz wave moving along it has a wavelength of 50 cm?

(b)   What mass must be hung from the end of the string over a pulley to produce this tension?

11.44   A wave of frequency 60 Hz and amplitude 0.020 m travels from left to right along a stretched string at 4.0 m/s. At $t = 0$ the displacement at $x = 0$ is 0.020 m. Sketch the shape of the string at a time $t = 3.0$ s, clearly indicating distance along the string from the point $x = 0$.

11.45   A transverse wave of amplitude 0.12 m and wavelength 1.5 m travels at 1.5 m/s from left to right along a long, horizontally stretched string. At $t = 0$, the left end of the string has a displacement $y = 0.12$ m.

(a)   Write an equation describing the motion of a particle at the left end of the string as a function of time.

(b)   Write an equation describing the motion of a particle 1.0 m from the left end of the string as a function of time.

(c)   What is the maximum transverse velocity of any particle in the string?

11.46   The equation of a transverse traveling wave on a string is $y = 0.10 \cos (40\pi x - 100\pi t)$, where all distances are in meters. Find (a) the amplitude, (b) the wavelength, (c) the frequency, (d) the period, (e) the speed of the wave. (f) What is the tension in the string, if its linear density is 0.60 kg/m?

*****11.47**   A transverse wave travels down a string with an amplitude of 0.010 m, a frequency of 120 Hz, and a wavelength of 0.10 m.

(a)   What is its speed?

(b)   If, when $t = 0$ and $x = 0$, the displacement $y = Y_0/4$, what is the phase angle ϕ in radians?

(c)   Write a general expression for this traveling wave.

(d)   Find the displacement of the string at $t = 3.0$ s and $x = 2.5$ m.

*****11.48**   (a)   Use the results of Example 11.4 to write a general expression for the wave discussed there.

(b)   If, when $t = 0$ and $x = 0$, $y = Y_0/2$, what is the phase angle ϕ?

(c)   Prove from the general expression for this wave that its speed is 2.0 m/s.

(d)   Find the value of the displacement of the string at $t = 10$ s and $x = 2.0$ m.

*****11.49**   Two transverse waves are traveling down a string in the same direction. They have a speed of 4.0 m/s, a frequency of 240 Hz, and an amplitude of 1.0 cm. The two waves can be described by the equations

$$y_1 = Y_0 \cos \left[2\pi \left(\frac{t}{T} - \frac{x}{\lambda} \right) \right]$$

$$y_2 = Y_0 \cos \left[2\pi \left(\frac{t}{T} - \frac{x}{\lambda} \right) + \phi \right]$$

When $t = 0$, the resultant displacement of the string at $x = 0$ is 0.50 cm.

(a)   What is the value of the phase angle difference ϕ between the two waves?

(b)   What is the resultant displacement of the string 1.5 m along the string at $t = 3.00$ s?

*****11.50**   Two waves travel in the same direction along a string. They have the forms

$$y_1 = 2.0 \cos [2\pi(x - 40t)]$$

$$y_2 = 2.0 \cos \left[2\pi(x - 40t) + \frac{\pi}{4} \right]$$

where x and y are in centimeters and t is in seconds. Determine the values of x at which, for $t = 0$, (a) there is total cancellation of the waves; (b) the displacement is a maximum.

11.51 A metal wire is under a tension of 200 N. The length of the wire is 1.00 m and its mass is 4.00 g.

(a) What is the speed of a transverse wave along it?

(b) What is the frequency of the fundamental vibration of this wave?

(c) What is the frequency of its second and third harmonics?

11.52 A banjo string is 30.0 cm long and has a fundamental frequency of 256 Hz.

(a) What is the speed of waves along it?

(b) What is the tension in the string if its linear density is 0.200 g/cm?

11.53 A wire stretched under tension has a fundamental frequency of 440 Hz. What is the fundamental frequency of a second wire made of the same material but with twice the cross-sectional radius, twice the tension, and twice as long as the first?

11.54 Two wires of the same cross section are stretched with equal tensions between the same two supports. One wire is made of steel (density 7.8×10^3 kg/m^3) and the other of nickel (density 8.9×10^3 kg/m^3). If the fundamental frequency of the steel wire is 220 Hz, what is the fundamental frequency of the nickel wire?

11.55 A string attached to supports at both ends vibrates at 440 Hz with four nodes between its two ends, not counting the nodes at the two ends. At what frequency will the number of nodes be reduced to two?

11.56 A 6.0-g steel piano wire 0.50 m long is stretched with a tension of 500 N.

(a) What is its fundamental frequency?

(b) What is its highest harmonic that could be heard by a person able to hear frequencies up to 15,000 Hz?

∗11.57 Assume that the fundamental frequency of a stretched string must depend on the mass, length, and tension in the string, according to the equation

$$f_1 = Cm^x L^y F^z \qquad \text{where } C \text{ is a dimensionless constant.}$$

Use dimensional analysis to determine x, y, and z. Then compare your result with the complete result in Eq. (11.28) and find the value of C.

11.58 A standing wave is set up on a string by combining two waves traveling in opposite directions with identical amplitudes of 0.015 m and the same wavelength of 0.12 m. What is its amplitude for the following values of x along the string, if $x = 0$ corresponds to a node?

    (a) 0   (b) 0.030 m   (c) 0.060 m

    (d) 0.090 m   (e) 0.18 m   (f) 0.20 m

∗11.59 Two waves travel in opposite directions along a string. They have the forms

$$y_1 = 5.0 \cos [2\pi(0.050x - 25t)]$$

$$y_2 = 5.0 \cos \left[2\pi(0.050x + 25t) + \frac{\pi}{2} \right]$$

where x and y are in centimeters and t in seconds.

(a) What is the distance between adjacent nodes in the standing wave pattern produced?

(b) What is the position of the node nearest the end of the string at which $x = 0$?

11.60 A stretched wire on a sounding board is clamped at both ends.

(a) At what point should the wire be plucked to make its fundamental note sound loudest?

(b) If it is desired to make the second harmonic louder, where should the wire be plucked? Would touching the wire at its center after it has been plucked help make the second harmonic louder than the fundamental?

(c) What would you suggest to make the third harmonic loud?

11.61 A string 3.5 m long with a linear density of 0.040 kg/m is attached to a mechanical vibrator of variable frequency. The other end of the string passes over a pulley to a 20-kg mass.

(a) What is the lowest frequency of the vibrator at which this string will resonate?

(b) What frequencies correspond to the second, third, and fourth harmonics of the string?

∗11.62 Use dimensional analysis to prove that the speed of waves on a stretched string can only depend on the tension F and the linear density μ in such a way that $v \propto \sqrt{F/\mu}$.

Additional Readings

Bascom, Willard: "Ocean Waves," *Scientific American*, vol. 201, no. 2, August 1959, pp. 74–84. Some interesting data on one of the most complicated kind of waves—those on the oceans of the world.

Boorse, D. M.: "Motion of the Ground in Earthquakes," *Scientific American*, vol. 237, no. 6, December 1977, pp. 68–78. Discusses how earthquakes are produced and detected, and the nature of seismic waves.

Crawford, Frank S., Jr.: *Waves* (Berkeley Physics Course, Vol. 3), McGraw-Hill, New York, 1968. A more advanced text that contains some excellent diagrams and examples of wave motion.

French, A. P.: *Vibrations and Waves*, Norton, New York, 1971. A textbook at a more advanced level mathematically than this one; it contains some very good illustrations and experimental results.

Kock, Winston E.: *Sound Waves and Light Waves*, Doubleday Anchor, Garden City, N.Y., 1965. A very well-illustrated account of the similarities between sound and light waves, with particular emphasis on applications to the communications industry.

CHAPTER 12

Sound Waves

For this was the first time that I came in closer contact with the world leaders in scientific research in those days—Helmholtz, above all the others. But I learned to know Helmholtz also as a human being, and to respect him as a man no less than I had always respected him as a scientist. For with his entire personality, integrity of convictions and modesty of character, he was the very incarnation of the dignity and probity of science. These traits of character were supplemented by a true human kindness, which touched my heart deeply.

Max Planck (1858–1947)

Many of our greatest pleasures—friendly conversation around the dinner table, the rustle of autumn leaves, the boom of the ocean surf, a song by a favorite recording artist, a Beethoven symphony—come to us by sound waves striking our ears. In this chapter we apply to sound waves the ideas about waves we developed in the preceding chapter. We also discuss the miracle of the human ear and its exquisite sensitivity to sound.

12.1 The Nature of Sound Waves

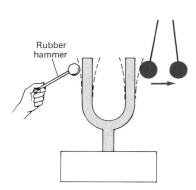

FIGURE 12.1 A tuning fork excited by a blow from a rubber hammer. The bouncing ball on the right shows that the tuning fork is vibrating.

Sound is the name given to a wave disturbance that produces the sensation of hearing. Sound waves are longitudinal waves in a medium like air with frequencies in the range 20 to 20,000 Hz—the frequency range to which the normal human ear is sensitive. At times, however, the word *sound* is used by physicists to include all compressional waves in a medium, no matter what the frequency.

Since sound is a mechanical wave, it requires a medium such as air, water, or steel for its propagation, and the properties of the medium determine the speed of the sound waves.

To understand better the nature of a sound wave let us consider what happens when a rubber hammer strikes a tuning fork, as in Fig. 12.1. The prongs of the tuning fork move back and forth in SHM at a frequency determined by the material and shape of the tuning fork. The amplitude of the vibration is very small, but that the fork moves is clear from its effect on a light cork ball, as in the figure. Suppose we put a small tube open at both ends next to one prong of the tuning fork so that the prong moves parallel to the length of the tube. When the prong moves toward the tube, it compresses the air inside the tube and this compression is passed on from one group of air particles* to another along the length of the tube. When the tuning fork prong moves away from the tube, it produces a partial vacuum and reduces the density of the air. This reduced density is called a *rarefaction*, and it follows the compression down the length of the tube, as in Fig. 12.2a. As the tuning fork continues to vibrate in SHM, it imparts a succession of compressions and rarefactions to the air. These disturbances travel down the tube and make up the sound wave. Just as the tuning fork executes SHM, so too the particles of the air execute SHM and pass on the sound to their neighbors by collisions.

The result is a wave in every way equivalent to the transverse wave of Fig. 11.14 except that, since this is a longitudinal wave, the motion of the air particles is *parallel to the direction of motion of the wave* instead of at right angles to it.

We can therefore define a sound wave as follows:

Sound wave: A longitudinal wave, transmitted by the air or some other medium, in the form of alternate compressions and rarefactions of the medium.

It is more difficult to diagram a compressional wave, even though the physics of its motion is basically the same as that of a transverse wave. In Fig. 12.2b we show as peaks the compressions, in which the particles are pushed closer together and the pressure and density take on their maximum values. The rarefactions, in which the particles are pulled farther apart and the pressure and density are lowered, are shown as troughs. Halfway between a peak and a trough is a point on the wave pattern where the air pressure and density retain their equilibrium values. (Don't be confused that we are plotting the pressure variations along the y direction even though they actually occur in the x direction; on a graph the significance of the x and y axes is what we choose to make it.)

The wavelength of a sound wave is the distance between any two consecutive equivalent points on the wave pattern for the sound wave. These points may be points of maximum compression, maximum rarefaction, or any other set of equivalent points, as in Fig. 12.2. The wavelength is related to the speed of the wave and to its frequency by the basic equation for all wave motion, $v = \lambda f$.

As the compressions and rarefactions move down the tube, each group of particles executes SHM along the wave direction. The displacement from its equilibrium position of any particle in the path of the wave can be represented by an equation similar to Eq. (11.23). For the displacement of the particle from its equilibrium position, we use the symbol X (to distinguish it from x, the distance along the horizontal axis), and for the amplitude, or maximum value of X, we use the symbol X_{max}. For sound waves X_{max} is very small, varying from about 10^{-5} to 10^{-11} m. Our wave equation for a sound wave moving in the positive x direction is then

$$X = X_{max} \cos \left[2\pi \left(\frac{x}{\lambda} - \frac{t}{T} \right) \right]$$
(12.1)

where, as usual, T is the period of the SHM, λ is the wavelength of the wave, and x is the distance along the x axis. This is shown in Fig. 12.3. A compressional wave moving in the opposite direction would then be represented by the equation

---

*The term *particle* is used deliberately to allow for the possibility that these particles are not individual molecules but clusters of molecules. For a gas the periodic motion of these clusters of molecules is superimposed on the random translational motion of the individual gas molecules.

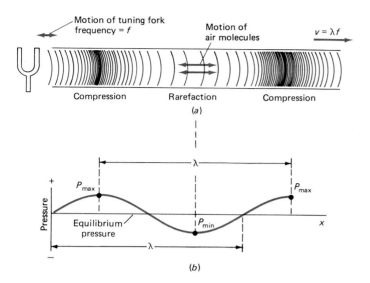

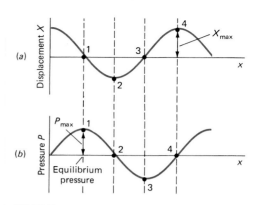

FIGURE 12.3 (a) The displacement of the particles in a gas as a function of distance for a sound wave passing through the gas. (b) The pressure variations for the same sound wave.

FIGURE 12.2 A sound wave moving through air. (a) The compressions and rarefactions caused by the wave moving through air. (b) The pressure of the air in the tube as a function of position along the tube.

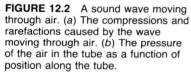

$$X = X_{max} \cos\left[2\pi\left(\frac{x}{\lambda} + \frac{t}{T}\right)\right]$$

(12.2)

where in both these cases we have assumed that the phase angle is zero.

It should be clear that, in the case just discussed, only *one* sound wave passes through the gas. But there are two different ways of looking at this one wave—either in terms of particle displacements or in terms of pressure variations. This leads to what are called the *displacement wave* and the *pressure wave*. Both waves are periodic, but they are out of phase with each other, since they describe different aspects of the one sound wave.

To see this, suppose the displacement of the particles transmitting the wave varies along the path of the wave, as shown in Fig. 12.3a at a particular instant of time. (Here, again, we plot displacements in the *x* direction along the vertical axis.) At this time the particles to the left of point 1 are moving in the positive *x* direction, and those to the right are moving in the negative *x* direction. Therefore a concentration of particles builds up at point 1, and the density and pressure are large at this point, as in Fig. 12.3b.

On the other hand, at this same instant, at point 2 in Fig. 12.3a the particles to either side of point 2 are moving in the same direction with the same displacement, and so the density of the gas does not change at that point. The pressure therefore remains at its equilibrium value, as shown on the pressure graph. Continuing this analysis for points 3 and 4, we find the values of the pressure in Fig. 12.3b. The pressure wave is seen to be 90° out of phase with the displacement wave in Fig. 12.3a, although they both describe exactly the same sound wave. This 90° phase difference corresponds to a shift of $[(\pi/2)/2\pi]\lambda = \lambda/4$ of the pressure wave with respect to the displacement wave along the direction of propagation, as seen in the figure.

The particle displacements and the pressure variations in a sound wave are 90° (λ/4) out of phase with each other.

For this reason, if the displacement wave is represented by a cosine function, the pressure wave will be represented by a sine function, as in Fig. 12.3.

Propagation of Sound Waves

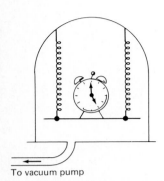

To vacuum pump

FIGURE 12.4 An alarm clock ringing in a chamber from which the air is evacuated.

Compressional waves propagate readily through all forms of matter—solids, liquids, and gases alike. In this way they differ from transverse mechanical waves, which cannot pass through gases and liquids, since fluids possess no elasticity in a direction perpendicular to the direction of wave propagation. Sound waves cannot propagate through a vacuum, since there are no molecules present to carry the wave. This can easily be seen by putting an alarm clock inside a vacuum chamber, setting it to alarm and then withdrawing the air, as in Fig. 12.4. As the air pressure decreases, the sound grows weaker and weaker and finally becomes inaudible. Makers of science fiction movies like to provide loud sounds when spacecraft explode in space; now you know enough not to believe them!

Sound propagates faster through solids than through gases, for in solids there are more molecules per unit volume to carry the sound, they are closer together, and solids have larger elastic moduli.

Example 12.1

Compressional waves have a speed of about 340 m/s in air at 20°C, and audible sound has a frequency range from about 20 to 20,000 Hz. **(a)** What is the wavelength of a sound wave of frequency 1.0×10^4 Hz? **(b)** Write an equation describing the longitudinal motion of the particles of the air through which this sound passes, as a function of time and distance along the positive x axis, which is the direction in which the sound is moving.

SOLUTION

(a) Since for any wave $\lambda = v/f$,

$$\lambda = \frac{340 \text{ m/s}}{1.0 \times 10^4/\text{s}} = 0.034 \text{ m} = \boxed{3.4 \text{ cm}}$$

(b) Our general equation for the motion of the particles in the path of a sound wave is Eq. (12.1):

$$X = X_{\max} \cos \left[2\pi \left(\frac{x}{\lambda} - \frac{t}{T} \right) \right]$$

Here

$$T = \frac{1}{f} = \frac{1}{10^4/\text{s}} = 1.0 \times 10^{-4} \text{ s} \quad \text{and} \quad \lambda = 0.034 \text{ m}$$

Hence

$$X = X_{\max} \cos \left[2\pi \left(\frac{x}{0.034} - \frac{t}{10^{-4}} \right) \right]$$

$$= \boxed{X_{\max} \cos \left[2\pi (29x - 10^4 t) \right]}$$

where times are in seconds and distances in meters.

Again we have assumed here that the phase angle ϕ is equal to zero, and so $X = X_{\max}$ when $t = 0$ and $x = 0$.

EXERCISE 1 Compare the wavelength found in part (*a*) with the average distance between the molecules in a gas at STP.

12.2 The Speed of Sound Waves

In the preceding chapter we saw that the speed of a transverse wave on a string is $v = (F/\mu)^{1/2}$, where μ is the mass per unit length, an *inertial* factor, and F is the tension in the string, an *elastic* factor. The tension F determines how tightly coupled one molecule of the string is to its neighbors and hence how elastically the string rebounds when it is displaced at right angles to its length. We might, for these reasons, expect the speed of sound waves to depend in the same way on the elastic and inertial properties of the medium through which they pass. This indeed turns out to be the case.

If the medium is a fluid, i.e., a liquid or a gas, the most important inertial property is the density of the undisturbed fluid ρ_0, for this is an inertial property independent of volume. Similarly, the only elastic property a fluid has is its resistance to compression, and this is measured by its bulk modulus, given by Eq. (10.20) as $B = -P/(\Delta V/V_0)$. By analogy with $v = (F/\mu)^{1/2}$ for waves on a string, we might expect the speed of sound in a fluid to take the form

$$v = \left(\frac{B}{\rho_0}\right)^{1/2} \tag{12.3}$$

This is the same result obtained by a more careful derivation of the speed of sound in fluids.

The speed of longitudinal waves in fluids depends directly on the square root of the bulk modulus and inversely on the square root of the equilibrium density of the fluid.

As might be expected, for a solid metal rod struck at one end, we must replace the bulk modulus by the elastic modulus for the length change in the solid, i.e., by Young's modulus Y. Then we obtain in the same way

$$v = \left(\frac{Y}{\rho_0}\right)^{1/2} \tag{12.4}$$

Generally sound travels more rapidly in liquids than in gases and more rapidly in solid metals than in liquids. This is because the ratio of elastic modulus to density is higher for liquids than for gases, as can be seen in Table 12.1. The same is true for solids compared to liquids. The agreement of experimental values for the sound velocity in a great variety of materials with the values calculated from

TABLE 12.1   Elastic and Inertial Properties and the Speed of Sound

			Speed of sound (m/s)	
Fluids	**Density ρ_0 (kg/m^3)**	**Bulk modulus B (N/m^2)**	**Calculated value $v = (B/\rho_0)^{1/2}$**	**Measured value**
Air (STP)	1.29	1.41×10^5	3.31×10^2	3.32×10^2
Carbon dioxide (STP)	1.98	1.41×10^5	2.66×10^2	2.58×10^2
Hydrogen (STP)	9.0×10^{-2}	1.41×10^5	1.25×10^3	1.27×10^3
Water	1.00×10^3	2.00×10^9	1.41×10^3	1.43×10^3
Mercury	1.35×10^4	2.5×10^{10}	1.36×10^3	1.45×10^3

			Speed of sound (m/s)	
Solids	**Density ρ_0 (10^3 kg/m^3)**	**Young's modulus Y (10^{10} N/m^2)**	**Calculated value $v = (Y/\rho_0)^{1/2}$**	**Measured value**
Aluminum	2.70	7.0	5.09×10^3	5.10×10^3
Copper	8.89	11.0	3.52×10^3	3.56×10^3
Iron	7.20	19.0	5.14×10^3	5.13×10^3
Nickel	8.90	21.4	4.90×10^3	4.97×10^3
Tungsten	19.0	36.0	4.35×10^3	4.32×10^3

the simple equations $v = (B/\rho_0)^{1/2}$ and $v = (Y/\rho_0)^{1/2}$ gives us confidence in the correctness of these equations. In physics, all theories are suspect unless they are confirmed by experiment.

Dependence of Sound Velocity on Temperature

The velocity of sound (or of nonaudible longitudinal waves) varies little with temperature for liquids or solids. In gases, however, the dependence is considerable. It can be shown (see Prob. 12.29) that the speed of sound v_1 at a temperature T_1 is related to the speed of sound v_2 at a temperature T_2 by the equation

$$\frac{v_1}{v_2} = \left(\frac{T_1}{T_2}\right)^{1/2}$$

(12.5)

where both T_1 and T_2 are Kelvin temperatures.

The speed of sound in a gas is proportional to the square root of the absolute (Kelvin) temperature.

Example 12.2

The density of hydrogen iodide (HI) gas is 5.79 kg/m³ at STP. What is the speed of sound in hydrogen iodide at 50°C, if we assume that the speed of sound in air at STP is 331 m/s?

SOLUTION

The speed of sound in a gas is given by Eq. (12.3), where ρ_0 is the equilibrium density of the gas. (The actual density varies about this value as the sound wave passes through the gas.) We use this equation to first determine the speed of sound in HI at STP, and then use Eq. (12.5) to calculate the speed at 50°C. We have

$$\frac{v_{HI}}{v_{air}} = \frac{(B_{HI}/\rho_{0,HI})^{1/2}}{(B_{air}/\rho_{0,air})^{1/2}}$$

But the bulk modulus for all ideal gases is the same, and so $B_{HI} = B_{air}$, and

$$v_{HI} = \left(\frac{\rho_{0,air}}{\rho_{0,HI}}\right)^{1/2} v_{air} = \left(\frac{1.29 \text{ km/m}^3}{5.79 \text{ km/m}^3}\right)^{1/2} v_{air}$$

$$= 0.472 \ (331 \text{ m/s}) = \boxed{156 \text{ m/s}}$$

This is the speed of sound in HI at STP. If we keep the pressure fixed at atmospheric and raise the temperature to 50°C, we have from Eq. (12.5) for the speed of sound in HI,

$$\frac{v_{50}}{v_0} = \left(\frac{323}{273}\right)^{1/2} \quad \text{or} \quad v_{50} = 1.09 \ (156 \text{ m/s}) = \boxed{170 \text{ m/s}}$$

12.3 Standing Sound Waves; Resonance

Sound waves are reflected from the end of an organ pipe similar to the way transverse waves on a string are reflected from the end of the string. If the organ pipe is closed at the top end away from the source, then no displacement of the particles is possible at the closed end, and the closed end must be a displacement node. If the organ pipe is open at the top end, on the other hand, the air moves with maximum displacement there and we have an antinode (loop). This sets the boundary condition for the displacement wave at the top end of the pipe. Since the bottom end is where the organ pipe is excited, that end is assumed to be an antinode.

Since the pressure wave is 90°, or $\lambda/4$, out of phase with the displacement wave, a displacement antinode corresponds to a pressure node, and a displacement node to a pressure antinode. At the end of a closed organ pipe, therefore, we have a pressure antinode; at the end of an open pipe, a pressure node.

Standing Waves

As we saw in the preceding chapter, a stretched string has certain fundamental or natural frequencies that depend on its length and on the speed of the transverse waves moving along it. Standing waves can be set up in the string by waves of the same frequency and speed moving in opposite directions. The form of these standing waves is given by Eq. (11.26):

$$y = 2Y_0 \cos \frac{2\pi x}{\lambda} \cos \frac{2\pi t}{T} \tag{12.6}$$

The same analysis can be applied to two sound waves traveling in opposite directions along the length of an organ pipe (or some similar tube). Standing waves again result, in which the displacement of a particle at a position x and a time t can be described by a similar equation:

$$X = 2X_{\text{max}} \cos \frac{2\pi x}{\lambda} \cos \frac{2\pi t}{T} \tag{12.7}$$

Here X_{max} is the displacement amplitude of each of the original waves, and X is the displacement resulting from the superposition of the two waves traveling in opposite directions. We see that X is periodic in time, but that it has an amplitude that varies with position x along the organ pipe. At some places this amplitude is zero; at other places it takes on its maximum value $2X_{\text{max}}$. Hence we have a standing compressional wave in the pipe.

In Fig. 12.5 we show the displacement nodes and antinodes for a closed organ pipe. Distances along the organ pipe are plotted along the x axis. Along the y axis we plot the displacement of the particles *in the x direction*. The result is a graph showing points of no displacement (nodes N) at the end of the pipe and at intervals of $\lambda/2$ from that end; and other points halfway between the nodes where the displacement has its maximum value of $2X_{\text{max}}$. These are the antinodes (A). This graph represents well, therefore, the physical content of Eq. (12.7).

We can also plot the pressure variations in a standing sound wave in a closed organ pipe. Since there is a phase shift of $\lambda/4$ along the direction of propagation between the pressure and displacement waves, we would expect that the cos $(2\pi x/\lambda)$ term in Eq. (12.7) would be replaced by a sine function to allow for this phase shift. The standing pressure wave can therefore be described by the equation

$$P = 2P_{\text{max}} \sin \frac{2\pi x}{\lambda} \cos \frac{2\pi t}{T}$$

This equation is plotted in Fig. 12.6, along with the similar equation for the

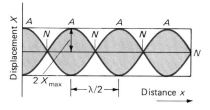

FIGURE 12.5 Displacement nodes and antinodes for a standing sound wave in a closed organ pipe. Here N are the nodes and A the antinodes. Compare the standing wave pattern in this case with that shown for transverse waves in Fig. 11.26.

FIGURE 12.6 Nodes and antinodes for the displacement and pressure in a standing sound wave in a closed-end organ pipe. Subscripts D and P are used for displacement and pressure. Note that pressure nodes (N_P) occur at the same positions as displacement antinodes (A_D), and vice versa.

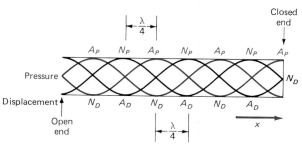

displacement. As time goes on, there is no pressure variation at the nodal points (N_P) in the diagram, whereas there is maximum variation at the antinodes (A_P).

We see from this figure that *the nodes of the pressure wave are displaced by a distance of $\lambda/4$ from the displacement nodes.* As a result, pressure antinodes coincide with displacement nodes, and vice versa, as seen in the figure.

Resonant Frequencies

An organ pipe, therefore, has a number of natural frequencies that are determined by the length of the pipe and the boundary conditions at the two ends. If a sound is produced at the bottom of the pipe, this end of the pipe is an antinode (we assume this in what follows). If the sound introduced is at one of the natural frequencies of the pipe, a standing wave pattern of large amplitude is set up, because all the reflected waves add together at the antinodes to give large amplitudes there. This resonant condition results in a loud sound.

The natural frequencies of open and closed organ pipes differ greatly. For an open pipe, with displacement antinodes at the two ends, the longest wavelength possible is $\lambda_1 = 2L$, where L is the length of the pipe (see Fig. 12.7). This corresponds to the fundamental frequency $f_1 = v/\lambda_1 = v/2L$. The second harmonic will have $\lambda_2 = L$, or $f_2 = v/L = 2f_1$. Similarly, the third harmonic has $f_3 = 3f_1 = 3v/2L$, and so forth. In this case (open pipe) all harmonics are present, and the fundamental frequency is $v/2L$. The arrangement of nodes and antinodes for an open organ pipe vibrating at these frequencies is shown in Fig. 12.7.

An organ pipe closed at the top is quite different. Here we still must have a displacement antinode at the open end to enable the pipe to be excited, but there is a node at the closed end. Since nodes and antinodes are only $\lambda/4$ apart, in this case the fundamental vibration must have a wavelength $\lambda_1 = 4L$, or $f_1 = v/\lambda_1 = v/4L$, which is half the fundamental for an open pipe. The next harmonic that can occur is shown in Fig. 12.8. The length of the pipe in this case must be $\frac{3}{4}\lambda$, and $f_3 = v/\lambda_3 = 3v/4L$. We have designated this frequency by f_3 since it is 3 times the fundamental and hence is the third harmonic. Similarly, the next possible harmonic is $\lambda_5 = 4L/5$, or $f_5 = 5v/4L = 5f_1$, which is the fifth harmonic.

A closed organ pipe, therefore, resonates at the fundamental frequency $v/4L$, where L is the length of the pipe, and only odd harmonics of this frequency occur. No even harmonics can be produced, since they cannot satisfy the boundary conditions set by the physical situation.*

When organ pipes are excited by an organist, not only the fundamental but many harmonics of that fundamental are excited. It is clear that two organ pipes of the same length, one open and one closed, emit very different sounds because both the fundamental frequency and the harmonic content are different.

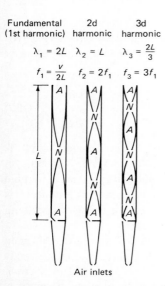

Fundamental (1st harmonic)	2d harmonic	3d harmonic
$\lambda_1 = 2L$	$\lambda_2 = L$	$\lambda_3 = \frac{2L}{3}$
$f_1 = \frac{v}{2L}$	$f_2 = 2f_1$	$f_3 = 3f_1$

Air inlets

FIGURE 12.7 Standing sound waves in an organ pipe open at the top end. The lines indicate particle displacements.

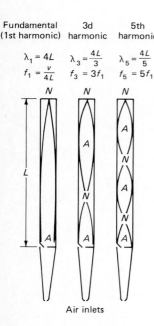

Fundamental (1st harmonic)	3d harmonic	5th harmonic
$\lambda_1 = 4L$	$\lambda_3 = \frac{4L}{3}$	$\lambda_5 = \frac{4L}{5}$
$f_1 = \frac{v}{4L}$	$f_3 = 3f_1$	$f_5 = 5f_1$

Air inlets

FIGURE 12.8 Standing sound waves in an organ pipe closed at the top end. The lines indicate particle displacements.

---

*A physician's *stethoscope* has a body-contact piece connected by hollow rubber tubing to earpieces that enable the doctor to hear sounds from inside the human body. Different contact pieces are used depending on the frequency of the sound to be detected, just as different-sized organ pipes resonate at different frequencies.

Example 12.3

The apparatus in Fig. 12.9 can be used to obtain the speed of sound in air. A glass tube is filled with water, the level of which can be adjusted by moving the water container up and down. A 1000-Hz tuning fork is set into vibration above the open top of the glass tube, and the sound intensity is large when the water level is 8.3, 24.8, and 41.3 cm below the open end of the tube. Find the speed of sound in air.

SOLUTION

The sound intensity is large when resonance occurs and standing waves are set up in the air column. The standing waves set up must have a displacement node at the air-water interface and a displacement antinode at the top of the tube, as shown in Fig. 12.10. Here we construct three standing waves that satisfy these boundary conditions, where the wavelength is in every case the same, since v and f are both fixed and $\lambda = v/f$.

$$\frac{\lambda}{4} = 8.3 \text{ cm} \qquad \text{or} \qquad \lambda = 33.2 \text{ cm}$$

$$\frac{3\lambda}{4} = 24.8 \text{ cm} \qquad \text{or} \qquad \lambda = 33.1 \text{ cm}$$

$$\frac{5\lambda}{4} = 41.3 \text{ cm} \qquad \text{or} \qquad \lambda = 33.0 \text{ cm}$$

Average $\lambda = 33.1$ cm $= 0.331$ m

1000-Hz tuning fork

FIGURE 12.9

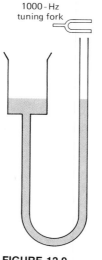

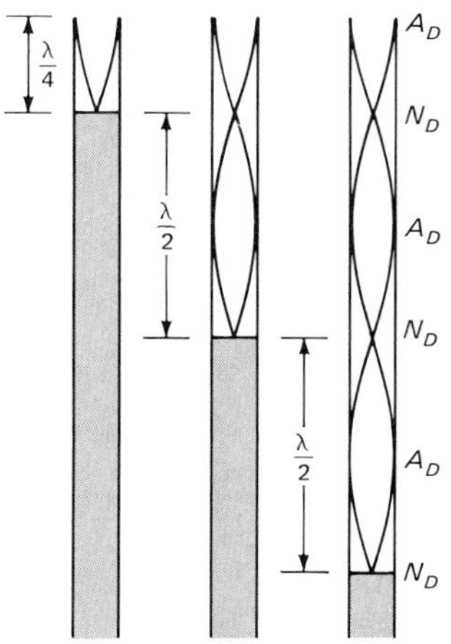

FIGURE 12.10 Diagram for Example 12.3. Lines indicate particle displacements.

Then the speed of sound in air under the conditions of the experiment is

$$v = \lambda f = (0.331 \text{ m})(1000 \text{ s}^{-1}) = \boxed{331 \text{ m/s}}$$

Example 12.4

A 1.00-m-long aluminum rod is mounted by a clamp at its center and set into an audible longitudinal vibration by rubbing it with resin. [This is referred to as a *Kundt's tube*, after the German physicist August Kundt (1839–1894).] One end of the rod is attached to a diaphragm that protrudes into a glass tube which has been dusted with talcum powder, as in Fig. 12.11.

The position of the piston at the other end of the tube can be adjusted so that the talcum powder forms little mounds 6.3 cm apart. What is the measured speed of sound in the air of the tube?

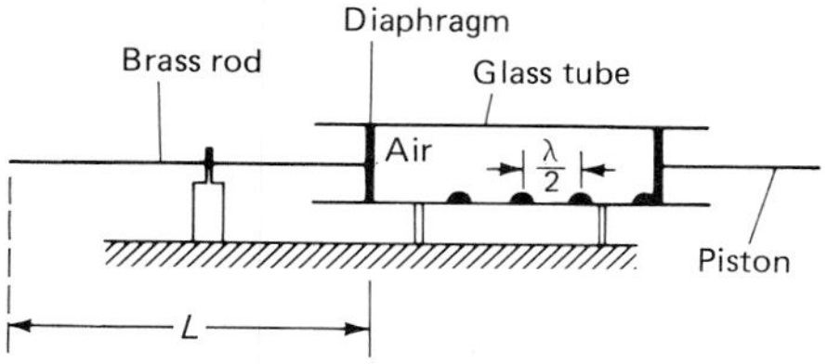

FIGURE 12.11 Kundt's tube experiment. The vibrating rod produces sound waves in the air of the tube. For the correct tube length, obtained by moving the piston, standing waves are produced and talcum powder accumulates in mounds ˙ ⁻ displacement nodes, which are $\lambda/2$ apart.

SOLUTION

This is an example of two different standing longitudinal waves, one in the aluminum rod and another in the air in the tube. The frequency of these two standing waves must be the same if resonance is to occur, as it does when the powder is sufficiently agitated to form the little mounds.

For the aluminum rod, which is supported at its center, the fundamental vibration is one with a node at the center and antinodes at the two ends. Hence the length L of the rod is $\lambda_R/2$, and so $\lambda_R = 2.00$ m.

Since no other information about the rod is given except the fact that it is aluminum, it can be assumed that we must obtain the speed of sound in aluminum from Table 12.1. We then have

$$f = \frac{v}{\lambda_R} = \frac{5.10 \times 10^3 \text{ m/s}}{2.00 \text{ m}} = 2.55 \times 10^3 \text{ Hz}$$

In the glass tube the powder accumulates at the displacement nodes to which it is driven by the agitated molecular motion at the displacement antinodes. Since these nodes are $\lambda/2$ apart, we have

$$\frac{\lambda}{2} = 6.3 \text{ cm} \qquad \text{or} \qquad \lambda = 12.6 \text{ cm} = 0.126 \text{ m}$$

The speed of sound in the air of the tube is then

$$v = \lambda f = (0.126 \text{ m})(2.55 \times 10^3/\text{s}) = \boxed{3.21 \times 10^2 \text{ m/s}}$$

12.4 Interference of Sound Waves of Different Frequencies; Beats

In the preceding section we discussed the production of standing waves by the superposition of sound waves of the same frequency traveling in opposite directions inside an organ pipe. In that case displacement nodes and antinodes occur as a function of *distance* along the length of the pipe. It is also possible to produce interference between two waves that results in amplitude variations not as a function of distance but as a function of *time*.

What we need to produce this effect are two sound waves of slightly different frequencies traveling through space in the same direction. An example might be the sound waves produced by two tuning forks, of very close frequencies, both traveling along the same path to a listener's ear. Here we concentrate on the motion of the air particles at some fixed distance x along the path of the two waves. The air is being subjected to the simultaneous displacements of the two waves, and its resultant motion as a function of time is just the superposition of these two motions. The displacement of the air at the desired position due to the first wave, of frequency f_1, is then

$$X_1 = X_{max} \cos \frac{2\pi t}{T_1} = X_{max} \cos 2\pi f_1 t$$

Similarly, for the second wave, which we assume has the same amplitude X_{max} but a different frequency f_2,

$$X_2 = X_{max} \cos 2\pi f_2 t$$

From the principle of superposition, the resultant motion of the air is

$$X = X_1 + X_2 = X_{max} (\cos 2\pi f_1 t + \cos 2\pi f_2 t) \tag{12.8}$$

Now, from trigonometry (Appendix 1.C), we have

$$\cos A + \cos B = 2 \cos \frac{A + B}{2} \cos \frac{A - B}{2} \tag{12.9}$$

Then, if we let $A = 2\pi f_1 t$ and $B = 2\pi f_2 t$, Eq. (12.8) becomes

$$X = X_{max} \left[2 \cos \frac{2\pi(f_1 + f_2)t}{2} \cos \frac{2\pi(f_1 - f_2)t}{2} \right]$$

or

$$X = \left[2X_{max} \cos \frac{2\pi(f_1 - f_2)t}{2} \right] \cos \frac{2\pi(f_1 + f_2)t}{2} \tag{12.10}$$

The resultant displacement of the air at the position under consideration occurs, therefore, at a frequency $(f_1 + f_2)/2$, which is just the average of the two frequencies. The amplitude of this displacement varies in time at a frequency $(f_1 - f_2)/2$, however. This results in what are called *beats*.

Beats: The regular increase and decrease in the amplitude of sound (or other) waves caused by two waves of slightly different frequencies being superimposed on the same medium.

Although we have derived Eq. (12.10) for only one distance from the source, it should be clear that it holds for every distance from the source. How the am-

FIGURE 12.12 Beats produced at a point in space by two sound waves of slightly different frequencies. (*a*) Displacements of the two waves traveling in the same direction as a function of time. (*b*) Resultant displacement obtained by superimposing the displacements of the two traveling waves.

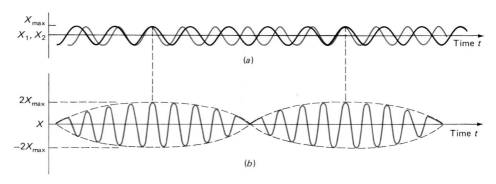

plitude of the resultant wave varies at one position x as a function of *time* is shown in Fig. 12.12. In looking at this figure, keep in mind that beats occur because we are constantly going from a situation where the two waves interfere constructively to one where they interfere destructively at a later time. This leads to the rise and fall in the amplitude of the resultant wave, and hence in the loudness of the sound heard.

When we speak of a beat, we mean a maximum amplitude for the resulting motion that produces a large pulse of sound. This occurs whenever $\cos [2\pi(f_1 - f_2)t/2]$ in Eq. (12.10) equals either 1 or -1. The amplitude is a maximum twice in each cycle, therefore, and so the number of beats per second is $2(f_1 - f_2)/2 = f_1 - f_2$. *The beat frequency is equal to the difference in the frequencies of the two sound sources.* This gives us a convenient way of tuning musical instruments to a desired frequency. That frequency is generated by some frequency standard; then the instrument is tuned to "zero beat" with it, i.e., tuned until the number of beats has been reduced exactly to zero. Then $f_1 - f_2 = 0$, or $f_2 = f_1$.

A piano tuner uses a zero-beat technique to adjust the frequency of a piano string to that of a standard tuning fork; the members of a symphony orchestra tune their instruments by listening for beats between their instruments and the standard tone produced by a violin or oboe at 440 Hz. The human ear can detect beats only up to about seven per second. Beyond this, however, various kinds of electronic pulse and frequency counters can be used to measure the beat frequency.

Example 12.5

Two open organ pipes excited simultaneously in their fundamental modes produce four beats per second. If the longer pipe is 2.30 m in length, find the length of the shorter pipe.

SOLUTION

The fundamental mode of an open organ pipe has antinodes at its two ends, and so its fundamental wavelength is $\lambda = 2L$ (see Fig. 12.7), where L is the length of the organ pipe. In this case we have

$$4.00 \text{ s}^{-1} = f_2 - f_1 = \frac{v}{\lambda_2} - \frac{v}{\lambda_1} = \frac{v}{2L_2} - \frac{v}{2L_1}$$

from which

$$\frac{1}{L_2} = \frac{2}{v}(4.00 \text{ s}^{-1}) + \frac{1}{L_1} = 0.459$$

and so

$$L_2 = \boxed{2.18 \text{ m}}$$

12.5 The Detection of Sound

Since sound waves in air consist of small variations in the displacement of the air particles and in the pressure of the air, any device sensitive to these changes can be used as a sound detector. In a microphone, for example, a thin diaphragm is set into vibration by the incident sound wave. The motion of the diaphragm is then converted into an electric signal by a *transducer,* a device that converts mechanical impulses to electric impulses. These electric impulses are then carried over telephone lines to a telephone receiver, or transmitted through space as electromagnetic waves to radio and television sets.

The Human Ear

The ear is an extremely sensitive detector of sound, as good as the most sensitive microphone. Thus a displacement of the eardrum by only 10^{-11} m (smaller than the average diameter of an atom, which is about 2×10^{-10} m) can give rise to a sound that the normal human ear can hear.

Figure 12.13 is a drawing of the human ear, which performs the same function as a microphone, but in a much more complicated and delicate fashion. The eardrum, which divides the outer ear from the middle ear, is set into vibration by an incident sound wave. The ear then amplifies this motion mechanically and converts it into electric signals that proceed to the brain.

The outer ear contains the ear canal, which is about 2.7 cm long and ends at the eardum. If we consider this canal to be a pipe closed at one end, vibrating in its fundamental mode, we have, from Fig. 12.8, $\lambda_1 = 4L = 4(0.027 \text{ m}) = 0.11$ m. For sound waves with a speed of 340 m/s in air, this corresponds to a frequency $f = v/\lambda = (340 \text{ m/s})/(0.11 \text{ m}) = 3.1 \times 10^3$ Hz. The ear is therefore most sensitive to frequencies around 3000 Hz because at that frequency resonance in the outer ear increases the pressure fluctuations on the eardrum. This resonance is rather broad, and so the ear is most sensitive over a range of a few thousand hertz on either side of 3000 Hz. The ear canal in women is a little shorter than in men. For this reason women are more sensitive to higher frequencies than are men.

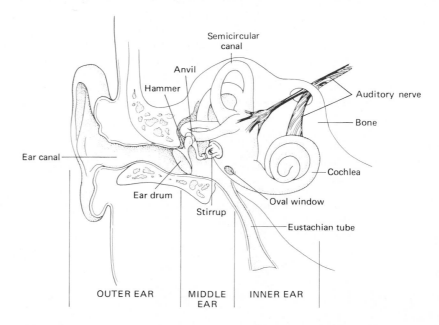

FIGURE 12.13 The human ear.

The middle ear transmits the sound from the eardrum to the oval window that separates the middle ear from the inner ear. The sound is transmitted by three tiny bones (ossicles), the hammer, anvil, and stirrup.

The inner ear contains the semicircular canals, which are important for maintaining balance, and the cochlea, a snail-shaped cavity filled with a liquid called lymph. The stirrup presses on the oval window and sets up sound vibrations in the lymph. Thirty thousand sensory cells located on the partition that divides the cochlea into two parts detect the sound vibrations in the lymph, convert them to electrical nerve impulses, and pass on the signal to the auditory nerve fibers, which carry them to the brain where they are perceived as sounds.

Defective hearing can be due to a breakdown at any point in the ear's complicated sound-transmission system. Hearing aids can sometimes help. These are basically combinations of a microphone and an amplifier that boost the sound level at the eardrum and thus compensate for defects in the ear structure itself.

12.6   Intensity and Loudness

By the intensity of a sound wave we mean the following:

Intensity: The amount of energy transported per second per unit area normal to the direction of wave propagation.

$$\text{Intensity } I = \frac{\mathcal{W}}{tA} \qquad\qquad (12.11)$$

Since *power* is energy per unit time, the dimensions of intensity are those of *power per unit area,* and the units of intensity are watts per square meter (W/m^2). (Note that $\mathcal{W}$ signifies *energy* and W signifies *watts*.) For example, if the sound generated by a rock group causes 1 J of sound energy per second to pass through a 1-m^2 area between them and their audience, the sound intensity is 1 W/m^2.

We have seen (Sec. 6.4) that for the SHM of a mass at the end of a spring, the total mechanical energy of the system is $\mathcal{W} = \frac{1}{2}kX_0^2$, where X_0 is the amplitude in the equation of motion $x = X_0 \cos 2\pi ft$. In a sound wave, each particle of the medium oscillates back and forth in SHM in a way described by the similar equation

$$X = X_{max} \cos 2\pi ft \qquad\qquad (12.12)$$

where $f = (1/2\pi)\sqrt{k/m}$ from Eq. (11.8), and the displacement X is parallel to the direction of the wave. Since $k = 4\pi^2 f^2 m$, the mechanical energy of each vibrating particle is $\mathcal{W}_1 = \frac{1}{2}kX_{max}^2 = 2\pi^2 f^2 m X_{max}^2$. As the sound wave moves through the material, it imparts this amount of energy to each particle it passes. The total energy imparted per second to all the particles in a volume V is then

$$\frac{\mathcal{W}}{t} = \frac{\mathcal{W}_1 Vn}{t} = \frac{\mathcal{W}_1 ALn}{t}$$

where n is the number of particles per unit volume, and the volume V is equal to AL, where L is the length of the volume. But L/t is the speed of the wave along the x axis, and so

$$\frac{\mathcal{W}}{t} = \mathcal{W}_1 Avn \qquad\qquad (12.13)$$

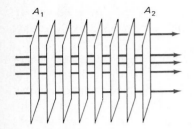

FIGURE 12.14 A plane sound wave. The intensity remains constant with distance, and the cross-sectional area of the wave front is the same for all distances.

Hence the total energy imparted to the particles in a volume of unit cross-sectional area per second is, from the above equations,

$$I = \frac{\mathcal{W}}{tA} = \frac{\mathcal{W}_1 A v n}{A} = 2\pi^2 f^2 X_{max}^2 m v n \qquad (12.14)$$

The intensity of a wave is proportional to the square of the amplitude of the vibrating particles in the medium and to the square of the frequency.

The presence of the area A in Eq. (12.11) means that the intensity falls off with distance from the source, unless the wave is a plane wave carrying energy only in one direction, as in Fig. 12.14. In a plane wave the sound travels in straight lines perpendicular to the planes of equal phase. If a sound is radiated in all directions from a point source such as an exploding shell, however, its intensity falls off as $1/A$, or $1/(4\pi R^2)$, where $4\pi R^2$ is the area of the surface of a sphere of radius R. As R increases, the intensity falls off as $1/R^2$, as shown in Fig. 12.15. Here the same amount of energy ($\mathcal{W}$) passes through the areas A_1 and A_2 per second. Since A_2 is larger than A_1, and $I = \mathcal{W}/tA$, the intensity at A_2 is *lower* than at A_1 by a factor A_1/A_2, and hence by a factor of R_1^2/R_2^2.

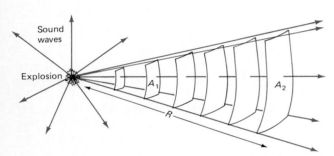

FIGURE 12.15 A spherical sound wave. The intensity falls off as $1/R^2$, since the area through which the wave passes increases as R^2.

Loudness

The intensity of a sound wave is an objective property that can be measured quantitatively by instruments. The loudness of a sound, on the other hand, is a *subjective property* that depends on the nature and condition of the ear doing the hearing. The frequency of a sound will, for example, affect the perceived loudness, even for the same intensity. (A dog can hear an intense sound of 20,000 Hz, even though most people would hear nothing at that frequency.) It is still possible, however, to make the general statement that the more intense a sound of a given frequency, the louder it will seem to any listener.

Decibel Scale

An individual does not perceive a sound B that has 10 times the intensity of A to be 10 times as loud as A. Rather, the human ear, like other sense organs, has a logarithmic response to changes in intensity.* By this we mean that if B has 10 times the intensity of A, and C has 100 times the intensity of A, then the ear perceives C to be as much louder than B as B is than A, because the ratios 100/10 and 10/1 are equal.

---

*This is an important psychological law, the *Weber-Fechner law,* named in honor of Ernst Weber (1795–1878), who discovered the law, and Gustav Fechner (1801–1887), who popularized its use.

When B has 10 times the intensity of A, the ratio of their intensities is 10, and the logarithm of 10 is 1. The difference in sound intensity between A and B is then said to be 1 bel, named in honor of Alexander Graham Bell (1847–1922), the inventor of the telephone. Similarly, if C is 10 times more intense than B, it is 1 bel more intense than B, and therefore 2 bels more intense than A. A sound 1000 times more intense than A has an intensity 3 bels above that of A, and so forth. This reduces multiplicative factors of 10 to additive factors of 1 and creates a *logarithmic scale*. This kind of scale is useful because it corresponds to the actual way the human ear responds to sound.

The bel is too large a unit to be convenient. Therefore we use another unit, one-tenth as large, called the *decibel* (dB). We define the *intensity level* of a sound, designated by the Greek letter β (*beta*) as follows:

$$\beta \text{ (in dB)} = 10 \log \frac{I}{I_0} \qquad \textbf{(12.15)}$$

The factor 10 arises simply because of our converting from bels to decibels. I is the intensity of the sound for which β is being calculated, and I_0 is some reference intensity, usually taken as the minimum audible intensity, or *threshold of hearing*, which is $I_0 = 1.0 \times 10^{-12} \text{ W/m}^2$.

For example, the intensity of the sound emitted by a jet plane just leaving the runway, which is about 100 W/m^2, has an intensity level

$$\beta = 10 \log \frac{I}{I_0} = 10 \log \frac{100 \text{ W/m}^2}{1.0 \times 10^{-12} \text{ W/m}^2} = 10 \log 10^{14} = 10(14) = 140 \text{ dB}$$

Every increase in intensity by a factor of 10 therefore corresponds to an increase of 10 dB. Since a jet engine has an intensity level of 140 dB, the intensity of the sound it emits is 10^{14} times the threshold of hearing.

Some intensities and intensity levels for commonly heard sounds are given in Table 12.2.

TABLE 12.2   Intensity and Intensity Level of Some Common Sounds

Sound	Intensity* (W/m^2)	Intensity level β (compared to $I_0 = 10^{-12} \text{ W/m}^2$) (dB)
Threshold of hearing	10^{-12}	0
Rustle of leaves	10^{-11}	10
Whisper	10^{-10}	20
Radio playing softly	10^{-8}	40
Ordinary conversation	10^{-6}	60
Busy traffic	10^{-5}	70
Jackhammer	10^{-2}	100
Motorcycle	10^{-1}	110
Drums of symphony orchestra at climax of *1812 Overture*	1	120
Loud indoor rock concert	1	120
Threshold of pain	10	130
Jet engine	10^2	140

*At a location very near the sound source.

Frequency Response of the Human Ear

Figure 12.16 shows how the threshold of hearing of the human ear varies with frequency. The ear is most sensitive to frequencies between about 2000 and 5000 Hz. Sound waves are often defined as compressional waves with frequencies between 20 and 20,000 Hz, but few, if any, ears can hear this full range of frequencies. As a matter of fact, for many people expensive stereo systems are a bad investment, since they cannot hear the very low and very high frequencies in the "full-frequency response" of the electronic components.

The *threshold of pain* for the human ear occurs at about 130 dB and varies very little with frequency, as is clear from Fig. 12.16. There is a slight dip in the curve at the frequencies where the ear is most sensitive, as might be expected. Clearly the 140-dB sound of a jet plane 30 m above the ground is, from Table 12.2, about 10 times greater than the threshold of pain for the ear, and continued exposure to such intense sounds can damage the ear. Protracted exposure to loud sounds below the threshold of pain may also degrade hearing. (Turn that stereo down!)

Figure 12.16 points up clearly the difference between intensity and loudness. A 40-dB sound would appear rather loud to a listener if its frequency were 3000 Hz, but would be barely audible at 80 Hz.

Example 12.6

An amplified steel guitar has an intensity level 5 dB higher than the same unamplified sound. How much more intense is the amplified sound than the unamplified sound?

SOLUTION

Intensity levels may be used to compare any two intensities, since

$$\frac{I_2/I_0}{I_1/I_0} = \frac{I_2}{I_1}$$

and so, from Eq. (12.15),

$$\beta = 10 \log \frac{I_2}{I_1} = 5 \text{ dB}$$

where I_2 is the intensity of the amplified sound and I_1 that of the unamplified sound. Therefore

$$\log \frac{I_2}{I_1} = \frac{5}{10} = 0.5$$

Taking antilogs, we have

$$\frac{I_2}{I_1} = \text{antilog } 0.5$$

and from a hand calculator we obtain

$$\frac{I_2}{I_1} = 3.2 \quad \text{or} \quad I_2 = 3.2 I_1$$

Hence the amplified sound is $\boxed{3.2 \text{ times}}$ more intense than the unamplified sound.

Example 12.7

At a frequency of about 1000 Hz a human ear can hear sounds very slightly above the *threshold of hearing*, which corresponds to an intensity I of 10^{-12} W/m² (see Fig. 12.16). What is the displacement amplitude corresponding to this intensity? (Assume that the sound is heard at about room temperature.)

FIGURE 12.16 Response of human ear to sound as a function of frequency. The upper curve indicates the intensity that produces pain in the ear; the lower curve the minimum audible intensity at various frequencies. Note that the frequency scale is logarithmic.

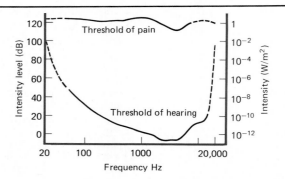

SOLUTION

We need an expression relating the amplitude of the sound wave to the known intensity of 10^{-12} W/m². This is given by Eq. (12.14),

$$I = 2\pi^2 f^2 X_{max}^2 mvn$$

where X_{max} is the displacement amplitude and v is the speed of sound. In this equation mn is the number of particles per unit volume times the mass of each particle, and thus $mn = \rho$, the density of the air. So we have

$$X_{max} = \left[\frac{I}{2\pi^2 f^2 \rho v} \right]^{1/2}$$

$$= \left[\frac{1.0 \times 10^{-12} \text{ W/m}^2}{2(9.87)(1000/\text{s})^2(1.29 \text{ kg/m}^3)(340 \text{ m/s})} \right]^{1/2}$$

$$= [1.15 \times 10^{-22} \text{ m}^2]^{1/2}$$

$$= \boxed{1.1 \times 10^{-11} \text{ m}}$$

Since an atom has a diameter of about 10^{-10} m, the maximum displacement is only about one-tenth the diameter of an atom, and still a sound with this displacement amplitude can be heard! This indicates the exquisite sensitivity of a human ear. (It is worth checking to make sure that the units in the answer actually correspond to a displacement.)

EXERCISE 2 What is the displacement amplitude corresponding to the threshold of pain of the human ear at 1000 Hz? Does this tell you how easy it is to damage the human ear by subjecting it to excessively loud sounds?

12.7 Quality of Sound and Pitch

Frequency is the fundamental physical property that indicates the number of oscillations of a string or air column per second. From a physiological point of view the corresponding attribute of sound is called the *pitch*.

Pitch: The perceived physiological response to the frequency of a sound wave.

Frequency and pitch are closely related but are not identical, since the loudness of a note of a particular frequency may change its perceived frequency (or its pitch) as heard by a listener. Thus a loud foghorn sounds lower in pitch than a sound of the same frequency but of less intensity.

Musicians make use of a variety of instruments with vibrating parts to produce sound. Many instruments like the violin, piano, cello, and guitar contain vibrating strings for this purpose. The length of the string, its linear density, and its tension can then be chosen to produce the desired fundamental note, as discussed previously. Instruments like flutes, French horns, trumpets, and pipe organs have vibrating air columns inside, and the players vary the length of these air columns and the method of excitation to select the desired note. How the particular fundamental is excited also determines the number and relative intensity of the higher harmonics generated.

Sounds produced by musical instruments are often referred to by subjective terms like *reedy, round, mellow.* The *quality* of sound, or *timbre,* is determined by the number of harmonics present, their relative intensities, and how quickly they die out. The timbre of an instrument is determined not merely by the size of the resonant cavity in the instrument but also by the shape of the instrument and the material from which it is made. Three sound ''spectra'' for a piano are shown in Fig. 12.17. Note how all the harmonics are integral multiples of one fundamental frequency. The relative intensities of the harmonics depend on the loudness of the note struck on the piano, and hence on the force applied to the fundamental string.

When the fundamental is very loud, the fourth harmonic is almost as loud as the fundamental. We can show by a simple experiment that harmonics are present in any note struck on the piano. If we release the dampers on the strings of the piano, and then strike one key, a number of other strings vibrate in addition to the

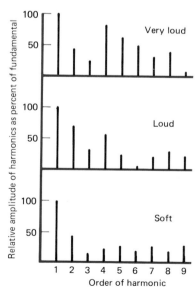

FIGURE 12.17 Sound spectra of a piano for a fundamental at 256 Hz sounded at different intensities: very loud, loud, and soft.

Hermann von Helmholtz (1821–1894)

FIGURE 12.18 Hermann von Helmholtz. This photograph was taken in his laboratory in Berlin on July 7, 1894, just a few months before his death. (*AIP Niels Bohr Library.*)

One of the most versatile scientists who ever lived, Hermann von Helmholtz was born in Potsdam, Germany, in 1821, the son of a high school teacher. From an early age he wanted to be a physicist but his family could not afford the money required for his studies. Instead, his father persuaded him to take up medicine, since he could receive a free medical education on condition that he serve as a doctor in the Prussian army after receiving his medical degree. This he did, studying at the Institute for Medicine and Surgery in Berlin from 1838 to 1842, and fulfilling his obligation as an army surgeon from 1843 to 1848. Helmholtz's real interest was always research; even in the army barracks he insisted on setting up a small laboratory for research in physiology and physics.

In 1847 Hemholtz published a mathematical formulation of the principle of conservation of energy that finally convinced scientists of the validity of this principle. On the basis of this and other research, he was appointed professor of physiology at the University of Königsberg. There he devoted himself to the physiology of the eye, first explaining the mechanism of lens accommodation in the eye. Then in 1851 he invented the ophthalmoscope (see Sec. 25.8). This instrument, still the basic tool used by eye doctors to peer into the eye's interior, immediately made Helmholtz famous. In 1852 he also became the first experimentalist to measure the speed of nerve impulses in the human body. He was particularly happy about this, because one of his former professors, an eminent biologist, had said that it could not be done.

Helmholtz also studied a second important sense organ, the ear. He advanced the theory that the ear detected differences in pitch through the action of the cochlea, the spiral organ in the inner ear; he was the first to point out that the quality of a sound depends on the nature, number, and relative intensities of the harmonics present; and he applied the principles of physics to the understanding of music.

In 1858 he was appointed to a chair in anatomy at Heidelberg, and while there devoted most of his time to studies in optics and sound. In 1867 he published his famous *Treatise on Physiological Optics* in three volumes. His days in Heidelberg, from 1858 to 1871, enabled him to work with other great scientists like the physicist Kirchhoff (see Sec. 18.5) and the chemist Bunsen in what has been called an era of brilliance "such as has seldom existed for any university and will not readily be seen again."

In 1871 Helmholtz abandoned physiology in favor of physics. He accepted the chair of Physics at the University of Berlin and spent the rest of his life there.

As he grew older, he became more and more interested in the mathematical side of physics and made noteworthy contributions to fluid mechanics, thermodynamics, and electrodynamics. He devoted the last decade of his life to an attempt to unify all of physics under one fundamental principle, the principle of least action. This attempt, while evidence of Helmholtz's lifelong interest in philosophy, was no more successful than Einstein's quest for a unified field theory. Helmholtz died in 1894 as the result of a fall suffered on board ship while on his way back to Germany from the United States.

It is difficult to exaggerate the influence Helmholtz had on nineteenth-century science, not merely in Germany but throughout the world. It was during his lifetime that Germany gained its preeminence in science, which it was not to lose until the Second World War. His own research contributions, together with the impetus he gave to other scientists and would-be scientists by his teaching, research administration, and popular lectures, had much to do with the scientific renaissance that Germany experienced during his lifetime. At that time Helmholtz was, next to Bismarck and the old emperor, the most illustrious man in the German empire.

Helmholtz was a sensitive and sickly man all his life, plagued by severe migraine headaches and fainting spells. He sought relief from his pain in music and the arts and in mountain climbing in the Alps. It is intriguing to imagine what he might have accomplished if he had been in good health for all his 73 years! On Helmholtz's death Lord Kelvin summed up his accomplishments: "In the historical record of science the name of Helmholtz stands unique in grandeur, as a master and leader in mathematics, and in biology, and in physics."

one struck. These are strings corresponding to harmonics of the fundamental note. This note must therefore contain these frequencies to an extent sufficient to excite the other strings by resonance.

Analysis of Musical Sounds

The Greeks were the first to understand something of the nature of music. After the Greeks the greatest progress was made by Hermann von Helmholtz (see Fig. 12.18 and accompanying biography), who devoted much of his career to a study of sound and light, particularly to their physiological effects. He designed a set of spherical cavities, called *Helmholtz resonators,* which resonated at certain fixed frequencies determined by their dimensions. By playing a complex musical sound at a frequency whose harmonics correspond to the resonant frequencies of these cavities, he was able to set them into oscillation and so determine the harmonic content of the musical note. Today this can be done using oscilloscopes and electronic filters to exclude all harmonics but the one of interest. That harmonic can then be observed on an oscilloscope screen (similar to a TV screen) and its amplitude measured. Since the intensity of the harmonic is proportional to the square of its amplitude, the percentage of the sound made up by this particular harmonic can be determined. In this way sound spectra like those in Fig. 12.17 can be plotted.

Helmholtz also succeeded in synthesizing musical sounds by combining harmonics in different proportions. He set up electrically driven tuning forks of the proper frequency and amplitude to duplicate any desired complex sound. Modern electronic music synthesizers are improved versions of Helmholtz's device.

12.8 The Doppler Effect in Sound

We have all probably observed that if a fire engine is approaching our street, the frequency of its siren appears higher than when the same engine is leaving our neighborhood. This illustrates how the frequency of a sound changes depending on the relative motion of the sound source and the hearer. This effect has been named the *Doppler effect*, after Christian Doppler (1803–1853), the Austrian physicist who first explained it in 1842.

Source at Rest; Observer in Motion

Suppose a vibrating tuning fork at a fixed point in space emits a frequency f. This sound moves through the air in all directions with a speed v determined by the properties of the air, and a wavelength determined by the fundamental equation $v = \lambda f$. If, now, an observer moves toward the tuning fork at a speed v_0, as in Fig. 12.19, he or she hears a higher frequency. This may be explained as follows.

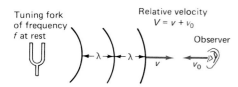

FIGURE 12.19 The Doppler effect: observer in motion toward a stationary sound source. The frequency of the observed sound is increased.

We assume that the air is at rest with respect to the tuning fork. The number of complete wavelengths passing the observer per second if he or she were at rest would be $f = v/\lambda$. If the observer is in motion at a speed v_0 toward the source of sound, the observer's relative speed with respect to the sound wave becomes $V = v + v_0$. The wavelength is not affected by the observer's motion, but the frequency heard is. The apparent frequency observed is

$$f' = \frac{V}{\lambda} = \frac{v + v_0}{\lambda} = f + \frac{v_0}{\lambda} = f\left(1 + \frac{v_0}{f\lambda}\right) = f\left(1 + \frac{v_0}{v}\right)$$

Hence the frequency heard by the observer is not f but a higher frequency

$$f' = f\left(1 + \frac{v_0}{v}\right) \tag{12.16}$$

that depends on the ratio of the observer's speed (v_0) to the speed of sound (v). Since the velocity of sound in air at room temperature is about 340 m/s, the observer would have to travel at 3.4 m/s to increase the frequency heard by 1 percent.

If the observer is moving away from the sound source,

$$f' = \frac{v - v_0}{\lambda} = f\left(1 - \frac{v_0}{f\lambda}\right) = f\left(1 - \frac{v_0}{v}\right) \tag{12.17}$$

In this case the frequency heard would be lower than the tuning fork frequency f.

In the case of a moving observer, then, the frequency heard shifts because the number of complete wavelengths reaching the observer's ear per second changes.

Source in Motion; Observer at Rest

This is a case very different from the one just discussed. Here the motion of the source with respect to the air changes the *wavelength* of the sound wave in the air. This sound wave still travels through the air at a speed v determined by the physical properties of the air. When it reaches the observer, however, a different frequency is heard because the wavelength has been changed.

Consider, as in Fig. 12.20, a tuning fork initially at S_1 emitting sound at a frequency f and moving through the air at a velocity v_s toward an observer who is at rest. We consider the *wave fronts* marked 1, 2, and 3. These are surfaces on which the phase is the same at all points on the surface, and these waves are produced at points S_1, S_2, and S_3. The wavefronts selected correspond to a one-

FIGURE 12.20 The Doppler effect: Source in motion toward observer. Wave front 1 is emitted when the source is at S_1, wave front 2 when the source is at S_2, and so forth. This leads to a decrease of the wavelength in the forward direction and an increase in the wavelength in the backward direction. The result is an increase in the frequency heard by the observer when the sound source is approaching.

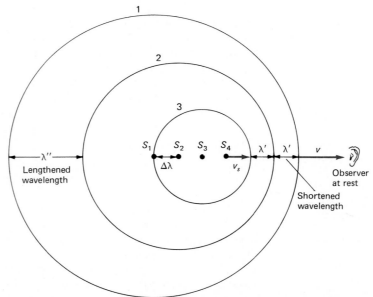
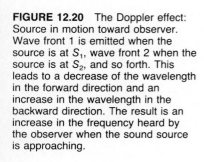

wavelength difference between two consecutive wave fronts. In this case the motion of the source pushes together the waves it emits in the direction in which it is moving, as shown in the figure. In the periodic time $T = 1/f$, the source moves a distance $\Delta\lambda = v_s/f$ and the sound wave travels a distance $\lambda = v/f$. Hence the actual wavelength produced in the air by the moving source is no longer λ but

$$\lambda' = \lambda - \Delta\lambda = \frac{v}{f} - \frac{v_s}{f} = \frac{v - v_s}{f}$$

This sound wave then reaches the observer at its original velocity v with respect to the air, since the observer is at rest with respect to the air. The frequency heard is therefore

$$f' = \frac{v}{\lambda'} = \frac{v}{(v - v_s)/f} = \frac{1}{1 - v_s/v} f \qquad (12.18)$$

The frequency that is heard is increased in this case by the motion of the source toward the observer. If the source moves away from the observer, the wavelength is increased as shown by λ'' in the figure, and we can show in the same way that the new (lower) frequency is

$$f' = \frac{1}{1 + v_s/v} f \qquad (12.19)$$

The results of Eqs. (12.16) to (12.19) can be combined into the following single equation for the *Doppler effect*:

$$\boxed{f' = \frac{1 \pm v_0/v}{1 \mp v_s/v} f} \qquad (12.20)$$

where the upper signs in numerator and denominator apply to motion of source and observer toward each other and the lower signs apply to motion away from each other.

Notice that, even though relative motion of source and observer toward each other always raises the frequency, and relative motion away from each other always lowers the frequency, the actual change differs depending on whether the source or the observer does the moving. This is because $1 + v_0/v$ is not exactly equal to $1/(1 - v_s/v)$, even when $v_0 = v_s$.

The Doppler effect also occurs with light (and other electromagnetic) waves, although there are differences because light, unlike sound, requires no medium for its propagation. It remains true, however, that the frequency of light is reduced if the source and observer are moving apart, and increased if they are coming together. If light source and observer are moving at a relative speed v_r, the formula for the optical Doppler shift in frequency is similar to Eqs. (12.16) and (12.17) above. We have $f' = f(1 \pm v_r/c)$, where c is the speed of light. Then

$$\Delta f = f' - f = \pm \frac{v_r}{c} f$$

and so

$$\frac{\Delta f}{f} = \pm \frac{v_r}{c} \qquad (12.21)$$

Since c is 3×10^8 m/s, the relative shift in frequency ($\Delta f/f$) is very small unless we are dealing with stars or other objects whose speeds approach that of light.

The Doppler effect for light waves explains why the frequency of visible light from the stars is shifted to lower frequencies (toward the *red*), indicating that the stars are moving away from the earth. This "red shift" is the main evidence we have for the commonly accepted theory of the expanding universe. "Doppler radar" techniques are also used for the more mundane tasks of catching speeders on the nation's highways and for determining the speed of a baseball pitcher's fastball.

Example 12.8

A physics teacher is demonstrating the Doppler effect in class by moving a tuning fork of frequency 440 Hz very rapidly toward a blackboard at the front of the classroom, thus producing beats between the sound heard directly by the class from the moving tuning fork and that reflected from the blackboard. If the students are to hear a beat frequency of 2 Hz, how rapidly must the teacher move the tuning fork?

SOLUTION

The speed with which the tuning fork moves must be such as to produce one beat with a stationary source at the blackboard. Motion of the tuning fork at this speed away from the class lowers the direct frequency heard by the students from 440 to 439 Hz. Motion of the tuning fork toward the blackboard at this same speed increases the frequency reflected by the blackboard from 440 to 441 Hz, since the frequency reflected is the same as that which would be heard by a stationary listener at the blackboard. The students then hear frequencies of 439 and 441 Hz and a beat note between them of $441 - 439 = 2$ Hz.

The speed that will lower the frequency by 1 Hz is then, since the source is moving,

$$f' = \frac{f}{1 + v_s/v} \quad \text{or} \quad \frac{v_s}{v} f' = f - f' = 1 \text{ Hz}$$

and so

$$\frac{v_s}{v} = \frac{1}{f'} = \frac{1}{439}$$

and

$$v_s = \frac{1}{439} (340 \text{ m/s}) = \boxed{0.775 \text{ m/s}}$$

Motion of the tuning fork at this speed leads to a beat note of 2 Hz between the direct and reflected sound waves.

12.9 Supersonic Speeds and Ultrasonic Waves

Although the terms *supersonic* and *ultrasonic* may appear to have similar meanings, technically they have quite different significances. *Supersonic* is used for *speeds* greater than the speed of sound; *ultrasonic* refers to *frequencies* of sound waves higher than those of audible sound, i.e., above 20,000 Hz.

Supersonic Speeds

The elastic and inertial properties of the air molecules determine the speed of sound waves in air. The faster the air molecules rebound after being compressed, the faster the sound wave is propagated. The speed with which air molecules can rebound from a flying aircraft also determines whether they can get out of the way of the plane. As the plane approaches the speed of sound, it is increasingly more difficult for the air molecules to get out of its way, since the plane is going almost as fast as the speed at which the molecules can rebound. At the speed of sound, the plane piles up all the air molecules in front of it into one large compressional crest (since they cannot move fast enough to escape). This creates what was once called a "sound barrier," since it was believed that any airplane would be destroyed by turbulence if it tried to fly through the layer of molecules it had piled up in front of it.

On October 24, 1947, a plane piloted by Charles ("Chuck") Yeager broke through the sound barrier by traveling at a speed greater than the speed of sound in the surrounding air. This is referred to as "exceeding Mach 1," where Mach 1 means flight at the speed of sound, i.e., about 340 m/s near the earth's surface. Similarly, Mach 2 means twice the speed of sound, or about 680 m/s. For twenty years the SR-71 Blackbird spy plane flew routinely at Mach 3.2, until it was retired in 1990. Speeds greater than five times the speed of sound are referred to as *hypersonic*.

Mach number: The ratio of the speed of a moving airplane or other object (v_P) to the speed of sound (v) at the flight altitude.

$$\text{Mach number} = \frac{v_P}{v}$$ (12.22)

When a plane is flying at supersonic speeds (i.e., greater than Mach 1), it compresses the air ahead of it and sets up a sharp dividing line between this region of strong compression and the surrounding air. This compression region extends in a cone behind the plane, as shown in Fig. 12.21, and is called a shock wave.

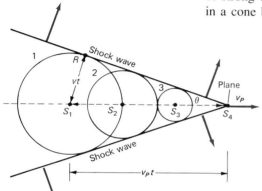

FIGURE 12.21 Shock wave generated by a plane in supersonic flight at approximately Mach 3. The circles are the sound wave fronts emitted by the plane at S_1, S_2, and S_3. In the time it takes the plane to move from S_1 to S_4 at a speed v_P, the sound wave moves from S_1 to R at a speed v.

The angle of the shock wave cone depends on the Mach number. From Fig. 12.21 it can be seen that the cone angle θ is an angle in the right triangle $S_1 S_4 R$ and has a sine given by

$$\sin \theta = \frac{vt}{v_P t} = \frac{v}{v_P} = \frac{1}{\text{Mach number}}$$ (12.23)

Hence the higher the Mach number, the smaller the angle of the cone produced by the shock wave.

When a plane passes overhead traveling at supersonic speeds, the piled-up compressions of the air spread out from the shock wave produced by the plane. The shock wave gradually changes into an ordinary sound wave as it moves away from the plane. When this wave reaches the ground, as in Fig. 12.22, an observer hears a *sonic boom*. This boom is heard after the plane has passed overhead, since the plane is traveling more rapidly than are the sound waves reaching the ears of the observer. A sonic boom is therefore due to the constructive interference of a large number of compressional pulses at the ground that can shatter windows and create noise levels harmful to human beings and animals. Much of the destructive power of nuclear or conventional explosives comes from the shock waves they generate. Figure 12.23 shows a shock wave produced by a rifle bullet moving through air at supersonic speed.

Sonic booms associated with the SST, the supersonic transport that Air France and British Airways fly across the Atlantic to the United States, have led to the banning of supersonic flights by SSTs over landmasses in the United States. They now fly at supersonic speeds only over oceans.

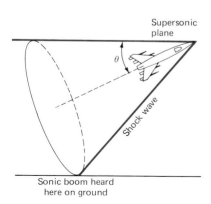

FIGURE 12.22 Sonic boom connected with a shock wave.

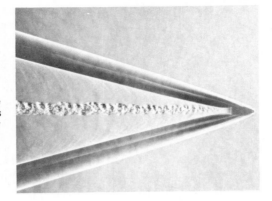

FIGURE 12.23 A shock wave produced by the bullet fired from a hunting rifle. The bullet has a Mach number of 3.4 and a Reynolds number of 10^6. An aircraft flying at the same speed at high altitude generates a similar shock wave. (*Photo courtesy of Professor Peter P. Wegener, Yale University.*)

Ultrasonic Frequencies

As noted before, the frequency range of the human ear is roughly from 20 to 20,000 Hz. Compressional waves of frequencies below and above this range can be produced and detected by a variety of electronic techniques. "Sound" waves with frequencies below 20 Hz are called *infrasonic* waves (i.e., below the audible range). They are produced by earthquakes, thunderstorms, and very heavy vibrating machinery. Infrasonic waves are also produced by the muscles and fluids of the human body, so the fact that we do not perceive these "sounds" is evidence of sensible planning in the design of humans. Sound waves with frequencies above 20,000 Hz are called *ultrasonic* waves (i.e., above the audible range), and are of great importance from both a theoretical and a practical point of view.

Ultrasonic waves can be generated by using quartz crystals as sources. Such crystals are *piezoelectric*, which means that if an alternating electric voltage is applied to them, they vibrate mechanically and generate sound waves at the same frequency as the applied voltage. In this way sound waves of very short wavelength can be generated. For example, at a frequency of 100 kHz, the wavelength is only about 3 mm for sound waves in air, and at 1 GHz, it is only 3×10^{-4} mm. Ultrasonic waves have been produced at frequencies as high as 600 GHz, at which frequency the wavelength is only 5×10^{-10} m, or approximately the size of an atom.

Normal audible sound waves have such long wavelengths that they simply bend around small objects without much reflection occurring, in the same way

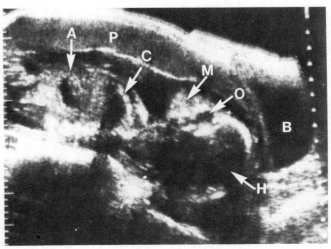

FIGURE 12.24 Ultrasound photograph of a child in the mother's womb. The letters on the photo have the following significance: H = head; O = eye socket; M = mouth; C = chest; A = abdomen; P = placenta; B = maternal bladder. (*Courtesy of Steven A. Mervis, Division of Ultrasound and Radiologic Imaging, Thomas Jefferson University Hospital, Philadelphia, Pennsylvania.*)

water waves flow around the support pillars of a pier. If the wavelength is short enough, however, sound waves behave more like light waves and are reflected from objects larger than their wavelengths. For this reason, ultrasonic waves can be used to find flaws or imperfections in metals, to locate tumors in the human body, and to detect intruders in homes or suspicious articles in baggage. These are examples of nondestructive testing using ultrasound. In the form of what is now called *sonar* (sound *navigation and ranging*), ultrasonic waves can also be used underwater to locate submarines or schools of fish.

Because it is less dangerous to living tissue, ultrasound has gradually replaced x-rays for many medical applications. Ultrasound can even be used to look at the development of a fetus without risk to mother or fetus (see Fig. 12.24).

Example 12.9

An earth satellite must move at a speed of about 18,000 mi/h to stay in a stationary orbit 22,400 mi above the earth. If a jet plane could achieve the same speed (It can't!) at a height in the earth's atmosphere where the speed of sound is 300 m/s,

(a) What would be the Mach number at which the plane would be flying? **(b)** What would be the angle of the shock wave produced by such a plane?

SOLUTION

(a) $18,000 \text{ mi/h} = (18,000 \text{ mi/h}) \left(0.447 \dfrac{\text{m/s}}{\text{mi/h}} \right)$

$= 8.0 \times 10^3 \text{ m/s}$

Then the Mach number of the plane's flight is

$\text{Mach number} = \dfrac{8.0 \times 10^3 \text{ m/s}}{3.0 \times 10^2 \text{ m/s}} = \boxed{27}$

(b) $\sin \theta = \dfrac{1}{\text{Mach number}} = \dfrac{1}{27} = 0.037$

$\theta = \boxed{2.1°}$

Summary: Important Definitions and Equations

Sound wave: A longitudinal wave, transmitted by the air or some other medium, in the form of alternate compressions and rarefactions of the medium and in the frequency range 20 to 20,000 Hz.

Equation for the displacement of particles in a sound wave moving along the positive x axis:

$$X = X_{max} \cos \left[2\pi \left(\frac{x}{\lambda} - \frac{t}{T} \right) \right] \qquad (12.1)$$

For sound waves moving in the negative x direction:

$$X = X_{max} \cos \left[2\pi \left(\frac{x}{\lambda} + \frac{t}{T} \right) \right] \qquad (12.2)$$

Speed of sound in gases or liquids: $v = \sqrt{\dfrac{B}{\rho_0}}$ $\qquad (12.3)$

Speed of sound in solids: $v = \sqrt{\dfrac{Y}{\rho_0}}$ $\qquad (12.4)$

Standing sound waves: $X = 2X_{max} \cos \dfrac{2\pi x}{\lambda} \cos \dfrac{2\pi t}{T}$

$\qquad (12.7)$

Reflection of sound waves:
1 At closed end of a pipe: Phase change of π rad, or 180°, in the displacement.
2 At open end of a pipe: No phase change in the displacement.

Resonant frequencies of organ pipes:
1 Open: $f_1 = \dfrac{v}{2L}$; $f_2 = 2f_1$; $f_3 = 3f_1$; ... (All harmonics present)

2 Closed: $f_1 = \dfrac{v}{4L}$; $f_3 = 3f_1$; $f_5 = 5f_1$; ... (No even harmonics)

Interference of two waves of different frequencies traveling in the same direction:

$$X = \left[2X_{max} \cos \frac{2\pi(f_1 - f_2)t}{2} \right] \cos \frac{2\pi(f_1 + f_2)t}{2}$$

$\qquad (12.10)$

Beats: The regular increase and decrease in sound intensity caused by two waves of slightly different frequencies being superimposed on the same medium. The beat frequency is the difference in the frequencies of the sound sources that produce the two waves.

Intensity of a sound wave: The amount of energy transported per second per unit area normal to the direction of the wave:

$$I = \frac{\mathcal{W}}{tA} \propto X_{max}^2 f^2 \qquad \textbf{(12.11; 12.14)}$$

Loudness: The perceived strength of a sound.

Decibel (db): A unit on a logarithmic scale used to measure sound intensities. Every increase in intensity by a factor of 10 corresponds to an increase of 10 dB (or 1 bel).

Intensity level of a sound: β (in dB) $= 10 \log \dfrac{I}{I_0}$ **(12.15)**

Threshold of hearing: $I_0 = 1.0 \times 10^{-12}$ W/m^2 (weakest sound that can be heard by human ear).

Pitch: The perceived physiological response to the frequency of a sound wave.

Physical and physiological properties of sounds

Physical property	Corresponding physiological property
Intensity	Loudness
Frequency	Pitch
Harmonic content	Quality, timbre

Doppler effect for sound

$$f' = \frac{1 \pm v_0/v}{1 \mp v_s/v} f \qquad \textbf{(12.20)}$$

Supersonic speeds: Velocities greater than the speed of sound in the medium.

$$\text{Mach number} = \frac{v_P}{v} \qquad \textbf{(12.22)}$$

Shock wave: An intense compressional pulse produced, for example, by planes flying at supersonic speeds.

Angle of shock wave cone:

$$\sin \theta = \frac{v}{v_P} = \frac{1}{\text{Mach number}} \qquad \textbf{(12.23)}$$

Ultrasonic waves: Sound waves with frequencies above 20,000 Hz.

Questions

1 Why are transverse mechanical waves unable to pass through gases and liquids?

2 Can you suggest a way that waves could be used to test whether the core of the earth is a liquid or a solid?

3 Can you suggest how a sound wave could be used as a crude vacuum gauge to determine the gas pressure within a closed tube?

4 Why are steel wires rather than copper wires used for stringed musical instruments?

5 In a traveling sound wave, (*a*) are there any particles that are always at rest as the wave passes through? (*b*) Are there any particles that are sometimes at rest? Indicate which particles, and for how long they remain at rest.

6 (*a*) In a standing sound wave, are there any particles that are always at rest?

(*b*) Where are these particles located?

7 For a standing sound wave:

(*a*) On the basis of Eq. (12.7) describe in words the motion of a particle at the fixed position $x = x_1$ as a function of time.

(*b*) In the same way, describe, at a fixed instant of time $t = t_1$, the displacements of the particles as x varies along the length of the tube.

8 Why does your voice sound better when you sing in the shower?

9 Explain why doubling the intensity of a 70-dB sound does not raise it to 140 dB.

10 In the text, we have found the resonant frequencies for a tube open at both ends and for a tube closed at one end. What would be the nodal conditions for a tube closed at both ends, and what resonant frequencies would you expect?

11 Perhaps you have noticed that some people have a harder time than others in carrying an ice-cube tray filled almost to the top with water without spilling the water. If we presume that all the people involved exercise equal care to avoid spilling the water, what might be a reasonable physical explanation of the different results?

12 Would it be correct to say that the intensity of a sound wave is proportional to the square of the amplitude of the pressure variations in the wave? Why?

Multiple-Choice and Simple Exercises

(Unless otherwise stated, use 340 m/s for the speed of sound in air in all problems.)

12.1 The speed of sound in metals exceeds its speed in air by about:

(*a*) One order of magnitude
(*b*) Two orders of magnitude
(*c*) Three orders of magnitude
(*d*) Four orders of magnitude
(*e*) They are about equal

12.2 Longitudinal waves cannot be propagated through:
(a) Water vapor   (b) Liquids   (c) Solids
(d) Gases   (e) A vacuum

12.3 If the air in an organ pipe closed at one end is in resonance with a tuning fork so that the air column (including the two ends) contains two nodes and two antinodes, the length of the pipe is approximately:
(a) $\lambda/2$ (b) $\lambda/4$ (c) $3\lambda/4$ (d) λ
(e) 2λ

12.4 A standing sound wave is set up in a closed organ pipe at a frequency of 3400 Hz. At the position $x_1 = 0.050$ m there is a displacement node. A pressure node occurs at:
(a) 0.10 m   (b) 0.075 m   (c) 0.20 m
(d) 0.15 m   (e) 0.25 m

12.5 If the speed of sound in air is 340 m/s, the shortest air column closed at one end that will respond to a tuning fork of frequency 512 Hz has a length of approximately:
(a) 4.2 cm   (b) 0.40 cm   (c) 17 cm
(d) 33 cm   (e) 66 cm

12.6 A firecracker produces a loud sound when it explodes. At a distance x from the explosion, the sound intensity is I_1. The sound intensity at a distance $10x$ from the firecracker is:
(a) $I_1/10$ (b) $I_1/3.2$ (c) $I_1/100$ (d) I_1
(e) $10^{-4}I_1$

12.7 What is the ratio of the intensities of a 50-dB sound and a 10-dB sound?
(a) 40   (b) 10^5 (c) 5   (d) 10^4 (e) 4

12.8 We can distinguish between a note played on a violin and the same note sung by a tenor because the two notes differ in:
(a) Pitch   (b) Intensity   (c) Phase
(d) Number and loudness of harmonics
(e) Their fundamental frequencies

12.9 A fire engine passes an intersection at 20 m/s, while sounding its siren at 5.0×10^3 Hz. After the fire engine has passed, a person in a car stopped at the intersection hears a frequency of:
(a) 5.0×10^3 Hz   (b) 4.7×10^3 Hz
(c) 5.3×10^3 Hz   (d) 2.9×10^2 Hz
(e) None of the above

12.10 A plane is flying at Mach 2.5 at a height where the speed of sound is 300 m/s. The speed of the plane is:
(a) 120 m/s   (b) 300 m/s   (c) 250 m/s
(d) 750 m/s   (e) None of the above

12.11 A sound wave of displacement amplitude 1.0×10^{-12} m in air has a wavelength of 6.8 cm and a frequency of 5000 Hz.
(a) What is its speed?
(b) What is the maximum acceleration experienced by a particle in the air through which the wave moves?

12.12 The density of silver is 10.5×10^3 kg/m³, and Young's modulus for silver is 7.75×10^{10} N/m². What is the speed of sound in silver?

12.13 What is the frequency of the third harmonic of an organ pipe closed at one end and 1.40 m long.

12.14 A nickel rod 1.6 m long is mounted at a point 0.40 m from one end. It is then set into longitudinal vibration by being rubbed with resin. At what frequency will the rod vibrate?

12.15 Two violins are badly out of tune. When a frequency of 440 Hz is played on one, a beat frequency of 6 Hz is heard between the two violins. What are the possible frequencies of the second violin?

12.16 The sound emitted by a small radio carries 1.0×10^{-12} J of energy per second through an area of 1.0 cm² at right angles to the direction of the sound wave. What is the intensity of the sound?

12.17 What is the intensity of a sound that is 80 dB above the threshold of hearing?

12.18 A ship is approaching a dock at 7.8 m/s. The ship sounds its warning whistle at 10.0×10^3 Hz. What is the frequency heard by a person on the dock?

12.19 A plane is flying at supersonic speed at a height where the speed of sound is 310 m/s. The shock wave generated makes a cone angle of 45°. What is the Mach number corresponding to the plane's speed?

12.20 A plane is flying at Mach 0.85. Its engines produce a whining sound at 3000 Hz. What frequency is heard by an observer in a lookout tower as the plane approaches?

Problems

12.21 The string corresponding to middle C on a piano vibrates at 261.6 Hz when excited in its fundamental mode and has an amplitude of 1.5 mm. What are the maximum velocity and maximum acceleration of the midpoint of the string?

12.22 (a) A sound wave in water has a speed of 1.40×10^3 m/s. What must the frequency of this sound wave be for it to have a wavelength of 30.0 cm in water?
(b) Write an equation describing the longitudinal motion of a particle of the water as a function of time and distance, as the sound wave passes through (assume that the wave is moving in the positive x direction).

12.23 The speed of sound in tin is half the speed of sound in aluminum, and the density of tin is 7.2×10^3 kg/m³. Use these data and Table 12.1 to find Young's modulus for tin.

12.24 At the beginning of a 100-m dash the starting signal is a gunshot fired by a starter at one end of the line of runners. There are eight lanes, each 1.22 m wide.
(a) What is the time delay between the starting signal heard by the sprinter in lane 1 and by the sprinter in lane 8?
(b) Could this have any effect on the outcome of the race?

12.25 Use the data in Table 12.1 to determine the percentage decrease in the volume of the water contained in a cylinder and compressed by a piston, when the pressure on the water is increased by 2.0×10^5 N/m².

*12.26 Use dimensional analysis to show that the dependence of the speed of sound on the density and bulk modulus of a fluid must take the form $v \propto (B/\rho_0)^{1/2}$, where ρ_0 is the density of the undisturbed fluid and B is the bulk modulus $[B = -P/(\Delta V/V_0)]$.

∗12.27   What stress is required in a stretched nickel wire if the speed of transverse waves in it is to be 1/50 the speed of longitudinal waves in the same wire? (*Hint:* First solve algebraically to cancel out unknown quantities.)

12.28   A child drops a rock down a well and hears the sound of the rock hitting water 3.20 s after the rock is dropped. How deep is the well? (Ignore air resistance.)

∗12.29   Use the assumption that the speed of sound in a gas is proportional to the rms speed of the gas molecules to prove that the speed of sound v_1 at a temperature T_1 is related to the speed of sound v_2 at a temperature T_2 by

$$\frac{v_1}{v_2} = \left(\frac{T_1}{T_2}\right)^{1/2}$$

12.30   (*a*)   An organ pipe of length 1.60 m open at both ends is emitting a sound that includes a frequency of 425 Hz. To what harmonic does this frequency correspond?

(*b*)   If the air in the organ pipe is replaced with carbon dioxide, what will the frequency of this same harmonic be?

12.31   The lowest frequency that can be heard by the human ear is about 20 Hz. What is the smallest size room in which standing sound waves at this frequency can be set up?

12.32   A glass tube is filled with water whose height can be adjusted, as in Fig. 12.9. The tube above the water is kept filled with helium gas. A tuning fork, of fixed frequency 1760 Hz, is set into vibration above the open top of the tube. The sound intensity is large when the water level is 13.7, 41.5, and 69.2 cm from the open end of the tube. Find the speed of sound in helium.

12.33   A copper rod of length 1.6 m is clamped at a position 0.40 m from one end.

(*a*)   At what frequency will such a rod vibrate?

(*b*)   If this rod is used to agitate the air in a Kundt's tube like that in Fig. 12.11, how far apart will the mounds of talcum powder be?

12.34   A Kundt's tube (Fig. 12.11) is filled with air and set into vibration by a diaphragm at one end. The small mounds of powder are 15 cm apart on the average. The air is then replaced by an unknown gas, and the distance between the mounds averages 22 cm.

(*a*)   What is the speed of sound in the unknown gas?

(*b*)   Is this gas more or less dense than air?

12.35   Many of us enjoy singing in the shower because sound resonates between the walls of the shower stall. If a particular shower stall has dimensions 0.80 by 1.0 m, what are the three lowest-frequency sound waves that will produce standing waves in it?

12.36   The three lowest frequencies emitted by an organ pipe are 75.0, 225, and 375 Hz.

(*a*)   Is the pipe open or closed?

(*b*)   What are the wavelengths corresponding to these three frequencies?

(*c*)   What is the length of the organ pipe?

12.37   The fourth harmonic of an organ pipe of fundamental frequency 440 Hz is excited. This fourth harmonic is then compared with an unknown frequency produced by a tuning fork. Three beats are heard per second when the two are sounded at the same time.

(*a*)   What are possible frequencies of the tuning fork?

(*b*)   If the frequency of the tuning fork is lowered by sticking a bit of wax on the prongs, the beat frequency increases. What is the actual frequency of the tuning fork?

12.38   Two pure notes are sounded, one at 440 Hz and the other at 446 Hz.

(*a*)   What is the frequency of the resultant note that is heard?

(*b*)   What beat frequency is heard?

12.39   The frequency of a rotating mirror is measured by comparing the whistling frequency it produces when rotating with an audio-frequency oscillator generating a fixed frequency of 100 Hz. It is found that 5 beats are produced between these two sources, and that if the frequency of the rotating mirror is slightly increased, the number of beats is reduced. What is the frequency of the rotating mirror?

12.40   Two identical sound sources A and B of frequency 440 Hz and equal amplitude are located side-by-side 3.0 m from a microphone M. The two sources are turned on at the same time and are in phase.

(*a*)   Describe what happens as B is moved toward M. In particular, at what distances from M is the sound most intense, and at what distances from M is the sound least intense?

(*b*)   Is there any place on the straight line connecting the speakers and the microphone where there will be complete cancellation of the sounds received by the microphone?

12.41 B and C are two identical radio speakers, placed side-by-side at a distance of 1.5 m from a microphone at M. When B alone is emitting sound, the amplitude of the sound waves at M is X_{max}.

(*a*)   If B and C each emit the same volume of sound at the same frequency and in phase, what is the resulting amplitude of the sound produced by the two speakers together at M?

(*b*)   How does the sound intensity at M when both B and C are turned on compare with that when only B is emitting sound?

(*c*)   What is the increase in the intensity level in dB when speaker C is turned on?

∗12.42   Two small loudspeakers A and B face each other at the end of a 4.00-m laboratory bench. A small omnidirectional microphone is located at C, a point 1.00 m from A on the line joining A and B. A variable-frequency audio oscillator feeds the two loudspeakers, and they produce sounds of the same frequency, amplitude, and phase. At what frequencies will minimum sound be heard at C?

12.43   A noise-level meter indicates that the sound intensity level in rush-hour traffic is 75 dB. What is the corresponding sound intensity?

12.44   (*a*)   The sound of a crying baby reaches an intensity level of 10^{-3} W/m². How much energy passes in 1 min through an area 1 m on a side near the baby?

(*b*)   How high could this amount of energy lift a 1.0-g mass?

12.45   A woman is adjusting the balance of the speakers in her stereo set and decides that she must double the power output of the sound from one speaker. By how many decibels must she increase the intensity level of that speaker?

12.46 One speaker of a stereo set is putting out sound with an intensity of 5.0×10^{-8} W/m^2. The second speaker is putting out sound of intensity 8.0×10^{-9} W/m^2. By how many decibels does the first sound exceed the second in intensity?

12.47 The intensity level 40 m from a loudspeaker is 90 dB. What is the power output in watts, if we assume that the energy spreads uniformly in all directions?

*12.48 There is one intensity level for sound to which the following statement applies: "If the intensity level in decibels is quadrupled, the intensity of the sound is also quadrupled." What is this intensity level?

12.49 Show that the observed frequency for a tuning fork sounding a note of 500 Hz is different in cases (a) and (b).

(a) The tuning fork is moving away from a stationary observer at a speed of 34.0 m/s.

(b) The observer is moving away from a stationary tuning fork at a speed of 34.0 m/s.

(c) What would happen if the speeds were 340 m/s in parts (a) and (b)?

12.50 A car is traveling at 30 m/s directly toward a building. The driver sounds his horn, which emits a short burst of sound at 500 Hz. A few seconds later, the echo of the sound returns to the driver after being reflected from the building. At this time the car is still moving at 30 m/s toward the building. What frequency is heard by the driver?

12.51 Jennifer and Gregory each have a tuning fork that emits a pure tone at 440 Hz. If Gregory stands still and Jennifer runs toward him while holding her tuning fork, how fast must she run if he is to hear 2 beats per second between the two tuning forks?

12.52 A firefighter is riding on the back of a fire engine moving at 25 m/s. The truck's siren, which is at the front, is sounding a note of 3000 Hz.

(a) What is the wavelength of the sound wave in the air near the middle of the fire engine?

(b) What frequency does the firefighter hear?

12.53 A sound source at rest emits waves at a frequency of 5000 Hz in air.

(a) What is the wavelength reaching an observer at rest with respect to the source?

(b) What is the wavelength reaching the observer if the source moves with a speed of 34 m/s toward the listener?

(c) What frequency does the listener hear?

(d) What is the wavelength reaching an observer moving toward the stationary source at a speed of 34 m/s?

(e) What frequency is heard in this case?

12.54 A police car moving at 36 m/s is chasing a speeding motorist traveling at 30 m/s. The police car has a siren going at 3000 Hz. What frequency does the motorist hear?

12.55 Two trains are moving along in opposite directions on parallel tracks and are approaching a train station. The trains are each sounding their horns at 500 Hz. One train is moving at 35 m/s. What is the speed of the other train if a passenger on the platform hears a beat frequency of 4 Hz?

*12.56 Derive an expression for the Doppler effect for the two cases of moving source and moving observer in the situation where the air is also moving with a speed of u m/s with respect to the ground.

12.57 The flight of a plane is being simulated in a wind tunnel by air blown at very high speeds past the stationary model of an airplane. How fast does the air have to travel for a shock wave of cone angle 45° to develop?

12.58 A plane is flying at Mach 0.90 and carries a sound source that emits a loud 2000-Hz signal. After the plane passes an observer on the ground, what frequency does the observer hear?

12.59 An SST is flying at Mach 1 and producing a vibration at 1000 Hz. What frequency is heard by an observer directly in the path of the plane?

*12.60 Prove that, at constant temperature, the speed of sound in a gas is independent of the pressure.

Additional Readings

Crombie, A. C.: "Helmholtz," *Scientific American*, vol. 198, no. 3, March 1958, pp. 94–102. An excellent article by a well-known British historian of science.

Devey, Gilbert B., and Peter N. T. Wells: "Ultrasound in Medical Diagnosis," *Scientific American*, vol. 238, no. 5, May 1978, pp. 98–112. An excellent article on the use of ultrasound for medical diagnosis, containing many ultrasonic pictures of human organs and even of triplets in their mother's womb. Also, an account of how the Doppler effect can be used with ultrasonic waves in medical diagnosis.

Fletcher, Neville H., and Suzanne Thwaites: "The Physics of Organ Pipes," *Scientific American*, vol. 242, no. 1, January 1983, pp. 94–103. An interesting article on how organ pipes produce their majestic sounds.

Hudspeth, A. J.: "The Hair Cells of the Inner Ear," *Scientific American*, vol. 248, no. 1, January 1983, pp. 54–64. An account, including scanning electron micrographs, of the exquisitely sensitive transducers that enable us to hear.

The Physics of Music (readings from *Scientific American*, with an introduction by Carleen Maley Hutchins), Freeman, San Francisco, 1978. An excellent collection of articles by experts in the field of sound and acoustics.

Pierce, John R.: *The Science of Musical Sound*, Freeman, San Francisco, 1983. A *Scientific American* book on acoustics for the nonspecialist, written from the viewpoint of contemporary electronic and computer-generated music.

Roedderer, Juan: *The Physics and Psychophysics of Music*, Springer-Verlag, New York, 1979. This book contains much fascinating material on the physics and neuroscience behind the perception of sound.

Taylor, Charles: *Sounds of Music*, British Broadcasting Corporation, London, 1976. A popular account based on the Christmas lectures at the Royal Institution in London in 1971.

Wegener, Peter P.: "The Science of Flight," *American Scientist*, vol. 74, May–June 1986, pp. 268–278. A fascinating and extremely well-illustrated discussion of the flight of aircraft and the production of shock waves. The photograph of a shock wave in Fig. 12.23 is taken from this article.

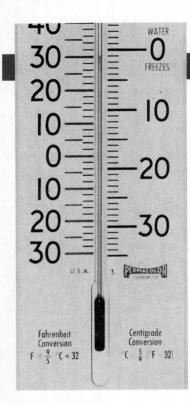

Fahrenheit
Conversion

$°F = \frac{9}{5}°C + 32$

Centigrade
Conversion

$°C = \frac{5}{9}(°F - 32)$

CHAPTER 13

Temperature and Heat

And, in reasoning on this subject, we must not forget to consider that more remarkable circumstance, that the source of Heat generated by friction, in these experiments, appeared evidently to be inexhaustible. . . . And it appears to me to be extremely difficult, if not quite impossible, to form any distinct idea of anything capable of being excited and communicated in the manner the Heat was excited and communicated in these experiments, except it be motion.

Benjamin Thompson, Count Rumford (1753–1814)

In the next three chapters we discuss the basic laws governing the conversion of heat to other forms of energy and the conversion of other forms of energy to heat. These laws are crucial to our health and comfort, which depend so sensitively on our body temperature. Likewise, a proper assessment of society's energy problems requires an understanding of the nature of temperature and heat.

There are two very different approaches to this important subject. The first, the science of *thermodynamics*, deals only with such measurable attributes of large-scale matter as volume, pressure, and temperature. The second approach relates these quantities to the behavior of the atoms and molecules that constitute matter. This molecular approach uses the basic quantities of mechanics, such as velocities and kinetic energies, to derive the properties of large-scale matter. Since the number of molecules is so large, statistical techniques must be used to obtain average values for these physical quantities. This approach is therefore called *statistical mechanics*.

We have already used the statistical approach in our discussion of the kinetic theory of gases (Sec. 9.6). In the next three chapters we will be using the insights of thermodynamics, and we will see how the insights of thermodynamics and those of statistical mechanics complement each other.

13.1   Temperature

We saw in Sec. 9.5 a rough definition of temperature as a measure of hotness or coldness. These are very subjective concepts, however, and we need a more objective and precise definition of temperature for scientific work. This we find in the kinetic theory of gases, which leads to Eq. (9.17): $\frac{3}{2}kT = \frac{1}{2}m\overline{v^2}$.

The absolute (Kelvin) temperature T of a gas is directly proportional to the average random translational kinetic energy of the gas molecules.

At a fixed temperature T, the molecules of all gases have the same average kinetic energy, no matter what the masses, pressures, or volumes of the gases are. This is the significance of temperature at the atomic or molecular level.

13.2   Gas Thermometers

It is very difficult to measure the average kinetic energy of the molecules of a substance directly. We must obtain the temperature *indirectly* by first measuring some property of the substance that changes with temperature, and then using this measured value to obtain the temperature. To do this we use a *thermometer*. This is a device that contains a solid, liquid, or gas with some physical property which varies with temperature, as, for example, the pressure of a gas at constant volume, the volume of a liquid, or the length of a solid rod.

For accurate temperature measurements constant-volume gas thermometers are used, in which the volume is held constant and the pressure measured as a function of temperature. The ideal gas law, $PV = nRT$ [Eq. (9.14)] predicts that, for V constant, P is directly proportional to the absolute temperature T, and so the absolute temperature can be obtained from a pressure measurement.

Even with a constant-volume gas thermometer, different gases give slightly different results for the temperature of an object, because the ideal gas law is not exact for a nonideal gas. If the density of the gas is sufficiently low, however, all gas thermometers yield exactly the same temperature. For this reason, the constant-volume gas thermometer is used to establish the temperature scale employed for scientific work throughout the world.

A constant-volume gas thermometer is shown in Fig. 13.1. Such a thermometer is far from convenient to use, since the temperature probe is a large bulb containing gas. For this reason simpler thermometers, such as the common mercury-in-glass thermometer, are more frequently used in routine temperature measurements. They are all ultimately checked (*calibrated*) against a standard constant-volume gas thermometer.

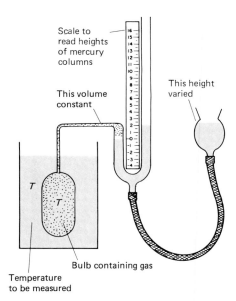

FIGURE 13.1 A constant-volume gas thermometer. By raising and lowering the bulb at the right, the volume of the gas in the bulb at the left can be kept constant. The manometer is then used to measure the pressure at constant volume. Since $PV = nRT$, the absolute temperature is proportional to the pressure.

Scale to read heights of mercury columns

This volume constant

This height varied

Bulb containing gas

Temperature to be measured

Absolute Zero

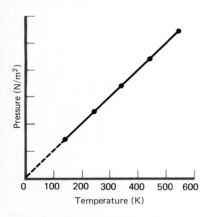

FIGURE 13.2 Graph of pressure against absolute temperature for a constant-volume gas thermometer.

A gas thermometer gives a precise physical meaning to the zero of temperature. If we use a constant-volume gas thermometer and gradually reduce the temperature, we obtain the experimental curve for pressure shown in Fig. 13.2. If the temperature could be lowered far enough, the pressure would go to zero at $T = 0$ (actually any real gas would liquefy before this could happen). Hence we can define absolute zero as follows:

Absolute zero: That temperature at which the pressure of an ideal gas at constant volume would fall to zero.

This temperature is chosen as the zero on the absolute (Kelvin) scale of temperature and has an "absolute" significance not shared by the zeros on the Celsius or Fahrenheit scales, which are arbitrarily defined points and represent no limit to the lowest temperature attainable. Absolute zero (i.e., zero Kelvin) corresponds to $-273.15°C$, as we have seen in Sec. 9.7.

The Quest for Absolute Zero

Over the years physicists have devoted much effort to producing lower and lower temperatures in order to investigate the properties of matter at these temperatures.

Temperatures as low as 4.2 K are easily attainable, since ordinary helium ($_2^4He$) can be liquefied at this temperature by using specialized refrigeration equipment. Every low-temperature physics laboratory contains a supply of liquid helium for experiments at 4.2 K. The temperature can be lowered even further by pumping away the helium vapor that constantly "boils" off the liquid helium. The evaporation of high-speed molecules from a liquid lowers the temperature of the liquid. In this way temperatures as low as 0.7 K have been achieved with $_2^4He$. Using the isotope $_2^3He$, which liquefies at 3 K, even lower temperatures of 0.3 K have been reached. Still lower temperatures, as low as 10^{-6} K, can be attained using the interaction of paramagnetic salts with large magnetic fields.

What physicists find in their quest for absolute zero, however, is that the closer they get, the more difficult it becomes to go further. They can always reduce the temperature somewhat below the lowest temperature previously attained, but this never brings them to absolute zero, no matter how many times they repeat the process. A mathematician would say that these experiments approach absolute zero "asymptotically," which means that they get very close but never actually arrive at the desired goal. For this reason it is generally accepted as a basic law of nature that zero on the Kelvin scale is unattainable in a finite number of operations. This is sometimes referred to as *the third law of thermodynamics*.

13.3 Other Kinds of Thermometers

In addition to gas thermometers, which are the ultimate standards for all temperature measurements, there are other, more convenient types such as mercury-in-glass thermometers, bimetallic-strip thermometers (Sec. 13.5), and resistance thermometers (Sec. 18.4).

Mercury-in-Glass Thermometers

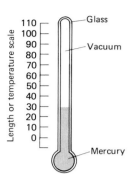

FIGURE 13.3 A mercury-in-glass thermometer. The width of the column is exaggerated for clarity.

Suppose we fill a small glass bulb with mercury metal, which is a liquid at ordinary temperatures, and attach to the bulb a long, thin, hollow glass tube with a uniform bore, as in Fig. 13.3. If we apply heat, the mercury expands and rises in the glass tube (usually called a *capillary tube*). The glass also expands, but this effect is small, since the coefficient of volume expansion for Pyrex glass is only about one-twentieth that for mercury. If we assume that the volume of mercury changes linearly with the temperature, as is true for most substances, we have

$$\Delta V = k \, \Delta T \qquad \text{or} \qquad \pi R^2 \, \Delta L = k \, \Delta T \tag{13.1}$$

where k is a constant of proportionality, R is the radius of the capillary tube, and ΔL is the change in length of the mercury column for a change in temperature ΔT. Since ΔT is proportional to ΔL, the temperature can be determined from a length measurement on the mercury column.

A mercury-in-glass thermometer with an extremely narrow bore can be a very accurate and sensitive temperature-measuring instrument in the range from somewhat below 0°C to a few hundred degrees Celsius. Its range is limited because mercury becomes a solid at -39°C and a vapor at 357°C.

Example 13.1

Mercury increases in volume by 18×10^{-5} times its original volume for each 1°C increase in temperature. A mercury-in-glass thermometer has a quartz bulb of volume 0.500 cm³ and an attached quartz tube with a bore of 0.10 mm. Initially the bulb is just filled with mercury, and there is no mercury in the straight tube. If the temperature in the room changes from 20 to 25°C, what is the increase in the height of the mercury in the tube?

SOLUTION

The increase in volume of the mercury is

$$\Delta V = \frac{18 \times 10^{-5}}{°C} \, V \, \Delta T$$

$$= \frac{18 \times 10^{-5}}{°C} \, (0.500 \text{ cm}^3)(25°C - 20°C)$$

$$= 4.5 \times 10^{-4} \text{ cm}^3$$

Now, the increased volume of mercury must raise the height of the mercury column since the bulb is completely filled. The change in height ΔL of a mercury column of cross-sectional area πR^2 is related to the increased volume ΔV by

$$\Delta V = \Delta L \, \pi R^2$$

Hence

$$\Delta L = \frac{\Delta V}{\pi R^2} = \frac{4.5 \times 10^{-4} \text{ cm}^3}{3.14(0.0050 \text{ cm})^2} = \boxed{5.7 \text{ cm}}$$

13.4 Temperature Scales

The temperature of any system in thermal equilibrium (Sec. 9.5) can be represented by a number. We now want to see how to assign such numbers, in other words, how to set up a workable temperature scale. This is done by selecting certain *fixed points* that correspond to easily reproducible temperatures and assigning numbers to them on a chosen temperature scale. Good examples of fixed points are the boiling point and the freezing point of water at standard atmospheric pressure.

Absolute (Kelvin) Temperature Scale

On the absolute scale, zero is the temperature at which the pressure of an ideal gas goes to zero at constant volume. On this scale the freezing point of water, as determined by a gas thermometer, is 273.15 K, and the boiling point of water is 373.15 K. The difference in temperature between the freezing and boiling points is exactly 100 kelvins, just as it is 100 Celsius degrees on the Celsius scale.

Celsius Temperature Scale

This temperature scale, first proposed in 1742 by the Swedish astronomer Anders Celsius (1701–1744), is much used both for scientific and practical purposes. On this scale the freezing point of water is assigned the number 0°C, and the boiling point of water, 100°C. This scale is sometimes called the centigrade scale (*centi-* $= \frac{1}{100}$) because there are 100 equal divisions between these two fixed points on this scale, just as there are 100 kelvins between these same two points. The size of a degree on the absolute scale is therefore exactly the same as the size of a degree on the Celsius scale. Consequently *temperature differences* have the same numerical values on these two scales, as shown in Fig. 13.4.

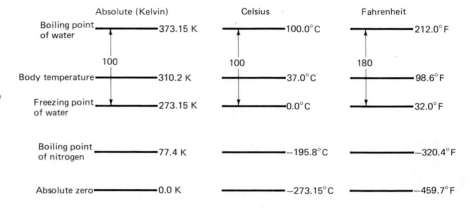

FIGURE 13.4 A comparison of three temperature scales.

Fahrenheit Temperature Scale

The German physicist Gabriel Daniel Fahrenheit (1686–1736) was the first to use mercury as a thermometric liquid. He set 100°F as the approximate temperature of the human body, and 0°F as the lowest temperature reached by an ice-salt mixture. This made the boiling point of water 212°F and the temperature of an ice-water mixture (the freezing point of pure water) 32°F. He then divided the distance between these two fixed points on a mercury-in-glass thermometer into 180 equal divisions, or Fahrenheit degrees, and thus created the Fahrenheit temperature scale. We now know that the normal temperature of the human body is actually closer to 98.6°F than to 100°F.

Relations between Temperature Scales

An interval of one degree on the absolute scale is of exactly the same size as on the Celsius scale, and so to obtain the absolute temperature from a given Celsius temperature all we must do is add 273.15 to the Celsius value:

$$T(\text{K}) = T(°\text{C}) + 273.15 \qquad \textbf{(13.2)}$$

In the range between the freezing and boiling points of water, there are 180 Fahrenheit degrees but only 100 Celsius degrees. Hence the Celsius degree is *larger* than the Fahrenheit degree by the factor $180/100 = 9/5$. There are therefore *fewer* Celsius degrees than Fahrenheit degrees in any given temperature interval.

For example, suppose we have a Celsius temperature of 45°. This is 45 Celsius degrees above the freezing point of water, which occurs at 0°C. On the Fahrenheit scale it is $\frac{9}{5}(45) = 81$ Fahrenheit degrees above the freezing point of water. The

TABLE 13.1 Important Physical Temperatures on the Fahrenheit, Celsius, and Kelvin Temperature Scales

Physical quantity	Temperature		
	(°F)	(°C)	(K)
Deuterium-deuterium fusion	4×10^8	2×10^8	2×10^8
Surface of hottest known star	4×10^5	2×10^5	2×10^5
Surface of sun	10,000	5500	5800
Melting point of tungsten	6,170	3410	3683
Light-bulb filament	4,600	2500	2800
Bunsen burner flame	3,400	1870	2150
Steam temperature in a typical fossil-fuel power plant	930	500	770
Boiling point of water	212.0	100.0	373.15
Human body temperature	98.6	37.0	310.2
Room temperature	70	21	294
Triple point of water	32.02	0.01	273.16
Freezing point of water	32.0	0	273.15
Solid CO_2 (dry ice)	-110	-74	194
Liquid nitrogen (boiling point)	-320	-196	77
Liquid helium (boiling point)	-452	-269	4.2
Absolute zero	-459.67	-273.15	0

freezing point of water is, however, not 0°F but 32°F, and so 45°C corresponds to a Fahrenheit temperature of 81°F + 32°F = 113°F.

This result can be put in the form of an equation:

$$T'(°F) = \tfrac{9}{5}T(°C) + 32 \qquad\qquad (13.3)$$

This yields

$$T(°C) = \tfrac{5}{9}[T'(°F) - 32] \qquad\qquad (13.4)$$

There is only one temperature at which Fahrenheit and Celsius scales yield exactly the same temperature: -40°C, which is equal to -40°F. This is sometimes helpful in remembering the conversion formulas given above.

The relationship of these three temperature scales is shown in Fig. 13.4. Values of some important temperatures on the Celsius, Fahrenheit, and absolute scales are given in Table 13.1. Note that they range all the way from absolute zero to hundreds of millions of degrees.

The difference between the freezing and boiling points of water is 100 Celsius degrees, which is usually written as 100 C°. This notation is used to distinguish a *temperature range* (or temperature difference) of 100 C° from the *individual temperature* of the boiling point, which is written as 100°C. We will adopt this convention in what follows.

Example 13.2

A pot of coffee is heated from 20 to 70°C. What is this change in temperature on **(a)** the Kelvin scale and **(b)** the Fahrenheit scale?

SOLUTION

(a) The change on the Celsius scale is $\Delta T_C = 70°C - 20°C = 50\ C°$. (Note the subtle change from °C to C° to indicate that this is a *temperature difference*.) Since the change in the temperature on the Kelvin scale is the same as on the Celsius scale (for one degree has the same size on the two scales), we have

$$\Delta T = \boxed{50\ K}$$

(b) From Eq. (13.3), we have

$$\Delta T_F = [\tfrac{9}{5}(70°C) + 32] - [\tfrac{9}{5}(20°C) + 32]$$

$$= \tfrac{9}{5}(50\ C°) = \boxed{90\ F°}$$

Therefore 90 F° = 50 C°, where the notation applies to *temperature differences, not to individual temperatures*.

The same result can be obtained more easily by using the fact that Fahrenheit degrees are only five-ninths as large as Celsius degrees. Therefore there are nine-fifths as many of them in the same temperature interval. For this reason

$$50\ C° = \tfrac{9}{5}(50\ F°) = \boxed{90\ F°}$$

13.5 Expansion of Solids and Liquids with Temperature

Many thermometers for the scientific measurement of temperature rely on the fact that a solid changes its length and a liquid its volume *linearly* with temperature. Solids and liquids are therefore convenient to use in thermometers. As we saw in Chap. 10, the molecules of a metal vibrate about fixed positions as if they were connected by springs. Heating a metal causes the molecules to vibrate more vigorously and therefore to move farther apart on the average. If the metal is in the form of a long, thin rod of original length L_0 at some temperature T_0, then we can denote by ΔL the change in length due to a change in temperature ΔT, as in Fig. 13.5. From experiment we find that, for most materials at ordinary temperatures, the increase in length is proportional both to the original length of the rod and to the change in temperature,

$$\Delta L \propto L_0\ \Delta T \qquad \text{or} \qquad \Delta L = \alpha L_0\ \Delta T \tag{13.5}$$

Here α is called the *coefficient of linear expansion* and is defined as follows:

Coefficient of linear expansion α: The increase in length of a solid per unit length per unit change in temperature:

$$\alpha = \frac{\Delta L}{L_0 \Delta T} \tag{13.6}$$

The new length of the rod is then

$$L = L_0 + \Delta L = L_0 + \alpha L_0\ \Delta T$$

$$\text{or} \qquad \boxed{L = L_0(1 + \alpha\ \Delta T)} \tag{13.7}$$

Since the change in length, $\Delta L = L - L_0$, is proportional to the change in temperature, it can be used to measure an unknown temperature by relating it to some known temperature.

The coefficient of linear expansion varies from substance to substance, and its value must be obtained by experiment. Table 13.2 gives values of α for a variety of useful substances. Note that α has units $(C°)^{-1}$, since $\Delta L/L_0$ is dimensionless, and so $\alpha\ \Delta T$ must also be dimensionless.

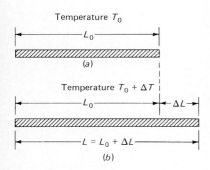

FIGURE 13.5 Increase in length of a metal rod with temperature: (a) Length at the original temperature T_0; (b) length at the new temperature $T = T_0 + \Delta T$. The change in length is greatly exaggerated.

TABLE 13.2 Thermal Coefficients of Linear Expansion

Solid substances	Coefficient of linear expansion $\alpha[10^{-6}(C°)^{-1}]$
Aluminum	23
Brass	19
Copper	17
Glass (Pyrex)	3.2
Invar	0.70
Lead	29
Quartz (fused)	0.4
Steel	11

TABLE 13.3   Thermal Coefficients of Volume Expansion

Substance	Coefficient of volume expansion $\beta[10^{-5}(C°)^{-1}]$
Solids:	
Aluminum	7.2
Brass	6.0
Copper	4.2
Glass (Pyrex)	0.96
Invar	0.27
Lead	8.7
Quartz (fused)	0.12
Steel	3.6
Liquids:	
Ethyl alcohol	112
Glycerin	51
Mercury	18
Turpentine	97

If we consider a two-dimensional object like a flat rectangular plate of original length L_0 and width W_0, then both L_0 and W_0 increase with temperature to new values $L_0 + \Delta L$ and $W_0 + \Delta W$. Hence the new area is

$$A = (L_0 + \Delta L)(W_0 + \Delta W) = (L_0 + \alpha L_0 \, \Delta T)(W_0 + \alpha W_0 \, \Delta T)$$

$$= L_0 W_0 (1 + 2\alpha \, \Delta T + \alpha^2 \, \Delta T^2) \simeq A_0 (1 + 2\alpha \, \Delta T)$$

where we have dropped $\alpha^2 \, \Delta T^2$ because it is usually small compared with unity (for brass, if ΔT is 100°C, $\alpha \, \Delta T = 1.9 \times 10^{-3}$ and $\alpha^2 \, \Delta T^2 = 3.6 \times 10^{-6}$). Hence we obtain, to a good approximation,

$$A = A_0(1 + 2\alpha \, \Delta T) \tag{13.8}$$

The change in area is therefore also proportional to the change in temperature, the proportionality factor being in this case 2α rather than α.

A similar analysis for volumes (try to derive it yourself) shows that

$$V = V_0(1 + 3\alpha \, \Delta T) \tag{13.9}$$

It is also possible to define a *thermal coefficient of volume expansion* for solids in a fashion similar to the way we defined a coefficient of linear expansion in Eq. (13.7):

$$\boxed{V = V_0(1 + \beta \, \Delta T)} \tag{13.10}$$

where β is the coefficient of volume expansion of the solid.

Comparing Eq. (13.9) with (13.10), we see that for solids β ought to be equal to 3α. Some values of α and β for important solids are given in Tables 13.2 and 13.3. Note that for these solids β is indeed approximately equal to 3α.

Bimetallic Strips and Thermostats

A bimetallic strip consists of two thin strips of different metals cemented together along their flat surfaces. Since the two metals expand at different rates as the temperature increases, the bimetallic strip bends further as the temperature increases, as shown in Fig. 13.6. This is the only way the two metals can expand without cracking the cement holding them together. Such strips can be wound into a coil and attached to a pointer that indicates the temperature on a calibrated scale. This makes a crude, but simple thermometer, as illustrated in Fig. 13.7.

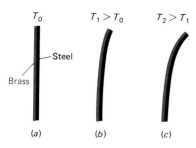

FIGURE 13.6   A bimetallic strip consisting of two flat metal strips cemented together (we are looking at the edge in the figure).

FIGURE 13.7   A thermometer employing a bimetallic strip.

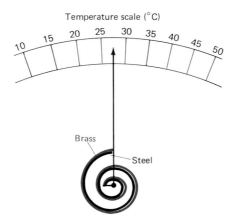

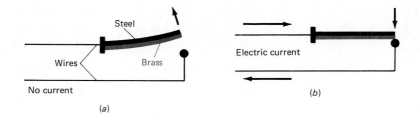

FIGURE 13.8 A thermostat consisting of a bimetallic strip that turns a heating unit on and off.

Some thermostats contain a bimetallic strip that controls a heating (or cooling) unit to keep the temperature of a region of space (a room or an oven, for example) within preset limits. If the temperature exceeds the desired limit, the bimetallic strip bends enough to break the electrical contact, as shown in Fig. 13.8a. The electric current ceases and the heating unit turns off. As the bimetallic strip cools down, it bends in the opposite direction and makes contact again, as in Fig. 13.8b. The current is reestablished and the heater turns back on. A thermostat like this is the simplest kind of *feedback mechanism*, since it constantly turns the furnace on and off in response to a signal that depends on the temperature.

Liquids

Because their shape is so easily changed, only volume changes are significant for liquids. Here again we find that the change in volume is proportional to the change in temperature, the change being given by

$$V = V_0(1 + \beta \, \Delta T) \tag{13.11}$$

where β is the thermal coefficient of volume expansion of the liquid. Some values for β for important liquids are included in Table 13.3.

Most liquids change volume linearly with temperature. Water, our most common liquid, is an exception and is therefore not very useful as a thermometer liquid. The volume of a fixed amount of water is smallest at 4°C, and its density is therefore a maximum, as shown in Fig. 13.9. Above 4°C the expansion of water is not linear, although it is reasonably close to it. But from 4 down to 0°C, where it freezes, water again expands, and so ice at 0°C is *less dense* than water is at 4°C. It is for this reason that ice floats at the top of a lake or river. If this were not the case—if ice were *denser* than cold water—lakes and rivers would freeze from the bottom up and never completely unfreeze even in summer.

Volume expansion is the same in principle for solids, liquids, and gases, but very different quantitatively. As Table 13.3 shows, the coefficient of volume expansion is about 10 times larger for liquids than for solids. For gases, the coefficient of volume expansion is about 300 times that for solids.

FIGURE 13.9 Density of water in the vicinity of 0°C. As can be seen, water is densest and hence has its smallest volume at 4°C. Water expands both above and below 4°C.

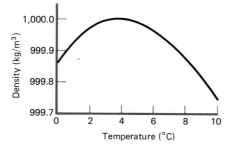

Example 13.3

Invar (an alloy of iron and nickel in a 5/3 ratio) is a material with a very low coefficient of linear expansion frequently used in clocks and precise scientific instruments, in which slight changes in temperature might otherwise introduce errors.

(a) Calculate the increase in length of an Invar pendulum rod 1.0 m long when the temperature changes from 20 to 30°C.
(b) What is the percentage change in the period of the pendulum produced by this change in temperature?

SOLUTION

(a) The increase in length is given by Eq. (13.5), $\Delta L = \alpha L_0 \Delta T$. Table 13.2 tells us that α here is 0.70×10^{-6} $(C°)^{-1}$, and the change in temperature is $30°C - 20°C = 10$ C°. Then

$$\Delta L = \frac{0.70 \times 10^{-6}}{C°}(1.0 \text{ m})(10 \text{ C}°) = \boxed{0.70 \times 10^{-5} \text{ m}}$$

This is a change of less than $\frac{1}{100}$ mm.
(b) We assume for purposes of calculation that the clock pendulum approximates a simple pendulum of length exactly 1.0 m at 20°C. The *period* is $T = 2\pi(L/g)^{1/2}$. For a length of 1.000000 m at 20°C, we have

$$T_{20} = 2\pi(1.000000/g)^{1/2}$$

and for a length of 1.000007 m, we have

$$T_{30} = 2\pi(1.000007/g)^{1/2}$$

Then
$$\frac{T_{30}}{T_{20}} = \left(\frac{1.000007}{1.000000}\right)^{1/2} = (1.000007)^{1/2}$$

$$T_{30} = 1.000004\, T_{20}$$

$$\frac{\Delta T}{T_{20}} = \frac{T_{30} - T_{20}}{T_{20}} = \frac{0.000004}{1.000000} = 0.000004$$

The change in the period of the pendulum for a change of 10 C° in temperature is therefore only $\boxed{0.0004 \text{ percent,}}$ or 4 parts in a million.

EXERCISE 1   Repeat this calculation for a pendulum made of brass.

13.6   Heat as a Form of Energy

We have seen that, at the microscopic level, the temperature of a gas is a measure of the average random translational kinetic energy of the gas molecules. The higher the temperature, the greater the molecular kinetic energy, and therefore the greater the average molecular speed. When we measure the temperature of a gas, we measure the average random translational kinetic energy of its molecules, and nothing else. This temperature is independent of how much gas is present or of what kind of gas it is.

What then do we mean by *heat*, and how do we distinguish the concept of heat from that of temperature? To answer this question we must first clarify what we mean by the thermal energy of a gas like He at a temperature T. By *thermal energy* we mean the sum of all the energies associated with the random motions of the molecules in the substance. Thermal energy depends on the temperature, but it also depends on the number of molecules present. Since the molecules have an average kinetic energy determined by the temperature T, the more molecules present, the greater the total thermal energy of the gas, even if T remains fixed.

For more complicated systems, there are other kinds of motions that make additional contributions to the thermal energy even for the same values of the temperature and the same number of molecules. Thus in a sample of nitrogen gas, an N_2 molecule can rotate like a dumbbell in space, or the two N atoms can vibrate back and forth as if they were connected by springs (Fig. 13.10). Hence, while the *temperature* depends only on the translational kinetic energy of the gas molecules, the *thermal energy* depends not only on the temperature but also on the number and kinds of atoms and molecules in the gas. In other words:

Thermal energy: The sum of all the random mechanical energies of the atoms *or* molecules in a substance.

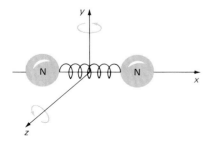

FIGURE 13.10   Vibration and rotation of a nitrogen molecule (N_2). The two N atoms can vibrate back and forth as if connected by a spring, and the whole molecule can rotate like a dumbbell around the y and z axes.

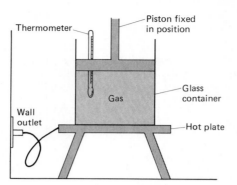

FIGURE 13.11 A gas in a container on a hot plate.

What then is heat? The term *heat* is usually reserved for the *transfer* of thermal energy from one substance to another.

Heat: The thermal energy transferred between a system and its surroundings as a result of temperature differences only.

Thermal energy and heat are the same kind of physical quantity, just as mechanical energy and work are in mechanics. In the same way that we use the term *work* to refer to the amount of mechanical energy transferred from one part of a system to another, so too we use the term *heat* for the amount of thermal energy transferred from one part of a system to another. This distinction will become important when we discuss the first law of thermodynamics.

To clarify the distinction between heat and temperature a bit more, consider what happens when we increase the temperature of the hot plate in Fig. 13.11 to a temperature T_2 higher than the initial temperature T_1 of the gas, while keeping the volume fixed. Heat flows from the hot plate through the bottom of the glass into the gas molecules in the container. On a molecular level this means that the agitated molecules in the hot plate pass on some of their random molecular motion to the glass molecules, which pass it on to the gas molecules inside. We say that a *flow of heat* has increased the translational kinetic energy of the gas molecules and thus has increased the thermal energy and hence the temperature of the gas. The amount of heat ΔQ required to change the temperature by an amount $\Delta T = T_2 - T_1$ obviously depends on how large ΔT is. We can therefore write $\Delta Q \propto \Delta T$.

But the amount of heat, that is, the amount of thermal energy, required to raise the temperature by an amount ΔT depends also on how many gas molecules are present: the amount of heat required to change the temperature of 10^{12} molecules from T_1 to T_2 is 10 times greater than the amount of heat required to raise the temperature of 10^{11} identical molecules by the same amount ΔT. In other words, the amount of heat will also be proportional to the mass of gas present, and so, for ΔT constant, $\Delta Q \propto m$. We must then combine these two results and obtain

$$\Delta Q \propto m\,\Delta T \qquad \text{or} \qquad \boxed{\Delta Q = cm\,\Delta T} \qquad\qquad \textbf{(13.12)}$$

where c is a constant of proportionality that differs from gas to gas or from substance to substance. It depends on the kinds of molecules present in the substance and is called the *specific heat capacity*. The specific heat capacity of a substance is usually obtained from measurements made in the laboratory, although in some cases it is possible to calculate specific heat capacities theoretically.

13.7 Specific Heat Capacities

Since heat is a form of energy, it could well be measured in joules, the unit of energy in the SI system. Initially, however, the science of heat grew up separately from the science of mechanics and developed its own units, the kilocalorie in countries using the metric system and the British thermal unit in the British engineering system. Both these units are still much used today, and for this reason we treat them here.

Kilocalorie (kcal or Cal): The amount of heat required to raise the temperature of one kilogram of water one degree Celsius (from 14.5 to 15.5°C).

British thermal unit (Btu): The amount of heat required to raise the temperature of one pound of water one degree Fahrenheit (from 63 to 64°F).

There is a smaller unit, the calorie (cal), which is 1/1000 of the kilocalorie. When we talk about the number of calories in a potato or a candy bar, we are really talking about the kilocalorie (the "big calorie"), not the calorie ("small calorie").

Note that in the two definitions above, the kilocalorie and the British thermal unit are so defined that the specific heat capacity of water is unity in both systems. For water $c = 1$ kcal/(kg·C°) in the metric system, and $c = 1$ Btu/(lb·F°) in the British system. This leads to the result that 1 kcal = 3.96 Btu ≃ 4 Btu.

The basic equation needed for problems involving the heating or cooling of substances is, from above,

$$\Delta Q = cm \, \Delta T \tag{13.12}$$

Here the symbol c designates the *specific heat capacity* of the substance being heated or cooled.

Specific heat capacity c: The quantity of heat that must be supplied to a unit mass of a substance to raise the temperature by one degree.

For water (H_2O in its liquid phase), $c = 1.00$ kcal/(kg·C°), if the mass of water is given in kilograms, the amount of heat in kilocalories, and the temperature change in either Celsius degrees or kelvins. Thus it takes 10 kcal to raise the temperature of 5 kg of water 2 C°, as is clear from substituting in Eq. (13.12).

For other substances the specific heat capacity is different from that of water and almost always less than 1 kcal/(kg·C°).

In many practical problems involving specific heat capacities, two bodies at different temperatures come in contact, as when we pour cold milk into hot coffee. The coffee loses heat and the milk gains heat until the two come to an equilibrium temperature. The equilibrium temperature can be found, if we know the amounts of each substance present and their specific heat capacities, by equating the heat lost by the coffee to the heat gained by the milk (see Example 13.4).

In Table 13.4 the specific heat capacities of some important substances are given. Since specific heat capacities vary slightly with temperature, the values given are for temperatures in the immediate vicinity of room temperature (about 20°C).

TABLE 13.4 Specific Heat Capacities

Substance	Specific heat capacity [kcal/(kg·C°)]
Metals:	
Aluminum	0.22
Beryllium	0.47
Copper	0.093
Iron	0.11
Lead	0.031
Mercury	0.033
Silver	0.056
Nonmetals:	
Body tissue	0.80
Glass	0.12
Human blood	0.92
Human body (average)	0.85
Ice	0.55
Steam	0.48
Sugar	0.27
Water	1.00

Example 13.4

Suppose we drop a 1.0-kg piece of aluminum at 200°C into 5.0 kg of water at room temperature (20°C). If no heat is exchanged with the surroundings, what is the final temperature of the water and the aluminum?

SOLUTION

The aluminum will lose heat and the water will gain heat until they reach a final equilibrium temperature T_f. The amounts of heat gained by the water and lost by the aluminum must be equal, according to the principle of conservation of energy, and can be calculated from Eq. (13.12). Thus the heat given out by the aluminum is

$$\Delta Q_{Al} = c_{Al} m_{Al} \, \Delta T_{Al}$$

$$= \left(0.22 \, \frac{kcal}{kg \cdot C°}\right) (1.0 \text{ kg})[(200 - T_f)C°]$$

$$= 0.22(200 - T_f) \text{ kcal} = (44 - 0.22T_f) \text{ kcal}$$

Similarly, the heat taken in by the water is

$$\Delta Q_w = c_w m_w \, \Delta T_w$$

$$= \left(1.0 \, \frac{kcal}{kg \cdot C°}\right) (5.0 \text{ kg})[(T_f - 20)C°]$$

$$= 5.0(T_f - 20) \text{ kcal} = (5T_f - 100) \text{ kcal}$$

Since no heat is lost or gained by the system, $\Delta Q_{Al} = \Delta Q_w$,

and so $44 - 0.22T_f = 5T_f - 100$, from which $T_f =$ $\boxed{28°C}$

The answer is only given to two significant figures, since the values of c in Table 13.4 are only to two significant figures.

Note that the final temperature of the mixture is within 8C° of the initial temperature of the water, even though the aluminum was at 200°C. This is because there is five times as much water as aluminum, and also because the water has a specific heat capacity five times that of the aluminum. The water, because of its large specific heat capacity, "hides" the effect of the heat given to it by the aluminum.

In general, all problems of this kind can be solved by setting up equations in the form

$$c_1 m_1 (T_1 - T_f) = c_2 m_2 (T_f - T_2)$$

where T_f is the final equilibrium temperature. The units used must be consistent. These are kilocalories, kilograms, and Celsius degrees or joules, kilograms, and Celsius degrees. Absolute temperatures may also be used, since temperature *differences* like $T_1 - T_f$ are the same on the Celsius and absolute temperature scales.

13.8 The Mechanical Equivalent of Heat (Joule's Equivalent)

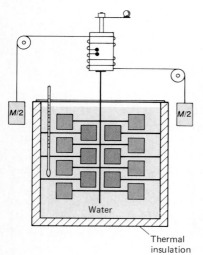

FIGURE 13.12 Paddle-wheel apparatus used by Joule in measuring the mechanical equivalent of heat (Joule's equivalent).

The fields of mechanics and heat developed independently, and as a result work was for many years measured in mechanical units (joules) while heat was measured in thermal units (calories or kilocalories). The conclusive experimental work that established the exact quantitative relationship between a quantity of work (in joules) and a quantity of heat (in kilocalories) was performed by James Joule (see Fig. 13.13 and accompanying biography). In his most famous experiment Joule used the paddle-wheel apparatus of Fig. 13.12, which consisted of two heavy masses attached to a rope wound around a pulley. When the masses were allowed to fall under gravity, they unwound the rope and turned a set of paddles immersed in a tank of water thermally insulated from its surroundings. The churning paddles did work on the water and heated it up. In this way Joule could compare the work done by the falling mass with the heat required to produce the measured temperature change in the water.

If each mass is $M/2$ and if each is raised to a height h off the ground, the gravitational potential energy of the two masses with respect to the ground is Mgh. When the two masses fall slowly through a distance h back to the ground, their combined potential energy is converted to an amount of work

$$\mathcal{W} = Mgh$$

where $\mathcal{W}$ is in joules.

As the rope unwinds, the moving paddles stir up the water and increase its temperature by an amount ΔT that can be measured with a thermometer. The amount of heat in kilocalories absorbed by the water is, then, from Eq. (13.12),

$$\Delta Q = cm \, \Delta T$$

where, if m is the mass of the water in kilograms and c is equal to 1 kcal/(kg·C°), ΔQ is in kilocalories.

James Prescott Joule (1818–1889)

FIGURE 13.13 Sir James Joule. (*NBS Archives; courtesy of AIP Niels Bohr Library.*)

Of the four men usually associated with the development of the first law of thermodynamics—Mayer, Rumford, Helmholtz, and Joule—the last was the true experimental physicist. It was his lifelong dedication to confirming the conservation-of-energy principle in the laboratory which finally convinced his contemporaries of that principle's universal validity.

Joule was born in 1818 at Salford, near Manchester, England, the son of a wealthy brewery owner. He was educated at home and for 3 years had the eminent chemist John Dalton, then in his seventies, as his tutor. Dalton imparted to Joule his great love for science and his passion for sound numerical data on which to base scientific theories and laws. Unfortunately Joule had little mathematical training,

and this prevented him from making even more significant contributions to physics in his later life.

Joule had no real profession and no job except for his involvement in running his father's brewery. Up until 1854, when the brewery was sold, Joule worked there and did his experiments before or after work. After 1854 he had the time and the funds to continue his physics experiments in a laboratory he built at his home. Later in his life Joule suffered financial reverses and needed a government subsidy from Queen Victoria to continue his research.

From 1837 to 1847 Joule devoted all his available time to a variety of experiments on the conversion of various forms of energy—mechanical, electric, chemical—into heat. He developed thermometers capable of measuring temperatures to 1/200 of a Fahrenheit degree, and this gave his work a precision unattainable by other physicists up to that time. In 1840 he proposed that the rate at which heat was generated by an electric current I passing through a wire of resistance R was $\Delta Q / \Delta t = I^2 R$. This is now called *Joule's law*.

In June 1847 Joule presented a paper to the British Association meeting in Oxford which contained the most accurate experimental value for the mechanical equivalent of heat (Joule's equivalent) obtained up to

that time. It represented a significant achievement even when judged by today's higher standards of precision. Joule's peers were sleepy and unimpressed until a young man in the audience, William Thomson (later Lord Kelvin), pointed out to his fellow scientists the significance of Joule's work. This was the turning point in Joule's career.

In 1850 Joule was elected to the Royal Society on the basis of his work on energy conversion, and he became very influential in physics circles. He received many scientific honors, serving as president of the British Association for the Advancement of Science in 1872 and again in 1887, and having the SI unit of energy named after him.

Joule had excellent laboratory skills and a passion for careful, precise work. Even on his honeymoon he took time out to measure the temperature of the water at the top and bottom of a scenic waterfall to see if the difference agreed with the value predicted by the conservation-of-energy principle! He believed that nature was simple, and strove to find the simple relationships (like Joule's law in electricity) which he was convinced must exist between important physical quantities. His discovery of just two such relationships made a major contribution to the development of the concept of energy as we know it today.

Joule compared the values of $\mathcal{W}$ and ΔQ obtained in his paddle-wheel experiment and in this way found a numerical value for the mechanical equivalent of heat, or *Joule's equivalent*:

$$\text{Joule's equivalent} = \frac{\mathcal{W}}{\Delta Q} \tag{13.13}$$

where $\mathcal{W}$ is in joules and ΔQ is in kilocalories. Joule's efforts to determine the mechanical equivalent of heat culminated in a paper he wrote in 1847. It contained an overwhelming amount of experimental data from five series of experiments using water, sperm oil, and mercury as fluids and churning paddle wheels, frictional heating, and electricity as heat sources. Joule obtained 5 percent agreement among his results from the five series. Translated into SI units, his average result corresponds to a value of 4.15×10^3 J/kcal for the mechanical equivalent of heat. The best modern value of Joule's equivalent is 4.184×10^3 J/kcal, and so Joule's work was accurate to better than 1 percent.

Joule's equivalent: $\boxed{1 \text{ kcal} = 4.184 \times 10^3 \text{ J}}$

13.9 Change of Phase: Heats of Fusion and Vaporization

By the term *phase* we here mean the state in which a substance exists, i.e., solid, liquid, or gas. Thus the chemical substance H_2O can exist in the gaseous phase as steam, in the liquid phase as ordinary water, and in the solid phase as ice. All substances can normally exist in any one of these three phases under the proper conditions of pressure and temperature. For example, water at atmospheric pressure becomes steam at temperatures above 100°C and ice at temperatures below 0°C.

From our discussion of specific heat capacities we might think that whenever we add heat to a substance, we necessarily raise its temperature. A simple experiment shows us that this is not always the case. Consider a glass beaker containing water and ice cubes, as in Fig. 13.14. We insert a thermometer and find that the temperature of the water-ice mixture is 0°C. Then we begin to heat the ice-water mixture. We might expect the temperature to rise immediately, but it does not! Instead the temperature remains fixed at 0°C until all the ice has melted, as shown in Fig. 13.15. Only then does the temperature begin to rise.

The explanation of this behavior is that we are dealing here not with a simple substance like water or oxygen gas but with a mixture of ice and water, i.e., with H_2O in *two different phases*. To convert ice to water at 0°C, that is, to produce a *phase change*, requires an input of energy to break the bonds holding the H_2O molecules together in the ice crystal. As long as any ice is present, all the heat we put in is used in breaking bonds and converting ice to water. Only after all the ice has been changed to water does the applied heat increase the random kinetic ener-

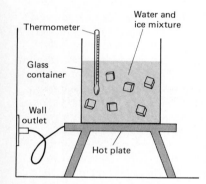

FIGURE 13.14 A beaker filled with ice and water at 0°C and resting on a hot plate.

FIGURE 13.15 Variation of temperature of an ice-water mixture with heat input. Note the break in the scale for the boiling part of the curve. Since the heat of vaporization is 540 kcal/kg and the heat of fusion is about 80 kcal/kg, the boiling part of the curve should be about seven times as long as the melting part. Note also that, since the specific heats of ice and of steam are less than the specific heat of water, the rising straight lines for ice and steam have steeper slopes than does the line for water.

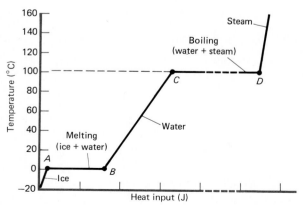

gies of the water molecules and show up as a change in temperature of the water. Again we see from this example that temperature and heat are two very different concepts. At its melting point the amount of heat added to the ice increases in time, but the temperature is unchanged.

Heat of fusion: The amount of heat required to convert 1 kg of a substance from solid to liquid.

The word *fusion* refers to the reverse process to melting. For water the heat of fusion is 80 kcal. In other words, when ice melts, each kilogram absorbs 80 kcal and, when ice freezes (or ''fuses''), each kilogram releases 80 kcal. The numerical value of the heat of fusion is different for different substances. We may, then, write in general

$$Q = L_f m \qquad \text{(13.14)}$$

where Q is the heat in kilocalories, m is the mass in kilograms, and L_f is the heat of fusion of the substance in kilocalories per kilogram.

If, after all the ice is melted, we continue to heat the water, its temperature increases constantly, as shown in Fig. 13.15, until 100°C is reached. Then boiling occurs, and all the applied heat is used to break the H_2O molecules free from the liquid and allow them to escape as vapor molecules. To do this requires a ''heat of vaporization'' of 540 kcal/kg. While boiling is taking place, the temperature remains fixed at 100°C. It is this constancy of temperature at the point where a change of phase occurs that makes melting and boiling points so useful in defining temperature scales.

No matter how vigorously water boils, it remains at the same temperature (100°C at atmospheric pressure). This is why double boilers are useful for heating substances that scorch easily. In a double boiler such substances are subjected only to the boiling temperature of the water regardless of how hot the burner on the stove may be.

For vaporization we can write an equation similar to Eq. (13.14),

$$Q = L_v m \qquad \text{(13.15)}$$

where L_v is the *heat of vaporization* of the substance involved.

Heat of vaporization: The heat required to convert 1 kg of a substance from liquid to vapor.

After all the water has been converted into steam, we can collect the steam in a closed vessel and continue to heat it. Its temperature will increase above 100°C according to Eq. (13.12), where c is now the specific heat capacity of *steam*, 0.48 kcal/(kg·C°). This is shown in Fig. 13.15.

Values of the heats of fusion and vaporization for various substances are given in Table 13.5. Notice how much larger the values are for water than for other common substances.

The specific heat of water is also larger than that of most common substances, as Table 13.4 shows. This explains the moderating effect on the climate of large bodies of water like the oceans, since large amounts of heat are required to change the temperature of the oceans by small amounts.

TABLE 13.5 Heats of Fusion and Vaporization

Substance	Heat of fusion (kcal/kg)	Heat of vaporization (kcal/kg)
Ethyl alcohol	16.4	204
Helium	· · ·	5
Hydrogen	14.0	108
Mercury	2.82	70.6
Methyl alcohol	16.4	262.8
Nitrogen	6.09	47.6
Oxygen	3.30	50.9
Water	79.7	539.6

Example 13.5

An ice storm leaves a 5.1-cm layer of ice on the driveway of a professor's house. The next day the temperature is exactly 0°C, and the professor wants to know whether to break up the ice or let the sun melt it away. If the driveway, which has an area of 80 m², is directly exposed to the sun and if the average solar energy falling on the driveway is 200 W/m², how long will it take to melt all the ice? (The density of ice was given in Table 10.3 as 0.92×10^3 kg/m³.)

SOLUTION

The volume of ice in an area of 1.0 m² is

$$V = (0.051 \text{ m})(1.0 \text{ m}^2) = 0.051 \text{ m}^3$$

and the mass of this much ice is $m = \rho V = (0.92 \times 10^3$ kg/m³)(0.051 m³) = 47 kg.

The amount of heat required to melt this ice to water at 0°C is

$$Q = mL_f = (47 \text{ kg})(80 \text{ kcal/kg}) = 3.8 \times 10^3 \text{ kcal}$$

or, converting to joules,

$$Q = (3.8 \times 10^3 \text{ kcal})(4.18 \times 10^3 \text{ J/kcal}) = 1.6 \times 10^7 \text{ J}$$

Now, 200 W/m² is equal to 200 J/s falling on 1.0 m², and so the time required to melt all the ice is

$$t = \frac{Q}{P} = \frac{1.6 \times 10^7 \text{ J}}{200 \text{ J/s}} = 8.0 \times 10^4 \text{ s} = \boxed{22 \text{ h}}$$

If the professor wants the ice removed, then, she had better start chopping away at it, since it will be dark before half the ice melts.

Note: The area of the driveway is of no importance in this problem, since the ice is assumed to have the same thickness over the whole driveway and the energy input from the sun is the same over the whole driveway. Hence the result would be the same, whether calculated for 1 m² or 80 m².

Vapor Pressure and Boiling

The heat of vaporization is usually given for the normal boiling temperature of liquids, for example, 100°C for water at atmospheric pressure. At different pressures, however, liquids boil at different temperatures and also have different heats of vaporization at these temperatures.

If a liquid substance partially fills a closed container, some molecules will evaporate from the liquid into the gaseous state and other vapor molecules will return from the gaseous to the liquid state. When the number of molecules leaving the liquid is exactly equal to the number returning to the liquid, the vapor is said to be *saturated*, and the vapor phase is in *equilibrium* with the liquid phase.

Vapor pressure: The pressure exerted by a vapor when in equilibrium with its own liquid phase, i.e., the pressure of the saturated vapor.

As the temperature rises, the number of molecules leaving the liquid increases, and as a result the vapor pressure also increases. If a liquid is heated continuously, some evaporation occurs even at temperatures well below the boiling point and increases as the temperature increases. Bubbles of vapor form within the liquid, but if the external pressure is greater than the vapor pressure of the liquid at the given temperature, the bubbles collapse as they are formed. If the temperature is increased to the point where the vapor pressure of the liquid becomes equal to the external pressure, then these bubbles rise up through the liquid and *boiling* occurs. For water at atmospheric pressure, this occurs at 100°C.

Boiling point: That temperature at which the vapor pressure of a liquid is equal to the external pressure.

While boiling, the liquid is rapidly converted to a vapor at the boiling point temperature, as long as for each kilogram of liquid boiled off an amount of energy equal to the heat of vaporization is provided by some source external to the liquid.

For water under atmospheric pressure (760 mmHg, or 760 torr), boiling occurs at 100°C with a heat of vaporization of 540 kcal/kg. At Denver, Colorado, which

TABLE 13.6 Melting and Boiling Temperatures for Some Important Substances

Substance	Melting point (°C)	Boiling point (°C)
H_2O	0	100
Aluminum	660	2057
Copper	1083	2300
Gold	1063	2600
Lead	327	1620
Nickel	1455	2900
Platinum	1774	4300
Silver	961	1950
Tungsten	3370	5900
Uranium	1133	4740

is approximately 1 mi above sea level, atmospheric pressure is reduced to about 610 torr, and water boils at about 94°C with a heat of vaporization of 543 kcal/kg. On the other hand, in a pressure cooker the boiling temperature of water is raised by increasing the pressure on the water. In this way it is possible to cook food at a higher temperature.

Changes in pressure also affect the freezing point of liquids but to a much more limited extent. Thus an increase in pressure of 1 atm lowers the freezing point of water by only about 0.007°C. Table 13.6 gives the melting and boiling temperatures of some important substances at atmospheric pressure.

Sublimation

Under conditions of reduced pressure, ice changes directly to steam without going through the liquid state. This is called *sublimation*, and the heat required to cause this transformation is called the *heat of sublimation*. Sublimation is not normally observed for water, because the required pressure is so low, below 4.58 torr.

A pressure-temperature phase diagram for water is shown in Fig. 13.16a. It includes the *triple point* of water at 0.01°C and 4.58 mmHg.

Triple point: That temperature and pressure at which a substance can exist in all three phases (solid, liquid, and vapor) simultaneously.

Note on the phase diagram for water that, by keeping the pressure constant below 4.58 mmHg and increasing the temperature, it is possible to have the substance *sublime*, as indicated by the line $A \rightarrow B$ on the diagra n.

Figure 13.16b shows a similar phase diagram for carbon dioxide (CO_2), on which the triple point occurs at 5.11 atm and -56.6°C. Hence at atmospheric pressure CO_2 sublimes, passing directly from the solid phase (''dry ice'') to the vapor phase along the line AB. The fusion curve for CO_2 has a positive slope, which is typical of most substances and unlike the negative slope shown for H_2O (a most untypical substance) in Fig. 13.16a. Notice the similar forms of the curves describing the phase transitions for these two substances, even though the pressures and temperatures differ greatly.

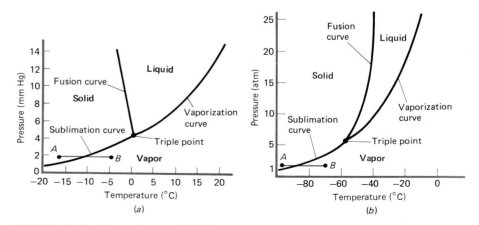

FIGURE 13.16   Phase diagrams for (a) water (H_2O) and (b) carbon dioxide (CO_2). Note the different units on the pressure scales in the two cases.

Example 13.6

Tissue ablation (removal of diseased tissue) is often accomplished by laser surgery. Because body tissue is 70 to 90% water, thermal ablation using lasers may, to a first approximation, be considered simply as the heating and vaporization of water. Approximately how long would it take for a YAG laser, which can deliver 10 W of power on a continuous basis, to destroy 1 cm³ of diseased tissue?

SOLUTION

We assume, as an average value, that the tissue is 80% water. The water content of the tissue is then 0.80 cm³, and its mass is 0.80 g = 8.0×10^{-4} kg. The thermal energy required first to raise the temperature of the water from body temperature (37°) to 100°C and then to vaporize it is

$$W = cm\,\Delta T + mL_v$$

$$= \left(\frac{1\ \text{kcal}}{\text{kg}\cdot\text{C}°}\right)(8.0 \times 10^{-4}\ \text{kg})(100°\text{C} - 37°\text{C})$$

$$+ (8.0 \times 10^{-4}\ \text{kg})(540\ \text{kcal/kg})$$

$$= 5.0 \times 10^{-2}\ \text{kcal} + 0.432\ \text{kcal}$$

$$= 0.48\ \text{kcal} = 2.0 \times 10^3\ \text{J}$$

If the YAG laser has a power output of 10 W, then

$$W = Pt \quad \text{or} \quad t = \frac{W}{P} = \frac{2.0 \times 10^3\ \text{J}}{10\ \text{J/s}} = \boxed{2.0 \times 10^2\ \text{s}}$$

About 3 min would therefore be required to destroy the diseased tissue.

Note that most of this energy is required to vaporize the water in the tissue; much less thermal energy is needed to raise the temperature of the water from 37 to 100°C.

Summary: Important Definitions and Equations

Temperature: The absolute (Kelvin) temperature T of a gas is directly proportional to the average random translational kinetic energy of the gas molecules.
Absolute zero: That temperature at which the pressure of an ideal gas would fall to zero at constant volume; 0 K = $-273.15°$C.
Third law of thermodynamics: Zero on the Kelvin scale is unattainable in a finite number of operations.
Temperature scales (Kelvin, Celsius, and Fahrenheit):

$$T(\text{K}) = T(°\text{C}) + 273.15 \qquad \textbf{(13.2)}$$

$$T'(°\text{F}) = (\tfrac{9}{5})T(°\text{C}) + 32 \qquad \textbf{(13.3)}$$

Thermal expansion of a rod:

$$L = L_0\,(1 + \alpha\,\Delta T) \qquad \textbf{(13.7)}$$

where α is the coefficient of linear expansion.

Volume expansion of a solid or liquid:

$$V = V_0(1 + \beta\,\Delta T) \qquad \textbf{(13.10)}$$

where β is the coefficient of volume expansion.

Thermal energy: The sum of all the random mechanical energies of the atoms or molecules in a substance.

Heat: The thermal energy transferred between a system and its surroundings as a result of temperature differences only.

$$\Delta Q = cm\,\Delta T = cm(T_2 - T_1) \qquad \textbf{(13.12)}$$

Specific heat capacity (c): The quantity of heat that must be supplied to a unit mass of a substance to raise its temperature by one degree.

Heat units
One kilocalorie (kcal): The amount of heat required to raise the temperature of one kilogram of water one degree Celsius (from 14.5 to 15.5°C).
One British thermal unit (Btu): The amount of heat required to raise the temperature of one pound of water one degree Fahrenheit (from 63 to 64°F).
Mechanical equivalent of heat (Joule's equivalent): 1 kcal = 4.184×10^3 J.

Phase: The state in which a substance exists, i.e., solid, liquid, or gas.
Phase transition: A change in the phase of a substance accompanied by the absorption or liberation of heat without any change in temperature. $Q = Lm$, where L is the heat absorbed or liberated per kilogram of substance.
Heat of fusion (L_f): The amount of heat required to convert 1 kg of a substance from solid to liquid.
Heat of vaporization (L_v): The amount of heat required to convert 1 kg of a substance from liquid to vapor. (For water L_f = 80 kcal/kg; L_v = 540 kcal/kg.)
Heat of sublimation: The amount of heat required to change 1 kg of a substance directly from solid to vapor.
Vapor pressure: The pressure exerted by a vapor when in equilibrium with its own liquid phase, i.e., the pressure of the saturated vapor.
Boiling point: That temperature at which the vapor pressure of a liquid is equal to the external pressure.
Triple point: That temperature and pressure at which a substance can exist in all three phases (solid, liquid, and vapor) simultaneously.

Questions

1 Does the question, "Do you have a temperature?" often asked by a parent of a sick child, make sense physically? What is really meant?

2 Name three properties of substances, other than the thermal expansion of gases, liquids, and solids, that might be used to measure temperature.

3 Why in practice cannot a single thermometer be used to measure temperatures all the way from 0 to 10^8 K?

4 Is there any temperature that is the same in degrees Fahrenheit and in kelvins?

5 If it proves difficult to open a jar with a screw-on metal lid, the conventional wisdom is that you should put the lid under hot running water. Why would you expect such a technique to help in opening the jar?

6 A flat piece of metal with a round hole in it and initially at room temperature is heated to 1000°C. Does the hole get larger, smaller, or remain the same size? Why?

7 Does the temperature of a gas depend on how much gas is present?

8 Show, by analogy with the flow of water, why heat always flows of its own accord from a higher to a lower temperature and never the reverse.

9 Heats of fusion and vaporization were once referred to as "latent heats." Can you explain why this term was introduced in connection with the processes of freezing and boiling?

10 Explain why in a state like Maine less snow and higher temperatures occur in winter along the coast than inland.

11 The vapor pressure of a liquid does not depend on the external pressure, i.e., the pressure of the atmosphere pressing down on the liquid. Why, then, does the boiling point of a liquid depend on the external pressure?

12 You want to boil water but have no way to heat it. You do have available, however, a vacuum chamber from which the air can be pumped out. Explain how you might make water boil at room temperature (about 20°C) using this apparatus.

13 Why is water a particularly good substance to use in home heating systems, whether they be hot-water or steam heating systems?

14 Explain why in a house whose furnace breaks down in the middle of winter there is danger that the water pipes will burst. Can you suggest a way to prevent this from happening?

15 Count Rumford once remarked that he was amazed at how long apple pies retain their heat after being taken out of the oven. What explanation can you provide for this?

16 Which is more effective in cooling down a cup of boiling coffee to a drinkable temperature: 20 g of water at 0°C or 20 g of ice at 0°C? Why?

Multiple-Choice and Simple Exercises

13.1 If the pressure of the hydrogen in a constant-volume hydrogen gas thermometer is 54.6 cmHg when the temperature is 0°C, a pressure of 84.8 cmHg indicates a temperature of:
(a) 42°C   (b) 122°C   (c) 151°C
(d) 212°C   (e) 524°C

13.2 A nurse finds that a patient's temperature is 102°F. On the Celsius scale, this is:
(a) 102°C   (b) 39°C   (c) 35°C   (d) 375°C
(e) 312°C

13.3 When a metal is heated, the order of magnitude of its change in length per unit length per Celsius degree is:
(a) 10^{-3} (b) 10^{-5} (c) 10^{-7} (d) 10^{-9}
(e) 10^{-12}

13.4 If a brass rod is exactly 1.0 m long at 25°C and 1.00135 m long at 100°C, its average coefficient of linear expansion [in units of $(C°)^{-1}$] between these two temperatures is:
(a) 0.000018   (b) 0.000054   (c) 0.000135
(d) 0.00135   (e) 1.00135

13.5 If the coefficient of linear expansion of iron is $1.00 \times 10^{-5}/C°$, an iron cube 5.00 cm on an edge heated from 10.0 to 60.0°C has its volume increased by:
(a) 0.0375 cm³   (b) 0.188 cm³   (c) 0.225 cm³
(d) 0.500 cm³   (e) 0.625 cm³

13.6 It takes 10 kcal to raise the temperature of a certain mass of water 10 C°. The amount of heat required to raise the temperature of the same mass of ice 10 C° is about:

(a) 10 kcal   (b) 1.0 kcal   (c) 20 kcal
(d) 5.0 kcal   (e) 2.0 kcal

13.7 The thermal energy of 1 mol of an ideal monatomic gas at 300 K is about:
(a) 4×10^3 J   (b) 4×10^3 kcal
(c) 4×10^6 J   (d) 4×10^3 cal   (e) 4 J

13.8 The specific heat capacity of beryllium metal is about twice that of aluminum. Identical masses of aluminum and beryllium at room temperature (20°C) are dropped together into a can of hot water. When the system has come to equilibrium:
(a) The two masses are at the same temperature.
(b) The final temperature of the two masses depends only on the initial temperature of the water.
(c) The aluminum is at a higher temperature than the beryllium.
(d) The final temperature depends on the shape of the two masses.
(e) The beryllium is at a higher temperature than the aluminum.

13.9 When water freezes:
(a) Heat is absorbed from the surroundings.
(b) Heat is given off to the surroundings.
(c) There is no heat exchange with the surroundings.
(d) The temperature of the water increases.
(e) The temperature of the water decreases.

13.10 The principal advantage of a pressure cooker is that:
(*a*) Less heat is lost than in an open vessel.
(*b*) The high pressure generates heat.
(*c*) Increase in pressure raises the boiling point of water.
(*d*) Only a small amount of water is necessary.
(*e*) Less heat is required to boil the water.

13.11 Express 0, 300, and 600 K in degrees Fahrenheit.

13.12 Express −40, 68, and 98.6°F in degrees Celsius.

13.13 Waterford crystal is made by first melting a mixture of silica, potash, and red lead oxide in a furnace at 1200°C. What is this temperature in degrees Fahrenheit and in kelvins?

13.14 (*a*) In kelvins, what is the ratio of the melting temperature of platinum to that of lead?
(*b*) What is this ratio in degrees Celsius? (See Table 13.6.)

13.15 A steel bridge is 600 m long at 20°C. By how much does it shorten when the temperature falls to 0°C?

13.16 If the earth were a steel ball, what temperature change would make its radius increase 35 m?

13.17 How much heat is required to produce a 10 C° temperature change in 2.0 kg of (*a*) aluminum, (*b*) ice, and (*c*) water?

13.18 How many kilocalories are required to heat 1.0 kg of water in a 0.10-kg aluminum container (*a*) from 5.0 to 50°C? (*b*) From 50 to 95°C?

13.19 What is the specific heat capacity of mercury in J/(kg·K)?

13.20 (*a*) How much heat is required to convert 2.0 kg of solid mercury at −38.87°C (its freezing point) to liquid mercury at the same temperature?
(*b*) How much heat is required to convert 0.50 kg of liquid mercury at 356.6°C (its boiling point) to mercury vapor?

13.21 How much heat must be extracted from 0.50 kg of hydrogen gas at 20.4 K (its boiling point) to convert it to liquid at the same temperature?

13.22 Healthy human bodies radiate heat at a rate of about 120 W. If all this heat could be used to melt ice, how many kilograms of ice per second would be melted?

13.23 A pot of water is heated on a stove from 20 to 90°C. How much is this change in temperature on the Fahrenheit and the Kelvin scales?

13.24 The temperature of the sun is about 6.0×10^3 K. What is this temperature in degrees Fahrenheit and in degrees Celsius?

Problems

13.25 At what temperature is the reading of a thermometer in degrees Fahrenheit equal to twice the same temperature in degrees Celsius?

13.26 The bulb of a constant-volume gas thermometer is placed in an ice-water mixture and the pressure recorded as 250 mmHg. The bulb is then immersed in a liquid, and the pressure rises in the gas thermometer to 350 mmHg. What is the temperature of the liquid in degrees Celsius?

****13.27** Consider a mercury-in-glass thermometer. If the cross section of the capillary is a constant A and if the mercury just fills the bulb of volume V_0 at 0°C, show that the length of the mercury in the capillary at temperature t°C is $L = (V_0/A)(\beta − 3\alpha)t$, where β is the volume coefficient of expansion of mercury and α is the linear coefficient expansion of the glass.

13.28 One mole of CO_2 vapor is heated at constant pressure from 20 to 55°C. What is the ratio of the final volume to the initial volume?

13.29 A Pyrex flask in a refrigerator at −10°C is filled to the top and holds exactly 1000 cm³ of ethyl alcohol. If the refrigerator fails, how much alcohol spills by the time the temperature reaches 22°C?

13.30 A solid brass block at exactly 20°C has a circular hole in it of radius 2.000 cm. If the brass block is thrown into a pot of boiling water at 100°C, what is the cross-sectional area of the hole in the block when the temperature of the block reaches 100°C?

13.31 A steel tape measure gives the length of a brass rod to be 125.00 cm when both are at 20°C. What would the tape measure read when the temperature increases to 40°C? (*Note:* Five significant figures are required in your answer.)

****13.32** Two rods, each of 5.0-cm² cross-sectional area and 15 cm long, are placed end to end and held rigidly in clamps at the two free ends. One rod is made of steel and the other of brass. If, after the rods are put in place, they are heated until their temperature increases by 100 C°, what is the stress set up in each rod?

****13.33** When the outside temperature is 10°C, a steel beam of cross-sectional area 120 cm² is installed in a building with the ends of the beam bolted securely to two immovable pillars. In the summer the temperature of the beam rises to 35°C. What is the compressional force on the beam?

13.34 An aluminum meterstick of mass 0.40 kg reads correctly at 20°C. How much longer will the meterstick be if it absorbs 7.0 kcal of heat?

13.35 A 13.6-kg sample of liquid mercury occupies 1.00×10^3 cm³ at its melting point (−38.9°C).
(*a*) What is its volume just as it reaches its boiling point (356.9°C)?
(*b*) How many kilocalories of heat does this take?

13.36 (*a*) Suppose a 75-kg person consumes food providing 2500 kcal of energy in a day. If all that energy went into heat energy in the person's body, what would the person's final temperature be? [Assume $c = 0.83$ kcal/(kg·C°).]
(*b*) What is the total number of kilocalories that must be taken in to get the person to the boiling point of water? (Neglect any heat losses from the body.)

13.37 Suppose air has density 1.3 kg/m³ and specific heat capacity 0.20 kcal/(kg·C°). If the volume of air in the passenger compartment of an automobile is 2.0 m³, how long would it take the body heat of one person (120 W) to raise the air temperature from 15 to 37°C, if no air enters or leaves the car?

13.38 A 0.70-kg cup [$c = 0.22$ kcal/(kg·C°)] is at 20°C. If 0.25 kg of water at 95°C is poured into the cup, what is the final temperature?

13.39 Suppose we drop a 1.0-kg piece of iron at 200°C into 5.0 kg of water in a 0.20-kg aluminum container at room temperature (20°C). If no heat is exchanged with the surroundings, what is the final temperature of the iron?

13.40 A 5.0×10^{-2} kg iron vessel and the 0.20 kg of water in it are at 20°C. If 8.0×10^{-2} kg of metal shot at 100°C is dropped into it, the final temperature is 24°C. What is the specific heat capacity of the metal?

13.41 A 0.21-kg brass calorimeter (an insulated container used for heat measurements) contains 0.10 kg of water. If 1.20 kcal is required to raise the temperature of the water and container 10 C°, what is the specific heat capacity of brass?

13.42 According to the law of Dulong and Petit, most metals have a molar heat capacity of 6 cal/(mol·C°) at room temperature. A metal has molecular mass 197. If it obeys the Dulong-Petit law, what is its specific heat capacity?

13.43 The temperatures of three different liquids are maintained at 10, 20, and 30°C, respectively. It is found that mixing equal masses of the first two liquids leads to a final temperature of 16°C and that mixing equal masses of the second and third liquids leads to a final temperature of 24°C. What is the final temperature if equal masses of the first and third liquids are mixed together?

13.44 An aluminum can of mass 60 g is used as a calorimeter in which heat experiments are performed. The can contains 120 g of a water-ice mixture. An aluminum bar of mass 120 g is taken out of a steam bath, where it has reached a temperature of 100°C, and dropped into the can. The temperature of the contents rises to 6.0°C. How much ice was originally in the can?

13.45 In a student experiment intended to measure Joule's equivalent, a mass of 5.00 kg falls through a height of 3.00 m and rotates a paddle wheel that stirs up 0.150 kg of water: If the water is initially at 20.0°C and no heat is lost to the container or the surroundings, what is the final temperature of the water?

13.46 A machinist is drilling a hole in a 1.0-kg block of iron with an electric drill. Electric power is supplied to the drill at a rate of 0.50 kW for 3.2 min to complete the hole.

(a) How much heat is produced?

(b) If it all goes to warm the iron block, how much does the temperature of the block rise?

13.47 Water flows over Niagara Falls at the United States-Canada border and crashes to the river 50 m below the top of the falls. If all the energy in the water is converted into heat in the water at the bottom of the falls, how much does the temperature of the water rise?

13.48 A 2.5-g lead bullet is fired at a speed of 250 m/s into the 4.0-kg wooden mass of a ballistic pendulum similar to that in Fig. 7.13.

(a) How much energy is converted into heat in this inelastic collision?

(b) How much does the temperature of the bullet increase, if we assume it absorbs all the heat generated?

13.49 A mixture of 100 g of ice and 200 g of water is heated by adding 5.0 kcal of heat to the mixture with a heating coil. What is the final temperature of the ice-water mixture?

13.50 One kilogram of water gives up 80 kcal upon freezing. If another 80 kcal is removed, what is the final temperature of the ice?

13.51 A thermos bottle contains 100 g of ice at 0°C.

(a) How much water at 100°C must be added to just melt the ice?

(b) What mass of steam at 100°C is just sufficient to melt the ice?

13.52 A 5.0-kg iron ball at 1200°C is put into a 2.0-kg iron pail containing 12 kg of water at 20°C. What mass of steam is produced?

13.53 A 1.0-mm-thick crust of ice covers a windshield of area 0.50 m².

(a) How much energy is required to melt the ice?

(b) If the job is to take exactly 3 min, what must be the output of the car's heater (in watts)?

13.54 Air conditioners are sometimes rated in tons. Thus a 2-ton air conditioner removes heat at the same rate as would the melting of 2 tons (1 ton = 2000 lb) of ice per day. What is the rate (in watts) at which a 2.0-ton air conditioner removes heat from a house?

13.55 To compare the energy required per molecule for the melting or vaporization of three very different substances:

(a) Calculate the heats of fusion per mole for water (H_2O), mercury (Hg), and hydrogen (H_2).

(b) Do the same for the heats of vaporization per mole.

(c) Comment on any notable trends in these data.

13.56 If 0.50 kg of ice at $-10°C$ is put into a large container of water at 0°C, how much water changes to ice?

13.57 From what height must an ice cube fall to acquire sufficient energy to melt itself?

13.58 For simplicity, assume that the heat of vaporization of water is independent of temperature. If so, how many kilograms of perspiration (water) are needed to change the temperature of a 75-kg person by 1 C°?

13.59 A marathon runner in the middle of a race generates heat at a rate of 10^3 W (as compared with 120 W for a person at rest).

(a) How many kilograms of water per hour must the runner lose by evaporation, with the skin at a temperature of 35°C, to keep the runner's body temperature from rising? (Assume that the heat of vaporization of water at 35°C is 577 kcal/kg.)

(b) What percentage of the runner's total body mass of 80 kg does this loss represent?

13.60 Compare the amount of steam that must be circulated in a steam heating system to produce 10^3 kcal of heat with the amount of hot water needed to achieve the same heating effect. Assume that the hot water leaves the furnace at 60°C and returns at 35°C.

13.61 A closed vessel contains liquid water in equilibrium with its vapor at 100°C and 1 atm. Compare the mean kinetic energy of a vapor molecule with the energy required to transfer one molecule from the liquid to the vapor phase.

13.62 In 1986 a study of 17,000 Harvard alumni showed that moderate physical exercise in adult life can significantly increase life expectancy. Men who participated in activities like walking, stair climbing, and active sports that burned up over 2000 kcal/week had death rates one-half to one-third lower than those in the study group who were least active.

(a) How far would a 90-kg man have to climb vertically at a slow rate to burn up 2000 kcal?

(b) How many flights of stairs (each flight about 4 m high) would the man have to climb to burn up this much energy?

Additional Readings

Holton, Gerald: *Introduction to Concepts and Theories in Physical Science*, 2d ed. (revised and with new material by Stephen G. Brush), Addison-Wesley, Reading, Mass., 1973. Chapter 17 in particular is excellent with respect to the historical and philosophical aspects of the material in this and the next chapter.

MacDonald, D. K. C.: *Near Zero: An Introduction to Low Temperature Physics*, Doubleday/Anchor, Garden City, N.Y., 1961. A popular book which includes discussions of magnetic cooling, superconductivity, the properties of liquid helium, and some interesting applications.

Scientific American, vol. 191, no. 3, September 1954. This issue contains nine articles on heat, thermal energy, and temperature. Especially good is the one entitled "What Is Heat?" by Freeman J. Dyson.

Shamos, Morris H. (ed.): *Great Experiments in Physics*, Holt, Rinehart and Winston, New York, 1959. Each chapter of this book is devoted to an important experiment in the history of physics, as described by the originator of the experiment and annotated by the editor. Chapter 12 is devoted to Joule's work on the mechanical equivalent of heat.

Steffens, Henry John: *James Prescott Joule and the Concept of Energy*, Science History Publications, New York, 1979. A useful account not merely of the work of Joule but also of the contributions of Mayer, Kelvin, Helmholtz, and Clausius to the development of thermodynamics.

Zemansky, M. W.: *Temperatures Very Low and Very High*, Van Nostrand, New York, 1964. An interesting account of how matter behaves at extreme temperatures.

The two books on Count Rumford by Sanford Brown referred to in Chap. 6 are also relevant here.

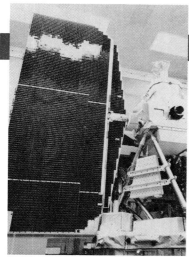

CHAPTER 14

The Transfer of Heat and the First Law of Thermodynamics

I shall lose no time in repeating and extending these experiments, being satisfied that the grand agents of nature are, by the Creator's fiat, indestructible; and that, whenever mechanical force is expended, an exact equivalent of heat is always obtained.

Sir James Joule (1818–1889)

Many of us have used a hand pump to put air into a bicycle or automobile tire and have noticed that both the tire and the pump became hot in the process. Where did this heat come from? No obvious source of heat was present, since the air in the tire and the air outside the tire were probably initially at the same temperature. We did work, however, to operate the pump, and hence we might suspect that there was some relationship between the work we did and the thermal energy which raised the temperature of the air in the tire. In this chapter, after first discussing the heat transfer processes of conduction, convection, and radiation, we consider how mechanical work can increase the thermal energy of a substance. Then, by combining these ideas on the transfer of heat and the performance of work, we arrive at the first general principle of the science of heat, the first law of thermodynamics.

14.1 The Conduction of Heat

The molecules in a metal hot plate are in violent, agitated motion, since they are at a high temperature. These molecules collide with the glass molecules in the bottom of a gas-filled glass flask set on top of the hot plate and transfer some of their energy to the glass molecules. These in turn increase their motion, and, even though they cannot move freely through the glass, they collide with and pass on some of their excess energy to neighboring molecules. These in turn pass energy on to their neighbors, etc., until finally the gas molecules in the container receive the heat that has passed through the glass.

This process, in which heat is carried from one molecule to the next through a series of collisions, is called *conduction. In heat conduction, molecules, while remaining at relatively fixed positions in space, absorb and pass on heat in the form of random mechanical motion.* We describe this process by saying that a certain amount of heat ΔQ is conducted through a substance from a higher temperature T_2 to a lower temperature T_1. The amount of heat conducted per second depends on both the number and the kind of molecules involved.

In Fig. 14.1 we consider the conduction of heat through a solid metal rod of

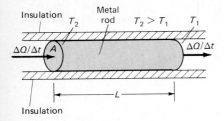

Insulation   Metal rod

T_2 $T_2 > T_1$ T_1

$\Delta Q/\Delta t$ A $\Delta Q/\Delta t$

L

Insulation

FIGURE 14.1 The conduction of heat along a solid metal rod of length L and cross-sectional area A. Heat flows from the higher temperature T_2 to the lower temperature T_1.

length L whose ends are kept at temperatures T_2 and T_1 and whose sides are insulated so that no heat can escape. We might expect the amount of heat transferred to be directly proportional to the cross-sectional area A of the rod normal to the direction of heat flow (for this determines the number of molecules involved) and to the length of time Δt the heat flows. We might also expect the amount of heat transferred to be proportional to the temperature gradient $(T_2 - T_1)/L$, just as the flow of water in a pipe is proportional to the pressure gradient according to Poiseuille's law, Eq. (10.11). We can therefore write

$$\Delta Q \propto \frac{A(T_2 - T_1)\,\Delta t}{L}$$

The rate at which heat flows is then

$$\frac{\Delta Q}{\Delta t} \propto \frac{A(T_2 - T_1)}{L}$$

where $\Delta Q/\Delta t$ has the dimensions of *power*. We can convert this proportionality into an equality by introducing a constant of proportionality called the *coefficient of thermal conductivity* k. We then have

$$\boxed{\frac{\Delta Q}{\Delta t} = k\,\frac{A(T_2 - T_1)}{L}} \tag{14.1}$$

In this equation, which has been well confirmed by experiment, the rate of heat conduction $\Delta Q/\Delta t$ is usually expressed in kilocalories per second, A in meters squared, L in meters, and $T_2 - T_1$ in Celsius degrees, or kelvins. The units for thermal conductivity k are therefore

$$\frac{\text{kcal}}{\text{s}} \cdot \frac{\text{m}}{\text{m}^2} \cdot \frac{1}{\text{C}°} \qquad \text{or} \qquad \text{kcal}/(\text{s}\cdot\text{m}\cdot\text{C}°)$$

TABLE 14.1 Thermal Conductivities

Material	Thermal conductivity k [W/(m·C°)]*
Metals:	
Aluminum	2.4×10^2
Brass	1.1×10^2
Copper	3.8×10^2
Silver	4.2×10^2
Steel	0.50×10^2
Nonmetals:	
Air	0.024
Concrete	~1.0
Glass (average)	0.84
Ice	2.1
Rock wool	0.041
Styrofoam	0.010
Water (liquid)	0.58
Wood	0.084–0.17

*1 W/(m·C°) $= 0.239 \times 10^{-3}$ kcal/(s·m·C°).

The rate of heat flow, since it has the dimensions of power, can also be expressed in joules per second (watts), in which case k has to be in $W/(m \cdot C°)$. The conversion factor is $1\ W/(m \cdot C°) = 0.239 \times 10^{-3}\ kcal/(s \cdot m \cdot C°)$.

The thermal conductivity k varies greatly from material to material, as Table 14.1 shows, being large for a good heat conductor and small for a poor one. Metals have large thermal conductivities and are very good conductors of heat just as they are good conductors of electricity. Gases are poor heat conductors, or good heat insulators. Materials like Styrofoam and rock wool are good insulators because they are porous and contain large amounts of air in their interstices. Similarly, clothing insulates the body by trapping a layer of warm air that prevents heat flow from the body to the cold air outside.

Carpeting in a bathroom is more comfortable than tile to bare feet on a winter morning because the carpeting conducts heat poorly. As a consequence, it heats up quickly to the temperature of our bodies and feels relatively warm to the touch. A bare tile bathroom floor, on the other hand, quickly conducts heat away from our feet and for this reason feels much colder.

Example 14.1

(a) Calculate the rate of heat flow through a glass window 1.0 m wide and 1.5 m high if the glass is 2.0 mm thick. Assume that the temperatures of the inner and outer surfaces of the glass are 10 and 5°C, respectively. **(b)** The window is replaced by a wall with rock wool insulation 15 cm thick. Assume that the inner and outer surfaces of the wall remain at the same temperatures as the inner and outer surfaces of the window, and calculate the rate of heat flow through the $1.0 \times 1.5\ m^2$ area of wall.

SOLUTION

(a) From Eq. (14.1),

$$\frac{\Delta Q}{\Delta t} = k \frac{A(T_2 - T_1)}{L}$$

$$= \left(\frac{0.84\ W}{m \cdot C°}\right)(1.0\ m)(1.5\ m)\frac{(10 - 5)\ C°}{2.0 \times 10^{-3}\ m}$$

$$= \boxed{3.2 \times 10^3\ W}$$

(b) $\frac{\Delta Q}{\Delta t} = \left(\frac{0.041\ W}{m \cdot C°}\right)(1.0\ m)(1.5\ m)\frac{(10 - 5)\ C°}{0.15\ m}$

$$= \boxed{2.1\ W}$$

The loss of heat is reduced by a factor of 1.5×10^3 when the window is replaced by a wall, under the conditions stated in the problem.

This rough calculation shows that a major source of heat loss in houses is by conduction through windows. Note that we have not said that the temperatures inside and outside the house were 10 and 5°C. It is likely that in this case the inside temperature might be close to 20°C and the outside air temperature well below freezing. Much of the fall in temperature occurs in the thin layers of air that are immediately adjacent to the glass on the two sides. How effective these layers are in reducing the conduction of heat depends on how much they are stirred up by the motion of the surrounding air. A strong wind, for example, continuously replaces the layer of air on the outside of the window pane with colder air and increases the heat loss through the window. Storm windows and thermal-pane windows reduce this heat loss by introducing a layer of trapped air, or a vacuum, between the inner and outer windows. This trapped air layer reduces the heat losses by both conduction and convection (Sec. 14.2).

Heat Conduction in Houses

Understanding Eq. (14.1) can help us save energy in our homes. Since most heat losses in a house are by conduction, insulation is very important. This means making k, A, and $T_2 - T_1$ as small as possible and L as large as possible. We have little control over T_1, which in this case is the outside temperature, but turning the thermostat down a few degrees and thus lowering T_2 can provide real savings. The surface area A in contact with the outside is a function of the kind of house in which we live. Townhouses and apartments are much easier to keep warm in winter than detached houses because the latter have more surface area in direct contact with the elements. The crucial factor in insulating a house (in addition to caulking cracks and holes) is to use materials with low thermal conductivities and to make L as large as possible by having the walls as thick as possible. Modern building practice calls for relatively thin outer and inner walls, with the space between them filled with large thicknesses of insulating materials of low k. This contrasts with older building styles, in which very thick stone walls were used.

14.2   The Convection of Heat

If we set a beaker of water on a hot plate, we find that once the heat is conducted through the glass bottom of the beaker, a new kind of heat transfer process occurs. The glass molecules conduct heat to the water molecules touching the glass. The heated water molecules then rise to the surface of the water and are replaced by cooler water molecules, which are then heated and repeat the cycle, as in Fig. 14.2.

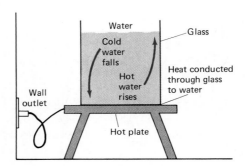

FIGURE 14.2   The convection of heat in a beaker of water.

Convection: The process whereby heated molecules carry thermal energy with them as they move from one part of a fluid to another.

In this case convection occurs because hot water, being lighter than cold water, spontaneously rises to the surface. A similar process occurs in a room heated by a radiator. Air molecules are constantly being heated by contact with the radiator. They rise and are replaced by cooler air molecules, which are in turn heated, rise, and repeat the cycle.

Convection is a difficult topic to handle quantitatively, because it involves both a hydrodynamic problem (the complicated flow of a fluid) and a heat transfer problem. In fact, it is impossible to derive any general equation to describe convection. The best we can do is to define a convection coefficient h by analogy with the coefficient of thermal conductivity k previously introduced. We therefore write for the rate at which heat is transferred by convection,

$$\frac{\Delta Q}{\Delta t} = hA(T_2 - T_1) \tag{14.2}$$

where A is the area of the hot surface (at temperature T_2) with which the fluid is in contact and T_1 is the temperature of the main body of the fluid. The convection constant h, which must have units of $W/(m^2 \cdot C°)$ or $kcal/(s \cdot m^2 \cdot C°)$ to make Eq. (14.2) dimensionally correct, has to be determined from experiment in each case, since it depends on the properties of the fluid, including its rate of flow, and on the size, shape, and orientation of the hot object.

14.3 Heat Transfer by Radiation

The third method by which heat can be transferred from one object to another is by *radiation*. In this chapter, when we speak of radiation, we always mean *electromagnetic radiation*. This kind of radiation includes a great variety of electromagnetic waves—such as radio waves, light waves, and x-rays—that are produced and detected in different ways depending on their wavelengths. (This is discussed more fully in Chap. 22).

Radiation: The transfer of electromagnetic energy from one object to another without any transfer of matter and without the need of a material medium.

For example, when we are stretched out on the beach in the summertime, we are absorbing energy from the sun, but we can detect no material substance being transported from the sun to our bodies. Similarly, the warmth we derive from a fireplace is mostly from radiation, since most of the heated air is convected up the chimney and not into the room. In this chapter we concentrate on one particular kind of electromagnetic radiation:

Thermal radiation: The electromagnetic radiation emitted by hot objects because of their temperature.

Any object that is hotter than its surroundings loses heat to its surroundings. Some of this loss occurs by conduction and convection. But much of this loss occurs by thermal radiation, especially at high temperatures.

Thermal radiation is produced by the random motion of the atoms and molecules in a hot object like the sun. The motion of charged particles in the sun produces electromagnetic waves that are radiated out into space in the same way that the motion of electric charges in the transmitting antenna of a radio station sends radio waves out into space. Since the sun is so hot (6000 K), the electromagnetic waves radiated from it are partly in the visible region of the spectrum, where we perceive them as light, and partly in the infrared region of the spectrum, where we perceive them by their heating effects. For objects at temperatures below 1000 K, most of the electromagnetic radiation emitted is in the infrared region of the spectrum, with wavelengths from about 0.7×10^{-6} m (the red end of the visible region) to about 1 mm. For this reason infrared radiation produced by hot objects is often referred to as thermal radiation, or heat radiation, although it is nothing but electromagnetic radiation in a particular wavelength range.

The frequency range spanned by infrared radiation includes the vibrational and rotational frequencies of many molecules in the human body. Hence when radiation from the sun or from an infrared lamp strikes our bodies, resonance occurs between frequencies in the radiation and the natural frequencies of the molecules in our bodies. This sets the molecules into forced motion, increases their random kinetic energies, and thus increases our body temperature.

Laws Governing Thermal Radiation

Stefan's radiation law, first proposed by the Austrian physicist Josef Stefan (1835–1893) to explain experimental data he had collected, states that the rate at which energy is radiated by an object at a Kelvin *temperature T* is proportional to the surface area A of the object, to the fourth power of the absolute temperature, and to the *emissivity* (e) of the substance emitting the radiation:

$$\frac{\Delta Q}{\Delta t} = e\sigma A T^4$$

(14.3)

where the constant of proportionality σ (Greek *sigma*) is a universal constant called the *Stefan-Boltzmann constant*, with a value $\sigma = 5.67 \times 10^{-8}$ W/(m^2·K^4). The emissivity e is a number between 0 and 1 that depends on the substance emitting the radiation and its surface condition. Dark, rough surfaces have emissivities close to 1, the value for a perfect radiator, while white, shiny surfaces have values closer to 0. For example, the value of σ for a polished copper surface is about 0.3, for light human skin about 0.6, and for dark human skin about 0.8.

A second law governing thermal radiation was proposed by the German physicist Wilhelm Wien (1864–1928) in 1893, and is called the *Wien displacement law*. Wien observed that the wavelength at which the emitted radiation had maximum intensity is displaced toward the red as the absolute temperature of the source decreases, as in Fig. 14.3. He was able to fit his experimental data by the following simple equation:

$$\lambda_{max}T = w = 2.898 \times 10^{-3} \text{ m·K} \tag{14.4}$$

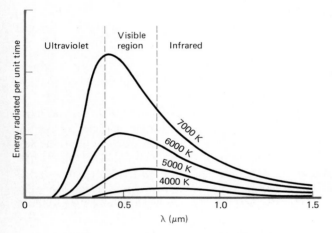

FIGURE 14.3 Experimental radiation curves for different temperatures of the emitter. Note that the total amount of energy emitted per unit time increases rapidly with the temperature and that the peaks move to shorter wavelengths as the temperature increases.

This equation relates the wavelength at which the maximum intensity radiated by a hot object occurs to the Kelvin temperature. The constant w is called *Wien's constant*. This law predicts, for example, that if the sun emits radiation as a blackbody at 6000 K (which is approximately correct), the wavelength at which maximum emission occurs is

$$\lambda_{max} = \frac{w}{T} = \frac{2.898 \times 10^{-3} \text{ m·K}}{6000 \text{ K}} = 0.48 \times 10^{-6} \text{ m} = 0.48 \ \mu\text{m} = 480 \text{ nm}$$

This wavelength is in the green region of the visible spectrum, which is indeed where solar radiation has its greatest intensity. Figure 14.3 shows that at 6000 K, all visible wavelengths are roughly equally intense in the output of the sun. This leads to the sun's yellowish-white appearance.

Wien's displacement law also suggests why a piece of steel in a blast furnace will, as it heats up, first glow a dull red, then bright red, then yellow, and finally look almost pure white. The steel appears white because most of the radiation being emitted is in the visible region of the spectrum, and all the visible colors are present to a sufficient extent to produce an overall white effect. Table 14.2 shows how the color of a star can be used to estimate its approximate temperature. The temperature can be calculated from Eq. (14.4) using observations on the color of the star.

TABLE 14.2 Color Temperature of Stars

Color	Temperature (K)
Reddish stars	2000–3000
Orange stars	3000–5000
Yellow stars (sun)	5000–8000
White stars	8000–12,000
Bluish stars	>12,000

14.4   Radiation Balance

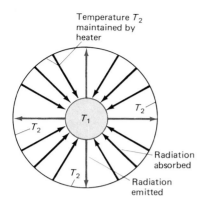

Temperature T_2 maintained by heater

T_2

T_1

T_2

T_2

Radiation absorbed

Radiation emitted

FIGURE 14.4   A metal sphere surrounded by an outer spherical shell maintained at a temperature T_2. As time goes on, the temperature of the metal sphere increases from T_1 to T_2.

Suppose we put a metal sphere at an initial temperature T_1 inside an evacuated spherical shell that is maintained at a higher temperature T_2, as in Fig. 14.4. Both the sphere and the shell radiate energy at a rate determined by Stefan's law, but since the outside shell is hotter, it radiates at a faster rate. As time goes on, the temperature of the metal sphere gradually increases and comes to a final temperature T_2 that is the same as that of the outer shell. The sphere is now in thermal equilibrium with the outer shell, but it is radiating energy at a rate $\Delta Q/\Delta t = e\sigma A T_2^4$, according to Stefan's law, since its new temperature is T_2. Because the sphere's temperature is not changing, it must also be *absorbing* radiation from the outer shell at exactly the same rate, $\Delta Q/\Delta t = e\sigma A T_2^4$, at which it is *radiating* energy. This is an example of what is called *radiation balance*.

The rate of absorption of energy by the sphere is the same as its rate of emission at the same temperature. A good emitter is therefore a good absorber, and a poor emitter is a poor absorber, the emissivity e being the same for both processes.

In the situation just discussed, when the inner sphere was at temperature T_1, it was emitting radiation at a rate $(\Delta Q/\Delta t)_1 = e\sigma A T_1^4$ and absorbing radiation at a rate $(\Delta Q/\Delta t)_2 = e\sigma A T_2^4$. The net flow of energy from the hot outer shell to the cooler inner sphere was therefore

$$\left(\frac{\Delta Q}{\Delta t}\right)_{\text{net}} = e\sigma A(T_2^4 - T_1^4) \tag{14.5}$$

Why do not all hot objects eventually radiate all their energy away as heat and cool down to absolute zero? Only then will the radiated power be equal to zero in Eq. (14.3). This is exactly what would happen if energy were not continuously supplied to these objects to enable them to radiate without cooling down. The sun, for example, has fusion processes going on in its interior that continuously convert mass into energy. This is the source of the energy the sun radiates. The earth, for its part, is continuously receiving some of the sun's energy to replenish the energy the earth radiates into space. It is the radiation balance between the energy falling on the earth from the sun and the energy the earth radiates into space that determines the earth's temperature at any time. That temperature continuously adjusts itself until the earth is emitting exactly the same amount of energy it receives from the sun. Any drastic change in the amount of energy radiated by the sun or in its distance from the earth could destroy this balance and bring an end to life on earth by making its temperature either too high or too low to sustain life.

A similar balance occurs with the human body. Since the skin temperature is about 33°C, and the walls of a room on a cold day may be at 18°C, the human body is continuously losing thermal energy by radiation to the walls. This, together with other sources of heat loss like conduction and convection, would be expected to make the temperature of the body fall. The body responds, however, by increasing the rate at which it converts food or body fat into thermal energy to keep its temperature at the normal interior value of 37°C. This heat is then carried to the surface of the body by *convection*, with the heart acting as the pump and the blood as the circulating fluid.

A person who maintains constant weight must take in energy in the form of food at an average rate of about 120 J/s, or 120 W. This same person loses heat by conduction, convection, and (mostly) radiation at the same rate and therefore converts stored energy to heat at the same rate that a 120-W light bulb would convert electrical energy to light and heat.

Example 14.2

(a) Compare the rate at which energy is radiated by 1 m² of the sun's surface at 6000 K with the rate at which energy is radiated by 1 m² of the earth's surface at 290 K. Assume that the emissivity e is the same for both bodies. **(b)** If the radius of the sun is 6.96×10^8 m and its emissivity is about 0.93, find the rate at which it radiates energy at 6000 K.

SOLUTION

(a) From Eq. (14.3), we have for the ratio of the radiation rates for the sun and earth,

$$R = \frac{e\sigma A T_S^4}{e\sigma A T_E^4} = \frac{T_S^4}{T_E^4} = \frac{(6000 \text{ K})^4}{(290 \text{ K})^4} = \boxed{1.8 \times 10^5}$$

Hence in 1 s, 1 m² of the sun's surface radiates 1.8×10^5 times the energy radiated by the same surface area of the earth in the same time. Since the surface area of the sun is 1.2×10^4 times that of the earth, the overall ratio of the rates of radiation is 2.2×10^9. The sun therefore radiates energy at a rate approximately 2 billion times that of the earth.

(b) The rate at which the sun radiates energy is

$$\left(\frac{\Delta Q}{\Delta t}\right)_s = e\sigma A T^4 = e\sigma(4\pi R_S^2)T^4$$

$$= 0.93 \left(5.67 \times 10^{-8} \frac{\text{W}}{\text{m}^2 \cdot \text{K}^4}\right)$$

$$\times (4\pi)(6.96 \times 10^8 \text{ m})^2 (6000 \text{ K})^4$$

$$= \boxed{4.2 \times 10^{26} \text{ W}}$$

EXERCISE 1 At what wavelength does the earth emit radiation most strongly?

Example 14.3

A swimmer clad in a tight-fitting bathing suit is standing near an indoor pool. The surface temperature of the swimmer's body is 33°C, and the temperature of the walls of the room is 20°C. **(a)** If the surface area of the swimmer's body is 1.3 m² and its average emissivity is 0.65, find the net rate at which energy is radiated by the swimmer's body. **(b)** At what wavelength is the energy radiated a maximum?

SOLUTION

(a) From Eq. (14.5),

$$\left(\frac{\Delta Q}{\Delta t}\right)_{net} = (0.65)\left(5.67 \times 10^{-8} \frac{\text{W}}{\text{m}^2 \cdot \text{K}^4}\right)$$

$$\times (1.3 \text{ m}^2)(306^4 - 293^4) \text{ K}^4$$

$$= (4.8 \times 10^{-8})(1.4 \times 10^9) \text{ W} = \boxed{67 \text{ W}}$$

The rate at which heat is radiated from the swimmer's body is roughly equivalent to the power consumed by a 60-W light bulb.
(b) From Eq. (14.4) we have

$$\lambda_{max} = \frac{w}{T} = \frac{2.898 \times 10^{-3} \text{ m} \cdot \text{K}}{306 \text{ K}} = \boxed{9.47 \text{ } \mu\text{m}}$$

This is in the infrared region of the spectrum, the approximate limits of which are 0.7 μm and 1 mm.

14.5 Some Applications of Heat Transfer Processes

Let us consider a few applications of the heat transfer processes discussed in this chapter to get some feeling for their practical importance.

Newton's Law of Cooling

Among the many discoveries in physics made by Sir Isaac Newton was a simple law obeyed by objects on cooling down from a higher temperature to room temperature. If the initial temperature of the object is T_2 and the temperature of its surroundings T_1, as in Fig. 14.5, Newton found that the rate of heat loss is

$$\frac{\Delta Q}{\Delta t} = K(T_2 - T_1) \tag{14.6}$$

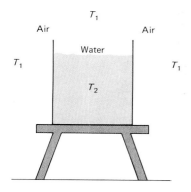

FIGURE 14.5 Newton's law of cooling. Hot water at a temperature T_2 in a room where the air temperature is T_1 cools down at a rate $K\,(T_2 - T_1)$, if $T_2 > T_1$.

Here K is an empirical constant, i.e., one that must be obtained from experiment for each particular physical situation. Equation (14.6) is called *Newton's law of cooling*. It predicts that the rate of cooling slows down as T_2 is reduced, and goes to zero when T_2 becomes equal to T_1. This is illustrated in Fig. 14.6.

It can be shown that Newton's law of cooling is an immediate consequence of the laws of conduction, convection, and radiation, as long as T_2 is not too much larger than T_1 (see Prob. 14.45).

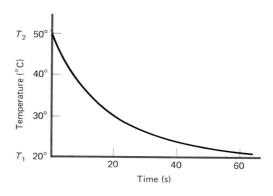

FIGURE 14.6 The temperature of a hot object initially at temperature T_2 as a function of time. The temperature decreases exponentially and in the limit approaches room temperature (T_1).

Thermos Bottles

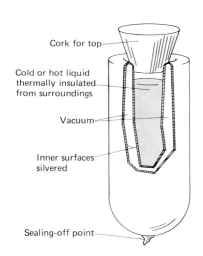

FIGURE 14.7 A thermos bottle, or dewar.

A thermos bottle [or *dewar*, after its inventor, the British physicist Sir James Dewar (1842–1923)] is able to maintain the temperature of hot and cold liquids for long periods of time. Why does the hot coffee in a thermos not cool down quickly to room temperature? The thermos makes heat transfer from the hot coffee to the cooler air as difficult as possible by drastically reducing the rate at which conduction, convection, and radiation occur.

A typical thermos bottle is shown in Fig. 14.7. Evacuating the space between the inner and outer walls leaves few molecules in this space to convey heat by conduction or convection. Hence the only way heat can be lost (except for inevitable small losses near the neck of the bottle) is through radiation. Radiation loss can be greatly reduced by silver-plating the two inner surfaces of the vacuum jacket. These silvered surfaces reflect back any radiation trying to escape from the liquid and any radiation trying to enter from outside. They also radiate very little energy, because of their low emissivity.

Thermography

Infrared radiation from an object can be detected by a camera equipped with infrared-sensitive film, or by a more sensitive type of infrared scanner that converts small variations in the temperature of radiating objects into pictures called *thermograms*, in which each scanned region is displayed with a color shading characteristic of its temperature.

Thermography is very useful in medicine. The human body, with a skin temperature of about 33°C, radiates in the infrared region of the spectrum, according to the Wien displacement law. Temperature differences in the body can be recorded on thermograms, as in Fig. 14.8. This can be quite helpful in detecting tumors, vascular disorders, arthritis, and other diseases. Tumors are often found

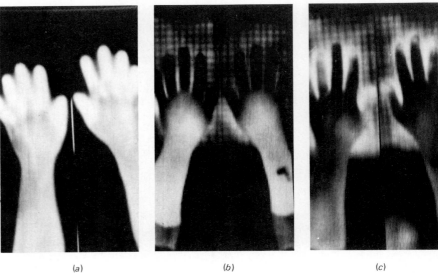

(a)   (b)   (c)

FIGURE 14.8 Three thermograms of the hands of a 24-year-old woman who smokes a pack of cigarettes a day. (a) The presmoking thermogram demonstrates normal warm (white) hands; (b) 1 min after she smokes a single filtered cigarette, her hands become 6C° cooler; (c) 30 min later the temperature is starting to rise again (the fingers are becoming white in the thermogram. (*Courtesy of American Cancer Society.*)

in areas where the temperature is higher due to increased metabolic activity in the vicinity. This leads to more intense infrared radiation. Such a technique is especially useful in finding breast cancers, since thermograms can find temperature differences of less than 0.07°C, and the temperature differences associated with breast cancers are usually at least a few tenths of a Celsius degree. Growths as small as 1 cm in diameter can be detected with this valuable technique, which also avoids the problems associated with the use of x-rays.

Solar Radiation

The sun is the ultimate source of 99.98 percent of all the energy incident on the earth's surface, with the remaining 0.02 percent coming from the earth's interior. The sun radiates energy into space at a rate of 4.2×10^{26} J/s, or 4.2×10^{26} W, as we found in Example 14.2. The source of this enormous energy supply is the conversion of the sun's mass to energy at a rate of 4.6×10^9 kg/s.* This is a fantastic rate of decrease in mass, but since the sun's total mass is 2×10^{30} kg, the sun could continue to exist for some 10^{17} years even at its present rate of converting mass into energy.

Figure 14.9 shows that the sun radiates a continuous spectrum with maximum intensity in the green region at a wavelength of about 0.5 μm, or 500 nm. About one-half the total energy radiated by the sun is concentrated in the visible range, from 0.4 to 0.7 μm, with the remainder mostly in the infrared, as the figure shows.

Although the sun radiates energy at a rate of 4.2×10^{26} W, this radiation is spread out in all directions over a full sphere. Only one two-billionth of it falls on the earth, but this still amounts to about 2×10^{17} W, which is 500,000 times the electric power generating capacity of the United States!

When we consider that this radiant power is spread out over the whole earth, that much of it is absorbed or reflected by the earth's atmosphere, that cloud cover obscures the sun on many days of the year, and that half the earth is in darkness at any one time, the average intensity of solar energy falling on a house in the central United States turns out to be only about 200 W/m². In other words, if all

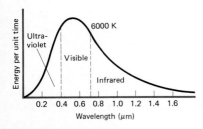

FIGURE 14.9 The distribution of the energy radiated by the sun at 6000 K as a function of wavelength. Most of the radiation is in the visible and infrared regions of the spectrum (1 μm = 10^3 nm).

---

*This conversion is governed by Einstein's famous equation $E = mc^2$ (Sec. 26.7).

the solar energy falling on 1 m² of the earth's surface could be collected and fully converted to electric energy, on the average it would just suffice to light two 100-W electric light bulbs on a continuous basis. This diffuseness is one of the main problems associated with using solar energy to meet our energy needs.

Example 14.4

Assume that the average rate at which solar energy falls on a house in the central United States is 200 W/m². Consider a one-story house of 140 m² floor space, and assume that it requires 10^9 J of energy per day to provide heat and hot water on a cold day in January. **(a)** How many joules of solar energy are incident per day on the flat roof of the house? **(b)** If this energy could be collected and used with 100 percent efficiency (which is not possible), what fraction of the roof area of the house would be required for heat collectors? **(c)** If a solar-electric power plant capable of producing electric energy at a rate of 1 GW (or 10^9 W) were to be built in the central United States, what is the *minimum* land area such a power plant would require?

SOLUTION

(a) The total incident power is

$$P = \left(200 \frac{W}{m^2}\right)(140 \ m^2) = 28 \ kW$$

The energy incident in 1 day is

$$\mathcal{W} = Pt = \left(28 \frac{kJ}{s}\right)(8.64 \times 10^4 \ s) = \boxed{2.4 \times 10^9 \ J}$$

This is more than twice the energy needs of the house, even in January.
(b) The fraction of the roof area needed for collectors at 100 percent efficiency is

$$\frac{1 \times 10^9 \ J}{2.4 \times 10^9 \ J} = 0.42 = \boxed{42\%}$$

(c) To produce 10^9 W when the average incident power is 200 W/m² would require an area

$$A = \frac{10^9 \ W}{200 \ W/m^2} = 5.0 \times 10^6 \ m^2 = \boxed{5.0 \ km^2}$$

This would be a square area about 2.2 km (or 1.4 mi) on a side. This is the *minimum* area needed, because solar energy cannot be collected and converted into usable heat with 100 percent efficiency. Fossil-fuel or nuclear plants providing this same amount of power, which is the output of a typical large electric power plant, occupy much smaller land areas.

Radiation Cooling

On a *clear* night the earth loses large amounts of energy by radiation, since the earth's temperature is high compared to that of the cold night sky. As a consequence of this radiation loss, the earth's temperature falls. This is called *radiation cooling*. Objects that do not receive much heat by conduction from the ground to replace that lost by radiation—car windows, for example—cool down rapidly. The water vapor from the air may then condense as frost on such objects, even when the ground temperature is above freezing. This is why we find our car windows frosted over on autumn mornings, although there is no frost on the cement driveway on which the car is parked.

However, on a *cloudy* night, some of the energy the earth loses by radiation is absorbed by the clouds, and the clouds reemit this radiant energy back toward the earth. Because of this return of energy to the earth, the earth's temperature does not fall as rapidly as it would on a clear night.

The next time you hear TV weather people predict that, on a night when there is little cloud cover, the temperature will be lower than it would be on a cloudy night, you will understand why.

The Greenhouse Effect

Carbon dioxide in the earth's atmosphere can play the same role as the glass walls of a florist's greenhouse. The CO_2 allows the visible radiation from the sun to pass through to the earth, but absorbs and reradiates the infrared wavelengths the earth emits at its 290 K temperature [Eq. (14.4)]. This infrared (thermal) radiation is trapped between the earth and the CO_2 in the atmosphere and increases the earth's temperature the same way the temperature is increased inside a greenhouse.

Scientists and government officials worry a great deal at the present time about the increased heating of the earth caused by growing amounts of atmospheric CO_2 from the emissions of factories, power plants, and automobiles. The problem is a complicated and difficult one, however, and much research needs to be done before we understand all its facets. For example, even the analogy with a florist's greenhouse may be misleading, since recent studies have shown that the lack of convection as a mechanism for removing heat from a greenhouse may be more important as the cause of greenhouse heating than the trapping of infrared radiation.

14.6 Heat and Work

The experiments of Sir James Joule (Sec. 13.8) showed conclusively that work can be converted to heat and that the amount of heat produced from a given amount of work is always the same.* We also know that heat can be converted to work. For example, if a gas is heated while its temperature is held constant, it will expand and do work on a piston, as in Fig. 14.10. In this case, the heat transferred to the gas is converted into work. It would therefore appear that heat and work are both forms of *energy*, and that this is the reason one can be converted into the other.

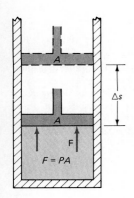

FIGURE 14.10 An expanding gas doing work on a movable piston. The work done is $\Delta W = F\,\Delta s \doteq P\,\Delta V$.

Work Done by a Gas

Let us consider a gas contained in a cylinder and maintained at constant pressure by a piston at one end, as in Fig. 14.10. If the gas expands a distance Δs against the pressure P exerted by the piston, which has a cross-sectional area A, then the work done by the gas on expanding is

$$\Delta W = \mathbf{F} \cdot \mathbf{\Delta s} = PA\,\Delta s = P\,\Delta V \qquad \text{or} \qquad \boxed{\Delta W = P\,\Delta V} \qquad (14.7)$$

since $F = PA$ and $\Delta V = A\,\Delta s$.

The work done by the gas is considered *positive* if the volume increases, as in this case. If the gas contracts, then the force exerted by the gas is in the opposite direction to the displacement $\mathbf{\Delta s}$, and the work done by the gas is negative. In this case work is done *on the gas* by the piston. In the SI system, the work will be in joules if the pressure is in newtons per square meter and the volume in cubic meters, since $1\ \text{N·m} = 1\ \text{J}$.

If the pressure remains constant and equal to P_0, then the work done by the gas in expanding from V_1 to V_2 is

$$\Delta W = P_0(V_2 - V_1)$$

---

*In the quotation from Joule at the beginning of this chapter, he uses the term *mechanical force* to describe what we now call *work*.

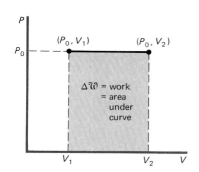

FIGURE 14.11 The work done by an expanding gas at constant pressure P_0. The work done is the area shown: $\Delta W = P_0 (V_2 - V_1)$.

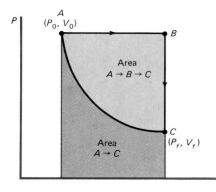

FIGURE 14.12 The work done by a gas in going from point $A(P_0, V_0)$ to point $C(P_f, V_f)$ depends on the path taken. The work done along the path $A{\rightarrow}B{\rightarrow}C$ is greater than along the curved path $A{\rightarrow}C$ since for $A{\rightarrow}B{\rightarrow}C$ the area under the curve is all that between the line AB and the x axis.

FIGURE 14.13 The work done in a cyclic process. The work done by a gas in going from A to B to C and then back to A again, by the path shown, is the area indicated on the diagram.

Heat Flow

In Fig. 14.11, $P_0(V_2 - V_1)$ is simply the area under the PV curve between the two volumes V_1 and V_2. We therefore say that *the work done by a gas on expanding is equal to the area under its PV curve.* We sometimes call this PV *work.*

Suppose we have a gas in an initial state specified by its pressure and volume, P_0 and V_0, and we want to change the gas to a new state specified by P_f and V_f. There are a number of ways to do this, as indicated in Fig. 14.12. We can simply let the gas expand at constant temperature according to Boyle's law (Sec. 9.7) from point A to point C. We can also let the gas expand at constant pressure from point A to point B and then reduce the pressure at constant volume to bring the gas to point C, where the pressure and volume are the desired P_f and V_f. Since the amount of gas remains constant, and n is therefore constant in the ideal gas equation $PV = nRT$, and since $P_0V_0 = P_fV_f$ for the two processes, the temperature at the end of the two processes is the same and is equal to T_0.

We could also carry out this expansion in a variety of other ways. Clearly in each case the work done would be different, since the area under the PV curve would be different, as Fig. 14.12 shows. We see then that the work done by a gas on expanding depends not merely on the initial and final states but also on the *path* (on a PV diagram) the gas takes in going from its initial to its final state. *The work done on or by the gas is dependent on the path.* Thus in Fig. 14.13 if we take the gas from the initial state to the final state by path ABC and then back to the initial state by the path CA, the gas has done a net amount of work in the process, although the final state of the system is the same as it was initially. The amount of work done in going from one state to another can be determined, therefore, only if the path taken is fully specified.

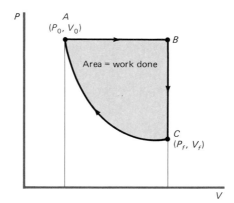

Heat flow is similarly dependent on the path followed by the process. In Fig. 14.12 a greater amount of heat must be supplied to take the system from the initial to the final state by path ABC than directly by path AC, if the initial and final temperatures are to be the same (T_0) for the two processes. More work is done along the path ABC, and heat is required to do that work. Hence the flow of heat into or out of the gas is also dependent on the path taken and not merely on the initial and final states. For this reason the phrase *the heat content of a system* is meaningless. For example, if we could assign an arbitrary value to the heat content of a gas in some standard state, the heat content in some other state would be the heat content in the standard state plus whatever heat was added in going from that standard state to the new state. But the amount of added heat depends on the path taken. Since there are an infinite number of paths, there are an infinite number of values for the heat content, and hence this term is physically meaningless.

Both heat and work therefore depend on path and are not conserved quantities in the sense that the energy of a particle in a gravitational field is conserved. In the gravitational case the work done in moving the particle from one point to another depends only on the two endpoints and not on the path taken. Neither heat nor work share this conservative property; both are dependent not merely on the coordinates (i.e., for a gas, P, V, and T) of the endpoints but also on the physical *path* taken to go from one endpoint to the other.

Example 14.5

In the *PV* diagram of Fig. 14.14 a gas expands at constant pressure from point A to point B, and the pressure is then changed at constant volume to take the gas from B to C. The gas is finally compressed back to A along the path CA. How much work has been done *by* the gas in going **(a)** from A to B; **(b)** from B to C; **(c)** from C to A? **(d)** What is the net work done by the gas in the complete process?

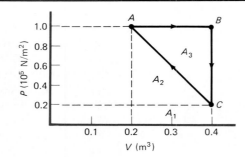

FIGURE 14.14

SOLUTION

(a) The work done here is the constant pressure times the change in volume:

$$\Delta W_{AB} = P(V_B - V_A)$$

$$= (1.0 \times 10^5 \text{ N/m}^2)(0.40 \text{ m}^3 - 0.20 \text{ m}^3)$$

$$= 0.20 \times 10^5 \text{ N·m} = \boxed{0.20 \times 10^5 \text{ J}}$$

which is the area between the x axis and the line AB that is bounded by $V = 0.20$ m^3 and $V = 0.40$ m^3.

(b) $\Delta W_{BC} = P(V_C - V_B) = \boxed{0}$ since $V_B = V_C$

(c) $\Delta W_{CA} = P(V_A - V_C)$

Here the pressure varies along with the volume, but the work done is still the area under the *PV* curve. This is the sum of the two areas A_1 and A_2, with the latter being the area of a triangle. Hence the total area A under the curve is

$$A = A_1 + A_2$$

$$= (0.20 \times 10^5 \text{ N/m}^2)(0.20 \text{ m}^3 - 0.40 \text{ m}^3)$$

$$+ \tfrac{1}{2}[(1.0 - 0.20) \times 10^5 \text{ N/m}^2](0.20 \text{ m}^3 - 0.40 \text{ m}^3)$$

$$= -0.12 \times 10^5 \text{ J}$$

and so $\Delta W_{CA} = \boxed{-0.12 \times 10^5 \text{ J}}$

This work done *by the gas* is *negative*, since work is being done *on* the gas to compress it.

(d) The net work done by the gas is

$$\Delta W = \Delta W_{AB} + \Delta W_{BC} + \Delta W_{CA}$$

$$= 0.20 \times 10^5 \text{ J} + 0 - 0.12 \times 10^5 \text{ J} = \boxed{8.0 \times 10^3 \text{ J}}$$

with the work *positive*, indicating that net work is done *by* the gas. This work is equal to the area A_3 inside the triangle ABC on the *PV* diagram.

14.7   The First Law of Thermodynamics

Thus far we have used the phrase *thermal energy* to describe the random motion of the atoms and molecules in a gas or other substance. This thermal energy can be increased by doing work on the gas or decreased by having the gas do work on an external object like a piston. A physical system has other forms of energy, however, such as the energy in chemical bonds, the binding energy of atoms, and the energy stored in the atomic nucleus—energy that cannot under normal circumstances be released to do work. To state the first law of thermodynamics precisely, therefore, we need to introduce a broader term to include all the forms of energy that a physical system like a gas can have. This we call *internal energy*.

Internal energy (U): The sum of the kinetic and potential energies of all the individual particles making up a physical system.

A system with internal energy U can interact with its surroundings in two ways. First the system can have *work done on it*, in which case *U increases*, or it can *do work* on its surroundings, in which case *U decreases*. Secondly, *heat* can flow *into* the system and thus *increase* the internal energy U, or heat can flow *out* of the system and thus *decrease U*.

The great difference between internal energy and work or heat is that internal energy is a *state function*. By this we mean that U depends only on the values of certain physical quantities (for a gas, P, V, and T), and that once these quantities are determined, the value of U is completely determined. The quantities P, V, and T are called *thermodynamic state variables*. No matter how complicated the interaction of the gas with its surroundings, once it returns to the equilibrium state specified by its original values of P, V, and T, its internal energy is exactly what it was originally.* This is true of all thermodynamic state functions. Internal energy is therefore a quantity similar to potential energy in mechanics. Its value is independent of the processes (or paths) by which a substance arrives at a particular state, in the same way that potential energy in a gravitational field is a function only of a particle's position in the field and not of the path taken to get there. This contrasts with heat and work, which (as we saw in Sec. 14.6) both depend on the *path* taken by the substance in any thermodynamic process.

We can now state the first law of thermodynamics in terms of the internal energy of a system. Suppose that the initial internal energy of the system is U_i, that an amount of heat ΔQ flows into the system, and an amount of work ΔW is done by the system, as in Fig. 14.15. Then, if U_f is the final internal energy of the system, we can state the first law as follows:

First law of thermodynamics: Every thermodynamic system in equilibrium can be described by a state function U, called the *internal energy*, whose changes are governed by the equation

$$\Delta U = \Delta Q - \Delta W$$

(14.8)

where $\Delta U = U_f - U_i =$ *increase* in internal energy of the system
$\Delta Q =$ heat flow *into* the system
$\Delta W =$ work done *by* the system

The minus sign on ΔW is needed because when work is done *by* the system, energy flows *out* of it.

All three quantities in Eq. (14.8) must be expressed in the same units, e.g., joules or kilocalories. (If a problem is stated in inconsistent units, you can use Joule's equivalent to convert joules to kilocalories, or vice versa, as Example 14.6 shows.)

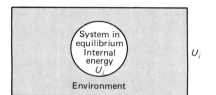
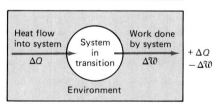
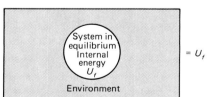

FIGURE 14.15 The first law of thermodynamics. A system in equilibrium with its environment and with internal energy U_i interacts with its environment by absorbing heat ΔQ from the environment and doing work ΔW on the environment. The final internal energy of the system is $U_f = U_i + \Delta Q - \Delta W$.

---

*Note that if P, V, and n are known for a gas, then its temperature is fully determined by the gas law $PV = nRT$.

Example 14.6

Determine Joule's equivalent from the following data: 4.00 kcal of heat is supplied to a system; the system does 6.70×10^3 J of external work; at the end of the process the internal energy of the system has increased by 1.00×10^4 J.

SOLUTION

From the first law of thermodynamics,

$$\Delta U = \Delta Q - \Delta W$$

and so

$$1.00 \times 10^4 \text{ J} = 4.00 \text{ kcal} - 6.70 \times 10^3 \text{ J}$$

or

$$4.00 \text{ kcal} = 1.00 \times 10^4 \text{ J} + 6.70 \times 10^3 \text{ J}$$
$$= 1.67 \times 10^4 \text{ J}$$

and thus

$$\text{Joule's equivalent} = \frac{1.67 \times 10^4 \text{ J}}{4.00 \text{ kcal}} = \boxed{4.18 \times 10^3 \text{ J/kcal}}$$

Conservation of Energy

When we are dealing with an isolated system, i.e., a system that does not exchange either heat or work with its surroundings, both ΔQ and ΔW are zero in Eq. (14.8), and so

$$U_f - U_i = \Delta U = \Delta Q - \Delta W = 0, \qquad \text{or} \qquad U_f = U_i \qquad \text{(14.9)}$$

Since the internal energy consists of the potential and kinetic energies of the atoms and molecules making up the system, the *first law of thermodynamics* represents an extension of the mechanical principle of conservation of energy to include thermodynamical systems:

> In any isolated system the total internal energy remains constant, even though it may be transformed from one kind of energy to another.

This is the most general form of the principle of conservation of energy. The internal energy of an isolated system cannot be changed by any process (including a chemical or biological process) taking place *inside* the system. If the system is not isolated, then heat can flow into or out of it, and the system can do work or have work done on it. In this case, as we have seen, $\Delta U = \Delta Q - \Delta W$.

Example 14.7

In Fig. 14.16, when an ideal gas is taken from state A to state B along the path ACB, 1.91×10^{-2} kcal of heat flows into the system and 30 J of work is done by the system. What is the change in the internal energy of the gas **(a)** along path ACB; **(b)** along path ADB? **(c)** How much heat flows into the system along path ADB if the work done by the system along ADB is 10 J?

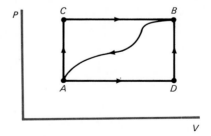

FIGURE 14.16

SOLUTION

To make the units consistent, we use joules throughout. The heat flowing into the system is

$$\Delta Q = (1.91 \times 10^{-2} \text{ kcal})(4.18 \times 10^3 \text{ J/kcal}) = 80 \text{ J}$$

(a) From Eq. (14.8), we have for the change in the internal energy of the gas along path ACB,

$$\Delta U_{AB} = \Delta Q - \Delta W = 80 \text{ J} - 30 \text{ J} = \boxed{50 \text{ J}}$$

(b) For path *ADB* the change in internal energy must be the same, since ΔU *is independent of the path*, and so

$$\Delta U_{AB} = \boxed{50 \text{ J}}$$

(c) Since $\Delta U_{AB} = \Delta Q - \Delta W = 50$ J,

$$\Delta Q = \Delta U_{AB} + \Delta W = 50 \text{ J} + 10 \text{ J} = \boxed{60 \text{ J}}$$

EXERCISE 2 When the system is returned from *B* to *A* along the curved path in Fig. 14.16, the work done is 20 J. Does the system absorb or liberate heat, and how much?

Perpetual-Motion Machines of the First Kind

Equation (14.9) makes clear that a device known as a perpetual-motion machine of the first kind cannot work.

Perpetual-motion machine of the first kind: A machine designed to do work continuously either with no input of energy or with an input of energy less than the work output.

For example, suppose we have a steam engine that operates in a closed cycle, which leaves the internal energy *U* unchanged. If such an engine could produce more work than the heat taken in, it would be a perpetual-motion machine of the first kind (we discuss perpetual-motion machines of the second kind in Chap. 15). An alternative statement of the first law of thermodynamics is that *a perpetual-motion machine of the first kind is impossible*, since it would violate the first law of thermodynamics. We cannot emphasize too strongly that this limitation is *not* due to engineering problems in designing or building such machines; it is due to the nature of the physical universe and is therefore not subject to change. We cannot get more work out of a machine than the energy we put in, no matter how complicated or ingenious the machine.

Efficiency

The *efficiency* (*e*) of a physical process is defined as follows:

$$e = \frac{\text{work done (or energy produced) by a system}}{\text{energy input to the system}} \tag{14.10}$$

The maximum possible value of *e* is unity, since we can never get more work out of a system than the energy we put in. Thus an oil furnace in a home is said to be 60 percent efficient ($e = 0.60$) if 60 percent of the oil's heat of combustion is delivered to the house as heat.

Example 14.8

In a coal-burning steam-electric power plant generating electricity at a rate of 1.0 GW (10^9 W), coal is burned to provide steam for the turbines at a rate of 7.2×10^5 kcal/s. What is the efficiency of such a plant?

SOLUTION

In this case by the efficiency we mean

$$e = \frac{\text{electric energy out}}{\text{thermal energy in}} = \frac{W_{el}}{W_{th}}$$

Since we are given rates of energy production rather than absolute amounts of energy, we need to select a certain time interval, say, 1 s, over which to compare the energies. The electric energy produced in 1 s is then

$$W_{el} = (1.0 \text{ GW})(1.0 \text{ s}) = (10^9 \text{ J/s})(1.0 \text{ s}) = 1.0 \times 10^9 \text{ J}$$

The thermal energy put in is

$$W_{th} = \left(7.2 \times 10^5 \frac{\text{kcal}}{\text{s}}\right)\left(4.18 \times 10^3 \frac{\text{J}}{\text{kcal}}\right)(1.0 \text{ s})$$
$$= 3.0 \times 10^9 \text{ J}$$

Then the efficiency is

$$e_1 = \frac{W_{el}}{W_{th}} = \frac{1.0 \times 10^9 \text{ J}}{3.0 \times 10^9 \text{ J}} = 0.33 = \boxed{33\%}$$

This is a typical efficiency for large steam-electric power plants.

14.8 Applications of the First Law of Thermodynamics

Let us consider the implications of the first law of thermodynamics in the form of Eq. (14.8) for a number of important physical processes. For simplicity we confine our applications to ideal gases.

Isothermal Process

An *isothermal* ("same-temperature") *process is one that takes place at constant temperature*. In order for the temperature to remain constant when a gas expands, the changes in pressure and volume must be carried out very slowly so that every state in the process is an equilibrium state. For an ideal gas, the internal energy does not change during an isothermal expansion, since the internal energy depends only on the temperature, which does not change. The quantity ΔU is therefore zero, and so from Eq. (14.8), $\Delta Q = \Delta W$. Hence an amount of heat flows into the gas that is just sufficient to enable the gas to do the work required for the expansion without changing its internal energy.

Adiabatic Process

An *adiabatic* (from the Greek word meaning "impassable") *process is one in which no heat enters or leaves the system*. Here $\Delta Q = 0$, and so from Eq. (14.8), $\Delta U = -\Delta W$. If work is done by the system, then the system's internal energy *decreases* by an amount equal to the work done.

An adiabatic process can be achieved either by thermally insulating the system or by carrying out the process so rapidly that heat has no chance to flow into or out of the system. Thus the compressions and rarefactions of a gas in a sound wave are good approximations to adiabatic processes, since they take place so rapidly that heat cannot flow away from the compressions or into the rarefactions. The measured velocity of sound in gases confirms this interpretation. Adiabatic processes play an important role in mechanical engineering, because many processes in internal-combustion engines take place adiabatically.

Refrigerators work because of a special adiabatic process called a *throttling process*, in which a gas, originally at high pressure, flows through a needle valve or other tiny opening into a region of lower pressure. Since the process is assumed to be adiabatic, $\Delta Q = 0$, and so $\Delta U = -\Delta W$, where ΔW is the work done by the gas in expanding. If the gas is an ideal gas, and if it expands into a vacuum, then $\Delta W = P \Delta V$. Since for a vacuum $P = 0$, we must have $\Delta W = 0$, and so $\Delta U = 0$. The internal energy of an *ideal gas* does not change in this case, and so the temperature also does not change.

For most *real gases*, however, the gas molecules exert small forces on one another. Therefore, on expanding, the gas does work against these forces, and ΔW is finite and positive. Hence $\Delta U = -\Delta W$ is negative, and the temperature falls as the internal energy decreases. Such a throttling process is used in the gas refrigerator shown in Fig. 14.17.

When a liquid such as ammonia or freon is subjected to a throttling process, it vaporizes when it expands into the region of lower pressure because the reduced pressure lowers the boiling point below the actual temperature of the liquid. An amount of heat equal to the heat of vaporization of the liquid must be supplied to evaporate each kilogram of liquid. Since $\Delta Q = 0$, this energy can come only from the internal energy of the fluid, and so the temperature falls, as it does when the liquid in an aerosol can is converted to vapor. Refrigeration systems almost always involve changes of state because the associated large heats of vaporization make the cooling more effective.

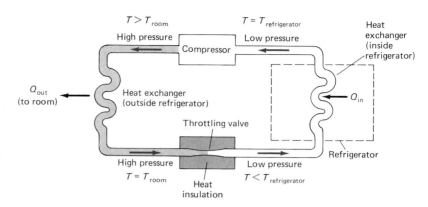

FIGURE 14.17 Throttling process. Such a process causes the temperature to fall and plays an essential role in the gas refrigerator shown here. The temperatures indicated are the temperatures of the gas as it moves through the refrigeration cycle. As the gas flows through the throttling valve, its temperature falls from room temperature to below the temperature inside the refrigerator.

Isochoric Process

An *isochoric* ("same-volume") *process is one in which the volume of the system remains unchanged.* If heat is added to a gas and the volume is unchanged, then no work is done and $\Delta W = 0$. As a consequence $\Delta U = \Delta Q$ and all the heat added goes into the internal energy of the gas. The very sudden increase in temperature and pressure when a gasoline-air mixture ignites in an automobile engine is close to an isochoric process. Here the heat added to the system increases the pressure and the internal energy of the mixture but does not change its volume. This isochoric process is immediately followed by the power stroke of the engine, which is basically an adiabatic process. This performs useful work (by turning the car's wheels) at the expense of the internal energy of the system, which had just been increased by the isochoric combustion process.

Isobaric Process

An *isobaric* ("same-pressure") *process is one that takes place at constant pressure.* Here the work done is $\Delta W = P\,\Delta V$, and so

$$\Delta U = \Delta Q - P\,\Delta V \tag{14.11}$$

When water is heated and boiled in a pot open to atmospheric pressure, the heating of the water and the conversion of water to steam all take place isobarically. In this case,

$$\Delta Q = cm\,\Delta T + mL_v$$

where m is the mass of the water and L_v is the heat of vaporization. At the boiling point the volume changes from V_{liq} to V_{vap}, and so

$$P\,\Delta V = P(V_{vap} - V_{liq})$$

where P is atmospheric pressure. Then

$$\Delta U = U_{vap} - U_{liq} = \Delta Q - P\,\Delta V = cm\,\Delta T + mL_v - P(V_{vap} - V_{liq}) \tag{14.12}$$

Because $\Delta Q = \Delta U + P\,\Delta V$ for an isobaric process, from Eq. (14.11), any heat added to a gas at constant pressure goes partly into the internal energy of the gas and partly into PV work.

Living Systems and the First Law

A living human body is *not* an isolated system, and energy is *not* transferred into the system only in the form of heat or work. Rather, the supply of internal energy in the human body is constantly being replenished by the food we eat. For a healthy adult, the energy taken in as food minus the energy lost in the form of waste products is exactly equal to the sum of the work done by the individual and the heat lost to the surroundings. The first law of thermodynamics remains valid because food provides a very important source of energy. If the intake of food exceeds the sum of the work done by the individual and the heat lost to the surroundings, the excess energy is stored in the body. If the person is physically active, this energy is stored as muscle tissue; if not, it is stored as fat. This excess energy can then be made available to the body at a later time when food is lacking.

14.9 Molar Heat Capacities of an Ideal Gas

A very important application of the ideas of this and the previous chapter is to the molar heat capacities of an ideal gas. Historically this application has contributed greatly to our understanding of molecular processes and to the transition from classical to modern quantum physics.

Let us consider an ideal, monatomic gas like He in a closed container with a piston at one end, as in Fig. 14.10. We assume that the collisions of the gas molecules among themselves and with the walls of the container are perfectly elastic and that no other forces act except those during collisions. From Eq. (9.16) we know that, under such assumptions, the kinetic energy per mole of gas is

$$\text{KE} = \tfrac{1}{2}M\overline{v^2} = \tfrac{3}{2}RT \tag{14.13}$$

where R is the universal gas constant [in J/(mol·K)] and M is the mass of 1 mol of gas. Since the only energy the monatomic gas possesses is kinetic, its internal energy is

$$U = \tfrac{3}{2}nRT \tag{14.14}$$

where n is the number of moles of gas. Equations (14.13) and (14.14) show that the internal energy of an ideal monatomic gas depends only on its temperature.

We now heat the gas shown in Fig. 14.10 to determine its specific heat capacity. Since we are here dealing with n mol of the gas, we use the molar heat capacity C. This is *the amount of heat required to raise the temperature of 1 mol of the gas 1 C° (or 1 K)*. We can add heat to the gas in many different ways, however, two of the most important being at constant volume and at constant pressure.

1 *Constant volume:* Here we do not allow the piston in Fig. 14.10 to move. The amount of heat ΔQ required to change the temperature by ΔT is then

$$\Delta Q = nC_V\,\Delta T \tag{14.15}$$

where C_V is the molar heat capacity *at constant volume*. Since the volume of the gas does not change, the heat added must all go into the internal energy of the gas. The increase in internal energy is therefore

$$\Delta U = \Delta Q = nC_V\,\Delta T \tag{14.16}$$

2 *Constant pressure:* Here we allow the piston to move up, *maintaining* the gas at a *constant pressure P*. The gas does an amount of work $\Delta W = P\,\Delta V$ in moving the piston of area A through a distance Δs. In this case the heat input is

$$\Delta Q = nC_P\,\Delta T \tag{14.17}$$

where C_P is the molar heat capacity at constant pressure. Part of the heat added is used to perform the work ΔW, the rest is stored in the gas as internal energy ΔU.

We have, therefore,

$$\Delta Q = \Delta W + \Delta U \qquad \text{or} \qquad nC_P \, \Delta T = P \, \Delta V + \Delta U \qquad \text{(14.18)}$$

Now the internal energy of the gas depends only on the temperature, as Eq. (14.14) shows. The change in internal energy is given by Eq. (14.16) as $\Delta U = nC_V \, \Delta T$ and *has the same numerical value for any particular change in temperature* ΔT, irrespective of whether the change was carried out at constant pressure, constant volume, or in some more complicated fashion, since the internal energy depends only on the temperature. Combining Eqs. (14.16) and (14.18), we have, therefore,

$$nC_P \, \Delta T = P \, \Delta V + nC_V \, \Delta T \qquad \text{(14.19)}$$

But, from the ideal gas law [Eq. (9.14)], $PV = nRT$, and so, at constant pressure, $P \, \Delta V = nR \, \Delta T$. We have, then, on substituting in Eq. (14.19),

$$nC_P \, \Delta T = nR \, \Delta T + nC_V \, \Delta T$$

or $\qquad \boxed{C_P - C_V = R} \qquad \text{(14.20)}$

Since $R = 8.31$ J/(mol·K) $= 1.99$ cal/(mol·K), as we saw in Sec. 9.8, we expect the difference between the molar heat capacities of gases at constant pressure and constant volume to be about 2 cal/(mol·K). This is indeed what we find from the experimental values in Table 14.3, not merely for monatomic gases like He and Ne but also for a variety of other, more complicated, molecular gases such as O_2 and CO_2.

TABLE 14.3 Experimental Values of Molar Heat Capacities and γ for Selected Gases

Gas	Molar heat capacities [cal/(mol·K)]		$C_P - C_V$ [cal/(mol·K)]*	$\gamma = C_P/C_V$
	C_P	C_V		
Monatomic:				
He	4.97	2.98	1.99	1.67
Ne	4.97	2.98	1.99	1.67
Ar	4.97	2.98	1.99	1.67
Diatomic:				
H_2	6.87	4.88	1.99	1.41
N_2	6.95	4.96	1.99	1.40
O_2	7.03	5.03	2.00	1.40
Cl_2	8.29	6.15	2.14	1.35
Polyatomic:				
CO_2	8.83	6.80	2.03	1.30
H_2O	8.20	6.20	2.00	1.32
C_2H_6	12.35	10.30	2.05	1.20

*Or kcal/[(kg mol)·K].

Value of Molar Heat Capacities C_P and C_V

If the temperature of an ideal monatomic gas changes by ΔT, we have, from Eq. (14.14),

$$\Delta U = \tfrac{3}{2}nR \, \Delta T$$

But, from Eq. (14.16) for an expansion at constant volume,

$$\Delta U = nC_V \, \Delta T$$

and so $\quad \frac{3}{2}nR \, \Delta T = nC_V \, \Delta T$

or $\quad C_V = \frac{3}{2}R$ $\qquad\qquad\qquad\qquad$ **(14.21)**

The molar heat capacity at constant volume of an ideal monatomic gas is therefore $\frac{3}{2}R$, or about 3 cal/(mol·K).

Since, from Eq. (14.20), $C_P = C_V + R$, $C_P = \frac{3}{2}R + R = \frac{5}{2}R$, and, for a monatomic gas, the molar heat capacity at constant pressure is about 5 cal/(mol·K).

A very useful quantity in thermodynamics is the ratio of the two molar heat capacities. This ratio is designated by the symbol γ (Greek *gamma*), where

$$\gamma = \frac{C_P}{C_V}$$ $\qquad\qquad\qquad\qquad$ **(14.22)**

Since C_P and C_V have the same units, their ratio γ is dimensionless. For a monatomic gas $\gamma = C_P/C_V = \frac{5}{2}R/\frac{3}{2}R = \frac{5}{3} = 1.67$. The data in Table 14.3 show that these theoretical values for C_P, C_V, and γ are all indeed verified by experiment for monatomic gases.

It can be seen that, for the diatomic gases listed in Table 14.3, γ is *not* equal to 1.67 but is close to 1.40. This can be explained by the fact that a diatomic molecule resembles a rigid dumbbell which can rotate in space. Such a molecule can rotate freely about two perpendicular axes of rotation. It is therefore said to have two other *degrees of freedom* in addition to its three degrees of translational freedom for motion along the x, y, and z axes. Each degree of freedom has associated with it a kinetic energy of $\frac{1}{2}RT$ per mole. Hence, instead of C_V being equal to $\frac{3}{2}R$, it is equal to $\frac{5}{2}R$, and $C_P = \frac{5}{2}R + R = \frac{7}{2}R$. Then $\gamma = C_P/C_V = \frac{7}{2}R/\frac{5}{2}R = 1.40$. This agrees with the value of γ for diatomic gases in Table 14.3. This is an application of the classical principle of the *equipartition of energy*, which states that each degree of freedom has associated with it an energy of $\frac{1}{2}RT$ per mole. The failure of this principle to give results in agreement with experiment for other substances contributed significantly to the development of quantum physics.

Example 14.9

The temperature of 2.00 mol of He gas is raised from 20 to 150°C isobarically. Find **(a)** the increase in internal energy and **(b)** the external work done by the gas. **(c)** Find the heat that must be added to the gas in this isobaric process and also **(d)** in an isochoric process between the same two temperatures.

SOLUTION

From Table 14.3 the molar heat capacities of He gas are $C_P = 4.97$ cal/(mol·C°) $= 20.8$ J/(mol·C°), and $C_V = 2.98$ cal/(mol·C°) $= 12.5$ J/(mol·C°).

For an isobaric (constant-pressure) process, we have

$$\Delta Q = \Delta U + \Delta W \qquad \text{or} \qquad nC_P \, \Delta T = \Delta U + \Delta W \qquad (1)$$

where n is the number of moles of gas. Both ΔU and ΔW are, however, unknown, and we must use another approach to obtain one of them. Let us consider the same temperature change for an isochoric (constant-volume) process, for which, since $\Delta W = 0$,

$$nC_V \, \Delta T = \Delta U \qquad (2)$$

Now ΔU is the same for the isobaric and the isochoric processes, since ΔU depends *only on the change in temperature*, which is the same in the two cases. [Remember that, from Eq. (14.14), $\Delta U = \frac{3}{2}nR \, \Delta T$, no matter the process that brings about the change in temperature.]

(a) Since ΔU is the same for the isobaric and for the isochoric processes, we can use the simpler expression (2) to find it.

$$\Delta U = nC_V \, \Delta T = (2.00 \text{ mol}) \left(\frac{12.5 \text{ J}}{\text{mol·C}°} \right) (130 \text{ C}°)$$

$$= \boxed{3.25 \times 10^3 \text{ J}}$$

(b) To find ΔW for the isobaric process, we subtract Eq. (2) from Eq. (1) and obtain

$$n(C_P - C_V)\,\Delta T = \Delta W$$

or $\quad \Delta W = (2.00 \text{ mol}) \left[\dfrac{(20.8 - 12.5)\text{ J}}{\text{mol·C°}}\right] (130 \text{ C°})$

$$= \boxed{2.16 \times 10^3 \text{ J}}$$

(c) For the isobaric process,

$$(\Delta Q)_P = \Delta U + \Delta W = \boxed{5.41 \times 10^3 \text{ J}}$$

(d) For the isochoric process $\Delta W = 0$, and so from the first law of thermodynamics.

$$(\Delta Q)_V = \Delta U = \boxed{3.25 \times 10^3 \text{ J}}$$

Summary: Important Definitions and Equations

The transfer of heat

1 *Conduction:* A process in which molecules, while remaining at relatively fixed positions in space, absorb and pass on heat in the form of random mechanical motion.

$$\frac{\Delta Q}{\Delta t} = \frac{kA(T_2 - T_1)}{L} \tag{14.1}$$

2 *Convection:* A process in which heated molecules carry thermal energy with them physically as they move from one part of a fluid to another.

$$\frac{\Delta Q}{\Delta t} = hA(T_2 - T_1) \tag{14.2}$$

3 *Radiation:* The transfer of electromagnetic energy from one object to another without any transfer of matter or the need of a material medium.
Thermal radiation: The electromagnetic radiation emitted by hot objects because of their temperature.
Stefan's radiation law:

$$\frac{\Delta Q}{\Delta t} = e\sigma AT^4 \tag{14.3}$$

where the Stefan-Boltzmann constant $\sigma = 5.67 \times 10^{-8}$ W/(m²·K⁴).

Wien displacement law:

$$\lambda_{max}T = w = 2.898 \times 10^{-3} \text{ m·K} \tag{14.4}$$

Newton's law of cooling:

$$\frac{\Delta Q}{\Delta t} = K(T_2 - T_1) \tag{14.6}$$

Heat and work
Heat (Q): Thermal energy that flows from one object to another because of a temperature difference between them.

Work (W): Mechanical energy that is transferred when one object exerts a force on another object and moves it through a distance in the direction of the force.
Work done by a gas: The work done by a gas on expanding is equal to the area under its pressure-volume curve:

$$\Delta W = P\,\Delta V \tag{14.7}$$

Both the work done and the heat transferred in any physical process are dependent on the path followed by that process.

Internal energy (U): The sum of the kinetic and potential energies of all the individual particles making up a physical system. The internal energy is independent of the path followed by the physical process and depends only on the endpoints. Internal energy is a *state function.*

First law of thermodynamics

$$\Delta U = \Delta Q - \Delta W \tag{14.8}$$

where ΔU is the increase in internal energy of the system, ΔQ is the heat flow into the system, and ΔW is the work done by the system.

Principle of conservation of energy: In any isolated system the total internal energy remains constant, even though it may be transformed from one kind to another; therefore, a perpetual-motion machine of the first kind is impossible.

Efficiency

$$e = \frac{\text{work done (or energy produced) by a system}}{\text{energy input to the system}} \tag{14.10}$$

Thermodynamic processes
Isothermal (constant temperature):

$$\Delta U = 0 \quad \text{and so} \quad \Delta Q = \Delta W$$

Adiabatic (no heat enters or leaves the system):

$$\Delta Q = 0 \quad \text{and so} \quad \Delta U = -\Delta W$$

Isochoric (constant volume):

$$\Delta W = 0 \quad \text{and so} \quad \Delta U = \Delta Q$$

Isobaric (constant pressure):

$$\Delta U = \Delta Q - \Delta W = \Delta Q - P\,\Delta V \tag{14.11}$$

Molar heat capacities of an ideal gas:

$$C_P - C_V = R \tag{14.20}$$

where C_P = molar heat capacity at constant pressure
C_V = molar heat capacity at constant volume
R = the universal gas constant
Ratio of molar heat capacities:

$$\gamma = \frac{C_P}{C_V} \tag{14.22}$$

Questions

1 Why is it more difficult to keep a house warm on a winter's day if a very strong wind is blowing?

2 Why is the term *radiator* for the heating fixtures used in houses inappropriate?

3 Cool ocean breezes, produced by convection currents, are one of the delights of the seashore in the summer. Show why these breezes blow off the ocean toward the land on a hot summer's day, while at night in the summer they blow off the land toward the water.

4 It is sometimes said that, if you want to keep your coffee warm for a longer period of time, it is advisable to put the cream in immediately after pouring the coffee and not wait to do so. Can you justify such an opinion from the viewpoint of physics?

5 What are the basic problems associated with the use of solar energy to produce large amounts of electric energy in a convenient and economical way?

6 If both $\Delta\mathcal{W}$ and ΔQ are dependent on the path taken in any thermodynamic process, how can the change in internal energy ΔU be independent of path?

7 Does the first law of thermodynamics apply to living systems? Does it have to be modified in any way to be so applied?

8 Many modern refrigerators have been made more attractive by reducing the thickness of their walls. If the thickness of all the insulating walls of a refrigerator is cut in half, what will be the effect on the cost of running the refrigerator for a day, all other factors remaining the same? Why?

9 Argon atoms (Ar) and oxygen molecules (O_2) have roughly the same masses. The specific heat capacity of O_2 is, however, about twice that of Ar. Explain from the structure of the two gases why this is so.

10 Does the earth absorb exactly the same amount of energy from the sun as the earth radiates into space? What role is played by the heat flowing from the hot interior of the earth to its surface?

11 The plunger of a bicycle pump is pushed down quickly, compressing the air inside the cylinder. Why do the air and the pump's cylinder become warm?

12 Distinguish as carefully as you can between *heat* and *temperature*, using examples to do so.

13 Why do sleeping animals curl up to keep warm in winter and stretch out to keep cool in summer?

14 Doctors have reported that a number of people have died of heat stroke after remaining in hot tubs at about 45°C for several hours. Can you suggest what happens physically to raise the body temperature excessively under such circumstances?

15 (*a*) Indicate the roles played by conduction, convection, and radiation in cooling down a lake to near freezing on a cold winter's day when the air temperature is $-10°C$.

(*b*) What happens to the convection process when the temperature of the water at the bottom of the lake reaches $+4°C$?

(*c*) Draw a temperature profile of the lake water as a function of depth when the water surface is frozen solid.

(*d*) Does a thick layer of ice at the lake surface and the snow that accumulates on it prevent the water from freezing more rapidly than it would otherwise? Why?

Multiple-Choice and Simple Exercises

14.1 The amount of heat conducted per second through a cylindrical copper rod of cross-sectional area 0.10 m², whose ends are 10 m apart and kept at 50 and 30°C, is:
(*a*) 1.8×10^{-2} kcal   (*b*) 7.6×10^2 kcal
(*c*) 7.6×10^2 J   (*d*) 1.8×10^{-2} J
(*e*) None of the above

14.2 A piece of metal is at 27°C. This temperature is then increased to 327°C. The radiation emitted by the hot metal increases by a factor of about:
(*a*) 2.0×10^4 (*b*) 12   (*c*) 4   (*d*) 8
(*e*) 16

14.3 The absolute temperature of a blast furnace is increased by one order of magnitude. The amount of radiation emitted per second by the furnace increases by:
(*a*) Four orders of magnitude
(*b*) Three orders of magnitude
(*c*) Two orders of magnitude
(*d*) One order of magnitude
(*e*) None of the above

14.4 The earth radiates like a blackbody at about 290 K. Blackbody radiation from the earth therefore has a maximum intensity at a wavelength of about:
(*a*) 10 m   (*b*) 10 μm   (*c*) 1 μm
(*d*) 0.5 μm   (*e*) 10^{-7} m

14.5 Water flows over a waterfall that is 50 m high. The increase in temperature of the water at the bottom of the fall compared with the top should be about:
(*a*) 0.01C°   (*b*) 0.1C°   (*c*) 1.0C°
(*d*) 10C°   (*e*) 50C°

14.6 The efficiency of an electric power plant that takes in 1.5×10^9 J of heat in the form of steam each second, and produces electricity at the rate of 0.40×10^9 W, is:
(*a*) 2.7%   (*b*) 27   (*c*) 0.27   (*d*) 37%
(*e*) 3.7

14.7 A gas expands isothermally and does an amount of work equal to 50 J. The heat added to the gas during the expansion is:
(*a*) -50 J   (*b*) 100 J   (*c*) -100 J
(*d*) 50 J   (*e*) 0

14.8 A gas is compressed adiabatically from a volume of 2.0×10^{-3} to 1.0×10^{-3} m³. The work done on the gas during the compression is 50 J. The change in internal energy is:
(*a*) 0   (*b*) $+50$ J   (*c*) -50 J   (*d*) $+100$ J
(*e*) -100 J

14.9 A gas is compressed at a constant pressure of 10^5 N/m² from a volume of 2.0×10^{-3} to 1.0×10^{-3} m³. The work done by the gas is:

(a) -10^2 J   (b) 10^2 J   (c) 0   (d) 10^2 kcal
(e) -10^2 kcal

14.10   In an isochoric process 20 J of heat is added to a gas whose pressure is 1.0×10^5 N/m² and whose volume is 1.0×10^{-4} m³. The change in internal energy of the gas is:
(a) +20 J   (b) -20 J   (c) 0   (d) +10 J
(e) -10 J

14.11   Consider a window of area 1.0 m² and thickness 1.0×10^{-3} m. If the temperatures of the inner and outer surfaces are 10 and 5°C, respectively, how much heat per second flows through such a "window" of (a) copper, (b) glass, and (c) Styrofoam?

14.12   The inside of the steel wall of a boiler is at 110°C and the outside is at 62°C. If the wall is 1.0×10^{-2} m thick, how much heat per second comes through a 1.0-m² section?

14.13   The air temperature at the upper surface of the ice layer on a pond is -20°C. The temperature of the bottom surface is 0°C.
(a)   If the ice layer is 1.0×10^{-2} m thick, how much heat per second comes through 1.0 m² of surface?
(b)   Does ice form more rapidly when the layer is thick or thin?

14.14   How much heat per second is radiated by a 1.0-m² surface with emissivity $e = 1.0$ at 10^4 K?

14.15   At what wavelength is the radiation intensity highest for a radiating object at 0°C?

14.16   What would the temperature of a body have to be if its peak radiation intensity were at $\lambda = 10^{-10}$ m (i.e., an x-ray)?

14.17   If all the energy radiated by the sun on the earth could be used to heat the oceans, how long would it take to change the temperature of the oceans from 0 to 100°C, i.e., from freezing to boiling? Assume that the mass of the oceans is about 1.5×10^{21} kg.

14.18   A gas expands adiabatically, doing 200 J of work.
(a)   How much heat is added?
(b)   Calculate ΔU for the gas.

14.19   A gas expands isothermally, doing 150 J of work.
(a)   How much heat must be added?
(b)   Calculate ΔU for the gas.

14.20   During a constant-volume process, the internal energy of a gas increases by 300 J.
(a)   How much heat must be added?
(b)   How much work is done by the gas?

14.21   You blow up a balloon to a volume 0.10 m³.
(a)   How much work does the surface of the balloon do against the atmosphere?
(b)   How high would this amount of work lift a 90 kg person?

14.22   During a physics demonstration, a metal can is crushed by removing the inside air with a pump. The initial volume is 9.0×10^{-3} m³ and the final volume is 1.0×10^{-4} m³. How much work does the atmosphere do while crushing the can?

Problems

14.23   The submerged area of a 1.5×10^{-2} m thick wooden boat is 4.0 m². For the wood the coefficient of thermal conductivity $k = 0.12$ W/(m·C°). If the inside of the boat is 7.0 C° warmer than the water, at what rate does heat flow from the boat into the water?

14.24   A wooden shack is in the form of a cube 2.0 m on a side. The planks are 1.5×10^{-2} m thick. The temperature difference across the wall is 5.0 C°. At 120 W per person, at least how many people must be in the shack? Take $k = 0.12$ W/(m·C°).

14.25   (a)   What is the temperature gradient across the window in Example 14.1?
(b)   Thirty-five miles below the earth's surface the temperature of the melting rock is about 1400°C. If the surface temperature is 20°C, what is the temperature gradient?
(c)   Since the thermal conductivity of glass and granite are roughly the same, what conclusion do you draw?

14.26   The surface area of a Styrofoam ice chest is 2.0 m². How thick are its walls if it takes 24 h to melt 5.0 kg of ice inside the chest on a day when the temperature is 35°C?

14.27   In Prob. 14.26 if the Styrofoam were replaced by steel of the same thickness, how long would the ice last?

14.28   The inside wall temperature of a steel oven is 300°C and the outside is 120°C.
(a)   If the surface area of the oven is 1.5 m² and the walls are 5.0×10^{-3} m thick, how much heat per second is given off by conduction?
(b)   What thermal conductivity would make the result 5.0×10^3 W?

14.29   A copper bar is 2.0 m long. The cross-sectional area is 4.0×10^{-4} m². It is insulated so that heat may flow only from one end to the other. If one end of the bar is in boiling water and the other is in ice, at what rate does heat flow along the bar?

14.30   Since the specific heat of the human body is close to that of water, let us assume that the body's thermal conductivity equals that of water, 0.58 W/(m·C°). Using $T_{internal} - T_{surface} = 4.0$ C° and an effective thickness of 10^{-1} m, calculate how much heat per second the swimmer of Example 14.3 loses by conduction.

***14.31**   The thickness and thermal conductivities of a two-layer wall are denoted by L_1, L_2, k_1, and k_2. For a temperature difference $T_2 - T_1$ across the combination, show that the heat per second transferred across an area A is

$$\frac{\Delta Q}{\Delta t} = \frac{A(T_2 - T_1)}{L_1/k_1 + L_2/k_2}$$

(For three layers, a term L_3/k_3 would appear in the denominator, and similarly for more layers.)

14.32   (a)   Calculate $\Delta Q/\Delta t$ for a 1-m² glass window 0.66 cm thick. Assume $T_2 - T_1 = 5.0$ C°.
(b)   Repeat for a 1-m² window composed of two 0.33-cm-thick glass panes sandwiching a 0.33-cm-thick layer of air. Assume $T_2 - T_1 = 5.0$ C° (see Prob. 14.31).

***14.33**   A cooking utensil is to be 3 mm thick, one layer being 2 mm and the other 1 mm thick. If aluminum and brass are used, which combination will pass more heat per second? (See Prob. 14.31.)

*14.34 Two flat metal plates are firmly soldered together. Each plate has a surface area of 400 cm² and a thickness of 5.0 mm. One plate is made of copper and the other of brass. The plates are laid flat on a hot metal plate whose surface temperature is 150°C. The upper (brass) surface of the metal plates is at room temperature (20°C).

(a) Find the rate at which heat flows through the plates to the air. (Neglect any heat loss at the edges.)

(b) Find the temperature of the surface where the copper and brass plates are soldered together.

14.35 Using the results of Example 14.3 and Prob. 14.30, calculate the convection coefficient for the swimmer if the total heat loss is 120 W.

14.36 (a) What power, in watts, is radiated by a 1.0×10^{-4}-m² object ($e = 1$) at 20°C? What would this be if (b) the Celsius temperature were doubled, (c) the Kelvin temperature were doubled?

14.37 Molten iron is in a blast furnace at 1600°C. If the surface area of the iron is 10 m², how much heat is lost by radiation each second? (Take $e = 0.83$.)

14.38 A light bulb filament is 2.5×10^{-1} m long and has a cross-sectional diameter 5.0×10^{-4} m.

(a) If the filament ($e = 0.3$) runs at 1100°C, what power is radiated?

(b) Is all this in the visible region?

14.39 A person sits perfectly still in a bathtub. The body's surface temperature is 33°C, and the body's surface area is 1.3 m². What should the water temperature be if the bather is to receive a net 20 W of radiation from the water? (Take $e = 0.6$.)

14.40 From our discussion of sound (Table 12.2) we know that the thresholds of pain and barely audible sound correspond to intensities of 1.0 W/m² and 1.0×10^{-12} W/m², respectively. At what temperatures would a perfect radiator ($e = 1$) radiate with these intensities?

*14.41 A kilogram of water at 0°C is ejected from an orbiting satellite into surroundings which are at a temperature of 3.0 K.

(a) How much time elapses before the water turns into ice?

(b) Why doesn't 2 kg take exactly twice as long to turn to ice? (Assume that the water takes on a spherical shape and that the emissivity of the water is $e = 1$.)

14.42 An incandescent light bulb has a tungsten filament of surface area 0.50 cm² and operates at a temperature of 2000°C. Assume that the emissivity of the tungsten is 0.90 and that all the energy furnished to the bulb is radiated from it. How much power must be furnished to the light bulb when it is operating?

14.43 A furnace in Minneapolis breaks down at 9 P.M., when the outside temperature is 0°C. During the 3 h from 9 P.M. to midnight the house temperature falls from 20 to 17°C. If the outside temperature remains right at the freezing point all night, approximately what is the total time it will take for the house temperature to fall to (a) 14°C; (b) 11°C (c) 8°C? (Hint: Use Newton's law of cooling.)

14.44 The average rate at which solar energy is incident in the southwestern United States is about 250 W/m². Of this energy about one-third can be conveniently collected and, of this, about 30 percent can be converted into electrical energy.

Use these facts to calculate the ground area required by a solar power plant intended to produce 1 GW of electrical power.

*14.45 Show that Newton's law of cooling follows directly from the equations governing the conduction, convection, and radiation of heat so long as the two temperatures involved are not too far apart. [Hint: Show that for radiation $T_2^4 - T_1^4 \simeq K(T_2 - T_1)$.]

*14.46 When a person sleeps in pajamas, covered only by a sheet, 53 percent of the heat loss from the person's body to the room is by way of radiation, 28 percent by evaporation, and 19 percent by convection, with conduction contributing an insignificant amount. If the person's skin temperature is 33°C and room temperature is 20°C, what is the rate at which the person is losing heat to the room by all possible processes? Assume that the body surface area is 1.3 m² and that the emissivity of the skin is 0.65 in this case.

14.47 A cylinder contains 0.15 m³ of neon gas at 20°C and 5.0 atm. How much heat is required to raise the temperature of this gas to 100°C?

*14.48 A balloon is filled with 3.0 m³ of helium at STP. How much heat is needed to raise the temperature of the helium by 25 C°, if the balloon expands at atmospheric pressure?

14.49 One mole of hydrogen gas (diatomic H_2) is heated at constant volume from 20 to 100°C.

(a) How much heat is required to produce this change?

(b) How much heat would be required if this process were carried out at constant pressure instead of constant volume?

*14.50 Oxygen gas (O_2) is confined inside a vertical metal cylinder of cross section 0.010 m² by a piston of weight 100 N that is free to move up and down. The equilibrium position of the piston is 0.30 m from the bottom of the container when the gas is at 20°C, and the pressure outside the tank is 1 atm.

(a) How many moles of oxygen gas are in the cylinder?

(b) How much heat is required to raise the temperature of the gas to 200°C?

14.51 One mole of oxygen gas (O_2) is heated at a constant pressure starting at 0.00°C. How much heat must be added to the gas to triple its initial volume?

14.52 The cycle in Fig. 14.18 is for an ideal gas.

(a) Find the work done during each of the processes AB, BC, CD, and DA, specifying whether work is done by or on the gas.

(b) What is the total work done by the gas on going around the path ABCDA?

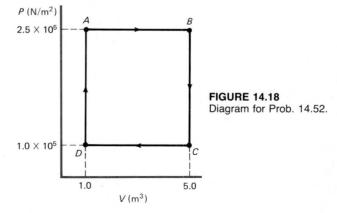

FIGURE 14.18
Diagram for Prob. 14.52.

14.53 If the cycle in Fig. 14.18 is for helium and the temperature at D is 27°C:

(a) Find the temperature at A, B, and C.

(b) Calculate ΔQ for AB, BC, CD, and DA.

*__14.54__ Two moles of N_2 gas are at 100°C.

(a) What is v_{rms} for a nitrogen molecule?

(b) If the gas does 1.0×10^3 J of work during an adiabatic expansion, what is v_{rms} at the end of the process?

14.55 In Prob. 14.54 what is the final v_{rms} if 1.0×10^3 J of work is done on the gas during an adiabatic compression?

14.56 One kilogram of copper is heated from 20 to 60°C.

(a) How much work does the copper do by pushing back the atmosphere? Take $P = 1$ atm.

(b) Calculate $\Delta W / \Delta Q$ for this process. Does the approximation $\Delta W = 0$ seem justified?

*__14.57__ The temperature of 5.0 mol of nitrogen (N_2) gas is raised from 20 to 200°C at constant atmospheric pressure.

(a) Find the increase in internal energy of the gas.

(b) Find the external work done by the gas.

(c) Find the heat added to the gas by an external source. (*Hint:* Use the values of C_P and C_V given in Table 14.3.)

*__14.58__ A cylinder containing He gas is closed off at the top by a 10-kg movable piston of surface area 100 cm². The piston is subjected to atmospheric pressure at its top surface. The He is first heated from 20 to 120°C at constant pressure, and the piston rises 30 cm. Then the piston is held fast in its new position while the gas is cooled back down to 20°C.

(a) How much heat was added to the gas during the heating process?

(b) How much heat did the gas lose as it cooled back down to 20°C?

14.59 One kilogram of iron is heated from 20 to 60°C. For purposes of this problem, the expansion of the iron can be considered negligible over this temperature range.

(a) How much heat is added to the system?

(b) What is the increase in internal energy of the iron?

(c) If all the increased internal energy could be converted into useful work, for example, to lift another 1.0-kg iron mass, how high could the 1.0-kg mass be lifted?

*__14.60__ At atmospheric pressure, 1.0 kg of water, with a volume of 1.0×10^{-3} m³, becomes 1.67 m³ when boiled.

(a) How much of the heat supplied to vaporize the water goes into external work?

(b) How much of the heat goes into internal energy?

14.61 How much does the internal energy of 5.0 kg of ice at 0°C increase as it is melted completely to water at 0°C? (Neglect the small difference in volume.)

14.62 Find the change in internal energy of a gas, (a) when it expands isothermally from 0.10 m³ at STP to 0.30 m³; (b) when it expands adiabatically from 0.10 m³ at STP to 0.30 m³ and does 8.0 J of work while expanding.

14.63 One mole of argon (Ar) gas at 20°C is in a small cylinder of constant volume. The cylinder is placed in an oven at 200°C. How much heat flows into the cylinder as the gas temperature rises to 200°C?

14.64 How many kilograms of water at 100°C could be evaporated per hour by the heat conducted through a steel plate 3.0 mm thick and of cross-sectional area 0.15 m², if one side of the plate is at 20°C and the other at 100°C?

*__14.65__ Assume that the internal energy of the gas in Fig. 14.16 is 20 J at A and 60 J at D, that the work done by the system along path ADB is 10 J, and that the difference in internal energy between A and B is $\Delta U_{AB} = 50$ J, as in Example 14.7.

(a) Find the heat absorbed in the expansion from A to D.

(b) Find the heat absorbed in the constant-volume process in which the gas goes from D to B.

14.66 One mole of He gas is heated at constant volume from 27 to 97°C.

(a) What is the change in internal energy of the gas?

(b) What is its molar heat capacity at constant volume?

(c) What is its molar heat capacity at constant pressure?

(d) What is the ratio of the specific heats $\gamma = C_P / C_V$?

14.67 A sample of He gas of initial volume 1.0×10^{-2} m³ and initial pressure 1.0×10^5 N/m² is heated from 300 to 400 K. Show by numerical calculation that the change in internal energy of the gas is the same and equal to $nC_V \Delta T$ whether the process is carried out at (a) constant volume (isochoric process) or (b) constant pressure (isobaric process).

Additional Readings

Angrist, Stanley W.: "Perpetual Motion Machines," *Scientific American*, vol. 218, no. 1, January 1968, pp. 114–122. A fascinating article relating attempts made over the past 400 years to escape the implications of the first and second laws of thermodynamics. It contains many interesting photographs and sketches of proposed perpetual-motion machines.

Feynman, Richard: *The Character of Physical Law*, MIT Press, Cambridge, Mass., 1965. Chapter 3, "The Great Conservation Principles," is particularly relevant.

Holton, Gerald: *Introduction to Concepts and Theories in Physical Science*, Addison-Wesley, Reading, Mass., 1973. Chapter 17 is very good on the first law of thermodynamics.

Kelly, James B.: "Heat, Cold, and Clothing," *Scientific American*, vol. 194, no. 2, February 1956, pp. 109–116. A brief article on the function of clothing in keeping us warm and cool.

Tierney, John: "Perpetual Commotion," *Science 83*, vol. 4, no. 4, May 1983, pp. 30–39. A good account, accompanied by photographs, of attempts to evade the implications of the laws of thermodynamics.

CHAPTER 15

The Second Law of Thermodynamics

In order to go further it is necessary to talk quantitatively. We must measure heat precisely in terms of numbers. . . . First it is clear that to specify heat we must use at least two numbers: one to measure the quantity of energy, the other to measure the quantity of disorder. The quantity of energy is measured in terms of a practical unit called the calorie. . . . The quantity of disorder is measured in terms of the mathematical concept called entropy.

Freeman J. Dyson (b. 1923)

Science and technology move forward together. Sometimes advances in physics lead to giant steps forward in technology. The development of the transistor in research physics laboratories, for example, revolutionized the electronics industry. At other times, technological developments lead to great progress in pure science. One noteworthy case was the development of radar electronics during World War II. The radar equipment developed for the detection of enemy aircraft turned out to be almost perfectly adapted to studies of atomic and molecular structure. Thus the years immediately following the war, when physicists returned to their laboratories and brought their radar gear with them, saw the rapid development of new fields such as

FIGURE 15.1 An early form of steam engine. This sketch by Theodore R. Davis shows President Ulysses S. Grant and Emperor Dom Pedro II of Brazil starting the gigantic Corliss Engine at the International Centennial Exposition in Philadelphia in 1876. (*The Granger Collection.*)

microwave spectroscopy, electron spin resonance, and atomic and molecular beam spectroscopy.

Thermodynamics has profited from technology in a special way. Around 1800 the first truly practical steam engine was put into operation in Great Britain, marking the beginning of the industrial revolution. Throughout the nineteenth century the steam engine (see Fig. 15.1) took over the operation of pumps and industrial machines, the propulsion of boats and trains, and onerous tasks previously performed by people and draft animals. Physicists were anxious to develop the steam engine to its maximum efficiency. This was the basis for much of the progress in thermodynamics in the nineteenth century, and in particular for Carnot's pioneering work on the theory of heat engines, which led directly to the second law of thermodynamics.

15.1   Reversible and Irreversible Processes

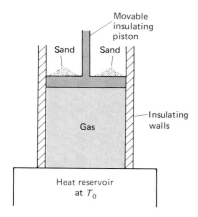

FIGURE 15.2   A gas in a container with insulating walls and a movable insulating piston at the top. The gas is in contact with a heat reservoir at temperature T_0.

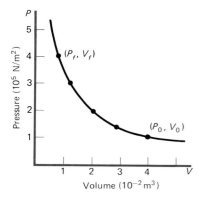

FIGURE 15.3   *PV* diagram for an isothermal (reversible) compression of the gas in Fig. 15.2, an isotherm.

Steam engines and all other kinds of heat engines are inefficient. Such inefficiency is inevitable in light of a fundamental principle of physics called the second law of thermodynamics. To understand heat engines and the second law we must first clarify the difference between reversible and irreversible thermodynamic processes, for this difference is crucial to understanding the second law.

Consider a gas in thermal equilibrium, like the one in Fig. 15.2, with its thermodynamic state defined by its volume V_0, pressure P_0, and temperature T_0. We surround the container with perfectly insulating walls at the sides and an insulating piston at the top, and set the container on a heat reservoir which allows heat to flow into or out of the gas to keep its temperature at T_0. We then reduce the volume of the gas to $V_f = V_0/4$ by carrying out either of the following processes:

1   *Reversible compression.* We gradually increase the pressure P by dropping sand, a few grains at a time, onto the piston and allowing the gas to remain in thermal equilibrium with the reservoir as the gas is compressed. In this way the volume is slowly decreased to $V_0/4$, while the pressure increases to $4P_0$ and the temperature remains constant at T_0. Each intermediate state is, therefore, an equilibrium state, and the change can be graphed on a *PV* diagram as in Fig. 15.3. This slow compression through a very large number of equilibrium states is an example of a *reversible process*. Such a process can be made to go in the reverse direction by an infinitesimal change in the operating conditions, say, by removing a few grains of sand at a time from the piston.

2   *Irreversible compression.* We can also push the piston in Fig. 15.2 down very suddenly and change the pressure from P_0 to $4P_0$. The temperature initially increases, then decreases, and the pressure and volume fluctuate in unpredictable ways during the compression. Finally, however, the temperature returns to T_0 because the gas must come into thermal equilibrium with the reservoir that is maintained at T_0. At the end of the process the pressure is $P_f = 4P_0$, the volume $V_f = V_0/4$, and the temperature $T_f = T_0$, just as in the case of the reversible compression. During the actual process of compression, however, we have no way of knowing the values of P, V, and T, since the temperature fluctuates, and Boyle's law for an isothermal process does not apply. *The intermediate states in the process do not correspond to states in which the gas is in thermal equilibrium.* The system therefore passes from one equilibrium state specified by (P_0, V_0, T_0) to another equilibrium state specified by $(4P_0, V_0/4, T_0)$ through a series of unknown non-

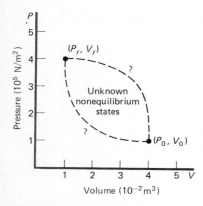

FIGURE 15.4 *PV* diagram for an irreversible compression of the gas in Fig. 15.2. The question marks are intended to indicate that we do not know the path followed by the gas.

Isothermal and Adiabatic Processes

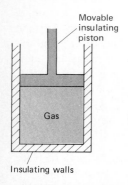

FIGURE 15.5 A reversible adiabatic process, in which a gas is surrounded on all sides by insulating walls and a movable insulating piston and is then slowly compressed.

equilibrium states, as suggested in Fig. 15.4. Such a process is called *irreversible* because it cannot be made to go in the reverse direction by a small change in the operating conditions. The sudden compression discussed above is irreversible because we do not even know the path followed by the gas in going from V_0 to V_f and hence there is no way to reverse it. Any process that is not *completely* reversible is said to be *irreversible*.

Any macroscopic physical process in the real world is irreversible. On a cold day heat always flows from the warm air inside the house to the cold air outdoors, never the reverse. When hot coffee and cold milk are mixed, they eventually reach some common, intermediate temperature; once mixed, however, they never separate out into hot coffee and cold milk. The mixing process is *irreversible*.

A reversible process is an idealized case that is extremely important from a theoretical point of view but *can never be achieved in practice*. The slow expansion or compression of an ideal gas is, however, a very close approximation to a reversible process, and we will use it as our model of a reversible process in the discussion that follows.

All isothermal processes are reversible, because the only way the temperature can be kept constant is to carry out the process very slowly. In the case of the isothermal expansion or compression of an ideal gas, we have already seen (Sec. 9.7) that the product of the pressure and the volume must remain constant, according to Boyle's law:

Isothermal process: $\boxed{PV = \text{constant}}$ **(9.11)**

The graph of such a process on a *PV* diagram is called an *isotherm*. It takes the form of the hyperbola shown in Fig. 15.3.

Adiabatic processes can be either reversible or irreversible. Here we consider a *reversible adiabatic process*, such as that in Fig. 15.5, in which a gas is surrounded by insulating walls and an insulating piston, and then slowly compressed. In this case no heat flows out of the gas, and so the temperature must rise, because the work done on the gas in the compression increases its internal energy. Hence $PV\ (= nRT)$ can no longer be a constant, since T is not a constant. Rather we find experimentally that, if the compression is carried out slowly so that the gas is in thermal equilibrium at every stage of the process, the PV graph is as shown in Fig. 15.6. Such a graph of a reversible adiabatic process is called an *adiabat*. *This has a steeper slope than an isotherm* and crosses a series of isotherms as the gas is compressed, since the temperature continually rises in such an adiabatic compression. As T increases, the PV product must increase according to the gas law $PV = nRT$. The pressure P is therefore higher than in the isothermal case for the same value of V, and the slope of the graph is steeper. This explains the difference between the isotherms and the adiabat in Fig. 15.6. It is found that the adiabat in Fig. 15.6 satisfies the following equation:

Adiabatic process: $\boxed{PV^\gamma = \text{constant}}$ **(15.1)**

where γ (Greek *gamma*) is the ratio of the molar heat capacities (C_P/C_V) defined in Sec. 14.9, and the constant depends on the amount of gas present. Equation (15.1) describes the behavior of the pressure and volume in a reversible adiabatic process, just as Boyle's law, $PV = $ constant, describes the same behavior for an isothermal process.

Since, for any ideal gas $PV = nRT$, we have from Eq. (15.1) for an adiabatic process,

$$\frac{nRT}{V} V^\gamma = \text{constant} \qquad \text{or} \qquad \boxed{TV^{\gamma-1} = \text{constant}} \qquad \textbf{(15.2)}$$

for n and R are both constants for any sample of gas, and have no effect on the functional form of the equation. The two basic equations governing reversible adiabatic processes in an ideal gas are therefore

$$P_1 V_1^\gamma = P_2 V_2^\gamma \qquad \textbf{(15.3a)}$$

and $\qquad T_1 V_1^{\gamma-1} = T_2 V_2^{\gamma-1} \qquad$ **(15.3b)**

These equations can be used to find how any one of the three fundamental quantities for an ideal gas—P, V, and T—varies when the other quantities vary in an adiabatic process, just as $PV = $ constant can be used in the isothermal case. Note that the temperatures must be expressed on the absolute scale, since the temperature in the ideal gas law is the absolute temperature.

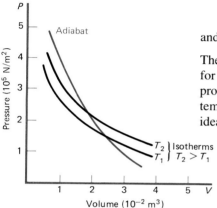

FIGURE 15.6 *PV* diagram for a reversible adiabatic process; an adiabat. Notice how the adiabat crosses the isotherms, since adiabats have steeper slopes.

Example 15.1

Suppose that 1.00×10^{-3} m³ of neon gas at room temperature and atmospheric pressure is compressed adiabatically to twice atmospheric pressure. What are **(a)** the final volume and **(b)** the final temperature of the gas?

SOLUTION

(a) For an adiabatic process, we have $P_1 V_1^\gamma = P_2 V_2^\gamma$. Here $V_1 = 1.00 \times 10^{-3}$ m³ and $P_2 = 2P_1$. Also, since neon is a monatomic gas, $\gamma = 1.67$ (Table 14.3). We have, therefore,

$$\left(\frac{V_2}{V_1}\right)^\gamma = \frac{P_1}{P_2} = \frac{1}{2}$$

or $\qquad \dfrac{V_2}{V_1} = \left(\dfrac{1}{2}\right)^{1/\gamma} = \left(\dfrac{1}{2}\right)^{1/1.67} = \left(\dfrac{1}{2}\right)^{(0.60)} = 0.660$

and so $\qquad V_2 = 0.660\, V_1 = \boxed{0.660 \times 10^{-3} \text{ m}^3}$

(b) Here $T_1 V_1^{\gamma-1} = T_2 V_2^{\gamma-1}$, and so

$$\frac{T_2}{T_1} = \left(\frac{V_1}{V_2}\right)^{\gamma-1} = \left(\frac{1.00 \times 10^{-3} \text{ m}^3}{0.660 \times 10^{-3} \text{ m}^3}\right)^{0.67}$$

$$= (1.52)^{0.67} = 1.32$$

$$T_2 = 1.32T_1 = 1.32(293 \text{ K}) = 387 \text{ K} = \boxed{114°\text{C}}$$

Note on the numerical calculation: The unusual powers to which numbers are raised in a problem of this sort can be calculated by use of logarithms. However, it is much easier to use the y^x function on a hand calculator.

Work Done in Isothermal and Adiabatic Processes

We saw in Sec. 14.6 that the work done by a gas in expanding its volume by an amount ΔV at a pressure P is $\Delta W = P\, \Delta V$. For an *isobaric* process, the total work done during the expansion at pressure P_0 is $W = P_0(V_f - V_0)$. This result does not help us to find the work done in isothermal and adiabatic processes, however, since the pressure is *not constant* in such processes. To obtain the work in such

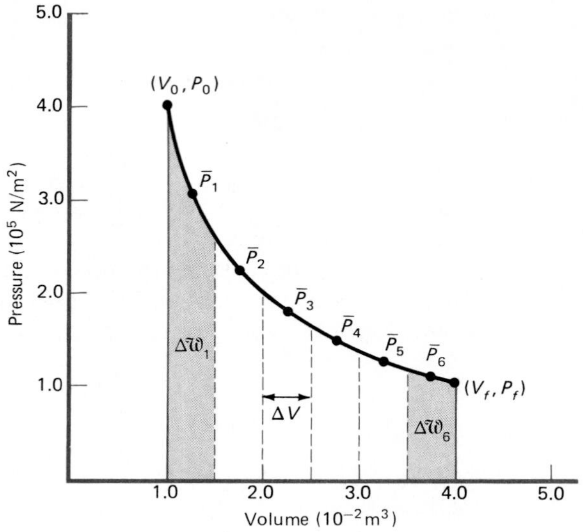

FIGURE 15.7 Calculation of the work done during an isothermal expansion.

cases, calculus is needed, but useful results can also be obtained graphically, as we will now show for the work done by a gas in an isothermal expansion.

The graph of an isothermal expansion, in which the volume of a gas increases from $V_0 = 1.00 \times 10^{-2}$ m^3 to $V_f = 4.00 \times 10^{-2}$ m^3, is shown in Fig. 15.7. Since the temperature remains constant, the pressure and the volume must satisfy the equation

$$PV = P_0V_0 = (4.00 \times 10^5 \text{ N/m}^2)(1.00 \times 10^{-2} \text{ m}^3) = 4.00 \times 10^3 \text{ J}$$

and so $P = (4.00 \times 10^3 \text{ J})/V$. The values of P obtained from this equation are given with the corresponding values of V in Table 15.1.

The work done in this expansion is the area under the curve between V_0 and V_f. To obtain an approximate value for this area graphically, we break up the area into six vertical strips, each of width $\Delta V = 0.500 \times 10^{-2}$ m^3. We then calculate the average value of the pressure $(\overline{P})$ for each strip by simply averaging the values of P at the two edges of each strip. The area of each strip is then, to a reasonable approximation, the work done by the gas during the volume change covered by the strip. The total work done by the gas is the sum of the areas of the six strips:

TABLE 15.1 Work Done by a Gas in an Isothermal Expansion

Volume V (10^{-2} m^3)	Pressure P (10^5 N/m^2)	Average pressure $\overline{P}$ for vertical strip (10^5 N/m^2)	Width of strip ΔV (10^{-2} m^3)	Work done $\Delta W = \overline{P}\,\Delta V$ (10^3 J)
1.00	4.00			
1.50	2.70	3.35	0.50	1.68
2.00	2.00	2.35	0.50	1.18
2.50	1.60	1.80	0.50	0.90
3.00	1.30	1.45	0.50	0.73
3.50	1.10	1.20	0.50	0.60
4.00	1.00	1.05	0.50	0.53

Total work done $W = 5.62 \times 10^3$ J

$$\mathcal{W} = \sum_{i=1}^{6} \Delta \mathcal{W}_i = \sum_{i=1}^{6} \overline{P}_i \, \Delta V$$

This summation is carried out in Table 15.1 and leads to the result $\mathcal{W} = 5.62 \times 10^3$ J.*

For an adiabatic process, since $PV^\gamma = P_0 V_0^\gamma$, we have $P = \text{constant}/V^\gamma$, where the constant is $P_0 V_0^\gamma$. This equation can be used to plot P versus V, and the work done obtained from the area under the curve, as in the above example.

This graphical method can be used to find the work done by a gas in any kind of physical process, as long as it is possible to construct a continuous PV graph for the process.

15.2 Entropy

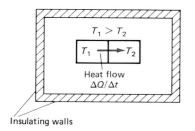

Insulating walls

FIGURE 15.8 Flow of heat between two bodies at different temperatures in a thermally insulated box. Since $T_1 > T_2$, heat flows from 1 to 2. The process is an irreversible one, and the entropy increases.

Suppose we have two blocks of metal, one at a temperature T_1, the other at temperature T_2, where $T_1 > T_2$. The blocks are initially insulated from each other. In such a condition they represent an *ordered system*, for we can clearly distinguish two groups of molecules, those at temperature T_1, with a higher average kinetic energy, and those at T_2, with a lower average kinetic energy, and we can therefore rank-order them according to their temperatures. (These two blocks could be used as the hot and cold reservoirs in a heat engine to produce useful work, as we will see.)

Suppose, now, that the two blocks are put in contact with each other, as in Fig. 15.8. Heat conduction from the hotter block to the cooler block soon brings them to the same intermediate temperature T_f. Now we can no longer distinguish one group of molecules from the other or use the blocks to do useful work. Order has thus given way to *disorder*, or randomness, in an irreversible way. To separate the original two groups of hot and cold molecules from each other once they have come to the same average temperature would require the flow of heat from a colder temperature to a hotter one—an event that has never been experimentally observed.

Why does the first process, the flow of heat from a higher to a lower temperature, occur so easily, whereas the reverse process never occurs? The answer given by physics is that in the first process a quantity called *entropy* increases, and therefore the process can occur in practice, whereas the second process never occurs because it would require a *decrease* in entropy.[†]

What is this mysterious quantity called entropy? The above example indicates that entropy is a measure of *disorder*. An increase in entropy means an increase in disorder. Since, as we all know from experience, things spontaneously become more disordered (consider the clothes in your room or the records in your record collection), in any spontaneous process the entropy always increases.

---

*From calculus it can be found that the exact value for the work done by the gas in this isothermal expansion from V_0 to V_f is

$$\mathcal{W} = P_0 V_0 \ln \frac{V_f}{V_0} = (4.00 \times 10^5 \text{ N/m}^2)(1.00 \times 10^{-2} \text{ m}^3) \ln \frac{4}{1}$$

$$= (4.00 \times 10^3 \text{ J})(1.39) = 5.55 \times 10^3 \text{ J}$$

Our graphical result agrees with this exact result to two significant figures, which is quite good considering the crudeness of the averaging process used to obtain $\overline{P}$.

[†]The German physicist Rudolf Clausius (1822–1888) introduced the word *entropy* in 1865, basing it on a Greek word signifying *transformation*, and deliberately making it as similar as possible to the word *energy*.

Disorder and Probability

The reason *all* physical systems tend spontaneously to move from an ordered state to a disordered state is simply that disordered states are *more probable* because there are more of them. For example, consider a collection of 100 identical coins. Let us characterize this collection by specifying the number of *heads* that turn up when the coins are tossed. The most ordered state of this system is one with 100 heads. If the coins are tossed, however, it is most improbable that 100 heads will turn up. The reason is that there is only one way this can be achieved. On the other hand, a situation with 50 heads and 50 tails resulting from tossing the coins is the most probable, since there are a very large number of ways in which this state can be achieved. But this state is also the most disordered state. We therefore conclude that disordered states are highly probable but that the one highly ordered state of all heads is most improbable. Since this result depends on the theory of probability, the more coins there are, the more likely it is that half of them will come up heads when tossed (we assume that the number of coins is always even).

The entropy of any isolated system therefore always tends to increase, because disorder (which entropy measures) is always more probable than order. The relationship between entropy and disorder is usually expressed in the form of an equation derived by Ludwig Boltzmann (1844–1906):

$$S = k \ln W \tag{15.4}$$

where S is the entropy, k is Boltzmann's constant (Sec. 9.8), W is the thermodynamic probability, and ln is the natural logarithm (base e). This equation connects a thermodynamic, or macroscopic, quantity, the *entropy*, with a statistical, or microscopic, quantity, the *thermodynamic probability*, which is the number of possible arrangements leading to the same observed state of a physical system. As an example, let us return to our 100 coins. The thermodynamic probability of all 100 coming up heads when tossed is 1, since there is only one way this configuration can be achieved. For this highly improbable state, $S = k \ln 1 = 0$. But for 50 heads and 50 tails, it can be shown that $W = 10^{29}$, and so $\ln W = 67$. We see, therefore, that for this highly probable and therefore disordered state, the entropy S is very much larger than for the state with 100 heads, for which $S = 0$.

Example 15.2

If you toss 3 pennies, what is the relative probability that the pennies will come up **(a)** 3 heads; **(b)** 2 heads, 1 tail; **(c)** 1 head, 2 tails; **(d)** 3 tails?

SOLUTION

The easiest way to solve this problem is simply to make a table of the possibilities for each of the 3 pennies:

	3 heads	2 heads, 1 tail	1 head, 2 tails	3 tails
Penny 1	H	H H T	T T H	T
Penny 2	H	H T H	T H T	T
Penny 3	H	T H H	H T T	T
Possible ways	1	3	3	1

Since the number of possible different configurations of the three pennies is 8, the relative probability of:

(a) 3 heads is $\boxed{1/8}$

(b) 2 heads and 1 tail is $\boxed{3/8}$

(c) 1 head and 2 tails is $\boxed{3/8}$

(d) 3 tails is $\boxed{1/8}$

EXERCISE 1 Calculate the entropy corresponding to (a) all 3 coins turning up the same; (b) 2 coins the same, 1 different.

15.3   Entropy Changes

To understand better the difficult concept of entropy, we define entropy in a measurable way first introduced by Rudolf Clausius. Suppose heat flows into a gas slowly and reversibly so as to keep the temperature T constant while the gas does work on a piston. We define the change in entropy (ΔS) of the gas in this reversible process as the heat added to the gas divided by the absolute temperature at which the process takes place:

Change in entropy:

$$\Delta S = \frac{\Delta Q}{T}$$

(15.5)

where ΔQ is positive if heat flows into the system and negative if heat flows out of the system. Common units for entropy are joules per kelvin (J/K), or kilocalories per kelvin (kcal/K).

We have already seen that Q is not a state function. On the other hand, it turns out that entropy, as defined above, *is* a state function (like the internal energy U), since it always has the same value for the same values of P, V, and T. Because entropy is a state function, the change in entropy of a system in any closed cyclic process is zero, just as the change in the internal energy of a thermodynamic system in a cyclic process is zero, as discussed in Sec. 14.7. Therefore,

$$\sum_{\text{cycle}} \Delta S = \sum_{\text{cycle}} \frac{\Delta Q}{T} = 0$$

Reversible Process

Suppose an ideal gas expands isothermally from a volume V_0 to a volume V_f, as in Fig. 15.7. In so doing the gas must take in an amount of heat ΔQ from a heat reservoir to keep the temperature constant at its initial value T_0. The change in the entropy of the gas (in this case the gas is the *system*) is therefore $\Delta S_{\text{sys}} = \Delta Q/T_0$. The heat reservoir (which we call the *environment*) must lose an equal amount of heat ΔQ at the same temperature. Therefore the change in the entropy of the environment is $\Delta S_{\text{env}} = -\Delta Q/T_0$. The total change in the entropy of the system plus environment is therefore

$$\Delta S = \Delta S_{\text{sys}} + \Delta S_{\text{env}} = \frac{\Delta Q}{T_0} - \frac{\Delta Q}{T_0} = 0$$

The sum of the entropy of the system and the entropy of its environment remains unchanged in this and in all *reversible processes*.

Irreversible Process

As an example of an irreversible process, consider our previous example of the flow of heat between two blocks of metal, one at T_1 and the other at T_2, with $T_1 > T_2$. We put these two blocks in contact with each other inside a thermally insulated box, as in Fig. 15.8, and they eventually come to a final intermediate temperature T_f. This is an irreversible adiabatic process.

It is difficult to calculate entropy changes in irreversible processes. We can, however, easily show that entropy must *increase* in this irreversible process. The hotter metal's temperature falls from T_1 to T_f, and hence its average temperature T_H during this process is certainly greater than T_f. Similarly, the colder metal's temperature rises from T_2 to T_f, and so its average value T_C is certainly less than T_f. If the heat lost by the hotter metal and gained by the colder metal is ΔQ, then the change in entropy of the system is

$$\Delta S_{\text{sys}} = -\frac{\Delta Q}{T_H} + \frac{\Delta Q}{T_C}$$

Since $T_H > T_f > T_C$, we must have $\Delta Q/T_C > \Delta Q/T_H$ and

$$\Delta S_{\text{sys}} = -\frac{\Delta Q}{T_H} + \frac{\Delta Q}{T_C} = \Delta Q \left(\frac{T_H - T_C}{T_H T_C}\right) > 0$$

In this irreversible process the entropy of the system increases. The entropy of the environment is unchanged, since the system is thermally insulated from the environment, i.e., it is an *isolated system*. This result turns out to be valid for all irreversible processes. We therefore conclude:

In an irreversible process the entropy of any physical system plus its environment always increases.

We can combine our two results in the following summary statement about entropy:

> The entropy of any physical system plus its environment increases in all irreversible processes and remains constant in all reversible processes. Hence *the total entropy of any isolated system never decreases.*

Since, as we have seen, all real macroscopic processes are irreversible, the entropy of any isolated physical system is continually increasing. This is one statement of the *second law of thermodynamics*, which we will discuss in the next section. Just as the first law states that the total energy in a closed system is conserved, so the second law tells us that in any real physical process the entropy increases and the energy available to do work (*the available energy*) decreases. The ultimate justification of the second law is found in experiment.

In solving problems involving entropy changes, it is important to note that in a thermodynamical process, whether reversible or irreversible, the entropy of the *system* (say, a gas) may either increase, decrease, or remain unchanged. But the corresponding entropy change in the *environment* must always be such that the total entropy change of *system plus environment* remains constant in reversible processes and increases in irreversible processes.

Example 15.3

In an isothermal expansion at 300 K, the volume of 2.00 mol of argon gas increases from V_0 to $3V_0$. What is the change in the entropy of the gas?

SOLUTION:

We can use the result given in the footnote on page 383 for the work done by a gas in an isothermal expansion from V_0 to V_f:

$$\mathcal{W} = P_0 V_0 \ln \frac{V_f}{V_0} = P_0 V_0 \ln \frac{3V_0}{V_0} = P_0 V_0 \ln 3$$

But, from the ideal gas law,

$$P_0 V_0 = nRT_0$$
$$= (2.00 \text{ mol}) \left(8.31 \frac{\text{J}}{\text{mol·K}}\right) (300 \text{ K})$$
$$= 4.99 \times 10^3 \text{ J}$$

and so $\mathcal{W} = (4.99 \times 10^3 \text{ J}) \ln 3 = (4.99 \times 10^3 \text{ J})(1.10)$

$$= 5.48 \times 10^3 \text{ J}$$

For an isothermal process, $\Delta U = 0$, and so $\Delta Q = \Delta \mathcal{W}$, where $\Delta \mathcal{W}$ is the work done by the gas, equal to 5.48×10^3 J, and thus

$$\Delta S = \frac{\Delta Q}{T} = \frac{5.48 \times 10^3 \text{ J}}{300 \text{ K}} = \boxed{18.3 \text{ J/K}}$$

The entropy of the gas has increased and the gas is now more disordered, for it occupies three times its original volume. The flow of an amount of heat ΔQ into the gas keeps its *internal energy* constant, since the gas does an amount of work $\Delta \mathcal{W} = \Delta Q$. At the same time the flow of heat *into* the gas increases its *entropy* by an amount $\Delta S = \Delta Q/T$, while the flow of heat out of the environment lowers the entropy of the environment by the same amount. The total entropy change of system plus environment therefore remains constant, as it must in any isothermal process, which is necessarily a reversible process.

Example 15.4

In a thermally isolated system, a sample of 2.00 kg of water at 28.0°C is mixed with 2.00 kg of water at 20.0°C. **(a)** Calculate the change in entropy of this system in the mixing process. **(b)** Does the entropy of the system plus its environment increase or decrease?

SOLUTION

This is a more difficult example than 15.3 because all mixing processes of this kind are irreversible (as we will show). Ordinary thermodynamics deals only with equilibrium states, and therefore we can make calculations only for reversible processes. There are ways around this limitation, but we will be content with an approximate calculation that leads to a result of the same sign and order of magnitude obtained by more sophisticated methods.

(a) Since the masses of the two samples of water are equal, the final temperature of the water will be 24.0°C. An amount of heat ΔQ therefore leaves the hotter water and flows into the cooler water to bring them both to the same equilibrium temperature. The amount of heat that flows is

$$\Delta Q = cm \, \Delta T = \left(1.00 \, \frac{\text{kcal}}{\text{kg·C}°}\right)(2.00 \text{ kg})(4.0 \text{ C}°)$$

$$= 8.00 \text{ kcal} = 3.35 \times 10^4 \text{ J}$$

The change in entropy of the hot water is then $\Delta S_H = -\Delta Q/T$, where we take T as approximately equal to the average temperature of the hot water as it cools from 28 to 24°C, that is, 26°C or 299 K. Then

$$\Delta S_H = -\frac{3.35 \times 10^4 \text{ J}}{299 \text{ K}} = -112 \text{ J/K}$$

For the cold water,

$$\Delta S_C = +\frac{3.35 \times 10^4 \text{ J}}{295 \text{ K}} = +114 \text{ J/K}$$

The overall entropy change of the water is therefore

$$\Delta S = \Delta S_H + \Delta S_C = \boxed{+2 \text{ J/K}}$$

(b) The entropy of system plus environment increases, since the entropy of the environment is unchanged, for the system is thermally insulated from its environment. The reason that the overall entropy increases in the mixing process is that the process is irreversible. The final state of the system is more disordered than the initial state, and *entropy* increases along with *disorder*.

Some Practical Consequences of the Concept of Entropy

Entropy must increase in any *closed* system. An embryo is less ordered than a mature adult, and as a result human beings become more ordered as they grow. The explanation of this seeming anomaly is that a human body is not a closed system. In the growth process food that is highly ordered and therefore of low entropy is taken into the body, while waste products that are disordered and therefore of high entropy are eliminated. As a consequence, the entropy of the total closed system—human body plus environment—continually increases, in accordance with the second law of thermodynamics.

This increase in entropy with time has been called *time's arrow*, for it gives us a way to distinguish clearly the direction in which time is flowing: time is always flowing in the direction of *increasing entropy*. If, for example, we see a film of Humpty-Dumpty sitting on a wall and then falling off and breaking into thousands of little pieces, we judge that the film is moving forward in time, since order is being replaced by disorder, and entropy is increasing. If, on the other hand, we see a film in which many thousands of little pieces fly together to form the original

Humpty-Dumpty, we can say immediately that this does not correspond to the real world, since entropy is decreasing, which means that time must be running backward. Thus the increase in entropy—time's arrow—gives us certain knowledge of time's direction.

Since the entropy of the universe is continuously increasing, it will some day reach a maximum. This will happen when everything in the universe is at the same temperature. All physical, chemical, and biological processes will then cease, and there will be no energy available for doing useful work. For example, heat engines, which require hot- and cold-temperature reservoirs for their operation, will be unable to perform useful work. This condition is sometimes referred to as the *heat death of the universe*. Fortunately it will be some billions of years before such a condition is likely to occur.

15.4 The Second Law of Thermodynamics

The second law of thermodynamics is unique in physics in that it is not a statement of what is or what can be but rather of what *cannot* be. It is a *statement of impossibility*. The many different forms in which this impossibility can be expressed make the law appear more complex than it really is.

The most basic and far-reaching statement of the second law is the entropy statement discussed in the last section: *The total entropy of any isolated system never decreases*. This is based on the spontaneous tendency for nature to proceed from a state of greater order to one of lesser order (greater disorder). Gases expand to fill any available space; two objects at different temperatures when put in contact finally reach the same temperature. The explanation for this tendency is found ultimately in a statistical analysis of the behavior of the huge number of molecules that constitute matter as we know it. Systems move spontaneously from less probable states to more probable states, and in so doing increase their entropy. Movement from more probable to less probable states never occurs, and entropy therefore never decreases.

An immediate consequence of this entropy statement is that heat can never flow spontaneously from a colder object to a hotter one. If it did, the entropy of the system and its environment would decrease (prove this!), and we have just excluded this as a possibility.

There are other important statements of the second law of thermodynamics based on the behavior of heat engines. We will discuss these statements later in this chapter.

Clausius' Summary of Thermodynamics

Probably the most terse statement of the laws of thermodynamics has been provided by Rudolf Clausius:

First law: The *energy* of the universe is constant.

Second law: The *entropy* of the universe tends always toward a maximum.

In more colloquial terms, the first law tells us that we cannot win, since we can never get something for nothing in the energy business. The second law tells us that we cannot break even, because we must always lose some useful energy in any energy-conversion process.

The entropy statement of the second law implies that any isolated system will spontaneously go from an ordered state to a disordered state. The greater the disorder, the less work the system will be able to do, because less energy *useful* for doing work will be available. Even though energy is always conserved, an increase of entropy in a system means that the system will be able to do less work than if the system had a lower entropy. Degradation of the quality of energy therefore goes hand in hand with an increase in entropy—and entropy is always increasing in any isolated system. This is important for understanding both thermodynamics and the world's energy problems.

15.5 Heat Engines; the Carnot Engine

Our technological civilization depends on the efficient and economical conversion of the energy in fossil and nuclear fuels into useful mechanical or electrical energy. In a heat engine, the process followed in doing this always includes the following steps:

Internal energy (in fuel) → heat → mechanical or electrical energy

Even though the first step can be carried out with nearly 100 percent efficiency, the second is a very inefficient process: only 5 to 40 percent of the thermal energy is converted to mechanical or electrical energy, as we will see. The rest of the energy is lost in the form of heat exhausted at a lower temperature.

Heat engine: A device that, operating in a closed cycle, converts heat into useful work.

A heat engine takes a *working substance* (often water) through a series of operations known as a *cycle*. In all heat engines the basic process is the same. The internal energy of a fuel like gas or coal is used to heat the working substance to a high temperature in a boiler. The working substance then does some useful work and discharges the remainder of the original heat at a lower temperature. Such a heat engine can be represented by the flow diagram in Fig. 15.9.

In this process the working substance is used over and over again. It is always restored to its original condition at the end of the cycle by contact with the heat reservoir at temperature T_H, which receives its energy from the fuel being burned. Since the working substance is unchanged in a complete cycle, its internal energy is also unchanged, and so, from the first law,

$$\Delta U = \Delta Q - \Delta W = 0$$

or $$\Delta W = \Delta Q = \Delta Q_H - \Delta Q_C$$

The total work done by the engine is thus the difference between the heat ΔQ_H taken in from the hot reservoir and the heat ΔQ_C given out to the cold reservoir.

The efficiency e of such an idealized heat engine is defined as the ratio of the work it produces to the energy it takes in as heat. Therefore

$$e = \frac{\text{work output during cycle}}{\text{heat input during cycle}} = \frac{\Delta W}{\Delta Q_H} \tag{15.6}$$

or $$e = \frac{\Delta Q_H - \Delta Q_C}{\Delta Q_H} = 1 - \frac{\Delta Q_C}{\Delta Q_H} \tag{15.7}$$

Hence, for maximum efficiency ΔQ_C must be as small as possible and ΔQ_H as large as possible. However, no heat engine has ever been built that makes ΔQ_C zero, and so no heat engine is 100 percent efficient.

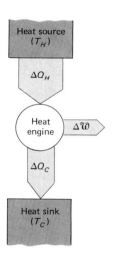

FIGURE 15.9 A generalized heat engine. The engine takes in heat ΔQ_H at a temperature T_H, converts part of it into useful work ΔW, and discards the remainder (ΔQ_C) as heat to a cold reservoir at temperature T_C. Here $\Delta W = \Delta Q_H - \Delta Q_C$.

Nicolas Léonard Sadi Carnot (1796–1832)

FIGURE 15.10 Nicolas Léonard Sadi Carnot in 1813 when he was 17. He is wearing the uniform of the École Polytechnique in Paris, a school founded to train engineers for the French army. (*Courtesy of AIP Niels Bohr Library.*)

Carnot was a remarkable member of a very distinguished French family. He was born in Paris, the eldest son of Lazare Carnot, who served as minister of war under Napoleon and wrote treatises on military strategy, mechanics, geometry, and calculus. The young Carnot was educated at home by his father in mathematics, physics, languages, and music. At the age of 16 he entered the École Polytechnique, of which his father was a founder, and continued his studies there until 1814. In that year Carnot petitioned Napoleon to allow the Polytechnique students to help defend France against an alliance of several European armies. Napoleon granted this request, and in March 1814 Carnot and his fellow students fought bravely but unsuccessfully to keep the attacking armies out of Paris. He then took a course in military engineering at Metz, and from 1816 to 1818 served in the French army as a second lieutenant in charge of planning fortifications.

Carnot did not have the temperament of a soldier and in 1818 took a permanent leave of absence from the military to pursue his real interest, the application of physics and mathematics to the improvement of the steam engines that were then becoming popular in France as part of the industrial revolution. This passion occupied him until his premature death from cholera in 1832 at the age of 36, except for two interruptions: a trip to Magdeburg in the summer of 1821 to visit his then-exiled father, and another tour of active duty with the military from 1826 to 1828.

In 1824 Carnot published his great and only work, *Reflections on the Motive Power of Heat*. In it he addressed questions such as whether there is a limit to the amount of work that can be produced from a given quantity of heat, whether steam is the best substance to use in engines, and, if so, what can be done to improve the efficiency of steam engines. His approach to these problems was in great part derived from his father. Although he used the caloric theory of heat in his book, it seems clear that he broke away from it in his later years and embraced the kinetic theory of heat, which was ultimately to win out.

Because Carnot was the first physicist to treat the interconversion of work and heat quantitatively, he deserves to be called the father of thermodynamics. He showed that what is important in producing work from heat are the high and low temperatures between which the heat engine operates. It is possible to use Carnot's results to obtain the second law of thermodynamics, even though he never made this clear in his work. Kelvin and Clausius did so years later, and in so doing paid tribute to Carnot's pioneering work, Kelvin referring to him as "the profoundest thinker in thermodynamic philosophy in the first thirty years of the nineteenth century."

Carnot was a quiet, scholarly individual, who lived by the avowed principle: "Say little about what you know, nothing about what you do not know." He was by no means a narrow scientist; throughout his short life he continued to study economics, French literature, and music, and to cultivate gymnastics, fencing, and dancing as relaxations from his scientific work. He was also decidedly practical, a prime motivation in his work being the hope that he could improve people's lives by developing cheaper and more efficient steam engines to reduce their burden of work.

Carnot Cycle

A *Carnot engine* is a particular type of idealized heat engine that Sadi Carnot (see Fig. 15.10 and accompanying biography) proved was the most efficient possible heat engine operating between any two given temperatures. Carnot proved that all engines of this type operating between the same two temperatures have the same efficiency regardless of the nature of the working substance. In our case we choose, for simplicity, an ideal gas as the working substance.

The distinguishing feature of a Carnot engine is that the cycle is defined by

FIGURE 15.11 (Right) The four steps in a Carnot cycle: (*a*) Isothermal expansion; (*b*) adiabatic expansion; (*c*) isothermal compression; and (*d*) adiabatic compression. In a Carnot cycle all four processes are reversible.

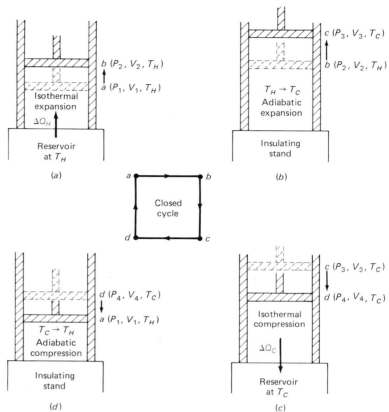

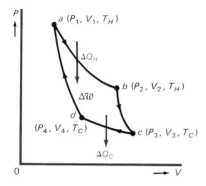

FIGURE 15.12 (Above) *PV* diagram for the Carnot cycle of Fig. 15.11. The enclosed area (= ΔW) is bounded by two isotherms and two adiabats. (For clarity the difference between the two slopes has been exaggerated.)

two *reversible* isothermal processes and two *reversible* adiabatic processes. The gas is confined to a cylinder with insulating walls and a movable, insulating piston at one end. At the other end the cylinder can be put in contact with heat reservoirs or insulating stands as needed. The Carnot cycle consists of the four steps illustrated in Fig. 15.11:

(a) *Isothermal expansion:* The gas is put in contact with the reservoir at T_H and, after achieving temperature equilibrium, is expanded isothermally. Its initial volume V_1 changes to a new volume V_2, with the pressure falling from P_1 to P_2. The gas takes in an amount of heat ΔQ_H from the reservoir at T_H, an amount that is just sufficient to keep the temperature constant at T_H throughout the expansion. On the *PV* diagram in Fig. 15.12, the gas moves from point *a* to point *b*.

(b) *Adiabatic expansion:* The heat reservoir is replaced by an insulating stand and the gas is expanded adiabatically and reversibly. The quantities P_2, V_2, T_H change to P_3, V_3, T_C, with $T_C < T_H$; that is, the temperature falls. In Fig. 15.12 the gas goes from *b* to *c*.

(c) *Isothermal compression:* The insulating stand is replaced by another heat reservoir at a temperature T_C, and the gas is compressed isothermally at T_C. An amount of heat ΔQ_C is given off to the reservoir at the temperature T_C so as to keep the gas temperature constant. The quantities P_3, V_3 change to P_4, V_4, and the gas goes from *c* to *d* on the *PV* diagram.

(d) *Adiabatic compression:* An insulating stand again replaces the heat reservoir. The gas is compressed adiabatically back to its initial state. The quantities P_4, V_4, T_C change to P_1, V_1, T_H (the original values), and the gas goes from *d* to *a* in Fig. 15.12.

The gas has therefore been taken around a closed, reversible cycle. As Fig. 15.12 shows, heat ΔQ_H flows into the gas at temperature T_H, and a lesser amount of heat ΔQ_C flows out of the gas at T_C. Since the internal energy of the gas is unchanged in the cycle, the difference $\Delta Q_H - \Delta Q_C$ has been converted to an amount of work equal to the area enclosed by the two isotherms and the two adiabats in Fig. 15.12.

The entropy change of the gas in this Carnot cycle is

$$\Delta S = \frac{\Delta Q_H}{T_H} - \frac{\Delta Q_C}{T_C}$$

Since entropy is a state function, the change in entropy of the gas is zero for the complete cycle, and so

$$\frac{\Delta Q_H}{T_H} - \frac{\Delta Q_C}{T_C} = 0 \quad \text{or} \quad \boxed{\frac{\Delta Q_C}{\Delta Q_H} = \frac{T_C}{T_H}} \tag{15.8}$$

The overall efficiency of the Carnot cycle is therefore, from Eq. (15.7),

$$e_{\text{max}} = 1 - \frac{\Delta Q_C}{\Delta Q_H} = 1 - \frac{T_C}{T_H} \quad \text{or} \quad \boxed{e_{\text{max}} = 1 - \frac{T_C}{T_H}} \tag{15.9}$$

We call e_{max} the Carnot efficiency. It is the *maximum possible efficiency* for a heat engine operating between temperatures T_H and T_C, since Carnot proved that a Carnot engine is the most efficient heat engine operating between any two given temperatures.

For a Carnot engine operating between 600 and 100°C, this efficiency is

$$e_{\text{max}} = 1 - \frac{273 + 100}{273 + 600} = 1 - 0.43 = 0.57 = 57\%$$

Any heat engine that is not a Carnot engine has a lower efficiency than this. One reason is that, since all processes in a Carnot cycle are reversible, it represents an idealized case. All real heat engines involve some irreversible processes, like friction or turbulence, which dissipate energy and reduce the efficiency of the engine below the Carnot value. Since Carnot had proved that the efficiency was independent of the nature of the working substances, he was able to conclude:

> The efficiency of all reversible heat engines operating between the same two temperatures is the same, and no irreversible engine working between the same two temperatures can have as great an efficiency.

Thermodynamics determines the maximum possible efficiency of all heat engines, and nothing can be done by human ingenuity to improve this efficiency except to make T_H as high as possible and T_C as low as possible in Eq. (15.9).

A Carnot engine is not a practical engine, since the cyclic process must be carried out reversibly and therefore slowly. For this reason the power output of such an engine is too small to be practical. The Carnot engine is, however, of great importance from a theoretical point of view because it sets limits on the efficiency that can be reached by any practical heat engine.

Example 15.5

A nuclear reactor on a submarine produces steam at 350°C to drive a steam-electric turbine. The temperature on the low side of the turbine is 27°C. **(a)** What is the maximum possible efficiency of the steam turbine? **(b)** If 1.0×10^{12} J of heat is supplied by the nuclear fuel, how much of this is converted into useful work (assuming maximum possible efficiency)? **(c)** How much energy is dissipated as waste heat?

SOLUTION

(a) The maximum possible efficiency is that of a Carnot engine operating between these two temperatures:

$$e_{max} = 1 - \frac{T_C}{T_H} = 1 - \frac{300}{623} = 0.52 = \boxed{52\%}$$

(b) From Eq. (15.6), we have

$$e_{max} = \frac{\Delta W}{\Delta Q_H} = 0.52$$

and so

$$\Delta W = e_{max}\,\Delta Q_H = (0.52)(1.0 \times 10^{12}\ \text{J}) = \boxed{5.2 \times 10^{11}\ \text{J}}$$

(c) The waste heat is, from the principle of conservation of energy,

$$\Delta Q_C = \Delta Q_H - \Delta W$$

$$= 1.0 \times 10^{12}\ \text{J} - 0.52 \times 10^{12}\ \text{J} = \boxed{4.8 \times 10^{11}\ \text{J}}$$

The 52 percent efficiency found in (a) is, of course, the *maximum* possible efficiency. This is reduced, for a typical nuclear reactor, to about 30 to 35 percent by energy losses in the operation of the reactor.

15.6 Practical Heat Engines

One of the practical inventions on which the industrial revolution was built was the steam engine.

Steam Engines

This type of engine utilizes the energy in high-pressure steam to do useful work. A reciprocating steam engine consists of a cylinder with two valves A and B, one at each end, as in Fig. 15.13. These valves serve alternately as input and exhaust valves for the steam. First, hot steam is injected from a steam boiler into the cylinder through valve A. As the steam drives the piston to the right, valve A is

FIGURE 15.13 A model of a reciprocating steam engine. Hot gases from a steam boiler enter at A and B and drive the piston P back and forth. The crosshead converts this back-and-forth motion into the rotary motion of a flywheel, which does useful work.

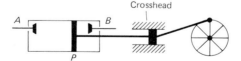

closed and valve B is opened to exhaust the steam to the right of the piston. When the steam has cooled because of the work it has done in moving the piston, valve A is again opened, but now to serve as an exhaust valve. At the same time, hot steam is injected from a steam boiler through valve B. This drives the piston back to the left, and the remaining spent steam is exhausted from valve A. The process continues in this way, with successive bursts of high-pressure steam driving the piston first to the right and then to the left. The piston's motion is converted to the rotary motion of a flywheel by the action of the connecting rods and the "crosshead" shown in the figure. The flywheel can then do useful work.

Steam Turbines

A steam turbine is a special kind of engine in which the steam produced in a boiler impinges on a set of blades attached to a rotating wheel. Some of the internal energy of the steam is converted into rotational energy of the wheel.

A prime example of a practical steam turbine is the typical fossil-fuel steam-electric power plant shown in Fig. 15.14. (The nature of the electric generator will be discussed in Chap. 20.) The boiler converts water to steam, which impinges on the blades of a turbine and causes the turbine to rotate. This rotating turbine then produces electricity. The steam loses some of its energy in driving the turbine and is condensed back to water in a set of pipes cooled by flowing river water or blown air. The overall function of the steam turbine is to take heat from the steam at temperature T_H, convert part of it into useful work ΔW, and deliver the rest of it as waste heat to the condenser at the temperature of the cooling water T_C.

The theoretical Carnot efficiency of such an engine is $e_{max} = 1 - T_C/T_H$, and the actual efficiency even less than this. Figure 15.15 shows the theoretical and practical efficiencies of a typical fossil-fuel power plant as a function of T_H, when $T_C = 300$ K. We can see that even for $T_H = 973$ K, the actual efficiency is only about 50 percent and has leveled off so much as a function of T_H that higher temperatures would not be justified in terms of the increased efficiency to be expected. For this reason most fossil-fuel plants today use steam at about 800 K and have theoretical efficiencies of about 60 percent and overall practical efficiencies of about 40 percent. This means that 60 percent of the input energy is thrown away as waste heat (resulting in what is called *thermal pollution*).

Nuclear-powered steam-electric plants are even less efficient than fossil-fuel-powered plants, because safety considerations require that the steam be produced at a lower temperature (≈ 625 K). As a consequence, the practical efficiency of such nuclear plants is only about 30 percent.

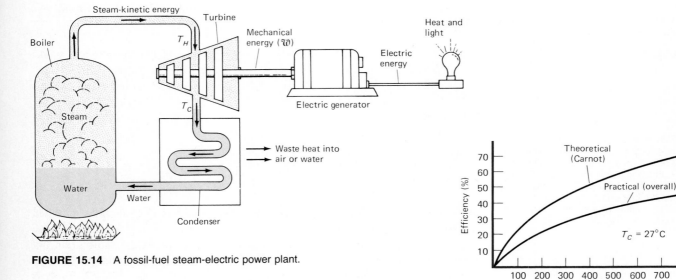

FIGURE 15.14　A fossil-fuel steam-electric power plant.

FIGURE 15.15　Theoretical and practical efficiencies of a typical fossil-fuel power plant.

Internal-Combustion Engines

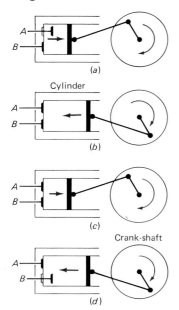

FIGURE 15.16 An internal-combustion engine. A complete cycle involves the following processes: (a) intake stroke, (b) compression stroke, (c) power stroke, and (d) exhaust stroke.

Such engines differ from steam engines (external-combustion engines) in that the heat is produced by combustion inside the engine and not in a separate boiler outside, as in a steam engine. A good example is the gasoline engine used in automobiles. In each cylinder of such an engine, gasoline is burned and the expanding hot gases from the combustion do work on the piston to drive the wheels of the car. Only part of the energy from the gasoline is converted into useful work; the remainder is lost in the form of heated air and exhaust products.

The four processes that make up one complete cycle of the piston are illustrated in Fig. 15.16. On the *intake stroke* (a) the piston is pulled back by the rotating crankshaft. The intake valve A opens and a mixture of gasoline vapor and air is drawn into the cylinder. This is followed by the *compression stroke* (b), in which the piston compresses this fuel mixture to about one-seventh of its original volume (called the *compression ratio*). A spark from the spark plug ignites the fuel mixture, and the temperature and pressure in the cylinder increase greatly. This increased pressure drives the piston outward in the *power stroke* (c). When the volume of gas in the cylinder reaches its maximum value, determined by the position of the piston, the *exhaust stroke* (d) starts. Valve B opens and the exhaust products of the combustion are forced out. The piston is now back where it started and the cycle begins again. Work is done during the power stroke to keep the crankshaft rotating, and the rotating crankshaft moves the car.

Diesel engines are the most efficient of all internal-combustion engines because the combustion chamber becomes hot enough to ignite the fuel without the need for a spark. As a result, T_H is higher and the efficiency improves. The prices paid for this increased efficiency are the added initial cost of the sturdier metals required to stand the higher temperatures and the greater weight of the engine per horsepower produced. Ordinary automobile engines are 20 to 25 percent efficient, whereas Diesel engines have efficiencies in the range of 26 to 38 percent.

Example 15.6

Consider two electric power plants, one a fossil-fuel plant with an efficiency of 40 percent, the other a nuclear plant with an efficiency of 30 percent. If each generates the same amount of energy, how much more waste heat is produced by the second plant than by the first?

SOLUTION

Since the work done by either plant is $\Delta W = e\,\Delta Q_H$, and since $e = 1 - \Delta Q_C/\Delta Q_H$, we have

$$\frac{\Delta Q_C}{\Delta Q_H} = 1 - e \qquad \text{or} \qquad \Delta Q_H = \frac{\Delta Q_C}{1 - e}$$

Hence

$$\Delta W = \frac{e}{1 - e}\,\Delta Q_C$$

Since ΔW is the energy generated by the plants in some fixed time interval and is the same for the two plants, we have, using subscripts n and f for the nuclear and fossil-fuel plants, respectively,

$$\Delta W = \frac{e_n}{1 - e_n}\,\Delta Q_{C,n} = \frac{e_f}{1 - e_f}\,\Delta Q_{C,f}$$

But $e_n = 0.30$ and $e_f = 0.40$, and so

$$\frac{0.30}{0.70}\,\Delta Q_{C,n} = \frac{0.40}{0.60}\,\Delta Q_{C,f}$$

or

$$\frac{\Delta Q_{C,n}}{\Delta Q_{C,f}} = \frac{0.70(0.40)}{0.30(0.60)} = \boxed{1.6}$$

Hence 60 percent more waste heat is produced by the nuclear plant, even though its efficiency is only 10 percent lower than the fossil-fuel plant.

Nuclear plants produce more thermal pollution than fossil-fuel plants, but they produce no chemical pollution like that produced by coal-, oil-, and gas-burning power plants. Safety and waste-disposal problems are, however, much more acute for nuclear than for fossil-fuel power plants.

15.7 The Absolute Thermodynamic Temperature Scale

We have seen in Sec. 15.5 that the heat transferred between a Carnot engine and a heat reservoir with which it is in contact is proportional to the temperature of that reservoir. This fact was expressed in Eq. (15.8) in the form $\Delta Q_C / \Delta Q_H = T_C / T_H$. This equation is at the core of thermodynamics in the same way that Newton's second law is at the core of mechanics. In 1848 the British physicist William Thomson [Lord Kelvin (1824–1907)] proposed that the heat ΔQ_H taken in and the heat ΔQ_C given out by a Carnot engine be used to define a new thermodynamic scale of temperature. This is now called the *absolute thermodynamic scale*, since it is independent of the working substance. It is also called the *Kelvin scale* after its founder. The ratio of two temperatures on this scale is defined as

$$\frac{T_C}{T_H} = \frac{\Delta Q_C}{\Delta Q_H} \tag{15.10}$$

Even though ΔQ_C and ΔQ_H depend on the substance used in the Carnot engine, their ratio does not. This ratio depends only on the two temperatures involved and can therefore be used to define a temperature scale.

This absolute scale is still not complete, for it yields only ratios of temperatures. We must also fix one point on the scale by assigning a particular numerical temperature to a well-defined physical system. We choose to take 273.16 K as the temperature of the triple point of water (see Fig. 13.16a). Since the triple point corresponds to 0.01 C° above the freezing point of water, this makes the freezing point of water 273.15 K, and absolute zero -273.15°C.

Such a scale has advantages over the similar scale previously defined in terms of measurements with a constant-volume gas thermometer. The problem with such a thermometer is that measured temperatures differ slightly with the gas used, and even with the density of a particular gas, as we have seen. In the limiting case of very low density gases, however, the temperature measured with a constant-volume gas thermometer agrees perfectly with the absolute thermodynamic temperature. In what follows we will always consider these two scales as equivalent and use kelvins to denote the temperature on both scales.

In practice no gas thermometer works well at low temperatures. However, Eq. (15.10) can still be used. If we carry out a Carnot cycle using He between temperatures of, say, 20 K and some unknown temperature, as in Fig. 15.17, then

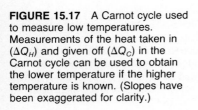

FIGURE 15.17 A Carnot cycle used to measure low temperatures. Measurements of the heat taken in (ΔQ_H) and given off (ΔQ_C) in the Carnot cycle can be used to obtain the lower temperature if the higher temperature is known. (Slopes have been exaggerated for clarity.)

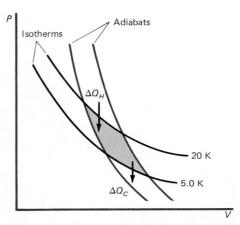

$$T_C = \frac{\Delta Q_C}{\Delta Q_H} T_H = \frac{\Delta Q_C}{\Delta Q_H} (20 \text{ K})$$

If we find experimentally that $\Delta Q_C = \frac{1}{4} \Delta Q_H$, then we can conclude that $T_C = 20 \text{ K}/4 = 5.0 \text{ K}$.

As the temperature falls, the amount of heat ΔQ_C given off in a Carnot cycle at the lower temperature T_C gets smaller and smaller, according to Eq. (15.10). In the limit of absolute zero, ΔQ_C also goes to zero. Another definition of absolute zero is therefore the following:

Absolute zero: The temperature of a reservoir into which a Carnot engine would eject no heat.

The Carnot Cycle and the Second Law of Thermodynamics

Another statement of the second law is based on the extension of Carnot's work by Lord Kelvin. For a 100 percent efficient Carnot engine, T_C in Eq. (15.9) would have to be zero. But we have seen (Sec. 13.2) that absolute zero cannot be reached in a finite number of steps. Kelvin therefore concluded that a 100 percent efficient heat engine is impossible:

> It is impossible to construct a heat engine that, operating in a cycle, produces no effect except the extraction of heat from a source and the conversion of this heat completely into work.

A perpetual-motion machine of the *first kind* [Sec. 14.7] is one that could run forever without the input of any energy, or with less energy input than the work output. Such a machine would violate the *first law* of thermodynamics. By analogy a machine that would violate the *second law* of thermodynamics is called a perpetual-motion machine of the *second kind*.

Perpetual-motion machine of the second kind: A machine that would take heat from a reservoir and convert it entirely into work.

Such a machine would be a *100 percent efficient heat engine*, which the Kelvin statement of the second law does not allow. We therefore conclude:

> Since a heat engine with an efficiency of 100 percent is impossible, a perpetual-motion machine of the second kind is also impossible.

An example of a perpetual-motion machine of the second kind would be a ship that could take in water from the surface of the ocean, remove heat from it, use the heat to power an engine to move the ship, and then dump the water back into the surface waters of the ocean from which it originally came. Carnot's work shows that this is impossible because the efficiency of a Carnot engine working between two identical temperatures T_0 is zero, from Eq. (15.9), and such a machine could perform no work. Note that if the intake and output waters are from different depths and so have different temperatures, in principle a heat engine could be driven by this temperature difference. The efficiency of such an engine would be very low, however, because of the small temperature differences available in the oceans (see Example 15.7).

Example 15.7

In recent years there has been renewed interest in a process for generating energy from the oceans first proposed in 1881 by Jacques d'Arsonval (1851–1940) and now usually called OTEC (ocean thermal energy conversion). This plan would use the temperature difference between the surface waters and the deep waters of the ocean to generate electric energy. In the tropics the surface temperature of the ocean is about 25°C and at depths of 1000 m the temperature falls to about 5°C. What is the efficiency of a Carnot engine working between these two temperatures?

SOLUTION

Since $e_{max} = 1 - T_C/T_H$, we have

$$e_{max} = 1 - \frac{278 \text{ K}}{298 \text{ K}} = 0.067 = \boxed{6.7\%}$$

This is, of course, the *maximum* efficiency. The practical efficiency would be closer to 2 to 3 percent. Hence very large flow rates of water through the engine will be required to produce the same power output provided by more efficient engines operating with large temperature differences. As a result the pumps and pipes required are very large and costly, and it is unclear at this time whether OTEC will ever be a commercially viable system. While it does not violate the second law of thermodynamics, its efficiency is just too low to be useful.

15.8   Some Applications: Heat Pumps and Refrigerators

We now apply the principles developed in the preceding section to some practical devices commonly used in homes and businesses.

Heat Pumps

According to the second law of thermodynamics, heat never flows spontaneously from a low-temperature reservoir to a high-temperature one, since that would result in a decrease in the entropy of the system and its environment. For this reason we can never expect to heat a house by opening a window on a cold day and hoping that heat will flow into the house from the outside cold air. The reverse is obviously what happens. The second law, however, in no way prevents us from *pumping* heat from the outside air into the house to warm it. In the process the pump must do work and consume energy, but *the heat transferred can be much greater than the energy consumed by the pump*. This is why the *heat pump* is becoming increasingly popular for the heating of houses in moderate climates.

In Fig. 15.18 we compare a heat pump with a heat engine from a thermodynamic viewpoint. In the heat engine shown in Fig. 15.18a an amount of heat ΔQ_H is taken from a hot reservoir at temperature T_H. Some of this heat is converted into useful work ($\Delta W = \Delta Q_H - \Delta Q_C$), and the remaining ΔQ_C is exhausted to the cold reservoir at temperature T_C. In a heat pump we reverse this process. As shown in Fig. 15.18b, we take an amount of heat ΔQ_C from a cold reservoir at temperature T_C and pump it up, by doing work on it, to a higher-temperature reservoir at temperature T_H. The amount of heat delivered to the hot reservoir is then $\Delta Q_H = \Delta Q_C + \Delta W$, since we assume that no energy is lost in the process. In a practical heat pump T_C is the temperature of the outside air, T_H is the temperature of the inside of the house, and ΔW is the energy contributed by the pump, which is usually a compressor run by electric power.

As stated previously, the practical value of a heat pump lies in the fact that the amount of heat transferred to the interior of the house can be much greater than the amount of energy consumed in the transfer process. As an example of this, let us suppose that in Fig. 15.18b an amount of heat ΔQ_C is taken from outside the house, where the temperature is T_C, and an amount ΔQ_H is delivered to the house at the warmer temperature T_H inside, during each cycle of the heat pump. Then

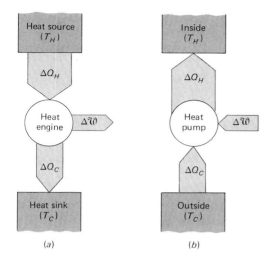

FIGURE 15.18 Comparison of (a) a heat engine with (b) a heat pump.

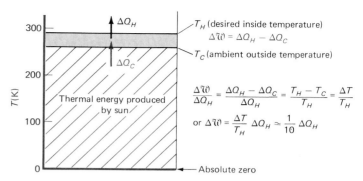

FIGURE 15.19 An illustration of why a Carnot heat pump has such a high ideal coefficient of performance. The work that must be done to transfer the heat needed to heat a house is only about 10 percent of the heat energy transferred. This is because the ambient temperature produced by the sun's radiation is 10 times greater (on the Kelvin scale) than the change in temperature required to make the house comfortable (*Adapted from a diagram by H. Socolow,* Physics Today, *Jan., 1986, p. 66, with permission*).

we have, from the first law, $\Delta W = \Delta Q_H - \Delta Q_C$. For an ideal Carnot cycle we can also make use of Eq. (15.8). This leads to the result

$$\frac{\Delta Q_H}{\Delta W} = \frac{\Delta Q_H}{\Delta Q_H - \Delta Q_C} = \frac{1}{1 - \Delta Q_C/\Delta Q_H} = \frac{1}{1 - T_C/T_H} = \frac{T_H}{T_H - T_C}$$

The amount of heat transferred to the inside of the house is therefore

$$\Delta Q_H = \frac{T_H}{T_H - T_C} \Delta W \qquad \text{(15.11)}$$

Note the remarkable result here. If T_H and T_C are close together (i.e., if it is not too cold outside), then ΔQ_H can be much greater than ΔW; that is, much more heat can be transferred than the energy input required by the pump (ΔW). Of course, this does not mean that energy is being created; the heat brought into the house is still just the sum of the work done and the heat extracted from the outside air.

Figure 15.19 shows the key to the heat pump's success. Even though it may be freezing outside, the air still contains considerable thermal energy, since it is at a temperature some hundreds of degrees above absolute zero. This ambient temperature is maintained by the sun's radiation. In a heat pump we convert some of the relatively useless energy stored by the sun in the outside air into useful heat to warm our houses. As Fig. 15.19 indicates, we can do this with an expenditure of energy that is as small as 10 percent of the heat transferred to the inside of the house, because the increase in temperature required to heat the house is only about 10 percent of the Kelvin temperature of the air.

Coefficient of Performance of a Heat Pump Since $\Delta Q_H/\Delta W$ for a heat pump can be much greater than unity (or 100 percent efficiency), and since efficiencies are usually defined so as to have a maximum value of 100 percent, the quantity $\Delta Q_H / \Delta W$ is not called the efficiency, but rather the *coefficient of performance* (CP) of the system. For an ideal, or Carnot heat pump, we have for the CP of the system, from Eq. (15.11),

$$CP_{max} = \frac{\Delta Q_H}{\Delta W} = \frac{T_H}{T_H - T_C} \qquad \text{(15.12)}$$

Note that this is the maximum possible CP and is correct only for a Carnot heat pump; the CP is considerably less than CP_{max} for real heat pumps but usually in the range from 3 to 10. Example 15.8 shows how sensitive the CP of a heat pump is to the temperature difference $T_H - T_C$.

The great advantage of a heat pump is evident when we pay our electric bills. We pay only for ΔW, the work done, and not for the heat extracted from the outside air, which is free.

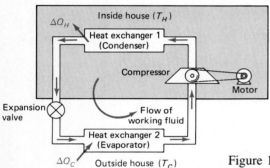

FIGURE 15.20   A practical heat pump. The working fluid flows around in a closed cycle, constantly taking heat from the outside air and carrying it inside the house.

Figure 15.20 shows how a practical heat pump works. The outdoor source of heat for the pump may be a water reservoir buried in the ground or may be the earth itself, but in practice the outside air is most frequently used. This makes for a simpler, less costly system but reduces the efficiency of the heat pump when it is very cold outside. For this reason an auxiliary heating system must be used to supplement the heat pump when the outside temperature falls much below freezing. One way to avoid this problem is to use solar heating to increase the temperature of the air providing the thermal input to the heat pump.

Example 15.8

Suppose we want to keep the inside of a house at 70°F (21°C). Compute the required coefficient of performance of a heat pump working to do this **(a)** when the outside temperature is 10°C and **(b)** when the outside temperature is −10°C.

SOLUTION

(a) When $T_C = 10°C = 283$ K, the maximum coefficient of performance is

$$CP_{max} = \frac{(273 + 21) \text{ K}}{(294 - 283) \text{ K}} = 27$$

Note that in such calculations we could use Celsius degrees *in the denominator* since the difference between two temperatures is the same whether they be expressed in kelvins or Celsius degrees, but we *must* use *Kelvin* temperatures in the numerator.

(b) When $T_C = -10°C = 263$ K,

$$CP_{max} = \frac{294 \text{ K}}{(294 - 263) \text{ K}} = 9.5$$

Thus the CP of the heat pump is reduced by a factor of 3 when the temperature outside the house falls 20 C°.

Note that even when the temperature is below freezing outside, a heat pump can, in the ideal order, transfer to the interior of a house an amount of heat which is almost 10 times greater than the work performed by the pump. This is consistent with Fig. 15.19.

Air Conditioners

An air conditioner is nothing but a heat pump run backward; i.e., the outside and inside of the house are simply exchanged in Fig. 15.20. The working fluid in this case (usually freon) is cooled by a throttling process as it enters the house, picks up heat inside the house, is compressed as it leaves the house so that its temperature rises above that of the outside air, and then delivers heat to the outside air before starting the cycle all over again.

The coefficient of performance of an air conditioner is slightly different than for a heat pump, because the important thing here is not the amount of heat (ΔQ_H) delivered to the outside air but rather the heat (ΔQ_C) removed from inside the house. We define the coefficient of performance for an ideal air conditioner as

$$\mathrm{CP}_{max} = \frac{\Delta Q_C}{\Delta W} = \frac{\Delta Q_C}{\Delta Q_H - \Delta Q_C} = \frac{T_C}{T_H - T_C} \qquad (15.13)$$

Real CPs are somewhat less than this ideal value because of friction and other irreversible factors in the operation of the unit. The amount of heat removed from the house when an amount of work ΔW is performed is then

$$\Delta Q_C = \mathrm{CP}_{max}\,\Delta W = \frac{T_C}{T_H - T_C}\,\Delta W \qquad (15.14)$$

Again the heat removed can be much larger than the work done.

Refrigerators

A refrigerator is a special kind of air-conditioning unit that removes heat from the inside of the refrigerator box and deposits it in the kitchen. (Feel the condensing coils at the rear of your refrigerator and notice how warm they are.) The maximum coefficient of performance for a refrigerator is also given by Eq. (15.13). The use of a throttling process for cooling in a gas refrigerator was illustrated in Fig. 14.17.

Summary: Important Definitions and Equations

Reversible process: A thermodynamic process carried out so slowly that every intermediate state is an equilibrium state; a process that can be reversed by an infinitesimal change in the operating conditions.

Irreversible process: A thermodynamic process for which the intermediate states are not equilibrium states; any process that is not completely reversible.

Equations for reversible adiabatic process:

$$P_1 V_1^\gamma = P_2 V_2^\gamma \qquad (15.3a)$$

$$T_1 V_1^{\gamma-1} = T_2 V_2^{\gamma-1} \qquad (15.3b)$$

$$\gamma = C_P/C_V \qquad (14.22)$$

Entropy: A state variable that measures the disorder of a physical system. For any reversible process the change in entropy is

$$\Delta S = \frac{\Delta Q}{T} \qquad (15.5)$$

Boltzmann equation for entropy:

$$S = k \ln W \qquad (15.4)$$

where W is the *thermodynamic probability*, the number of possible arrangements leading to the same observed state of a physical system.

Entropy statement of second law of thermodynamics: The entropy of any physical system plus its environment increases in all irreversible processes and remains constant in all reversible processes. Hence the total entropy of any isolated system never decreases.

Clausius' summary of thermodynamics

First law: The *energy* of the universe is constant.

Second law: The *entropy* of the universe tends always toward a maximum.

Heat engine: A device that, operating in a closed cycle, converts heat into useful work.

Carnot engine: The most efficient possible heat engine operating between any two given temperatures. The efficiency of a Carnot engine is

$$e_{max} = \frac{\Delta W}{\Delta Q_H} = 1 - \frac{\Delta Q_C}{\Delta Q_H} = 1 - \frac{T_C}{T_H} \qquad \text{(15.6–15.9)}$$

Absolute thermodynamic (Kelvin) temperature scale: A temperature scale based on the fact that for a Carnot cycle, $\Delta Q_C / \Delta Q_H = T_C / T_H$, and the definition of the triple point of water as 273.16 K.

Absolute zero (thermodynamic definition): The temperature of a reservoir into which a Carnot engine would eject no heat.

Perpetual-motion machine of the second kind: A 100 percent efficient engine that takes heat from a reservoir and converts it entirely into work (such a machine is impossible according to the second law of thermodynamics).

Coefficient of performance of an ideal heat pump:

$$CP_{max} = \frac{\Delta Q_H}{\Delta W} = \frac{T_H}{T_H - T_C} \qquad \text{(15.12)}$$

Coefficient of performance of an ideal air conditioner (or refrigerator):

$$CP_{max} = \frac{\Delta Q_C}{\Delta W} = \frac{T_C}{T_H - T_C} \qquad \text{(15.13)}$$

Questions

1 Distinguish clearly between a perpetual-motion machine of the first kind and a perpetual-motion machine of the second kind. Give a few examples of each.

2 Discuss all the energy transformations that take place in a steam-electric turbine (see Fig. 15.14), beginning with the energy in the fossil or nuclear fuel and ending with the waste heat delivered to the cooling system.

3 Why cannot the theoretical efficiency of a Carnot engine be achieved in any practical engine?

4 One way to help solve the problem of shortages developing in our supplies of metals like aluminum and tin is to recycle these metals, i.e., collect old tin cans and aluminum containers and convert them to useful forms in a reprocessing plant. Why cannot we solve our present energy problems in a similar fashion by recycling the energy we have available, since we know from the first law of thermodynamics that energy is never destroyed but is merely converted to a different form?

5 Is it possible to cool down the kitchen on a hot summer's day by opening the door of the refrigerator? Why? What will the result be? Would the result be the same if a window air conditioner were placed in the middle of a room and turned on?

6 Why cannot the huge amount of energy stored in the core of the earth and in the oceans be used to solve our energy problems? Can these sources of energy contribute in some way to our energy needs? How?

7 Name a few spontaneous processes in which an ordered state changes to a disordered (or less-ordered) state.

8 Explain what is meant by the statement: "The flow of heat down a temperature gradient is always associated with an increase of entropy and a degradation of energy."

9 A high-entropy state has a greater probability of occurring than a low-entropy state. From your knowledge of the physical significance of entropy, explain why you would expect this to be the case.

10 Which has the greater entropy, 1 mol of liquid mercury, or 1 mol of solid mercury?

11 (a) In an isothermal expansion of an ideal gas, does the entropy of the gas increase, decrease, or stay the same?

(b) In an adiabatic expansion of an ideal gas, does the entropy of the gas increase, decrease, or stay the same?

12 Which is the stronger statement of the first and seconds laws of thermodynamics, that in terms of an isolated system, or Clausius' statement in terms of the energy and entropy of the universe? Why?

13 Since the earth remains at a reasonably constant temperature, it must radiate into space roughly the same amount of energy it receives from the sun. Compare the entropy changes in the earth due to the thermal radiation it receives from the sun and the thermal radiation it emits into space. Does this help explain why the entropy of the earth also remains roughly constant over time?

14 In living organisms order increases and hence entropy appears to decrease. Explain why living organisms do not constitute a violation of the law of increasing entropy.

15 Why cannot a gas at 40°C spontaneously split into two pieces, one half at 30°C and the other half at 50°C?

16 A sign on a light switch says: *Conserve energy! Turn off light when not in use.* But, according to the first law of thermodynamics, energy is *always* conserved. Can you use the second law of thermodynamics to reconcile these two statements?

Multiple-Choice and Simple Exercises

(For some of the problems in this chapter the values of γ given for selected gases in Table 14.3 are needed.)

15.1 A gas is expanded irreversibly from a volume V to a volume $3V$. At the end of the process, the temperature is the same as it was at the beginning and the pressure of the gas has changed from P to $P/3$. When the volume was equal to $2V$ during the expansion, the pressure was:

(a) $P/2$ (b) $2P$ (c) P (d) $P/3$

(e) It is impossible to say

15.2 If 0.50 m³ of argon gas at room temperature and atmospheric pressure is compressed adiabatically to a pressure 5 times atmospheric, the final volume is:
(a) 2.5 m³   (b) 0.10 m³   (c) 0.38 m³
(d) 0.19 m³   (e) 2.5 m³

15.3 A gas expands adiabatically and reversibly. In such an expansion:
(a) The internal energy of the gas remains constant.
(b) The entropy of the gas is unchanged.
(c) The pressure of the gas is unchanged.
(d) The volume of the gas is unchanged.
(e) No work is done on or by the gas.

15.4 Two kilograms of water at 0°C is frozen to ice at the same temperature. The change in entropy of the ice-water system is:
(a) $+2.4 \times 10^3$ J/K   (b) $+6.3 \times 10^3$ J/K
(c) -2.4×10^3 J/K   (d) -6.3×10^3 J/K
(e) 0

15.5 The Carnot efficiency of a heat engine operating between 27 and 627°C is:
(a) 0   (b) 67%   (c) 33%   (d) 4.3%
(e) 96%

15.6 The ratio of the energy thrown away as waste heat to the energy consumed as fuel in a practical heat engine is about:
(a) 0   (b) 1.0   (c) 0.20   (d) 0.40
(e) 0.70

15.7 In a low-temperature physics experiment a Carnot cycle is carried out between a warmer reservoir at 4.0 K and a colder reservoir whose temperature is unknown. In the cycle the heat taken from the upper reservoir is 4.0 J and the heat delivered to the lower reservoir is 0.20 J. What is the temperature of the colder reservoir?
(a) 2.0 K   (b) 0.20 K   (c) 4.0 K
(d) 0.020 K   (e) 1.0 K

15.8 A heat pump operates between 0°C outside a house and 20°C inside. If it requires 10^9 J of heat to keep the house warm for a day, in the ideal order the electrical input to the heat pump has to be:
(a) 1.0×10^9 J   (b) 6.8×10^7 J
(c) 15×10^9 J   (d) 5.0×10^8 J   (e) 0

15.9 The coefficient of performance of a heat pump working between $-10°C$ outside and 10°C inside a house is:
(a) 0.92   (b) 0.64   (c) 1.55   (d) 14
(e) 1.1

15.10 The coefficient of performance of an electric freezer working between $-10°C$ and room temperature (20°C) is:
(a) 8.8   (b) 9.8   (c) 1.0   (d) 0.10
(e) 0.33

15.11 A sample of nitrogen gas is at $T = 400$ K and $V = 0.10$ m³. After an adiabatic expansion, the final volume is 0.50 m³. What is the final temperature?

15.12 A sample of argon gas is at $P = 3.0 \times 10^5$ N/m² and $V = 1.0$ m³. After an adiabatic expansion, the final pressure is 1.0×10^5 N/m². What is the final volume?

15.13 Calculate the change in entropy for the melting of 1.0 kg of solid hydrogen at $-259°C$ (see Table 13.5).

15.14 An engine does 10^4 J of work for each 10^5 J of heat added. What is its efficiency?

15.15 During every cycle 1.0×10^5 J of heat is added to an engine and 1.0×10^4 J of heat is thrown away by the engine. What is its efficiency?

15.16 What is the efficiency of a Carnot engine that operates between:
(a) 100 and 200 K   (b) 200 and 300 K
(c) 300 and 400 K?

15.17 What would the temperature of the low-temperature reservoir of a Carnot engine have to be in order that the efficiency be 100 percent?

15.18 A Carnot device operates between 0 and 300°C.
(a) Calculate the efficiency if it is run as an engine.
(b) Calculate the coefficient of performance if it is run as a heat pump.

15.19 A Carnot air conditioner operates between 20 and 38°C. What is its coefficient of performance?

15.20 A bathysphere is 1000 m below the ocean surface, where the water temperature is 5°C. The interior of the bathysphere is to be at 20°C. What is the coefficient of performance of a Carnot heat pump that will maintain this temperature difference?

Problems

15.21 A 1.0×10^{-3}-m³ sample of neon gas at 300 K and atmospheric pressure is allowed to expand adiabatically to a final pressure of 0.50 atm.
(a) What is the final volume of the gas?
(b) What is its final temperature?
(c) What is the change in its internal energy?

*15.22 A monatomic gas for which $P_0 = 1.0 \times 10^6$ N/m² and $V_0 = 1.0$ m³ expands so that the final pressure is 1.0×10^5 N/m².
(a) What is the final volume if the expansion is isothermal?
(b) What is the final volume if the expansion is adiabatic?
(c) Calculate the work done in both expansions by drawing the process on a PV diagram.
(d) In which process does the gas do more work?

15.23 At the base of a mountain, the air pressure is 1.0 atm and the temperature is 300 K.
(a) If air rises adiabatically to the top of the mountain $(P = 0.95$ atm), what is the temperature? (Use $\gamma = 1.4$ since air is mostly oxygen and nitrogen.)
(b) If some of the water vapor condenses, will the temperature tend to go back up?

15.24 A cylinder contains 0.100 m³ of He gas at 1.00×10^5 N/m² and 27°C. The gas is heated first at constant volume to a pressure of 2.00×10^5 N/m² and then at constant pressure to a temperature of 427°C, as in Fig. 15.21.
(a) Calculate the total heat input during these processes (a to b and b to c). The gas is next cooled at constant volume to its original pressure and then at constant pressure to its original volume.

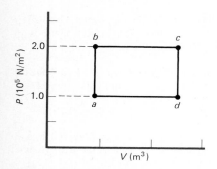

FIGURE 15.21 Diagram for Prob. 15.24.

(b) Find the total heat output during these processes (c to d and d to a).

(c) Find the total work done by the gas in the entire cyclic process (abcda). (*Hint:* Use Table 14.3.)

15.25 A reversible heat engine carries 1 mol of an ideal monatomic gas around the cycle shown in Fig. 15.22. Process 1-2 takes place at constant volume, process 2-3 is adiabatic, and process 3-1 takes place at constant pressure. The temperatures are $T_1 = 300$ K, $T_2 = 600$ K, $T_3 = 455$ K.

(a) Compute the heat taken in by the gas (ΔQ), the change in internal energy (ΔU), and the work done by the gas (ΔW) for each of the three processes and for the cycle as a whole.

(b) If the initial pressure at point 1 is 1.0×10^5 N/m², find the pressure and the volume at points 2 and 3.

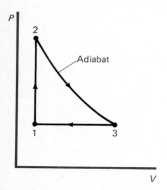

FIGURE 15.22 Diagram for Prob. 15.25.

*15.26 A mole of ideal gas goes from temperature T_1 to temperature T_2 during an adiabatic process.

(a) Show that the change in internal energy of the gas is given by

$$\frac{3}{2} RT_1 \left[\left(\frac{V_1}{V_2} \right)^{\gamma - 1} - 1 \right]$$

(b) If $2V_2 = V_1$, would the energy increase be larger for Ne or for CO_2?

*15.27 Three moles of neon gas is initially at 20°C in a cylinder closed by a piston at one end. The initial volume is 0.80 m³. The neon is expanded first at constant pressure until

the volume is 2.0 m³, and then adiabatically until the temperature returns to its initial value.

(a) What is the final volume of the gas?

(b) Draw a PV diagram for the process.

(c) What is the net change in the internal energy of the gas?

(d) What is the net heat supplied in the two expansions?

(e) What is the total work done by the gas?

15.28 In a dice game using two dice, a player wants to know the relative chances of his throwing a 7 as compared to an 11. What would you tell him?

15.29 A system is made up of four coins that are randomly tossed.

(a) In how many ways can the system have all four heads up?

(b) In how many ways can the system have two out of four coins with heads up?

(c) What is the entropy of the system in each case?

15.30 A boy tosses four pennies and counts the number of heads and tails.

(a) What is the most probable result if we average the results over a large number of tosses?

(b) What are the least probable results? How frequently will they occur?

(c) In how many different ways can the following results be obtained: four heads; three heads and one tail; two heads and two tails?

(d) What is the relationship between the number of different ways a result can be achieved and the probability that this result will occur when the coins are tossed?

15.31 Four kilograms of water is slowly heated from 20 to 100°C at atmospheric pressure. What is an approximate value of the entropy change in the water?

15.32 Two kilograms of water at 20°C is mixed with 4 kg of water at 90°C in a thermally insulated container.

(a) What is the final temperature of the system?

(b) What is an approximate value of the entropy change in the system?

15.33 A gas at 1.0×10^5 N/m² expands adiabatically and reversibly from 0.20 to 0.50 m³. What is its change in entropy?

15.34 Three moles of helium gas expands isothermally at 0°C. The gas volume is increased to 2.5 times its initial volume V_0.

(a) What is the change in entropy of the gas?

(b) What is the change in entropy of the environment?

(c) What is the change in entropy of gas plus environment?

15.35 Two containers, one at 80°C and the other at 40°C, each hold 1.5×10^{-3} kg of the same gas [$c = 0.20$ kcal/(kg·C°)]. The two containers are insulated from their surroundings and put in contact with each other so that heat flows from the hot gas to the cooler gas.

(a) Find the entropy change of each of the two gases when the temperature of the hot gas falls to 78°C.

(b) Does the total entropy of the system increase, decrease, or remain the same when the temperature of the hot gas falls to 78°C? How much is the change, if any?

15.36 Five kilograms of water at 0°C is frozen to ice at 0°C.

(*a*) Calculate the change in entropy of the ice-water system.

(*b*) Show that the result is consistent with the entropy statement of the second law of thermodynamics.

15.37 (*a*) If 500 J of heat is extracted from a gas during an isothermal compression, how much work is done?

(*b*) Is the work done on the gas or by the gas?

(*c*) What is the entropy change for the gas if the temperature is 27°C?

15.38 A gas does 10^4 J of work during an isothermal expansion. What is the increase in entropy of the gas if the expansion takes place at (*a*) 0.0°C and (*b*) −150°C?

*15.39** In Fig. 15.23, *ab* is an isotherm, *bc* an isochor, and *ca* an adiabat; and $P_a = 2.0 \times 10^5$ N/m². (Assume 1 mol of an ideal monatomic gas and remember that the entropy is a state variable.)

(*a*) Calculate ΔS_{ab} for 1 mol of an ideal gas.

(*b*) Calculate ΔS_{ac}, noting that path *ac* is an adiabat.

(*c*) Calculate ΔS_{bc}.

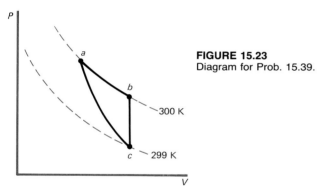

FIGURE 15.23
Diagram for Prob. 15.39.

*15.40** Starting with Eq. (14.11) for an isobaric process, $\Delta Q = \Delta U + P\,\Delta V$, show that

$$\Delta S = C_V \frac{\Delta T}{T} + R \frac{\Delta V}{V} = R\left(\frac{3}{2}\frac{\Delta T}{T} + \frac{\Delta V}{V}\right)$$

for 1 mol of a monatomic ideal gas. (Note that the universal gas constant has dimensions of entropy per mole. Note also that we can have $\Delta S \neq 0$ even if $\Delta T = 0$ so long as there is a change in volume.)

15.41 One kilogram of water at 10°C is mixed with one kilogram of water at 0°C.

(*a*) What is the final temperature?

(*b*) Calculate the entropy decrease of the warm water between 10 and 9°C, using 9.5°C for the average temperature. Do the same for 9 to 8°C, etc., down to 5°C.

(*c*) Calculate the entropy increase of the cold water between 0 and 1°C, using 0.5°C for the average temperature. Do the same for 1 to 2°C, etc., up to 5°C.

(*d*) Add the five entropy increases and the five entropy decreases together. Is the result positive, negative, or zero? Does this show that mixing is an irreversible process?

15.42 An engine operates between 500 and 300 K. If it takes in 5.0×10^5 J of heat and does 1.5×10^5 J of work each cycle, how does its efficiency compare with the corresponding Carnot engine?

15.43 Assume that the turbines in a power plant are as efficient as Carnot engines. They operate between 450 and 27°C.

(*a*) Calculate the efficiency.

(*b*) If 1.0×10^{12} J of work must be done, how much heat must be added at the high temperature and rejected at the low temperature?

15.44 For a Carnot cycle similar to that of Fig. 15.12, the entropy increase of the gas from point *a* to point *b* is 10^5 J/K. If $T_a = T_b = 400$ K,

(*a*) How much heat is given to the gas along *ab*?

(*b*) How much work is done by the gas along *ab*?

(*c*) How much heat is given to the gas along *bc*?

15.45 A 90-kg person consumes 2.0×10^3 kcal. If this person were a Carnot engine operating between 37 and 33°C, how high could he or she be lifted with the work produced?

15.46 The cycle shown in Fig. 15.24 consists of two isotherms and two isochors. The numbers of joules listed indicate how much heat is taken in or expelled in each process.

(*a*) How much work is done per cycle?

(*b*) What is the efficiency?

(*c*) What is the efficiency of a Carnot engine operating between these isotherms?

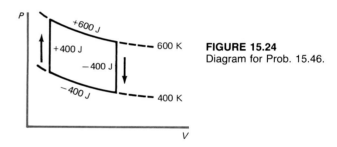

FIGURE 15.24
Diagram for Prob. 15.46.

*15.47** In Fig. 15.25 curves *AB*, *FC*, and *ED* are isotherms. *AFE* and *BCD* are adiabats.

(*a*) What is the efficiency of cycle *ABCDEFA*?

(*b*) In an attempt to improve the efficiency, cycles *ABCFA* and *FCDEF* are operated in such a way that the heat rejected by the former is the heat input to the latter. What is the efficiency of this tandem setup?

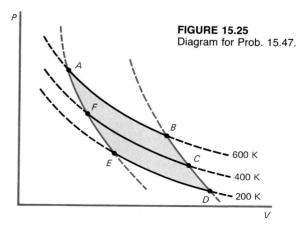

FIGURE 15.25
Diagram for Prob. 15.47.

*15.48   Prove that a 100 percent efficient heat engine would lead to a *decrease* in entropy of the heat engine plus the environment, which would violate the entropy statement of the second law.

15.49   A Carnot engine performs work at the rate of 100 kW while taking in heat at a rate of 100 kcal/s. If the temperature of the heat source is 550°C, at what temperature is the waste heat exhausted?

15.50   A Carnot engine whose low-temperature reservoir is at 275 K has an efficiency of 30 percent. To increase this efficiency to 40 percent:

(a)   At what temperature must the high-temperature reservoir be if the low-temperature reservoir stays at 275 K?

(b)   At what temperature must the low-temperature reservoir be if the high-temperature reservoir remains fixed at its original temperature?

15.51   Physics laboratories often contain apparatus to liquify helium gas at 4.2 K.

(a)   If the gas has been cooled to 4.5 K, and room temperature is 20°C, how does the heat delivered to the room compare with the heat removed from the helium gas?

(b)   If the liquifier operates in a closed Carnot cycle, how does the work done by the liquifying machine compare with the actual heat removed from the gas?

15.52   Five moles of He gas is compressed at 20°C to a volume half the initial volume.

(a)   What is the change in entropy of the gas?

(b)   If the resultant entropy change is negative, why does this not constitute a violation of the second law of thermodynamics?

*15.53   Two large, solid-metal spheres are of such great specific heat capacity that they can supply or receive small amounts of heat without changing temperature. The hotter of these two spheres is at $T_H = 320$ K and the cooler at $T_C = 300$ K. Prove from the entropy changes involved that, if these two spheres are put in contact with each other, heat will flow from the hotter to the colder sphere, and not the reverse.

*15.54   Clausius stated the second law of thermodynamics as follows:

> There is no such thing as a perfect refrigerator, i.e., a machine which transfers heat from a cold temperature to a hot temperature in a closed cyclical process without any other change in the system.

Prove that, if a perfect refrigerator existed, its entropy plus that of its environment would *decrease* in such a refrigeration process, which would violate the entropy statement of the second law.

15.55   A Carnot heat pump runs between 20 and 0°C.

(a)   If 2000 J of heat is absorbed from the outside air to heat the room, how many joules is supplied to the room?

(b)   How much work must the compressor do?

(c)   What is the coefficient of performance?

15.56   A Carnot air conditioner runs between 20 and 37°C.

(a)   If 2000 J of heat is extracted from the inside, how many joules is given to the outside?

(b)   How much work must the compressor do?

(c)   What is the coefficient of performance?

15.57   Before an ideal Carnot heat pump is turned on, the temperature inside and outside a cabin is 4°C.

(a)   If 200 J per cycle is delivered by the compressor (see Fig. 15.20), how much heat per cycle does the cabin receive when the inside temperature has reached 5°C?

(b)   How much when it is 32°C?

(c)   What is the coefficient of performance in the ideal order in each case?

15.58   (a)   A Carnot engine operates between 320 and 260 K. If it absorbs 1000 J of heat at the hot reservoir, how much work does it deliver?

(b)   If the same engine, working in reverse, functions as a refrigerator between the same two reservoirs, how much work must be supplied to remove 3000 J of heat from the cold reservoir?

Additional Readings

Atkins, P. W.: *The Second Law*, Freeman, San Francisco, 1984. An interesting, comprehensive, and well-illustrated discussion of the second law of thermodynamics.

Bennett, Charles H.: "Demons, Engines and the Second Law," *Scientific American*, vol. 257, no. 5, November 1987, pp. 108–116. An interesting account of attempts to create a Maxwell demon, a creature that seems to violate the second law of thermodynamics.

Dyson, Freeman J.: "What Is Heat?," *Scientific American*, vol. 191, no. 3, September 1954, pp. 58–63. The quotation at the beginning of this chapter is from page 59 of this thoughtful article (with permission).

Fenn, John B.: *Engines, Energy, and Entropy: A Thermodynamics Primer*, Freeman, San Francisco, 1982. An attempt to make thermodynamics clear and enjoyable to readers with minimal technical backgrounds.

Mendelssohn, Kurt: *The Quest for Absolute Zero*, 2d ed., Wiley, New York, 1977. A more technical discussion of low-temperature physics than in McDonald's book (Chap. 13).

Ord-Hume, Arthur W. J. G.: *Perpetual Motion: The History of an Obsession*. St. Martin's Press, New York, 1977. An interesting, well-illustrated account of perpetual-motion machines.

Saperstein, Alvin M.: *Physics: Energy in the Environment*, Little, Brown, Boston, 1975. Chapters 10 and 11 of this book contain a detailed discussion of entropy.

Wilson, S. S.: "Sadi Carnot," *Scientific American*, vol. 245, no. 2, August 1981, pp. 134–145. One of the best short accounts of the little-known life of Carnot.

Zener, Clarence: "Solar Sea Power," *Physics Today*, vol. 26, no. 1, 1973, pp. 48–53. A good article on the prospects for extracting useful energy from the oceans.

CHAPTER 16

Electric Charges and Electric Fields

In going on with these experiments, how many pretty systems do we build, which we soon find ourselves obliged to destroy! If there is no other use discovered of electricity, this, however, is something considerable, that it may help to make a vain man humble.

Benjamin Franklin (1706–1790)

We live in an electrical age. We use electrical devices—refrigerators, computers, television sets, VCRs—every day and would be lost without them. Modern industry and business rely to an ever-increasing extent on electric power to drive machines and process data. In the next seven chapters we examine the basic facts and laws of electricity to understand better the electrical world in which we live. We start with the basic ideas governing charges at rest (electrostatics) and build through the study of charges in motion (electrodynamics) and the magnetic fields they produce (electromagnetism) to the grand synthesis of electromagnetic theory contained in Maxwell's equations.

16.1 Elementary Electric Charges and Their Interaction

As we saw in Chap. 9, particles such as electrons and protons have a property called electric charge by which they exert electrical forces on each other. It is this electrical force that keeps the negatively charged electrons in atoms, since they are attracted by the positively charged protons in the nucleus of the atom.

Many everyday phenomena can be explained by the existence of electric charges. When we walk on a wool rug on a winter's day, we separate positive and negative charges by frictional rubbing. This leads to a slight electric shock and sparking when our bodies, now containing an excess of one kind of charge, touch a metal doorknob or water pipe. On a large scale, lightning is produced when large amounts of positive and negative charge build up on clouds, eventually leading to the violent flow of charge through the atmosphere (Fig. 16.1).

A great many electrostatic phenomena like these can be explained once we understand the structure of atoms. In the eighteenth century, however, when most of the original work on electrostatics was done, scientists not only did not understand the structure of atoms, they were not sure that atoms even existed.

FIGURE 16.1 Lightning at the John F. Kennedy Space Center at Cape Canaveral, Florida, on August 30, 1983, almost struck the space shuttle STS-8 on the launchpad. The storm passed and the shuttle was launched on schedule. (*Courtesy of NASA.*)

Benjamin Franklin (1706–1790)

FIGURE 16.2 Benjamin Franklin at his workbench. Notice the Leyden jar (used for electrostatic experiments) near Franklin's right elbow. (*Courtesy of the Burndy Library, Norwalk, Connecticut.*)

One of the founding fathers of the United States and an author of the Declaration of Independence, Benjamin Franklin is also world-famous as the man who tamed and explained lightning, introduced the idea of positive and negative electric charges, and made sense of the many puzzling phenomena of electrostatics.

Franklin was born in Boston in 1706, the fifteenth of 17 children. When he was 17 years old, he went to Philadelphia to work as a printer. He had only 2 years of formal schooling, but he loved books and studied foreign languages, philosophy, and science on his own. He became so involved in scientific study that in 1748 he turned over to his foreman what was then his own printing business, intending to devote the rest of his life to science.

From 1747 to 1753 Franklin car-ried out his most important work on the nature of electricity, the conservation of charge, and the properties of Leyden jars (Fig. 17.8). This work culminated in the publication in 1751 of his *Experiments and Observations on Electricity*, which went through five editions in English, three in French, one in Italian, and one in German.

One of Franklin's discoveries in electricity was the important role played by pointed metal objects. He found that a grounded metal rod with a pointed end could cause a charged conductor to lose its charge if the point was brought within 6 in of the charged object, but that a flat or rounded conductor had no such effect. Franklin used this discovery in the design of lightning rods to protect buildings. A lightning rod is simply a sharp-pointed piece of metal that is connected by a very good conducting path from the roof of a building to the ground. A lightning rod allows charge to leak off thunder clouds before it builds up sufficiently to create destructive lightning bolts. Also, if lightning does strike, it hits the lightning rod, not the house, and the charge passes through the rod to the ground without damaging the house.

For at least 50 years before Franklin's time, European scientists had speculated that lightning was electrical in nature, but he was the first scientist with the insight and the courage to prove this by his famous lightning-kite experiment.

By 1753 Franklin had become so involved in the political life of colonial America that he was forced to abandon most of his scientific work. His later success in winning the support of France for the American Revolution was as much due to his stature in Europe as a scientist and an intellectual as it was to his ability to charm both the men and women of France by his simplicity, urbanity, and wit. His experiments on lightning especially endeared him to the French people because he helped them understand the real nature of lightning. The resulting relief from the fear of the unknown earned Franklin his almost divine status in France.

His good humor is reflected in the comment he made after trying to kill a turkey with an electric shock and nearly killing himself in the process: "I meant to kill a turkey, and instead I nearly killed a goose."

Among Franklin's many inventions were the rocking chair, the Franklin stove, and bifocal glasses. Franklin had an attitude toward life that reflected the youthfulness of his spirit. In 1787, at the age of 81, after 50 years of public service, he wrote to a Dutch friend asking him to come to America, where "in the little remainder of my life . . . we will make plenty of experiments together."

Benjamin Franklin was the first great American scientist, and has been called the "wisest American," a well-deserved title.

Franklin's Contributions

The scientist who did much of the original work to clarify our understanding of the behavior of electric charges at rest was the great American statesman Benjamin Franklin (see Fig. 16.2 and accompanying biography). In the early 1740s Franklin carried out a variety of experiments with static charges produced by frictional rubbing or by electrostatic machines. He introduced the idea of *positive* and *nega-*

tive charges and concluded that the total charge was always exactly the same after the rubbing as it was before. This idea is now known as the *principle of conservation of electric charge*. The concepts of positive and negative charge, together with the principle of conservation of electric charge, are Franklin's most significant contributions to *electrostatics*, the study of charges at rest.

Modern Version of Franklin's Theory

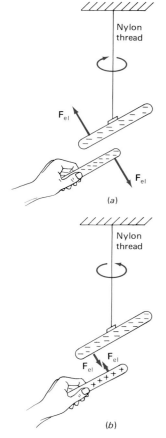

FIGURE 16.3 (a) The repulsion of two amber rods that contain excess negative charge. (b) The attraction of a glass rod for an amber rod; the glass rod is charged positively and the amber rod negatively.

Franklin's theory, combined with our modern understanding of atomic structure, leads to the following summary of the behavior of electric charges.

There are two kinds of electric charge: the kind of charge on the proton, which is designated as positive, and the kind of charge on the electron, which is designated as negative. Charges of like sign repel each other, and charges of unlike sign attract each other.

Many electrostatic phenomena require only these simple laws for an explanation. For example, if two amber rods are rubbed with fur, they repel each other. This can be shown by mounting one rod as shown in Fig. 16.3a and bringing the other rod near it. Also, if two glass rods are rubbed with silk, one rod will repel the other in the same way. If, however, a charged glass rod is brought near a charged amber rod, they attract each other, as in Fig. 16.3b. We explain these cases by noting that the amber rods are negatively charged, the glass rods positively charged: like charges repel each other, and unlike charges attract.

Franklin's law of *conservation of charge* can be stated as follows:

> In any closed system the algebraic sum of all the electric charge remains constant.

By *algebraic sum* we mean that each charge must be given its proper sign before the addition of charges is carried out.

As an example of this conservation law, consider what happens in frictional rubbing. Before a glass rod is rubbed with silk, both the rod and the silk are electrically neutral; each has zero net charge. If, however, we rub the glass rod vigorously with the silk, electrons are rubbed off the glass rod onto the silk, and the glass rod becomes positively charged and the silk negatively charged. In such a process, billions of electrons may be transferred from the glass to the silk. But the algebraic sum of all the electric charge on the silk and glass combined is still what it was before the rubbing: zero. No electric charge has been created; the charge has merely been redistributed.

This principle of conservation of electric charge joins the principle of conservation of mass, of energy, and of linear and angular momentum as the foundation stones on which physics is built. The unity and coherence of physics as a discipline is based in great part on the general validity and wide applicability of these conservation laws.

16.2 Conductors and Insulators

Suppose two metal spheres are connected together by a long wooden rod, as in Fig. 16.4a. If one sphere is then charged by touching it with an amber rod rubbed with fur, which allows electrons from the amber to collect on the sphere, the second

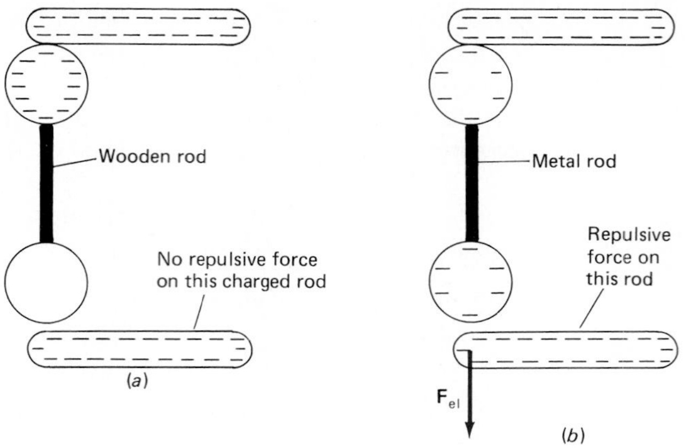

FIGURE 16.4 (*a*) Demonstration that a wooden rod is an electric insulator. (*b*) Demonstration that a metal rod is an electric conductor.

Wooden rod

No repulsive force on this charged rod

(*a*)

Metal rod

Repulsive force on this rod

F_{el}

(*b*)

sphere exerts no force on a charged amber rod brought near it. If, however, the two spheres are connected by a metal rod, as in Fig. 16.4*b*, the second sphere does exert a repulsive force on another charged amber rod. It appears that charge can flow from the first sphere to the second through the metal rod but is unable to flow through the wood. Thus metals are called electric *conductors*, while substances like wood, glass, wax, and plastics are called electric *insulators*.

Metals are good conductors of electricity because they contain many free electrons. In metals at least one electron from each atom is free to move through the crystal lattice. Since 1 mol of a metal contains 6.02×10^{23} atoms, there are in 1 mol of a metal at least 6.02×10^{23} electrons free to carry charge from one place to another. The lattice is still electrically neutral, since there are just as many positive charges as negative charges in it. In wood or glass, on the other hand, almost all the electrons are tightly bound to atoms or molecules and cannot move about freely. In a perfect insulator an electric current cannot exist; i.e., no electrons can flow through the material.

At room temperature the best conductors known are silver and copper. Since silver is a precious metal, and therefore expensive, most electric wiring is made of copper. Good insulators are quartz, glass, and synthetic materials like Teflon.

16.3 Electroscopes and Electrometers

Thus far our exploration of electrostatics has been descriptive and qualitative, as it was in the eighteenth century. But physics is a quantitative science, and eighteenth-century physicists badly needed an instrument to measure electric charge. Such an instrument is the gold-leaf electroscope, first developed by Abraham Bennet in 1787. This consists of a metal shaft with a round metal knob on top and two pieces of very light gold foil at the bottom, as in Fig. 16.5. The shaft of the electroscope passes through an insulating support in the case, which is enclosed in glass to protect the delicate gold leaves from air currents. When the electroscope is charged by touching its knob with a negatively charged rod, excess electrons flow to the gold leaves, driven by the repulsion exerted on negative charges by the charged rod. The two leaves acquire negative charges and therefore repel each other; their deflection can be used as a rough measure of the charge on the electroscope. For more quantitative measurements the two gold leaves can be replaced by a flat metal plate and a single gold leaf mounted on the shaft in such

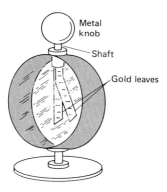

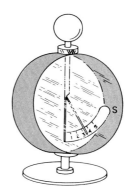

FIGURE 16.5 A gold-leaf electroscope.   **FIGURE 16.6** A simple gold-leaf electrometer.

a way that the mutual repulsion between the two causes the gold leaf to pull away from the plate, as in Fig. 16.6. The angular deflection of the gold leaf with respect to the brass plate is then a measure of the charge on the instrument, which can be read off the scale S. Such an instrument is called an *electrometer*, i.e., an instrument for measuring electric charge.

Forces Exerted by Electric Charges at a Distance

Suppose we touch the knob of an electroscope with a charged amber rod. Some of the excess electrons from the amber will flow to the gold leaves, which will then diverge, as in Fig. 16.7a. If we now remove the amber rod, charge it up again, and bring it close to the knob without touching it, the gold leaves separate even more (Fig. 16.7b). The excess electrons of the charged rod repel the free electrons in the knob and shaft of the electroscope. The repelled free electrons move as far from the charged rod as possible, ending up on the gold leaves. With more excess electrons on the gold leaves, the leaves deflect more, and the size of this increase depends on how close the charged rod is to the knob. If the rod is removed, the excess deflection disappears and the leaves return to the same position as in Fig. 16.7a. On the other hand, if a *positively* charged glass rod is brought near the knob of the electroscope, as in Fig. 16.7c, electrons are drawn *away* from the gold leaves by the attractive force of the positive charges and the electroscope deflection is *decreased* as long as the glass rod is held near the electroscope knob. Hence an electroscope containing a charge of known sign can be used to determine the sign of the charge on objects brought near it.

FIGURE 16.7 Effect of a charged rod on a charged electroscope. (a) Electroscope charged negatively. (b) A negatively charged rod increases the separation of the leaves of the electroscope. (c) A positively charged rod decreases the separation of the leaves.

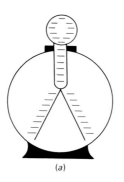

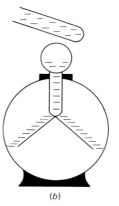

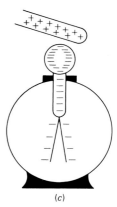

(a)   (b)   (c)

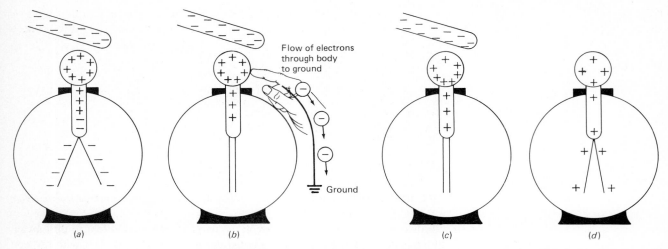

FIGURE 16.8 Charging an electroscope by induction. (*a*) A negatively charged amber rod is brought near the knob of an uncharged electroscope. (*b*) The knob is touched with the free hand. In an effort to get as far away as possible from the charged rod, electrons flow from the electroscope to ground through the body. (*c*) The free hand is then removed, breaking the connection to ground and leaving the electroscope positively charged. (*d*) The charged rod is removed, and the separation of the leaves indicates that the electroscope contains a net positive charge.

Charging an Electroscope by Induction

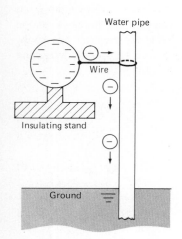

FIGURE 16.9 Grounding a charged sphere. The excess negative charge flows off the sphere to ground and leaves the sphere uncharged.

We can charge an electroscope without touching it with a charged rod or other source of charge, as in Fig. 16.8. We say that the electroscope has been charged by *induction*, i.e., without any charged body making direct contact with it.

Charging by induction is achieved by providing a path for the excess negative charge to flow off the electroscope to ground through the human body, as described in the caption to Fig. 16.8. The word *ground* is used here to mean the *earth*, which is an essentially infinite source of electric charge and a reasonably good conductor. Hence if any charged object is connected to the earth by a conducting path (such as the human body), any net negative charge on the object will flow from the object to ground, and any net positive charge on the object will be canceled by a flow of electrons from ground to the object. This is illustrated in Fig. 16.9. The symbol ⏚ designates a ground in electrical diagrams.

Charges can be induced even on good insulators. For example, a charged amber rod will attract small bits of paper. The negative charges on the rod will repel the electrons in the atoms of the paper and attract the positively charged nuclei. Even though paper is an insulator and the electrons are not free to move through the paper, the atoms near the surface become *polarized*; i.e., the positive charges are displaced slightly in one direction and the negative charges in the opposite direction. This results in a layer of positive charge near the surface where the paper touches the rod, with the electrons pushed back slightly from the surface. The attraction of the electrons in the amber for the positive charges in the paper is greater than the repulsion between the electrons on the rod and the electrons in the paper, since the latter electrons are farther away from the rod. A net attractive

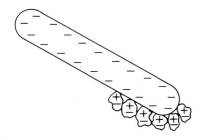

force results, and the amber picks up the bits of paper, as in Fig. 16.10. The electrostatic precipitators used to remove coal ash, sulfur oxides, and other pollutants from the chimneys of coal-burning power plants work in a similar fashion.

FIGURE 16.10 Attraction of a charged amber rod for small pieces of paper. The positive and negative charges of the paper are slightly separated from each other, and the result is a net attractive force.

16.4 Coulomb's Law

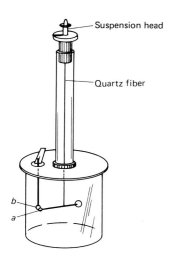

FIGURE 16.11 Coulomb's electrostatic torsion balance. The charged sphere *b* exerts a force on charged sphere *a*, which twists the quartz fiber about a vertical axis. The force between the two spheres can be measured in terms of the twist of the suspension.

In 1788 Charles-Augustin Coulomb (1736–1806), a French engineer and scientist, performed the first quantitative electrostatic experiments. Coulomb used a torsion balance (Fig. 16.11) and a technique similar to the one used by Henry Cavendish to measure the gravitational constant (see Sec. 4.1). Coulomb concluded that *the electrostatic force between two charges at rest is directly proportional to the product of the two charges and inversely proportional to the square of the distance between them*. In the form of an equation (see Fig. 16.12), we have

$$\textit{Coulomb's law:} \qquad F_{\text{el}} = \frac{k_e q q'}{r^2} \qquad\qquad (16.1)$$

where q and q' are the magnitudes of the two charges (with the proper signs), r is the separation between them, F_{el} is the electrostatic force, and k_e is a proportionality constant.

The unit of charge is the coulomb (C), which we tentatively define as the magnitude of the charge on 6.24×10^{18} electrons.* In SI units F_{el} is measured in newtons and r in meters. Experiments indicate that the force between two 1.00-C charges separated by 1.00 m is 8.99×10^9 N, and so $k_e = 8.99 \times 10^9$ N·m^2/C^2.

Coulomb's law is analogous to Newton's law of universal gravitation, both in the inverse-square dependence on distance and in the direct dependence on the magnitude of the charges (or masses, in the gravitational case). The major difference is that the electrostatic force is repulsive ($F_{\text{el}} > 0$) when q and q' are of the same sign, and attractive ($F_{\text{el}} < 0$) when they are of opposite signs, whereas the gravitational force between two masses is always attractive, since masses do not have positive and negative signs. The electrostatic force between two electrons is also larger than the gravitational force by 42 orders of magnitude (Prob. 16.19). For this reason, in electrostatic problems we ignore the gravitational forces between the charges, which are negligible compared with the electrostatic forces between even very small charges.

The coulomb is very large compared with the charges on the electron and proton. The charge on the electron was first measured by the American physicist Robert A. Millikan (see Fig. 16.13). Millikan measured the electron's charge by balancing the electrostatic force on charged droplets of oil against the gravitational

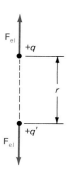

FIGURE 16.12 The Coulomb's law force between two point charges: $F_{\text{el}} = k_e q q'/r^2$.

---

*The coulomb is not a fundamental SI unit but a derived unit more properly defined in terms of the unit of electric current, the *ampere*. This is discussed more fully in subsequent chapters.

force acting on the same droplets. The apparatus he used for this purpose is shown in Fig. 16.14.

Millikan and his coworkers made measurements on thousands of oil drops in the years 1906 to 1910, sometimes observing through the telescope the same oil drop for days at a time. They found that the measured charge q was always an integral multiple of an elementary charge e; that is, $q = e, 2e, 3e, 4e, \ldots, ne$ with n an integer. Even when an x-ray source was used to change the charge on an oil drop by ionization, *the change in charge was always found to be an integral multiple of e*. Also, no charge was ever observed to be less than e, which was found to be approximately 1.60×10^{-19} C.

Millikan's interpretation of these results was that e was the magnitude of the charge on the electron, and that each oil drop measured had either one, two, three, or n extra electrons, and hence that the magnitude of the charge was an integral multiple of the so-called *elementary charge*—the smallest charge ever experimentally observed.

Elementary charge: $\boxed{e = 1.60 \times 10^{-19} \text{ C}}$

In terms of this elementary charge e, the charge of the proton is $+e$, and of the electron $-e$.

In recent years strong evidence has accumulated for the existence of particles called *quarks* inside protons and neutrons. Quarks have charges of $+\frac{2}{3}e$ and $-\frac{1}{3}e$, but no physicist has yet been able to detect a free quark (Sec. 31.6).

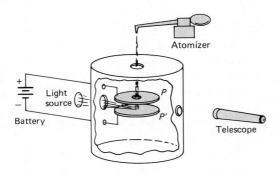

FIGURE 16.14   Millikan's apparatus for measuring the charge on the electron. An atomizer sprays tiny oil droplets into the region between the charged plates P and P'. The motion of the oil droplets under the combined gravitational and electrostatic fields is then observed through the telescope.

Example 16.1

A comb drawn through a person's hair on a dry day causes 10^{12} electrons to leave the person's hair and stick to the comb. **(a)** Is the force between the comb and the hair attractive or repulsive? **(b)** What is the magnitude of this force when the comb is 10 cm from the person's hair?

SOLUTION

(a)   The force is attractive because the charge on the comb is negative and that on the hair is positive. Unlike charges attract.
(b)   That the force is attractive results from a calculation using Coulomb's law, in which we label positive charges as + and negative charges as −. If the resulting force has a negative sign, the force is attractive; if positive, it is repulsive.

Since charge is conserved, if 10^{12} electrons leave the person's hair, the hair must contain 10^{12} net positive charges. Then

$$F_{\text{el}} = \frac{k_e q_1 q_2}{r^2} = [(9.0 \times 10^9 \text{ N·m}^2/\text{C}^2)(10^{12})(+1.6 \times 10^{-19} \text{ C})$$
$$\times (10^{12})(-1.6 \times 10^{-19} \text{ C})]/(0.10 \text{ m})^2$$
$$= \boxed{-2.3 \times 10^{-2} \text{ N}}$$

where the minus sign means that the force is attractive. This is approximately equal to the force of gravity on a 2.3-g mass.

16.5   Electrostatic Problems Involving Point Charges

Coulomb's law can be used to solve a variety of problems involving point charges at rest. By a *point charge* we mean that the object containing the charge is so small that all the charge may be considered concentrated at one point in space. In applying Coulomb's law, three basic ideas should be kept in mind.

1   *Newton's third law.* To every action there is equal and opposite reaction. If a charge q_A exerts an electrostatic force $\mathbf{F}_{AB}$ on q_B, then q_B exerts an equal and opposite force $\mathbf{F}_{BA}$ on q_A. Note that this equality is true even if q_A and q_B are very different in magnitude. Thus q_A could be 100 μC and q_B only 2.0 μC, and it would still be true that $\mathbf{F}_{BA} = -\mathbf{F}_{AB}$; that is, the two forces are equal in magnitude but opposite in direction. This must be true since the equation for Coulomb's law is unchanged when q and q' are interchanged.

2   *The principle of superposition.* Each electrostatic force acts independently of all other forces present. The electrostatic force on a charge is, therefore, the resultant of the individual forces exerted on that charge by the other charges.

3   *Vector addition.* Electrostatic forces are, like all other forces, *vectors*. If two charges q_A and q_B exert forces on q_C, as in Fig. 16.15, then the forces $\mathbf{F}_{AC}$ and $\mathbf{F}_{BC}$ must be added *vectorially* to obtain the resultant force $\mathbf{F}_{el}$ on q_C. The direction of each force depends on whether the interaction of the charges is attractive or repulsive, and this is determined by the signs of the two charges.

Using these basic ideas and Coulomb's law, it is possible to solve all electrostatic problems involving point charges, no matter how many charges are involved.

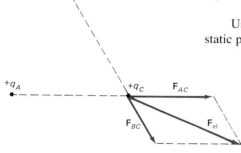

FIGURE 16.15   Application of Coulomb's law to a configuration of three charges, indicating how the electrostatic forces must be added as *vectors* to obtain correct results.

Example 16.2

(a) The force required to give a 1000-kg car an acceleration of 2.50 m/s² is 2.50×10^3 N. What identical charges must two spheres 1.00 m apart possess to produce the same force? **(b)** If the charge on each sphere is negative, how many excess electrons are there on each sphere?

SOLUTION

(a)   From Coulomb's law, with $q' = q$, we have

$$F_{el} = \frac{k_e qq'}{r^2} = \frac{k_e q^2}{r^2}$$

Hence

$$q^2 = \frac{r^2 F_{el}}{k_e} = \frac{(1.00 \text{ m})^2(2.50 \times 10^3 \text{ N})}{8.99 \times 10^9 \text{ N·m}^2/\text{C}^2}$$

$$= 2.78 \times 10^{-7} \text{ C}^2$$

and so $q = 5.27 \times 10^{-4}$ C = $\boxed{527 \ \mu\text{C}}$

Two charges of 527 μC separated by a distance of 1 m would therefore produce this relatively large force of 2.50×10^3 N.

(b)   The magnitude of the charge of the electron is 1.60×10^{-19} C, and so the number N of electrons in 527 μC is

$$N = \frac{527 \times 10^{-6} \text{ C}}{1.60 \times 10^{-19} \text{ C/electron}} = \boxed{3.29 \times 10^{15} \text{ electrons}}$$

Example 16.3

Three charges are arranged at the corners of a right triangle, as shown in Fig. 16.16, where q_B and q_C are located on the x axis. Find the magnitude and direction of the resultant force acting on q_C.

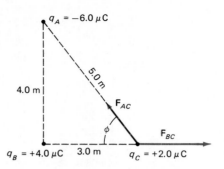

FIGURE 16.16

SOLUTION

The magnitude of the force exerted on q_C by q_A is

$$F_{AC} = \frac{k_e q_A q_C}{r^2}$$

$$= \left(9.0 \times 10^9 \frac{\text{N·m}^2}{\text{C}^2}\right) \frac{(-6.0 \times 10^{-6}\ \text{C})(+2.0 \times 10^{-6}\ \text{C})}{(5.0\ \text{m})^2}$$

$$= -4.3 \times 10^{-3}\ \text{N}$$

This is an attractive force (the sign is $-$) and is directed from q_C toward q_A along the line joining q_A and q_C.

$$F_{BC} = \left(9.0 \times 10^9 \frac{\text{N·m}^2}{\text{C}^2}\right) \frac{(+4.0 \times 10^{-6}\ \text{C})(+2.0 \times 10^{-6}\ \text{C})}{(3.0\ \text{m})^2}$$

$$= +8.0 \times 10^{-3}\ \text{N}$$

This is a repulsive force directed along the positive x axis and away from q_B at q_C.

The two forces on q_C are as shown in Fig. 16.17. To find the resultant, we first find the x and y components of the two forces acting on q_C. The angles in the triangle can be found from our knowledge of the three sides of the triangle: since $\sin \phi = \frac{4}{5}$, $\phi = 53°$.

$$F_x = 8.0 \times 10^{-3}\ \text{N} - (4.3 \times 10^{-3}\ \text{N}) \cos 53°$$

$$= 5.4 \times 10^{-3}\ \text{N}$$

$$F_y = (4.3 \times 10^{-3}\ \text{N}) \sin 53° = 3.4 \times 10^{-3}\ \text{N}$$

Then we have, since $F_{el} = \sqrt{F_x^2 + F_y^2}$,

$$F_{el} = \sqrt{(5.4)^2 + (3.4)^2} \times 10^{-3}\ \text{N} = \boxed{6.4 \times 10^{-3}\ \text{N}}$$

The direction of the force is given by

$$\tan \theta = \frac{F_y}{F_x} = \frac{3.4\ \text{N}}{5.4\ \text{N}} = 0.63 \quad \text{and} \quad \boxed{\theta = 32°}$$

Hence the resultant force on q_C is of magnitude 6.4×10^{-3} N at an angle of $32°$ with respect to the positive x axis, as shown in Fig. 16.17.

EXERCISE 1 Find the magnitude and direction of the force on the charge q_B in the same problem.

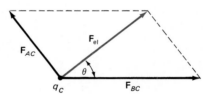

FIGURE 16.17

16.6 The Electric Field

Coulomb's law describes the electrostatic force that one charge exerts on another. Only two objects are involved in this force: the two charges. This force exists even in a vacuum. But, you may ask, how can two charges exert forces on each other when there is nothing but empty space in between?

Physicists have answered this question by considering the interaction of two charges in a different but equivalent way. A charge creates an *electric field*, which is some alteration in the character of the space around the charge, and another charge experiences a force because of the *electric field* created by the first. Although this is a more abstract view of the interaction between the two charges, it gives us a model for visualizing the nature of the interaction between two charges at a distance. We will find this concept of an *electric field* to be crucial in explaining the nature of electromagnetic waves, including radio waves and visible light.

Suppose we have a negative point charge $-q$ at some point in space. We can explore the nature of the force exerted by this charge by bringing into its neighborhood a so-called test charge, i.e., a positive point charge that we use to find the magnitude and direction of the force which would be produced by $-q$ on the test charge when positioned at various points in space. If the test charge is $+q'$ and it is located a distance r from $-q$, then q' will experience a force given by Coulomb's law as $F_{el} = -k_e qq'/r^2$, where the negative sign indicates that the force on $+q'$ is directed toward $-q$. If we move the test charge around, we find that the force is always directed radially toward the charge $-q$, and that its magnitude falls off as $1/r^2$. We say that the charge $-q$ produces a *field* at every point in space, since a test charge always experiences an electric force when placed in the vicinity of the charge $-q$. This is similar to what happens with the gravitational field of the earth. We know such a field exists because any object near the surface of the earth experiences a force directed toward its center.

We now want to introduce a quantity independent of the magnitude of the test charge. This is a vector quantity called the *electric field* $\mathbf{E}$, defined as follows:

Electric field: The electrostatic force $\mathbf{F}_{el}$ experienced by a positive test charge q' at some position in space, divided by the magnitude of that test charge.

$$\boxed{\mathbf{E} = \frac{\mathbf{F}_{el}}{q'}} \qquad (16.2)$$

The electric field $\mathbf{E}$ has dimensions of force per unit charge; its units are newtons per coulomb (N/C).

In the case of the point charge above, the magnitude of the electric field is $k_e qq'/q'r^2 = k_e q/r^2$, and the direction of the electric field is toward the source charge $-q$, as in Fig. 16.18. If the source charge were positive, the field would be of the same magnitude but directed away from the positive source charge.

We now go a step further and assert that an electric field surrounds any isolated charge or group of charges, whether we use a test charge to explore that field or not. An electric field therefore exists at any point in space surrounding a source charge. The magnitude of the field is just $k_e q/r^2$, and its direction is toward or away from the source charge depending on whether the source charge is negative or positive. A test charge is a convenient way to explore an electric field, but we are convinced that the nature of an electric field is independent of any test charge used to explore it.

FIGURE 16.18 A test charge q' at point A experiences a force $\mathbf{F}_{el}$ due to the charge $-q$. The electric field at point A is then in the direction of $\mathbf{F}_{el}$ and of magnitude $E = F_{el}/q'$.

Example 16.4

In Example 16.3 what is the magnitude and direction of the electric field due only to charges q_A and q_B at the position occupied by charge q_C in Fig. 16.16?

SOLUTION

We found that the resultant force on q_C was $F_{el} = 6.4 \times 10^{-3}$ N at an angle of $32°$ with respect to the positive x axis. The electric field at q_C is in this same direction, and its magnitude is

$$E = \frac{F_{el}}{q_C} = \frac{6.4 \times 10^{-3}\ \text{N}}{2.0 \times 10^{-6}\ \text{C}} = \boxed{3.2 \times 10^3\ \text{N/C}}$$

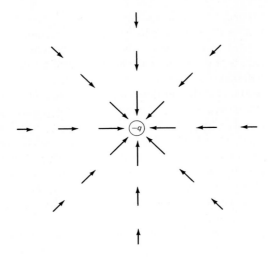

FIGURE 16.19 One way of diagraming the electric field of a point charge. Here the direction of the arrows indicates the direction of the electric field at that point in space, and the length of the arrows indicates the magnitude of the electric field.

Electric Field Lines

Suppose we calculate the electric field at various positions in space for a stationary source charge $-q$. This will lead to a series of vectors pointing toward $-q$, where the direction of each vector is the direction of the electric field and the length of the vector is the magnitude of the electric field at that place, as in Fig. 16.19. If we connect these vectors together to form continuous lines, we obtain a representation of the electric field at various positions in space. We can also adopt an arbitrary convention that the number of electric field lines passing through a given region of space is proportional to the magnitude of the source charge producing the field. Thus the electric field lines not only tell us the direction of the electric field; they also indicate its magnitude, since that magnitude can be measured in terms of the number of field lines passing through a unit cross-sectional area perpendicular to the field. Where the lines are closer together, the electric field is stronger; where the lines are further apart, the field is weaker. This method allows us to diagram the electric field lines for various arrangements of source charges. Here we show the results for only a few important types of charge distributions. Figures 16.20 and 16.21 show the field lines of single negative and positive point charges. We see that the field lines are always directed toward a negative charge and away from a positive charge.

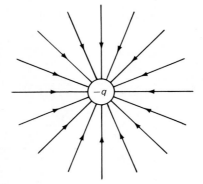

FIGURE 16.20 The electric field of a negative point charge. These lines actually fill up an entire sphere in three dimensions, but in the diagram only a two-dimensional cross section is shown.

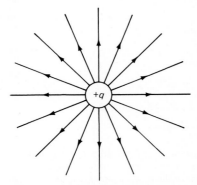

FIGURE 16.21 The electric field of a positive point charge.

In Figs. 16.20 and 16.21, which show only the two-dimensional cross sections of a three-dimensional electric field, it can be seen that the number of electric field lines falls off as $1/r^2$, in accordance with Coulomb's law, since the number of electric field lines depends on the magnitude of the source charge, but the area normal to the field lines increases as r^2.

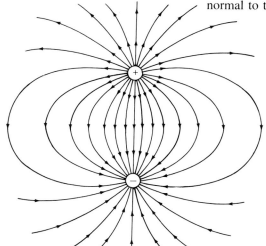

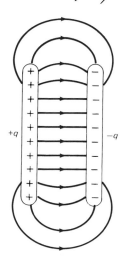

FIGURE 16.22   The electric field of an electric dipole. These lines actually fill up a three-dimensional space. The complete picture can be obtained by rotating this figure through 180° about the line connecting the two charges.

A very important charge distribution is that of an *electric dipole*, shown in Fig. 16.22. This consists of two charges of equal magnitude but opposite sign placed very close together. Some molecules like HCl are natural electric dipoles. Other molecules can be polarized in strong electric fields, leading to a separation of the centers of positive and negative charge, and to what we call an *induced dipole*. The pattern of electric field lines shown in Fig. 16.22 is appropriate to either kind of electric dipole.

Figure 16.23 shows the electric field between two flat metal plates, with total charge $+q$ on the left plate and $-q$ on the right plate. Here the charge is distributed uniformly over each of the plates. In this case the electric field lines are perpendicular to the plates in the region inside the plates and not too close to its edges. Since the number of electric field lines passing through a unit area perpendicular to the field lines remains unchanged as we move from one plate to the other, we see that the magnitude of the electric field **E** is *constant* at any point between the plates. The electric field lines always begin on positive charges and end on negative charges. This is illustrated in Figs. 16.20 to 16.23. It is impossible for an electric field line, which originates on a positive charge, to close on itself; to do this, the field line would have to point *toward* the positive charge, which contradicts our definition of the electric field. In many cases, as in Fig. 16.21, the negative charges on which the field lines end may be a very large distance away from the positive charges originating the field. Despite this, we can be confident that the field lines for static charges end on negative charges somewhere in space.

FIGURE 16.23   Cross-sectional view of the electric field between two flat plates containing equal and opposite electric charges. At the top and bottom of the plates the field lines fringe, indicating that the electric field is weaker at these places.

Example 16.5

Points A and B are at the end of a 1.00-m-long straight line, as in Fig. 16.24. A $+50$-μC point charge is placed at A and a $+100$-μC point charge at B. **(a)** What is the electric field at D, halfway between A and B? **(b)** What would be the force on a $+20$-μC charge placed at point D? **(c)** Where is the electric field zero on the line between A and B?

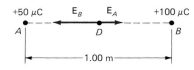

FIGURE 16.24

SOLUTION

(a) At point D the magnitude of the electric field due to the charge at A is

$$E_A = \frac{k_e q_A}{(0.50 \text{ m})^2} = \left(9.0 \times 10^9 \frac{\text{N·m}^2}{\text{C}^2}\right) \frac{50 \times 10^{-6} \text{ C}}{(0.50 \text{ m})^2}$$

$$= 1.8 \times 10^6 \text{ N/C}$$

The direction of $\mathbf{E}_A$ is along the positive x axis, which we choose as our positive direction. Similarly $E_B = -3.6 \times 10^6$ N/C, where the minus sign means that $\mathbf{E}_B$ is directed along the negative x axis. Since the electric field is a vector, we must add $\mathbf{E}_A$ and $\mathbf{E}_B$ vectorially. We obtain

$$\mathbf{E} = \mathbf{E}_A + \mathbf{E}_B = \boxed{-1.8 \times 10^6 \text{ N/C}}$$

where the minus sign means that the resultant electric field at D is directed toward point A.
(b) Since $\mathbf{F}_{el} = q'\mathbf{E}$,

$$\mathbf{F}_{el} = (20 \times 10^{-6} \text{ C})(-1.8 \times 10^6 \text{ N/C}) = \boxed{-36 \text{ N}}$$

$\mathbf{F}_{el}$ is in the same direction as $\mathbf{E}$, that is, toward A in the diagram.
(c) At the point P where the two fields cancel, the fields must be in opposite directions and have equal magnitudes. Hence

$$\frac{k_e q_A}{r_A^2} = \frac{k_e q_B}{r_B^2} = \frac{k_e q_B}{(1.00 \text{ m} - r_A)^2}$$

where r_A is the distance from A to P. We have, therefore,

$$\frac{(1.00 \text{ m} - r_A)^2}{r_A^2} = \frac{q_B}{q_A} = \frac{100 \ \mu\text{C}}{50 \ \mu\text{C}} = 2$$

Taking square roots, $(1.00 \text{ m} - r_A)/r_A = 2^{1/2} = 1.41$, from which

$$r_A = \boxed{0.41 \text{ m}}$$

The resultant field is zero at a point 0.41 m from charge A.

Example 16.6

Two charged metal plates are 0.25 m apart, with an electric field between the plates of magnitude 2000 N/C in the direction shown in Fig. 16.25. An electron of charge -1.6×10^{-19} C and mass 9.1×10^{-31} kg, is released from rest at the negative plate. **(a)** What is the acceleration of the electron? **(b)** How long will it take for the electron to reach the positive plate? **(c)** What is the speed of the electron just before it strikes the positive plate?

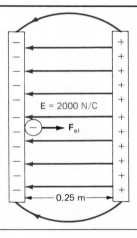

FIGURE 16.25

SOLUTION

(a) The field between the plates is directed from the positive to the negative plate. But since the electron is negatively charged, the force on it is directed from the negative to the positive plate. In other words, the negatively charged electron is attracted by the positively charged plate and therefore moves to the right. The magnitude of the force on the electron is

$$F_{el} = eE = (-1.6 \times 10^{-19} \text{ C})(-2.0 \times 10^3 \text{ N/C})$$

$$= +3.2 \times 10^{-16} \text{ N}$$

The electron's acceleration by this force is also toward the positive plate and has a magnitude, from Newton's second law,

$$a = \frac{F_{el}}{m} = \frac{3.2 \times 10^{-16} \text{ N}}{9.1 \times 10^{-31} \text{ kg}} = \boxed{3.5 \times 10^{14} \text{ m/s}^2}$$

since
$$1 \frac{\text{N}}{\text{kg}} = \frac{1 \text{ kg·m/s}^2}{\text{kg}} = 1 \text{ m/s}^2$$

(b) This is now a problem in uniformly accelerated motion, in which $v_0 = 0$ and $x = 0.25$ m. Then, from Eqs. (2.18) and (2.19), we have

$$x = v_0 t + \tfrac{1}{2}at^2$$

or
$$t = \sqrt{\frac{2x}{a}} = \left[\frac{2(0.25 \text{ m})}{3.5 \times 10^{14} \text{ m/s}^2}\right]^{1/2} = \boxed{3.8 \times 10^{-8} \text{ s}}$$

(c) $v = v_0 + at = (3.5 \times 10^{14} \text{ m/s}^2)(3.8 \times 10^{-8} \text{ s})$

$$= \boxed{1.3 \times 10^7 \text{ m/s}}$$

This is a very high speed, about one-thirtieth the speed of light. As a result, an exact calculation requires correction for relativistic effects, as we will see in Chap. 26.

16.7 Gauss' Law

One of the great contributions to physics made by the eminent German mathematician Karl Friedrich Gauss (1777–1855) is called *Gauss' law* and describes an important property of electrostatic fields. Gauss' law provides a very useful technique for the solution of many practical problems in electrostatics.

Consider a positive point charge q whose electric field is that shown in Fig. 16.26, with N electric field lines drawn from the charge to represent its magnitude q (for our purposes N can be any arbitrarily chosen number proportional to the value of q). Suppose we draw two spherical surfaces centered on the charge, one with radius R, the other with radius $2R$. These two surfaces are everywhere normal to the radial lines of the electric field **E**. The surface area of the first sphere is $4\pi R^2$, and that of the second is $4\pi(2R)^2 = 16\pi R^2$. The magnitude of the electric field at R is $E_R = k_e q/R^2$, and at $2R$ it is $E_{2R} = k_e q/(2R)^2 = k_e q/4R^2$. Hence, on multiplying the magnitude of the electric field by the surface area perpendicular to the field lines in each case, we obtain

$$E_R(4\pi R^2) = \frac{k_e q}{R^2}(4\pi R^2) = 4\pi k_e q$$

and

$$E_{2R}(16\pi R^2) = \frac{k_e q}{4R^2}(16\pi R^2) = 4\pi k_e q$$

We obtain exactly the same result ($4\pi k_e q$) for the product of electric field and normal area, whether the radius of the sphere is R or $2R$. In fact, we would also obtain the same result for any sphere centered on q no matter what its radius.

This is Gauss' law, which we have proved only for the simple case of a point charge and spherical surfaces. It states that, if we multiply each element of the area of any closed surface by the component of **E** normal to that surface element and then sum over the entire surface, the result is a constant ($4\pi k_e$) times the total charge q inside the surface:

$$\sum E_\perp \, \Delta A = 4\pi k_e q$$

The quantity $\sum E_\perp \, \Delta A$ is called the *electric flux* Φ_E through the surface. It can be shown that Gauss' law is valid for *any* number of charges inside *any* closed surface. The more general form of Gauss' law can then be written

Gauss' law $\boxed{\Phi_E = \sum E_\perp \, \Delta A = 4\pi k_e \sum q}$ (16.3)

where $E_\perp$ is the component of the electric field perpendicular to an element of area ΔA and the sums are taken over the total area of the surface and the total charge within the surface.

Electric flux $\Phi_E = \Sigma E_\perp \, \Delta A$ has units of N·m²/C and so does the right side of Eq. (16.3). We adopt the convention that the electric flux is positive if the electric field points outward through the surface surrounding the volume; electric flux is negative if the electric field points inward.

The surface used in applying Gauss' law is called a *gaussian surface*. A gaussian surface is an arbitrary, imaginary, mathematical surface, and Gauss' law is valid for any gaussian surface enclosing electric charge. However, you might wonder why we are concerned with imaginary surfaces. The answer is this: Gauss' law greatly simplifies calculations of electric fields in certain situations. Without Gauss' law many calculations of electric fields would be extremely difficult; with it, they can be extremely simple.

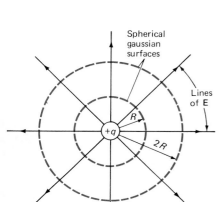

FIGURE 16.26 Gauss' law: The electric field of a positive point charge. Note that the lines drawn here as circles really represent spheres.

For Gauss' law to be useful, however, we must be able to choose a gaussian surface with a symmetry appropriate to the problem we want to solve. The symmetry of a gaussian surface will be closely related to the symmetry of the charge distribution producing the electric field we desire to calculate. We will show here just a few examples that indicate the usefulness and power of Gauss' law.

Electric Field of a Charged Metal Sphere

If a hollow metal sphere is charged negatively with a charge $-Q$, the excess electrons will repel one another and assume a uniform distribution over the whole surface of the sphere. The electric field lines in this case point toward the surface of the metal and end on the excess electrons. They are also normal to the surface of the metal, as in Fig. 16.27. If they were not, the electric field would have a component parallel to the metal surface. Since the electrons are free to move, they would therefore move to a position where that parallel component would be reduced to zero. Hence *the only components of the electric field that remain are perpendicular to the surface of the metal.*

Let us construct a spherical gaussian surface of radius R just outside the charged metal sphere and concentric with the sphere. This gaussian surface has the proper spherical symmetry for our problem and therefore simplifies the calculation of the electric flux. Since the lines of the electric field are everywhere perpendicular to the gaussian surface, Gauss' law tells us that

$$\Phi_E = EA = 4\pi k_e \sum q = 4\pi k_e(-Q)$$

where E is the electric field strength on the surface of area A and has the same value at every point on the surface, from the symmetry of the problem. And so

$$E = \frac{-4\pi k_e Q}{A} = \frac{-4\pi k_e Q}{4\pi R^2} = \frac{-k_e Q}{R^2}$$

But this is just the result that would be obtained from Coulomb's law for a point charge of magnitude $-Q$ at the center of a sphere of radius R. Hence *the electric field of a charged metal sphere at any point outside the sphere is the same as if all the charge were concentrated at its center.*

This is extremely useful in solving problems involving uniformly charged spherical surfaces. We can treat all the charge as located at the center of the sphere, just as for the motion of a planet about the sun we can consider all the mass as lo-

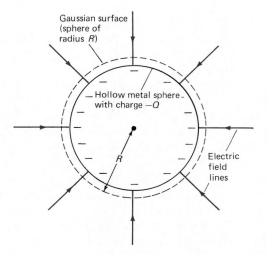

FIGURE 16.27 Field of a negatively charged hollow metal sphere.

Gaussian surface
(sphere of
radius R)

Hollow metal sphere
with charge $-Q$

R

Electric
field
lines

cated at the center of the planet. The reason is the same in the two cases: we have spherical symmetry and a field (gravitational or electrostatic) that falls off as $1/r^2$.

Location of the Charge on Metal Objects

Metal object

Gaussian surface

FIGURE 16.28 Charge distribution on an irregularly shaped solid metal object. Gauss' law predicts that all the excess charge must be on the surface of the metal object.

Suppose we put a negative charge on an irregularly shaped solid metal object like the one in Fig. 16.28. Where does the excess negative charge end up after the very brief time required for the charges to move to equilibrium positions and come to rest?

To answer this question, we construct a gaussian surface just *inside* the outer boundary of the metal, as in the figure. We do this because we know the electric field inside the metal and can therefore hope to find the charge distribution inside the metal. Then we apply Gauss' law:

$$\Phi_E = \sum E_\perp \, \Delta A = 4\pi k_e \sum q \tag{16.3}$$

But the electric field strength inside any metal must be zero, since the charges are free to move and will continue to move until no net field remains. Hence the left side of Eq. (16.3) is zero, and therefore the right side must also be zero. This can only be true if $\Sigma q = 0$. Therefore there can be no excess electric charge in the metal's interior. *All the excess charge on a metal is on its outer surface.* This insight from Gauss' law is of great practical importance when we try to accelerate particles by creating lightninglike discharges between large charged spheres, as in the Van de Graaff generator (see Sec. 17.5).

We have here obtained some very important insights into the behavior of charges in metals by using Gauss' law. To obtain the same results by using Coulomb's law would be much more difficult.

Example 16.7

Find the magnitude of the electric field **E** at a distance R from a long, straight wire that contains a total charge q spread uniformly over a length L. Assume that R is small compared with L.

SOLUTION

The charge per unit length on the wire is $\lambda = q/L$. Since a wire can be considered to be a very thin cylinder, the wire has cylindrical symmetry. This gives us a hint that, if we apply Gauss' law, the gaussian surface we use should be a cylindrical surface, as in Fig. 16.29. We choose the gaussian surface to have a radius R and a length ΔL. The surface area of the curved sides of this cylinder is then $2\pi R \, \Delta L$.

We have seen that the electric field near a metal is everywhere perpendicular to the surface of the metal, in this case, the wire. The electric field lines spread out symmetrically from the wire so as to be perpendicular to the curved surface of the cylinder at every point on it. For the same reason no lines of the electric field pass through the two flat ends of the cylinder, and the electric field is radial and equal to E_R. Hence we have, from Gauss' law,

$$\sum E_\perp \, \Delta A = 4\pi k_e \sum q \quad \text{or} \quad E_R(2\pi R \, \Delta L) = 4\pi k_e \lambda \, \Delta L$$

since the amount of charge within the gaussian surface is $\lambda \, \Delta L$, and so

$$E_R = \frac{4\pi k_e \lambda \, \Delta L}{2\pi R \, \Delta L} = \boxed{\frac{2k_e \lambda}{R}}$$

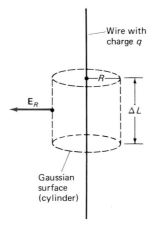

Wire with charge q

E_R

Gaussian surface (cylinder)

FIGURE 16.29 Diagram for Example 16.7. The charge per unit length on the wire is $\lambda = q/L$.

The field of a long, straight wire therefore falls off as $1/R$. This example shows the power of Gauss' law in solving a problem that otherwise would be very complicated.

It is worth summarizing some interesting results here. From Coulomb's law we know that the electric field of a point charge falls off as $1/R^2$. Using Gauss' law we have just found that the electric field of a long wire falls off as $1/R$. For a solid plane of charge, like that of Fig. 16.23, we saw that the electric field near the plane is constant and does not vary at all with distance from the plane. [This can also be proved using Gauss' law. (See Prob. 16.46.)] Thus, as we move from a point charge (zero dimensions) to a line charge (one dimension) to a plane charge (two dimensions), the electric field dependence changes from $1/R^2$ to $1/R$ to 1.

Summary: Important Definitions and Equations

Electric charge: The two kinds of electric charge are called positive ($+$) and negative ($-$). Charges of like sign repel each other and charges of unlike sign attract each other.

Principle of conservation of charge: In any closed system the algebraic sum of all the electric charges remains constant.

Coulomb's law:

$$F_{\text{el}} = \frac{k_e qq'}{r^2} \quad \text{where } k_e = 8.99 \times 10^9 \text{ N·m}^2/\text{C}^2 \quad \textbf{(16.1)}$$

Elementary charge: The magnitude of the charge on the electron, equal to 1.602×10^{-19} C.

Coulomb (preliminary definition): The charge on 6.24×10^{18} electrons.

Electric field: The electrostatic force $\mathbf{F}_{\text{el}}$ experienced by a positive test charge q' at some position in space, divided by the magnitude of the test charge.

$$\mathbf{E} = \mathbf{F}_{\text{el}}/q' \quad \textbf{(16.2)}$$

Electric field lines: Lines drawn to describe an electrostatic field. The direction of the lines is that of the field, and the number passing through unit area perpendicular to the field indicates the strength of the field.

Gauss' law:

$$\sum E_\perp \, \Delta A = 4\pi k_e \sum q \quad \textbf{(16.3)}$$

where $E_\perp$ is the component of the electric field perpendicular to an element of area ΔA and the summations are taken over the area of the surface and the total charge within that surface.

Electric flux Φ_E: The quantity $\Sigma E_\perp \, \Delta A$ evaluated for every point on a Gaussian surface.

Gaussian surface: An imaginary, mathematical surface chosen to fit the symmetry of a particular electrostatic problem to which Gauss' law is being applied.

Dependence of electric fields on charge distributions
Uniformly charged flat surface: $E = $ constant
Uniformly charged straight wire: $E \propto 1/R$
Point charge (or charged sphere at points outside the sphere): $E \propto 1/R^2$

Questions

1 Explain why distilled water is a very poor conductor of electricity, but if ordinary table salt (NaCl) is poured into the water, it becomes a good conductor.

2 Suggest some ways in which the electrostatic field resembles the gravitational field, and some ways in which the electrostatic field differs.

3 Draw the electric field lines surrounding two small spherical charges, each of magnitude $+q$, separated by a distance R that is much larger than the radius of the charged spheres.

4 How could you test to see if an electric field existed at some given point in space?

5 Is it possible for two lines of an electric field to cross each other? Why?

6 If you comb your hair repeatedly on a dry winter day, the comb becomes electrically charged. Indicate how you would determine whether this charge is positive or negative. What would be the resulting charge on your hair?

7 Why do woolen socks coming out of a clothes dryer tend to stick together? What does a "cling-free" sheet do to stop this condition?

8 Why do electrostatic experiments not work well on humid days?

9 An uncharged cork sphere at the end of a string is attracted by a charged amber rod, collides with the rod, and then jumps violently away from it. From that time on the sphere is repelled by the amber rod. Explain.

10 Is the force that one charge exerts on a second charge changed in any way when other charges are brought into the immediate neighborhood of the first two charges?

11 (*a*) A metal sphere is charged positively by touching it with a charged glass rod. Does the mass of the sphere increase or decrease, and by how much?

(*b*) A metal sphere is charged negatively by touching it with a charged amber rod. Does the mass of the sphere increase or decrease, and by how much?

Multiple-Choice and Simple Exercises

16.1 The electrostatic force between a proton and an electron separated by 10^{-10} m is:
(a) 2.3×10^{-8} N   (b) 2.3×10^{-4} N
(c) 2.3×10^8 N   (d) 2.3×10^4 N   (e) 2.3 N

16.2 The ratio of the electric force between two protons to the gravitational force between them is about:
(a) 10^{42} (b) 10^{39} (c) 10^{36} (d) 10^{45}
(e) 10^{48}

16.3 A 1.0-μC point charge is 2.0 cm away from a 2.0-μC point charge. If the magnitude of the force on the larger charge is F, the magnitude of the force on the smaller charge is:
(a) $F/4$ (b) $F/2$ (c) F (d) $2F$
(e) $4F$

16.4 Two charged bodies separated by a distance d repel each other with a force F. This force will change to $9F$ when the two bodies are separated by a distance:
(a) $3d$ (b) $9d$ (c) $d/3$ (d) $d/9$
(e) $d/81$

16.5 At the vertices of an equilateral triangle, 5.0 cm on a side, are placed charges $+Q$, $-Q$, and $+Q'$. The net force on the $+Q'$ charge is directed:
(a) Toward $-Q$ (b) Toward $+Q$
(c) Directly away from $+Q$
(d) At right angles to the line joining $+Q$ and $-Q$
(e) Parallel to the line joining $+Q$ and $-Q$

16.6 How many excess electrons are required to give a charged sphere a negative charge of 5.0 μC?
(a) 5.0×10^6 (b) 1.6×10^{19}
(c) 3.1×10^{13} (d) 3.1×10^{19}
(e) 3.1×10^{16}

16.7 The electric field **E** at a point 0.30 m to the right of a charge $Q = 3.0 \times 10^{-4}$ C is (where a minus sign on E means that the field is directed toward Q, a plus sign means away from Q):
(a) -3.0×10^7 N/C   (b) $+3.0 \times 10^7$ N/C
(c) -3.0×10^7 N   (d) $+3.0 \times 10^7$ N
(e) -3.0×10^7 J

16.8 A long, straight wire carries a static electric charge of 10^{-4} C/m. The magnitude of the electric field 2.0 cm from the wire is:
(a) 9.0×10^7 N/C   (b) 4.5×10^9 J/C
(c) 9.0×10^9 N/C   (d) 9.0×10^7 N/m
(e) 4.5×10^9 N/C

16.9 Two point charges, one of $+20$ μC, the other of -10 μC, are separated by 40 cm in air.
(a) What are the magnitude and direction of the electric force on the 20-μC charge?
(b) If this positive charge is on a small plastic sphere of mass 1.0 g, what is the acceleration of this sphere caused by the electrostatic force?

16.10 A point charge of 100 μC repels another point charge with a force of 2.0 N when placed 0.50 m away from the second charge. Find the sign and magnitude of the second charge.

16.11 Two metal spheres are charged by induction so that one has a surplus of 2.0×10^{10} electrons and the other has a deficiency of 1.0×10^{10} electrons.
(a) If the two spheres are separated by 1.0 m, what are the magnitude and direction of the force on the negatively charged sphere?
(b) What are the magnitude and direction of the force on the positively charged sphere?

16.12 Two cork balls contain electrostatic charges of $+2.5$ and -1.5 μC. If they are separated by 20 cm:
(a) What are the magnitude and direction of the force on the negatively charged ball?
(b) What are the magnitude and direction of the electric field at the position of the negatively charged ball?

16.13 What are the magnitude and direction of the electric field **E**:
(a) 20 cm from a 30-μC positive point charge?
(b) 20 cm from a 30-μC negative point charge?

16.14 What are the magnitude and direction of the electric field 1.0 m from a long, thin copper wire containing a charge of 10 μC per meter of length?

16.15 (a) What is the acceleration of a proton in an electric field of magnitude 400 N/C?
(b) How does this acceleration compare with the acceleration of an electron in the same field?
(c) How does the acceleration of the electron compare with the acceleration due to gravity?

16.16 A $+5.0$-μC charge experiences a force of 2.4×10^{-3} N in the positive y direction at a certain place in a room.
(a) What is the electric field at that place?
(b) What force would a -15.0-μC charge experience at the same place?

Problems

16.17 Three positive charges are at the corners of an equilateral triangle of sides 20 cm, with $q_A = 50$ μC, $q_B = 25$ μC, and $q_C = 100$ μC.
(a) What is the magnitude of the resultant force on q_C?
(b) What angle does this force make with respect to the line joining q_A and q_B?

16.18 In Fig. 16.16, what are the magnitude and direction of the resultant force on q_A?

16.19 Calculate the ratio of the electric force between two electrons to the gravitational force between them for the same distance r between the electrons in the two cases.

16.20 Three point charges are placed at the following positions on the x axis: $+3.0$ μC at the origin; -4.0 μC at $x = +30$ cm; -5.0 μC at $x = +90$ cm. What is the force (a) on the $+3.0$-μC charge and (b) on the -5.0-μC charge?

16.21 Four point charges, each of magnitude $+5.0$ μC, are placed at the corners of a square 30 cm on a side. Find the magnitude and direction of the force on any one of the charges.

16.22 Repeat Prob. 16.21 for the situation where one of the four charges is negative. What is the force on this negative charge?

16.23 Charges of $+3.0$, $+5.0$, and -8.0 μC are placed at the vertices of an equilateral triangle of sides 3.0 cm. What are the magnitude and direction of the force (a) on the -8.0-μC charge; (b) on the $+3.0$-μC charge?

*16.24 Two identical small metal spheres are given charges of $+4.0$ and -9.0 μC.

(a) If the spheres are separated by 30 cm, what is the force of attraction between them?

(b) If they are touched together and then separated to their original 30-cm distance, what is the force between them?

16.25 In the Bohr theory of the hydrogen atom an electron circles the hydrogen nucleus (a proton) in an orbit of radius 0.53×10^{-10} m. The electrostatic attraction of the proton for the electron furnishes the centripetal force needed to hold the electron in its circular orbit.

(a) Find the force of electrical attraction between the proton and the electron.

(b) Find the speed of the electron.

16.26 Suppose all the electrons in a gram of carbon ($^{12}_6$C) could be moved 1.0 m away from the carbon nuclei. What would be the force of attraction between the electrons and the nuclei?

*16.27 Two small cork balls, each of mass 10.0 g, are hung at the ends of two thin threads, each of lengths 20.0 cm, from a common support. The balls are given identical electric charges, repel each other, and come to equilibrium with an angle of $10.0°$ between the strings. What is the magnitude of the charge on each ball?

16.28 Two small conducting balls, each of mass 1.0 g, hang from a stand on parallel insulating threads of length 75 cm so that the two balls just touch. A positively charged rod then touches the two balls and gives them equal positive charges. They repel each other and take up new positions in which each hangs at an angle of $30°$ with respect to the vertical. What is the charge on each ball?

*16.29 Two small spheres carry a total charge of -50μC. When they are separated by a distance of 25 cm, they repel each other with a force of 20 N. What is the charge on each sphere?

16.30 A $+10.0$-μC charge is placed 50 cm away from a -2.5-μC charge.

(a) Where can a third charge (of either sign) be placed so that it experiences no net force?

(b) Does this position correspond to stable or unstable equilibrium?

16.31 (a) Determine the electric field midway between two positive charges of 10 μC each, if the charges are 5.0 cm apart.

(b) What is the electric field if one charge is positive and the other negative?

16.32 A square $ABCD$ has sides 20 cm in length. Positive charges of 50 μC are placed at corners A and B, and negative charges of 50 μC are placed at corners C and D. Calculate the magnitude and direction of the electric field at the center of the square.

16.33 Two point charges, $q_1 = +2.0$ μC and $q_2 = -5.0$ μC, are separated by 2.0 m.

(a) At what place on the line joining the charges, or on its extension, is the electric field zero?

(b) If both charges were positive, would there be a place on the line joining the charges where the electric field would be zero? Where would it be located?

16.34 For the three charges shown in Fig. 16.30 calculate the electric field **E** (a) halfway between charges q_A and q_B and (b) at a point 5.0 cm above the charge q_B and equidistant from the charges q_A and q_C.

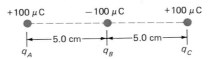

FIGURE 16.30 Diagram for Prob. 16.34.

16.35 (a) What is the electric field at the center of the square in Prob. 16.21?

(b) What is the electric field at the center of the square in Prob. 16.22?

16.36 A 10.0-g pendulum bob is charged with $+150$ μC. It is suspended by a cord that holds it in equilibrium in an electric field of magnitude 500 N/C. What is the tension in the cord if the electric field is directed (a) downward and (b) upward?

*16.37 A circular metal hoop of radius 0.40 m carries a charge of $+100$ μC.

(a) What is the electric field at the center of the hoop, if the 100-μC charge is symmetrically distributed around the hoop?

(b) A 1.0 cm piece is cut out of the hoop so that a gap of 1.0 cm is introduced. The hoop is then charged to the same $+100$ μC as in part (a). What is the electric field at the center of the hoop in this case?

16.38 A proton is released from rest in a uniform electric field and moves 1.0 m in the direction of the field in 1.0×10^{-3} s. What is the magnitude of the electric field?

16.39 A Van de Graaff accelerator is used to accelerate a beam of protons from rest. It produces a uniform electric field of 1.2×10^7 N/C over the 10-cm distance through which the protons move.

(a) What is the kinetic energy of the protons after they have been accelerated?

(b) What is their speed?

16.40 An electron is shot out of an electron gun at an initial speed of 2.0×10^6 m/s along the $+x$ axis. It slows down and stops because of the action of a uniform electric field. If the electron stops in a distance of 1.0 m, what are the magnitude and direction of the field?

16.41 An electron is released from rest at the surface of a filament in a uniform electric field and travels 10.0 cm in 5.0×10^{-6} s. What is the magnitude of the electric field?

16.42 An alpha particle has a charge $+2e$, where e is the elementary charge. Find the electric flux (a) through a sphere of radius 2.0×10^{-10} m surrounding the alpha particle and (b) through a sphere of radius 2.0×10^{-9} m surrounding the alpha particle.

*16.43 Positive charge is uniformly distributed throughout a spherical volume of radius R, the volume charge density being ρ in units of C/m^3.

(a) Use Gauss' law to prove that the electric field inside the volume, at a distance r from the center, is $\rho r/3\epsilon_0$.

(b) Prove that the electric field at the surface of the sphere is the same as that produced by a point charge located at the center and carrying the same total charge as the sphere.

*16.44 Two concentric spheres, one of radius 10.0 cm, the other of radius 12.0 cm, are uniformly charged, the inner sphere with a charge of $+200$ μC, and the outer sphere with a charge of $-200/\mu C$. What is the electric field (a) 5.0 cm from the common center of the spheres; (b) 11.0 cm; (c) 15.0 cm?

*16.45 Two long concentric cylinders, one of radius 6.0 cm, the other of radius 10.0 cm, are charged, the inner cylinder with a charge of $+150$ μC per meter of length, and the outer cylinder with a charge of -150 μC per meter of length. What is the electric field at a distance of (a) 3.0 cm from the common axis of the two cylinders; (b) 8.0 cm from the axis; (c) 12.0 cm?

*16.46 A flat slab of glass of very large surface area is rubbed on one side with a silk cloth until there is a uniform charge on the glass of $+0.20$ $\mu C/cm^2$ (The surface charge density σ is therefore $+0.20$ $\mu C/cm^2$).

(a) Use Gauss' law to find an expression for the magnitude of the electric field near the charged surface of the glass slab.

(b) What is the magnitude of the electric field 2.0 cm from the charged glass surface and near the center of the large slab?

(c) What is the direction of the electric field at this place?

Additional Readings

Cohen, I. Bernard: "Benjamin Franklin," in *Lives in Science*, a *Scientific American* book, Simon and Schuster, New York, 1957. An excellent brief article by the man who has done much to restore Franklin to his rightful place in the history of physics. Cohen sees Franklin as "a major transitional figure between the physics of Newton and the physics of Faraday and Clerk Maxwell."

———: *Benjamin Franklin, Scientist and Statesman,* Scribner, New York, 1975. A well-illustrated account based on Cohen's article on Franklin in the *Dictionary of Scientific Biography*.

Fletcher, Harvey: "My Work with Millikan on the Oil-Drop Experiment," *Physics Today*, vol. 35, no. 6, June 1982, pp. 43–47. An interesting article suggesting that many of the best ideas in Millikan's work were provided by Fletcher when he was a graduate student.

Heilbron, J. L.: "Franklin's Physics," *Physics Today*, vol. 29, no. 7, July 1976, pp. 32–37. A short, well-illustrated discussion of Franklin's contributions to physics.

Kargon, Robert H.: *The Rise of Robert Millikan,* Cornell University Press, Ithaca, N.Y., 1982. The author uses Millikan's life to illustrate the great changes that took place in science during Millikan's impressive career as a physicist.

Roller, Duane, and Duane H. D. Roller: *The Development of the Concept of Electric Charge*, Harvard University Press, Cambridge, Mass., 1954. This is one of the best of the Harvard Case Studies in Experimental Science, and considers the development of electricity from the Greeks to Coulomb.

Shamos, Morris H.: *Great Experiments in Physics*, Holt, Rinehart and Winston, New York, 1959. Coulomb's work is described in his own words in Chap. 5 of this very useful book.

Stewart, Ian: "Gauss," *Scientific American*, vol. 237, no. 1, July 1977, pp. 122–131. A discussion of the life and accomplishments of one of the greatest mathematicians who ever lived, who was also the discoverer of the famous law of electrostatics discussed in this chapter.

Williams, Earle R.: "The Electrification of Thunderstorms," *Scientific American*, vol. 259, no. 5, November 1988, pp. 88–99. An up-to-date account of scientists' endeavors to understand the mechanisms producing lightning discharges in the atmosphere.

CHAPTER 17

Electric Potential Energy, Potential Difference, and Capacitance

Through and through the world is infected with quantity. To talk sense, is to talk in quantities. It is no use saying that the nation is large—How large? It is no use saying that radium is scarce—How scarce? You cannot evade quantity. You may fly to poetry and to music, and quantity and number will face you in your rhythms and your octaves. Elegant intellects which despise the theory of quantity are but half developed. They are more to be pitied than blamed. . . .

Alfred North Whitehead (1861–1947)

In this chapter we continue our discussion of electrostatics by showing how changing the position of a charge in an electrostatic field changes its electric potential energy. We then introduce the concepts of electric potential and potential difference. Finally we discuss the storage of electric charge in capacitors and the relationships among electric charge, potential difference, capacitance, and energy.

17.1 Electric Potential Energy and Electric Potential

Let us consider the electric field between two oppositely charged flat metal plates, as in Fig. 17.1, where the two charges are of the same magnitude. The electric field between the plates is uniform, as we saw in Sec. 16.6. Suppose the distance between points A and B is d, and we introduce a test charge $+q'$ and move it slowly from B to A. Since we are moving a positive charge in the direction opposite to the electric force $\mathbf{F}_{el}$ exerted on it by the electric field $\mathbf{E}$, we must do work against this force. The amount of work we do is (just as it was in mechanics) the product of the force we exert and the distance moved in the direction of the force [Eq. (6.1)]:

$$\mathcal{W} = \mathbf{F} \cdot \mathbf{d} = Fd = q'Ed \tag{17.1}$$

since the force $\mathbf{F}$ we exert is equal in magnitude to the force $\mathbf{F}_{el} = q'\mathbf{E}$.

The electric force is a conservative force similar to the gravitational force discussed in Sec. 6.3, as might be expected from the similarity between Newton's

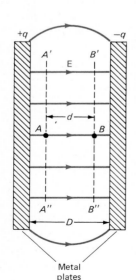

Metal plates

FIGURE 17.1 Cross-sectional view of the electric field between two charged flat metal plates. Here $A'A''$ and $B'B''$ are two flat surfaces parallel to the metal plates. (As we will see later in this section, these two surfaces are equipotential surfaces.) We must do work to move a positive test charge from B to A.

law of universal gravitation and Coulomb's law. The work done in moving the charge from B to A must therefore increase the electric potential energy of the charge, just as lifting a rock increases the gravitational potential energy of the rock. Since the charge is moved so slowly that it can be assumed no work goes into kinetic energy, and since the force is conservative so that no energy is lost as heat, all the work done must be converted into electric potential energy:

$$\mathcal{W} = \Delta PE = PE_A - PE_B \tag{17.2}$$

Here PE_A means the electric potential energy of the charge q' at point A, and PE_B is the same at point B. On dividing Eq. (17.2) by q', we have

$$\frac{\mathcal{W}}{q'} = \frac{PE_A}{q'} - \frac{PE_B}{q'}$$

The potential energy per unit charge at point P in an electric field is called the electric potential (or simply the potential) V_P at that point.

$$\boxed{V_P = \frac{PE_P}{q'}} \tag{17.3}$$

In Fig. 17.1 the potential at A is $V_A = PE_A/q'$ and that at B is $V_B = PE_B/q'$. We then have, from above,

$$\boxed{\frac{\mathcal{W}}{q'} = V_A - V_B = V_{AB}} \tag{17.4}$$

where $V_{AB} = V_A - V_B$ is called the electric *potential difference* between points A and B. The potential difference (V_{AB}) between two points (A and B) in an electric field is therefore numerically equal to the work done in moving a unit charge from one point (B) in the field to the other (A). This result turns out to be completely general, even though derived for a special case here.

The dimensions of both electric potential and potential difference are work per unit charge, and the units are joules per coulomb (J/C). If 1 J of work is done to move 1 C of charge from point B to point A in an electrostatic field, then the potential difference V_{AB} is said to be $+1$ volt (V), since $V_A > V_B$. We can therefore define a *volt* as follows:

$$1 \text{ V} = 1 \text{ J/C}$$

For example, if a charge of 2.0 μC is moved from B to A through a potential difference of 3.0 V, the work done is, by Eq. (17.4),

$$\mathcal{W} = q'V_{AB} = (2.0 \ \mu C)(3.0 \ V) = (2.0 \ \mu C)(3.0 \ J/C) = 6.0 \times 10^{-6} \ J$$

Since, in Fig. 17.1, a positive charge q' would have a higher potential energy at A than at B, we say that there is a *drop in potential* from A to B, and that a positive charge will, if released, accelerate "downhill" from A to B. Similarly, if we want to move a positive charge "uphill" from B to A, we must do an amount of work $\mathcal{W} = q'V_{AB}$ against the field, which is directed from A to B. On the other hand, a *negative* charge released at B will accelerate from B to A under the action of the field; that is, it will accelerate *up* a potential hill, because in moving from B to A, it *loses* electric potential energy.

Example 17.1

For the charged metal plates of Fig. 17.1, suppose that the plate separation D is 5.00 cm, the distance d from A to B is 3.50 cm, and the electric field is 1.40×10^3 N/C. **(a)** How much work must I do to move a charge of $+1.20$ μC from B to A? **(b)** What is the potential difference V_{AB} between A and B? **(c)** What is the potential difference between the two metal plates? **(d)** How much work must I do to move a charge from A to B?

SOLUTION

(a) The work I must do to move the charge from B to A is $\mathcal{W} = \mathbf{F} \cdot \mathbf{d} = q'Ed$, since $\mathbf{F}$ and $\mathbf{d}$ are in the same direction, for I must exert a force in a direction opposite to that of the electric field $\mathbf{E}$.

$$\mathcal{W} = (1.20 \times 10^{-6} \text{ C})(1.40 \times 10^3 \text{ N/C})(3.50 \times 10^{-2} \text{ m})$$

$$= \boxed{5.88 \times 10^{-5} \text{ J}}$$

This work goes into increased PE of the charge q' at its new position (A) in the field.

(b) The potential difference between A and B is

$$V_{AB} = V_A - V_B = \frac{\mathcal{W}}{q'} = \frac{5.88 \times 10^{-5} \text{ J}}{1.20 \times 10^{-6} \text{ C}}$$

$$= 49.0 \text{ J/C} = \boxed{49.0 \text{ V}}$$

A is at a higher potential than B, since the electric field is directed from A to B.

(c) Since the electric field between the plates is constant, the potential difference is directly proportional to distance. Therefore $V_{AB}/d = V/D$, where V is the potential difference between the plates and D is the distance between the plates. Therefore

$$V = \frac{D}{d} V_{AB} = \frac{5.00 \text{ cm}}{3.50 \text{ cm}} (49.0 \text{ V}) = \boxed{70.0 \text{ V}}$$

(d) Since the electrostatic field is conservative, the work done in taking a charge around a closed path must be zero. Therefore the work in taking the charge from A to B must be the negative of the work required to move the charge from B to A, as found in part (a). Therefore the work is $\boxed{-5.88 \times 10^{-5} \text{ J}}$.

17.2   Potential Difference and Voltage

The terms *potential difference* and *voltage* are often used interchangeably. They have the same dimensions and are expressed in the same unit, the *volt* (V), where 1 V = 1 J/C. Both are equal to the difference in electric potential energy per unit charge between two points in an electric field. For clarity, we will use the term *potential difference* for the remainder of our discussion of electrostatics, but we will switch to the more commonly used term *voltage* when we discuss current electricity.

A battery (to be discussed more at length in Chap. 18) is a very convenient way to apply a desired potential difference to an electric system. The potential difference between the positive and negative terminals of an automobile storage battery is usually 12 V. This means that there is an electric field in the region around the battery terminals, and that if we move a positive charge from the negative to the positive terminal, the work we do is 12 J for each coulomb of positive charge moved. Since the electric force is conservative (Sec. 17.1), this amount of work is the same no matter what path is taken.

The Electronvolt: A Unit of Energy

We have just seen that if the potential difference between two points is 1.00 V, then it takes 1 J of work to move 1 C of charge between these two points. Suppose, however, we do not want to move 1 C of charge, but only a single electron with a charge 1.60×10^{-19} C. The numerical value of the work in this case is

$$\mathcal{W} = (1.60 \times 10^{-19} \text{ C})(1.00 \text{ V}) = 1.60 \times 10^{-19} \text{ J}$$

This amount of work (or energy) is called an *electronvolt* (eV).

$$\boxed{1 \text{ eV} = 1.6 \times 10^{-19} \text{ J}}$$

Electronvolt: The work required to move one electronic charge through a potential difference of one volt: 1 eV $= 1.60 \times 10^{-19}$ J.

The electronvolt is much used as an energy unit in atomic and nuclear physics.

Example 17.2

An electron in a TV set is accelerated toward the screen by a potential difference of 1000 V. The screen is 35 mm from the electron source. **(a)** How much work is done by the electric field in accelerating the electron? **(b)** What is the speed of the electron when it strikes the screen?

SOLUTION

(a) Since the work done per unit charge by the field is, from Eq. (17.4), equal to the potential difference V_{AB}, the work done on a charge q is

$$\mathcal{W} = qV_{AB} = (1.60 \times 10^{-19} \text{ C})(1000 \text{ V})$$

$$= \boxed{1.60 \times 10^{-16} \text{ J}}$$

Another approach is to note that an electron moving through a potential difference of 1000 V has an energy of 1000 eV. But 1000 eV is $1000(1.60 \times 10^{-19} \text{ J}) = 1.60 \times 10^{-16} \text{ J}$, as found above.

(b) Since no external forces act, by the principle of conservation of energy all this work must be converted into kinetic energy of the electron:

$$KE = \tfrac{1}{2}mv^2 = 1.60 \times 10^{-16} \text{ J}$$

and

$$v = \left[\frac{2(1.60 \times 10^{-16} \text{ J})}{9.11 \times 10^{-31} \text{ kg}} \right]^{1/2} = \boxed{1.87 \times 10^7 \text{ m/s}}$$

Note that the distance between source and screen is of no importance here. The crucial factor is the potential difference between source and screen; this completely determines the work done on the electron, and hence its final KE and speed.

17.3   Electric Potential and the Electric Field

There are several important differences between the concepts of electric potential and electric field. First of all, the electric field $\mathbf{E}$ is a *vector*, $\mathbf{E} = \mathbf{F}_{el}/q'$, and can be mapped by electric field lines that give the magnitude and direction of the electric field at every point in space. Second, the units of the electric field are clearly newtons per coulomb. Since

$$1 \frac{\text{N}}{\text{C}} = \frac{1 \text{ N·m}}{1 \text{ C·m}} = \frac{1 \text{ J/C}}{1 \text{ m}} = \frac{1 \text{ V}}{\text{m}}$$

the electric field may be expressed either in newtons per coulomb or in volts per meter.

The dimensions of the electric potential at a point P, on the other hand, are work per unit charge rather than force per unit charge, and its units are joules per coulomb or volts. Secondly, the electric potential is a *scalar*, without direction, whose magnitude may be mapped by the use of *equipotential surfaces*.

Equipotential Surfaces

In understanding and diagraming electric fields, equipotential surfaces are an important tool.

Equipotential surface: An imaginary surface in space, all parts of which are at the same electric potential.

Since there is no potential difference between any two points on such a surface, no work need be done in moving a charge from one place to another on that surface. We therefore conclude that the electric field has no components parallel to the equipotential surface, for if it did, work would have to be done in moving a charge along that surface against the field. Therefore the *electric field must be everywhere perpendicular to an equipotential surface*. (This is similar to the situation for the earth's gravitational field: the gravitational equipotentials are parallel to the earth's surface, and the gravitational field is perpendicular to the earth's surface.)

For example, in Fig. 17.1 $A'A''$ and $B'B''$ are flat surfaces parallel to the two charged flat plates. Any two points on the surface $A'A''$ are at the same electric potential because they are the same distance from the plates. $A'A''$ is therefore an equipotential surface, as is $B'B''$. These surfaces, as can be seen from the diagram, are everywhere perpendicular to the lines of the electric field. This remains true, however complicated the charge distribution.

Relationship between E and V

We now want to establish a relationship between the electric field and the rate at which the electric potential changes with distance.

The work done *by the electric field* in moving a charge from A to B in Fig. 17.1 is

$$\mathscr{W} = \mathbf{F}_{el} \cdot \mathbf{d} = q'Ed$$

since $\mathbf{F}_{el}$ and $\mathbf{d}$ are in the same direction. But, from Eq. (17.4), if the charge moves from A to B, then the work done on the charge by the field must be $\mathscr{W}/q' = V_B - V_A = V_{BA} = -V_{AB}$, and so $\mathscr{W} = -q'V_{AB}$.

Combining these two results for the work, we obtain

$$\mathscr{W} = q'Ed = -q'V_{AB}$$

and so the magnitude of the electric field is

$$E = -\frac{V_{AB}}{d} \qquad\qquad\qquad \textbf{(17.5)}$$

This was derived for the uniform electric field of Fig. 17.1. It is, however, valid for all electrostatic situations, no matter how complicated. Since in many cases $\mathbf{E}$ changes in both magnitude and direction from point to point in the field, to calculate the magnitude of $\mathbf{E}$ at a point we must consider a small change in distance Δs normal to the equipotentials and a corresponding small change in the electric potential, which we designate by ΔV. We then obtain instead of Eq. (17.5),

$$\boxed{E = -\frac{\Delta V}{\Delta s}} \qquad\qquad\qquad \textbf{(17.6)}$$

where $\Delta V/\Delta s$ is the rate at which the electric potential changes with distance normal to the equipotentials. This quantity is called the *gradient* of the potential V. Here the minus sign indicates that E is positive when $\Delta V/\Delta s$ is negative, since $\mathbf{E}$ always points in the direction in which V *decreases*. Thus when ΔV has a minus sign, E must have a plus sign. We therefore say that *the electric field is the negative gradient of the potential*. Both the potential gradient and the electric field are measured in volts per meter or newtons per coulomb.

Example 17.3

Show that the results of Example 17.1 are consistent with Eq. (17.5).

SOLUTION

In Example 17.1 we found that the potential difference between the plates was 70 J/C, and the distance between the plates was 5.00 cm. The field is directed from the positive to the negative plate, and the electric potential therefore *decreases* as we move from the positive to the negative plate. Therefore

$$\frac{V_{AB}}{d} = -\frac{70 \text{ J/C}}{5.00 \times 10^{-2} \text{ m}} = -1.40 \times 10^3 \text{ N/C}$$

and so $E = -V_{AB}/d = \boxed{+1.40 \times 10^3 \text{ N/C}}$. This result is consistent with the statement of Example 17.1.

Metals

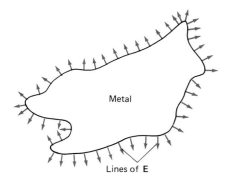

FIGURE 17.2 Electric field lines for an irregularly shaped metal object. The lines of **E** are perpendicular to the surface at every point.

For a solid piece of metal charged with excess electrons, every point in the metal must be at the same potential. If it were not, an electric field would exist because of the change in the potential with distance [$E = -\Delta V/\Delta s$, from Eq. (17.6)]. This field would cause the free electrons to move until the electric field within the metal is reduced to zero. The electrons, as we have learned from Gauss' law, all wind up on the surface of the metal. Every point on this surface is at the same potential, no matter how complicated its shape. The surface of the metal is therefore an equipotential, and the electric field lines are everywhere perpendicular to the surface of the metal, as in Fig. 17.2. An electric field exists, of course, just outside the metal, and the equipotentials outside and near the metal surface have roughly the shape of the metal surface. This conclusion is also true if the metal is positively charged.

17.4 Electric Potential in the Vicinity of a Point Charge

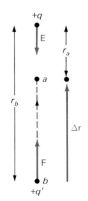

FIGURE 17.3 Two point charges q and q' initially separated by a distance r_b. Work must be done on q' to move it to a new position a distance r_a from q.

Suppose we have two point charges $+q$ and $+q'$ initially separated by a distance r_b, as in Fig. 17.3. If we move $+q'$ to a new position a distance r_a from $+q$, the work we do is

$$\mathcal{W} = \mathbf{F} \cdot \Delta\mathbf{r} = F\,\Delta r$$
$$= q'E(r_b - r_a)$$

or, since the magnitude of the field produced by q is $E = k_e q/r^2$ (Sec. 16.6), we have

$$\mathcal{W} = q'\,\frac{k_e q}{r^2}\,(r_b - r_a)$$

Here r is the distance between the two charges, which varies continuously from r_b to r_a. This means that the force varies continuously along the path, making the calculation difficult. To simplify it, we introduce the approximation that r is the geometrical average of r_a and r_b; that is, $r^2 = r_a r_b$. Then we have

$$\mathcal{W} = \frac{k_e qq'}{r_a r_b}\,(r_b - r_a) = k_e qq'\left(\frac{1}{r_a} - \frac{1}{r_b}\right)$$

The increase in the electric potential energy of the system is thus

$$\Delta\mathrm{PE} = \mathcal{W} = k_e qq'\left(\frac{1}{r_a} - \frac{1}{r_b}\right)$$

and the electric potential difference between the two points is

$$V_{ab} = \frac{\Delta\mathrm{PE}}{q'} = k_e q\left(\frac{1}{r_a} - \frac{1}{r_b}\right) \tag{17.7}$$

where V_{ab} is the potential difference (i.e., the difference in the potential energy per unit charge) between point a and point b in the electric field.

Even though we have derived Eq. (17.7) in an approximate fashion, an exact calculation (using calculus) leads to the same result. Just as an electric field is a characteristic of the space in the vicinity of a source charge, so too is the electric potential. Each is independent of any test charge.

Absolute Potential

As we saw in Sec. 6.3, the zero of gravitational potential energy is completely arbitrary; the only quantity of interest is the *difference* in potential energy between two points at different distances from the earth's center. Similarly, what matters for electric potential energy is the *difference* in potential energy between two points in space. In certain electrical problems, however, it is useful to choose a particular zero for the potential energy in an experimentally reproducible fashion.

For electric circuits it is customary to connect one part of the circuit to ground and take the electric potential energy of a charge at ground potential to be zero. In this way the so-called absolute electric potential at point a (also called simply the potential V_a at point a), is defined as the potential difference between point a and ground potential.

Another way to choose a zero for the electric potential energy is particularly useful in dealing with atoms and molecules. A charge q' in the field of a point charge q has an electric potential energy that depends on its distance from q. We can define the *absolute electric potential energy* of q' as *the work required to bring q' from a very large distance away* (which we usually call infinity) *to the point at which it is located.* This is equivalent to choosing $r_b = \infty$ as the place of zero potential in Fig. 17.3. Then from Eq. (17.7),

$$V_{ab} = V_a - V_b = V_a - 0 = k_e q \left(\frac{1}{r_a} - \frac{1}{\infty} \right) = \frac{k_e q}{r_a}$$

since, as $r_b \to \infty$, $1/r_b \to 0$.

The absolute electric potential V_a at a distance r_a from a point charge q is numerically equal, therefore, to the work done in bringing a unit positive test charge from infinity to that point in the field, and is

$$\boxed{V_a = \frac{k_e q}{r_a}} \tag{17.8}$$

Equation (17.8) is valid whether q is positive or negative. If q is negative, no external work must be done to move $+q'$ toward q, but instead the electric field does work on $+q'$. The electric potential energy therefore decreases, and V_a is negative, as predicted by Eq. (17.8). The total energy in the field is conserved, and, as q' moves toward q, its KE increases as its electric PE decreases.

While the field of a point charge varies as $1/r^2$, the absolute potential in the vicinity of a point charge varies as $1/r$. Since the absolute potential has the same units as potential difference, V_a has units of joules per coulomb or volts.

As an example of the usefulness of equipotential surfaces, consider the field of the point charge in Fig. 17.4. Suppose successive pairs of equipotential surfaces vary in electric potential by a constant ΔV, which might be, say, 10 V. Then the closer together the equipotential surfaces are, the smaller the value of Δs and therefore the greater the magnitude of the electric field, since $E = -\Delta V/\Delta s$. Hence the electric field is large where the equipotential surfaces are crowded together and weak where the equipotential surfaces are far apart. This can be compared to the way contour maps show elevations on the surface of the earth. A

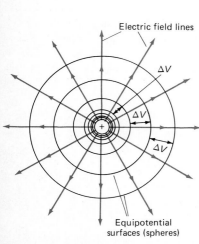

Electric field lines

ΔV

ΔV

ΔV

Equipotential surfaces (spheres)

FIGURE 17.4 Cross-sectional drawing of the field lines and equipotential surfaces for a positive point charge.

crowding together of gravitational equipotential surfaces means a large change in height over a small distance along the earth's surface and thus indicates the presence of mountains or cliffs.

Example 17.4

Consider two point charges $q_1 = +1.00 \ \mu C$ and $q_2 = +2.00 \ \mu C$ separated by 10.0 m, as in Fig. 17.5. **(a)** At what point on the line joining the charges or on its extension is the electric field zero? **(b)** At what point on the same line is the absolute electric potential zero? **(c)** Is there any point in the space surrounding these two charges at which the absolute electric potential is zero? Why?

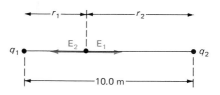

FIGURE 17.5

SOLUTION

(a) Since the electric field is a *vector* and the two charges are positive, the only place where the fields due to the two charges can cancel is on the line between the two charges, for only there are the two electric field vectors in opposite directions.

Let r_1 be the distance from q_1 to the point where the field vanishes, and let $r_2 = 10.0 - r_1$ be the distance from q_2 to the same point. Then we have, from the definition of the electric field,

$$E_1 = \frac{k_e q_1}{r_1^2} \qquad E_2 = \frac{k_e q_2}{r_2^2} = \frac{k_e q_2}{(10.0 - r_1)^2}$$

At the point where the field vanishes, $E_1 = E_2$ in magnitude, and they are in opposite directions. We have, therefore,

$$\frac{k_e(1.00 \ \mu C)}{r_1^2} = \frac{k_e(2.00 \ \mu C)}{(10.0 - r_1)^2} \quad \text{and} \quad (10.0 - r_1)^2 = 2.00 r_1^2$$

from which $2.41 r_1 = 10.0$ m and $r_1 = \boxed{4.15 \ \text{m}}$. Hence the

point where the electric field vanishes is on the line between the two charges and 4.15 m from the smaller charge. There is no other point on the line joining the two charges or on its extension where the field vanishes, since outside the charges the two fields are in the same direction. Of course, as the distance from the charges goes to ∞, $E \to 0$.

(b) The electric potential is a *scalar*. In this case it is equal to $k_e q_1/r_1$ for the first charge and $k_e q_2/r_2$ for the second charge. Since both charges are positive, the sum of the potentials will never be zero. It is only because of the *vector* nature of the electric field that there is one point where the *field* vanishes. This points up the clear distinction between the *vector* electric field and the *scalar* electric potential. It also demonstrates why it is easier to solve electrostatic problems by using potentials (scalars) than by using electric fields (vectors).

(c) From the argument just presented there is no point in space at which the electric potential due to two positive charges is zero, except for a point infinitely far from the two charges, since in that case both potentials go to zero.

Example 17.5

An electron is released from rest at a distance of 0.40 nm from a proton, which is initially also at rest. What is the elec-

tron's kinetic energy (in electronvolts) when it is 0.10 nm from the proton?

SOLUTION

Since the proton has a mass about 2000 times that of the electron, it will move very little in the time the electron moves 0.30 nm. Hence we assume that the proton remains fixed in position and that all the kinetic energy is acquired by the electron.

There are no *external* forces acting on this system of charges, since the electrostatic attraction is *internal* to the system. The electric force is a conservative force, and so, since no external forces act, the total energy (in this case, KE plus electric PE) is conserved:

$$KE_f + PE_f = KE_i + PE_i$$

or

$$KE_f = 0 + PE_i - PE_f$$

$$= \frac{k_e q q'}{0.40 \ \text{nm}} - \frac{k_e q q'}{0.10 \ \text{nm}}$$

$$= k_e q q'(2.5 - 10)\left(\frac{10^9}{\text{m}}\right) = -7.5 \left(\frac{10^9}{\text{m}}\right) k_e q q'$$

$$= -7.5 \left(\frac{10^9}{\text{m}}\right)\left(9.0 \times 10^9 \ \frac{\text{N·m}^2}{\text{C}^2}\right)$$

$$\times (1.6 \times 10^{-19} \ \text{C})(-1.6 \times 10^{-19} \ \text{C})$$

$$= +1.7 \times 10^{-18} \ \text{J} \left(\frac{1.0 \ \text{eV}}{1.6 \times 10^{-19} \ \text{J}}\right) = \boxed{11 \ \text{eV}}$$

17.5   The Storage of Electric Charge

Consider a hollow metal sphere mounted on an insulating stand. We initially ground the metal so that no excess charge remains on it, and then we remove the ground. Next we introduce into the center of the sphere a silk thread with a metal ball at the end, and then fill in the sphere with a metal plug, as in Fig. 17.6. The ball has been previously given a charge $+Q$. What happens to the large hollow sphere as a result of the introduction of the positive charge into the cavity?

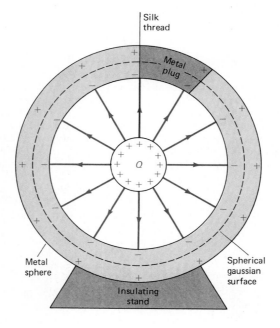

FIGURE 17.6   The introduction of a charge $+Q$ into the interior of a hollow metal sphere.

To answer this question, we construct a spherical gaussian surface confined everywhere to the interior of the metal sphere, as in the figure. Then we have, from Gauss' law,

$$\Phi_E = \sum E_\perp \, \Delta A = 4\pi k_e \sum q$$

But the electric field inside a metal, as we have seen, is everywhere zero, and so the left side of this equation is zero. Hence the right side is also zero, and so $\Sigma q = 0$. Since we know we have a charge $+Q$ on the small ball inside the gaussian surface, we must have a charge $-Q$ distributed over the inner surface of the large sphere to make the total charge within the gaussian surface zero. Electrons are therefore pulled away from the outer surface of the hollow sphere, which is left positively charged, with a charge $+Q$, as shown. The effect of introducing the charge $+Q$ into the interior of the hollow sphere has been to induce a charge $+Q$ on the outer surface of the hollow sphere and a charge $-Q$ on its inner surface.

This interpretation is verified by attaching the outer surface of the sphere to an electroscope. When the charge Q is introduced, the leaves of the electroscope diverge. Moving the charge Q around inside the hollow sphere does not affect the electroscope deflection. Finally, if we touch the ball to the interior of the large sphere, there is still no change in the electroscope's response. This is because the

FIGURE 17.7 A schematic diagram of a Van de Graaff generator. A nonconducting belt runs over the pulleys P and P'. These pulleys are made of different materials so that the belt picks up a positive frictional charge from pulley P and a negative frictional charge from pulley P'. The left belt carries the positive charge to electrode B, which is connected to the interior of the large sphere S. The sphere S therefore develops a large positive charge as the belt rotates. A large voltage is built up between point C on the sphere's surface and a target D at ground potential. Atomic particles can then be accelerated from C to D by this high voltage.

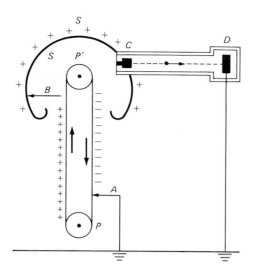

charge $+Q$ on the ball cancels the charge $-Q$ on the interior of the large sphere, leaving a charge $+Q$ still on the outside of the sphere. Since the electroscope responds to this charge on the outer surface, and this charge does not change, there is no change in the electroscope deflection.

The effect of touching the charge $+Q$ to the inside of the sphere is therefore to transfer that charge completely to the outside of the sphere. If we repeat this process over and over again, we can transfer large amounts of charge to the sphere and thus build up a large potential with respect to ground. Since a charged sphere produces an electric field at any point outside the sphere identical to that of a point charge at the center of the sphere, it also produces a potential similar to that of a point charge. Equation (17.8) is therefore valid, and the absolute potential of the sphere is proportional to its charge.

This idea of transferring charge from one conductor to another by internal contact is put to good use in the *Van de Graaff generator* (Fig. 17.7), which in this way builds up a high potential on a large sphere. This high potential can then be used to accelerate atomic particles.

17.6 Capacitors and Capacitance

In the early days of electricity, a crucial piece of apparatus for electrical experiments was the Leyden jar.* This is a glass jar with both inner and outer surfaces covered with metal foil, as in Fig. 17.8. A metal knob at the top is attached through a metal shaft mounted on a nonconducting plug in the neck of the jar to a metal chain making contact with the inside metal foil at the bottom of the jar. A Leyden jar therefore consists of two electric conductors separated by the insulating glass of the jar. It was the first example of what we now call a *capacitor* (sometimes called a *condenser*), which is a device used for storing electric charge.

If the knob of a Leyden jar is repeatedly touched with a negatively charged rod, a large negative charge builds up on the inside metal face. If the outside of

FIGURE 17.8 A Leyden jar. (*Courtesy of Smithsonian Institution.*)

---

*The name *Leyden jar* comes from the University of Leyden in Holland, at which the first capacitor of this kind was used for electrical experimentation.

the jar is grounded, electrons will flow to ground and leave a positive charge on the outer face. This positive charge will be exactly equal to the negative charge on the inside of the jar. Such a device, in which we have equal amounts of positive and negative charge separated by an insulator (glass in this case) is a typical capacitor.

Parallel-Plate Capacitor

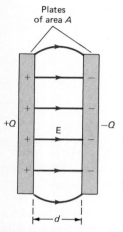

FIGURE 17.9 A parallel-plate capacitor.

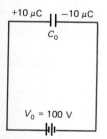

FIGURE 17.10 A capacitor connected across a battery.

Consider a very simple type of capacitor, a parallel-plate capacitor in a vacuum. In this case we have two metal plates, one with a charge $+Q$, the other with a charge $-Q$, at a distance d apart in a vacuum, as in Fig. 17.9. If we vary the amount of charge on the two plates, we find that the potential difference V_0 between the plates increases as the charge on each plate increases, so that

$$Q \propto V_0 \quad \text{or} \quad Q = C_0 V_0$$

where the subscripts on V_0 and C_0 refer to a capacitor with a vacuum between the plates. The constant of proportionality C_0 is called the *capacitance*.

Capacitance: A measure of the ability of a system to store electric charge; the ratio of the charge on a capacitor to the potential difference between its plates.

$$C_0 = \frac{Q}{V_0} \tag{17.9}$$

In Eq. (17.9) the capacitance C_0 depends on the dimensions of the capacitor and the material between the plates (in this case a vacuum). Since $V_0 = Q/C_0$, a large C_0 indicates the ability of the system to store large charges Q without the potential difference becoming too great. The units of capacitance are coulombs per volt, now called *farads* (F) after the British physicist Michael Faraday (1791–1867). Thus 1 F = 1 C/V.

In Fig. 17.10 the vacuum capacitor, indicated by the symbol —||—, has a capacitance

$$C_0 = \frac{Q}{V_0} = \frac{10 \ \mu C}{100 \ V} = 0.10 \times 10^{-6} \ F = 0.10 \ \mu F$$

Gauss' law is useful in determining the value of the capacitance in terms of the dimensions and material of a capacitor. We use Gauss' law to find the electric field at one point between the plates of a parallel-plate capacitor. Since the field is uniform in Fig. 17.9 (if we neglect the fringing at the edges), the electric field is the same at all points between the plates. To find the field at one point, we construct a cylindrical gaussian surface of cross-sectional area ΔA that starts within the left plate and ends at some point between the plates, as in Fig. 17.11. Within the metal plate, where there can be no electric field, the electric flux $\Phi_E = E_\perp \Delta A = 0$. The electric flux is equal to zero for the curved sides of the cylinder, since $\mathbf{E}$ is parallel to the surface, and so $E_\perp \Delta A = 0$. The only place $E_\perp \Delta A$ is not zero is therefore at the right end of the cylinder, where it has the value $E \Delta A$, since the electric field is perpendicular to this end. We have, therefore, from Gauss' law,

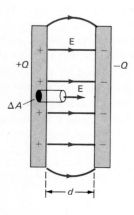

FIGURE 17.11 Gaussian surface for a parallel-plate capacitor. A cylinder of cross-sectional area ΔA, with one end inside the left plate and the other end between the plates, contains an amount of charge $\Delta Q = \sigma \ \Delta A$.

$$\Phi_E = \sum E_\perp \Delta A = E \Delta A = 4\pi k_e \sum q$$

where E is the magnitude of the electric field between the plates. But the charge ΔQ within the gaussian cylinder is $\sigma \Delta A$, where σ is the surface density of charge (in coulombs per square meter) on the right surface of the left plate, and so

$$E \Delta A = 4\pi k_e \sigma \Delta A \qquad \text{or} \qquad E = 4\pi k_e \sigma$$

We now define a new constant

$$\epsilon_0 = \frac{1}{4\pi k_e} \tag{17.10}$$

and we obtain $E = \sigma/\epsilon_0$. Since Q is the total charge on the plate of area A, σ must equal Q/A, and so

$$\boxed{E = \frac{Q}{\epsilon_0 A}} \tag{17.11}$$

Also, since the field between the plates is uniform, $V_0 = Ed$, from Eq. (17.5). On substituting this value for V_0 we obtain

$$C_0 = \frac{Q}{V_0} = \frac{Q}{Ed} = \frac{Q}{Qd/\epsilon_0 A} = \frac{\epsilon_0 A}{d} \tag{17.12}$$

The capacitance of a parallel-plate capacitor is directly proportional to the area of the plates and inversely proportional to the distance between the plates. In SI units A must be in square meters and d in meters to obtain C_0 in farads. Note that C_0 depends only on the size and shape of the capacitor and on ϵ_0; later we will show that because of this dependence on ϵ_0, the capacitance also depends on the material between the plates. It does *not* depend on either Q or V_0.

The constant ϵ_0, the permittivity of a vacuum (sometimes called the *permittivity of free space*), has a value

$$\epsilon_0 = \frac{1}{4\pi k_e} = \frac{1}{4\pi(8.99 \times 10^9 \text{ N·m}^2/\text{C}^2)} = 8.85 \times 10^{-12} \frac{\text{C}^2}{\text{N·m}^2}$$

Since $\epsilon_0 = C_0 d/A$, equivalent units for permittivity are farads per meter. For this reason the value of ϵ_0 in tables is usually given as $\epsilon_0 = 8.85 \times 10^{-12}$ F/m.

The permittivity is an electrical property of the material filling the region between the plates of the capacitor. Thus far we have assumed that this is a vacuum. In Sec. 17.7 we take up the discussion of insulating materials whose electrical properties differ greatly from those of a vacuum.

The permittivity is simply related to the constant k_e in the Coulomb law equation by Eq. (17.10) above. This means that Coulomb's law may be written

$$F = \frac{k_e q q'}{r^2} = \frac{q q'}{4\pi\epsilon_0 r^2}$$

This shows us that the larger the permittivity of the material between the charges (which is ϵ_0 only for a vacuum), the smaller the force between the two charges. This is because the electric field *polarizes* the material (see Sec. 16.3) filling the space between q and q', and this induced polarization reduces the force between q and q'. This happens, for instance, when an insulating material like glass or paper is used between the plates of a capacitor.

Example 17.6

A parallel-plate capacitor consists of two metal plates, each of area 150 cm^2, separated by a vacuum space 0.60 cm thick. **(a)** What is the capacitance of such a device? **(b)** What potential difference must be applied to the plates if the capacitor is to hold a charge of 1.00×10^{-3} μC?

SOLUTION

(a) The capacitance is given by Eq. (17.12):

$$C_0 = \frac{\epsilon_0 A}{d} = \left(8.85 \times 10^{-12} \frac{C^2}{N \cdot m^2}\right)\left(\frac{150 \times 10^{-4} \text{ m}^2}{0.60 \times 10^{-2} \text{ m}}\right)$$

$$= 2.2 \times 10^{-11} \text{ C}^2/(\text{N} \cdot \text{m}) = 22 \times 10^{-12} \text{ F} = \boxed{22 \text{ pF}}$$

since $\dfrac{1 \text{ C}^2}{\text{N} \cdot \text{m}} = \left(\dfrac{1 \text{ C}^2}{\text{N} \cdot \text{m}^2}\right) \text{m} = \left(\dfrac{1 \text{ F}}{\text{m}}\right) \text{m} = 1 \text{ F}$

(b) Since $C_0 = Q/V_0$, $V_0 = Q/C_0$, and

$$V_0 = \frac{1.00 \times 10^{-9} \text{ C}}{22 \times 10^{-12} \text{ F}} = \boxed{45 \text{ V}}$$

17.7 Dielectrics

We now want to see what happens when the material between the plates of a capacitor, like the one in Fig. 17.11, is not a vacuum but an insulating material with permittivity ϵ. Such a material, when used in capacitors, is commonly called a *dielectric*.

Suppose we have the parallel-plate capacitor of Fig. 17.11 with its plates charged but disconnected from any battery or other source of charge. We first measure the voltage between the plates with an electrometer. Then we push in a slab of dielectric material such as glass to fill the region between the plates. The voltage is observed to decrease, although the charge on the plates remains exactly what it was before. The presence of the dielectric *increases* the capacitance of the capacitor over its value for a vacuum, since $C = Q/V$. Whereas for a vacuum we had $C_0 = Q/V_0$, we now have a larger capacitance

$$C = \frac{Q}{V} = KC_0 \tag{17.13}$$

where $K = C/C_0$, *the ratio of the capacitance with a dielectric between the plates to the capacitance with vacuum between the plates*, is called the *dielectric constant of the material*. The process just outlined gives us an operational way of defining exactly what we mean by the dielectric constant. For a vacuum $K = 1$, and for air it is only 0.06 percent greater than for a vacuum. Values for some common dielectrics are given in Table 17.1.

The capacitance of our new capacitor with dielectric inserted is then

$$C = KC_0 = \frac{K\epsilon_0 A}{d} = \frac{\epsilon A}{d} \tag{17.14}$$

where *the permittivity of the dielectric* is

$$\epsilon = K\epsilon_0 \tag{17.15}$$

Since K is always greater than 1, the presence of the dielectric *always increases* the capacitance over that of a vacuum.

From a slightly different viewpoint, *the effect of a dielectric is to increase the amount of charge that can be stored in a capacitor without increasing the capac-*

TABLE 17.1 Dielectric Constants of Various Materials*

Material	Dielectric constant $K = \epsilon/\epsilon_0$
Vacuum	1.00000
Air	1.00059
Water	80
Paper	3.5
Porcelain	6.5
Bakelite	4.8
Quartz	4.3
Pyrex glass	4.5
Amber	2.7
Polystyrene	2.6
Teflon	2.1
Transformer oil	2.2
Titanium dioxide	~100
Barium titanate	~10,000

*For 20°C and unvarying or very low frequency electric fields.

itor's voltage, since $C = Q/V$. This is the reason the Leyden jar was so useful as a charge-storage device. The glass in the jar is thin and has a dielectric constant of about 5; hence it is a good way to store charge.

Molecular View of Dielectrics

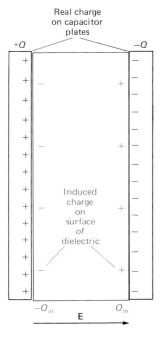

How do we explain the reduction of voltage that occurs when we insert a dielectric between the plates of a parallel-plate vacuum capacitor? To understand this, we must introduce the molecular structure of dielectric materials. Some dielectrics (water) consist of molecules that are permanent electric dipoles; i.e., the centers of positive and negative charge are slightly displaced in the molecule. Other dielectrics (glass) have no permanent dipoles, but dipoles can be induced by electric fields. In either case, when a dielectric is placed between the plates of a charged capacitor, these molecular dipoles line up in the field of the charged plates, and the ends of the dielectric near the plates become charged with induced charges Q_{in} of the sign shown in Fig. 17.12. (The interior of the dielectric remains electrically neutral. It is only at the ends that the induced charges become important.)

These induced charges therefore produce a potential difference V_{in} that is the reverse of the potential difference caused by the free charges on the capacitor plates. The effective potential difference across the capacitor is therefore reduced to $V = V_0 - V_{in}$. More careful analysis leads to the conclusion that $V = V_0/K$. We see that the molecular structure of the dielectric fully explains our observations of what happens when a dielectric is inserted between the plates of a capacitor.

FIGURE 17.12   A dielectric between the charged plates of a parallel-plate capacitor. The induced charge Q_{in} on the dielectric reduces the effective charge on the capacitor plates from Q to $Q - Q_{in}$. This reduces the effective voltage across the plates from V_0 to V_0/K.

Practical Capacitors

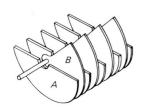

FIGURE 17.13   A tuning capacitor for a radio. Plates *B* rotate with respect to plates *A* to change the effective area of the plates.

Dielectrics also make it possible to bring the plates of a capacitor very close together, thus increasing the capacitance, and reduce the likelihood that high voltages on the plates will produce ionization of the air and an electric discharge from one plate to the other.

Commercial capacitors often consist of sheets of metal foil separated by thin sheets of plastic or other dielectric material. These can be rolled into cylinders with large surface areas A and plates very close together so that d is small. They therefore have large capacitances, although the capacitance usually remains in the microfarad (μF) range.

Capacitors are often used as circuit elements in ac circuits (Chap. 21), for example, in radios. Capacitors in radios can be used in the receiving circuit to "tune" the receiver to resonance with the incoming signal. Tuning capacitors vary the capacitance by moving one set of charged plates with respect to a second, stationary set of plates, as in Fig. 17.13. This changes the effective area of the plates and therefore the capacitance. The capacitance of radio capacitors is usually between 1 microfarad (10^{-6} F) and 1 picofarad (10^{-12} F).

Example 17.7

A parallel-plate capacitor has a plate area of 50 cm^2 and a plate separation of 1.0 cm. A potential difference $V_0 = 200$ V is applied across the plates with no dielectric present. The battery is then disconnected and a piece of Bakelite inserted that fills the volume between the plates. What is **(a)** the capacitance, **(b)** the charge on the plates, and **(c)** the potential difference between the plates before and after the dielectric is inserted?

SOLUTION

(a) Before the dielectric is inserted, air fills the space between the plates. Since the dielectric constant of air is very close to unity,

$$C_0 = \frac{\epsilon_0 A}{d} = \frac{(8.85 \times 10^{-12} \text{ C}^2/\text{N·m}^2)(50 \times 10^{-4} \text{ m}^2)}{0.010 \text{ m}}$$

$$= 4.4 \times 10^{-12} \text{ F} = \boxed{4.4 \text{ pF}}$$

After the dielectric is inserted, using the value $K = 4.8$ from Table 17.1, we have

$$C = KC_0 = 4.8(4.4 \text{ pF}) = \boxed{21 \text{ pF}}$$

(b) Before the dielectric is inserted,

$$Q_0 = C_0 V_0 = (4.4 \times 10^{-12} \text{ F})(200 \text{ V}) = 880 \times 10^{-12} \text{ C}$$

$$= \boxed{8.8 \times 10^{-10} \text{ C}}$$

After the dielectric is inserted the charge *must remain the same*, since the battery is disconnected and the charge cannot

leave the plates. Hence, since electric charge is always conserved,

$$Q = Q_0 = \boxed{8.8 \times 10^{-10} \text{ C}}$$

(c) The potential difference before the dielectric is inserted is given as

$V_0 = \boxed{200 \text{ V}}$. The potential difference after the Bakelite is inserted is then

$$V = \frac{Q}{C} = \frac{8.8 \times 10^{-10} \text{ C}}{21 \times 10^{-12} \text{ F}} = \boxed{42 \text{ V}}$$

The same result may be found by using the relationship $Q = CV = C_0 V_0$, since charge is conserved. Therefore $V/V_0 = C_0/C = 1/K$, from Eq. (17.13), and so

$$V = \frac{V_0}{K} = \frac{200 \text{ V}}{4.8} = 42 \text{ V}$$

as above.

17.8 Capacitors in Series and in Parallel

In practical electric circuits combinations of capacitors often occur. For this reason it is useful to be able to find the equivalent capacitance corresponding to various electrical combinations of single capacitors. In this section we consider two capacitors and the ways in which they can be electrically connected: in *series* and in *parallel*.

Capacitors in Parallel

Figure 17.14 shows two capacitors connected *in parallel*, i.e., with the positively charged plates of the two connected by a metal wire and their negatively charged plates also connected by a wire. In this case the potential difference across the two capacitors is the same, since the same battery is connected across the two. The charge Q, however, is divided between the two capacitors because it must distribute itself on the plates of the capacitors in such a way that the voltage is the same across the two. Since the two capacitors may have different capacitances, their charges Q_1 and Q_2 may also be different.

We now desire to find the equivalent capacitance of this parallel combination, i.e., the capacitance of a single capacitor that would produce the same electrical effect as the parallel combination. This is simply $C_{eq} = Q/V$, where V is the potential difference applied by the battery and $Q = Q_1 + Q_2$, since we want the equivalent capacitor to store the same amount of charge as the two capacitors separately would. We have, therefore, for the equivalent capacitance,

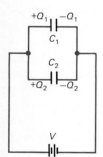

FIGURE 17.14 Two capacitors in parallel: $C = C_1 + C_2$.

$$C_{eq} = \frac{Q}{V} = \frac{Q_1 + Q_2}{V} = \frac{Q_1}{V} + \frac{Q_2}{V} = C_1 + C_2$$

or

$$\boxed{C_{eq} = C_1 + C_2}$$

(17.16)

For capacitors in parallel, the equivalent capacitance is the sum of the individual capacitances.

For n capacitors in parallel, therefore, $C = \sum_n C_n$.

Capacitors in Series

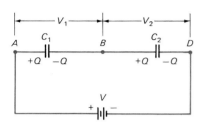

FIGURE 17.15 Two capacitors in series: $1/C = 1/C_1 + 1/C_2$.

Figure 17.15 shows two capacitors, of capacitance C_1 and C_2, combined *in series*, i.e., in a line so that the positive plate of one is connected to the negative plate of the other. We desire to find the equivalent capacitance of this series combination.

In this case the potential difference between the two terminals of the battery is, from Fig. (17.15),

$$V = V_{AD} = V_A - V_D = (V_A - V_B) + (V_B - V_D) = V_1 + V_2$$

So we see that the potential difference provided by the battery is the sum of the potential differences across the two capacitors. In this case V_1 and V_2 may be different, but the charge on each capacitor must be the same, since charge flows until the left plate of capacitor 1 and the right plate of capacitor 2 have the same final charge. These charged plates induce equal and opposite charges on the interior plates of the two capacitors, since the net charge must be conserved, so that each capacitor ends up with an identical charge $Q = Q_1 = Q_2$.

Since, from above, $V = V_1 + V_2 = Q/C_1 + Q/C_2$, and the equivalent capacitance of the series combination is $C_{eq} = Q/V$, we have

$$\frac{V}{Q} = \frac{1}{C_{eq}} = \frac{1}{C_1} + \frac{1}{C_2}$$

and so
$$\boxed{\frac{1}{C_{eq}} = \frac{1}{C_1} + \frac{1}{C_2}}$$
  (17.17)

For capacitors in series, the reciprocal of the equivalent capacitance is the sum of the reciprocals of the individual capacitances.

For n capacitors in series, therefore, $1/C = \sum_n (1/C_n)$.

Example 17.8

(a) What is the equivalent capacitance between points A and B in Fig. 17.16? **(b)** What is the charge on the 4.00-μF capacitor? **(c)** What is the charge on each 1.00-μF capacitor?

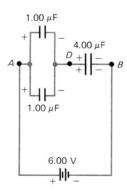

FIGURE 17.16

SOLUTION

(a) The equivalent capacitance of the two 1.00-μF capacitors in parallel is $C_{eq} = C_1 + C_2 = 2.00\ \mu F$. When C_{eq} is combined in series with the 4.00-μF capacitor, we have for the equivalent capacitance between points A and B,

$$\frac{1}{C_{AB}} = \frac{1}{2.00\ \mu F} + \frac{1}{4.00\ \mu F} = \frac{3}{4.00\ \mu F}$$

and so $\quad C_{AB} = \dfrac{4.00}{3}\ \mu F = \boxed{1.33\ \mu F}$

(b) The charge delivered by the 6.00-V battery is then

$$Q = C_{AB}V = (1.33\ \mu F)(6.00\ V) = \boxed{8.00\ \mu C}$$

This is the charge on the 4.00-μF capacitor, since the negative terminal of the battery is connected directly to the capacitor.

(c) The total charge on the left plates of the two 1.00-μF capacitors must also be 8.00 μC, because the battery provides equal amounts of positive and negative charge. Since the two 1.00-μF capacitors are identical, each must have a charge of

$$\boxed{4.00\ \mu C.}$$

17.9 The Energy Stored in a Capacitor

Work must be done in charging a capacitor, since charges must be moved from one plate to the other against the force set up by the charges already on the two capacitor plates. This work becomes stored electric energy in the capacitor. We want to find how much energy is stored in a capacitor when its final charge is Q (that is, $+Q$ on one plate and $-Q$ on the other) and when the potential difference between the plates is V.

The work done in charging the capacitor is the work required to move the total charge Q from one plate to the other against the potential difference between the plates. Since the potential difference between the plates varies linearly from 0 to V as the charge on the plates changes from 0 to Q, the average value of the potential difference is $V/2$. The work done is, therefore, from Eq. (17.4),

$$\mathcal{W} = Q(\tfrac{1}{2}V) = \tfrac{1}{2}QV \tag{17.18}$$

This then is the energy stored in the capacitor. Equation (17.18) turns out to be true in general for any capacitor.

Since $C = Q/V$, we may also write *the energy stored in a capacitor* in the following equivalent ways:

$$\boxed{\mathcal{W} = \tfrac{1}{2}QV = \tfrac{1}{2}CV^2 = \frac{1}{2}\frac{Q^2}{C}} \tag{17.19}$$

Energy in the Electric Field

It is possible to use Eq. (17.19) to find the energy stored in an electric field. Let us concentrate on a parallel-plate capacitor with a substance of dielectric constant K between the plates. Then its capacitance is, from Eq. (17.14), $C = \epsilon A/d$. Also the potential difference between the plates of a parallel-plate capacitor is $V = Ed$, where E is the magnitude of the electric field and d is the distance between the plates. Hence the energy stored in the capacitor is, from Eq. (17.19),

$$\mathcal{W} = \tfrac{1}{2}CV^2 = \frac{1}{2}\frac{\epsilon A}{d}E^2 d^2 = \tfrac{1}{2}\epsilon E^2 Ad$$

But Ad is the volume between the plates, and so the energy stored per unit volume, or the *energy density w*, is

$$w = \frac{W}{Ad} = \tfrac{1}{2}\epsilon E^2 \qquad\qquad (17.20)$$

The electric energy stored per unit volume in a region between the plates is therefore $\tfrac{1}{2}\epsilon E^2$, where $\epsilon = K\epsilon_0$ is the permittivity of the material in which the field is located. In a vacuum $w = \tfrac{1}{2}\epsilon_0 E^2$.

This expression for the energy density, while derived here for a special case, can be shown to be valid for any region of space in which an electric field exists. It will turn out to be of considerable importance when we discuss the propagation of electromagnetic energy through space.

Example 17.9

A parallel-plate capacitor with air between its plates has a capacitance of 5.0 pF. A potential difference of 100 V is applied by a storage battery across the plates, which are 1.0 cm apart. **(a)** What is the energy stored in the capacitor? **(b)** The battery is disconnected and one plate is moved until it is 2.0 cm from the other plate. What is the energy stored in the capacitor now? **(c)** What is the energy stored in the capacitor when the battery is left connected and one plate is again moved to a position 2.0 cm from the other plate?

SOLUTION

(a) The energy stored in the air capacitor is

$$W = \tfrac{1}{2}CV^2 = \tfrac{1}{2}(5.0 \times 10^{-12} \text{ F})(100 \text{ V})^2 = \boxed{2.5 \times 10^{-8} \text{ J}}$$

(b) The energy stored in a capacitor is also given by $Q^2/2C$. Hence if the battery is disconnected so that the charge remains unchanged, the energy varies inversely with the capacitance. In this case C is reduced by a factor of 2 [since d in Eq. (17.12) is doubled when the plate is moved]. We therefore have

$$W' = \frac{Q^2}{2C'} = \frac{Q^2}{2C/2} = \frac{Q^2}{C} = 2W$$

The stored energy has been doubled by pulling the plates apart, and

$$W' = 2(2.5 \times 10^{-8} \text{ J}) = \boxed{5.0 \times 10^{-8} \text{ J}}$$

The increase in the energy is equal to the work done in separating the plates against the electric attraction pulling them together.

(c) If the battery is left connected, $V' = V$, and so

$$W' = \tfrac{1}{2}C'V^2 = \frac{1}{2}\left(\frac{C}{2}\right)V^2 = \tfrac{1}{2}(\tfrac{1}{2}CV^2) = \tfrac{1}{2}W$$

In this case the stored energy is *reduced* to

$$\tfrac{1}{2}(2.5 \times 10^{-8} \text{ J}) = \boxed{1.3 \times 10^{-8} \text{ J}}$$

Even though the stored energy in the capacitor has been reduced, positive work has been done to pull the plates apart against the force pulling them together. Where does this work go? Both this work and half the original stored energy in the capacitor must be returned to the battery if energy is to be conserved. This energy can be used to increase the chemical potential energy of the battery. This is called *charging* the battery, a term that is somewhat inappropriate, since the battery's net charge remains zero.

Summary: Important Definitions and Equations

Electric potential: The potential energy per unit charge of a charge q' at a point P in an electric field:

$$V_P = \frac{\text{PE}_P}{q'} \qquad\qquad (17.3)$$

Potential difference (also called **voltage**): The difference in potential energy per unit charge between two points in an electric field. If W is the work done by the field in moving a charge q from A to B in the field, then the potential difference between A and B is

$$V_{AB} = V_A - V_B = \frac{\text{PE}_A}{q'} - \frac{\text{PE}_B}{q'} = \frac{W}{q'} \qquad (17.4)$$

Volt (unit of potential difference or voltage): If 1 J of work is done to move 1 C of charge from one point to another in an electric field, then the potential difference between these two points is 1 V; that is, $1 \text{ V} = 1 \text{ J/C}$.
Electronvolt: The work required to move one elementary electric charge through a potential difference of one volt. $1 \text{ eV} = 1.60 \times 10^{-19} \text{ J}$.

Equipotential surfaces: Imaginary surfaces in space that are at every point perpendicular to the electric field, and all points of which are therefore at the same potential.

Relationship between the magnitude of the electric field E and the potential V

$$E = -\frac{\Delta V}{\Delta s} \tag{17.6}$$

Absolute potential of a point charge: The work done in bringing a unit positive test charge from infinity up to a distance r_a from the source charge q:

$$V_a = \frac{k_e q}{r_a} \tag{17.8}$$

Capacitor: A device for storing electric charge, consisting of two charged conductors separated by a dielectric.

Capacitance: The ratio of the charge on a capacitor to the potential difference between its plates:

$$C = \frac{Q}{V} \tag{17.13}$$

The unit of capacitance is the farad (F), where 1 F = 1 C/V.

Capacitance of a parallel-plate capacitor

In vacuum: $C_0 = \dfrac{\epsilon_0 A}{d}$ \qquad (17.12)

With dielectric: $C = \dfrac{\epsilon A}{d}$ \qquad (17.14)

Dielectric constant: The ratio of the permittivity of a material to the permittivity of a vacuum:

$$K = \frac{\epsilon}{\epsilon_0} \tag{17.15}$$

Electric field between the plates of a parallel-plate vacuum capacitor of charge Q and plate area A

$$E = \frac{Q}{\epsilon_0 A} \tag{17.11}$$

where $\epsilon_0 = 1/4\pi k_e$, is the permittivity of a vacuum (or free space) and is equal to 8.85×10^{-12} F/m.

Capacitors in series

$$\frac{1}{C_{eq}} = \frac{1}{C_1} + \frac{1}{C_2} \tag{17.17}$$

Capacitors in parallel

$$C_{eq} = C_1 + C_2 \tag{17.16}$$

Energy stored in a capacitor

$$\mathcal{W} = \tfrac{1}{2}QV = \tfrac{1}{2}CV^2 = \frac{1}{2}\frac{Q^2}{C} \tag{17.19}$$

Energy density in an electric field

$$w = \frac{\mathcal{W}}{Ad} = \tfrac{1}{2}\epsilon E^2 \tag{17.20}$$

Questions

1 Two point charges are a distance d apart. The electric field is zero at one point on the line connecting the two charges. What conclusion can you draw about the sign and magnitude of the charges?

2 Two point charges are a distance d apart. The electric potential is zero at one point on the line connecting them. What conclusion can you draw about the magnitude and sign of the two charges?

3 If the electric field is zero at a particular point in space, does this mean that the electric potential V is also zero at that point? Give an example to prove your point.

4 If the electric potential in a given region of space is constant, what is the electric field in that region?

5 A large hollow metal sphere is charged to a potential of 1000 V.
  (a) What is the electric potential inside the sphere?
  (b) What is the electric field inside the sphere?

6 Can two different equipotential surfaces intersect? Why?

7 Is the surface of the earth an equipotential surface for the gravitational force? If this is true, is it exactly true or is it only an approximation?

8 A physicist decides to call the potential of the earth (i.e., ground potential) $+2000$ V instead of zero. Is this possible without invalidating the basic laws of electrostatics? What effect would such an assumption have on measured values of absolute potentials and of potential differences?

9 If a capacitor is connected directly across a battery, does each of its plates receive exactly the same charge? Why? Is this true even if the plates are of very different sizes?

10 (a) What happens to the charge, the voltage, and the capacitance when a slab of glass is inserted between the plates of a charged parallel-plate capacitor with the voltage source disconnected? Assume that the plate separation is equal to the thickness of the glass, which has the same area as the plates.
  (b) Repeat part (a) with the glass slab only half as thick as the distance between the plates.

11 A capacitor is charged by a battery, which is then disconnected. If a dielectric slab is inserted between the plates, must work be done to push the slab into its final position? Justify your answer.

12 A capacitor is charged by a battery that remains connected to the plates. If a dielectric slab is inserted between the plates, must work be done to push the slab into its final position? Justify your answer.

13 Is the resulting capacitance of two capacitors connected in series greater or less than the capacitance of the larger of the two capacitors? Than the capacitance of the smaller of the two capacitors?

Multiple-Choice and Simple Exercises

17.1 The radius of a spherical gold nucleus is about 6.6×10^{-15} m, and the nucleus contains 79 protons. The electric potential at the surface of the nucleus is:

   (a) 1.7×10^7 J   (b) 1.7×10^7 C
   (c) 2.6×10^{21} V   (d) 1.7×10^7 V
   (e) 1.9×10^{-3} V

17.2 The potentials for two equipotential surfaces between the charged plates of a parallel-plate capacitor are $V_A = 30$ V and $V_B = 20$ V, where A is 4.0 cm from the left plate and B is 5.0 cm from the same plate. With a minus sign meaning that the field is directed toward the left plate, the electric field in the region between the plates is:

   (a) $+25$ V/m   (b) $+10^3$ V/m   (c) -10^3 V
   (d) $+10^3$ V   (e) -10^3 V/m

17.3 A potential difference of 120 V drives 10 C of electric charge through an electric heater each second. The work done by the voltage source in 1 s is:

   (a) 1.2×10^3 J   (b) 1.2×10^3 eV
   (c) 2.4×10^3 J   (d) 0.60×10^3 J
   (e) 1.2×10^{-3} J

17.4 Which of the following units could be properly used to specify a voltage of 1 V (C here means coulomb)?

   (a) 1 eV   (b) 1 J   (c) 1 N/C   (d) 1 C/J
   (e) 1 C/F

17.5 Which of the following is *not* an energy unit (C here means coulomb)?

   (a) Calorie   (b) Electronvolt
   (c) Volt   (d) N·m   (e) C·V

17.6 To increase the capacitance of a parallel-plate capacitor by two orders of magnitude, the diameter of its round plates must be increased by:

   (a) One order of magnitude
   (b) Two orders of magnitude
   (c) Four orders of magnitude
   (d) Three orders of magnitude
   (e) None of the above

17.7 The capacitance of an air capacitor with plates of area 5.0 cm² separated by a distance of 1.0 mm is:

   (a) 4.4 pF   (b) 4.4 μF   (c) 4.4×10^{-10} F
   (d) 1.8 pF   (e) 1.8 μF

17.8 A 3.0-μF capacitor is put in series with a 4.0-μF capacitor. The resulting capacitance of the combination is:

   (a) 7.0 μF   (b) 1.0 μF   (c) 0.58 μF
   (d) 1.71 μF   (e) 0.14 μF

17.9 A parallel-plate capacitor with a vacuum between the plates has a capacitance of 3.0 μF. A second parallel-plate capacitor has plates of twice the area of the first, spaced four times as far apart, and the same capacitance. The dielectric constant of the material between the plates of the second capacitor is:

   (a) 1   (b) 2   (c) 4   (d) 8   (e) 16

17.10 A 100-pF capacitor is charged to a potential difference of 5000 V. The energy stored in the capacitor is:

   (a) 2.5×10^{-7} J   (b) 1.25×10^{-3} J
   (c) 2.5×10^{-3} J   (d) 1.25×10^3 V
   (e) 1.25×10^{-3} V

17.11 How much work does a 6.0-V battery do in moving a single electron through the battery from its positive to its negative terminal?

17.12 (a) If 10 J of work must be done to move 0.050 C of positive charge from point A to point B in an electrostatic field, what is the difference in potential between A and B?

   (b) Which point, A or B, is at the higher potential?

17.13 The positive terminal of a dry-cell battery is $+1.5$ V with respect to the negative terminal. The two terminals are 4.0 cm apart. What are the magnitude and direction of the electric field along a straight line connecting the two terminals?

17.14 How many joules are there in 1.0 MeV?

17.15 A 20-pF capacitor is charged by a voltage of 400 V. What is the charge on the capacitor?

17.16 What is the capacitance of a parallel-plate capacitor with air between the plates, a plate area equal to 0.010 m², and a plate separation of 1.0 mm?

17.17 Two capacitors, one of capacitance 20 μF and the other of capacitance 30 μF, are connected in series. What is the capacitance of the combination?

17.18 Two capacitors, one of capacitance 15 μF and the other of capacitance 25 μF, are connected in parallel. What is the capacitance of the combination?

17.19 A capacitor has a capacitance of 2.0 μF and is charged to 1000 V. How much energy is stored in the capacitor?

17.20 What is the energy density between the plates of a parallel-plate capacitor when the voltage across the plates is 500 V and the distance between the plates is 1.0 mm (in air)?

Problems

17.21 A hydrogen atom consists of an electron moving in a circular orbit about a nucleus, which is a single proton. The electron's distance from the proton is 0.53×10^{-10} m.

   (a) Find the electric potential energy of the electron in the field of the proton.

   (b) What is the absolute electric potential at 0.53×10^{-10} m from the proton?

17.22 The following point charges are located along the x axis: $+3.0$ μC at $x = +20$ cm, -4.0 μC at $x = +35$ cm, $+5.0$ μC at $x = +50$ cm. What is the absolute potential at the point $x = 100$ cm?

17.23 Two charged spheres, one of charge $+100$ μC and the other of charge -100 μC, are held in fixed positions with their centers separated by 20.0 cm.

   (a) What is the electric potential energy of the two charges?

   (b) What is the absolute electric potential at the -100-μC charge?

17.24 Two point charges, $q_1 = +2.0\ \mu C$ and $q_2 = -5.0\ \mu C$, are separated by 2.0 m.

(a) At what point on the line joining the charges, or on its extension, is the electric field zero?

(b) At what point on the same line is the absolute electric potential zero?

(c) If both charges were positive, would there be a point on the line joining the charges where the electric field would be zero? Where would it be located?

(d) If both charges were positive, would there be a point where the absolute electric potential would be zero?

17.25 In a modern version of Millikan's experiment, polystyrene spheres of mass 1.60×10^{-14} kg were used. It was found that a potential difference of 4000 V across two horizontal plates 4.00 mm apart was able to balance out the force of gravity on a sphere. If we assume that the sphere contained one electronic charge, what is the value for the electronic charge?

17.26 The sphere of a Van de Graaff generator is to be charged to a potential of 10^6 V. The dielectric strength of air is 3.0×10^6 V/m, which means that an electric field greater than 3.0×10^6 V/m will cause the air to ionize and allow charge to escape from the sphere. What is the minimum radius the sphere must have to retain its charge?

17.27 (a) How much energy is required to separate completely the two ions in the potassium iodide molecule, K^+I^-, where the ions are initially 3.0×10^{-10} m apart in air?

(b) How would this energy differ if the potassium iodide were immersed in water?

17.28 An electron is located midway between two parallel metal plates 2.0 cm apart. One of the plates is at a potential 300 V higher than the other.

(a) What is the potential gradient between the plates?

(b) What is the magnitude of the force on the electron?

(c) What is the direction of this force?

17.29 An electron starts from rest and is accelerated by a potential difference of 500 V. What is its final kinetic energy in (a) joules and (b) electronvolts? (c) What is its final speed?

17.30 A positive charge of magnitude $+12\ \mu C$ is at point A.

(a) What is the absolute potential at point B 20 cm from A?

(b) What is the absolute potential at point C 45 cm from A?

(c) How much work is required to carry a $+0.10\text{-}\mu C$ charge from C to B?

(d) Does it make any difference what path is chosen in going from C to B? Why?

17.31 In laboratory courses in college physics the oil drops in Millikan's experiment are frequently replaced by tiny polystyrene spheres of known density and radius. In this way the mass of the spheres can be accurately obtained, since $m = \rho V = \frac{4}{3}\pi R^3 \rho$. In one such experiment, the mass of a particular sphere was determined in this way to be 1.60×10^{-14} kg. The sphere was first balanced in an electric field by applying a voltage of 2035 V across two horizontal plates 2.00 mm apart, with the voltage positive on the upper plate. After this measurement, a radioactive source was used to change the charge on the sphere; each time the charge was changed, the new voltage required for equilibrium was measured. It was found that the sphere could be held at rest by voltages of the following size: 995, 503, 390 V.

Use these experimental data to calculate the best possible value for the elementary charge e.

17.32 A parallel-plate capacitor with large plates 2.5 cm apart is connected to a 500-V power supply. After the connection to the power supply has been broken:

(a) Find the electric field between the plates.

(b) Find the force on an electron in the region between the plates.

(c) Find the change in the electric potential energy of the electron as it moves from the negative to the positive plate.

(d) Find the kinetic energy of the electron after it has traversed the 2.5 cm distance between the plates.

17.33 (a) Calculate the electric capacitance of the earth.

(b) How many excess electrons must there be on the earth to raise its potential by 1.0 V?

17.34 (a) What is the capacitance of the moon?

(b) What would the electric potential of the moon be if it were charged to 100 μC?

17.35 Prove that 1 F = 1 $C^2/(N \cdot m)$.

17.36 The plates of a parallel-plate capacitor are of area 0.50 m^2 and are 4.0 mm apart. If a voltage of 10^5 V is applied across the capacitor, which has air between the plates; what are (a) the capacitance, (b) the charge on each plate, and (c) the electric field between the plates?

17.37 A parallel-plate capacitor has two plates, each of area 0.030 m^2, separated by a 0.30-cm air gap.

(a) Find the capacitance.

(b) If the capacitor is connected across a 600-V source, find the charge on its plates.

(c) Find the electric field between the plates.

17.38 (a) Calculate the capacitance of a parallel-plate capacitor with a 0.50-cm-thick layer of Bakelite between the plates, each of which has an area of 0.010 m^2.

(b) If this capacitor is connected to a 600-V source, calculate the charge on each plate of the capacitor.

17.39 A parallel-plate capacitor has a plate area of 100 cm^2 and a plate separation of 4.0 mm. A potential difference of 1000 V is applied across the plates with only air between the plates. The battery is then disconnected, and a piece of Pyrex glass inserted to fill completely the space between the plates. Both before and after the dielectric is inserted, what are (a) the capacitance, (b) the charge on the plates, and (c) the potential difference across the plates?

*17.40 A 10.0-μF capacitor is charged to 100 V, and a 5.0-μF capacitor is charged to 150 V. The capacitors are then disconnected from the voltage source. The positive plates are connected together, and the negative plates are also connected together.

(a) What is the potential difference across each capacitor?

(b) What is the charge on each capacitor? (*Hint:* Charge must be conserved.)

*17.41 A spherical capacitor consists of an inner metal sphere of radius R_a supported on an insulating stand at the center of a hollow metal sphere of inner radius R_b. The inner sphere has a charge $+q$ and the outer sphere a charge $-q$. Prove that the capacitance of the two spheres is

$$C = \frac{1}{k_e}\frac{R_b R_a}{R_b - R_a}$$

17.42 A voltage of 50 V is connected across three capacitors in series. The charge delivered by the battery to the three capacitors is 5.0 μC. If two capacitors have capacitances of $C_1 = 0.50$ μF and $C_2 = 0.25$ μF, what is the value of C_3?

17.43 A 0.30-μF capacitor is connected in parallel with a 0.15-μF capacitor. This combination is then connected in series with a 0.40-μF capacitor. What is the equivalent capacitance of the combination?

17.44 Two capacitors, one of capacitance 5.0 μF and the other of capacitance 15.0 μF, are connected in series across 1500 V.

(*a*) What is the equivalent capacitance C of the series combination?

(*b*) What is the charge on each capacitor?

(*c*) What is the potential difference across each capacitor?

(*d*) What is the energy stored in each capacitor?

17.45 Repeat Prob. 17.44 for the case where the two capacitors are connected in parallel instead of series.

17.46 What is the equivalent capacitance between points A and B in Fig. 17.17?

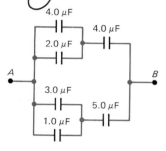

FIGURE 17.17 Diagram for Prob. 17.46.

17.47 We have two capacitors, one of 10 μF with a charge of 50 μC, the other uncharged with a capacitance of 20 μF. If the two capacitors are connected in parallel, what is the final charge on each capacitor?

17.48 A parallel-plate capacitor in air has a capacitance of 12 μF. It is charged up by a 120-V power supply, which is then disconnected.

(*a*) What is the charge on the capacitor?

(*b*) If a slab of Bakelite is slid between the plates and fills the entire volume of the capacitor, what are the new capacitance, voltage across the plates, and charge on the plates?

17.49 A parallel-plate capacitor with plates separated by air receives 10 μC of charge when connected to a 300-V battery. The plates, still connected to the battery, are then immersed in transformer oil. How much additional charge must flow from the battery to the plates of the capacitor?

17.50 (*a*) A 15-μF capacitor with air between the plates is charged to a potential difference of 2000 V. What is the energy stored in the capacitor?

(*b*) If the voltage source remains connected to the capacitor, but the air is replaced by a piece of quartz that fills the entire region between the plates, what is the new value of the energy stored in the capacitor?

*17.51 A parallel-plate capacitor, which has a capacitance of 5.0 μF with a slab of Pyrex glass filling the region between the plates, is charged to a potential difference of 1000 V, and the voltage source is then disconnected. How much work must be done to pull the Pyrex slab out from between the plates? (*Hint:* Use energy considerations to calculate the work.)

*17.52 Two metal spheres, one of radius 10 cm and charge $+2.0$ μC, and the other of radius 20 cm and charge $+10.0$ μC, are initially some distance apart. If the two spheres are brought together and allowed to touch, what will be the final charge on each sphere?

Additional Readings

Anderson, David L.: *The Discovery of the Electron*, D. Van Nostrand (Momentum Books paperback), Princeton, N.J., 1964. An interesting account of the experimental and theoretical developments that led to our present ideas about static electricity and current electricity.

Moore, A. D.: "Electrostatics," *Scientific American*, vol. 226, no. 3, March 1972, pp. 46–59. A brief article that stresses modern applications of electrostatic phenomena, including Xerox copying machines and Cottrell precipitators.

Moore, A. D.: *Electrostatics: Exploring, Controlling and Using Static Electricity*, Doubleday Anchor Books, Garden City, N.Y., 1968. A fascinating account of practical electrostatics by an electrical engineer who has devoted his life to the subject.

McDonald, D. K. C.: *Faraday, Maxwell and Kelvin*, Doubleday Anchor Books, Garden City, N.Y., 1964. This book discusses the great revolution in physics consequent on Faraday's introduction of electric field lines.

CHAPTER 18

Direct Current Electricity

The most startling result of Faraday's laws is perhaps this. If we accept the hypothesis that the elementary substances are composed of atoms, we cannot avoid concluding that electricity also, positive as well as negative, is divided into definite elementary portions, which behave like atoms of electricity.

Hermann von Helmholtz (1821–1894)

The phenomena of electrostatics are basic to our understanding of the physical universe, but electric currents are more important for practical purposes. (Exceptions are the Xerox dry copying process and the Cottrell dust precipitator, which depend completely on electrostatics.) To power our machines and computers, to heat our toasters and stoves, charges at rest are of little value. Instead, we need a continuous flow of electric charge—in other words, an electric current—to accomplish such tasks. In this chapter we begin with a discussion of direct currents (dc). In practice most electrical machines run on alternating current (ac), but many of the ideas developed for direct current are applicable to alternating current (which we discuss in Chap. 21).

18.1 EMF's and Direct Currents

When charge flows around a closed electric circuit, energy must be supplied to keep the charge flowing. This energy is supplied by a device that we call a *source of emf** (pronounced *ee-em-ef*). Such sources include batteries, electric generators, and silicon solar cells.

Source of emf: A device (such as a battery) that transforms nonelectric energy into electric energy.

All sources of emf produce a potential difference between two conductors, such as those labeled *A* and *B* in Fig. 18.1. These conductors are called the electric *terminals* of the device and become positively and negatively charged, as shown. In the diagram there is no complete circuit, and the device is said to be on *open circuit*. In this case there are electrostatic forces on the positive charges at *A* that are directed from *A* to *B*, but no charge flows between *A* and *B* because the source of emf produces a force in the opposite direction (due to chemical or mechanical

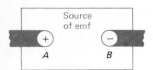

FIGURE 18.1 A source of emf on *open circuit*. Terminal *A* becomes positively charged and terminal *B* negatively charged because of nonelectric processes going on inside the source of emf.

---

*Emf stands for "electromotive force," an unfortunate term that arose in the formative years of electrical theory. Since an emf is not a force (it is a *potential difference*), we refer to it simply as an *emf*.

activity) to prevent this charge flow. On open circuit these two forces are equal and opposite, and no net force acts on the charges.

We define the strength of the source of emf in a fashion similar to the way we defined electrostatic potential difference; that is, V_{AB} is numerically equal to the work done against the electrostatic force in moving a unit positive charge from B to A [Eq. (17.4)]. In the present case we define the emf ($\mathscr{E}$) as numerically equal to the work done by a source of emf in moving a unit positive charge from the negative to the positive terminal inside the source:

$$\mathscr{E} = \frac{\mathscr{W}}{q'} \tag{18.1}$$

$\mathscr{W}$ is, for example, the work done to move a positive test charge q' from B to A in Fig. 18.1.

EMF's are measured in joules per coulombs or volts, since they are dimensionally work done per unit charge. We denote the magnitude of an emf by $\mathscr{E}$ (not E, which we continue to reserve for the magnitude of the *electric field*).

A battery is a source of emf that converts chemical energy into electric energy on a continuous basis. In a 1.5-V battery, each coulomb of charge that flows through the battery gains 1.5 J of electric potential energy at the expense of the chemical energy in the battery. Since $\mathscr{E}$ is the work done per unit charge, the work done by a battery in moving a charge Q through the battery is $\mathscr{W} = Q\mathscr{E}$. The value of the emf $\mathscr{E}$ is determined by the chemical processes going on inside the battery and has a different numerical value for each kind of battery.

The concept of emf is similar to that of voltage, and both have the same dimensions and units (usually volts). The term *source of emf*, however, is applied only to parts of a circuit in which forms of energy other than electric energy are continuously converted into electric energy. Also, emf differs from potential difference (voltage) conceptually in that the emf is a constant, intrinsic property of the source of emf. The potential difference between the terminals of a battery when current is being delivered, however, varies with the amount of current drawn by the external circuit, as we will see (Sec. 18.6).

Electric Currents

If the terminals of a source of emf are connected by a wire, we have a complete, or closed circuit, as in Fig. 18.2. Positive charge then flows through the external circuit from A to B. This decreases the electrostatic force between A and B within the source below the value of the nonelectrostatic force produced by the source of emf. As a result positive charges are driven from B to A to replenish those flowing through the external circuit. The circuit finally settles down to a steady state in which the same amount of charge flows through the circuit both inside and outside the source of emf.

When a current flows, the source of emf does work on a positive charge to move it from B to A inside the source. This charge then flows down a potential hill from A to B in the external circuit and uses the energy provided to it by the source of emf to do useful work, for example, to drive an electric motor.

In this chapter we assume that our source of emf always produces a *direct current* in the external circuit.

Electric current: The rate at which electric charge passes a point in the circuit.

Direct current (dc): A current that is always in the same direction in the circuit carrying the current.

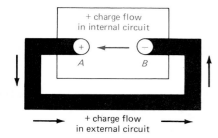

FIGURE 18.2 A source of emf on *closed circuit.* In this case charge flows through the external circuit from terminal A to terminal B.

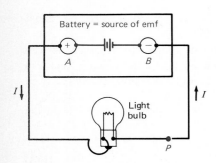

FIGURE 18.3 An electric current I in a wire. The source of emf is a battery with terminal A positive and terminal B negative. Charge flows from A to B in the external circuit. The electric current (in amperes) is the amount of charge (in coulombs) passing point P in the wire in 1 s.

Suppose, for example, that we have a conductor in which positive electric charge is flowing in the external circuit from A to B, as in Fig. 18.3. The magnitude of the direct current I is the amount of charge in coulombs that passes any point in the wire in 1 s. Thus, if an amount of charge ΔQ passes point P, say, in time Δt, then by definition the current I is

$$I = \frac{\Delta Q}{\Delta t} \tag{18.2}$$

In the SI system of units, a current of *one coulomb per second* is called an *ampere* (A), and so

$$1 \text{ A} = 1 \text{ C/s}$$

The direction of the current is taken to be the direction in which a *positive* charge would flow if free to move around the electric circuit, since we have defined the electric field in terms of the force exerted on a *positive* charge. For the electric current to be in the direction of the electric field in a wire, we therefore take the direction of the current as that in which a *positive* charge would flow. This is only a *convention*, or arbitrary decision, and for this reason such a current is sometimes called the *conventional current*. (In metals all the current is carried by electrons moving in a direction *opposite* to that of the conventional current.) In this book, by the direction of a direct current we always mean the direction of the *conventional current*, as in Fig. 18.4.

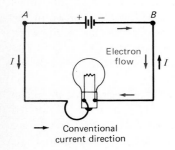

Conventional current direction

FIGURE 18.4 The conventional current I in a wire. Its direction is that in which a positive charge would flow and is therefore in the opposite direction to that of the electrons which actually carry the current in a wire.

Practical Batteries

Because of chemical reactions within a battery, a positive charge has a higher potential energy at the positive terminal than at the negative terminal. This difference in potential energy is exactly equal to the work done by the chemical forces that move the positive charge from the negative to the positive terminal inside the battery. Hence if the two terminals are connected to an electric motor, charge flows in the external circuit from the positive to the negative terminal of the battery and this current powers the motor, which can then do useful work. This work has its ultimate source in the chemical energy provided by the battery.

There are many varieties of batteries in common use. The most popular are the *dry cells* used in flashlights, automobile storage batteries, cameras, hand calculators, and watches. Storage batteries and some of the newer batteries can be *recharged*, which means that after some of the chemical energy in the battery has been converted into electric energy, an electric current can be forced through the battery in the reverse direction to convert electric energy back into chemical energy. In an automobile both discharging and recharging of the car's storage batteries go on automatically while the car is running.

In electric circuits a battery is designated by the symbol ⊣ ⊢, where the long vertical line represents the positive terminal of the battery, i.e., the terminal at the higher potential. A repeated symbol like ⊣⊪⊢ indicates that the battery contains more than one simple cell. If the cells are connected *in series*, i.e., with

FIGURE 18.5 (*a*) Two batteries connected in a series-aiding circuit. The voltages have the same sign and add. (*b*) Two batteries connected in a series-opposing circuit. The voltages have opposite signs and subtract.

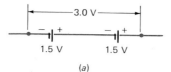

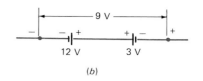

(*a*)

(*b*)

the positive terminal of one cell connected to the negative terminal of the next cell, then the voltages of the two cells add and the cells are said to be *in series aiding*. Thus two 1.5-V dry cells connected in this way yield a 3.0-V battery, as in Fig. 18.5*a*. If two cells are connected in series but with two positive (or two negative) terminals together, then the cells are said to be *in series opposing* and the effective emf is the difference of the two individual emf's, as in Fig. 18.5*b*.

Example 18.1

A dry cell with an emf of 1.5 V is connected to two flat metal plates inside a vacuum chamber, as in Fig. 18.6. How much kinetic energy would **(a)** a proton gain in moving from plate *D* to plate *C*; **(b)** an electron gain in moving from plate *C* to plate *D*? **(c)** Where does this energy come from?

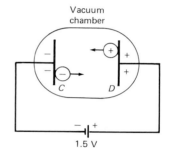

FIGURE 18.6

SOLUTION

(a) In Fig. 18.6, plate *D* is kept at a potential 1.5 V higher than plate *C* by the battery. Hence the electric potential energy of a proton, with charge $q = +1.6 \times 10^{-19}$ C, is higher at *D* than at *C* by an amount

$$\Delta PE = qV_D - qV_C = q(V_D - V_C) = (1.6 \times 10^{-19} \text{ C})(1.5 \text{ V})$$
$$= 2.4 \times 10^{-19} \text{ J}$$

This electric potential energy is converted into kinetic energy when the proton moves from *D* to *C*, attracted by the negatively charged plate *C*. The proton gains 2.4×10^{-19} J of kinetic energy. Its velocity just before it strikes plate *C* can be obtained from the fact that $\frac{1}{2}mv^2 = \boxed{2.4 \times 10^{-19} \text{ J}}$, if it

starts from rest and if there is no resistive force opposing the proton's motion.

Another way to look at this problem is to note that, if a single elementary charge moves through a potential difference of 1.5 V, it gains a kinetic energy of 1.5 eV. But this is $(1.5 \text{ eV})(1.6 \times 10^{-19} \text{ J/eV}) = 2.4 \times 10^{-19}$ J.

(b) An electron will gain just as much kinetic energy in moving from *C* to *D* as a proton does in moving from *D* to *C*, since the charge on the electron is negative and of the same magnitude as the proton's charge. Its kinetic energy will also be the same, although its speed will be quite different because of the difference between the two masses.

(c) Ultimately this kinetic energy comes from the chemical energy in the battery.

EXERCISE 1 What will the difference be in the final velocities of the two particles if they both start from rest?

18.2 Electric Currents in Electrolytes

Distilled water, like many other pure liquids, is a poor electric conductor, since it lacks free electrons or ions to move through the liquid. If some sodium chloride (ordinary table salt) is added, however, the water becomes a good conductor. Such solutions of salts in water are termed *electrolytes*. If a battery is connected to two metal plates immersed in the salt solution, as in Fig. 18.7, the Na^+ ions will migrate to the negative plate and the Cl^- ions will migrate to the positive plate,

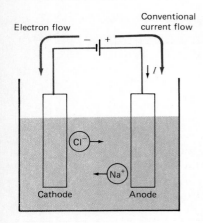

Electron flow

Conventional current flow

$-$ $+$

I

Cl$^-$ →

← Na$^+$

Cathode   Anode

FIGURE 18.7 An electrolytic cell containing salt (NaCl) in solution. Inside the cell the current is carried by both positive Na ions and negative Cl ions. Outside the cell the current is carried by electrons in the wire.

resulting in an electric current. In this case the current is carried by both positive and negative charges moving in opposite directions through the liquid, although in the wire connected to the battery only electrons carry the current.

Faraday's Laws of Electrolysis

In 1833 Michael Faraday (see Fig. 20.2) showed that the study of currents in electrolytes can cast much light on the structure of matter and the nature of electricity. Faraday found in his experiments on what we now call *electrolysis* that the passage of a current through a NaCl solution liberates chlorine gas at the anode (the positive, or higher-potential, plate) and deposits solid sodium on the cathode (the negative, or lower-potential, plate). Careful study of this effect led Faraday to the following conclusions that are called *Faraday's laws of electrolysis*:

1 The mass of substance deposited or liberated from a solution undergoing electrolysis is directly proportional to the amount of charge flowing through the solution.

2 One faraday of charge ($F = 96,500$ C) deposits or liberates an amount of substance equal to the mass of one mole of the substance divided by its valence.

Atomicity of Charge

Faraday's laws are capable of a very simple atomic interpretation. For monovalent substances like Na and Cl, each ion in solution carries a net electric charge of either plus or minus one elementary charge. Every time a Cl$^-$ ion reaches the anode, it gives up its excess electron to the anode, becomes a neutral atom, and bubbles up to the surface as a gas. Similarly, a Na$^+$ ion picks up an electron from the cathode and attaches itself to the cathode as neutral sodium metal. Suppose, for purposes of illustration, that the electron picked up by the Na$^+$ is the same one deposited by the Cl$^-$ at the anode. For each molecule of NaCl coming out of solution, one electron thus moves through the circuit outside the electrolytic cell. The more electrons that flow, the larger the current in the circuit and the more Na and Cl liberated.

If this atomic interpretation is correct, a charge of 96,500 C corresponds to Avogadro's number (N_A) of electrons flowing in the circuit, for this amount of charge releases 1 mol of Na$^+$ and 1 mol of Cl$^-$, or 1 mol of NaCl in all. Corresponding to each pair of ions released there is one electron flowing through the external circuit, as we have just seen, and so we can write

$$F = 96,500 \text{ C} = N_A e \tag{18.3}$$

where F is the faraday, e is the charge on the electron, and N_A is Avogadro's number. We therefore obtain

$$e = \frac{96{,}500 \ \text{C/mol}}{(6.02 \times 10^{23})/\text{mol}} = 1.60 \times 10^{-19} \ \text{C}$$

If we assume that Avogadro's number is known, electrolysis experiments therefore yield a value for the charge on the electron in good agreement with the one found by Millikan in the oil-drop experiment discussed in Sec. 16.4.

Example 18.2

A current of 15 A is maintained for 30 min through an electrolytic cell containing $CuSO_4$. The salt breaks up into Cu^{2+} and SO_4^{2-} in solution. How much copper is deposited at the cathode?

SOLUTION

The amount of charge flowing through the cell is

$$Q = It = (15 \ \text{A})(30 \ \text{min}) = \left(15 \ \frac{\text{C}}{\text{s}}\right)(1800 \ \text{s})$$

$$= 2.7 \times 10^4 \ \text{C}$$

Now, from Faraday's laws, 1 faraday (96,500 C) will plate out 63.5/2 g of Cu^{2+} since the mass number of Cu (from Table F.6) is 63.5, and each ion carries two elementary charges. To determine the mass of the copper deposited, we can set up a proportion:

$$\frac{x}{2.7 \times 10^4 \ \text{C}} = \frac{\frac{1}{2}(63.5) \ \text{g}}{9.65 \times 10^4 \ \text{C}} \qquad \text{or} \qquad x = \boxed{8.9 \ \text{g}}$$

Notice that we have obtained our answer in *grams* because we are dealing with *gram* moles.

18.3 Electric Currents in Metals

Unlike the current in electrolytes, which is carried by both positive and negative ions, the current in a metal wire is carried solely by electrons. The metal atoms, which lose electrons and become positive ions, are bound to fixed positions in the crystal lattice and cannot move. Only the outermost (valence) electrons from these atoms can move freely through the wire, moving in a direction opposite that of the conventional current.

Ohm's Law

Suppose we attach a voltage source across the two ends of a piece of thin copper wire of length L. What happens? If the left end of the wire is at a potential V_A and the right end at a potential V_B, as in Fig. 18.8, the work done by the voltage source in moving a charge q' from A to B in the wire is, from Eq. (17.4),

$$\mathscr{W} = (V_A - V_B)q'$$

The work done per unit charge is therefore

$$\frac{\mathscr{W}}{q'} = \frac{F_{el}L}{q'} = V_A - V_B$$

where F_{el} is the electric force on the charge q'. But F_{el}/q' is the magnitude of the electric field strength **E** in the wire, [Eq. (16.2)], and so

$$E = \frac{V_A - V_B}{L}$$

In a wire with a potential difference between its two ends there is an electric field of magnitude $E = (V_A - V_B)/L$, which is directed parallel to the wire at any point along its length. Note that, unlike the situation in electrostatics, an electric field *can* exist inside a conductor when a current exists inside the conductor. This is

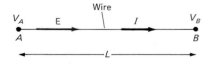

FIGURE 18.8 Electric field **E** and current in a wire when a potential difference is applied across it.

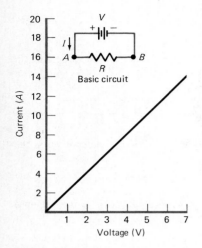

FIGURE 18.9 Ohm's law for the relationship of the current to the applied voltage in a simple circuit containing only a resistance R.

because the applied voltage does not allow the charges in the metal to come to rest at positions where they could cancel the field.

Let $V = V_A - V_B$ be the voltage across the wire. If we now increase V, the field inside the wire increases, and experiment shows that the current through the wire also increases in direct proportion to V, as shown in Fig. 18.9. We have, therefore,

$$I \propto V \quad \text{or} \quad V \propto I$$

and we can write, on introducing a constant of proportionality R,

Ohm's law: $\boxed{V = IR}$ (18.4)

For any applied voltage V, the current I decreases as R increases: R measures the resistance of the wire to the flow of charge. It is called the *resistance* of the wire and is denoted by the symbol ⋀⋀⋁ on circuit diagrams.

Electrical resistance: The ratio of the voltage applied across any part of a dc circuit to the current through that part of the circuit.

Equation (18.4) is called *Ohm's law* after its discoverer, the German physicist Georg Simon Ohm (1787–1854). The unit of resistance is also named after him.

Ohm: The resistance of a wire across which the application of a voltage of one volt produces a current of one ampere ($1 \ \Omega = 1 \ \text{V/A}$).

The symbol for the ohm is capital *omega*, the last letter in the Greek alphabet.
Note that there is a voltage *drop* across a resistor in a dc circuit, like that of Fig. 18.9, so that $V_B < V_A$ if the current is in the direction from A to B. This is because the battery must do work to move the charge from A to B. Energy is therefore lost in the process and appears as heat in the resistor R.
Ohm's law also applies to other kinds of resistors, such as carbon resistors, that conduct electricity but are nonmetallic. The equation $R = V/I$ is always valid as the definition of R, even when Ohm's law is no longer valid, that is, even when R is not constant but varies with I.

Example 18.3

The resistance of a graphite rod is $3.0 \times 10^{-1} \ \Omega$. **(a)** What voltage must be applied across the two ends of the rod to produce a current of 10 A through the rod? **(b)** If the required voltage is supplied by a battery that delivers current for 10 min, how much work does the battery do?

SOLUTION

(a) From Ohm's law,

$$V = IR = (10 \ \text{A})(3.0 \times 10^{-1} \ \Omega) = \boxed{3.0 \ \text{V}}$$

(b) Since $\mathcal{W} = (V_A - V_B)q' = (3.0 \ \text{V})q'$, and

$$q' = It = (10 \ \text{A})(10 \ \text{min}) = (10 \ \text{A})(600 \ \text{s}) = 6.0 \times 10^3 \ \text{C}$$

we have $\quad \mathcal{W} = (3.0 \ \text{V})(6.0 \times 10^3 \ \text{C}) = \boxed{1.8 \times 10^4 \ \text{J}}$

Simple Electric Circuits

A simple kind of electric circuit is a flashlight containing a small bulb lighted by two D cells, each with an emf of 1.5 V, as in Fig. 18.10. In this case the path of the conventional electric current is from the positive terminal of the top battery through wires to the filament of the light bulb and then back to the negative terminal

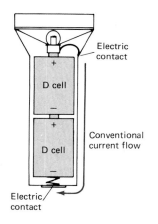

FIGURE 18.10 A flashlight: a simple electric circuit. The return circuit is through the metal case of the flashlight.

of the lower battery by way of the metal case of the flashlight (or a connecting wire).

The circuit diagram corresponding to the flashlight is shown in Fig. 18.11. The filament of the light bulb is represented by the resistance R. For a flashlight R is about 10 Ω. The connecting wires do have some resistance, but it is so small that it can be neglected or lumped in with the resistance of the filament wire to make up the resistance R. In circuit diagrams straight lines like those connecting the battery to the resistance R are therefore assumed to have zero resistance. As a result, in Fig. 18.11 points like A and A' or B and B' are at exactly the same potential, since, by Ohm's law, $V_A - V_{A'} = IR' = 0$, if $R' = 0$, and so $V_A = V_{A'}$. *Only wires (or other objects) in which an electric current encounters sufficient resistance to generate a measurable amount of heat are called resistors and indicated by the resistance symbol on a circuit diagram.* The electric current is the same at every point in the circuit of Fig. 18.11. The electrons act as an incompressible fluid which flows around the circuit in the same way that water flows through a pipe. An equation similar to the hydrodynamic equation of continuity is therefore valid. For a closed loop the amount of charge passing any point in the circuit per second is exactly the same as the amount passing any other point in the circuit per second, since electric charge can neither be created nor destroyed.

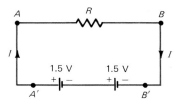

FIGURE 18.11 Circuit diagram for the flashlight. The resistance R includes the resistance of the flashlight bulb and all other resistances in the circuit.

18.4 Resistance and Resistivity

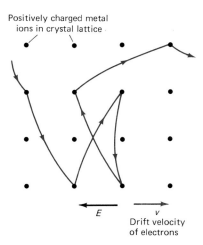

FIGURE 18.12 Drift velocity of electrons through a wire. The electrons move in a direction opposite to the electric field in the wire at a rate limited by the number of collisions they make with the positive ions in the metal crystal.

When a potential difference is applied across a wire, the electrons do not move very rapidly along the length of the wire. An electron is accelerated by the voltage, since it experiences a force of magnitude eE, where e is one elementary charge and E is the electric field strength in the wire. The acceleration of the electron is then, by Newton's second law, $a = eE/m$. Because the electron mass m is so small, this acceleration is very large even for a small applied voltage. The electron therefore picks up speed very rapidly but almost immediately collides with one of the positively charged metal ions in the wire. It loses a little of its energy in the collision, bounces off the ion, is accelerated again, collides again, and thus makes its way by a very irregular path along the length of the wire, as in Fig. 18.12. As a result the *drift velocity* of the electrons through the wire (that is, the average velocity at which they progress along the length of the wire) is only on the order of 1 mm/s. This is in contrast to the very large speeds the electrons can acquire during the brief periods of time between collisions.

The magnitude of the current through a wire is proportional to the drift velocity of the electrons through the wire. Hence the effect of the collisions the electrons undergo is to reduce the current, much in the same way that rocks, fallen trees, and other debris in a river reduce the flow rate of the river water. The smaller the current for any given applied voltage, the larger the resistance.

**Factors Determining
Electric Resistance**

To find the factors that determine the electric resistance of a wire, let us do a simple experiment. Suppose we take a piece of copper wire 1 m long, apply a voltage of 6 V across it with a storage battery, and measure the current which flows. (This measurement can be made with an *ammeter*, to be discussed in the next chapter.) Suppose we find that the current is 1 A. Then, from Ohm's law, the resistance of the wire is $R = V/I = 6\ V/1\ A = 6\ \Omega$. If we repeat this experiment with a piece of the same wire 2 m long, we find that this time R is 12 Ω; if the wire is 3 m long, its resistance is 18 Ω; and so on. The resistance of the wire is thus found by experiment to be proportional to its length. The reason is that the electric field inside the wire is given by $E = V/L$, where V is the applied voltage and L is the length of the wire. Hence increasing L decreases E and therefore decreases the current, since the electric field E provides the force to move the charges through the wire.

Similarly, if we use copper wires of the same length but different cross-sectional areas, we find by experiment that the resistance varies inversely with the area of the wire. This is reasonable, since the larger the wire's cross section, the more free electrons there are in any given length to carry the current.

Putting these experimental observations together, we may write for the resistance $R \propto L/A$ or

$$R = \rho \frac{L}{A} \qquad\qquad \textbf{(18.5)}$$

Here L and A are the length and cross-sectional area of the wire used; they are geometric factors that are independent of the material out of which the wire is made. The dependence of the resistance on the material of the wire is contained in the proportionality constant ρ (Greek *rho*, the equivalent of the English r), which

TABLE 18.1   Electric Resistivity and Its Temperature Coefficient at 0°C

Material	Resistivity ρ ($\Omega \cdot$m)	Temperature coefficient of resistivity α ($10^{-3}/C°$)
Conductors:		
Silver	1.5×10^{-8}	4.1
Copper	1.7×10^{-8}	4.1
Gold	2.4×10^{-8}	4.0
Aluminum	2.6×10^{-8}	3.9
Tungsten	5.5×10^{-8}	4.7
Iron	8.85×10^{-8}	6.2
Platinum	9.83×10^{-8}	3.7
Lead	22×10^{-8}	3.9
Mercury	94×10^{-8}	0.88
Nichrome	100×10^{-8}	0.4
Semiconductors (see Secs. 30.3 and 30.4):		
Germanium	~0.5*	−50
Silicon	20–2000*	−70
Nonconductors (insulators):		
Amber	~5×10^{14}	
Fused quartz	~10^{17}	

*Depending on purity.

is called the *resistivity* of the material. The resistivity must be obtained by laboratory measurement; it is small for good conductors and large for poor conductors, as Table 18.1 shows. The units of resistivity are, from Eq. (18.5), $\Omega \cdot m^2/m = \Omega \cdot m$ (ohm-meters). Thus copper has a resistivity of $1.7 \times 10^{-8} \ \Omega \cdot m$ at 0°C.

Temperature Coefficient of Resistivity

Like many physical properties, resistivity depends on temperature and therefore so does resistance. The resistivity of pure metals increases linearly with temperature because raising the temperature increases the amplitude of vibration of the ions in the metal lattice. This increases the likelihood of collisions and decreases the current through the wire. (It is harder to cross a dance floor when people are moving around in all directions than when they are all standing still.) The expression for the increase of resistivity with temperature is similar to that for the increase in length of a metal rod with temperature [Eq. (13.7)]. We have

$$\rho_T = \rho_0(1 + \alpha \, \Delta T) \tag{18.6}$$

where ρ_T is the resistivity of the metal at a temperature T°C, ρ_0 is its resistivity at 0°C, and α is called the *temperature coefficient of resistivity* and is a measure of *the fractional increase in resistivity per 1 C° rise in temperature*. Since $\alpha \, \Delta T$ must be dimensionless, α has units $(C°)^{-1}$. Values of α for some important metals are given in Table 18.1. The dependence of the resistance of a copper wire on temperature is shown in Fig. 18.13.

From Tables 13.2 and 18.1 it can be seen that the temperature coefficients of resistivity for metals are at least 100 times larger than the coefficients of linear expansion for the same metals, and so temperature effects on L and A can be reasonably ignored in calculating the resistance R as a function of temperature.

Since the resistance R for a metal is proportional to its resistivity, Eq. (18.6) is equivalent to

$$\boxed{R_T = R_0(1 + \alpha \, \Delta T)} \tag{18.7}$$

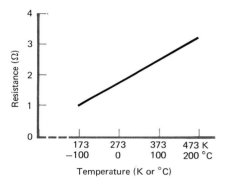

FIGURE 18.13 The resistance of a piece of copper wire, 100 m long and 1 mm² in cross-sectional area, as a function of temperature.

Because the resistance of a metal is a function of temperature, *resistance thermometers* are frequently used to measure high temperatures at which ordinary thermometers will not function. (How would you expect a resistance thermometer to work?) A platinum resistance thermometer is particularly useful for this purpose, since platinum has a very high melting point (1774°C) and is readily available in highly purified form.

Resistance of the Human Body

The tissues and fluids just beneath the skin of humans contain a large number of ions and hence conduct electricity almost as well as many metals. By producing a small current between two points near the surface of the body, it is therefore possible to measure the electric resistance between these points. Often one electrode is attached to a patient's leg and the other is moved over the body, with a voltage of 60 V or so applied between the two electrodes. Since, for constant voltage, the current varies inversely with the resistance of the path, the electric resistance between the two electrodes can be determined in this way. (This kind of electrical experiment should not be attempted except under skilled supervision.)

It is found that nerve damage or tumors beneath the skin greatly increase the electric resistance near the site of the abnormality. This technique has been found to be very effective in detecting cancers. The measured increase in resistance even provides a good indication of how advanced the cancer is.

Example 18.4

(a) What is the resistance at 0°C of a 1.0-m-long piece of no. 5 gauge copper wire that has a cross-sectional area of 16.8 mm²? (b) What would the resistance of the same piece of wire be at room temperature (20°C)?

SOLUTION

(a) Using the basic equation $R = \rho L/A$ and the value of ρ for copper from Table 18.1, we have

$$R = \frac{(1.7 \times 10^{-8}\ \Omega\cdot m)(1.0\ m)}{(16.8\ mm^2)(10^{-3}\ m/mm)^2} = \boxed{1.0 \times 10^{-3}\ \Omega}$$

(b) Since, from Table 18.1, $\alpha = 4.1 \times 10^{-3}/C°$, we have $R_T = R_0(1 + \alpha T)$, or

$$R_{20°} = R_{0°}[1 + (4.1 \times 10^{-3}/C°)(20\ C°)] = R_{0°}(1 + 0.08)$$

$$= (1.0 \times 10^{-3}\ \Omega)(1.08) = \boxed{1.1 \times 10^{-3}\ \Omega}$$

Hence the resistance changes by about 10 percent for a 20 C° change in temperature.

18.5 Kirchhoff's Rules

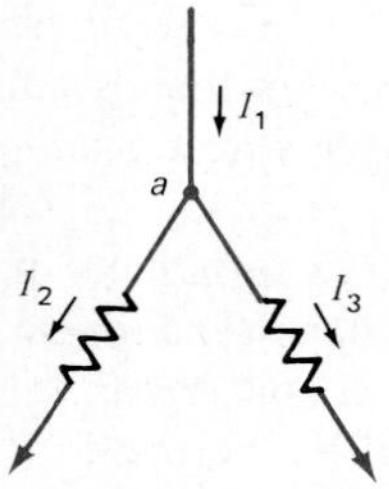

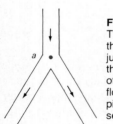

FIGURE 18.14 Kirchhoff's junction rule for three currents: $I_1 = I_2 + I_3$.

FIGURE 18.15 The behavior of the current at a junction point is the same as that of a fluid that flows from one pipe into two separate pipes.

In 1845 Gustav Kirchhoff (see Fig. 18.17 and accompanying biography) proposed two extremely useful rules relating to electric circuits.

Kirchhoff's first rule is based on the law of conservation of electric charge and applies to *junction points* in a circuit, i.e., to points at which three or more wires come together.

Junction rule: The sum of all the currents entering any junction point is equal to the sum of all the currents leaving that junction point.

For example, as applied to the junction point a in Fig. 18.14, Kirchhoff's first rule states that, if the currents are in the directions shown in the figure, then $I_1 = I_2 + I_3$. In writing down this equation we must often assume the directions of the three currents, and our assumptions may be incorrect. If so, we may find on solving the circuit problem that one or more of the currents are negative. This means that our original assumption about the direction of that particular current was wrong and that the actual direction is the reverse of our original conjecture.

If we consider water flowing in a pipe instead of charges flowing in a wire, as in Fig. 18.15, then clearly all the water that flows into point a in one pipe must flow away from a in the other two pipes. The same is true of electric charge and hence of electric current. The amount of charge flowing into a per second (I_1) must be equal to the amount of charge leaving a per second ($I_2 + I_3$), if charge is to be conserved. Kirchhoff's junction rule is a statement of this fact.

Kirchhoff's second rule is based on the principle of conservation of energy applied to complete circuits, or *loops*.

Loop rule: The algebraic sum of the changes in electric potential encountered in a complete traversal of any closed circuit is equal to zero.

When there is a steady current in a closed circuit, there is a fixed value for the electric potential at every point in that circuit (i.e., for the potential energy per unit charge at each point). If we take a charge at point H in Fig. 18.16 and follow it around the circuit, returning finally to point H, then its potential energy must be the same at the end as it was at the beginning, since electrical forces are conservative. Hence the algebraic sum of the changes in potential it encounters as it goes around the complete circuit must be zero—which is the loop rule.

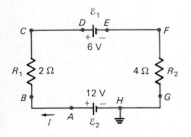

FIGURE 18.16 Kirchhoff's loop rule: Since the algebraic sum of the changes in electric potential around the complete loop is zero, the sum of the emf's is equal to the sum of the IR drops around the circuit.

FIGURE 18.17 (*Courtesy of AIP Niels Bohr Library; Meggers Gallery of Nobel Laureates.*)

One of the group of university professors who led Germany to preeminence in physics at the end of the nineteenth century, Kirchhoff was born in Königsberg. The son of a law counselor who had many friends on the faculty at the University of Königsberg, he married the daughter of one of his professors, and he spent his life doing research and teaching at the universities of Breslau, Heidelberg, and Berlin. While in Heidelberg, he worked in close association with R. Bunsen, the famous chemist, and H. von Helmholtz (Fig. 12.18), the physiologist-physicist, during what has been called Heidelberg's golden age of science.

Kirchhoff's first important contribution to physics, the analysis of electric networks, was made while he was still a university student and only 21 years old.

When he went to Heidelberg in 1854, Kirchhoff found Bunsen trying to analyze salts on the basis of the distinctive colors they imparted to the flame of what we now call a Bunsen burner. Kirchhoff pointed out that a much more reliable test for the presence of various elements would be the line spectrum they emit, which could be detected using the spectroscope which he proceeded to construct with Bunsen's help. Together they determined the line spectra of many of the elements, and in the process they discovered the elements cesium and rubidium.

Kirchhoff's work on the emission spectra of the elements led him to consider the relationship of their spectral lines to the dark lines observed by Fraunhofer in the sun's spectrum. Kirchhoff showed that the famous yellow D lines in the emission spectrum of sodium corresponded perfectly to two dark lines in the solar spectrum. This made him conclude that the sun's atmosphere contained sodium and that this sodium absorbed characteristic wavelengths out of the continuous radiation emitted by the sun. He verified that these were indeed the sodium lines by means of a laboratory experiment involving a sodium light source and NaCl salt ignited in a Bunsen burner.

His study of the relationship between the emission and absorption of radiation then led Kirchhoff to investigate the radiation from hot objects and to propose a hollow metal cavity (a blackbody) as an ideal source for such radiation. His radiation law became the key to the study of radiation. Max Planck, who succeeded Kirchhoff in the chair of theoretical physics at the University of Berlin, took up Kirchhoff's ideas and used them as the basis for his quantum theory, one of the truly great developments in the history of physics (see Chap. 27).

Despite his enormous research ability, Kirchhoff was a somewhat dull lecturer. Still, he worked assiduously on his theoretical physics lectures at Berlin in the years after 1875, when his poor health prevented further experimental work. Many of the physicists who made Germany the mecca of theoretical physics from 1900 to 1935 learned physics from Kirchhoff's published lectures.

For many years, as the result of an accident, Kirchhoff was confined to a wheelchair or crutches. Yet he retained his cheerful disposition and great enthusiasm for his work. One day his banker, unimpressed by Kirchhoff's ability to locate elements in the sun with his spectroscope, asked him, "Of what use is gold in the sun if I cannot bring it down to earth?" Some years later, when Great Britain presented Kirchhoff with a gold medal and a large cash prize in gold sovereigns for his research, Kirchhoff handed it over to the same banker, with the sly remark: "Here is your gold from the sun."

Another way to express the loop rule is that, in going around a closed loop, the sum of the emf's is equal to the sum of the *IR* drops. Let us see if this is true for the circuit of Fig. 18.16. Finding the single current in the circuit of Fig. 18.16 is not difficult. The net emf is 12 V − 6 V = 6 V, since the two emf's are opposed and the resultant emf drives current in a clockwise direction around the

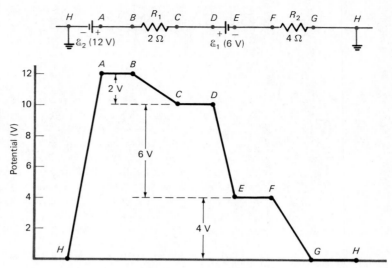

FIGURE 18.18 Circuit of Fig. 18.16 unfolded into a straight line, with values of the potential at various points throughout the circuit indicated on the vertical scale. Since point *H* is at ground potential, the circuit is completed through ground.

circuit (we assume here that the emf's and terminal voltage of the batteries are the same). The resistance of the circuit is $R = 2\Omega + 4\Omega = 6\Omega$, since the voltage drop across the two resistors in series (as they appear in this circuit) is the sum of the voltage drops across the two resistors separately. We have $I = (6 \text{ V})/(6 \ \Omega) = 1$ A. There is therefore a current of 1 A in a clockwise direction in the circuit.

To consider the changes in potential as we go around this circuit, let us unfold the circuit into a straight line, as in Fig. 18.18, where points *H* at the beginning and end of the unfolded circuit are identical electrically. This diagram enables us to indicate below each circuit element exactly how the potential varies as we move through that element. For simplicity, we choose to ground point *H* to put it at zero potential.

As we move through $\mathscr{E}_2$ from *H* to *A*, the potential increases by 12 V, the emf of the battery. Between *A* and *B* there is no change in potential, since there is no resistance between *A* and *B*. When we pass through the resistor R_1 the potential falls by 2 V, since $1 \text{ A} \times 2 \ \Omega = 2$ V, and *C* is at a lower potential than *B*. Again there is no change in potential between *C* and *D*. When we move through $\mathscr{E}_1$, however, the potential *decreases* by another 6 V, since we are going from a higher potential to a lower potential. Finally, in the resistor R_2 the potential drops again from 4 V to zero, and we return to point *H* at the same (zero) potential at which we started.

The changes in potential in going around the circuit are $12 \text{ V} - 2 \text{ V} - 6 \text{ V} - 4 \text{ V} = 0$, as required by the loop rule. The sum of the emf's $(12 \text{ V} - 6 \text{ V} = 6 \text{ V})$ is therefore equal to the sum of the *IR* drops (the products of the current and the resistances), which in this case are $4 \text{ V} + 2 \text{ V} = 6$ V. This calculation makes clear two points of great importance in assigning signs to the changes in potential when applying the loop rule.

1 If a source of emf is traversed in the direction of increasing potential, the change in potential is $+\mathscr{E}$; if in the opposite direction, it is $-\mathscr{E}$. (Look carefully at Fig. 18.18 and be sure you understand how these sign conventions are used.)

2 If a resistor is traversed in the same direction as the current in any branch of a circuit, the change in potential is $-IR$ because there is a *drop* in potential in moving through a resistor in the direction of the current; if a resistor is traversed in the opposite direction to that of the current, the change in potential is $+IR$.

In applying the loop rule, the assumed directions of the currents must be the same as those assumed in applying the junction rule. These may turn out to be the wrong directions, but the correct answer will still be obtained as long as the current directions are taken consistently throughout the problem, i.e., in all applications of both the junction rule and the loop rule.

The application of Kirchhoff's two rules to a circuit in which all the emf's and resistances are given leads to a number of equations in which all terms except the currents are known. In general, there will be a sufficient number of independent equations to solve algebraically for all the unknown currents. This is illustrated in the following example and in Sec. 18.6.

Example 18.5

Solve the circuit of Fig. 18.19 for the three currents I_1, I_2, and I_3.

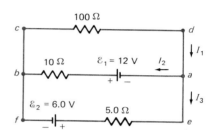

FIGURE 18.19

SOLUTION

Applying the junction rule to point a and assuming the currents to be in the directions shown in Fig. 18.19, we have

$$I_1 = I_2 + I_3 \tag{1}$$

We now apply the loop rule, going around each of the three loops in a clockwise direction, and noting that there is a different current in each branch. For loop *abcd* we have, using the sign convention previously discussed,

$$\mathcal{E}_1 - I_2(10\ \Omega) - I_1(100\ \Omega) = 0 \tag{2}$$

For the lower loop *aefb* we have (being careful of sign conventions)

$$-I_3(5.0\ \Omega) - \mathcal{E}_2 + I_2(10\ \Omega) - \mathcal{E}_1 = 0 \tag{3}$$

For the full loop *defc* we have

$$-I_3(5.0\ \Omega) - \mathcal{E}_2 - I_1(100\ \Omega) = 0 \tag{4}$$

Then (4) becomes, on replacing I_1 by $I_2 + I_3$ from (1),

$$-5.0I_3 - 6.0 - (I_2 + I_3)100 = 0$$

and so

$$100I_2 = -105I_3 - 6.0 \tag{5}$$

From (3) we have $-5.0I_3 - 6.0 + 10I_2 - 12 = 0$, or

$$10I_2 = 5.0I_3 + 18 \tag{6}$$

On multiplying (6) by 10 and combining with (5), we have

$$100I_2 = 50I_3 + 180 = -105I_3 - 6.0$$

which leads to

$$I_3 = \frac{-186}{155} = \boxed{-1.2\ \text{A}}$$

Then, from (6), $10I_2 = 5.0(-1.2\ \text{A}) + 18$, or

$$I_2 = \boxed{1.2\ \text{A}}$$

and so

$$I_1 = I_2 + I_3 = 1.2\ \text{A} - 1.2\ \text{A} = \boxed{0}$$

Our results indicate that $I_1 = 0$, $I_2 = 1.2$ A and is in the originally assumed direction, and $I_3 = 1.2$ A but is in a direction *opposite* to that shown in the diagram. Hence there is a counterclockwise current of 1.2 A in the lower loop and no current at all in the upper loop. The reason for this is that the potential difference between points a and b due to $\mathcal{E}_1$ is 12 V $-$ $(10\ \Omega)(1.2\ \text{A}) = 0$. Since $V_{ab} = V_{dc}$ there is no voltage between d and c to drive current through the 100-Ω resistor. Therefore $I_1 = 0$, as found from Kirchhoff's rules.

Note that we did not use Eq. (2) at all in this solution. Actually it is redundant, since only two of the three loop equations are independent, and so the two loop equations and the junction equation must be used to solve the problem.

18.6 Some Applications of Kirchhoff's Rules

EMF and Terminal Voltage of a Battery

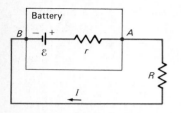

FIGURE 18.20 Terminal voltage and emf for a battery: $V_{AB} = \mathcal{E} - Ir$.

In a battery chemical energy is continuously being converted into electric energy when the battery is in use. No battery is perfect, however; some energy is always lost in the form of heat inside the battery. This loss depends on the amount of current through the battery. The battery therefore behaves as if it had an *internal resistance r* that functions in exactly the same way an ordinary resistance would function in the circuit external to the battery.

The correct circuit diagram for a battery delivering a current I to an external resistance R is shown in Fig. 18.20, where we have included the internal resistance r of the battery. Applying Kirchhoff's loop rule, we find, on going around the circuit in the direction of the conventional current,

$$\mathcal{E} - Ir - IR = 0 \quad \text{and} \quad \mathcal{E} - Ir = IR$$

Now, $\mathcal{E} - Ir$ is the *terminal voltage* of the battery, which is the voltage V_{AB} between points A and B in the circuit. This voltage is measured by attaching a voltmeter (an instrument for measuring voltage) to the two terminals of the battery while it is delivering current. Hence

$$V_{AB} = \mathcal{E} - Ir \tag{18.8}$$

The terminal voltage of a battery is its emf reduced by the Ir drop across its internal resistance.

Note that the terminal voltage is equal to the emf only when $I = 0$, that is, when no current is being drawn from the battery. The more current drawn, the lower the terminal voltage, according to Eq. (18.8). If the battery were a storage battery being charged, the terminal voltage would be *larger* than the emf (prove this).

In starting a car very large currents (~ 100 A) must be provided by the car's battery for a short period of time. Thus if you start your car with the headlights on, the lights dim while the battery is delivering this large starting current, since the terminal voltage is reduced by Ir. This means that, in Fig. 18.21, the voltage across the headlights is momentarily decreased, and so the current through the headlights and the light they give is reduced.

The internal resistance of a fresh battery is very small, perhaps about 0.05 Ω for a small D cell. However, as the battery ages, its internal resistance increases to many ohms. Finally the internal resistance becomes so large that the battery ceases to be useful, for as soon as it delivers any measurable current its terminal voltage falls almost to zero.

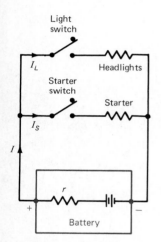

FIGURE 18.21 Connections of the starter circuit and the headlights to an automobile battery. I is the current delivered by the battery, which has internal resistance r. I_L is the current through the headlights, I_S the starter current.

Example 18.6

In Fig. 18.22, a 12-V battery has an internal resistance of 0.10 Ω. **(a)** If the battery is connected to two light bulbs whose resistances when hot are 50 and 100 Ω, respectively, as shown in the figure, what are the currents in the three branches of the circuit? **(b)** By how much does the terminal voltage of the battery differ from its emf?

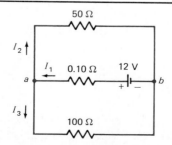

FIGURE 18.22

SOLUTION

Let us solve the problem using Kirchhoff's rules. At the junction point a we have $I_1 = I_2 + I_3$ from the junction rule.

From the loop rule we have, on going around the loops in a clockwise direction,

For the upper loop: $12 \text{ V} - (0.10 \ \Omega)I_1 - (50 \ \Omega)I_2 = 0$

For the lower loop: $-12 \text{ V} + (0.10 \ \Omega)I_1 + (100 \ \Omega)I_3 = 0$

On adding these two equations, we obtain

$$100I_3 = 50I_2 \quad \text{or} \quad I_2 = 2I_3$$

From which $I_1 = I_2 + I_3 = 2I_3 + I_3 = 3I_3$

On substituting this in our second loop equation, we have

$$-12 \text{ V} + (0.10 \ \Omega)(3I_3) + (100 \ \Omega)I_3 = 0$$

from which $I_3 = \dfrac{12 \text{ V}}{100.3 \ \Omega} = \boxed{0.12 \text{ A}}$

Hence $I_2 = 2I_3 = 2(0.12 \text{ A}) = \boxed{0.24 \text{ A}}$

and $I_1 = 3I_3 = 3(0.12 \text{ A}) = \boxed{0.36 \text{ A}}$

In this case all the currents come out positive, which means that our original assumption about directions was correct.

(b) The terminal voltage differs from the emf by $I_1 r = (0.36 \text{ A})(0.10 \ \Omega) = 0.036 \text{ V}$.

Resistors in Series

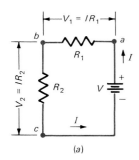

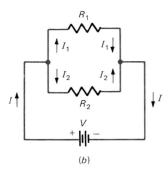

FIGURE 18.23 (a) Two resistors in series: $R_1 = R_2 + R_3$. (b) Two resistors in parallel: $1/R = 1/R_1 + 1/R_2$.

Kirchhoff's rules can be used to obtain some very practical equations for combining resistors in electric circuits.

There are two ways to combine resistors. The first way is to connect them *in series*, as in Fig. 18.23a. Here the current is the same in the two resistors R_1 and R_2. By Kirchhoff's loop rule we have

$$V - IR_1 - IR_2 = 0$$

where V is the terminal voltage of the battery, and so

$$V = I(R_1 + R_2)$$

But the resistance of a circuit that draws the same current I when the same voltage V is applied to it, the so-called equivalent resistance, is

$$R_{eq} = \frac{V}{I} \quad \text{or} \quad V = IR_{eq}$$

and so we have

$$IR_{eq} = I(R_1 + R_2)$$

or $$\boxed{R_{eq} = R_1 + R_2} \tag{18.9}$$

For *three* resistors in series,

$$R_{eq} = R_1 + R_2 + R_3 \quad \text{or, in general,} \quad R_{eq} = \sum_n R_n$$

We therefore conclude:

Resistors in series: The equivalent resistance is the sum of the individual resistances.

There is a practical problem with any series circuit of the kind in Fig. 18.23a. For example, suppose R_1 is the filament resistance of a light bulb and R_2 is the resistance of a toaster's heating element. Then if the light bulb burns out, the filament wire breaks and the toaster will not work if the two are connected together in series, for there is no longer a complete circuit from a to c. This was a great nuisance in the series-wired Christmas-tree lights used before World War II. Every time one bulb burned out, the whole set went dark and considerable time was wasted in finding the burnt-out bulb. Series connections are now rarely used in house wiring. The exception is a fuse put in series with an appliance precisely to cut off the current to the appliance if a malfunction blows the fuse.

Resistors in Parallel

The second way to combine resistors is to connect R_1 and R_2 in parallel, as in Fig. 18.23b. Here the current I splits into two parts, a current I_1 passing through R_1 and a current I_2 passing through R_2.

By Kirchhoff's loop rule applied to the upper branch,

$$V - I_1R_1 = 0 \quad \text{or} \quad V = I_1R_1$$

Similarly, for the lower branch:

$$V - I_2R_2 = 0 \quad \text{or} \quad V = I_2R_2$$

$$\text{Hence} \quad I_1 = \frac{V}{R_1} \quad \text{and} \quad I_2 = \frac{V}{R_2}$$

But, from Kirchhoff's junction rule, $I = I_1 + I_2$, and so

$$I = \frac{V}{R_{eq}} = \frac{V}{R_1} + \frac{V}{R_2}$$

where R_{eq} is the equivalent resistance of the circuit. If we now cancel V on both sides of the preceding equation, we obtain

$$\boxed{\frac{1}{R_{eq}} = \frac{1}{R_1} + \frac{1}{R_2}}$$

(18.10)

For three resistors in parallel,

$$\frac{1}{R_{eq}} = \frac{1}{R_1} + \frac{1}{R_2} + \frac{1}{R_3} \quad \text{or, in general,} \quad \frac{1}{R_{eq}} = \sum_n \frac{1}{R_n}$$

Resistors in parallel: The reciprocal of the equivalent resistance is the sum of the reciprocals of the individual resistances.

Note that in this case the equivalent resistance is *less* than any of the individual resistances. The reason is that by providing alternate, parallel paths we are in a sense increasing the cross-sectional area of the conductor and thus decreasing the overall resistance.

In most household circuits, appliances are connected in parallel. Each appliance therefore has the full 120 V applied across it, and the current it draws is determined by its resistance. Parallel circuits have the great advantage that the burnout of one light bulb or other device does not interrupt the current to the other appliances connected in parallel with it.

Example 18.7

Find the current delivered by a 6.0-V battery when it is connected to two light bulbs, each of resistance 50 Ω and connected **(a)** in parallel; **(b)** in series.

SOLUTION

(a) The circuit is as shown in Fig. 18.24a, with $R_1 = R_2 = 50\ \Omega$. Since the two resistors are in parallel,

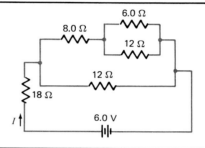

FIGURE 18.24

(a)

(b)

$$\frac{1}{R_{eq}} = \frac{1}{50\ \Omega} + \frac{1}{50\ \Omega} = \frac{2}{50\ \Omega} \quad \text{or} \quad R_{eq} = 25\ \Omega$$

and so $\quad I = \dfrac{V}{R_{eq}} = \dfrac{6.0\ \text{V}}{25\ \Omega} = \boxed{0.24\ \text{A}}$

(b) In this case, the circuit is as in Fig. 18.24b, and $R_{eq} = R_1 + R_2 = 100\ \Omega$. Then

$$I = \frac{V}{R_{eq}} = \frac{6.0\ \text{V}}{100\ \Omega} = \boxed{0.060\ \text{A}}$$

Example 18.8

(a) In the circuit of Fig. 18.25 find the current I in the 18-Ω resistor. The terminal voltage of the battery is 6.0 V.
(b) Find the currents in the individual branches of the circuit.

FIGURE 18.25

SOLUTION

(a) Here we must first find the equivalent resistance of the circuit because this is required to solve for the current delivered by the battery. In such complicated circuits the best approach in applying the laws for combining resistances in series and in parallel is to start with the innermost combination of resistances. This is the combination of the 6.0-Ω and the 12-Ω resistors in parallel. We find

$$\frac{1}{R_{eq}} = \frac{1}{6.0\ \Omega} + \frac{1}{12\ \Omega} = \frac{3.0}{12\ \Omega} \quad \text{or} \quad R_{eq} = 4.0\ \Omega$$

as in Fig. 18.26a. We can then add this in series with the 8.0-Ω resistor to obtain

$$R'_{eq} = 8.0\ \Omega + R_{eq} = 8.0\ \Omega + 4.0\ \Omega = 12\ \Omega$$

as in Fig. 18.26b.

Next we must combine R'_{eq} with the 12-Ω resistor in parallel to obtain R''_{eq}. We have

$$\frac{1}{R''_{eq}} = \frac{1}{R'_{eq}} + \frac{1}{12\ \Omega} = \frac{1}{12\ \Omega} + \frac{1}{12\ \Omega} = \frac{1}{6.0\ \Omega}$$

and so $R''_{eq} = 6.0\ \Omega$, as in Fig. 18.26c. This must then be combined in series with the 18-Ω resistor to yield a final equivalent resistance for the circuit of $R = 24\ \Omega$, as in Fig. 18.16d.

Our circuit has now been reduced to a voltage of 6.0 V driving a current through an equivalent resistance of 24 Ω. Hence, from Ohm's law,

$$I = \frac{V}{R} = \frac{6.0\ \text{V}}{24\ \Omega} = \boxed{0.25\ \text{A}}$$

This is the current in the 18-Ω resistor.

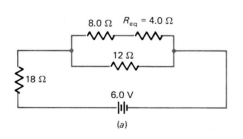

(a)

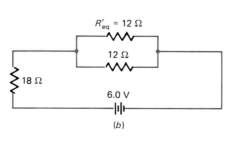

(b)

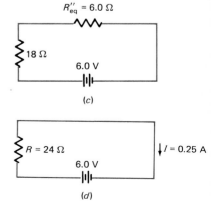

(c)

(d)

FIGURE 18.26   Steps in combining resistors in Example 18.8.

(b)   Since $R'_{eq} = 12\ \Omega$ and is in parallel with another 12-Ω resistor, the 0.25-A current must split in two, with half going through the 12-Ω resistor and half going through the other branch. The 0.125 A going through the upper branch must also split so that twice as much current goes through the 6.0-Ω resistor as through the 12-Ω resistor, since the voltage across the two is the same. Hence $\boxed{0.042\ \text{A}}$ goes through the

12-Ω resistor and $\boxed{0.084\ \text{A}}$ through the 6.0-Ω resistor, as in Fig. 18.27.

EXERCISE 2   Calculate the voltages across all elements of this circuit, and show that they all add correctly to the 6.0 V provided by the battery, as they must according to Kirchhoff's loop rule. The results are shown in Fig. 18.27.

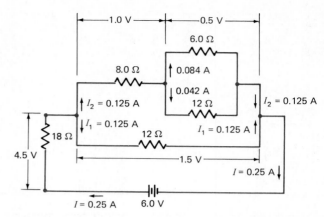

FIGURE 18.27   Voltages across circuit elements in Example 18.8.

DC Circuits Containing Resistors and Capacitors

Capacitors do not usually play important roles in dc circuits because no direct current can flow through a capacitor. An applied voltage will merely charge up a capacitor in a dc circuit, and then the current will stop. For example, consider the circuit of Fig. 18.28. Suppose the switch K is closed and charge flows. Charge will build up on the plates of the capacitor very quickly until the voltage between points a and b is equal to the applied voltage V, that is, until $V_{ab} = Q/C = V$. After the charge reaches a value Q, determined by $Q = CV$, the capacitor behaves as if it were no longer in the circuit. All the current is now through the resistor R, with its value being given by Ohm's law, $I = V/R$.

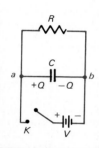

FIGURE 18.28   A capacitor and a resistor in parallel.

Example 18.9

Consider the circuit of Fig. 18.29, where the emf's are assumed to have zero internal resistance. **(a)** Find the currents I_1, I_2, and I_3. **(b)** Find the voltage across the capacitor. **(c)** Find the charge on the capacitor.

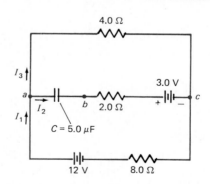

FIGURE 18.29

SOLUTION

(a)   No charge can flow through a capacitor, and so once the capacitor is charged, there is no current through the whole middle branch. We have, therefore,

$$\boxed{I_2 = 0}\qquad \text{and so}\qquad I_1 = I_3$$

We then have a simple series circuit of only one branch,

with a voltage of 12 V driving a current through the 4.0- and 8.0-Ω resistors in series:

$$I_1 = \frac{\sum \mathscr{E}}{R_{eq}} = \frac{12\ \text{V}}{4.0\ \Omega + 8.0\ \Omega} = \frac{12\ \text{V}}{12\ \Omega} = \boxed{1.0\ \text{A}}$$

$$I_3 = I_1 = \boxed{1.0\ \text{A}}$$

(b) Let us apply Kirchhoff's loop rule to the upper branch of the circuit, going clockwise around the circuit starting at c. Then we have, since there is no current through the 2.0-Ω resistor,

$$3.0 \text{ V} + V_{ba} - (4.0 \ \Omega)(1.0 \text{ A}) = 0$$

or $\quad V_{ba} = 4.0 \text{ V} - 3.0 \text{ V} = \boxed{1.0 \text{ V}}$

where side a of the capacitor is 1.0 V more positive than side b.

(c) Since $C = Q/V$,

$$Q = CV = (5.0 \times 10^{-6} \text{ F})(1.0 \text{ V})$$

$$= 5.0 \times 10^{-6} \text{ C} = \boxed{5.0 \ \mu\text{C}}$$

18.7   Energy and Power in DC Circuits

Electric energy is particularly useful to us because it is so easily converted into other forms of energy—heat in a heating element, light in a light bulb, mechanical motion in a motor. In all these cases the electric energy produced in a seat of emf is eventually converted into some other form of energy.

Electric Power

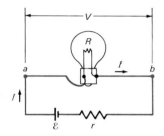

FIGURE 18.30   Electric light bulb connected to a battery. The applied voltage is shown as the emf of the battery reduced by the Ir drop across its internal resistance.

In dc circuits like that in Fig. 18.30, when a charge q flows from point a to point b through a light bulb of resistance R, the work done by the battery in moving the charge from a to b is

$$\mathcal{W} = \text{PE}_a - \text{PE}_b = qV$$

where V is the voltage drop between a and b. If this work $\mathcal{W}$ is done in time t, then the power delivered by the battery to the light bulb is

$$P = \frac{\mathcal{W}}{t} = \frac{qV}{t} = IV \qquad \text{or} \qquad \boxed{P = VI} \tag{18.11}$$

Power in a dc electric circuit is the product of the voltage and the current.

This is true for the circuit as a whole or for any individual part of the circuit. In SI units, in which voltage is measured in volts and current in amperes, the unit of power is the watt (W), or joule per second, so that

$$\boxed{1 \text{ W} = 1 \text{ V} \cdot 1 \text{ A}}$$

This equation shows the advantage of the SI system of units. Using this system, electric power can be expressed in the same unit as mechanical power, the watt, if the voltage is measured in volts and the current in amperes.

Home Usage of Electricity

The *rate* at which a home uses electric energy is therefore 120 V (the commonly used house voltage) times the current drawn at any particular time.* This is the *power*, or *the rate at which energy is consumed*, and is measured in watts or kilowatts (10^3 W). If electric energy is used for 1 h at the rate of 1 kW, the amount of energy consumed is 1 kW · 1 h = 1 kWh. This is

---

*Even though almost all house wiring now carries alternating current (ac) rather than direct current (dc), Ohm's law and other basic equations of this chapter remain valid as long as the circuits contain only resistances. This is true of many practical appliances such as electric heaters, electric stoves, toasters, light bulbs, etc. In such cases, I refers to the *effective value* of the ac current, i.e., the dc current that would have the same heating effect as the ac being considered, and V refers to the *effective value* of the ac voltage. (These ideas are clarified and expanded in Chap. 21.)

$$1 \text{ kWh} = \left(10^3 \frac{J}{s}\right)(3600 \text{ s}) = 3.6 \times 10^6 \text{ J} \quad \text{or} \quad \boxed{1 \text{ kWh} = 3.6 \times 10^6 \text{ J}}$$

The kilowatt is a unit of *power*, but the *kilowatthour* is a unit of *energy*. Kilowatts and kilowatthours are frequently confused in newspaper and magazine articles. Monthly electric bills are based on the number of kilowatthours used, not on the number of kilowatts used, since the latter indicates only the *rate* at which electric energy is used and not the *amount* of energy used. When you fill up your gasoline tank at a service station, you pay not for the rate at which gasoline is pumped into your car but for the amount of gasoline put into the tank. Similarly, you pay the utility company not for the rate (in kilowatts) at which they deliver electric energy to your house but for the total electric energy (in kilowatthours) delivered.

Example 18.10

A 150-W light bulb is connected to a 120-V line. **(a)** What is the current drawn from the line? **(b)** What is the resistance of the light bulb while it is burning? **(c)** How much energy is consumed if the light is kept on for 6.0 h? **(d)** What is the cost of this energy at 8.0 cents/kWh?

SOLUTION

(a) Since the power is equal to IV, we have

$$I = \frac{P}{V} = \frac{150 \text{ W}}{120 \text{ V}} = \boxed{1.25 \text{ A}}$$

(b) The resistance of the light bulb is

$$R = \frac{V}{I} = \frac{120 \text{ V}}{1.25 \text{ A}} = \boxed{96 \ \Omega}$$

(c) $W = Pt = (150 \text{ W})(6.0 \text{ h})\left(3600 \frac{s}{h}\right) = 3.2 \times 10^6 \text{ J}$

Since $1 \text{ kWh} = 3.6 \times 10^6 \text{ J}$,

$$W = (3.2 \times 10^6 \text{ J})\left(\frac{1 \text{ kWh}}{3.6 \times 10^6 \text{ J}}\right) = \boxed{0.90 \text{ kWh}}$$

The same result could be obtained by using kilowatthours as the energy unit from the beginning, so that

$$W = Pt = (150 \text{ W})(6.0 \text{ h}) = 900 \text{ Wh} = \boxed{0.90 \text{ kWh}}$$

(d) $\text{Cost} = \left(\frac{8.0 \text{ cents}}{1 \text{ kWh}}\right)(0.90 \text{ kWh}) = \boxed{7.2 \text{ cents}}$

18.8 Joule Heating

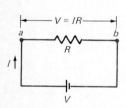

FIGURE 18.31 Joule heating in a resistor: $P = I^2R$.

We can now combine two previous results into an equation that James Joule (Sec. 13.8) used to help establish the first law of thermodynamics. Consider the simple circuit of Fig. 18.31, in which a battery of terminal voltage V drives a current I through a coiled wire of resistance R. In the battery, chemical energy is being converted into electric energy. In the resistor, this electric energy is being converted into thermal energy, since the moving electrons collide with the positive ions in the metal and give up some of their kinetic energy to these ions. The random motion of the metal ions is increased, and so their temperature (and therefore their thermal energy) increases. The energy imparted to the electrons by the battery is ultimately used to heat the metal. This is called *Joule heating*.

Since the voltage across R is V, the power delivered to the resistor is $P = VI$. But, from Ohm's law, $V = IR$, and so

$$P = (IR)I = I^2R \quad \text{or} \quad \boxed{P = I^2R} \tag{18.12}$$

The rate at which thermal energy is generated in a resistor is proportional to its resistance and to the square of the current through the resistor.

The total energy converted into heat in time t in the resistor is then

$$\boxed{\mathcal{W} = Pt = I^2Rt}$$ (18.13)

Electrical Equivalent of Heat

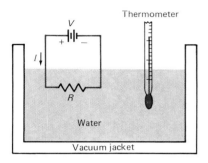

Thermometer

R

Water

Vacuum jacket

FIGURE 18.32 Experiment to determine the electrical equivalent of heat. From a measurement of V and I, the electric energy delivered in time t can be determined. From a measurement of the increase in temperature of the water of mass m, the heat input to the water can be determined.

To show the validity of the first law of thermodynamics we can perform an experiment first performed by Joule. We take the resistor of Fig. 18.31 and immerse it in water inside a thermally insulated container, as in Fig. 18.32. In time t the amount of electric energy delivered to the resistor is I^2Rt. This energy is all converted into heat in the resistor, the heat is conducted from the resistor into the water, and the water temperature increases. If the mass of the water is m, then the increase in thermal energy of the water is $cm\,\Delta T$, where c is the specific heat capacity of the water. In this way we can measure an amount of electric energy in joules (I^2Rt) and the equivalent amount of thermal energy in kilocalories ($cm\,\Delta T$).

When Joule performed this experiment, he consistently found the same relationship between the joule and the kilocalorie:

$$1 \text{ kcal} = 4.18 \times 10^3 \text{ J}$$

This is the same basic result Joule obtained from his paddle-wheel experiment (Sec. 13.8) in which mechanical energy was converted into heat.

Here is another demonstration that energy is neither created nor destroyed but only changes form. In this case energy is conserved but is degraded in the conversion process, since the ordered chemical energy stored in the battery is ultimately converted into the disordered, random motion of the water molecules. Thus in this electrical version of Joule's experiment we find confirmation of both the first and second laws of thermodynamics: energy is conserved and entropy increases.

Example 18.11

A 50-Ω resistor is attached to a 120-V dc voltage source and immersed in 1.5 kg of water. If current is drawn from the line for 15 min, what is the increase in temperature of the water, assuming that all the electric energy goes into heating the water?

SOLUTION

The current drawn from the circuit is

$$I = \frac{V}{R} = \frac{120 \text{ V}}{50 \,\Omega} = 2.4 \text{ A}$$

The electric energy going into heating the water is then

$$\mathcal{W} = I^2Rt$$
$$= (2.4 \text{ A})^2(50 \,\Omega)(15 \text{ min})\left(60\,\frac{\text{s}}{\text{min}}\right)$$
$$= 2.6 \times 10^5 \text{ J}$$

and the heat produced is

$$\Delta Q = (2.6 \times 10^5 \text{ J})\left(\frac{1 \text{ kcal}}{4.18 \times 10^3 \text{ J}}\right) = 62 \text{ kcal}$$

Then, from our equation for the specific heat capacity (from Sec. 13.7),

$$\Delta T = \frac{\Delta Q}{cm} = \frac{62 \text{ kcal}}{[1.0 \text{ kcal/(kg·C}°)](1.5 \text{ kg})} = \boxed{41 \text{ C}°}$$

Summary: Important Definitions and Equations

Source of emf ($\mathscr{E}$): A device (such as a battery) that transforms nonelectric energy into electric energy.

$$\mathscr{E} = \frac{W}{q'} \tag{18.1}$$

where $\mathscr{E}$ is in volts if W is in joules and q' in coulombs.

Battery: A device that converts chemical energy into electric energy on a continuous basis.
Terminal voltage of battery:

$$V_{AB} = \mathscr{E} - Ir \tag{18.8}$$

where r is the internal resistance of the battery.

Electric current: The rate at which electric charge passes a point in the circuit.

$$I = \frac{\Delta Q}{\Delta t}; \ 1 \text{ ampere (A)} = 1 \text{ C/s} \tag{18.2}$$

Direct current: A current always in the same direction.

Conventional current: A current in the direction in which positive charges would flow in an electric circuit.

Faraday (F): The amount of charge (96,500 C) that deposits or liberates an amount of substance from an electrolytic solution equal to the mass of one mole of the substance divided by its valence.

$$F = N_A e \tag{18.3}$$

Ohm's law (for currents in metals):

$$V = IR \tag{18.4}$$

where the resistance R (in ohms) is

$$R = \frac{\rho L}{A}, \tag{18.5}$$

and ρ is the resistivity (in ohm-meters) of the material in the wire.

1 ohm (Ω) = 1 V/A

Temperature dependence of resistance

$$R_T = R_0(1 + \alpha \, \Delta T) \tag{18.7}$$

where α is the temperature coefficient of resistivity, or the fractional increase in resistivity per Celsius degree rise in temperature.

Combinations of resistors

In series: $\quad R_{eq} = R_1 + R_2 \tag{18.9}$

In parallel: $\quad \dfrac{1}{R_{eq}} = \dfrac{1}{R_1} + \dfrac{1}{R_2} \tag{18.10}$

Kirchhoff's rules

1 Junction rule: The sum of all the currents entering any junction point is equal to the sum of all the currents leaving that junction point.

2 Loop rule: The algebraic sum of the changes in potential encountered in a complete traversal of any closed circuit is equal to zero.

Sign conventions: If a resistor is traversed in the same direction as the current through the resistor, the change in potential is $-IR$; if in the opposite direction, the change in potential is $+IR$. If a seat of emf is traversed in the direction of the emf, the change in potential is $+\mathscr{E}$; if in the opposite direction, it is $-\mathscr{E}$.

Electric power

$$P = VI \tag{18.11}$$

$$1 \text{ W} = 1 \text{ V} \cdot 1 \text{ A}$$

Joule heating

$$P = I^2 R \tag{18.12}$$

$$W = I^2 R t \tag{18.13}$$

Questions

1 When a dry cell is driving current through an external circuit, electrons inside the battery are moving from the positive to the negative terminal. How can this happen, since the electrons should be attracted to the positive terminal and repelled by the negative terminal?

2 Give some examples of situations in which charges flow through a closed circuit but Ohm's law is not obeyed. Is it proper to call a relationship like $V = IR$ a law of physics if it has only limited applicability?

3 A 6-V battery is connected in series with a capacitor.
(*a*) If the wires leading to the capacitor have a very low resistance, what is the voltage across the capacitor plates after all the charge has ceased to flow?
(*b*) If the wires leading to the capacitor had a very high resistance, how would the situation change? Would the battery have expended the same amount of chemical energy as in part (*a*)?

4 Compare the formulas for the equivalent resistance of resistors in series and parallel with the same formulas derived in the last chapter for capacitors in series and parallel. Can you give a reason why these formulas turn out to be different in the two cases?

5 As the temperature of a wire is increased, the average kinetic energy of the electrons increases in proportion to the absolute temperature. Would you not expect, therefore, that the drift velocity of the electrons would increase with the temperature? Why, then, does the resistance of a wire increase with increasing temperature?

6 Would you expect the internal resistance of a dry cell to be a constant? Why?

7 Can the terminal voltage of a battery be zero? Why?

8 In a flashlight battery containing zinc and carbon electrodes, both the zinc and the carbon, if left by themselves, are electrically neutral. How then can a flashlight produce a potential difference between the zinc and the carbon electrodes?

9 If you measure the resistance of a 100-W light bulb, you will find that it is smaller than for a 60-W bulb. Is this not inconsistent with the fact that $P = I^2R$?

10 Why is it dangerous to replace a burnt-out 20-A fuse that is constantly blowing with a 30-A fuse in the hope that the fuse will no longer blow?

11 What happens when a light bulb burns out? Why is there a momentary flash of light just before the bulb goes dark?

12 Car batteries are often rated in ampere-hours (A·h). Is this a unit of current, power, voltage, energy, or charge? Explain.

13 The potential difference across a connecting wire in an electric circuit is usually assumed to be zero, since the wire has no resistance, and so $V = IR = 0$. How then can there be a current through the wire? Can you suggest a good analogy between this situation and the flow of a liquid through a pipe?

14 An Associated Press release on December 16, 1988, discussed the deactivation of the first commercial nuclear reactor in the United States, dedicated by President Eisenhower at Shippingport, Pennsylvania, in 1957. The article states that between 1957 and 1982 this reactor generated about 6.5 billion kilowatts of electricity. Comment on the accuracy of this statement.

Multiple-Choice and Simple Exercises

18.1 The work done by a 6.0-V battery is moving a single electron through the battery from its positive to its negative terminal is:
(a) 6.0 J (b) 6.0 V (c) 1.6×10^{-19} J
(d) 9.6×10^{-19} J (e) 1.6×10^{-19} W

18.2 A 6.0-V battery has an internal resistance of 0.50 Ω. To reduce the terminal voltage to 3.0 V, the battery will have to deliver to the external circuit a current of:
(a) 3.0 A (b) 6.0 A (c) 0.50 A
(d) 1.0 A (e) 12.0 A

18.3 A storage battery delivers a 10-A current for 10 min to a heating coil. The amount of charge that flowed through the heating coil in the 10 min is:
(a) 6.0×10^3 C (b) 100 C (c) 10 C
(d) 6.0×10^3 J (e) 1.0 C

18.4 The amount of sodium deposited on the cathode of an electrolytic cell by the flow of 1 C of charge through the cell is about:
(a) 1 mol (b) 10^{-6} mol (c) 10^{-5} mol
(d) 0.5 mol (e) 10^{-3} mol

18.5 A piece of metal wire is cut into three equal lengths, and the pieces are then put together side by side to form a thicker wire. The resistance of the new wire is less than that of the original wire by a factor of:
(a) 3 (b) 6 (c) 9 (d) 27 (e) 81

18.6 If the temperature of a copper wire increases by 250 C°, the resistance of the wire increases by a factor of about:
(a) 2 (b) 10^2 (c) 250 (d) 10^3
(e) 10

18.7 A 200-Ω and a 100-Ω resistor are connected in parallel. The equivalent resistance of the combination is:
(a) 300 Ω (b) 20 Ω (c) 150 Ω
(d) 67 Ω (e) 100 Ω

18.8 In Fig. 18.33, $R_1 = R_2 = R_3$. What happens to the current (I_3) through R_3 when the circuit is broken at point O?

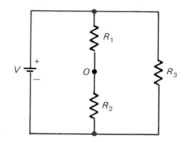

FIGURE 18.33 Diagram for Exercises 18.8 and 18.9.

(a) I_3 increases to three-halves its original value.
(b) I_3 decreases to two-thirds its original value.
(c) I_3 remains the same.
(d) I_3 increases to twice its original value.
(e) I_3 is reduced to one-half its original value.

18.9 In Fig. 18.33, $R_1 = R_2 = R_3$. If resistor R_1 is removed from the circuit, leaving R_2 and R_3 in parallel, the following is true:
(a) The current through R_3 increases.
(b) The current through R_3 decreases.
(c) The current through R_2 remains the same as when R_1 was in the circuit.
(d) The current through R_2 is half what it was with R_1 in the circuit.
(e) The currents through R_2 and R_3 are now the same.

18.10 A power line carries 1000 A and has a resistance of 0.10 Ω. The rate at which electric energy is converted into heat in the line is:
(a) 100 W (b) 1.0×10^5 W
(c) 1.0×10^5 J (d) 1.0×10^4 W
(e) 10 W

18.11 A 12-V automobile battery is being charged with a current of 20 A. The charging process lasts 6.0 h. How much electric charge has passed through the battery?

18.12 A current of 10 A is maintained for 45 min through an electrolytic cell containing Cu^{2+} ions. What mass of copper is deposited?

18.13 A voltage of 120 V is applied to a light bulb of resistance 200 Ω.

(a) What is the current through the light bulb?

(b) How much charge flows through the light bulb every hour?

18.14 A sparrow stands on a 240 kV high-tension wire carrying 1000 A. The line has a resistance of 2.0×10^{-5} Ω/m. What voltage does the bird feel across its body if its feet are 5.0 cm apart?

18.15 What is the resistance, at 0°C, of a piece of ni-chrome wire 0.50 m long and of cross-sectional area 1.0 mm²?

18.16 Two resistors, $R_1 = 40.0$ Ω and $R_2 = 60.0$ Ω, are connected in parallel to a 120-V source.

(a) What is the equivalent resistance of the circuit?

(b) What is the current delivered by the battery?

(c) What is the current through R_1?

18.17 A 6.0-V battery that has an internal resistance of 0.50 Ω is delivering 2.0 A to the external circuit. What is the terminal voltage of the battery?

18.18 A 60-W bulb is connected to a 120-V line.

(a) What is the current in the circuit?

(b) What is the resistance of the light bulb?

18.19 A power line of resistance 1.0 Ω carries 100 A of current for a day. How much energy has the power company lost in the form of heat in the line?

18.20 The current through a 100-W light bulb is 0.83 A.

(a) What is the bulb's resistance?

(b) What is the line voltage?

Problems

18.21 How much work is done by a storage battery, of emf 12 V and negligible internal resistance, if it delivers a 50-A current for 10 s?

18.22 An electron is accelerated from rest by a voltage of 15×10^3 V between the cathode and the anode of an x-ray tube.

(a) When the electron crashes into the anode, what energy does it possess?

(b) What is the electron's speed just before it hits the anode?

18.23 In the video tube of a TV set the electron beam is accelerated toward the screen by a voltage of 2.0×10^4 V. With what speed does an electron strike the screen?

18.24 A voltage of 10 V is applied across the ends of a piece of copper wire 10 cm long.

(a) What force is experienced by an electron in the wire?

(b) What is the electron's acceleration?

(c) What is the speed of the electron after traveling 10^{-11} m, if it does not collide with a copper ion over this distance?

18.25 Chlorine gas (Cl_2) can be produced from a solution containing Cl^- ions, e.g., from Na^+Cl^-, by electrolysis.

(a) What must the current through an electrolytic cell be to produce 1.0 kg of Cl_2 gas in 9.0 h?

(b) What is the volume of this gas, stored as Cl_2 at 0°C and 50 atm?

18.26 Objects made of iron are often plated with cadmium to make them rustproof. If the cadmium is present in an electrolytic cell in the form of Cd^{2+} ions, how much cadmium will be plated on an iron object in 5.0 h by a 20 A current?

18.27 Objects can be gold- or silver-plated using electrolysis. It is desired to gold-plate a metal object to a thickness of 20 μm (that is, 20×10^{-6} m). The surface area of the object is 0.010 m², and it is placed in an electrolytic cell containing gold (Au^{3+}) ions. If a current of 5.0 A is driven through the cell, how long does the plating take?

18.28 In starting a car a 12-V battery is required to de-liver an extremely large current, perhaps as much as 200 A, for a few seconds. There is always some resistance in the wires leading to the motor from the battery and in the lugs attaching the wires to the battery terminals. What is the maximum allow-able resistance in each of the two leads if a 200-A current is to be delivered by the battery? (This provides a good reason for keeping the battery lugs on a car clean to reduce the contact resistances as much as possible.)

18.29 What is the resistance at 100°C of a piece of no. 5 gauge copper wire of length 4.5 m and cross-sectional area 16.8 mm²?

18.30 Figure 18.34 is a diagram of an ohmmeter, a de-vice used to measure electric resistance. The meter measures the current in the circuit, where V is the applied voltage, R_{ser} is a fixed series resistance, and R_x is the unknown resistance. The scale on the meter can be marked to read resistance directly, even though the instrument is highly nonlinear and not very precise.

(a) Show that the needle deflection of the meter is roughly inversely proportional to the resistance R_x, if R_{ser} is small compared to R_x.

(b) What is the value of the current when $R_x \rightarrow \infty$?

(c) When $R_x \rightarrow 0$ but the circuit is still complete between points a and b?

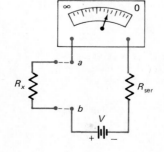

FIGURE 18.34
Diagram for Prob. 18.30.

18.31   A platinum resistance thermometer has a resistance of 50 Ω at 0°C. It is then immersed in a mixture of solid and liquid tin at 2260°C. What resistance will the thermometer now show?

18.32   A 100-W incandescent bulb has a resistance of 16 Ω when it is at room temperature and 144 Ω when it is hot and delivering light to the room. What is the approximate temperature of the bulb when in use, if the filament is made of tungsten?

18.33   If you are given some nichrome wire, with $R = 24$ Ω, what is the better way to get more heat from this wire—by winding one heating coil out of it, or by cutting it in half, winding two separate coils, and connecting them in parallel? In each case, the coils are connected across a 120-V line.

*18.34   Copper is a slightly better conductor than aluminum, but its density is greater. Show that for two electric power lines 1000 m in length an aluminum line will have a lower resistance than a copper line of the same weight.

18.35   In Fig. 18.35, $R_1 = 5.0$ Ω, $R_2 = 7.0$ Ω, $R_3 = 8.0$ Ω, $\mathscr{E}_1 = 9.0$ V, $\mathscr{E}_2 = 6.0$ V, with the emf equal to the terminal voltage for each battery,

(a)   What is the current in the circuit?

(b)   Is the energy of battery 1 increasing or decreasing when a current exists in the circuit? Why? (Assume that both batteries are rechargeable.)

(c)   Is the energy of battery 2 increasing or decreasing when a current exists? Why?

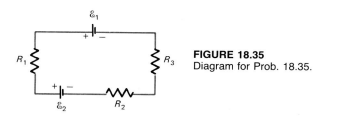

FIGURE 18.35
Diagram for Prob. 18.35.

18.36   Find I_1, I_2, and I_3 in Fig. 18.36. Assume that the batteries have zero internal resistance.

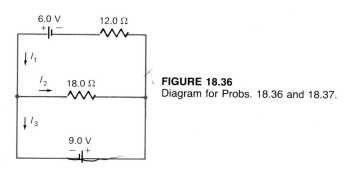

FIGURE 18.36
Diagram for Probs. 18.36 and 18.37.

18.37   Solve the circuit of Fig. 18.36 for I_1, I_2, and I_3 under the assumption that each battery has an internal resistance of 1.0 Ω.

18.38   Prove that when a storage battery with internal resistance r is being charged, the voltage that must be applied to charge it is $V = \mathscr{E} + Ir$, where I is the charging current and $\mathscr{E}$ is the emf of the battery.

*18.39   (a)   Determine I_1, I_2, and I_3 in Fig. 18.37 under the assumption that the internal resistance (r) of each battery is 1.0 Ω.

(b)   What is the terminal voltage of the 18-V battery?

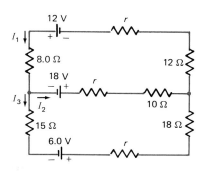

FIGURE 18.37   Diagram for Prob. 18.39.

18.40   In Fig. 18.38, $\mathscr{E}_1 = 12$ V, $\mathscr{E}_2 = 6.0$ V, and both have zero internal resistance. Also $R_1 = 24$ Ω, $R_2 = 12$ Ω, and $R_3 = 6.0$ Ω. Find I_1, I_2, and I_3.

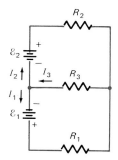

FIGURE 18.38   Diagram for Prob. 18.40.

18.41   (a)   Calculate the current delivered by each emf in Fig. 18.39.

(b)   Calculate the current through the 5.0-Ω resistor.

(c)   What is the terminal voltage of each battery?

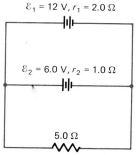

FIGURE 18.39   Diagram for Prob. 18.41.

*18.42 Calculate the potential difference between P and P' in the circuit of Fig. 18.40. Assume that the three emf's have negligible internal resistance and that $\mathcal{E}_1 = 40$ V, $\mathcal{E}_2 = 20$ V, and $\mathcal{E}_3 = 60$ V.

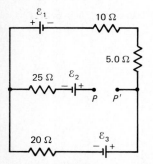

FIGURE 18.40 Diagram for Prob. 18.42.

18.43 (a) What is the potential difference between B and C in Fig. 18.41? (b) Which point is at the higher potential?

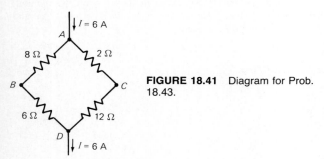

FIGURE 18.41 Diagram for Prob. 18.43.

18.44 Figure 18.42 is the diagram of a Wheatstone bridge, a device used to measure unknown resistances to high accuracy by a "null" method that compares them with standard resistors of precisely known values. In the diagram R_1 and R_2 are fixed resistances, R is a variable standard resistor, and R_x is the unknown. Prove, using Kirchhoff's rules, that, if R is adjusted until the meter M indicates no current when the switch K is closed, then $R_x = (R_2/R_1)R$.

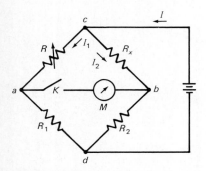

FIGURE 18.42 Diagram for Prob. 18.44.

*18.45 Two cells, one of emf 1.2 V and internal resistance 0.50 Ω and the other of emf 2.0 V and internal resistance 0.10 Ω, are connected in parallel, as shown in Fig. 18.43. This combination of two emf's is then connected in series with an external resistance of 5.0 Ω.
   (a) What is the current through the external resistor?
   (b) Prove that the power generated by the two emf's is equal to the power dissipated as heat in the circuit.

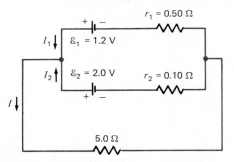

FIGURE 18.43 Diagram for Prob. 18.45.

18.46 What is the equivalent resistance when a 4.0- and an 8.0-Ω resistor are connected (a) in series and (b) in parallel?

18.47 Calculate the equivalent resistance of the following combinations of resistors:
   (a) A 3.0-Ω resistor, a 4.0-Ω resistor, and two 5.0-Ω resistors in series
   (b) A 4.0-Ω resistor, a 2.0-Ω resistor, and a 6.0-Ω resistor in parallel

18.48 In Fig. 18.44 calculate (a) the equivalent resistance of the circuit and (b) the current through each resistor.

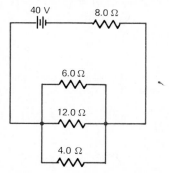

FIGURE 18.44 Diagram for Prob. 18.48.

*18.49 In Fig. 18.45, in which the internal resistance of the battery is 0.40 Ω, calculate (a) the equivalent resistance of the circuit and (b) the current through each resistor.

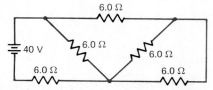

FIGURE 18.45 Diagram for Prob. 18.49.

18.50 Compute the current through the 12-Ω resistor in Fig. 18.46.

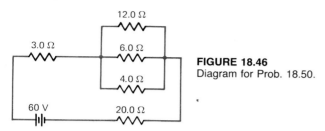

FIGURE 18.46
Diagram for Prob. 18.50.

18.51 Calculate the equivalent resistance between points A and B in the circuit of Fig. 18.47.

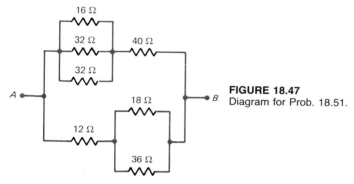

FIGURE 18.47
Diagram for Prob. 18.51.

*18.52 (a) In Fig. 18.48, find the currents I_1, I_2, and I_3. (b) Find the charge on the capacitor.

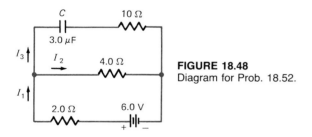

FIGURE 18.48
Diagram for Prob. 18.52.

*18.53 In the circuit of Fig. 18.49, what is the charge on the 10-μF capacitor?

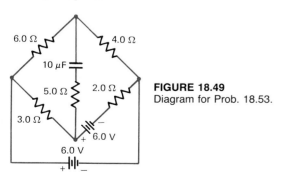

FIGURE 18.49
Diagram for Prob. 18.53.

18.54 In Fig. 18.50 find the charge on the capacitor if the battery has an internal resistance of 1.0 Ω.

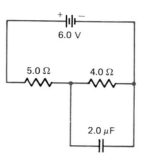

FIGURE 18.50 Diagram for Prob. 18.54.

18.55 If an electric air-conditioning unit draws 15 A from a 120-V line and is used 24 h a day during July, how much will the electricity to run this unit cost for the month if the local electrical rate is 8.0 cents/kWh?

18.56 Suppose you burn five 100-W bulbs, each for 6.0 h a day.
(a) How much electric energy do you consume each day?
(b) What is the cost of this energy at 8.0 cents/kWh?

*18.57 Some light bulbs have two separate filaments and allow three different wattages depending on how the filaments are connected, for example, 50-100-150 W.
(a) What resistances would be required in the two filaments for the bulb to consume 50 and 100 W of electric power when connected to a 120-V line?
(b) Show that the resistance needed to make the bulb function as a 150-W bulb is merely the sum of the two filament resistances in parallel. (Assume that Ohm's law remains valid, even though the current and voltage here are alternating rather than direct.)

18.58 Batteries are often rated in terms of ampere-hours, i.e., the number of hours the battery would last if current were drawn from it at the rate of 1 A. Consider a simple 1.5-V D cell used in flashlights. If it is rated for 3.0 A·h and the current drawn by the flashlight is 0.15 A:
(a) How long will the battery last?
(b) What will be the total energy delivered by the battery during its lifetime?
(c) If the battery costs 50 cents, what is the cost per kilowatthour for the electric energy used?
(d) How does this compare with the rate of about 8 cents/kWh at which utilities sell electric energy?

18.59 A student leaves her car in the parking lot of the university at 8 A.M., forgetting to turn off the car's lights. Assume that the two front lights each have 40-W bulbs and the two rear lights 6-W bulbs. The battery is rated at 100 A·h. How long will it take to completely run down the car's fresh 12-V battery?

18.60 In 1983 the Public Service Electric and Gas Co. of New Jersey installed two batteries costing $100 million each and occupying an acre of land. These batteries are used to store electric energy at off-peak periods and then deliver it when it is needed, e.g., to mass transit trains at rush hours. These batteries are zinc chloride batteries that can produce 500 kWh over a 5.0-h period before having to be recharged. At what rate does one of these batteries deliver energy to the electric-grid distribution system when it is discharging?

18.61 Two resistors are connected in series across a 60-V battery. One resistor has a resistance of 12 Ω.

(a) If the potential difference across the 12-Ω resistor is 20 V, what is the resistance of the second resistor?

(b) What is the power dissipated in this second resistor?

18.62 In Fig. 18.36 (assuming that the batteries have zero internal resistance), calculate:

(a) The power produced by the two batteries.

(b) The power dissipated as heat in the two resistors.

(c) Do your results indicate that energy is conserved?

18.63 An electric space heater draws 7.5 A from a 120-V line.

(a) If this heater is kept on 14 h/day, how much electric energy is used?

(b) If all the electric energy is converted into heat, how much heat is produced in kilocalories?

18.64 A chain of 16 identical 15-W Christmas tree bulbs are connected in series across a 115-V line. What is the resistance of each bulb?

Additional Readings

Asimov, Isaac: *Asimov's Biographical Encyclopedia of Science and Technology*, 2d rev. ed., Doubleday, Garden City, N.Y., 1982. This book contains brief biographies of some 1510 famous scientists, written in Asimov's lucid and entertaining style. Among the physicists included are Kirchhoff, Ohm, Volta, Galvani, and others who developed the physics presented in this chapter.

Bordeau, Sanford P.: *Volts to Hertz . . . The Rise of Electricity*, Burgess Publishing Co., Minneapolis, 1982. This popular account of the development of electricity contains brief biographies of 16 physicists (including Ampère, Faraday, Oersted, Henry, Maxwell, and Hertz) together with some excellent illustrations.

Healy, Timothy J.: *Energy, Electric Power and Man*, Boyd and Fraser, San Francisco, 1974. Included are good discussions of electric power, electric automobiles, and alternative energy sources.

Magie, William Francis: *A Source Book in Physics*, Harvard University Press, Cambridge, Mass., 1963. In this useful volume are a brief account of Ohm's life and a long excerpt from the paper in which he first proposed his famous law. It also contains material on the discovery of the battery by Galvani and Volta.

Rosenfeld, L.: "Gustav Robert Kirchhoff," in *Dictionary of Scientific Biography*, Scribner, New York, 1970, vol. 7, pp. 379–382. One of the few accounts available in English of Kirchhoff's life and contributions to physics.

Shamos, Morris H.: *Great Experiments in Physics*, Holt, Rinehart and Winston, New York, 1959. Faraday's experiments on electrolysis are described in his own words in Chap. 10 of this collection of important research papers in the history of physics.

Shepard, G. M.: "Microcircuits in the Nervous System," *Scientific American*, vol. 238, no. 2, February 1978, pp. 92–103. An interesting discussion of the microcircuits in the brain which provide a deeper understanding of the neural mechanisms underlying behavior.

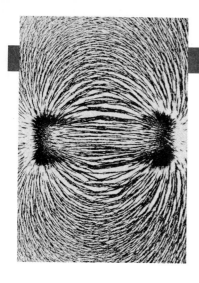

CHAPTER 19

Magnetism

The experimental investigation by which Ampère established the laws of the mechanical action between electric currents is one of the most brilliant achievements in science. The whole, theory and experiment, seems as if it had leaped full grown and full armed from the brain of the "Newton of Electricity."

James Clerk Maxwell (1831–1879)

The year 1820 was a notable one in the history of physics. In that year Oersted discovered that an electric current deflected a compass needle—the first clear evidence that magnetism and electricity were interrelated. A week after Oersted's discovery was reported to the French Academy of Sciences in Paris, Ampère had extended Oersted's work to show that two current-carrying conductors exerted forces on each other through the magnetic fields they produced. Thus began the branch of physics called *electrodynamics*, a name given to the field by Ampère himself.

Immediately scientists all over the world rushed to their laboratories to carry out experiments involving magnets and electric currents, driven not only by a desire to understand nature better but also by the sense that such remarkable discoveries might have very important practical consequences—as they indeed did. Witness the motors, generators, electric meters, telephones, and other more sophisticated electromagnetic devices that surround us today. As Isaac Asimov once remarked, nothing like the discoveries of Oersted and Ampère were seen again "until the announcement of nuclear fission a century later."

19.1 Magnets and the Magnetic Field

Most of us played with bar magnets and compass needles when we were young. Observations made during such experimentation lead to the following basic facts about the behavior of magnets:

1 *There are two kinds of magnetic poles*, or places on a magnet where the magnetic effects are particularly strong. These are called north (N) and south (S) poles because, when placed in the earth's magnetic field, the north pole of a magnetic needle will point north and the south pole will point south. The two kinds of magnetic pole (N and S) are analogous to the two kinds of electric charge (+ and −).

2 *Like magnetic poles repel each other and unlike poles attract each other.* This again is similar to the behavior of electric charges. (See Fig. 19.1.)

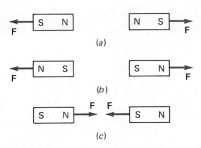

FIGURE 19.1 The interaction between magnetic poles. (a) Two north poles repel each other. (b) Two south poles repel each other. (c) A north pole attracts a south pole.

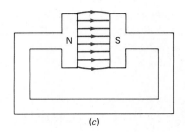

FIGURE 19.2 Cutting a bar magnet in two. The result is two smaller bar magnets.

3 *It is impossible to produce an isolated north or south pole in a magnet.** The production of one kind of magnetic pole is always accompanied by the simultaneous production of an opposite pole on the other end of the magnet. As an example, north and south poles cannot be separated by cutting a bar magnet in half. All that happens is that two smaller bar magnets result, each with N and S poles at the two ends, as shown in Fig. 19.2.

4 *Either pole of a magnet will attract previously unmagnetized materials containing iron, cobalt, or nickel, which are the primary magnetic substances.* (See Fig. 19.3.)

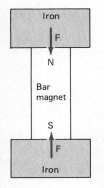

FIGURE 19.3 Either pole of a magnet will attract unmagnetized materials that contain iron, cobalt, or nickel.

The Magnetic Field

We have seen that surrounding any stationary electric charge is an electric field, which can be represented by electric field lines whose direction indicates the direction of the electric field at any point in space. Similarly, surrounding any magnetic pole is a magnetic field, which we represent by magnetic field lines. By analogy with the electric field case, we can define the magnetic field as follows:

Magnetic field (**B**): A vector whose direction at any point in space is the direction of the force which would be exerted on a small, isolated north pole at that point, and whose magnitude is proportional to the force experienced by that north pole.

According to this convention, the magnetic field lines always begin at the N pole of a magnet and end at a S pole, as can be seen in Fig. 19.4 for a variety of permanent magnets. An isolated N pole, if such existed, would be pushed by a

---

*The history of physics is marked by an almost continuous effort to isolate magnetic ''monopoles'' in the laboratory, but it is at present unclear whether such odd ''beasts'' will ever be captured and tamed.

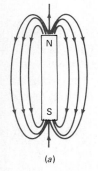

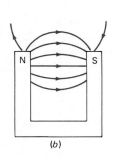

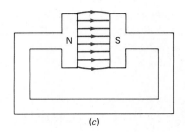

FIGURE 19.4 Some permanent magnets, with magnetic field lines drawn to show that the lines of the field start on north poles and end on south poles. (a) A bar magnet; (b) a horseshoe magnet; (c) a laboratory magnet with large, flat pole pieces.

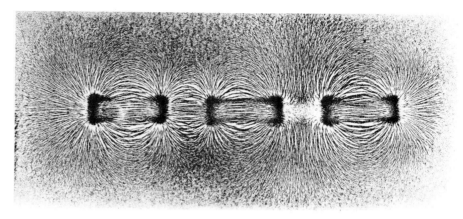

FIGURE 19.5 Magnetic field lines made visible by the use of iron filings that align themselves in the direction of the magnetic field. The adjacent poles of the first two magnets attract each other, while those of the second and third magnets repel each other. Notice the similarity of these magnetic field patterns to those of an electric dipole shown in Fig. 16.22. (*Courtesy of Omikron/Photo Researchers.*)

magnetic force from the N pole of the magnet to the S pole and would therefore move along the field lines in the direction shown by the arrows.

The lines of the magnetic field are illustrated by the use of iron filings in Fig. 19.5. The iron filings become magnetized in a magnetic field and line up along the lines of the magnetic field, just as tiny compass needles would.

19.2 Force on a Current-Carrying Conductor in a Magnetic Field

A stationary magnetic field has no effect on *stationary* electric charges, but it does exert a force on *moving* electric charges and therefore on electric currents.

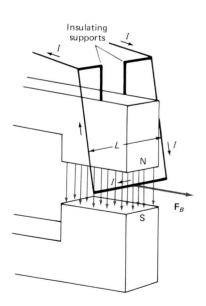

FIGURE 19.6 Apparatus to measure the force on a current-carrying conductor in a magnetic field. When a current *I* passes through the wire of length *L* in the magnetic field **B**, the wire experiences a force $F_B = B_\perp IL$ in the direction shown.

One of the easiest ways to obtain the magnitude of the magnetic field **B** is to measure the force exerted on a wire carrying a known current through a magnetic field. Suppose we have a magnetic field **B** in the vertical direction provided by a permanent magnet, as in Fig. 19.6. We use two light wires to suspend a heavier straight length of wire in the horizontal plane and at right angles to the magnetic field. When a current passes through the wire of length L, the magnetic field exerts a force $\mathbf{F}_B$ on the wire, pushing the wire out, as shown. This force can be measured in terms of the angle the wire makes with the vertical when it comes to rest, for this angle depends on the relative sizes of the magnetic (horizontal) and gravitational (vertical) forces acting on the wire (see Example 19.2).

Further experimentation leads to the conclusion that the magnitude of $\mathbf{F}_B$ depends directly on the length L of the wire in the magnetic field, on the magnitude of the current I through the wire, and on the component of the magnetic field at right angles to the wire, which we designate by $B_\perp$:

$$F_B \propto B_\perp IL$$

As usual, we can convert this proportion into an equality by introducing a constant of proportionality. In this case we choose to define the units for magnetic field in such a way as to make the proportionality constant equal to unity so that

$$\boxed{F_B = B_\perp IL} \tag{19.1}$$

Since in the SI system F_B is in newtons, I in amperes, and L in meters, the unit for $B_\perp$ must be N/(A·m) to make Eq. (19.1) dimensionally correct. In the SI system 1 N/(A·m) is called a *tesla* (T).

Tesla: The magnitude of that magnetic field in which a current-carrying conductor of length one meter oriented perpendicular to the magnetic field and carrying a current of one ampere experiences a force of one newton [1 T = 1 N/(A·m)].

An older, non-SI unit still frequently used for magnetic field strength is the *gauss* (G), where $1\ G = 10^{-4}\ N/(A\cdot m) = 10^{-4}\ T$. A field of 1 T, or 10^4 G, is a rather large magnetic field. The magnetic field of the earth is on the order of 0.5 G, or 5×10^{-5} T.

The direction of the force on the current-carrying conductor can be found by experiment. The results conform to the following *right-hand force rule* (Fig. 19.7). (Check the direction of deflection shown in Fig. 19.6 against this rule.)

Right-hand force rule: If the fingers of the outstretched right hand are placed in the direction of the magnetic field and the thumb in the direction of the conventional current I, then the palm of the hand faces in the direction of the force on the current-carrying conductor.

FIGURE 19.7 The right-hand force rule. If the fingers are in the direction of the magnetic field and the thumb in the direction of the current, then the palm of the hand faces in the direction of the force produced.

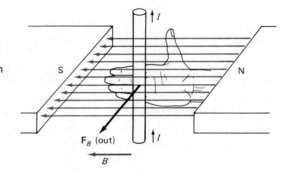

Note that the magnetic field must have a component normal to the current or else no force arises. If the angle between **B** and the current-carrying wire is θ, then $B_\perp$ equals $B \sin \theta$. Hence the force varies as the sine of the angle between **B** and the wire. The sine varies from 0 to 1 as θ goes from 0 to 90° (Fig. 19.8).

Note the clear distinction Eq. (19.1) makes between the magnetic force $\mathbf{F}_B$ and the magnetic field **B**. The magnetic force $\mathbf{F}_B$ differs from **B** not only in magnitude and units but also in *direction*.

FIGURE 19.8 How the magnetic force on a wire varies with the angle between the current direction and the magnetic field. (a) Current is parallel to lines of **B**; no force results. (b) Current makes an angle θ with **B**; force is $BIL \sin \theta$. (c) Current is perpendicular to **B**; force is BIL.

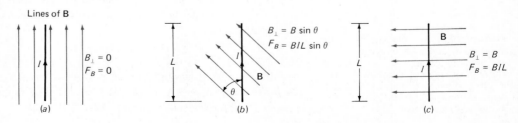

Example 19.1

A straight wire carries a 1.5-A current at right angles to a magnetic field of magnitude $B = 0.010$ T, as in Fig. 19.9. If the length of the wire in the magnetic field is 5.0 cm, what are **(a)** the magnitude of the force on the wire and **(b)** its direction? (Assume that the magnetic field of the magnet is confined to the 5.0-cm width of its pole pieces.)

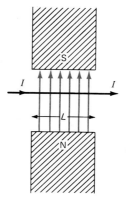

SOLUTION

(a) The magnitude of the magnetic force is $F_B = B_\perp IL$. Since the field is perpendicular to the direction of the current,

$$F_B = (0.010 \text{ T})(1.5 \text{ A})(5.0 \times 10^{-2} \text{ m}) = \boxed{7.5 \times 10^{-4} \text{ N}}$$

(b) If the situation is as in Fig. 19.9 and we apply the right-hand rule, we find that the force is directed *out of the plane of the paper*.

FIGURE 19.9

Example 19.2

In the setup of Fig. 19.6, the horizontal wire has a mass of 100 g, is 10.0 cm long, and carries current out of the page. The magnetic field is 0.500 T. What angle do the massless supporting wires make with the vertical when a current of 4.00 A passes through the wire?

SOLUTION

Since the magnetic field is perpendicular to the wire, the magnitude of the horizontal force on the length L of wire carrying the current in the magnetic field **B** is

$$F_B = B_\perp IL = (0.500 \text{ T})(4.00 \text{ A})(0.100 \text{ m}) = 0.200 \text{ N}$$

The direction of this force is shown in the FBD of Fig. 19.10.

The solution of the problem then involves merely the addition of vectors. The magnetic and gravitational forces must add vectorially to make the supporting wires assume an angle θ with the vertical. The gravitational force on the wire is its weight acting vertically downward:

$$w = mg = (0.100 \text{ kg})(9.80 \text{ m/s}^2) = 0.980 \text{ N}$$

and so $$\tan \theta = \frac{F_B}{w} = \frac{0.200 \text{ N}}{0.980 \text{ N}} = 0.204$$

$$\theta = \boxed{11.5°}$$

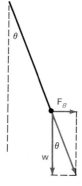

FIGURE 19.10   Free-body diagram for Example 19.2.

19.3   Force on a Charged Particle in a Magnetic Field

In atomic and nuclear physics, we are more interested in the force on individual charged particles in motion through a magnetic field than we are in forces on current-carrying wires.

Let us consider positive charges moving in a beam with a velocity **v** at right angles to a uniform magnetic field **B**, as in Fig. 19.11. Since moving charges constitute a current, we expect that the magnitude of the magnetic force on a length L of a beam will be $F_B = B_\perp IL$, as given by Eq. (19.1). The current is the amount of charge passing any point along the path of the beam in 1 s and is therefore

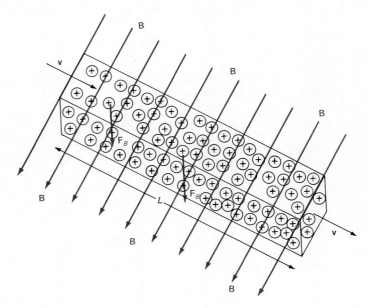

FIGURE 19.11 Forces exerted by a magnetic field on charged particles moving in a beam through a magnetic field. The force is at right angles to both **B** and **v**. In this case, the right-hand force rule predicts that the force is directed into the plane of the paper.

$$I = \frac{nq}{t} \tag{19.2}$$

where n is the number of charges passing that point in time t and q is the magnitude of each charge. We assume that the charge density in the beam is constant so that I is the same at any point along the beam.

For convenience we choose t to be the time it takes for a charge to move the distance L at a speed v (assumed the same for all charges in the beam), where L is the length of the beam in the magnetic field. Then

$$L = vt \tag{19.3}$$

But for the time t, as chosen in this way, n is equal to the total number of charges in the length L. This is because the charges at the beginning of L will just reach the end of L in time t, for they move at a speed $v = L/t$. Hence the magnitude of the magnetic force on the beam, which is the force on n individual charges, is, from Eqs. (19.1) to (19.3).*

$$F_B = B_\perp IL = B_\perp \frac{nq}{t} vt = B_\perp nqv$$

The force on a single charge q is then of magnitude

---

*This derivation shows that we can replace IL by nqv in the expression $F_B = B_\perp IL$. This same kind of substitution turns up in many other fields of physics, e.g., in the kinetic theory of gases in Sec. 9.6. As another example, if a stretch of highway 1000 m in length contains n automobiles (q here stands for an automobile), each moving at 25 m/s, and if the number of cars passing any point on the road per second is 10 (this is equivalent to I being the "car current"), then the total number of cars on the 1000-m stretch is

$$n = \frac{IL}{qv} = \frac{(10 \text{ cars/s})(1000 \text{ m})}{(1 \text{ car})(25 \text{ m/s})} = 400$$

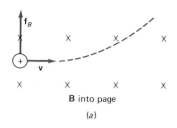

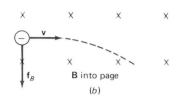

$$f_B = \frac{F_B}{n} = B_\perp qv \tag{19.4}$$

A charge q moving with a velocity **v** *in a magnetic field* **B** *experiences a force at right angles to both* **B** *and* **v** *of magnitude* $B_\perp qv = Bqv \sin \theta$.

For a positive charge the right-hand force rule again gives us the direction of the force $\mathbf{f}_B$:

If the fingers of the right hand are placed in the direction of **B** *and the thumb in the direction of the velocity* **v**, *then the palm of the hand faces in the direction of the force* $\mathbf{f}_B$ *on a moving positive charge.**

FIGURE 19.12 The direction of the forces on (a) positive and (b) negative charges moving through a magnetic field. The effect of these forces is to deflect the charged particles in the way shown. The ×'s indicate the tails of magnetic field vectors directed into the page.

If the charge q is negative, as it would be for an electron, f_B in Eq. (19.4) is also negative, so that the direction of the force on a negative charge is opposite to that given by the right-hand rule for a positive charge. The directions of the forces in these two cases are shown in Fig. 19.12.

Example 19.3

An electron in a vacuum tube is accelerated by a voltage of 2000 V and enters a region in which there is a uniform magnetic field of 1.0 T at right angles to the direction of the electron's motion. What is the force on the electron due to the magnetic field?

SOLUTION

To find the magnetic force on the electron, we first need to know its speed. Since the electron has been accelerated by 2000 V, the work done on it is $\mathcal{W} = qV = \frac{1}{2}mv^2$, for in this case all the work done must go into kinetic energy. Hence

$$v = \sqrt{\frac{2qV}{m}} = \left[\frac{2(1.6 \times 10^{-19} \text{ C})(2.0 \times 10^3 \text{ V})}{9.1 \times 10^{-31} \text{ kg}}\right]^{1/2}$$

$$= 2.7 \times 10^7 \text{ m/s}$$

Then the magnitude of the magnetic force is

$$f_B = Bqv = \left(1.0 \; \frac{\text{N}}{\text{A·m}}\right)(1.6 \times 10^{-19} \text{ C})\left(2.7 \times 10^7 \; \frac{\text{m}}{\text{s}}\right)$$

$$= \boxed{4.3 \times 10^{-12} \text{ N}}$$

The direction of the force on the electron can be found from the fact that the force on a positive charge would be in the direction given by the right-hand force rule. The force on the electron is the reverse of this because of its negative charge.

Hall Effect

Suppose we take a thin, flat ribbon of a metal like copper with the flat side perpendicular to a magnetic field **B** (into the paper in Fig. 19.13) and pass a current I along the length of the wire ribbon.

As we have seen, a positive charge flowing in one direction is completely equivalent to a negative charge flowing in the opposite direction. Therefore, when

---

*Both the magnitude and direction of the magnetic force on a moving charge are given by the cross product (introduced in the footnote on p. 126): $\mathbf{f}_B = q(\mathbf{v} \times \mathbf{B})$. The magnitude of the force $\mathbf{f}_B$ is $Bqv \sin \theta$, where θ is the angle between **v** and **B**. This is $B_\perp qv$, as in Eq. (19.4). The direction of $\mathbf{f}_B$ is the direction in which a right-handed screw would move if rotated from **v** into **B**, which is the same direction given by the right-hand force rule.

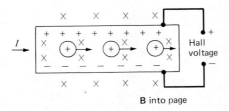

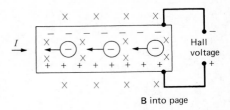

FIGURE 19.13   The Hall effect: Current carried by positive charges moving in the direction of the conventional current *I*.

FIGURE 19.14   The Hall effect: Current carried by negative charges moving in the opposite direction to that of the conventional current.

we establish a current in the ribbon, we do not know whether positive or negative charges carry the current. A phenomenon called the *Hall effect* enables us to decide unambiguously between these two possibilities. Suppose that in Fig. 19.13 the conventional current is to the right. Then, if the current were carried by positive charges, the magnetic field would deflect the charges upward, and the upper edge of the ribbon would become positively charged and the lower edge negatively charged, according to our right-hand force rule. The accumulated charges would produce a voltage called the *Hall voltage* across the ribbon, with the voltage being directed from the top to the bottom.

If, on the other hand, the current is carried by *electrons* moving to the *left*, then, because of their negative signs the electrons would also be deflected *upward* in the magnetic field. In this case the upper edge of the wire would become negatively charged and the lower edge positively charged so that the Hall voltage would be from the bottom to the top of the ribbon, as in Fig. 19.14.

To find out exactly what kind of charge carries the current in metals, all we need do is measure the Hall voltage across the ribbon. When we do this, we find that the voltage is as in Fig. 19.14. Therefore we may conclude that *the electric current in metals is carried by electrons moving in a direction opposite to that of the conventional current*. This is not the case in some semiconductors, where the Hall voltage shows that the current is carried by positive charges (whose nature is discussed in Chap. 30).

The magnitude of the force on a charge q moving at right angles to a magnetic field **B** at a drift velocity $\mathbf{v}_D$ is, from Eq. (19.4), $f_B = Bqv_D$. When this force moves a charge q across the width w of the ribbon, it does work $\mathcal{W} = Bqv_Dw$. The work per unit charge (or the potential difference from top to bottom) is then

$$V_H = \frac{\mathcal{W}}{q} = Bv_Dw \tag{19.5}$$

where V_H is the Hall voltage. The magnetic field is therefore directly proportional to the magnitude of the Hall voltage. This relationship is used in an instrument called a *Hall-effect gaussmeter** for the measurement of magnetic fields. Such gaussmeters are very convenient instruments calibrated to convert measurements of the Hall voltage directly into values of the magnetic field.

---

*Since the gauss is a unit of magnetic field, the name *gaussmeter* is given to instruments designed to measure magnetic fields. Similarly, a *degausser* is a device for removing magnetic fields induced in submarines, watches, or other steel or iron objects after they have been in a magnetic field like that of the earth.

Example 19.4

A Hall-effect gaussmeter has a measuring probe that contains a thin ribbon of conducting material 1.2 mm thick and 5.6 mm wide. A current of 0.25 A flows along this ribbon, which has its flat side perpendicular to the magnetic field to be measured. The volume density of the electrons that carry the current in the material is $1.0 \times 10^{26}/m^3$. Find the Hall voltage corresponding to a magnetic field of 1.4 T.

SOLUTION

The statement of the problem gives the magnetic field B and the width w of the ribbon in Eq. (19.5). The only other quantity needed to solve the problem is the drift velocity v_D of the electrons. If n is the volume density of the electrons in the metal, then the number of electrons in a length L of the ribbon is $nV = ndwL$, where d is the thickness and w the width of the current-carrying ribbon.

The current is therefore $I = ndwLq/t$, where we choose the time t to be the time it takes for a charge q to move the distance L at a velocity v_D. The drift velocity is, from Eq. (19.3), $v_D = L/t = I/ndwq$, and the Hall voltage becomes, from Eq. (19.5),

$$V_H = Bv_D w = B\frac{I}{ndwq}w = \frac{BI}{ndq}$$

$$= \frac{(1.4 \text{ T})(0.25 \text{ A})}{(1.0 \times 10^{26}/m^3)(1.2 \times 10^{-3} \text{ m})(1.6 \times 10^{-19} \text{ C})}$$

$$= 1.8 \times 10^{-5} \text{ V} = \boxed{18 \ \mu\text{V}}$$

EXERCISE 1   What is the drift velocity of the electrons in this case?

Paths of Charged Particles in a Magnetic Field

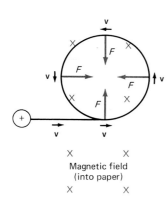

Magnetic field
(into paper)

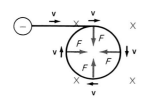

FIGURE 19.15   Charged particles moving at right angles to a magnetic field directed into the page. They are bent by the magnetic field into closed circles, with (in this case) negatively charged particles rotating clockwise, positively charged particles counterclockwise.

If an atomic particle of charge q is shot into a magnetic field with a velocity **v** at right angles to the direction of the field **B**, as in Fig. 19.15, it is deflected at right angles to its original direction by the magnetic field. As it changes its direction slightly, it continues to be deflected at right angles to its new direction, and so on. Since the force $\mathbf{F}_B$ continues to be at right angles to the velocity **v**, it acts like a string keeping a ball rotating in a circle. As a result the path of the particle is a *circle*. The radius of the circle can be found by noting that the centripetal force needed for the circular motion is here provided by the magnetic field, so that, from Eqs. (19.4) and (4.31),

$$Bqv = \frac{mv^2}{r} \quad \text{or} \quad r = \frac{mv}{Bq} \qquad \textbf{(19.6)}$$

(In most of the following examples the arrangement of the apparatus ensures that **B** is perpendicular to **v**. Thus we will write simply B instead of $B_\perp$.) If the charge of the particle (usually one or two elementary charges) and the magnitude of the magnetic field are known, the momentum mv of the particle can be obtained from a measurement of the radius of the circle in which the particle moves, since $mv = Bqr$. For electrons and protons moving at the same speed, the electrons circle with much smaller radii. This makes it easy to identify electrons in the photographs of particle tracks taken in cloud chambers and other kinds of nuclear detectors. From the direction in which the particle is bent in the magnetic field it is also possible to tell whether its charge is positive or negative, as in Fig. 19.15.

If the speed v of the particle is known, for example, from the voltage used to accelerate it to that speed, then from Eq. (19.6),

$$\frac{q}{m} = \frac{v}{Br} \qquad \textbf{(19.7)}$$

and the charge-to-mass ratio can also be obtained from a measurement of r.

The charge-to-mass ratio is a very important quantity for any atomic particle, since, once q/m is known, a measurement of the charge leads directly to a value for the particle's mass.

Example 19.5

In Example 19.3, what path does the electron follow in the magnetic field? What is the size of this path?

SOLUTION

The direction of the electron's velocity will constantly change as the particle is continuously deflected at right angles to its original direction. It will therefore move in a *circle* at a constant speed v, already found to be 2.7×10^7 m/s. The radius of the circle is given by Eq. (19.6):

$$r = \frac{mv}{Bq} = \frac{(9.1 \times 10^{-31} \text{ kg})(2.7 \times 10^7 \text{ m/s})}{(1.0 \text{ T})(1.6 \times 10^{-19} \text{ C})}$$

$$= \boxed{1.5 \times 10^{-4} \text{ m}}$$

Note that the force produced by the magnetic field on the electron is always perpendicular to the electron's velocity and hence can never change the *magnitude* of that velocity.

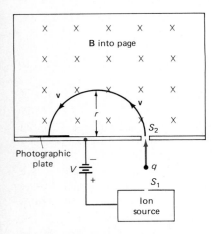

FIGURE 19.16 A magnetic mass spectrometer. The mass of the ion can be determined from the position at which it strikes the photographic plate.

Mass Spectrometer A mass spectrometer is an instrument used to separate gaseous ions in a magnetic field according to their masses, and to measure atomic and molecular masses. In this way the molecules making up the gas and the concentration of each kind of molecule can be determined.

Charged ions are first produced by electron bombardment of the gas. Positive ions emerge in a beam through slit S_1 in Fig. 19.16 and are accelerated toward slit S_2 by a voltage applied between the two slits. An ion with a charge $+q$ will have a kinetic energy qV, where V is the voltage accelerating the ion, and a speed $v = (2qV/m)^{1/2}$. After passing through slit S_2, the ion moves at constant speed into a magnetic field perpendicular to $\mathbf{v}$. The ion is deflected in a circular path of radius r, given by Eq. (19.6) as

$$r = \frac{mv}{Bq} = \frac{m}{Bq}\sqrt{\frac{2qV}{m}} = \sqrt{\frac{2Vm}{B^2 q}}$$

and so

$$m = \frac{B^2 q r^2}{2V} \tag{19.8}$$

The ion strikes a photographic plate or electronic detector after moving through a semicircle, as shown in the figure, and in this way r can be measured. Since q, V, and B are known, the mass m of the ion can be determined.

Cyclotron Large particle accelerators are used in national laboratories like Brookhaven on Long Island and Fermilab near Chicago to study the collision of atomic particles at very high speeds and in this way to learn about the ultimate constituents of matter. Many of these accelerators use magnetic fields to control the paths of the particles. The *cyclotron*, one of the earliest particle accelerators, was developed in 1930 by Ernest O. Lawrence (Fig. 19.17) at the University of California at Berkeley. It used a large magnetic field to keep protons or other charged particles moving in circles.

A cyclotron has two D-shaped cavities called *dees*, as in Fig. 19.18. Each

FIGURE 19.17 Ernest O. Lawrence (1901–1958) supplied the key idea needed to accelerate charged particles to very high speeds around circular paths. For his development of the cyclotron, he received the Nobel Prize in physics in 1939. This photograph shows Lawrence with one of the cyclotrons he built at the University of California, Berkeley. (*Lawrence Radiation Laboratory, Berkeley, California, photo; courtesy of AIP Niels Bohr Library*).

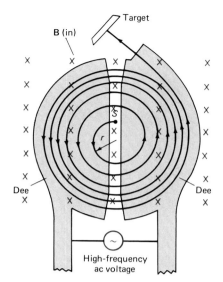

B (in)

Target

Dee

Dee

High-frequency
ac voltage

FIGURE 19.18 Schematic diagram of a cyclotron. Protons are accelerated by electric fields between the dees and kept moving in a circle by the strong magnetic field. The protons spiral outward, moving more rapidly as the radius of their orbit increases, and finally crash into a target outside the cyclotron.

time the protons cross the region between the dees, a voltage is applied to accelerate the protons. They therefore move with increasing speed and in paths of increasing radius as time goes on. Finally the protons acquire the desired high energy and are directed out of the cyclotron to an external target where their interaction with other nuclei is studied.

The voltage applied to the dees must be in resonance with the frequency of the circular motion of the protons if large energies are to be developed, for every time the protons pass through the gap, the voltage must accelerate them again. The voltage must reverse itself, therefore, in the time it takes for the protons to make half a revolution. This is accomplished by using an alternating voltage across the dees that is in perfect synchronism with the particles' motion. The protons are therefore accelerated twice in each revolution by the electric field produced by the voltage on the dees.

The magnetic field supplies the force to keep the protons moving in a circle of radius r. The speed v changes each time the protons are accelerated by the electric field in the gap. The radius r therefore changes along with v.

The time required for the protons to make a complete revolution is the period, which is, using Eq. (19.6),

$$T = \frac{\text{distance}}{\text{speed}} = \frac{2\pi r}{v} = \frac{2\pi mv}{Bqv} = \frac{2\pi m}{qB}$$

and so the frequency of revolution is

$$f = \frac{1}{T} = \frac{qB}{2\pi m} \qquad\qquad \textbf{(19.9)}$$

What makes the cyclotron work is the fact that *the frequency f is independent of either v or r*, as is clear from Eq. (19.9). From a knowledge of B and q/m for the particle, we can find the frequency f to be used for the accelerating voltage. This frequency is unchanged throughout the experiment. Even though v and r increase rapidly, they increase in such a way that v/r remains constant and so the frequency also remains constant. This constant frequency is readily provided with standard electronic oscillators.

19.4 J. J. Thomson's Determination of e/m for the Electron

Sir Joseph John Thomson is usually given credit for the discovery of the electron. (See Fig. 19.19 and accompanying biography.) Although there had been electron theories before Thomson, it was his work in 1896–1897 that pinned down the nature of the electron. Thomson performed many careful, ingenious experiments on the deflection of cathode rays in electric and magnetic fields. (*Cathode ray* is the name given to the glow that spreads out from the cathode of an evacuated tube when a high voltage is applied between the anode, or positive terminal, and the cathode, or negative terminal.)

Figure 19.20 shows the evacuated glass tube used by Thomson in his most famous experiment, in which he showed that the cathode rays were deflected by both electric and magnetic fields, proving that they were material particles. Also, the direction of their deflection indicated that their charge was negative. He then clarified the nature of these particles by measuring the ratio of their charge e to their mass m.

To measure e/m, Thomson first determined the position of the beam on the

FIGURE 19.19
Ernest Rutherford showing J. J. Thomson to his seat for the annual photograph of the Cavendish Laboratory staff, probably in 1930. (*Courtesy of AIP Niels Bohr Library.*)

The British physicist J. J. Thomson is famous both as the discoverer of the electron and as the teacher and mentor of many of this century's greatest physicists. He became a physicist almost by accident. His father wanted him to be an engineer, but when the time came, the family did not have the money to pay for his engineering apprenticeship. A friend suggested to the elder Thomson that he send his son to Owens College in Manchester. At Owens young Thomson had excellent instructors in mathematics and physics who kindled his interest in a physics career.

In 1876 J. J. (as he was always called by everyone, including his son, Nobel Laureate George Thomson) went to Trinity College at Cambridge University to study mathematics and science—and spent the next 64 years of his life there! In 1884, at the age of 28, he was elected to the Cavendish Professorship at Cambridge and became director of the Cavendish Laboratory, the outstanding physics laboratory in England at that time. He was elected Master of Trinity College at Cambridge in 1918 and gave up the Cavendish Chair in 1919 in favor of Ernest Rutherford. He remained Mas-

ter of Trinity until shortly before his death in 1940.

Thomson's most famous work in physics was the clarification of the nature of cathode rays and the proof that they were what we now call electrons. He did other important work on the behavior of positive ions in electric and magnetic fields, and on the structure of atoms. He was neither a high-powered theoretical physicist nor a gifted experimentalist but rather an unusually adept "broker" between the two. In 1906 he received the Nobel Prize in physics "for his theoretical and experimental investigation of the conduction of electricity in gases" and 2 years later was knighted by the British Crown.

Thomson did magnificent work in stimulating and encouraging the research of the outstanding group of physicists who flocked from all over the world to work with him at the Cavendish. He trained eight Nobel Laureates, 27 Fellows of the Royal Society, and many other highly successful physics teachers and researchers. In the 1920s it was said that all physics professors in the United States were either Thomson's students or students of his students. He was a good, if

somewhat penurious, administrator, and much beloved by his students. Their respect and admiration for him are reflected in the parodies of popular songs they wrote in his honor to be sung at the annual meetings of the Cavendish Physical Society, rather raucous occasions when old stories were retold, much wine consumed, and ditties sung about the Cavendish and its professors.

No one has captured J. J.'s boyish enthusiasm and remarkable genius better than his former Cavendish student, Nobel Laureate F. W. Aston, in this statement on the occasion of Thomson's death in 1940:

> Working under him never lacked thrills. When results were coming out well his boundless, indeed childlike, enthusiasm was contagious and occasionally embarrassing. Negatives just developed had actually to be hidden away for fear he would handle them while they were still wet. Yet when hitches occurred, and the exasperating vagaries of an apparatus had reduced the man who had designed, built and worked with it to baffled despair, along would shuffle this remarkable being, who, after cogitating in a characteristic attitude over his funny old desk in the corner, and jotting down a few figures and formulae in his tiny tidy handwriting, on the back of somebody's Fellowship thesis, or on an old envelope, or even the laboratory cheque book, would produce a luminous suggestion, like a rabbit out of a hat, not only revealing the cause of trouble, but also the means of cure. This intuitive ability to comprehend the inner working of intricate apparatus without the trouble of handling it appeared to me then, and still appears to me now, something verging on the miraculous, the hallmark of a great genius.*

The London Times, Sept. 4, 1940.

FIGURE 19.20 The evacuated glass tube used by Thomson in his work on the charge-to-mass ratio *e/m* of the electron. The electrons originate at *C*, are accelerated by the positive voltage on the ring electrode at *F*, pass through the slits *A* and *B* and the deflecting plates *D* and *G*, and strike the screen at the end of the tube.

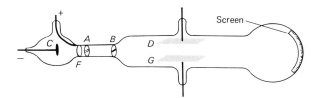

screen in the absence of electric and magnetic fields. Then he applied a voltage *V* between plates *D* and *G* so that an electric field was established between the plates of magnitude $E = V/d$, where *d* is the distance between the plates. This field exerted a force on the electrons equal to $F_y = Ee$, which produced an upward deflection of the electron beam from its original position on the screen.

Next, Thomson established a uniform magnetic field at right angles to the electric field. He chose a magnetic field that would deflect the particles downward and varied this field until the "crossed" electric and magnetic fields restored the beam to its original, undeflected position. Then the magnetic force was equal in magnitude and opposite in direction to the electric force so that, using Eq. (19.4),

$$Ee = Bev_x \qquad \text{or} \qquad v_x = \frac{E}{B} \tag{19.10}$$

Using this value of v_x and the fact that the particle was acted on by the electric field only while it was between the plates, i.e., during a time *t* equal to L/v_x, where *L* is the length of the deflecting plates, Thomson obtained the following result (see Prob. 19.41):

$$\boxed{\frac{e}{m} = \frac{E \tan \theta}{LB^2}} \tag{19.11}$$

Here $\tan \theta$ is just the ratio y/x, where *y* is the deflection on the screen in the absence of the magnetic field and *x* is the distance from the center of the plates to the screen, as shown in Fig. 19.21. Since *x* and *y* can be measured, as can *B*, *E*, and *L*, a value of *e/m* can be obtained.

Thomson found that, if the charge on his cathode-ray particles was equal to the minimum charge on H^+ ions in Faraday's electrochemical experiments, then the mass of the cathode-ray particles was less than $\frac{1}{1000}$ that of a hydrogen atom. The cathode-ray particles were thus much lighter (and presumably much smaller) than a hydrogen atom. Thomson's work thus opened up the whole field of subatomic particles.

FIGURE 19.21 Diagram showing how the angle θ in Thomson's experiment may be obtained by measurements of the deflection *y* of the beam on the screen and the distance *x* between the center of the deflecting plates and the screen.

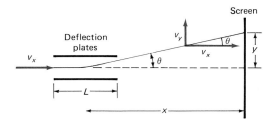

In 1891 George Stoney had proposed the name *electron* for the hypothetical minimum charge involved in carrying electric currents. After Thomson's clarification of the nature of the cathode-ray particles, Lorentz applied this name (over Thomson's objections) to these particles rather than to their charges. Since Thomson was the first to determine the approximate mass and size of the particles and their charge-to-mass ratio, he is rightly called the discoverer of the electron.

The presently accepted value of 1.76×10^{11} C/kg for e/m can be combined with the charge on the electron, as determined by the Millikan oil-drop experiment, $e = 1.60 \times 10^{-19}$ C, to yield a value for the electron mass:

$$m = \frac{e}{e/m} = \frac{1.60 \times 10^{-19} \text{ C}}{1.76 \times 10^{11} \text{ C/kg}} = 9.11 \times 10^{-31} \text{ kg}$$

19.5 Magnetic Field of a Straight, Current-Carrying Wire

Thus far we have considered how moving charges, whether in wires or in space, react to the presence of magnetic fields; we have not asked how magnetic fields are produced.

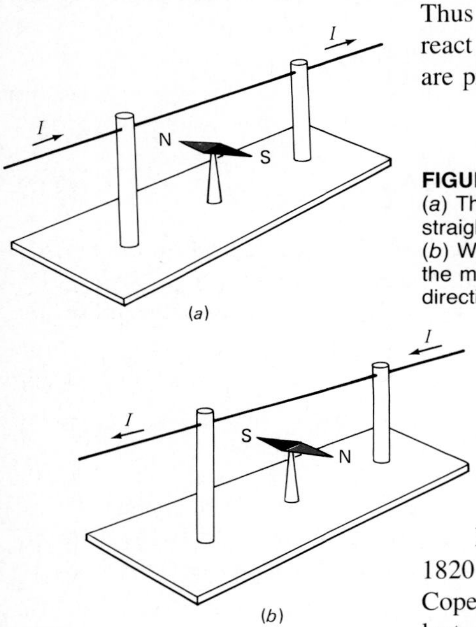

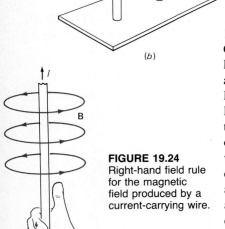

FIGURE 19.22 Oersted's experiment. (a) The magnetic field near a long, straight wire carrying a current *I*. (b) When the current reverses direction, the magnetic field also reverses direction.

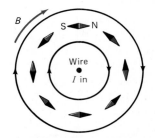

FIGURE 19.23 Magnetic field lines around a current-carrying wire. The compass needles show that the magnetic field lines are circles about the wire. Here the current in the wire is flowing into the paper.

FIGURE 19.24 Right-hand field rule for the magnetic field produced by a current-carrying wire.

Let us consider the simple, but revolutionary, experiment first performed in 1820 by Hans Christian Oersted (1777–1851), a professor at the University of Copenhagen. According to Oersted's own account, he was giving a course of lectures on electricity and magnetism and had set up a current-carrying wire above a compass needle to see if there was any interaction between the two. During the lecture he passed current through the wire and observed a feeble force acting to line up the needle in a direction *at right angles* to the wire. The effect was so weak that his audience was not impressed, but in July 1820 Oersted repeated his demonstration with a more powerful battery and with a group of notables present to witness his discovery. When Oersted reversed the direction of the current, the compass needle reversed its direction but still lined up perpendicular to the wire, as in Fig. 19.22. As Oersted moved the needle around the wire, it always pointed at right angles to the wire. Hence he concluded that the magnetic field must form closed circles around the wire, as in Fig. 19.23. The direction of the magnetic field can be specified by the following right-hand rule (see Fig. 19.24):

Right-hand field rule: If a wire is grasped with the right hand with the thumb pointed in the direction of the current, the magnetic field is circular about the wire and in the direction in which the fingers curl.

FIGURE 19.25 André-Marie Ampère (1775–1836), the founder of "electrodynamics," a name he coined himself. The unit of electric current is named after him. (*Courtesy of the Burndy Library, Norwalk, Connecticut.*)

Oersted was the first to demonstrate the connection between electricity and magnetism that many physicists had suspected for years. His discovery marked the beginning of the study of *electrodynamics*. News of it reached Paris in September 1820, and André-Marie Ampère (Fig. 19.25) immediately set out to verify and extend Oersted's discovery. Within a week, Ampère presented important new results to the French Academy. Not only do magnets and electric currents exert forces on each other, electric currents exert forces on other electric currents. We would say that a steady direct current in a straight wire produces a magnetic field about the wire, which in turn exerts a force on another current-carrying wire placed in that field. Ampère concluded from his experiments that all magnetism is produced by electric currents and that even in permanent magnets there must be some kind of electric current to produce the magnetism.

With a Hall-effect gaussmeter or other magnetic-field-measuring device, we can measure $\mathbf{B}$ as a function of the current I in a straight wire and the distance r from the wire to the point at which the field is measured. We find that the magnitude of B varies as follows:

$$B \propto \frac{I}{r} \quad \text{or} \quad B = \frac{k'I}{r} \tag{19.12}$$

We now proceed to determine k' by considering the force between two current-carrying wires.

Force between Two Long, Straight Current-Carrying Conductors

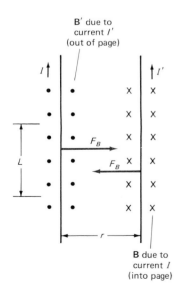

B' due to current I' (out of page)

I

I'

F_B

F_B

L

r

B due to current I (into page)

If we have two long parallel wires, one carrying a current I and the other carrying a current I' in the same direction, as in Fig. 19.26, we find, as Ampère did, that they attract each other. The reason is that the wire with current I produces a magnetic field which is circular around the wire and of magnitude $B = k'I/r$, from Eq. (19.12). This field is perpendicular to the current I' in the other wire at any point along the length L of the wire. Hence, from Eq. (19.1) we have

$$F_B = B_\perp I'L = k'\frac{II'L}{r} \quad \text{or} \quad \frac{F_B}{L} = k'\frac{II'}{r} \tag{19.13}$$

This is the magnitude of the force per unit length between the two wires. By first applying the right-hand field rule and then the right-hand force rule, you can see that the two wires attract each other. If I and I' are in opposite directions, the two wires repel each other (prove this also).

FIGURE 19.26 Force between two wires carrying electric current in the same direction. The two wires attract each other.

Definition of the Ampere

If, in Eq. (19.13), we set $I = I' = 1$ A, and $r = 1$ m, then

$$\frac{F_B}{L} = k'\frac{(1 \text{ A})(1 \text{ A})}{1 \text{ m}}$$

This enables us to define the *ampere* as follows:

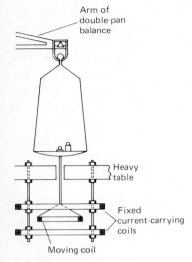

FIGURE 19.27 A current balance of the type used to determine the standard ampere at the National Institute of Standards and Technology in Gaithersburg, Maryland. The same current is established in three circular loops, two of which are fixed rigidly to the table, and the middle one, which is hung from one arm of an equal-arm balance, like that in Fig. 1.10. Here, because of the direction of the current, the moving coil is attracted by the upper coil and repelled by the lower coil. This force can be balanced by adding weights to the pan of the balance. The standard ampere is then defined in terms of the weights added and the coil dimensions.

Ampere: The magnitude of the current that, when present in each of two long parallel wires one meter apart, results in a force between the wires of exactly 2×10^{-7} N per meter of length.

In terms of this definition, then, k' must have exactly the value

$$k' = 2 \times 10^{-7} \text{ N/A}^2 = 2 \times 10^{-7} \text{ T·m/A}$$

since $\quad 1 \text{ T·m/A} = \left(1 \dfrac{\text{N}}{\text{A·m}}\right)\left(1 \dfrac{\text{m}}{\text{A}}\right) = 1 \text{ N/A}^2$

Frequently k' is written as $\mu_0/2\pi$, where $\mu_0 = 4\pi \times 10^{-7}$ T·m/A. μ_0 is called the *permeability of free space* (or of a vacuum) and plays in magnetism a role similar to that played by ϵ_0, the permittivity of free space, in electrostatics. Equation (19.12) for the magnetic field at a distance r from a long, straight wire carrying a current I then becomes

$$B = k'\frac{I}{r} = \frac{\mu_0 I}{2\pi r} = 2 \times 10^{-7}\frac{I}{r} \qquad (19.14)$$

The preceding definition of the ampere is an *operational definition* that allows us to obtain 1 A precisely in terms of *operations* to be performed using a *current balance*. This is an instrument in which the repulsive force between two equal currents in coils of the same dimensions can be balanced against the gravitational force required to restore the wires to their original separation. Figure 19.27 shows a current balance of the kind used to determine the standard ampere at the National Institute of Standards and Technology in Gaithersburg, Maryland.

Once the ampere is defined in this way, the coulomb can be rigorously defined as the amount of charge passing any point in a wire per second when the wire is carrying a current of one ampere; that is, $1 \text{ C} = (1 \text{ A})(1 \text{ s}) = 1 \text{ A·s}$. This definition is perfectly consistent with the tentative definition of the coulomb in Sec. 16.4—a definition that is no longer needed.

Example 19.6

Two long, straight wires, each of length 0.50 m and separated by a distance of 1.0 cm, are connected to a battery, as in Fig. 19.28. The resistance of the circuit external to the battery is 3.0 Ω, and the applied voltage is 6.0 V. **(a)** What is the current in the circuit? **(b)** What is the magnitude of the force between the wires? **(c)** Is this force one of attraction or repulsion?

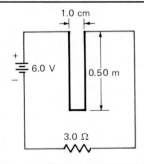

FIGURE 19.28

SOLUTION

(a) From Ohm's law, $I = \dfrac{V}{R} = \dfrac{6.0 \text{ V}}{3.0 \text{ Ω}} = \boxed{2.0 \text{ A}}$

(b) From Eq. (19.13) and our definition of k',

$$F_B = \frac{\mu_0 II'L}{2\pi r} = \left(2.0 \times 10^{-7}\frac{\text{T·m}}{\text{A}}\right)\left[\frac{(2.0 \text{ A})(2.0 \text{ A})(0.50 \text{ m})}{(0.010 \text{ m})}\right]$$

$$= 400 \times 10^{-7} \text{ N} = \boxed{4.0 \times 10^{-5} \text{ N}}$$

(c) Since the currents are in opposite directions in the two wires, the force is one of *repulsion*.

19.6 Ampère's Circuital Law

Equation (19.14) resulted from an experimental determination of the magnetic field **B** produced by the current in a long, straight wire. What about the fields produced by more complicated configurations of current-carrying wires? In 1826 Ampère generalized Eq. (19.14) to obtain the magnetic field produced by any configuration of currents. This equation is called *Ampère's circuital law* (or simply *Ampère's law*).

The basis of this law is that work must be done to carry a magnetic pole in a closed path about any wire or group of wires carrying electric current. Because the amount of work done depends on the strength of the magnetic field produced by the current, a calculation of the work can, in principle, lead to a value for the magnetic field.

Suppose we have a number of wires carrying currents through a plane, as in Fig. 19.29. If we make a complete circuit, or path, around the wires in the plane, and along each short piece Δl of this path evaluate the product of Δl and the component of **B** parallel to Δl (we label this parallel component $B_\parallel$), then the following equation is valid:

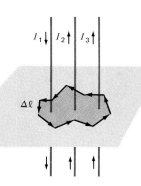

FIGURE 19.29 Ampère's circuital law. Three wires carry currents through the plane *ABCD*. If any continuous path is taken around these three wires in this plane, with Δl a distance along the path and $B_\parallel$ the value of the magnetic field **B** parallel to Δl, then $\Sigma B_\parallel \Delta l = \Sigma \mu_0 I$ according to Ampère's law. Here $\Sigma I = I_2 + I_3 - I_1$, since I_1 is in the opposite direction to I_2 and I_3.

Ampère's circuital law: $\boxed{\sum B_\parallel \, \Delta l = \sum \mu_0 I}$ (19.15)

The first sum is taken around any closed path, and the second sum over all the currents (taken with the correct sign) passing through the surface bounded by this closed path. Ampère proved that his circuital law is valid for *any* closed path taken around the wires.

Ampère's law is difficult but extremely useful. It enables us to find the magnetic field produced by complicated systems of currents in the same way that Gauss' law enables us to find the electric field produced by complicated charge distributions. In most cases symmetry must exist in the physical situation to make these laws useful in practice. They are, however, also of great conceptual importance in building up a consistent theory of electromagnetism.

Let us apply Ampère's law to the example already discussed, the magnetic field at a distance r from a long, straight wire.

In this case we choose as our path a circle of radius r around the wire. Then **B** is circular around the wire, is everywhere parallel to Δl, and has the same magnitude at every point on the circle, since each point is the same distance from the wire. Also there is only one current I through the plane bounded by the path we have chosen. We have

$$\sum B_\parallel \, \Delta l = B \sum \Delta l = B(2\pi r) \qquad \text{and} \qquad \sum \mu_0 I = \mu_0 I$$

and so Eq. (19.15) becomes

$$2\pi r B = \mu_0 I \qquad \text{or} \qquad B = \frac{\mu_0 I}{2\pi r}$$

which is Eq. (19.14).

We make limited use of Ampère's law in solving problems in this chapter, but its theoretical importance will become clear when we discuss electromagnetic waves in Chap. 22.

19.7 Circular Coils and Solenoids

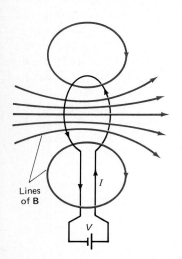

Lines of B

FIGURE 19.30 Magnetic field of a current loop.

A very important kind of magnetic field is that produced by a single circular coil of wire, which carries a current I, as in Fig. 19.30. We can apply the right-hand field rule to find the direction of the magnetic field **B** shown in the figure. At the center of the coil, the magnetic field is perpendicular to the plane of the coil over a significant distance. Since the center of the coil is at the same distance r from every point on the coil, we might expect that the magnitude of **B** would vary directly with the current through the coil and inversely with the distance r from the wire to the center of the coil. This is what we found in Eq. (19.14) for the field of a long, straight wire. Here the same dependence is found from theory and by experiment except that the constant of proportionality is different in this case. The magnetic field at the center of a current loop of radius r turns out to be:

Magnetic field at the center of a current loop: $$B = \frac{\mu_0 I}{2r}$$ **(19.16)**

The magnetic field produced by a single loop of wire, as shown in Fig. 19.30, is very similar to the electric field of an electric dipole, as shown in Fig. 16.22. A current-carrying loop of wire is therefore called a *magnetic dipole*, and the quantity IA, where A is the area of the circular loop, is called the *dipole moment* of the current loop. The direction of this magnetic dipole is defined to be perpendicular to the plane of the loop in the same direction as the magnetic field at the loop's center. If there are N turns of wire instead of one turn, the magnetic field becomes $B = \mu_0 NI/2r$ and the dipole moment becomes NIA. When placed in a magnetic field, a current loop tends to line up so that its dipole moment is in the direction of the applied magnetic field, and so the plane of the current loop is perpendicular to the direction of the applied magnetic field.

The Solenoid

A *solenoid* is a tight cylindrical coil of wire with a diameter small compared with its length, as in Fig. 19.31. Let us apply the circuital law to such a solenoid containing N turns of wire in a length L, each turn carrying a current I. The magnetic field in the center of the solenoid (or the *core*) is directed along the axis of the solenoid, as can be seen from the diagram for a single loop in Fig. 19.30. Outside the solenoid the magnetic field is very much weaker, as the rough sketch of the magnetic field in Fig. 19.31 indicates. We choose as the path for our application of Ampère's law the rectangular path *abcda* in Fig. 19.32. Along *ab* the magnetic field is very weak and $B_\parallel \, \Delta l$ is nearly zero. Along *bc* and *da* the field is very weak outside the solenoid and is at right angles to Δl inside the solenoid, and so $\Sigma B_\parallel \, \Delta l$ is also zero for these pieces of the path. This leaves the path *cd*, which is of length L, along which **B** has a constant magnitude B parallel to L. Ampère's law in the form $\Sigma B_\parallel \, \Delta l = \Sigma \mu_0 I$ becomes, therefore,

$$BL = \mu_0 NI$$

since in the length L there are N turns of wire that carry current I through the surface bounded by the path *abcd*. Hence

$$B = \mu_0 \frac{NI}{L} = \mu_0 nI \qquad \text{or} \qquad B = \mu_0 nI \qquad \textbf{(19.17)}$$

In this expression for the magnetic field produced by a solenoid on axis near its

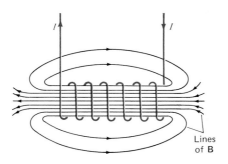

FIGURE 19.31 A solenoid, carrying a current I, and the magnetic field produced.

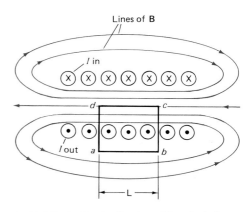

FIGURE 19.32 Application of Ampère's law to obtain the field of a solenoid near its center and along its axis. The figure shows a plane cut through the three-dimensional solenoid.

center, n is the number of turns per unit length of the solenoid (N/L) and I is the current in each turn. B is constant inside the solenoid except near the ends, where it begins to "fringe," or bend away from the solenoid axis, as shown in Fig. 19.32.

Example 19.7

A solenoid with air in its central core is wound with 50 turns of wire for every 10 cm of length. A current of 1.0 A exists in the windings of the solenoid. What is the magnetic field inside the solenoid at a point on its axis?

SOLUTION

$$B_0 = \mu_0 n I = \left(4\pi \times 10^{-7}\ \frac{\text{T·m}}{\text{A}} \right) \left(\frac{50}{0.10\ \text{m}} \right)(1.0\ \text{A}) = \boxed{6.3 \times 10^{-4}\ \text{T}}$$

Electromagnets

The magnetic field of the solenoid in Fig. 19.31 is almost identical to that of the bar magnet shown in Fig. 19.33. The solenoid is therefore an *electromagnet* that behaves basically like a bar magnet when there is a current in the solenoid, but has no magnetic field when the current is turned off. To increase the magnetic field strength of electromagnets, the central core of the solenoid is often filled with an iron alloy. The magnetic dipole moments of the iron atoms line up with the applied magnetic field and increase greatly the final magnetic field produced.

Both permanent magnets and electromagnets attract objects containing iron, cobalt, or nickel by inducing magnetic dipole moments in the material, in the same way that an electric field induces electric dipole moments in some materials. For

FIGURE 19.33 The magnetic field produced by a bar magnet.

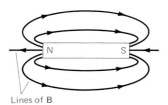

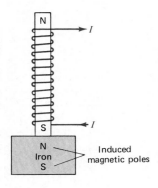

example, if an iron-core solenoid is brought near a piece of iron, as in Fig. 19.34, the magnetic field of the solenoid will so modify the iron that a north magnetic pole will appear near the solenoid's south pole and a south magnetic pole will appear at the other end of the iron. Since the iron's north pole is closer to the solenoid's south pole, a net attractive force results. Electromagnets are often used on the ends of derrick cables to lift large pieces of steel or scrap iron, because the magnetic field can be turned on or off as desired merely by turning the electric current on or off.

FIGURE 19.34 The use of a solenoid with an iron core to induce a magnetic moment in a piece of iron, resulting in an attractive force between the solenoid and the iron.

19.8 Magnetic Properties of Materials

An air-core solenoid makes a very weak electromagnet. If we place a soft-iron* core inside the solenoid, however, the magnetic field inside the solenoid can be increased by a factor of a thousand or more. The iron greatly enhances the magnetic field and changes the expression for B from $B_0 = \mu_0 nI$, where μ_0 is the magnetic permeability of a vacuum (or of air, which is very close to a vacuum magnetically), to $B = \mu nI$, where μ is the magnetic permeability of the iron or other material filling the core of the solenoid. The ratio B/B_0 is called the *relative magnetic permeability* and is represented by the symbol K_m, where

$$K_m = \frac{B}{B_0} = \frac{\mu nI}{\mu_0 nI} = \frac{\mu}{\mu_0} \qquad \text{(19.18)}$$

so that

$$\boxed{\mu = K_m \mu_0} \qquad \text{(19.19)}$$

We can distinguish three kinds of materials with very different magnetic properties that depend on the value of K_m.

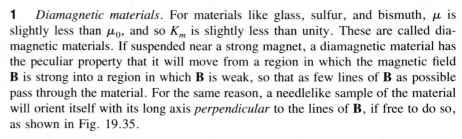

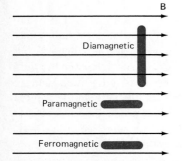

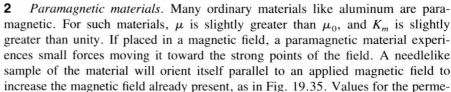

FIGURE 19.35 Orientation of needlelike samples of diamagnetic, paramagnetic, and ferromagnetic materials in a magnetic field.

1 *Diamagnetic materials.* For materials like glass, sulfur, and bismuth, μ is slightly less than μ_0, and so K_m is slightly less than unity. These are called diamagnetic materials. If suspended near a strong magnet, a diamagnetic material has the peculiar property that it will move from a region in which the magnetic field **B** is strong into a region in which **B** is weak, so that as few lines of **B** as possible pass through the material. For the same reason, a needlelike sample of the material will orient itself with its long axis *perpendicular* to the lines of **B**, if free to do so, as shown in Fig. 19.35.

2 *Paramagnetic materials.* Many ordinary materials like aluminum are paramagnetic. For such materials, μ is slightly greater than μ_0, and K_m is slightly greater than unity. If placed in a magnetic field, a paramagnetic material experiences small forces moving it toward the strong points of the field. A needlelike sample of the material will orient itself parallel to an applied magnetic field to increase the magnetic field already present, as in Fig. 19.35. Values for the perme-

---

*"Soft" iron refers to iron that is easily magnetized but does not hold a permanent magnetic field. "Hard" iron is an alloy that is difficult to magnetize but, once magnetized, makes a good permanent magnet.

TABLE 19.1   Relative Magnetic Permeabilities at Room Temperature

Material	Relative permeability $K_m = \mu/\mu_0$
Ferromagnetic:	
Iron	5,500*
Permalloy[†]	25,000*
Mumetal[‡]	100,000*
Paramagnetic:	
Aluminum	1.000023
Magnesium	1.000012
Titanium	1.000071
Uranium	1.00040
Diamagnetic:	
Bismuth	0.999834
Copper	0.999990
Gold	0.999964
Mercury	0.999971

*Maximum values; actual values depend on the applied field.
[†](55% Fe, 45% Ni)
[‡](77% Ni, 16% Fe, 5% Cu, 2% Cr)

abilities of some diamagnetic and paramagnetic materials are given in Table 19.1. None of them differs much from unity.

3   *Ferromagnetic materials.* Ferromagnetic materials are mainly the elements iron, cobalt, and nickel, and their alloys. They have very large values of μ and hence relative permeabilities of 1000 to 100,000, as shown in Table 19.1. As we know from experience, ferromagnetic materials are subject to large forces pulling them into the strongest part of a magnetic field. Similarly, iron needles and bar magnets line themselves up in the direction of an applied magnetic field. Ferromagnetic materials are much used in permanent magnets.

Atomic Explanation of Magnetism

Ampère first suggested that all objects contain a multitude of small current-carrying loops. We now know that Ampère's idea was basically correct. The elementary current loops are thought to consist of (1) the motion of electrons in orbits around atomic nuclei and (2) the spin of each electron about an axis through its center, just as the earth rotates on its axis while moving in its orbit around the sun. This leads to magnetic dipole moments for the electrons and to associated magnetic fields. In diamagnetic and paramagnetic materials these dipole moments tend to cancel each other out and produce almost zero net magnetic dipole moment.

Ferromagnetic Materials

Ferromagnetic materials have electrons with magnetic dipole moments that do not cancel out. These dipole moments can be aligned in an applied magnetic field and can increase that field greatly. This is the phenomenon of *ferromagnetism* (from the Latin word *ferrum*, meaning ''iron'').

Iron contains regions where all the magnetic dipoles align themselves in the same direction, even in the absence of an external magnetic field. Such regions are called *magnetic domains*, and they have dimensions of less than 1 mm on a side. The domains are oriented randomly with respect to one another in the absence of an applied magnetic field, as in Fig. 19.36. When an external field is applied, however, whole domains rotate and line up in the direction of the field (Fig. 19.37) and produce a greatly enhanced field **B**. Also, the size of some domains is increased by the action of the field. This makes μ greater than μ_0 by orders of magnitude, and the substance is therefore ferromagnetic. The value of μ is not constant but depends on the applied field and the previous magnetic history of the sample. For large currents all magnetic domains line up and *saturation* occurs, making it impossible to increase the magnetic field any further.

Domains can be made visible by putting some finely divided iron oxide powder

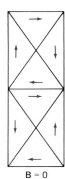

FIGURE 19.36
Magnetic domains oriented in all directions with no preferred orientation in absence of a magnetic field. (The situation is oversimplified to make comparisons with Fig. 19.37 clearer.)

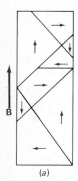
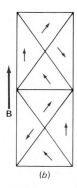

(a)   (b)

FIGURE 19.37 Orientation of magnetic domains in the direction of an applied magnetic field. (a) *Domain growth:* Domains in the direction of the magnetic field grow, while domains in the opposite direction decrease in size. (b) *Domain rotation:* Individual domains rotate in such a way that they have a larger component of their magnetic moments in the direction of the applied field.

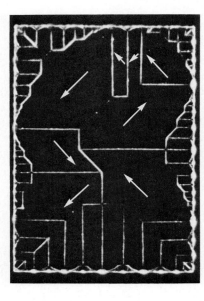

FIGURE 19.38 A photograph of ferromagnetic domain boundaries in a single-crystal nickel platelet of dimensions 71 by 99 μm. The arrows indicate the direction of the magnetic field in each domain. (*Photograph courtesy of Ralph W. DeBlois, General Electric Research Center.*)

on the iron. The powder tends to collect on the boundaries between adjacent domains and make them visible, as in Fig. 19.38.

The existence of magnetic domains also explains the existence of permanent magnets. In ferromagnetic substances some of the domains remain aligned in the original field direction even after the external field is removed, thus producing a permanent magnet.

19.9 The Earth's Magnetic Field

The earth acts as if it had a relatively short bar magnet, some few hundred kilometers long, buried near its center, as shown in Fig. 19.39. This produces a dipole field similar to that of a bar magnet. The axis of this magnet does not correspond

FIGURE 19.39 Magnetic field of the earth, showing that it behaves as if it had a bar magnet some few hundred kilometers long buried near its center.

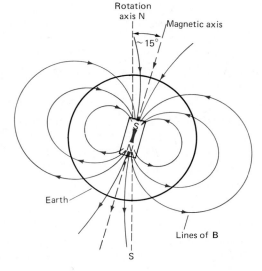

to the axis of rotation of the earth but is tilted about 15° with respect to it, as shown in the figure. Since the north pole of a compass needle is, by definition, the end of the needle that points *north* when free to rotate in the earth's field, it must be attracted by a *south* pole. Hence the earth's south magnetic pole is situated near its geographic north pole.

The angle made by a compass needle with true geographic north is called the *magnetic declination* and varies from about 20°E to 20°W for most of the United States. Figure 19.39 shows that the direction of **B** at most places on the earth is not parallel to the earth's axis. In addition, the lines of the earth's magnetic field are not horizontal at any place on the earth's surface except near the equator. This means that in the northern hemisphere a compass needle will dip below the horizontal if pivoted around a horizontal axis at the earth's surface. The angle of dip below the horizontal is called the *magnetic inclination*. In the United States the dip angle varies from a low of 39° for Hawaii to 75° for the state of Maine.

In practice the magnitude of the earth's magnetic field is measured in two steps. First a compass needle, free to move only in a horizontal plane, is used to find the direction of the earth's magnetic field at the point of interest. The magnitude of the horizontal component of **B** is then measured with a gaussmeter. Then a magnetic dip needle (see Fig. 19.40), whose axis of rotation in the horizontal plane is perpendicular to the horizontal component of the earth's field, is used to measure the angle made by the earth's field with the horizontal.

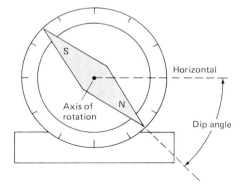

FIGURE 19.40 A magnetic dip needle used to measure *magnetic inclination.* The dip needle's axis of rotation is at its center and is horizontal.

It is noteworthy that in 1989 the spaceship *Voyager 2* found that the magnetic field of Neptune points at an angle of more than 30° with the planet's rotation axis. *Voyager 2* had previously found in 1986 that for Uranus this angle was 60°. These angles are considerably larger than those for the earth, and their large values are surprising and as yet unexplained. If currents of electrically charged material deep within the planets produce their magnetism, as is commonly believed, it would be expected that the magnetic axis would not deviate this much from the planet's rotation axis.

Example 19.8

The magnetic inclination at Ponce in Puerto Rico is 49.5°. If the *vertical* component of the earth's magnetic field is measured with a gaussmeter and found to be 0.35 G, what is the magnitude of the earth's magnetic field at Ponce?

SOLUTION

Since the magnetic inclination is the angle between the horizontal and the direction of the earth's magnetic field, the situation is as shown in Fig. 19.41. Then $\sin \theta = B_y/B$, or $B = B_y/\sin \theta$, and

$$B = \frac{0.35 \text{ G}}{\sin 49.5°} = \frac{0.35 \text{ G}}{0.76} = \boxed{0.46 \text{ G}}$$

This value, close to 0.5 G, is typical of magnetic field values on the earth's surface. This is a very small field relative to the tesla fields $(1 \text{ T} = 10^4 \text{ G})$ that can be produced in the laboratory.

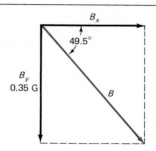

FIGURE 19.41

19.10 Galvanometers; Electric Meters

An important application of the ideas of this chapter is to the development of electric meters to measure currents and voltages. Suppose we suspend a rectangular coil of wire in a magnetic field **B**, as in Fig. 19.42a. The two sides of the coil are of length L and are at right angles to **B**. When a current I is established in the coil, side 1 experiences a force $\mathbf{F}_B$ out of the plane of the paper, according to our right-hand force rule. Similarly side 2 experiences a force equal in magnitude but opposite in direction, i.e., into the plane of the paper. Therefore two torques (Sec. 5.4) are set up to twist the coil about its axis, the magnitude of the total torque being the sum of the force times the lever arm for each force (Fig. 19.42b):

FIGURE 19.42 A rectangular coil of wire in a uniform magnetic field **B**. (a) Front view: The forces on the two sides of the coil exert a torque that rotates the coil. (b) Top view: The magnitude of the counterclockwise torque is $2(F_B D/2) = F_B D = B_\perp IA$. (c) Top view: The coil rotates to a new position where the magnetic torque is balanced by the mechanical torque exerted by the twisted suspending wires. The angular displacement θ of the coil is proportional to the current I through the coil and can be used to measure this current.

$$\tau = 2 \left(F_B \frac{D}{2} \right) = F_B D = (B_\perp IL)D = B_\perp IA \qquad (19.20)$$

where A is the area of the coil. If the coil consists of N turns of wire, then $\tau = B_\perp NIA$. Note that the quantity NIA is the *magnetic dipole moment* μ of the loop, which is directed perpendicular to the plane of the coil, and that the field tries to orient this dipole moment in the direction of the field, just as it would a compass needle.

This torque twists the loop to an equilibrium position where the restoring torque of the twisted suspension just balances out the torques on the current-carrying conductors in the magnetic field (Fig. 19.42c). Since the angular displacement of the coil is proportional to the torque τ, this displacement is also proportional to the current I, from Eq. (19.20). By putting known currents through the coil, the

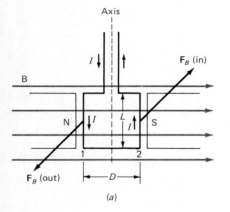

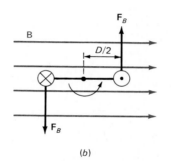

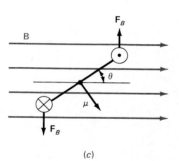

(a)   (b)   (c)

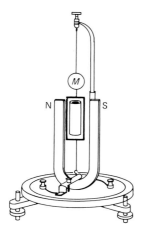

instrument can be calibrated; i.e., the readings corresponding to these known currents can be marked on the scale. By interpolation, the values of unknown currents can then be found. Such an instrument is called a *galvanometer*, a very sensitive instrument for measuring small currents (Fig. 19.43).

Galvanometers can measure small currents directly but must be modified if they are to be used to measure large currents or voltages (see Example 19.9).

FIGURE 19.43   A galvanometer of the type developed by Jacques d'Arsonval (1851–1940). The rectangular coil is suspended by very fine metal wires that also serve to carry current to the coil. The coil moves between the poles of a horseshoe magnet, with the magnetic field concentrated by a fixed soft iron cylinder inside the coil. The angular displacement of the mirror *M* is proportional to the current, and a light beam reflected off the mirror may be used to indicate the value of the current on a linear scale.

Example 19.9

A galvanometer has a resistance of 40 Ω and deflects full-scale for a current of 10 μA through its coil. **(a)** What resistance must be connected in parallel with the galvanometer to convert it into a 1.0-A ammeter? **(b)** What resistance must be connected in series with the galvanometer to convert it into a 1000-V voltmeter?

SOLUTION

(a)   Since the galvanometer can take a current of only 10 μA before deflecting off-scale, we must in some way bypass most of the 1.0-A current around the galvanometer. To do this we place a resistor R_P in parallel with the galvanometer coil, which has a resistance of 40 Ω. The circuit is therefore that of Fig. 19.44, in which R_P is called a *shunt resistor*.

Since the voltage between a and b is the same for the two current paths,

$$I_G R_G = I_P R_P \qquad \text{or} \qquad R_P = \left(\frac{I_G}{I_P}\right) R_G$$

But we want the galvanometer to deflect full-scale when the current $I = I_G + I_P = 1.0$ A. Since this full-scale deflection occurs when $I_G = 10$ μA, we have

$$I_P = I - I_G = 1.0 \text{ A} - 10^{-5} \text{ A} = 0.99999 \text{ A}$$

and so $\qquad R_P = \dfrac{10^{-5} \text{ A}}{0.99999 \text{ A}} (40 \ \Omega) = \boxed{4.0 \times 10^{-4} \ \Omega}$

(b)   To convert the same galvanometer into a 1000-V voltmeter, we put a resistor R_S in series with the galvanometer, and choose R_S in such a way that the galvanometer deflects full-scale when 1000 V is placed across the combination $R_G + R_S$.

We have, from Ohm's law, applied to the circuit in Fig. 19.45,

$$V_{ab} = I(R_G + R_S)$$

where I must be 10 μA for full-scale deflection of the galvanometer. Hence

$$1000 \text{ V} = 10^{-5} \text{ A}(40 \ \Omega + R_S)$$

from which $\qquad R_S = 10^8 \ \Omega - 40 \ \Omega = \boxed{10^8 \ \Omega}$

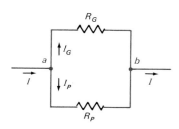

FIGURE 19.44   Conversion of a galvanometer into an ammeter by putting a shunt resistor R_P in parallel with the galvanometer (R_G).

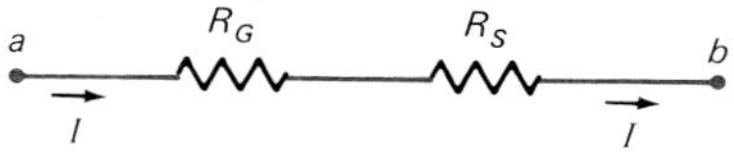

FIGURE 19.45   Conversion of a galvanometer into a voltmeter by putting a resistor R_S in series with the galvanometer (R_G).

Summary: Important Definitions and Equations

Magnetic field (B): A vector whose direction at any point in space is the direction of the force which would be exerted on a small, isolated north pole at that point, and whose magnitude is proportional to the force experienced by that north pole.

Force on a current-carrying conductor of length L in a magnetic field

Magnitude: $F_B = B_\perp IL$ **(19.1)**

where $B_\perp$ is in teslas (T); 1 tesla = 1 N/(A·m) = 10^4 G
Direction (right-hand force rule): If the fingers of the outstretched right hand are placed in the direction of the magnetic field and the thumb in the direction of the conventional current I, then the palm of the hand faces in the direction of the force on the current-carrying conductor.

Force on a charged particle in a magnetic field

Magnitude: $f_B = B_\perp qv$ **(19.4)**

Direction: Given by right-hand force rule for a *positive charge.*

Magnetic field produced by a straight, current-carrying wire

Magnitude: $B = \dfrac{\mu_0 I}{2\pi r} = 2 \times 10^{-7} \dfrac{I}{r}$ **(19.14)**

where r is the distance from the wire and $\mu_0 = 4\pi \times 10^{-7}$ T·m/A is the permeability of free space.
Direction (right-hand field rule): If a wire is grasped with the right hand with the thumb pointed in the direction of the current, the magnetic field is circular about the wire and in the direction in which the fingers curl.

Force between two long, straight, parallel current-carrying conductors

Magnitude: $\dfrac{F_B}{L} = \dfrac{\mu_0 II'}{2\pi r}$ **(19.13)**

Direction: Attraction if two currents are in same direction; repulsion if in opposite directions.

Ampere (official definition): The magnitude of the current that, when present in each of two long parallel wires one meter apart, results in a force between the wires of exactly 2×10^{-7} N per meter of length.

Ampère's circuital law

$$\sum B_\parallel \Delta l = \sum \mu_0 I \qquad \textbf{(19.15)}$$

Magnetic field of a circular current loop

$$B = \frac{\mu_0 I}{2r} \qquad \textbf{(19.16)}$$

Magnetic field of a solenoid

In vacuum (or air): $B = \mu_0 n I$ **(19.17)**

where n is the number of turns per unit length.
In material of magnetic permeability μ: $B = \mu n I$

Relative magnetic permeability

$$K_m = \frac{\mu}{\mu_0} \qquad \textbf{(19.19)}$$

Magnetic materials

Diamagnetic: μ is slightly less than μ_0; K_m is slightly less than 1.

Paramagnetic: μ is slightly greater than μ_0; K_m is slightly greater than 1.

Ferromagnetic: μ is very much greater than μ_0; K_m is very much greater than 1; also μ is not constant but varies with both the magnetizing current and the previous history of the material.

Atomic explanation of magnetism: Magnetism is caused by current loops at the atomic level. These current loops act as magnetic dipoles whose behavior in magnetic fields determines the magnetic behavior of the material.

Questions

1 What would happen if the north pole of a magnet were brought near a positively charged light plastic ball? The south pole?

2 Why does either pole of a permanent magnet attract unmagnetized iron?

3 Why can you form a chain of paper clips at the end of a small bar magnet? What has happened to each clip?

4 You have an electron at rest and want to accelerate it to a high speed. Can you do this with a static electric field? With a static magnetic field? Explain your answer.

5 If a particle with charge q enters a magnetic field **B** with a velocity **v** that has a component in the direction of **B**, describe the path of the particle in the magnetic field.

6 Two ions have the same mass, but one has a single positive charge and the other a double positive charge. How will their final positions vary as measured by a mass spectrometer like that in Fig. 19.16?

7 Does the magnetic field do any work on a charged particle moving at right angles to the field? If not, how can it deflect the particle? If so, where does the energy come from?

8 What conclusions can you draw from the fact that J. J. Thomson repeated his e/m determination for vacuum tubes with electrodes of different metals like aluminum, platinum, and iron, and containing different gases like hydrogen and carbon dioxide in addition to air, and found, within the limited accuracy of his experiment, always the same value for e/m?

9   J. J. Thomson performed another measurement of e/m for the electron; the experiment involved the collision of a beam of electrons with the inside of a metal can. Thomson measured with an electrometer the total charge on the can, and obtained the total kinetic energy of the electron beam by measuring the increase in temperature of the can. Would you expect this method to be as accurate as the method for e/m discussed in the text? Why?

10   Would you expect the earth's magnetic field to introduce a sizable error in J. J. Thomson's measurement of e/m? How could you determine the approximate size of the error introduced?

11   What is the evidence that cathode rays are really electrons and not some kind of electromagnetic radiation?

12   The picture on your TV screen is rather blurry, and a friend suggests that it is probably due to the deflection of the electron beam by the earth's magnetic field. How would you go about deciding whether this could possibly be the cause of the poor picture?

13   Discuss all the laws of physics that are illustrated in the operation of a cyclotron.

14   In the interaction of two wires carrying current in the same direction, how can you be sure that the interaction is not an electrostatic one? (*Hint:* What would happen if you had three wires all carrying current in the same direction?)

15   Two electrons at rest repel each other. Do two electrons moving along parallel lines attract each other? Why or why not?

16   A bright student suggests that, since a long wire experiences a force in a magnetic field, we could have airplanes land and take off vertically by passing large currents through metal rods running the length of the plane. The force provided by the earth's magnetic field on the metal rod would then provide the needed upward or downward force. Discuss this suggestion from the point of view of the physics involved.

17   What would you expect the direction of the magnetic inclination on the earth to be in the southern hemisphere?

Multiple-Choice and Simple Exercises

$e = -\mathcal{I}$

19.1   The north pole of a bar magnet is held in a fixed position near a positively charged Ping-Pong ball suspended at the end of a long thread. Which of the following accurately describes what happens?

(*a*)   The ball moves toward the north pole of the magnet.
(*b*)   The ball moves away from the north pole.
(*c*)   The ball moves in a horizontal plane at right angles to the line connecting the ball and the north pole.
(*d*)   The ball moves in a vertical plane at right angles to the line connecting the ball and the north pole.
(*e*)   Nothing will happen.

19.2   Consider the three-dimensional rectangular coordinate system shown in Fig. 19.46. A uniform magnetic field **B** points downward along y into the xz plane. A wire is stretched along the x axis and carries a current I in the negative x direction. The force on the wire is in the direction of:

(*a*) $-y$ (*b*) $-z$ (*c*) $+y$ (*d*) $+z$
(*e*) $-x$

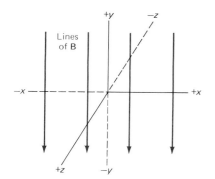

FIGURE 19.46   Diagram for Exercises 19.2 and 19.3.

19.3   An electron moves in the negative z direction in Fig. 19.46. The force on the electron is in the direction of:

(*a*) $+x$ (*b*) $-x$ (*c*) $+z$ (*d*) $-z$
(*e*) $+y$

19.4   If a current of 1.0 A is carried by a wire of length 10 cm at right angles to a magnetic field **B** and the wire experiences a force of 1.0×10^{-2} N, the strength of the magnetic field is:

(*a*)   1.0 T   (*b*)   10 T   (*c*) 1.0×10^{-1} T
(*d*) 1.0×10^{-3} T   (*e*) 1.0×10^{-2} T

19.5   A proton and an electron enter the same magnetic field with identical velocities at right angles to the field. The ratio of the radius of the circular orbit of the proton to the radius of the electron's orbit will be about:

(*a*)   1   (*b*)   1/2000   (*c*)   2000/1
(*d*) $(4 \times 10^6)/1$ (*e*) $1/(4 \times 10^6)$

19.6   An electron of speed 2.0×10^6 m/s enters a region of space in which there are crossed electric and magnetic fields. If the electric field is 2.0×10^4 V/m and the electron emerges from the fields without being deflected from its original path, the magnitude of the magnetic field must be

(*a*) 4.0×10^{10} T   (*b*) 2.0×10^4 T
(*c*) 1.0×10^2 T   (*d*) 1.0×10^{-1} T
(*e*) 1.0×10^{-2} T

19.7   A particle of velocity 2.0×10^6 m/s is shot into a 2.0-T magnetic field at right angles to **v**. The particle moves in a circle of radius 1.0 cm. The particle is a:

(*a*)   Electron   (*b*)   Neutron   (*c*) ^{3}He nucleus
(*d*) ^{4}He nucleus   (*e*)   Proton

19.8   A solenoid with an air core has 10 turns per centimeter and carries 1.0 A of current. The magnetic field inside the solenoid is about:

(*a*)   1 T   (*b*) 10^{-2} T   (*c*) 10^2 T
(*d*) 10^{-3} T   (*e*) 10^{-5} T

19.9 The permeability of a particular sample of iron is 2.5×10^{-3} T·m/A. The relative magnetic permeability (K_m) of the iron sample is about:

(a) 2.5×10^{-3} T   (b) 2.0×10^3 T·m/A
(c) 2.0×10^3 (d) 2.5×10^{-3}
(e) 5.0×10^{-4}

19.10 In a laboratory experiment on the optical pumping of rubidium, a magnetic field of 50 G is used. The error introduced in such an experiment by the neglect of the earth's magnetic field would be about one part in:

(a) 10^4 (b) 10   (c) 10^2 (d) 10^3
(e) 10^5

19.11 In the situation shown in Fig. 19.9, the length L of the wire is 10 cm, the current is 8.0 A, and the magnetic field **B** has a strength of 0.50 T in the direction shown. What is (a) the magnitude and (b) the direction of the force on the wire?

19.12 A rectangular galvanometer coil of area 5.0 cm² is suspended in a 0.10-T magnetic field with the plane of the coil parallel to the magnetic field. What is the torque on the coil when the current through the coil is 0.15 mA?

19.13 What is the magnitude of the force on an electron moving with a speed of 2.0×10^5 m/s at right angles to a magnetic field of strength 0.55 T?

19.14 How many complete revolutions does an electron make per second in a cyclotron in which the magnetic field **B** is 50×10^{-4} T?

19.15 What is the magnetic field 2.0 cm from a straight wire carrying a current of 1.5 A?

19.16 An electric power line carries a dc current of 500 A in a straight line at a height of 20 m above the ground.

(a) What is the magnitude of the magnetic field produced at the ground?

(b) How does this field compare with the magnetic field of the earth?

19.17 Two long parallel wires are 20.0 cm apart and carry currents of 2.0 and 5.0 A in the same direction. Find the magnitude and direction of the force per unit length between the wires.

19.18 What is the magnetic field of a solenoid that is 50 cm long and contains 500 turns, when there is a current of 12 A through the solenoid?

19.19 Approximately what is the maximum magnetic permeability of a sample of permalloy?

19.20 The horizontal component of the earth's magnetic field at Tucson, Arizona, is 0.26 G, and the magnetic inclination is 59°. Find the strength of the earth's magnetic field at Tucson.

Problems

19.21 A long electric cable is suspended above the earth and carries a 100-A current parallel to the surface of the earth. The earth's magnetic field has a value of 0.56×10^{-4} T and makes an angle of 70° with respect to the cable. What is the force exerted on a 10-m length of the cable due to the earth's magnetic field?

19.22 A galvanometer coil consists of a single turn rectangular in shape and of length 4.0 cm and width 2.0 cm. It is mounted with its two long sides perpendicular to a magnetic field of 0.010 T, with its two short sides initially parallel to the magnetic field. If the current through the coil is 100 µA:

(a) What is the force on one of the long sides?

(b) What is the torque on one of the long sides about an axis parallel to the long side and passing through the middle of the short sides?

(c) If the coil rotates through 90° so that its plane is now perpendicular to the magnetic field, what is the force on one of its long sides?

(d) What is now the torque on one of the long sides?

19.23 A rectangular galvanometer coil of length 3.0 cm and width 1.5 cm, containing 10 turns of light wire, is suspended in a magnetic field with the plane of the coil parallel to the direction of the field. If the torque on the coil is 1.8×10^{-7} N·m when the current through the coil is 0.20 mA, what is the strength of the magnetic field?

19.24 A rectangular coil of 20 turns is suspended in a magnetic field of magnitude 0.20 T, with its plane parallel to the direction of the field. The coil has a height of 12 cm and a width of 6.0 cm. What must be the current in the coil to produce a torque of 5.0×10^{-6} N·m?

19.25 A 0.40-T magnetic field is directed along the $+x$ axis. A particle of speed 3.0×10^6 m/s is shot into the magnetic field at an angle of 35° above the magnetic field direction. Describe the path followed when the particle is (a) a proton, (b) an electron, and (c) a neutron.

19.26 An electron is accelerated from rest by a voltage of 3000 V. If it then enters a region where a 6.5×10^{-4} T magnetic field exists perpendicular to its velocity, what is the radius of the electron's orbit?

19.27 What is the radius of the circular orbit of a 0.50-MeV proton in a 1.0-T magnetic field?

19.28 In a cyclotron a proton moves in a circle of radius 0.40 m. The magnitude of the magnetic field perpendicular to the orbit is 1.5 T. What are (a) the cyclotron frequency and (b) the kinetic energy of the proton?

19.29 An electron in a vacuum is first accelerated by a voltage of 1.4×10^4 V and then enters a region in which there is a uniform magnetic field of 0.60 T at right angles to the direction of the electron's motion.

(a) What is the force on the electron due to the magnetic field?

(b) What is the speed of the electron after it has been in the magnetic field for 10 s? Why?

19.30 Crossed electric and magnetic fields can be used as a "velocity selector" to select particles of a particular speed from a beam containing charged particles with a variety of speeds. If the electric field in a particular region of space is 6.0×10^7 V/m and the magnetic field at right angles to the electric field is 1.5 T, what will be the speed of protons that can pass perpendicular to these fields without being deflected?

19.31 A beam of electrons passes undeflected through "crossed" electric and magnetic fields, where $E = 8.0 \times 10^3$ V/m and $B = 2.0 \times 10^{-3}$ T. If the electric field is turned off, the electrons move in the magnetic field in circular paths of radius 1.2 cm. Determine e/m for the electrons from these data.

19.32 Protons in a cyclotron are moving with a speed of 2.0×10^7 m/s in a circle of radius 0.80 m.

(a) What is the strength of the magnetic field required to keep them moving in this circle?

(b) What is the magnitude of the force exerted on the protons by this magnetic field?

19.33 A proton is moving with a speed of 2.0×10^6 m/s in a magnetic field whose direction is unknown. When moving along the positive z axis in Fig. 19.47, the proton experiences no force. When moving along the positive x axis, it feels a force of 1.0×10^{-13} N in the positive y direction. What are the magnitude and direction of the magnetic field?

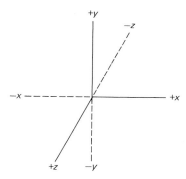

FIGURE 19.47   Diagram for Prob. 19.33.

19.34 A proton is accelerated by a voltage of 50,000 V and then enters a region in which a 1.0-T uniform magnetic field exists.

(a) What is the force on the proton?

(b) What is the radius of the circle in which the proton moves?

19.35 In a repetition of J. J. Thomson's experiment, a beam of electrons is first accelerated through a voltage of 1.2×10^4 V. The beam then enters into a region between two deflection plates, as in Fig. 19.21, where a voltage of 0.90×10^3 V is applied to the two plates separated by a distance of 2.0 cm. What must be the value of the magnetic field that, when applied at right angles to the electric field, allows the electrons to pass through the plates without being deflected?

19.36 Singly charged uranium ions, some of mass number 235 and the others of mass number 238, make up the beam in a mass spectrometer.

(a) What is the percentage difference in the radii of the two isotopes in the mass spectrometer, if they enter the magnetic field with the same speed?

(b) What conclusions can you draw from this about the feasibility of using electromagnetic separation to produce the ^{235}U needed for nuclear power plants and nuclear weapons?

19.37 Prove that the momentum of a charged particle moving at right angles to a uniform magnetic field is directly proportional to the radius of its circular orbit.

19.38 A proton is shot at 5.0×10^6 m/s at an angle of $30°$ with respect to a 0.30-T magnetic field directed along the positive x axis. Describe the path followed by the proton. (*Hint*: Resolve the motion of the proton into its x and y components.)

19.39 An electron of speed 3.0×10^6 m/s is shot into a magnetic field of 50×10^{-4} T. It then moves in a circle of 3.5-mm radius. What is the value of e/m for the electron?

19.40 An electron is moving in a straight line at 5.0×10^6 m/s. Find (a) the magnitude and direction of the magnetic field that causes the electron to move in a clockwise circular path of radius 2.0 cm, as viewed from above; (b) the time it takes the electron to move once around its circular orbit.

*__19.41__ Derive J. J. Thomson's result for e/m for the electron [Eq. (19.11)]. (*Hint*: Use Fig. 19.21 and the values of the electric and magnetic fields acting on the electron.)

19.42 Two types of mass spectrometers are used for isotope analysis. In one the charged isotopes first pass through a velocity selector and enter the magnetic field with the same speeds. In the second type, the charged isotopes are first accelerated by the same electric field before entering the magnetic field. What is the ratio of the radii of the circular paths followed by two He$^+$ ions, one of mass $3u$ and the other of $4u$ (a) in the first type of mass spectrometer and (b) in the second type?

19.43 The starting motor of an automobile draws a current of 200 A from the car's battery for a few seconds while the car is starting.

(a) What is the strength of the magnetic field 1.5 cm from one wire leading to the battery? If the two battery wires are at one place parallel and 6.0 cm apart, as in Fig. 19.48, what are the magnitude and direction of the magnetic field (b) at A; (c) at D?

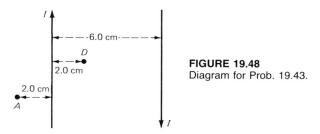

FIGURE 19.48
Diagram for Prob. 19.43.

19.44 A 12-V storage battery is connected across two long wires in parallel, one of resistance 4.0 Ω and the other of resistance 3.0 Ω. A 1.0-m length of one wire is placed parallel to a 1.0-m length of the second wire, and 10 cm from it. What is the force between the two 1.0-m lengths?

19.45 Three long, straight wires carry identical parallel currents of $I = 10$ A, as shown in Fig. 19.49.

(a) What is the resultant force on the center wire?

(b) What is the force on the wire at the left?

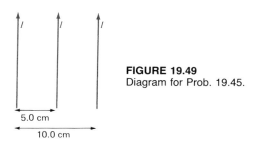

FIGURE 19.49
Diagram for Prob. 19.45.

*19.46 Two long parallel wires are hung by cords of 3.0-cm length from a common axis, as in Fig. 19.50. The wires have a mass of 0.010 kg/m and carry the same current in opposite directions. What is the current if each cord makes an angle of 30° with the vertical?

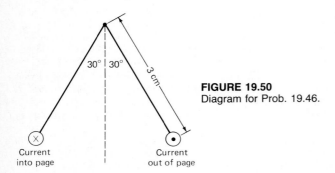

FIGURE 19.50
Diagram for Prob. 19.46.

*19.47 A long, straight wire carries a current of 2.0 A. An electron travels at 5.0×10^4 m/s parallel to the wire, 20 cm from the wire, and in the same direction as the current. What force does the magnetic field of the current exert on the moving electron?

*19.48 A toroid is a long solenoid that has been bent into a circle of radius r. The average length of the toroid is then $2\pi r$. If the toroid contains N turns in all, use Ampère's law to find the field inside the coils of the toroid when it carries a current I.

*19.49 A coaxial cable consists of a small, solid central conductor, separated from a cylindrical outside conductor by insulating disks. If the two conductors carry equal currents in opposite directions, use Ampère's circuital law to find the magnetic field outside the cable.

19.50 What current is required in an air-core solenoid with 100 turns per meter in order to produce a magnetic field equal to that of the earth, i.e., about 5×10^{-5} T?

19.51 What is the maximum possible value of the magnetic field produced by an iron-core solenoid of 100 turns and length 20 cm when a current of 1.0 A is established in it?

19.52 According to Bohr's model of the hydrogen atom (Chap. 28), the electron in hydrogen travels with a speed of 2.2×10^6 m/s in a circle of radius 5.3×10^{-11} m. What are the magnitude and direction of the magnetic field at the nucleus due to the electron's motion?

19.53 A long, straight wire carries a current of 25 A along the axis of a long solenoid. The solenoid has 200 turns per meter and carries a current of 10 A. Find the magnitude of the resultant field at a point inside the solenoid and 4.0 mm from its axis.

19.54 If the horizontal component of the earth's magnetic field at Cambridge, Massachusetts, is 0.17 G and the magnetic inclination is 73°, find the vertical component of the earth's magnetic field at Cambridge.

19.55 In Florida the average horizontal component of the earth's magnetic field in 1912 was 28 μT and the average dip angle was 57.5°. What was the magnitude of the earth's magnetic field in Florida at that time?

*19.56 Estimate the effect of the earth's magnetic field on the electron beam in a TV set. Assume that the accelerating voltage is 2.0×10^4 V, and calculate the approximate deflection of the beam over its travel distance of 45 cm to the screen of the tube.

*19.57 A solenoid of diameter 5.0 cm has 250 turns in its 50-cm length and carries a current of 8.0 A. A small circular coil of diameter 1.0 cm is located at the center of the solenoid with its axis perpendicular to the solenoid axis. If a current of 1.5 A is passed through the circular coil, what mechanical torque must be applied to keep the circular coil from changing its orientation in the magnetic field of the solenoid?

19.58 A galvanometer has a circuit sensitivity of 100 μA and a resistance of 15 Ω. It is desired to use this galvanometer as a 30-A full-scale ammeter.
(a) How can the galvanometer be modified to accomplish this goal?
(b) What is the equivalent resistance of the galvanometer after it is thus modified?

19.59 A galvanometer of current sensitivity 50 μA and resistance 50 Ω is to be used as a high-voltage voltmeter to measure voltages up to 10^4 V.
(a) How can the galvanometer be modified to accomplish this goal?
(b) What is the equivalent resistance of the galvanometer after it is thus modified?

19.60 (a) In part (a) of Example 19.9, what is the equivalent resistance of the ammeter constructed from the 10-μA galvanometer?
(b) What advantages does an ammeter with such a resistance have when used to measure current?

19.61 (a) In part (b) of Example 19.9, what is the equivalent resistance of the voltmeter constructed from the 10-μA galvanometer?
(b) What advantage does such a voltmeter have when used in a circuit to measure voltage?

Additional Readings

Bitter, Francis: *Magnets: The Education of a Physicist*, Double-day Anchor, Garden City, N.Y., 1959. An unusual book that combines an abundance of useful information about magnetism with the story of the author's career as a distinguished physicist.

Bloxham, Jeremy, and David Gubbins: "The Evolution of the Earth's Magnetic Field," *Scientific American*, vol. 261, no. 6, December 1989, pp. 68–75. An explanation of the earth's magnetic field in terms of the churning of molten iron in the outer core around the solid inner core of the earth.

Carrigan, Richard A., Jr., and W. Peter Trower: "Superheavy Magnetic Monopoles," *Scientific American*, vol. 246, no. 4, April 1982, pp. 106–118. Many research laboratories are searching for the type of monopole discussed in this article.

Dibner, Bern: *Oersted and the Discovery of Electromagnetism*, Blaisdell, New York, 1962. A historical account of Oersted's life and work.

Kunzler, J. E., and M. Tannenbaum: "Superconducting Magnets," *Scientific American*, vol. 206, no. 6, June 1962, pp. 60–67. Describes the development of the first generation of superconducting magnets.

Rayleigh, Lord, *The Life of Sir J. J. Thomson*, Dawsons of Pall Mall, London, 1969. An interesting biography by a physicist who knew Thomson well.

Runcorn, S. K.: "The Earth's Magnetism," *Scientific American*, vol. 193, no. 3, September 1955, pp. 152–162. Geophysical explanation of the earth's magnetic field and its changes over time.

Thomson, George Paget: *J. J. Thomson, Discoverer of the Electron*, Doubleday Anchor Books, Garden City, N.Y., 1966. A popular account of J. J.'s life and scientific accomplishments by his son, also a Nobel Laureate in physics.

Weinberg, Steven: *The Discovery of Subatomic Particles, Scientific American* Library, Freeman, San Francisco, 1983. Contains a good account of Thomson's measurements and of the work of Oersted and Ampère.

Williams, L. Pearce: "André-Marie Ampère," *Scientific American*, vol. 260, no. 1, January 1989, pp. 90–97. An interesting account of the French physicist who first quantified the magnetic effects of electric currents and who constructed the first galvanometer.

CHAPTER 20

Electromagnetic Induction

And if the path indicated was a false one, warning could only come from an intellect of great freshness—from a man who looked at phenomena with an open mind and without preconceived notions, who started from what he saw, not from what he had heard, learned, or read. Such a man was Faraday.

Heinrich Hertz (1857–1894)

Just as the cathedrals of Chartres and Mont-Saint-Michel were the symbols of medieval western culture, so the dynamo—the electric generator—was the symbol of the industrial revolution and the technological age that followed in Europe and North America. Although a more appropriate symbol for present-day civilization might be the computer chip or the industrial robot, electric generators and motors remain the workhorses of today's world. In this chapter we develop the basic theory that enables us to understand these industrial "movers and shakers."

20.1 Motional EMF's

We have seen that a wire carrying a current in a magnetic field experiences a force which tends to set the wire in motion. Physicists like Michael Faraday suspected that the reverse process might also occur, i.e., that the motion of a conductor through a magnetic field might cause a current to flow in the conductor.

This does indeed happen. Consider the circuit of Fig. 20.1, in which a metal rod of length l between points a and b slides along a U-shaped wire frame in a

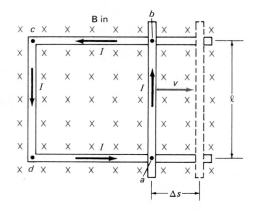

FIGURE 20.1 A motional emf induced in a rod moving to the right in a magnetic field directed into the paper.

magnetic field **B** directed into the paper. Suppose we move the rod to the right at a constant velocity **v** at right angles to the lines of **B**. Positive charges in the rod experience forces in the direction from *a* to *b*, as given by the right-hand force rule, and negative charges experience forces in the direction from *b* to *a*. Since there is a complete circuit, electrons will flow in a clockwise direction around the circuit as long as the rod continues to move. We have therefore produced (or *induced*) a *conventional current* in the *counterclockwise* direction *abcda*. The motion of the rod has induced an emf and in this case also a current. For this reason the emf is called a *motional emf*, which is one kind of *induced emf*. For a motional emf, the work we do in moving the rod is converted into the electric energy of moving charges.

From Eq. (19.4) the force on each charge has a magnitude

$$f_B = B_\perp q v$$

and the work done to move the charge from *a* to *b* is

$$\mathcal{W} = f_B l = B_\perp q v l$$

The motional emf $\mathcal{E}$ is then the work done per unit charge, or

$$\mathcal{E} = \frac{\mathcal{W}}{q} = B_\perp v l \qquad \text{(20.1)}$$

The magnitude of the motional emf is thus the product of the magnetic field, the length of the rod, and the speed with which the rod moves at right angles to the magnetic field. As usual, $\mathcal{E}$ is in volts if B is in teslas [N/(A·m)], v is in meters per second, and l is in meters.

If B is not perpendicular to the plane of the circuit, then $B_\perp = B \cos \theta$, and Eq. (20.1) becomes

$$\mathcal{E} = B v l \cos \theta \qquad \text{(20.2)}$$

where θ is the angle between B and the normal to the plane.

In Fig. 20.1 the induced emf produces a conventional current that is counterclockwise, as we find by applying the right-hand force rule. Example 20.1 shows that the current must be in this direction if energy is to be conserved.

Example 20.1

The length of the rod resting on the metal frame in Fig. 20.1 is 0.20 m, and its speed *v* to the right is 0.10 m/s at right angles to the magnetic field. If the resistance of the loop is 0.020 Ω and if $B = 1.0$ T [= 1.0 N/(A·m)], find **(a)** the motional emf; **(b)** the current that flows; **(c)** the magnitude and direction of the force on the rod because of the current flowing through it; **(d)** the power needed to move the rod against this force. **(e)** Is this power equal to the rate at which electric energy is being produced? **(f)** What is the rate at which energy is consumed in the resistance of the circuit?

SOLUTION

(a) The motional emf is, from Eq. (20.1),

$$\mathcal{E} = B_\perp v l = \left(1.0 \ \frac{\text{N}}{\text{A·m}}\right)\left(0.10 \ \frac{\text{m}}{\text{s}}\right)(0.20 \text{ m})$$

$$= 0.020 \text{ N·m/C} = 0.020 \text{ J/C} = \boxed{0.020 \text{ V}}$$

(b) The current is, from Ohm's law,

$$I = \frac{\mathcal{E}}{R} = \frac{0.020 \text{ V}}{0.020 \ \Omega} = \boxed{1.0 \text{ A}}$$

(c) Since the rod is carrying a current of 1.0 A in a magnetic field **B**, it experiences a force equal, from Eq. (19.1), to

$$F_B = B_\perp I l = \left(1.0 \ \frac{\text{N}}{\text{A·m}}\right)(1.0 \text{ A})(0.20 \text{ m}) = \boxed{0.20 \text{ N}}$$

The direction of this force can be found from the right-hand force rule and is to the left in the diagram. Hence the current produces a force (in the opposite direction to **v**) that opposes the motion producing the emf.

(d) To move the rod requires work by some external agent against this opposing force. The rate at which this work is done is, from Eq. (6.18),

$$P = \frac{\Delta \mathcal{W}}{\Delta t} = \frac{F \Delta s}{\Delta t}$$

where Δs is the distance that the rod moves to the right in time Δt so that $\Delta s / \Delta t = v$. We have, therefore,

$$P = Fv = (0.20 \text{ N})(0.10 \text{ m/s}) = \boxed{0.020 \text{ W}}$$

(e) The rate at which electric energy is produced is

$$P = \mathcal{E}I = (0.020 \text{ V})(1.0 \text{ A}) = \boxed{0.020 \text{ W}}$$

The mechanical power put into the circuit by the external force moving the rod is equal to the electric power generated. Energy is therefore conserved, as it must be.

(f) Since $P = I^2 R$ [Eq. (18.12)] we have

$$P = (1.0 \text{ A})^2 (0.020 \text{ }\Omega) = \boxed{0.020 \text{ W}}$$

The work done therefore ends up as heat in the resistance of the circuit. The work done in moving the rod is first converted into electric energy. This is, in turn, converted into thermal energy in the resistance of the circuit. Energy is transformed, but no energy is lost.

20.2 Faraday's Induction Law

Equation (20.1) is one expression of a more general law called *Faraday's induction law* after the great British physicist Michael Faraday (see Fig. 20.4 and accompanying biography), who is credited with discovering electromagnetic induction.*

Magnetic Flux

In Fig. 20.1 the lines of the magnetic field **B** are perpendicular to the area A bounded by the four sides of the electric circuit. We call the product BA the *magnetic flux* Φ (Greek capital *phi*). In the general case, **B** may not be perpendicular to A, and we write instead

$$\Phi_B = B_\perp A = BA \cos \theta \tag{20.3}$$

as is clear from Fig. 20.3. As we will see, it is the rate of change in time of the magnetic flux through a circuit that induces an emf.

If we have a small area ΔA over which B is constant and if the magnetic field lines are all perpendicular to ΔA, then the magnetic flux is $\Delta \Phi_B = B \Delta A$, and

$$B = \frac{\Delta \Phi_B}{\Delta A} \tag{20.4}$$

FIGURE 20.2 Joseph Henry (1797–1878), the first American scientist to do any serious work in electricity after Benjamin Franklin's great contributions made around 1750. Henry's life and scientific discoveries paralleled those of Faraday in many ways. Henry is credited with developing the principles of the telegraph (before Morse), of electromagnetic induction (independently of Faraday), of self-induction, and of the electric motor. The unit of inductance, the henry, is named after him. *(Courtesy of National Portrait Gallery, Smithsonian Institution.)*

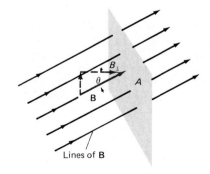

Lines of **B**

FIGURE 20.3 Definition of magnetic flux. The lines of the magnetic field **B** pass through the area A in a direction that is not perpendicular to A. The magnetic flux Φ is defined as $\Phi_B = AB_\perp = AB \cos \theta$, where $B_\perp = B \cos \theta$ is the component of **B** perpendicular to A.

---

*Joseph Henry (Fig. 20.2), the first head of the Smithsonian Institution in Washington, D.C., actually did some of the key experiments on electromagnetic induction in 1830 before Faraday. But heavy teaching responsibilities at the Albany Academy kept Henry from completing and publishing his work.

Michael Faraday (1791–1867)

FIGURE 20.4 Michael Faraday. Portrait painted in 1842 by Thomas Phillips, R.A.; now in National Portrait Gallery, London. *(Courtesy of AIP Niels Bohr Library.)*

Born into a poor family that could not afford to pay for his education, Michael Faraday was almost completely self-taught. Despite this initial handicap, his scientific curiosity and laboratory skills were so great that he made outstanding contributions to both chemistry and physics, and is considered by many to be the greatest experimental physicist who ever lived. His chief competitor for this title was his compatriot, Ernest Rutherford (Sec. 9.2).

Faraday was saved from oblivion by a felicitous combination of hard work and good fortune. As a youth he was apprenticed to a bookbinder and spent his lunch hours reading some of the books sent to the shop for binding. An article on electricity in the *Encyclopedia Britannica* particularly captured his fancy, and he decided to devote

his life to science. He applied to Sir Humphrey Davy, the director of the Royal Institution in London, for a position and eventually was taken on as a valet and laboratory assistant to Davy.

Faraday's laboratory skills developed so quickly that he soon outshone his scientific sponsor—a source of lasting tension between them. Faraday became known as one of the best analytical chemists in England, much sought after by the growing chemical industry of the day, especially after he succeeded in isolating benzene. Later he moved on to a study of ions in solution and discovered the laws governing electrolysis (Sec. 18.2). His most famous contributions to science, however, were his induction law and his development of a prototype electric generator.

Faraday believed that, because a current-carrying wire experiences a force and tends to move in a magnetic field, the reverse should also be true—a magnet should be able to produce a current in a wire. For many years he was unable to prove this experimentally, but one day he plunged a magnet down into a coil and observed that a current did exist for an instant. The essential thing was the *relative motion* of magnet and coil, something that had escaped him before.

Faraday was the first physicist to develop the idea of electric and magnetic *fields*. He did not have the mathematical tools needed to treat field problems theoretically, and his monu-

mental work, *Experimental Researches in Electricity*, contains not a single equation! James Clerk Maxwell (Fig. 22.1) took Faraday's ideas and used them as the basis for his famous equations describing the electromagnetic field. For this reason Albert Einstein once said that Faraday had the same relationship to Maxwell in the development of electromagnetism that Galileo had to Newton in the development of mechanics.

Despite his many successes as a scientist, Faraday always remained "as simple, charming and unaffected as a child," in the words of H. Helmholtz. He was truly one of the saints of science. He rejected worldly honors and possible wealth because of his devotion to science and to the religious ideals of the obscure Sandemanian sect to which he remained faithful throughout his life. He turned down the presidency of both the Royal Institution and the Royal Society of London and refused to be knighted. He requested a "gravestone of the most ordinary kind" when he died, and that is exactly what he received. His grave in Highgate Cemetery, London, contains a small tombstone inscribed simply: "Michael Faraday, Born 22 September 1791. Died August 25 1867."

The true memorials to Faraday's genius are the laws of electrolysis and of electromagnetic induction that now bear his name, as do the unit for capacitance, the farad, and the unit for a mole of electric charge, the faraday.

The magnetic field is therefore sometimes called the *magnetic flux density*, since it can be measured in terms of the number of lines of magnetic flux passing through a given area at normal incidence.

The unit of magnetic flux is the weber (Wb), named in honor of Wilhelm Weber (1804–1890), a German physicist and collaborator of Gauss. Since
$\Phi_B = BA$,

$$1 \text{ Wb} = \left(1 \frac{N}{A \cdot m}\right)(1 \text{ m}^2) = 1 \text{ N} \cdot m/A$$

Since $B = \Phi_B/A$, the units of the magnetic field are often given as webers per square meter (Wb/m²), where

$$1 \frac{Wb}{m^2} = 1 \frac{N}{A \cdot m} = 1 \text{ T}$$

Here again different units are used to describe the same physical quantity, as happens frequently in physics. In what follows we confine ourselves to the use of the tesla and the weber per square meter as units for the magnetic field and will choose between these two to simplify calculations as much as possible.

Example 20.2

If the length *da* in Fig. 20.1 is 0.30 m and if all other quantities are as in Example 20.1, what is the magnetic flux through the circuit *abcd*?

SOLUTION

The magnetic field $B = 1.0$ T $= 1.0$ Wb/m². Also the length *da* is equal to 0.30 m, and the length *ab* is 0.20 m. The area enclosed by the circuit *abcd* is, then,

$$A = (0.30 \text{ m})(0.20 \text{ m}) = 0.060 \text{ m}^2$$

Since the magnetic field is perpendicular to A, we have

$$\Phi_B = BA = \left(1.0 \frac{Wb}{m^2}\right)(0.060 \text{ m}^2) = \boxed{0.060 \text{ Wb}}$$

Faraday's Law

Now that we have introduced the concept of magnetic flux, let us look at Fig. 20.1 from a different point of view. We found in Eq. (20.1) that the induced emf is $\mathscr{E} = B_\perp vl$. If the rod moves a distance Δs in time Δt, however, $v = \Delta s/\Delta t$. Now, as the rod moves, $l \Delta s$ is simply the change in area (ΔA) of the closed loop in time Δt. Using these results in Eq. (20.1), we have

$$\mathscr{E} = B_\perp \frac{\Delta s}{\Delta t} l = B_\perp \frac{\Delta A}{\Delta t} = \frac{\Delta \Phi_B}{\Delta t} \tag{20.5}$$

The induced emf in the circuit is numerically equal to the time rate of change of magnetic flux through the circuit.

Equation (20.5) gives us the magnitude of the induced emf when magnetic flux changes through an electric circuit. We show in the next section that a negative sign is needed to account for the direction of the emf, and so the correct form of Faraday's law is $\mathscr{E} = -\Delta\Phi_B/\Delta t$.

If, instead of a single turn of wire, we have N turns connected in series, an emf is induced in each of them. The total emf induced in the complete circuit is then

Faraday's induction law: $$\boxed{\mathscr{E} = -N \frac{\Delta \Phi_B}{\Delta t}} \tag{20.6}$$

This is *Faraday's induction law*, one of the most important equations in electromagnetism and the basis for the operation of generators, motors, transformers, and other electric devices. It is a completely general law. The flux Φ_B

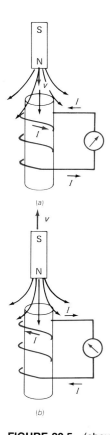

(a)

(b)

FIGURE 20.5 (above) A change in magnetic flux through a circuit induces an emf and, if the circuit is complete, a current. The direction of the current in (b) is opposite that in (a), since in (a) the flux through the circuit is increasing while in (b) it is decreasing.

need not be changed by moving part of a circuit, as in the example above. No matter how the flux through a circuit changes, there is an induced emf in the circuit. For example, if a bar magnet is pushed into or pulled out of the coil of wire in Fig. 20.5, an emf is induced as long as the magnet is moving, since it is changing the magnetic flux through the circuit. Once the magnet stops moving, the induced emf vanishes. In this case a moving magnetic field produces an electric field, just as in the case of a moving charge a moving electric field (attached to the charge) produces a magnetic field. This is the "symmetry" between electric and magnetic fields that physicists like Faraday expected to find in nature.

Another example of the Faraday induction law is shown in Fig. 20.6. Here we have two solenoids, one connected to a battery, the other to an electric meter. If we throw the switch and pass a current through the upper coil, there is an induced emf in the lower coil while the magnetic field in the upper coil is increasing from zero to its final value. During this time $\Delta\Phi_B/\Delta t$ is not zero in the lower coil, because some field lines from the upper coil pass through the lower coil, thus linking the two coils. Once the magnetic field in the upper coil reaches its final value, the induced emf vanishes. If, now, we open the switch and allow the magnetic field in the upper coil to fall to zero, an emf is again induced in the lower coil, but in the opposite direction from the first case, as shown in Fig. 20.7.

If we change the flux through the lower coil by first increasing the current through the upper coil and then decreasing it, and we do this repeatedly, an emf and a changing current will be induced in the lower coil as long as the current in the upper coil is changing. Hence we have a way of inducing a current in a coil that has *no direct physical connection* with a power source or with a coil carrying a current. This is how a *transformer* works.

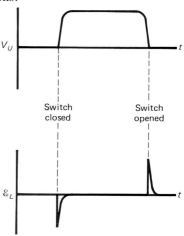

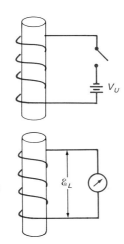

FIGURE 20.6 (right) An emf is induced in the lower coil by a changing current in the upper coil, since some of the magnetic flux from the upper coil passes through the lower coil.

FIGURE 20.7 (a) The applied voltage V_U as a function of time in the upper coil of Fig. 20.6. (b) The induced emf $\mathcal{E}_L$ in the lower coil of Fig. 20.6 as a function of time. Note that an emf is only induced in the lower coil when the current through the upper coil is *changing*.

Example 20.3

In the circuit of Fig. 20.1 the rod *ab* is at rest at a position where the length *da* is 0.30 m, as in Example 20.2. The magnetic field **B**, produced by an electromagnet, is suddenly turned off

and falls to zero in 0.25 s. What is the magnitude of the induced emf?

SOLUTION

As found in Example 20.2, the flux through the coil in this case is initially 0.060 Wb. Since there is only one turn, $N = 1$, and Faraday's law becomes

$$\mathscr{E} = -N\frac{\Delta\Phi_B}{\Delta t} = -N\frac{\Phi_{final} - \Phi_{init}}{t_{final} - t_{init}}$$

$$= -\frac{0 - 0.060 \text{ Wb}}{0.25 \text{ s}} = 0.24 \frac{\text{Wb}}{\text{s}} = \boxed{0.24 \text{ V}}$$

EXERCISE 1 Prove that $1 \text{ Wb/s} = 1 \text{ V}$.

20.3 Lenz' Law

The Russian physicist Heinrich Lenz (1804–1865) duplicated many of the results of Faraday and Henry without any prior knowledge of their work. The law that goes by his name accounts for the negative sign in Faraday's law.

Lenz' law: If an induced emf sets up a current in a complete circuit, the direction of the induced current is always such as to *oppose the change in magnetic flux* that produces it.

The cause of an induced emf and current may be the motion of a conductor in a magnetic field. If we move the rod in Fig. 20.1, to the right, the induced current sets up a force to the left to oppose the motion, as we found in Example 20.1. This is in accordance with Lenz' law. It is also in accordance with the principle of conservation of energy, for if the force produced by the induced current were to the right, we would be able to generate an electric current without doing any work, and that would violate the first law of thermodynamics.

Another illustration of Lenz' law is shown in Fig. 20.5. If we push the north pole of the magnet into the coil, the induced emf must cause a current in such a direction as to make the end of the solenoid near that north pole also a north pole, so the two will repel each other. This means that we must do work to push the magnet into the coil, and this work is converted into electric energy in the coil. Similarly, if we pull the north pole away from the coil, the induced current is in such a direction as to make the near end of the coil a south pole, so the two poles attract each other and work must be done to separate them. In this case the induced current is in the opposite direction from the current produced when the north pole was pushed into the coil. The minus sign in Eq. (20.6) is therefore required.

Lenz' law is easy to visualize and apply; it is therefore the best way to determine the direction of the induced emf predicted by Faraday's law. Note that Lenz' law does *not* state that the induced current opposes the flux itself; the induced current opposes the *change in flux* that induces the emf.

Example 20.4

A long solenoid with an air core has 400 turns per meter and a cross-sectional area A of 10 cm^2. The current through the solenoid increases from 0 to 50 A in 2.0 s in the direction shown in Fig. 20.8. A loop of wire consisting of 10 turns and of cross-sectional area 100 cm^2 and resistance 0.050 Ω is put in place around the solenoid near its center. What is the magnitude of **(a)** the induced emf and **(b)** the current in the wire loop? **(c)** What is the direction of the induced emf and current?

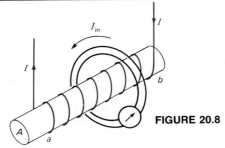

FIGURE 20.8

SOLUTION

(a) We must first find the flux through the solenoid. In this case all the flux is confined to the inside of the solenoid, and so the cross-sectional area of the outer coil is of no importance, for the flux through it is always the same, no matter what its area. Since B is perpendicular to the area A, we have, from Eqs. (20.3) and (19.17),

$$\Phi_B = BA = \mu_0 n I A$$

The rate of change of the flux with time is then, since n and A are constant,

$$\frac{\Delta\Phi_B}{\Delta t} = \mu_0 n A \frac{\Delta I}{\Delta t}$$

$$= \left(4\pi \times 10^{-7}\,\frac{\text{Wb}}{\text{A·m}}\right)\left(\frac{400}{1\text{ m}}\right)(10\text{ cm}^2)$$

$$\times \left(\frac{10^{-4}\text{ m}^2}{1\text{ cm}^2}\right)\left(\frac{50\text{ A}}{2.0\text{ s}}\right)$$

$$= 1.3 \times 10^{-5}\text{ Wb/s}$$

Then, by Faraday's induction law, the induced emf is

$$\mathcal{E} = -N\frac{\Delta\Phi_B}{\Delta t} = -10(1.3 \times 10^{-5}\text{ Wb/s})$$

$$= \boxed{-1.3 \times 10^{-4}\text{ V}}$$

(b) The induced current is then

$$I_{\text{in}} = \frac{\mathcal{E}}{R} = \frac{-1.3 \times 10^{-4}\text{ V}}{0.050} = -2.6 \times 10^{-3}\text{ A}$$

$$= \boxed{-2.6\text{ mA}}$$

and this can be measured by the meter shown in the diagram.
(c) The current through the solenoid is clockwise and increasing as we look along its axis from its lower end in the figure. The magnetic field inside the solenoid is therefore from a to b, and the flux of this field through the outer coil is also increasing from a to b. The induced current must therefore produce a magnetic field in the opposite direction, i.e., from b to a. A *counterclockwise* induced current accomplishes this. The direction of the induced current is therefore opposite the direction of the current increasing the flux, as predicted by Lenz' law. The minus sign on the current found in (*b*) indicates the same result, that the induced current is counterclockwise as we look along the solenoid from a to b.

Eddy Currents

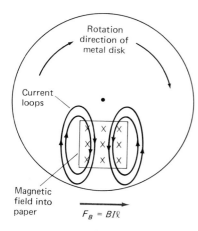

FIGURE 20.9   Eddy currents in a rotating disk.

An interesting illustration of Lenz' law is found in *eddy currents*, which are induced currents in metals placed in a region of space where magnetic flux is changing. Consider Fig. 20.9, in which a metal disk rotates in a magnetic field produced by a horseshoe magnet whose poles cover only a small rectangular portion at the bottom of the disk. According to our right-hand force rule, positive charges moving with the disk to the left through the magnetic field experience a downward force. This causes a conventional current directed toward the bottom of the disk. This current must complete a closed circuit and hence is directed upward in regions of the disk outside the magnetic field, as shown in the figure. According to Lenz' law, the downward current produces a force $F_B = BIl$ to the right that acts to slow down the wheel (apply the right-hand force rule). The upward current, on the other hand, is outside the magnetic field and creates no force on the disk. As a consequence the net effect of the eddy currents is to *oppose* the motion of the disk. This is an example of *eddy-current damping*, often used to damp galvanometers and other oscillatory systems and to provide braking action on subway cars.

20.4   Mutual Induction and Self-Induction

In Fig. 20.10 two coils are wound together on the same core. An induced emf is produced in coil 2 by a change in the current in coil 1, since some flux produced by coil 1 links coil 2. It is often useful to express the induced emf in terms of this change in current rather than in terms of the change in the magnetic flux. We will

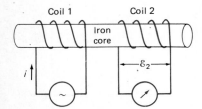

FIGURE 20.10 Two coils wound together on the same core so that magnetic flux through coil 1 produces flux through coil 2.

indicate the value of the *changing current* in coil 1 at any instant of time by the symbol i.

Let Φ_B be the flux produced in coil 2 by a current i in coil 1. This flux is proportional to i, and so if i changes by Δi, then the flux in coil 2 will change by $\Delta\Phi_B$. Since there are N turns of wire on coil 2 and N is a constant, we can write

$$N\,\Delta\Phi_B \propto \Delta i$$

If we call the constant of proportionality M, this becomes

$$N\,\Delta\Phi_B = M\,\Delta i \qquad (20.7)$$

$$\text{or} \qquad M = N\frac{\Delta\Phi_B}{\Delta i}$$

M is called the *mutual inductance* and is a measure of how the magnetic flux through the N turns of coil 2 is influenced by a changing current in coil 1. M is assumed to be a constant for any set of coils, since the flux Φ_B through coil 2 increases in direct proportion to the current in coil 1, and so $\Delta\Phi_B/\Delta i$ is a constant.

Now, from Faraday's law [Eq. (20.6)], we have for the induced emf

$$\mathscr{E} = -N\frac{\Delta\Phi_B}{\Delta t} = -M\frac{\Delta i}{\Delta t} \qquad (20.8)$$

The mutual inductance M is the induced emf in coil 2 per time rate of change of current in coil 1. M is therefore a measure of the coupling of the coils, i.e., of how large an emf is induced in coil 2 by a current change in coil 1. Notice that there is no *direct* coupling, either mechanical or electrical, of the two coils. Rather, they are coupled through the magnetic field, whose reality is strongly confirmed by induction effects of this kind.

The mutual inductance M has units of $V/(A/s)$, or *henrys* (H), where

$$1\text{ H} = \frac{1\text{ V}}{1\text{ A/s}} = 1\text{ V·s/A}$$

This unit takes its name from the American physicist Joseph Henry (Fig. 20.2).

Example 20.5

Two coils of wire are placed next to each other on a common axis. The mutual inductance of the pair is 50 mH. If the current in one coil increases from 0 to 10 A in 0.10 s, what is the induced emf in the other coil?

SOLUTION

As we have just seen,

$$\mathscr{E} = -M\frac{\Delta i}{\Delta t}$$

$$= -(50 \times 10^{-3}\text{ H})\left(\frac{10\text{ A}}{0.10\text{ s}}\right) = -5.0\,\frac{\text{V·s}}{\text{A}}\frac{\text{A}}{\text{s}} = \boxed{-5.0\text{ V}}$$

The minus sign indicates that the induced emf drives current in a direction *opposite* to that of the increasing current which induces the emf, according to Lenz' law. If the current in the first coil were *decreasing* from 10 A to 0, then the induced emf would drive current in the *same* direction as the original current, again according to Lenz' law.

Self-Induction

We really do not need two coils to produce inductive effects. If we have a single coil and vary the current in it, the magnetic field produced changes in time. Since this field exists in the region inside the coil, the magnetic flux through the coil changes. This results in a *self-induced emf* opposing the change in the current, according to Faraday's induction law. This induced emf delays the change in current, whether the current is increasing or decreasing, and the amount of delay depends on the self-inductance of the circuit. By analogy with mutual induction, we define the *coefficient of self-inductance* of the circuit as follows:

$$L = N \frac{\Delta\Phi_B}{\Delta i}$$

(20.9)

so that $L\,\Delta i = N\,\Delta\Phi_B$

Note that Φ_B is the flux produced by the current i in the same coil. The self-inductance L (usually called simply the *inductance*) depends on geometric factors like the size and shape of the coil, as well as on the permeability of the material filling the core of the coil. It is much greater for an iron core, for example, than it is for air. For ferromagnetic materials it is also not constant but varies with the current. (Throughout this section, however, we will assume that both M and L are constant.)

Since $L\,\Delta i = N\,\Delta\Phi_B$, Faraday's law becomes

$$\mathcal{E} = -N \frac{\Delta\Phi_B}{\Delta t} = -L \frac{\Delta i}{\Delta t}$$

(20.10)

where the unit for L is again the henry.

Any element of a circuit having a self-inductance is called an *inductor* and is represented by the symbol ⌒⌒⌒ on circuit diagrams. Since coils are made of wire, they also have a resistance R associated with the inductance L. Such coils can therefore be considered as a resistance R in series with an inductance L.

20.5 Energy Stored in an Inductor

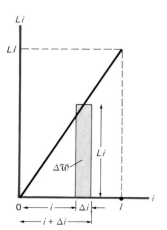

FIGURE 20.11 Energy stored in an inductor. The energy is the area under the curve (in this case, a straight line), when we plot Li against i, leading to the result $\mathcal{W} = \frac{1}{2}LI^2$.

Suppose we have an inductor of inductance L through which we establish a current by connecting a variable dc voltage across it. We increase the voltage and thus increase the current i from zero to its final value I. An induced emf $\mathcal{E} = -L\,\Delta i/\Delta t$ is produced across the inductor, and this emf acts to oppose the increase in current, by Lenz' law. Hence work must be done *against this emf* by the battery to establish the current. The work done by the battery during time Δt is, using Eq. (20.10),

$$\Delta\mathcal{W} = P\,\Delta t = \mathcal{E}\,i\,\Delta t = L \frac{\Delta i}{\Delta t}\,i\,\Delta t = Li\,\Delta i$$

To find the total work done in establishing the current I, we must sum Δi over all the values of i from zero to I. [Note the similarity here to the equation for the work done by the expansion of an ideal gas, $\Delta\mathcal{W} = P\,\Delta V$ (Eq. 14.7).] Here the work done is the area under the curve when we plot Li against i, as in Fig. 20.11, just as in the PV case the work was the area under the curve when we plotted P versus V.

On the straight-line graph of Li versus i, the work done in changing the current from i to $i + \Delta i$ is $\Delta\mathcal{W} = Li\,\Delta i$, which is the area of the small vertical rectangle shown in color. Just as for the ideal gas, the total work done is the sum of all such small rectangles, or the total area under the curve. Since this total area is triangular, its area is one-half the base of the triangle times its altitude, or

$$\mathscr{W} = \tfrac{1}{2}I(LI) = \tfrac{1}{2}LI^2$$

The energy stored in an inductor, which is equal to the energy needed to establish a current I through the inductor, is

$$\mathscr{W} = \tfrac{1}{2}LI^2 \tag{20.11}$$

Once the current is established, no further work is required to maintain the current in a "pure" inductor. If the inductor has resistance, of course, work must be done to replenish the energy lost as heat in the resistance.

If the circuit is broken after the current I is established, then the energy $\mathscr{W} = \tfrac{1}{2}LI^2$ is returned to the circuit when the magnetic field collapses and is dissipated as heat in the resistive elements of the circuit.

Energy in the Magnetic Field The magnetic field of a typical inductor is the same as that for a solenoid and is given, for an air-core solenoid carrying a current i, by Eq. (19.17),

$$B = \mu_0 ni = \mu_0 \frac{N}{l} i$$

where we assume that there are N turns in length l of the solenoid.* The magnetic flux Φ_B is thus, from Eq. (20.3),

$$\Phi_B = BA = \mu_0 \frac{N}{l} iA \qquad \text{and so} \qquad \Delta\Phi_B = \mu_0 \frac{N}{l} A \, \Delta i$$

and, from Eq. (20.9),

$$L = N \frac{\Delta\Phi_B}{\Delta i} = \mu_0 \frac{N^2}{l} A \tag{20.12}$$

This is the equation for the self-inductance of a solenoid of length l and cross-sectional area A, containing N turns of wire.

Using this value for L in our expression for the energy in an inductor [Eq. (20.11)] and using Eq. (19.17) to replace i, we have

$$\mathscr{W} = \tfrac{1}{2}LI^2 = \tfrac{1}{2}\mu_0 \frac{N^2}{l} A \left(\frac{Bl}{\mu_0 N}\right)^2 \qquad \text{or} \qquad \mathscr{W} = \frac{1}{2}\frac{B^2}{\mu_0} lA$$

This represents the energy stored in the magnetic field of the inductor. But the interior volume of the solenoid is lA, and so the energy density, or the energy per unit volume (J/m^3), is

$$w = \frac{\mathscr{W}}{V} = \frac{\mathscr{W}}{lA} = \frac{1}{2}\frac{B^2}{\mu_0}$$

Energy density in magnetic field:
$$w = \frac{B^2}{2\mu_0} \tag{20.13}$$

If the inductor has a core with permeability μ, then Eq. (20.13) becomes $w = B^2/2\mu$.

---

*We avoid using L for length here because of the danger of confusion with the self-inductance L.

We therefore see that it is possible to have energy stored in free space where there is no matter present. Wherever a magnetic field exists, energy is present.

Energy Density in Combined Electric and Magnetic Fields

If electric and magnetic fields exist together in space, then the total energy stored in unit volume of the combined electric and magnetic fields is, from Eqs. (17.20) and (20.13),

$$w = \tfrac{1}{2}\epsilon_0 E^2 + \frac{1}{2}\frac{1}{\mu_0} B^2 \qquad\qquad (20.14)$$

and depends on the square of the magnitude of both the electric and magnetic fields. Note the symmetry that exists in these two expressions; that is, E and B both appear squared in the expression for w, and the constants multiplying E^2 and B^2 are of similar form. The existence of such fields in free space, and the possibility that energy can be transferred back and forth between the electric and magnetic fields, form the basis for electromagnetic waves, as we will see in Chap. 22.

Example 20.6

In a certain region of space the magnetic field has a value of 1.0×10^{-2} T and the electric field a value of 2.0×10^6 V/m. What is the density of the energy stored in the combined electric and magnetic field?

SOLUTION

We have, from Eq. (20.14),

$$w = \tfrac{1}{2}\epsilon_0 E^2 + \frac{1}{2}\frac{1}{\mu_0} B^2$$

For the electric field,

$$\tfrac{1}{2}\epsilon_0 E^2 = \tfrac{1}{2}\left(8.85 \times 10^{-12}\ \frac{C^2}{N \cdot m^2}\right)(2.0 \times 10^6\ V/m)^2$$

$$= \boxed{18\ J/m^3}$$

For the magnetic field,

$$\frac{1}{2}\frac{1}{\mu_0} B^2 = \frac{1}{2}\left(\frac{1}{4\pi \times 10^{-7}}\ \frac{A}{T \cdot m}\right)(1.0 \times 10^{-2}\ T)^2$$

$$= \boxed{40\ J/m^3}$$

Hence the energy density for the two fields together is

$$w = 18\ J/m^3 + 40\ J/m^3 = \boxed{58\ J/m^3}$$

20.6 Circuit with Inductance and Resistance: The *RL* Circuit

Figure 20.12 shows an inductor in series with a resistor. The resistance R includes the resistance of the wire coils of the inductor in addition to any other resistors in the circuit. The result is an idealized combination of a noninductive resistor and a nonresistive inductor connected to a source of constant voltage V by throwing the switch shown in the figure to cd so that a is connected to c and b to d.

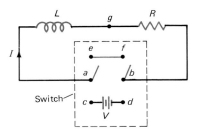

FIGURE 20.12 An *RL* circuit. The dashed lines indicate a special kind of switch (a so-called double-pole, double-throw switch) that allows points a and b to be connected either to e and f or to c and d.

Once a constant current I is established in the inductance coil, this current can be reduced to zero by throwing the switch to ef so that a is connected to e and b to f, and e and f are shorted together by a short, low-resistance wire. Although the current in the circuit has no applied voltage to sustain it, it will not decrease instantly to zero; rather the current will fall to zero in a time determined by the inductance and resistance of the circuit.

Decreasing Current

Initially the current is constant and clockwise. When the switch is thrown to ef, the current starts to decrease. An emf is induced in the inductor that acts to oppose this reduction in the current.

We can apply Kirchhoff's loop rule by traversing the circuit clockwise from a back to a. We obtain, since there is no applied voltage,

$$V_{ag} + V_{gb} = 0 \qquad \text{or} \qquad -L\frac{\Delta i}{\Delta t} - Ri = 0$$

and so

$$\frac{\Delta i}{\Delta t} = -\frac{Ri}{L} \tag{20.15}$$

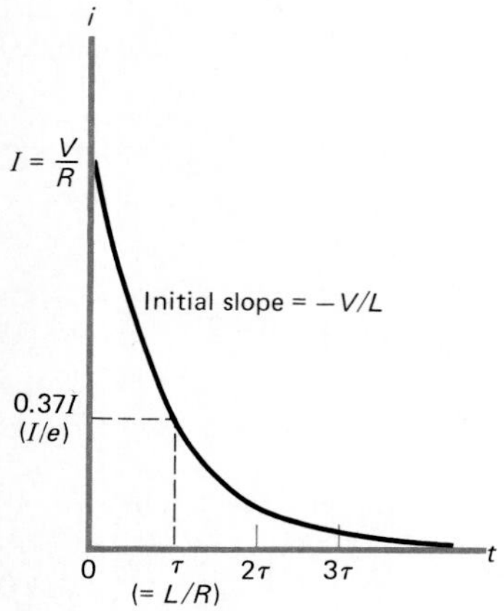

Here $\Delta i/\Delta t$ is the rate at which the current is changing, or the slope of the graph of i against t. In this case the current is decreasing, but the rate of decrease becomes less as the current becomes smaller, since $\Delta i/\Delta t$ is proportional to i. If we plot an experimental curve of i against t in this case, we obtain a *decreasing exponential* curve of the type shown in Fig. 20.13. The actual rate of decrease depends on the values of R and L, being rapid for R large and L small and more gradual for R small and L large.

The solution to Eq. (20.15) can be shown (using calculus, which is beyond the scope of this book) to be

$$i = Ie^{-Rt/L} \tag{20.16}$$

where e is the *natural base* of the exponential function and has a value $e = 2.718$. (When this base is used, the value of the slope of a graph of e^x against x is equal to the value of e^x itself at any point, that is, $\Delta e^x/\Delta x = e^x$.*)

In Eq. (20.16), when $t = L/R$, $i = I/e = I/2.718 = 0.37\ I$. Hence the current has fallen from I to $0.37I$ in L/R s. For this reason L/R is called the *time constant* τ of the circuit.

Time constant: A quantity that determines the rate at which a physical quantity (the current in this case) approaches its final, steady-state value.

Time constants are very useful for describing complicated electric circuits.

FIGURE 20.13 Decay of current in the circuit of Fig. 20.12 as a function of time after the switch is thrown to ef, removing the battery from the circuit. In L/R, one time constant τ, the current falls to $0.37I$.

Increasing Current

If we now throw the switch back to position cd, the current rises but does not immediately reach its final value because of the effect of the inductor. At any instant of time the potential difference across the inductor is $V_{ag} = L\ \Delta i/\Delta t$, where a is at a higher potential than g because the induced emf is opposing the increase of current. Also the voltage across the resistor is $V_{gb} = iR$. By Kirchhoff's loop rule we therefore have

---

*For a further discussion of exponential functions see Appendixes 1.E and 2.C, and Clifford Schwartz, *Prelude to Physics*, Wiley, New York, 1983.

$$V = V_{ag} + V_{gb} = L\frac{\Delta i}{\Delta t} + iR \qquad \text{and so} \qquad \frac{\Delta i}{\Delta t} = \frac{V - iR}{L} = \frac{V}{L} - \frac{iR}{L} \quad \textbf{(20.17)}$$

As the current i increases, the right side of this equation gets smaller and so $\Delta i/\Delta t$ also becomes smaller.

The graph of i versus t will therefore be a line starting with a slope V/L when $t = 0$, but that slope will get smaller and approach zero as i approaches its final value I. The resulting experimental curve is shown in Fig. 20.14. When the current reaches its final value, $\Delta i/\Delta t$ is zero, and so

$$\frac{\Delta i}{\Delta t} = 0 = \frac{V}{L} - \frac{R}{L}I \qquad \text{or} \qquad I = \frac{VL}{LR} = \frac{V}{R}$$

which is what we would expect from Ohm's law if there were no inductor in the circuit. *An inductance has an effect on a circuit only when the current through it is changing.*

Equation (20.17) can be solved for i as a function of time, and the solution takes the form

$$i = \frac{V}{R}(1 - e^{-Rt/L})$$

Here again the solution contains a decreasing exponential term $e^{-Rt/L}$, which falls from an initial value 1 to 0 as t goes from 0 to ∞. As a result the current i increases gradually from 0 to its final value $I = V/R$. The time constant of the circuit is again that value of t for which the exponential term is equal to $1/e$; that is, $t = L/R$, as in the decreasing-current case. Here, when $t = L/R$, $i = (V/R)(1 - 1/e) = (V/R)(1 - 0.37)$, or $i = 0.63V/R = 0.63I$, and the current has reached 63 percent of its final steady-state value when $t = L/R$.

The time constant τ is therefore the same for *LR* circuits whether the current is increasing or decreasing. It is

$$\boxed{\tau = \frac{L}{R}} \qquad\qquad\qquad \textbf{(20.18)}$$

RL circuits with short time constants allow the current to reach its final value very quickly after the voltage is applied. Circuits with large time constants, i.e., large inductances and small resistances, take much longer for the current to reach its final, steady-state value. An inductor therefore acts as an inertial term in an electric circuit. We will see in the next chapter how the value of the inductance helps determine the frequency of an electric oscillator, just as the mass determines the frequency of a mechanical oscillator.

Note that the units of the time constant L/R are

$$\frac{H}{\Omega} = \frac{V \cdot s}{A} \frac{A}{V} = s$$

which shows that τ is indeed dimensionally a time.

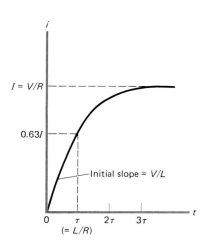

FIGURE 20.14 Increase of the current through the circuit of Fig. 20.12 as a function of the time after the switch is thrown to *cd* connecting the battery to the circuit. When $t = L/R$, $i = 0.63I$.

Example 20.7

A coil has a resistance of 5.0 Ω and an inductance of 100 mH. At a particular instant of time after a battery is connected to the coil, the current is 2.0 A and is increasing at a rate of 20 A/s. What are **(a)** the voltage applied to the coil; **(b)** the time constant of the circuit; **(c)** the final value of the current?

SOLUTION

(a) From Kirchhoff's loop rule

$$V = L\frac{\Delta i}{\Delta t} + iR$$

where V is the applied voltage and i is the current at some particular instant of time. When $i = 2.0$ A,

$$V = (100 \text{ mH})\left(20\,\frac{A}{s}\right) + (2.0 \text{ A})(5.0\,\Omega) = \boxed{12 \text{ V}}$$

(b) The time constant of the circuit is L/R, and so

$$\tau = \frac{L}{R} = \frac{100 \times 10^{-3} \text{ H}}{5.0\,\Omega} = 20 \times 10^{-3} \text{ s} = \boxed{0.020 \text{ s}}$$

(c) When $\Delta i/\Delta t = 0$, $V = IR$, and

$$I = \frac{V}{R} = \frac{12 \text{ V}}{5.0\,\Omega} = \boxed{2.4 \text{ A}}$$

20.7 Circuit with Capacitance and Resistance: The *RC* Circuit

Because of the parallelism between *RL* and *RC* circuits, we here consider a circuit containing a resistor and a capacitor in series. Consider the *RC* circuit of Fig. 20.15, with a switch to allow us to connect a voltage across the *RC* combination or to complete the circuit with no applied voltage in the circuit.

Charging the Capacitor

If we throw the switch so that a is connected to c and b to d, the battery provides a current which charges the capacitor C. The initial current is $I = V/R$, but it decreases to zero as time goes on, because no direct current can flow through the capacitor. We have, from Kirchhoff's loop rule,

$$V = V_{ah} + V_{hb} = \frac{q}{C} + Ri$$

where i is the current through R at some instant of time t and q is the charge on the capacitor at the same instant of time. Hence

$$i = \frac{V - q/C}{R} = \frac{V}{R} - \frac{q}{RC} = \frac{V}{R}\left(1 - \frac{q}{CV}\right) = \frac{V}{R}\left(1 - \frac{q}{Q}\right)$$

As time goes on, q increases until it reaches its final value Q, when the voltage V is completely across the capacitor. When this happens, $i = 0$. The current has therefore decreased exponentially from I to zero, as q increased from zero to Q, as shown in Figs. 20.16 and 20.17. (Note the similarity of these figures for a capacitor to Figs. 20.13 and 20.14 for an inductor.) The exponential decrease in i can be written

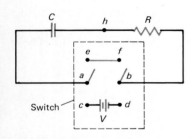

FIGURE 20.15 An *RC* circuit.

$$i = \frac{V}{R}e^{-t/RC} = Ie^{-t/RC} \tag{20.19}$$

In this case the *time constant* τ is

$$\tau = RC \tag{20.20}$$

for when $t = RC$ the current has fallen to $I/e = 0.37I$ and the charge has risen to $0.63Q$.

Note that the units of the time constant RC are

$$(\Omega)(F) = \left(\frac{V}{A}\right)\left(\frac{C}{V}\right) = \left(\frac{C}{C/s}\right) = s$$

as must be the case for a time constant.

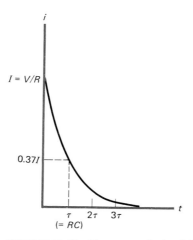

FIGURE 20.16 The current in the *RC* circuit of Fig. 20.15 as a function of time. After the switch is thrown to *cd*, the current falls exponentially to zero with a time constant $\tau = RC$.

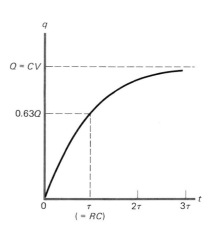

FIGURE 20.17 The charge on the capacitor in the *RC* circuit of Fig. 20.15 as a function of time. After the switch is thrown to *cd*, the charge approaches its final value with a time constant $\tau = RC$.

Discharging the Capacitor

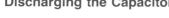

FIGURE 20.18 The variation with time of the charge on the capacitor of Fig. 20.15 when it discharges through a resistor *R*. The charge decays exponentially with a time constant $\tau = RC$.

Once the capacitor has its full charge Q, we can throw the switch to position *ef* in Fig. 20.15 and let the capacitor discharge through the resistor. Here both q and i fall gradually to zero. We have, from Kirchhoff's loop rule,

$$0 = V_{ah} + V_{hb} \qquad \text{or} \qquad 0 = +\frac{q}{C} - iR$$

since the direction of i is now from h to a. Hence $i = q/RC$, and initially, when $i = I$ and $q = Q$, we have

$$I = \frac{Q}{RC} = \frac{V}{R}$$

where V is the original voltage used to charge the capacitor. Here

$$q = Qe^{-t/RC} \tag{20.21}$$

and

$$i = Ie^{-t/RC} \tag{20.22}$$

The charge q decreases to $0.37Q$ within one time constant $\tau = RC$, as shown in Fig. 20.18, and the same is true for the current.

The shorter the time constant, the more rapidly a capacitor can be charged or discharged. The longer the time constant, the more slowly the capacitor charges or discharges. This can have disastrous results if someone, feeling confident because the voltage source has been disconnected, touches both ends of a large capacitor that has been charged to a high voltage (Example 20.8).

Example 20.8

A discharge of more than 10 J of energy through the human body is extremely hazardous and often lethal. **(a)** To what voltage would an 80-μF capacitor have to be charged to store 10 J of energy? **(b)** Suppose this capacitor is charged to 1000 V and the voltage source then disconnected and the capacitor allowed to discharge through a resistance of 1 MΩ. If a TV repairer is working on a TV set and touches both terminals of such a high-voltage capacitor 1 min after the voltage has been disconnected, how much energy would be discharged through the repairer's body?

SOLUTION

(a) The energy stored in the capacitor is, from Eq. (17.19), $\mathcal{W} = \frac{1}{2}CV^2 = 10$ J, and so

$$V = \sqrt{\frac{2(10\ J)}{80 \times 10^{-6}\ F}} = \boxed{5.0 \times 10^2\ V}$$

(b) In this case we have an RC series circuit with $R = 10^6\ \Omega$ and $C = 80 \times 10^{-6}$ F. The time constant of this circuit is

$$\tau = RC = (10^6\ \Omega)(80 \times 10^{-6}\ F) = 80\ s$$

The charge q on the discharging capacitor as a function of time, with t taken as zero when the voltage is disconnected, is given by Eq. (20.21) as

$$q = Qe^{-t/RC}$$

where in this case $Q = CV = (80 \times 10^{-6}\ F)(1000\ V) = 8.0 \times 10^{-2}$ C. After 1 min the charge on the capacitor is

$$q = (8.0 \times 10^{-2}\ C)(e^{-(60\ s/80\ s)}) = (8.0 \times 10^{-2}\ C)\ e^{-0.75}$$
$$= (8.0 \times 10^{-2}\ C)(0.47) = 3.8 \times 10^{-2}\ C$$

The energy still stored in the capacitor is therefore

$$\mathcal{W} = \frac{q^2}{2C} = \frac{(3.8 \times 10^{-2}\ C)^2}{2(80 \times 10^{-6}\ F)} = \boxed{9.0\ J}$$

The energy discharged through the repairer's body is therefore still almost a lethal 10-J dose, even though 1 min has elapsed since the voltage source was disconnected from the capacitor. In other words, always short-circuit large capacitors before handling them! Otherwise you may live (or not live!) to regret it.

Thyratron Sweep Circuit

An interesting and practical circuit using a capacitor and a resistor in series is the sweep circuit (Fig. 20.19) used to sweep the electron beam back and forth across TV screens or oscilloscopes. In the figure R is a much larger resistor than R'. The voltage source charges the capacitor C through the resistor R, with the charging process being confined to the lower portion of the charging curve shown in Fig. 20.17, so that the charging curve approximates a straight line. The voltage V_{ab} across the capacitor is also present across a gas-filled vacuum tube called a *thyratron*, since there is no voltage drop across R' as long as there is no current through R'. When the voltage V_{ab} reaches a preselected value V', it is sufficient to ionize the gas in the thyratron tube so that it conducts. The capacitor therefore

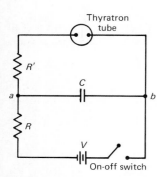

FIGURE 20.19 A thyratron sweep circuit.

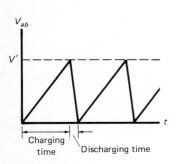

FIGURE 20.20 The sawtooth voltage produced by a thyratron sweep circuit of the kind shown in Fig. 20.19. V_{ab} is the voltage across the capacitor C. This is the type of voltage used to sweep an electron beam back and forth across a TV or oscilloscope screen.

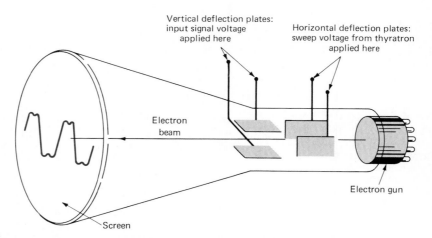

FIGURE 20.21 Diagram of a cathode-ray tube of the type used in oscilloscopes. The electron beam is swept horizontally across the screen by a sweep voltage (like that in Fig. 20.20) applied to the horizontal deflection plates. These plates exert electrostatic forces on the electrons and move them back and forth horizontally on the screen. The signal to be studied is applied at the same time to the vertical deflection plates. This enables variations in the signal as a function of time to be studied.

discharges very rapidly through the ionized gas, because R' is so small that the time constant is also very small. Once the discharge ends, the positive ions in the gas recombine with the electrons, and the thyratron no longer conducts. This allows the voltage source to take over and charge up the capacitor again. This process repeats itself continuously.

The resulting *sawtooth voltage* across the capacitor is shown in Fig. 20.20. In a TV set this voltage is used to deflect the electron beam. The relatively slow rise of the V_{ab}-versus-t curve corresponds to the sweep of the electron beam from left to right on the TV screen. The much more rapid fall of the voltage from V' back to zero corresponds to the almost instantaneous flyback of the beam from right to left. In oscilloscopes (Fig. 20.21) and other instruments that display electric signals, such as electrocardiographs, the proper selection of C and R fixes the appropriate time constant for any desired application.

Example 20.9

The electron beam in a TV set sweeps across the screen at a rate of 30 lines per second. If the sweep mechanism is a thyratron circuit, provided by charging and discharging a 10-μF capacitor through a resistor, what must be the approximate values of R and R' in Fig. 20.19 in order for the sweep circuit to function properly?

SOLUTION

Since the time of one sweep is $\frac{1}{30}$ s, the time constant for the charging of the capacitor must be approximately $\frac{1}{30}$ s. Since the time constant $\tau = RC$, we have

$$R = \frac{\tau}{C} = \frac{\frac{1}{30}\text{ s}}{10 \times 10^{-6}\text{ F}} = \boxed{3.3 \times 10^3\ \Omega}$$

For the flyback of the electron beam, a time constant much shorter than this is needed. Let us take 10^{-4} s as a reasonable value. Then we have

$$R' = \frac{10^{-4}\text{ s}}{10 \times 10^{-6}\text{ F}} = \boxed{10\ \Omega}$$

20.8 Electric Generators

Faraday's induction law is the basis for the operation of electric generators, machines that convert mechanical energy into electric energy. A simple model of an electric generator, or dynamo, is shown in Fig. 20.22. A rectangular coil *abcd* is mechanically forced—for example, by connection to a steam engine or steam turbine—to rotate about an axis OO' perpendicular to a uniform magnetic field of strength B. This rotation converts mechanical energy into electric energy and produces an emf that drives current through a closed circuit. The ammeter in the circuit is used to study the form of the current through the load.

Figure 20.23 gives an end-on view along the axis OO' in Fig. 20.22. The width of the coil is w, and its sides extend inward (into the paper) to a depth l. The two sides *ad* and *bc* have motional emf's and currents induced in them as they rotate through the magnetic field.

If the coil rotates with an angular velocity ω, then the linear speed of the side *ad* is $v = \omega(w/2)$, since $w/2$ is the radius of the coil. The motional emf induced in side *ad* is $\mathcal{E} = B_\perp vl$, where l is the length *ad*. The direction of $\mathcal{E}$ is from d to a, by the right-hand rule. But in this case

$$B_\perp v = Bv \sin \theta$$

where θ is the angle between the lines of B and the *normal to the plane of the coil*, as shown in Fig. 20.23. When v is parallel to B, $\theta = 0$, $\sin \theta = 0$, and no emf is induced. When v is at right angles to B, $\theta = 90°$, $\sin \theta = 1$, and the induced

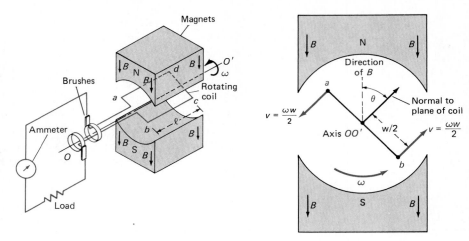

FIGURE 20.22 (left) A simple electric generator. Mechanical motion of the coil in a magnetic field produces a current through the load resistor. This current can be measured as a function of time by the ammeter.

FIGURE 20.23 (right) End-on view of the generator of Fig. 20.22. The coil is rotating with an angular velocity ω, and the sides of the coil ad and bc cut the magnetic field lines and produce an emf.

emf is a maximum. Hence $Bv \sin \theta$ correctly represents the physical situation. Thus we have

$$\mathscr{E} = Bvl \sin \theta = B\omega \frac{w}{2} l \sin \theta = \frac{BA\omega}{2} \sin \theta$$

since the area of the coil $A = wl$. (Note that this expression for $\mathscr{E}$ shows that the flux $\Phi_B = BA$ through the coil is changing as the coil rotates, in accordance with Faraday's induction law.)

This is the induced emf only for the side ad. For side bc, an identical expression is obtained. Since this second emf is in the direction from b to c, these two emf's reinforce each other. Also, let us assume that there are N windings on the coil. Then the total emf is

$$\mathscr{E} = 2N \frac{BA\omega}{2} \sin \theta = NBA\omega \sin \theta = NBA\omega \sin \omega t \qquad \textbf{(20.23)}$$

since $\theta = \omega t$ and $\omega = 2\pi f$, where f is the frequency, or the number of complete rotations per second.

The maximum emf occurs when $\theta = \omega t = 90°$, and $\mathscr{E}_{max} = NBA\omega$. Hence we have

$$\boxed{\mathscr{E} = \mathscr{E}_{max} \sin \omega t = \mathscr{E}_{max} \sin 2\pi ft} \qquad \textbf{(20.24)}$$

As time goes on, therefore, the rotating coil produces a sinusoidal voltage of the kind shown in Fig. 20.24. As the coil rotates through one complete cycle, or 360°, the emf goes from a maximum to zero, to a maximum in the opposite direction, to zero, and finally back to maximum in the original direction, as shown. If the circuit is complete, a current results, which is constantly changing in time according to Eq. (20.24). This is called an *alternating current* (ac) *generator*.

The output of the generator can be fed to an external circuit by attaching the two ends of the coil to rotating rings that make contact with metal brushes, as in Fig. 20.22. The "load" resistors in Figs. 20.22 and Fig. 20.25 represent light bulbs, motors, or other devices to which the generator supplies electric energy.

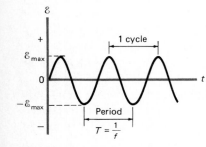

FIGURE 20.24 The emf produced by the generator of Fig. 20.22 as a function of time. Notice that the emf is "alternating" in nature and produces an alternating current when the circuit is complete.

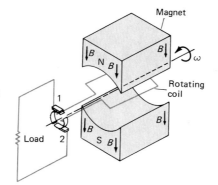 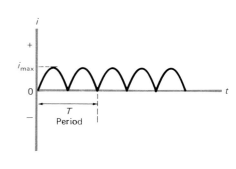

FIGURE 20.25 (left) A generator using a split-ring commutator to produce direct current. The commutator reverses the current every half cycle so as to keep it always in the same direction, although changing in magnitude.

FIGURE 20.26 (right) The pulsating *dc* current produced by the generator of Fig. 20.25.

DC Generators

The current in Fig. 20.24 is obviously positive in direction half the time and negative the other half. If we want direct current, we can obtain it from the same generator by attaching the two ends of the coil to a single ring that is split in half and makes contact with two metal brushes, as shown in Fig. 20.25. In this case the split ring, called a *commutator*, reverses the external contacts just at the instant the current reverses. As a result, the current through the load is always in one direction but is pulsating in magnitude, as shown in Fig. 20.26. Such a pulsating current can be smoothed out by using more than one coil rotating about the same axis or by other electrical techniques to give a good imitation of the direct current delivered by a battery.

Practical Electric Generators

In a typical fossil-fuel steam-electric power plant, coal, oil, or natural gas is burned to heat water and produce steam. The high-energy steam molecules then strike the blades of a turbine (a modern version of an old-fashioned water wheel) and set it into rotational motion (see Fig. 15.14). This rotational motion is conveyed through a shaft to the rotating coil of an electric generator, which converts mechanical energy into electric energy, as described above, and feeds it into transmission lines for the use of consumers.

The above discussion does not indicate how the load (e.g., the light bulb or toaster in Fig. 20.27) affects the operation of the generator. If a generator is running with no load attached, as in Fig. 20.27a, once the rotor is moving at full speed the only mechanical power the turbines must provide to keep the shaft turning is that needed to overcome friction and other mechanical losses in the system. If, now, we connect a light bulb (or a whole city of light bulbs) to the generator, the bulb draws current from the generator. That current is in such a direction that, by Lenz' law, it sets up a magnetic force which opposes the shaft's rotation. To overcome this turning resistance, additional mechanical power must be provided by the steam turbine, which means that more fuel must be burned to produce the steam to drive the turbine. If we now add a toaster (or another city) in parallel with the light bulb, the current increases even more, as does the opposing force, and still more mechanical power is needed.

FIGURE 20.27 Loading the generator. (a) No load—open circuit. (b) Closed circuit—load attached. The addition of a light bulb and a toaster to the generator output increases the required energy input to the generator.

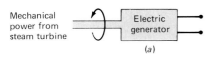

(a)

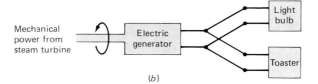

(b)

The greater the load, therefore, the more mechanical energy required and the more fuel consumed. This is another example of the *principle of conservation of energy*, since the energy consumed by the light bulb and the toaster must ultimately come from the energy provided by the fossil fuel or other energy source.

Example 20.10

The rotating coil (called the *armature*) of a small 60-Hz ac generator has 200 coils of wire wound on it. The area of each coil is 0.25 m². **(a)** What must be the magnitude of the magnetic field in which this armature rotates if the peak voltage output of the generator is to be 170 V? **(b)** Write a general expression for the emf induced in this armature.

SOLUTION

(a) From Eq. (20.23), the maximum emf produced by the rotating armature is $\mathcal{E}_{max} = NBA\omega$, and so

$$B = \frac{\mathcal{E}_{max}}{NA(2\pi f)} = \frac{170 \text{ V}}{200(0.25 \text{ m}^2)(120\pi/\text{s})} = \boxed{0.0090 \text{ T}}$$

(b) From Eq. (20.24),

$$\mathcal{E} = \mathcal{E}_{max} \sin 2\pi ft = \boxed{(170 \text{ V})(\sin 120\pi t)}$$

20.9 Electric Motors

An electric motor is a device that converts *electric energy* into rotational kinetic energy (a form of *mechanical energy*) which can then be used to do work. For example, the shaft connected to the rotating coil can be coupled to some external device like a fan blade or a helicopter propeller. A motor is nothing but a generator run in reverse. Whereas in a generator a current is produced by a rotating coil, in a motor a current driven through the coil by an external voltage source causes a coil to rotate.

As we saw in our discussion of the galvanometer in Sec. 19.10, a current-carrying coil in a magnetic field experiences a rotational torque. If the current is direct and the plane of the coil is originally parallel to the magnetic field, the torque will be a maximum and will rotate the coil to a position where its plane is at right angles to the magnetic field. There, however, the coil will stop, because in Fig. 20.28 the torques at coil positions 1 and 2 are in opposite directions and tend to drive the coil to position 3, where it "freezes."

To make the coil rotate continuously, we must make the current through the coil reverse every time it moves through position 3. Then the torque will be a continuous one and the coil will pick up speed as it rotates, as in Fig. 20.29. This

FIGURE 20.28 (left) A coil carrying a direct current in a fixed magnetic field. The torques on the current-carrying wires normal to the paper at positions 1 and 2 are in opposite directions and drive the coil to position 3, where it "freezes."

FIGURE 20.29 (right) Current through the coil reverses every half cycle so that the torques at positions 1 and 2 are in the same direction and the coil is set into rapid rotation about the axis at O.

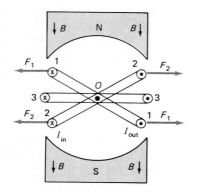

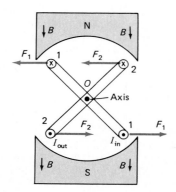

can be accomplished by using an ac voltage to drive current through the coil.

Motors can also be made to operate off dc voltages using split-ring commutators to reverse the current.

Back EMF in a Motor

The constant torque on the coil of a motor does not accelerate it to greater and greater rotational speeds because, as the coil turns, the emf generated in it opposes its rotation according to Lenz' law. This is called a *back emf*. As the rotor speeds up, the back emf increases until a constant rotational speed is reached. The back emf depends on $B_\perp vl$, [Eq. (20.1)], and thus the faster the coil rotates, the larger the back emf and the larger the torque opposing the rotation of the coil.

This situation is quite similar to a ball falling through a viscous fluid. The resistive force is proportional to the ball's velocity, and when this force equals the force of gravity, the ball ceases to be accelerated and falls at a constant terminal velocity, as shown in Sec. 3.7. Similarly in the case of a motor, the coil comes to a constant terminal rotational speed, because the back emf opposes the applied voltage.

When the coil is rotating at constant speed, the voltage across the coil must be equal to the back emf, for otherwise the coil would speed up or slow down. The power delivered to the coil is therefore

$$P = \mathcal{E}_{back}I \qquad\qquad (20.25)$$

where $\mathcal{E}_{back}$ is the back emf. This is the useful power produced by the motor and is equal to the rate at which electric energy is converted into mechanical energy in the motor.

We show schematically a motor circuit in Fig. 20.30, in which an external voltage source of 120 V drives a current through the coil of a motor that has a resistance of 20 Ω. Suppose that, when the motor is running at constant speed, the back emf is 80 V. Then we have, from the loop rule,

$$V = \mathcal{E}_{back} + IR$$

and so $\quad I = \dfrac{V - \mathcal{E}_{back}}{R} = \dfrac{(120 - 80)\ \text{V}}{20\ \Omega} = 2.0\ \text{A}$

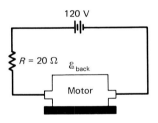

120 V

$R = 20\ \Omega$ $\quad \mathcal{E}_{back}$

Motor

FIGURE 20.30 A circuit in which an external dc voltage of 120 V drives a current through an electric motor. In this diagram R includes all the resistance in the motor circuit.

The current is therefore limited not only by the resistance R but also by the back emf. The useful power produced by the motor is the product of this current and the back emf.

Since the back emf limits the current flowing through the motor, overloading a motor slows down the rotor and reduces the back emf. Hence more current is drawn from the source. If this current becomes too large, the motor burns out.

When a motor is just starting, there is little back emf, and the motor draws a larger current when it is starting than when it is up to full speed. This explains why house lights dim when a large motor like that in an air conditioner is first turned on. The large starting current drawn by the air conditioner causes an increased IR drop in the house wiring that reduces the voltage at electrical outlets in the same circuit as the air conditioning unit.

Example 20.11

A dc motor has a resistance of 4.0 Ω. When running at full speed on a 120-V line, it draws a current of 5.0 A. **(a)** What is the back emf produced by the motor? **(b)** What power is delivered to the motor by the line? **(c)** At what rate is energy lost as heat in the motor? **(d)** How much useful power is produced by the motor? **(e)** What is the efficiency of such a motor? **(f)** What current does such a motor draw when it is just starting up?

SOLUTION

(a) If the applied voltage is V and the back emf is $\mathscr{E}_{back}$, then the applied voltage must be equal to the sum of the voltage drops across the motor, or $V = \mathscr{E}_{back} + IR$. Then

$$\mathscr{E}_{back} = V - IR = 120 \text{ V} - (5.0 \text{ A})(4.0 \text{ } \Omega) = \boxed{100 \text{ V}}$$

(b) The power delivered by the line is, from Eq. (18.11),

$$P_1 = VI = (5.0 \text{ A})(120 \text{ V}) = \boxed{600 \text{ W}}$$

(c) Energy is lost as heat at a rate [Eq. (18.12)]

$$P_2 = I^2 R = (5.0 \text{ A})^2 (4.0 \text{ } \Omega) = \boxed{100 \text{ W}}$$

(d) The useful power produced by the motor is the difference between (b) and (c):

$$P = P_1 - P_2 = 600 \text{ W} - 100 \text{ W} = \boxed{500 \text{ W}}$$

The same result could have been obtained by taking the product of the back emf and the current, using Eq. (20.25):

$$P = \mathscr{E}_{back}I = (100 \text{ V})(5.0 \text{ A}) = \boxed{500 \text{ W}}$$

(e) The efficiency is the ratio of the useful power out to the total power in, or

$$\text{Efficiency} = \frac{P}{P_1} = \frac{500 \text{ W}}{600 \text{ W}} = 0.83 = \boxed{83\%}$$

(f) When the motor is starting up, the back emf is zero since v is zero, and the starting current is

$$I = \frac{V}{R} = \frac{120 \text{ V}}{4.0 \text{ } \Omega} = \boxed{30 \text{ A}}$$

Note how much larger this current is than the 5.0-A current when the motor is running at full speed.

Motors, Generators, and the Laws of Electromagnetic Induction

In concluding this chapter let us reemphasize that the operation of generators and motors is based on the laws of Faraday and Lenz. That we may end where we began, we consider our original sliding-rod device of Sec. 20.1, which is a primitive kind of electric generator (Fig. 20.31). In this case our generator is a linear generator rather than a rotational generator.

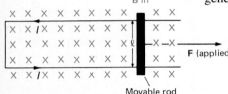

FIGURE 20.31 An electric generator. The motion of the rod to the right, produced by the applied force, induces a current I in the closed circuit.

If we exert a force **F** to move the rod to the right at a constant speed v, the magnetic flux into the paper increases as a result of the motion, since the area of the magnetic field inside the coil increases. By Faraday's induction law, this induces an emf in the rod and, since the circuit is complete, a current also. By Lenz' law, this current is in a direction to reduce the flux through the circuit and is therefore counterclockwise. This induced current, in turn, produces a force on the rod given by $F = B_\perp Il$ in a direction opposite the applied force. Because of this opposing force, mechanical work must be done to move the rod, and this work is numerically equal to the electric energy produced by this primitive generator.

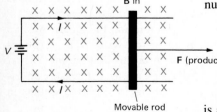

FIGURE 20.32 An electric motor. The current I produced by the voltage source V leads to the motion of the rod in the direction of the force **F**.

In the case of the simplified motor shown in Fig. 20.32, an external voltage is applied to drive current clockwise around the loop. This produces a force on the rod that moves it to the right, thus increasing the magnetic flux through the circuit by making the area enclosed by the circuit greater. This changing magnetic flux induces an emf, and by Lenz' law the direction of the emf is such as to oppose the increase in magnetic flux through the circuit by producing magnetic flux in a direction opposite that of the original flux. The direction of the emf is therefore counterclockwise. This is the back emf that opposes the applied voltage and decreases the current drawn by a motor as it comes up to speed.

We see, therefore, that Faraday's induction law governs the general behavior of both generators and motors, while Lenz' law is particularly useful in predicting the *direction* of the induced emf's and currents.

Summary: Important Definitions and Equations

Induced emf's

Motional emf: The emf induced by the motion of a conductor through a magnetic field: $\mathcal{E} = B_\perp vl$ (20.1)

Faraday's induction law: $\mathcal{E} = -N \dfrac{\Delta \Phi_B}{\Delta t}$ (20.6)

where Φ_B = magnetic flux = $B_\perp A$.   (20.3)
Unit of magnetic flux: 1 Wb (weber) = 1 N·m/A.

Lenz' law: The direction of an induced emf is always such as to oppose the change in magnetic flux producing it.
EMF due to mutual induction:

$\mathcal{E} = -M \dfrac{\Delta i}{\Delta t}$ (20.8)

EMF due to self-induction:

$\mathcal{E} = -L \dfrac{\Delta i}{\Delta t}$ (20.10)

Inductors
Unit of inductance (M or L): 1 henry (H) = 1 V·s/A
Energy stored in an inductor: $\mathcal{W} = \frac{1}{2}LI^2$ (20.11)
Energy density in a magnetic field B (energy stored per unit volume):

$w = \dfrac{B^2}{2\mu_0}$ (20.13)

Self-inductance of a solenoid:

$L = \dfrac{\mu_0 N^2 A}{l}$ (20.12)

Energy density in combined electric and magnetic fields:

$w = \frac{1}{2}\epsilon_0 E^2 + \dfrac{1}{2}\dfrac{1}{\mu_0} B^2$ (20.14)

RL circuit: Time constant $\tau = L/R$ (20.18)

Time constant: A time that determines the rate at which a physical quantity approaches its final, steady-state value.
RC circuit: Time constant $\tau = RC$ (20.20)

Electric generators: Machines that convert mechanical energy into electrical energy.

EMF produced by an ac generator:

$\mathcal{E} = \mathcal{E}_{max} \sin \omega t = \mathcal{E}_{max} \sin 2\pi f t$ (20.24)

where $\mathcal{E}_{max} = NBA\omega$ (20.23)

Electric motors: Machines (equivalent to electric generators run in reverse) that convert electric energy into mechanical energy.

Back emf: The emf produced by an electric motor because of Faraday's induction law. It opposes the external emf applied to the motor.
Useful power produced by a motor:

$P = \mathcal{E}_{back} I$ (20.25)

Questions

1   What would happen if a straight rod that is not part of a complete circuit moved at right angles to a magnetic field? Would an emf be produced? Would a current be produced? If so, how long would the current last?

2   Is it correct to argue that, since a static electric charge can be used to induce another static charge on a metal, so too an electric current should be able to induce another electric current?

3   If, in Fig. 20.1, the rod *ab* is at rest and the magnetic field is suddenly turned off, is the induced emf clockwise or counterclockwise? Explain your answer.

4   Discuss the relationship between Faraday's induction law and the operation of an electric generator.

5   A copper ring is mounted on a ring stand 1 m above the floor. A bar magnet is held above the ring with its north pole toward the floor. If the magnet is then dropped through the ring to the floor, what are the forces acting on the magnet (*a*) before it reaches the ring; (*b*) while it is passing through the ring; (*c*) after it has passed through the ring?

6   Sensitive galvanometers (see Fig. 19.43) use closed coils of wire mounted adjacent to the deflecting coils of the meter to damp the motion of the galvanometer coil after it is displaced. Explain how such a magnetic damping mechanism operates.

7   A galvanometer has an eddy-current damper attached to it somewhat like that in Fig. 20.9. Does this damping system reduce the energy supplied to the galvanometer by the current through it? If so, where does the energy go?

8   Can you suggest a way to decrease inductive effects

in the winding of resistance coils for use in standard resistance boxes? (Resistance coils are tightly wound spools of wire of a thickness and length chosen to provide the desired resistance.)

9 How could you decide whether the earth is moving through a magnetic field that fills all space or is carrying its own magnetic field along with it as it moves?

10 Why is any work required to keep a steam-electric generator turning when no load is attached? What happens when a load is attached?

11 Are nuclear steam-electric power plants any more efficient than coal-burning plants? Explain.

12 A motor forced to do too much work will frequently blow a fuse or circuit breaker before any damage is done to the motor itself. Explain what happens in this case. When is the fuse most likely to blow?

13 Explain how a combination of strong magnetic fields and eddy-current effects could be used to separate ground-up solid waste into three groups: ferrous metals, nonferrous metals, and nonmetallic substances.

Multiple-Choice and Simple Exercises

20.1 A conductor consisting of a single long straight rod is moved at right angles to a uniform magnetic field. Which of the following statements is false?

(a) The faster the motion of the rod, the greater the induced voltage.

(b) There is a continuous current from one end of the rod to the other.

(c) The stronger the magnetic field, the greater the induced voltage if the speed at which the rod moves remains constant.

(d) The effect is the same whether the magnetic field is produced by a permanent magnet or an electromagnet.

(e) The induced voltage is independent of the conducting material out of which the rod is made.

20.2 In Fig. 20.1 the rod ab, which is 0.20 m long, is moved to the right with a speed of 3.0 m/s. The magnetic field strength is 1.5 T and is directed at right angles to the plane of the current loop. If the current loop has a break in it, the induced emf is:

(a) 0   (b) 0.90 V   (c) 0.90 T   (d) 1.0 V
(e) 0.10 V

20.3 The Wb/m^2 is a unit of:

(a) Magnetic flux   (b) Area
(c) Magnetic field   (d) Force
(e) Pressure

20.4 A circular coil of wire rests on your desk. A magnetic field is directed at right angles to the plane of your desk and passes through the coil. For a clockwise current (as you look down on your desk) to be generated, the magnetic field must be directed:

(a) Downward toward the desk and be increasing in time
(b) Downward toward the desk and be constant in time
(c) Upward away from the desk and be increasing in time
(d) Upward away from the desk and be decreasing in time
(e) Upward away from the desk and be constant in time

20.5 Lenz' law states that the current induced in a loop of wire when the magnetic flux through the loop changes is always:

(a) Clockwise   (b) Counterclockwise
(c) In such a direction as to oppose the change in the original flux
(d) In a direction opposite that of the current producing the original flux
(e) In the same direction as that of the current producing the original flux

20.6 A circular coil of wire consists of 20 turns, and has an area of 0.10m^2. The magnetic field through the coil changes by 2.0 T in 0.10 s. The magnitude of the induced emf is:

(a) 40 T   (b) 4.0 V   (c) 400 V   (d) 40 J
(e) 40 V

20.7 The unit 1 Wb/s is the same as:

(a) 1 J   (b) 1 V   (c) 1 T   (d) 1 N
(e) 1 Wb/m^2

20.8 A coil has a self-inductance of 100 mH. If the current through it is increasing at the rate of 5.0 A/s, the induced voltage in the coil is:

(a) −0.50 V   (b) +50 V   (c) + 0.50 V
(d) −50 V   (e) −0.020 V

20.9 An air-core coil of length 0.10 m has a cross-sectional area of 0.025 m^2 and contains 100 turns. The inductance of such a coil is:

(a) 0.25 H   (b) 0.25 T   (c) 3.14×10^{-7} T
(d) 3.14×10^{-7} H   (e) 3.14×10^{-3} H

20.10 A circuit contains an inductance coil with $L = 50$ mH and $R = 100 \ \Omega$. A voltage of 100 V is placed across the coil. The time it takes for the current to reach 0.63 A is:

(a) 5.0×10^3 s   (b) 5.0×10^{-3} s   (c) 5.0 s
(d) 5.0×10^{-5} s   (e) 5.0×10^{-4} s

20.11 A circuit consists of a 10-μF capacitor in series with a 500-Ω resistor. The current 5.0×10^{-3} s after a voltage of 100 V is connected across this combination is:

(a) 0.20 A   (b) 0.063 A   (c) 0.13 A
(d) 0.037 A   (e) 0.074 A

20.12 A coil of 50 turns has a cross-sectional area of 5.0×10^{-3} m^2. It rotates at a frequency of 60 Hz in a magnetic field of 2.0 T. The maximum induced emf is:

(a) 30 V   (b) 1.9×10^2 V   (c) 1.9×10^4 V
(d) 60 V   (e) 3.8×10^2 V

20.13 If a generator has an emf of 120 V when its armature rotates at 1800 rev/min, an armature rotating at 1200 rev/min will produce an emf of:

(a) 180 V   (b) 53 V   (c) 80 V   (d) 60 V
(e) 110 V

20.14 A rod slides along a metal frame similar to that in Fig. 20.1. If the length of the rod is 0.50 m, its speed is 0.30 m/s, and a magnetic field of 0.10 T is directed at right angles to the plane of the metal frame, find the induced emf in the rod.

20.15 What is the flux produced by a magnetic field directed perpendicular to a coil of area 0.20 m^2, if B is 0.50 Wb/m^2?

20.16 (*a*) The magnetic field at right angles to a coil of 50 turns and area 0.025 m^2 increases at a rate of 0.10 T/s. What is the induced emf in the coil?

(*b*) How would the result change if the magnetic field were decreasing at a rate of 0.10 T/s?

20.17 Two coils are wound on a common iron core, as in Fig. 20.10. If a current of 4.0 A in coil 1 produces a magnetic flux of 1.2 × 10^{-3} Wb/m^2 in coil 2, which has 150 turns, what is the mutual inductance between the two coils?

20.18 The current through one coil of a transformer is increasing at a rate of 6.0 A/s.

(*a*) If the self-inductance of the coil is 200 mH, what is the magnitude of the induced emf in the coil?

(*b*) What is the direction of the induced emf?

20.19 How much energy is stored in the magnetic field of a 200-mH inductor when the current through it is 5.0 A?

20.20 What is the time constant of a circuit consisting of an 80-mH inductance in series with a 2000-Ω resistor?

20.21 What is the time constant of a circuit consisting of a 2.0-μF capacitor and a 10-MΩ (10 × 10^6 Ω) resistor in series?

20.22 It is desired to produce an alternating voltage of maximum value 5.0 × 10^3 V with a generator containing a coil of 200 turns and cross-sectional area 1.0 m^2 rotating at 60 Hz. What must be the strength of the magnetic field in which the coil rotates?

20.23 A dc motor has a resistance of 8.0 Ω and draws a current of 10 A when running at full speed on a 120-V line. What is the back emf produced?

Problems

20.24 A train is moving at a speed of 10 m/s in a northerly direction. If the vertical component of the earth's magnetic field in the vicinity is 5.0 × 10^{-5} T, find the magnitude and direction of the emf induced in a 2.0-m-long steel axle of the train.

20.25 An airplane is flying along the surface of the earth at a place where the earth's magnetic field is 0.47 × 10^{-4} T and has a dip angle of 37° with respect to the vertical. If the plane is flying at a speed of 200 m/s, what is the motional emf between the two ends of its steel wings, which are separated by 40 m?

20.26 A rod slides along a metal frame of the shape shown in Fig. 20.33. The earth's magnetic field has a strength of 0.35 × 10^{-4} T at right angles to the plane of the frame. If the rod is of length 0.20 m inside the frame and moves at 1.0 m/s:

(*a*) What is the motional emf produced by the rod?

(*b*) If the resistance of the frame and rod combined is 0.050 Ω, what current results?

(*c*) What are the magnitude and direction of the force on the rod?

(*d*) Show that the mechanical power needed to move the rod is exactly equal to the electric power consumed in the resistance of the frame and rod.

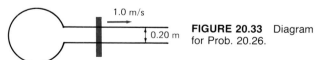

1.0 m/s

0.20 m

FIGURE 20.33 Diagram for Prob. 20.26.

20.27 A 6.0-cm-diameter circular coil of wire has a resistance of 2.0 Ω. It is located in a magnetic field of 0.30 T directed at right angles to the plane of the coil. If the coil is removed from the field in 0.10 s:

(*a*) What is the induced emf in the coil?

(*b*) How much electric energy is lost in the process?

(*c*) Where does this energy go?

*20.28 The moving rod in Fig. 20.1 is assumed to have negligible resistance, but the U-shaped conductor contains a resistance *R*.

(*a*) Prove that the rate at which mechanical work must be done to move the rod is $P = B^2l^2v^2/R$.

(*b*) Prove that this is equal to the electric power dissipated as heat in the circuit.

20.29 An air-core solenoid contains 100 turns of wire per meter and has a cross-sectional area of 0.12 m^2. A current of 2.5 A is established in the solenoid and is then reduced to zero in 3.0 s by opening a switch. A pickup coil of 50 turns circles the solenoid near its center.

(*a*) What are the magnitude and direction of the induced emf in the pickup coil?

(*b*) If the resistance of the loop is 0.0080 Ω, what is the induced current?

*20.30 A rod 0.50 m long is pivoted so that it can rotate about a fixed point *P*, as in Fig. 20.34. If it rotates at 4.0 rev/s and a magnetic field of 0.20 T exists parallel to its axis of rotation and covering the whole area swept out by the rod, what is the potential difference between the two ends of the rod?

P

FIGURE 20.34 Diagram for Prob. 20.30.

*20.31 A "search coil" combined with a ballistic galvanometer (whose deflection is proportional to the charge flowing through it) is often used to measure magnetic fields. If a search coil in a magnetic field of strength *B* is suddenly pulled completely out of the field, show that the magnetic field is given by $B = RQ/NA$, where *N* is the number of turns, *A* is the area of the coil, *R* is the resistance of the coil and galvanometer combined, and *Q* is the total charge that flows through the galvanometer.

20.32 One meter of no. 18 copper wire (diameter = 0.10 cm) is formed into a circular loop. It is then placed in a uniform magnetic field directed at right angles to the plane of the loop. The magnetic field increases in time at a constant rate of 0.020 T/s. At what rate is heat generated in the loop?

20.33 The magnetic field perpendicular to a single-turn loop of wire of radius 0.20 m and resistance 0.10 Ω changes with time, as shown in Fig. 20.35.

(*a*) Plot the induced emf as a function of time.

(*b*) Plot the rate of energy dissipation as a function of time.

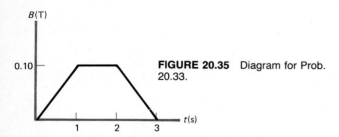

FIGURE 20.35 Diagram for Prob. 20.33.

20.34 A rectangular coil of length 10 cm and width 6.0 cm has 100 turns and a resistance of 0.050 Ω. It is initially placed in a magnetic field directed perpendicular to the coil. It is then pulled along its length out of the field at a speed of 6.0 m/s, and a clockwise current of 0.10 A is observed in the coil. What are (*a*) the magnitude and (*b*) the direction of the magnetic field?

20.35 A transformer consists of two coils of wire wound on a common toroidal iron core. If the mutual inductance of the pair is 200 mH and the current in the first coil decreases from 20 A to 0 in 0.50 s, what is the induced emf in the second coil?

20.36 An inductance coil of self-inductance 100 mH is connected to a circuit in which the current is changing with time in a sawtooth pattern, as in Fig. 20.36. Draw a graph indicating how the induced emf in the circuit varies in time, showing the actual numerical value of $\mathscr{E}$ at each instant.

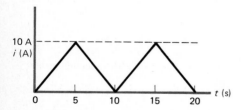

FIGURE 20.36 Diagram for Prob. 20.36.

20.37 An air-core solenoid has a length of 30 cm, a cross-sectional area of 0.020 m², and contains 100 turns.

(*a*) What is its inductance?

(*b*) If the current through the solenoid increases by 5.0 A, how does the magnetic flux through the solenoid change?

20.38 Two coils of wire C and C' have 500 and 1000 turns, respectively. A current of 2.0 A in C produces a flux of 2.0×10^{-4} Wb through C and a flux of 1.5×10^{-4} Wb through C'. Find:

(*a*) The self-inductance of C

(*b*) The mutual inductance of C and C'

(*c*) The average induced emf in C' when the current in C falls to zero in 0.50 s

20.39 A coil in the form of a solenoid is to have a 15-mH self-inductance. It is wound on an air-core cylinder of 5.0 cm diameter and 80 cm in length.

(*a*) How many turns will have to be wound on the cylindrical form?

(*b*) If a second coil of 90 turns is wound directly over the first coil, what is the mutual inductance between the two coils?

20.40 An air-core solenoid consists of 200 turns of wire wound on a form that is 40 cm long and has an inner diameter of 4.0 cm. If a current of 25 A is established in this solenoid:

(*a*) What is its self-inductance?

(*b*) What is the magnetic field at its center?

(*c*) How much energy is stored in it?

(*d*) What is the energy density in the solenoid?

20.41 Prove that the expression $\sqrt{L/C}$ has the units of *ohms*. Here L is an inductance and C is a capacitance.

20.42 (*a*) What must the magnitude of a magnetic field be to store as much energy per unit volume as is stored in an electric field of strength 10^4 V/m?

(*b*) What is the numerical ratio of E (in V/m or N/C) to B (in T) in this case?

20.43 A coil has a self-inductance of 2.5 mH. If the current through this coil increases uniformly from zero to 2.0 A in 0.10 s:

(*a*) What is the magnitude of the induced emf?

(*b*) What is its direction with respect to that of the current?

(*c*) How much energy has been stored in the inductor when the current has reached its maximum value?

20.44 (*a*) What must the current through a 50-mH inductor be if the energy stored in the inductor is to be 10 J?

(*b*) What happens to the energy when the battery producing the current is disconnected?

20.45 A circuit consists of a 20-Ω resistor and a 500-mH inductor connected in series to a 12-V battery. What is the value of the current when the current is increasing at the rate of 10 A/s?

20.46 A current of 10 A exists in a circuit consisting of a 10-mH inductor in series with a 100-Ω resistor. A switch in the circuit is opened at $t = 0$. How long will it take for the current to fall to 3.7 A?

20.47 A long iron-core solenoid is connected across a 12-V battery, and the current rises to 0.63 of its final value in 0.80 s. This procedure is then repeated with the iron core removed, and the time required for the current to rise to 0.63 of its final value is 0.0010 s.

(*a*) Calculate the relative permeability of the iron.

(*b*) Find the inductance of the air-core solenoid if the maximum current is 0.60 A.

*20.48 An inductance coil of resistance 20 Ω and inductance 0.50 H is connected to a 120-V dc source. At what rate will the current in the coil rise (*a*) at the instant the coil is connected to the power source; (*b*) at the instant the current reaches 75 percent of its maximum value?

20.49 A circuit consisting of a 100-μF capacitor in series with a 2.0-MΩ resistor is connected across a 6.0-V battery.

(a) What is the time constant of the circuit?

(b) What is the initial value of the current when a switch in the circuit is closed?

(c) What is the charge on the capacitor when the current i is 2.0×10^{-6} A?

20.50 A 5.0-μF capacitor is attached to a 100-V battery and a 4.0×10^{4} Ω resistor.

(a) What is the final charge on the capacitor?

(b) If the battery is shorted out of the circuit so as to allow the capacitor to discharge, how long does it take for the charge on the capacitor to fall to 37 percent of its initial value?

(c) Consider the situation where the capacitor is being charged. Prove that at the instant of time when $q = 0.37Q$, where Q is the final charge on the capacitor, the voltages across C and R add to give the 100 V provided by the battery.

20.51 A 6.0-μF capacitor is charged to a potential of 100 kV. After being disconnected from the power source, it is connected across a 10-MΩ resistor.

(a) At what rate does the capacitor discharge initially?

(b) How long does it take for the voltage across the capacitor to fall to 37 kV?

20.52 A generator coil rotates in a magnetic field. If the coil contains 100 turns, has an area of 1.5 m^2, and makes 60 rev/s in a magnetic field of 0.10 T:

(a) What is the maximum emf induced in the coil?

(b) Draw a graph of emf as a function of time for one complete cycle.

20.53 An electric generator delivers 20 A of current at 120 V to run a large piece of machinery. If the generator's armature rotates at 1000 rev/min and 400 W is lost as heat in the generator, what torque must be applied to the generator to drive it at this rotational speed?

20.54 A 100-turn coil 30 cm long and 20 cm wide is rotated about its long axis at 25 rev/s in a uniform magnetic field of 0.36 T. If the magnetic field is perpendicular to the axis of rotation:

(a) What is the maximum value of the emf produced?

(b) Write an expression for the emf as a function of time.

20.55 A motor has a back emf of 110 V and a current through its armature of 40 A when running at 900 rev/min.

(a) What power does the motor produce?

(b) What torque does it develop?

20.56 An ac motor develops an emf of 120 V when rotating at 1500 rev/min. At what speed must it rotate to generate 240 V?

20.57 A dc motor with a resistance of 1.5 Ω draws a current of 12 A when running at full speed off a 60-V battery.

(a) What is the starting current for this motor?

(b) What is the back emf when the motor is up to full speed?

*20.58 For the motor of Prob. 20.57 prove that the electric power delivered by the battery is equal to the sum of the power delivered by the motor and the power lost as heat in the motor resistance (a) at the instant the motor is starting up; (b) when the motor is running at full speed.

20.59 A motor has a resistance of 4.0 Ω and is connected to a 120-V power line. The motor draws a current of 5.0 A when running at full speed. What current will the motor draw if the speed is reduced to 90 percent of full speed by application of a load?

Additional Readings

Coulson, Thomas: *Joseph Henry, His Life and Work*, Princeton University Press, Princeton, N.J., 1950. The standard biography of a great American physicist.

Kaplan, Joseph: "Michael Faraday (1791–1896)," in *Great Men of Physics: The Humanistic Element in Scientific Work*, Tinnon-Brown, Los Angeles, Calif., 1969. A popular lecture on Faraday's contributions to physics.

Kondo, Herbert: "Michael Faraday," in *Lives in Science, a Scientific American* book, Simon and Schuster, New York, 1957, pp. 127–140. A brief and useful account of Faraday's life and contributions to physics.

Shiers, George: "The Induction Coil," *Scientific American*, vol. 224, no. 5, May 1971, pp. 80–87. A fascinating history of the development of the induction coil as a source of high voltage and of the role it played in important physics discoveries by Roentgen, Hertz, and J. J. Thomson.

Williams, L. Pearce: "Michael Faraday and the Physics of 100 Years Ago," in Jerry B. Marion, *A Universe of Physics: A Book of Readings*, Wiley, New York, 1970, pp. 85–97. An account of how Faraday broke with the mechanistic concepts of his day. Williams has also written the definitive biography of Faraday: *Michael Faraday*, Basic Books, New York, 1964.

Wilson, Mitchell: "Joseph Henry," in *Lives in Science, a Scientific American* book, Simon and Schuster, New York, 1957, pp. 141–153. A good short description of the life of Joseph Henry.

CHAPTER 21

Alternating Current Circuits

Any notice about the progress of physics in the latter part of the last century would be like the play of Hamlet *without the Prince of Denmark if it did not deal with the part played in it by Lord Kelvin, who for more than forty years before his death in 1907 had been the most potent influence in British physics.*

J. J. Thomson (1856–1940)

At the end of the nineteenth century a heated controversy erupted over whether the electric power to be delivered to consumers in the United States should be in the form of alternating or direct current. George Westinghouse embraced the ideas of Nikola Tesla and strenuously supported alternating current. Thomas Edison, on the other hand, fought furiously and often unscrupulously for direct current. He lobbied the New York State Legislature to adopt alternating current for the electric chair and then used the electric chair as an example of the deadly nature of alternating current to advocate the adoption of the "less dangerous" direct current for commercial use! In 1893 the proponents of alternating current won two crucial victories: Tesla's ac system was chosen to light the World Columbian Exposition in Chicago, and Westinghouse received the contract to deliver hydroelectric power from Niagara Falls in ac form to American homes and factories. Since that time ac power has assumed almost complete dominance in the electric utility industry in the United States.

21.1 AC Circuit Containing Only Resistance

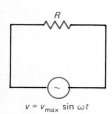

$v = v_{max} \sin \omega t$

FIGURE 21.1 An ac voltage of form $v = v_{max} \sin \omega t$ applied across a simple resistor, with $\omega = 2\pi f$.

As we have seen in Sec. 20.8, the emf of an ac generator takes the form $\mathscr{E} = \mathscr{E}_{max} \sin 2\pi ft$, where f is the frequency, or the number of complete rotations of the generator coil per second, and $\mathscr{E}$ is the emf at any instant of time t. This emf has the form of a sine wave that continually changes direction and is referred to as *alternating* or *sinusoidal*. If this emf is applied across a simple resistor, as in Fig. 21.1, the terminal voltage of the source of emf may be reduced slightly from the open-circuit emf, but the form of the voltage will still be sinusoidal: $v = v_{max} \sin 2\pi ft = v_{max} \sin \omega t$. Experiment shows that the current through the resistor, $i = v/R$, is also sinusoidal. It takes the form $i = i_{max} \sin \omega t$. The current reaches its maximum value i_{max} at the same time as the voltage does and passes through zero at the same instant as the voltage, as shown in Fig. 21.2. We therefore conclude:

In an ac circuit containing only resistance, the current and the voltage are in phase.

FIGURE 21.2 The ac voltage and current as a function of time for a purely resistive circuit. The current is in phase with the voltage.

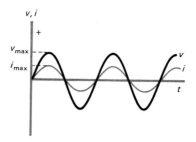

(To refresh your mind on phase and phase relationships to which we will be referring frequently in what follows, review Sec. 11.8).

In dealing with ac currents and voltages it is important to distinguish a number of different terms used to describe ac quantities:

i *(or v):* The *instantaneous value* of an ac quantity, i.e., its value at some particular instant of time.

i_{max} *(or v_{max}):* The *amplitude* in the expression for an ac quantity, for example, $i = i_{max} \sin \omega t$. The amplitude is also referred to as the *peak value* of the quantity.

For a purely resistive circuit, Ohm's law [Eq. (18.4)] applies, and we have for the instantaneous current and voltage

$$i = \frac{v}{R}$$

Since the peak values i_{max} and v_{max} are just the values of i and v at particular instants of time, and since the current and the voltage reach their peaks at the same instant of time, we also have

$$i_{max} = \frac{v_{max}}{R}$$

Effective Values of Current and Voltage

From Fig. 21.2 we can see that the *average value* of an ac voltage or current is zero, since it is positive for as much time as it is negative. In many cases, however, such as in supplying power to a light bulb, an alternating current can be of considerable use even though its average value is zero. The effectiveness of the current in producing heat in the filament of the light bulb does not depend on the direction of the current. The rate at which heat is generated is determined by Joule's law, $i^2 R$ [Eq. (18.12)], and this quantity is *not* zero when averaged over a cycle, even though the current i is.

To find the *effective value* of an ac current, we average the heating effect of the current over one cycle; we then equate it to the constant current that would produce the same effect. We have for the power,

$$P = \overline{i^2 R} = \overline{i^2} R = I^2 R$$

where I is the *effective current,* or the *constant current that would produce the same heating effect* as does i averaged over a complete cycle. Here $\overline{i^2}$ is the mean-square alternating current averaged over one cycle, i.e., the average squared current for the cycle. Then, since $i = i_{max} \sin \omega t$, we have, on canceling R on the two sides,

$$I^2 = \overline{i_{max}^2 \sin^2 \omega t} = i_{max}^2 \ \overline{\sin^2 \omega t}$$

FIGURE 21.3 The effective, or rms, value (*V* or *I*) and the peak, or maximum, value (v_{max} or i_{max}) of (*a*) an ac voltage and (*b*) an ac current.

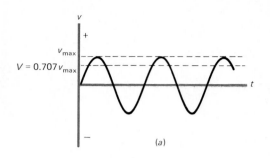

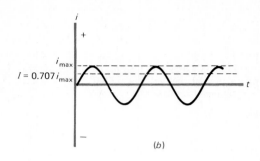

Now the average value of $\sin^2 \omega t$ is $\frac{1}{2}$,* and so

$$I^2 = \frac{i^2_{max}}{2} \quad \text{and} \quad I = \frac{i_{max}}{\sqrt{2}} = 0.707 i_{max} \tag{21.1}$$

Similarly, $\quad V = \dfrac{v_{max}}{\sqrt{2}} = 0.707 v_{max} \tag{21.2}$

The effective values of ac currents and voltages are thus 0.707 times their peak values. This is shown in Fig. 21.3. Since $i_{max} = v_{max}/R$,

$$0.707 i_{max} = \frac{0.707 v_{max}}{R} \quad \text{and} \quad I = \frac{V}{R}$$

In a purely resistive ac circuit, the effective current and the effective voltage satisfy Ohm's law.

Effective currents and voltages are also called *root-mean-square* (rms) currents and voltages since they are obtained by first *squaring* the quantity, then taking the *mean* or average of the squared quantity over a period, and finally taking the square *root* of this average, as we did for rms speeds in Sec. 9.8:

$$i_{rms} \equiv I = 0.707 i_{max} \quad \text{and} \quad v_{rms} \equiv V = 0.707 v_{max}$$

For ordinary ac house voltages, which have effective values of about 120 V, the peak voltage is $V/0.707 = 170$ V. Most ac voltmeters and ammeters are calibrated to read effective (rms) values directly.

In this chapter we always use the symbols I and V to denote the *effective* values of ac currents and voltages. The basic relations between the various kinds of ac currents and voltages are listed in Table 21.1.

TABLE 21.1   AC Quantities

Current	
Instantaneous:	$i = i_{max} \sin \omega t$
Effective (rms):	$I = i_{rms} = 0.707 i_{max}$

Voltage	
Instantaneous:	$v = v_{max} \sin \omega t$
Effective (rms):	$V = v_{rms} = 0.707 v_{max}$

---

*To show this, we use a trigonometric identity from Appendix 1.C, with $x = \omega t$.

$$\cos 2\omega t = 1 - 2\sin^2 \omega t \quad \text{or} \quad 2\sin^2 \omega t = 1 - \cos 2\omega t$$

Taking time averages of all quantities, we have

$$\overline{2\sin^2 \omega t} = 1 - \overline{\cos 2\omega t}$$

But the average value of any sine or cosine function over one period is zero, since it is positive just as much as it is negative. Hence

$$\overline{2\sin^2 \omega t} = 1 \quad \text{or} \quad \overline{\sin^2 \omega t} = \tfrac{1}{2}$$

Example 21.1

An ac voltage of effective value 240 V is connected across a 62.5-Ω resistor. What are **(a)** the effective, **(b)** rms, **(c)** maximum, and **(d)** peak currents?

SOLUTION

(a) Since Ohm's law is valid for an ac voltage across a resistor,

$$I = \frac{V}{R} = \frac{240 \text{ V}}{62.5 \ \Omega} = \boxed{3.84 \text{ A}}$$

(b) $\boxed{3.84 \text{ A.}}$ RMS current is just another name for the effective current.

(c) $i_{max} = \dfrac{I}{0.707} = 1.41(3.84 \text{ A}) = \boxed{5.41 \text{ A}}$

(d) $\boxed{5.41 \text{ A.}}$ Peak current is another name for maximum current.

21.2 AC Circuit Containing Only Inductance

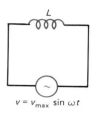

$v = v_{max} \sin \omega t$

FIGURE 21.4 An ac voltage applied to a purely inductive circuit.

Next we consider what happens when we apply an ac voltage across an inductance coil, as in Fig. 21.4. Initially we assume that the resistance of the coil is negligible, but in a later section we consider more realistic coils having both resistance and inductance.

Even if the coil has no resistance, it limits the ac current just as the resistor limits the current in Fig. 21.1. An equation similar to Ohm's law for a pure-resistance circuit applies in this case and takes the form

$$V = IX_L \tag{21.3}$$

where X_L is called the *inductive reactance* and V and I are the *effective* values of the voltage and current. X_L has the dimensions of resistance and is measured in ohms, just like an ordinary resistance. The *reactance* of a circuit is a measure of the opposition presented by an inductor or capacitor to alternating current.

Phase of the Current

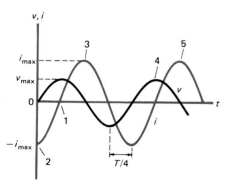

FIGURE 21.5 Current and voltage for an ac circuit containing only inductance (Fig. 21.4). The current lags the voltage by 90°.

When an ac voltage is applied across a coil, the alternating current in the coil produces a changing magnetic flux through the coil. This changing flux produces an induced emf $- L \, \Delta i/\Delta t$. Hence Kirchhoff's loop rule (Sec. 18.5) gives

$$v - L\frac{\Delta i}{\Delta t} = 0 \quad \text{or} \quad v = L\frac{\Delta i}{\Delta t} \tag{21.4}$$

The applied voltage v is therefore proportional to the time rate of change of the current, or, in other words, to the slope of the current curve. Since the applied voltage is represented by a sine curve of the form $v = v_{max} \sin \omega t$, we want to find a function to represent the current i, which has the property that its slope is a sine curve. To find this function, let us first plot a few important values of i.

When $\sin \omega t = 1$, $v = v_{max}$, $\Delta i/\Delta t$ has its maximum positive value, and so i is increasing most rapidly in time. This corresponds to point 1 in Fig. 21.5, at which point $i = 0$ and $\Delta i/\Delta t = v_{max}/L$. From Eq. (21.4), when $v = 0$, $\Delta i/\Delta t = 0$, which means that there is no change in i with time, and so i must be either at its maximum positive value or at its maximum negative value. Since we already know that the slope of i versus t is upward at point 1, points 2 and 3 must be as shown on the graph.

If we plot a whole series of points in this way, the current curve shown in Fig. 21.5 results. The graph for i is clearly not a sine curve, since i is not equal to 0 when $t = 0$. Rather it turns out to be the negative of a cosine curve, since the current value for $t = 0$ is not i_{max}, as it would be for a cosine function, but $-i_{max}$, as expected for the negative of a cosine curve. We can therefore write the current through the inductor as a function of time in the following form:

$$i = -i_{max} \cos \omega t$$

Both v and i therefore follow sinusoidal curves, but *the current lags behind the voltage in time*. That is, the current reaches its maximum value (at point 5) at a *later time* than the voltage reaches its maximum value (at point 4). The time lag between v and i is seen from the graph to be one-fourth of a cycle $(T/4)$, which is equivalent to 90°, or $\pi/2$ rad.

The current in an ac circuit containing only inductance lags the applied voltage by 90°, or $\pi/2$ rad.

Dependence of X_L on Frequency

It can be shown (using calculus) that X_L in Eq. (21.3) takes the form

$$X_L = \omega L = 2\pi f L \qquad \textbf{(21.5)}$$

where f is the frequency of the applied voltage and $\omega = 2\pi f$.

To see whether Eq. (21.5) makes good physical sense, let us see what it predicts for very low and very high values of the frequency. If we connect an inductor with zero resistance to a dc voltage, $f = 0$, $X_L = 0$, and $I = V/X_L \rightarrow \infty$. In this case there is no induced emf and hence no opposition to the current, since $L \, \Delta i/\Delta t = 0$. Any real coil, of course, has some resistance that prevents the current from reaching an infinite value.

At the other extreme, if f is large (and L is nonzero), then X_L is large and the effective current I is small, no matter how large the applied voltage. An inductor therefore acts as a low resistance at low frequencies and as a high resistance at high frequencies. Experimental results confirm these predictions of Eq. (21.5). We therefore conclude:

In an inductive ac circuit, the factor that is analogous to the resistance in a purely resistive circuit is the inductive reactance $X_L = 2\pi f L = \omega L$.

The fact that X_L is proportional to both f and L is also consistent with the requirement that X_L have the dimensions of resistance and be measured in ohms, since $X_L = V/I$. The units of fL are

$$\frac{1}{s} \cdot H = \frac{1}{s} \cdot \frac{V \cdot s}{A} = \frac{V}{A} = \Omega$$

as required.

Example 21.2

An inductance coil has an inductance of 400 mH and a resistance of 2.0 Ω. Find the current in the coil if the applied voltage is **(a)** 120 V dc and **(b)** 120 V ac at 60 Hz. **(c)** What is the peak current in case (b)?

SOLUTION

(a) For direct current the inductance has no effect, since $f = 0$ and so $X_L = 0$. We have, then, from Ohm's law,

$$I = \frac{V}{R} = \frac{120 \text{ V}}{2.0 \ \Omega} = \boxed{60 \text{ A}}$$

(b) For alternating current the inductive reactance is

$$X_L = 2\pi fL = 2\pi(60/\text{s})(0.40 \text{ H}) = 151 \ \Omega$$

Since this is almost 100 times larger than the 2.0-Ω resistance, we omit the resistance for now (we will see shortly how to include it) and obtain for the rms ac current:

$$I = \frac{V}{X_L} = \frac{120 \text{ V}}{151 \ \Omega} = \boxed{0.79 \text{ A}}$$

Notice the tremendous difference between the ac and dc currents in these two cases, even though the circuits are identical and the effective ac voltage applied in (b) is the same as the dc voltage in (a).

(c) $i_{\text{peak}} = i_{\text{max}} = 1.41I = 1.41(0.79 \text{ A}) = \boxed{1.1 \text{ A}}$

21.3 AC Circuit Containing Only Capacitance

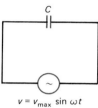

$v = v_{\text{max}} \sin \omega t$

FIGURE 21.6 An ac circuit containing only a capacitor.

Let us consider the circuit of Fig. 21.6, in which an ac voltage is connected to a capacitor C. In this case the ac current is limited by the capacitor in a way similar to that in which a resistor limits the current in a pure-resistance circuit. Here we have

$$V = IX_C \tag{21.6}$$

where X_C is called the *capacitive reactance*.

Phase of the Current

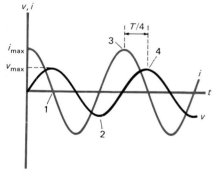

FIGURE 21.7 Current and voltage for an ac circuit containing only capacitance (Fig. 21.6). The current leads the voltage by 90°.

If the applied voltage is given by $v = v_{\text{max}} \sin \omega t$, then the charge on the capacitor at any instant of time is, from Eq. (17.9), $q = Cv = Cv_{\text{max}} \sin \omega t$. The *charge* on the capacitor is *in phase* with the applied voltage.

The current in the circuit, however, is $i = \Delta q/\Delta t = C \ \Delta v/\Delta t$, and so the current is not in phase with the applied voltage; rather i is proportional to the slope of the curve $v = v_{\text{max}} \sin \omega t$.

We therefore desire to find a mathematical function that represents the slope of a sine curve. To find this function, we plot a few values of the slope of the voltage curve in Fig. 21.7. At point 1, $v = v_{\text{max}}$, its slope is zero (the graph is horizontal), and i must be equal to zero. At point 2, $v = -v_{\text{max}}$, the slope of v is again 0, and so i at point 2 is also equal to zero. Also, since $i = C \ \Delta v/\Delta t$, the current is positive when v is increasing in time and negative when v is decreasing in time. Using this information we can plot the current curve shown in Fig. 21.7. It is a cosine curve, and so we have found that the slope of a sine curve is a cosine curve. We can therefore write the current in the capacitive circuit as a function of time in the following form:

$$i = i_{\text{max}} \cos \omega t$$

In this case it can be seen that *the current leads the voltage by 90°*. By this we mean that the current attains any particular value like that at point 3 in the diagram one-fourth of a cycle $(T/4)$ *ahead of* the voltage (point 4). This is equivalent to 90°, or $\pi/2$ rad.

In an ac circuit containing only capacitance, the current leads the voltage by 90°, or $\pi/2$ rad.

Dependence of X_C on Frequency

For a capacitor in an ac circuit, it is possible to find (using calculus) that

$$X_C = \frac{1}{\omega C} = \frac{1}{2\pi f C} \tag{21.7}$$

where C is the capacitance and f the frequency of the applied voltage in Fig. 21.6. This result agrees with our previous finding that $i = \Delta q/\Delta t = C\,\Delta v/\Delta t$. The current should therefore *increase* with both C and $\Delta v/\Delta t$, and so X_C should *decrease* as both C and $\Delta v/\Delta t$ increase.

For zero frequency (dc), $X_C \to \infty$, and no current flows, since a dc current cannot pass through a capacitor. At a very high frequency, on the other hand, X_C becomes very small and the current is limited very little by the capacitor in the circuit. A capacitor therefore acts as a high resistance at low frequencies and as a low resistance at high frequencies.

Experimental results confirm these predictions of Eq. (21.7). We therefore conclude:

In a capacitive ac circuit, the factor that is analogous to the resistance in a purely resistive circuit is the capacitive reactance $X_C = 1/(2\pi f C)$.

The fact that X_C is inversely proportional to both f and C is also consistent with the requirement that X_C have the dimensions of resistance, since $X_C = V/I$. The units of X_C are, from Eq. (21.7), those of $1/fC$:

$$\frac{1}{(1/s)(F)} = \frac{1}{(1/s)(C/V)} = \frac{V}{A} = \Omega$$

as required.

A summary of the major results of the preceding three sections is given in Table 21.2.

TABLE 21.2   Comparison of AC Circuits Containing R, L, and C

Circuit	$\dfrac{V}{I}$	Phase relations	Power loss in circuit (from Sec. 21.5)
R only	R	Current and voltage are *in phase*	I^2R
L only	$X_L = 2\pi f L$	Current *lags* voltage by 90°, or $\pi/2$ rad	0
C only	$X_C = \dfrac{1}{2\pi f C}$	Current *leads* voltage by 90°, or $\pi/2$ rad	0

Example 21.3

A 10.0-μF capacitor is connected to a 120-V, 60-Hz ac source. **(a)** What is the effective value of the current? **(b)** How does this current change if the frequency is increased to 6.00 MHz?

SOLUTION

(a) We have just seen that the capacitive reactance is

$$X_C = \frac{1}{2\pi fC} = \frac{1}{2\pi(60/s)(1.00 \times 10^{-5} \text{ F})} = 265 \ \Omega$$

$$I = \frac{V}{X_C} = \frac{120 \text{ V}}{265 \ \Omega} = \boxed{0.453 \text{ A}}$$

(b) In this case, where $f = 6.00 \times 10^6$ Hz,

$$X_C = \frac{1}{2\pi fC} = \frac{1}{2\pi(6.00 \times 10^6/s)(1.00 \times 10^{-5} \text{ F})}$$
$$= 2.65 \times 10^{-3} \ \Omega$$

$$I = \frac{V}{X_C} = \frac{120 \text{ V}}{2.65 \times 10^{-3} \ \Omega} = \boxed{4.53 \times 10^4 \text{ A}}$$

Notice the very large ac currents that can exist in a capacitive circuit at high frequencies, even if the applied voltage is relatively low, as in this case.

21.4 The *LC* Circuit

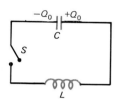

FIGURE 21.8 An *LC* circuit in series with a switch *S*. The capacitor is initially charged with a charge Q_0, and then the switch *S* is closed.

Let us next consider a circuit containing both an inductor L and a capacitor C in series with a switch, as in Fig. 21.8, usually referred to as an *LC* circuit. We assume initially that there is no resistance and no battery or other source of emf in the circuit. The capacitor is given an initial charge Q_0 so that the initial energy stored in the capacitor is $Q_0^2/2C$ [Eq. (17.19)]. We want to study the behavior of this circuit once the switch S is closed.

The best approach to this problem is to note that the total energy (the sum of electric and magnetic) is conserved over time, since there is no resistance present to dissipate energy. We can therefore write an expression for the constant energy in the system at any instant of time. Once charge flows from the capacitor, the energy initially stored in the capacitor is partially converted to energy in the magnetic field of the inductor, and we have, using Eqs. (17.19) and (20.11),

$$\frac{Q_0^2}{2C} = \frac{q^2}{2C} + \tfrac{1}{2}Li^2$$

where Q_0 is the original charge on the capacitor and q and i are the *instantaneous* values of the charge on the capacitor and the current. Then, solving for i, we have

$$i = \pm\sqrt{\frac{1}{LC}} \ \sqrt{Q_0^2 - q^2} \qquad\qquad \textbf{(21.8)}$$

This equation is very similar in form to the equation for the velocity of a particle of mass m moving in simple harmonic motion at the end of a spring, as given by Eq. (11.11):

$$v = \pm\sqrt{\frac{k}{m}} \ \sqrt{X_0^2 - x^2}$$

Just as $v = \Delta x/\Delta t$ in the mechanical case, so too $i = \Delta q/\Delta t$ in the electrical case. Also, just as X_0 is the maximum value of x, or the amplitude of the mechanical motion, so too Q_0 is the maximum value of the charge q on the capacitor.

Hence, just as the displacement was $x = X_0 \cos \omega t$ for the mechanical case, with $\omega = \sqrt{k/m}$, we have, in the electrical case,

$$q = Q_0 \cos \omega t = Q_0 \cos 2\pi ft$$

where $\qquad \omega = 2\pi f = \sqrt{\dfrac{1}{LC}} \quad$ or $\quad \boxed{f = \dfrac{1}{2\pi} \sqrt{\dfrac{1}{LC}}} \qquad\qquad \textbf{(21.9)}$

FIGURE 21.10 William Thomson, Lord Kelvin (1824–1907). Kelvin is famous for his work in thermodynamics (Kelvin temperature scale; second law) and electrodynamics and for his many scientific contributions to the first 1900-mi underwater transatlantic cable laid between Ireland and Newfoundland from 1850 to 1866. *(Courtesy of AIP Niels Bohr Library.)*

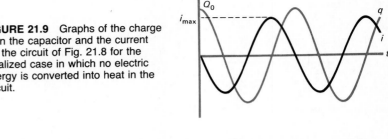

FIGURE 21.9 Graphs of the charge q on the capacitor and the current i in the circuit of Fig. 21.8 for the idealized case in which no electric energy is converted into heat in the circuit.

This is the *natural frequency* of oscillation of an *LC* circuit, and it is seen to depend only on the values of L and C. Note that the inertial factor m in the mechanical case corresponds to a similar inertial factor L in the electrical case, since the inductance acts to oppose any change in the current through the circuit. Also the elastic constant k, which determines the restoring force in the system, is replaced by $1/C$ in the electrical case, since the smaller C is, the more rapidly the capacitor must charge and discharge and hence the higher the frequency must be. Graphs of q and i in Fig. 21.9 parallel perfectly those for the displacement x and the velocity v in Fig. 11.6.

Equation (21.9) was first derived in 1853 by the British physicist William Thomson, or Lord Kelvin, as he was later called (Fig. 21.10). This simple equation makes it possible to calculate the natural frequency of different electrical systems, and it explains why we can tune radio transmitters and receivers to the same frequency by varying L and C. J. J. Thomson once referred to Eq. (21.9) as the basic equation of "modern wireless telegraphy, telephony, and broadcasting."

Energy Transfer in an *LC* Circuit

In an *LC* circuit the energy surges back and forth between the electric field ($\frac{1}{2}q^2/C$) and the magnetic field ($\frac{1}{2}Li^2$) in the same way that the energy changes back and forth from elastic potential ($\frac{1}{2}kx^2$) to kinetic energy ($\frac{1}{2}mv^2$) in the mechanical case. What happens is shown in Fig. 21.11. Initially the capacitor is charged to a voltage v_{max} and has a charge Q_0 as in (a). When the switch is closed, the capacitor discharge produces a current. The flow of charge increases until $q = 0$ and i takes its maximum value in Eq. (21.8) of $i_{max} = \sqrt{1/LC}\,\sqrt{Q_0^2 - 0} = \sqrt{1/LC}\,Q_0$, as in (b). At this point all the energy is in the magnetic field of the inductor. The current then continues but decreases in magnitude as the charge builds up until it has the same maximum value but the opposite sign to what it had originally, as in (c). The capacitor then discharges again and the current builds up to a maximum value in the opposite direction, as in (d). One complete cycle corresponds to q returning to its original value of $+Q_0$ and i returning to its initial value of zero, as shown in (e). In Fig. 21.11a, c, and e the energy is all in the electric field of the capacitor; in (b) and (d) it is all in the magnetic field of the inductor.

FIGURE 21.11 Oscillation of energy between electric and magnetic fields in the *LC* circuit of Fig. 21.8. In (e), at the end of one complete cycle or period, the circuit is back in its original condition (a).

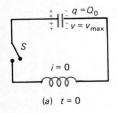

(a) $t = 0$

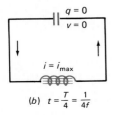

(b) $t = \dfrac{T}{4} = \dfrac{1}{4f}$

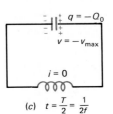

(c) $t = \dfrac{T}{2} = \dfrac{1}{2f}$

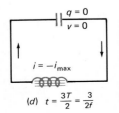

(d) $t = \dfrac{3T}{2} = \dfrac{3}{2f}$

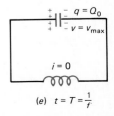

(e) $t = T = \dfrac{1}{f}$

TABLE 21.3   Comparison of Mechanical and Electrical Oscillations

Mechanical: SHM of a mass on a spring (*mk* "circuit")	Physical quantity	Electrical: *LC* circuit
m	Inertial factor	L
k	Elastic factor	$\dfrac{1}{C}$
$x(x = X_0 \cos \omega t)$	Displacement—Charge	$q(q = Q_0 \cos \omega t)$
$v = \dfrac{\Delta x}{\Delta t}$	Velocity—Current	$i = \dfrac{\Delta q}{\Delta t}$
$\left(v = \pm \sqrt{\dfrac{k}{m}} \sqrt{X_0^2 - x^2} \right)$		$\left(i = \pm \sqrt{\dfrac{1}{LC}} \sqrt{Q_0^2 - q^2} \right)$
$\omega = \sqrt{\dfrac{k}{m}}$	Angular frequency of oscillation	$\omega = \sqrt{\dfrac{1}{LC}}$
$f = \dfrac{1}{2\pi} \sqrt{\dfrac{k}{m}}$	Linear frequency (oscillations per second)	$f = \dfrac{1}{2\pi} \sqrt{\dfrac{1}{LC}}$
$\tfrac{1}{2}mv^2$	Kinetic energy—Magnetic energy	$\tfrac{1}{2}Li^2$
$\tfrac{1}{2}kx^2$	Elastic PE—Electric energy	$\dfrac{1}{2}\dfrac{1}{C}q^2$
$\tfrac{1}{2}kx^2 + \tfrac{1}{2}mv^2 = \tfrac{1}{2}kX_0^2$	Conservation of energy	$\dfrac{1}{2}\dfrac{1}{C}q^2 + \dfrac{1}{2}Li^2 = \dfrac{1}{2}\dfrac{1}{C}Q_0^2$

We see, therefore, that there is a very good analogy between the oscillations of an *mk* mechanical circuit and of an *LC* electric circuit. The same oscillation of energy at a frequency determined by L and C (or m and k) is found, and mechanical quantities in one case correspond to electrical quantities in the other. This is evident from Table 21.3, in which these correspondences are explicitly shown.

This striking parallelism illustrates the symmetry that connects many fields of physics. In the present case we can describe an *LC* circuit with mathematical equations that are similar in structure to those for the motion of a mass at the end of a spring. The solutions must therefore also be similar. Once we understand the behavior of the mechanical system, the behavior of the corresponding electric system becomes much easier to understand.

Effect of Resistance

If we add a series resistance R to an *LC* circuit, we obtain the *RLC* circuit of Fig. 21.12. This is equivalent to a *damped* simple harmonic oscillator. Just as frictional damping dissipates mechanical energy, so here resistance dissipates electric energy, turning it into thermal energy. Energy still oscillates back and forth between the electric field of the capacitor and the magnetic field of the inductor, but the total amount of electromagnetic energy decreases with each cycle. Hence the amplitude of the current decreases with time, as shown in Fig. 21.13.

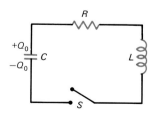

FIGURE 21.12 A circuit containing a resistance R in series with an inductance L and a capacitance C, that is, an *RLC* series circuit. The capacitor is charged with a charge Q_0, and then the switch S is closed.

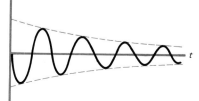

FIGURE 21.13 Current in an *RLC* series circuit as a function of time.

Any real *LC* circuit has some resistance. To produce an electrical oscillation of constant amplitude over a long period of time, therefore, energy must be constantly fed into the circuit to compensate for the energy being lost as heat in the resistor.

Example 21.4

A 10.0-μF capacitor is connected in series with an inductance of 100 mH and a switch. The capacitor is first charged to a voltage of 100 V. The charging unit is then removed, and the switch is closed. Determine **(a)** the natural frequency of oscillation of this circuit; **(b)** the maximum charge on the capacitor; **(c)** the maximum energy stored in the capacitor; **(d)** the maximum energy stored in the inductor; **(e)** the maximum current in the circuit.

SOLUTION

(a) $f = \dfrac{1}{2\pi}\sqrt{\dfrac{1}{LC}} = \dfrac{1}{2\pi}\dfrac{1}{\sqrt{LC}} = \dfrac{1}{2\pi}\dfrac{1}{\sqrt{(1.00 \times 10^{-5}\text{ F})(0.100\text{ H})}}$

$= \boxed{159\text{ Hz}}$

(b) The maximum charge on the capacitor is the initial charge

$Q_0 = CV = (1.00 \times 10^{-5}\text{ F})(100\text{ V}) = \boxed{1.00 \times 10^{-3}\text{ C}}$

(c) The maximum energy stored in the capacitor is

$\mathcal{W} = \tfrac{1}{2}CV^2 = \tfrac{1}{2}(1.00 \times 10^{-5}\text{ F})(100\text{ V})^2 = \boxed{0.0500\text{ J}}$

(d) From the energy-conservation principle the maximum energy stored in the inductor is also 0.500 J, since the energy oscillates back and forth between the electric field of the capacitor and the magnetic field of the inductor.

(e) The maximum energy in the inductor is $\tfrac{1}{2}Li^2_{max} = 0.0500$ J, and so

$i^2_{max} = \dfrac{2(0.500\text{ J})}{L} = \dfrac{0.100\text{ J}}{0.100\text{ H}} = 1.00\ \dfrac{\text{V}\cdot\text{A}\cdot\text{s}}{\text{V}\cdot\text{s/A}} = 1.00\text{ A}^2$

or $i_{max} = \boxed{1.00\text{ A}}$

21.5   *RLC* Series Circuit with Applied AC Voltage

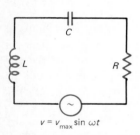

FIGURE 21.14 An *RLC* series circuit with an applied ac voltage.

We next consider the "driven" *RLC* series circuit of Fig. 21.14. An ac voltage of form $v = v_{max}\sin \omega t$ drives current through a resistor, inductor, and capacitor in series. We assume that the effective current I in the circuit is proportional to the effective voltage V applied, so

$$\boxed{V = ZI} \tag{21.10}$$

where the proportionality factor Z is called the *impedance* of the circuit and is some combination, still to be determined, of the resistances and reactances in the circuit. Equation (21.10) is the generalization of Ohm's law for ac circuits.

To find both the numerical and the phase relationships between current and voltage, we use the vector diagram of Fig. 21.15. We make use of the fact that the current is the same in all three elements of the circuit, and therefore we take the effective current I, drawn along the positive x axis, as the reference line with respect to which the phases of all the voltages are determined. V_R, V_L, and V_C are the effective voltages across R, L, and C, respectively. $V_R = IR$ is in phase with the current and hence is plotted along the positive x axis in the direction of I. On the other hand, the current I lags the voltage V_L across the inductor by 90°, and so V_L is drawn along the positive y axis; and the current leads the voltage across the capacitor by 90°, and so V_C is drawn along the negative y axis, as shown in the figure. (Note that all angles are measured *counterclockwise* from the positive x axis.)

The effective voltage V applied to the circuit is then the sum of V_R, V_L, and

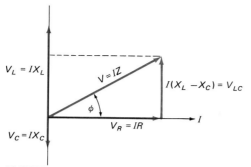

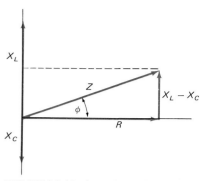

FIGURE 21.15 Relationship of the current and the voltages across the resistance, capacitance, and inductance in a driven *RLC* series circuit. ϕ is the phase angle between the applied voltage V and the current I.

FIGURE 21.16 Impedance diagram for *RLC* circuit: $Z^2 = R^2 + (X_L - X_C)^2$. The angle ϕ is the phase angle.

V_C. Since these three have different phases, they must be added as *vectors* to take account of their phase relationships. From Fig. 21.15 we have for the magnitude of the resultant of these three vectors,

$$V^2 = V_R^2 + (V_L - V_C)^2 \quad \text{or} \quad I^2 Z^2 = I^2 R^2 + I^2 (X_L - X_C)^2$$

from which

$$Z = \sqrt{R^2 + (X_L - X_C)^2} \tag{21.11}$$

Hence the impedance Z in any series *RLC* circuit is the square root of the sum of the squares of the total resistance and the total reactance in the circuit, as shown in Fig. 21.16. The units of Z are ohms, since X_L, X_C, and R are all in ohms.

The effective (rms) voltage and current in an RLC series circuit are related by the equation $V = ZI$, where the impedance $Z = \sqrt{R^2 + (X_L - X_C)^2}$.

Power Factor

From Fig. 21.16,

$$\cos \phi = \frac{R}{Z} \tag{21.12}$$

where ϕ is the *phase angle* between the applied voltage V and the current I. The rate at which energy is converted into joule heat is then

$$P = I^2 R = I^2 (Z \cos \phi) = I(IZ \cos \phi)$$

and, from Eq. (21.10),

$$P = IV \cos \phi \tag{21.13}$$

Here $\cos \phi$ is called the *power factor* of the *RLC* circuit.

For a pure resistance we have $Z = R$, $\cos \phi = R/Z = 1$, and $P = IV$, as in the dc case [Eq. (18.11)]. For a pure inductance or a pure capacitance, on the other hand, $\phi = \pm 90°$, $\cos \phi = 0$, and so $P = IV(0) = 0$. Hence *no power is dissipated as heat in a pure inductance or capacitance*. Electric energy simply flows in and out of the capacitor as it charges and discharges, and magnetic energy flows in and out of the coil as the current builds up and decays, as discussed in Sec. 21.4.

> *In an RLC series circuit, all the power delivered is dissipated as I^2R heat in the resistor R.*

These results are included in the last column of Table 21.2.

Example 21.5

A 1000-Ω resistor is connected in series with a 200-mH inductor and a 3.00×10^{-7} F capacitor. The applied voltage is $v = 170 \sin 2\pi ft$, where $f = 2000$ Hz. Find **(a)** the impedance of the circuit; **(b)** the effective current in the circuit; **(c)** the effective voltages across the resistor, capacitor, and inductor; **(d)** the power factor of the circuit; **(e)** the power dissipated by the circuit. **(f)** Explain how the voltages across R, L, and C add to give the voltage applied to the circuit.

SOLUTION

(a) $Z^2 = R^2 + (X_L - X_C)^2$

$$= (1.00 \times 10^3\ \Omega)^2 + \left[2\pi(2000)(0.200) - \frac{1}{2\pi(2000)(3.00 \times 10^{-7})} \right]^2 \Omega^2$$

$$= 6.04 \times 10^6\ \Omega^2$$

$$Z = \boxed{2.46 \times 10^3\ \Omega}$$

(b) Since the maximum voltage is 170 V, the effective voltage is 0.707(170 V) = 120 V. Then

$$I = \frac{V}{Z} = \frac{120\ \text{V}}{2.46 \times 10^3\ \Omega} = \boxed{48.8\ \text{mA}}$$

(c) $V_R = IR = (48.8\ \text{mA})(1000\ \Omega) = \boxed{48.8\ \text{V}}$

$V_L = IX_L = (48.8\ \text{mA})(2\pi)(2000/\text{s})(2.00 \times 10^{-3}\ \text{H})$

$$= \boxed{123\ \text{V}}$$

$$V_C = IX_C = \frac{I}{2\pi fC} = \frac{48.8\ \text{mA}}{2\pi(2000/\text{s})(3.00 \times 10^{-7}\ \text{F})}$$

$$= \boxed{12.9\ \text{V}}$$

(d) $\cos\phi = \frac{R}{Z} = \frac{1000\ \Omega}{2.46 \times 10^3\ \Omega} = \boxed{0.407}$

ϕ is therefore 66.0°. Since X_L is greater than X_C, the current *lags* the voltage by 66° in this circuit.

(e) $P = VI\cos\phi = (120\ \text{V})(48.8 \times 10^{-3}\ \text{A})(0.407)$

$$= \boxed{2.38\ \text{W}}$$

The same result can be obtained from $P = I^2R = (48.8 \times 10^{-3}\ \text{A})^2(1000\ \Omega) = 2.38$ W.

(f) The result of direct addition would be V = (49 + 123 + 13) = 185 V, which is more than the effective voltage applied, and so is clearly wrong. The reason is that in adding the voltages, *we must take account of their phase differences*. We then obtain

$$V^2 = V_R^2 + (V_L - V_C)^2 = (48.8\ \text{V})^2 + (123 - 13)^2\ \text{V}^2$$

$$= 14.5 \times 10^3\ \text{V}^2$$

and so $V = 120$ V

which is equal to the effective applied voltage, as it must. This is shown in Fig. 21.17.

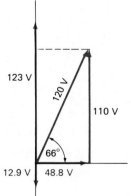

FIGURE 21.17

21.6 Series Resonance

We have seen that the phase angle ϕ is the angle between the current and the voltage in an ac circuit, where, from Fig. 21.15,

$$\tan\phi = I\frac{X_L - X_C}{IR} = \frac{X_L - X_C}{R} \tag{21.14}$$

If $X_L = X_C$, $\tan \phi = 0$, $\phi = 0$, and the current and the voltage are *in phase*, as in a purely resistive circuit. When $X_L = X_C$, we have for the impedance [Eq. (21.11)],

$$Z = \sqrt{R^2 + (X_L - X_C)^2} = R$$

Hence $V = IR$ and the current takes on its maximum value, which is limited only by the resistance in the circuit. In this case the circuit is said to be in *series resonance*.

Series resonance: The condition in an *RLC* series circuit in which the inductive reactance X_L and the capacitive reactance X_C are equal; the current is then a maximum and in phase with the applied voltage.

Figure 21.18 shows the behavior of X_L, X_C, and I as a function of the frequency of the voltage applied to the circuit, where the amplitude of the voltage is constant with frequency. When $X_L = X_C$, the current I is a maximum.

Since the resonant frequency is that frequency at which X_L is equal to X_C, we have

$$2\pi f_{\text{res}} L = \frac{1}{2\pi f_{\text{res}} C} \qquad \text{or} \qquad \boxed{f_{\text{res}} = \frac{1}{2\pi} \sqrt{\frac{1}{LC}}} \qquad \textbf{(21.15)}$$

This is, of course, the equation previously obtained for the natural frequency of a simple *LC* circuit [Eq. (21.9)]. *Resonance occurs when the frequency of the applied voltage is equal to this natural frequency.*

The quality factor Q (or Q factor) of a resonant circuit is defined as the ratio of the voltage across the capacitor (or inductor) at resonance to the voltage across the resistor at resonance, V_C/V_R. The larger the Q factor, the sharper the resonance curve, and therefore the sharper the tuning. It can be shown (Prob. 21.52) that the quality factor may be expressed as

$$Q = \frac{1}{R} \left(\frac{L}{C} \right)^{1/2} \qquad \textbf{(21.16)}$$

so that Q varies inversely with the resistance R. Since Q is the ratio of two voltages, it is a pure number without dimensions or units.

The dependence of the resonance curves on R is shown in Fig. 21.19. The larger R, the greater the energy loss per cycle and the smaller the quality factor.

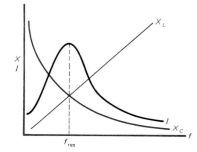

FIGURE 21.18 Capacitive reactance X_C, inductive reactance X_L, and effective current I in a driven *RLC* series circuit as a function of the frequency of the applied voltage. Notice how X_C and X_L are equal and therefore cancel (since they have opposite phases) at the resonant frequency f_{res}, at which the current attains its maximum value.

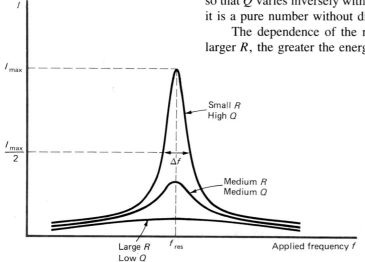

FIGURE 21.19 Series resonance curves for three circuits with the same values of L and C but different values of R. The Q factor, and therefore the sharpness of resonance, vary inversely with R. The higher the Q factor, the smaller the frequency width Δf at half height for the resonance curve.

Importance of Resonance

Resonance phenomena occur in all branches of physics, from mechanics and classical electromagnetism to lasers and elementary particle physics.

The mechanical analogue of an *RLC* circuit is a mass vibrating at the end of a spring and damped by friction or some other resistive force. If this mass is driven by an external force at a varying frequency, there will be one frequency at which the motion of the mass will build up to very large amplitudes (Sec. 11.4). This is the natural mechanical frequency of the system. A common example of such behavior is the pushing of a child on a swing. If large amplitudes are to develop, the pushing frequency must be equal to the natural frequency of the swing, and the pushes must be in phase with the motion of the swing.

A very practical use of electric resonance occurs in radio and TV sets, where series resonance is used to select a desired station or channel from the multitude of frequencies arriving at the scene from different transmitters. The tuning capacitance in an *RLC* series circuit is varied until the antenna circuit is in series resonance with the carrier frequency of the desired broadcasting station. At this frequency, the emf induced in the antenna by the incident wave causes a large current in the antenna circuit. This current can then produce an audio or video signal. Other stations or channels have frequencies not in resonance with the antenna circuit and hence produce negligibly small currents in that circuit. The selectivity of the receiver depends on the *Q* factor of the receiving circuit.

Danger of Resonance Circuits

It is important to realize that in a series circuit at resonance the voltage across the capacitor and the inductor individually may be many times larger than the 120-V line voltage or other applied voltage. This is true even though the total voltage across the combination of *L* and *C* together is zero because of the opposite phases of the voltages across each. This can present dangers both to the circuit elements and to anyone foolhardy enough to touch the leads to either the inductor or the capacitor. That very large voltages can arise in this way when a circuit is at resonance is shown in Example 21.6.

Example 21.6

An *RLC* series circuit is connected to a variable-frequency ac generator whose effective voltage output is maintained at 120 V. The circuit elements are $L = 100$ mH, $C = 20.0 \times 10^{-9}$ F, $R = 50.0$ Ω.

What are **(a)** the resonant frequency of the circuit; **(b)** the current in the circuit at resonance; **(c)** the voltage across the capacitor at resonance; **(d)** the voltage across the inductor at resonance; **(e)** the voltage across the resistor at resonance; **(f)** the *Q* factor for this circuit?

SOLUTION

(a) For resonance,

$$f_{res} = \frac{1}{2\pi}\sqrt{\frac{1}{LC}} = \frac{1}{2\pi}\frac{1}{\sqrt{(0.100\text{ H})(20.0 \times 10^{-9}\text{ F})}}$$

$$= \boxed{3.56 \times 10^3 \text{ Hz}}$$

(b) The effective current at resonance is

$$I = \frac{V}{R} = \frac{120\text{ V}}{50.0\text{ }\Omega} = \boxed{2.40\text{ A}}$$

(c) $V_C = IX_C = \dfrac{I}{2\pi fC}$

$$= \frac{2.40\text{ A}}{2\pi(3.56 \times 10^3/\text{s})(20.0 \times 10^{-9}\text{ F})}$$

$$= \boxed{5.36 \times 10^3 \text{ V}}$$

(d) $V_L = IX_L = (2.40\text{ A})(2\pi)(3.56 \times 10^3/\text{s})(0.100\text{ H})$

$$= \boxed{5.36 \times 10^3 \text{ V}}$$

This is the same as V_C, indicating that the circuit is indeed resonant at 3.56×10^3 Hz.

(e) $V_R = IR = (2.40\text{ A})(50.0\text{ }\Omega) = 120$ V

as expected.

Note in (c) and (d) that the voltages across the capacitor and inductor are each over 5000 V, which is almost 50 times the applied voltage! Since these voltages are always opposite in phase and since they have the same value at resonance, the voltage across the capacitor and inductor in series at resonance is zero, and all the 120 V applied is across the resistor. Despite this fact, the 5000 V across either capacitor or inductor by itself is enough to kill a person under the right (or wrong!) circumstances.

(f)   By definition $Q = V_C/V_R = 5.36 \times 10^3$ V/120 V $= 45$. The same result can be obtained from Eq. (21.16). Note that high-Q circuits are the most dangerous because of the high voltages across L and C individually at resonance. These voltages are Q times the applied voltage.

21.7   Transformers

Almost 500,000 km of overhead high-voltage electric cables line the United States. Any transmission of electric power over electric cables loses energy in the form of Joule heat, $W = I^2Rt$, where I is the effective ac current and R is the resistance of the line. Since resistance is directly proportional to the length of the line, the longer the power line, the greater the energy loss. Consequently electric power transmission over long distances can be very wasteful; in some cases 10 percent of the energy transmitted is lost. This loss can be avoided by transmitting the power at a high voltage and thus reducing the current I and hence the I^2R losses. If the voltage is increased by a factor of 10^3, for example, the power loss is decreased by a factor of 10^6. It is for this reason that many recently installed power lines in the United States carry power at voltages as high as 765,000 V.

The advantage of ac over dc voltages in power transmission is mainly the ease with which ac voltages can be "stepped up" or "stepped down." In electric power distribution systems it is desirable at both the generating and the receiving end to deal with relatively low voltages for safety and convenience. No one wants a child's electric train to operate off 100,000 V. On the other hand, power companies, for economic reasons, want to transmit power at high voltages. They therefore step up (i.e., increase) the voltage at the generator, transmit this high-voltage power over the electric lines, and then step down the voltage at the house or factory where the power is to be used. This is done by a *transformer*.

Transformer: An electric device that uses electromagnetic induction to step up voltages and simultaneously to step down currents, or vice versa, in an ac circuit.

A simplified version of a transformer is shown in Fig. 21.20. It consists of two coils of wire, a primary coil (1) and a secondary coil (2), wound on the same iron core. The core concentrates the magnetic flux and guarantees that it is the same in the primary, which has N_1 turns of wire, as in the secondary, which has N_2 turns of wire.

In Fig. 21.20 the primary coil of the transformer is connected to an ac generator whose voltage output is $v_1 = v_{max} \sin \omega t$. Initially we assume that switch

FIGURE 21.20 A simplified version of a transformer. For such a transformer, $V_2/V_1 = N_2/N_1 = I_1/I_2$.

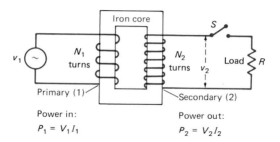

Power in:
$P_1 = V_1 I_1$

Power out:
$P_2 = V_2 I_2$

S is open, so there is no current in the secondary circuit. The changing current in the primary produces a changing magnetic flux $\Delta \Phi_B$ with time in the primary. This changing magnetic flux induces an emf in the primary. By Kirchhoff's loop rule,

$$v_1 - L \frac{\Delta i}{\Delta t} = 0$$

and so, from Eq. (20.9),

$$v_1 = L \frac{\Delta i}{\Delta t} = N_1 \frac{\Delta \Phi_B}{\Delta t}$$

But $\Delta \Phi_B / \Delta t$ is the same, by assumption, in both the primary and secondary coils. The induced emf in the secondary, which has N_2 turns, is therefore obtained from

$$\frac{\Delta \Phi_B}{\Delta t} = \frac{v_1}{N_1} = \frac{v_2}{N_2} \quad \text{or} \quad \frac{v_2}{v_1} = \frac{N_2}{N_1}$$

If we take rms averages of v_1 and v_2, this becomes

$$\boxed{\frac{V_2}{V_1} = \frac{N_2}{N_1}} \tag{21.17}$$

N_2/N_1 is called the *turns ratio* of the transformer. When $N_2 > N_1$, the transformer is a *step-up transformer*. When $N_2 < N_1$, it is a *step-down transformer*, where *step-up* and *step-down* refer to the *voltage* change. The great advantage of using a transformer is that, without any moving parts, we can transform voltages merely by selecting the right values for N_1 and N_2. Transformers also consume very little energy. The ease with which transformers operate on alternating current, and the lack of any equivalent device for dc operation, are prime reasons for the use of alternating rather than direct current in household and industrial applications.

Delivery of Power

We now close the switch S in Fig. 21.20 and deliver power to a load resistance R. Then the power dissipated in the load is $I_2^2 R$. An equal amount of power must be fed into the primary from the generator so that, if there are no losses in the system, we have

$$P_1 = P_2 \quad \text{or} \quad V_1 I_1 = V_2 I_2$$

and so

Transformer equation: $\quad \boxed{\dfrac{I_1}{I_2} = \dfrac{V_2}{V_1} = \dfrac{N_2}{N_1}} \tag{21.18}$

The transformer therefore *steps down the current* at the same time and by the same ratio it *steps up the voltage*, and vice versa.

Transformers are used not merely in the transmission of electric power but for a variety of other useful purposes. They are, for example, used to step down 120 V to 6 V for doorbells in homes, and to step up 120-V power to 10^4 V to accelerate the electron beam in television sets.

Impedance Matching

Another very important use of transformers is as impedance-matching devices. It can be shown (using calculus) that for maximum power delivery from a source to a load, the load must have the same resistance as the internal resistance of the source. This can be accomplished by using a transformer to match the two resistances. If the load resistance is R_2, then $R_2 = V_2/I_2$. But, if a transformer is inserted between the source and the load, from the transformer equation we have

$$V_2 = \frac{N_2 V_1}{N_1} \quad \text{and} \quad I_2 = \frac{N_1 I_1}{N_2}$$

so that

$$R_2 = \frac{(N_2/N_1)V_1}{(N_1/N_2)I_1} = \frac{V_1}{(N_1/N_2)^2 I_1}$$

and so

$$I_1 = \frac{V_1}{(N_1/N_2)^2 R_2} = \frac{V_1}{R'}$$

Hence the load resistance R_2 appears to the source to be an effective load resistance

$$R' = \left(\frac{N_1}{N_2}\right)^2 R_2 \tag{21.19}$$

where N_2/N_1 is the turns ratio of the transformer. By choosing the proper turns ratio, therefore, the effective load resistance R' can be made equal to the internal resistance R_1 of the source. This is called *impedance matching*, since in more complicated cases the load may include both reactances and resistances.

Example 21.7

An electric generating plant produces electric energy at a rate of 1 GW and a voltage of 12 kV. **(a)** If it is desired to transmit this power at 720 kV, what must be the turns ratio of the step-up transformer? **(b)** What current would be sent out over the power lines if the transmission were at 12 kV? **(c)** What current would be transmitted if the voltage were stepped up to 720 kV? **(d)** What would be the ratio of the power lost as heat in (b) and (c)?

SOLUTION

(a) $\dfrac{N_2}{N_1} = \dfrac{V_2}{V_1} = \dfrac{720 \text{ kV}}{12 \text{ kV}} = \boxed{60/1}$

(b) $I_1 = \dfrac{P}{V_1} = \dfrac{1.0 \times 10^9 \text{ W}}{12 \text{ kV}} = \boxed{8.3 \times 10^4 \text{ A}}$

(c) Since the power remains unchanged because we assume there are no losses in the transformer,

$$I_2 = \frac{P}{V_2} = \frac{1.0 \times 10^9 \text{ W}}{7.2 \times 10^5 \text{ V}} = \boxed{1.4 \times 10^3 \text{ A}}$$

(d) The ratio of the power lost in the two cases would be

$$\frac{P_1}{P_2} = \frac{I_1^2 R}{I_2^2 R} = \frac{(8.3 \times 10^4 \text{ A})^2}{(1.4 \times 10^3 \text{ A})^2} = (60)^2 = \boxed{3.6 \times 10^3}$$

Hence more than 3000 times the power lost as heat at 720 kV would be lost at 12 kV.

Example 21.8

An audio amplifier with an internal impedance of 2.0 kΩ is used to drive a loudspeaker with an impedance of 5.0 Ω. What must the turns ratio of the transformer in Fig. 21.21 be to match the impedances?

FIGURE 21.21 Using an impedance-matching transformer to match the output of a high-impedance audio amplifier to a low-impedance speaker system.

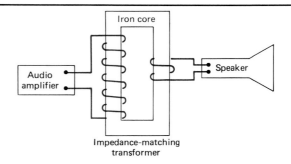

SOLUTION

We want the transformer to convert the impedance of the loudspeaker from $R_2 = 5.0 \ \Omega$ to an effective resistance, as seen by the audio amplifier, of $R' = 2.0 \ k\Omega$, where, from Eq. (21.19),

$$R' = \left(\frac{N_1}{N_2}\right)^2 R_2 \quad \text{or} \quad \left(\frac{N_1}{N_2}\right)^2 = \frac{R'}{R_2} = \frac{2.0 \ k\Omega}{5.0 \ \Omega} = 4.0 \times 10^2$$

and so

$$\frac{N_1}{N_2} = \sqrt{4.0 \times 10^2} = \boxed{20}$$

Hence the primary of the transformer must have 20 turns for each turn of the secondary to match impedances and deliver maximum power to the loudspeaker.

21.8 Practical Aspects of Electric Power

Figure 21.22 shows how transformers are typically used in the transmission of electric power. A generating plant produces ac power at 12,000 V. Before leaving the plant, this voltage is stepped up to 240,000 V by using a 20/1 step-up transformer. The power is then transmitted at 240,000 V over high-voltage power lines to a substation where a second transformer steps down the voltage to 12,000 V. The power is then distributed at this voltage to sites where it is needed. At a large factory or in residential areas additional smaller transformers step down the voltage again by a ratio of 50/1 to yield the 240 V normally supplied to homes and factories.

This transmission process avoids the large power losses that would be incurred in the transmission of electric power over large distances at low voltage (and hence high current) levels, and at the same time provides the large amount of current needed at 120 or 240 V ac to run most electric machines and appliances. Another advantage of such a procedure is that it keeps excessively high voltages out of homes and factories, where they could be dangerous to human life.

Home Electricity

The power company uses a three-wire power line to supply the needed power at 120 and 240 V to consumers. One of these wires is a ground wire. The other two each carry 120 V ac with respect to ground but are adjusted in phase so that the effective voltage difference between the two "hot" or "live" wires is 240 V. Large appliances and machinery are connected to this 240-V supply, whereas smaller home appliances are connected to the 120 V between one of these wires and the ground wire. In modern homes a number of independent 120-V circuits each carry a maximum of 15 or 20 A of current. An individual circuit may include three or four double outlets in the walls of a room. Each time we plug in a new light or appliance we draw more current from that particular circuit, for the appliances are all connected in parallel. In most cases little danger exists of exceeding the 20-A capacity of a circuit, since we would have to plug twenty-four 100-W bulbs into the same 120-V circuit to exceed the capacity of the line. When a number of electric heating appliances are being used, however, as in the kitchen, it is quite easy to exceed this 20-A capacity.

Suppose, for example, that we have the situation shown in Fig. 21.23. We have two double outlets above a kitchen counter, both part of the same circuit. We plug into the outlets an electric toaster requiring 800 W of power, an electric frying pan rated at 1200 W, and an automatic coffee maker rated at 1000 W. The toaster draws a current of $I = P/V = (800 \ W)(120 \ V) = 6.7 \ A$; the frying pan $(1200 \ W)/(120 \ V) = 10 \ A$; and the coffee maker $(1000 \ W)/(120 \ V) = 8.3 \ A$. Since the wiring in the circuit is meant to carry only 20 A, we have overloaded the circuit, and there is danger of overheating the wires and starting a fire. To prevent this, safety devices must be included in the circuit.

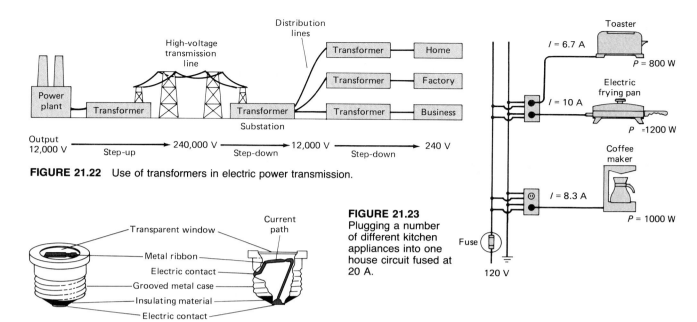

FIGURE 21.22 Use of transformers in electric power transmission.

FIGURE 21.23 Plugging a number of different kitchen appliances into one house circuit fused at 20 A.

FIGURE 21.24 An electric fuse. The current path is from the bottom tip of the fuse through a fine metal ribbon to the metal side of the fuse body to complete the circuit. An overload causes the ribbon to melt and open the circuit.

Fuses, Circuit Breakers, and GFIs Building codes now require protection of each house circuit by fuses or circuit breakers. A *fuse* is merely a screw-in device that puts a piece of metal ribbon in series with the circuit, as in Fig. 21.24. If the current is more than 20 A, the ribbon melts and cuts off all further current. If one of the appliances is removed from the line and a new fuse put in place, everything should again function normally.

In most modern homes *circuit breakers* are used in place of fuses. A circuit breaker is also wired in series with the rest of the circuit. Two kinds are in common use—a thermal and a magnetic device. The thermal circuit breaker is a bimetallic strip (see Sec. 13.5) that becomes so hot when it carries excessive current that it bends enough to open the circuit. In a magnetic circuit breaker, excessive current produces a magnetic field large enough to pull open a switch and interrupt the current. The advantage of circuit breakers over fuses is that they can be reset easily and need not be replaced.

Another way that a fuse may be blown or a circuit breaker tripped is if a *short circuit* occurs. A short circuit occurs when two wires, one at a high potential and the other at ground potential, cross or touch. As a result, the current path to ground has a very low resistance (since the wires have low resistance) and a very large current is drawn from the line. A short circuit may occur either in the house wiring itself or in an appliance plugged into the wiring.

A short circuit is indicated if fuses or circuit breakers are frequently blown on the same circuit. All appliances should then be removed from the line and plugged in again one at a time. The one that blows the fuse is clearly the culprit and needs repair. If the circuit itself is defective, the fuse will blow even when nothing is plugged into the line.

Some electric circuits now contain a *ground fault interrupter* (GFI) as a protection against electric shocks. This device detects small leakage currents in electric equipment. Such leakage currents indicate a present or incipient short circuit that could be a danger to life. As soon as the GFI detects a leakage current above the safety limit (about 5 mA), it automatically throws a switch and opens the circuit.

Such devices have saved many lives by alerting people to small leakage currents before they could become lethal.

Grounding of Appliances   Most 120-V appliances now have a third (round) prong on the power plug. This connects the metal case of the appliance to ground. If the live wire inside the appliance should by accident touch the case of the appliance, a short circuit will result and a fuse will blow (as in Fig. 21.25). If the ground wire were absent, as in older two-pronged plugs, once the high-voltage wire touched the case of the appliance the case would be at 120 V above ground. If a person touched it, there would be 120 V between points A and B in the figure. This voltage could drive a large current through the person's body, producing an electric shock that could have fatal consequences.

FIGURE 21.25   The use of a three-pronged plug on an electric appliance to prevent electric shocks. The plug grounds the case of the appliance so that no shock will be received from the case even if high-voltage wires inside the appliance should make contact with the case.

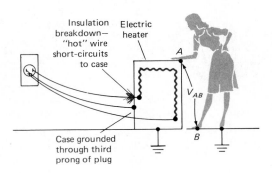

Electric Shocks

The most important factors in determining the consequences of an electric shock are the amount of current through the body and the length of time it persists.

Currents of less than 5 mA are usually considered safe, although they give a slight shock. Currents of 50 mA cause more severe shocks to the body, and currents above 100 mA can cause ventricular fibrillation, i.e., a fluttering of the heart that cuts off the blood supply to the brain and often results in death. Currents in the ampere range both stop the heart and produce severe external burns on the body.

Of course, the actual current through the body depends on the voltage across the body and the resistance of the body between the voltage source and ground. In most household accidents the voltage arises from an electric fault in an appliance that puts 120 V across the body. Contact with live power lines may mean voltages thousands of times greater than this.

Inside the human body the electric resistance from head to toe is between 500 and 1000 Ω. This is increased greatly by the skin resistance where the body touches the voltage source and by the resistance of the path from the feet or other part of the body to ground. Dry skin may have a resistance of 10^6 Ω, but this can be reduced by a factor of 100 or more by moisture. This is the reason that electrical accidents in a bathtub or swimming pool can be fatal, since both the skin resistance and the resistance to ground are greatly reduced by the presence of water.

If low skin resistance and a good ground connection reduces the overall body resistance to, say, 2000 Ω, then the expected current is 120 V/2000 Ω = 60 mA. This is dangerously close to the 100 mA that can cause the heart to spasm.

To conclude, let us return to Edison's statement (referred to in our introduction) that alternating current is more dangerous than direct current. The fact of the matter is that an alternating current of effective value 100 mA through the human body is neither more nor less dangerous than 100 mA dc. Both produce the same heating effects and the same destructive impact on the body. The only possibility of increased danger from alternating current would be if the frequency of the ac voltage produced resonance with some natural frequency in the human body. A frequency of 60 Hz is sufficiently different from all known natural frequencies of the organs or molecules in the body to make this an unlikely possibility.

Summary: Important Definitions and Equations

AC currents and voltages

Instantaneous currents and voltages of the form

$$i = i_{max} \sin \omega t = i_{max} \sin 2\pi f t$$

$$v = v_{max} \sin \omega t = v_{max} \sin 2\pi f t$$

where f is the frequency of the current or voltage.

Effective (or rms) values:

$$I = 0.707 i_{max} \tag{21.1}$$

$$V = 0.707 v_{max} \tag{21.2}$$

Resistive circuit (R only): $V = IR$; I and V are in phase.

Inductive circuit (L only):

$$V = IX_L; I \text{ lags } V \text{ by } 90°, \text{ or } \pi/2 \text{ rad.} \tag{21.3}$$

Inductive reactance: $X_L = \omega L = 2\pi f L \tag{21.5}$

Capacitive circuit (C only):

$$V = IX_C; I \text{ leads } V \text{ by } 90°, \text{ or } \pi/2 \text{ rad.} \tag{21.6}$$

Capacitive reactance: $X_C = \dfrac{1}{\omega C} = \dfrac{1}{2\pi f C} \tag{21.7}$

LC circuit

Natural frequency: $f = \dfrac{\omega}{2\pi} = \dfrac{1}{2\pi}\sqrt{\dfrac{1}{LC}} \tag{21.9}$

RLC series circuit with applied ac voltage

$$V = ZI \tag{21.10}$$

where impedance $Z = \sqrt{R^2 + (X_L - X_C)^2} \tag{21.11}$

Power factor: $\cos \phi = R/Z \tag{21.12}$

Power delivered to an RLC circuit:

$$P = IV \cos \phi \tag{21.13}$$

In an *RLC* series circuit all the power delivered is dissipated as I^2R heat in the resistance, none in the inductor or capacitor.

Series resonance: That condition in an *RLC* series circuit when X_L and X_C are equal; the current is then a maximum and in phase with the applied voltage. This occurs when the applied frequency is equal to the natural frequency of the *RLC* circuit, $f = (1/2\pi)\sqrt{1/LC}$.

Quality factor: The ratio of the voltage across the capacitor (or inductor) to the voltage across the resistor at resonance.

$$Q = \frac{1}{R}\left(\frac{L}{C}\right)^{1/2} \tag{21.16}$$

Transformer: An electric device that uses electromagnetic induction to step up voltages and simultaneously step down currents, or vice versa, in an ac circuit.

Transformer equation: $\dfrac{I_1}{I_2} = \dfrac{V_2}{V_1} = \dfrac{N_2}{N_1} \tag{21.18}$

Questions

1 Would you expect ac voltages and currents to be more dangerous than dc voltages and currents of the same effective value? Why?

2 Can you suggest how to wind an electric resistor so that it has no inductance?

3 Why would you expect a coil with inductance L and no resistance to be an idealization, never achievable in the real world?

4 Why does a capacitor behave like a high resistance at low frequencies and like a low resistance at high frequencies?

5 An ac circuit contains an inductor that has a resistance R. Is there any electric power lost in the inductor? If so, how much?

6 What are the maximum and minimum possible values of the power factor for an ac circuit? Is the power factor characteristic of the applied voltage, or of the load, or of some combination of the two?

7 A doorbell transformer is designed for a primary voltage of 120 V and a secondary voltage of 6 V. Suppose by mistake you mix up the primary and secondary connections when you install the transformer. What would you expect to happen?

8 Why does not a transformer violate the first law of thermodynamics, since it enables a voltage to be increased to any desired amount?

9 Will the length of a copper cable carrying power-line currents at high voltages increase or decrease as the temperature rises? Will its resistance increase or decrease as the temperature rises? What conclusion can you draw from these facts?

10 Why is 240 V ac instead of 120 V used for clothes dryers, electric stoves, and other appliances that draw large electric currents?

11 Point out some practical jobs for which direct current is absolutely essential and others for which either direct or alternating current will do.

12 When an added load is attached to the output of a transformer, e.g., an air conditioner, how is the primary able to adjust so that it always draws from the power line the exact amount of power needed by the load?

13 The advice is often given to technicians or physicists working with high-voltage electronic circuits that they should always keep one hand in a pocket when they are taking measurements in the vicinity of high-voltage wires. Why is this good advice?

Multiple-Choice and Simple Exercises

21.1 A circuit contains only a 50-Ω resistor. An ac voltage of peak value 340 V and frequency 60 Hz is applied to this circuit. The rms current is:

(a) 6.8 A   (b) 3.4 A   (c) 0.15 A
(d) 0.21 A   (e) 4.8 A

21.2 In an ac circuit containing only an inductor:
(a) The current lags the voltage by 90°.
(b) The current lags the voltage by 180°.
(c) The voltage lags the current by 90°.
(d) The voltage lags the current by 180°.
(e) The current and voltage are in phase.

21.3 A circuit contains only a 0.20-H inductor. An ac voltage of peak value 170 V and frequency 60 Hz is applied to the circuit. When the ac voltage is 170 V, the ac current is:
(a) 0   (b) 1.6 A   (c) 2.3 A   (d) −1.6 A
(e) −2.3 A

21.4 An ac voltage of peak value 85 V and frequency 50 Hz is applied to a 20-μF capacitor. The rms current is:
(a) 0.54 A   (b) 0.38 A   (c) −0.54 A
(d) −0.38 A   (e) 0

21.5 A circuit containing resistance only is connected to a 110-V, 60-Hz line. If a capacitor is connected in series with the resistance, the current:
(a) Lags the emf   (b) Is in phase with the emf
(c) Is decreased   (d) Is increased
(e) Remains constant

21.6 If the frequency of the voltage source applied to a purely capacitive circuit is increased by 5 orders of magnitude, the current through the circuit:
(a) Increases by 10^5 (b) Decreases by 10^5
(c) Increases by 10^{10} (d) Decreases by 10^{10}
(e) Remains the same

21.7 An LC circuit oscillates at a frequency of 1000 Hz. If it is desired to raise this frequency by three orders of magnitude (i.e., by a factor of 1000), the values of the inductance L and the capacitance C must be changed as follows:
(a) Both L and C decreased by three orders of magnitude
(b) Both L and C decreased by six orders of magnitude
(c) Either L or C decreased by three orders of magnitude
(d) Either L or C increased by six orders of magnitude
(e) None of the above

21.8 An LC circuit with $L = 0.50$ H and $C = 0.20$ μF has an initial charge on the capacitor of 100 μC. The maximum current is:
(a) 1.0×10^2 μA   (b) 0.32 A   (c) 50 A
(d) 1.0 A   (e) 0.01 A

21.9 An RLC series circuit has $L = 10$ mH, $C = 1.0$ μF, $R = 20$ Ω. A voltage is applied at 1000 Hz. The power factor of the circuit is:
(a) 1.0   (b) 0.090   (c) 0.90   (d) 0.20
(e) 0.020

21.10 It is desired to step 120 V up to 96.0 kV. The turns ratio N_2/N_1 of the transformer used must be:
(a) 28/1   (b) 800/1   (c) 1/800   (d) 1/28
(e) $(64 \times 10^4)/1$

21.11 An audio amplifier with an internal impedance of 2000 Ω is used to drive a loudspeaker with an impedance of 5.0 Ω. What must be the turns ratio (N_2/N_1) of the transformer used to match the impedances?
(a) 4.5/1   (b) 1/20   (c) 1/4.5
(d) 400/1   (e) 1/400

21.12 An ac voltage of rms value 120 V and frequency 50 Hz is connected across a light bulb with a resistance of 100 Ω. What are (a) the rms current and (b) the maximum current?

21.13 A 60-Hz ac voltage of effective value 120 V is applied across a pure inductor of inductance 200 mH.
(a) What is the effective current?
(b) When the voltage has its maximum value of 170 V, what is the current?

21.14 A 0.10-μF capacitor is connected to an ac voltage source of rms value 8.0 V at 60 Hz. What are (a) the rms current and (b) the maximum current through the capacitor?

21.15 A circuit consists of a 200-Ω resistor, a 0.15-μF capacitor and a 100-mH inductor with a resistance $R_L = 50$ Ω in series with a 60-Hz voltage source. What are (a) the inductive reactance of the circuit; (b) the capacitive reactance of the circuit; (c) the impedance of the circuit?

21.16 What is the natural frequency of an LC circuit with $L = 0.20$ mH and $C = 2.0$ μF?

21.17 What is the resonant frequency of an RLC circuit with $L = 0.20$ mH, $C = 100$ μF, and $R = 0.10$ Ω?

21.18 An RLC circuit is in resonance with a 60-Hz applied voltage. If the inductance in the circuit is 60 mH and the resistance is 40 Ω, what is the capacitance in the circuit?

21.19 An electric-train transformer steps down the 120-V house current to 6.0 V. If the current supplied to the transformer by the house circuit is 0.20 A, what is the current output of the transformer?

21.20 An electric generator produces ac power at an effective voltage of 19,125 V. It is desired to transmit this power over high-voltage lines at a voltage of 765,000 V by using a step-up transformer. If the primary consists of 100 turns, how many turns must the secondary have?

Problems

21.21 An ac voltage of the form $v = v_{max} \sin 2\pi ft$, with $f = 60$ Hz and $v_{max} = 170$ V, is applied across a 60-W light bulb.

(*a*) When $t = 0$, that is, when the voltage is first applied, what is the current through the circuit?

(*b*) When $t = T/4$, where T is the period (equal to $1/f$) of the applied voltage, what is the current in the circuit?

(*c*) What is the rms current in the circuit?

21.22 An ac voltage of the form $v = v_{max} \sin 2\pi ft$, with $f = 60$ Hz and $v_{max} = 340$ V, is applied to a light bulb of resistance 100 Ω. Draw graphs of the applied voltage and the current as a function of time for one voltage cycle. Indicate the numerical values of the maximum voltage and current.

21.23 A voltage of the form $v = v_{max} \sin 2\pi ft$, with $f = 60$ Hz and $v_{max} = 170$ V, is applied to a pure inductor with inductance 0.30 H. What is the current through the circuit (*a*) when $t = T/4$, where T is the period; (*b*) when $t = T/2$; (*c*) when $t = T$?

21.24 An ac voltage of the form $v = v_{max} \sin 2\pi ft$, with $f = 2.5 \times 10^6$ Hz and $v_{max} = 0.10$ V, is applied to a pure inductor with $L = 0.15$ H.

(*a*) On the same graph, plot the voltage v and the current i as a function of time.

(*b*) What is the effective current in the circuit?

21.25 A radio capacitor of capacitance 0.020 μF is attached to a voltage source of the form $v = v_{max} \sin 2\pi ft$, where $f = 60$ Hz and $v_{max} = 170$ V. What is the current through the circuit (*a*) at the instant the voltage is applied ($t = 0$); (*b*) when $v = 170$ V; (*c*) when $t = T/2$, where T is the period?

21.26 For the situation of Prob. 21.25:

(*a*) Plot on the same graph the voltage and current as a function of time.

(*b*) What is the effective value of the current?

(*c*) What is the maximum charge on the capacitor?

21.27 A 60.0-μF capacitor is connected in series with a 10.0-mH inductance and a switch. The capacitor is first charged to a voltage of 170 V. The charging battery is then removed, and the switch is closed. What are (*a*) the maximum charge on the capacitor; (*b*) the maximum current in the circuit; (*c*) the resonant frequency of the circuit?

21.28 In the situation given in Prob. 21.27 plot on the same graph the energy stored in the magnetic field of the inductor and the energy stored in the electric field of the capacitor, both as a function of time. Show that the total energy in the circuit remains constant in time.

*21.29 A capacitor is placed in parallel across a resistive load R in the circuit of Fig. 21.26. Suppose that R is 250 Ω and that C is 0.20 μF. What fraction of the incoming current will pass through the capacitor to ground rather than through the load when the frequency is (*a*) 100 Hz; (*b*) 1.0 MHz?

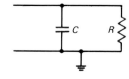

FIGURE 21.26 Diagram for Prob. 21.29.

21.30 A 25-mH coil with a resistance of 9.0 Ω is connected in series with a capacitor C and a 60-Hz voltage source.

(*a*) What must C be if the current and voltage are to be in phase?

(*b*) If the source delivers power at an effective voltage of 120 V, what is the current through the capacitor?

21.31 A circuit consists of an idealized 100-mH inductor with no resistance, and a 10.0-μF capacitor.

(*a*) What is the natural frequency of this circuit if the capacitor is first charged to 200 μC and then allowed to discharge through the inductor?

(*b*) What is the maximum energy stored in the capacitor?

(*c*) What is the maximum energy stored in the inductor?

(*d*) Plot the energy stored in the capacitor and the inductor individually as a function of time for one oscillation cycle.

21.32 A 15-μF capacitor is put in series with a 50-Ω resistor and connected to a 120-V, 60-Hz power line. Find (*a*) the current in the circuit; (*b*) the phase angle between the current and the applied voltage; (*c*) the power factor; (*d*) the power loss in the capacitor.

21.33 What is the resistance of a coil if its impedance is 50 Ω and its inductive reactance is 25 Ω?

21.34 A 100-mH inductor with $R_L = 10$ Ω is connected to a 120-V 60-Hz line. Find (*a*) the current in the circuit; (*b*) the phase angle between current and applied voltage; (*c*) the power factor; (*d*) the power loss in the inductor.

21.35 A capacitor is connected in series with a 25-Ω resistance.

(*a*) If the impedance of the circuit at 60 Hz is 75 Ω, what is the capacitive reactance of the capacitor?

(*b*) What is C?

21.36 Find the impedance of an RLC circuit with $L = 50$ mH, $C = 20$ μF, and $R = 60$ Ω (*a*) at 60 Hz; (*b*) at 600 Hz.

21.37 If an effective ac voltage of 120 V at 60 Hz is applied to an RLC circuit with $L = 10$ mH, $C = 0.10$ μF, and $R = 100$ Ω, what is the effective current?

21.38 In Prob. 21.37:

(*a*) What is the power factor of the circuit?

(*b*) What is the power consumed in the circuit?

(*c*) In which element of the circuit is the power consumed?

21.39 An inductance coil with $L = 0.15$ H and resistance 20 Ω is connected across a 120-V, 60-Hz line. Find (*a*) the current through the coil; (*b*) the phase angle between the current and the applied voltage; (*c*) the power factor; (*d*) the power lost in the coil in the form of heat.

21.40 A 300-mH inductance coil has a resistance of 20 Ω. Find the current in the coil if the applied voltage is (*a*) 120 V dc; (*b*) 120 V ac at 10^6 Hz.

21.41 A 100-Ω resistor is connected in series with a 100 mH inductor and a 0.300 μF capacitor. The applied voltage has the form $v = 170 \sin 2\pi ft$, where $f = 1000$ Hz.

(*a*) Find the impedance of the circuit.

(*b*) Find the effective current.

(*c*) Find the effective voltages across R, L, and C.

(*d*) Show by means of a vector diagram how the voltages across R, L, and C add to give the applied voltage.

*21.42 A "black box" contains two circuit elements in series, but it is not known whether they are capacitors, inductors, or resistors. When connected to a 120-V, 60-Hz circuit, the effective current is 6.0 A, and this current lags the voltage by 40°. What are the two elements in the black box and what are their numerical values?

21.43 An inductance coil draws 2.0 A dc when connected to a 12-V battery. When connected to a 120-V 60-Hz power source the effective current drawn is 5.0 A. What are the (*a*) resistance and (*b*) inductance of the coil?

*21.44 A 12-V battery is connected in series with two unknown electric components, and the current is 600 mA. A 60-Hz, 12-V ac source then replaces the battery, and the current becomes 300 mA.

(*a*) What are the circuit elements and what are their numerical values?

(*b*) How does the current change if the frequency is increased to 1.0 MHz?

*21.45 When a cathode-ray tube has its electrostatic deflection plates connected across a 90-V battery, as in Fig. 21.27*a* the spot on the fluorescent screen is deflected through 10 cm. If the plates are now connected across a resistance of 20 Ω in parallel with an ac voltmeter and in series with a 60-Hz ac generator, as in Fig. 21.27*b*, the voltmeter reads 45 V and the length of the trace on the screen becomes 14 cm.

(*a*) Can you explain these seemingly contradictory data? (*Hint:* What is the beam displacement per volt in the two cases?)

(*b*) What is the rms current through the 20-Ω resistor?

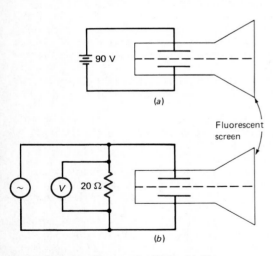

FIGURE 21.27 Diagram for Prob. 21.45.

*21.46 A 5.0-H inductance coil with resistance 200 Ω is connected to a 120-V, 60-Hz source. A capacitor is connected in series with the coil to bring the power factor up to 1.0.

(*a*) What is the current in the circuit after the capacitor has been inserted?

(*b*) What capacitance is needed to bring the power factor to 1.0?

(*c*) What is the peak voltage the capacitor must be able to sustain?

*21.47 Figure 21.28 shows an ac bridge that is similar to a Wheatstone bridge for direct current (see Fig. 18.42) but which in this case is used to measure frequencies. When the bridge is balanced, *a* is at the same potential as *b* and there is no current through the earphones connected between *a* and *b*. If $R_1 = R_2$, $L = 100$ mH, and $C = 1.00$ μF, at what frequency will the bridge be balanced? (Assume that the resistance of the coil is negligible.)

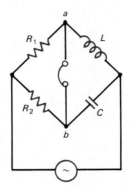

FIGURE 21.28 Diagram for Prob. 21.47.

21.48 A radio is tuned to a chosen station by a circuit consisting of a fixed 0.25-mH inductance and a variable capacitor set to 20 pF.

(*a*) What is the frequency of this station?

(*b*) What is the wavelength of the waves it transmits?

21.49 A 120-V voltage is applied at the resonant frequency of a circuit consisting of a 10-Ω resistor, a 20-mH inductor, and a 10-μF capacitor in series. What are (*a*) the voltages across the inductor and the capacitor individually at resonance; (*b*) the voltage across the resistor at resonance; (*c*) the total voltage across the inductor, capacitor, and resistor in series at resonance?

21.50 An *RLC* series circuit is connected to a variable-frequency ac generator of rms voltage 120 V. In the circuit $L = 100$ mH, $C = 2.00 \times 10^{-6}$ F, and $R = 150$ Ω. What are (*a*) the resonant frequency of the circuit; (*b*) the effective current at resonance; (*c*) the voltages across the capacitor, inductor, and resistor individually at resonance; (*d*) the power delivered to the circuit at resonance?

21.51 A voltage of 120 V is applied across a circuit consisting of a 10-Ω resistor, a 50-mH inductor, and a 30-μF capacitor in series. What are (*a*) the resonant frequency of this circuit; (*b*) the power factor at resonance?

*21.52 (*a*) Prove that the quality factor for an *RLC* series circuit is $Q = (1/R)(L/C)^{1/2}$. (*Hint:* Start with the definition of the *Q* factor).

(*b*) What is the *Q* factor for an *RLC* series circuit with $R = 25$ Ω, $L = 30$ mH, and $C = 12$ μF?

21.53 An *RLC* series circuit resonates at 5000 Hz.

(*a*) If its capacitance is 0.0050 μF, what must its resistance be for the circuit to have a *Q* factor at resonance of 500?

(*b*) What is *L* in this circuit?

21.54 An electric generating plant produces electric energy at a rate of 0.50 GW and a voltage of 10,000 V.

(*a*) If it is desired to transmit this power at 400,000 V, what must be the turns ratio of the transformer used?

What current would be sent out over the power lines if it were transmitted at (*b*) 10,000 V; (*c*) at 400,000 V?

21.55 A stereo amplifier has an impedance of 1500 Ω. What must be the turns ratio (N_2/N_1) of the transformer used to match this impedance to (*a*) a loudspeaker of impedance 3.0 Ω; (*b*) a public-address system of impedance 50 Ω?

21.56 An ac source of internal resistance 10,000 Ω is to feed current to a load of resistance 20 Ω.

(*a*) How should the source be matched to the load?

(*b*) When the impedances are matched, what is the ratio of the current through the source to the current through the load?

21.57 A step-down transformer at the end of a transmission line reduces the voltage from 2400 to 120 V. The power output of the transformer is 10.0 kW, and the overall efficiency of the transformer is 92 percent. The primary coil has 1000 turns.

(*a*) How many turns does the secondary have?

(*b*) What is the power input to the transformer?

(*c*) What is the current in each transformer coil?

21.58 How many 150-W light bulbs can be used in a 120-V circuit fused for 20 A before the fuse is blown, when the bulbs are connected (*a*) in parallel; (*b*) in series?

21.59 An audio amplifier has output connections for 4.0-, 8.0-, and 16-Ω outputs. If two 8.0-Ω speakers are connected to this amplifier, to which output terminals should they be connected if they are connected (*a*) in parallel; (*b*) in series?

21.60 An rms current of 100 mA from a 120-V, 60-Hz power line can be lethal if it passes through the human body for as little as 1.0 s.

(*a*) What would the resistance of the body have to be for such a 100-mA current to pass through the body?

(*b*) How much energy would be delivered to the body in this 1 s?

Additional Readings

Cheney, Margaret: *Tesla: Man Out of Time*, Laurel-Dell, New York, 1983. A very interesting and more complete life than O'Neill's, but a bit weak on the scientific side.

Coltman, John W.: "The Transformer," *Scientific American*, vol. 258, no. 1, January 1988, pp. 86–95. An interesting account of one of technology's unsung wonders, and the key element in the distribution of electrical power.

Gray, Andrew: *Lord Kelvin: An Account of His Scientific Life and Work*, Chelsea, New York, 1973. A delightful book, a reprint of the original 1908 edition by a former student and assistant of Kelvin at the University of Glasgow.

Hammond, Allen L., William D. Metz, and Thomas H. Maugh II: *Energy and the Future*, American Association for the Advancement of Science, Washington, D.C., 1973. This book contains four brief but excellent chapters on new approaches to the transmission of electric power.

Morgan, M. Granger: Book review of Paul Brodeur's *Currents of Death. Scientific American,* vol. 262, no. 4, April 1990, pp. 118–123. A fair-minded, balanced account of Brodeur's controversial book on the public health hazards of electric power lines, electric blankets, and computer terminals.

O'Neill, John J.: *Prodigal Genius: The Life of Nikola Tesla*, Ives Washburn, New York, 1944. A lively study of the life and mind of an eccentric scientific genius by a personal friend.

Schwartz, Brian B., and Simon Foner: "Large-Scale Applications of Superconductivity," *Physics Today*, vol. 30, no. 7, July 1977, pp. 34–43. Includes applications of superconductivity to power transmission, and to electric motors and generators, superconducting magnets, and magnetic levitation of trains.

Sharlin, Harold Issadore: *Lord Kelvin, The Dynamic Victorian*, Pennsylvania State University Press, University Park, Pa., 1979. The most recent account of Kelvin's life and scientific accomplishments.

CHAPTER 22

Electromagnetic Waves

A Clerk Maxwell was required, a second man of the same power and independence of intellect, to build up by the normal methods of science the great structure, whose plan Faraday had conceived and which he had tried to make clear to his contemporaries.

Hermann von Helmholtz (1821–1894)

In the preceding six chapters we have introduced the basic physical laws governing the behavior of charges at rest and in motion, and electric and magnetic fields and their interactions. Now we can form a complete picture of electromagnetic theory. This was first done in the years between 1864 and 1873 by James Clerk Maxwell (see Fig. 22.2 and accompanying biography). The four equations he put forward, now called simply Maxwell's equations, unified elements of electricity and magnetism once thought entirely distinct and independent. In addition they integrated electromagnetic theory with light by predicting the existence of electromagnetic waves traveling at the same speed as light. As a consequence, Maxwell's equations provide the theoretical foundation on which today's radio, radar, and television communication systems are built. In the realm of electromagnetics, Maxwell's equations play the same important role that Newton's laws of motion play in mechanics.

22.1 Maxwell's Equations

Maxwell showed that all the experimental facts of electromagnetism can be summarized in four basic equations which describe the behavior of electric and magnetic fields in space. Of these four equations three are restatements of results previously obtained by Coulomb, Gauss, Oersted, Faraday, Henry, and others—results we have already seen in previous chapters. The fourth equation, while based on the work of Ampère, introduced Maxwell's key idea of *displacement current*, the last piece needed to complete the intricate mosaic of electromagnetic theory.

Because precise statements of Maxwell's equations require calculus, we will confine ourselves here to stating in words the physical content of each equation and to showing how Maxwell completed them by introducing the idea of displacement current.

Table 22.1 shows the physical content of Maxwell's four equations and indicates the physical facts and equations on which they are based.

Maxwell was bothered by the lack of symmetry between Faraday's induction law and Ampère's circuital law. If a changing magnetic field produces an electric

TABLE 22.1   Maxwell's Equations

Content of each equation	Basis in theory and experiment
1 All lines of an electrostatic field begin and end on electric charges.	Coulomb's law: $$F_{\text{el}} = \frac{k_e q q'}{r^2} \qquad \textbf{(16.1)}$$ Gauss' law: $$\Phi_E = \sum E_\perp \, \Delta A = 4\pi k_e \sum q \qquad \textbf{(16.3)}$$
2 All lines of a static magnetic field are continuous and neither start nor stop in space.	No magnetic poles exist. $$\Phi_B = \sum B_\perp A = 0$$
3 A changing magnetic field produces an electric field.	Faraday's induction law: $$\mathscr{E} = -N \frac{\Delta \Phi_B}{\Delta t} = -NA \frac{\Delta B_\perp}{\Delta t} \qquad \textbf{(20.6)}$$
4A (Incomplete form—before Maxwell): A conventional electric current produces a magnetic field.	Ampere's circuital law: $$\sum B_\parallel \, \Delta l = \sum \mu_0 I \qquad \textbf{(19.15)}$$
4B (Complete form—after Maxwell): Either a conventional electric current or a changing electric field produces a magnetic field.	Ampere's circuital law *plus* Maxwell's displacement current: $$\sum B_\parallel \, \Delta l = \mu_0 I + \mu_0 \epsilon_0 A \frac{\Delta E_\perp}{\Delta t}$$ $$= \mu_0 I + \mu_0 \epsilon_0 \frac{\Delta \Phi_E}{\Delta t} \qquad \textbf{(22.6)}$$

field, why does not a changing electric field produce a magnetic field? Maxwell was able to show that Ampère's circuital law [Eq. (19.15)] is really incomplete, and that when properly completed it does indeed restore the expected symmetry between electric and magnetic fields.

22.2   Maxwell's Displacement Current

Maxwell's great achievement was the introduction of the idea of a displacement current, which exists whenever an electric field changes in time. A displacement current produces a magnetic field in the same way that conventional currents do but does not consist of the flow of electric charges through wires or through space. Instead, a displacement current consists of a changing electric field. In this section, we present a somewhat simplified version of Maxwell's argument that a varying electric field is equivalent to an electric current.

We consider a capacitor of capacitance C in series with a resistance R, as in Fig. 22.1, with the separation between the capacitor plates being d. At time $t = 0$, the switch S is closed and the battery delivers a current i, which falls from its initial value to zero in a time determined by the time constant of the RC circuit (Fig. 20.16). At the same time, the charge q on the capacitor increases gradually from zero to its final value Q_0, as in Fig. 20.17. As the charge q varies, so does the voltage v across the capacitor, since $v = q/C$ [Eq. (17.9)].

The electric field in the region between the plates is $E = v/d$, and so

$$q = Cv = CdE \qquad \textbf{(22.1)}$$

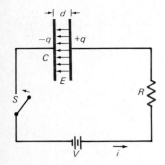

FIGURE 22.1 Maxwell's displacement current. When a capacitor C is being charged through a resistor R, the electric field between the plates changes at the rate $\Delta E/\Delta t$. Maxwell showed that this was equivalent to a displacement current between the plates of magnitude $I_D = \epsilon_0 A\, \Delta E_\perp/\Delta t$, where A is the area of the capacitor plates.

In this equation C and d are both constants. As the charge on the plates changes in time, therefore, the electric field must change according to the equation

$$\frac{\Delta q}{\Delta t} = Cd\,\frac{\Delta E}{\Delta t} \tag{22.2}$$

Here $\Delta q/\Delta t$ is just the instantaneous current i in the wire, and from Eq. (17.12) the capacitance of a parallel-plate capacitor in a vacuum is just $C = \epsilon_0 A/d$. Hence

$$i = \frac{\Delta q}{\Delta t} = Cd\,\frac{\Delta E}{\Delta t} = \epsilon_0 A\,\frac{\Delta E}{\Delta t}$$

Since the field E is here assumed to be normal to the area A, this equation can be written as

$$i = \epsilon_0 A\,\frac{\Delta E_\perp}{\Delta t} \tag{22.3}$$

Equation (22.3) relates the change in the electric field between the plates of the capacitor to the current in the rest of the circuit. If we assume that there is a current equal to $\epsilon_0 A\,\Delta E_\perp/\Delta t$ between the capacitor plates when the field $E_\perp$ changes with time, then we have a continuous current through the whole circuit, including the region between the plates. As long as the electric field between the plates is changing, we can say that there is an equivalent current present in the capacitor, and this equivalent current is numerically equal, from Eq. (22.3), to the real current in the circuit. Maxwell called this current the *displacement current*:

Maxwell's displacement current:
$$\boxed{I_D = \epsilon_0 A\,\frac{\Delta E_\perp}{\Delta t}} \tag{22.4}$$

Just as Maxwell's third equation (Faraday's induction law) states that a changing magnetic field produces an emf and therefore also an electric field, so Eq. (22.4) states that a changing electric field produces a displacement current, which in turn creates a magnetic field, since all electric currents produce magnetic fields.

Even though Eq. (22.4) has been derived for the field between the plates of a capacitor, it is valid for all situations in which an electric field changes with time. In all such situations, since $A\,\Delta E_\perp/\Delta t$ is equal to the electric flux Φ_E, the displacement current can be written

$$\boxed{I_D = \epsilon_0\,\frac{\Delta \Phi_E}{\Delta t}} \tag{22.5}$$

If this displacement current is treated as an additional current in Ampère's circuital law, we obtain

$$\sum B_\parallel\,\Delta l = \mu_0(I + I_D) = \mu_0 I + \mu_0 \epsilon_0\,\frac{\Delta \Phi_E}{\Delta t} \tag{22.6}$$

which is Maxwell's complete fourth equation. His introduction of this displacement current was the crucial step needed to introduce symmetry between the electric and magnetic fields, and thus to complete electromagnetic theory.

If there were a dielectric material between the plates of the capacitor, then

FIGURE 22.2 *(Courtesy of AIP Niels Bohr Library.)*

James Clerk Maxwell (1831–1879)

The greatest theoretical physicist of the nineteenth century was born in 1831, the year in which Faraday discovered electromagnetic induction, and died in 1879, the year in which Einstein was born. James Clerk Maxwell was a true link between these two great physicists, for it was he who put Faraday's experimental results and far-reaching insights into useful mathematical equations, and it was on the foundation of Maxwell's equations that Einstein built his theory of relativity.

Maxwell was born in Edinburgh of a well-known Scottish family. His mother died when he was 9 years old, and he was raised by his father and an aunt. As a youth he revealed great mathematical genius, which earned him the nickname "Dafty" from his schoolmates. Maxwell had not only the ability to handle mathematical ab-

stractions but also a remarkable curiosity about how things worked and a fascination with constructing working models of practical devices. He constantly bombarded his elders with questions such as "What does it do?" or "What's the particular go of that?"

In 1847 Maxwell enrolled at the University of Edinburgh, and in 1850 moved on to Cambridge University, from which he graduated in 1854, second in his class in mathematics. In 1856 he accepted an appointment as a professor at Marischal College in Aberdeen, Scotland, mainly to be near his father, who unfortunately died just before his son took over his new post. After 5 years (1860 to 1865) at King's College in London, Maxwell went into semiretirement at his family estate, Glenlair, near the village of Parton in Scotland, to devote as much time as possible to research.

In 1871 he agreed to come out of retirement to organize the newly created Cavendish Laboratory at Cambridge, the first laboratory for experimental physics at that university. There Maxwell began the great tradition that has marked physics at Cambridge ever since.

Maxwell made great contributions to many fields of physics. Among these were his work on kinetic theory and the statistical theory of gases, his theory of color vision, his work in thermodynamics and astrophysics, and in particular his development of the equations of the electromagnetic field (discussed in this chapter). Heinrich

Hertz once referred to Maxwell's equations as behaving as if they had "an independent life and an intelligence of their own, as if they were wiser than ourselves, indeed wiser than their discoverer, as if they gave forth more than he had put into them." Today, even after the revolutionary discoveries in this century of relativity and quantum mechanics, Maxwell's equations remain as valid and useful as ever.

This remarkable scientist was a gentle, religious, selfless man who loved children (he and his wife had none of their own), family, and good friends. He once wrote: "Work is good, and reading is good, but friends are better." He also had a fine sense of humor, once delivering a lecture about the telephone, then newly developed, in which he referred to "the perfect symmetry of the whole apparatus—the wire in the middle, the two telephones at the ends of the wire, and the two gossips at the ends of the telephones."

Maxwell's career came to a sudden, sad end when he succumbed to cancer at the age of 48. The Rev. H. M. Butler, in a sermon after Maxwell's death, remembered him in these terms: "[he] was even more loved than he was admired, retaining after twenty years of fame that mirth, that simplicity, that child-like delight in all that is fresh and wonderful which we rejoice to think of as some of the surest accompaniments of true scientific genius."

the electric field would *displace* the centers of positive and negative charge in the dielectric with respect to each other, as discussed in Sec. 17.7. Thus we have the name *displacement current*. Maxwell's theory gives a vacuum the same standing as a dielectric, though a vacuum has a dielectric constant of unity, whereas dielectrics have larger dielectric constants than this, as shown in Table 17.1.

FIGURE 22.3 Magnetic field **B** produced by (a) a conventional electric current *I* in a wire and (b) a change in the electric field **E** between the plates of a capacitor, i.e., a *displacement current*. Since these currents have the same numerical value, the two magnetic fields are identical.

Figure 22.3 illustrates how a magnetic field can be produced either by a conventional electric current or by a changing electric field, as predicted by Maxwell's theory of displacement currents. Note that a displacement current does not involve charges in motion but consists of a changing electric field which has the same physical effects as those produced by real charges in motion. For this reason we refer to a displacement current as an *equivalent current*.

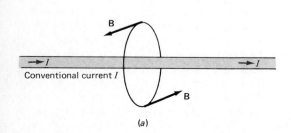

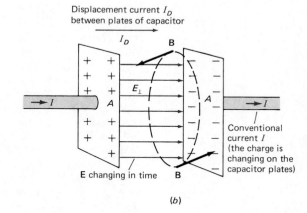

Example 22.1

A capacitor with plates of area 400 cm² and an air gap between the plates of 1.0 mm is being charged by a dc power supply at a constant rate of 5.0 nA. **(a)** At what rate is the electric field between the plates changing? **(b)** What is the dis-placement current while the capacitor is being charged? **(c)** What is the magnetic field 50 cm from the wire that carries the charge to the capacitor? **(d)** What is the magnetic field 50 cm from the center of the capacitor?

SOLUTION

(a) To find the rate at which the electric field is changing, we need to know the rate at which the voltage across the plates is changing. Since $v = q/C$, this is

$$\frac{\Delta v}{\Delta t} = \frac{1}{C}\frac{\Delta q}{\Delta t}$$

Here $\Delta q/\Delta t = I = 5.0 \times 10^{-9}$ A, and the capacitance is that of a parallel-plate capacitor:

$$C = \frac{\epsilon_0 A}{d} = \left(8.85 \times 10^{-12}\, \frac{\text{C}^2}{\text{N·m}^2}\right)\left(\frac{0.040\ \text{m}^2}{0.0010\ \text{m}}\right)$$

$$= 3.5 \times 10^{-10}\ \text{F}$$

and so

$$\frac{\Delta v}{\Delta t} = \frac{5.0 \times 10^{-9}\ \text{C/s}}{3.5 \times 10^{-10}\ \text{F}} = 14\ \text{V/s}$$

Then, since $E = v/d$, we have

$$\frac{\Delta E}{\Delta t} = \frac{1}{d}\frac{\Delta v}{\Delta t} = \frac{14\ \text{V/s}}{0.0010\ \text{m}} = \boxed{1.4 \times 10^4\ \text{V/(m·s)}}$$

(b) The displacement current I_D is, from Eq. (22.4),

$$I_D = \epsilon_0 A \frac{\Delta E}{\Delta t}$$

$$= \left(8.85 \times 10^{-12}\, \frac{\text{C}^2}{\text{N·m}^2}\right)(0.040\ \text{m}^2)\left(1.4 \times 10^4\, \frac{\text{V}}{\text{m·s}}\right)$$

$$= \boxed{5.0 \times 10^{-9}\ \text{A}}$$

which is numerically equal to the current in the external circuit, as we expect from Eq. (22.3).

(c) Since the magnetic field around a wire is, from Eq. (19.14), $B = \mu_0 I/2\pi r$ and since $I = 5.0 \times 10^{-9}$ A, we have

$$B = \left(\frac{4\pi \times 10^{-7}}{2\pi}\, \frac{\text{N}}{\text{A}^2}\right)\left(\frac{5.0 \times 10^{-9}\ \text{A}}{0.50\ \text{m}}\right) = \boxed{2.0 \times 10^{-15}\ \text{T}}$$

(d) Since the displacement current I_D is numerically equal in magnitude to the conventional current, the magnetic field B around the capacitor at a distance of 50 cm from its center is the same as that calculated for the field around the wire,

$\boxed{2.0 \times 10^{-15}\ \text{T}}$, as in part (c). This is quite a small magnetic field because the rate at which the capacitor is being charged is very low.

EXERCISE 1 Prove that in part (b) above the units of the answer are really amperes (A).

22.3 The Nature of Electromagnetic Waves

Once we understand the content of Maxwell's equations, the nature of electromagnetic waves becomes clearer. Time-varying electric fields moving through space produce magnetic fields that also change in time. These time-varying magnetic fields in turn produce electric fields. Thus, even in regions of space in which there are no charges, no magnets, and no real currents, electric and magnetic fields can exist. A "leap-frogging" effect of changing electric fields producing magnetic fields, which in turn produce electric fields, occurs continuously, as in Fig. 22.4. As time goes on, the whole package of electric and magnetic fields moves through space. This combination of moving electric and magnetic fields is called an *electromagnetic wave*. The energy in the electric and magnetic fields is thus transported through space by electromagnetic waves, even through regions in which there is no matter.

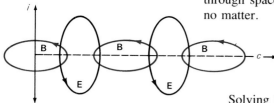

FIGURE 22.4 The propagation of an electromagnetic wave. An oscillating current *i* produces a changing magnetic field **B** that produces a changing electric field **E**, which in turn produces a changing magnetic field **B**, and so forth. As time goes on, this leap-frogging effect carries electromagnetic energy through space in the direction shown.

FIGURE 22.5 A plane electromagnetic wave moving along the *x* axis. The electric field is confined to the *xy* plane and the magnetic field to the *xz* plane, so the wave is transverse.

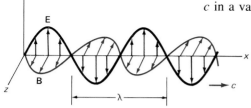

Solving Maxwell's equations for the electric and magnetic fields in a region of space where there are no charges leads to the following important results:

1 Electromagnetic waves in a vacuum (or in "free space") are all *transverse*. The magnetic field **B** and the electric field **E** are at right angles to each other and to the direction of propagation. This is illustrated in Fig. 22.5 for the usual case where both **E** and **B** are sinusoidal, with **E** being confined to the *xy* plane and **B** to the *xz* plane, and the wave moves along the positive *x* axis.

2 All electromagnetic waves travel with the same speed in a vacuum. (We show in the next section that this speed is $c = 1/\sqrt{\mu_0\epsilon_0}$.)

3 The different kinds of electromagnetic waves (radio, visible, x-rays, etc.) differ in frequency f (the number of oscillations of **B** and **E** per second) and wavelength λ [the distance between any two consecutive points along the direction of the wave at which the phases of **E** (or of **B**) are the same]. The wavelength and frequency are connected by the usual equation for waves, $\lambda f = c$, where the speed c in a vacuum is the same for all frequencies and wavelengths.

22.4 The Predicted Speed of Electromagnetic Waves

It is possible to use Maxwell's equations to obtain a value for the speed of electromagnetic waves in a vacuum in terms of known constants. We confine ourselves here to outlining the method and stating the important results obtained.

We first consider an electromagnetic wave made up of a sinusoidal electric field in the *y* direction and an associated sinusoidal magnetic field in the *z* direction, as in Fig. 22.5. The wave moves along the positive x axis at a speed c. We apply Maxwell's third equation to determine the emf induced as the magnetic field **B** changes in time. This emf leads to an electric field **E** in space that also changes sinusoidally in time. Most importantly, we find that the magnitude of the electric and magnetic fields must be related by

$$E = cB \tag{22.7}$$

where c is the speed of the electromagnetic wave along the x axis.

Speed of Electromagnetic Waves in Free Space

We now go a step further and apply Maxwell's fourth equation to find the magnetic field produced by the changing electric field in Fig. 22.5. Since the wave is moving through a vacuum, there can be no conventional electric currents, but there is a displacement current. This is caused by the changing electric field and produces in turn a changing magnetic field. This magnetic field must be equal to the original magnetic field whose change in time produced the electric field. Maxwell's fourth equation then leads to the result that the magnetic field produced by the changing electric field must have a magnitude

$$B = \mu_0 \epsilon_0 E c \tag{22.8}$$

where μ_0 is the permeability of free space and ϵ_0 is the permittivity of free space. But we have already seen [Eq. (22.7)] that $E = cB$, and thus

$$B = \mu_0 \epsilon_0 c^2 B \qquad \text{or} \qquad c^2 = \frac{1}{\mu_0 \epsilon_0}$$

and so

$$c = \frac{1}{\sqrt{\mu_0 \epsilon_0}} \tag{22.9}$$

This is the speed of electromagnetic waves in free space (vacuum). This speed is seen to depend on two constants, one (ϵ_0) that tells us how strong an electric field is produced by a given electric charge and a second (μ_0) that tells us how strong a magnetic field is produced by a given current.

The postulated configuration of electric and magnetic fields is therefore consistent with the laws of electromagnetism only if the wave moves with a speed $c = 1/\sqrt{\mu_0 \epsilon_0}$. This turns out to be a completely general result that follows from Maxwell's equations for all electromagnetic waves, no matter what their frequency. The numerical value of c is, therefore, using the values of μ_0 and ϵ_0 in Table B.2,

$$c = \frac{1}{\sqrt{[8.854 \times 10^{-12} \ C^2/(N \cdot m^2)](4\pi \times 10^{-7} \ N \cdot s^2/C^2)}} = 2.998 \times 10^8 \ \text{m/s*}$$

At the time Maxwell did his work (1864) it had already been determined by the optical experiments of Fizeau (1849) and Foucault (1850) that light traveled at a speed very close to 3.0×10^8 m/s. The agreement of this experimental result with the above prediction of Maxwell's equations suggests strongly that light is an electromagnetic wave. Subsequent work has confirmed this suggestion. Maxwell had thus achieved a true synthesis of light and electromagnetism.

---

*The symbol c is universally used for the speed of electromagnetic waves in a vacuum. The c stands for *celeritas*, a Latin word meaning "speed."

Example 22.2

An electromagnetic wave in free space consists of a sinusoidal electric field whose rms value is 5.00×10^3 N/C and a magnetic field perpendicular to this electric field. What is the rms value of the magnetic field?

SOLUTION

Since for all electromagnetic waves the relationship between the magnitudes of the electric field **E** and the magnetic field **B** is $E = cB$, we have

$$B = \frac{E}{c} = \frac{5.00 \times 10^3 \text{ N/C}}{3.00 \times 10^8 \text{ m/s}} = \boxed{1.67 \times 10^{-5} \text{ T}} = 0.167 \text{ G}$$

The units are correct because

$$\frac{\text{N/C}}{\text{m/s}} = \left(\frac{\text{N}}{\text{C}}\right)\left(\frac{\text{s}}{\text{m}}\right) = \text{N/(A·m)}$$

But, since $F = BIL$, 1 T = 1 N/(A·m). The answer is therefore correctly given in teslas.

The Experiments of Hertz

FIGURE 22.6 Heinrich Hertz (1857–1894) was a protégé of Helmholtz at the University of Berlin. Although he died very young, at the age of only 36, Hertz made many outstanding contributions to physics, in particular his experimental confirmation of the predictions of Maxwell's electromagnetic theory. His work foreshadowed the development of radio by Marconi. *(Photo from Deutsches Museum, Munich; courtesy of AIP Niels Bohr Library.)*

FIGURE 22.7 Hertz's apparatus for transmitting and detecting electromagnetic waves. A high voltage across the air gap of the transmitter causes the air to ionize and produces an oscillatory spark discharge across the gap. The frequency of this discharge is determined by the inductance and capacitance of the transmitting circuit. Some of the energy in the discharge is converted into electromagnetic waves in space, which produce a tiny spark in the air gap of the detector.

Maxwell predicted, on the basis of his equations, that electromagnetic radiation with essentially the same properties as light should exist at frequencies far below and far above those of light. Almost 10 years after Maxwell's death, the German physicist Heinrich Hertz (Fig. 22.6) used an induction coil and spark gap to produce electromagnetic radiation at frequencies near 10^8 Hz, which are now called *radio* frequencies. He detected these waves with a detector coil that could be adjusted to resonance with the transmitted frequency (see Fig. 22.7). Sparks jumping across the narrow air gap of the detector coil indicated that radiation from the transmitter was reaching the detector.

In 1887, at the University of Karlsruhe, Hertz produced standing electromagnetic waves in air and measured the distance between adjacent nodes to obtain their wavelength. Using this wavelength and the frequency of the waves, which he obtained from the measured inductance and capacitance of the oscillating circuit, he calculated the speed of the waves to be 3.2×10^8 m/s. He cautiously stated that his value "only holds good as far as the order of magnitude is concerned," but it was sufficiently accurate to show that light and radio waves traveled with essentially the same speed in air.

Hertz also proved that the electric field was transverse to the direction of propagation of the radio waves and that all the electric vectors were in the same direction. In addition he was able to show that radio waves could be reflected and refracted (bent) in the same way as is light. His work proved conclusively that electromagnetic waves of wavelengths near 1 m and light waves with wavelengths 10^6 times shorter had the same fundamental properties. Thus Hertz provided the experimental confirmation needed to establish Maxwell's theory.

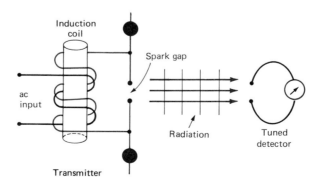

Speed of Electromagnetic Waves in Dielectric Materials

Electromagnetic waves do not propagate very far into conductors, but they do propagate through insulators, or dielectrics. In this case the speed of the wave is

$$v = \frac{1}{\sqrt{\epsilon\mu}} \qquad \text{(22.10)}$$

where ϵ, the permittivity of the dielectric, replaces the permittivity of free space ϵ_0, and μ, the magnetic permeability of the dielectric, replaces the permeability of free space μ_0 in Eq. (22.9). The ratio ϵ/ϵ_0 is the dielectric constant K [Eq. (17.15)], and the ratio μ/μ_0 is the relative permeability K_m [Eq. (19.19)], and so

$$v = \frac{1}{\sqrt{\epsilon\mu}} = \frac{1}{\sqrt{KK_m\epsilon_0\mu_0}} = \frac{c}{\sqrt{KK_m}}$$

For all but ferromagnetic materials K_m is very close to unity, and so*

$$\boxed{v = \frac{c}{\sqrt{K}}} \qquad \text{(22.11)}$$

As we will see when we discuss light, this is sometimes written as $v = c/n$, where $n = \sqrt{K}$ is called the *index of refraction* of the material.

Example 22.3

For some of his experiments Heinrich Hertz used electromagnetic waves whose wavelength he measured to be 30 cm. **(a)** What is the frequency of such waves? If such waves pass from air into a block of quartz, for which $K = 4.3$ when $\lambda = 30$ cm, what are their new **(b)** speed; **(c)** frequency; **(d)** wavelength?

SOLUTION

(a) Since $c = f\lambda$, we have

$$f = \frac{c}{\lambda} = \frac{3.0 \times 10^8 \text{ m/s}}{0.30 \text{ m}} = \boxed{1.0 \times 10^9 \text{ Hz}}$$

(b) The dielectric constant of quartz is given as 4.3, while the dielectric constant of air is very close to that of free space (it is 1.00059 compared with 1.00000). Hence we have

$$v = \frac{c}{\sqrt{K}} = \frac{3.0 \times 10^8 \text{ m/s}}{\sqrt{4.3}} = \boxed{1.4 \times 10^8 \text{ m/s}}$$

(c) The frequency of an electromagnetic wave is determined by the source producing it and does not change when the medium changes.

(d) What does change is the wavelength, which varies in proportion to the speed in the medium, and so

$$f = \frac{c}{\lambda_{air}} = \frac{v}{\lambda_{quartz}}$$

or

$$\lambda_{quartz} = \frac{v}{c}\lambda_{air} = \frac{1.4}{3.0}(30 \text{ cm}) = \boxed{14 \text{ cm}}$$

22.5 Energy in Electromagnetic Waves; Intensity

The modern equivalent of Hertz's apparatus is a television transmitter sending electromagnetic waves through space to the antennas on our houses. These waves are then detected and converted into sound and pictures on TV sets. No *matter* is transported through space; what is transported is pure *energy*. In this section we investigate how the amount of energy transported by a wave depends on the strength of the electric and magnetic fields making up the wave.

---

*The dielectric constant K is frequency-dependent. Its value in Eq. (22.11) must therefore be chosen to fit the frequency involved. For example, at very low frequencies K for water is about 80, but at visible frequencies it is 1.78, so $n = 1.33$.

In Sec. 20.5 we found that the total energy stored in unit volume of the combined electric and magnetic fields was

$$w = \tfrac{1}{2}\epsilon_0 E^2 + \tfrac{1}{2}\frac{1}{\mu_0} B^2 \tag{22.12}$$

Since for an electromagnetic wave $B = E/c$ [Eq. (22.7)], and since $c = 1/\sqrt{\epsilon_0\mu_0}$, we can write Eq. (22.12) as

$$w = \tfrac{1}{2}\epsilon_0 E^2 + \frac{1}{2}\frac{E^2}{c^2\mu_0} = \tfrac{1}{2}\epsilon_0 E^2 + \tfrac{1}{2}\epsilon_0 E^2$$

or $w = \epsilon_0 E^2$ (22.13)

Comparing Eq. (22.12) with (22.13), we see that half the energy in the wave is stored in the electric field and half in the magnetic field.

As an electromagnetic wave moves through space, it carries along with it the energy in the electromagnetic field. Let us denote by P the *incident power per unit area*, i.e., *the energy transported per unit time across a unit cross-sectional area perpendicular to the direction in which the wave is traveling.*

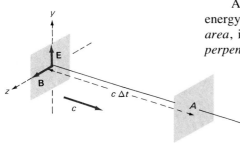

FIGURE 22.8 The transport of energy through the area A by a plane electromagnetic wave. The intensity of the wave is the amount of energy carried by the wave per second through a 1-m² area perpendicular to the direction of propagation of the wave.

In Fig. 22.8 a plane wave front that is part of a continuous wave moving along the positive x axis travels a distance $c\,\Delta t$ in time Δt. If we consider a cross-sectional area A at right angles to the direction in which the wave is traveling, then in time Δt the wave moves through a volume ΔV of space, where $\Delta V = Ac\,\Delta t$. The amount of energy filling this volume is, since w is the energy density,

$$\Delta W = w\,\Delta V = \epsilon_0 E^2 Ac\,\Delta t$$

The incident power *per unit area* is, then, from the definition given above,

$$P = \frac{\Delta W}{A\,\Delta t} = \frac{\epsilon_0 E^2 Ac\,\Delta t}{A\,\Delta t}$$

so that $\boxed{P = \epsilon_0 E^2 c}$ (22.14)

This is the rate at which energy is transported across a unit area at right angles to the direction of propagation of the wave. It is measured in J/(s·m²), or W/m².

Since half the energy is in the electric field and half is in the magnetic field, it is useful to use Eqs. (22.7) and (22.9) to convert this to a form including both the electric and magnetic fields:

$$P = \epsilon_0 cE(cB) = \epsilon_0 c^2 EB = \frac{EB}{\mu_0}$$

so that $\boxed{P = \frac{1}{\mu_0} EB}$ (22.15)

This equation shows how the power delivered by the electromagnetic wave depends symmetrically on E and B.

Equation (22.15) gives the power per unit area delivered by the electromagnetic wave at any instant of time when E and B are the *instantaneous* magnitudes of the electric and magnetic fields. To find the *average* value of P, we must take averages of E and B over time. If the electric field takes the form $E = E_0 \sin \omega t$, the average value of E^2 is $\overline{E^2} = E_0^2/2$ (as for the effective value of an ac current or voltage; see Sec. 21.1), and so

$$S = \overline{P} = \epsilon_0 \frac{E_0^2}{2} c = \frac{E_0 B_0}{2\mu_0} = \frac{1}{\mu_0} \frac{B_0^2}{2} c \qquad (22.16)$$

In this equation S, which is the average value of P, is called the *intensity* of the wave and is defined as the average power transmitted across a unit area normal to the direction in which the wave is traveling. The intensity S of an electromagnetic wave therefore depends directly on the square of the amplitudes of the E and B fields, according to Eq. (22.16). The electric and magnetic fields each contribute exactly the same amount to the energy transported.

Example 22.4

Find the intensity of the wave described in Example 22.2 using **(a)** the value given there for the rms value of the electric field; **(b)** the result calculated there for the rms value of the magnetic field.

SOLUTION

(a) Since, from Eq. (22.14), $P = \epsilon_0 E^2 c$, the intensity is $S = \overline{P} = \epsilon_0 \overline{E^2} c$. But, by definition,

$$E_{rms}^2 = \overline{E^2} \qquad \text{and so} \qquad S = \epsilon_0 E_{rms}^2 c$$

and we have

$$S = \left(8.85 \times 10^{-12} \frac{C^2}{N{\cdot}m^2}\right)\left(5.00 \times 10^3 \frac{N}{C}\right)^2 (3.00 \times 10^8 \text{ m/s})$$

$$= 6.6 \times 10^4 \frac{N}{s{\cdot}m} = 6.6 \times 10^4 \frac{J}{s{\cdot}m^2} = \boxed{6.6 \times 10^4 \text{ W/m}^2}$$

(b) Since, from Eq. (22.16),

$$S = \frac{c}{\mu_0} \frac{B_0^2}{2} = \frac{c}{\mu_0} \overline{B^2} = \frac{c}{\mu_0} B_{rms}^2$$

we have

$$S = \left(\frac{3.00 \times 10^8 \text{ m/s}}{4\pi \times 10^{-7} \text{ N/A}^2}\right)(1.67 \times 10^{-5} \text{ T})^2$$

$$= \left(6.6 \times 10^4 \frac{m/s}{N/A^2}\right)\left(1 \frac{N}{A{\cdot}m}\right)^2 = \boxed{6.6 \times 10^4 \text{ W/m}^2}$$

This is the same result obtained in part (a), as must be the case.

22.6   The Complete Electromagnetic Spectrum

All electromagnetic waves have the same basic nature and travel at the same speed c in a vacuum, but different kinds of electromagnetic waves have different frequencies and wavelengths, which are related by the equation $\lambda f = c$. Electromagnetic waves in different frequency ranges are produced in different ways, are detected in different ways, and interact with matter in different ways. This is one example of a frequently occurring situation in physics in which quantitative changes (in this case in frequency and wavelength) lead to important qualitative differences. Figure 22.9 shows the *complete electromagnetic spectrum*, i.e., the complete range of electromagnetic waves. It gives the wavelengths, frequencies, and energies corresponding to each part of the spectrum and some indication of how electromagnetic waves of a particular frequency are produced and detected.

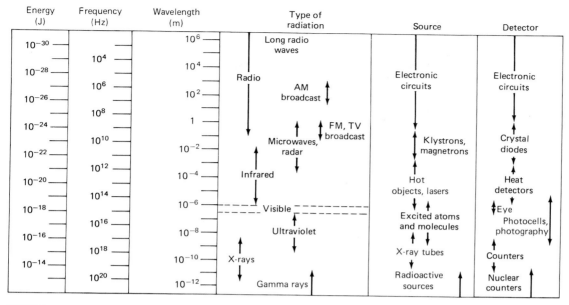

Energy (J)	Frequency (Hz)	Wavelength (m)	Type of radiation	Source	Detector

FIGURE 22.9 The complete electromagnetic spectrum.

The complete electromagnetic spectrum stretches all the way from long radio waves with frequencies below 10 Hz to gamma rays with frequencies above 10^{22} Hz. The names given to the various regions of the electromagnetic spectrum are based on the way a particular kind of wave is produced and detected. There is considerable overlap in the designations used at the boundaries between the various spectral regions. For example, short-wavelength radio waves with wavelengths below about 1 m are usually called *microwaves* or radar waves and are produced in electron tubes known as *klystrons* or *magnetrons*. In the region below 1 cm they blend into the *infrared* region of the spectrum, where the electromagnetic radiation is produced by the vibration and rotation of molecules inside very hot objects.

Similarly, electromagnetic waves with wavelengths of about 10^{-11} m are called either x-rays or gamma rays depending on their origin. If such radiation is produced by the sudden stopping of an electron beam, it is called x-radiation (or x-rays); if it is emitted from the nucleus of an atom, it is called gamma radiation (or gamma rays).

22.7 Measurements of the Speed of Electromagnetic Waves

The precise measurement of the speed of electromagnetic waves constitutes one of the most interesting and important chapters in the history of experimental physics. The first successful measurement of the speed of light over a light path confined to the earth was carried out in 1849 by the French physicist Armand Fizeau (1819–1896). Fizeau set up a rapidly rotating toothed wheel on one hilltop and a mirror on another 8 km away. Light from a strong source passed through a narrow gap between two adjacent teeth of the wheel, as in Fig. 22.10, traveled the 8 km to the mirror, and was reflected back. If the wheel was turning at the right speed, the reflected light pulse returned just in time to pass through another gap (1 or 2 in the figure) between the teeth of the disk. From the wheel's speed of rotation Fizeau was able to calculate the time required for light to travel the 16 km. His

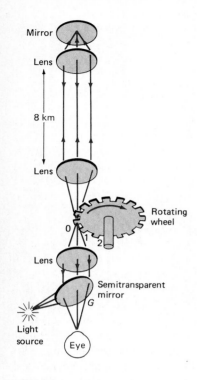

value for c was some 5 percent too high, but his success encouraged other physicists to improve on his basic method.

Measurements like Fizeau's are called *time-of-flight measurements*, i.e., measurements of the time required for a beam of electromagnetic radiation to travel a known distance. More accurate values of c can be obtained by *standing wave measurements*, i.e., by measuring the wavelength and the frequency of electromagnetic waves and then calculating c from the equation $c = \lambda f$. Almost all determinations of c are made in one of these two ways.

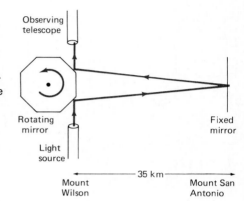

FIGURE 22.10 (left) Fizeau's rotating toothed-wheel apparatus for measuring the speed of light c. The semitransparent mirror reflects the light from the source through the rotating wheel but allows the returning light to reach the eye.

FIGURE 22.11 (right) Michelson's eight-sided rotating mirror for measuring c.

Time-of-Flight Measurements

The most famous measurements of the speed of light in the United States were those carried out by the American physicist A. A. Michelson (see Fig. 25.28 and accompanying biography) in the years 1878 to 1930, using a rotating mirror in place of Fizeau's toothed wheel. In 1923 Michelson had a 35-km path between two California mountain peaks, Mount Wilson and Mount San Antonio, surveyed to an accuracy of a few centimeters. He used an eight-sided revolving mirror, shown in Fig. 22.11, to time the travel of a light pulse over the 70-km round-trip path.

Michelson directed a beam of light at the rotating mirror. Some of this light struck one face of the mirror at an angle so that it was reflected toward the distant mirror and then reflected back to the rotating mirror. If, in the time the pulse of light traveled the distance $2d$, the mirror rotated one-eighth of a turn, another mirror face would be in the right position to reflect light into an observing telescope. The speed of rotation of the mirror was varied until bright light was observed in the telescope. Then the time for light to travel $2d$ was one-eighth of the period of rotation of the eight-sided mirror. Michelson's 1926 result for c was 299.796 ± 4 km/s, a value that agrees within his stated accuracy with the value accepted today.

An improvement on Michelson's method was introduced by E. Bergstrand in 1950. Bergstrand used a *geodimeter*, which replaced Michelson's rotating mirror with an electronic device that pulsed both the light beam and the detector on and off at a variable frequency. By varying this on-off frequency until maximum light intensity was obtained at the detector, Bergstrand was able to time the passage of light to a distant reflector and back. He improved on both the accuracy and the precision with which c was known, his final result being $c = 299,793.1 \pm 0.4$ km/s. The geodimeter is now used routinely in land surveys, since, once the speed of light is known accurately, measuring the time required for light to travel an unknown distance leads to a precise value for that distance.

**Standing Wave
Measurements**

The technique used for standing wave measurements of c is similar to that used to measure the speed of transverse waves on strings. We set up a standing wave pattern, measure the distance (d) between nodes to obtain the wavelength ($\lambda = 2d$), and from that and the known frequency obtain c, since $c = \lambda f$. This was the same technique used by Hertz in his verification of Maxwell's theory.

Before the development of lasers it was not possible to apply this technique to infrared or visible radiation, since it was impossible to measure directly the frequency of such radiation. Important measurements were, however, made in the microwave and radio wave regions of the spectrum, where frequencies could be measured more easily.

For example, in 1958 K. D. Froome in Great Britain measured c using a microwave interferometer. His apparatus, which is shown in Fig. 22.12, consisted of a four-horn* microwave interferometer operating at 72 GHz ($\lambda = 4.2$ mm). Froome generated a beam of microwaves whose frequency could be measured to very high accuracy using well-known microwave or radar techiques. This radiation was split into two beams and radiated from two horns toward a detector mounted on a cart that could be moved along a track. As the cart was moved along, positions of constructive and destructive interference were observed between the two beams. The distance moved by the cart between nulls was $\lambda/2$, since the motion of the cart through a distance of one-half wavelength decreased one path by $\lambda/2$ and increased the other path by the same amount. Since many nulls could be detected over the 8-m movement of the cart, a very precise value of λ could be obtained. From $c = \lambda f$ Froome found that c was equal to 299,792.5 $\pm$ 0.1 km/s, which improved on both the accuracy and the precision of Bergstrand's work.

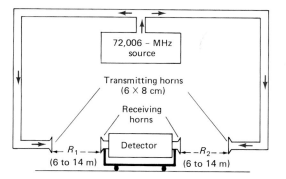

FIGURE 22.12 Froome's microwave interferometer for measuring the speed of light c.

The most accurate measurements of c ever made were those carried out at the National Bureau of Standards in the United States and at the National Physical Laboratory in England in the early 1970s. After Froome's measurement it became clear to many physicists that the ideal way to determine c was to extend Froome's method from the microwave to the infrared and visible regions of the spectrum and simultaneously measure frequency and wavelength for a monochromatic light source in a vacuum. A high-precision wavelength measurement could be made in

---

*A horn is merely a piece of flanged hollow metal for guiding the microwaves out into the air from inside a metal pipe called a *waveguide*.

TABLE 22.2 Results of Measurements of the Speed of Electromagnetic Waves

Date	Experimenter	Method	Velocity of electromagnetic waves in vacuum (km/s)	Experimental error (km/s)
1862	Foucault	Rotating mirror	298,000	±500
1925	Michelson	Rotating mirror	299,796	±4
1950	Essen	Cavity resonator	299,792.5	±3
1950	Bergstrand	Geodimeter	299,793.1	±0.4
1958	Froome	Microwave interferometer	299,792.5	±0.1
1973	Evenson et al.	Wavelength and frequency measurements of 3.39-μm He–Ne laser line	299,792.4574	±0.0011
1974	Blaney et al.	Wavelength and frequency measurements of 9.32-μm CO_2 laser line	299,792.4590	±0.0008

this region, using well-established interferometric techniques (to be discussed in Sec. 25.8).

The real scientific breakthrough came when it became possible to measure the *frequency* of infrared and visible laser radiation to very high accuracy. This is done by taking microwave frequencies (at about 10^{11} Hz), which are accurately known in terms of the Cs (cesium) frequency standard (see Sec. 1.5), and multiplying them into the near-infrared region of the spectrum, using a variety of advanced electronic techniques. This results in the measurement of the frequency of near-infrared radiation (at about 10^{14} Hz) to almost the accuracy of the Cs frequency standard (a few parts in 10^{14}).

Some of the values for the speed of electromagnetic waves in vacuum obtained from the above measurements are presented, in chronological order, in Table 22.2. Note the improvement in accuracy of these measurements as physics experiments became more sophisticated and better apparatus became available. The laser measurements in the early seventies were among the most precise and accurate determinations of a physical constant ever made and led to the value $c = 299,792.458 \pm 0.001$ km/s. As indicated in Sec. 1.5, this result was considered so certain that the speed of light (and therefore of all electromagnetic waves) in vacuum has been *defined* to be exactly 299,792,458 m/s. The meter is then equal to the distance traveled by light in 1/299,792,458 s.

Example 22.5

In the measurement of c using an eight-sided rotating mirror, Michelson found that the mirror rotated $\frac{1}{8}$ rev in the time it took light to travel to a mirror 35.0 km away and return to the rotating mirror. From the known value of c, what was the frequency of revolution of the mirror in Michelson's experiment?

SOLUTION

The time for one complete revolution of the eight-sided mirror is $T = 1/f$, where f is the number of revolutions per second of the mirror. The time for a rotation of $\frac{1}{8}$ rev is therefore $t = 1/8f$, and so

$$c = \frac{d}{t} = \frac{70.0 \times 10^3 \text{ m}}{1/8f} = (70.0 \times 10^3 \text{ m})(8f)$$

or $\quad f = \dfrac{2.998 \times 10^8 \text{ m/s}}{560 \times 10^3 \text{ m}} = \boxed{535 \text{ rev/s}}$

22.8   The Generation of Radio Waves

Let us now consider how electromagnetic waves are generated and detected. These processes differ from one region of the electromagnetic spectrum to another, but as a useful example we consider here the generation of waves in the *radio-frequency* region of the spectrum, i.e., waves with wavelengths of about 100 m.

We have already seen (Sec. 21.5) that an *RLC* circuit has a resonant frequency $f = (1/2\pi)(\sqrt{1/LC})$. If we feed energy into such a circuit at this frequency, we build up large oscillations of energy between the electric field of the capacitor and the magnetic field of the inductor. These oscillations still do not constitute an electromagnetic wave, since the electric and magnetic fields are largely confined to separate regions of space (the regions near the capacitor and the inductor) and do not overlap to any extent. To produce electromagnetic waves we need overlapping electric and magnetic fields to allow a changing electric field to produce a magnetic field, and a changing magnetic field in turn to produce an electric field, as required to sustain the waves.

The Electric Field

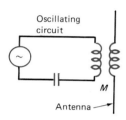

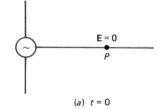

FIGURE 22.13   A dipole antenna for transmitting radio waves.

To produce overlapping electric and magnetic fields we use an antenna. The type show in Fig. 22.13 is called a *dipole antenna*. It consists of a long rod connected through a mutual inductance M to an oscillating circuit feeding in energy at a frequency f.

The net distribution of electric charge on such an antenna varies periodically in time with a period $T = 1/f$. Figure 22.14 shows the charge distribution at intervals of $T/4$ from $t = 0$ to $t = T$. At $t = 0$ the entire antenna is assumed to be electrically neutral and the electric field at point P is zero. A time $T/4$ later, however, the oscillator has moved electrons from the bottom to the top of the antenna so that the top becomes electrically negative and the bottom electrically positive, as in Fig. 22.14*b*. The electric field at P is therefore the vector sum of the electric fields produced by the net positive and negative charges. This field is directed vertically upward, as shown in the figure. A time $T/4$ later the oscillator has driven a sufficient number of electrons to the bottom of the rod to make both top and bottom electrically neutral again, and the field zero, as in Fig. 22.14*c*. This process then repeats itself in the opposite direction, and at time $3T/4$, **E** has the same magnitude as in (*b*) but is in the opposite direction. When $t = T$, the oscillator is as it was initially, and the electric field is again zero.

The crucial point here is that the electric field is not static. As the oscillator drives charge up and down the antenna, these electric field lines move away from

FIGURE 22.14   Charge distribution on a dipole antenna at intervals of $t = T/4$. The electric field **E** at point P oscillates in time as shown.

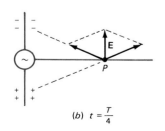

(*b*) $t = \dfrac{T}{4}$

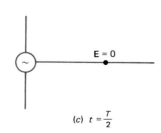

(*c*) $t = \dfrac{T}{2}$

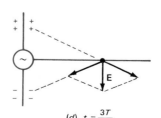

(*d*) $t = \dfrac{3T}{4}$

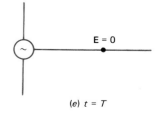

(*e*) $t = T$

the antenna with a speed c. Hence the electric field configuration along the positive x axis looks like that in Fig. 22.15, where the five points marked correspond to the five times discussed in relation to Fig. 22.14. It is clear that such an electric field distribution is identical to that shown in Fig. 22.5, with the wavelength being the distance between points a and e in Fig. 22.15.

Note that the electric field lines no longer begin and end on charges once the wave is radiated into space. Rather the field lines form continuous curves in space, as indicated in Fig. 22.16. The electric field shown in Fig. 22.15 is only the strength of the electric field along the positive x axis. The dashed lines in Fig. 22.16 represent the electric field lines in space. Notice that all the lines of **E** form closed curves.

FIGURE 22.15 (left) Electric field **E** produced by a dipole antenna as a function of distance along the x axis. The wave pattern moves along the x axis with the speed of electromagnetic waves in air.

FIGURE 22.16 (right) Complete lines of the electric field for a traveling electromagnetic wave. The lines of **E** form closed loops in space.

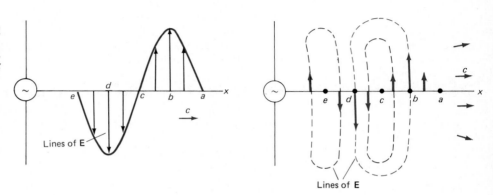

The Magnetic Field

While the oscillator is moving charge up and down the antenna in the y direction, it is also producing a changing magnetic field in the z direction. As the oscillator goes through one period T, the conventional current goes from a maximum in one direction to zero to a maximum in the other direction and then back to zero again, and these currents produce magnetic fields. For this reason the magnetic field **B** also goes through a complete cycle in one period. The lines of the magnetic field form closed circles in the xz plane around the antenna rod and are always at right angles to the lines of the electric field. This magnetic field moves through space at a speed c and is diagramed in Fig. 22.17. Since the electric field is in the y direction and the magnetic field in the z direction when the wave is moving along the positive x axis, the electric field **E**, the magnetic field **B**, and the direction of propagation are mutually perpendicular to one another, as must be the case for electromagnetic waves in free space.

Phase Relations

Although the electric and magnetic fields are not in phase with each other near the antenna, they come into phase in the ''far-field'' region away from the transmitter. We can therefore combine the electric field of Fig. 22.15 with the magnetic field of Fig. 22.17 to obtain Fig. 22.18, which is identical with Fig. 22.5.

The accelerated motion of electrons up and down a straight-rod dipole antenna therefore produces electromagnetic waves that satisfy the predictions of Maxwell's equations: the electric and magnetic fields are periodic in both time and space, are transverse and mutually perpendicular, and move with the speed of light c, as in Fig. 22.18.

Electromagnetic radiations in the microwave and x-ray regions of the spectrum are also produced by the acceleration of electrons. It is the *acceleration* that is the

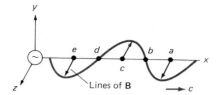

FIGURE 22.17 Magnetic field **B** in the *z* direction produced by a dipole antenna as a function of distance from the antenna along the *x* axis.

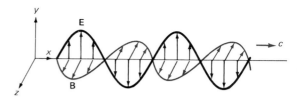

FIGURE 22.18 Combination of the electric and magnetic fields shown for an electromagnetic wave moving along the positive *x* axis in Figs. 22.15 and 22.17. The result is the electromagnetic wave of Fig. 22.5.

real source of the radiated electromagnetic energy. To accelerate the electrons requires a force, and this force does work in moving the electrons. For radio waves this work is provided by the oscillator, and the energy input to the oscillator ultimately determines the energy radiated.

22.9 The Detection and Modulation of Radio Waves

Electromagnetic waves in the radio region can be detected using an antenna similar to the transmitting antenna of Fig. 22.13. We set up a dipole receiving antenna in the direction of the electric field radiated by the transmitting antenna, as in Fig. 22.19. The incident ac field sets electrons oscillating in the receiving antenna, which generates a current in the tunable *LC* circuit connected to it. The capacitance *C* of the circuit can be tuned to resonance and large currents produced. These can then be amplified and made to produce the audio or video signals required for radio or television.

Magnetic Detection

Radio waves can also be detected by using a loop antenna to pick up the signal from the oscillating magnetic field of the wave, as in Fig. 22.20. If the magnetic field is in the *xz* plane and the loop is in the *xy* plane, so as to be normal to the lines of **B**, then as the wave passes the loop, the magnetic flux through the loop changes with time and induces an emf in the loop. Again in this case a tunable ac circuit connected to the loop can lead to sizable currents, which can be amplified further and used to produce sound or visible signals.

Both the dipole and loop antennas are highly directional because, for maximum signal reception, the dipole antenna must be lined up parallel to the electric field and the loop antenna must be at right angles to the magnetic field.

FIGURE 22.19 (left) A dipole receiving antenna. The dipole *ab* is oriented in the direction of the electric field **E**, which in this case is in the *y* direction.

FIGURE 22.20 (right) A loop antenna for receiving radio waves. The plane of the loop is at right angles to the direction of the magnetic field **B**, which in this case is in the *z* direction.

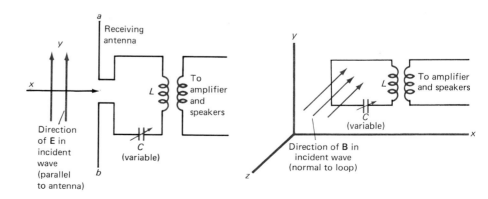

Modulation

Much of the information transmitted by radio consists of voices and music at audio frequencies in the range 20 to 20,000 Hz. The latter frequency corresponds to a wavelength for electromagnetic waves of 1.5×10^4 m. For good transmission and reception, dipole antennas must be about $\lambda/2$ in length; i.e., for audio frequencies of 20,000 Hz we would need antenna lengths of about 7500 m. This is clearly impractical. To overcome this difficulty information at audio frequencies is used to *modulate* higher-frequency electromagnetic waves at the transmitting end.

Modulation: The superposition of an audio (or video) signal on a higher-frequency carrier wave so that the information contained in the signal can be transmitted by the carrier wave.

In practice the carrier frequency for an FM radio station is about 10^8 Hz, corresponding to a wavelength of about 3 m. At the receiving end the audio signal is extracted from the higher-frequency carrier wave and converted to sound or video form. This means that the transmitting and receiving antennas can be designed for wavelengths much shorter than those in the audio range, making them more compact and practical.

There are two kinds of modulation used to transmit information at low frequencies using a higher-frequency carrier wave: amplitude modulation and frequency modulation.

Amplitude Modulation   With amplitude modulation (AM) the *amplitude* of a fixed-frequency carrier wave is modified, or "modulated," to transmit the desired information. For example, the dots and dashes of Morse code can be sent in this way by transmitting pulses of electromagnetic radiation of different lengths, as in Fig. 22.21. In this case the amplitude of the modulation varies between zero and some constant value.

Frequency Modulation   With frequency modulation (FM) the amplitude of the carrier wave is not changed, but its *frequency* is expanded or contracted to reflect the information being transmitted. To transmit dots and dashes in this way, the carrier frequency might be switched from 100 to 150 MHz and kept there for 1.5 s to indicate a dash, or for 0.5 s to indicate a dot, as in Fig. 22.22. The same technique can be used, with some modifications, to transmit music or the human voice.

FIGURE 22.21 Amplitude modulation (AM). The amplitude of the carrier wave is varied over time to carry the desired information. Here amplitude modulation is being used to transmit the dots and dashes of Morse code.

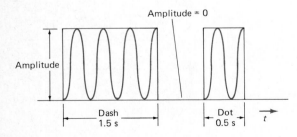

FIGURE 22.22 Frequency modulation (FM). The frequency of the carrier wave is varied over time to carry the desired information. Here frequency modulation is being used to transmit the dots and dashes of Morse code.

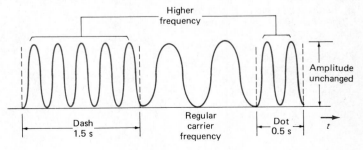

22.10 Practical Radio and Television

Figure 22.23 shows the various steps required to transmit and receive voice or music by means of radio. A microphone first converts the sound into an audio-frequency (AF) electric signal, which is then amplified and fed into a mixer. The mixer then superimposes the AF signal on the fixed-frequency, fixed-amplitude output of a radio-frequency (RF) oscillator. This modulation may be either AM or FM. The modulated RF signal is then amplified further and fed to a transmitting antenna from which it radiates outward at speed c.

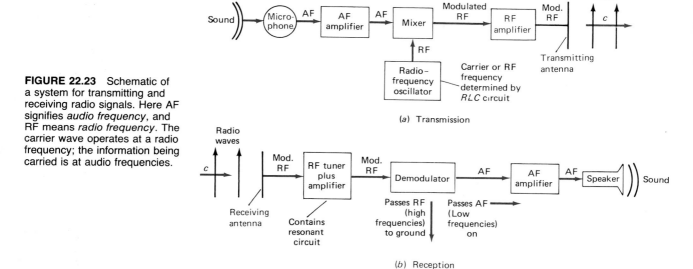

FIGURE 22.23 Schematic of a system for transmitting and receiving radio signals. Here AF signifies *audio frequency*, and RF means *radio frequency*. The carrier wave operates at a radio frequency; the information being carried is at audio frequencies.

TABLE 22.3 Frequencies Used in Radio and Television

Type of electromagnetic wave	Frequency range (MHz)
AM radio	0.54–1.6
CB (citizens band)	About 27
TV: channels 2 to 6	54–88
FM radio	88–108
TV: channels 7 to 13	174–216
UHF (ultrahigh frequency)	470–890

Expansion of Communication Channels

At the receiving end the modulated RF signal is picked up by a receiving antenna and fed to an RF tuner and amplifier. The tuner is a variable-frequency RLC circuit whose frequency can be tuned to resonance with the incoming RF frequency. This RF signal is then amplified and fed to a demodulator, which functions in a reverse fashion from the mixer. The demodulator passes low frequencies (the AF information) on to an AF amplifier, but bypasses the high-frequency RF to ground since it is no longer needed. The AF signal is then further amplified and fed to a speaker that converts it back into audible sound.

The process for picture transmission by television is basically the same, with a TV camera and TV receiver replacing the radio microphone and speaker. Television sound is usually transmitted by FM, while the video is transmitted by AM.

The approximate frequency ranges of radio and TV bands are shown in Table 22.3.

Every television channel has its own distinctive carrier frequency that, to avoid interference, must be well separated from all other allowed TV frequencies. For this reason the assigned width of a standard TV channel is 4 MHz. If we assume that this channel width remains fixed throughout the electromagnetic spectrum, then as we go to higher frequencies, one octave will contain a larger number of possible transmission frequencies (an *octave* is the frequency interval between a wave of frequency f and one of frequency $2f$). Thus in the octave from 60 to 120 MHz there is room for 15 TV channels, each 4 MHz wide. Three octaves up the scale, in the frequency range 480 to 960 MHz, one octave can contain 120 conventional TV channels, or eight times as many, of the same 4-MHz width.

The number of TV channels included in one octave of the electromagnetic spectrum thus doubles with the doubling of frequency that occurs as we go from one octave to the next. Since the frequencies of visible light (about 10^{14} Hz) are about a million times higher than those now used for TV transmission (about 10^8 Hz), this means that a million times more information channels are available in one octave of the visible electromagnetic spectrum than in one octave of the TV frequency range.

With the advent of lasers to provide the needed carrier waves, the possibility of using visible light to transmit voices and other information has become a reality. This opens up vast numbers of new channels of communication, so vast that interference between channels becomes extremely unlikely.

However, even laser light cannot penetrate buildings, heavy clouds, or other obstacles. But this sort of interference can be avoided by using optical fibers (light pipes) to transmit laser light through cables either above or below ground (see Sec. 23.2). The huge increase in the number of messages that can be sent simultaneously over such cables is revolutionizing the communications industry.

Summary: Important Definitions and Equations

Maxwell's equations: Four equations containing most of the important laws of electricity and magnetism and leading to the prediction that electromagnetic waves exist and travel at a speed c.

Maxwell's displacement current

$$I_D = \epsilon_0 \frac{\Delta \Phi_E}{\Delta t} = \epsilon_0 A \frac{\Delta E_\perp}{\Delta t} \qquad \text{(22.4, 22.5)}$$

where $E_\perp$ is the component of the electric field perpendicular to the area A.

Electromagnetic waves: Transverse waves in which the sinusoidal electric and magnetic fields are at right angles to each other and to the direction of propagation, and which travel through free space at a speed

$$c = 1/\sqrt{\mu_0 \epsilon_0} = 2.998 \times 10^8 \text{ m/s} \qquad \text{(22.9)}$$

For all electromagnetic waves in a vacuum $c = \lambda f$, where λ is the wavelength in meters and f is the frequency in hertz.

Relationship between magnitudes of E field and B field in an electromagnetic wave:

$$E = cB \qquad \text{(22.7)}$$

Speed of electromagnetic waves in dielectric materials:

$$v = \frac{c}{\sqrt{K}} \qquad \text{(22.11)}$$

where K is the dielectric constant.

Intensity of an electromagnetic wave: The average power transported across unit cross-sectional area perpendicular to the direction in which the wave is moving.

$$\text{Intensity } S = \overline{P} = \epsilon_0 \frac{E_0^2}{2} c = \frac{E_0 B_0}{2\mu_0} = \frac{1}{\mu_0} \frac{B_0^2}{2} c \qquad \text{(22.16)}$$

where E_0 and B_0 are the amplitudes of the periodic **E** and **B** fields.

Complete electromagnetic spectrum: All the electromagnetic waves from radio waves (low frequency, long wavelength) to gamma rays (high frequency, short wavelength).

Radio waves: Electromagnetic waves of frequencies between 10^1 and 10^9 Hz produced by the accelerated motion of electrons up and down a straight-rod dipole antenna.
Modulation: The superposition of an audio (or video) signal on a higher-frequency carrier wave so that the information contained in the signal can be transmitted by the carrier wave at a speed c.
Amplitude modulation (AM): The amplitude of the carrier wave is varied to carry the desired information.
Frequency modulation (FM): The frequency of the carrier wave is varied to carry the desired information.

Questions

1 A very long, thin bar magnet has its north pole 0.70 m from its south pole. A spherical surface is constructed that surrounds the north pole and intersects the magnet 0.30 m from its end, as in Fig. 22.24. What is the net flux of the magnetic field through this surface? Why?

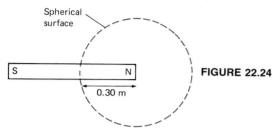

FIGURE 22.24

2 What was the key idea first introduced by Maxwell in putting forth his equations to describe electricity, magnetism, and their interactions? Why was this idea so important?

3 Point out three ways in which sound waves differ from electromagnetic waves and three ways in which a wave traveling along a stretched string differs from electromagnetic waves.

4 Heinrich Hertz, in an 1889 report on his experiments with electromagnetic waves, stated: "I am here to support the assertion that light of every kind is itself an electrical phenom-enon—the light of the sun, the light of a candle, the light of a glowworm." Discuss how the work of Maxwell and Hertz led to the general acceptance by scientists of this statement.

5 Electromagnetic waves transport energy through space. Would you expect them also to transport momentum? Why?

6 Can you suggest why gamma rays differ so radically from radio waves in their properties, even though they are both part of the same electromagnetic spectrum?

7 X-rays are produced when electrons moving at high speeds collide with a metal target. Can you suggest the cause of the continuous x-ray spectrum produced?

8 Why should you expect standing wave measurements to be more accurate than time-of-flight measurements of the speed of light in a vacuum?

9 Why would you expect less "static" and other kinds of interference on FM radio than on AM radio?

10 A radio transmitter has a vertical antenna, i.e., one directed along the y axis, as in Fig. 22.13. At the receiving end it is desired to detect the signal magnetically by placing a loop of wire in the direction that will yield maximum signal. If the radio wave is being propagated along the positive x axis, should the loop of wire be placed in the xy plane, the xz plane, or the yz plane?

11 Why are the "rabbit ears" on some television sets about a meter long when fully extended?

Multiple-Choice and Simple Exercises

22.1 The electric field between the plates of an air capacitor of plate area 0.050 m^2 is changing at a rate of 10^4 V/(m·s). Maxwell's displacement current is, in this case:
(a) 0   (b) 1.0×10^4 A   (c) 4.4×10^{-11} A
(d) 4.4×10^{-10} A   (e) 4.4×10^{-9} A

22.2 A state trooper's car sends out a radar signal at a frequency of 10.5 GHz. The wavelength of this signal is about:
(a) 3 m   (b) 30 m   (c) 0.3 m
(d) 0.003 m   (e) None of the above

22.3 When a light wave passes from air into water:
(a) Both its wavelength and its frequency remain unchanged.
(b) Its frequency remains unchanged, but its wavelength decreases.
(c) Its frequency remains unchanged, but its wavelength increases.
(d) Its wavelength remains unchanged, but its frequency increases.
(e) Its wavelength remains unchanged, but its frequency decreases.

22.4 An electromagnetic wave is traveling through free space. At a particular instant of time the electric field at a point along the path of the wave is 4.2×10^3 V/m. At the same time and at the same place the magnetic field is:
(a) 1.4×10^{-5} G   (b) 1.4×10^{-5} T
(c) 1.4×10^{-7} T   (d) 4.2×10^{-7} T
(e) 1.3×10^{12} T

22.5 At low frequencies the dielectric constant of Pyrex glass is 4.5. The speed of low-frequency radio waves in Pyrex should be about:
(a) 3.0×10^8 m/s   (b) 14×10^8 m/s
(c) 0.67×10^8 m/s   (d) 6.4×10^8 m/s
(e) 1.4×10^8 m/s

22.6 The speed of light in a piece of light flint glass is 1.90×10^8 m/s. The wavelength of the red light ($\lambda = 633$ nm in a vacuum) emitted by a He–Ne laser and passing through the flint glass is:
(a) 633 nm   (b) 999 nm   (c) 401 nm
(d) 78 nm   (e) 5.14 μm

22.7 Which of the following lists electromagnetic radiations in the correct order of *increasing wavelength*?
(a) Radio, ultraviolet, visible light, x-rays
(b) Microwaves, radio, visible light, infrared
(c) Gamma-rays, microwaves, infrared, visible light
(d) Radio, infrared, visible light, x-rays
(e) X-rays, infrared, microwaves, radio

22.8 The sodium atom has a spectrum with two very close spectral lines (a doublet) in the yellow region with a wavelength at about 589 nm. The third harmonic of these two lines occurs in the following region of the spectrum:
(a) Yellow   (b) Infrared   (c) Microwave
(d) Ultraviolet   (e) X-ray

22.9 An air capacitor with plates of area 150 cm^2 and a gap of 2.0 mm between the plates is being charged by a battery at a rate of 1.0×10^{-8} C/s. What is the rate at which the electric field between the plates is changing?

22.10 In Exercise 22.9, what is the displacement current while the capacitor is being charged?

22.11 A radar wave has associated with it a magnetic field of rms value 8.0×10^{-6} T. What is the rms value of its electric field?

22.12 A radio wave traveling through air consists of a sinusoidal electric field of rms value 4.0×10^{2} V/m and a magnetic field perpendicular to this electric field. How much energy is stored on the average in a unit volume of space as this wave passes through it?

22.13 An electromagnetic wave is traveling through free space. The rms value of the electric field associated with this wave is 5.6×10^{3} V/m. What is the intensity of this wave?

22.14 For light waves of 500-nm wavelength the index of refraction $n \; (= \sqrt{K})$ of flint glass is 1.58. What is the speed of 500-nm light in this glass? Is it greater or less than in air?

22.15 WGMS-FM in Washington, D.C., transmits at a frequency of 103.5 MHz. To what wavelength does this correspond?

22.16 What is the frequency band corresponding to the microwave region of the spectrum, extending from wavelengths of about 1 mm to 1 m?

22.17 How long did it take light to travel the 70-km round trip between Mount Wilson and Mount San Antonio in Michelson's 1923 measurement of the speed of light?

22.18 What is the wavelength range covered by TV channels 7 to 13?

Problems

22.19 An air capacitor with plates of area 100 cm^2 and a 0.50-mm air gap is being charged by a battery at a rate of 2.0×10^{-9} C/s. What are (a) the capacitance of the capacitor; (b) the current in the circuit while the capacitor is being charged; (c) the displacement current; (d) the magnetic field 20 cm from a wire carrying current from the battery; (e) the magnetic field 20 cm out from the central line connecting the two plates?

22.20 A 5.0-μF parallel-plate air capacitor is charged by a 24-V battery. After being fully charged, the capacitor is discharged through a 500-Ω resistor. What is the average displacement current during a time equal to the time constant of this circuit?

22.21 A parallel-plate air capacitor is part of an LC resonant circuit with natural frequency 5.0 MHz. The inductance of the coil in the circuit is 0.20 mH. The capacitor, which has a plate area of 20 cm^2, is first charged to 150 V and then connected to the inductance coil. What is the average value of the displacement current during the time it takes the voltage across the capacitor to change from $+150$ to -150 V?

*22.22 A parallel-plate capacitor consisting of two circular plates of area $A = \pi R^2$ separated by air is being charged by a constant current I.

(a) Show that the magnetic field between the plates at a distance r from the line connecting the centers of the two plates is $B = \mu_0 Ir/2A$ for r less than R. (Hint: Use Ampère's circuital law and Maxwell's displacement current.)

(b) Show that for r equal to R this expression reduces to the usual expression for the magnetic field at a distance R from a long, straight wire.

22.23 The maximum value for the electric field very close to a radar transmitter is 2.5×10^{-3} V/m.

(a) What is the maximum value of the magnetic field at the same place?

(b) How does this compare with the earth's magnetic field?

22.24 A straight wire 2.0 m long is oriented in the direction of the electric field vector **E** in an electromagnetic wave.

(a) To produce a voltage difference between the ends of the wire of maximum value 10^{-2} V, how large must the maximum electric field be?

(b) For the same wave, what is the maximum magnetic field?

22.25 A ruby laser emits intense red light at a wavelength of 0.694 μm. Crystals like KDP can be used to produce higher harmonics because of the laser radiation's great intensity. For example, the second harmonic of the 0.694-μm ruby laser line falls in the ultraviolet region of the spectrum. What are (a) the frequency of the second harmonic; (b) the wavelength of the second harmonic; (c) the speed of the second harmonic?

22.26 In Bohr's model of the hydrogen atom the electron circles the proton 3.6×10^{15} times per second.

(a) If this were the source of the radiation emitted by the hydrogen atom, what would the wavelength emitted be?

(b) In what spectrum region does this wavelength fall?

(c) How could such a model explain the fact that the hydrogen atom emits not one wavelength, but many different ones?

22.27 (a) From the expression $c = 1/\sqrt{\mu_0 \epsilon_0}$, prove that c has the dimensions of a velocity.

(b) From the expression $c = E/B$, prove that c has the dimensions of a velocity.

22.28 A 60-Hz radio wave passes from air into a block of Teflon of dielectric constant 2.1. What are (a) the speed of the radio wave in the Teflon; (b) the wavelength in the Teflon; (c) the frequency of the wave in the Teflon?

22.29 An electromagnetic wave is moving through a material for which $K = 4.3$ and $K_m = 1.0$ at the frequency of the wave.

(a) Find the speed of the wave through the material.

(b) If the frequency is 2.0 kHz, find the wavelength of the wave.

22.30 A plane electromagnetic wave has a wavelength of 10 cm and an electric field with an amplitude of 20 V/m. What are (a) the amplitude of the magnetic field associated with the wave; (b) the frequency of the wave; (c) the intensity of the wave?

22.31 The intensity of a radio wave is 2.5 W/m^2. What are (a) the rms value of the electric field associated with this wave; (b) the amplitude of the magnetic field associated with this wave?

22.32 A light wave traveling through air consists of a sinusoidal electric field of rms value 20 V/m and a magnetic field perpendicular to this electric field. Find the intensity of this wave, using (a) the rms value of the electric field; (b) the rms value of the magnetic field.

22.33 How much energy is transported across a 10-cm^2 area in 1 min by an electromagnetic wave whose electric field **E** has an rms value of 200 V/m?

22.34 (a) Show that the expression $\epsilon_0 E^2$ has the dimensions of energy per unit volume.

(b) Show that the expression $\epsilon_0 E^2 c$ has the proper dimensions for the intensity of an electromagnetic wave.

*22.35 The solar constant is the rate at which electromagnetic energy radiated by the sun falls on the earth's atmosphere. Its value, at high noon, for radiation falling on the atmosphere at right angles to the earth's surface, is 1.35 kW/m^2.

(a) Find the rms values of the electric and the magnetic fields needed to produce this energy flow from the sun to the earth.

(b) If the distance from the sun to the earth is 1.5×10^{11} m, find the total power radiated by the sun.

22.36 A 100-W light bulb emits about 10 percent of its 100-W input as visible light that is radiated uniformly in all directions. Calculate the rms values of the electric and magnetic fields for the light wave 3.0 m from the bulb.

*22.37 A radio transmitter emits electromagnetic waves at 98 MHz. At 150 m from the transmitter, the rms value of the electric field is 10 V/m. If we assume that the antenna is a point source radiating its energy equally in all directions, what are (a) the rms value of the magnetic field 1.5 km from the transmitter; (b) the intensity of the radio waves at 1.5 km from the transmitter; (c) the power radiated by the transmitter?

*22.38 It has been suggested that, to reduce energy losses in electric power cables, electricity should be transmitted through space in the form of electromagnetic waves. Suppose that we wanted to transmit an amount of electric power comparable to that handled by modern transmission lines, say, 1000 A at 750 kV, and to do this using a beam of 50-m^2 cross-sectional area.

(a) What would the rms values of E and B in the electromagnetic wave have to be?

(b) Does this strike you as a realistic scheme?

22.39 Helium-neon lasers emit very intense red light at a wavelength of 0.633 μm.

(a) What is the frequency of this light?

(b) What is the frequency of the second harmonic of this radiation?

(c) In what region of the spectrum does this second harmonic occur?

22.40 On the Fourth of July a large fireworks display is set off 1.5 km from an observer. How long after the observer sees the visual display will the sound of the initial explosion be heard? (Assume that the speed of sound is 340 m/s.)

22.41 Who will hear the voice of a radio announcer first, a woman in the studio audience 40 m away from the announcer whose voice reaches her directly, or her husband who is listening to the radio in their home 160 km from the radio studio? By how much? (*Hint:* See Prob. 22.40.)

22.42 "Ghosts" on TV screens arise from reflected signals that travel longer paths to the TV set than the main signal. We assume that, in a particular case, signals arriving 1.0 μs later than the main signal lead to a secondary image at a place 1.0 cm to the right of the main image on the screen. If a ghost image is 0.60 cm to the right of the main image, what is the difference in path length between direct and reflected signals?

22.43 An amateur radio operator desires to construct a tuner to receive wavelengths in the neighborhood of 20 m. She has available a capacitor variable between 20 and 80 pF. If she decides to wind an inductance coil to use in the tuning circuit, what must its inductance be?

22.44 An RLC circuit with $L = 1.0$ mH, $C = 10$ μF, and $R = 0.50$ Ω, resonates at its natural frequency. If this circuit is attached to a dipole antenna, what is the wavelength of the radiation produced?

22.45 An FM receiver has a dial marked from 88 to 108 MHz. If the fixed inductance in the circuit is 1.0 μH, what range of values must the variable capacitor cover to tune the receiver over the entire frequency range?

*22.46 An electromagnetic wave of frequency 10 MHz passes through a circular loop antenna that has a cross-sectional area of 0.050 m^2 and consists of 200 turns. If the maximum value of the electric field in the wave is 5.0×10^{-2} V/m, what is the approximate amplitude of the emf generated in the loop antenna?

Additional Readings

Hertz, Johanna (ed.): *Heinrich Hertz: Memoirs, Letters, Diaries*, San Francisco Press, San Francisco, 1977. A fascinating account of the education and development of a great physicist as told mostly in his own words.

Jackson, Charles Lee: "The Allocation of the Radio Spectrum," *Scientific American*, vol. 242, no. 2, February 1980, pp. 34–39. An account of the difficulty in deciding who gets to use what frequencies.

Jaffe, Bernard: *Michelson and the Speed of Light*, Doubleday Anchor, Garden City, N.Y., 1960. A brief biography of Michelson, emphasizing his lifelong obsession with the measurement of c.

MacDonald, D. K. C.: *Faraday, Maxwell and Kelvin*, Doubleday Anchor, Garden City, N.Y., 1964. Contains some very interesting biographical material and photographs relating to Maxwell's life.

Morrison, Philip, and Emily Morrison: "Heinrich Hertz," *Scientific American*, vol. 197, no. 6, December 1957, pp. 98–106. A brief biography of a man who accomplished a great deal in a short time.

Mulligan, Joseph F.: "Heinrich Hertz and the Development of Physics," *Physics Today*, vol. 42, no. 3, March 1989, pp. 50–57. An account of Hertz's life and his many contributions, both experimental and theoretical, to physics.

Newman, James R.: "James Clerk Maxwell," in *Lives in Science*, a *Scientific American* book, Simon and Schuster, New York, 1957, pp. 155–180. A brief account of Maxwell and his contributions to physics.

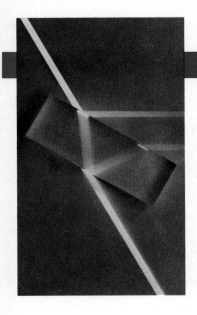

CHAPTER 23

Geometrical Optics: Mirrors and Lenses

To obtain information for myself and for others directly from nature gives me so much more satisfaction than to be always learning it from others and for myself alone—so much more that I can scarcely express it.

Heinrich Hertz (1857–1894)

In the previous chapter we presented strong evidence that light consists of electromagnetic waves moving through space at a speed c. The only difference between light and other forms of electromagnetic radiation is that light can be detected by the retina of the human eye and so can convey visual information to the human brain.

The *wave model* of light, to be discussed in the next chapter, works extremely well; for certain properties of light like diffraction and interference, such a wave model is absolutely necessary to explain experimental results. For other properties, however, such as reflection and refraction, a simpler *ray model* works equally well. In this chapter we confine ourselves to this simpler ray model.

23.1 Ray (Geometrical) Optics

The ray model of light assumes that light travels in straight lines along "rays." In Fig. 23.1 we see a plane electromagnetic wave moving along the positive x axis. We choose as our wave fronts those positions on the wave where the electric field **E** has its maximum value, as in Fig. 23.2. Consecutive wave fronts like AA' and BB' are then planes normal to the x axis and separated by one wavelength λ, as in the figure. We can then define wave fronts and rays as follows:

FIGURE 23.1 (left) A plane electromagnetic wave. The planes represent wave fronts spaced a wavelength λ apart, and the arrows indicate rays.

FIGURE 23.2 (right) The variation of the electric field **E** for the wave in Fig. 23.1. The wave fronts are shown separated by the distance between the maximum positive values of **E**, a distance equal to the wavelength λ.

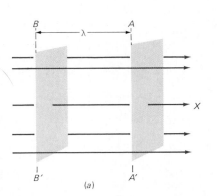

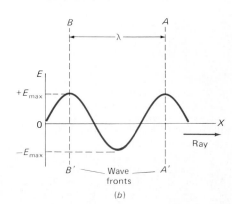

Wave front: A surface joining all adjacent points on a wave that have the same phase.

Ray: A perpendicular to the wave fronts at each point along the path of the wave.

For the plane wave in Fig. 23.1, the rays are all directed along the positive x axis, i.e., in the direction in which the wave is moving and energy is being propagated.

According to this ray model, opaque objects cast sharp shadows whose dimensions can be calculated by geometry, as in Fig. 23.3. For this reason *geometrical optics* is sometimes used as a synonym for *ray optics*. Similarly, if light passes through an aperture, the area of a screen illuminated by the light can be found (with the aid of a little geometry) by assuming that light travels in perfectly straight lines. This assumption is only valid as long as the wavelength of the light is small compared with the smallest dimension of the object or aperture on which the light is incident. We will see in the next chapter that this is often not the case.

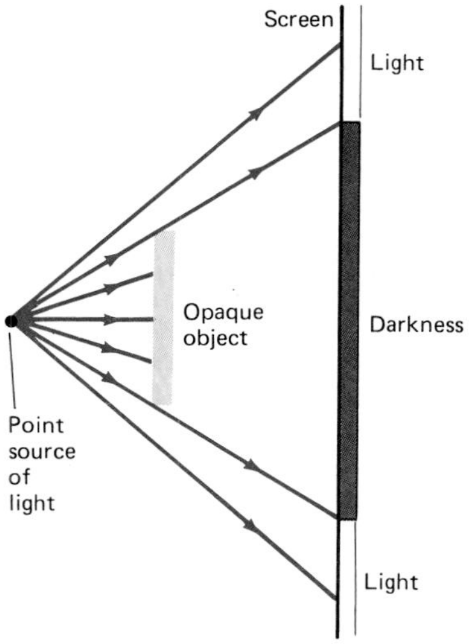

FIGURE 23.3 The shadow cast by an opaque object on a screen, on the assumption that light travels along rays in perfectly straight lines.

23.2 Reflection and Refraction on a Ray Model

The ray model of light is especially useful in dealing with reflection and refraction, and therefore for mirrors, lenses, and many optical instruments.

Reflection of Light

Consider a plane wave front incident on a plane, horizontal mirror, as in Fig. 23.4. The planes of the mirror and of the incident wave fronts are all drawn normal to the plane of the paper. A ray, which is by definition perpendicular to the wave front, is then in the plane of the paper. So too is the normal to the mirror. Hence the reflection of light can be represented either by the wave front diagram of Fig. 23.4 or by the ray diagram of Fig. 23.5.

Experimentally we find that the reflection of light is governed by the following two rules:

1 *The incident ray, the reflected ray, and the normal to the surface all lie in the same plane* (the plane of the paper in Fig. 23.5).

2 *The angle of reflection r is equal to the angle of incidence i, where both angles are measured with respect to the normal to the mirror.*

FIGURE 23.4 (left) Reflection of a plane wave front. The wave fronts shown are at right angles to the direction of wave propagation.

FIGURE 23.5 (right) Ray diagram for the reflection of light by a mirror. Experimentally the incident ray, the reflected ray, and the normal to the mirror are all in the same plane.

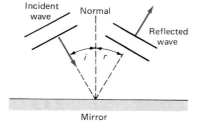

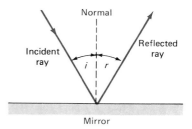

FIGURE 23.6 A ball bouncing off the ground at an angle. Here too the angle of reflection is equal to the angle of incidence.

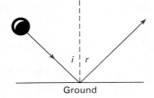
Ground

FIGURE 23.7 The diffuse reflection of light. The angle of reflection is still equal to the angle of incidence for each ray, but since the surface is rough, the reflected rays go off in many different directions.

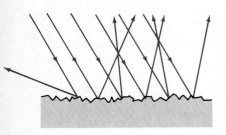

These results for the reflection of light are the same as would be obtained for the path of a spinless ball making a perfectly elastic collision with the floor at an angle of incidence i, as in Fig. 23.6. The ball would be reflected at an angle $r = i$, and its motion would be in the plane defined by its initial path and the normal to the floor, since no force acts on the ball to move it out of that plane.

For rough surfaces, the law of reflection still holds, but different rays are reflected in different directions because they strike parts of the surface oriented at different angles with respect to the incident ray. This we call *diffuse reflection*, which is illustrated in Fig. 23.7.

Example 23.1

A plane wave from a spotlight strikes a mirror at an angle of 37.5° with respect to the normal and is reflected by the mirror. What is the angle between the incident and the reflected waves?

SOLUTION

Since the angle of reflection is equal to the angle of incidence, we have for the angle between the incident and reflected waves

$$i + r = i + i = 2i = 2(37.5°) = \boxed{75.0°}$$

Example 23.2

Two plane mirrors are arranged at right angles to each other. A light ray is reflected off the two mirrors, as in Fig. 23.8, in such a way that the plane in which the light ray moves is normal to the surface of both mirrors. Show that, no matter what the angle of incidence on the first mirror, the final angle of reflection off the second mirror is such as to send the light back in the direction from which it came.

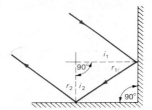

FIGURE 23.8

SOLUTION

Let the initial angle of incidence be i_1. The light is reflected off the first mirror at an angle $r_1 = i_1$, since the angle of reflection is equal to the angle of incidence. The light is then incident on the second mirror at an angle i_2 and reflected at an angle r_2 with respect to the normal, where $r_2 = i_2$. But, from Fig. 23.8,

$$r_1 + i_2 = 90° \qquad \text{and so} \qquad i_1 + r_2 = 90°$$

On adding these two equations, we obtain

$$i_1 + r_1 + i_2 + r_2 = 180°$$

But the left side of this equation is the total angle through which the incident ray has been deviated by the two reflections. It has

therefore been deviated by 180° and is moving in a direction opposite to its original direction. It is clear from this derivation that our result is independent of the angle of incidence.

A very practical application of this idea consists of *three* mirrors that are mutually perpendicular. Such a device, called for obvious reasons a *corner-cube reflector*, or *retroreflector*, reflects any incident ray through exactly 180°, no matter what the initial direction of the light. A small panel of corner-cube reflectors left behind on the moon's surface by *Apollo 11* astronauts in 1969 has provided 20 years of precise data for astronomers and astrophysicists. It has enabled the distance between the centers of the earth and its moon to be measured to within 3 cm in 3.84×10^8 m, or to one part in 10^{10}!

Refraction of Light

TABLE 23.1 Indices of Refraction*

Material	n
Air (STP)	1.00029
Hydrogen (STP)	1.00013
Water	1.33
Benzene	1.50
Ice	1.31
Glass:	
Borosilicate crown	1.52
Light flint	1.58
Heavy flint	1.65
Heaviest flint	1.89
Diamond	2.42

*For yellow light, with $\lambda = 589$ nm.

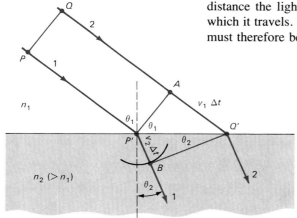

Refraction: The change in direction of a light ray when it travels from one medium to another.

The crucial point to remember about refraction is that the speed of light is different in different materials. This speed is determined by Eq. (22.11), $v = c/\sqrt{K} = c/n$, where c is the speed of light in a vacuum and n is the index of refraction of the material for light of the particular wavelength used. The speed of light is highest in a vacuum, somewhat smaller in gases, and much smaller in solids, as is clear from Table 23.1. (Refraction occurs for all kinds of waves, but we confine our discussion here to light waves.)

We consider two parallel light rays (1 and 2) incident at an angle θ_1 with respect to the normal to the interface between two different media, one of index of refraction n_1 and the other of index of refraction n_2, with n_2 greater than n_1, as in Fig. 23.9.

Light travels along the two rays from Q to A and from P to P' in the same time. But, as the light moves from A to Q' along ray 2 in a time Δt, the light originally moving along ray 1 travels a shorter distance, because light moves less rapidly in the second medium.

To find the new direction of ray 1 as it moves from the first medium to the second, we draw a circle of radius $R_2 = v_2 \Delta t$ about P' as a center, for this is the distance the light would travel in time Δt, and we do not know the direction in which it travels. The new wave front BQ' must be at right angles to ray 1 at B and must therefore be tangent to the circle at B, as shown in the figure.

FIGURE 23.9 The refraction of light on the basis of a ray model. This leads to Snell's law:

$$n_1 \sin \theta_1 = n_2 \sin \theta_2$$

Then, we have, from the diagram,

$$\sin \theta_1 = \frac{AQ'}{P'Q'} = \frac{v_1 \Delta t}{P'Q'} \qquad \text{and} \qquad \sin \theta_2 = \frac{P'B}{P'Q'} = \frac{v_2 \Delta t}{P'Q'}$$

But from Eq. (22.11), the speed of light in a medium of index of refraction n is $v = c/n$, and so

$$\frac{\sin \theta_1}{\sin \theta_2} = \frac{v_1 \Delta t}{v_2 \Delta t} = \frac{c/n_1}{c/n_2} = \frac{n_2}{n_1}$$

and

$$\boxed{n_1 \sin \theta_1 = n_2 \sin \theta_2} \tag{23.1}$$

In Fig. 23.9 θ_1 is the angle between the incident ray and the normal to the surface, and θ_2 is the angle between the refracted ray and the normal to the surface.

Equation (23.1) is called *Snell's law of refraction* after W. Snell (1591–1626), a Dutch mathematician. It predicts that a light ray will always be bent toward the normal in the optically denser material, i.e., the material with the larger index of refraction. If n_2 is greater than n_1, θ_2 must be smaller than θ_1.

Figure 23.10 shows a photograph of the reflected and refracted rays produced at air-glass interfaces.

FIGURE 23.10 The reflection and refraction of a light ray by a block of glass. The light enters the glass from above and is partially reflected and partially refracted at the upper surface. At the lower surface the ray is again partially reflected and partially refracted, as shown. Since the ray is refracted toward the normal on passing from air to glass and away from the normal on passing from glass to air, and the two faces of the block are parallel, the emerging ray has the same direction as the incident ray but has been displaced sideways. *(Courtesy of Educational Development Center, Newton, Massachusetts.)*

Example 23.3

A searchlight is being used at the edge of a swimming pool to locate a lost key at the bottom of the pool. The key is 4.0 m below the surface and 4.0 m from the edge of the pool. At what angle will the searchlight have to point with respect to the normal to locate the key, if the light enters the water 1.0 m from the edge of the pool?

SOLUTION

The situation is as in Fig. 23.11. Light is bent toward the normal in the water, and so angle θ_1 is greater than θ_2. From Snell's law we have

$$\sin \theta_1 = \frac{n_2}{n_1} \sin \theta_2 = \frac{1.33}{1} \sin \theta_2$$

Now, in the figure AC is the hypotenuse of the 3-4-5 right triangle ABC and therefore has a length of 5 m, and so

$$\sin \theta_2 = \frac{BC}{AC} = \frac{3}{5} = 0.60$$

and

$$\sin \theta_1 = 1.33(0.60) = 0.80$$

so that

$$\theta_1 = \boxed{53°}$$

This is the angle at which the searchlight would have to point with respect to the normal to the surface of the pool to locate the key at the bottom of the pool. Someone looking along the direction of the searchlight beam would think that the pool was

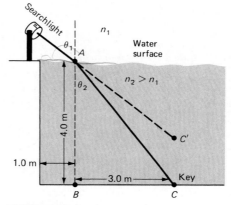

FIGURE 23.11

shallower than it really is and that the key was located at C', since the eye (and brain) would assume that the light reaching the eyes from the key had traveled in a straight-line path.

Total Internal Reflection

According to Snell's law,

$$\sin \theta_2 = \frac{n_1}{n_2} \sin \theta_1$$

If light is traveling from a denser (n_1) to a less dense (n_2) optical medium (from glass to air, for example), as in Fig. 23.12, n_1 is greater than n_2. For small angles

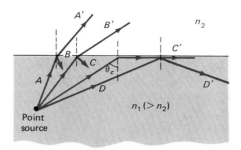

FIGURE 23.12 Total internal reflection. Light travels from a medium of index of refraction n_1 to a medium of index of refraction n_2, where $n_1 > n_2$. For incident angles greater than θ_c, where $\sin \theta_c = n_2/n_1$, no light passes into the upper medium; the light is totally reflected back into the lower medium.

FIGURE 23.13 Total internal reflection. Light rays enter a prism from the left at different angles with respect to the normal to the prism surface. The top four rays are partially reflected and partially refracted, as can be seen. The two lowest rays, however, are totally reflected at the glass-air interface, since their angle of incidence exceeds the critical angle. These two rays are much more intense than the other rays emerging from the bottom of the prism. *(Courtesy of Educational Development Center, Newton, Massachusetts.)*

of incidence, part of the light is refracted into the less dense medium, and part is reflected back into the denser medium. When the angle of incidence θ_1 is such that the angle of refraction $\theta_2 = 90°$ and thus $\sin \theta_2 = 1$, the light ray (C in the diagram) is refracted along the interface between the two materials. This particular angle of incidence is known as the *critical angle* θ_c. For any angle greater than θ_c, as that for ray D, *only reflection* (and no refraction) occurs; i.e., the light is no longer partially reflected but is *totally* reflected at the interface. This is called *total internal reflection* and occurs for all angles greater than the critical angle θ_c. Since $\sin \theta_2 = 1$, substitution in Eq. (23.1) yields for the critical angle:

Critical angle:
$$\sin \theta_c = \frac{n_2}{n_1}$$
(23.2)

Figure 23.13 shows two light rays undergoing total internal reflection.

When total internal reflection occurs, the interface acts as a perfect reflector. This makes it convenient, for example, to use 45° prisms (i.e., solid triangular blocks of glass, with two 45° angles and one 90° angle) in place of mirrors to reflect light in binoculars. This is shown in Fig. 23.14 in which all angles of incidence on the prism sides are 45°, which is greater than the critical angle of 41° for crown glass. Since ordinary mirrors always absorb some light, corner-cube reflectors use total internal reflection rather than reflection from plane mirrors to reverse the direction of a light beam.

Some crystal chandeliers and glassware are made from very heavy lead glass, for which n_1 is about 1.89 and the critical angle is only 32°. The crystal is therefore cut at such angles that much of the incident light is reflected back and forth inside

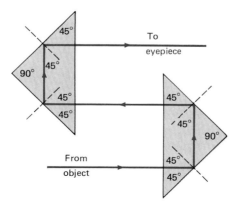

FIGURE 23.14 The use of 45° prisms to reflect light in binoculars. Since the critical angle for total internal reflection is about 41°, the light is totally reflected twice in each prism, as shown.

the crystal many times by successive total internal reflections before it escapes. This gives the crystal its uniquely bright, dazzling appearance. This sparkling effect is even greater in diamonds, for which $n = 2.42$ and θ_c is only 24°.

Example 23.4

Light passes from a prism of heavy flint glass directly into a second prism of borosilicate crown glass that is in contact with it. If the light incident on the interface makes an angle of 60.0° with respect to the normal, as in Fig. 23.15, what happens to the light after it strikes the interface?

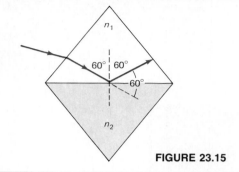

FIGURE 23.15

SOLUTION

Usually we would expect that light would be partially reflected and partially refracted at the interface. To see if that happens in this case, let us calculate the critical angle for light passing from the denser to the less dense glass. From Table 23.1, the critical angle is given by

$$\sin \theta_c = \frac{n_2}{n_1} = \frac{1.52}{1.89} = 0.804$$

and so $\theta_c = 53.5°$

Since the light is incident at an angle of 60°, which is greater than the critical angle, no refraction occurs, but *the light is totally reflected*.

For the reflected light, the angle of incidence is equal to the angle of reflection. The reflected ray therefore makes an angle of 60° with the direction of the incident ray just before it hit the interface, as can be seen from Fig. 23.15.

Fiber Optics

Total internal reflection enables light to be transmitted inside thin glass fibers, since the light is internally reflected off the sides of the fiber and therefore follows the fiber's contour. In this way light can be carried around corners as long as the bends are not too sharp so that the light always strikes the sides at angles greater than the critical angle.

In practice, round glass fibers of uniform thickness and diameters about 2×10^{-6} m are used. These are made of glass of high refractive index coated with a thin layer of glass or plastic of lower refractive index (called *cladding*; see Fig. 23.16a) to provide a suitable boundary and to reduce energy losses due to surface imperfections and contamination. These fibers are combined in bundles that are flexible enough to bend easily without breaking. If the end of a fiber bundle of this sort is polished flat, almost all the light entering one end will emerge from the other (Fig. 23.16b). These fiber bundles are therefore frequently referred to as *light pipes*.

Light pipes are now used routinely to take pictures inside the human body with fiber-optic endoscopes and to transmit high-intensity laser light inside the body for use in surgery and cancer therapy.

The most important future applications of optical fibers lie in the area of voice or video communications and data transmission. The information-carrying capacity of light waves is much greater than for microwaves and radio waves because of the higher frequency and the resulting greater *bandwidth* (See Sec. 22.10).

AT&T now has a 776-mi glass fiber link between Massachusetts and Virginia in place and a second cable linking Sacramento and San Diego, California, under construction. Such fiber-optic cables are capable of transmitting 2 billion "bits"

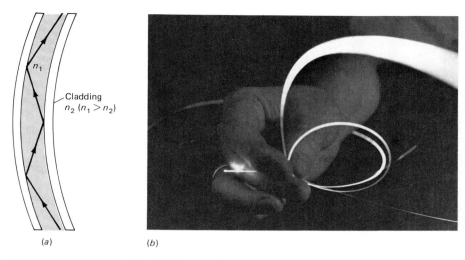

FIGURE 23.16 (a) Bending of light around curves by optical fibers. (b) Glass fibers used in light communication systems. *(Courtesy of AT&T Bell Laboratories.)*

of information per second over fiber lengths exceeding 100 km before amplification to boost the signal is required. At this rate the information in 500 novels could be transmitted in 1 s.

On December 14, 1988, the first fiber-optic telephone cable across the Atlantic went into service, vastly increasing the number of phone calls that can be placed at one time to Europe. The new cable can carry 40,000 two-way conversations simultaneously, whereas the three existing underwater copper cables together can carry only 12,000 calls. Because of this greatly increased load capacity, it is expected that by the end of the century all long-distance telephone and telegraph lines in the United States will consist of optical fibers rather than electric cables.

The use of fiber optics in medicine and communications illustrates how a simple physical principle like total internal reflection can have far-reaching practical consequences.

23.3 Dispersion of Light by a Prism

The index of refraction of any material varies with wavelength, as shown in Fig. 23.17 for three different kinds of glass. This is called *dispersion*.

Dispersion: The dependence of the index of refraction of a material on wavelength.

It is dispersion that causes a glass prism to break up white light into all the colors of the rainbow. Since the different colors correspond to different wavelengths, they are refracted by different amounts and emerge from the prism going in different directions.

Since the index of refraction n for the glass of the prism is greater for violet light (short wavelength) than for red (long wavelength), violet light is bent most and red light least by a prism. Hence, when white light is incident on a prism, as in Fig. 23.18, we obtain a continuous spectrum of colors in the deviated light, running from red, which is least deviated, through orange, yellow, green, and blue to violet, which is most deviated. In 1666 Sir Isaac Newton made the first recorded observation of this dispersion of sunlight by a prism.

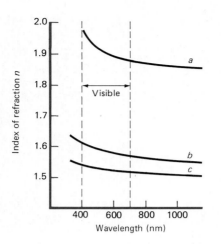

FIGURE 23.17 Dependence of index of refraction on wavelength for three kinds of glass: (*a*) heavy flint glass; (*b*) light flint glass; (*c*) zinc crown glass.

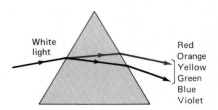

FIGURE 23.18 A prism used to break up white light into its component colors. The red light is deviated least and the violet most, leading to a complete visible spectrum.

If the light source emits not white light but light characteristic of a particular element or molecule, the prism produces images of a slit placed between the light source and the prism. These line images of the slit occur at different angles, depending on the characteristic wavelengths emitted by the element or molecule. The prism spectrometer is a device based on this principle. Such a spectrometer can be used for the qualitative and quantitative chemical analysis of materials.

23.4 Spherical Mirrors

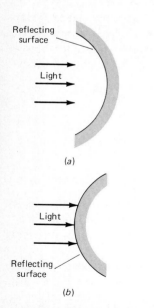

FIGURE 23.19 Types of spherical mirrors: (*a*) concave; (*b*) convex.

One of the most important applications of the laws governing the reflection of light is to mirrors. Most of us look at ourselves in a plane or concave mirror at least once a day, but probably never wonder how such a mirror works. An ordinary plane mirror is made by applying a thin metallic backing to a flat, polished glass surface. A spherical mirror has the shape of a piece cut out of a spherical surface, where the radius of the sphere *R* is called the *radius of curvature* of the mirror. Mirrors can be made to reflect almost 100 percent of the incident light, but some small amount of light is always absorbed.

If the side of a spherical mirror facing the incoming light corresponds to the *inside* portion of a sphere, the mirror is a *concave* mirror (the "cave" in *concave* has the same meaning as *cavity*). If the side facing the incoming light corresponds to the *outside* portion of the sphere, then the mirror is a *convex* mirror (see Fig. 23.19). Concave mirrors are often used as shaving or makeup mirrors, since they magnify objects placed near them. Convex mirrors are often found in stores and on automobiles, because a small convex mirror provides a large field of view.

In Fig. 23.20 the line *AV* drawn normal to the reflecting surface of a concave mirror at its center is called the *principal axis of the mirror*. (Note that, although all our diagrams of mirrors and lenses are only in two dimensions, a concave mirror is a three-dimensional object. If such diagrams are rotated through 180° around the principal axis, they provide a complete picture of the actual behavior of the mirror.) The point *V* at which the principal axis meets the mirror is called the *vertex*.

It is found by experiment that rays striking a spherical concave mirror parallel to its principal axis, and not too far off that axis, are reflected by the mirror so that they all pass through the same point *F*, called the *focal point* (or *focus*), on the principal axis, as in Fig. 23.20.

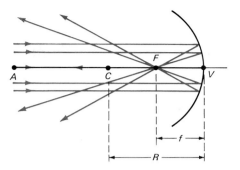

FIGURE 23.20 A concave mirror. Rays parallel to the principal axis and not too far off that axis all pass through the focal point *F* after reflection.

Focal point: The point through which all rays close to and parallel to the principal axis pass after reflection from a concave mirror.

Focal length (f): The distance of the focal point from the vertex.

All rays from an object a large distance away, such as a distant star, are parallel when they strike the mirror and are therefore brought to a focus at *F*.

Spherical Aberration

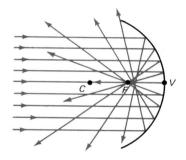

FIGURE 23.21 Spherical aberration. A spherical mirror does not focus off-axis rays at the focal point *F* but at points nearer to the mirror.

FIGURE 23.22 A light beam from an automobile headlight is produced by a small light bulb at the focus of a parabolic concave mirror.

The result of the preceding discussion is only an approximation, since as the rays depart farther from the principal axis of the mirror, they focus not at *F* but closer to the mirror, as shown in Fig. 23.21. This lack of perfect focusing by a spherical mirror is called *spherical aberration*. It can be overcome by using a parabolic mirror instead of a spherical one.

Since a parabolic mirror will bring all parallel rays incident on the mirror to a focus at *F*, a small light bulb at *F* will produce, after reflection, a beam of rays parallel to the principal axis of the mirror (this is referred to as the *principle of the reversibility of light rays*). For this reason automobile headlights have an inner surface consisting of a parabolic mirror, with a small incandescent bulb at its focus. Consequently such headlights cast very straight beams of light ahead of the car, as in Fig. 23.22.

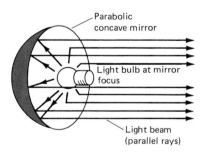

Parabolic
concave mirror

Light bulb at mirror
focus

Light beam
(parallel rays)

Image Formation by Concave Mirrors

In Fig. 23.23 a concave mirror produces an image *S'T'* of an object *ST*. To discuss the position and form of the image, the following definitions are useful:

Object distance p: The distance of the object from the mirror along the principal axis.

Image distance q: The distance of the image from the mirror along the principal axis.

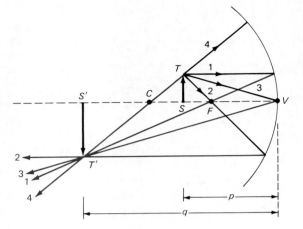

FIGURE 23.23 Rays used to locate the image formed by a concave mirror graphically. In this case the image $S'T'$ is real, inverted, and larger than the object.

There are two different ways of locating the image produced by a concave mirror—one graphical, the other analytical.

Graphical Method The graphical method consists of drawing the rays emanating from key points on the object and locating the points at which these rays are focused by the mirror, as in Fig. 23.23. For this purpose four kinds of rays are of particular importance:

1 A ray that is parallel to the principal axis and is reflected through the focal point F of the mirror

2 A ray that passes through the focal point first and is reflected parallel to the principal axis (the reverse of ray 1)

3 A ray that strikes the mirror at its vertex and is reflected so that the angle of incidence with respect to the principal axis is equal to the angle of reflection

4 A ray that passes through the center of curvature and is reflected back through the center of curvature, since it strikes the mirror normal to its surface

It is seen that all four rays originating at the tip T come together (*converge*) at T' to form the tip of the image (actually only two rays are needed, but it is helpful to draw at least three as a check). Similarly, all other points of the object are imaged in the same way. The bottom of the arrow at S is reflected by the mirror, which produces an image that must be on the principal axis, since a ray from S strikes the mirror normal to its surface and is reflected back on itself. Therefore an observer looking directly at the mirror from a distance greater than the image distance sees an image that in this case is real, inverted, and larger than the original (see Fig. 23.23).

By a *real image* we mean an image formed by light rays actually coming together (converging) to form the image, as in the figure. Whether the image is larger or smaller than the object depends on whether the object is inside or outside the center of curvature C.

Figure 23.24 shows the formation of a *virtual image* of an arrow by a concave mirror. In this case the four rays discussed above *diverge* after striking the mirror, and so no real image is formed. Rather the rays *appear to come* from a point T'

behind the mirror, although there are, of course, no real light rays behind the mirror. In this case of *diverging* rays, the image is said to be *virtual*. The image in Fig. 23.24 is upright, larger than the original, and virtual.

In dealing with mirrors, remember that, if the reflected (image) rays are *converging*, then the image is *real*; if they are *diverging*, then the image is *virtual*. Drawing a ray diagram at the beginning of each problem will clarify this distinction.

Analytical Method   We seek an equation relating the image distance, the object distance, the focal length, and the radius of curvature of a spherical mirror.

Consider an object ST placed on the principal axis of a concave mirror, as in Fig. 23.25. To find the image $S'T'$ produced by the mirror, we draw two rays from T to the mirror. The first ray travels from T to the vertex of the mirror and is reflected so that its angle of incidence (θ) is equal to its angle of reflection (θ). The second ray passes through the center of curvature of the mirror, strikes the mirror at B, and is reflected directly back on itself. These two rays meet at point T', and so $S'T'$ is the image of ST, since S' must fall on the principal axis.

In the triangle STV, we have $\tan \theta = h/p$, and in $S'T'V$ we have $\tan \theta = -h'/q$, where p and q are the object and image distances. A minus sign is used with h' because the image is *inverted* with respect to the object. We therefore find that $\tan \theta = h/p = -h'/q$, and so

$$M = \frac{h'}{h} = \frac{-q}{p} \tag{23.3}$$

Here M is the magnification of the mirror, the ratio of image size to object size.

The next step is to find a relation between p, q, and R, the radius of curvature of the mirror. From triangles STC and $S'T'C$ in Fig. 23.25, we have

$$\tan \alpha = \frac{h}{p - R} \qquad \text{and} \qquad \tan \alpha = \frac{-h'}{R - q}$$

from which

$$\frac{h}{p - R} = \frac{-h'}{R - q} \tag{23.4}$$

Combining Eqs. (23.3) and (23.4), we obtain

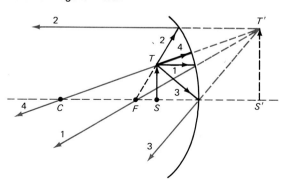

FIGURE 23.24   Formation of a virtual image by a concave mirror. In this and subsequent diagrams dashed lines mean that the light rays never actually pass over these paths. The dashed lines for the image $S'T'$ mean that the image is *virtual*.

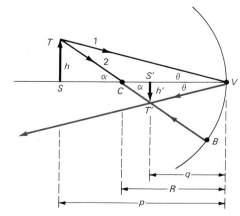

FIGURE 23.25
Obtaining the image of an object ST analytically. Rays 1 and 2 cross at T' to produce the real, inverted image $S'T'$.

$$\frac{-h'}{h} = \frac{R - q}{p - R} = \frac{q}{p} \quad \text{and so} \quad pR - pq = pq - Rq$$

or $\quad qR + pR = 2\,pq$

On dividing by pqR, we have, finally,

$$\frac{1}{p} + \frac{1}{q} = \frac{2}{R} \tag{23.5}$$

For an object very far away ($p \to \infty$), such that the rays from the object are parallel to the mirror axis, we have seen that the image is formed at F, the focal point of the mirror. Therefore in this case $q = f$, the focal length, and Eq. (23.5) becomes

$$0 + \frac{1}{f} = \frac{2}{R} \quad \text{or} \quad f = \frac{R}{2} \tag{23.6}$$

Our mirror equation is then

$$\boxed{\frac{1}{p} + \frac{1}{q} = \frac{1}{f}} \tag{23.7}$$

(We will see in Sec. 23.6 that an identical expression is obtained for the image formed by a thin lens.)

We now introduce sign conventions that apply to all cases of the formation of images by single mirrors, whether the mirror be concave, convex, or plane. The conventions we use are the following:

When the object, image, or focal point is on the front side of the mirror, i.e., the side from which light is coming toward the mirror, the corresponding distance is positive (+). If any of these is behind the mirror,* the corresponding distance is negative (−). If the magnification M (= − q/p) is positive, the image is upright; if negative, the image is inverted.

In the situation shown in Fig. 23.23 the object is on the reflecting side of the mirror and p is positive. The mirror is concave, and so f is on the reflecting side of the mirror and is positive. Graphically the image is on the reflecting side of the mirror, and so q is also positive in this case. Therefore the magnification is $M = -q/p$, which is negative, and so the image is *inverted*. If we want to solve any similar problem using Eq. (23.7), we substitute values for p and f and solve for q. If q is positive, the image is on the reflecting side of the mirror and the reflected rays *converge* to a *real* image; if q is negative, as in Fig. 23.24, the rays appear to *diverge* from an image on the nonreflecting side of the mirror and the image is *virtual*.

The analytical method is in most cases easier to use and more accurate than the graphical method, if only we stick faithfully to the adopted sign conventions.

---

*A situation in which an object is on the nonreflecting side of a mirror occurs in complex lens and mirror systems where the image produced by one element of the system becomes the object for another element of the system.

It is always important to draw a ray diagram to make sure that our results agree with the analytical solution with respect to the nature of the image (real or virtual, inverted or upright) and its approximate size and location. Mistakes can be caught easily in this way. These techniques will be clearer after we work through Examples 23.5 and 23.6.

Example 23.5

A chess pawn 3.00 cm in height is located 40.0 cm from a concave mirror with radius of curvature $R = 50.0$ cm, as in Fig. 23.26. Find the location and size of the image of the pawn, and determine whether it is inverted or upright.

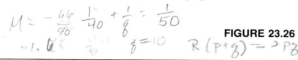

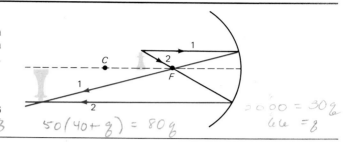

FIGURE 23.26

SOLUTION

From the mirror equation with $p = +40.0$ cm, $f = R/2 = (+50.0 \text{ cm})/2 = +25.0$ cm (since the mirror is concave), and $h = 3.00$ cm, we have

$$\frac{1}{q} = \frac{1}{f} - \frac{1}{p} = \frac{1}{25.0 \text{ cm}} - \frac{1}{40.0 \text{ cm}} = \frac{1}{66.7 \text{ cm}}$$

or $q = \boxed{66.7 \text{ cm}}$

The magnification is

$$M = -\frac{q}{p} = -\frac{66.7 \text{ cm}}{40.0 \text{ cm}} = -1.67$$

and so $h' = Mh = (-1.67)(3.00 \text{ cm}) = \boxed{-5.00 \text{ cm}}$

where the minus sign means that the image is inverted.

Hence the image is real, since q is positive; inverted, since M is negative; larger than the object; and located 66.7 cm from the reflecting side of the mirror. Figure 23.26 shows a ray diagram for this problem, which indicates that our analytical solution is correct. Only two rays are used here to find the image, but it is clear from the diagram that these two rays *converge* to form a *real* image, as was found from the analytical solution.

Example 23.6

This example is similar to Example 23.5, but the chess pawn is now at a distance 12.5 cm from the mirror, which is inside the focal point of the mirror. Find the location and size of the image, and determine whether it is inverted or upright.

SOLUTION

In this case we have

$$\frac{1}{q} = \frac{1}{f} - \frac{1}{p} = \frac{1}{25.0 \text{ cm}} - \frac{1}{12.5 \text{ cm}} = -\frac{1}{25.0 \text{ cm}}$$

and so $q = \boxed{-25.0 \text{ cm}}$

and the image is on the nonreflecting side of the mirror. It is therefore *virtual*. The magnification is

$$M = -\frac{q}{p} = -\frac{-25.0 \text{ cm}}{12.5 \text{ cm}} = +2.00$$

and the image size is

$$h' = Mh = 2.00(3.00 \text{ cm}) = \boxed{6.00 \text{ cm}}$$

In this case the concave mirror produces an upright, enlarged, virtual image of the type produced by a shaving or makeup mirror. Figure 23.27 shows a ray diagram for this case.

The rays arising at the top of the pawn and striking the mirror appear to an observer to *diverge* from the top of a larger image of the pawn behind the mirror, but no light rays actually penetrate behind the mirror to that image. The image is therefore virtual. It is useful to compare the ray diagram in Fig. 23.27 with that in Fig. 23.26 to see how the situation changes when the object is moved from outside F to inside F.

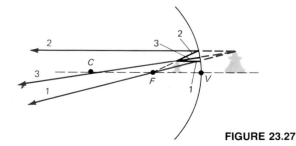

FIGURE 23.27

Convex Mirrors

When parallel rays fall on a convex mirror, as in Fig. 23.28*a*, they are reflected at angles that make it appear as if they came from a single point just behind the mirror. This point is called the *virtual focal point* of the convex mirror. The focal length is again equal to half the radius of curvature of the mirror, as shown.

The equations and sign conventions used for concave mirrors remain valid for convex mirrors, with the one change that the focal length *f* (and hence the radius of curvature *R*) must be taken as *negative*, since the focal point occurs on the nonreflecting side of the mirror, i.e., on the back side.

In constructing ray diagrams for convex mirrors we proceed as for concave mirrors except that the four rays of importance are the following:

1   A ray that is parallel to the axis and is reflected as if it came from the virtual focal point of the mirror

2   A ray that is directed toward the virtual focal point and is reflected parallel to the principal axis (the reverse of ray 1)

3   A ray that is directed toward the vertex of the mirror and is reflected so that the angle of reflection with respect to the principal axis is equal to the angle of incidence

4   A ray that is directed toward the center of curvature of the mirror and is reflected directly back on itself, since it strikes the mirror normal to its surface

These four rays enable the image of any point on an object to be located for a convex mirror; they are used to locate a virtual image in Fig. 23.28*b*.

FIGURE 23.28   (*a*) A convex mirror. Rays parallel to the principal axis are reflected so that they appear to come from a virtual focus behind the mirror. (*b*) Graphical method of obtaining the image formed by a convex mirror.

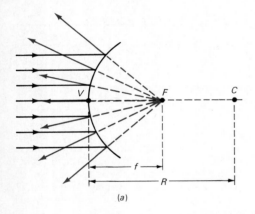

(a)

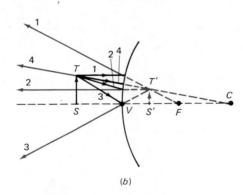

(b)

Example 23.7

A security mirror in a department store is a convex mirror of focal length 25.0 cm. What height does a man 1.80 m tall and 20.0 m from the mirror appear to be in the mirror, and what is the location of his image?

SOLUTION

Here $f = -25.0$ cm, since the mirror is convex, and $p = 20.0$ m. The mirror equation becomes

$$\frac{1}{q} = \frac{1}{f} - \frac{1}{p} = \frac{1}{-0.250 \text{ m}} - \frac{1}{20.0 \text{ m}} = -\frac{20.25}{5.00 \text{ m}}$$

and so $q = \dfrac{-5.00 \text{ m}}{20.25} = -0.247 \text{ m} = \boxed{-24.7 \text{ cm}}$

The reflected rays therefore diverge, and the image of the man is *virtual* and approximately 25 cm behind the mirror.

In this case the magnification is

$$M = -\frac{q}{p} = -\frac{-0.247 \text{ m}}{20.0 \text{ m}} = +0.0124$$

and so the image is upright. The image size is

$$h' = Mh = 0.0124(1.80 \text{ m}) = 0.0223 \text{ m} = \boxed{2.23 \text{ cm}}$$

In this case the image is nearly 100 times smaller than the object. It is because of this drastic reduction in size that a small convex mirror can cover such a large field of view. A rough graphical solution (not to scale) of this problem is shown in Fig. 23.29.

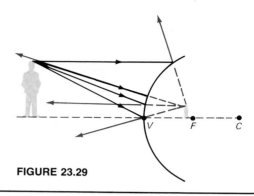

FIGURE 23.29

23.5 Plane Mirrors

A plane mirror can be treated as a special case of a concave (or convex) mirror, with the radius of curvature taken as infinite, since as R approaches infinity, a piece out of a spherical surface of radius R approaches a flat plane. If $R \rightarrow \infty$, $f = R/2 \rightarrow \infty$, $1/f \rightarrow 0$, and the mirror equation becomes

$$\frac{1}{p} + \frac{1}{q} = \frac{1}{f} = 0 \quad \text{and so} \quad q = -p$$

For a plane mirror the image distance is equal to the object distance. The image is virtual, since q is negative, and is as far behind the mirror as the object is in front of the mirror. This is shown graphically in Fig. 23.30. All the light rays originating at S and striking the mirror appear to come from a point S' behind the mirror. The law governing the reflection of light and a little geometry lead to the conclusion that the image and object distances are the same in the diagram.

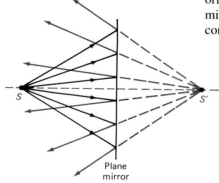

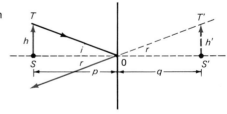

FIGURE 23.30 (left) Image formation by a plane mirror. The image always appears as far behind the mirror as the object is in front of the mirror.

FIGURE 23.31 (right) Formation of the image of an extended object by a plane mirror.

Size and Nature of the Image

Suppose we have an extended object like the arrow in Fig. 23.31, and we want to know not merely where the image is but whether it is upright or inverted and how large it is. We have already found that $q = -p$ for a plane mirror. The magnification of the object is then

$$M = -\frac{q}{p} = -\frac{-p}{p} = +1$$

The image is therefore upright and the same size as the object. This agrees with our experience with plane mirrors. Since the image is behind the mirror, it is also *virtual*. This means that a piece of white paper put at point S' will never have an image formed on it, and a photographic plate at S' will never record an image.

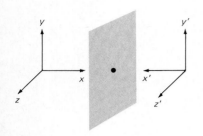

FIGURE 23.32 Reflection in a plane mirror changes a right-handed coordinate system into a left-handed coordinate system.

Figure 23.32 shows a reflection in a plane mirror. In this case the object is a right-handed set of coordinate axes, characterized by the fact that if the $+x$ axis is turned in the direction of the $+y$ axis, a *right-handed* screw (an ordinary screw) would move along the $+z$ axis; that is, it would move *out* of the plane of the paper in the figure (see Fig. 8.14). On reflection in the plane mirror the three axes have the images shown. In the image, however, the coordinate axes are *left-handed*, for if a right-handed screw were turned from $+x'$ into $+y'$, it would no longer move in the direction of $+z'$ but in the opposite direction, that is, *into* the plane of the paper in the figure. The image is reversed with respect to the plane of the mirror and is said to be *perverted*.

Just as a right-handed coordinate system is converted into a left-handed system by reflection in a plane mirror, so the right hand of a girl brushing her hair seems, on reflection in a plane mirror, to be her left hand. This change of right-handed systems into left-handed systems on reflection in a plane mirror has considerable theoretical importance in elementary-particle physics (Sec. 31.7).

23.6 Thin Converging Lenses

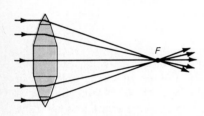

FIGURE 23.33 A glass lens considered as a series of prisms. The drawing shows the bending occurring at the center of each prism, although it actually occurs at the two faces of each prism.

Just as the laws of reflection determine the behavior of mirrors, so Snell's law of refraction determines the behavior of lenses. A glass lens can be considered to be a series of glass prisms, as in Fig. 23.33, with the light passing through the lens being refracted by the prisms as shown. In practice the surfaces of lenses are polished curved surfaces, of course, with none of the sharp breaks evident in the figure.

There are two basic kinds of lenses, *converging* and *diverging*.

Converging lens: A lens that brings all incident light rays parallel to its principal axis together at a point called the focal point of the lens.

Diverging lens: A lens that spreads out all incident light rays parallel to its principal axis so that they appear to diverge from a focal point on the object side of the lens.

The behavior of these two kinds of lenses for incident rays parallel to the principal axis (i.e., for an object very far from the lens) is shown in Fig. 23.34. Sketches of some frequently used types of converging and diverging lenses are shown in Fig. 23.35.

In what follows we limit our discussion to *thin lenses*, i.e., to lenses whose thicknesses are small compared with their focal lengths. For such lenses we draw ray diagrams as if the refraction produced by the lens occurs at the central plane of the lens, as in Fig. 23.34.

FIGURE 23.34 Types of lenses: (a) converging; (b) diverging. The focal length f of each is shown.

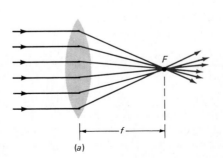
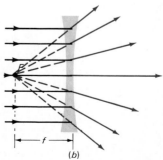

(a)   (b)

Converging lenses:

Double
convex

Plano-
convex

Convex
meniscus

Diverging lenses:

Double
concave

Plano-
concave

Concave
meniscus

FIGURE 23.35   Some frequently used types of converging and diverging lenses. Note that converging lenses are always thicker at the center than at the rim of the lens, while the opposite is true for diverging lenses.

Converging lenses behave very much like concave mirrors, as shown in Fig. 23.36, and the definitions, sign conventions, and ray diagrams used are also very similar. For lenses we use the following definitions, as illustrated in Fig. 23.37.

Object distance p: The distance of the object from the center of the lens as measured along or parallel to the principal axis.

Image distance q: The distance of the image from the center of the lens as measured along or parallel to the principal axis.

Focal length f: The image distance for an object at infinity:

Sign conventions

The object distance p is positive $(+)$ if the object is on the front side of the lens, i.e., the side from which the light is coming. In this case the rays coming from the object are diverging from a real object.*

The image distance q is positive $(+)$ if the image is on the back side of the lens. In this case the rays are converging toward a real image. If the image is on the front side of the lens, the image distance q is negative.

The focal length is positive for a converging lens and negative for a diverging lens.

If the magnification M $(= -q/p)$ is positive, the image is upright; if negative, the image is inverted.

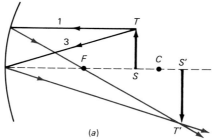
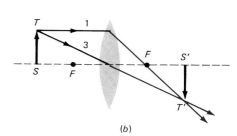

(a)   (b)

FIGURE 23.36   Comparison of a converging lens with a concave mirror: (a) a concave mirror; (b) a converging lens. Both form inverted real images of objects outside their focal points. The numbers on the rays are those used for the rays in the discussion of mirrors and lenses in the text.

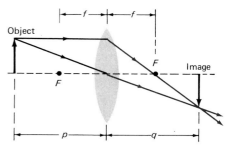

FIGURE 23.37   Definition of object distance p, image distance q, and focal length f for a converging lens. Note that any lens has two focal points, one on each side of the lens and equidistant from the lens center.

*The object distance is negative only when we are dealing with systems of lenses. In that case the image produced by one lens may be on the other side of a second lens than the side from which the light is coming. See Example 23.10.

Ray Diagrams

In locating graphically the images formed by lenses, it is useful to draw rays from a point on the object that come together to form a point on the image. The rays we use are the following (Fig. 23.38):

1 A ray that is parallel to the principal axis and is refracted through the focal point of the lens

2 A ray that passes through the focus on the object side of the lens and is refracted parallel to the principal axis (the reverse of ray 1)

3 A ray that passes through the exact center of the lens and is not refracted at all by the lens. (Actually such a ray, while not bent, is displaced slightly to one side. For a thin lens, however, this displacement is very small and can be neglected.)

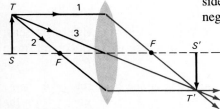

FIGURE 23.38 Graphical location of the real image formed by a converging lens.

These three rays are used in Fig. 23.38 to locate the image of an arrow that serves as the object for a converging lens. We obtain the real, inverted image shown in the figure.

Thin-Lens Equation

In Fig. 23.39 the right triangles SOT and $S'OT'$ are similar, and so

$$\frac{-h'}{h} = \frac{OS'}{OS} = \frac{q}{p}$$

The magnification of a thin converging lens is therefore

$$\boxed{M = \frac{h'}{h} = \frac{-q}{p}} \tag{23.8}$$

This is the same result obtained for a mirror.

In Fig. 23.39 the right triangles OPF and $S'T'F$ are also similar, and so

$$\frac{-S'T'}{OP} = \frac{FS'}{OF} \quad \text{or} \quad \frac{-h'}{h} = \frac{q-f}{f}$$

FIGURE 23.39 Diagram used for deriving the lens equation: An object ST at a distance p from the lens produces an image $S'T'$ at a distance q from the lens.

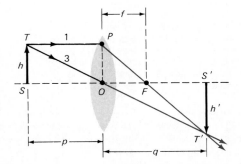

Note that we have taken h' as negative since it is in the opposite direction from h. But, from Eq. (23.8), $h'/h = -q/p$, and so

$$\frac{q - f}{f} = \frac{q}{p} \qquad \text{or} \qquad pq - pf = qf$$

Finally, on dividing by pqf and rearranging, we have

Lens equation: $\boxed{\dfrac{1}{p} + \dfrac{1}{q} = \dfrac{1}{f}}$ **(23.9)**

This is the equation for a thin converging lens. It is identical in form to the mirror equation, Eq. (23.7).

The sign conventions introduced above must be followed in applying the lens equation. The use of these sign conventions will be clarified in Example 23.8.

Example 23.8

A book 20.0 cm wide and 25.0 cm high is placed 1.50 m from the center of a large converging lens of 50.0 cm focal length and at right angles to the lens axis, as in Fig. 23.40. **(a)** What is the position and nature of the image produced? **(b)** What is the size of the image of the book?

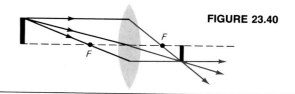

FIGURE 23.40

SOLUTION

(a) Using the lens equation, we have

$$\frac{1}{q} = \frac{1}{f} - \frac{1}{p} = \frac{1}{50.0 \text{ cm}} - \frac{1}{150 \text{ cm}} = \frac{2}{150 \text{ cm}} = \frac{1}{75.0 \text{ cm}}$$

and so $q = \boxed{75.0 \text{ cm}}$

Since this image distance is positive, the image is *real*.
(b) The magnification is

$$M = -\frac{q}{p} = -\frac{75.0 \text{ cm}}{150 \text{ cm}} = -\frac{1}{2}$$

The image is therefore reduced in size and inverted. The width of the image is reduced to

$$w' = Mw = -\tfrac{1}{2}(20.0 \text{ cm}) = \boxed{-10.0 \text{ cm}}$$

and the height to

$$h' = Mh = -\tfrac{1}{2}(25.0 \text{ cm}) = \boxed{-12.5 \text{ cm}}$$

where the minus signs merely indicate the inverted nature of the real image.

The image of the book is therefore 10.0 by 12.5 cm in area. It would so appear on a screen placed 75.0 cm from the lens.

A ray diagram of how the image is produced is shown in Fig. 23.40. It shows only how the height is reduced in the image; the width is changed in the same proportion.

EXERCISE 1 Repeat this example with the book only 15.0 cm from the lens. Figure 23.41 is a ray diagram against which to check your answers.

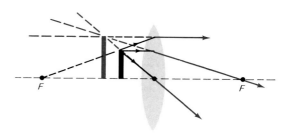

FIGURE 23.41

23.7 Thin Diverging Lenses

A diverging lens behaves in a fashion similar to the way a convex mirror behaves, as can be seen from Fig. 23.42. A diverging lens produces an upright, virtual image of a real object. This can be seen from the ray diagram in Fig. 23.43 or by applying the thin-lens equation. Remember that the focal length of a diverging lens

FIGURE 23.42 Comparison of (a) a diverging lens with (b) a convex mirror. Both form virtual, upright images of objects.

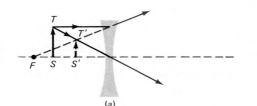

(a)

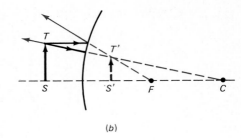

(b)

FIGURE 23.43 Ray diagram for image formation by a diverging lens. The image S'T' is virtual and upright and, in this case, smaller than the object ST.

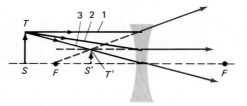

is always negative, according to our sign conventions. In drawing ray diagrams for diverging lenses the three best rays to use are (1) a ray that is parallel to the principal axis and is refracted so that it appears to come from the focus of the lens on the object side; (2) a ray that is directed at the focus on the other side of the lens and is refracted parallel to the principal axis; (3) a ray that passes through the center of the lens and whose path is unchanged. These three rays are shown in Fig. 23.43.

The image distance q for a diverging lens can also be found from the lens equation

$$\frac{1}{p} + \frac{1}{q} = \frac{1}{f} \tag{23.10}$$

For a diverging lens, since $1/q = 1/f - 1/p$, where f is negative, it can be seen that the right side of this equation is *always* negative if $p > 0$, and so the image distance q must also be negative. For this reason the image of a real object produced by a diverging lens is always virtual and upright with respect to the original object. The magnification can then be obtained from $M = h'/h = -q/p$.

Example 23.9

An arrow is 24.2 cm from a double-concave (diverging) lens whose focal length is 36.4 cm, as in Fig. 23.43. **(a)** What will be the nature and location of the image of the arrow?

(b) How large will the image be compared with the size of the original arrow?

SOLUTION

(a) Since a diverging lens has a negative focal length, we have $f = -36.4$ cm. Our lens equation therefore becomes

$$\frac{1}{q} = \frac{1}{f} - \frac{1}{p} = \frac{1}{-36.4 \text{ cm}} - \frac{1}{24.2 \text{ cm}} = -0.0688$$

and so $q = \boxed{-14.5 \text{ cm}}$

This means that the image is 14.5 cm from the lens on the same side of the lens as the object, and is therefore *virtual*.

(b) The magnification is

$$M = -\frac{q}{p} = -\frac{-14.5 \text{ cm}}{24.2 \text{ cm}} = +0.599$$

The image is therefore upright and three-fifths the size of the object. The ray diagram in this case is similar to that in Fig. 23.43.

23.8 Combinations of Lenses

Most practical optical systems such as those used in microscopes and telescopes use more than one lens or mirror. It is therefore important to be able to locate the final image produced by a system of two or more lenses. Here we do this for a few general cases. More specific applications will be taken up in Chap. 25.

In dealing with combinations of lenses, we take the lenses one at a time in the order in which the light strikes them, while ignoring for the moment all other lenses. The image produced by the first lens, as determined using the lens equation and the sign conventions already introduced, is then taken as the object for the second lens. If there is a third lens, the image produced by the second lens is used as the object for the third lens, and so on, until the final image is obtained. Example 23.10 illustrates this technique.

Example 23.10

The telephoto lenses much used in photography and television allow larger images of distant objects to be formed without requiring excessively long cameras. Figure 23.44 is a diagram of a telephoto lens combination consisting of a converging lens L_1 followed by a diverging lens L_2. L_1 has a focal length of 125 mm and L_2 has a focal length of -70 mm (diverging), and the two lenses are separated by a distance of 75 mm in the camera. **(a)** If this camera is used to take a close-up shot of a football player on the field 100 m away, what is the position of the final image formed by this lens combination? **(b)** What is the magnification produced by the two lenses? **(c)** To accomplish the same task, what would the length of a single-lens cam-

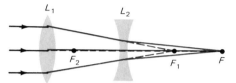

FIGURE 23.44 Diagram for Example 23.10. F_1 is the position where the image formed by the converging lens would fall if L_2 were not present. The actual image formed by the two lenses in combination falls at F.

era have to be? How does this compare with the length of the camera using the telephoto lens?

SOLUTION

(a) For the first lens,

$$\frac{1}{q_1} = \frac{1}{f_1} - \frac{1}{p_1} = \frac{1}{0.125 \text{ m}} - \frac{1}{100 \text{ m}}$$

and so $q_1 = 0.125$ m

and $M_1 = -\dfrac{q_1}{p_1} = -\dfrac{0.125 \text{ m}}{100 \text{ m}} = -0.00125$

The image would therefore be real and inverted, if the light were not intercepted by the second lens.

The image produced by the first lens then serves as the object for the second lens. The object distance is in this case

$$p_2 = 0.075 - 0.125 = -0.050 \text{ m}$$

where the negative sign means that the object is on the back side of the lens, i.e., on the *opposite side* of the lens from where the light is coming. Hence the lens equation for the second lens becomes

$$\frac{1}{q_2} = \frac{1}{f_2} - \frac{1}{p_2} = \frac{1}{-0.070 \text{ m}} - \frac{1}{-0.050 \text{ m}}$$

and so $q_2 = \boxed{+0.175 \text{ m}}$

and the image is 17.5 cm to the right of lens L_2 and is a real image, since light rays actually converge to that point.

(b) The magnification produced by the second lens is

$$M_2 = -\frac{q_2}{p_2} = -\frac{0.175 \text{ m}}{-0.050 \text{ m}} = +3.5$$

The overall magnification is then

$$M = M_1 M_2 = (-0.00125)(+3.5) = \boxed{-0.0044}$$

and the final image is inverted and greatly reduced in size.

Note that the overall magnification *cannot* be obtained by taking $-q_2/p_1 = -0.175/100 = -0.00175$. The overall magnification of the lens combination is always the product of M_1 and M_2: $M = M_1 M_2 = -0.0044$, as found above.

(c) The length of the above lens system from lens L_1 to the final image is

$$\text{Length of camera} = 7.5 \text{ cm} + 17.5 \text{ cm} = \boxed{25 \text{ cm}}$$

For a single-lens camera to produce the same image size for an object 100 m away, we would need

$$p = 100 \text{ m}$$

$$q = Mp = 0.0044(100 \text{ m}) = 0.44 \text{ m} = \boxed{44 \text{ cm}}$$

The length of the camera from lens to film would therefore be 44 cm, which is about twice as long as required by the telephoto lens system, for the same size image of an object 100 m away.

Diopters

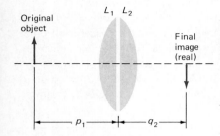

Original object
L_1 L_2
Final image (real)
p_1
q_2

FIGURE 23.45 Combination of two plano-convex lenses placed very close together. They form a real, inverted image of the object.

When testing a patient's eye for glasses, an optometrist places several different lenses in front of the eye at the same time, and from this combination of lenses determines the single lens the patient needs to see clearly with glasses. How does the optometrist arrive at the focal length of the single lens needed to produce the same effect as the correct combination of lenses?

This is a simple problem in the combination of lenses, but here the lenses being combined are all very close together so that it is reasonable to neglect their separation compared with the object and image distances involved. The situation is therefore as in Fig. 23.45 for two plano-convex lenses very close together, where we measure all distances with respect to the center of the two-lens system.

We proceed as in Example 23.10. We first find the image produced by lens L_1 while ignoring lens L_2. We have

$$\frac{1}{p_1} + \frac{1}{q_1} = \frac{1}{f_1} \tag{23.11}$$

In this case the image is clearly on the other side of the lens from the object.

We next apply the lens equation to the second lens L_2. But here the object is the image produced by the first lens, and this image is to the right of the lens. The object distance for the second lens is therefore *negative* and is $p_2 = -q_1$. The lens equation for the second lens is then

$$\frac{1}{p_2} + \frac{1}{q_2} = \frac{1}{f_2} \quad \text{and so} \quad \frac{1}{-q_1} + \frac{1}{q_2} = \frac{1}{f_2} \tag{23.12}$$

If we add Eqs. (23.11) and (23.12), we obtain

$$\frac{1}{p_1} + \frac{1}{q_2} = \frac{1}{f_1} + \frac{1}{f_2} \tag{23.13}$$

Here p_1 is the distance of the original object from the lenses, and q_2 is the distance of the final real image from them. The two lenses act together, therefore, as if they were a single lens of focal length f, where

$$\frac{1}{p_1} + \frac{1}{q_2} = \frac{1}{f} \tag{23.14}$$

On comparing Eqs. (23.13) and (23.14), we see that

$$\frac{1}{f} = \frac{1}{f_1} + \frac{1}{f_2} \tag{23.15}$$

The reciprocal of the focal length for the combination of two lenses located very close together is the sum of the reciprocals of the focal lengths of the individual lenses. (It might be helpful to remember that this is exactly the way two electric resistors combine when connected in parallel.)

We define the *power* of a lens as follows:

Power (P) of a lens [in diopters* (D)]: The reciprocal of its focal length in meters.

$$P = \frac{1}{f} \quad (1 \text{ D} = 1 \text{ m}^{-1}) \tag{23.16}$$

---

*Diopter is derived from two Greek words meaning "see through."

The power of two lenses expressed in diopters can be added directly to give the power of the lens combination. As is clear from Eq. (23.15),

$$P = P_1 + P_2 \tag{23.17}$$

The power of a converging lens is positive when expressed in diopters, since the focal length is positive. Similarly, the power of a diverging lens is negative. Converging and diverging lenses are therefore often referred to as *positive* and *negative* lenses, respectively.

Example 23.11

Two converging lenses with focal lengths 0.205 and 0.307 m, respectively, are combined with a third diverging lens of focal length 0.504 m to form a single compound lens, where all three lenses are very close together. What is the focal length of this compound lens?

SOLUTION

The diverging lens must be assigned a negative focal length according to our sign convention. We have

$$\frac{1}{f} = \frac{1}{f_1} + \frac{1}{f_2} + \frac{1}{f_3}$$

$$= \frac{1}{0.205 \text{ m}} + \frac{1}{0.307 \text{ m}} + \frac{1}{-0.504 \text{ m}} = 6.15 \text{ m}^{-1}$$

and $\quad f = \dfrac{1}{6.15} \text{ m} = \boxed{0.163 \text{ m}}$

The same result could have been obtained using the power lenses in diopters. We have

$$P = P_1 + P_2 + P_3$$

$$= \frac{1}{0.205} \text{ D} + \frac{1}{0.307} \text{ D} - \frac{1}{0.504} \text{ D} = 6.15 \text{ D}$$

Then, using the definition of lens power, we have

$$f = \frac{1}{P} = \frac{1}{6.15 \text{ D}} = \boxed{0.163 \text{ m}}$$

as before.

Example 23.12

One way to measure the focal length of a diverging lens is to place a converging lens of known focal length in contact with the diverging lens and find the focal length of the combination. If parallel light from the sun is focused at a point 30.5 cm from the center of the lens combination, and the focal length of the converging lens is $f_1 = 15.2$ cm, what is the focal length of the diverging lens?

SOLUTION

Since $1/f = 1/f_1 + 1/f_2$, we have

$$\frac{1}{0.305 \text{ m}} = \frac{1}{0.152 \text{ m}} + \frac{1}{f_2}$$

where f_2 is the focal length of the diverging lens. From this we obtain

$$f_2 = \boxed{-0.303 \text{ m}}$$

The focal length is about 30 cm and is negative, as required for a diverging lens.

EXERCISE 2 Solve the same problem by using the powers of the two lenses.

23.9 Lens Aberrations

The theory we have developed for image formation by lenses assumes that a simple lens would focus all rays arising at one point on an object to one point on the image regardless of the location of the point on the object or the wavelength of the light used. This oversimplifies the complicated process of image formation by a simple lens. No simple lens forms a perfect image of any given object. Imperfections, which we call *lens aberrations*, lead to lack of clarity and sharpness in the image. There are five principal kinds of such lens aberrations, but for our purposes we will discuss only two: spherical aberration and chromatic aberration.

Spherical Aberration

If we use a simple spherical lens to form an image of a point source of light, we find that rays passing through the outer portion of the lens surface are focused closer to the lens than rays passing near the center of the lens (the same thing happens with a spherical mirror). As a result there is no single sharp image of the point object, but rather a tiny circular patch of light. The diameter of this circular patch will vary with distance from the lens. In practice, the best place to put a photographic film to record the image is where this circle has its smallest diameter. This is called the *circle of least confusion* and is shown in Fig. 23.46.

Just as with mirrors spherical aberration can be corrected by using parabolic mirrors, so too spherical aberration can be eliminated by using nonspherical lens surfaces. These are very difficult and expensive to produce, however, and it is easier to use combinations of lenses chosen to reduce spherical aberration. Most compound lenses used in research equipment are corrected for spherical aberration in this way.

Even with a simple lens, spherical aberration can be reduced by stopping down the lens with a diaphragm. This, of course, greatly reduces the light passed by the lens and the intensity of the image produced.

FIGURE 23.46 Spherical aberration. Rays passing through the center of the lens are focused at a different distance from the lens than are rays passing through the outside portions of the lens; a point object does not produce a point image.

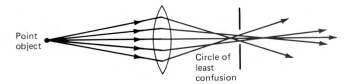

Point object

Circle of least confusion

Chromatic Aberration

Spherical aberration occurs even with the perfectly monochromatic light from a laser. There is another aberration, however, that occurs because the index of refraction of the glass in a lens is different for different wavelengths. We have seen that a prism refracts violet light more than it refracts red light. The same is true of a lens, since a lens can be considered to be an array of small prisms, as in Fig. 23.33. As a result a simple lens will focus violet or blue light closer to the lens than it focuses red light. Hence a white light source produces a blurred image of an object, and this image has colored edges, as shown in Fig. 23.47. This lens defect is called *chromatic aberration* (*chroma* is the Greek word for "color"). The human eye has the same problem in focusing on both red and blue colors at the same time, since the focal length of the eye for the two colors is slightly different.

In 1758 John Dollond (1706–1761), an English optician, discovered a way to eliminate chromatic aberration. He combined two lenses, one converging, the other diverging, to make an *achromatic doublet*, i.e., two lenses showing no color. The two lenses are made of different kinds of glass with indices of refraction such that the combination brings any two chosen colors to the same sharp focus.* The light paths through an achromatic doublet designed for blue and red light are shown in

---

*Dollond's discovery was particularly important for the development of high-grade microscopes. Newton's discovery of reflecting telescopes had eliminated the problem of chromatic aberration from astronomical work by replacing lenses with mirrors. This was not possible, however, for microscopes.

FIGURE 23.47 Chromatic aberration. Because the index of refraction of glass is different for different colors, images in these colors fall at different distances from the lens. A blue image with red edges is formed near the lens, and a red image with blue edges slightly farther away from the lens.

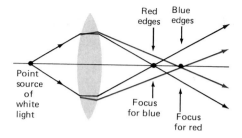

FIGURE 23.48 An achromatic doublet. The first element in the compound lens is made from low-dispersion glass. The second element, a diverging lens, is weaker than the first element but is made from high-dispersion glass. The combination cancels out all dispersion and focuses all colors at the same focal point. (The dispersion is greatly exaggerated for clarity.)

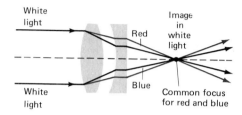

Fig. 23.48. In many cases achromatic triplets or more complicated compound lenses are used to correct chromatic aberration. Often they can be designed to correct for spherical aberration at the same time.

The design of compound lenses to reduce aberrations has been greatly improved in recent years by the use of high-speed computers. The usual technique is to combine two lenses, or a lens and a mirror, in such a way that the aberration introduced by one just cancels the aberration caused by the other. Even with computers, not all aberrations can be drastically reduced at the same time, but it is possible to determine which aberrations are most troublesome for a particular application and to choose a lens system that reduces these aberrations to within tolerable limits.

Summary: Important Definitions and Equations

Ray (or geometrical) optics: A model of light which assumes that light travels in perfect straight lines, or along "rays," which are at right angles to the wave fronts.

Reflection of light
1 The incident ray, the reflected ray, and the normal to the reflecting surface all lie in the same plane.
2 The angle of incidence (i) is equal to the angle of reflection (r).

Refraction of light
Snell's law: $n_1 \sin \theta_1 = n_2 \sin \theta_2$ (**23.1**)
Total internal reflection: When light passes from a denser optical medium to a less dense optical medium, total reflection occurs for light incident at an angle greater than the critical angle θ_c, where $\sin \theta_c = n_2/n_1$. (**23.2**)

Dispersion: The dependence of the index of refraction of a material on wavelength.

Mirrors

Object distance (p): The distance of the object from the mirror along the principal axis.

Image distance (q): The distance of the image from the mirror along the principal axis.

Focal length (f): The distance of the focal point F from the mirror along the principal axis.

Focal point: The point through which all rays parallel to the mirror's principal axis are reflected by a mirror;

$$f = \frac{R}{2} \qquad (\textbf{23.6})$$

where R is the radius of curvature of the mirror.

Real image: An image formed by light rays actually coming together to form the image.

Virtual image: An image formed by light rays which appear to come from a particular point but never actually pass through that point.

Thin lenses

Converging lens: A lens which brings all incident light rays parallel to its principal axis together at the focal point of the lens.

Diverging lens: A lens which spreads out all incident light rays parallel to its principal axis so that they appear to come from a focal point on the other side of the lens.

(The definitions given for mirrors of object distance, image distance, focal length, focus, real image, and virtual image are the same for lenses except that *lens* replaces *mirror* in each definition. All important equations and sign conventions for mirrors and lenses are summarized in Table 23.2.)

Power of a lens (in diopters): The reciprocal of its focal length in meters.

$$P = \frac{1}{f} \qquad 1 \text{ diopter (D)} = 1 \text{ m}^{-1} \qquad \textbf{(23.16)}$$

Combination of two lenses close together

$$\frac{1}{f} = \frac{1}{f_1} + \frac{1}{f_2} \qquad \textbf{(23.15)}$$

$$P = P_1 + P_2 \qquad \textbf{(23.17)}$$

Lens aberrations

Spherical aberration: Lack of sharpness in focusing by a lens because off-axis rays parallel to the principal axis have a different focus from those passing through the center of the lens.

Chromatic aberration: Lack of sharpness and the presence of colored fringes in the image caused by variation of focal length of a lens with the wavelength of the light.

TABLE 23.2   Important Equations and Sign Conventions for Single Mirrors and Lenses

Mirrors and lenses: $\dfrac{1}{p} + \dfrac{1}{q} = \dfrac{1}{f}$ **(23.7, 23.9)**

Quantity	Positive sign (+)	Negative sign (−)
Mirrors:		
Object distance p	If object is on front side* of mirror (real object)	If object is on back side of mirror (virtual object)
Image distance q	If image is on front side of mirror (real image)	If image is on back side of mirror (virtual image)
Focal length $f = R/2$ **(23.6)**	If focal point is on front side of mirror (concave mirror)	If focal point is on back side of mirror (convex mirror)
Lateral magnification: $M = h'/h = -q/p$ **(23.3)**	If image is upright	If image is inverted
Lenses:		
Object distance p	If object is on front side of lens (real object)	If object is on back side of lens (virtual object)
Image distance q	If image is on back side of lens (real image)	If image is on front side of lens (virtual image)
Focal length f	If lens is converging	If lens is diverging
Lateral magnification: $M = h'/h = -q/p$ **(23.8)**	If image is upright	If image is inverted

*By *front side* is meant the side from which light is coming toward the mirror or lens.
For a single mirror or lens: Real images are always *inverted*; virtual images are always *upright*.

Questions

1 Give a few examples of optical observations made in everyday life that *cannot* be explained by a ray theory of light.

2 What advantage can you see in using prisms at angles greater than the critical angle to reflect light instead of using plane mirrors for the same purpose? What disadvantages can you see in the use of such prisms?

3 (a) Why is the index of refraction of any actual medium always larger than unity?

(b) What kind of hypothetical medium would have an index of refraction less than unity?

4 Can a light ray in air be totally reflected at a glass surface if the angle of incidence is just right? Why?

5 Show by ray diagrams and from the mirror equation that the magnification of a concave mirror is less than 1 if the object is outside the center of curvature and greater than 1 if it is inside the center of curvature. What is the situation when the object is placed at the center of curvature?

6 A concave mirror produces a real image. Is the image necessarily inverted?

7 (a) Indicate the similarities between converging lenses and concave mirrors.

(b) Do the same for diverging lenses and convex mirrors.

8 Can you suggest any circumstances in which the image formed by a diverging lens would be a real image? Draw a ray diagram to explain your answer.

9 Some newspapers in an apartment ignite and the fire department is called to put out the fire. The fire captain claims that the fire was caused by a fishbowl full of water sitting just inside a sunlit window. The bowl focused the sun's rays and set the newspapers on fire. Do you think this is a possible explanation of the fire?

10 Is it possible for the same glass lens to be either converging or diverging depending on the medium in which it is embedded? Explain your answer, using a ray diagram.

11 (a) How would you determine the focal length of a converging lens in the laboratory?

(b) How would you determine the focal length of a diverging lens in the laboratory?

Multiple-Choice and Simple Exercises

23.1 Light enters a plate of borosilicate crown glass at an angle of 43° with respect to the normal to the air-glass surface. The angle made by the refracted beam with the normal is:
(a) 0° (b) 27° (c) 63° (d) 32°
(e) 90°

23.2 Light travels down a light pipe made of flint glass ($n = 1.66$) coated on the outside by borosilicate crown glass. The critical angle for total internal reflection inside the light pipe is:
(a) 66.3° (b) 23.7° (c) 0° (d) 41.1°
(e) 37.0°

23.3 The image of a real object formed in a convex mirror is:
(a) Real, inverted, magnified
(b) Real, inverted, reduced in size
(c) Virtual, erect, magnified
(d) Virtual, erect, reduced in size
(e) Virtual, inverted, reduced in size

23.4 A concave mirror has a radius of curvature of 2.0 m. A bouquet of roses is placed 1.0 m from the mirror on its reflecting side. The roses will be focused sharply on a screen that is a distance from the mirror of:
(a) 2.0 m (b) 1.0 m (c) 0.50 m
(d) 4.0 m (e) None of the above

23.5 A convex mirror has a focal length of 1.0 m. An upright arrow is located 0.60 m from the reflecting side of the mirror. The image produced is:
(a) Virtual and 0.63 times as tall as the object
(b) Real and 0.63 times as tall as the object
(c) Inverted and 4.4 times as tall as the object
(d) Upright and 4.4 times as tall as the object
(e) None of the above

23.6 The distance of the lens in the human eye from the retina, on which the image is focused, is about 1.7 cm. To focus on a book 0.50 m from the eye, the focal length of the eye must be about:
(a) 50 cm (b) 0.59 cm (c) 1.6 cm
(d) 1.6 m (e) 3.4 cm

23.7 The image formed by a single converging lens is real, inverted, and larger than the object. From this information it may be concluded that the object is:

(a) Inside the focal point of the lens
(b) At the focal point of the lens
(c) A distance between f and $2f$ from the lens
(d) A distance $2f$ from the lens
(e) A distance greater than $2f$ from the lens

23.8 It is not possible with a single lens, which may be either converging or diverging, to form an image that is:
(a) Real, erect, and larger than the object
(b) Real, inverted, and larger than the object
(c) Real, inverted, and smaller than the object
(d) Virtual, erect, and larger than the object
(e) Virtual, erect, and smaller than the object

23.9 A vase is located 40 cm from a diverging lens of focal length 40 cm. The location of the image will be:
(a) 20 cm from the lens on the same side as the object
(b) 20 cm from the lens on the opposite side from the object
(c) 40 cm from the lens on the same side as the object
(d) 40 cm from the lens on the opposite side from the object
(e) None of the above

23.10 A converging lens of focal length 20 cm and a diverging lens of focal length 50 cm are combined into a compound lens. The focal length of the compound lens is:
(a) −33 cm (b) 14 cm (c) 33 cm
(d) −14 cm (e) 70 cm

23.11 Two lenses are combined to form a compound lens. One of the lenses has a power of +10 D and the other of −5.0 D. The focal length of the resulting lens is:
(a) 100 cm (b) 50 cm (c) 20 cm
(d) 10 cm (e) None of the above

23.12 The critical angle for total internal reflection of a light beam passing from diamond to air is 24°. What is the index of refraction of diamond?

23.13 If an object 0.60 m from a concave mirror produces a real image 0.20 m from the mirror, what is the radius of curvature of the mirror?

23.14 An object is placed 0.20 m from the reflecting surface of a convex mirror of focal length 0.60 m. What is the location, size, and nature of the image produced?

23.15 What is the magnification produced by a converging lens of focal length 0.15 m for an object 0.20 m from the lens?

23.16 A converging lens of focal length 0.20 m forms a virtual image of an object. The image appears to be 0.90 m from the lens on the same side as the object. Where was the object located?

23.17 A double-concave (diverging) lens of focal length 0.20 m is used to view a bowl of fruit 2.0 m from the lens.

(*a*) What is the location of the image?

(*b*) What is the nature of the image and the magnification?

23.18 (*a*) Two lenses, one of power $+2.0$ D and the other of power -1.5 D, are combined into one compound lens. What is the power of the compound lens in diopters?

(*b*) What is the focal length of the compound lens?

23.19 Two diverging lenses with focal lengths 0.20 and 0.25 m, respectively, are combined with a converging lens of focal length 0.60 m to form a single compound lens.

(*a*) What is the focal length of the combination?

(*b*) What is the power of the compound lens?

Problems

23.20 Two plane mirrors are in contact at one edge and make an angle of 45.0° with each other. A beam of light falls on one mirror at an angle of incidence of 20.5°. It is reflected off the first mirror and then strikes the second mirror.

(*a*) What is the angle of reflection off the second mirror?

(*b*) What is the final direction of the light with respect to its original direction?

23.21 What must be the minimum height of a vertical plane mirror to allow a woman 1.70 m tall to see herself completely in the mirror? Assume that her eyes are 12.0 cm below the top of her head.

23.22 A light beam falling on the mirror of a Cavendish balance (see Sec. 4.1) is initially reflected directly back on itself. When two heavy lead spheres are brought near the two small lead balls at the ends of the balance arm, the mirror is deflected through an angle of 1.50°. If a straight meterstick is mounted 20.0 m from the mirror, what is the deflection of the reflected light beam on the meterstick?

23.23 A light beam travels from air into a block of ice and falls on the ice at an angle of incidence of 42.5°.

(*a*) What angle does the refracted beam make with the normal to the surface?

(*b*) What is the speed of light in the ice?

23.24 (*a*) What is the critical angle for the total internal reflection of a light beam passing from a diamond into very heavy flint glass?

(*b*) What is the critical angle for a light beam passing in the opposite direction, from very heavy flint glass into diamond?

23.25 If a light pipe is made of very heavy flint glass fibers coated with borosilicate crown glass, what is the largest bend in the pipe which can be sustained without losing some of the light by refraction?

23.26 Light is incident on one end of an optical fiber at an angle of incidence θ_1 and is refracted into the fiber at an angle θ_2, as in Fig. 23.49. It then strikes the side of the fiber at an angle i. If the index of refraction of the fiber is 1.40, what is the largest angle of incidence θ_1 that the ray can have and still be totally reflected from the wall of the fiber?

23.27 An optical fiber consists of an interior glass fiber of index of refraction $n_1 = 1.65$ coated with a thin layer of glass of index of refraction $n_2 = 1.40$.

(*a*) What is the critical angle for total reflection of a ray inside the fiber?

(*b*) If light is incident on the end of the fiber, as in Fig.

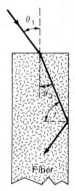

FIGURE 23.49 Diagram for Prob. 23.26.

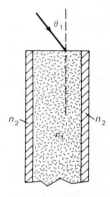

FIGURE 23.50 Diagram for Prob. 23.27.

23.50, what is the largest angle of incidence θ_1 that the ray can have and still be totally reflected from the fiber wall?

*23.28 A small rock lies at the bottom of a large water tank, 8.0 m below the surface of the water. Determine the diameter of a circular piece of wood which, when floating on the surface of the water directly above the rock, will totally hide the rock, regardless of the angle of the line of sight.

*23.29 A light ray passes downward into a block of transparent plastic with an angle of incidence of 78.4°, as in Fig. 23.51. If total internal reflection is to occur when the light strikes the left edge of the block at point A, what must the index of refraction of the plastic be?

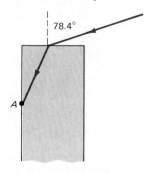

FIGURE 23.51 Diagram for Prob. 23.29.

*23.30 A beam of white light is traveling through a block of heavy flint glass ($n = 1.65$) and strikes a glass-air interface at an angle of incidence θ.

(a) Can you suggest how the light emerging into the air could be made to look reddish?

(b) What angle of incidence would be needed to do this?

(c) How would it be possible to obtain bluish light with the same arrangement?

23.31 Red light at 633 nm from a He–Ne laser enters an equilateral (all angles = 60°) prism made of light flint glass in such a way that, after reflection at the first surface, the light travels parallel to the base of the prism, as in Fig. 23.52.

(a) What must the angle of incidence of the light be for this to occur?

(b) What will be the angle of incidence of the light on the second prism face?

(c) What will be the angle of refraction of the light at the second prism face?

(d) What will be the total deviation of the beam from its original direction? (See Table 23.1 for index of refraction.)

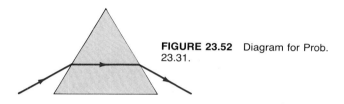

FIGURE 23.52 Diagram for Prob. 23.31.

*23.32 (a) Repeat Prob. 23.31 for blue light of wavelength 436 nm (use 1.60 as the index of refraction for this wavelength).

(b) If both the 633- and 436-nm radiation are moving together in the same direction when they strike the prism face, what is the angular deviation between them when they leave the prism?

23.33 A concave mirror has a radius of curvature of 1.55 m. Calculate the position and kind of image produced when an object is placed at the following positions:

(a) 2.00 m from the mirror.

(b) 1.00 m from the mirror.

(c) 0.50 m from the mirror.

(d) Describe how the image changes as the object is gradually moved nearer to the mirror as in parts (a) to (c).

23.34 (a) A concave mirror has a radius of curvature of 0.60 m. A vase is placed 0.50 m from the mirror and on its principal axis. Use graphical methods to locate the image and obtain its approximate size.

23.35 The real image produced by a concave mirror is found to be one-third the size of the object. If the distance from the mirror to the screen on which the image appears is 0.852 m, what is the focal length of the mirror?

23.36 A convex mirror with a 100-cm radius of curvature is used to reflect the light from an object placed 66.3 cm in front of the mirror. Find the location of the image and its relative size.

23.37 A concave mirror of radius of curvature 1.50 m is placed behind a chess piece which is 0.205 m from the reflecting side of the mirror along its principal axis:

(a) What is the nature of the image produced, and where does it occur?

(b) What is the magnification?

23.38 A convex mirror of radius of curvature 1.50 m is placed 0.205 m from a chess piece which is on the reflecting side of the mirror.

(a) What is the nature of the image produced, and where does it occur?

(b) What is the magnification?

23.39 A convex mirror has a radius of curvature of 1.00 m. Calculate the position and kind of image produced when an object is placed at the following points:

(a) 1.50 m from the mirror.

(b) 0.751 m from the mirror.

(c) 0.353 m from the mirror.

(d) Describe how the image changes as the object is gradually moved nearer to the mirror as in parts (a) to (c).

23.40 (a) A convex mirror has a radius of curvature of 0.75 m. A vase is placed 0.60 m from the mirror on its principal axis. Use graphical methods to locate the image and obtain its approximate size.

(b) Solve the problem by using the mirror equation and compare the two results.

*23.41 Prove that for a convex mirror, no matter where an object is placed on the reflecting side of the mirror, the image will always be virtual and erect.

*23.42 A spherical mirror has both sides reflecting. A woman looks into one side of the mirror and sees an image of her face 42.6 cm in back of the mirror. When she looks into the other side of the mirror, she sees an image 14.2 cm in back of the mirror.

(a) How far is her face from the mirror, assuming that the distance is the same in both cases?

(b) What is the focal length of each side of the mirror?

23.43 A mirror forms an image of a hammer on a screen 15.0 cm from its vertex. The image of the head of the hammer is at the top of the screen. The real hammer is 25.5 cm from the mirror.

(a) Is the mirror convex or concave?

(b) What is its focal length?

(c) What is the orientation of the original object (the hammer)?

23.44 You desire to make a photographic self-portrait. If you hold your camera close to your body and 1.53 m from a plane mirror, for what distance should the camera be focused to give a sharp picture of yourself?

23.45 The magnification produced by a converging lens is found to be -2.5 for an object placed 0.203 m from the lens. What is the focal length of the lens?

23.46 A double-convex lens (converging) has a focal length of 0.400 m. Calculate the position and kind of image (including its size) produced when an object is placed at the following points:

(a) 1.55 m from the lens.

(b) 0.600 m from the lens.

(c) 0.200 m from the lens.

(d) Describe how the image changes as the object is moved gradually nearer to the mirror as in parts (a) to (c).

23.47 (a) A converging lens has a focal length of 0.55 m. A vase is placed 0.65 m from the lens on its principal axis. Use graphical methods to locate the image and obtain its magnification.

(*b*) Solve the same problem analytically by using the lens equation, and compare the two results.

23.48 Repeat both parts of Prob. 23.47 for the same lens and an object distance of 0.30 m.

23.49 How far apart are the object and image if a lens of focal length 30.0 cm forms a real image five times as high as the object?

****23.50** Prove that for a glass lens, with n_g as its index of refraction, immersed in a liquid of index of refraction n_l, the focal length of the lens in the liquid is not f but

$$f' = \frac{n_l(n_g - 1)f}{n_g - n_l}$$

where f is the focal length in air.

23.51 Show that the formula in Prob. 23.50 correctly predicts the behavior of a glass lens immersed in a liquid for the two limiting cases (*a*) $n_l \rightarrow n_g$; (*b*) $n_l \rightarrow 1$.

****23.52** A light bulb and a screen are placed at a fixed distance D apart.

(*a*) Show that if a converging lens of focal length f, where $f < D/4$, is inserted between them, it will produce a real image of the object on the screen for two positions separated by a distance $d = \sqrt{D(D - 4f)}$.

(*b*) Prove that at the two positions found in part (*a*) the ratio of the two image sizes is $(D - d)^2/(D + d)^2$.

(*c*) Why have the focal lengths been limited to values such that $f < D/4$?

23.53 Consider two converging lenses, one of focal length 25.5 cm and a second of focal length 50.0 cm, with their centers 75.5 cm apart. If an object is placed 20.2 cm from the center of the first lens on the side away from the second lens, what is the nature of the final image produced and where does it fall?

23.54 A man 1.98 m tall is observed through a diverging lens of focal length 1.12 m. The man is a distance of 3.05 m from the lens.

(*a*) What will be the nature and location of the image of the man?

(*b*) How large will the man appear to be?

23.55 (*a*) A diverging lens has a focal length of 1.6 m.

A rose in a bud vase is placed 1.9 m from the lens on its principal axis. Use graphical methods to locate the image and to obtain its magnification.

(*b*) Solve the same problem analytically and compare the two results.

23.56 Repeat both parts of Prob. 23.55 for the same lens and an object distance of 1.0 m.

****23.57** If a diverging lens is to be used to form an image that is half the size of the object, where must the object be placed?

23.58 A converging lens has a focal length of 25.0 cm, and a second converging lens has a focal length of 50.5 cm. An object is placed 25.0 cm from the first lens and on the side away from the second lens.

(*a*) If the lenses are 50.0 cm apart, where does the final image appear?

(*b*) If the lenses are 2.00 m apart, where does the image appear?

23.59 A converging lens of focal length 35.2 cm is put into contact with a diverging lens of unknown focal length. This compound lens focuses sunlight at a point 40.4 cm from the center of the lens combination. What is the focal length of the diverging lens?

23.60 An object is placed 24.4 cm in front of a converging lens with $f = 16.0$ cm. At a distance of 72.3 cm beyond the first lens is a diverging lens of focal length 12.0 cm.

(*a*) Find the position and size of the final image produced.

(*b*) Is the image real or virtual, erect or inverted?

23.61 An object is placed 40.6 cm in front of a converging lens which has a focal length of 20.0 cm. This lens is 50.8 cm from a plane mirror. Find the final image produced by this system (assume that the light from the object passes through the lens twice).

23.62 A flint glass lens, with $n = 1.66$, has a power in air of $+3.00$ D. What would be its power if submerged in water? (*Hint:* See Prob. 23.50.)

23.63 Light from a neon sign passes through a converging lens of focal length 20.9 cm and forms a sharp upright image of the letter T on a screen 30.5 cm from the lens.

(*a*) What is the position of the neon sign with respect to the lens?

(*b*) Is the letter T in the sign upright or inverted?

Additional Readings

Fraser, Alistair B., and William H. Mach: "Mirages," *Scientific American*, vol. 234, no. 1, January 1976, pp. 102–111. A discussion of an interesting optical phenomenon caused by the refraction of light.

Greenler, Robert: *Rainbows, Halos and Glories*, Cambridge University Press, New York, 1980. A discussion of optical and meteorological phenomena in the atmosphere.

Katzir, Abraham: "Optical Fibers in Medicine," *Scientific American*, vol. 260, no. 5, May 1989, pp. 120–125. An up-to-date account that bridges the gap between physics and medicine.

Miller, Stewart E.: "Lightwaves and Telecommunication,"

American Scientist, vol. 72, no. 1, January-February 1984, pp. 66–71. An interesting account of recent developments in the use of light pipes for long-range communication.

Minnaert, M.: *Light and Color in the Open Air*, Bell, London, 1959. A classic discussion of some commonly observed optical phenomena.

Smith, F. Dow: "How Images Are Formed," *Scientific American*, vol. 219, no. 3, September 1968, pp. 96–108. How wave theory and computer programs are needed to supplement ray optics in designing high-quality lenses. This issue of *Scientific American* also contains some other excellent articles on light.

CHAPTER 24

The Wave Nature of Light: Interference, Diffraction, and Polarization

He [Thomas Young] was one of the most acute men who ever lived, but had the misfortune to be too far in advance of his contemporaries. They looked on him with astonishment, but could not follow his bold speculations, and thus a mass of his most important thought remained buried and forgotten in the Transactions of the Royal Society *until a later generation by slow degrees arrived at the rediscovery of his discoveries, and came to appreciate the force of his arguments and the accuracy of his conclusions.*

Hermann von Helmholtz (1821–1894)

Before the middle of the nineteenth century, heated controversies about the nature of light marked the history of physics. Some scientists, like Newton, thought that light consisted of particles; others believed it was a wave. When Maxwell proposed his electromagnetic theory in 1865 and predicted a speed $c = 2.998 \times 10^8$ m/s for all electromagnetic waves, a value equal to the then-known speed of visible light, it appeared that light had finally been proved to be an electromagnetic *wave*.

The place of visible light in the electromagnetic spectrum is shown in Fig. 24.1. The wavelengths of visible light range from about 3.9×10^{-7} to 7.6×10^{-7} m. Often light wavelengths are given in units of micrometers (1 μm $= 10^{-6}$ m), nanometers (1 nm $= 10^{-9}$ m), or angstroms (1 Å $= 10^{-10}$ m). Thus the visible spectrum runs from 390 to 760 nm or from 3900 to 7600 Å. (Remember that most atoms are 1 to 2 Å in diameter.)

Although Maxwell's work was far from the last shot in the battle over the nature of light, as we will see in Chap. 27, a wave theory does explain most observed optical phenomena including reflection, refraction, diffraction, interference, and polarization. In this chapter we apply a wave theory of light to these phenomena.

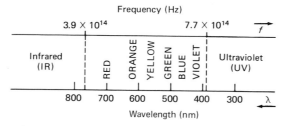

FIGURE 24.1 Place of visible light in the electromagnetic spectrum. It is bounded by the infrared on the low-frequency (long-wavelength) side and by the ultraviolet on the high-frequency (short-wavelength) side.

619

24.1 Huygens' Principle

In 1670, long before Maxwell's day, Christiaan Huygens (see Fig. 7.8) had proposed a wave theory of light. Although his work has now been both justified and replaced by Maxwell's more comprehensive theory, it contains one key idea that is very useful in predicting how light (or any wave) will behave in particular physical situations. This is *Huygens' principle*:

Every point on a wave front may be considered a source of secondary spherical wavelets that spread out in the forward direction at the speed of light in the medium. The new wave front is then the surface tangent to all these secondary wavelets.

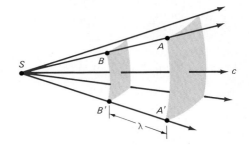

FIGURE 24.2 A spherical electromagnetic wave. The wave fronts are spherical in shape and are spaced a wavelength apart, while the rays are radial. Far away from the source, however, small sections of the wave fronts become approximately planar.

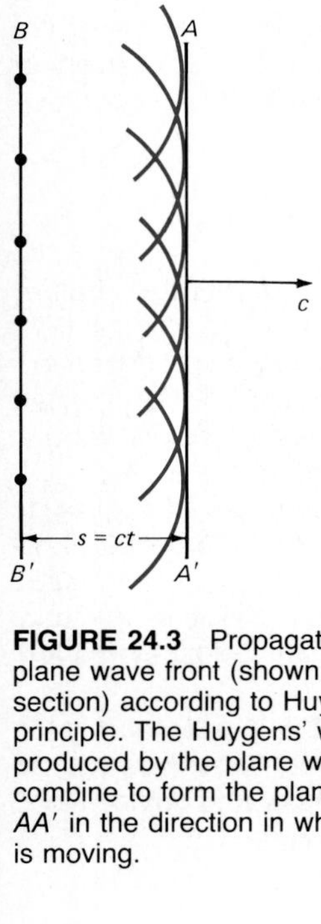

For a plane wave, the wave fronts are planes normal to the direction of propagation, and separated by one wavelength λ, as in Fig. 23.1. For a spherical electromagnetic wave the wave fronts AA' and BB' traveling in any given direction are parts of spherical surfaces, again separated by a distance λ, as shown in Fig. 24.2. In both these cases *rays* can be drawn at right angles to the wave fronts to indicate the direction in which the wave is actually traveling. For the plane waves in Fig. 23.1, the rays are directed along the positive x axis; for the spherical waves in Fig. 24.2, they are directed radially outward from the point source of the wave. Experiment shows that plane waves remain planar and spherical waves remain spherical as they move through space. Huygens' principle shows us why this is so.

For the plane wave shown in Fig. 24.3, if we take one plane wave front and consider that all points on it produce Huygens' wavelets in the forward direction, we obtain the result shown in the figure. All the spherical wavelets travel exactly the same distance in the same time interval. Therefore at some later time the tangents to all the wavelets form another plane wave front that has moved a distance s along the x axis, where $s = ct$. In this case Huygens' principle correctly predicts the observed motion of the plane wave along the positive x axis.

The same consistent results are obtained when Huygens' principle is applied to a spherical wave front, as in Fig. 24.4. In this case the tangents to the wavelets arising at a spherical wave front form a spherical surface similar to the original wave front but of larger radius. The spreading of the wave from the original point source is again correctly predicted.

Huygens' principle is particularly useful when we deal with light waves moving from one medium to another or falling on various obstacles and apertures.

FIGURE 24.3 Propagation of a plane wave front (shown in cross section) according to Huygens' principle. The Huygens' wavelets produced by the plane wave front BB' combine to form the plane wave front AA' in the direction in which the wave is moving.

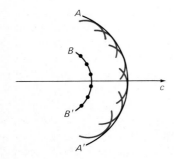

FIGURE 24.4 Propagation of a spherical wave front (shown in cross section) according to Huygens' principle. The Huygens' wavelets produced by the spherical wave front BB' combine to form the spherical wave front AA'.

Example 24.1

Use Huygens' principle to prove that, for the reflection of a plane light wave by a flat mirror, the angle of reflection is equal to the angle of incidence.

SOLUTION

In Fig. 24.5 we show the reflection of light by a mirror. Three light rays are drawn in the direction of the wave motion, and three wave fronts are drawn at right angles to the rays. We assume the experimental fact that the incident and reflected rays and the normal to the mirror are all in the same plane (called the *plane of incidence*).

Let us assume that wavefront AA' is the source of Huygens' wavelets. Then in the time it takes one wavelet to go from A' to R', a wavelet at A'' travels from A'' to B and is then reflected back to B'. A wavelet originating at A travels the distance AR in the same time. The new reflected wave front is therefore the surface tangent to these three wavelets and is RR' in the diagram.

Now $A'R' = AR$, since these are distances traveled by light in equal times in the same medium. The right triangles

$AA'R'$ and ARR' are therefore congruent, since they have three equal sides. Hence angle $A'R'A$ is equal to angle RAR'. But, from Figs. 24.5 and 24.6,

$$\angle A'R'A = 90° - i \qquad \text{and} \qquad \angle RAR' = 90° - r$$

and so $\qquad 90° - i = 90° - r \qquad$ and $\qquad \boxed{r = i}$

On reflection *the angle of reflection is therefore equal to the angle of incidence*, as discussed in the last chapter.

EXERCISE 1 Use a similar wave approach to derive Snell's law of refraction. (*Hint:* Use Fig. 23.9 together with Huygens' principle.)

FIGURE 24.5 (left) Diagram used to prove that the angle of reflection is equal to the angle of incidence when light is reflected from a mirror. Here the distance AR is equal to the distance $A'R'$.

FIGURE 24.6 (right) Relationship between angles for the reflection of light. Since $90° - i = 90° - r$, we have $i = r$.

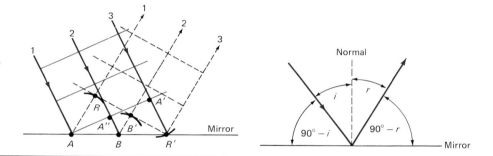

Coherent and Incoherent Light Sources

In the following sections we discuss the superposition and interference of light waves, for which the distinction between coherent and incoherent light sources is crucial.

In many cases electromagnetic radiation is produced by oscillating electric charges with the frequency of oscillation equal to the frequency of the omitted radiation. If, in a given source, the charges all oscillate *in step*, or *in phase*, the source is said to be *coherent*. If the charges oscillate independently and randomly, the source is said to be *incoherent*. Coherent sources produce coherent waves; incoherent sources produce incoherent waves.

Ordinary sources of visible light—incandescent light bulbs, fluorescent tubes, carbon arcs—are incoherent. Laboratory sources of radio waves and microwaves are normally coherent, because they amplify the output of coherent oscillators such as *LC* circuits. The development of the *laser*—which is basically an optical amplifier—has extended the range of coherent sources into the optical region of the electromagnetic spectrum. (Both the laser and coherence are discussed more fully in Sec. 28.6.)

Superposition and interference are easily demonstrated with light from a laser, because the superimposed waves are all from the same coherent source. Suppose two light waves leave the laser in phase but travel different distances to a screen or to the eye. They will produce constructive or destructive interference depending on the phase difference between the two waves, as discussed for waves on a string in Sec. 11.8.

With a conventional light source the interference of light can be demonstrated only by selecting a very small region of the source. The light emitted from such a small region has sufficient coherence to produce interference. A broad, incoherent light source will not, however, produce an observable interference pattern because waves from different parts of the source will have random phase relationships. The superposition of such waves of random phase produces no observable constructive or destructive interference.

24.2   Interference: Young's Double-Slit Experiment

One of the crucial experiments that established the wave nature of light was first performed in 1803 by a remarkable British medical doctor, Thomas Young (see Fig. 24.10 and accompanying biography). As a physician, Young was interested in the workings of human ears and eyes and dedicated much time to a study of sound and light. He knew that if two sound waves of equal intensity reach the ear 180° out of phase, they cancel each other out and no sound is heard. It occurred to him that a similar interference effect should result with two beams of light if, like sound, light consisted of waves. This led Young to an experiment now commonly referred to as *Young's double-slit experiment.*

Consider the physical situation shown in Fig. 24.7, in which we use a laser for convenience. Two very narrow slits separated by a distance d between their centers are cut in a piece of metal. Light from the laser is directed at the two slits, passes through, and strikes a screen at a large distance L from the slits. Huygens' wavelets produced at the slits combine to form cylindrical wave fronts (because of the long, narrow slits), and these wave fronts from the two slits are superimposed on the screen. This superposition leads to constructive or destructive *interference* depending on the *path difference* between the wavelets from the two slits. Therefore bright and dark fringes appear on the screen.

A bright fringe will occur at point P, which is exactly opposite the center of the two slits, because the path lengths for the Huygens' wavelets from the two slits to P are the same, and the wavelets from the two slits are initially in phase. Consider next the fringes produced at distances y_m to either side of P on the screen. Constructive interference occurs when the wavelets from the two slits travel distances to the screen that differ by $\Delta = m\lambda$, with m integral. From Fig. 24.7 we have $\sin \theta_m = \Delta/d$, and so the condition for *constructive interference* (bright fringes) is

$$\sin \theta_m = \frac{m\lambda}{d} \qquad (m = 0, 1, 2, \ldots)$$ **(24.1)**

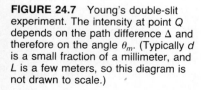

FIGURE 24.7   Young's double-slit experiment. The intensity at point Q depends on the path difference Δ and therefore on the angle θ_m. (Typically d is a small fraction of a millimeter, and L is a few meters, so this diagram is not drawn to scale.)

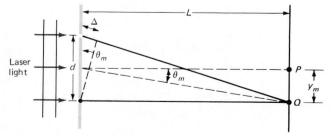

For *destructive interference* (dark fringes), the path difference must be

$$\sin \theta_m = \frac{(m + \frac{1}{2})\lambda}{d} \qquad (m = 0, 1, 2, \ldots) \qquad \textbf{(24.2)}$$

We next find the distance of the bright and dark fringes from the center of the interference pattern. The angle θ_m is always very small, since $\sin \theta_m = m\lambda/d$, and the wavelength λ is usually very small compared to the slit separation d in a practical experiment. We therefore set $\sin \theta_m = \tan \theta_m = y_m/L$, where y_m is the distance of the fringe in question from the center of the pattern, and obtain

Bright fringes: $\quad y_m = L \sin \theta_m = \dfrac{m\lambda L}{d} \qquad$ (*m* integral) $\qquad \textbf{(24.3)}$

For destructive interference, on the other hand, the path difference for the two waves must be $\Delta = (m + \frac{1}{2})\lambda$, and so

Dark fringes: $\quad y'_m = \dfrac{(m + \frac{1}{2})\lambda L}{d} \qquad$ (*m* integral) $\qquad \textbf{(24.4)}$

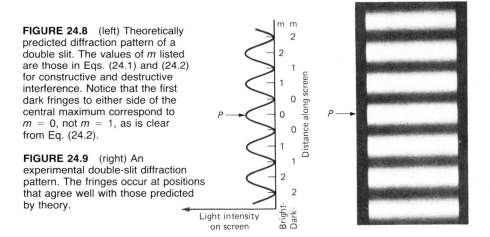

FIGURE 24.8 (left) Theoretically predicted diffraction pattern of a double slit. The values of *m* listed are those in Eqs. (24.1) and (24.2) for constructive and destructive interference. Notice that the first dark fringes to either side of the central maximum correspond to $m = 0$, not $m = 1$, as is clear from Eq. (24.2).

FIGURE 24.9 (right) An experimental double-slit diffraction pattern. The fringes occur at positions that agree well with those predicted by theory.

Figure 24.8 shows what we would expect on the basis of the above theory. If Young's interpretation of his double-slit experiment is valid, we expect to find a series of alternating bright and dark fringes on the screen. Figure 24.9 provides ample confirmation of Young's ideas. Here we see that the fringes are equally spaced as we go out from the central image at *P*, exactly as predicted by theory.

Equations (24.3) and (24.4) can be used to obtain a value for the wavelength λ of the light used, since the slit separation d can be measured with a traveling microscope and y_m and L with a measuring tape or meterstick. The results confirm the fact that the wavelength of visible light is about 500 nm. As a result d must be small or the fringes would be so close together that it would be impossible to distinguish them.

As the slits are brought together, the interference pattern spreads out, according to Eq. (24.3). If the wavelength increases, the pattern also spreads out.

Example 24.2

Coherent light of wavelength 633 nm from a He-Ne laser falls on a double slit with a slit separation of 0.103 mm. An interference pattern is produced on a screen 2.56 m from the slits. Calculate the separation on the screen of the two fourth-order bright fringes on either side of the central image.

SOLUTION

The easiest way to handle this problem is to calculate the distance y_4 of the fourth-order bright fringe on one side from the central image and then to double this value to obtain the distance between the two fourth-order fringes. From Eq. (24.3) we have

$$y_4 = \frac{4\lambda L}{d} = \frac{4(633 \times 10^{-9} \text{ m})(2.56 \text{ m})}{0.103 \times 10^{-3} \text{ m}}$$

$$= 6.29 \times 10^{-2} \text{ m} = 6.29 \text{ cm}$$

The distance between the two fourth-order fringes on either side is therefore

$$2y_m = \boxed{12.6 \text{ cm}}$$

EXERCISE 2 If, in a similar experiment, a different slit is used, and the separation between adjacent maxima is 1.25 cm on the screen 2.56 m from the slits, what is the new slit separation d?

Example 24.3

In a modern version of Young's experiment, a double slit is illuminated by light from a He-Ne laser at 633 nm and an interference pattern is formed. Then a piece of flint glass with index of refraction 1.58 at 633 nm is placed between the laser and one slit. Experiment shows that the locations of the light and dark fringes are interchanged in the pattern. What is the smallest thickness of glass that will produce this result?

SOLUTION

Remember (Sec. 22.4) that the speed of light in a medium of index of refraction n is $v = c/n$, where c is the speed of light in a vacuum (very close to its speed in air). Light therefore takes longer to travel a distance d in glass than it does in air, and so the introduction of the piece of glass shifts the phase of the light from one slit with respect to that from the other slit. In the time that it takes the light to travel a distance d in the glass, it would have traveled a distance nd in air (or vacuum), since the light travels at a slower speed c/n in the glass.

The difference in optical paths introduced between the two beams is therefore $\Delta = nd - d$. But inserting the glass is observed to change constructive interference to destructive interference and therefore must change the optical path by $m\lambda/2$, where m is an integer and λ is the wavelength in air. The minimum thickness of glass must correspond to $m = 1$, and so

$$\Delta = nd - d = (n - 1)d = \frac{\lambda}{2}$$

or $d = \dfrac{633 \text{ nm}}{2(1.58 - 1)} = 546 \text{ nm} = \boxed{0.55 \times 10^{-6} \text{ m}}$

Notice that we have given only two significant figures in the answer, since, when we subtract 1.00 from 1.58, only two significant figures remain.

24.3 Multislit Interference: The Diffraction Grating

If the number of slits producing interference fringes is gradually increased from 2 to 3 to 4, and so on, while keeping d the same, we find that the bright fringes become narrower and brighter. In the limit of many slits, we have a *diffraction grating*, invented in 1820 by Joseph von Fraunhofer (1787–1826), and consisting of as many as 10,000 or more slits per centimeter. Such a device is called a *diffraction* grating because diffraction effects (to be discussed in Sec. 24.6) are important in its full explanation. Transmission gratings are produced by highly specialized engines that use a diamond stylus to rule very closely spaced, parallel lines on glass. The unruled regions of glass then act as the slits. Reflection gratings are made by ruling closely spaced lines on a polished metal surface. Light reflected by the metal surface then shows interference effects. Such diffraction gratings produce extremely narrow, bright interference fringes and are much used in spectrometers for the analysis of the spectra of atoms and molecules.

Thomas Young (1773–1829)

FIGURE 24.10 *(Courtesy of AIP Niels Bohr Library.)*

Few men or women have ever contributed so much first-rate scholarly work to so many different fields of learning as did Thomas Young.

Born in Somerset, England, in 1773, Young learned to read at the age of 2 and by the age of 6 is said to have read the complete Bible twice. By 19 he had learned on his own to read Latin, Greek, Italian, Hebrew, Arabic, Persian, Turkish, and Ethiopian, and had time to master calculus, chemistry, and physics on the side! No wonder his fellow students at Cambridge nicknamed him "Phenomenon Young."

From 1792 to 1799 Young was engaged in his medical studies at London, Edinburgh, Göttingen, and Cambridge, receiving his M.D. degree from Göttingen in 1796. For most of his life he practiced medicine, but without great success. Constantly torn between the practice of medicine and his love for pure research, Young lacked sufficient interest in the routine problems of patients to build up a successful medical practice. Fortunately he had inherited a large fortune from his uncle and did not have to rely on medicine for a livelihood.

During his medical studies Young became interested in the physiology of the human eye. He was the first to explain how the human eye can focus on objects at different distances (accommodation). He also first explained, by experimenting with his own eyes, the nature of astigmatism (see Sec. 25.2). Finally, research on the nature of color blindness led him to put forth his three-color theory of color vision, which proposed that only three different kinds of color receptors are present in the human eye, one for blue, one for green, and one for red light. All other colors are merely combinations of these three basic colors. This theory, now called the Young-Helmholtz theory, is the basis of modern color photography and color television.

Young's work on vision naturally led him to explore the nature of light. His experiments with double-slit interference, Newton's rings, and thin films convinced him that light was a wave motion of some kind. Young never developed the mathematics to explain his wave model of light properly, and so the French physicist Fresnel, who did, is often given more credit than Young for the wave theory.

Young also made contributions to other fields of physics, e.g., to the stretching of metals (Young's modulus), to the understanding of the concept of energy, and to the theory of the tides.

In his latter years Young devoted most of his time to his first love, languages. He devoted years of intense study to deciphering the Egyptian hieroglyphics found on the Rosetta stone discovered in the Nile Delta in 1799. He never achieved complete success, but his ideas are the basis for our understanding of this important ancient form of communication. As a consequence, the memorial tablet on his tomb in Westminster Abbey reads: "[He] first established the undulatory theory of light, and first penetrated the obscurity which had veiled for ages the hieroglyphics of Egypt."

Young was a brilliant intellectual, but had little understanding of human sensitivities. He was honest—sometimes too honest—in appraising the scholarly work of his contemporaries, and made enemies as a result. Much of his scientific work was published anonymously because he did not want his patients to think that his heart was not in his medical practice.

Thomas Young was never a modest man and liked to boast: "I never spent an idle day in my life." Given the number and importance of his contributions to knowledge, we can well believe him.

The theory of the diffraction grating is identical with that of the double slit. We consider a large number of Huygens' wavelets moving from the slits in a direction such that the path difference between the wavelets from two adjacent slits is Δ, as in Fig. 24.11. For *constructive interference* adjacent wavelets must satisfy

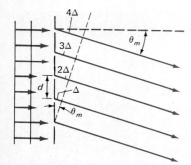

FIGURE 24.11 Interference produced by light passing through a diffraction grating. The intensity of light at any point on a screen to the right of the grating depends on the value of Δ.

the condition $\Delta = m\lambda$, where m is an integer. Hence, if d is the distance between two adjacent slits, we have

Grating equation (bright fringes): $\boxed{d \sin \theta_m = m\lambda}$ **(24.5)**

[Notice that this equation is identical with Eq. (24.1) for two slits.] Here θ_m is the angle corresponding to the mth-order bright fringe, where m is 0 for the central, undeviated image, $m = 1$ for the first-order fringes to either side of the central image, $m = 2$ for the second-order fringes, etc. The slit separation d, in meters, is the reciprocal of the number of lines per meter ruled on the grating.

In practice a diffraction grating is mounted as in Fig. 24.12. A *collimator* (which uses a lens of proper focal length) takes diverging light from a light source and converts it into a beam of plane waves. These plane waves strike a diffraction grating normal to its surface. The parallel waves emerging from all the slits at any particular angle are then collected and the image observed with a telescope. This telescope is free to move in a circle around an axis parallel to the grating rulings. The angles corresponding to bright fringes can then be read from a scale over which the telescope moves. In this way θ can be measured and the wavelength λ of the light obtained from the grating equation. This apparatus (Fig. 24.12) is usually called a *grating spectrometer*.

The main difference between the fringe patterns from a double slit and a grating is the remarkable increase in intensity and sharpness of the fringes with the grating. Figure 24.13 illustrates the situation in which d, the distance between slits, is the same in the two cases. The increased intensity and sharpness occurs because, with so many sources of light provided by the slits in the grating, even a slight change in the angle θ_m produces a sufficient amount of destructive interference to

FIGURE 24.12 A grating spectrometer. A diffraction grating produces images of the collimator slit at angular positions that depend on the wavelengths emitted by the light source.

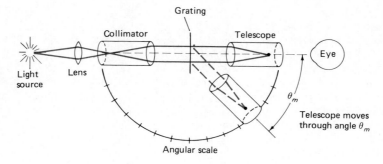

FIGURE 24.13 Comparison between interference fringes produced by a double slit and those produced by a diffraction grating. The grating fringes are much more intense and sharp.

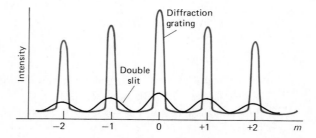

cancel out the light at all angles except those satisfying Eq. (24.5). All the original light is therefore concentrated into very intense, narrow fringes in the interference pattern.

If white light is incident on a grating, and the collimator contains a narrow slit through which the light enters the system, then a bright white line is seen with the telescope for light passing straight through the grating. This is the zero-order ($m = 0$) fringe. To either side of this central image we find a complete spectrum like a rainbow stretching from the violet, which has the shortest wavelength and is bent least, to the red, which has the longest wavelength and is bent most. This is the first-order ($m = 1$) spectrum, in this case a continuous spectrum because the white light contains all visible wavelengths. Beyond this first-order spectrum there are weaker, higher-order spectra.

If the light source is a mercury arc lamp (such as those used to light highways), we find a central image that is a bluish-purple line, and discrete slit images to either side of it in the first order, ranging outward from 405 nm (violet) through 436 nm (blue) and 546 nm (green) to 579 nm (yellow). We call this a discrete spectrum, or a *line spectrum*. (Some line spectra may be found in the colored insert on Optical Spectra in Chap. 28.) Beyond these first-order lines are higher-order images, the number depending on the value of the grating space d. Since $\sin \theta_m$ cannot exceed 1, there is a limit on the order m possible for any given wavelength and grating space.

Note that the spectrum produced by the prism spectrometer discussed in Sec. 23.3 differs from a grating spectrum in two ways: (1) The arrangement of wavelengths (or colors) in the prism spectrum is the reverse of that in the grating spectrum, and (2) a prism spectrum has no higher orders; all the light is concentrated in a single spectrum.

Example 24.4

A spectrometer uses a grating ruled with 8000 lines per centimeter. **(a)** How many orders of the sodium D line at 589 nm can be observed with such a spectrometer? **(b)** How many orders of the mercury violet line at 405 nm can be observed?

(c) A second-order orange line from an unknown element is observed to fall very close to the third-order mercury violet line of wavelength 405 nm. What is the wavelength of the unknown line?

SOLUTION

(a) 8000 lines per centimeter corresponds to a grating space

$$d = \frac{1}{8.00 \times 10^5/\text{m}} = 1.25 \times 10^{-6} \text{ m}$$

From the grating equation we have

$$\sin \theta_m = \frac{m\lambda}{d} = \frac{m(589 \times 10^{-9} \text{ m})}{1.25 \times 10^{-6} \text{ m}} = \frac{m}{2.12}$$

Now, the maximum value of $\sin \theta_m$ is 1, for which θ_m is 90°, and so the largest possible integral value of m is $\boxed{m = 2}$, and only the first- and second-order spectra show the sodium D lines at 589 nm.

(b) For the mercury 405-nm line, we find in the same way,

$$\sin \theta_m = \frac{m(405 \times 10^{-9} \text{ m})}{1.25 \times 10^{-6} \text{ m}} = \frac{m}{3.09}$$

and the largest possible value of m is $\boxed{m = 3}$. The third-order mercury line at 405 nm is therefore also visible.

(c) From the grating equation, with θ_m the same for the two lines, since they fall very close together, we have

$$m_2\lambda_2 = m_1\lambda_1$$

Here $m_1 = 3$ and $\lambda_1 = 405$ nm for the mercury violet line. Also $m_2 = 2$ for the unknown line of the second order. Hence

$$2\lambda_2 = 3(405 \text{ nm}) \quad \text{or} \quad \lambda_2 = \frac{3}{2}(405 \text{ nm}) = \boxed{608 \text{ nm}}$$

This method can be used to find the wavelengths of unknown lines in experimental spectra and is referred to as the *method of overlapping orders*.

24.4   Interference in Thin Films

A. A. Michelson once observed that the brilliant colors exhibited by the diamond beetle are caused by a diffraction grating built into the beetle's wings by nature. The beautiful colors seen in soap bubbles or in the feathers of peacocks, however, are produced by interference in the light reflected by thin films.

For example, suppose that a very thin film of air is trapped between two pieces of flat glass, as in Fig. 24.14. If a beam of yellow light from a sodium arc shines down almost normal to the glass surfaces, some of the light will be reflected from the interface between the bottom of the upper plate and the air, and some will be reflected from the interface between the air and the top of the lower plate. The eye can focus these two parallel beams at one spot on the retina, and these two beams will produce destructive or constructive interference depending on whether their path difference is equal to an odd or an even number of half wavelengths.

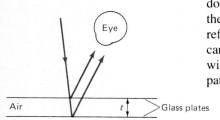

FIGURE 24.14   Interference produced by a thin air film of thickness t. (The thickness of the film and the separation of the two reflected beams are exaggerated for clarity.)

Let the thickness of the air film be t. Then the difference in the paths of the two beams shown in the figure is $\Delta = 2t$. We would expect that constructive interference and bright light would result if $\Delta = m\lambda$, where m is an integer, and destructive interference and darkness would result if $\Delta = (m + \frac{1}{2})\lambda$, with m again an integer.

This is not the entire picture, however, since an additional phase difference is introduced between the two beams on reflection. The first beam is reflected where the optically denser glass meets the less dense air. There is therefore no phase change on reflection, just as there is no phase change when a wave on a string is reflected from a free end of the string (see Sec. 11.9). But the other beam that is reflected from the interface where the air film meets the top surface of the lower piece of glass undergoes a phase change of 180° on reflection, just as does a transverse wave on a string when reflected from a fixed end, as in Fig. 11.25. An additional phase difference of 180° is therefore introduced between the beams on reflection, which is equivalent to an additional path difference of $\lambda/2$. When this added phase change is taken into account, the condition for *constructive interference* becomes not $2t = m\lambda$, but

Constructive interference: $\boxed{2t = (m + \frac{1}{2})\lambda}$ **(24.6)**

Similarly, for destructive interference and dark fringes we have, when the phase change on reflection is allowed for,

Destructive interference: $\boxed{2t = m\lambda}$ **(24.7)**

For white light this means constructive interference for some wavelengths and destructive interference for others, and so white light reflected back from a thin film of air will exhibit those colors for which constructive interference occurs.

Films of Materials Other Than Air

If the thin film consists of water, oil, or some other material between glass plates, the observed results are basically the same as those observed for air, except that the wavelength of the light in the film is reduced from λ to λ/n, since the speed of light in the film is reduced to $v = c/n$. For *destructive interference* we must have (allowing for the phase change on reflection)

$$t = \frac{m\lambda}{2n} \qquad \text{or} \qquad nt = \frac{m\lambda}{2} \tag{24.8}$$

The product nt of the index of refraction of the material and its actual thickness is called the *optical thickness* of the film. If the optical thickness is used in place of t, Eqs. (24.6) and (24.7) remain valid, with λ the vacuum wavelength for the light, as usual. This explains the colors seen in thin oil films spread out on wet surfaces.

If the thin film is not of constant thickness, but wedge-shaped, as in Fig. 24.15, then as the thickness increases, a series of alternating bright and dark fringes appear in monochromatic light. This is because the thickness satisfies the conditions first for constructive interference, then for destructive interference, and so forth, given by Eqs. (24.6) and (24.7). If both the plates are optically flat, i.e., ground flat to within a small fraction of a wavelength, the fringes will be straight. If the plates are not perfectly flat, fringes like those in Fig. 24.16 appear. One glass plate that is known to be optically flat is often used to test the flatness of another glass plate. If the fringes produced by the two plates are not straight, the second plate is polished until they become straight. In this way the optically flat mirrors and glass plates needed for today's research laboratories are produced.

Wedges

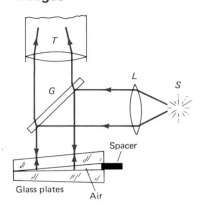

FIGURE 24.15 Experimental setup for observing fringes produced by a wedge-shaped, thin air film.

FIGURE 24.16 Observed fringes for a wedge-shaped air film between two glass plates that are not flat. Each dark fringe corresponds to a region of equal thickness in the film; between two adjacent fringes the change in thickness is $\lambda/2n$. The fringes indicate that the plates are approximately flat only near their left edge.

Newton's Rings

A striking example of thin-film interference called *Newton's rings* occurs when the film is the air trapped between a flat glass plate and the convex surface of a lens with a very large radius of curvature, as in Fig. 24.17. This is virtually the same situation as for the wedge discussed above except that the fringes now are circular, because the points at which the film has the same thickness form a series of concentric circles about the point of contact of lens and plate. In this case the distance between the two glass surfaces does not increase linearly, as it did for two flat plates. Rather, because of the convex curvature of the lens, the thickness of the air film increases more rapidly the farther out we go from the center of the lens. As a result the fringes are spaced farther apart near the center of the lens and closer together as we go out from the center.

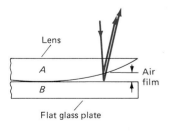

FIGURE 24.17 Newton's rings. The diagram indicates how reflections from the top and bottom surfaces of the air film between the long-focal-length lens and the flat glass plate lead to interference fringes.

The center spot, where the lens makes contact with the glass plate, is not bright, but dark. Here the optical path difference is much smaller than a wavelength (since the plates are very close together), so this is good evidence that the phase of one of the reflected beams has been shifted 180° relative to the other. Figure 24.18 shows the experimental arrangement for producing Newton's rings, and Fig. 24.19 shows the resulting pattern.

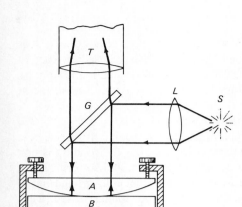

FIGURE 24.18 (left) Experimental arrangement for producing interference rings. The air wedge has circular symmetry, leading to the circular fringes called Newton's rings. In the diagram, *S* is the light source, *T* is the observing microscope, and *G* is a beam splitter.

FIGURE 24.19 (right) A Newton's rings interference pattern made by Dr. William P. Meggers using a mercury source containing only the single isotope $^{198}_{80}$Hg. *(Photo courtesy of National Institute of Standards and Technology, and Omikron/Photo Researchers, Inc.)*

Soap Bubbles

A soap bubble consists of a thin film of soapy water solution. When looked at in white light, a soap bubble shows bright colors. Let us suppose that for some particular wavelength of light the optical thickness of the bubble wall (*nt*) is exactly $\lambda/2$. Then the optical path difference for light reflected from the inside and outside of the bubble wall will be exactly λ. But the beam reflected at the front surface of the film undergoes a 180° phase change, while the beam reflected at the back experiences no phase change. Hence the two are 180° out of phase and cancel. The light corresponding to this wavelength is therefore not reflected by the bubble. For other wavelengths, however, constructive interference occurs and brightly colored reflected light is seen. As the bubble grows in size, its wall thins out and the colors seen by reflected light change. When the thickness of the wall becomes much less than the wavelength of any visible light, the bubble appears black, just as in a Newton's rings pattern the central fringe by reflected light is black.

Example 24.5

A soap bubble 250 nm thick is illuminated by white light. The index of refraction of the soap film is 1.36. **(a)** What colors are *not* seen by reflected light? **(b)** What colors appear strong in reflected light? **(c)** What color does the soap bubble appear to be for normal incidence?

SOLUTION

In this case our treatment of films of materials other than air in the preceding section, and in particular Eq. (24.8), are applicable.

(a) For *destructive interference* we must have $nt = m\lambda/2$. The wavelengths that are *not* reflected must satisfy the equation

$$\lambda_m = \frac{2nt}{m} \quad \text{with } m = 1, 2, 3, \ldots$$

and so

$$\lambda_1 = \frac{2nt}{1} = 2(1.36)(250 \text{ nm}) = \boxed{680 \text{ nm}}$$

$$\lambda_2 = \frac{2nt}{2} = (1.36)(250 \text{ nm}) = 340 \text{ nm}$$

These are the only wavelengths even close to the visible region of the spectrum for which destructive interference occurs: 680 nm is right in the middle of the red region of the spectrum, while 340 nm is in the ultraviolet and cannot be seen in any case. The only nonreflected color is *red*.

(b) For *constructive interference* we must have, using Eq. (24.6) modified for the proper optical thickness of the soap film,

$$2nt = (m + \tfrac{1}{2})\lambda \quad \text{or} \quad \lambda_m = \frac{2nt}{m + \tfrac{1}{2}}$$

$$\lambda_0 = \frac{2nt}{1/2} = 4(1.36)(250 \text{ nm}) = 1360 \text{ nm}$$

$$\lambda_1 = \frac{2nt}{3/2} = \frac{4}{3}(1.36)(250 \text{ nm}) = \boxed{453 \text{ nm}}$$

$$\lambda_2 = \frac{2nt}{5/2} = \frac{4}{5}(1.36)(250 \text{ nm}) = 272 \text{ nm}$$

A wavelength of 272 nm is in the ultraviolet region and 1360 nm is in the infrared. The only strong reflection is therefore in the blue-violet region at 453 nm.

(c) The reflected light will be weak in the red region of the spectrum and strong in the blue-violet region. Its color will therefore be a pronounced blue.

24.5 Some Applications of Interference

Nonreflecting Films

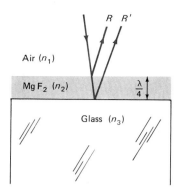

FIGURE 24.20 A nonreflective coating for glass. If the two reflected beams R and R' are 180° out of phase, they cancel and no light is reflected. Here $n_1 < n_2 < n_3$, and the phase changes at the two surfaces are the same. (The thickness of the glass and the separation of the rays are exaggerated for clarity.)

The thin films used as antireflection coatings on camera lenses and other optical components are an interesting and practical application of interference. Suppose we want all light of a particular wavelength to pass through a glass plate without any loss due to reflection. We deposit a thin film of a transparent substance like magnesium fluoride (MgF_2) on the outer surface of the glass and make the optical thickness (nt) of the film equal to $\lambda/4$, where λ is the wavelength of the light in air and n is the index of refraction of the film. The film material is usually chosen to have an index of refraction between that of air and glass so that the same phase change of 180° occurs on reflection at each interface. Then, as shown in Fig. 24.20, the light reflected at the film-glass interface will have traveled a distance $\lambda/2$ more than the light reflected from the air-film interface. The eye focuses the two parallel rays R and R' to a common point. Destructive interference occurs, and so no light is reflected.

Almost all optical parts of high quality are now coated to reduce reflection. Coated lenses and mirrors often have a purplish hue by reflected light. This is because perfect cancellation of the two beams can only occur for one wavelength, usually chosen to be in the middle of the visible spectrum in the green. Some red and some violet light is still reflected, leading to the observed purplish color.

Modern antireflection coatings for good photographic lenses contain typically five or more layers of inert materials like titanium oxide ($n = 2.45$) and silicon dioxide ($n = 1.45$). Such coatings reduce reflected light to less than 0.1 percent compared to the usual 4 percent reflected light for uncoated lenses. These multilayer films are effective across the whole visible spectrum.

Example 24.6

Magnesium fluoride (MgF_2) has an index of refraction $n = 1.38$. How thick a coating of MgF_2 is needed on a store window to produce minimum reflection from the glass in the green region of the visible spectrum near 550 nm? Assume that light strikes the coating almost at normal incidence.

SOLUTION

For zero reflection we want the optical film thickness nt to be equal to $\lambda/4$ for green light:

$$t = \frac{\lambda}{4n} = \frac{550 \text{ nm}}{4(1.38)} = \boxed{99.6 \text{ nm}}$$

EXERCISE 3 What other values of film thickness t will produce minimum reflection for light of 550 nm wavelength?

Holography

A more recent application of light interference is the storage of three-dimensional images on two-dimensional photographic plates by a process called *holography*. Holography is a two-step process in which an object is first illuminated by coherent light (like that from a laser) and an interference pattern is produced on a photographic plate between the light arising from the coherent source and the light

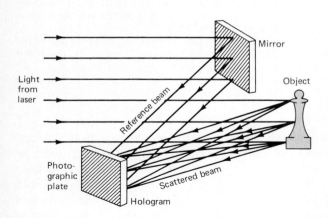

FIGURE 24.21 Making a hologram. An interference pattern is produced on the photographic plate between the light coming directly from a laser and the light scattered from the object of which a hologram is desired, in this case, a chess pawn.

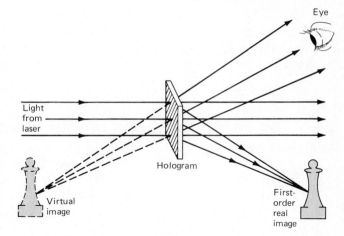

FIGURE 24.22 Viewing a hologram. Laser light is incident on the hologram at the same angle at which the plate was exposed to the reference beam. The recorded interference pattern on the hologram produces a virtual image, which is an exact replica of the original object.

scattered by the object, as in Fig. 24.21. Then the developed photographic plate (called a *hologram*) is illuminated by light from the same coherent source with the same alignment with the laser beam as when the hologram was made. An image of the original object results, as in Fig. 24.22. Amazingly, the holographic image appears truly three-dimensional. A viewer changing position with respect to the image experiences the same changes in perspective experienced in moving around the original three-dimensional object.

Holography is a very active research field at the present time. Holographic microscopes and moving pictures are being produced. Not only visible-light holograms but holograms using infrared, ultraviolet, and microwave radiation, and even sound waves at ultrasonic frequencies, are being made. Acoustic holography has been used, for example, to obtain images of the embryo in a mother's womb in which all the baby's organs and their relative sizes and positions can be seen clearly, without any known danger to the mother or infant.

24.6 The Diffraction of Light

One of the common objections once raised against the wave theory of light was that familiar waves like water waves and sound waves bend around obstacles, whereas light appears to travel in perfect straight lines and to cast sharp shadows, as would a beam of particles. But in the middle of the seventeenth century the Italian Jesuit priest Francesco Grimaldi (1618–1663) showed that this was not the whole story. Grimaldi demonstrated that when a beam of light passes through two small holes, one behind the other, and then falls on a dark surface, the band of light on that surface is a little wider than the original beam and has colored edges. Grimaldi surmised that the beam had been bent outward by a slight amount at the edges of the hole. This bending he called *diffraction*.

Diffraction: The bending of waves into the shadow region when they pass through apertures or around obstacles comparable in size to the wavelength.

All types of waves exhibit diffraction, but it becomes most apparent when the wavelength is large compared with the size of the aperture or obstacle. Diffraction is much less apparent when the wavelength is very small compared to the aperture or obstacle. Sound waves, with wavelengths measured in centimeters or in meters, diffract readily around small obstacles, but light, with wavelengths of only a few hundred nanometers, does not. For this reason the diffraction of light is not easily observed in everyday life.

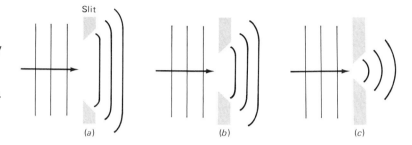

FIGURE 24.23 Diffraction of light by slits of varying width: (a) very wide slit; (b) narrower slit; (c) very narrow slit. The designations of width are all relative to the wavelength of the light.

FIGURE 24.24 Diffraction of water waves of different wavelengths by a slit. In (a) the wavelength λ of the water waves is 0.3w, where w is the slit width. A very clear shadow of the slit can be seen to the right of the slit. In (b) λ has been increased to 0.5w, and some bending into the shadow region can be seen. In (c) λ is 0.7w, and the plane wave passing through the opening has been almost completely converted into a cylindrical wave. No shadow of the slit can be seen. *(Courtesy of Educational Development Center, Newton, Massachusetts.)*

The diffraction of light can be explained by Huygens' principle. Suppose a plane beam of light falls on a very wide slit, as in Fig. 24.23a. Using Huygens' construction we find that the wave fronts remain planar across the slit, but that there is a slight bending of the wave front at the edges, since there are no additional Huygens' wavelets beyond the edges to create the same kind of plane wave front which is produced in the interior of the slit. As long as the wavelength of light (5×10^{-7} m) is small compared with the width of the slit (say, 10 cm), this bending or diffraction is not easily noticeable. If we narrow the slit down considerably, however, as in Fig. 24.23b, there is a more noticeable bending of light into the shadows at the two edges of the slit. Finally, if the slit becomes of the same relative size as the wavelength of the light, for example, 10^{-6} m, or 10^{-3} mm, then there is room for only a few Huygens' wavelets across the slit width and the plane wave is converted into a cylindrical wave, as shown in Fig. 24.23c. The corresponding behavior for water waves is shown in Fig. 24.24.

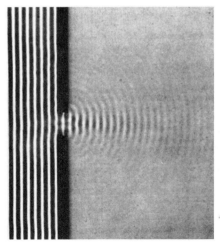

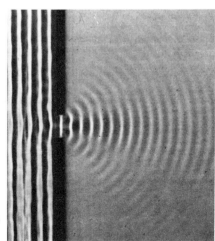

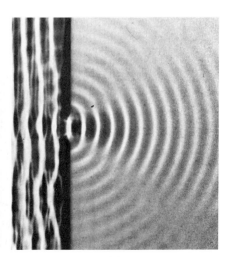

(a)   (b)   (c)

Single-Slit Diffraction

For simplicity we use a He-Ne laser as a light source, as in Fig. 24.25. The laser produces intense, highly monochromatic (one-color), coherent red light of wavelength 633 nm. This light falls on a narrow slit of width w and then on a screen a very large distance from the slit so that $w \ll L$. The observed pattern on the screen is referred to as the *diffraction pattern* of the single slit. It is found experimentally to consist of alternating bright and dark fringes parallel to the slit.

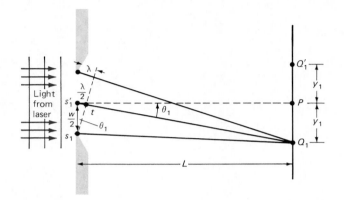

FIGURE 24.25 Calculation of a single-slit diffraction pattern. Light from the two halves of the slit are 180° out of phase and cancel at Q_1.

To explain the observed diffraction pattern we need to apply the ideas of superposition and interference to Huygens' wavelets originating at the slit. The Huygens' wavelets that go straight through the slit will arrive at position P on the screen in phase and produce a bright central fringe. To find the width of this central fringe we must find the positions to either side of P on the screen at which the Huygens' wavelets from different parts of the slit cancel out because they travel *different distances* and arrive at the screen 180° out of phase. This leads to darkness at the points Q_1 and Q_1' on either side of P on the screen.

To find the distance y_1 from P to Q_1 we consider light traveling in a direction from slit to screen so that the path difference between a Huygens' wavelet arising at the upper edge of the slit in Fig. 24.25 and the wavelet coming from the lower edge of the slit is $\Delta = \lambda$, where λ is the wavelength of the light being used. Then we divide the slit in Fig. 24.25 into two halves (or zones), a top half and a bottom half, and consider the superposition of pairs of Huygens' wavelets, one from the top half and one from the bottom half, arising at points separated by a distance $w/2$ across the slit. (Remember that we are looking at this slit parallel to its long axis. It actually extends some distance into the plane of the paper.) If point Q_1 is chosen as indicated above, then a wavelet from s_1 cancels a wavelet from s_1' at Q_1, and so on, as we take pairs of Huygens' wavelets from points a distance $w/2$ apart, since the two paths always differ by $\lambda/2$ for these wavelets. The difference in the paths of these pairs is $s_1't$ in Fig. 24.25. Now, from Fig. 24.26,

$$s_1't = \frac{w}{2} \sin \theta_1$$

Also for $L \gg y_1$ (which we assume), θ_1 is a very small angle, and so

$$\sin \theta_1 \simeq \tan \theta_1 = \frac{y_1}{L}$$

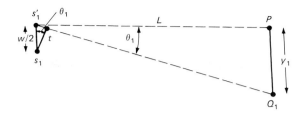

FIGURE 24.26 Angles used in calculating single-slit diffraction.

For *dark fringes* we must have

$$s_1't = \frac{\lambda}{2} = \frac{w}{2} \sin \theta_1 \qquad \text{or} \qquad \sin \theta_1 = \frac{\lambda}{w} \tag{24.9}$$

and so $\qquad y_1 = L \sin \theta_1 \qquad$ or $\qquad y_1 = \frac{\lambda L}{w}$ (24.10)

The first dark fringe therefore occurs at Q_1, a distance $\lambda L/w$ from P. A similar dark fringe is found at Q_1' on the other side of P. Note that the width of the central image is $2y_1 = 2\lambda L/w$ and that it gets wider as the wavelength increases and the slit gets narrower. Since $\sin \theta_1 = \lambda/w$, when λ is equal to w, $\sin \theta_1 = 1$, $\theta_1 = 90°$, and the image of the slit spreads to cover the entire hemisphere to the right of the slit.

Beyond the first dark fringe on either side of the central maximum there is another much less intense bright fringe. This is followed by a second dark fringe at a distance y_2 from P. To find y_2 we consider wavelets from the two edges of the slit in Fig. 24.25 that have path differences of 2λ. Then we break up the slit into four zones, and consider the superposition of Huygens' wavelets arising from points on the slit $w/4$ apart. In a fashion similar to that used for the first dark fringe, we find that in this case dark fringes occur, according to Eq. (24.9), when

$$\sin \theta_2 = \frac{2\lambda}{w} \qquad \text{and so} \qquad y_2 = \frac{2\lambda L}{w}$$

If this process is repeated, we find that *dark fringes* occur when

Dark fringes: $\quad \boxed{\sin \theta_m = \frac{m\lambda}{w} \qquad y_m = \frac{m\lambda L}{w} \qquad (m \text{ integral})}$ (24.11)

The distance between dark fringes is then $\lambda L/w$, which is half the width of the central bright fringe. Between adjacent dark fringes, bright fringes of decreasing brightness occur. Since Eq. (24.11) contains the wavelength, the single-slit diffraction pattern varies with the wavelength of the light used. If white light is used, a series of colored fringes is obtained, since y_m varies with λ. These are the colored diffraction fringes first observed by Grimaldi.

The predicted diffraction pattern on the screen is shown in Fig. 24.27*a*, and a photograph of an actual pattern is shown in Fig. 24.27*b*. For $w \gg \lambda$, $y_1 \rightarrow 0$ and we have almost perfect straight-line propagation of light. As a consequence, a *ray theory of light* (Chap. 23) should be valid as long as there are no obstacles or openings as small as the wavelength of the light used.

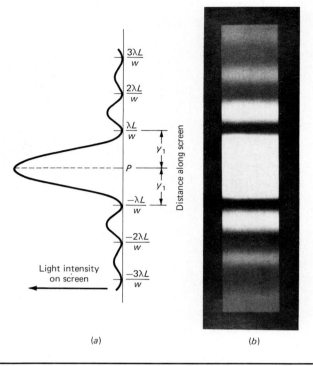

FIGURE 24.27 (a) Theoretically predicted diffraction pattern of a single slit. (b) An experimental diffraction pattern for a single slit. Note that the width of the central maximum is indeed twice the width of the secondary maxima, as predicted by theory.

Example 24.7

Light from a He-Ne laser ($\lambda = 633$ nm) is incident on a single slit of width 0.10 mm. **(a)** Describe the diffraction pattern formed on a screen 2.0 m away from the slit. **(b)** Describe what happens if the slit width is reduced to 0.010 mm.

SOLUTION

(a) By Eq. (24.11) the dark fringes occur at distances $y_m = m\lambda L/w$ from the center of the pattern. The first dark fringe falls at

$$y_1 = \frac{1(633 \times 10^{-9} \text{ m})(2.0 \text{ m})}{0.10 \times 10^{-3} \text{ m}} = 1.3 \times 10^{-2} \text{ m} = \boxed{1.3 \text{ cm}}$$

The central image is therefore 2(1.3 cm) = 2.6 cm wide, and the dark fringes on either side of the central image are spaced 1.3 cm apart.

(b) If the slit is narrowed to 0.010 mm, we have

$$y_1 = \frac{1(633 \times 10^{-9} \text{ m})(2.0 \text{ m})}{0.01 \times 10^{-3} \text{ m}} = \boxed{13 \text{ cm}}$$

In this case the central image is 10 times as wide, and the fringes are spaced 10 times as far apart as in (a).

Diffraction around a Circular Obstacle

If a plane light wave from a laser is incident on a penny, the penny will throw a shadow on a screen behind it, but there will be weak fringes around the shadow arising from the superposition of Huygens' wavelets diffracted around the edge of the penny. In addition there will be a very tiny, bright spot at the exact center of the shadow, as shown in Fig. 24.28. This spot is expected according to Huygens' principle. Suppose the light in the plane wave hitting the penny is coherent; i.e., all the waves are perfectly in phase. Then the Huygens' wavelets produced at the edge of the penny will be in phase at a point on the screen directly opposite the center of the penny, since all the wavelets travel exactly the same distance to reach that point. We would therefore expect a bright spot of light at the center. A photograph of a diffraction pattern obtained in this way is shown in Fig. 24.29.

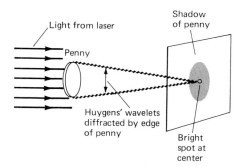

FIGURE 24.28 Prediction of bright spot at the exact center of the diffraction pattern of a penny.

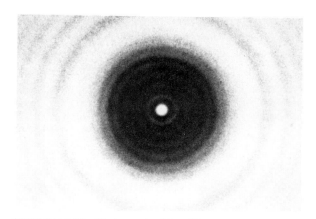

FIGURE 24.29 Photograph of the diffraction pattern of a circular opaque object in laser light, indicating the predicted bright spot at the center of the shadow.

When the French physicist Augustin Fresnel (1788–1827) first presented his wave theory of light in 1816, it was ridiculed by many French scientists. One of Fresnel's strongest critics was S. Poisson (1781–1840), who pointed out that Fresnel's theory led to the "absurd" conclusion that a bright spot should appear at the center of the diffraction pattern of an opaque circular object. Much to Poisson's chagrin, in 1818 another French physicist, D. F. Arago (1786–1853), actually carried out the experiment and found the bright spot! To commemorate Poisson's unintentional contribution to this landmark proof of Fresnel's wave theory of light, the bright spot is now called *Poisson's spot.*

Interference and Diffraction

Interference and diffraction are similar in that they are both basically *superposition* effects which depend on the addition of wave disturbances at a given point, taking account of the phase differences between the superimposed waves. The distinction usually made between interference and diffraction is that *interference* arises from the superposition of a finite number of waves coming from *different* coherent sources, as in Young's double-slit experiment. *Diffraction,* on the other hand, comes from the superposition of a very large (really an infinite) number of Huygens' wavelets arising from different places on the *same* original wave front, as in single-slit diffraction. In many cases, as with double slits and diffraction gratings, both interference and diffraction occur.

For example, Fig. 24.30*a* shows the interference fringes produced by a double-slit system in which each slit also produces a single-slit diffraction pattern with

FIGURE 24.30 (*a*) Combined interference and diffraction for a double-slit system. (*b*) If one of the slits is covered, the interference fringes disappear and we see only the diffraction pattern of a single slit.

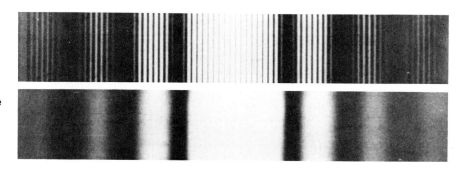

spacings about 10 times larger than the interference-fringe spacings. It can be seen that the equally spaced interference fringes have superimposed on them an intensity profile due to single-slit diffraction. Figure 24.30*b* justifies this interpretation: if one of the slits is covered, the interference fringes disappear, and we see only the diffraction pattern of a single slit, similar to that shown previously in Fig. 24.27*b*.

24.7 The Polarization of Light

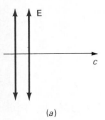

(a)

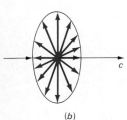

(b)

FIGURE 24.31 (a) Linearly polarized light. All electric vectors in the light are in the same direction and transverse to the direction of the light. (b) Unpolarized light. All electric vectors are still transverse but are in all possible directions in a plane at right angles to the direction of the light.

Diffraction and interference are clear examples of wave properties of light. They do not, however, indicate whether light is a transverse or a longitudinal wave. Sound, after all, is a longitudinal wave and still exhibits diffraction and interference effects. The most convincing experimental proof that light is a *transverse* wave comes from the phenomenon called *polarization*.

Just as the word *polarity* applied to a battery refers to the *direction* of the emf produced by the battery, so too the word *polarization* refers to the *direction* of the electric field vector **E** in a light wave or other kind of electromagnetic wave.

The electromagnetic waves discussed in Chap. 22 are transverse waves. For a wave moving along the positive x axis, as in Fig. 22.5, the electric field **E** and the magnetic field **B** are perpendicular to the direction of propagation and to each other. If, for example, the **E** field is along the y axis, the **B** field must be along the z axis, as shown in the figure. Such a wave is called a *linearly polarized* wave (Fig. 24.31*a*).

Linearly polarized light*: A light wave in which all the electric field vectors are oriented in the same direction.

Unpolarized light: A light wave in which the electric field vectors are distributed in all possible directions normal to the direction of propagation of the light.

Since the light emitted by most light sources is produced by the superposition of the light from a very large number of atoms oriented at random with respect to one another, light as normally found in nature consists of many independent transverse waves with randomly oriented **E** vectors. Such light is *unpolarized*.

A beam of unpolarized light in the plane of the paper can therefore be represented with its electric field vectors oriented in all possible directions normal to the direction of the beam, as in Fig. 24.31*b*. It is possible to change such an unpolarized beam of light into a beam linearly polarized along any desired direction at right angles to the direction of propagation by using a piece of *Polaroid.*[†]

Polaroid sheets have a preferred transmission axis (TA) determined by the arrangement of the molecules in the sheet. Light polarized parallel to this axis can pass through the Polaroid sheet; light polarized at right angles to this axis is absorbed and does not pass through. Light whose **E** vector makes an angle θ with the TA of the Polaroid has a component of **E** parallel to the axis of the Polaroid sheet, $E \cos \theta$, that is passed by the Polaroid, whereas the component normal to the axis, $E \sin \theta$, is absorbed.

---

*There are other important kinds of polarized light, such as *circularly polarized* light, which we do not discuss here.

[†]Polaroid is the trade name given to a plastic material capable of polarizing light, and invented in 1934 by Edwin H. Land (born 1909).

Malus' Law

When unpolarized light such as that in Fig. 24.31 passes through a piece of Polaroid, it emerges linearly polarized along the direction determined by the TA of the Polaroid, as in Fig. 24.32. Used in this way, a piece of Polaroid is called a *polarizer*. Once the light has been polarized, a second piece of Polaroid, with its axis rotated through an angle of 90° with respect to the polarizer, cuts off the light completely, as in Fig. 24.33, since all electric vectors are now at right angles to the TA of the Polaroid. A second piece of Polaroid, used in this way, is called an *analyzer*, for it can be used to determine (analyze) the polarization of the light incident on the Polaroid sheet. Note that a polarizer and an analyzer are identical. The different names arise from their different functions dictated by their different positions in the light path.

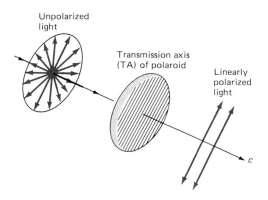

FIGURE 24.32 The conversion of unpolarized light into linearly polarized light by a Polaroid sheet (a polarizer).

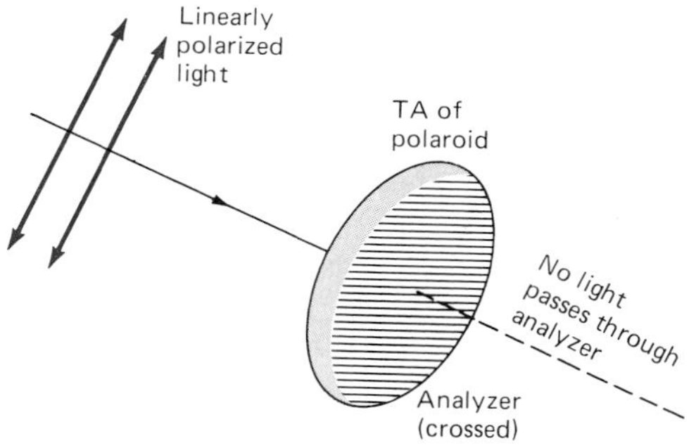

FIGURE 24.33 Use of a second sheet of Polaroid (an analyzer) to cut off a beam of polarized light.

If polarized light is incident on an analyzer, the amplitude of the transmitted wave depends on the component of the electric field vector parallel to the axis of the analyzer and therefore on the cosine of the angle θ between **E** and the TA of the Polaroid. For the amplitudes E'_{max} and E_{max} of the transmitted and incident waves we have

$$E'_{max} = E_{max} \cos \theta$$

Since the intensity of the waves is proportional to the square of their amplitude (Sec. 22.5), we have

$$\boxed{I' = I \cos^2 \theta}$$

(24.12)

an expression called *Malus' law*, after Étienne Louis Malus (1775–1812) who first introduced the term *polarized light*. The intensity of the light transmitted by an analyzer depends therefore on the square of the cosine of the angle between the polarization direction of the incoming wave and the TA of the Polaroid.

For example, if $\theta = 0$, $I' = I$ and the intensity is not reduced at all; if $\theta = 90°$, or the analyzer is "crossed" with respect to the polarization direction of the incoming light, $I' = 0$ and no light is transmitted.

Before the discovery of Polaroid, *Nicol prisms* made of calcite were used as polarizers and analyzers for polarized light. These are still used today for precision work in physics research laboratories.

Example 24.8

Linearly polarized light is incident on a piece of Polaroid at an angle of 60° with respect to the axis of the Polaroid. What is the intensity of the light passing through the Polaroid compared with the intensity of the incident light?

SOLUTION

Since $I' = I \cos^2 \theta$, we have

$$\frac{I'}{I} = \cos^2 60° = 0.25 = \boxed{\frac{1}{4}}$$

The intensity of the light passed by the Polaroid is one-fourth that of the incident light.

Polarization by Reflection

An unpolarized beam of light is equivalent to two equally intense beams polarized along any two mutually perpendicular axes, as in the incident beam in Fig. 24.34a, in which arrows indicate polarization in the plane of the paper, and dots polarization normal to the plane of the paper.

If unpolarized light strikes a material like glass or water at any but normal incidence, the beam becomes partially polarized by reflection. Thus in Fig. 24.34a an unpolarized light beam is incident on a sheet of glass at an angle θ_1 with respect to the normal. It is found by experiment that the reflected light is partially linearly polarized—i.e., the **E** vectors point in all directions perpendicular to the direction of propagation, but the amplitude of the electric-field vibrations is larger for **E** vectors parallel to the glass surface (the dots) than for **E** vectors at right angles to both these vectors and the direction of propagation (the arrows). Reflection therefore produces a partially polarized beam of light.

For one particular angle of incidence, θ_B, called *Brewster's angle*, 100 percent linear polarization is produced in the reflected light. This occurs when all the components of the electric field not parallel to the surface of the glass are refracted through the glass, and only the parallel components are reflected. It is found by experiment that this occurs when the reflected and refracted beams are at right angles, so that from Fig. 24.34b we have

$$\theta_B + \theta_2 = 90°$$

FIGURE 24.34   (a) Partial polarization of originally unpolarized light by reflection from a glass plate. (b) Complete polarization of light by reflection from a glass plate at Brewster's angle θ_B, where tan $\theta_B = n_2/n_1$. Here the reflected and refracted waves are at right angles, and all **E** vectors in the reflected light are parallel to the surface of the plate.

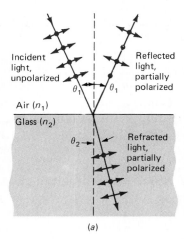

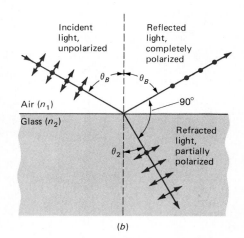

(a)

(b)

But, from Snell's law,

$$n_1 \sin \theta_B = n_2 \sin \theta_2$$

so that $\qquad n_1 \sin \theta_B = n_2 \sin (90° - \theta_B) = n_2 \cos \theta_B$

from which $\qquad \tan \theta_B = \dfrac{n_2}{n_1}$

This can be written as

$$\boxed{\tan \theta_B = n}$$

(24.13)

where θ_B is Brewster's angle, i.e., the angle for which complete polarization by reflection is achieved, and $n \, (= n_2/n_1)$ is the index of refraction of medium 2 with respect to medium 1.

The complete polarization produced when light is incident on a dielectric material at Brewster's angle is easily verified by rotating a piece of Polaroid in the path of the reflected beam.

It is not hard to understand the physical reason complete polarization occurs when light is reflected at Brewster's angle. The incident light sets the electrons in the atoms of the glass into oscillation, and these vibrating electrons produce the reflected beam. When the light is reflected perpendicular to the refracted beam, only vibrations parallel to the air-glass interface can contribute. This is because any vibration perpendicular to the air-glass interface would be, for reflection at Brewster's angle, parallel to the direction of the reflected beam, as shown in Fig. 24.34b. Since light is a *transverse* wave, no such longitudinal **E** vectors can occur in the reflected light, for **E** vectors must be perpendicular to the direction of propagation.

Sunlight reflected from a flat horizontal reflecting surface such as a lake, river, highway, ocean, or desert is partially polarized with its electric vector horizontal, as just discussed. Polaroid sunglasses employ sheets of Polaroid with their TA vertical to cut out most of this reflected light. The light passing through the glasses is then the vertically polarized part of the diffusely scattered and unpolarized light from the trees, boats, hills, or houses we are looking at. Polaroid sunglasses reduce the amount of light getting through to our eyes, but we see more clearly because much of the glare is removed.

Example 24.9

At what angle must light be incident on a piece of borosilicate crown glass, with index of refraction 1.52, to produce 100 percent linearly polarized light by reflection?

SOLUTION

From our discussion above, the light must be incident at Brewster's angle, given by

$$\tan \theta_B = n = \dfrac{n_2}{n_1}$$

Here n_1 is the index of refraction of air, which is equal to 1 to four significant figures, and so we have

$$\tan \theta_B = \dfrac{1.52}{1.00} = 1.52 \qquad \text{and} \qquad \theta_B = \boxed{56.7°}$$

For light incident on crown glass at 56.7°, therefore, the reflected light is 100 percent linearly polarized.

EXERCISE 4 What is Brewster's angle for light passing from borosilicate crown glass into air?

In conclusion, it may be worthwhile to emphasize two important conclusions from the material in this chapter:

1 *Diffraction* and *interference* demonstrate the *wave nature* of light.

2 *Polarization* demonstrates that light waves are *transverse*.

Summary: Important Definitions and Equations

Huygens' principle: Every point on a wave front may be considered a source of secondary wavelets that spread out in the forward direction at the speed of light in the medium. The new wave front is then the surface tangent to all these secondary wavelets.

Interference of light: The superposition of two or more beams of light so that they reinforce each other (constructive interference—bright fringes) or cancel each other out (destructive interference—dark fringes), depending on their relative phases.

Young's double-slit experiment
For bright fringes:

$$\sin \theta = \frac{m\lambda}{d} \qquad y_m = \frac{m\lambda L}{d} \qquad \textbf{(24.1, 24.3)}$$

For dark fringes:

$$\sin \theta = \frac{(m + \frac{1}{2})\lambda}{d} \qquad y_m' = \frac{(m + \frac{1}{2})\lambda L}{d} \qquad \textbf{(24.2, 24.4)}$$

Diffraction grating

Grating equation (bright fringes):

$$d \sin \theta_m = m\lambda \qquad \textbf{(24.5)}$$

Thin films:
1 *Air film between glass plates:*

Bright fringes in reflected light: $2t = (m + \frac{1}{2})\lambda$ **(24.6)**

Dark fringes in reflected light: $2t = m\lambda$ **(24.7)**

2 *Fluid film between glass plates (or other identical materials):*

Bright fringes in reflected light: $2nt = (m + \frac{1}{2})\lambda$

Dark fringes in reflected light: $2nt = m\lambda$ **(24.8)**

where nt is the optical thickness of the film and n the index of refraction.

Holography: The use of the interference of light to store three-dimensional images on two-dimensional photographic plates.

Diffraction of light: The bending of waves into the shadow region when they pass through apertures or around obstacles comparable in size to the wavelength.

Single-slit diffraction (dark fringes):

$$\sin \theta_m = \frac{m\lambda}{w} \qquad y_m = \frac{m\lambda L}{w} \qquad \textbf{(24.11)}$$

Polarization of light:

Linearly polarized light: A light wave in which all the electric vectors are oriented in the same direction.

Malus' law: $I' = I \cos^2 \theta$ **(24.12)**

Polarization by reflection: For 100 percent polarization of reflected light

$$\tan \theta_B = n \qquad \text{and} \qquad n = \frac{n_2}{n_i}$$

where θ_B = Brewster's angle **(24.13)**

Diffraction and *interference* demonstrate the *wave nature* of light.
Polarization demonstrates that light waves are *transverse*.

Questions

1 Why is a light beam reflected better by a plane mirror than by a piece of cardboard?

2 If light is incident normal to the surface of an air-glass interface, no refraction occurs. Show from Snell's law why this is to be expected.

3 Explain why an oar that is half in and half out of the water appears to be broken at the point where it enters the water.

4 If a source of sound can be heard when it cannot be seen because of obstacles in the way, how can both sound and light be propagated as waves?

5 Explain why coherent light diffracted around a penny will produce a bright spot at the center of the penny on a screen.

6 Why does a soap bubble appear dark (and not bright) just before it bursts as it grows larger and thinner?

7 An oil film spreads out on wet pavement and produces colors by interference in reflected light. Near the edges, where the film is thinnest, the oil slick appears bright.

(*a*) Why does the oil slick appear bright, rather than dark, where it is very thin?

(*b*) What limits can you set on the value of the index of refraction of the oil from the information given?

8 For Young's double-slit experiment explain what happens to the bright fringes as:

(*a*) The wavelength of the light is increased.

(b) The distance to the screen is increased.
(c) The distance between the slits is increased.
(d) The width of each individual slit is increased.

9 Give a theoretical justification for the experimental fact that in a Newton's rings interference pattern the fringes get closer together as they move out from the point of contact of lens and glass plate.

10 What is the difference between diffraction and interference?

11 (a) If light as it occurs in nature is "unpolarized," does this mean that the electric and magnetic vectors are not confined to a plane perpendicular to the direction of propagation?
(b) What then do we mean by unpolarized light?
(c) How is unpolarized light produced?

12 Why are the diameters of some microwave transmitting antennas so large? (These are the large dish-shaped antennas you often see on ships, at airports, and at radio relay stations.)

Assume that the microwaves used are from 1 to 10 cm in length.

13 Explain how one might use polarized light to measure the index of refraction of something opaque, like a lump of coal.

14 Unpolarized light is incident on a crossed polarizer and analyzer, and no light passes through the combination. A student claims to be able to make some light pass through the analyzer by placing a third piece of Polaroid between the polarizer and the analyzer. Will this work? If so, explain how a third piece of polarizing material can increase the transmission of the light. For what angle of the Polaroid axis with respect to the crossed axes of polarizer and analyzer, would the transmission be greatest?

15 (a) What phenomena in optics provide strong proof that visible light has wave properties?
(b) What phenomenon in optics provides strong proof that visible light consists of transverse waves?

Multiple-Choice and Simple Exercises

24.1 The wavelength of visible light is shorter than the wavelength of an FM radio wave of frequency 100 MHz by about:
(a) 1 order of magnitude
(b) 3 orders of magnitude
(c) 7 orders of magnitude
(d) 10 orders of magnitude
(e) 14 orders of magnitude

24.2 The separation of white light into its component colors is called:
(a) Diffraction (b) Reflection
(c) Polarization (d) Dispersion
(e) Refraction

24.3 To measure the wavelength of light by Young's interference method, one must know, among other things:
(a) The time required for the light to travel from the source to the screen
(b) The speed of light in the medium
(c) The size of the two slits
(d) On which side of the central fringe the selected maximum lies
(e) The distance between the two slits

24.4 A double-slit interference experiment uses a screen 1.00 m from the double slit, whose width is 0.500 mm. If the third-order bright fringe falls a distance of 4.00 mm from the center of the pattern, the wavelength of the light used must be:
(a) 670 nm (b) 396 nm (c) 472 nm
(d) 525 nm (e) 590 nm

24.5 A diffraction grating is ruled with 12,000 lines to the centimeter. It is illuminated with mercury light. The angular displacement of the first-order green line at 546 nm with respect to the central image will be:
(a) 41° (b) 3.7° (c) 0° (d) 49°
(e) None of the above

24.6 A thin film of water is trapped between two glass plates. On looking straight down on this water "sandwich," a blue color is observed, indicating that red light has not been reflected. One reasonable value for the thickness of the water film is:

(a) 0.26 μm (b) 0.35 μm (c) 0.15 μm
(d) 0.20 μm (e) None of the above

24.7 A soap bubble 350 nm thick is illuminated by white light. The index of refraction of the soap film is 1.36. The most intense color in the reflected light from the bubble will be:
(a) Red (b) Yellow (c) Green (d) Blue
(e) Violet

24.8 Diffraction effects become important for visible light when the light encounters slits or holes with dimensions of the order of:
(a) 1 m (b) 10^{-3} m (c) 10 cm
(d) 1 mm (e) 10^{-6} m

24.9 A single slit is 1.5 m from a screen. Monochromatic light of wavelength 633 nm falls on the slit and the third dark fringe falls at a distance of 3.0 cm from the center of the diffraction pattern. The width of the single slit is about:
(a) 1 mm (b) 0.1 mm (c) 1 cm
(d) 0.01 mm (e) None of the above

24.10 When light is polarized by reflection at Brewster's angle, the refracted ray:
(a) Is at right angles to the reflected ray
(b) Is completely polarized
(c) Makes any arbitrary angle with the incident ray
(d) Is at right angles to the incident ray
(e) Is completely unpolarized

24.11 Unpolarized light is incident on a piece of glass at an angle of 59.0° and it is found that the reflected light is 100 percent linearly polarized. The index of refraction of the glass with respect to the air is:
(a) 0.85 (b) 1.05 (c) 1.53 (d) 1.27
(e) 1.66

24.12 At time $t = 0$ a plane wave front of light is at a position 50 m from the origin along the x axis. What is the position of the same wave front 30×10^{-9} s later, if the light is moving through a vacuum along the $+x$ axis?

24.13 A point source of light emits a spherical light wave in air. At time $t = 0$ the radius of the wave front is 100 m. What is the radius of the wave front when $t = 8.0 \times 10^{-6}$ s?

24.14 A plane wave front in air is incident on a flat piece of plate glass ($n = 1.50$) so that the normal to the wave front makes an angle of $40.0°$ with respect to the normal to the glass surface.

(*a*)   What is the angle of reflection?

(*b*)   What is the angle of refraction in the glass?

24.15   Light of wavelength 579 nm from a mercury arc falls on a double slit, where the distance between the centers of the two slits is 0.080 mm. On a screen 2.0 m from the slit what is the location in the interference pattern (*a*) of the first dark fringe; (*b*) of the first bright fringe?

24.16   Light of wavelength 633 nm falls on a diffraction grating normal to its surface. The grating is ruled with 7500 lines per centimeter. What is the angle corresponding to the first bright fringe produced by the grating?

24.17   A soap bubble of index of refraction 1.36 is illuminated by white light. No red light in the region around 700 nm is reflected from the walls of the bubble. What is the minimum thickness (not zero) of the bubble walls that will explain this observation?

24.18   A magnesium fluoride film, of index of refraction 1.38, is used to coat a prism to reduce the light reflected from the front face of the prism. What must be the thickness of the film to reduce the reflection of the red light ($\lambda = 633$ nm) from a He-Ne laser to zero?

24.19   Light from a He-Ne laser ($\lambda = 633$ nm) is incident on a narrow slit of width 0.020 mm. What is the position of the fourth-order dark fringe produced by the slit on a screen 2.8 m away from the slit?

24.20   Unpolarized light in air falls on a glass surface at an angle of $56°$. It is found that the reflected beam is 100 percent linearly polarized with its electric vectors parallel to the surface of the glass. What is the index of refraction of the glass?

24.21   Linearly polarized light falls on a piece of Polaroid with the direction of polarization of the light making an angle of $45°$ with the transmission axis of the Polaroid. If the intensity of the light striking the Polaroid is 60 mW/cm^2, what is the intensity of the light emerging from the Polaroid?

Problems

24.22   What must be the separation between two slits to produce a fourth-order bright fringe at a distance of 2.00 cm from the center of the interference pattern on a screen 1.00 m from the slit? Assume that:

(*a*)   The light used is red at a wavelength of 700 nm.

(*b*)   The light used is blue at a wavelength of 400 nm.

24.23   A double slit has a slit separation of 0.010 mm. For red light at 633 nm what is the highest-order bright fringe that can be observed on a screen?

24.24   Light from a tunable dye laser is incident on a double slit. If the slit separation is 0.050 mm, and the first bright fringe occurs at a distance of 2.5 cm from the center of the interference pattern on a screen 2.0 m from the slit, what is the wavelength of the light from the laser?

24.25   An adjustable double slit is used in Young's experiment. Initially the slit separation is 0.10 mm and the distance between the fifth-order bright fringes to either side of the central image on the screen is 16 cm. What is the distance between the same fringes if the slit separation is reduced to 0.020 mm?

24.26   Light of wavelength 633 nm from a He-Ne laser falls on a double slit where each slit has a width of 0.010 mm and the distance between the centers of the two slits is 0.10 mm. Compare the widths of the single-slit diffraction pattern and the double-slit interference pattern on a screen 2.0 m away.

***24.27**   In the Young's double-slit experiment of Prob. 24.26 a piece of glass is placed in front of one of the slits. Then a thin film of MgF$_2$, with $n = 1.38$, is deposited on the glass to a thickness of 1.0×10^{-4} cm. How far does the first bright fringe move sideways when the thin film is deposited on the glass?

***24.28**   In a Young's double-slit experiment a pair of slits separated by 0.60 mm is illuminated by coherent light at 633 nm. The interference pattern is then observed on a screen 1.55 m from the slits.

(*a*)   At what distance does the third maximum fall from the center of the interference pattern?

(*b*)   A piece of glass with index of refraction 1.55 is placed in front of one of the slits. If a minimum is now seen to replace the third maximum, what is the minimum thickness of the glass that will produce this result?

24.29   A double slit is illuminated with coherent light at 633 nm. It is found that when a thin plate of glass of index of refraction 1.67 is placed between the source and one of the slits, the interference pattern shifts so that the fourth maximum to one side of the center now appears at the location previously occupied by the central maximum of the pattern. Determine the thickness of the glass plate.

24.30   Light of 546 nm from a mercury arc falls on a diffraction grating ruled with 30,000 lines per inch. What is the angular separation between the first-order images to either side of the central maximum?

24.31   How many orders of the 436-nm blue line of mercury can be observed when this light is incident on a grating ruled with 5000 lines per centimeter?

24.32   A third-order violet line from an unknown element is observed to fall at the same place as the second order of the mercury 579-nm yellow line. What is the wavelength of the unknown line?

24.33   An astrophysicist is studying the infrared spectrum of the sun with a grating spectrograph and finds that the fourth-order image of the visible hydrogen line at 656 nm falls at an angle of $32°$ with the direction of the incident light. He also finds, using an infrared thermal detector, that a first-order infrared line makes an angle of $24°$ with the incident light. What is the wavelength of the infrared line?

24.34   An ultraviolet grating spectrometer is being used to study the fine structure of the mercury line at 254 nm. The second-order image of this line occurs at an angle of $30°$ from the original direction of the light. How many lines per centimeter are ruled on the grating?

24.35   The yellow light from a sodium arc consists of two very close wavelengths of 589.0 and 589.6 nm, respectively. This light falls normally on a plane diffraction grating with 500 lines per centimeter.

(a) What is the angular separation of these two lines in the first order?

(b) In the second order?

(c) If the human eye can distinguish lines with an angular separation of at least 1 minute of arc, will it be able to see these two lines distinctly?

24.36 The wavelength limits of the visible spectrum are 390 and 760 nm. If white light falls at normal incidence on a plane diffraction grating ruled with 5000 lines per centimeter, what is the angular spread of (a) the first-order spectrum; (b) the second-order spectrum; (c) the third-order spectrum?

24.37 An oil film of index of refraction 1.33 is trapped between two pieces of glass. No light is reflected by such a film when 536-nm light falls on it at normal incidence. What is the minimum thickness of the oil film that will satisfy these conditions?

24.38 A metal shim is used to separate the ends of two long pieces of flat glass, while the other two ends remain in direct contact. The plates are 50.0 cm long. If mercury light with $\lambda = 436$ nm is incident on the plates, the distance between two consecutive dark fringes is found to be 1.00 cm. What is the thickness of the shim?

∗24.39 A transmission grating is used with light incident normal to the grating. The width of each slit on the grating is one-third the spacing between slits. Show, by considering the effects of single-slit diffraction, that the third-order maxima are missing from the diffraction pattern produced by the grating.

24.40 Metals like aluminum can be given a bright metallic color by a process called *anodization*. By this electrolytic process a thin transparent film is deposited on the metal surface to a thickness that gives the desired color by interference. If a transparent film of aluminum oxide of thickness $t = 250$ nm and index of refraction $n = 1.76$ is deposited on an aluminum cup, what is the color of the cup when observed in white light falling at normal incidence on the cup's surface (see Fig. 24.1)?

24.41 Two flat glass plates rest on a table. They touch at one edge and are separated by a metal shim at the other end, as in Fig. 24.35. When a mercury arc, with a filter in place that passes only the 436-nm line, illuminates this wedge directly from above, four dark fringes are seen, with dark fringes (D) appearing at both ends. What is the thickness t of the shim?

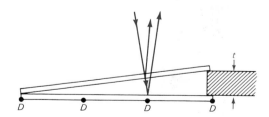

FIGURE 24.35 Diagram for Prob. 24.41.

∗24.42 In a Newton's rings experiment like that of Fig. 24.18, the light used is from a sodium arc, with $\lambda =$ 589 nm. The radius of the twelfth dark ring is 10.0 mm. What is the radius of curvature of the convex lens that rests on the glass plate?

∗24.43 A long-focal-length lens is supported horizontally a short distance above the flat end of a steel cylinder. The cylinder is 10.0 cm high and is rigidly clamped at its base. Newton's rings are produced between the curved surface of the lens and the flat surface of the top of the steel cylinder. The light used is that from a sodium lamp at 589 nm. A microscope is used to observe the Newton's rings. When the temperature of the cylinder is raised 20 C°, 80 fringes are observed to move past the cross hairs of the microscope. What is the coefficient of linear expansion of steel?

24.44 A magnesium fluoride film of index of refraction 1.38 has the minimum thickness required to reduce the reflection of green light ($\lambda = 546$ nm) almost to zero. What wavelengths will such a coated window reflect strongly?

24.45 (a) Calculate the minimum thickness of a coating of MgF_2 ($n = 1.38$) that must be applied to a light flint glass lens to cause minimum reflection of red light at 700 nm.

(b) What greater thicknesses would produce the same result?

24.46 It is desired to coat a lens so that it has maximum transmission of light in the center of the solar spectrum at 500 nm.

(a) What minimum thickness of MgF_2 ($n = 1.38$) should be used to coat the lens?

(b) What wavelengths will be transmitted least well by such a coated lens?

24.47 A soap film appears red ($\lambda = 650$ nm) when viewed with white light incident perpendicular to the film. If the index of refraction of the film is 1.45, what is the minimum thickness of the film?

24.48 (a) What would be the angular deviation of visible light from a straight-line path when it passes through a slit 1.0 mm in width?

(b) How does this result change for a 1.0-μm slit?

24.49 Calculate the approximate ratio of wavelength to obstacle size for a shed 10 m long by 10 m high for the following kinds of waves: (a) AM radio waves. (b) FM radio waves; (c) microwaves (radar waves); (d) light waves. (e) Point out how you would expect these four kinds of radiation to differ in their interaction with the shed.

24.50 What is the width of the central image of a diffraction pattern produced by a 1.06-mm slit on a screen 10.0 m from the slit, (a) if the light is that from a He-Ne laser at 633 nm; (b) if the radiation falling on the slit is a radio wave of 100-MHz frequency?

24.51 Mercury light at 436 nm is incident on a narrow slit of width 0.105 mm. Find the width of the central bright fringe on a screen 2.54 m away.

24.52 Light from a hydrogen discharge tube falls on a narrow slit 0.050 mm in width. On a scale 2.0 m from the slit what is the distance between the third-order dark fringe produced by the blue-green 486-nm line and that produced by the blue 434-nm line?

24.53 Light from a sodium arc, with $\lambda = 589$ nm, is incident on a slit of width 0.152 mm.

(a) Describe the diffraction pattern on a screen 1.0 m away from the slit.

(b)   Describe what happens if the slit width is reduced to 0.051 mm.

24.54   A slit of width w is illuminated by white light.

(a)   For what value of w will the first minimum for red light, with $\lambda = 650$ nm, fall at an angle of $30°$ to the normal?

(b)   For the same slit what is the wavelength λ' of the light whose first diffraction maximum (not counting the central maximum) coincides with the first minimum for red light? (Use the approximation that the diffraction maxima fall halfway between minima.)

(c)   If the slit width w is made smaller, will this coincidence of maximum for λ' with minimum for 650 nm remain?

(d)   What happens to the diffraction pattern as the slit width w is made smaller?

24.55   (a)   Unpolarized white light is incident on a polarizer and then falls on an analyzer whose TA makes an angle of $30°$ with the polarizer. What is the fraction of the original light transmitted through both the polarizer and analyzer?

(b)   If the analyzer is rotated until the transmitted intensity is 20 percent of the original light intensity, what is the new angle of the analyzer with respect to the polarizer?

24.56   Unpolarized light of intensity I_0 is incident on a crossed polarizer-analyzer combination:

(a)   If the analyzer is now rotated through $40°$, what percentage of the light incident on the analyzer is passed by the analyzer?

(b)   What percentage of the original light is passed by the analyzer?

24.57   At what angle is the sun above the horizon when the sunlight reflected from a swimming pool can be completely obscured with a piece of Polaroid?

24.58   The index of refraction of a plate of flint glass is 1.66. At what angle of incidence on this plate is the reflected light 100 percent linearly polarized if the plate is immersed (a) in water; (b) in benzene?

24.59   A light source is at the bottom of a small aquarium tank.

(a)   At what angle would the light have to strike the water-air interface at the top of the tank to be completely polarized by reflection?

(b)   At what angle would the light have to strike the water-glass interface at the side of the tank to be completely polarized by reflection (Assume that the wall of the tank is made of light flint glass).

(c)   At what angle would sunlight have to strike the water at the top of the tank to be completely polarized by reflection?

24.60   Unpolarized white light passes from air into a flat, heavy-flint glass plate.

(a)   Use the values of the index of refraction from Fig. 23.17 to calculate Brewster's angle for the limiting wavelengths of the visible spectrum, 390 nm (blue-violet) and 760 nm (red).

(b)   What is the difference in Brewster's angle for these two wavelengths?

(c)   If light passes from the heavy flint glass into air, what is Brewster's angle for 500-nm light?

*24.61   The captain of a naval vessel desires to send a message to a Coast Guard lighthouse on shore by using a light beam which is 100 percent linearly polarized parallel to the ocean surface. The captain decides to polarize the beam by reflecting it from the ocean surface. If the transmitter is 20 m above the ocean surface and the receiver is on a cliff 150 m above the ocean surface, how far from the cliff should the captain locate the ship to carry out this mission? (Assume that the index of refraction of the water at the frequency used is 1.33.)

Baumeister, Philip, and Gerald Pincus: "Optical Interference Coatings," *Scientific American*, vol. 233, no. 6, December 1970, pp. 58–75. Describes the making of multilayer interference films and their use for a variety of practical purposes.

Cagnet, M., M. Françon, and J. C. Thrierr: *Atlas of Optical Phenomena*, Springer-Verlag, Heidelberg, 1962. Contains a magnificent selection of photographs illustrating important optical phenomena, including diffraction and interference. Figures 24.9, 24.16, 24.27, 24.29, and 24.30 are taken from this reference (with permission of Springer-Verlag, Heidelberg).

Können, G. P.: *Polarized Light in Nature,* Cambridge University Press, New York, 1985. A charming book, with some outstanding photographs of polarization phenomena in the world about us.

Light and Its Uses: Making and Using Lasers, Holograms, Interferometers, and Instruments of Dispersion, readings from *Scientific American* (introduction by Jearl Walker), Freeman, New York, 1980. Contains useful articles on holography and a good bibliography.

Moore, A. D.: "Henry A. Rowland," *Scientific American*, vol. 246, no. 2, February 1982, pp. 150–161. An account of the success of the famous Johns Hopkins University professor in producing high-precision diffraction gratings.

Shamos, Morris H. (ed.): *Great Experiments in Physics*, Holt, Rinehart and Winston, New York, 1959. Chapter 7 is devoted to Young's work on the interference of light, and chap. 8 to Fresnel's work on diffraction.

Shurcliff, William A., and Stanley S. Ballard: *Polarized Light*, Van Nostrand, New York, 1964. The nature and applications of polarized light.

Van Heel, A. C. S., and C. H. F. Velzel: *What Is Light?*, McGraw-Hill, New York, 1968. Much of this introductory college text is devoted to well-illustrated discussions of diffraction, interference, and polarization.

Wehner, Rudiger: "Polarized Light Navigation by Insects," *Scientific American*, vol. 235, no. 1, July 1976, pp. 106–115. An account of experiments showing that bees and ants find their way home by detecting the polarization of scattered sunlight from the sky.

Wood, Alexander, and Frank Oldham: *Thomas Young, Natural Philosopher, 1773–1829*, Cambridge University Press, Cambridge, 1954. One of the few biographies of a surprisingly neglected figure in the history of science.

CHAPTER 25

Optical Instruments

Where the telescope ends, the microscope begins. Who is to say of the two, which is the grander view?

Victor Hugo (1802–1885)

Cameras, microscopes, telescopes, and more specialized optical instruments are among the triumphs of modern technology. Without them we would neither know as much as we do about the world nor be able to enjoy its wonders as fully as we now can. All optical instruments consist of combinations of the lenses and mirrors discussed in Chap. 23, sometimes in combination with prisms and diffraction gratings. The behavior of such instruments can be understood using the theory developed in the preceding two chapters.

25.1 Cameras

The simplest optical device for viewing images on a screen or for recording images on a photographic plate is the *pinhole camera*.*

Pinhole Camera

The pinhole camera uses neither lenses nor mirrors but is still capable of producing sharp images with good *depth of field*; that is, the camera can form sharp images of objects at very different distances from the camera. It consists of a single small pinhole in the front of a light-tight box. If an illuminated object like the light bulb filament in Fig. 25.1 is placed opposite the pinhole, a ray from point A on the filament passes through the pinhole and strikes the screen at the rear of the box at point A'. Similarly, a ray from B strikes the screen at B'. If the pinhole is small enough, each point on the object is in the same way imaged at a single point on the screen. A real, inverted image of the light bulb filament is therefore produced on the screen. The magnification M produced by this simple camera is equal to q/p, where q is the image distance and p the object distance, as for mirrors and lenses.

Excellent photographs of stationary objects can be taken with a pinhole camera and photographic film. However, very little light gets through the tiny pinhole, and so long exposure times are needed to collect sufficient light to form an image bright enough to expose the film (how long depends on the film speed). During this exposure time the object, pinhole, and film must remain perfectly motionless.

FIGURE 25.1 A pinhole camera. Its magnification is q/p.

---

*The pinhole camera was originally called a *camera obscura*, the Latin words for "dark room." (When a judge is in his chambers, he is said to be *in camera*.)

If the hole is increased in size to reduce the exposure time, the image begins to blur, because rays from different parts of the object now overlap at the same point on the screen.

Cameras with Lenses

For most practical purposes the pinhole camera has been replaced by cameras employing converging lenses in place of pinholes. In the old-fashioned box camera the photographic film was placed in the focal plane of the lens, that is, the plane of the focal point of the lens. Distant objects would then be focused sharply on the film. Since the lens had a fixed focal length, objects closer to the lens would be out of focus. This difficulty was overcome by adding a bellows to the camera that allows the image distance to be adjusted until a sharp image of the desired object is obtained on the photographic plate or film. This process of moving the lens with respect to the photographic plate is called *focusing*. For any given object distance p, the image distance is adjusted to conform to the lens equation: $1/q = 1/f - 1/p$. A sketch of a simple bellows camera is shown in Fig. 25.2.

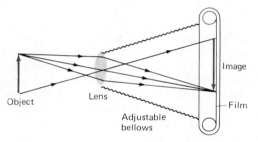

FIGURE 25.2 A bellows-type camera. The distance between the converging lens and the photographic film can be adjusted with the bellows to bring into focus objects at different distances from the lens.

f-Number

A very important characteristic of a camera is the brightness of the image it forms; the brighter the image, the shorter the exposure time required. Short exposure times usually result in sharper images, since the image is relatively unaffected by motion of the camera or object during the exposure. The brightness of the image depends on the amount of light passing into the camera per second. This, in turn, depends on the area of the lens through which the light enters, which is $A = \pi R^2 = \pi D^2/4$, where D is the diameter of the lens. We may therefore write

$$\text{Image brightness} \propto D^2 \qquad \qquad \textbf{(25.1)}$$

The brightness also depends on the focal length of the lens. For a simple lens we have for the magnification, from Eq. (23.8),*

$$M = \frac{h'}{h} = \frac{q}{p}$$

For a camera, q is adjustable only within narrow limits and is approximately equal to the focal length f of the lens, and so

$$\frac{h'}{h} \simeq \frac{f}{p} \qquad \text{or} \qquad h' \simeq \frac{hf}{p}$$

The height of the image is therefore proportional to f, and so the area of the image is proportional to f^2,

---

*In this chapter we will abandon the convention of inserting a minus sign in this equation to predict whether the image is upright or inverted. This will be clear from the ray diagrams drawn for the optical instruments discussed.

$$A \propto f^2 \tag{25.2}$$

The larger the area A of the image, the lower the brightness, since the same amount of light is spread over a larger area. We have therefore

$$\text{Image brightness} \propto \frac{1}{A} \propto \frac{1}{f^2} \tag{25.3}$$

Putting Eqs. (25.1) and (25.3) together, we obtain

$$\text{Image brightness} \propto \left(\frac{D}{f}\right)^2 \tag{25.4}$$

This leads to the following definition:

f-number: The ratio of focal length to diameter (f/D) for a lens.

A lens of diameter 5 cm and f-number equal to 2 (often written as f/2) has a focal length given by

$$f = D(\text{f-number}) = (5 \text{ cm})(2) = 10 \text{ cm}$$

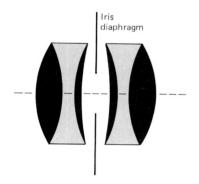

FIGURE 25.3 A typical corrected compound camera lens. The adjustable iris diaphragm removes the aberrations that would be introduced by the outer portion of the lens.

The brightness of an image varies inversely with the square of the f-number, according to Eq. (25.4), and so *the smaller the f-number, the brighter the image.* ''Fast'' lenses, which have large diameters and short focal lengths, often have f-numbers as low as 1.2. Because fast lenses are large and have short focal lengths, they also have larger lens aberrations than slower lenses and so need considerable correction before they can be used in good cameras. Since the smaller the aperture, the greater the depth of field (see the discussion of the pinhole camera), fast lenses with large apertures have greatly reduced depths of field.

A sketch of a corrected compound camera lens is shown in Fig. 25.3.

Modern cameras contain adjustments to change both the effective f-number of the lens and the shutter speed. The f-number is controlled by a ''stop,'' an iris diaphragm that determines the brightness of the image by varying the effective diameter D of the lens. Most cameras have f-stop markings of 1.0, 1.4, 2.0, 2.8, 4.0, 5.6, 8, 11, 16, 22, and 32. Each number in this series is approximately $\sqrt{2}$ times the preceding one. Since the f-number is equal to f/D, increasing the f-number from 1 to 1.4 means that the effective lens diameter D has been decreased by a factor of 1.4. Therefore the open lens area and the brightness are both decreased by a factor of 2, since $A \propto D^2$. Hence the image brightness is reduced by a factor of 2 for each successive f-stop marking.

Shutter Speed

The shutter controls the time during which light enters the camera and falls on the photographic film. This may vary from time exposures of seconds or minutes to flash exposures of 10^{-3} s or less, depending on the brightness of the object and the film speed. To photograph moving objects, very fast shutter speeds are required. Normally, to avoid the fuzziness in the image that results from inadvertent motion of the camera by the photographer, shutter speeds of 1/50 s or less are used.

The shutter speed and f-stop determine the amount of light entering the camera during an exposure. Too little light leads to *underexposure*, where none but the brightest parts of the image can be seen; too much light leads to *overexposure*, in which everything has a washed-out appearance. Modern cameras contain automatic devices to adjust the shutter and lens diaphragm for optimal exposures.

Example 25.1

(a) An f/2 camera lens is first used wide open, i.e., without a diaphragm to stop down the lens. How will the brightness of the image differ if the lens is then stopped down to f/8?

(b) How will the shutter speed have to change with the f/8 stop if the amount of light entering the camera is to remain the same?

SOLUTION

(a) The brightness depends on $(D/f)^2$ or on $(1/\text{f-number})^2$, and so the brightness of the f/8 lens is proportional to $1/64$. The brightness of the wide-open f/2 lens, on the other hand, is proportional to $1/4$. The relative brightness therefore is

$$\frac{1/4}{1/64} = \boxed{16}$$

The unstopped f/2 lens therefore produces an image 16 times brighter than the same lens stopped down to f/8.

(b) To compensate for the 16-fold reduction in brightness on changing to the f/8 stop, the shutter speed must be changed to increase the exposure time by a factor of 16. This allows the same amount of light to enter the camera.

25.2 The Human Eye

The human eye in many ways resembles a simple camera. The lens in the eye takes the place of the camera lens and forms a real, inverted image of an object on the *retina*, which is equivalent to the photographic film in a camera. Rods and cones in the retina act as transducers to convert light energy into electric signals, which travel along the optic nerve to the brain, where the final image is constructed.

Just as a bellows camera is focused by changing the lens-film distance, so too the eye is automatically focused by a process called *accommodation*. The lens in the eye is fixed in position about 1.7 cm from the retina. Since it is made of flexible tissue, however, its shape can be modified by the ciliary muscles of the eye. In this way a normal eye can be made to focus on objects from infinity to about 25 cm from the eye. Objects closer than 25 cm, called the *near point of the eye*, cannot be seen clearly with the unaided eye, since the lens of the normal eye cannot be made thick enough to focus at these short distances. Figure 25.4 shows how the shape of the lens changes as the object distance varies. The thicker the eye lens, the shorter its focal length and the greater its power.

Figure 25.5 is a diagram of the human eye. The *cornea* is the transparent membrane which acts as the outer window of the eye. The *iris* is the colored part—usually blue or brown—of the eye, and the *pupil* appears as the black circle at the

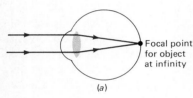

(a)

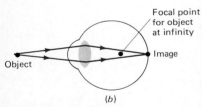

(b)

FIGURE 25.4 The change in the shape of the eye lens to focus on objects at different distances. (a) Eye focused on an object at infinity. (b) Eye focused on an object near the eye. The eye lens increases its thickness and therefore its power to focus on close objects. When its thickness is increased in this way, the eye focuses objects at infinity not on the retina but considerably in front of it, as shown. (For reasons of clarity, none of the drawings of the human eye in this chapter is to scale.)

FIGURE 25.5 Diagram of the human eye. Compare this with the bellows camera in Fig. 25.2.

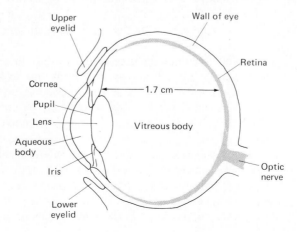

center. The pupil is black because it is a hole and no light is reflected from it. The iris acts as a diaphragm that adjusts the size of the pupil spontaneously to allow the proper amount of light into the eye for clear vision.

The *cornea* produces most of the refraction needed for the eye to converge the light incident on it. The *cornea* has a large index of refraction compared with air, and hence a large amount of refraction occurs at the air-cornea interface. On the other hand, the refractive indices of the cornea, the lens, and the aqueous and vitreous bodies adjacent to the lens are not too different, and relatively little refraction occurs once the light rays pass through the cornea. The optical power of the cornea is about 40 diopters (D), whereas all the other eye components, including the lens, have a total power of only 20 to 24 D (see Sec. 23.8). The overall power of the eye is therefore between 60 and 64 D. The amount of accommodation possible amounts to only about 4 D of this total focusing power.

If the eye lens is completely removed, as in a cataract operation, the patient can still see. Eyeglasses can make up for the converging power of the removed lens, but accommodation is no longer possible and objects at different distances from the eye appear fuzzy.

Example 25.2

(a) If a relaxed human eye, that is, the eye focused on an object far away, has a lens power of 60 D, what is its focal length? **(b)** How does the focal length of the eye change when, by accommodation, its power changes to 64 D? **(c)** To what object positions do these two focal lengths correspond, if we assume that the distance from eye lens to retina is 1.70 cm?

SOLUTION

(a) Since for any lens $f = 1/P$, we have

$$f = \frac{1}{60 \text{ D}} = \frac{1}{60} \text{ m} = 0.0167 \text{ m} = \boxed{1.67 \text{ cm}}$$

The relaxed human eye acts like a converging lens of 1.67 cm focal length.

(b) If the power changes to 64 D, we have

$$f = \frac{1}{64 \text{ D}} = \frac{1}{64} \text{ m} = 0.0156 \text{ m} = \boxed{1.56 \text{ cm}}$$

Hence the full range of accommodation of the average human eye changes its focal length by only about 0.1 cm, or 1 mm.

(c) For $f = 1.67$ cm, as in part (a).

$$\frac{1}{p} = \frac{1}{f} - \frac{1}{q} = \frac{1}{1.67 \text{ cm}} - \frac{1}{1.70 \text{ cm}} = 0.011 \text{ cm}^{-1}$$

and so

$$p = \frac{1}{0.011} \text{ cm} = \boxed{91 \text{ cm}}$$

For $f = 1.56$ cm, as in part (b),

$$\frac{1}{p} = \frac{1}{f} - \frac{1}{q} = \frac{1}{1.56 \text{ cm}} - \frac{1}{1.70 \text{ cm}} = 0.053 \text{ cm}^{-1}$$

and so

$$p = \frac{1 \text{ cm}}{0.053} = \boxed{19 \text{ cm}}$$

This range of focal lengths corresponds, therefore, to object distances from about 20 cm to 1 m.

Defects of the Eye

Nearsightedness Nearsightedness (myopia) allows a person to see *near* objects clearly but makes it impossible for the eye to focus on objects at a distance, even when the eye is totally relaxed. Correcting for myopia requires eyeglasses containing a diverging lens. This lens in combination with the eye lens enables distant objects to be focused clearly on the retina. This is illustrated in Fig. 25.6.

Farsightedness Farsightedness (hyperopia) allows a person to see *distant* objects clearly but makes it impossible to focus nearby objects on the retina. A perfectly normal eye in an adult cannot focus on objects closer than about 25 cm from the eye. Correcting for farsightedness requires glasses with converging lenses, as shown in Fig. 25.7. Farsightedness often occurs in older people as their ciliary

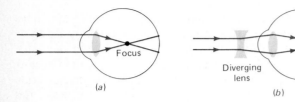

FIGURE 25.6 (*a*) Nearsightedness: The eye cannot focus on distant objects. (*b*) Correction for nearsightedness using diverging eyeglass lenses (not to scale).

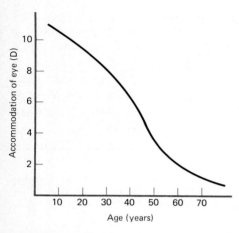

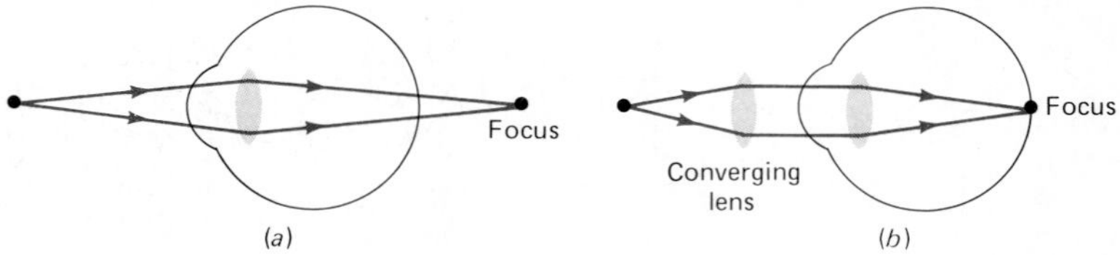

FIGURE 25.7 (*a*) Farsightedness: The eye cannot focus on near objects. (*b*) Correction for farsightedness using converging eyeglass lenses (not to scale).

FIGURE 25.8 Accommodation of the human eye as a function of age.

muscles lose the power to accommodate to near objects. The eye's power to accommodate changes from as much as 11 D at age 10 to 1 D at age 70, as shown in Fig. 25.8. Note particularly the rapid falloff during the years from 30 to 50.

Most prescriptions for eyeglasses range from 1 to 4 D, corresponding to focal lengths from 1 m to 25 cm. The wearer of *bifocal glasses* looks through the top diverging lenses at objects far away and through the bottom converging lenses at objects up close. *Trifocals* add a third lens, for middle-distance vision, between these two.

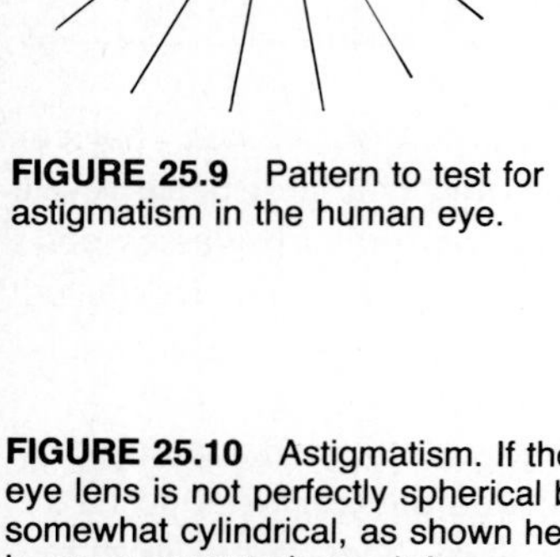

Astigmatism Astigmatism results from a lack of perfect spherical symmetry in the cornea and lens of the eye. In this case a person looking at the pattern of Fig. 25.9 sees some of the spokes of the wheel as darker and clearer than the others. This happens because the lenses of the person's eyes are not perfectly spherical but somewhat cylindrical. As a result the eye focuses rays from a point in a plane *perpendicular* to the cylinder axis to a point; but it focuses rays from a point in a plane *parallel* to the cylinder axis not to a point but to a line, as shown in Fig. 25.10. For this reason, lines drawn at different angles will look different to a person with astigmatism, and this enables an optometrist to locate the cylindrical axis in the astigmatic eye.

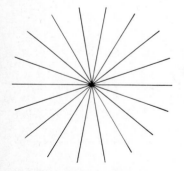

FIGURE 25.9 Pattern to test for astigmatism in the human eye.

FIGURE 25.10 Astigmatism. If the eye lens is not perfectly spherical but somewhat cylindrical, as shown here in an exaggerated way, it focuses point objects not into a point but into a line.

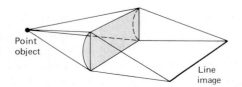

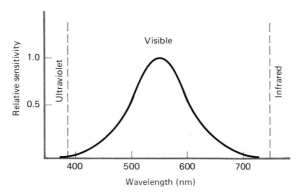

FIGURE 25.11   Sensitivity of human eye to wavelengths in the visible region of the spectrum. The eye is especially sensitive in the yellow-green region of the spectrum at about 550 nm.

Correcting for astigmatism requires eyeglasses with cylindrical lenses that cancel out the cylindrical components in the eye lens.

Chromatic aberration also exists in the human eye, but it is greatly reduced by the yellowish pigment called *macula lutea* in the fovea or color-sensitive part of the retina. This makes the eye much more sensitive in the yellow-green region of the spectrum, as shown in Fig. 25.11, and filters out much of the red and blue light that would make chromatic aberration more noticeable.

Example 25.3

An elderly man can read a newspaper only when it is at least 60 cm from his eyes. What kind of reading lenses does he need in his glasses? Assume that the glasses will be worn very close to his eyes.

SOLUTION

The near point of the human eye for a normal adult is 25 cm from the eye. Since the man is farsighted, he needs a converging lens in his glasses that will produce a virtual image of the newspaper 60 cm from his eyes, when the paper is held at the ordinary reading distance of 25 cm. Hence we have

$$\frac{1}{f} = \frac{1}{p} + \frac{1}{q} = \frac{1}{25 \text{ cm}} + \frac{1}{-60 \text{ cm}} = +\frac{1}{43 \text{ cm}}$$

Note that the sign on the image distance is negative, since it is on the same side of the lens as the object; i.e., it is a virtual image. The power of the required lens is thus

$$P = \frac{1}{f} = \frac{1}{+0.43 \text{ m}} = \boxed{+2.3 \text{ D}}$$

A lens of $+2.3$-D power is needed, i.e., a converging lens with power 2.3 D.

EXERCISE 1   A girl can read a book only when it is no more than 15 cm from her eyes. What power lenses does she need in her glasses?

25.3   The Magnifying Glass (Simple Magnifier)

The closer an object is to the eye, the larger the image on the retina and the more detail to be seen in it, as shown in Fig. 25.12. The actual size of an object is of little consequence in determining how big it appears. What is important is its *angular size*. An object h cm high at a distance d cm from the eye subtends an angle given by $\tan \phi = h/d$. Since h/d is usually small, we assume that $\phi \simeq \tan \phi = h/d$ and take ϕ (in radians) as a measure of the angular size of the object.

For example, a quarter (coin) of diameter 2.5 cm held at arm's length (60 cm) subtends an angle $\phi \simeq h/d = 2.5/60 = 0.042$ rad, whereas the sun, which has a diameter of 1.39×10^9 m, subtends an angle at the earth of only $\phi = (1.39 \times 10^9 \text{ m})/(1.49 \times 10^{11} \text{ m}) = 0.009$ rad—an angle five times smaller. The quarter therefore appears *bigger* than the sun under these conditions.

FIGURE 25.12 The size of the image formed on the retina of the eye. The object is of the same size in both (a) and (b), but the eye forms a much larger image on the retina in (b) because the object is closer to the eye.

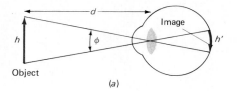

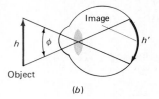

FIGURE 25.12 The size of the image formed on the retina of the eye. The object is of the same size in both (a) and (b), but the eye forms a much larger image on the retina in (b) because the object is closer to the eye.

The usefulness of a magnifying glass lies in its ability to increase the angular size of a viewed object. *Angular magnification* is defined as follows:

M_ϕ = angular magnification (or magnifying power)

$$= \frac{\text{angular size of image (with magnifier present)}}{\text{angular size of object (without magnifier)}} \qquad \textbf{(25.5)}$$

The simplest way to increase the angular magnification of an object is to bring it closer to the eye, as shown in Fig. 25.12. However, since most human eyes cannot focus at distances less than 25 cm, an additional lens is needed to view a closer object. A *magnifying glass* is a single converging lens that produces a large virtual image at infinity of an object placed at the focal point of the lens. The relaxed eye can then focus on this image at infinity and produce a real image on the retina that is larger than is possible without the magnifying lens.

Magnifying Power of a Simple Magnifier The angular magnification is given by Eq. (25.5) as

$$M_\phi = \frac{\phi'}{\phi} \qquad \textbf{(25.6)}$$

where ϕ and ϕ' are as shown in Fig. 25.13.

Now, from Fig. 25.13a,

$$\phi = \frac{h}{25 \text{ cm}}$$

FIGURE 25.13 A magnifying glass (simple magnifier). The magnifying glass produces an enlarged, virtual image of an object. The angular magnification is $M_\phi = \phi'/\phi$, where ϕ is the angle subtended by the object at the eye lens in (a) and ϕ' is the angle subtended by the object at the eye lens in (b).

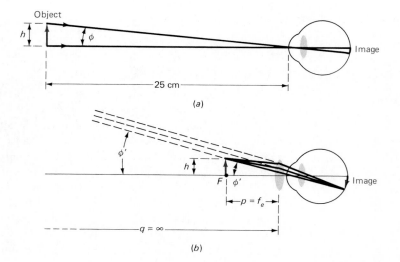

and, from Fig. 25.13b,

$$\phi' = \frac{h}{f_e}$$

where f_e is the focal length of the magnifying lens. (We assume that the eye is relaxed and sees the image at infinity, so that the object is precisely at the focal point of the lens.) Then

$$M_\phi = \frac{\phi'}{\phi} = \frac{h/f_e}{h/25} = \frac{25}{f_e} \qquad (25.7)$$

where both numerator and denominator are in centimeters. If the focal length of the magnifier is 4.0 cm and the eye is relaxed and focused on the virtual image at infinity, the magnification is therefore about six times.

A lens of high magnifying power must be a strongly positive lens, i.e., a lens with a short focal length, since M_ϕ depends inversely on f_e. Aberrations in simple converging lenses limit the achievable angular magnification M_ϕ to considerably less than 10. If these aberrations are corrected by using compound lenses, magnifications up to about 20 are possible, but no higher.

Example 25.4

A pencil 6.0 mm thick is viewed by a woman with normal vision using a magnifying glass of focal length 5.0 cm. Calculate the apparent thickness of the pencil when its virtual image is at infinity.

SOLUTION

When the virtual image is at infinity,

$$\frac{1}{p} = \frac{1}{5.0 \text{ cm}} - \frac{1}{\infty} \quad \text{and} \quad p = 5.0 \text{ cm}$$

In this case the object is exactly at the focal point of the lens.

The virtual image formed by the lens is now at infinity and is infinitely large. But its angular size is that of a 6.0-mm object at the focal distance of 5.0 cm (see Fig. 25.13b). We have, then, using Eq. (25.7),

$$M_\phi = \frac{25 \text{ cm}}{5.0 \text{ cm}} = 5$$

The pencil therefore appears to be $\boxed{3.0 \text{ cm}}$ thick.

25.4 The Compound Microscope

To increase the magnification over that possible with a simple magnifier, an additional lens may be added to form a *compound microscope*. In this case the magnifying lens is referred to as the *eyepiece lens* (or ocular) and the added lens as the *objective lens*. The distance between the lenses is L, as shown in Fig. 25.14. When an object of length h is placed just outside the focal point F_0 of the objective lens, this lens forms an enlarged real image of length h' at the focal point F_e of the eyepiece. The eyepiece then acts as a simple magnifier and produces a greatly enlarged virtual image at infinity of the real image formed by the objective lens. This image is inverted, since the image formed by the objective lens is inverted. This virtual image becomes the object for the eye itself, which forms the final real image on the retina, as shown.

The magnification of a compound microscope is simply the product of the magnifications produced by each of the two lenses. For the lateral magnification of the objective lens we have, from the lens formula,

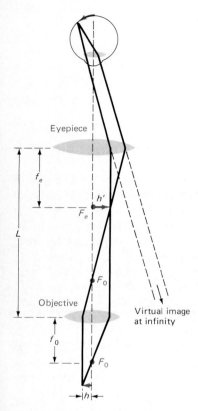

Eyepiece

f_e

F_e

h'

L

F_0

Objective

Virtual image at infinity

f_0

F_0

h

FIGURE 25.14 The compound microscope. It combines an objective lens with a simple magnifier, which in this case is called the eyepiece, or ocular.

$$M_0 = \frac{h'}{h} = \frac{q_0}{p_0} \tag{25.8}$$

But $p_0 \simeq f_0$ and $q_0 = L - f_e \simeq L$, since in working microscopes f_e is very small compared to L, the distance between the lenses. Hence

$$M_0 \simeq \frac{L}{f_0} \tag{25.9}$$

The eyepiece then acts as a simple magnifier, with magnification $M_\phi = 25/f_e$, from Eq. (25.7).

The overall magnification of the compound microscope is therefore

$$M = M_\phi M_0 \simeq \frac{25L}{f_e f_0} \tag{25.10}$$

where L is the distance between the two lenses. This equation and Eq. (25.11) below are only approximations, and their validity depends on how well the assumptions made in the above derivation are verified in a particular case.

In most practical microscopes the two lenses are separated by about 18 cm for the convenience of the user. With $L = 18$ cm, the overall magnification is approximately

$$M \simeq \frac{25(18)}{f_e f_0} = \frac{450}{f_e f_0} \tag{25.11}$$

where f_e and f_0 are the focal lengths, in centimeters, of the two lenses. The magnification is, of course, a dimensionless number.

Again we notice that the shorter the two focal lengths f_e and f_0, the larger the magnification. Of course, larger magnifications mean thicker lenses and hence greater lens aberrations. For this reason most microscope lenses are complex, multilens structures designed to reduce aberrations as much as possible.

Equation (25.11) shows that, if the two lenses in a compound microscope have focal lengths of 1 cm each, the overall magnification is only about 450. With most light microscopes no additional detail is seen for magnifications above about 400 because of limitations on the resolving power of lenses, to be discussed in Sec. 25.6.

Example 25.5

A compound microscope consists of a $10\times$ eyepiece (i.e., an eyepiece with a 10-fold magnification) and a $40\times$ objective lens at the ends of a microscope barrel 18 cm long. Find **(a)** the total magnification of the microscope; **(b)** the position at which the object must be placed so that the final image is at infinity; **(c)** the focal length of each lens.

SOLUTION

(a) The total magnification is

$$M = M_\phi M_0 = 10 \times 40 = \boxed{400}$$

(b) For the final object to be at infinity the image produced by the objective lens must be at the focal point of the eyepiece. (The diagram of Fig. 25.14 should be used here.) The image distance for the objective lens is

$$q_0 = L - f_e$$

where $L = 18$ cm, and

$$f_e = \frac{25 \text{ cm}}{M_\phi} = \frac{25 \text{ cm}}{10} = 2.5 \text{ cm}$$

Hence, $q_0 = 18 - 2.5 = 15.5$ cm. Now, from Eq. (25.8), $p_0 = q_0/M_0$, and so the object distance for the objective lens is

$$p_0 = \frac{15.5 \text{ cm}}{40} = \boxed{0.39 \text{ cm}}$$

(c)   We have found in part (b) that

$$f_e = \boxed{2.5 \text{ cm}}$$

We can find f_0 from the lens equation

$$\frac{1}{f_0} = \frac{1}{p_0} + \frac{1}{q_0} = \frac{1}{0.39 \text{ cm}} + \frac{1}{15.5 \text{ cm}}$$

or $$f_0 = \frac{0.39(15.5) \text{ cm}}{15.9} = \boxed{0.38 \text{ cm}}$$

Hence the object is just 0.01 cm outside the focal point of the objective lens, which is about what we would expect.

25.5   Telescopes

Two features are particularly important in a telescope: (1) *resolving power*, or the ability to form distinct images of objects very close together, and (2) *light-gathering power*, or the ability to collect large amounts of light to form the final image. Since both resolving power and light-gathering power improve with the size of the objective lens (or mirror), telescopes with very large lenses (and mirrors) are found at the great astronomical observatories of the world.

Refracting Telescopes

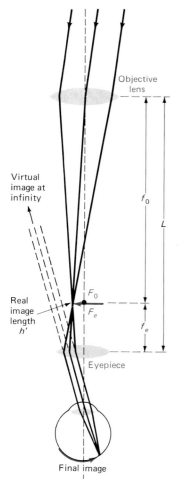

Objective lens

Virtual image at infinity

f_0

L

Real image length h'

F_0
F_e

f_e

Eyepiece

Final image

In 1611 Kepler took a refracting telescope sent to him by Galileo and modified it into one that employed two convex lenses in much the same positions as in a compound microscope. This is shown in Fig. 25.15. An objective lens forms a real inverted image of a distant object. Since the object is so far away, the light rays entering the telescope are parallel and the image is produced at the focal plane of the objective lens. This image is much smaller than the object and serves, in turn, as the new object for the eyepiece lens. The eyepiece produces an enlarged virtual image of this object. If this virtual image is to be seen by the relaxed eye at infinity, the object being viewed through the eyepiece must be at the focal point of the eyepiece lens. Hence the length of the telescope is fixed by the fact that the focal points of the objective and the ocular lenses must coincide. The barrel length L is then the sum of the two focal lengths, as shown in Fig. 25.15.

To obtain the magnifying power of a refracting telescope, or *refractor*, of this kind, let us consider a ray that comes from one edge E of a distant object, passes through the center of the objective lens, and forms a real image E' in the focal plane of an objective lens of focal length f_0, as in Fig. 25.16. The object is of length h at a very large object distance p, and so its angular size as seen with the unaided eye is $\phi \simeq \tan \phi = h/p$. The image size h' is much smaller than h, but the angular size of the image as seen with the eyepiece is $\phi' \simeq \tan \phi' = h'/f_e$, which is larger than ϕ if f_e is small. The magnifying power is, from Eq. (25.6),

$$M_\phi = \frac{\phi'}{\phi} = \frac{h'/f_e}{h/p}$$

But from similar triangles in Fig. 25.16, we have

$$\frac{h'}{f_0} = \frac{h}{p}$$

FIGURE 25.15   An astronomical telescope using two converging lenses.

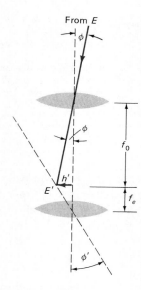

and so

$$M_\phi = \frac{h'/f_e}{h'/f_0} = \frac{f_0}{f_e} = \frac{\text{focal length of objective}}{\text{focal length of eyepiece}} \qquad (25.12)$$

The magnifying power of a refracting telescope is the ratio of the focal length of the objective lens to the focal length of the eyepiece.

For large magnifications a refractor should therefore have an objective lens of long focal length and an eyepiece of very short focal length. A refracting telescope produces an inverted image. This is not troublesome in astronomical work but is particularly bothersome for terrestrial observations. For this reason the spyglasses popular with the pirates of old contained a third lens to make the final image right side up. This made the spyglasses uncomfortably long.

FIGURE 25.16 Light from a distant object entering a refracting telescope. The angular magnification is $M_\phi = \phi'/\phi = f_0/f_e$.

Reflecting Telescopes

Most large astronomical telescopes in use today are *reflectors* rather than refractors. Astronomical telescopes need great light-gathering power, and the objective lenses must be large. Since a lens can only be supported at its edges and must be mounted at the top of the telescope tube, large lenses begin to sag under their own weight and thus distort the images they produce. For this reason large concave mirrors, which can be supported over their entire back surface, are now used almost exclusively in large research telescopes. A mirror also has only one surface instead of two that must be ground, and is free from the chromatic aberration that exists in all lenses.* By using parabolic mirrors spherical aberration may also be eliminated.

Since the focal plane of a reflecting telescope is inside the main tube of the telescope, many schemes have been developed to divert the image to a place outside the tube, where it can be observed and recorded. Two possible configurations—the newtonian focus and the Cassegrainian focus—are shown in Fig. 25.17.

The two largest telescopes in the world are the 5-m (200-in) Hale reflector at Mount Palomar in California (see Fig. 25.18) and the 6-m (236-in) Bolshoi alt-

FIGURE 25.17 Two types of reflecting astronomical telescopes: (a) the newtonian focus; (b) the Cassegrainian focus.

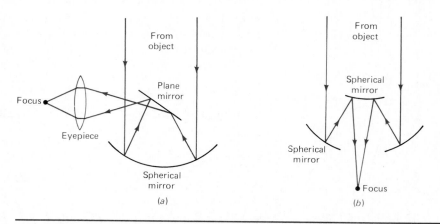

*It was to solve the problem of chromatic aberration that Newton first developed the reflecting telescope.

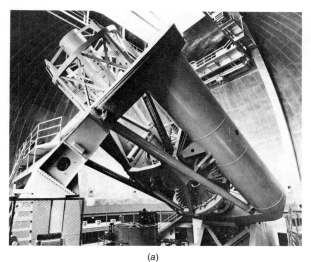

(a)

(b)

FIGURE 25.18 (a) The 200-in Hale reflecting telescope at Mount Palomar, California. *(Courtesy of Palomar Observatory.)* (b) The back of the 200-in mirror. Notice the honeycomb structure used to reduce the weight of the mirror. The hole in the mirror (required for the Cassegrainian focus) is approximately equal to the aperture of the 40-in Yerkes refractor. *(Photo by A. A. Stevens, Palomar Observatory, Courtesy of AIP Niels Bohr Library.)*

azimuth reflector (BAT) in the Soviet Union. The largest refractor in the world, the 40-in refractor at the Yerkes Observatory in Williams Bay, Wisconsin, is considerably smaller than these reflectors, for the reasons indicated above.

New Ideas in Telescope Design

After 40 years of relative quiescence in the construction of very large telescopes, astronomers and engineers are now feverishly designing reflecting telescopes with effective diameters larger than 6 m. These will have both greater resolution (see Sec. 25.6) and greater light-gathering power than any telescope now in operation.

The first step toward bigger and better telescopes was taken with the Multiple-Mirror Telescope on Mount Hopkins in Arizona, which uses six 72-in mirrors to achieve the effective aperture of a single mirror 176 in across. A follow-up to this multiple-mirror approach is the W. M. Keck Observatory 10-m (394-in) reflector now under construction atop Hawaii's Mauna Kea. This gigantic mirror is made up of 36 hexagonal segments, each 1.8 m across at its widest diameter, that focus light to the same point with a conventional Cassegrainian focus (Fig. 15.17b).

Among other telescopes now being designed are three using 8-m (315-in) mirrors to be built using a "spincasting" technique, in which molten metal is rotated at a speed that forces it into the particular paraboloid shape needed for the final mirror surface.

In addition to these huge, land-based telescopes, the Hubble Space Telescope (Fig. 25.19) was launched on April 24, 1990. This $1.5 billion telescope will orbit above the earth's obscuring atmosphere. It is expected that it will be able to detect celestial objects 50 times fainter with 10 times better resolution than any existing telescope.

The Space Telescope will use a 94.5-in concave mirror covered with a magnesium fluoride coating 1/1,000 the thickness of a human hair. It is expected to uncover astronomical secrets comparable to those revealed to Galileo in 1609 when he first directed his telescope at the skies.

It is likely that the Space Telescope and the new generation of land-based telescopes will provide an abundance of new information about stars, galaxies, pulsars, and black holes, and help answer astronomers' three big questions about the universe: its age, its density, and its rate of expansion.

FIGURE 25.19 The NASA Hubble Space Telescope. *(Courtesy of NASA.)*

Example 25.6

The large refracting telescope at Yerkes Observatory in Wisconsin has a 40-in-diameter objective lens of 63.5-ft focal length and an eyepiece of focal length 10.0 cm. What is the approximate magnifying power of this telescope?

SOLUTION

$$M_\phi = \frac{f_0}{f_e} = \frac{(63.5 \text{ ft})(12.0 \text{ in/ft})(2.54 \text{ cm/in})}{10.0 \text{ cm}} = 194\times$$

25.6   The Resolution of Telescopes and Microscopes

The fundamental element in a telescope is the large-aperture objective lens or mirror used to collect light from distant objects. Some diffraction effects occur even with the largest objective lenses or mirrors, although these effects may be quite small for large apertures.

In Sec. 24.6 we saw that, on passing through a rectangular aperture of width w, light of wavelength λ is diffracted so that the first diffraction minimum occurs at an angle θ given by $\sin \theta = \lambda/w$. In the case of a telescope lens we are dealing not with a rectangular aperture of fixed width w and height h but with a circular opening of diameter d. This introduces a numerical factor of 1.22 (obtained from analysis that we will not describe) so that the first minimum for a circular aperture occurs at $\sin \theta = 1.22\lambda/d$. The larger the diameter d of the lens, the smaller θ and the less important diffraction effects. This is one reason astronomical telescopes use such large-diameter objective lenses or mirrors.

Because of diffraction, the image of a star, as formed by a telescope lens, consists of a bright central disk (known as an *Airy disk*) surrounded by a number of fainter diffraction rings, as shown in Fig. 25.20. This makes it difficult to distinguish two stars whose angular separation is small. If the stars are too close together, their diffraction patterns overlap to such an extent that it is impossible to distinguish one star from the other. We say that the two stars cannot be *resolved*. Figure 25.21 shows what happens for two stars of the same intensity when the distance between the stars is gradually reduced. In Fig. 25.21a the two stars can be clearly distinguished, even though their diffraction rings overlap; in (b) it is barely possible to distinguish one star from the other; in (c) where the stars are even closer together, the two images fuse and the stars can no longer be resolved.

FIGURE 25.20 Image of a star, as formed by a telescope lens, showing the diffraction rings around the image.

Resolution of a Telescope Objective

Figure 25.22 shows the light intensity on a photographic plate of the central image of the two stars. In Fig. 25.22b the central maximum of one star's diffraction pattern falls directly on top of the first diffraction minimum of the other. Since the two stars are just barely resolvable under these conditions, Lord Rayleigh (1842–1919) suggested that this be taken as the criterion for resolution, a criterion now called the *Rayleigh criterion*. Since the angular separation of the two stars in this case is the same as the angular separation of the first minimum from the central image in the diffraction pattern of either star, we have for the Rayleigh criterion

$$\sin \theta_R = \frac{1.22\lambda}{d} \tag{25.13}$$

where θ_R is the minimum angular separation for resolution. Since θ_R is always small, θ_R (in radians) $\simeq \sin \theta_R$, and so

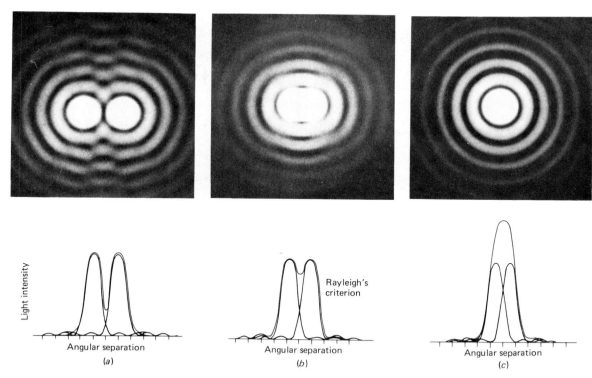

FIGURE 25.21 (top) Variation in overlap of the images of two adjacent stars as the distance between the stars is reduced.

FIGURE 25.22 (bottom) Rayleigh criterion for the resolution of a telescope or microscope. The angular separation in (b) is taken as the minimum angle that can be resolved. The images of the two stars in (c) overlap to form a single image.

Rayleigh criterion:
$$\theta_R \simeq \frac{1.22\lambda}{d}$$
(25.14)

This equation sets the theoretical limit on the ability of a telescope to resolve images of objects very close together. Stars with angular separations greater than θ_R in radians can be resolved, as in Fig. 25.22a; those with angular separations less than θ_R cannot be resolved, as in Fig. 25.22c. Note that the Rayleigh criterion applies to perfect lenses (and mirrors) whose performance is limited only by diffraction and therefore by the wave nature of light. Lens aberrations in imperfect lenses drastically reduce the achievable resolution. For example, the resolution of a high-quality camera lens is usually limited by diffraction for small lens openings and by lens aberrations for large lens openings.

Example 25.7

Experiments on the human eye indicate that its limit of resolution (sometimes called *visual acuity*) is about 2.0×10^{-4} rad (less than 1 min of arc). **(a)** If the finite size of the eye opening limits the resolution, what would the aperture of the eye's pupil have to be to limit the resolution to 2.0×10^{-4} rad? **(b)** For larger eye apertures the ultimate limitation on the resolution of the eye is a physiological one, the spacing of the sensitive photoreceptor cells on the retina. If the distance between the eye lens and the retina is 1.7 cm, what is the distance between the positions at which two just-resolvable light rays strike the retina?

SOLUTION

(a) Diffraction effects in the eye become important when the iris makes the pupil diameter small enough that the Rayleigh criterion leads to angles smaller than 2.0×10^{-4} rad. When this happens, we must have

$$\theta_R = \frac{1.22\lambda}{d} = 2.0 \times 10^{-4} \text{ rad}$$

or, taking 500 nm as the most sensitive response region of the eye,

$$d = \frac{1.22(500 \times 10^{-9} \text{ m})}{2.0 \times 10^{-4}} = 3.0 \times 10^{-3} \text{ m} = \boxed{3.0 \text{ mm}}$$

It is well known that the eye's pupil is varied by the iris from about 2 to 6 mm to control the amount of light entering the eye. Our answer of 3.0 mm is therefore quite reasonable.

(b) For two light rays subtending an angle of 2.0×10^{-4} rad at the eye lens, we have

$$\tan \theta = \frac{d}{1.7 \times 10^{-2} \text{ m}}$$

where d is the separation of the two light rays on the retina. If we assume that $\tan \theta \simeq \theta$, we have

$$d = (1.7 \times 10^{-2} \text{ m})(\theta) = (1.7 \times 10^{-2} \text{ m})(2.0 \times 10^{-4} \text{ rad})$$

$$= 3.4 \times 10^{-6} \text{ m} = \boxed{3.4 \text{ }\mu\text{m}}$$

It is known from the physiology of the eye that the average retinal photoreceptor cell has a diameter of about 1 μm.

We have found, therefore, that the eye can resolve light falling on two cells separated by one additional photoreceptor. This seems reasonable, since it takes more than one cell to determine the presence or absence of a boundary in an image.

For larger eye apertures, spherical and chromatic aberration, and ultimately the size of the photoreceptors, limit the eye's resolving power. At smaller apertures (less than about 3 mm) diffraction is the limiting factor, as we have just seen. These two effects tend to compensate and leave the resolution of the eye roughly constant at something less than 1 min of arc. This corresponds to the diameter of a human hair held at arm's length!

Resolution of Microscopes

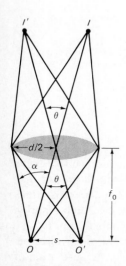

FIGURE 25.23 Resolution of a microscope. A point O on the object produces an image at I; a point O' on the object produces an image at I'. Since the radius of the lens subtends an angle α at the object, the full lens diameter subtends an angle 2α.

Unlike telescopes, whose objects are usually far away, microscopes look at objects very close to the objective lens. We assume that the objective lens subtends an angle 2α at the object, which is located close to the focal point of the objective, as in Fig. 25.23. We wish to find the smallest distance s between two points O and O' on the object that will produce images I and I' that can just be resolved. The resolution is, of course, limited by the diffraction occurring at the lens aperture, which will lead to overlapping diffraction patterns at I and I'. The criterion for resolution is again the Rayleigh criterion $\theta_R \simeq 1.22\lambda/d$.

In this case two rays from O and O' passing through the center of the lens arrive at the image points I and I'. The angle θ subtended by these rays at the lens must be greater than or equal to θ_R for O and O' to be resolved. But the angle θ is approximately equal to s/f_0. For resolution we must have

$$\theta = \theta_R \simeq \frac{1.22\lambda}{d} \cong \frac{s}{f_0} \quad \text{or} \quad s \simeq \frac{1.22\lambda f_0}{d} \tag{25.15}$$

This is the smallest separation of two objects that produces clearly distinct images. Notice that f_0/d is just the f-number of the objective lens, and so the smaller the f-number, the finer the detail that can be resolved.

If the object is approximately at the focal point of the lens, and if the angle α is small, we have

$$\sin \alpha \simeq \tan \alpha = \frac{d/2}{f_0} \tag{25.16}$$

or

$$d \simeq 2f_0 \sin \alpha \tag{25.17}$$

By combining Eqs. (25.15) and (25.17), we obtain

$$s = \frac{1.22\lambda f_0}{2f_0 \sin \alpha} = \frac{1.22\lambda}{2 \sin \alpha} \tag{25.18}$$

This is the minimum separation that can be resolved by the microscope. It shows that the larger α (and hence the larger the aperture of the lens, or the shorter its focal length), the better the resolution.

In microscopes of high magnification the space between the object and the objective lens is often filled with oil. This improves the resolution, since it changes the wavelength of the light in the oil from λ to λ/n, and so Eq. (25.18) becomes

$$s \simeq \frac{1.22\lambda}{2n \sin \alpha} \tag{25.19}$$

This is the minimum distance that can be resolved by an *oil-immersion microscope*. The product $n \sin \alpha$ is called the *numerical aperture* (NA) of the objective lens. The NA is usually specified on the side of the microscope housing, along with the magnification. The greater the NA, the better the resolution.

Objective lenses in air have numerical apertures of between 0.12 and 0.85, while oil-immersion lenses have numerical apertures as high as 1.3. The limit of resolution of an oil-immersion microscope using light of wavelength 500 nm is then approximately

$$s = \frac{1.22(500 \times 10^{-9} \text{ m})}{2(1.3)} = 2.3 \times 10^{-7} \text{ m}$$

We see, therefore, that the resolution of a good optical microscope (230 nm) is of the same order of magnitude as the wavelength of the light used (500 nm). This leads to the important practical conclusion that *with an optical microscope it is not possible to resolve structure which is much smaller than the wavelength of the light used.*

Example 25.8

What is the minimum separation that a microscope objective lens stopped down to $f/16$ is able to resolve in green light ($\lambda = 500$ nm)?

SOLUTION

The minimum separation that the lens can resolve depends on the diffraction produced by its finite aperture. This distance is given by Eq. (25.15):

$$s = 1.22 \frac{f_0\lambda}{d}$$

Here f_0/d is just the f-number of the lens, which is given as 16 in this case, and so

$$s \simeq 1.22(16)(500 \text{ nm}) = 10^4 \text{ nm} = 10^{-5} \text{ m} = \boxed{10^{-2} \text{ mm}}$$

Example 25.9

(a) What must be the numerical aperture of an oil-immersion microscope designed to resolve two bacteria separated by 2.0×10^{-4} mm, if the bacteria are illuminated by violet light at 400 nm? (b) What is the angle subtended by the objective lens of the microscope at the object? (Assume that $n = 1.5$ for oil.)

SOLUTION

(a) Since the numerical aperture NA $= n \sin \alpha$ and, from Eq. (25.19), the smallest distance that can be resolved is $s = 1.22\lambda/2$NA, we have,

$$\text{NA} = \frac{1.22\lambda}{2s} = \frac{1.22(400 \times 10^{-9} \text{ m})}{2(2.0 \times 10^{-7} \text{ m})} = \boxed{1.2}$$

(b) We have, from the definition of the NA,

$$\sin \alpha = \frac{NA}{n} = \frac{1.2}{1.5} = 0.80 \qquad \text{and} \qquad \alpha = 53°$$

The full angle subtended at the object is then

$$2\alpha = 2(53°) = \boxed{106°}$$

As this example indicates, α is often not a small angle, and so the assumption made in the derivation of Eq. (25.19) that $\sin \alpha \simeq \tan \alpha$ is only a rough approximation. A more rigorous derivation, however, leads to the same result.

25.7 Specialized Kinds of Microscopes

In biology and medicine a great variety of specialized microscopes are used for two basic purposes: (1) to decrease diffraction effects and thus improve the *resolution* of the instrument; (2) to increase the *contrast* between the object and the background.

Improving Resolution

We have already seen that one important way to improve resolution is to shorten the wavelength of the radiation used. An *oil-immersion microscope* accomplishes this to a limited degree by reducing the effective wavelength from λ to λ/n, where n is the index of refraction of the medium. If $n = 1.5$, this leads to a 50 percent increase in resolving power.

Another useful technique is to use only the short-wavelength components of visible light in a microscope to reduce diffraction and increase resolution; thus many microscope illuminators use blue or violet light.

An *ultraviolet (UV) microscope* reduces the wavelength used to about 100 nm and improves the resolution of the microscope by a factor of about 5. The problem is that glass does not transmit UV nor can the human eye detect it (although it can be seriously burned by it), and so quartz or fluorite lenses are needed, together with special photographic or photoelectric recording. This means added expense and reduced convenience. For this reason UV microscopes are only used for special purposes. When employed, however, they have the added advantage of detecting objects transparent to visible light, since many molecules, such as proteins and nucleic acids, absorb strongly in the UV region and show strong contrast under UV light. In cellular biology, UV microscopes have been used to observe the behavior and determine the function of DNA and RNA in living cells.

X-rays are so energetic that it is difficult to find ways to focus or control them, and so *x-ray microscopes* are not in common use, although they have been used successfully in some specialized research projects.

Electron microscopes, which have greater resolving power than any of the above types, will be discussed in Sec. 27.8, after we know more about the wave properties of electrons.

Increasing Contrast

Two important ways to increase contrast in microscopes are exemplified in the phase-contrast microscope and the polarizing microscope.

Phase-Contrast Microscope   The phase-contrast microscope, developed in 1935 by the Dutch physicist F. Zernike (1888–1966), converts *phase changes* produced by microscope specimens into *amplitude changes* in the final image. The human eye can easily detect amplitude changes since they vary the intensity of the light, but it does not respond directly to phase changes in the light. In a phase-contrast microscope, transparent objects which differ only slightly from their surroundings in refractive index can be made visible by using light interference to vary the

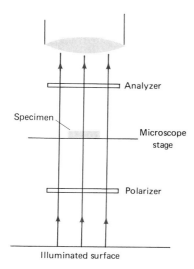

FIGURE 25.24 A polarizing microscope. The analyzer can be rotated to increase contrast between image and background.

brightness of the background compared with that of the object being studied.

Phase-contrast microscopes are very useful in the study of bacteria, which tend to be transparent and invisible when viewed with an ordinary microscope.

Polarizing Microscope   A polarizing microscope contains a polarizer below the microscope stage and an analyzer above it, as shown in Fig. 25.24. When the polarizer and analyzer are crossed in the absence of a specimen, no light enters the eyepiece and the field is dark. If a specimen is introduced that is optically active (or *birefringent**), it will rotate the plane of polarization and will appear bright against a dark background. Fortunately many important biological substances like nucleic acids and proteins are birefringent and thus make apt specimens for study with a polarizing microscope.

Both the phase-contrast microscope and the polarizing microscope have great advantages over the older light microscope, which requires that specimens be stained to make them visible. Staining usually kills the specimen, whereas the use of phase-contrast and polarizing microscopes does not. It is therefore possible to use living cells as specimens and to observe important biological processes like *mitosis* (cell division) as they actually occur in living cells.

25.8   Other Important Optical Instruments

Other optical instruments employing lenses and mirrors range from everyday devices like slide projectors to high-precision research instruments like interferometers. We will discuss only a few of these.

The Ophthalmoscope

The *ophthalmoscope* is one of the many contributions to science and medicine made by Hermann von Helmholtz (see Fig. 12.18). It is the basic instrument for examining the inner structure of the eye. Helmholtz' original ophthalmoscope, made in 1851, consisted of a flat glass plate held at an angle of 45° to the vertical between the subject's eyes and the doctor. A light was placed above the glass plate, which reflected some of the light into the subject's eye. The doctor could then look through the plate and see the illuminated interior of the eye. The lens and cornea of the eye act as a simple magnifying glass to form a large, virtual, upright image of the patient's eye at a large distance behind the eye. In this way the doctor could examine the retina under high magnification, at no risk to the patient.

Such a device is called a *direct ophthalmoscope*. A modern version of Helmholtz' device is shown in Fig. 25.25. In it the glass plate is replaced by a small mirror with a hole in its center to allow the doctor to see into the patient's eye.

The newest version of Helmholtz' invention is the *indirect ophthalmoscope*, shown in Fig. 25.26. This consists of a concave mirror and a battery-powered light source within a tubular handle, with the doctor sighting through a single or binocular eyepiece. The patient's eye is still used as a magnifying glass, but in addition there is an added convex glass lens that functions much as does an eyepiece in a

---

*A birefringent (or double refracting) crystal breaks up light into two beams polarized at right angles to each other. When these two beams combine on emerging from the crystal, the plane of polarization of the incident light is found to be rotated.

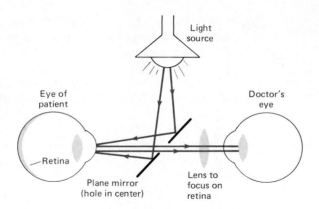

FIGURE 25.25   A modern version of the original ophthalmoscope of H. von Helmholtz. The doctor can see into the patient's eye, using the light reflected into the eye by the mirror.

microscope. Modern ophthalmoscopes have a rotating disk system that allows the convex lens to be changed to enable the doctor to observe the eye at varying depths and magnifications. Such eye examinations are often enhanced by administering eyedrops to dilate the pupil and thus allow more light into the eye.

Interferometers

One of the most precise and powerful research instruments is the *interferometer*, which uses the interference of two light beams to measure wavelengths of light, determine distances to high accuracy, and measure the index of refraction of gases.

Michelson Interferometer   The interferometer developed by A. A. Michelson (see Fig. 25.28 and accompanying biography) played a crucial role in the development of Einstein's special theory of relativity.

A sketch of Michelson's two-beam interferometer is shown in Fig. 25.27. It consists of a light source S, which sends monochromatic light to a beam-splitting mirror B that is lightly coated with a reflecting metal so that half of the light is reflected to the stationary mirror M_1 and the other half passes through the beam splitter to the movable mirror M_2. The light striking M_1 is reflected back on itself, and some of it passes through B and enters the eye at E. The light striking M_2 is also reflected back on itself, and some of it is reflected by B and also enters the eye. These two beams of light differ in that one has traveled twice over the distance D_1 from B to M_1 and the other has traveled twice over the distance D_2 from B to M_2. (The plate C is a compensating plate introduced to ensure that the path of light through glass is the same for both beams.)

Since the light in the two beams originated at a common point source S and

FIGURE 25.26   An indirect ophthalmoscope. *(Courtesy of Reichert Ophthalmic Instruments, Cambridge Instruments Inc.)*

FIGURE 25.27   The Michelson interferometer. Bright and dark fringes pass before the eye as the mirror M_2 is moved.

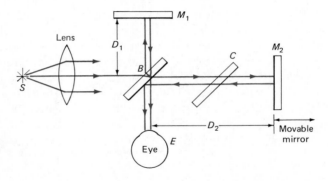

Albert Abraham Michelson (1852–1931)

FIGURE 25.28 Albert A. Michelson at his desk in the Ryerson Physical Laboratory at the University of Chicago in 1927. *(Courtesy of AIP Niels Bohr Library.)*

Although born in Europe, A. A. Michelson is usually considered the first American to receive a Nobel Prize in science, since he came to this country at the age of 4 and became a naturalized citizen. He received the 1907 Nobel Prize in physics "for his optical instruments of precision and the spectroscopic and metrologic investigations which he carried out by means of them."

Michelson was born in 1852 at what is now Strzelno in Poland. His parents were Polish Jews who immigrated to the United States in 1856. Here his father became a merchant selling supplies to the gold and silver miners of California and Nevada in the

gold rush days. Albert did very well in school, especially in mathematics and science, and, after failing to win an appointment to the Naval Academy at Annapolis, made a trip across the country to plead his case with President Grant. Owing to the kindness of the president and the authorities at Annapolis, he was finally given a special appointment to the Naval Academy.

There he did very well in science and athletics (he was a boxing champion) but less well in the humanities and seamanship. After graduating in 1873, he became a science instructor at the Naval Academy and there began his lifetime passion, the measurement of the speed of light to evergreater precision (see Sec. 22.7). Feeling the need for better training in optics, Michelson spent the years 1880 to 1882 in Europe, studying and doing research at the great research laboratories of Berlin, Heidelberg, and Paris. He profited particularly from working with Helmholtz in Berlin and there designed and built the first of his many interferometers.

After 12 years of naval service, Michelson resigned his commission in 1881 to devote himself completely to physics. He became professor of physics at the new Case School of Applied Science in Cleveland and collaborated with E. W. Morley, a professor of chemistry at Western Reserve University, in the famous Michelson-Morley experiment (see Sec. 26.3).

In 1892 Michelson was made

head of the department of physics at the University of Chicago. He remained there until 1930, making it a world-famous center for research in precision optics and spectroscopy, and ruling diffraction gratings that set new standards for size and resolving power. In his latter years he spent a great deal of time in California trying to improve on his already highly accurate determinations of the speed of light. He died in 1931 after several strokes suffered during preparations for a measurement of c in a mile-long evacuated metal tube.

Michelson was married twice and had three children by each wife. He does not appear to have been an easy person to live with. He disliked social functions and felt that university administrative tasks, even the direction of graduate students, took too much time from his beloved research. Still, during the First World War, his patriotism drove him to abandon research and, at the age of 65, to serve his country as a reserve officer. He mellowed considerably toward the end of his life, admitting in a letter to a friend that "I've also found that human nature is not so abominable as I have sometimes thought."

Michelson loved nothing more than experimental research. When asked by reporters in 1882 why he was attempting to measure the speed of light, he answered: "Because it's such great fun." Fifty years later Einstein asked Michelson a similar question. His reply was still the same.

was therefore originally in phase, the relative phase of the two beams at E depends on the difference in the paths traveled by the two beams. Thus, if the paths differ by a whole number of wavelengths of the light used, i.e., if

$$2D_1 = 2D_2 + m\lambda \qquad (m \text{ integral})$$

(25.20)

FIGURE 25.29 Photograph of fringes produced with a Michelson interferometer. The straight fringes are deformed in the neighborhood of a candle flame due to the change in the density of the air in the heated region.

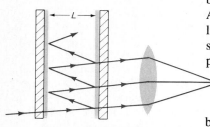

FIGURE 25.30 The Fabry-Perot interferometer.

constructive interference results. If, on the other hand, we have

$$2D_1 = 2D_2 + (m + \tfrac{1}{2})\lambda \qquad (m \text{ integral}) \qquad \textbf{(25.21)}$$

the two beams are 180° out of phase and cancel. The eye sees light or darkness depending on the phase difference between the two beams.

For perfect alignment of the mirrors at 90° to each other the interference pattern consists of a series of dark and bright circles corresponding to destructive and constructive interference between the two beams, because the path difference varies with the angle of vision. If one mirror is inclined at a slight angle with respect to the other, straight fringes similar to those produced by a wedge of air between two pieces of glass are produced (see Fig. 25.29). As the mirror M_2 is moved through a distance $\lambda/2$, a dark fringe changes to a light fringe and then back again to a dark fringe, and the bright and dark fringes appear to move horizontally past a reference marker. These fringes can be counted either manually or by using photo-counters, and the distance moved by the mirror can thus be measured in terms of wavelengths of light.

Thus, if 216 bright fringes pass the reference marker, it means that mirror M_2 has moved a distance of $216\,(\lambda/2) = 108\lambda$. For red light at 633 nm from a He–Ne laser, this corresponds to 6.84×10^{-5} m, or 0.0684 mm. In this way distances can be converted into wavelengths of light.

Fabry-Perot Interferometer   Another type of interferometer much used in research laboratories is the multibeam Fabry-Perot interferometer* shown in Fig. 25.30.

The Fabry-Perot interferometer consists of two perfectly flat glass plates coated on their facing surfaces with a thin film of metal that reflects 99 percent and transmits only 1 percent of the incident light. A beam of light is incident on these two interferometer plates from the left, as shown. Most of the beam is trapped between the plates and is reflected back and forth a very large number of times. At each reflection by the right-hand plate, however, about 1 percent of the incident light passes through. As a result a large number of beams of light emerge at the same angle at which they entered and can be focused by a converging lens to a point P.

Any two adjacent beams emerging normal to the plates have paths differing by $2L$, where L is the distance between the reflecting surfaces of the plates. If $2L$ is equal to a whole number (m) of wavelengths for some particular wavelength of light, constructive interference occurs and a bright spot appears at P. The spot is very bright because a large number of beams combine to form the image and they are all perfectly in phase, since each one differs from its immediate neighbors in path length by exactly the same amount, $2L$. If, on the other hand, the path difference $2L$ is not exactly equal to a whole number of wavelengths of the light used, destructive interference results because there are so many different beams, with each one somewhat out of phase with all the others. The two plates therefore serve as a highly selective wavelength filter. For a fixed distance between the plates, only those wavelengths that satisfy the equation $2L = m\lambda$ (m integral) are strongly transmitted.

---

*This interferometer is named after the two French physicists, Charles Fabry (1867–1945) and Alfred Pérot (1863–1925), who invented it in 1899.

FIGURE 25.31 Interference fringes produced by a Fabry-Perot interferometer using the 546-nm green line of mercury (Hg). The left side of the pattern was made with a natural Hg source, and the structure present is due to the many different isotopes found in natural Hg. The clearer and sharper fringes on the right were produced with a single-isotope $^{198}_{80}$Hg source. *(Photo courtesy of U.S. National Institute of Standards and Technology.)*

If the light comes from a point source, the path difference depends on the angle made by the incoming ray with the surface of the mirrors. As a result we would expect a circular interference pattern centered on a line drawn normal to the mirrors from the point source of light. This is indeed observed, as shown in Fig. 25.31. Fabry-Perot fringes are much sharper than those produced by a Michelson interferometer. (See color photograph of Fabry-Perot fringes on cover.)

Wavelength differences of less than 1 part in a million can be detected with a Fabry-Perot interferometer. It is therefore extremely useful for measuring the wavelengths of spectral lines and for measuring distances in terms of spectral wavelengths.

Example 25.10

A Michelson interferometer is illuminated with red light from a He–Ne laser. An accurate screw with an attached scale measures the distance traveled by the movable mirror to the nearest 10^{-3} mm, or 10^{-6} m. It is found that moving the mirror through a distance of 0.316 mm causes a shift of 1000 fringes, as recorded by a photocell and counter. What is the wavelength of the laser light?

SOLUTION

The passage of 1000 fringes means that the total (two-way) light path has been changed by 1000 wavelengths. Hence we have

$$1000\lambda = 2(0.316 \text{ mm})$$

or

$$\lambda = \frac{2(0.316 \times 10^{-3} \text{ m})}{1000} = \boxed{632 \text{ nm}}$$

Example 25.11

In the expression for the bright spot at the center of the interference pattern of a Fabry-Perot interferometer, $2L = m\lambda$, the integer m is called the order of *interference*. What is the order of interference for an interferometer with $L = 10.0$ cm that uses green light with $\lambda = 500$ nm?

SOLUTION

Since $2L = m\lambda$,

$$m = \frac{2L}{\lambda} = \frac{2(0.100 \text{ m})}{500 \times 10^{-9} \text{ m}} = \boxed{4.00 \times 10^5}$$

The path difference between two adjacent beams therefore corresponds to 400,000 wavelengths of the light used.

Summary: Important Definitions and Equations

Cameras: Optical instruments consisting of a lens or system of lenses to focus light onto a photographic plate. Cameras produce real, inverted images.

f-Number of lens: The ratio of focal length to lens diameter (f/D).

Human eye: Resembles a simple camera; forms a real, inverted image of an object on the retina, which is the light-sensitive part of the eye.

Accommodation of eye: The process by which the focal length of the eye lens is automatically varied to produce a sharp image on the retina.

Defects of eye:
   Nearsightedness (myopia): The inability to focus on distant objects; requires diverging lens to compensate.
   Farsightedness (hyperopia): The inability to focus on near objects; requires converging lens to compensate.
   Astigmatism: A defect due to lack of perfect spherical symmetry in the eye lens; requires cylindrical lens to compensate.

Simple Magnifier (magnifying glass): An optical instrument using a single converging lens to increase the angular magnification of an object.

Angular magnification, or magnifying power:

$$M_\phi = \frac{\text{angular size of image (with magnifier present)}}{\text{angular size of object (without magnifier)}}$$

$$= \frac{25}{f_e} \tag{25.7}$$

with f_e the focal length in centimeters.

Compound microscope: A combination of two converging lenses, an objective lens that forms an enlarged, real, inverted image, and an eyepiece (or ocular) that produces a greatly enlarged virtual image of the real image produced by the objective.

Magnification: $M \simeq \dfrac{25L}{f_0 f_e}$ **(25.10)**

where L is the distance between the two lenses.

Telescope: An optical instrument for collecting radiation from a distant object and producing an image that can then be magnified by an eyepiece.

Refractor: Uses a converging lens to collect the radiation.

$$\text{Magnification } M_\phi = \frac{f_0}{f_e} \tag{25.12}$$

Reflector: Uses a concave mirror to collect the radiation.

Resolution: The ability of an optical instrument to form distinct images of two objects that are very close together.

Rayleigh's criterion for resolution
   Telescope: The minimum angle that can be resolved is

$$\theta_R \simeq \frac{1.22\lambda}{d} \tag{25.14}$$

   Microscope: The minimum linear distance that can be resolved is

$$s \simeq \frac{1.22\lambda}{2n \sin \alpha} \tag{25.19}$$

where $n \sin \alpha$ is the numerical aperture (NA) of the objective lens, n is the index of refraction of the medium, and 2α is the angle subtended by the objective lens at the object.

Specialized microscopes

Oil-immersion microscope: Uses oil between the object and the objective lens to decrease the effective wavelength and increase the resolution.

Ultraviolet (UV) microscope: Uses shorter-wavelength UV light to increase the resolution.

Phase-contrast microscope: Converts phase changes produced by microscope specimen to intensity changes to improve contrast.

Polarizing microscope: Converts changes in polarization into intensity changes to improve contrast.

Interferometers: Optical instruments employing the interference of light beams to measure wavelengths of light and linear distances to very high accuracy.

Michelson interferometer: Uses the interference between two light beams that have traveled over two different paths at right angles to each other.

Fabry-Perot interferometer: Uses the multibeam interference of parallel light beams reflected back and forth many times between two parallel reflecting plates.

Questions

1 Prove from geometry that the magnification of a pinhole camera is q/p.

2 Why does a camera produce a sharper image when it is stopped down to a smaller aperture?

3 Compare the human eye with a simple camera with respect to:

(*a*)   Focusing mechanism.
(*b*)   Control of amount of light used to produce image.
(*c*)   Type of image produced.
(*d*)   Aberrations leading to distorted images.

4 (*a*)   When a swimmer's eyes are open under water, distant objects seem blurred and out of focus. Why?

(b) Why will wearing goggles, which keeps the water away from the swimmer's eye, solve this problem?

5 Why are bifocals worn mostly by older men and women rather than by those in their thirties?

6 Discuss the characteristics of an eye that could see electromagnetic radiation whose wavelengths varied by a factor of 20, rather than 2, from one end of the visible spectrum to the other.

7 Does the use of white light in a microscope reduce the resolution over that which could be achieved with monochromatic light? Why?

8 Prove that the distance a telescope can see into space is directly proportional to the diameter of the telescope's objective lens.

9 Why can you not use a negative lens as a simple magnifier?

10 Compare the optical system of a compound microscope with that of a refracting telescope. How do they differ, and in what respect are they similar?

11 (a) Atoms and molecules have dimensions of roughly 10^{-10} m. Would you expect to be able to see them with visible light? Why?

(b) What wavelength radiation would be needed to see an atom? In what region of the spectrum would this radiation fall?

12 Describe how a Michelson interferometer might be used to measure the index of refraction of a gas.

13 Which of the following optical instruments normally produces images that can be recorded on photographic film: (a) camera, (b) compound microscope, (c) projection lantern, (d) reflecting telescope with a newtonian focus, (e) Michelson interferometer?

Multiple-Choice and Simple Exercises

25.1 In a darkened room we place a lighted tungsten bulb in a box that has one end covered with tinfoil. If we make a pinhole in the tinfoil, an image of the filament appears on a screen facing the tinfoil. If we now make a second pinhole just above the first, there will appear on the screen:
(a) A brighter image of the filament
(b) A larger image of the filament
(c) A smaller image of the filament
(d) A second image of the filament above the first
(e) A second image of the filament below the first

25.2 The bellows of an adjustable camera can be extended so that the largest distance from lens to film is 1.5 times the focal length of the lens. If the focal length of the lens is 6.0 cm, the nearest object that can be sharply focused on the film must be at a distance from the lens of:
(a) 6.0 cm   (b) 12 cm   (c) 18 cm
(d) 24 cm   (e) 36 cm

25.3 A boy used the correct exposure time in photographing a distant object. For a second exposure he makes the diameter of the diaphragm opening just twice its initial value, but keeps everything else except the exposure time the same. Compared with the first exposure time, the second exposure time should be:
(a) One-half as long   (b) Twice as long
(c) The same   (d) Four times as long
(e) One-quarter as long

25.4 A converging lens with f-number equal to 2.0 and focal length 5.0 cm has a diameter of:
(a) 2.0 cm   (b) 0.40 cm   (c) 10 cm
(d) 2.5 cm   (e) None of the above

25.5 A photograph is taken with a camera that is stopped down to f/11. Another photograph is taken with the same camera, with an f/1.4 lens, wide open. The ratio of the brightness of the image in the first photograph to that in the latter is:
(a) 62   (b) 0.016   (c) 0.18   (d) 7.9
(e) 0.13

25.6 The magnifying power of a magnifying glass of focal length 2.0 cm is about:
(a) 2.0   (b) 4.0   (c) 50   (d) 13
(e) None of the above

25.7 The magnification of a compound microscope containing an eyepiece of focal length 2.0 cm and an objective lens of focal length 1.5 cm is about:
(a) 3.0×10^2 (b) 2.3×10^2 (c) 8.0
(d) 6.0   (e) None of the above

25.8 An astronomical telescope has an objective lens of focal length 300 cm and an eyepiece of focal length 2.00 cm. The magnifying power of this telescope is:
(a) 600   (b) 150 cm   (c) 1/150   (d) 150
(e) 300

25.9 The resolution of a lens can be improved by using:
(a) A smaller aperture in front of the lens
(b) A slower lens
(c) Longer-wavelength light
(d) Shorter-wavelength light
(e) A smaller object

25.10 The smallest angular separation that can be resolved by a stellar telescope of lens diameter 25 cm, using a blue filter at 450 nm, is:
(a) 2.2×10^{-6} rad   (b) 2.2×10^{-6} degree
(c) 7.9×10^{-3} s of arc   (d) 2.2×10^{-6} s of arc
(e) None of the above

25.11 In a Michelson interferometer using red light at 633 nm, a difference in the two optical paths of the following amount will lead to destructive interference of the two beams:
(a) 633 nm   (b) 2216 nm   (c) 1266 nm
(d) 1899 nm   (e) None of the above

25.12 A pinhole camera is used to photograph a child who is 1.0 m tall. If the child is 2.5 m from the pinhole and the pinhole is 10 cm from the photographic film, how tall will the child be in the picture?

25.13 How great a change in brightness is produced by a change in the f-stop used on a camera lens from f/16 to f/4?

25.14 A camera with a lens of 12-cm focal length is used to photograph a man 2.0 m tall who stands 4.0 m from the camera. How tall is the man's image?

25.15 A 25-diopter lens is used as a simple magnifier. What is the angular magnification of the lens for an image at infinity?

25.16 What is the magnifying power of a refracting telescope with an objective lens of 15-m focal length and an eyepiece of focal length 5.0 cm?

25.17 A small airplane with a 25-m wingspread is flying toward an airport with a light on each end of its wing. How close must the plane be before an air-traffic controller in the tower can distinguish these two lights as separate light sources? Assume that the human eye can resolve 1 min of arc.

25.18 What is the numerical aperture of the lens in an oil-immersion microscope if the objective lens used has a diameter of 10 cm and a focal length of 7.0 cm, and the oil between the objective lens and the object has an index of refraction of 1.5?

25.19 A Michelson interferometer is being used with red He–Ne laser light at 633 nm to measure a distance of 2.00 cm in terms of wavelengths of red light. When the movable mirror is moved from one end of the 2.00-cm standard to the other, how many bright fringes pass the cross hairs of an observing telescope?

25.20 A Michelson interferometer illuminated with krypton orange-red light at 606 nm is used to measure the distance between two points. If 240 dark fringes pass the telescope cross hairs as the mirror is moved from one of these points to the other, what is the distance between the two points?

Problems

25.21 A camera has a 4.0-mm lens opening. A good photograph is taken with it at a shutter speed of 1/60 s. If the camera lens is now stopped down to a 0.40-mm aperture, what would be the proper exposure time using the same object and the same film?

25.22 A camera takes a good picture with an f-stop of f/2 and a 1/50-s shutter speed. If the f-stop is changed to f/16, what must the shutter speed be to take an equally good picture?

25.23 In a simple box camera the distance from the lens to the film is 12.0 cm. If the lens has a focal length of 10.0 cm, for what position of the object will the image be sharpest?

25.24 A football trophy 30 cm wide and 50 cm high is to be photographed with a camera from a distance of 15 m.

(a) If the camera has a lens of 5.0 cm focal length, what is the size of the image on the film?

(b) If the camera lens is replaced with a 50.0-cm-focal-length telephoto lens, what area on the film is now occupied by the image?

25.25 A camera has a lens of 5.5 cm focal length. Objects at positions from 2.0 m to infinity can be focused on the camera film by using a bellows to move the lens. Over what distance must the lens be moved if the focus is changed over the full range from 2.0 m to infinity?

25.26 (a) What is the power of the eyeglasses required for reading purposes by a man whose near point is at 200 cm?

(b) The far point of a nearsighted woman is at 30 cm. What is the power of the spectacles she needs for distant vision?

25.27 A 70-year-old woman can read a book only when it is held at full arm's length, in her case at 50 cm from her eyes. If she uses reading glasses to enable her to read more comfortably with the book at the near point of the eye, what kind of lenses does she need and how strong must they be?

25.28 A man is not able to see objects clearly unless they are closer than about 1.60 m from his eyes.

(a) Is he nearsighted or farsighted?

(b) What type of correcting lenses does he need and how strong should they be?

25.29 (a) What is the focal length of the reading glasses required by a woman whose near point is 100 cm from her eyes?

(b) How far from her eyes would she have to hold a book to read it while wearing reading glasses of power 2.0 D?

25.30 A child in the first grade is seen to hold a comic book 10 cm from her eyes in order to see the book clearly.

(a) Is she nearsighted or farsighted?

(b) What power lens will she need in her glasses to enable her to read normally?

25.31 A woman can see objects clearly only if they lie between 50 and 100 cm from her eyes. Calculate the power of the lenses (in diopters) needed in bifocal glasses which will enable the woman to see distant objects clearly through the top half of her glasses and read a newspaper at a distance of 25 cm through the bottom half of her glasses.

*25.32 A man 40 years of age is prescribed +2.0-D lenses for reading at a normal distance of 25 cm. Fifteen years later he finds that he must hold the book or newspaper 40 cm from his eyes to see the type clearly with the same +2.0-D lenses. What should be the power of his new reading lenses?

25.33 A boy uses a magnifying glass to set a piece of paper on fire and finds that putting the paper 5.0 cm from the center of the lens provides the best focus for the sun's rays. If this lens were then used to read some small newsprint, what would be the magnification for a virtual image at infinity?

25.34 A small, round bug of diameter 3.0 mm is being observed under a magnifying glass, which is a simple converging lens of focal length 4.0 cm. Calculate the angular magnification of the magnifier when the virtual image is at infinity.

25.35 The length of a microscope tube is 12 cm. The focal length of the objective lens is 0.60 cm and the focal length of the eyepiece is 3.0 cm. What is the magnifying power of the microscope?

25.36 A microscope has an objective lens of 1.0-cm focal length and an eyepiece of 3.0-cm focal length. An object is in sharp focus when it is 1.1 cm from the objective.

(a) What is the best separation of the two lenses?

(b) What is the magnification produced by the microscope?

25.37 Find an expression for the magnification of a compound microscope in terms of the power of each of the two lenses and the distance between the lenses.

25.38 A compound microscope with an 18-cm barrel connecting the two lenses has an eyepiece of focal length 1.0 cm. If an overall magnification of 400 is desired, what must be the focal length of the objective lens?

25.39 A telescope has a magnification of $20\times$. The distance between eyepiece and objective lens is 25 cm.

(a)   What is the focal length of the objective lens?

(b)   What is the focal length of the eyepiece?

25.40 A compound microscope has a $5\times$ eyepiece and a $50\times$ objective lens at the end of a microscope barrel 18 cm long. Find (a) the total magnification of the microscope; (b) the position at which the object must be placed so that the final image is at 25 cm, the near point of the eye; (c) the focal length of each lens.

25.41 A refracting telescope has a 1.5-D objective lens and a 20-D eyepiece. What is its magnification?

25.42 You are given a 100-cm-long mailing tube and two lenses of focal lengths 60 and 12 cm that just fit inside the tube.

(a)   Describe how you could construct a telescope from these components.

(b)   What would be the magnification of such a telescope?

25.43 The 40-in refractor at Yerkes Observatory has an objective lens with a focal length of about 19 m (63.5 ft). If this telescope is being used to look at the moon, which is 3.8×10^8 m from the earth, how large will a distance of 1.0 km on the moon's surface appear in the image created by the objective lens?

25.44 The barrel of an optical instrument consists of an objective lens of focal length 25 cm and an eyepiece of focal length 1.5 cm separated by a distance of 27 cm. The device is to be used to view objects through the eyepiece with the eye relaxed, i.e., with parallel rays entering the eye.

(a)   Is this device intended to function as a microscope or a telescope?

(b)   For what approximate object distance is the device intended?

(c)   What is its magnifying power?

25.45 The lenses of an astronomical telescope are 96 cm apart when adjusted for viewing a distant object with the relaxed eye. The angular magnification of the telescope is 50. Compute the focal length of each lens.

25.46 A terrestrial telescope consists of three converging lenses in a row, as shown in Fig. 25.32. The focal lengths of the three lenses are $f_1 = 50$ cm, $f_2 = 5.0$ cm, $f_3 = 5.0$ cm.

(a)   Find the overall magnification of this telescope.

(b)   Describe the nature of the final image produced by this telescope.

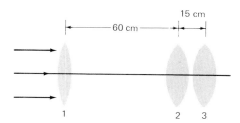

FIGURE 25.32 Diagram for Prob. 25.46.

*25.47 The first astronomical telescope invented by Galileo used a diverging lens in place of an eyepiece. Use a ray

diagram to show that by placing the diverging lens between the converging objective lens and its focal point, the telescope forms a virtual, enlarged, upright image of a distant object.

25.48 How close to a TV screen must a person sit to resolve two different color dots on the screen separated by a distance of 1.0 mm? (See Example 25.7.)

25.49 An oil-immersion microscope uses oil with $n = 1.5$. The objective lens can accept light scattered by the object up to 50° on either side of the vertical.

(a)   What is the numerical aperture of the objective lens?

(b)   What is the minimum linear distance this microscope can resolve for 450-nm light?

25.50 A telescope lens has an f-number of 4.0 and a diameter of 50 cm.

(a)   What is the minimum angular separation of two distant objects that such a telescope can resolve? (Assume that the wavelength of the light coming from the object is 600 nm.)

(b)   How far apart are the centers of the diffraction patterns of the two objects in the focal plane of the lens?

25.51 Two stars are 30 light-years away from earth. They can be just barely resolved by the 40-in refractor at Yerkes Observatory. If the telescope is using a filter that passes only 450-nm light, how far apart are the two stars?

25.52 An ophthalmologist uses a simple ophthalmoscope to examine the retina of a patient's eye. If the ophthalmologist uses the lens of the patient's eye, which has a focal length of 1.7 cm, as a simple magnifier, what is the angular magnification under which the doctor is able to examine the patient's eyes, if the virtual image is at infinity?

25.53 An oculist is using a simple ophthalmoscope to examine the retina of a patient's eye. If the power of the lens in the patient's eye is 60 D, how large will the spacings between two adjacent photoreceptors on the retina, which are actually 2.0 μm apart, appear when the oculist focuses on the virtual image produced by the lens at infinity?

25.54 How far must the movable mirror M_2 in a Michelson interferometer be moved if 750 bright fringes produced by 535-nm light are observed to move past the cross hairs of an observing telescope?

25.55 A gauge block (a metal block of accurately known length used by machinists) is measured with a Michelson interferometer, first using red light at 633 nm, and then green light at 535 nm. With the red light 6352 bright fringes are counted as the mirror M_2 is moved from one end of the gauge block to the other.

(a)   What is the length of the gauge block?

(b)   How many fringes would be counted when the green light at 535 nm is used in the interferometer?

25.56 A small evacuated glass cell of length 3.80 cm is inserted in the path of one light beam in a Michelson interferometer. A gas is slowly allowed to leak into the cell until the gas pressure is equal to atmospheric pressure, and the bright fringes passing the cross hairs of the observing telescope are counted. It is found that 16 fringes pass the cross hairs when neon light of wavelength 633 nm is used. What is the index of refraction of the gas?

*25.57 A problem occurs when a Michelson interferometer is used in the ordinary way with yellow sodium light. For one position of the movable mirror the fringes are very sharp and clear, but as the mirror is moved the fringes become more and more blurred and finally disappear altogether. If the mirror is moved even farther, the fringes become sharp again. The explanation for this is that the yellow line is actually a doublet with wavelengths 589.0 and 589.6 nm.

(*a*) Show how this fact explains the periodic variation in the nature of the observed fringes.

(*b*) How far would the mirror M_2 in Fig. 25.27 have to be moved for the fringes to change from sharp to nonexistent and then back to sharp again?

Additional Readings

Cagnet, M., M. Françon, and J. C. Thrierr: *Atlas of Optical Phenomena*, Springer-Verlag, Heidelberg, 1962. Source for the photographs of Figs. 25.20, 25.21, 25.29 (by permission of Springer-Verlag, Heidelberg).

Horridge, G. Adrian: "The Compound Eye of Insects," *Scientific American*, vol. 237, no. 1, July 1977, pp. 108–120. An interesting account of how insects see.

Jaffe, Bernard: *Michelson and the Speed of Light*, Doubleday Anchor, Garden City, N.Y., 1960. A brief, popular book on Michelson's experimental work. It is most successful in conveying the spirit that motivated Michelson, and therefore the spirit of modern science.

Livingston, Dorothy Michelson: *The Master of Light: A Biography of Albert Michelson*, University of Chicago Press, Chicago, 1973. A personal, nonscientific account by Michelson's daughter that contains some unusual insights into Michelson's character.

Maran, Stephen P.: "Beyond Galileo's Vast Crowd of Stars," *Smithsonian*, vol. 18, no. 3, June 1987, pp. 40–53. A beautifully illustrated account of recent developments in astronomical telescopes.

Michelson, Albert A.: *Light Waves and Their Uses*, University of Chicago Press, Chicago, 1903. Michelson's fascinating 1899 Lowell Lectures at Harvard.

Nelson, Jerry: "The Keck Telescope," *American Scientist*, vol. 77, March-April 1989, pp. 170–176. A good account of a revolutionary approach to the design of large-aperture telescopes.

Price, William H.: "The Photographic Lens," *Scientific American*, vol. 235, no. 2, August 1976, pp. 72–83. A discussion of lens aberrations and the use of modern computer techniques to improve photographic lenses.

Shankland, Robert S.: "Michelson and His Interferometer," *Physics Today*, vol. 27, no. 4, April 1974, pp. 36–43. A brief account of how Michelson applied his interferometer to important problems in astronomy, atomic spectra, and metrology.

Smith, Robert W.: *The Space Telescope: A Study of NASA, Science, Technology and Politics,* Cambridge University Press, New York, 1990. A fascinating, nontechnical book on the development and construction of the Hubble Telescope over the years 1968 to 1990. It is especially important as a case study of the nonscientific factors that had such an impact on the final shape and form of this spectacular research instrument.

Wald, George: "Eye and Camera," *Scientific American*, vol. 183, no. 2, August 1950, pp. 32–41. A comparison of the basic physics and chemistry of the eye and the camera by the eminent Harvard biologist and 1967 Nobel Laureate.

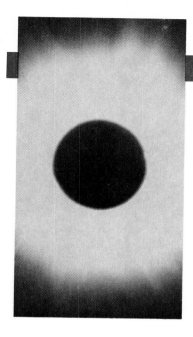

CHAPTER 26

The Theory of Relativity

It is no doubt that Michelson's experiment was of considerable influence upon my work insofar as it strengthened my conviction concerning the validity of the principle of the special theory of relativity. On the other side I was pretty much convinced of the validity of the principle before I did know this experiment and its result. In any case, Michelson's experiment removed practically any doubt about the validity of the principle in optics, and showed that a profound change of the basic concepts of physics was inevitable.

Albert Einstein (1879–1955)

In 1894, A. A. Michelson delivered an address at the dedication ceremonies for the new Ryerson Physical Laboratory at the University of Chicago. In this address Michelson echoed with favor the view of his old friend Lord Kelvin (see Fig. 21.10) that all great discoveries in physics had probably been made, and that the future of physics would be confined to measuring known quantities such as the velocity of light to greater and greater accuracy. Never have two renowned scientists been so completely wrong in their judgment! During the next 40 years Roentgen discovered x-rays and Becquerel radioactivity, atomic and nuclear physics developed and flourished, and the theories of relativity and quantum mechanics radically changed our view not only of physics but of philosophy as well. The remainder of this book is devoted to these developments, to what is often called *modern physics* as opposed to the classical physics we have been studying thus far.

In this chapter we take up Albert Einstein's famous theory of relativity. Most of our treatment is confined to special relativity, i.e., to the theory that applies to objects and phenomena in one reference system moving with constant velocity with respect to another. In the last section we discuss briefly the more complicated general theory of relativity, which applies to accelerated reference systems.

26.1 Galilean Relativity

Long before Einstein's day, physicists working in the field of mechanics were accustomed to dealing with the constant-velocity motion of one reference frame relative to another, something that has come to be called *Galilean relativity* (also called *Newtonian relativity*).

Consider a physical event such as the collision of two particles. The event must occur at some point in space and at some instant of time. We therefore specify the event by the three space coordinates x, y, and z, measured with respect to the origin of the coordinate system, and by the time t at which the event occurs.

Frame of reference: A system in which an event can be specified by three distances measured with respect to the origin of the coordinate system, and by the time elapsed from some chosen initial time.

Inertial frame of reference: A frame of reference in which Newton's first law of motion, i.e., the law of inertia, is valid.

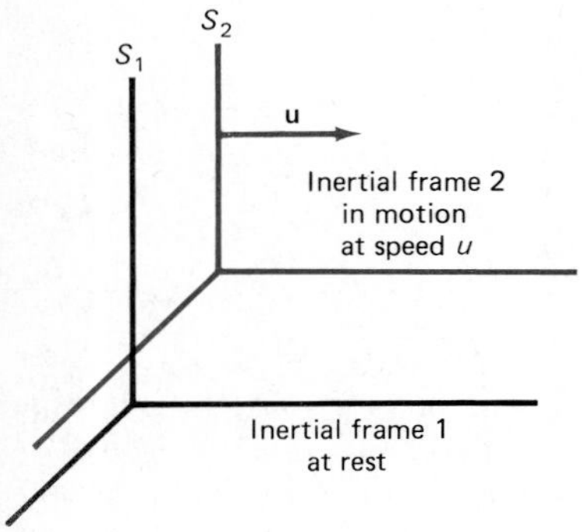

An inertial reference frame is an *unaccelerated* frame. Any object in this frame not acted on by a net force either remains at rest or continues to move with constant speed in a straight line. A rocket ship drifting freely in outer space, well away from gravitational forces and with its engines cut off, is an ideal inertial system. Newton assumed that the fixed stars made a good inertial frame. Once we have found an inertial frame, any other frame moving with constant speed in a straight line with respect to the first frame is also an inertial frame (Fig. 26.1).

Reference frames *accelerating* with respect to a chosen inertial frame are not inertial frames, because in such frames Newton's first law is no longer valid: the accelerating frame produces pseudoforces of the kind produced in a falling elevator, as discussed in Example 3.6. In practice we often neglect the small effects of the rotation and orbital motion of the earth and take the earth's surface as an inertial frame. Any reference frame moving with constant velocity with respect to the earth's surface is then also an inertial frame, but an accelerating car or a rotating merry-go-round is not, as illustrated in Fig. 26.2.

Consider two inertial frames of reference S_1 and S_2 consisting of rectangular coordinate systems in which distances along the three axes with respect to the origin are specified by x_1, y_1, z_1 and x_2, y_2, z_2, respectively. The second frame S_2 is moving in the positive x direction with a speed u with respect to the first frame, as in Fig. 26.3. At time $t = 0$ the two origins coincide and the clocks in the two frames are synchronized to read exactly the same time. We assume that the clocks remain synchronized throughout the motion, because the same measure of time prevails in the two reference frames, and so $t_2 = t_1$. (This may seem obvious, but in Einstein's theory it will prove to be an *invalid* assumption!)

Suppose that at time t_1 some event occurs at point P in space, say, a firecracker

FIGURE 26.1 Inertial frames of reference. If we have an inertial reference frame S_1, then any other frame S_2 moving with constant speed u in a straight line with respect to the first reference frame is also an inertial frame.

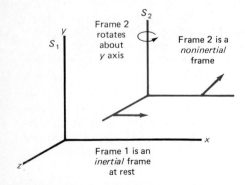

FIGURE 26.2 A frame of reference that rotates with respect to an inertial frame of reference. The rotating frame is a noninertial frame.

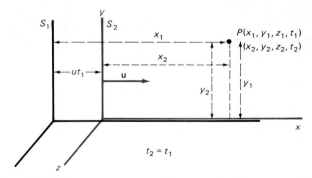

FIGURE 26.3 A Galilean transformation. Inertial frame S_2 moves with constant speed u along the x axis with respect to inertial frame S_1, which is at rest. From the diagram $x_2 = x_1 - ut_1$.

is exploded. The space and time coordinates of this event in frame S_1 are then (x_1, y_1, z_1, t_1) and in frame S_2 they are (x_2, y_2, z_2, t_2). From Fig. 26.3 it can be seen that the coordinates of this event in the two frames are related by the following equations, usually called the *Galilean transformation equations*:

$$x_2 = x_1 - ut_1 \qquad \textbf{(26.1)} \qquad\qquad z_2 = z_1 \qquad \textbf{(26.3)}$$

$$y_2 = y_1 \qquad \textbf{(26.2)} \qquad\qquad t_2 = t_1 \qquad \textbf{(26.4)}$$

The only coordinate affected is the x coordinate, since the motion is along the x axis. Also, since the distance between the origins of the two systems is equal to ut_1, the x coordinates of the particle in the two frames differ by ut_1, and so $x_1 = x_2 + ut_1$, or $x_2 = x_1 - ut_1$.

The Galilean transformation equations describe nonrelativistic transformations between any two inertial frames of reference. They express the commonsense assumption that space intervals and time intervals are absolute and completely independent of the motion of the observer. For example, when a firecracker explodes, an observer moving with the reference frame S_2 perceives that the event occurs at the same position in space and at the same time as does an observer in the stationary frame S_1.

If we want to transform from frame S_2 back to frame S_1, the inverse transformation is obtained by interchanging 1 and 2 and reversing the sign of u, since if S_2 is assumed at rest, then S_1 is moving with a speed u in the negative x direction with respect to S_2. The inverse transformation equations are the same as Eqs. (26.1) to (26.4) above except that Eq. (26.1) is replaced by

$$x_1 = x_2 + ut_2 \qquad\qquad \textbf{(26.5)}$$

These transformations show that even in classical Newtonian physics all mechanical motion is relative, i.e., that positions and velocities vary depending on the reference frame with respect to which they are measured. The speed of a person walking down the aisle of a moving train is different when measured with respect to the ground than it is when measured with respect to the floor of the train, as shown in Fig. 26.4. Since KE $= \frac{1}{2}mv^2$, the same is true of the person's kinetic energy.

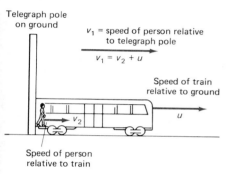

Telegraph pole on ground

v_1 = speed of person relative to telegraph pole

$v_1 = v_2 + u$

Speed of train relative to ground

u

v_2

Speed of person relative to train

FIGURE 26.4 The speed of a person on a train depends on the frame of reference in which it is measured. The person's speed relative to the train is v_2; the person's speed relative to a telephone pole is $v_1 = v_2 + u$.

Invariance of Classical Mechanics under Galilean Transformations

It can be shown that all the laws of *mechanics* (e.g., Newton's three laws of motion) remain valid under a Galilean transformation. They also remain *invariant*, by which we mean that the mathematical form of the equations expressing laws like the conservation of momentum and the conservation of energy is the *same* for any two inertial frames of reference. We prove this only for one simple case, the conservation of linear momentum.

Consider a head-on collision between two particles, as seen in two inertial frames S_1 and S_2, where S_2 is moving with a speed u along the positive x axis with respect to S_1 (see Fig. 26.5). We here use lowercase letters for the mass and velocity of the particle of mass m and capital letters for the mass and velocity of the particle of mass M. As seen by an observer in S_2, the law of conservation of momentum takes the form

$$m\mathbf{v}_2 + M\mathbf{V}_2 = m\mathbf{v}_2' + M\mathbf{V}_2' \qquad\qquad \textbf{(26.6)}$$

where the velocities are vector quantities and the unprimed velocities are those before the collision and the primed velocities those after the collision.

How does this collision appear to an observer in S_1?

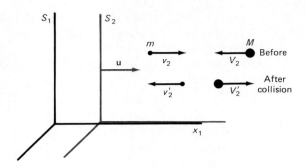

FIGURE 26.5 Conservation of linear momentum. The collision of two particles as seen in two inertial frames S_2 and S_1, where S_2 is moving with a constant speed u along the positive x axis with respect to S_1.

The observer in S_1 sees the velocities increased by the velocity $\mathbf{u}$ of S_2 with respect to S_1, and we have

$$\mathbf{v}_1 = \mathbf{v}_2 + \mathbf{u} \qquad \text{and so} \qquad \mathbf{v}_2 = \mathbf{v}_1 - \mathbf{u} \tag{26.7}$$

$$\mathbf{V}_1 = \mathbf{V}_2 + \mathbf{u} \qquad \text{and so} \qquad \mathbf{V}_2 = \mathbf{V}_1 - \mathbf{u} \tag{26.8}$$

Similarly, $\qquad \mathbf{v}_2' = \mathbf{v}_1' - \mathbf{u} \qquad$ and $\qquad \mathbf{V}_2' = \mathbf{V}_1' - \mathbf{u}$ $\qquad$ (26.9)

On substituting in Eq. (26.6), we have

$$m(\mathbf{v}_1 - \mathbf{u}) + M(\mathbf{V}_1 - \mathbf{u}) = m(\mathbf{v}_1' - \mathbf{u}) + M(\mathbf{V}_1' - \mathbf{u}) \tag{26.10}$$

where $\mathbf{u}$, the relative velocity of the reference frames, is unchanged by the collision. Then, on canceling $-m\mathbf{u}$ and $-M\mathbf{u}$ on both sides, we have

$$m\mathbf{v}_1 + M\mathbf{V}_1 = m\mathbf{v}_1' + M\mathbf{V}_1' \tag{26.11}$$

This equation expresses the conservation of linear momentum, as seen in reference frame S_1. Note that the total momentum measured in the two frames is *different*, but momentum is still conserved in both frames. The mathematical form of Eq. (26.11) is also *identical* with that of Eq. (26.6), with the subscripts merely changed from 2 to 1. We have therefore shown that the law of conservation of linear momentum is *invariant* under a Galilean transformation.

Similar methods show the Galilean invariance of the other basic laws of mechanics, like Newton's second law of motion or the principle of conservation of energy. We therefore conclude:

Principle of Galilean relativity: All the laws of classical mechanics are *invariant* under a Galilean transformation and are therefore the same in all inertial frames of reference.

For this reason it is impossible by means of mechanical experiments to detect uniform rectilinear motion from inside a moving system. Consider the motion of the balls on a billiard table below deck on an ocean liner moving with constant velocity across the ocean. The balls behave exactly as they would on a billiard table in your campus student union. The constant-velocity motion of the ship in no way affects the motion of the billiard balls.

The principle of Galilean relativity implies that the three basic quantities of mechanics—mass, length, and time—are all independent of the motion of the observer. We now show that this is no longer true when we move from mechanics into the realm of electromagnetism.

Example 26.1

A man is on a train that moves along a straight track with a constant speed of 40 m/s with respect to the ground. At time $t = 0$ the man begins to walk forward from his seat at the rear of the car at the instant the rear of the car passes a telegraph pole on the track outside. He walks up the aisle at a speed of 2.0 m/s. (It may be helpful to look at Fig. 26.4 to visualize what is happening.) **(a)** What is his position 10 s later with respect to the rear of the car? **(b)** What is his position 10 s later with respect to the telegraph pole? **(c)** What is his speed 10 s later with respect to the ground?

SOLUTION

It is convenient here to take reference frame S_2 as moving with the train and reference frame S_1 as fixed with respect to the ground.

(a)   Then $x_2 = v_2 t_2 = (2.0 \text{ m/s})(10 \text{ s}) = \boxed{20 \text{ m}}$

The man is therefore 20 m from the rear of the car after 10 s.
(b)   Here we want to transform from the moving frame S_2 back to the rest frame S_1. The inverse Galilean transformation yields

$$x_1 = x_2 + ut_2 = 20 \text{ m} + (40 \text{ m/s})(10 \text{ s}) = \boxed{420 \text{ m}}$$

At the end of 10 s the man is 420 m from the telegraph pole on the ground.
(c)   The speeds are related in this case by the fact that the speed of the man with respect to the ground is the sum of his speed with respect to the train and the speed of the train with respect to the ground, since all three are in the same direction. We have

$$v_1 = v_2 + u \qquad \text{[see Eq. (26.7)]}$$

and so $v_1 = 2.0 \text{ m/s} + 40 \text{ m/s} = \boxed{42 \text{ m/s}}$

The man's speed with respect to the ground is therefore 42 m/s in the direction in which he is walking.

26.2   The Aether and the Speed of Light

In 1900 light was believed to consist of transverse electromagnetic waves in a medium called the *aether* (sometimes spelled *ether*), just as sound waves are longitudinal waves in a medium called air. But we know from our consideration of the Doppler effect in sound (Sec. 12.8) that if a car moves at a speed v through the air toward a sound source which emits sound traveling with a speed u relative to the air, then the speed of the sound relative to the car is $u + v$, as in Fig. 26.6. If, now, light consists of waves in an aether, then the velocity of light waves, which presumably is fixed with respect to the aether, should change with the motion of an observer through the aether, just as in the case of sound. In classical physics there seemed to be little doubt about this. Light required some medium for its propagation ("you cannot have waves unless there is something waving"). Since the speed of light is fixed with respect to that medium by the properties of the medium, the speed of light should be different for observers moving at different speeds with respect to that medium.

According to this reasoning, there is one preferred frame of reference in which the speed of light is exactly 2.998×10^8 m/s in a vacuum (Sec. 22.7); this is the frame of reference at rest with respect to the aether. Observers in all other frames of reference, including all inertial frames, would measure a different value for the speed of light depending on how they move with respect to the aether. Therefore the speed of light would certainly not be invariant under a Galilean transformation.

FIGURE 26.6 Speed of sound in air as measured in a moving frame. The speed of the sound relative to the car is $v + u$.

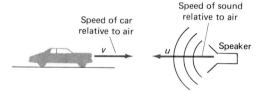

Maxwell's equations, as we have seen, led to the prediction that the speed of electromagnetic waves in a vacuum was related to two fundamental quantities of electricity and magnetism by the expression $c = 1/\sqrt{\epsilon_0\mu_0}$. If the measured speed of light differs for two inertial frames in relative motion, then Maxwell's equations, which are so intimately linked to the speed of light c, would change form for these two inertial frames. Thus we appear to have a situation in which the laws of mechanics are invariant under a Galilean transformation, but the laws of electromagnetism are not.

Albert Einstein (see Fig. 26.7 and accompanying biography) found it hard to accept this conclusion because it seemed to destroy the symmetry he expected to find in the laws of physics. Before we consider Einstein's great contributions, however, it will be helpful to discuss further the aether and the attempts to detect its presence.

The Luminiferous Aether

The word *aether* was first used by Aristotle to describe the substance of the heavens and celestial bodies. Nineteenth-century physicists postulated it as the medium for carrying light waves and also gravitational, electric, and magnetic forces through space. In case more than one aether actually existed, the name *luminiferous aether*, or light-carrying aether, specified that particular aether which transmitted *light*.

This aether had odd, and seemingly contradictory, properties. It had to be a very low density gas, since the heavenly bodies moved through it without being slowed down. Once Maxwell and Hertz had shown that electromagnetic waves were transverse waves, however, it was necessary to postulate that the aether acted like a very elastic solid, since only such a solid could transmit transverse waves at so high a speed (elastic objects such as ball bearings are very hard). The aether therefore had to have simultaneously the properties of both a very tenuous gas and a very hard solid. Physicists were not happy to accept an aether with such strange properties, but they felt they had no other choice.

The aether provided an ideal frame of reference, since it was thought to be at rest, so that all motions could be determined with respect to it. In nineteenth-century physics, therefore, the aether came to be just another name for "absolute space," a concept as dear to classical physicists as that of "absolute time."

It occurred to many physicists that the existence and properties of the aether might be clarified by measuring the speed of light in several inertial systems. If the aether theory were correct, then different values of the speed of light would be obtained for inertial systems moving through the aether with different speeds and in different directions. It was thought that the speed of the earth in its orbit around the sun, which is about $10^{-4}c$, could be used for such a test. Any measurable effects turn out, however, to depend on u^2/c^2, where u is the speed of one reference frame with respect to the other. For the earth's motion around the sun we then have $u^2/c^2 \simeq 10^{-8}$. Such a small effect is hard to measure, but A. A. Michelson (Fig. 26.8) thought he could do it by using one of his interferometers to convert velocity differences into measurable shifts in interference fringes.

26.3 The Michelson-Morley Experiment

In 1887 Michelson and his colleague E. W. Morley (1838-1923) set out to explore the mysterious aether. If an aether existed, the measured speed of light should change with the speed of the observer through the aether. The experiment Michelson devised is analogous to timing two trips by a rower in a flowing stream,

Albert Einstein (1879–1955)

FIGURE 26.7

The most revered scientist of the twentieth century, Albert Einstein, began his spectacular scientific career slowly. The first time around he failed the general knowledge section of his entrance examination at the Polytechnic Institute in Zurich, Switzerland, but was finally admitted in 1895.

There he cut most of his classes to devote himself to studying theoretical physics, which had become much more fascinating to him than pure mathematics. He succeeded in graduating only with the aid of excellent lecture notes taken by a friend. Once out of school he had trouble finding a teaching position, for he did not have a strong academic record and had the added disadvantages of being Jewish and not being a Swiss citizen.

After a few brief jobs as a tutor he was appointed an examiner in the Swiss Patent Office in Bern in 1902. His duties there were not onerous, and he had time for his beloved physics, to which he devoted most of his free time. In 1905 he published five research papers in the best German physics periodical, the *Annalen der Physik*. Three of these contributions were worthy of the Nobel Prize. One was the law of the photoelectric effect (see Sec. 27.3), which was mentioned explicitly in Einstein's 1921 Nobel Prize citation. The second was the theory of Brownian motion, the motion of tiny pollen grains when submerged in a liquid, and the third was his precedent-shattering special theory of relativity.

These contributions caused such a stir that in 1909 Einstein received his first academic appointment, a professorship at the University of Zurich, from which he had received his own Ph.D. degree in 1905. In 1911 he moved to the German university in Prague, Czechoslovakia, and in 1914 was made a professor at the University of Berlin. While in Berlin he published in 1915 his first paper on the general theory of relativity (see Sec. 26.9).

Einstein remained in Berlin until 1933. When conditions under Hitler became too difficult, he accepted a lifetime appointment as the first senior fellow at the newly formed Institute for Advanced Study in Princeton, New Jersey, where he remained until his death in 1955. The 30 years from 1925 to 1955 were devoted in great part to solitary work in developing a unified field theory to tie together the gravitational and electromagnetic fields—an endeavor which never quite succeeded.

In 1901 Einstein married a mathematician, Mileva Marič. After having two children they were divorced, and his wife took the two boys back to Switzerland. (According to one biographer, this was the only time anyone ever saw Einstein cry.) In 1919 he married his cousin Elsa, who took good care of him until her death in 1936.

Einstein once wrote: "The only thing that gives me pleasure apart from my work, my violin and my sailboat, is the appreciation of my fellow workers." He did not need much contact with other people and found the accolades and invitations he constantly received wearisome and distracting.

Dr. J. Robert Oppenheimer (see Fig. 1.3), who was the Director of the Institute for Advanced Study when Einstein was there, once described Einstein as follows: "He was almost wholly without sophistication and wholly without worldliness. . . . There was always with him a wonderful purity at once childlike and profoundly stubborn."

Einstein became a U.S. citizen in 1940, and was the physicist chosen by his colleagues to alert President Roosevelt to the need for developing the nuclear bomb during World War II. Given his deep humanitarian instincts and his longing for peace and a community of nations, he lived to regret the development of the bomb and to worry about what it would mean for the future of humanity.

one trip across the stream and back, the other trip the same distance down the stream and back. It turns out that the times for these two trips are not the same, even though the rower's speed with respect to the water is the same in all directions, and that the time difference increases with the speed of flow of the stream (see Example 26.2).

FIGURE 26.8 Photograph of A. A. Michelson (left front) with A. Einstein (front center) and R. A. Millikan (right front) at the California Institute of Technology, Pasadena, California, in 1931, shortly before Michelson's death. *(Photo courtesy of the California Institute of Technology Archives.)*

Example 26.2

Consider the situation shown in Fig. 26.9. Two rowers start out at the same time from a dock at the edge of a river. One rower rows across the river from A to B and back again to A. The second rower rows along the riverbank from A to C and back again to A. Both rowers row at identical speeds of $v = 2.00$ m/s with respect to the water. There is a constant current in the river directed downstream and equal to 1.00 m/s with respect to the riverbank. The distances AB and AC are each equal to 100 m. Which rower will arrive back at the dock A first, and how much later will the other rower arrive?

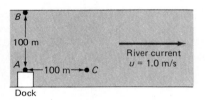

FIGURE 26.9

SOLUTION

The first rower must aim upstream on going from A to B in order that the resultant velocity be in the direction from A to B. The first rower's speed, as shown in Fig. 26.10, will actually be $(v^2 - u^2)^{1/2}$ from A to B. The same is true for the return trip from B to A. Hence the total time to go from A to B and back to A again will be

$$t_1 = \frac{2D}{(v^2 - u^2)^{1/2}} = \frac{2(100 \text{ m})}{[(2.00 \text{ m/s})^2 - (1.00 \text{ m/s})^2]^{1/2}} = 115 \text{ s}$$

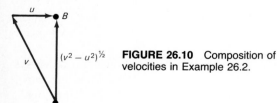

FIGURE 26.10 Composition of velocities in Example 26.2.

For the second rower who rows along the direction of the riverbank, the speed downstream will be $v + u$, where v is the rower's speed with respect to the water and u is the speed of the water with respect to the riverbank. Similarly, the rower's speed in rowing back from C to A will be $v - u$.

The total time for the round-trip will be

$$t_2 = \frac{D}{v + u} + \frac{D}{v - u}$$

$$= \frac{100 \text{ m}}{(2.00 + 1.00) \text{ m/s}} + \frac{100 \text{ m}}{(2.00 - 1.00) \text{ m/s}}$$

$$= 33 \text{ s} + 100 \text{ s} = 133 \text{ s}$$

The first rower, the one who rows across the river, arrives back at point A first, and does so $133 - 115 = \boxed{18 \text{ s}}$ before the second rower.

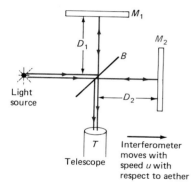

FIGURE 26.11 A Michelson interferometer used to compare the speed of light in different directions in the Michelson-Morley experiment. (For simplicity, the compensating plate shown in Fig. 25.27 is omitted.)

Michelson used his interferometer (Fig. 26.11) with the two mirrors M_1 and M_2 fixed in position and the path lengths D_1 and D_2 the same. Then, if the interferometer were at rest in the aether, the central image of the interference pattern, as observed in the telescope T, would be bright. If, now, the interferometer moves with a speed u through the aether in the initial direction of the light beam, then the times for light to travel the two perpendicular paths will no longer be equal, as should be clear from Example 26.2.

In this case it turns out that the time difference between the light beams traveling over the two perpendicular paths is

$$\Delta t = \frac{Du^2}{c^3}$$ (26.12)

where $D = D_1 = D_2$, since the paths are assumed to be the same length.

This is a very short time, even for the motion of the earth around the sun, where $u/c \simeq 10^{-4}$. With $D = 11$ m, a distance Michelson obtained by using multiple reflections of the beams back and forth in the interferometer, and $c = 3 \times 10^8$ m/s, we have

$$\Delta t = \frac{D}{c}\left(\frac{u}{c}\right)^2 = \frac{11\ m}{3 \times 10^8\ m/s}(10^{-4})^2 \simeq 4 \times 10^{-16}\ s$$

Despite the infinitesimal size of this time lag, Michelson felt confident that he could measure the fringe shift this time lag would produce in his interferometer.

Since light travels with a speed c, the path difference Δl corresponding to the time delay Δt is

$$\Delta l = c\ \Delta t = c\left(\frac{Du^2}{c^3}\right) = \frac{Du^2}{c^2}$$

and the number of complete wavelengths contained in this distance is

$$n = \frac{\Delta l}{\lambda} = \frac{Du^2}{\lambda c^2}$$

For each change in path length by λ, the fringes observed in the telescope change from light to dark and back to light again. Hence the number of fringes moving past the cross hairs in the telescope is $n' = Du^2/\lambda c^2$. Michelson then rotated the interferometer through 90°. This exchanged the two light paths and doubled the expected shift. Using sodium light at 590 nm, the expected fringe shift was then

$$2n' = 2\frac{Du^2}{\lambda c^2} = \frac{2(11\ m)}{590 \times 10^{-9}\ m}(10^{-8}) \simeq 0.37\ fringe$$

The expected fringe shift was only a little over one-third of a fringe! Michelson was confident, however, that he could observe a shift as small as one-hundredth of a fringe. In the actual experiment the shift was found to be *zero* within the limits of error of the experiment. Hence Michelson and Morley were forced to conclude, very reluctantly, that

$$\Delta t \simeq \frac{Du^2}{c^3} = 0$$

which means that $u = 0$! In other words, *the interferometer did not move at all with respect to the aether.*

The only reasonable conclusion was that either the aether was being carried along with the interferometer or *there was no aether*. The former conclusion was in conflict with two pieces of strong experimental evidence, that obtained by Fizeau and confirmed by Michelson for the speed of light in a moving liquid and that of Bradley on the aberration of light from the stars. Both experiments showed clearly that the aether is not dragged along by the earth. Many have repeated the Michelson-Morley experiment in the years since 1887. In every case the results have turned out to be negative.

The conclusion from the Michelson-Morley experiment and its modern equivalents is therefore clear and unambiguous: *There is no detectable aether*. In other words, there is no preferred frame of reference with respect to which light has a special speed c. Rather the measured speed of light in any two inertial systems is *always the same*, c, no matter how they are moving with respect to each other, as shown in Fig. 26.12.

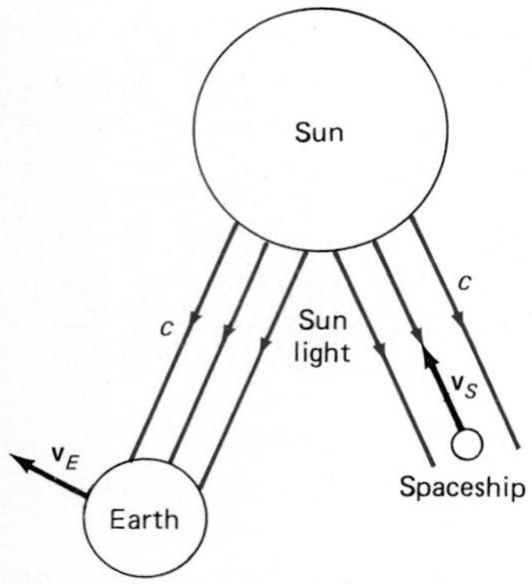

FIGURE 26.12 The speed of light as measured in two different inertial frames. An observer moving with the earth around the sun and an astronaut on a spaceship traveling directly toward the sun will measure exactly the same speed c for the light coming from the sun, even though the observer and the astronaut are moving with respect to each other.

26.4 Einstein's Special Theory of Relativity

The *special theory of relativity* was published by Albert Einstein in 1905. It represented a real break with classical modes of thought in physics and signaled the end of classical nineteenth-century physics. It applies to events in two inertial systems, one of which is moving with constant speed in a straight line with respect to the other.

Einstein's special theory is based on two deceptively simple postulates that have extraordinary consequences both for physics and for our perception of the universe. These *postulates of special relativity* are:

1 *Postulate of the constancy of the speed of light:* The speed of light in a vacuum is a constant c, independent of the inertial system in which it is measured or of the motion of the light source or observer.

2 *Principle of special relativity:* All the laws of physics, including those of mechanics and electromagnetism, have the same form in all inertial frames of reference (i.e., they are *invariant*). All inertial reference frames are therefore equivalent.

Postulate of Constancy of Speed of Light

Einstein's first postulate is completely consistent with the result of the Michelson-Morley experiment: Light travels with the same speed c both in the direction in which the interferometer is moving and at right angles to that direction. One possible conclusion from the Michelson-Morley experiment was that the aether did not exist. Einstein did not go quite that far but considered the aether a meaningless concept. Only motion relative to material bodies has physical significance; motion with respect to an aether that has never been observed makes no physical sense.

The first postulate therefore seems entirely consistent with experimental facts. Whether or not Einstein actually arrived at this postulate on the basis of the Michelson-Morley experiment is unclear.*

Principle of Special Relativity

Einstein's second postulate, the principle of special relativity, is not as closely tied to experiment as is Einstein's first postulate. Rather it is an extension of the principle of Galilean relativity from mechanics to electromagnetism and optics.

Einstein was led to assume the validity of this principle of special relativity on two counts:

1 If this principle were not valid, then different frames of reference would lead to different mathematical expressions for the laws of physics, some more complicated than others. The frame of reference in which the laws had their simplest expression could then be called *at rest*, and all the others *in motion*. This would destroy the equivalence of all inertial frames of reference. This would also lead to a dependence of the form of physical laws on the orientation and motion of physical systems in space. Thus empty space would have different properties in different directions—something which has never been observed.

2 The second argument in favor of Einstein's principle of special relativity is best expressed in Einstein's own words:

> There is a tremendous amount of truth in classical mechanics, since it supplied us with the actual motions of the heavenly bodies with a delicacy of detail little short of wonderful. The principle of relativity must therefore apply with great accuracy in the domain of *mechanics*. But that such a principle of such broad generality should hold with such exactness in one domain of phenomena [i.e., mechanics] and yet be invalid for another [i.e., electromagnetism] is *a priori* not very probable.

Einstein therefore postulated a principle of relativity that would be valid both in mechanics and in electromagnetism and optics, where Galilean relativity was certainly invalid. To do this he had to introduce more general transformation equations than those for a Galilean transformation. He did this by introducing into physics some revolutionary new ideas, in particular, new concepts of *time* and *simultaneity*. The need for these new ideas can be shown by a simple example.

Imagine that you measure the speed of light using a meterstick and a stopwatch. (Of course, this is not actually possible, but any measurement of a velocity requires the measurement of a length and a time.) Then the time required for light to travel the 1.0-m length of your meterstick is 3.3×10^{-9} s, yielding the speed $c = 3.0 \times 10^8$ m/s.

Suppose that a second observer, also with a meterstick and stopwatch, tries to measure the speed of light. This observer, however, is traveling at $0.60c$ in the same direction as the light. According to Galilean relativity, this observer would measure a speed of $0.40c$. In fact, this observer measures the same speed $c = 3.0 \times 10^8$ m/s that you obtained. This is to be expected from the negative result of the Michelson-Morley experiment and from Einstein's first postulate.

---

*Much ink has been spilled by writers trying to trace the exact historical relationship between Einstein's theory and the Michelson-Morley experiment. Even Einstein himself was not particularly consistent in discussing this relationship. Perhaps the most convincing statement is the one by Einstein quoted at the beginning of this chapter, a statement made by Einstein in a letter to Bernard Jaffe and quoted in Jaffe's book, *Michelson and the Speed of Light*, Doubleday, Garden City, N.Y., 1960, pp. 100–101.

But how is it possible for two observers, one moving with respect to the other at $0.60c$, to measure the *same* speed for light? The answer must be that the relative motion produces changes in the space intervals and the time intervals in the two frames of reference which preserve the constancy of the speed of light. The two observers measure *different* space intervals and *different* time intervals but the *same* speed of light. We now introduce the transformation equations needed to achieve this surprising result.

26.5 The Lorentz Transformation Equations

The more general transformation equations used by Einstein are called the *Lorentz transformation equations.*

It is possible to derive the Lorentz equations using simple algebra. We look for a set of equations similar to those for a Galilean transformation but which satisfy the two postulates of special relativity. We do not attempt this derivation here but merely state the result and show that the Lorentz equations do, indeed, satisfy the two postulates of special relativity, and reduce to the Galilean transformation equations when u/c is small.

We first write down the equations for the Galilean and Lorentz transformations for two reference frames S_1 and S_2, where S_2 moves along the x axis with a constant speed u with respect to S_1, as in Fig. 26.13. We have:

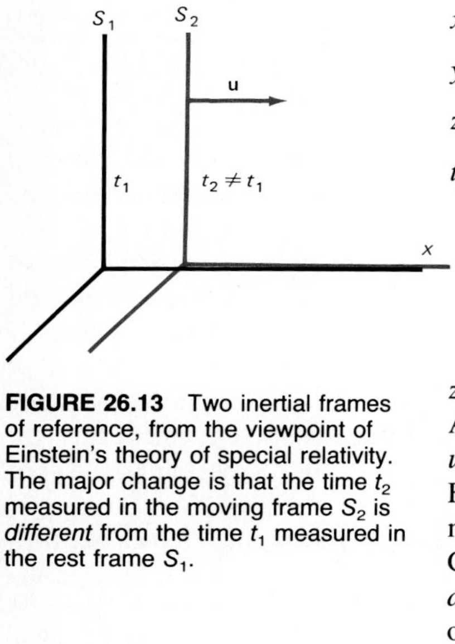

Galilean transformation:

$$x_2 = x_1 - ut_1 \qquad \text{(26.1)}$$

$$y_2 = y_1 \qquad \text{(26.2)}$$

$$z_2 = z_1 \qquad \text{(26.3)}$$

$$t_2 = t_1 \qquad \text{(26.4)}$$

Lorentz transformation:

$$x_2 = \gamma(x_1 - ut_1) \qquad \text{(26.13)}$$

$$y_2 = y_1 \qquad \text{(26.14)}$$

$$z_2 = z_1 \qquad \text{(26.15)}$$

$$t_2 = \gamma\left(t_1 - \frac{ux_1}{c^2}\right) \qquad \text{(26.16)}$$

$$\gamma = \left(1 - \frac{u^2}{c^2}\right)^{-1/2} \qquad \text{(26.17)}$$

FIGURE 26.13 Two inertial frames of reference, from the viewpoint of Einstein's theory of special relativity. The major change is that the time t_2 measured in the moving frame S_2 is *different* from the time t_1 measured in the rest frame S_1.

Since S_2 is moving along the x axis, we would expect no change in the y and z coordinates in the relativistic case, just as there was none in the Galilean case. An important change is the factor γ in the expression for x_2, where γ depends on u/c, the ratio of the relative speed of the coordinate systems to the speed of light. For $u/c \ll 1$, the Lorentz transformation for x_2 reduces to the Galilean transformation. The biggest change in the Lorentz transformation (compared with the Galilean transformation) is that *the times are not identical in the two frames S_1 and S_2*: t_2 depends not only on t_1 but also on the position x_1 at which an event occurs, and on the ratio u/c. This time transformation is essential if the two postulates of relativity are to be satisfied simultaneously. For u/c small, $\gamma = 1$ and Eq. (26.16) reduces to $t_2 = t_1$, as in classical physics.

Lorentz Transformation for a Light Pulse

We want to show that Einstein's two postulates are consistent if the Lorentz transformation equations are used to transform from one reference frame to another. Consider Fig. 26.13 and imagine that at the instant $t_1 = t_2 = 0$ we produce a bright spark at the common origin of the two frames of reference. As viewed by an observer in frame 1, the resulting light pulse travels out in all directions with the speed c. At any instant t_1, the light pulse from the spark forms a spherical wave front of radius $r_1 = ct_1$. This wave front must satisfy the equation

FIGURE 26.14 A plot of $\gamma = 1/\sqrt{1 - u^2/c^2}$ against u/c. As $u/c \rightarrow 1$, $\gamma \rightarrow \infty$.

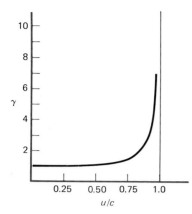

FIGURE 26.15 Hendrik Lorentz (1853–1928), professor of theoretical physics at the University of Leiden. He is best known for his theory of the electron and for the Lorentz transformation equations. He shared the 1902 Nobel Prize in physics with his student P. Zeeman for their work on the effect of magnetic fields on the radiation emitted by atoms. *(Courtesy of AIP Niels Bohr Library.)*

$$r_1^2 = x_1^2 + y_1^2 + z_1^2 = c^2 t_1^2 \quad \text{or} \quad x_1^2 + y_1^2 + z_1^2 - c^2 t_1^2 = 0 \qquad \textbf{(26.18)}$$

Suppose, now, that an observer at rest in a reference frame S_2, which is moving with constant speed u along the positive x axis with respect to S_1, observes the same wave front. The way it appears in this reference frame can be found by transforming from frame S_1 to S_2, using the Lorentz equations (26.13) to (26.17). When we do this, we obtain (the detailed algebra is left to Prob. 26.25)

$$x_2^2 + y_2^2 + z_2^2 - c^2 t_2^2 = 0 \qquad \textbf{(26.19)}$$

Since Eqs. (26.18) and (26.19) are identical in form, we conclude that Eq. (26.18) is invariant under a Lorentz transformation, and so Einstein's second postulate is satisfied in this case. *Note the paradox here:* Both observers see spherical wave fronts, even though one observer is moving in a straight line with respect to the other.

Also, since the speed of the light pulse is c in Eq. (26.18), which is valid in frame S_1, we conclude from Eq. (26.19) that the light pulse also travels with a speed c in frame S_2, even though S_2 is moving with a speed u with respect to S_1. Since this is true for all inertial frames of reference, it follows that the speed of light is always c, no matter the inertial system in which it is measured. This is Einstein's first postulate.

For this case, then, the Lorentz transformation equations do indeed allow both postulates of special relativity to be satisfied. This turns out to be true in every case.

Equations (26.13) to (26.17) are therefore the desired equations that must replace the Galilean transformation equations according to Einstein's special theory of relativity. In the limit of u small compared with c, the Lorentz equations reduce to the Galilean equations, as we would expect.

A plot of the factor $\gamma = (1 - u^2/c^2)^{-1/2}$ against the ratio u/c is shown in Fig. 26.14. Note that, for values of u less than $c/2$, γ has a value very close to 1. (For $u = c/2$, γ is equal to 1.125.) As u increases above the value $c/2$, however, γ grows very rapidly, as can be seen in the figure.

The French mathematician Henri Poincaré (1854–1912) first gave the name *Lorentz equations* to Eqs. (26.13) to (26.17) because the Dutch physicist Hendrik Lorentz (see Fig. 26.15) had proposed them in his work on the classical theory of electrons, which preceded Einstein's work. Since Einstein first put the derivation of the Lorentz equations on a sound basis and first saw their true significance, he deserves full credit for the special theory of relativity.

26.6 Consequences of the Lorentz Transformation Equations

The Lorentz equations lead to some very important consequences, the most surprising of which is that space and time intervals are no longer the absolutes they were in classical physics but vary with the motion of the observer measuring them. This variation is necessary if the speed of light is to remain constant.

Length Contraction

Let us consider a rod of length L_0 as measured in an inertial reference frame S_2 in which the rod is at rest, as in Fig. 26.16. Suppose, now, that reference frame S_2 moves with constant speed u in a straight line along the positive x axis with respect to another inertial frame S_1. What is the length of the rod as measured in S_1? This can be found by a simple application of the first Lorentz equation, Eq. (26.13). In the frame in which the rod is at rest, S_2, the ends of the rod are designated by x_2

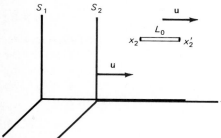

FIGURE 26.16 Lorentz contraction of a moving rod in the direction of its motion. Diagram used to derive the contraction equation $L = \sqrt{1 - u^2/c^2}\, L_0$.

and x_2' so that $L_0 = x_2' - x_2$. These coordinates are then related to the coordinates in the S_1 frame by the equations

$$x_2' = \gamma(x_1' - ut_1') \qquad \text{and} \qquad x_2 = \gamma(x_1 - ut_1)$$

To obtain the length of the rod in S_1, the observer in S_1 must measure the two ends at the same time so that $t_1 = t_1'$, and so

$$x_2' - x_2 = \gamma(x_1' - x_1) = L_0$$

or $\qquad x_1' - x_1 = \dfrac{x_2' - x_2}{\gamma} = \left(1 - \dfrac{u^2}{c^2}\right)^{1/2} L_0$

Now, $x_1' - x_1$ is the length of the rod L as measured in the frame with respect to which the rod is moving with a speed u, and so we have

$$\boxed{L = \left(1 - \frac{u^2}{c^2}\right)^{1/2} L_0} \qquad \qquad \textbf{(26.20)}$$

Here L_0 is called the *proper length* of the rod, or its length in a rest frame. L is the length *as measured in a frame with respect to which the rod is in motion at a speed u*. Thus we find that all length measurements are relative and depend on the motion of the object being measured with respect to the observer.

The rod is *contracted* in the direction of motion (L is smaller than L_0) as measured by an observer with respect to whom the rod is moving. This is called the *Lorentz-FitzGerald contraction* after the two men who had proposed (before Einstein) that such a contraction could explain the negative result of the Michelson-Morley experiment: Hendrik Lorentz (see Fig. 26.15) and the Irish physicist George Francis FitzGerald (1851–1901).

Example 26.3

An automobile 3.0 m long is moving at a speed of 25 m/s (55 mi/h). How much is the car contracted in length as measured by an observer at rest on the road over which the car moves?

SOLUTION

The observer sees the car contracted to a length

$$L = \left(1 - \frac{u^2}{c^2}\right)^{1/2} L_0 = \left[1 - \frac{(25 \text{ m/s})^2}{(3.0 \times 10^8 \text{ m/s})^2}\right]^{1/2} (3.0 \text{ m})$$

$$= [1 - (6.9 \times 10^{-15})]^{1/2}(3.0 \text{ m})$$

For x very much less than 1, the binomial expansion leads to $(1 - x)^{1/2} \simeq 1 - x/2$ (see Appendix 1.D). Hence

$$L = [1 - (3.5 \times 10^{-15})](3.0 \text{ m})$$

The contraction is therefore

$$(3.5 \times 10^{-15})(3.0 \text{ m}) \simeq \boxed{10^{-14} \text{ m}}$$

This shows how small, and therefore unobservable, the Lorentz contraction is for normal speeds in everyday life.

Example 26.4

A "flying saucer" moves past a meterstick with a speed relative to the meterstick of one-half the speed of light. What length do the creatures in the flying saucer measure for the meterstick?

SOLUTION

The flying saucer creatures see the meterstick contracted to a length given by

$$L = \left(1 - \frac{u^2}{c^2}\right)^{1/2} L_0$$

where $u/c = \frac{1}{2}$ and $u^2/c^2 = \frac{1}{4}$. Therefore

$$L = \left(1 - \frac{1}{4}\right)^{1/2} (1.0 \text{ m}) = \left(\frac{3}{4}\right)^{1/2} (1.0 \text{ m}) = 0.87 \text{ m} = \boxed{87 \text{ cm}}$$

In this case, unlike the previous example, the effect would be clearly observable because the relative velocity of the two inertial systems is so large.

Time Dilation

We now consider two events, the explosion of two firecrackers, at exactly the same place x_1 in a rest frame S_1, but at different times t_1 and t_1', as in Fig. 26.17. Then the time interval between t_1' and t_1 in the rest frame is

$$\Delta t_0 = t_1' - t_1$$

This time interval between two events that occur at the same place in the rest frame is called the *proper time interval*. Then, as seen in the moving frame S_2, the times for the two events are, from Eq. (26.16),

$$t_2' = \gamma\left(t_1' - \frac{ux_1'}{c^2}\right) \quad \text{and} \quad t_2 = \gamma\left(t_1 - \frac{ux_1}{c^2}\right)$$

But the firecrackers go off at exactly the same place in S_1 so that $x_1' = x_1$, and so

$$t_2' - t_2 = \gamma(t_1' - t_1)$$

If we designate the time interval as measured in the moving frame by $\Delta t = t_2' - t_2$, we have

$$\Delta t = t_2' - t_2 = \gamma(t_1' - t_1) = \gamma \Delta t_0$$

or

$$\boxed{\Delta t = \frac{\Delta t_0}{(1 - u^2/c^2)^{1/2}}} \tag{26.21}$$

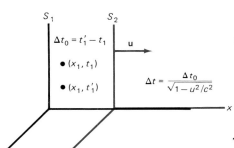

FIGURE 26.17 Time dilation. The time interval Δt_0 is expanded, or dilated, in the moving frame to $\Delta t = \Delta t_0/\sqrt{1 - u^2/c^2}$.

Thus the time interval measured in the moving frame is expanded, or *dilated*, and therefore we speak of a *time dilation*, meaning that time seems to be slowed down. Thus, if $u = 0.98c$ and we consider a clock at rest in S_1, then, as observed in S_2, the time interval corresponding to Δt_0 in S_1 is

$$\Delta t = \frac{\Delta t_0}{(1 - 0.96)^{1/2}} = \frac{\Delta t_0}{0.20} = 5.0 \Delta t_0$$

We would therefore see the second hand on a clock in the moving frame take 5 min to sweep through 1 rev requiring 1 min on a clock at rest. From another point of view, at $v = 0.98c$ the moving clock would only move 12 s while the clock in the rest frame moves 60 s. The moving clock runs very slow.

Example 26.5

Space travel from the earth to outside the sun's planetary system is difficult because the time to reach distant stars would exceed the lifetimes of the astronauts. One solution would be to have the astronauts travel so fast that the time for their trip would be greatly reduced in their reference frame because of time di-lation. If a star is 100 light-years away, i.e., if it would take 100 years, as measured on earth, to reach the star (traveling at the speed of light), how long would the trip take from the as-tronauts' point of view if they traveled at a speed of $0.998c$?

SOLUTION

Let the time for the trip, as measured on earth, be Δt, and, as measured by the astronauts in their rest frame, be Δt_0, where, from Eq. (26.21),

$$\Delta t = \frac{\Delta t_0}{(1 - u^2/c^2)^{1/2}}$$

Then $\Delta t_0 = \left(1 - \dfrac{u^2}{c^2}\right)^{1/2} \Delta t = (1 - 0.996)^{1/2}\, \Delta t$

$$= (0.004)^{1/2}\, \Delta t = 0.063(100 \text{ years}) = \boxed{6.3 \text{ years}}$$

For the astronauts, therefore, only 6.3 years would have elapsed, and they could hope to survive the trip, if the practical problem of achieving such speeds as $0.998c$ could be solved.

EXERCISE 1   Solve the same problem by calculating the con-traction in the earth-star distance measured by the moving astronauts.

Relativity of Simultaneity

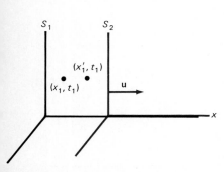

FIGURE 26.18   The relativity of simultaneity. Two events that are simultaneous but occur at different positions in S_1 are not simultaneous in S_2.

A result of the time dilation predicted by the Lorentz equations is that events which are simultaneous in one inertial frame may not be simultaneous in a second inertial frame moving with respect to the first. Suppose we consider two firecrackers going off at exactly the same time t_1 as measured in a reference frame S_1 in which the firecrackers are at rest but at two different positions x_1 and x_1' in that frame, as in Fig. 26.18. Then, as seen in another inertial frame S_2 moving with a speed u along the positive x axis with respect to S_1, the times of these two events are, from Eq. (26.16),

$$t_2 = \gamma\left(t_1 - \frac{ux_1}{c^2}\right) \quad \text{and} \quad t_2' = \gamma\left(t_1' - \frac{ux_1'}{c^2}\right)$$

In this case, t_1 and t'_1, are equal, since the events are simultaneous in S_1. But then we have

$$t_2 - t_2' = \gamma\left(\frac{-ux_1}{c^2} + \frac{ux_1'}{c^2}\right)$$

and so $\qquad \Delta t_2 = t_2 - t_2' = \gamma\dfrac{u}{c^2}(x_1' - x_1)$ $\qquad\qquad$ **(26.22)**

If the firecrackers go off at the same time but at different places in S_1, then, the explosions are *not simultaneous* as measured in S_2. *Simultaneity is therefore a relative concept; events simultaneous in one inertial system may not be simulta-neous in another inertial system.**

---

*This led to the famous bit of doggerel that appeared in *Punch* in 1923 and was a favorite of A. A. Michelson:

There was a young lady named Bright
Whose speed was far faster than light;
She set out one day
In a relative way
And returned home the previous night.

As Einstein has written, "Every reference body has its own particular time. Unless we are told the reference body to which the statement of time refers, there is no meaning in a statement of the time of an event."

To illustrate this by an example, consider a lightning flash that at the same instant strikes two points A and B along a railroad track, as in Fig. 26.19. If Jack is at rest at point P exactly halfway between A and B, then the lightning flashes from A and B will reach his eyes at exactly the same instant. He will therefore say that the two lightning flashes are simultaneous. But if his sister Jill is on a fast train traveling from B to A at a speed u and is at exactly the same point P when the lightning strikes, as judged by Jack, then the flashes will not appear simultaneous to Jill. Rather the light from A will reach Jill before the light from B because she is moving toward A and away from B, and the speed of light is finite. The effect will be small, since c is large, but the principle is still clear. Events simultaneous when measured in a rest frame are not simultaneous in another inertial system in motion with respect to that rest frame, if these events occur at different positions in space. This is the physical content of Eq. (26.22).

FIGURE 26.19 A lightning flash striking two points A and B at the same instant as observed by Jack in a rest frame. To an observer (Jill) in a moving frame these events are no longer simultaneous, but the lightning appears to strike A before B.

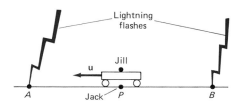

Example 26.6

Two events occur at the same time in an inertial frame S_1 and are separated by a distance of 10 km along the x axis. What is the time difference between the two events as measured by an observer in a frame S_2 moving with constant speed $0.50c$ along the positive x axis?

SOLUTION

The two events are no longer simultaneous, as seen in S_2, because of time dilation. The time delay is given by Eq. (26.22) as

$$\Delta t_2 = \gamma \frac{u}{c^2} (x_1' - x_1)$$

where here $\gamma = (1 - \tfrac{1}{4})^{-1/2} = 1.15$, and so

$$\Delta t_2 = 1.15 \left(\frac{0.50c}{c^2} \right) (10 \times 10^3 \text{ m}) = \boxed{19 \ \mu s}$$

Two events that appear simultaneous in S_1 therefore differ by a time interval of 19 μs as measured by an observer in S_2.

26.7 Mass, Energy, and Momentum

We have seen that the Lorentz transformation equations predict that a length L_0 measured in a rest frame becomes $L = (1 - u^2/c^2)^{1/2}L_0$ when measured in another inertial frame moving with a velocity u with respect to the rest frame. Now, u cannot be greater than the speed of light c, or L becomes the square root of a negative number. We therefore conclude:

The speed of light in a vacuum, c, is the limiting speed in the universe.

To understand why speeds greater than c are impossible, we return to Eq. (7.2), $\mathbf{F}_{net} = \Delta\mathbf{p}/\Delta t$, which is the most fundamental statement of Newton's second law of motion. Suppose $\mathbf{F}_{net}$ is a constant force applied to a particle in a fixed direction so that the change in momentum $\Delta\mathbf{p}$ of the particle is in the same direction as the force. Then

$$F_{net} = \frac{\Delta p}{\Delta t} = \frac{\Delta(mv)}{\Delta t}$$

If m is constant and the force is applied for a long time, v would increase without limit and eventually become greater than c—in contradiction to Einstein's theory. Einstein therefore concluded that at very high speeds m must vary in such a way as to prevent v from ever attaining the speed of light c. He was able to show that, according to the theory of relativity, the mass m of a particle moving at speed v is*

$$m = \frac{m_0}{\sqrt{1 - v^2/c^2}} \qquad (26.23)$$

where m_0 is the mass of the particle as measured in an inertial frame in which the particle is at rest; m_0 is called the *rest mass* of the particle.

As $v \to c$, $m \to \infty$, and the mass becomes so large that the applied force can increase the velocity very little. The predicted dependence of the mass on speed v is shown in Fig. 26.20. It is similar to the graph of γ against u/c in Fig. 26.14.

FIGURE 26.20 Values of m/m_0 as a function of the ratio v/c as calculated from Eq. (26.23). When $v = 0.87c$, the mass has doubled from m_0 to $m = 2m_0$.

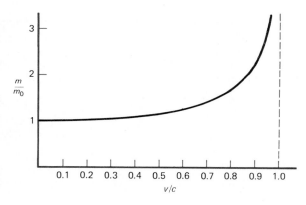

TABLE 26.1 Variation of Mass m of a Particle with Its Speed v (Theoretical Values)

$\dfrac{v}{c}$	$\dfrac{m}{m_0}$
0.10	1.005
0.20	1.021
0.30	1.048
0.40	1.091
0.50	1.155
0.60	1.250
0.70	1.400
0.80	1.667
0.90	2.294
0.95	3.203
0.98	5.025
0.99	7.089
0.995	10.013
0.999	22.366
0.9995	31.627
0.9999	70.712

Some values for m/m_0 as a function of the ratio v/c, as calculated from Eq. (26.23), are given in Table 26.1.

The theory of relativity shows us that our intuitive concept of mass, i.e., the quantity of matter or number of atoms in an object, is inadequate. For particles moving at speeds near that of light the amount of material substance does not change, but their relativistic mass m is very different from their rest mass m_0.

In our treatment of relativity we assume Eq. (26.23), which is confirmed every day in physics laboratories, and use it to obtain results of great theoretical and practical importance.

*Throughout this section we use v for the speed of a particle in an inertial frame to distinguish it from u, which we continue to use for the speed of one inertial frame with respect to another.

Einstein reasoned that, since the work a force does on a particle to increase its speed goes partially into increasing its mass [according to Eq. (26.23)], then in some way *mass and energy must be equivalent*. To show this we can take the expression for the mass m in Eq. (26.23) and expand the denominator by the binomial expansion (see Appendix 1.D). We obtain

$$m = \frac{m_0}{\sqrt{1 - v^2/c^2}} = m_0 \left(1 - \frac{v^2}{c^2} \right)^{-1/2}$$

and, on using the binomial expansion, we obtain

$$m = m_0 \left[1 - \frac{1}{2} \left(\frac{-v^2}{c^2} \right) + \cdots \right] = m_0 \left[1 + \frac{v^2}{2c^2} + \cdots \right]$$

or, neglecting higher-order terms, $m - m_0 = m_0 v^2/2c^2$, and so

$$mc^2 - m_0 c^2 = \tfrac{1}{2} m_0 v^2 = KE$$

or $$\boxed{mc^2 = m_0 c^2 + KE}$$ **(26.24)**

Dimensionally all the terms in this equation are *energies*. The term $m_0 c^2$, which contains the rest mass, must then be the energy possessed by the particle when it is at rest, a quantity we call the *rest energy* of a particle of mass m_0. This makes the term on the left the *total energy* of the particle, the sum of its *rest energy* and its kinetic energy. The total energy of a relativistic particle is therefore given by the famous Einstein equation

$$\boxed{E = mc^2}$$ **(26.25)**

and its kinetic energy is

$$KE = E - E_0 = mc^2 - m_0 c^2 = m_0 c^2 (\gamma - 1)$$ **(26.26)**

with $\gamma = (1 - v^2/c^2)^{-1/2}$. [Equation (26.26) turns out to be valid independent of the approximation used in the above derivation.]

For example, the rest energy of an electron is

$$m_0 c^2 = (9.11 \times 10^{-31} \text{ kg})(3.00 \times 10^8 \text{ m/s})^2$$

$$= \frac{8.19 \times 10^{-14} \text{ J}}{1.60 \times 10^{-19} \text{ J/eV}} = 0.511 \text{ MeV}$$

and the kinetic energy of an electron (in MeV) is, from Eq. (26.26),

$$KE = (0.511 \text{ MeV})(\gamma - 1)$$

For an electron at rest, $\gamma = 1$ and $KE = 0$, as expected. For γ small, the KE calculated from Eq. (26.26) is very close to the classical $KE = \tfrac{1}{2} mv^2$.

Mass-Energy Equivalence

When an object is at rest, it has a minimum (rest) energy $E_0 = m_0 c^2$. This is an enormous amount of energy, since $c^2 = 9 \times 10^{16} \text{ m}^2/\text{s}^2$, and so 1 kg of mass is the equivalent of 9×10^{16} J of energy. Because nuclear forces are so strong, mass changes in nuclear reactions are an appreciable fraction (about 0.1 percent) of the total mass of the involved nuclei. Since these mass changes are large, great amounts of energy can be released by nuclear reactions. It is such mass changes which produce the tremendous energies released by the sun and the stars and which fuel nuclear reactors and nuclear bombs.

Because of the possibility of converting mass into energy, the principles of conservation of mass and conservation of energy can be combined into one fundamental principle, that of the conservation of mass-energy, as indicated schematically in Fig. 26.21:

> ***Principle of conservation of mass-energy:*** The total energy (rest mass energy plus all other forms of energy) in a closed physical system is a constant.

The Einstein equation $E = mc^2$ has, therefore, both immense practical importance and great theoretical significance for our understanding of the fundamental laws of physics.

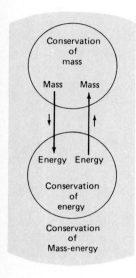

FIGURE 26.21 The principle of the conservation of mass-energy. Mass and energy need not be conserved individually, but the total amount of mass-energy in any closed system is constant. The two principles of conservation of mass and conservation of energy are absorbed into the one principle of conservation of mass-energy.

Example 26.7

(a) What is the kinetic energy of an electron of rest mass 9.11×10^{-31} kg that has a relativistic mass of 2.0×10^{-30} kg when in motion? Give your answer in both joules and electronvolts. **(b)** What is the speed of such an electron?

SOLUTION

(a) The KE of an electron, according to the theory of relativity [Eq. (26.26)], is

$$KE = mc^2 - m_0c^2$$

$$= (2.0 \times 10^{-30} \text{ kg} - 9.11 \times 10^{-31} \text{ kg})(3.0 \times 10^8 \text{ m/s})^2$$

$$= (1.1 \times 10^{-30} \text{ kg})(9.0 \times 10^{16} \text{ m}^2/\text{s}^2)$$

$$= \boxed{9.9 \times 10^{-14} \text{ J}}$$

In electronvolts this is

$$KE = (9.9 \times 10^{-14} \text{ J})(6.24 \times 10^{18} \text{ eV/J}) = \boxed{0.62 \text{ MeV}}$$

Note that the KE is slightly greater than the rest energy of the electron (0.51 MeV) and larger than the energies involved in molecular bonding by about five orders of magnitude.

Note also that in relativity theory the kinetic energy is *not* given by $\frac{1}{2}mv^2$, with m the relativistic mass—a mistake that students frequently make. Rather, the correct relativistic KE is given by Eq. (26.26).

(b) From Eq. (26.26),

$$\gamma - 1 = \frac{KE}{m_0c^2} = \frac{0.62 \text{ MeV}}{0.51 \text{ MeV}} = 1.22$$

and so $\gamma = 2.22 = (1 - v^2/c^2)^{-1/2}$, which leads to $v^2/c^2 = 0.80$, $v/c = 0.89$, and

$$v = 0.89c = \boxed{2.7 \times 10^8 \text{ m/s}}$$

This is a speed only about 10 percent lower than the speed of light in a vacuum.

Example 26.8

A proton has a total energy three times its rest energy. **(a)** What is its rest energy in electronvolts? **(b)** What is its KE in electronvolts? **(c)** What is its speed?

SOLUTION

(a) The rest energies of the proton and the electron are in the same ratio as their rest masses. Therefore

$$\frac{E_{0p}}{E_{0e}} = \frac{m_{0p}}{m_{0e}} = \frac{1.673 \times 10^{-27} \text{ kg}}{9.110 \times 10^{-31} \text{ kg}} = 1836$$

and so the proton rest energy is $E_{0p} = 1836 \ (0.511 \text{ MeV}) =$ $\boxed{938 \text{ MeV}}$.

(b) According to relativity theory, the KE may be obtained by substracting the rest energy from the total energy:

$$KE = 3m_0c^2 - m_0c^2 = 2m_0c^2 = 2(938 \text{ MeV}) = \boxed{1876 \text{ MeV}}$$

(c) From Eq. (26.26),

$$\gamma - 1 = \frac{KE}{m_0c^2} = \frac{1876 \text{ MeV}}{938 \text{ MeV}} = 2$$

and so $\gamma = 3 = (1 - v^2/c^2)^{-1/2}$, which leads to $1 - v^2/c^2 = \frac{1}{9}$ and $v/c = \sqrt{\frac{8}{9}} = 0.943$, or

$$v = (0.943)(3.00 \times 10^8 \text{ m/s}) = \boxed{2.83 \times 10^8 \text{ m/s}}$$

Relativistic Momentum

In relativity momentum plays the same important role it did in classical physics, and the principle of conservation of linear momentum remains valid. Relativistic momentum is defined in the same way as in classical mechanics, except that the mass varies with speed, as given by Eq. (26.23):

$$p = mv = \frac{m_0 v}{(1 - v^2/c^2)^{1/2}} \tag{26.27}$$

If $v \ll c$, then $p = m_0 v$, the classical result.

We can square Eq. (26.23) and rewrite it as follows:

$$m_0^2 = m^2 \left(1 - \frac{v^2}{c^2} \right) = m^2 - \frac{(mv)^2}{c^2} = m^2 - \frac{p^2}{c^2}$$

from which $\qquad m^2 = m_0^2 + \dfrac{p^2}{c^2} \tag{26.28}$

This equation enables us to calculate the extra mass a particle acquires when its momentum increases.

To find the relationship between the momentum and the energy, we multiply Eq. (26.28) by c^4 and write mc^2 as E, the total energy. We obtain

$$\boxed{E^2 = m_0^2 c^4 + c^2 p^2 = E_0^2 + c^2 p^2} \tag{26.29}$$

This important equation relates the total energy of a particle to its rest energy and momentum.

If the particle has no mass, then $m_0 = 0$ and $E = cp$, from which

$$\boxed{p = \frac{E}{c}} \tag{26.30}$$

This formula relates the momentum of a *massless* particle to its energy. It applies to particles of light (called *photons*) and to other particles like neutrinos that are believed to be massless. Maxwell's electromagnetic theory leads to the same relationship between energy and momentum as Eq. (26.30) for all forms of radiation in the electromagnetic spectrum, including visible light.

Example 26.9

What is the relativistic momentum of the electron in Example 26.7?

SOLUTION

The relativistic momentum of a particle is given by $\mathbf{p} = m\mathbf{v}$, where m is the relativistic mass and $\mathbf{v}$ is the velocity of the particle. In this case, the momentum is in the direction of $\mathbf{v}$ and has the magnitude $p = mv$. The mass is given in the statement of Example 26.7 as 2.0×10^{-30} kg, and the speed found in part (b) was 2.7×10^8 m/s. Therefore

$$p = (2.0 \times 10^{-30} \text{ kg})(2.7 \times 10^8 \text{ m/s})$$

$$= \boxed{5.4 \times 10^{-22} \text{ kg·m/s}}$$

Example 26.10

(a) What is the kinetic energy of a proton accelerated from rest through a potential difference of 100 MV? (b) What is its momentum? (c) Its speed?

SOLUTION

(a) By the principle of conservation of energy, all the energy supplied by the electric field must go to increase the total energy of the proton from $E_0 = m_0c^2$ to $E = mc^2$. Since, from Eq. (26.26), $mc^2 - m_0c^2$ is equal to the final KE of the proton, we have

$$\text{KE} = 100 \text{ MeV} = \boxed{1.60 \times 10^{-11} \text{ J}}$$

(b) The relativistic momentum is given by Eq. (26.29):

$$p^2 = \frac{E^2 - E_0^2}{c^2}$$

The rest energy of a proton is 938 MeV, as found in Example 26.8. Its total energy is therefore $E = E_0 + \text{KE} = 938 \text{ MeV} + 100 \text{ MeV} = 1038 \text{ MeV}$, and so

$$p = \frac{\sqrt{(1038)^2 - (938)^2} \text{ MeV}}{c} = \frac{445 \text{ MeV}}{c}$$

$$= \frac{445 \text{ MeV}(1.60 \times 10^{-13} \text{ J/MeV})}{3.00 \times 10^8 \text{ m/s}}$$

$$= \boxed{2.37 \times 10^{-19} \text{ kg·m/s}}$$

(c) Using the method of part (b) of Example 26.7, we find that $v = \boxed{1.29 \times 10^8 \text{ m/s}}$. (Work this out for yourself).

EXERCISE 2 Calculate the value of $\frac{1}{2}mv^2$ for the 100-MeV proton considered in this example, and show that it does *not* yield the correct value for the relativistic kinetic energy, which is 100 MeV, as found above.

26.8 Experimental Tests of the Special Theory of Relativity

In physics an elegant theory, such as that of special relativity, is a thing of beauty, but it does not endure unless it agrees with experiment. Strong experimental evidence exists for all the predictions of special relativity—the Lorentz contraction, time dilation, the change of mass with velocity, the conversion of mass into energy.

Length Contraction

There is no *direct* evidence for the length contraction predicted by special relativity because the only practical way to measure a length directly is to use a meterstick or a measuring tape in the reference frame of the object being measured. But if the object is in motion, its reference frame is also in motion and the meterstick will be shortened with respect to its length when at rest. Thus we would be attempting to measure a length change with a meterstick that shortens in exactly the same way as the object being measured, as shown in Fig. 26.22. Indirect confirmation of the Lorentz contraction comes, of course, from the negative result of the Michelson-Morley experiment. Additional confirmation comes from experiments on the lifetimes of mu mesons.

Time Dilation

A striking example of time dilation comes from the observed lifetimes of μ (mu) *mesons* (or *muons*) produced as secondary radiations from cosmic rays in the earth's upper atmosphere. Muons at rest have lifetimes of only about 2 μs on the average

FIGURE 26.22 Using a meterstick to measure the Lorentz contraction of a moving rod. The rod and the meterstick are moving at the same speed and hence contract in the same way.

FIGURE 26.23 Dependence of average lifetimes of muons on their speed. (The symbol u is used here for the muon speed because it is assumed that the muon is at rest in an inertial frame S_2 moving with speed u with respect to a rest frame S_1.)

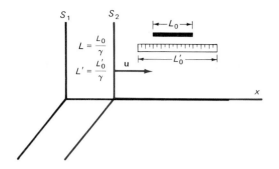

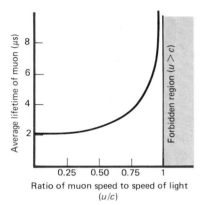

FIGURE 26.24 A commercial cesium atomic clock occupying its own seat on an airplane. This atomic clock was flown around the world on a jetliner to test Einstein's theory of relativity. *(Photo courtesy of U.S. National Institute of Science and Technology.)*

before they decay into other particles. Since the muons are created several thousand meters up in the atmosphere, we would expect few of them to reach earth. A muon moving with a speed of $0.995c$, for example, would only travel about 600 m in $2 \ \mu s$. Few muons would therefore survive and reach the ground if they started from a height of 2000 m. The muons are, however, moving with respect to an observer on the ground, and to that observer their lifetime is given by Eq. (26.21) as

$$\Delta T = \frac{\Delta T_0}{(1 - u^2/c^2)^{1/2}}$$

where ΔT_0 is their lifetime when at rest. In this case we have

$$\Delta T = \frac{2 \times 10^{-6} \ \text{s}}{[1 - (0.995)^2]^{1/2}} = 20 \ \mu s$$

Hence to an observer on earth the lifetime of the muons is increased by a factor of about 10, and we would expect them to survive through a distance 10×600 m, or about 6000 m. Thus the great majority of the muons would survive until they reached the ground. By measuring the number that reach the ground, the effective lifetime of the muons can be determined. The predicted variation of muon lifetime with speed is shown in Fig. 26.23.

Many experiments have been done on the lifetime of muons, and they all confirm the relativistic time dilation. In 1963 Frisch and Smith did an experiment in which they measured the number of muons incident at the top of Mount Washington, New Hampshire, at an altitude of 1911 m. Then they measured the number of muons incident on the ground at Cambridge, Massachusetts, where the muon flux from cosmic rays differs little from that at the foot of Mount Washington. They had predicted a time dilation of $\gamma = 8.4 \pm 2$; that is, the lifetimes were expected to be increased by a factor of about 8.4. The value obtained from their somewhat rough data was 8.8 ± 0.8, which agrees quite well within the limits of error of the experiment.

Another striking confirmation of time dilation has come from the work of two physicists, J. C. Hafele and R. E. Keating, who in 1971 carried four cesium atomic clocks around the world on commercial jet planes (see Fig. 26.24) and measured the time differences between their moving clocks at the end of the trip and a reference atomic clock that remained fixed at the U.S. Naval Observatory. They found excellent agreement with the predictions of relativity, as Table 26.2 shows. These results provide unambiguous proof of the reality of time dilation. A more recent version of this experiment has verified the predictions of relativity with an accuracy of 70 parts in a million.

TABLE 26.2 Measured and Predicted Values for Time Differences between Stationary and Moving Clocks*

	Time differences (10^{-9} s)	
	Eastward trip around earth	**Westward trip around earth**
Measured values (mean for four clocks)	-59 ± 10	273 ± 7
Prediction of relativity theory (including a gravitational effect)	-40 ± 23	275 ± 21

*From J. C. Hafele and Richard E. Keating, "Around-the-World Atomic Clocks," *Science*, vol. 177, July 14, 1972, pp. 166–170.

Dependence of Mass on Particle Speed

Figure 26.25 shows the data obtained by three different experimentalists on the change in the observed mass of high-speed electrons as their speed changed. Because of their light masses electrons are easily accelerated to speeds close to that of light. These electrons are deflected in a system of electric and magnetic fields, as in the experiments of J. J. Thomson discussed in Sec. 19.4. The electron's deflection in a magnetic field determines its momentum, and its deflection in an electric field determines its kinetic energy. In this way both the speed v and the ratio e/m can be determined for electrons of a variety of speeds.

It is clear that the data agree very well with the predictions of theory shown in Fig. 26.20. This is all the more impressive since the source of the electrons and the deflection techniques used differed considerably in the three experiments.

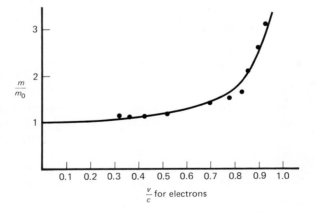

FIGURE 26.25 Experimental data on the dependence of the observed mass of high-speed electrons on their speed. The dots are the experimental points. Compare this experimental curve with the theoretical predictions shown in Fig. 26.20.

Mass-Energy Conversion

The conversion of mass into energy in conformity with Einstein's theory occurs every day in phenomena such as pair production, in which electron-positron pairs are produced from pure energy in the form of gamma rays, and pair annihilation (which is the reverse of pair production), and also in fission and fusion reactions. We will present the relevant experimental evidence for these cases in Chap. 29.

26.9 The General Theory of Relativity

Whereas the special theory of relativity finds its most important application at the atomic level, the general theory of relativity is most useful in what we usually call "the universe in the large"—the universe of planets, stars, and galaxies. The

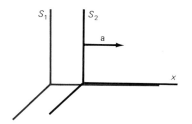

FIGURE 26.26 Reference frames in general relativity. Frame S_2 moves with an *acceleration* **a** along the positive x axis with respect to rest frame S_1.

general theory, first proposed by Einstein in 1915, applies to two reference frames one of which is in accelerated motion with respect to the other, as in Fig. 26.26. It is basically a theory of gravitation and is much more complicated and highly mathematical than is special relativity. All we can do here is indicate briefly the scope and status of this theory, which many physicists regard as the most elegant and perfect of all theories in physics.

There are two basic postulates of general relativity:

1   *Principle of invariance:* All the laws of physics have the same form in all frames of reference, no matter how they are moving with respect to one another.

This principle is merely an extension of the principle of special relativity to include not merely inertial frames of reference but *all* frames of reference, including *accelerated frames*.

2   *Principle of equivalence:* A uniformly accelerated reference frame is completely equivalent to a homogeneous gravitational field.

This principle is valid even in classical mechanics if the assumption is made that gravitational mass and inertial mass are identical. By *gravitational mass* we mean the mass m that occurs in Newton's law of universal gravitation, $f_G = Gmm_E/r^2$. Similarly, by *inertial mass* we mean the mass that occurs in Newton's second law, $f = ma$, as shown in Fig. 26.27. Measurements by the American physicist Robert H. Dicke (born 1916) have shown that the inertial and gravitational masses of any object are equal to better than 1 part in 10^{11}.

FIGURE 26.27 Distinction between (*a*) gravitational and (*b*) inertial mass. According to the general theory of relativity, the inertial and gravitational masses of any object are the same.

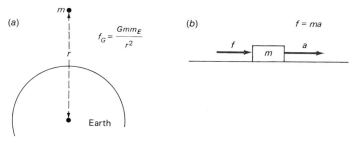

As long as gravitational and inertial masses are the same, no experiment in mechanics performed within an elevator or other closed compartment in space can distinguish whether the elevator is accelerating upward with an acceleration **a** or the objects in the elevator are being subjected to a gravitational force downward that produces a downward acceleration **g** = −**a**, as shown in Fig. 26.28. Einstein extended this principle of equivalence to all fields of physics, and assumed that there is no experiment of any kind which can distinguish uniformly accelerated motion from the presence of a gravitational field.

General relativity theory leads to three predictions that can be tested experimentally. The first is that a light beam should be deflected in a gravitational field, as shown in Fig. 26.29. Newton's theory of gravitation predicted that light skirting the sun should be bent by about 0.85 s of arc from its original direction. In 1915 Einstein predicted that the bending should, according to general relativity, be twice Newton's value, about 1.7 s of arc. Measurements made during the solar eclipse

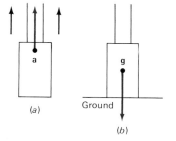

FIGURE 26.28 Equivalence of an accelerated frame of reference and a gravitational field. (*a*) The acceleration of an elevator upward with an acceleration **a** is completely equivalent to (*b*) a gravitational field producing an acceleration **g** downward inside the elevator, where **g** = −**a**.

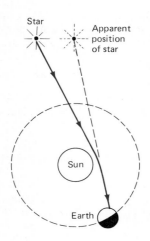

of May 29, 1919, proved conclusively that Einstein was right and Newton wrong. Overnight Einstein became famous in every corner of the world. Even though few could understand his general theory of relativity, all seemed to grasp the genius of the man and the importance of his work.

A second prediction of general relativity is that light emitted from regions in which strong gravitational fields exist should have its wavelength shifted toward the red. Third, general relativity predicts an added precession of the orbit of the planet Mercury over and above that caused by the gravitational attraction of the other planets. In recent years highly precise measurements have been made of these effects, and they all confirm the predictions of Einstein's theory. In addition, the development over the past 20 years of new, highly precise experimental techniques has confirmed general relativity in a variety of other ways, so there can no longer be any doubt about its validity.

FIGURE 26.29 Light from a star is deflected slightly by the gravitational field of the sun on its way to the earth. Measurements of this deflection can be made with visible light only at times of a total eclipse. Such measurements are now more conveniently made with radio waves from quasars.

Summary: Important Definitions and Equations

Galilean relativity

Frame of reference: A system in which an event can be specified by three distances measured with respect to the origin of the coordinate system, and by the time elapsed from some chosen initial time.

Inertial frame of reference: A frame of reference in which Newton's first law of motion is valid.

Galilean transformation equations

$$x_2 = x_1 - ut_1 \qquad z_2 = z_1$$
$$y_2 = y_1 \qquad t_2 = t_1$$

(26.1–26.4)

Invariance: If the equations expressing a physical law retain the same mathematical form after a coordinate transformation as they had before it, the equations are said to be *invariant* under the transformation.

Principle of Galilean relativity: All the laws of classical mechanics are invariant under a Galilean transformation. Hence all inertial systems are equivalent in classical mechanics.

Aether: A medium whose existence was postulated by classical physicists to carry light waves and gravitational, electric, and magnetic forces through space.

Michelson-Morley experiment: An experiment using a Michelson interferometer to detect the postulated influence of the aether on the speed of light.

Einstein's special theory of relativity

First postulate—Constancy of speed of light: The speed of light in a vacuum is a constant c, independent of the inertial system in which it is measured or of the motion of the light source or observer.

Second postulate—Principle of special relativity: All the laws of physics, including those of mechanics and electromagnetism, have the same form (i.e., they are invariant) in all inertial frames of reference.

Lorentz transformation equations (for S_2 moving along positive x axis with speed u with respect to S_1):

$$x_2 = \gamma(x_1 - ut_1) \qquad t_2 = \gamma\left(t_1 - \frac{ux_1}{c^2}\right)$$
$$y_2 = y_1 \qquad \gamma = \left(1 - \frac{u^2}{c^2}\right)^{-1/2}$$
$$z_2 = z_1$$

(26.13–26.17)

Consequences of the special theory of relativity

1 Length (Lorentz-FitzGerald) contraction:

$$L = (1 - u^2/c^2)^{1/2}L_0$$

(26.20)

where L_0 is the proper length

2 Time dilation: $\Delta t = \dfrac{\Delta t_0}{(1 - u^2/c^2)^{1/2}}$

(26.21)

where Δt_0 is the proper time interval.

3 Relativity of simultaneity: Events simultaneous in one inertial system may not be simultaneous in another inertial system.

4 Limiting speed: The speed of light in a vacuum, c, is the limiting speed in the universe.

5 Variation of mass with speed v:

$$m = \gamma m_0 = \frac{m_0}{(1 - v^2/c^2)^{1/2}}$$

(26.23)

where m_0 is the rest mass.

6 Equivalent of mass and energy: $E = mc^2$ (26.25)

   $KE = mc^2 - m_0c^2 = m_0c^2(\gamma - 1)$ (26.26)

7 Conservation of mass-energy

8 Relativistic momentum:

$$p = mv = \frac{m_0v}{(1 - v^2/c^2)^{1/2}}$$ (26.27)

$$E^2 = E_0^2 + c^2p^2$$ (26.29)

Einstein's general theory of relativity Applies to reference frames one of which is in accelerated motion with respect to the other.

First postulate—Principle of invariance: All the laws of physics have the same form in all frames of reference, no matter how they are moving with respect to one another.

Second postulate—Principle of equivalence: A uniformly accelerated reference frame is completely equivalent to a homogeneous gravitational field.

Questions

1 Can an experiment be performed in classical mechanics that will reveal whether a frame of reference is *accelerated* with respect to an inertial frame?

2 If you are shooting pool on a pool table below deck on an ocean liner and the ship has to slow down because of ocean fog, will this influence the motion of the balls on the pool table? Explain your answer.

3 In the phenomenon of time dilation, what exactly is "dilated"? Would "time retardation" be a more accurate term?

4 Two observers, one at rest in S_1 and the other at rest in S_2, which moves with a speed u along the positive x axis with respect to S_1, each carries a meterstick oriented along the x axis. Each observer finds, upon performing a measurement, that the other's meterstick is shorter than his or her meterstick. Who is right? Explain your answer.

5 Can a particle move through a medium at a speed greater than the speed of light in that medium (e.g., glass)? Explain your answer.

6 In optics it is well known that in certain crystals light rays travel at different speeds in different directions through the crystal. Why could not the same thing be true of light traveling through the aether?

7 Why is it easy to accelerate electrons to speeds approaching that of light, whereas it is much more difficult to do this for protons or alpha particles?

8 In e/m experiments, variations of e/m with the speed of the particle are normally taken as an indication that the mass m has changed with v. Why is it not possible that the charge e varies with v? (Consider what might happen in this case to heavy atoms, whose inner electrons travel at speeds close to that of light.)

9 Discuss changes that would occur in our everyday lives if the speed of light were 30 m/s instead of 3.0×10^8 m/s.

10 (*a*) If an astronaut on a spaceship is able to reach a speed of $0.6c$ away from the earth, would the mass, height, or heartbeat of the astronaut change as measured with instruments inside the spaceship?

(*b*) Suppose observers on earth were able to measure these properties through remote-sensing devices. Would they find any changes in the astronaut's mass, height, or heartbeat?

11 (*a*) Why are the effects of the special theory of relativity not normally observable in daily life?

(*b*) Suppose we lived in a universe in which everything on earth was reduced in size by a factor of 10^3. Would the effects of special relativity be any more evident than they are in our present world?

Multiple-Choice and Simple Exercises

26.1 Choose the *incorrect* statement from the five statements below.

(*a*) The velocity of light is independent of the velocity of the observer and the source.

(*b*) The velocity of light is constant irrespective of the medium through which it is traveling.

(*c*) The velocity of light in a vacuum is larger than the velocity of light in glass.

(*d*) The velocity of light in a vacuum does not depend on the wavelength of light.

(*e*) The velocity of light in a vacuum is the same as the velocity of x-rays in a vacuum.

26.2 A meterstick is at rest in an inertial frame S_2. If S_2 moves with a speed of $0.50c$ with respect to a rest frame S_1, the length of the meterstick as measured in frame S_1 is, according to classical physics:

(*a*) 1.15 m   (*b*) 1.00 m   (*c*) 0.87 m
(*d*) 0.75 m   (*e*) 1.33 m

26.3 A meterstick is at rest in an inertial frame S_2. If S_2 moves with a speed of $0.50c$ with respect to a rest frame S_1, the length of the meterstick as measured in frame S_1 is, according to the special theory of relativity:

(*a*) 1.15 m   (*b*) 1.00 m   (*c*) 0.87 m
(*d*) 0.75 m   (*e*) 1.33 m

26.4 Two firecrackers explode at the same place in a rest frame S_1 with a time separation of 10 s in that frame. The time between explosions, as measured in a frame moving with a speed $0.80c$ with respect to the rest frame is, according to classical physics:

(*a*) 6.0 s   (*b*) 16.7 s   (*c*) 8.0 s
(*d*) 12.5 s   (*e*) 10 s

26.5 Two firecrackers explode at the same place in a rest frame S_1 with a time separation of 10 s in that frame. The time between explosions, as measured in a frame moving with a speed of $0.80c$ with respect to the rest frame is, according to the special theory of relativity:

(a) 6.0 s   (b) 16.7 s   (c) 8.0 s
(d) 12.5 s   (e) 10 s

26.6 Two firecrackers explode at the same time, but one at a place with coordinate $x_1 = 10$ m in a rest frame S_1, and the other at a place with coordinate $x_1' = 110$ m in S_1. In a frame S_2 that moves with a speed $0.60c$ with respect to S_1, the time difference between the two explosions, according to the special theory of relativity is:

(a) 0.33 μs (b) 20 μs (c) 16 μs
(d) 0   (e) 0.25 μs

26.7 A particle of rest mass m_0 moves at a speed of $0.60c$ relative to an observer. Its observed mass is:

(a) m_0 (b) $1.25m_0$ (c) $0.80m_0$
(d) $1.33m_0$ (e) $0.60m_0$

26.8 A particle of rest mass m_0 moves at a speed $0.80c$ with respect to a rest frame. Its momentum, as measured in the rest frame is:

(a) $0.80m_0c$ (b) m_0c (c) $1.25m_0c$
(d) $1.33m_0c$ (e) $1.67m_0c$

26.9 To produce a change in mass of a particle by one order of magnitude, its speed as measured in a rest frame must be:

(a) $10c$ (b) c (c) $0.995c$ (d) $0.90c$
(e) $0.990c$

26.10 The energy associated with the rest mass of the proton is:

(a) 4.6×10^{-2} J   (b) 1.5×10^{-10} J
(c) 4.9×10^{-19} J   (d) 1.7×10^{-27} J
(e) 1.5×10^{-10} eV

26.11 An electron and a positron (which has the same mass as an electron) collide and annihilate each other. The energy produced in the form of gamma rays by the conversion of mass into energy is:

(a) 5.0×10^{-5} J   (b) 1.6×10^{-13} J
(c) 5.4×10^{-22} J   (d) 1.8×10^{-30} J
(e) 1.6×10^{-13} eV

26.12 The length of a 1.0-cm division on a meterstick is measured by an observer in a reference frame moving at a speed of $0.90c$ with respect to the meterstick. How long does the 1.0-cm division appear to the observer?

26.13 Calculate the Lorentz contraction of a tennis ball traveling at 100 mi/h (45 m/s) with respect to an observer at rest.

26.14 A physicist measures the lifetime of a neutron in a laboratory on earth and finds that it is 17 min. Another physicist has developed a new rocket that can travel at a speed of 1.5×10^8 m/s. If the second physicist moves at this speed with respect to the first physicist's laboratory, what lifetime does the second physicist measure for the neutron?

26.15 A particle called a pi meson, or pion, has a lifetime at rest on the earth of only about 1.8×10^{-8} s on the average.

(a) How long would such a pion live when approaching the earth at a speed of $0.90c$, as measured by an observer on earth?

(b) How long does an observer on the pion think the distance to earth is?

26.16 The star closest to our solar system is Alpha Centauri, which is 4.5 light-years away.

(a) How long would it take according to a clock on earth for a spaceship to make a round-trip to Alpha Centauri if the spaceship could travel at $0.995c$?

(b) How long would the trip take according to clocks on the spaceship?

26.17 Electrons in TV tubes are traveling at speeds of about $0.30c$ when they strike the fluorescent screen and produce the TV picture. What is their mass as measured in a frame in which the TV tube is at rest?

26.18 What is the fractional increase in mass of a jet plane flying at Mach 4 (that is, 4 times the speed of sound in air) as measured in a rest frame with respect to which the plane has this speed?

26.19 It takes 80 kcal at 0°C to melt 1 kg of ice to water, as we saw in Chap. 13. By how much does the mass of the water exceed the mass of the ice because of the energy added to melt it?

26.20 (a) How much energy would be obtained if the rest mass of 1.0 kg of coal could be converted entirely into energy?

(b) How long could this much energy provide electricity for a city that requires 10^6 kWh of electric energy per year, if the energy in the coal could be converted into electric energy with 100 percent efficiency?

26.21 What is the kinetic energy of an electron whose mass is five times its rest mass?

Problems†

*26.22** Show that it is possible to synchronize two clocks at different points in space so that they tell the same time. Assume that one clock is at the origin of the coordinate system of an inertial frame and that the second clock is a distance D away from the origin in the same inertial frame of reference.

*26.23** The Lorentz equations may be used to derive similar equations for transforming the components of the velocity v_1 in a rest frame S_1 to the corresponding velocity v_2 in a frame S_2. Show that the resulting equations are

$$v_{x2} = \frac{v_{x1} - u}{1 - uv_{x1}/c^2}$$

$$v_{y2} = \frac{v_{y1}/\gamma}{1 - uv_{x1}/c^2}$$

$$v_{z2} = \frac{v_{z1}/\gamma}{1 - uv_{x1}/c^2}$$

where frame S_2 is moving with a velocity u along the positive

---

†Additional problems on the conversion of mass into energy will be found at the end of Chap. 29.

x axis with respect to S_1. (*Hint:* Write $v_{x2} = \Delta x_2/\Delta t_2$, and use the Lorentz transformation equations for Δx_2 and Δt_2.)

26.24 A spaceship is moving away from a galaxy at $0.90c$, while the galaxy is moving away from the earth at a speed $0.90c$. What is the speed of the spaceship relative to the earth? (*Hint:* See Prob. 26.23.)

26.25 Consider a reference frame S_2 moving along the positive x axis at a speed u with respect to a rest frame S_1. Use the results of Prob. 26.23 to show that a light ray traveling along the x axis at a speed c in S_1 has the same speed c when measured in the moving frame S_2.

26.26 Rocket A travels to the right and rocket B to the left with speeds $v_A = 0.85c$ and $v_B = 0.70c$ with respect to an inertial frame at rest. What is the speed of rocket A as measured by an observer on rocket B? (See Prob. 26.23.)

*26.27 Prove that the Lorentz transformation equations are able to account for the negative results of the Michelson-Morley experiment by predicting a contraction of the interferometer in the direction of its motion.

26.28 A spaceship has a length L_0 in a rest frame. When in motion with a speed u with respect to that rest frame, its measured length is $L_0/2$.

(*a*) What is the speed of the spaceship?

(*b*) What is the dilation of time scale on the spaceship?

26.29 An airplane flies at a speed of 500 m/s. How long must it fly before its clock loses 1.0 s because of time dilation, as measured by a ground observer?

26.30 The radius of our galaxy is about 3×10^{20} m.

(*a*) Can a person, in principle, travel from the center of the galaxy to its edge in a normal lifetime by traveling very rapidly? Explain how he or she can accomplish this.

(*b*) What must the person's speed be to make the trip in 25 years?

26.31 Certain bacteria are observed to double in number every 20 days in a laboratory on earth. Two of these bacteria are placed on a spaceship and move in an orbit around the earth at a speed of $0.98c$ for a time of 1000 days as measured on earth. How many bacteria would be on the spaceship when it returns to earth?

26.32 Show that the same result can be obtained for Prob. 26.31 by using the Lorentz-FitzGerald contraction instead of time dilation to solve the problem.

26.33 An astronaut spends 20 days in orbit around the earth, traveling at an average speed of 7.8×10^3 m/s. Show that, when the astronaut splashes down, he or she is about 600 μs younger than would have been the case if there had been no space voyage.

26.34 A muon is 9000 m above the earth and moving directly toward the earth at a speed of $0.998c$. The muon's lifetime when at rest is 2.0 μs.

(*a*) What is the lifetime of the muon as observed by a physicist on the earth?

(*b*) What distance does the muon travel in this lifetime? Will it reach the earth before it decays?

(*c*) If an observer could travel along with the muon, what lifetime would she observe for the muon?

(*d*) What would be the distance from the muon's initial position to earth, as measured by the observer moving with the muon?

26.35 Two events occur at the same time in an inertial frame S_1 and are separated by a distance of 5.0 km along the x axis. What is the time difference between these two events as measured in frame S_2 moving with constant velocity u along x, if an observer in S_2 measures the spatial separation in S_1 to be 3.0 km?

*26.36 A cyclotron is designed to accelerate protons to kinetic energies of 400 MeV so that their relativistic mass becomes much larger than their rest mass. If the cyclotron's magnetic field remains fixed as the radius of the proton's circular orbit increases (see p. 488), the applied frequency to the dees must be varied to compensate for this mass change.

(*a*) Must the applied frequency be increased or decreased?

(*b*) What is the ratio of the final frequency applied to the initial frequency?

*26.37 In inertial system S_1, which is at rest, an event occurs at $x_1 = 50$ m, $y_1 = 0$, $z_1 = 0$. Exactly 10^{-6} s later a second event occurs at $x_1' = 550$ m, $y_1' = 0$, $z_1' = 0$.

(*a*) Does there exist another inertial frame S_2, moving with speed less than c along the x axis, such that the two events appear simultaneous as seen from S_2?

(*b*) If so, what is the magnitude and direction of the velocity of S_2 with respect to S_1?

(*c*) What is the spatial separation of x_1' and x_1 as seen in S_2?

(*d*) Answer parts (*a*), (*b*), and (*c*) for a situation where $x_1' = 150$ m.

26.38 A cube with all sides equal to a rests at the origin of an inertial frame S_1. The mass of the cube is m_0 in this rest frame. An observer moves along the positive x axis with a speed u with respect to S_1.

(*a*) What is the observed volume of the cube?

(*b*) What is the observed mass of the cube?

(*c*) What is the observed density of the cube?

(*d*) Show that, when u is much smaller than c, the density reduces to the usual value.

26.39 (*a*) A tennis ball is moving with a speed of 45 m/s with respect to the ground. What is the fractional change in its mass as measured by an observer in the tennis stands?

(*b*) An electron is moving at 98 percent of the speed of light with respect to a cathode-ray tube in which it is accelerated. What is the fractional change in its mass?

26.40 What is the speed of an electron whose kinetic energy is equal to its rest energy?

26.41 An electron is observed to move at 1.8×10^8 m/s with respect to a rest frame.

(*a*) What is the relativistic mass of the electron?

(*b*) What is its kinetic energy?

26.42 (*a*) Determine the voltage needed to accelerate electrons to a speed of $0.8c$.

(*b*) Determine the voltage needed to accelerate protons to $0.8c$.

26.43 An electron is accelerated from rest through a potential difference of 10^6 V and thus acquires a kinetic energy of 1.0 MeV.

(*a*) Find its speed and its mass when it has acquired this energy.

(*b*) What is the electron's total energy?

26.44 How much energy is required to accelerate an electron from rest to a speed of 0.99 that of light?

26.45 At the Stanford Linear Accelerator Center (SLAC), electrons are accelerated to such high speeds that their mass becomes 10^4 times their rest mass.

(a) What is the speed of the electrons when this happens?

(b) If the SLAC accelerating tube is 3.0 km long, how long is it in the reference frame of the moving electrons?

26.46 (a) If classical theory were valid, what accelerating voltage would be required to produce electrons with the speed of light?

(b) What is, according to relativistic mechanics, the actual velocity of the electrons for this potential?

(c) By what fraction has their mass increased?

*****26.47** The required resonance between the applied frequency and the circular motion of the charged particles in a cyclotron (Sec. 19.3) is destroyed when the energy of the electrons approaches 50 keV. Calculate the fractional change in the mass of the electron when this happens.

26.48 Two 0.10-kg masses with equal and opposite speeds of 50 m/s approach each other along a straight line, collide, and then stick together. What is the additional rest mass of the system after the collision?

26.49 The sun radiates energy at the rate of 3.9×10^{26} W, and this energy is produced by the conversion of the sun's mass into energy by fusion processes according to the equation $E = mc^2$.

(a) Find the rate at which the sun must convert mass into energy to produce its radiated energy.

(b) If the sun's total mass is 2.0×10^{30} kg, how long will the sun last if it continues to lose mass at this rate?

26.50 (a) How much energy is released in the explosion of a uranium bomb containing 5.0 kg of fissionable material, if we assume that $1/1000$ of the rest mass of the uranium is converted into energy?

(b) How much TNT would be required to produce the same amount of energy, if TNT liberates 820 kcal of energy per mole and each mole has a mass of 0.227 kg?

(c) What is the relative energy released in the two cases, for the same initial amounts of uranium and TNT?

26.51 An electron is accelerated from rest through a potential difference of 1.5 million volts and thereby acquires 1.5 MeV of energy. Find (a) the electron's final momentum; (b) the electron's final speed.

26.52 Find the total energy, kinetic energy, and momentum of a proton moving at a speed of $0.75c$.

26.53 An electron is accelerated through 10^6 V.

(a) What is its kinetic energy?

(b) What is its relativistic momentum?

*****26.54** Laboratory measurement shows that the speed of a charged particle is $0.71c$. The particle is then directed into a magnetic field $B = 1.0$ T, in which it moves in a circle of radius 3.1 m. Find the mass of the particle and identify it.

*****26.55** A proton describes an arc of 100 cm radius in a uniform magnetic field of 2.0 T. Find (a) the proton's momentum; (b) the proton's kinetic energy; (c) the ratio v/c, where v is the speed of the proton; (d) the factor $\gamma = (1 - v^2/c^2)^{-1/2}$.

Additional Readings

Bernstein, Jeremy: *Einstein*, Penguin, New York, 1976. Probably the best brief account of Einstein's personality and ideas.

Einstein, Albert: *Relativity: The Special and General Theory*, Crown, New York, 1961. A nontechnical book by the man who best understood relativity because he created it.

Epstein, Lewis C.: *Thinking Relativity*, Insight Press, San Francisco, 1985. A good popular book on relativity.

French, A. P.: *Special Relativity*, Norton, New York, 1968. An introductory textbook that contains a very good discussion of the historical origins of special relativity.

Gamow, George: *Mr. Tompkins in Paperback*, Cambridge University Press, New York, 1965. This collection of Gamow's popular books on physics includes the story of Mr. Tompkins, who lives in a world in which the speed of light is 10 mi/h and relativistic effects become quite important.

Geroch, Robert: *General Relativity from A to B*, University of Chicago Press, Chicago, 1978. Lectures on general relativity intended for nonscience undergraduates at the University of Chicago.

Hoffman, Banesh, and Helen Dukas: *Albert Einstein, Creator and Rebel*, Viking, New York, 1972. An account of Einstein's life and work by one of his collaborators and his former secretary.

Pais, Abraham: "How Einstein Got the Nobel Prize," *American Scientist*, vol. 70, no. 4, July-August 1982, pp. 358–365. The inside story on why Einstein received his Nobel Prize for the theory of the photoelectric effect and not for relativity theory. Pais, who was a colleague of Einstein's at Princeton, has also written a more technical book on Einstein: *'Subtle Is the Lord': The Science and the Life of Albert Einstein*, Oxford University Press, New York, 1982.

Shankland, Robert S.: "The Michelson-Morley Experiment," *Scientific American*, vol. 211, no. 5, November 1964, pp. 107–114. A discussion of the history and significance of the Michelson-Morley experiment. Shankland has another interesting article: "Conversations with Albert Einstein," *American Journal of Physics*, vol. 31, 1963, pp. 47–57.

Schwinger, Julian: *Einstein's Legacy*, Freeman, San Francisco, 1985. The story of one of the twentieth century's greatest achievements, as told by a Nobel Prize physicist.

Stachel, John: "Einstein and Ether-Drift Experiments," *Physics Today*, vol. 40, no. 5, May 1987, pp. 45–47. Recently discovered letters of Einstein shed new light on the origins of special relativity.

Swenson, Loyd S., Jr.: *Etherial Aether*. University of Texas Press, Austin, Texas, 1972. A detailed history of the Michelson-Morley and other aether experiments from 1880 to 1930.

Will, Clifford M.: *Was Einstein Right?* Basic Books, New York, 1986. An account of recent experimental confirmations of Einstein's theory of general relativity.

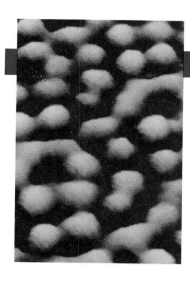

CHAPTER 27

The Experimental Basis of Quantum Mechanics

Almost half a century has elapsed since Max Planck's discovery of the quantum of action, a time sufficiently long to estimate its importance for science and, more generally, for the development of human thought. There is no doubt that it was an event of the first order, comparable with the scientific revolutions brought about by Galileo and Newton, Faraday and Maxwell.

Max Born (1882–1970)

Modern physics began at the turn of the century with two sharp breaks from nineteenth-century classical physics. The first was Einstein's special theory of relativity. The second and in some ways the more dramatic break is associated with the name of Max Planck. The quantum theory that Planck developed not only revolutionized physics but, as pointed out in the quotation above, produced one of the greatest intellectual upheavals in human history.

If the real world is governed by the laws of relativity and quantum theory, how then do we manage to survive in our daily lives while paying little attention to these theories? For particles on the earth, relativistic effects only become important for speeds approaching c, the speed of light. So too the quantum theory only becomes important for particles the size of atoms or smaller, as Fig. 27.1 attempts to show schematically. To describe particles of the size of golf balls or rockets moving at speeds low compared with 3×10^8 m/s, neither relativity nor quantum theory is needed. Relativity and quantum theory are still of crucial importance, however, insofar as they control the atomic and molecular structure of golf balls, rockets, and all other objects in our universe. These two theories are also needed to understand the properties of the universe on a cosmic scale and the universe's evolution to its present condition from the "big bang" believed to have been its origin.

FIGURE 27.1 A rough schematic outline of the domains of validity of various physical theories, depending on the size and speed of the particles involved.

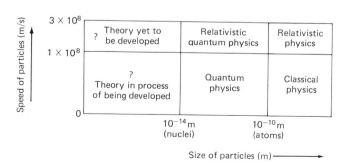

27.1 Particle Properties of Light; the Photon

In Chap. 24 we presented strong evidence that light consists of transverse electromagnetic waves traveling through space at a speed c. The phenomena of diffraction, interference, and polarization can all be explained not merely qualitatively but also quantitatively by such an assumption. But there are other experimental results—blackbody radiation, the photoelectric effect, the Compton effect—that cannot be explained on the basis of a classical wave model of light. These results also cannot be explained by the older corpuscular theories of light of the type proposed by Sir Isaac Newton. A new model for light is needed, and this was provided at the turn of the century by Planck, Einstein, and others who based their model on the concept of a *photon*.

Photon: A massless bundle (or ''quantum'') of electromagnetic energy that behaves like a particle.

As we will see in the next four sections, to explain the available experimental evidence it is necessary to postulate the existence of photons. The energy of a photon is directly related to its frequency by the equation $E = h\nu$, where E is the energy of the photon in joules, ν* is the frequency of the photon in hertz, and h is a constant that must have the value 6.63×10^{-34} J·s to give agreement with experiment. Only waves or other periodic motions can have a frequency, which is, after all, the number of *vibrations* per second. Inherent in the concept of photon, therefore, are both particle properties (momentum) and wave properties (frequency). This is our first exposure to the dual wave-particle nature of reality, which will turn up again and again in this and subsequent chapters.

We now consider the experimental evidence that led Planck and Einstein to the equation $E = h\nu$, which, despite its simplicity, is *the most important equation in all of modern physics*.

27.2 Blackbody Radiation

The first experimental evidence for the existence of photons came from the study of the radiation from hot objects, and particularly of *blackbody radiation*. Kirchhoff's invention of the prism spectrometer in 1859 led to a great interest in analyzing the visible light emitted by hot gases and solids, and to the accumulation of much important spectroscopic data. It was the attempt by Planck (who was a student of Kirchhoff's) to understand the data on the continuous radiation from heated solids that led to his theory of blackbody radiation and ultimately to the quantum, or photon, theory of light.

Blackbody: An ideal body which absorbs all the electromagnetic radiation that falls on it.

FIGURE 27.2 A blackbody cavity. Any radiation entering the hole is trapped inside by successive reflections from the walls.

A piece of black velvet is a good approximation to a blackbody for visible radiation. Kirchhoff pointed out that, for all forms of radiation, an even better approximation to a blackbody is a spherical cavity in a block of metal with one tiny entrance hole in it, as shown in Fig. 27.2. Any radiation entering the hole is trapped by multiple reflections inside the cavity and cannot escape. Such a cavity is black to all visible and near-visible electromagnetic radiation, since it reflects none of the radiation incident on the entrance hole but absorbs it all.

---

*As is customary in modern physics, for *frequency* we use the Greek letter ν (*nu*) instead of f in this and all subsequent chapters.

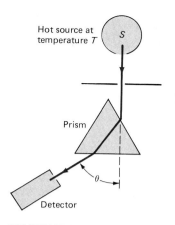

FIGURE 27.3 Experimental setup to measure the intensity of blackbody radiation as a function of temperature. The wavelength can be determined from the measured value of the angle θ.

FIGURE 27.4 Blackbody radiation curves. Each graph shows the intensity of the radiation emitted by a blackbody as a function of wavelength for one particular temperature. The four curves shown here are for the four different temperatures indicated.

Just as such a blackbody cavity is a nearly perfect absorber, so when heated it is also the most efficient emitter of radiation, since its emissivity e in Stefan's radiation law [Eq. (14.3)] is equal to unity. When heated, the molecules in the walls of the cavity vibrate and emit radiation into the interior of the cavity. Experiments on the radiation from blackbodies can thus be performed by putting such a cavity into a furnace and studying the intensity of the radiation emitted from the hole in the cavity as a function of furnace temperature and wavelength, as in Fig. 27.3. In the years between 1860 and 1900 physicists used spectrometers to obtain experimental curves for blackbody radiation and struggled to explain the curves they obtained.

Figure 27.4 shows the intensity of the radiation emitted by a blackbody as a function of wavelength at a series of temperatures. Note that as the temperature increases, the total amount of radiation emitted increases dramatically, as it must according to the fourth-power dependence on temperature in Stefan's law [Eq. (14.3)]. Also, we see that the peak of the curve moves to shorter and shorter wavelengths (higher frequencies) as the temperature increases, as predicted by the Wien displacement law [Eq. (14.4)], $\lambda_{max}T = w = 2.898 \times 10^{-3}$ m·K.

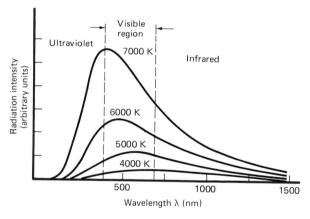

Wavelength λ (nm)

Although both the Stefan radiation law and the Wien displacement law were well known to physicists before 1900, no one had been able to derive either of these equations satisfactorily from the basic properties of radiation. This was accomplished in 1900 by Max Planck (see Fig. 27.5 and accompanying biography).

Planck's Quantum Theory

Planck set out to explain the form of blackbody radiation curves like those of Fig. 27.4 on the basis of first principles. There had been previous attempts to do this, but none had been very successful. One theory, due to the German physicist Wilhelm Wien (1864–1928), agreed well with experiments at short wavelengths but deviated noticeably from the experimental curves at long wavelengths. A second theory, due to the British physicists Lord Rayleigh (1842–1919) and Sir James Jeans (1877–1946), assumed that all wavelengths (and therefore all frequencies) would be radiated with equal probability. Since there are many more high frequencies than low frequencies possible, they concluded that the amount of energy radiated as a function of wavelength would take the form shown by the top colored line in Fig. 27.6. Most of the energy emitted would be in the violet and ultraviolet, in total disagreement with experiment. Physicists spoke about an "ultraviolet ca-

FIGURE 27.5 Max Planck is shown presenting the Planck Medal of the German Physical Society to Albert Einstein on June 28, 1929. On that day Planck received the first Planck medal, and then Einstein received the second from Planck's hands. *(Courtesy of AIP Niels Bohr Library, Fritz Reiche collection.)*

Max Planck (1858–1947)

The father of the quantum theory, Max Planck, was born in Kiel on April 23, 1858, into a well-known German family of lawyers, public servants, and scholars. Throughout his life he preserved a great devotion to his family, his friends, and, when possible, to his fatherland. He studied mathematics at the University in Munich, and then spent a year at the University of Berlin working in physics under Helmholtz and Kirchhoff. He received his doctorate from Munich in 1879 for a dissertation on the second law of thermodynamics. The key to his success in explaining blackbody radiation was his deep knowledge of thermodynamics. His book on this subject, first published in 1897, still influences the way physicists approach thermodynamics today.

In 1887 Planck received an appointment to the physics faculty at the University of Kiel. Two years later his reputation had so spread that he was offered the chair of theoretical physics in Berlin. Here he developed into one of the world's greatest theoretical physicists, aided by his association with Helmholtz, and in later years with Einstein, von Laue, and Born. He remained in Berlin for the rest of his active scientific life, enjoying the stimulating contacts with colleagues in physics, mathematics, chemistry, and philosophy at the university, the culture of the great capital city, and his walks in the forests near Berlin. He was an avid walker and hiker and took yearly mountain-climbing vacations in the Alps.

On October 19, 1900, Planck presented his paper on blackbody radiation to the Berlin Physical Society. The night after the meeting his colleague H. Rubens carried out additional measurements and the next morning informed Planck that his formula fitted all Rubens' data perfectly from very short to very long wavelengths. Thus was born the quantum theory of radiation.

Planck, while modest in public, realized fully the importance of his work. His son Erwin recalled that in 1900 his father took him for a walk through the Grünewald near Berlin, in the course of which his father astounded him by saying: "Today I have made a discovery as important as that of Newton." There is no doubt that Planck was correct in this judgment, although it took time and the work of Einstein, Compton, Bohr, and others to convince the physics establishment of this fact. For his pioneering work in quantum theory Planck received the 1918 Nobel Prize in physics.

In the years before the First World War, Planck and Einstein combined to make Berlin the world's great center for theoretical physics. Not only did these two great men work together, they played chamber music together, Planck at the piano and Einstein with his violin, both finding great joy and relaxation in the music they made. Unfortunately, this somewhat idyllic existence ended with the outbreak of World War I. Three of Planck's four children by his first wife died during the war, two daughters during childbirth and his son Karl in action in France.

The Hitler period in Germany brought increased personal tragedy to Planck. His son, Erwin, the only surviving child of his first marriage, was involved in the July 1944 plot to kill Hitler and was executed by the Nazis. Later Planck's home in Grünewald and his large personal library were destroyed in one of the allied air raids on Berlin. He took shelter in the woods near his home until rescued by American forces in 1945.

After the war Planck moved to Göttingen to live with a grandniece. He continued to write and lecture until 1947, when he died at the age of 89. His grave is marked by a simple rectangular headstone containing only his name and one additional line: $h = 6.62 \times 10^{-27}$ erg·sec. This nonzero value for Planck's constant, which distinguishes modern quantum physics from classical newtonian physics, assures Planck a unique place in the history of science. As Planck himself said in his 1911 address to the German Chemical Society, "the hypothesis of quanta will never vanish from the world."

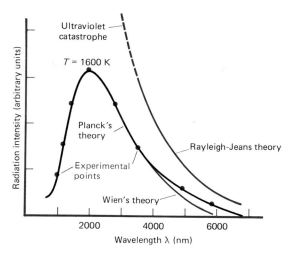

FIGURE 27.6 Comparison of the predictions of the Rayleigh-Jeans theory and of Wien's theory with the experimental curves for blackbody radiation at 1600 K. Note that at this temperature most of the radiation is in the infrared. The experimental points fall right on the curve calculated from Planck's theory [Eq. (27.1)].

tastrophe," since the total energy radiated by a blackbody would be infinite, for the curve in Fig. 27.6 goes to infinity at short wavelengths in the ultraviolet.

Planck looked for one equation that would fit the experimental data at all wavelengths. Since the wavelength distribution of the radiation emitted by the cavity is proportional to that inside the cavity, Planck set out to determine how the density of radiation inside the cavity depended on wavelength. After many unsuccessful attempts, he found, in 1900, the following equation for the dependence of the energy density on the wavelength:

$$U(\lambda) = \frac{8\pi hc}{\lambda^5} \frac{1}{e^{hc/\lambda kT} - 1} \tag{27.1}$$

where k is Boltzmann's constant, T is the absolute temperature, and c is the speed of light. This leaves the one new constant h to be explained. Planck was able to show that this constant had the dimensions of "action," where action is the product of energy and time. Planck's equation fitted the experimental data perfectly over the entire spectrum from the ultraviolet to the infrared if h were assigned the value 6.6×10^{-34} J·s.

To explain the experimental curve in Fig. 27.6, Planck had to understand the physical reason why the radiation intensity predicted from Eq. (27.1) fell off so rapidly at short wavelengths and thus avoided the ultraviolet catastrophe. Planck saw that this could happen only if the energy radiated into the cavity by the walls of the blackbody was not radiated continuously but came in discrete bundles which he called *quanta* (from the Latin *quantum*, meaning "how much"). In other words, energy was not radiated continuously from the walls like water from a hose but rather like bullets from an automatic machine gun. Planck assumed that each quantum had an energy proportional to its frequency. It was therefore much easier for lower-frequency oscillators to radiate than it was for higher-frequency oscillators, since they were more likely at any given temperature to have enough energy to emit a whole quantum of energy. If an oscillator did not have enough energy to emit a whole quantum at the frequency corresponding to its mode of vibration, it could not emit any radiation at all. (The situation is somewhat similar to one in which you need quarters for a parking meter that accepts only quarters. The fact that you may have a pocket full of nickels and dimes does you no good, since you do not have the necessary "quantum," the quarter.) At low temperatures there is only sufficient energy for the low-frequency oscillators to do most of the radiating (in other words, you have found a parking meter that takes nickels and dimes). As the temperature increases, however, the amount of energy available increases as T^4 (Stefan's radiation law), and high-frequency radiation becomes more likely.

On looking at his empirical equation [Eq. (27.1)], Planck saw that the quantity $hc/\lambda kT = h\nu/kT$ in the exponential term must be dimensionless, since all exponents are pure numbers. But we know from our treatment of thermodynamics that kT has the dimensions of energy. Hence $h\nu$ must also have the dimensions of energy, and so Planck for the first time wrote down the famous equation

$$E = h\nu \qquad \text{(27.2)}$$

This equation states the proportionality between the energy E and the frequency ν of a photon, where the proportionality constant is $h = 6.63 \times 10^{-34}$ J·s, a quantity now called *Planck's constant*.

With this value for h, Planck's blackbody equation leads to perfect agreement with experiment, as shown in Fig. 27.6. Planck's equation can also be shown to lead to the Stefan-Boltzmann T^4 law and to the Wien displacement law, and to reduce to the results of Wien and Rayleigh-Jeans in the proper limiting cases. These are additional proofs of its validity.

The real significance of Planck's work dawned on the physicists of the early twentieth century, however, only after Einstein, in another of his famous 1905 papers, carried Planck's ideas a step further and showed that radiation is not merely emitted and absorbed in the form of photons, it also travels through space as photons. Einstein's work, and that of Compton and others, made it clear that physicists must take the quantized nature of energy into account, i.e., that energy comes in discrete bundles of magnitude $E = h\nu$, whenever they deal with the interaction of matter and radiation. This quantum approach is the dividing line separating modern physics from classical physics and is basically due to the profound insight of one man, Max Planck.

Example 27.1

A 1.5-kg mass vibrates at the end of a spring of force constant $k = 25$ N/m with an amplitude of 2.0 cm. **(a)** What is the total mechanical energy of this system? **(b)** What is the smallest quantized energy change that can occur in the system, according to Planck's theory? **(c)** If the system changes energy by one quantum, what is its fractional change in energy? **(d)** What can you conclude about the classical or quantum nature of the system?

SOLUTION

(a) From Eq. (6.9) the total mechanical energy of the system is

$$\mathscr{W} = \tfrac{1}{2}kx_{max}^2 = \tfrac{1}{2}(25 \text{ N/m})(0.020 \text{ m})^2 = \boxed{5.0 \times 10^{-3} \text{ J}}$$

(b) The natural frequency of the oscillator is, from Eq. (11.8),

$$\nu = \frac{1}{2\pi}\left(\frac{k}{m}\right)^{1/2} = \frac{1}{2\pi}\left(\frac{25 \text{ N/m}}{1.5 \text{ kg}}\right)^{1/2} = 0.65 \text{ Hz}$$

and so the smallest energy change due to the emission or absorption of a photon is

$$E = h\nu = (6.6 \times 10^{-34} \text{ J·s})\left(\frac{0.65}{\text{s}}\right) = \boxed{4.3 \times 10^{-34} \text{ J}}$$

(c) Fractional change in energy $= \dfrac{E}{\mathscr{W}} = \dfrac{4.3 \times 10^{-34} \text{ J}}{5.0 \times 10^{-3} \text{ J}}$

$$= \boxed{8.6 \times 10^{-32}}$$

(d) For large-scale oscillators the quantized energy changes are so small compared to the total energy of the oscillator that quantum effects are of no importance. This is not true, however, of atomic and molecular oscillators, as we will see in the chapters ahead.

27.3 The Photoelectric Effect

The logical conclusion from Planck's work on blackbody radiation was that an oscillator in the walls of a cavity radiates electromagnetic energy into the cavity in the form of energy quanta. We might then expect that the interaction of electromagnetic radiation with matter would also be quantized rather than continuous.

Photoelectric effect: The release of electrons from a clean metal surface when electromagnetic radiation falls on it.

The photoelectric effect was discovered by Heinrich Hertz in 1887 during his attempts to detect electromagnetic waves from sparks produced across a small air gap. Apparatus that can be used for demonstrating the photoelectric effect is shown in Fig. 27.7. Light shines on the cathode C, which is at a negative potential with respect to the anode A. The light ejects electrons from the cathode, and these electrons are then collected by the anode. They flow through the complete circuit back to the cathode and thereby constitute an electric current that can be measured with the ammeter M. (This is one type of circuit used in burglar-alarm systems. An intruder cutting off the light beam stops the current and triggers the alarm.)

FIGURE 27.7 Apparatus used for experiments on the photoelectric effect.

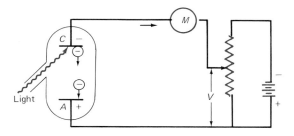

The kinetic energy of the emitted electrons can be measured by reversing the battery connections between C and A. If the voltage V (the *retarding potential*) is increased to a point where no electrons reach the anode, the initial kinetic energy of the most energetic electrons coming off the cathode is just reduced to zero by the electrons having to climb up the potential "hill" between C and A. If this happens for an applied voltage V_0, then we have

$$eV_0 = \tfrac{1}{2}mv_{\text{max}}^2 \tag{27.3}$$

where V_0 is called the *stopping*, or *cutoff*, *potential* and v_{max} is the speed of the most energetic electrons emitted by the cathode.

When experiments on the photoelectric effect are carried out, the results indicated in the first column of Table 27.1 are obtained.

Figure 27.9 shows experimental results for three metals. The value of the voltage V_0 required to cut off the electron flow (times the constant charge e on the electron) is seen to be proportional to the frequency ν of the incident light. For different materials the curves remain parallel to each other but are displaced so that the intercept on the x axis, ν_0, is different for different materials. These curves may be represented by the general equation for a straight line.

$$eV_0 = h(\nu - \nu_0) = h\nu - h\nu_0 \tag{27.4}$$

where h is a constant still to be determined. This equation rightly predicts that, when $\nu = \nu_0$, $eV_0 = 0$, as is clear from the figure. This frequency ν_0 is called the *threshold frequency*, since frequencies below ν_0 cannot eject electrons no matter how intense the light.

Turning Eq. (27.4) around, and replacing eV_0 by $\frac{1}{2}mv_{max}^2$ from Eq. (27.3), we have

$$h\nu = h\nu_0 + \tfrac{1}{2}mv_{max}^2 \tag{27.5}$$

TABLE 27.1 Experimental Results and Theoretical Predictions for the Photoelectric Effect

Experimental facts	Classical electromagnetic theory*		Einstein's quantum theory*	
1 Photocurrent flows almost instantaneously; delay less than 10^{-10} s.	⊖	For low light intensities delays of hours are to be expected.	⊕	Emission of electrons is produced by photon-electron collisions that occur almost instantaneously, no matter what the light intensity.
2 Photocurrent i is directly proportional to light intensity I (see Fig. 27.8).	⊕	The greater the light intensity, the greater the energy available to release electrons. Hence, more electrons are emitted and the current is greater.	⊕	The greater the light intensity, the greater the number of photons. More electrons are therefore emitted (*if frequency is above threshold*), and the current is greater.
3 For any particular material there is a cutoff frequency ν_0 below which no electrons are emitted by the light, no matter how great its intensity (see Fig. 27.9).	⊖	Any cutoff of the photocurrent should depend only on the intensity of the light and not on frequency.	⊕	The energy of a photon is $E = h\nu$. If this is less than the work function $\phi = h\nu_0$, the photon has insufficient energy to eject an electron, no matter what the intensity.
4 Maximum kinetic energy of electrons depends only on the frequency of the light, not on its intensity.	⊖	Maximum kinetic energy of electrons should depend only on the intensity of the light, not on the frequency.	⊕	$\frac{1}{2}mv_{max}^2 = h\nu - h\nu_0$; the intensity has no effect on $\frac{1}{2}mv_{max}^2$.

*⊕ indicates agreement of theory with experiment; ⊖ indicates disagreement.

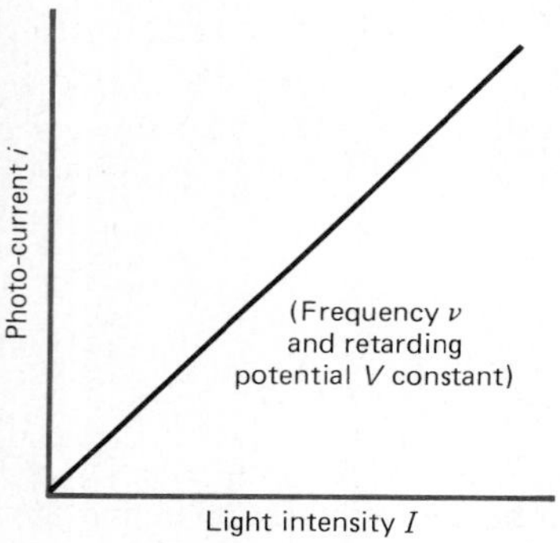

FIGURE 27.8 Dependence of photocurrent on light intensity for the photoelectric effect.

(Photo-current i vs. Light intensity I; Frequency ν and retarding potential V constant)

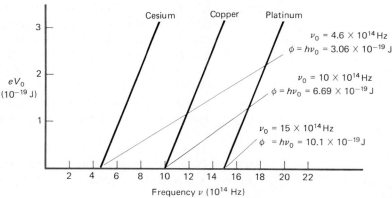

FIGURE 27.9 Experimental results on the photoelectric effect for cesium, copper, and platinum cathodes. The value of the retarding potential (times the charge on the electron) needed to cut off the electron flow is plotted as a function of the frequency ν of the incident light. In each case the slope of the curve is $h = eV_0/(\nu - \nu_0)$.

$\nu_0 = 4.6 \times 10^{14}$ Hz
$\phi = h\nu_0 = 3.06 \times 10^{-19}$ J

$\nu_0 = 10 \times 10^{14}$ Hz
$\phi = h\nu_0 = 6.69 \times 10^{-19}$ J

$\nu_0 = 15 \times 10^{14}$ Hz
$\phi = h\nu_0 = 10.1 \times 10^{-19}$ J

Classical electromagnetic theory, however, could not explain this experimental equation for the photoelectric effect, since its predictions agree with only one of the four experimental facts observed, as can be seen by comparing columns 1 and 2 in Table 27.1.

Einstein's Quantum Explanation of the Photoelectric Effect

In 1905 Albert Einstein explained the photoelectric effect in a very simple, elegant fashion. He assumed that the incident light consisted of photons of energy $E = h\nu$ and that individual photons collided with individual electrons in the metal and knocked them out of the metal by giving up to the electrons their *entire* energy. The principle of conservation of energy, applied to this collision of a photon and an electron, leads directly to *Einstein's photoelectric equation*:

$$h\nu = \phi + \tfrac{1}{2}mv_{max}^2$$

(27.6)

TABLE 27.2   Photoelectric Work Function ϕ for Clean Metal Surfaces

Metal	Photoelectric work function ϕ	
	(eV)	(10^{-19} J)
Cesium	2.14	3.42
Rubidium	2.10	3.36
Potassium	2.30	3.68
Aluminum	4.28	6.85
Lead	4.25	6.80
Copper	4.65	7.44
Tungsten	4.55	7.28
Silver	4.26	6.82
Gold	5.1	8.18
Platinum	5.65	9.04

Einstein interpreted this equation in the following way. An incident photon has an energy $E = h\nu$ that it gives up to a *single* electron in a collision process. Some of this energy goes into overcoming the attraction of the metal for the electron, and so the electron must do an amount of work ϕ, called the *work function* of the metal, in order to escape. After escaping, an electron near the surface has a kinetic energy $\tfrac{1}{2}mv_{max}^2$. Electrons in the interior of the metal must do additional work to escape from the metal, and they escape with kinetic energies smaller than $\tfrac{1}{2}mv_{max}^2$. If the work function ϕ is set equal to $h\nu_0$, then Eq. (27.6) can be interpreted as follows for the most energetic electrons:

$$h\nu \qquad\qquad = h\nu_0 \qquad\qquad\qquad + \tfrac{1}{2}mv_{max}^2$$

| Energy of incident photon | = | energy required to remove electron from metal | + | kinetic energy of emerging electron |

which is Eq. (27.5). Values of the work function, in both joules and electronvolts, for some important metals are given in Table 27.2.

Einstein's derivation of the photoelectric equation was a major step in the development of quantum theory. Table 27.1 makes clear the superiority of Einstein's theory over a classical theory for the photoelectric effect, since it accounts perfectly for all the observed facts.

Millikan's Experiments

The experimental work that established Einstein's theory of the photoelectric effect was performed by the American physicist Robert A. Millikan (see Figs. 16.13 and 26.8) during the years 1912 to 1916. The apparatus used for this work, for which Millikan received the Nobel Prize in 1923, is sometimes referred to as a "machine shop in a vacuum." Millikan continually had to slice off surface layers of atoms by remote control inside a complicated vacuum system to maintain the ultraclean metal surfaces he needed for his experiment. Millikan's results consisted of graphs similar to those of Fig. 27.9, where the slope of the curve is, from Eq. (27.4), $h = eV_0/(\nu - \nu_0)$. In this way h is obtained from the slope of these curves.

Millikan's value for h, as obtained in this way, was 6.569×10^{-34} J·s. The excellent agreement of this value for h with the value required to make Planck's equation fit the experimental data on blackbody radiation indicated the strong probability that both Planck's and Einstein's theories were correct and that light did indeed consist of photons of energy $E = h\nu$. A great variety of additional experimental evidence, some of which will be presented in subsequent sections, has confirmed this view and has led to a more accurate value of 6.6261×10^{-34} J·s for Planck's constant.

Example 27.2

(a) Blue light of wavelength 436 nm and intensity 1.5×10^3 W/m² falls on a piece of lithium metal in a vacuum. If the work function of lithium is 3.87×10^{-19} J, what is the maximum kinetic energy of the electrons emitted? **(b)** What is the maximum speed of the emitted electrons? **(c)** What is the longest wavelength that will eject electrons from the metal?

SOLUTION

The intensity of the light has no influence on the kinetic energy of the ejected electrons. The intensity of the light changes only the *number* of electrons emitted, as long as the photons have sufficient energy to eject any electrons at all from the metal.
(a) By Einstein's photoelectric equation,

$$h\nu = h\nu_0 + \tfrac{1}{2}mv_{max}^2$$

and so

$$\tfrac{1}{2}mv_{max}^2 = h\nu - h\nu_0 = \frac{hc}{\lambda} - \phi$$

$$= 4.56 \times 10^{-19} \text{ J} - 3.87 \times 10^{-19} \text{ J}$$

$$= \boxed{0.69 \times 10^{-19} \text{ J}}$$

(b) $v_{max} = \left(\dfrac{1.38 \times 10^{-19} \text{ J}}{9.11 \times 10^{-31} \text{ kg}} \right)^{1/2} = \boxed{3.89 \times 10^5 \text{ m/s}}$

(c) The longest wavelength which can eject electrons is that value of λ for which $\nu = \nu_0$, or $hc/\lambda = h\nu_0 = \phi$, and so

$$\lambda = \frac{hc}{\phi} = \frac{(6.63 \times 10^{-34} \text{ J·s})(3.00 \times 10^8 \text{ m/s})}{3.87 \times 10^{-19} \text{ J}}$$

$$= \boxed{514 \text{ nm}}$$

Hence only light of wavelength *less than* 514 *nm* (in the green) has photons with enough energy to eject electrons from lithium, irrespective of the intensity of the light.

EXERCISE 1 What is the cutoff wavelength for a silver cathode?

27.4 X-Rays

Other evidence for the photon nature of light (and other electromagnetic radiation) comes from the continuous x-ray spectrum.

Discovery of X-Rays

FIGURE 27.10 Wilhelm Roentgen (1845–1923), who received the first Nobel Prize in physics in 1901 for his discovery of x-rays. He refused to patent any aspect of x-ray production so that x-rays could be used freely for the good of humanity. *(Courtesy of AIP Niels Bohr Library, Landé collection.)*

In 1895 the German physicist Wilhelm Conrad Roentgen (see Fig. 27.10) was doing some experiments at the University of Würzburg on the effect of cathode rays (i.e., electron bombardment) on the luminescence of certain chemicals. He had a cathode-ray tube enclosed in black cardboard in a darkened room. When he turned on the voltage across the tube, he noticed a flash of light from a sheet of paper coated with barium platinocyanide at the other side of the room. This mystified him, since the cathode-ray beam was blocked off by the cardboard and could not possibly cause this light emission directly. A series of experiments led Roentgen to conclude that the cathode-ray tube must be producing a radiation that was invisible but very penetrating. Roentgen decided to call this radiation *x-radiation*, or *x-rays*, since x is the typical mathematical symbol for an unknown quantity.

Roentgen soon found that the x-rays could pass through paper and thin layers of metal but not through very thick layers of metal, and that they could ionize gases but were completely unaffected by electric and magnetic fields. This seemed to indicate that they were not charged particles.

On January 23, 1896, Roentgen gave his first lecture on x-rays. After his talk he asked a famous anatomist on the medical faculty at Würzburg for permission to photograph his hand. Roentgen took an x-ray photograph and passed it around the room. The audience broke into wild applause. Roentgen's discovery made him famous, for everyone saw the tremendous practical applications of these rays that could pass through the soft tissues of the body but cast shadows of bone or heavy foreign materials like bullets or pins embedded in human tissue.* Physics labora-

---

*Some members of the New Jersey legislature were so disturbed by the antiprivacy implications of x-rays that they tried to pass a law banning the use of x-ray opera glasses!

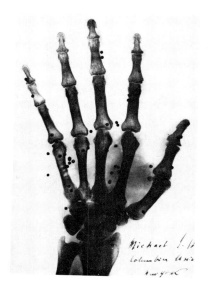

FIGURE 27.11 An x-ray photograph of a human hand. in 1896 Professor Michael I. Pupin of Columbia University first heard of Roentgen's discovery of x-rays from a physicist friend in Germany. He immediately set up equipment for x-ray experiments and in February 1896 was asked to take x-ray pictures of the hand of a New York lawyer who had been shot accidentally. This photo, which shows clearly the shotgun pellets in the hand, was one of the first x-ray photos taken in the United States. *(Courtesy of the Burndy Library.)*

tories all over the world turned their attention to x-rays, with resulting gains both for pure science and for diagnostic medicine. One of the first x-ray photographs taken in the United States is shown in Fig. 27.11.

The Continuous X-Ray Spectrum

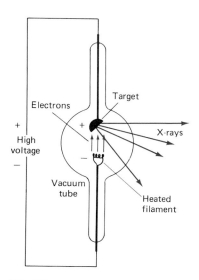

FIGURE 27.12 A Coolidge x-ray tube. Energy lost by the electrons on suddenly being brought to rest by the target is converted into x-ray photons.

FIGURE 27.13 (Right) The continuous x-ray spectrum arising from using different accelerating voltages for the electrons producing the x-rays.

An x-ray tube of the type developed in 1913 by the American physicist William D. Coolidge (1873–1975) is shown in Fig. 27.12. Electrons are accelerated by a high voltage from a heated filament, which acts as the cathode, to a metal target, which is the anode. The electrons plunge into the target at high speeds and produce x-rays in the collision process. The wavelength distribution of these x-rays can be determined using an x-ray spectrometer (see Fig. 27.22). The resulting plots of intensity against wavelength take the form of the continuous curves shown in Fig. 27.13. (There are, in fact, some sharp lines superimposed on this continuous spectrum, but we will defer a discussion of these lines until later.)

Experimentally the following facts have been established about the continuous x-ray spectrum:

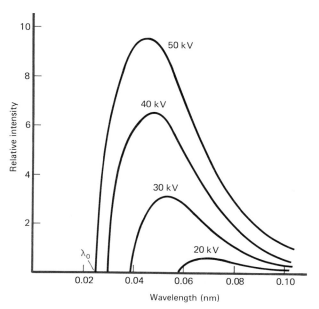

1 The wavelength characteristics of the *continuous spectrum* are completely independent of the target material and depend only on the voltage applied to the x-ray tube. The higher the voltage, the shorter the minimum (or cutoff) wavelength produced. The variation of cutoff wavelength with voltage can be seen in Fig. 27.13.

2 The intensity of the continuous spectrum depends both on the target material and thickness and on the applied voltage.

3 The cutoff wavelength λ_0 is related to the applied voltage V by the equation

$$\lambda_0 = \frac{1.24 \times 10^{-6} \text{ m}}{V} \tag{27.7}$$

where V is in volts. This is an empirical equation in which the constant has been chosen to fit the experimental observations. We now proceed to show that the same equation follows immediately from the principle of conservation of energy.

Explanation of Continuous X-Ray Spectrum

The sharp cutoff wavelength λ_0 is hard to explain on the basis of classical theory, but quantum theory provides a convincing explanation. The continuous x-ray spectrum is called *braking radiation* (in German, *Bremsstrahlung*), and it arises when the electrons plunge into the target, collide with atoms there, and rapidly decelerate. In this process much of the kinetic energy of the electrons is converted into heat, but some of it is radiated away in the form of x-ray *photons* of energy $h\nu = \Delta E$, where ΔE is the energy lost by a particular electron in a collision with a target atom. The amount of energy lost (ΔE) cannot exceed the total kinetic energy that the electron had before the collision, but it can have any value smaller than this, for multiple collisions lead to the emission of smaller quanta in each collision. Since the largest possible value of ΔE is set by the accelerating voltage V, we have for the cutoff frequency,

$$h\nu_0 = \Delta E = eV \quad \text{or} \quad \boxed{\nu_0 = \frac{eV}{h}} \tag{27.8}$$

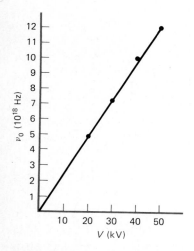

FIGURE 27.14 Graph of cutoff frequency ν_0 as a function of the voltage V applied to an x-ray tube.

where V is the voltage applied to the x-ray tube.

A plot of ν_0 versus V is a straight line with slope e/h, as shown in Fig. 27.14. From such a plot a value of e/h can be found and used as a check on the values of e and h already obtained from Millikan's oil-drop experiment and his work on the photoelectric effect.

If we use Eq. (27.8) in the expression for the cutoff wavelength λ_0, we find

$$\lambda_0 = \frac{c}{\nu_0} = \frac{hc}{h\nu_0} = \frac{hc}{eV} \tag{27.9}$$

or
$$\lambda_0 = \frac{(6.63 \times 10^{-34} \text{ J·s})(3.00 \times 10^8 \text{ m/s})}{(1.60 \times 10^{-19} \text{ C})V} = \frac{1.24 \times 10^{-6} \text{ m}}{V} \tag{27.10}$$

where again V is in volts. This agrees perfectly with our previous empirical equation [Eq. (27.7)]. Thus for a voltage of 50 kV applied to the x-ray tube, the cutoff wavelength is 2.5×10^{-11} m $= 0.025$ nm, in agreement with the experimental cutoff wavelength shown in Fig. 27.13. This wavelength is about 10^4 times shorter than that of visible light. Typical x-ray wavelengths fall in the range from 10^{-2} to 10 nm.

The sharp cutoff in the continuous x-ray spectrum at short wavelengths is therefore well explained by a quantum or photon theory, whereas classical physics was unable to provide a convincing explanation for this phenomenon.

Example 27.3

An x-ray tube has a voltage of 100 kV applied between the cathode and the steel plate that serves as the target producing the x-rays. **(a)** What is the highest-frequency photon to be expected in the x-ray beam? **(b)** What is the cutoff wavelength of the continuous x-ray spectrum?

SOLUTION

(a) Since the energy of a single photon cannot be greater than the energy of an electron crashing into the target, and this electron energy is determined by the applied voltage, we must have for the highest-energy photon $h\nu_0 = eV$, and so

$$\nu_0 = \frac{eV}{h} = \frac{(1.60 \times 10^{-19} \text{ C})(100 \times 10^3 \text{ V})}{6.63 \times 10^{-34} \text{ J·s}}$$

$$= \boxed{2.41 \times 10^{19} \text{ Hz}}$$

(b) The cutoff wavelength is the wavelength corresponding to the highest-energy photon produced in the braking radiation, and so

$$\lambda_0 = \frac{c}{\nu_0} = \frac{3.00 \times 10^8 \text{ m/s}}{2.41 \times 10^{19}/\text{s}} = \boxed{0.0124 \text{ nm}}$$

The same result could, of course, have been obtained directly from Eq. (27.10).

$$\lambda_0 = \frac{1.24 \times 10^{-6} \text{ m}}{V} = \frac{1.24 \times 10^{-6} \text{ m}}{100 \times 10^3} = \boxed{0.0124 \text{ nm}}$$

27.5   The Compton Effect

FIGURE 27.15   Arthur Holly Compton (1892–1962). For his discovery of the effect that now bears his name, he received the Nobel Prize in physics in 1927. *(Courtesy of AIP Niels Bohr Library.)*

A fourth piece of evidence on the nature of the photon was provided by another American physicist, Arthur Holly Compton (see Fig. 27.15) in 1922. Compton was studying the scattering of x-rays by a carbon target and directed a beam of x-rays with a discrete wavelength of 0.0710 nm at a carbon target. He found that, in addition to x-rays scattered with this original wavelength unchanged, other x-rays appeared with wavelengths of 0.0734 nm on the long-wavelength side of the original wavelength.

Compton effect: The increase in the wavelength of high-frequency electromagnetic radiation when scattered by free electrons.

Compton could not explain this result classically and therefore set out to see if some of the newer ideas of Planck and Einstein about photons might work. He assumed that photons could be treated as particles carrying energy and momentum and that the elastic collision between a photon and a free electron caused the photon to lose energy, thus decreasing its frequency and increasing its wavelength. (Notice that the binding energy of an electron in a carbon atom is so small compared with the energy of an x-ray photon that the electron can be considered *free*; note also that the photon's speed cannot change, since it always moves at the speed of light.) Compton then applied the laws of conservation of energy and momentum to the collision process* and found that the predictions of quantum theory did indeed

---

*A major difference between the photoelectric effect and the Compton effect should be emphasized. In the photoelectric effect, in which the electron is *bound* to the metal, both energy and momentum can be conserved when the incident light is completely absorbed by the metal. In the Compton effect, where the electron is essentially *free*, on the other hand, both energy and momentum *cannot* be conserved unless the photon remains in existence after the collision.

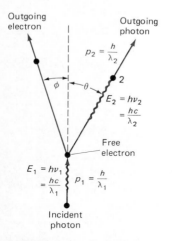

FIGURE 27.16. The Compton effect: The collision between an x-ray photon and an electron at rest. The initial frequency ν_1 of the photon is decreased to ν_2 after it is scattered.

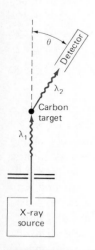

FIGURE 27.17 Sketch of Compton's apparatus for the study of the Compton effect. X-rays of wavelength λ_1 are scattered from the carbon target. The wavelength λ_2 of the scattered x-rays is measured by the detector.

agree with his experimental results.

To treat the photon as a particle colliding with an electron, we need formulas for the photon's energy and momentum. The photon energy is given simply by Planck's expression, $E = h\nu$, while the linear momentum of the photon is, from Eq. (26.30),

$$p = \frac{E}{c} = \frac{h\nu}{c} = \frac{h}{\lambda} \quad \text{or} \quad \boxed{p = \frac{h}{\lambda}} \tag{27.11}$$

The linear momentum of a photon is equal to Planck's constant divided by the photon wavelength [Notice that here again we are mixing together particle properties (momentum) and wave properties (wavelength)].

Compton then applied the laws of conservation of energy and conservation of linear momentum to the collision between a photon and an electron, as shown in Fig. 27.16, and obtained the equation

$$\boxed{\Delta\lambda = \lambda_2 - \lambda_1 = \frac{h}{m_0 c}(1 - \cos\theta)} \tag{27.12}$$

in which λ_1 is the wavelength of the incident x-ray photon and λ_2 is the wavelength of the scattered x-ray photon. This Compton effect equation predicts an increase in the wavelength of the scattered photon that is a function of the angle θ at which the photon is scattered. The shift in wavelength when $\theta = 90°$ and $\cos\theta = 0$ is

$$\Delta\lambda = \frac{h}{m_0 c}(1 - 0) = \frac{6.63 \times 10^{-34} \text{ J·s}}{(9.11 \times 10^{-31} \text{ kg})(3.00 \times 10^8 \text{ m/s})} = 0.00243 \text{ nm}$$

The quantity $h/m_0 c = 0.00243$ nm is called the *Compton wavelength of the electron.* The Compton shift is very small compared with the wavelength of visible light ($\sim$500 nm), but for x-rays of wavelength 0.0710 nm it amounts to a few percent of the original wavelength and can be easily measured with an x-ray spectrometer of sufficient resolution. When the calculated shift of 0.00243 nm is added to $\lambda_1 = 0.0710$ nm, the resulting value is $\lambda_2 = 0.0734$ nm, in excellent agreement with Compton's experimental results.

The experimental setup used by Compton is shown in Fig. 27.17. In practice both an unshifted line and a shifted line are observed. The unshifted line represents x-rays scattered from electrons whose binding energies are so great that they are not ejected from the atom. The change in the wavelength of the shifted line increases with the scattering angle for angles from 0 to 180°, in accordance with Compton's theoretical prediction in Eq. (27.12). Compton's results are shown in Fig. 27.18.

In his original paper Compton pointed out that "the wavelength and the intensity of the scattered rays are what they should be if a quantum of radiation bounced from an electron, just as one billiard ball bounces from another. . . . The obvious conclusion would be that x-rays, and so also light, consist of discrete units, proceeding in definite directions, each unit possessing the energy $h\nu$ and the corresponding momentum h/λ." This goes beyond the concept of the photon that emerged from the work of Planck and Einstein, which attributed energy to a photon but not momentum. The photon now is seen to have inertia and momentum and therefore many of the same properties as a billiard ball or a bullet.

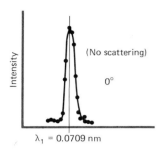

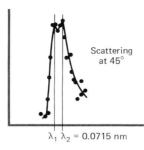

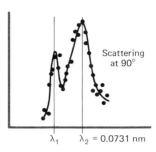

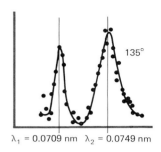

FIGURE 27.18 Experimental data on the Compton effect taken from Compton's original paper. Wavelength is plotted along the *x* axis and intensity along the *y* axis. The first graph shows the unscattered radiation, the other graphs the radiation scattered at 45, 90, and 135°. The wavelength of the scattered x-rays increases with angle, as predicted by Compton's theory.

> Some physicists believed the Compton effect sounded the death knell for the wave theory of light, but they underestimated the resiliency of that venerable theory.

Example 27.4

X-radiation of wavelength 0.112 nm is scattered from a carbon target. **(a)** Calculate the wavelength of the x-rays scattered at an angle of 90° with respect to the original direction. **(b)** Calculate the energy of the scattering electron after the collision. **(c)** Calculate the direction of the scattering electron after the collision.

SOLUTION

(a)   We presume that the scattering is due to collisions of the x-ray photons with electrons which are essentially free. Then the situation is well described by the equation for the Compton effect:

$$\Delta\lambda = \frac{h}{m_0 c}(1 - \cos\theta)$$

Here $\cos\theta = 0$, and so $\Delta\lambda = h/m_0 c = 0.0024$ nm, and

$$\lambda_2 = \lambda_1 + \Delta\lambda = 0.112 \text{ nm} + 0.0024 \text{ nm} = \boxed{0.114 \text{ nm}}$$

(b)   The energy given to an electron is equal to the energy lost by the x-ray photon, since energy must be conserved in the collision. This is

$$\Delta E = h\nu_1 - h\nu_2 = (6.63 \times 10^{-34} \text{ J·s})\left(\frac{c}{\lambda_1} - \frac{c}{\lambda_2}\right)$$

$$= (1.99 \times 10^{-25} \text{ J·m})\left(\frac{1}{0.112 \text{ nm}} - \frac{1}{0.114 \text{ nm}}\right)$$

$$= (1.99 \times 10^{-25} \text{ J·m})(0.16 \times 10^9 \text{ m}^{-1})$$

$$= \boxed{3.2 \times 10^{-17} \text{ J}}$$

Note that 3.2×10^{-17} J is equal to about 200 eV, which is roughly 200 times the binding energy of the valence electrons in carbon. The assumption that the electrons are essentially free is therefore justified.

(c)   The x-ray photon has an initial momentum h/λ_1 along the positive *x* axis; its final momentum is h/λ_2 along the positive *y* axis, since it is scattered through 90°, as shown in Fig. 27.19. To conserve momentum the electron must have a momentum after the collision with an *x* component $mv \cos\phi$ numerically equal to h/λ_1 along the positive *x* axis. The electron must also have a *y* component of momentum $mv \sin\theta$ numerically equal to h/λ_2 along the negative *y* axis, since the initial *y* component of momentum for the system was zero. Then, from the figure,

$$\tan\phi = \frac{mv \sin\phi}{mv \cos\phi} = \frac{h/\lambda_2}{h/\lambda_1} = \frac{\lambda_1}{\lambda_2} = \frac{0.112 \text{ nm}}{0.114 \text{ nm}} = 0.982$$

and so $\phi = \boxed{44.5°}$

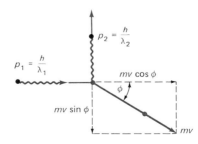

FIGURE 27.19

27.6   The Nature of Light

In 1920 physicists believed that the two aspects of reality represented by waves and particles were mutually exclusive and that no particle could exhibit wave properties and no wave could exhibit particle properties. The experimental results on blackbody radiation, the photoelectric effect, the continuous x-ray spectrum, and the Compton effect tended, therefore, to reduce physicists to a state of shock, since electromagnetic radiation now seemed to have both wave and particle properties.

We have seen that the phenomena of diffraction, interference, and polarization provide convincing evidence that light is a transverse wave. The work of Planck, Einstein, and Compton seemed to show, however, that light consists of beams of photons whose interactions with matter are correctly predicted only by a particle model. How can we reconcile these two seemingly contradictory models of radiation?

Our approach to this very difficult question is to postpone any attempt at an answer until we introduce a further complication with respect to particles such as electrons, protons, and hydrogen atoms. It is as if we are trying to piece together a gigantic jigsaw puzzle broken down into two separate sections. Some pieces from one section of the puzzle are mixed up with the other section, however, so that neither part can be solved by itself. If, however, the two sections are combined once again, the puzzle may be solvable. The nature of radiation and matter may be similar to our divided jigsaw puzzle. Perhaps by concentrating on radiation alone we are leaving out some key pieces of the puzzle. Perhaps we should pay more attention to the *particles* that make up matter.

27.7   Wave Properties of Matter: De Broglie Waves

As already mentioned, the concept of symmetry has played an important role in the development of physics. If, as Table 27.3 tries to indicate, radiation has both wave and particle properties, should not matter also have wave properties? If a photon of wavelength λ has a momentum $p = h/\lambda$, why should not a particle of momentum $p = mv$ have an associated wavelength $\lambda = h/p$? This was the reasoning of the French physicist L. de Broglie (see Fig. 27.20), who in 1923 was writing a doctoral dissertation at the Sorbonne in Paris. De Broglie's idea was so revolutionary that there was some skepticism on the part of his examiners about its validity. It was only after one of the examiners, Paul Langevin, sent de Broglie's thesis to Einstein and found him enthusiastic about it, that this skepticism disappeared.

TABLE 27.3   Wave and Particle Properties of Matter and Radiation

	Wave properties	Particle properties
Electromagnetic radiation (radio waves, light waves, x-rays, gamma rays)	Wavelength $\lambda\nu = c$ Diffraction Interference Polarization	$E = h\nu$ $p = \dfrac{h}{\lambda}$ Blackbody radiation Photoelectric effect Compton effect
Matter (electrons, protons, atoms, molecules)	$\lambda = \dfrac{h}{p}$ (?)	$KE = \frac{1}{2}mv^2$ $p = mv$

The name *de Broglie wave* is given to the wave associated with every particle having a momentum p. The associated wavelength is, by analogy with Eq. (27.11),

De Broglie wavelength:
$$\lambda = \frac{h}{p} = \frac{h}{mv}$$
(27.13)

There was no known experimental evidence for such an idea when de Broglie developed it, and for ordinary objects the predicted wavelength is much too short to have observable effects. Thus, for a 100-g ball moving at a speed of 25 m/s, the de Broglie wavelength is 2.6×10^{-34} m. To produce diffraction of such de Broglie waves, diffraction gratings with separations between rulings of about 10^{-34} m would be needed. These are impossible to obtain, since the required separations are smaller than the dimensions of the atoms making up the grating by about 24 orders of magnitude!

If, however, we consider an *electron* moving with a speed of 10^6 m/s, which is 1/300 of the speed of light, then its de Broglie wavelength is

$$\lambda = \frac{h}{mv} = \frac{6.6 \times 10^{-34} \text{ J·s}}{(9.1 \times 10^{-31} \text{ kg})(10^6 \text{ m/s})} = 7.3 \times 10^{-10} \text{ m}$$

This wavelength of 0.73 nm is of the same order of magnitude as the separation of planes of atoms in crystals, which are about 0.2 to 0.7 nm apart. Thus it should be possible to produce diffraction and interference of electrons with the proper wavelength by using crystals as diffraction gratings.

Diffraction and Interference of X-Rays

Before discussing the diffraction and interference of de Broglie waves, let us digress to consider the related phenomenon for x-rays. In 1912 the German physicist Max von Laue (1879–1960) predicted that x-rays could be diffracted by solid crystals with a spacing between atom planes of about the same size as the x-ray wavelength, in the same way that visible light can be diffracted by gratings. This was confirmed in 1912 by the British physicists William H. Bragg (1862–1942) and his son, William L. Bragg (1890–1971).

In a crystal such as sodium chloride (NaCl) the lattice spacing between adjacent atoms can be found from the density of the crystal and the atomic masses of sodium and chlorine. It turns out to be $d = 0.282$ nm for NaCl (see Example 27.5). If x-rays with wavelengths of 0.2 nm are incident on a salt crystal, each atom acts as a scattering center and radiates waves in all directions. The atoms in any one plane, however, radiate x-rays coherently in the same way a mirror reflects light. These lattice planes are called *Bragg planes*, and reflections from such planes are called *Bragg reflections*. If x-rays are reflected from two adjacent, parallel Bragg planes, constructive interference occurs for certain angles of incidence of the x-rays on the crystal planes.

Consider the Bragg planes marked 1 and 2 in Fig. 27.21. Let a beam of x-rays of wavelength λ fall on the crystal at an angle θ with respect to the Bragg planes, where θ is the angle between the direction of the x-ray beam and the Bragg

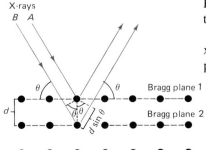

FIGURE 27.21 Reflection of x-rays by two adjacent Bragg planes in a crystal lattice.

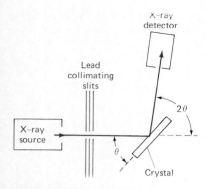

Lead
collimating
slits

X-ray
source

X-ray
detector

2θ

θ

Crystal

FIGURE 27.22 An x-ray
spectrometer.

plane (*not* the normal to the plane, as in the case of a mirror). Then waves scattered at the same angle by the two Bragg planes will be in phase and interfere constructively only if the path difference for the two is an integral number of x-ray wavelengths. From Fig. 27.21 we see that this condition is satisfied if

Bragg equation: $\boxed{2d \sin \theta = n\lambda}$ (27.14)

where n is a small integer (1, 2, 3, . . .) and λ is the x-ray wavelength. For x-rays incident at any other angle than one that satisfies this equation, the reflected x-rays will interfere destructively and no reflected beam will be observed.

The Bragg equation thus gives us a way of determining the wavelength of x-rays if the atomic separations are known or of determining d if the wavelength of the x-rays is known. An x-ray spectrometer similar to the one in Fig. 27.22 can be used for this purpose. The x-rays are collimated into a beam by slits in lead plates. The angle of incidence θ is gradually changed, and maximum intensities are then observed for angles θ satisfying the Bragg equation. In 1953 James D. Watson and Francis Crick worked out the famous double-helix structure of DNA by using x-ray diffraction photographs taken by Maurice Wilkens and Rosalind Franklin.

Example 27.5

The density of potassium chloride is 1.99 g/cm³ and its molecular mass 74.56 g. What is the distance between the atoms in the KCl crystal if it is a cubic structure with alternating K and Cl atoms?

SOLUTION

We first find the volume of a KCl crystal containing Avogadro's number of KCl molecules. This is, since the density $\rho = m/V$,

$$V = \frac{m}{\rho} = \frac{74.56 \text{ g}}{1.99 \text{ g/cm}^3} = 37.5 \text{ cm}^3$$

This volume contains Avogadro's number of KCl molecules and therefore 6.03×10^{23} potassium atoms and 6.03×10^{23} chlorine atoms. The total number of atoms present in the crystal is thus $2(6.03 \times 10^{23})$. If we consider each atom as being at the center of a cube in the crystal, the distance between atoms will

be the length of a side of this cube. To find the volume of such a cube we merely divide the total volume by the total number of atoms present. We have

$$V = \frac{37.5 \text{ cm}^3}{2(6.03 \times 10^{23})} = 3.11 \times 10^{-23} \text{ cm}^3$$

Then the distance d between atoms will be the length of the side of a cube of this volume, or

$$d = (3.11 \times 10^{-23} \text{ cm}^3)^{1/3} = \boxed{0.314 \text{ nm}}$$

Example 27.6

A beam of x-rays is reflected from a NaCl crystal whose lattice planes are separated by a distance $d = 0.282$ nm. The angle of incidence on the crystal planes that produces a strong reflected x-ray beam is 30°. **(a)** What are the possible wavelengths of the x-rays used? **(b)** How can we determine which of these possible wavelengths is the correct one?

SOLUTION

(a) The Bragg equation is applicable in this case, so

$$n\lambda = 2d \sin \theta = 2(0.282 \text{ nm})(0.500) = 0.282 \text{ nm}$$

The wavelength is not fully determined by this result unless the integer n, which specifies the order of interference, is known. For example, possible solutions are $\lambda = 0.282$ nm; 0.141 nm; 0.094 nm; and so on.

(b) To decide among the possible solutions found above, we could use the apparatus to see if any other strong reflections occur for values of θ less than 30°. If none is found, the presumption is that the beam incident at 30° corresponds to the first-order interference with $n = 1$ and that the x-ray wavelength is

$\boxed{0.282 \text{ nm}}$.

27.8 Electron Diffraction and Interference

De Broglie's suggestion of wave properties for particles immediately set physicists to thinking about the possibility of electron diffraction and interference. If electrons could be obtained with de Broglie wavelengths near 0.2 nm, then electrons should undergo diffraction and interference in crystalline solids, just as x-rays of the same wavelengths do. The wavelength of charged particles is also under the experimenter's control, for they can be accelerated to any desired speed and thus given any desired wavelength, since $\lambda = h/mv$.

For an electron accelerated by a voltage of 100 V, its final speed, if we neglect relativistic effects,* is 5.9×10^6 m/s. This corresponds to a de Broglie wavelength of 0.12 nm (Check this!), which is the same order of magnitude as the lattice separation in crystals.

Electrons can be easily produced with the proper wavelengths to make possible diffraction by crystals such as NaCl, Cu, or Ni. In this case the theory is exactly the same as for x-rays, and an equation identical with the Bragg equation results:

$$2d \sin \theta = n\lambda$$

(27.15)

In this equation d is the lattice spacing in the crystal, and λ is the de Broglie wavelength of the electrons.

Experimental Verification of de Broglie Waves

In 1927 American physicist Clinton J. Davisson (1881–1952), with the assistance of his colleague L. H. Germer at what are now the AT&T Bell Laboratories, obtained data on electron beams diffracted by a single crystal of nickel in excellent agreement with de Broglie's predictions. In the same year additional confirmation of de Broglie's ideas came from the somewhat different work of George P. Thomson (1892–1975), the son of J. J. Thomson, at the University of Aberdeen in Scotland. For their verification of de Broglie's speculations Davisson and Thomson shared the 1937 Nobel Prize for physics. The striking similarity of the patterns produced by a crystal for x-ray and electron diffraction is shown in Fig. 27.23. All this evidence points up the startling fact that, under the proper circumstances, electrons exhibit *wave* properties.

 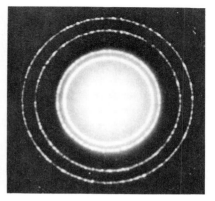

FIGURE 27.23 Comparison of experimental photographs obtained by x-ray and electron diffraction. (*a*) The diffraction pattern obtained by scattering 0.071-nm x-rays from an aluminum foil; (*b*) the diffraction pattern obtained by scattering 600-eV electrons off aluminum. Since the wavelengths are different, the scales of the two have been adjusted to highlight the similarities. *(Courtesy of Educational Development Center, Newton, Massachusetts.)*

---

*The rest energy of an electron is 0.51 MeV. If the accelerating voltage for an electron is much less than 0.51 MV, relativistic effects are small.

Example 27.7

Davisson and Germer used 54-V electrons in their experiment on the electron scattering by nickel crystals. If such electrons are scattered off a nickel crystal for which the separation of the crystal planes is 0.266 nm, at what angles will maxima occur for the scattered electrons?

SOLUTION

From the de Broglie wavelength equation, the wavelength of the incident electrons is

$$\lambda = \frac{h}{p} = \frac{h}{mv} = \frac{h}{m\sqrt{2eV/m}} = \frac{h}{\sqrt{2eVm}} = \frac{h}{\sqrt{2em}}\frac{1}{\sqrt{V}}$$

or $\lambda = \dfrac{1.23 \times 10^{-9} \text{ m}}{\sqrt{V}} = \dfrac{1.23 \text{ nm}}{\sqrt{V}}$

In this case $\lambda = 1.23 \text{ nm}/\sqrt{54} = 0.167 \text{ nm}$, and Eq. (27.15) yields for

$n = 1$: $\sin \theta_1 = \dfrac{\lambda}{2d} = \dfrac{0.167 \text{ nm}}{2(0.266 \text{ nm})} = 0.31$ $\theta_1 = 18°$

$n = 2$: $\sin \theta_2 = 2(0.31) = 0.62$ $\theta_2 = 38°$

$n = 3$: $\sin \theta_3 = 3(0.31) = 0.93$ $\theta_3 = 68°$

For $n = 4$ or higher, $\sin \theta > 1$, and the result is physically impossible. Hence strong electron beams are expected only for angles of incidence of 18, 38, and 68°.

The Electron Microscope

Electrons accelerated through 50,000 V have wavelengths 100,000 times shorter than green light. Since the resolution of any microscope is limited by the wavelength of the light used (Sec. 25.6), electron microscopes have the great advantage of achieving resolutions better by orders of magnitude than the best optical microscopes.

Transmission Electron Microscope   The transmission electron microscope (TEM) resembles an optical microscope in its lens system except that the lenses are electrostatic and magnetic arrays which deflect and focus electrons by electromagnetic interactions in the same way that glass lenses deflect and focus light, as shown in Fig. 27.24. An electron beam from a heated cathode is accelerated to a

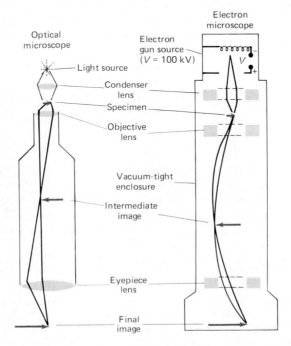

FIGURE 27.24   Comparison of a transmission electron microscope (TEM) with an optical microscope. The lenses in the TEM are magnetic lenses that use magnetic fields to focus the electrons to an image of the original object. To emphasize the similarity, the optical microscope is shown producing a real image, unlike the microscopes discussed in Chap. 25.

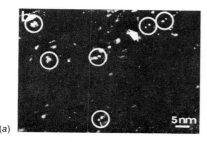

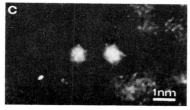

(a)

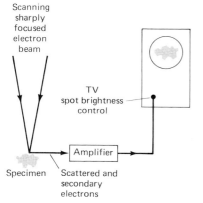

(b)

FIGURE 27.25 A very high resolution TEM photograph of barium atoms. (a) The circles show complicated organic molecules that have two barium atoms at their ends, separated by a distance of about 1.6 nm. The two dots inside the circles are the barium atoms. (b) Under even greater magnification the barium atoms can be clearly seen. According to the scale at the bottom of the photograph, their separation is indeed about 1.6 nm. The magnification here is about 10 million times! *(Courtesy of B. Jouffrey and D. Dorignac, Laboratoire d'Optique Electronique, Toulouse, France.)*

very high speed, condensed by such a lens system on a specimen, e.g., a virus or bacterium mounted on a very thin collodion film, and then focused by an objective lens to produce an intermediate image. This image is then the object for an eyepiece lens that produces the final image on a fluorescent screen, where it can be viewed by an observer, or on a photographic plate sensitive to electrons. Since the object attenuates the electron beam differently for different thicknesses and configurations, a very clear picture of the object under very high resolution can be obtained, as shown in Fig. 27.25. Modern transmission electron microscopes have resolved details as small as 0.23 nm compared with the 200 nm attainable with the best optical microscopes. The limitation on the resolution of the TEM is not the electron wavelength, which can be much shorter than 0.1 nm, but imperfections in the electrostatic and magnetic lenses that are used. Electron microscopes have the disadvantage, however, that a high vacuum is required throughout the entire instrument to reduce the scattering of electrons by air molecules.

Scanning Electron Microscope A more recent development is the scanning electron microscope (SEM), which is really not a true microscope at all. An SEM employs a well-focused beam of electrons (perhaps only 0.2 nm in diameter) to scan the surface of the specimen, as in Fig. 27.26. Some of these electrons are scattered by the specimen; others eject secondary electrons from the specimen. These electrons are then collected and amplified to produce a current that controls the brightness of the spot on a conventional TV screen. The TV electron beam is scanned in synchronism with the SEM beam and is modulated in intensity by the secondary electron beam. An image of the specimen is thus formed that varies with the shape, size, and contour of the object. The magnification of such a system can be controlled electronically by varying the sweep speed of the TV beam with respect to the SEM beam. The final image can also be transmitted immediately over conventional TV cables to distant receivers. The resolving power of an SEM is less than that of a TEM by a factor of about 10, but this is partially made up by the increased depth perception and almost three-dimensional quality of SEM pictures. It is also possible to convert brightness differences into colors to increase the contrast in SEM photographs.

The 1986 Nobel Prize in physics was shared by Ernst Ruska for his design of the first electron microscope in the early 1930s and by Gerd Binnig and Heinrich Rohrer of the IBM Research Laboratory in Zurich, Switzerland, for a new kind of SEM called the *scanning tunneling microscope*, built by them in the early 1980s (see Fig. 27.27).

FIGURE 27.26 (left) A scanning electron microscope (SEM). Here scattered electrons and secondary electrons knocked out of the target by the primary electron beam combine to produce the image on a TV screen.

FIGURE 27.27 (right) Scanning tunneling microscope photograph of the surface of a silicon crystal. *(Courtesy of Thomas J. Watson Research Center, IBM Corporation.)*

Scanning sharply focused electron beam

TV spot brightness control

Amplifier

Specimen

Scattered and secondary electrons

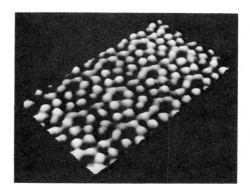

Wave Properties of Other Particles

The de Broglie relation $\lambda = h/mv$ attributes a wavelength to any particle that has momentum; *wave properties are intrinsic to all matter*. Research on neutrons indicates that uncharged particles as well as charged particles can be diffracted. Even atoms and molecules, which have internal structure, have been diffracted. The de Broglie equation is therefore a completely general relationship that applies to all matter; we apply it to electrons in atoms in the next chapter.

27.9 The Wave-Particle Duality

Because all energy is transported in the form of either particles or waves, the concepts of *particle* and *wave* take on great significance in physics. These two concepts provide *models* of reality that enable the physicist to understand the universe better. For example, a particle model, which considers an ideal gas as made up of infinitesimally small particles moving along random paths through the gas, provides a satisfactory description of most of the macroscopic properties of the gas—pressure, temperature, etc. Similarly, a wave model for radio waves adequately describes just about all the phenomena associated with the behavior of such waves—the speed of the waves, reflection, interference, etc. But in both these cases the models are highly idealized: no unstructured particles with all their mass concentrated at one point in space actually exist; neither do radio waves of a single, precisely known frequency, as we will see in the next section. Both these highly successful models are merely that—*models*, not definitive pictures of reality; the physical world about us is too complicated to be contained within, or perfectly described by, oversimplified models of this kind.

The situation becomes even more complicated when we discuss atomic systems and radiation. For example, we have a wave model for electromagnetic radiation that well describes phenomena like interference and diffraction. But we have just discovered that we need a photon—or particle—model to describe blackbody radiation, the photoelectric effect, the continuous x-ray spectrum, and the Compton effect. Similarly, to describe atomic particles like electrons and neutrons the Newtonian particle model is inadequate, for we need de Broglie's idea of matter waves to explain electron and neutron diffraction. This leads us to the following conclusion:

Wave-particle duality: An adequate description of reality at the atomic level requires the application of two different models, the wave model and the particle model, to encompass matter, radiation, and their interactions.

Bohr's Principle of Complementarity

There is an added problem in that the wave and particle models seem to be mutually incompatible. A wave must be spread out over a large region of space; a particle is confined to a very small region of space. How can we say that we need both models to describe an electron? The answer given by the Danish physicist Niels Bohr (see Fig. 28.2 and accompanying biography) in 1928 was that, if we look carefully at our interpretation of the experiments discussed in this and previous chapters, we find that we never apply the wave description and the particle description simultaneously, so the seemingly contradictory aspects of the two models never really come into conflict. Thus, in discussing Young's double-slit experiment, electromagnetic radiation is always treated completely on a wave model, and the concept of the photon is never introduced. Similarly, in dealing with the Compton effect we always apply particle mechanics to a collision between a photon and an electron, and the wave properties of the radiation never enter the picture. In other words, we need *both* a wave and a particle model to describe the atomic universe fully, but these models never introduce logical contradictions into our analysis because they are never applied simultaneously to the same physical situation.

This led Bohr to formulate the following *principle of complementarity* for atomic particles and radiation:

> *Bohr's principle of complementarity:*   The wave and particle descriptions of reality are *complementary*; both models are required to describe the behavior of electromagnetic radiation and atomic particles, but the two are never applied at the same time to the same physical experiment.

Newtonian physics was based on commonsense concepts like absolute space, time, and mass. Einstein's theory of relativity showed the inadequacy of these concepts. So too our ideas of particles and waves are borrowed from everyday experience with large-scale objects. Quantum theory has shown the inadequacy of these classical models as applied to electromagnetic radiation and to atomic particles. Both relativity and quantum theory represent revolutions not only in physics, therefore, but in philosophy as well.

27.10   The Heisenberg Uncertainty Principle

Further insight into the wave-particle duality inherent in the physical world may be obtained from a principle first enunciated in 1927 by the German physicist Werner Heisenberg (see Fig. 27.28) and now called the *Heisenberg uncertainty* (or indeterminacy) *principle*. It makes quantitative the implications of Bohr's principle of complementarity by showing to what extent the wave and particle pictures are incomplete. Heisenberg's work points out that any measurement of a physical system changes the system being measured to some extent and introduces a fundamental uncertainty into measurements of other properties of the system.

FIGURE 27.28   Werner Heisenberg (1901–1976), the originator of the uncertainty principle. Heisenberg received the 1932 Nobel Prize in physics for his contributions to the development of quantum mechanics. He also made substantial contributions to the theory of the atomic nucleus. *(Courtesy of AIP Meggers Gallery of Nobel Laureates.)*

Uncertainties in Position and Momentum of an Electron

Suppose we perform a "thought experiment" using an idealized microscope to determine the position of a single electron by illuminating it with a single photon. The electron is assumed to be at rest initially, and the photon approaches the electron from below, in the positive y direction, bounces off the electron, and enters the microscope lens, as in Fig. 27.29a. As we saw in Sec. 25.6, the minimum separation that can be resolved by a microscope is $1.22\lambda/(2n \sin \alpha)$, where λ is the wavelength of the light, α is the half angle subtended by the microscope lens, and n is in this case equal to 1. The uncertainty in the x position of the electron, as measured by the microscope, is then $\Delta x = 1.22\lambda/(2 \sin \alpha) \cong \lambda/(\sin \alpha)$, since we are interested only in the order of magnitude of the result.

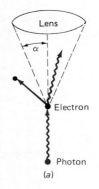

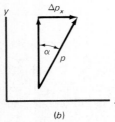

FIGURE 27.29 A thought experiment in which a single photon is used to locate an electron in an idealized light microscope. (a) A photon scattered by an electron and entering the lens of a microscope can have any angle from 0 to α with respect to its original direction. (b) As a result, after the collision the uncertainty in the momentum of the electron in the x direction is $\Delta p_x = p \sin \alpha$.

The photon has an initial momentum in the y direction of magnitude $p = h/\lambda$, where h is Planck's constant and λ is the photon wavelength. In the collision a small amount of momentum is transferred to the electron, and the photon moves off in a new direction with slightly decreased momentum. Depending on the direction in which the photon is scattered, it can have any x component of momentum from 0 to $p_x = p \sin \alpha$, where α is the half angle subtended by the microscope lens. Photons with x components of momentum greater than this do not enter the microscope and can be ignored. By the principle of conservation of momentum, the uncertainty in the x component of the electron's momentum after the collision must be equal to the uncertainty in the x component of the photon's momentum after the collision.

We have, therefore, for the electron,

$$\Delta p_x \simeq p \sin \alpha = \frac{h}{\lambda} \sin \alpha$$

Then, taking the product of these two uncertainties in the electron's position and momentum, we have

$$\Delta x \, \Delta p_x \simeq \frac{\lambda}{\sin \alpha} \frac{h \sin \alpha}{\lambda} \simeq h \qquad \text{(27.16)}$$

This is only an approximate result. Heisenberg's more careful calculation of the product of the uncertainties in position and momentum led to the more exact result

$$\boxed{\Delta x \, \Delta p_x \geq \frac{h}{4\pi}} \qquad \text{(27.17)}$$

The physical significance of this equation is the following:

> *Heisenberg's uncertainty principle:* It is impossible to measure the position and the corresponding momentum of a particle simultaneously with complete accuracy. The product of the uncertainties in the position and the corresponding momentum is greater than, or at best equal to, $h/4\pi$.

According to Newtonian mechanics the position and momentum of a particle can both be measured *exactly at the same time. This is no longer true in quantum physics.* If we try to locate the electron more precisely by decreasing the wavelength λ, we increase the momentum of the incident photon (since $p = h/\lambda$) and thus make the electron's momentum more uncertain. The same uncertainty exists for all other atomic particles.

The uncertainty principle has had profound consequences for both physics and philosophy. In classical physics it was assumed that, if we know the initial position and momentum of a particle at time $t = 0$, then we can calculate its trajectory at later times with any desired precision. The uncertainty principle tells us that this is *impossible* because we can *never* know the initial position and momentum of a particle exactly. This limitation has nothing to do with weaknesses in our measuring instruments or procedures; it is a direct logical consequence of quantum theory. From a philosophical point of view, therefore, it makes no sense to discuss the *exact* position and momentum of a particle at any instant of time, since they are in principle unknowable. This consequence of the Heisenberg principle has had a major impact on philosophical theories of knowledge.

Example 27.8

Classical mechanics assumes that objects have definite positions and momenta. In the light of the uncertainty principle, how can classical mechanics ever make correct predictions about the motion of billiard balls and rockets?

SOLUTION

The uncertainty introduced by the Heisenberg principle is too small to be observable for macroscopic objects.

As an example, consider an object of mass 100 g moving with a speed of 10 m/s. Suppose we know the speed to one part is a million. The uncertainty in the speed is therefore $\Delta v_x = 10^{-6}(10 \text{ m/s}) = 10^{-5}$ m/s, and the uncertainty in the momentum is

$$\Delta p_x = m \, \Delta v_x = (0.10 \text{ kg})(10^{-5} \text{ m/s}) = 10^{-6} \text{ kg·m/s}$$

For such an uncertainty in momentum, the uncertainty

principle states that the uncertainty in position must be greater than

$$\Delta x = \frac{h}{4\pi \, \Delta p_x} = \frac{6.6 \times 10^{-34} \text{ J·s}}{4\pi(10^{-6} \text{ kg·m/s})} = 5 \times 10^{-29} \text{ m}$$

This is less than one billion-billionth the diameter of an atom and too small to have any conceivable practical importance.

The Heisenberg principle, therefore, puts no practical limitations on our knowledge of the position and momentum of macroscopic objects.

Example 27.9

Suppose the speed of an electron in a hydrogen atom is measured to be 4.38×10^6 m/s with an uncertainty of 0.10 percent. What is the uncertainty in the simultaneous determination of the electron's position?

SOLUTION

From the Heisenberg uncertainty principle,

$$\Delta x \, \Delta p_x \geq \frac{h}{4\pi}$$

and so

$$\Delta x \, \Delta v_x \geq \frac{h}{4\pi m} \quad \text{and} \quad \Delta x \geq \frac{h}{4\pi m \, \Delta v_x}$$

We are told that $\Delta v_x = 10^{-3} v_x = 4.38 \times 10^3$ m/s, and so

$$\Delta x \geq \frac{6.6 \times 10^{-34} \text{ J·s}}{4\pi(9.1 \times 10^{-31} \text{ kg})(4.38 \times 10^3 \text{ m/s})}$$

$$\geq 1.3 \times 10^{-8} \text{ m} = 13 \text{ nm}$$

Since 13 nm is 100 times the diameter of a hydrogen atom, it is clear that, if the speed of this electron could be known to 0.10 percent, we would not even know that the electron was in the atom. This shows how important the Heisenberg principle is for small particles like the electron.

Uncertainties in Time and Energy

There is a second important version of Heisenberg's principle that relates uncertainties in time and in energy.

Consider a classical wave. To describe it fully we must know both its wavelength and its frequency exactly, for these quantities determine its speed. If we want to determine the frequency ν precisely, however, the wave must be infinitely long. To show this, suppose we choose to measure ν to high accuracy by beating it against the frequency of a standard oscillator. To be sure that the measurement is highly accurate, we might wait 1 min and judge that, if no beat occurs in that minute, then the two frequencies are exactly the same. But it is possible that, even if no beats were heard in the first minute, one might be heard in the second minute, or in an hour, a day, a year, or a century—which would indicate that the frequencies are not *exactly* the same. To be absolutely sure we would have to wait *forever*, in which time the wave would be spread out over an infinite distance.

Since none of us is patient enough to wait forever to measure a frequency, the question arises: What uncertainty is introduced into our knowledge of the frequency when we limit our measurement to a time Δt? Suppose that the unknown and the standard frequencies are ν_1 and ν_2 and the difference, or beat frequency, is $\Delta \nu = \nu_2 - \nu_1$ so that $\Delta \nu$ beats are observed per second. The time for one full beat is then $1/\Delta \nu$ s. Let us suppose that we count beats for a time about equal to the minimum possible time $\Delta t = 1/\Delta \nu$ which allows us to count one full beat. Then

$$\Delta t \simeq \frac{1}{\Delta \nu} \qquad \text{or} \qquad \Delta \nu \, \Delta t \simeq 1 \tag{27.18}$$

This is a purely *classical expression* for the uncertainty ($\Delta \nu$) in a frequency ν when measured over a time interval Δt. The shorter the time Δt, the larger the uncertainty in the frequency, $\Delta \nu$.

We can convert this classical result into another form of the Heisenberg principle by multiplying both sides of Eq. (27.18) by Planck's constant. Then we have $h \, \Delta \nu \, \Delta t \simeq h$, or, since $\Delta E = h \, \Delta \nu$,

$$\Delta E \, \Delta t \simeq h \tag{27.19}$$

Again, the more accurate calculation carried out by Heisenberg led to the result

$$\boxed{\Delta E \, \Delta t \geq \frac{h}{4\pi}} \tag{27.20}$$

The uncertainties in energy and time are therefore related in the same way as are the uncertainties in position and momentum.

Our conclusion is that, since we can never actually determine simultaneously the position and momentum (or speed) of any "particle" with complete accuracy, we are not justified in concluding that the object under observation is actually a particle in the usual sense of that term. Similarly, since we can never determine the frequency or wavelength of a "wave" with complete accuracy, we are not justified in concluding that the object under observation is actually a wave in the usual sense of the term. Rather we are dealing with more complicated entities requiring both particle and wave models to describe them fully. This is simply another statement of Bohr's principle of complementarity.

The Heisenberg principle in the form of Eq. (27.20) is very important for our knowledge of the energy levels of atoms and their spectra, which are the subjects of the next chapter.

Example 27.10

An electronics catalog advertises a frequency-measuring instrument that will automatically measure and display the frequency of any sinusoidal signal fed into it. To monitor the frequency over time the instrument averages the frequency over 2.0-s intervals and displays this updated value every 2 s. The catalog claims that the instrument is accurate to 0.1 Hz. Can you show why this is an exaggerated claim?

SOLUTION

The classical expression for the uncertainty in a frequency $\Delta \nu$ when measured over a time interval Δt is given by Eq. (27.18) as

$$\Delta \nu \, \Delta t \simeq 1$$

In this case, therefore,

$$\Delta \nu \simeq \frac{1}{2.0 \text{ s}} = 0.50 \text{ Hz}$$

We therefore see that an accuracy of 0.1 Hz is impossible with measurements which take only 2.0 s, no matter how good the instrument. Otherwise the classical version of the uncertainty principle discussed above would be violated.

Summary: Important Definitions and Equations

Particle properties of electromagnetic radiation

Photon: A massless bundle (or "quantum") of electromagnetic energy that behaves like a particle.

Energy of a photon: $E = h\nu$ **(27.2)**

where ν is the frequency of the photon and $h = 6.63 \times 10^{-34}$ J·s is Planck's constant.

Blackbody radiation

Blackbody: A perfect absorber which absorbs all the electromagnetic radiation that falls on it. A blackbody is a perfect emitter.

Planck's equation for blackbody radiation:

$$U(\lambda) = \frac{8\pi hc}{\lambda^5} \frac{1}{e^{hc/\lambda kT} - 1}$$ **(27.1)**

Photoelectric effect: The release of electrons from a clean metal surface when electromagnetic radiation falls on it.

Einstein's equation for photoelectric effect:

$$h\nu = h\nu_0 + \tfrac{1}{2}mv_{max}^2$$ **(27.5)**

where $h\nu_0 = \phi$ = work function of metal and $\tfrac{1}{2}mv_{max}^2 = eV_0$, where e is the electronic charge and V_0 is the cutoff potential.

Continuous x-ray spectrum

Cutoff frequency: $\nu_0 = \dfrac{e}{h} V$ **(27.8)**

Cutoff wavelength: $\lambda_0 = \dfrac{1.24 \times 10^{-6} \text{ m}}{V}$ **(27.7)**

where V is in volts.

Compton effect: The increase in the wavelength of electromagnetic radiation when scattered by free electrons. On a photon model the photon loses energy in the collision, and so its frequency is decreased and its wavelength increased.

Momentum of a photon:

$$p = \frac{h}{\lambda}$$ **(27.11)**

Equation for Compton effect:

$$\Delta\lambda = \lambda_2 - \lambda_1 = \frac{h}{m_0 c}(1 - \cos\theta)$$ **(27.12)**

where $h/m_0 c = 0.00243$ nm is called the Compton wavelength of the electron and $\Delta\lambda$ is the increase in the photon wavelength.

Wave properties of matter

De Broglie waves: Every particle of momentum p has associated with it a wave of wavelength

$$\lambda = \frac{h}{p} = \frac{h}{mv}$$ **(27.13)**

Electron diffraction: $2d \sin\theta = n\lambda$ **(27.15)**

where λ is the de Broglie wavelength of the electron.

Electron microscope: Microscopes using electrons with short de Broglie wavelengths to increase their resolution over that of optical microscopes.

The wave-particle duality: An adequate picture of reality at the atomic level requires the application of two different models, the wave model and the particle model, to describe fully matter, radiation, and their interactions.

Bohr's principle of complementarity: The wave and particle descriptions of reality are complementary; both models are required to describe the behavior of electromagnetic radiation and atomic particles, but the two are never applied at the same time to the same physical experiment.

Heisenberg uncertainty principle: It is impossible to measure the position and the corresponding momentum of a particle simultaneously with complete accuracy.

$$\Delta x \, \Delta p_x \geq \frac{h}{4\pi}$$ **(27.17)**

$$\Delta E \, \Delta t \geq \frac{h}{4\pi}$$ **(27.20)**

Questions

1 (a) Show that the Planck constant h has the dimensions of angular momentum.

(b) Does this mean that angular momentum must be quantized?

2 Try to suggest some everyday analogies with the situation where an atom only absorbs and emits energy in quantized bundles that we call photons.

3 In photographic darkrooms red, or "safety," lights are often kept on during the development process. Explain why this is possible, using the photon theory of light.

4 Explain why ultraviolet (UV) light is more dangerous to the human body (particularly to the eyes) than visible light.

5 Why could not a classical theory of light explain the results of the Compton effect?

6 Why is it impossible to observe the Compton effect using visible light?

7 Discuss changes that might occur in our daily lives if Planck's constant were four orders of magnitude larger than it is.

8 Why would gamma-ray microscopes not be particularly useful, even though they would have very high resolving powers?

9 Stanford physicist Wolfgang Panofsky is quoted as saying: "The smaller the objects, the bigger the microscope we must use to see them." Comment on the accuracy of this statement as applied to the electron microscope.

10 If the energies associated with the valence electrons in atoms and molecules are about 1 to 2 eV, why would you expect the quantum nature of visible light to be very important when light interacts with atoms and molecules?

11 Atoms vibrating in solid crystal lattices are found to have a small, "zero-point" energy at absolute zero (0 K). Explain why this is required by the Heisenberg uncertainty principle.

Multiple-Choice and Simple Exercises

27.1 The energy associated with a photon of red light is about:
(a) 2.8×10^{-19} eV   (b) 2.8×10^{-19} J
(c) 6.5×10^{-47} J   (d) 4.6×10^{-40} eV
(e) 4.6×10^{-40} J

27.2 A radio antenna radiates 10^4 W of power at 10^6 Hz. How many photons per second are emitted by the antenna?
(a) 1.0×10^2 (b) 2.4×10^{10}
(c) 2.4×10^{12} (d) 6.2×10^{23}
(e) 1.5×10^{31}

27.3 In Planck's equation for blackbody radiation the exponent $h\nu/kT$ is equal to 1 for visible light at a temperature of about:
(a) 2.5×10^2 K   (b) 25 K   (c) 2.5×10^{10} K
(d) 2.5×10^3 K   (e) 2.5×10^4 K

27.4 Which of the following is *not* an observed characteristic of the photoelectric effect?
(a) For a given material there is a minimum frequency below which electrons are not emitted.
(b) The number of photoelectrons emitted is proportional to the intensity of the light.
(c) The kinetic energy of the photoelectrons depends on the frequency of the incident light.
(d) The kinetic energy of the emitted photoelectrons depends on the intensity of the incident light.
(e) Electrons are immediately emitted when light is incident on the metal surface.

27.5 In a photoelectric effect experiment the intensity of the light is increased by a factor of 3, with the frequency held constant. The speed of the maximum-energy electrons being ejected from the cathode increases by a factor of:
(a) 9   (b) 2   (c) 3   (d) $\sqrt{3}$
(e) None of the above

27.6 An x-ray photon collides head-on with an electron and is scattered directly back at 180° to its original direction. The shift in the wavelength of the incident x-ray is:
(a) $+0.00243$ nm   (b) -0.00243 nm
(c) -0.00486 nm   (d) $+0.00486$ nm   (e) 0

27.7 The highest frequency that occurs in the continuous radiation from an x-ray tube to which a voltage of 1.0×10^5 V is applied is:
(a) 2.4×10^{12} Hz   (b) 2.4×10^{19} Hz
(c) 3.8×10^{18} Hz   (d) 2.4×10^{26} Hz
(e) 1.5×10^{20} Hz

27.8 To produce the diffraction of particles moving with a speed of 10^6 m/s with a diffraction grating ruled with 10,000 lines to the centimeter, the mass of the particles would have to be smaller than that of an electron by about:

(a) Six orders of magnitude
(b) Three orders of magnitude
(c) One order of magnitude
(d) It would be about equal to the electron mass.
(e) It would have to be greater than the electron mass.

27.9 For neutrons to be diffracted by crystals, the speed of the neutrons must correspond to a temperature of roughly:
(a) 600 K   (b) 0 K   (c) 6000 K
(d) 10^5 K   (e) None of the above

27.10 The de Broglie wavelength of an electron accelerated by a voltage V is:
(a) $\dfrac{h/(2em)^{1/2}}{V^{1/2}}$ (b) $\dfrac{(2em)^{1/2}h}{V^{1/2}}$ (c) $\dfrac{(2emh)^{1/2}}{V^{1/2}}$
(d) $\dfrac{h/(2em)^{1/2}}{V}$ (e) $\dfrac{h(2em)^{1/2}}{V}$

27.11 What is the uncertainty in the speed of an electron if it is desired to locate its position to within an uncertainty of 1.0 nm?
(a) 5.8×10^5 m/s   (b) 5.8×10^4 m/s
(c) 1.8×10^5 m/s   (d) 7.3×10^5 m/s
(e) 5.8×10^{-4} m/s

27.12 Find the energy associated with the photons emitted by the antenna of WGMS, a Washington, D.C., FM radio station that broadcasts at a frequency of 103.5 MHz.

27.13 During a chest x-ray a person's body normally absorbs about 1.5×10^{-3} J of energy. About how many x-ray photons are absorbed by a person having a chest x-ray? (Assume an x-ray wavelength of 0.10 nm.)

27.14 (a) Will red light cause electrons to be emitted from a clean aluminum surface? Why? (*Hint:* See Table 27.2.)
(b) What is the minimum frequency for incident radiation that will be able to eject electrons from aluminum?

27.15 It is found that light of wavelength shorter than 300 nm can eject electrons from a metal surface in a vacuum. What is the work function of the metal?

27.16 An electron beam is accelerated by a voltage of 250 V and then crashes into a tungsten target. What is the shortest wavelength in the x-radiation emitted by the target?

27.17 (a) What is the momentum of a gamma-ray photon of energy 1 MeV?
(b) What is the de Broglie wavelength of an electron moving with a speed of 3.0×10^5 m/s?

27.18 X-radiation of wavelength 0.095 nm is incident on a "free-electron gas," i.e., a collection of free electrons. Calculate the energy acquired by an electron that is struck by an x-ray photon, if the photon is deflected by an angle of 60° with respect to its original direction.

27.19 When x-rays are scattered from a KCl crystal with lattice spacing $d = 0.314$ nm, it is found that the smallest angle of incidence with respect to the lattice planes at which constructive interference occurs is 35.0°. What is the wavelength of the incident x-rays?

27.20 Some unstable nuclei have lifetimes in excited states of only 10^{-14} s. What is the limit on the accuracy with which the energy of such an excited state can be determined?

Problems

27.21 A rough value of Planck's constant can be obtained by making the reasonable assumption that, when $kT \simeq h\nu$ in Eq. (27.1), the energy density emitted by a blackbody at a wavelength $\lambda = c/\nu$ becomes large. Use the experimental fact that a blackbody glows a bright red at about 3000°C to obtain an order-of-magnitude value of Planck's constant h in this way.

27.22 A neon (Ne) lamp emits 40 W of red light at a wavelength of 680 nm. How many photons are emitted by the lamp per second?

27.23 Find the energies (in both joules and electron-volts) associated with (a) the infrared ($\lambda = 1000$ nm) and (b) the ultraviolet ($\lambda = 100$ nm) regions of the spectrum.

27.24 To cause sunburn the incident light must be able to break chemical bonds in the molecules of the human skin. To do this requires an energy of about 3.5 eV in the photons of the incident sunlight.

(a) To what wavelength does this energy correspond?

(b) In what region of the spectrum does this wavelength fall?

27.25 From each of the three experimental graphs shown in Fig. 27.9 calculate a value for Planck's constant h. From the average of your three results obtain the best possible value for h from the data given.

27.26 What cutoff voltage must be applied to stop the fastest photoelectrons emitted by a tungsten surface under the action of ultraviolet light of wavelength 250 nm? (*Hint:* See Table 27.2.)

27.27 It is found that violet light of wavelength 320 nm is able to eject from a metal electrons that have maximum speeds of 4.0×10^5 m/s. What is the work function of the metal?

27.28 Electrons with maximum speeds of 4.2×10^5 m/s are ejected from a clean metal surface by 200-nm ultraviolet radiation. Find (a) the work function of the metal; (b) the threshold wavelength of the metal for electron emission; (c) the retarding potential difference required to stop the electron flow.

27.29 For the metals given in Table 27.2 calculate enough threshold frequencies to determine from which metals electrons can be ejected by visible light.

***27.30** The wavelength of the light striking a metal surface is 200 nm less than the threshold wavelength. The measured maximum kinetic energy of the ejected photoelectrons is 2.5 eV. What is the threshold frequency for this metal?

27.31 An electron is deep inside a metal and requires 1.50 eV of energy to move to the surface of the metal before it can be ejected by an incident photon. The metal is potassium and the incident photon has a wavelength of 400 nm.

(a) What is the kinetic energy with which the electron emerges from the metal?

(b) What is the speed of this electron?

***27.32** Prove that it is impossible for a photon to give up all its energy and momentum to a *free* electron. As a result, the photoelectric effect can only take place when the electrons involved are *bound* to the metal lattice. (*Hint:* Write equations for conservation of energy and momentum.)

27.33 A dentist's x-ray machine has an accelerating voltage for the electron beam of 50,000 V. This beam strikes a lead target of work function 4.25 eV.

(a) What is the shortest wavelength present in the x-radiation produced?

(b) What is the highest-frequency photon produced?

27.34 An x-ray machine produces x-radiation of minimum wavelength 0.20 nm. What accelerating voltage does it employ?

27.35 (a) Derive a numerical relationship between the maximum frequency ν_{max} emitted by an x-ray tube and the voltage V applied to the tube.

(b) What is the minimum voltage that must be applied to a x-ray tube to produce x-rays of frequency 3.0×10^{18} Hz?

27.36 The potential difference used to accelerate the electron beam in a TV picture tube is usually about 2000 V. When the electrons strike the phosphor on the tube face, photons are emitted.

(a) What is the minimum wavelength of the emitted photons?

(b) Do you think these photons could be a source of danger to TV watchers?

27.37 An x-ray photon of wavelength 0.252 nm strikes a free electron and is scattered at an angle of 120°.

(a) What is the energy of the photon before the collision?

(b) What is the momentum of the photon?

(c) What is the wavelength of the x-ray photon after it is scattered?

(d) What is the energy of the scattering electron after the collision?

27.38 An x-ray photon of wavelength 0.200 nm is scattered by a free electron at an angle of 90.0°.

(a) What is the speed of the photon after the collision?

(b) What is the wavelength of the photon after the collision?

(c) What is the energy of the electron after the collision?

(d) What is its speed, if we neglect the effects of relativity?

***27.39** An x-ray photon of wavelength 0.120 nm is scattered from the electrons in a carbon target.

(a) Calculate the wavelength of the x-rays scattered at an angle of 60.0° with respect to the original direction.

(b) Calculate the direction of the velocity of the scattering electron after the collision.

27.40 What is the wavelength and the frequency of a 60-MeV photon?

27.41 An x-ray photon whose initial frequency was 1.60×10^{19} Hz emerges from a collision with an electron with a new frequency of 1.40×10^{19} Hz. How much kinetic energy was imparted to the electron?

∗27.42 Carry through the derivation of the equation for the Compton effect [Eq. (27.12)] from the equations describing the conservation of energy and momentum in a collision between an x-ray photon and an electron. (*Hint:* Equate the energies before and after the collision and the x and y components of momentum before and after the collision. The electron must be treated relativistically.)

∗27.43 A free proton at rest is struck by a photon in a Compton collision. The proton acquires a kinetic energy of 5.7 MeV in the collision. What is the minimum photon energy that can produce this result?

27.44 An inventor suggests using the momentum of photons to propel a rocket and recommends using a mercury lamp that emits 200 W of violet radiation at 356 nm. The rocket has a mass of 100 kg and is in free space.

(*a*) What force can be generated in this way?

(*b*) What will be the acceleration of the rocket?

27.45 (*a*) How fast would an electron have to travel to have the same wavelength as a 2.0-eV photon?

(*b*) How fast would a neutron have to travel to have the same wavelength as a 2.0-eV photon?

27.46 What wavelength must a photon have if its momentum is to be the same as that of an electron moving with a speed of 3.5×10^5 m/s?

27.47 What is the de Broglie wavelength of a 10-keV electron?

27.48 What is the energy and wavelength of a thermal neutron, i.e., a neutron at a temperature of about 300 K?

∗27.49 What is the wavelength of an electron with total energy 1.0 MeV? (Since the speed of such an electron is so high, relativistic formulas must be used.)

27.50 A beam of x-rays is reflected from the lattice planes of a KCl crystal, which are separated by a distance of 0.314 nm. The minimum angle of incidence with respect to the crystal planes that produces a strong reflected x-ray beam is 16.0°.

(*a*) What is the wavelength of the x-rays used?

(*b*) At what angle does the second-order reflection occur?

27.51 A beam of x-rays of wavelength 0.15 nm is reflected by the lattice planes in a crystal whose structure is being investigated. The minimum angle of incidence with respect to the lattice planes that produces a strong reflected x-ray beam is 26°.

(*a*) What is the lattice separation d in the crystal?

(*b*) At what angle would you expect to find the second-order diffraction of the x-ray beam?

27.52 A beam of 60-V electrons is incident on a KCl crystal, with lattice spacing between planes of 0.314 nm. For what angles of incidence with respect to the lattice planes will strong electron beams be detected?

27.53 A beam of electrons is incident on a NaCl crystal with lattice spacing $d = 0.282$ nm. The minimum angle of incidence for which a strong diffracted beam is observed is 4.6°. What is the energy of the electrons in the beam?

27.54 What potential difference would be required in an electron microscope to produce electrons with wavelengths of 0.016 nm? (Neglect the effects of relativity.)

27.55 A metal puck of mass 0.50 kg floats without friction on a layer of air blown through holes in the bottom of an air table. A physicist measures the position of this puck to an accuracy of 10^{-4} m.

(*a*) What speed is imparted to the puck by the act of measuring its position? (Set the speed equal to the uncertainty in the speed.)

(*b*) How long would it take for the puck to cross the 2.0-m table at this speed?

27.56 The energy of an electron state in a hydrogen atom is 3.4 eV. If it is desired to measure this energy to an accuracy of 2 percent, what is the minimum time required for the measurement?

27.57 (*a*) A proton is confined to a box of length 1.0 nm. What is the uncertainty in the proton's speed?

(*b*) An electron is confined to a box of length 1.0 nm. What is the uncertainty in the electron's speed?

27.58 Use the Heisenberg uncertainty principle to estimate the kinetic energy of a neutron in a nucleus of radius 7.0×10^{-15} m.

∗27.59 Calculate the uncertainties in position and momentum for a photon diffracted by a narrow slit of width w, and show that their product satisfies the Heisenberg uncertainty principle.

Additional Readings

Binnig, Gerd, and Heinrich Rohrer: "The Scanning Tunneling Microscope," *Scientific American*, vol. 253, no. 2, August 1985, pp. 50–56. An account of the work for which the authors received the 1986 Nobel Prize.

Born, Max: "Max Karl Ernst Ludwig Planck," *Obituary Notices of Fellows of the Royal Society of London*, vol. 6, 1948, pp. 161–180. A brief biography of Planck by a colleague and coworker who knew him well. This is reprinted in Henry A. Boorse and Lloyd Motz (eds.), *The World of the Atom*, vol. 1, Basic Books, New York, 1966, pp. 462–484.

de Broglie, Louis: *The Revolution in Physics: A Non-Mathematical Survey of Quanta* (trans. by Ralph W. Niemeyer), Noonday Press, New York, 1953. An especially lucid account of modern physics by a physicist who contributed greatly to its development.

Everhart, Thomas E., and Thomas L. Hayes: "The Scanning Electron Microscope," *Scientific American*, vol. 226, no. 1 January 1972, pp. 54–69. Contains some remarkable photographs of biological specimens made with the scanning electron microscope.

Gamow, George: "The Principle of Uncertainty," *Scientific*

American, vol. 198, no. 1, January 1958, pp. 51–57. A clear exposition of the background and significance of the Heisenberg uncertainty principle.

————: *Thirty Years That Shook Physics: The Story of Quantum Theory*, Doubleday Anchor, Garden City, N.Y., 1966. A clear and well-illustrated book on the development of the quantum theory.

Heilbron, John: *The Dilemmas of an Upright Man: Max Planck as a Spokesman for German Science*, University of California Press, Berkeley, 1986. A sensitive portrayal of the problems faced by Planck and other German physicists under Hitler.

Hey, Tony, and Patrick Walters: *The Quantum Universe*, Cambridge University Press, New York, 1987. A popular, well-illustrated account of quanta and their applications.

Kelves, D. J.: "Robert A. Millikan," *Scientific American*, vol. 240, no. 1, January 1979, pp. 142–151. A frank and interesting account of the best-known American scientist of his day.

Medicus, H. A.: "Fifty Years of Matter Waves," *Physics Today*, vol. 27, no. 2, February 1974, pp. 38–45. A historical discussion of how de Broglie waves were first predicted by de Broglie and then found in the laboratory.

Planck, Max: *Scientific Autobiography and Other Papers* (trans. by F. Gaynor), Philosophical Library, New York, 1949. Planck's very brief and often moving autobiography is included in this volume, along with other Planck lectures on a variety of subjects.

Segré, Emilio: *From X-Rays to Quarks*, Freeman, San Francisco, 1980. A personal account of modern physicists and their discoveries, starting with Roentgen's discovery of x-rays, by a physicist who contributed substantially to this exciting period.

CHAPTER 28

The Structure of Atoms

That this insecure and contradictory foundation was sufficient to enable a man of Bohr's unique instinct and tact to discover the major laws of the spectral lines and of the electron shells of the atoms together with their significance for chemistry, appeared to me like a miracle—and appears to me as a miracle even today. This is the highest form of musicality in the sphere of thought.

Albert Einstein (1879–1955)

The work of Planck and Einstein revolutionized our understanding of particles and radiation. But it indicated little about the structure of atoms and molecules, of those building blocks from which the universe is constructed. Ever since Kirchhoff invented the spectroscope in 1859, physicists had been accumulating data on the spectra of atoms. It had become clear that all atoms of an element emitted exactly the same series of spectral lines and that these lines could be used for the positive identification of that element. This suggested that all atoms of any given element probably had exactly the same structure. But what kind of a structure did atoms have? And why should all hydrogen atoms, for example, have the same spectrum, and therefore presumably the same structure?

The beginnings of answers to these questions were given by Niels Bohr, who carried the work of Planck and Einstein a giant step forward and provided our first important insight into the structure of atoms.

28.1 The Spectrum of the Hydrogen Atom

Let us begin with the simplest of atoms, hydrogen. This is the most important atom in physics because it contains only one electron and one proton. Thus it avoids the complications inherent in heavier, more complex atoms.

Suppose we excite *atomic* hydrogen (*not* H_2, which is *molecular* hydrogen) by an electric discharge, produced by a high voltage across electrodes in an evacuated glass tube, and observe the resulting hydrogen spectrum with a prism or grating spectrometer. This *emission spectrum* of atomic hydrogen consists of a number of sharp, discrete, bright lines on a dark background, the lines being images of the spectrograph slit. Such spectra are called *line* or *discrete spectra*.

The line spectrum of atomic hydrogen in the visible region is shown in Fig. 28.1 and on the color plate Optical Spectra facing p. 742. The lines are given their usual notation H_α, H_β, H_γ, . . . , with consecutive Greek letters referring to lines at higher frequencies (or shorter wavelengths). These lines approach a *series limit* where they bunch together. Beyond the series limit is a *continuum*, or continuous light pattern, in which discrete lines can no longer be seen.

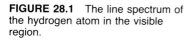

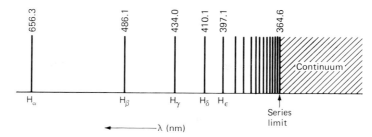

FIGURE 28.1   The line spectrum of the hydrogen atom in the visible region.

This visible spectrum of H is referred to as the *Balmer series* after the Swiss scientist J. J. Balmer (1825–1898), who in 1885 found a simple empirical formula for calculating the wavelengths of its lines. The *Balmer formula* is

$$\frac{1}{\lambda_n} = R\left(\frac{1}{2^2} - \frac{1}{n^2}\right) \tag{28.1}$$

where λ_n is the wavelength (in meters) of a particular line in the series, R is a constant now called the *Rydberg constant* after the Swedish physicist J. R. Rydberg (1854–1919), and n is an integer taking on values 3, 4, 5, . . . for the successive lines H_α, H_β, H_γ, ... A value for R of 1.0968×10^7 m^{-1} gives agreement with the wavelengths of the observed Balmer series lines. This Balmer formula explains nothing; it merely correlates the available experimental data in a convenient fashion. Thus, if we put $n = 4$ into Eq. (28.1), we find that the wavelength of H_β is 486 nm, in agreement with the experimental value in Fig. 28.1. Similar agreement is found for all lines in the visible spectrum. The lines get closer together as the frequency increases. The series limit corresponds to $n = \infty$ for which $\lambda_\infty = 4/R = 365$ nm, again in agreement with experiment.

Other Spectral Series in Hydrogen

In addition to the visible Balmer series, H has a similar series in the ultraviolet, the *Lyman series* [named after the American physicist Theodore Lyman (1874–1954)], and a number of additional spectral series in the infrared, the most important being the *Paschen series* [named after the German physicist L. Paschen (1865–1947)]. The Balmer, Lyman, and Paschen series can all be represented by one equation, similar to Eq. (28.1), called the *Rydberg equation*:

$$\frac{1}{\lambda} = R\left(\frac{1}{n_f^2} - \frac{1}{n_i^2}\right) \tag{28.2}$$

where the Rydberg constant R has the same value for all the series. Here n_f and

TABLE 28.1   Important Spectral Series for the Hydrogen Atom

| | | Values of | |
Region	Name of series	n_f	n_i
Ultraviolet	Lyman	1	2, 3, 4, . . .
Visible	Balmer	2	3, 4, 5, . . .
Infrared	Paschen	3	4, 5, 6, . . .

n_i are integers whose significance we will explain shortly. The three important series in H are characterized by the values of n_f and n_i given in Table 28.1.

Example 28.1

Calculate the wavelengths of the following lines in the spectrum of hydrogen, and indicate in what region of the spec-

trum they fall: **(a)** the first line in the Lyman series; **(b)** the second line in the Balmer series.

SOLUTION

Using the Rydberg equation and the data on the various spectral series in hydrogen from Table 28.1, we have

(a) $\dfrac{1}{\lambda} = R\left(\dfrac{1}{1^2} - \dfrac{1}{2^2}\right) = (1.0968 \times 10^7 \text{ m}^{-1})\left(\dfrac{3}{4}\right)$

or $\boxed{\lambda = 122 \text{ nm}}$ in the *ultraviolet*

(b) $\dfrac{1}{\lambda} = R\left(\dfrac{1}{2^2} - \dfrac{1}{4^2}\right) = (1.0968 \times 10^7 \text{ m}^{-1})\left(\dfrac{3}{16}\right)$

or $\boxed{\lambda = 486 \text{ nm}}$ in the *visible*

EXERCISE 1 Calculate the wavelength of the fourth line in the Paschen series.

28.2 The Bohr-Rutherford Atom

The first physicist to give a successful explanation of the observed spectrum and structure of hydrogen from first principles was Niels Bohr (see Fig. 28.3 and accompanying biography). In a daring paper published in 1913 Bohr tied together the results of classical physics with some highly original ideas to predict both the spectrum of the hydrogen atom and a precise value for the Rydberg constant R.

Among the classical ideas folded into Bohr's theory of the hydrogen atom were orbital motion, angular momentum, and centripetal force from classical mechanics, and Coulomb's law from electromagnetism. The basic model Bohr adopted for the hydrogen atom was Rutherford's nuclear model, which we must consider in more detail before we can understand how Bohr used it to explain the structure and spectrum of hydrogen.

Rutherford's Alpha-Particle Scattering Experiments

In the early years of this century Ernest Rutherford and his students were using alpha particles (that is, He atoms with their electrons stripped away) from radioactive materials to explore the structure of atoms. Suppose, for example, that we allow alpha particles from polonium, which have speeds of 1.6×10^7 m/s, to fall on gold foil, which is only 5×10^{-7} m thick, as in Fig. 28.2. On the other side of the foil from the polonium we place a screen painted with a material that scintillates when struck by an alpha particle (this is the forerunner of the scintillation counters used in nuclear physics today). Then by moving the screen and counting

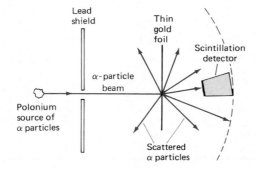

FIGURE 28.2 Alpha-particle scattering by a thin gold foil.

Niels Bohr (1885–1962)

FIGURE 28.3 Niels Bohr (right) and Ernest Rutherford with their wives (photograph by Mark Oliphant, whose wife appears in front of Rutherford). Mrs. Rutherford is just visible at the left. *(Courtesy of AIP Niels Bohr Library, Margrethe Bohr collection.)*

This prolific contributor to twentieth-century physics was the son of a physiology professor at the University of Copenhagen. While studying physics there, Bohr also developed into an excellent soccer player. The staunch experimentalist Lord Rutherford was once asked why he put more faith in Bohr's ideas than he did in those of other theoretical physicists. Rutherford answered: "Bohr's different. He's a football player!"

Bohr received his doctorate from Copenhagen in 1911. He resolved to go abroad for further study, since there was little physics research going on in Denmark at that time. He did research for a short time with J. J. Thomson in Cambridge, but became discouraged when Thomson could never find time to read his dissertation on the electron theory of metals. He moved to Manchester in 1912 to work with Rutherford, and there he developed the theory of atomic structure

and spectra that revolutionized physics and brought Bohr the 1922 Nobel Prize in physics.

He returned to Copenhagen as a lecturer in theoretical physics in 1913, then left to assume a similar post in Manchester during the years 1914 to 1916. He returned to Copenhagen as a professor of theoretical physics in 1916. He spent the rest of his life in Denmark except for the years 1943 to 1945, when he had to flee to Sweden to avoid arrest by the Nazis. He later moved to England and then to the United States to help with the atomic bomb project at Los Alamos, New Mexico. He and his family were happily reunited in Copenhagen in 1945.

In addition to his theory of the structure and spectra of hydrogen, which he extended to explain the shell structure of other atoms, Bohr made many other important contributions to physics. These include the principle of complementarity (Sec. 27.9), the correspondence principle, the liquid-drop model of the atomic nucleus, and the theory of nuclear fission.

In some ways Bohr made an even greater impact on physics by his work as director of the University Institute for Theoretical Physics, which was created for Bohr in Copenhagen in 1921 and supported by the Carlsberg Brewery. (As Isaac Asimov has pointed out, this was the greatest contribution made by beer to physics since the days of James Joule.) In the years between the two world wars Copenhagen became the mecca for the world's best theoretical physicists (some 600 over the years). They

came to discuss and argue physics with equally bright and dedicated colleagues from many different countries, always guided and stimulated by Bohr's penetrating understanding and deep interest in their work.

Niels Bohr provided another great service to physics by opening his institute to Jewish physicists when they lost their academic positions in Germany in 1933. He did everything possible to help them escape and to place them in academic positions in England and the United States.

While in the United States during the war Bohr tried to convince President Roosevelt and Prime Minister Churchill that they should reveal the existence of the nuclear bomb to the world and have nuclear weapons outlawed and put under international control. Roosevelt died before anything could come of Bohr's plan, and Churchill did not like Bohr's ideas. For his contributions to the peaceful use of atomic energy, Bohr received the Atoms for Peace award in 1957.

In 1912 Bohr had married Margrethe Norlund, who was an ideal spouse and companion over 50 years until Bohr's death in 1962. Bohr's personal life appears to have been an exceptionally happy one, except for the death in a boating accident in 1934 of the oldest of his five sons at the age of 19.

Niels Bohr was a simple, modest, generous human being, much admired by all who knew him. In addition to being a great physicist, he was, in the words of a colleague, "the wisest and most lovable of men."

the scintillations, we can study the angular distribution of the alpha particles scattered by the gold foil.

What we find is that most of the alpha particles pass right through the solid foil with little or no deflection. Only 1 in 1000 is deflected through more than 10°.

Occasionally, however, an alpha particle is deflected *through a very large angle*. Also, all the alphas, including the undeflected ones, are slowed down as they pass through the foil.

In 1911 Rutherford (see Fig. 9.1 and accompanying biography) proposed a model of the atom to explain this behavior. His model consisted of a very small, positively charged *nucleus*, containing almost all the mass of the atom, surrounded by a number of electrons equal to the number of elementary positive charges on the nucleus so that the atom was electrically neutral. Both the electrons and the nucleus were so small compared with the size of the atom that most of the atom was empty space. This would explain why most of the alphas passed through the foil undeflected, since it was only 2000 atoms thick. On a rare occasion, however, an alpha particle would hit a nucleus almost head-on and be deflected through a large angle, perhaps even having its direction reversed. Such reversals clearly implied the presence of a nucleus much more massive than the alphas being scattered. All the alphas also underwent collisions with electrons and swept them out of the atom but, since the alpha-particle mass was almost 10^4 times the electron mass, the alphas were not deflected but only slowed down by such collisions, in much the way a baseball would be on passing through a swarm of bees. On the basis of this model Rutherford calculated the expected angular distribution for the scattered alpha particles. These predictions were completely confirmed in a series of experiments carried out by Geiger and Marsden in Rutherford's laboratory in Manchester.

The Size of the Nucleus

In his nuclear model Rutherford had assumed that the force between the alpha particle and the positively charged nucleus was a repulsive Coulomb force of the form $F_{el} = k_e q_\alpha q_N / r^2$. This does not imply that the nucleus is a point charge, however. The nucleus could be a sphere of radius r_0 and the Coulomb force law would remain valid as long as the alpha particle did not penetrate the sphere. Since Geiger and Marsden's results indicated that the Coulomb force law remained valid even for head-on collisions, the alpha particle's distance of closest approach must have been greater than the radius of the nucleus r_0. The distance of closest approach for a head-on collision, therefore, sets an upper limit on the radius of the nucleus.

In a head-on collision an alpha particle moves in a straight line until its electric potential energy becomes exactly equal to its original kinetic energy, as in Fig. 28.4. Then it stops and retraces its path. (This is like throwing a ball up in the air; it stops at its highest elevation when its gravitational potential energy is equal to the kinetic energy it had when it left the thrower's hand.) Suppose r_0 is the distance of closest approach to the center of the nucleus. Then the electrostatic potential energy of the alpha particle at a distance r_0 from the gold nucleus is, from Eq. (17.8),

FIGURE 28.4 Distance of closest approach r_0 for a head-on collision of an alpha particle with a gold nucleus. The charge on the alpha particle is +2 elementary charges, and on the gold nucleus +79 elementary charges (drawing not to scale).

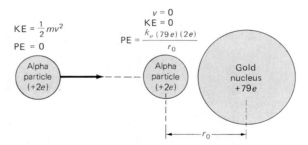

$$PE = \frac{k_e(79e)(2e)}{r_0} = \left(9.0 \times 10^9 \ \frac{\text{N·m}^2}{\text{C}^2}\right) \frac{79(2)(1.6 \times 10^{-19} \ \text{C})^2}{r_0}$$

$$= \frac{3.6 \times 10^{-26} \ \text{J·m}}{r_0}$$

The original kinetic energy of the polonium alpha particles, which have masses four times the proton mass and speeds of 1.6×10^7 m/s, was

$$KE = \tfrac{1}{2}(4m_p)v^2 = 8.6 \times 10^{-13} \ \text{J}$$

On equating this initial kinetic energy to the alpha's potential energy when it stops, we obtain

$$r_0 = \frac{3.6 \times 10^{-26} \ \text{J·m}}{8.6 \times 10^{-13} \ \text{J}} = 4.2 \times 10^{-14} \ \text{m}$$

The radius of a gold nucleus must therefore be less than 4.2×10^{-14} m! Since the radii of atoms are known to be closer to 10^{-10} m, the nucleus has a radius about 10^{-4} that of an atom and therefore a volume about $(10^{-4})^3 = 10^{-12}$ that of an atom. It is clear from this that the mass of an atom is concentrated almost exclusively in a very dense, positively charged nucleus at its center.

Rutherford's model of a hydrogen atom then consists of a nucleus made up of a single proton with charge $+e$, and of a single electron with charge $-e$ outside the nucleus. Bohr added the assumption that this electron moves in an orbit around the nucleus just as the earth moves in its orbit around the sun. On the basis of classical physics, however, such a model would predict that the *accelerated* electron (it is changing *direction*) would radiate energy (all accelerated charges do, according to Maxwell's electrodynamics) and spiral into the nucleus. This would happen in less than 10^{-6} s, and so hydrogen atoms would be extremely unstable. Bohr was therefore driven to some very nonclassical ideas and revolutionary postulates to explain the spectrum of hydrogen from first principles.

28.3 Bohr's Theory of the Hydrogen Atom

While in Manchester with Rutherford in 1912, Bohr thought of combining Rutherford's model of the atom with the quantum ideas of Planck and Einstein to explain the spectra of atoms. About this time he chanced to find Balmer's formula for the hydrogen spectrum and decided to apply his ideas to the spectrum of this simplest of all atoms, using what is now called the *Bohr-Rutherford model* of the atom.

Postulates of the Bohr Theory

In a paper published in 1913 Bohr was able to predict the spectrum of hydrogen from first principles on the basis of the following postulates:

1   The hydrogen atom exists in certain *stationary states* of discrete energies. While in such states the atom is stable and emits no radiation.

2   These stationary states are those for which the electron's classical angular momentum, $L = I\omega = mr^2\omega = mvr$, is quantized in units of $\hbar = h/2\pi$.

$$L = mvr = \frac{nh}{2\pi} = n\hbar \qquad\qquad \textbf{(28.3)}$$

where h is Planck's constant and n is an integer. (Note that the units of "action" and therefore of h, the quantum of action, are the same as the units of angular momentum. The quantity $\hbar = h/2\pi$ is called "h bar.")

3 Radiation is emitted by an atom when the atom undergoes a transition from a higher-energy stationary state (E_i) to a lower-energy state (E_f). The energy difference between these two states then appears as a single photon of energy $E = h\nu = E_i - E_f$, from which the photon frequency is

Bohr-Einstein frequency condition:
$$\nu = \frac{E_i - E_f}{h}$$
(28.4)

Bohr's first postulate contradicts classical electrodynamics, according to which an accelerated electron must radiate continuously. (As Bohr indicated in his paper, this postulate "is in obvious contrast to the ordinary ideas of electrodynamics, but appears to be necessary in order to account for experimental facts.") Bohr's second postulate that the angular momentum can only take on quantized values is again a daring new idea, with no counterpart in classical physics. A planet or earth satellite can have any value of angular momentum, so long as gravitational attraction can provide the centripetal force needed to keep the planet or satellite moving in its orbit. The actual angular momentum depends on the initial velocity of the planet or satellite when it goes into orbit. Bohr's postulate says that this is not the case for the hydrogen atom. No matter how the atom is formed, only certain discrete values of the angular momentum of the electron are possible. We will see in Sec. 28.10 that this is a consequence of the wave nature of the electron.

Application of Bohr's Postulates to Hydrogen

Bohr assumed that the centripetal force mv^2/r needed to hold the electron in a circular orbit about the nucleus must be provided by the Coulomb force, of magnitude $k_e e^2/r^2$, as shown in Fig. 28.5. We have, therefore,

$$\frac{mv^2}{r} = \frac{k_e e^2}{r^2} \quad \text{or} \quad mv^2 = \frac{k_e e^2}{r}$$
(28.5)

The total energy of the atom is then the sum of its kinetic and potential energies:

$$E = \frac{mv^2}{2} - \frac{k_e e^2}{r}$$

where $-k_e e^2/r$ is the potential energy of the electron at a distance r from the proton (see Sec. 17.4).

From Eq. (28.5) this becomes

$$E = \frac{k_e e^2}{2r} - \frac{k_e e^2}{r} = -\frac{k_e e^2}{2r}$$
(28.6)

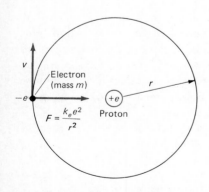

FIGURE 28.5 Bohr model of the hydrogen atom.

The total energy of the atom is *negative*, because the zero of potential energy is taken for the proton and electron infinitely far apart (Sec. 17.4). A negative sign therefore means that the electron is bound to the nucleus, and so the atom is stable.

This classical equation allows all possible values of r and hence all possible values of the energy E. Bohr now introduced his second postulate, according to which the angular momentum $L = mvr = n\hbar$. This postulate quantizes not merely the angular momentum but also the energy E. We specify the quantized values of

OPTICAL SPECTRA

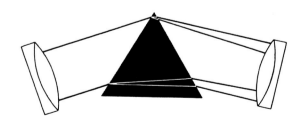

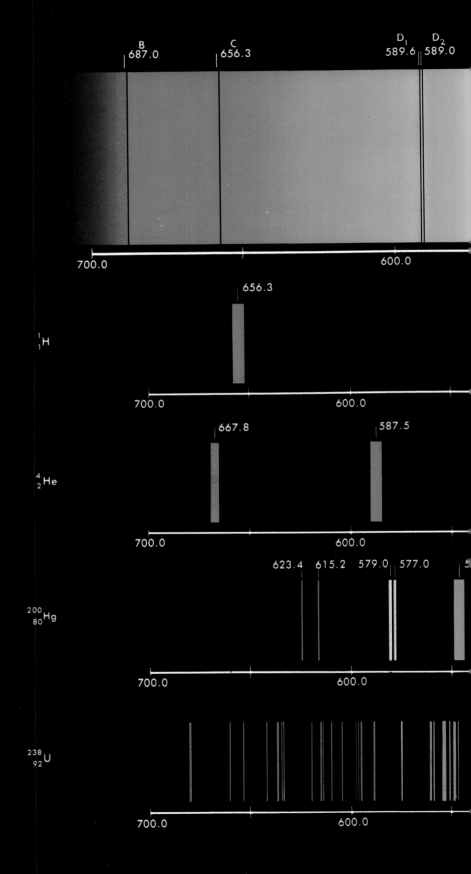

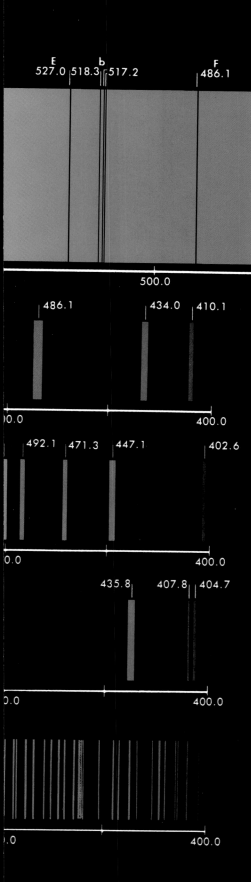

E 527.0 b 518.3 517.2 F 486.1 G 430.8

500.0 400.0

486.1 434.0 410.1

0.0 400.0

492.1 471.3 447.1 402.6

0.0 400.0

435.8 407.8 404.7

0.0 400.0

0.0 400.0

Such diverse and fundamental information on the nature of matter as the composition of distant stars and the structure of atoms and molecules has been obtained by analysis of the light emitted from substances heated to incandescence.

In the SPECTROSCOPE, such light, passed through a slit and a prism, is broken up into its component wavelengths, which are observed as colored lines (i.e., light of different energies) characteristic of the differences between the various electron energy levels of the atoms. This EMISSION SPECTRUM is CONTINUOUS when the images of the wavelengths are uninterruptedly overlapping; it is a LINE SPECTRUM when only certain specific wavelengths are emitted, as shown here for the elements hydrogen, helium, mercury, and uranium.

On the solar spectrum across the top of this plate appears a series of dark lines—FRAUNHOFER LINES—forming an ABSORPTION SPECTRUM. Some of the light from the intensely hot interior of the sun is absorbed by the cooler gases of its outer layers as the light energies raise the atoms in the cooler layers to higher energy states; bright lines are not, therefore, seen for these changes.

The spectra are calibrated in nanometers (1 nm = 10^{-9} m); the letters are arbitrary designations introduced by Fraunhofer for lines important in spectroscopy.

the energy by E_n, since they depend on the integer n, called the *principal quantum number* of the atom. We designate in the same way the values of the radius and the speed of the electron as r_n and v_n, since they are also quantized, as we will see.

From Bohr's second postulate we have $v_n = n\hbar/mr_n$, and so, from Eq. (28.5),

$$mv_n^2 = m\left(\frac{n\hbar}{mr_n}\right)^2 = \frac{k_e e^2}{r_n} \qquad \text{or} \qquad r_n = \frac{n^2\hbar^2}{mk_e e^2} \tag{28.7}$$

Discrete Energies in Hydrogen

To find the total energy of the system in the various energy states of H we use Eq. (28.6),

$$E_n = -\frac{k_e e^2}{2r_n} = -\frac{k_e e^2}{2}\frac{mk_e e^2}{n^2\hbar^2} = -\frac{1}{n^2}\frac{mk_e^2 e^4}{2\hbar^2} \tag{28.8}$$

Putting in numbers and keeping five significant figures (since spectral wavelengths can be measured to very high precision), we have

$$E_n = -\frac{1}{2n^2}\frac{(9.1095 \times 10^{-31}\text{ kg})(8.9876 \times 10^9\text{ N·m}^2/\text{C}^2)^2(1.6022 \times 10^{-19}\text{ C})^4}{(1.0546 \times 10^{-34}\text{ J·s})^2}$$

or
$$\boxed{E_n = \frac{-2.1799 \times 10^{-18}\text{ J}}{n^2} = \frac{-13.606\text{ eV}}{n^2} = \frac{-E_0}{n^2}} \tag{28.9}$$

where $E_0 = 13.606$ eV $= 2.1799 \times 10^{-18}$ J. These then are the energies of the quantized states of the hydrogen atom, as predicted by Bohr.

In Fig. 28.6 we plot the energy levels in hydrogen calculated from this equation. We also show the transitions between the various levels involved in the Lyman, Balmer, and Paschen spectral series in hydrogen. The combination of the Bohr-Einstein frequency condition with the energy values calculated from Eq. (28.9) leads to the wavelengths shown in the figure, which are all in excellent agreement with those observed spectroscopically.

The Rydberg Constant

From the Bohr-Einstein frequency condition (Bohr's third postulate) applied to a transition from a level with $n = n_i$ to a level with $n = n_f$, we have

$$h\nu = E_i - E_f \qquad \text{or} \qquad \frac{hc}{\lambda} = -\frac{E_0}{n_i^2} - \frac{-E_0}{n_f^2}$$

and so

$$\frac{1}{\lambda} = \frac{E_0}{hc}\left(\frac{1}{n_f^2} - \frac{1}{n_i^2}\right) \tag{28.10}$$

This has exactly the same form as the empirical Rydberg equation [Eq. (28.2)] used to describe the spectrum of hydrogen. According to Bohr's theory, the numerical value of the Rydberg constant should be

$$R = \frac{E_0}{hc} = \frac{2.1799 \times 10^{-18}\text{ J}}{(6.6262 \times 10^{-34}\text{ J·s})(2.9979 \times 10^8\text{ m/s})} \tag{28.11}$$

$$= 1.0974 \times 10^7 m^{-1}$$

This value is in reasonable agreement with the experimental value $1.0968 \times$

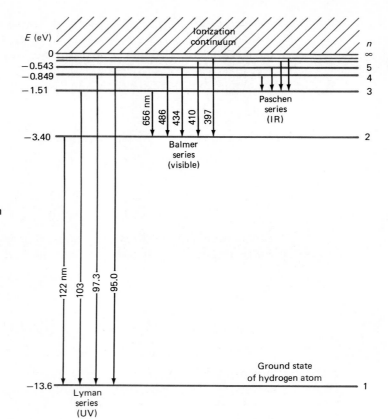

FIGURE 28.6 Energy-level diagram for atomic hydrogen.

10^7 m^{-1} for the hydrogen spectrum discussed in Sec. 28.1. Even better agreement is found when Bohr's assumption that the electron moves around a *stationary* nucleus is modified to allow for the fact that both electron and proton actually move about the center of mass of the two-particle system. Bohr's theory not only explained well the existence of discrete energy states and the observed spectrum of hydrogen; it was also able to calculate a precise numerical value for the Rydberg constant from first principles, using no numerical constants except those flowing naturally from his theory. This ability to predict so closely the results of experiment pointed to the basic correctness of Bohr's revolutionary ideas.

Electron Radius on Bohr Model

Although the idea of quantized energy levels, or states, is the most important and lasting result of Bohr's work, the Bohr theory also predicts that discrete radii are associated with the electron's motion in these energy states. Equation (28.7) shows that these Bohr radii have the discrete values

$$r_n = \frac{n^2\hbar^2}{mk_ee^2} \tag{28.12}$$

The radius of the electron orbits is quantized and increases as n^2. The smallest allowed radius is that for $n = 1$, called the *first Bohr radius*, for which

$$r_1 = \frac{\hbar^2}{mk_ee^2} = 5.28 \times 10^{-11} \text{ m} = 0.0528 \text{ nm}$$

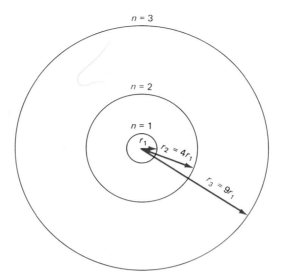

FIGURE 28.7   The first three Bohr orbits for the hydrogen atom.

This value agrees well as to order of magnitude with the known sizes of atoms, which have diameters from 0.1 to 0.4 nm. The other orbits then have radii $r_n = n^2 r_1$, as shown in Fig. 28.7.

Example 28.2

Calculate the values for the energy and the radius of the electron orbit, for the upper and lower energy states involved in

(a) the first line in the Lyman series; **(b)** the first line in the Balmer series.

SOLUTION

From Table 28.1 the energy levels involved in these cases are:

(a) Lyman series: $n_f = 1$ $n_i = 2$
(b) Balmer series: $n_f = 2$ $n_i = 3$

We need to find the values of E_n and r_n for only three separate levels, since the $n = 2$ level is both the upper level of the first line in the Lyman series and the lower level for the first line in the Balmer series.

From Eqs. (28.9) and (28.12), we have for these three levels,

$n = 1$: $E_1 =$ $\boxed{-13.6 \text{ eV}}$ $r_1 =$ $\boxed{0.0528 \text{ nm}}$

$n = 2$: $E_2 = \dfrac{-E_0}{n^2} = \dfrac{-13.6 \text{ eV}}{4} =$ $\boxed{-3.4 \text{ eV}}$

$r^2 = n^2 r_1 = 4r_1 = 4(0.0528 \text{ nm}) =$ $\boxed{0.211 \text{ nm}}$

In a similar way we obtain

$n = 3$: $E_3 =$ $\boxed{-1.5 \text{ eV}}$ $r_3 =$ $\boxed{0.475 \text{ nm}}$

28.4   Energy Levels in Hydrogen and Other Atoms

The main success of the Bohr theory was in explaining the presence of discrete energy levels in the hydrogen atom. All atoms have discrete energy levels similar to those shown for H in Fig. 28.6, but the Bohr theory is not able to calculate them as successfully as it did for H. The reason is that heavier atoms contain more electrons, and the calculation of energy levels becomes extremely hard. Despite this difficulty, Bohr's basic ideas about discrete energy levels and transitions between these levels as the source of the spectra of atoms are as valid for other atoms as they are for hydrogen.

Let us look at the *energy-level diagram* for H, Fig. 28.6, a bit more carefully. Using the Bohr-Einstein frequency condition, we have

$$\lambda = \frac{c}{\nu} = \frac{hc}{h\nu} = \frac{hc}{E_i - E_f} = \frac{hc}{\Delta E} = \frac{(6.63 \times 10^{-34} \text{ J}\cdot\text{s})(3.00 \times 10^8 \text{ m/s})}{\Delta E \text{ (eV)}(1.60 \times 10^{-19} \text{ J/eV})}$$

and so

$$\boxed{\lambda = \frac{1.24 \times 10^{-6} \text{ m}}{\Delta E}}$$

(28.13)

where ΔE is the energy difference (in *electronvolts*) between the two states of the atom involved in the transition. This equation provides a very convenient way to obtain quickly the wavelength corresponding to a transition between any two states of known energies.

There is one point about Fig. 28.6 that can cause confusion. The only important quantities in calculating wavelengths are the energy *differences* between levels; the absolute values of these energies are of no importance. Since we are free to choose our zero of energy in any convenient way, we choose as the zero of energy for the H atom that state in which the electron has been completely removed from the atom and is at rest at a great distance from the nucleus. This occurs at the series limit ($n = \infty$) in the diagram. Hence all the energy levels below the series limit are negative, indicating that the electron is bound to the nucleus and that the atom therefore has a lower energy than do a free proton and a free electron at rest. This is why all the energies in Fig. 28.6 have minus signs in front of them. The energy of the lowest, or ground, state ($n = 1$) is -13.6 eV. This means that 13.6 eV of energy must be provided from outside the atom to remove the electron from the atom, or to *ionize* the atom. The *ionization energy* for H is therefore 13.6 eV. The corresponding voltage (13.6 V) is called the *ionization potential*.

Note that the *continuum* above $n = \infty$ corresponds to electrons having kinetic energy after they have been removed from the atom by ionization. Since kinetic energies are not quantized for free particles, any energy is possible above the series limit, and a *continuum* results. We therefore have discrete energy states below the series limit and continuous energy states above the series limit. The same is true for the energy-level diagrams of other atoms.

Photon Absorption by Atoms

The process of emission can be reversed. If UV light of wavelength 122 nm shines on H gas, atoms in the ground state E_1 in Fig. 28.6 can absorb this photon and move up to state E_2 with an overall increase in energy of 10.2 eV for the atom. For this absorption to occur, however, it is found experimentally that the incident photon must have an energy *exactly* equal to the energy difference between the states E_2 and E_1. Otherwise the photon will not be absorbed and the atom will remain in its original state.

At room temperature most gas atoms are in their ground state. Since in hydrogen a transition from this ground state to the first excited state requires a photon of wavelength 122 nm, H does not absorb at all in the visible region, only in the ultraviolet. For this same reason most of the common gases like helium and neon are colorless since they do not absorb in the visible. In the ultraviolet, however, they show strong absorption.

If sunlight, with a continuous distribution of wavelengths, falls on sodium (Na) vapor, the atoms will absorb out of the sunlight two very close yellow lines that satisfy the Bohr-Einstein frequency condition for the Na atom. This leads to dark lines for Na and many other elements in the sun's spectrum, as shown in the colored plate included in this chapter. The Na lines are those marked D_1 and D_2 on the top spectrum. These lines are called *Fraunhofer lines* and can be used to identify the elements in the sun's atmosphere, where most of the absorption occurs.

Excitation of Atoms

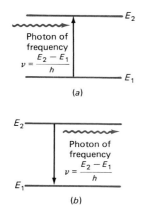

(a)

(b)

The absorption of light lifts atoms to excited states, from which they can then fall back to lower states by radiating photons of the proper frequency, as shown in Fig. 28.8. The absorption of light leads, therefore, to the excitation of atoms. The energy (in electronvolts) required to raise, or "excite," the atom from a lower energy state to a higher energy state is called the *excitation energy*. For example, the excitation energy for the $n = 2$ state in H is -3.4 eV $- (-13.6$ eV$) = 10.2$ eV. The voltage corresponding to this energy is called the *excitation potential*.

Atoms may also be excited thermally, although this is not a very effective process, since the KE of a gas atom at 3000 K is only 0.39 eV. (Compare this with the energies required to excite the higher-energy levels in hydrogen shown in Fig. 28.6.) A third method is to apply a high voltage across a tube containing atoms of the gas. An electric discharge then leads to atomic excitation by inelastic collisions.

FIGURE 28.8 Excitation of atoms by photon absorption. (a) The atom can absorb a photon of energy $h\nu = E_2 - E_1$. (b) Once in the excited state, it can fall back to the lower state, emitting a photon of exactly the same frequency ν.

Example 28.3

What must be the temperature of a gas for the average kinetic energy of a gas atom to be 1.0 eV?

SOLUTION

The kinetic energy of gas atoms in equilibrium at temperature T is described by a Boltzmann distribution of average energy $\frac{3}{2}kT$. Hence

$$\tfrac{3}{2}kT = 1.0 \text{ eV} = 1.6 \times 10^{-19} \text{ J}$$

or $$T = \frac{2}{3}\frac{1.6 \times 10^{-19} \text{ J}}{1.38 \times 10^{-23} \text{ J/K}} = \boxed{7.7 \times 10^3 \text{ K}}$$

The required temperature is close to 8000 K, at which temperature any container would melt.

28.5 The Franck-Hertz Experiment

In 1914 two German physicists, James Franck (see Fig. 28.9) and Gustav Hertz (1887–1975), the nephew of Heinrich Hertz, performed a crucial experiment in support of Bohr's ideas on atomic structure. They bombarded mercury (Hg) vapor with electrons of variable energy and studied how much energy was transferred from the electrons to the Hg atoms.

The apparatus used by Franck and Hertz is shown in Fig. 28.10. It consisted of a glass chamber containing mercury vapor at a gas pressure of about 1 torr (1 mmHg). A heated cathode C emitted electrons toward a grid G that was at a variable positive potential with respect to the cathode. Beyond the grid, which was a mesh structure through which electrons could pass, was a solid metal plate P maintained at about 0.5 V negative with respect to the grid. This provided the

FIGURE 28.9 James Franck (1882–1964; Nobel Prize in physics, 1925) at Princeton, N.J., in October 1954 with three other Nobel laureates in physics. Franck is the second from the left; the others are, left to right, Niels Bohr (1885–1962; Nobel Prize, 1922), Albert Einstein (1879–1955; Nobel Prize, 1921), and I. I. Rabi (1898–1988; Nobel Prize, 1944). *(Courtesy of AIP Niels Bohr Library, Margrethe Bohr Collection)*

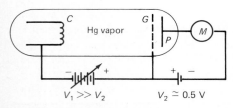

FIGURE 28.10 Apparatus for the Franck-Hertz experiment.

FIGURE 28.11 Results of the Franck-Hertz experiment: A plot of the current through the meter M in Fig. 28.10 against the applied voltage V_1.

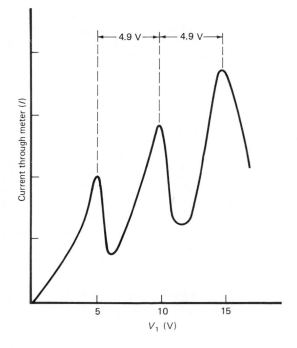

retarding potential needed to keep very low energy electrons from reaching the plate and contributing to the current measured by the ammeter M.

Franck and Hertz varied the voltage V_1 gradually and recorded the current through the meter as a function of this voltage. Their results are shown in Fig. 28.11. The current rose with the voltage until the voltage hit about 4.9 V. Then the current fell sharply to a low value. As the voltage V_1 between grid and cathode increased further, the current rose again, but at about 9.8 V it again fell sharply.

This behavior continued at higher voltages, with the current falling at intervals of 4.9 V. The graph of current against voltage V_1 obtained by Franck and Hertz consisted of current peaks with consecutive peaks separated by an average value of 4.9 ± 0.1 V, as can be seen in the figure.

How do we explain these experimental results? At voltages below 4.9 V the electrons have low kinetic energies and collide with the Hg atoms *elastically*. At about 4.9 V something new happens: an electron loses all its kinetic energy in a single *inelastic* collision with a Hg atom. The state of this Hg atom is therefore changed. It is excited from its normal (or ground) energy state to an excited state that has 4.9 eV more energy than the ground state. The electrons have therefore lost 4.9 eV of energy and cannot overcome the 0.5-V retarding potential V_2 to reach the plate. The current therefore falls sharply, as observed in Fig. 28.11.

If the voltage is increased above 4.9 V, the electrons again acquire enough energy to reach the plate after losing 4.9 eV in one inelastic collision, and the current increases again. But at 9.8 V the electrons have enough energy to make two consecutive inelastic collisions with Hg atoms, and so the current falls again. This falloff in current continues at intervals of 4.9 V as the voltage is increased further.

This seems like a reasonable physical explanation of what is going on in the mercury vapor, but how can we be sure that this interpretation is correct? We can look at the Hg gas with an ultraviolet grating spectrometer (see Sec. 24.3) as the voltage is increased. No spectrum at all appears until the voltage reaches 4.9 V. Then a single bright ultraviolet line at 253 nm stands out against the dark background. This corresponds to Hg atoms falling back to their ground state from the excited state to which they have been raised by a colliding electron. If we set this energy equal to $h\nu$, we can find the wavelength to be expected. It is, from Eq. (28.13),

$$\lambda = \frac{1.24 \times 10^{-6}\text{ m}}{\Delta E} = \frac{1.24 \times 10^{-6}\text{ m}}{4.9} = 253\text{ nm}$$

which agrees perfectly with the wavelength of the observed ultraviolet line.

The Franck-Hertz experiment therefore provides direct experimental proof of the existence of discrete energy levels in atoms, as predicted by Bohr, and of the way these energy levels lead to observed spectra. The electrons lose energy in their collisions with Hg atoms only in discrete energy chunks, or quanta, corresponding to a precise energy difference between two energy states in the atom, and these quanta are radiated away from the Hg vapor as photons of a single wavelength when the atom falls back to its ground state.

A rough energy-level diagram for Hg, as obtained from the work of Franck and Hertz, is shown in Fig. 28.12. Evidence from spectroscopy and from electron-impact studies like the Franck-Hertz experiment provides an answer to our initial question of why all atoms of a particular element are the same: All the atoms of any element have identical energy levels which determine their physical and chemical properties and distinguish that element from all other elements.

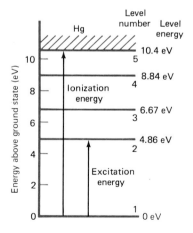

FIGURE 28.12 Partial energy-level diagram for the mercury atom (Hg).

Example 28.4

In an electron-impact experiment performed on atomic hydrogen gas, it is found that the atoms are excited from the ground state to excited states by electrons of energies 10.2 and 12.1 eV, respectively. **(a)** What wavelengths are to be expected in the radiation from such a gas after it has been excited in this way? **(b)** Are any of these wavelengths in the visible region of the spectrum? In what region of the spectrum are the other wavelengths?

SOLUTION

(a) The energy levels of the H atom, as determined from the experiment described in the problem, are shown in Fig. 28.13. There are therefore three possible transitions, one from state 3 to state 1, another from state 2 to state 1, and a third from state 3 to state 2. The corresponding wavelengths are given by Eq. (28.13) as $\lambda = (1.24 \times 10^{-6}\ m)/\Delta E$, and so

$$\lambda_{31} = \frac{1.24 \times 10^{-6}\ m}{E_{31}} = \frac{1.24 \times 10^{-6}\ m}{12.1} = \boxed{103\ nm}$$

$$\lambda_{21} = \frac{1.24 \times 10^{-6}\ m}{E_{21}} = \frac{1.24 \times 10^{-6}\ m}{10.2} = \boxed{122\ nm}$$

$$\lambda_{32} = \frac{1.24 \times 10^{-6}\ m}{E_{32}} = \frac{1.24 \times 10^{-6}\ m}{12.1 - 10.2} = \boxed{653\ nm}$$

(b) Of the three expected wavelengths only λ_{32} is in the visible region of the spectrum, for 653 nm is in the red. The other two wavelengths are much shorter and are in the ultraviolet. This red line near 653 nm can be seen on the colored spectrum of hydrogen facing p. 742.

   The red H_α line actually falls at 656 nm. The discrepancy between this and the 653 nm found here is caused by the loss of one significant figure on subtracting 10.2 from 12.1. As a consequence, our result for H_α is only good to two significant figures.

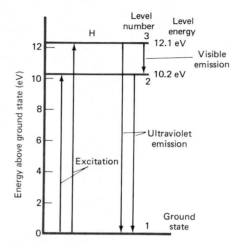

FIGURE 28.13

28.6   The Emission and Absorption of Radiation

Suppose we consider two energy levels E_1 and E_2 in an atom, as in Fig. 28.14. If the atom is originally in the lower state E_1, and a photon comes along of *exactly* the frequency $\nu = (E_2 - E_1)/h$, the atom can be excited from E_1 to E_2 by the *absorption* of the photon (Fig. 28.14a). Similarly, if the atom is in the excited state E_2, it can fall of its own accord from level 2 to level 1 and emit a photon of

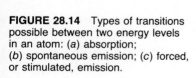

FIGURE 28.14   Types of transitions possible between two energy levels in an atom: (a) absorption; (b) spontaneous emission; (c) forced, or stimulated, emission.

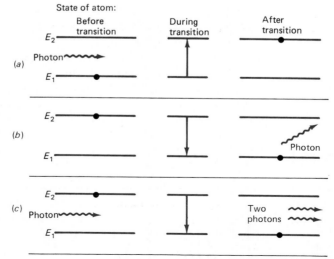

frequency $\nu = (E_2 - E_1)/h$. This is called *spontaneous emission* (Fig. 28.14*b*).

In addition to absorption and spontaneous emission there is a third possibility. Suppose the atom is in the upper state with energy E_2. If a photon comes along of *exactly* the frequency $(E_2 - E_1)/h$, it can *force* the atom to fall to the lower level and emit a photon of frequency $\nu = (E_2 - E_1)/h$. The result of this process of *forced* (or *stimulated*) *emission* is that the entering photon leaves the atom accompanied by a second photon of exactly the same frequency, as shown in Fig. 28.14*c*.

The Laser

The term *laser* is an acronym for *l*ight *a*mplification by the *s*timulated *e*mission of *r*adiation. The appropriateness of this name should be clear from the following discussion.

When a photon of the right energy is incident on an atom, it can produce either absorption or stimulated emission. Einstein showed that these two processes are equally probable and that net emission or net absorption results depending on whether there are more atoms in the upper excited state or in the lower state. Usually there will be more atoms in the lower state, since an atom will tend of its own accord to fall to a lower energy state by spontaneous emission.

A laser is a device which enables stimulated emission to win out over absorption to such an extent that atoms emit very strongly at one particular frequency. To accomplish this, two conditions must be satisfied:

1   The upper state involved in the transition must be *metastable*; i.e., the atom must remain in that state for a long time before it falls of its own accord to the lower state by spontaneous emission. For laser action to occur, the rate of stimulated emission must be far greater than the rate of spontaneous emission.

2   A *population inversion* must be produced. By this we mean that there must be more atoms in the upper state than in the lower state so that stimulated emission dominates absorption when light of the proper wavelength falls on the system. Such a population inversion must be artificially produced, since it is the reverse of what would occur naturally. A population inversion can be achieved in a variety of ways: by flooding the gas with high-intensity light and thus pumping atoms into excited states, by atomic collisions, or by other more sophisticated techniques.

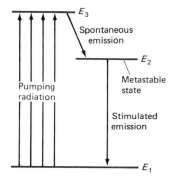

FIGURE 28.15   A typical scheme for producing a population inversion by optical pumping, using a three-level system. Since level E_2 is above the ground state, it is normally sparsely populated at room temperature. But this level is metastable, and so atoms excited to E_3 by the pumping radiation decay and accumulate in this metastable level, leading to a population inversion between states E_2 and E_1.

A typical scheme for inverting populations using "optical pumping" is shown in Fig. 28.15. Intense radiation from a flash lamp pumps an atomic system from the ground state into an excited state of energy E_3, from which it decays to the metastable state of energy E_2. Eventually level 2 has a higher population than level 1. Then photons at a frequency $\nu = (E_2 - E_1)/h$ can cause stimulated emission from state 2 to state 1. The pumping radiation is either kept on continuously (continuous lasers) or provided in short, repeated pulses (pulsed lasers) to replenish the population of the upper state.

Helium-Neon Gas Laser

A common type of laboratory laser is the He–Ne gas laser shown in Fig. 28.16. A high voltage placed across a mixture of 15% He and 85% Ne produces electrical breakdown of the gas into ions and electrons. Fast-moving electrons excite the He atom to the energy level E_2 in Fig. 28.17. This level is almost identical in energy to that of the Ne atom in the state E_2', so by inelastic collisions the He atoms can transfer their energy to the Ne atoms. In this way a large population is built up in this excited metastable state of Ne. Photons at frequencies $\nu = (E_2' - E_1')/h$ then

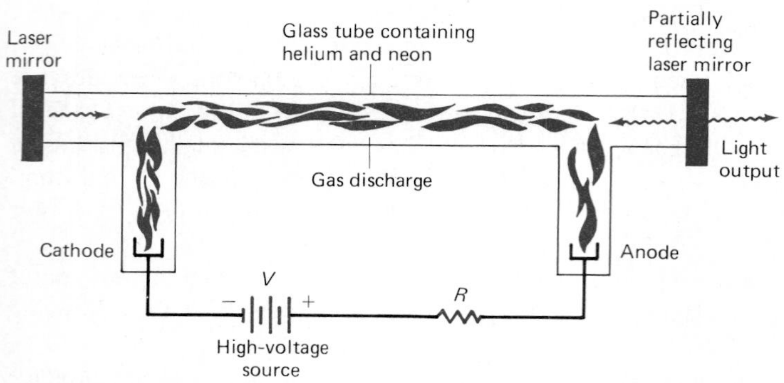

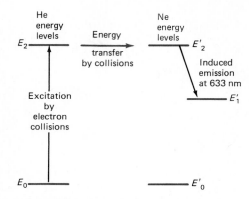

FIGURE 28.16 A schematic diagram of a He–Ne laser.

FIGURE 28.17 Energy-level diagram for He and Ne atoms showing how energy is transferred by collisions from the E_2 level of He to the E_2' level of Ne. This leads to a population inversion between the E_2' and E_1' states of Ne.

stimulate transitions to state E_1', with the emission of radiation at a wavelength of 633 nm. This explains why the He–Ne laser produces a very intense line in the red region of the spectrum.

In the He–Ne laser shown in Fig. 28.16, one photon of the proper frequency $\nu = (E_2' - E_1')/h$ moving along the axis of the tube causes forced emission of a second photon by a Ne atom. These two photons, which are at exactly the same frequency and moving in the same direction, then move along together and stimulate the emission of additional photons by other Ne atoms. A multiplication process results, and a very large number of photons of exactly the same frequency are produced in a small fraction of a second. This effect is increased even more by the mirrors at the ends of the tube, which keep the photons moving back and forth along the axis, stimulating more emissions as they go. Some of the light emerges through the partially transparent ("leaky") mirror at one end, and a very intense beam of single-frequency, highly directional light results.

A laser beam is highly *directional* because any photons not moving exactly perpendicular to the mirrors at the two ends are lost from the beam that finally emerges from one end of the tube. Laser light is also highly *monochromatic*, since the photon causing stimulated emission and the one resulting from stimulated emission must have exactly the same frequency. Laser radiation is also *coherent*, which means that all the photons in the beam of light emerging from the end of the laser are perfectly *in phase*. This was discussed in Sec. 24.1.

Uses of Lasers

Lasers have many practical uses because of their extreme monochromaticity, coherence, and directionality. They have been used to survey a perfect straight line for the laying of long pipes and tunnels, and for monitoring the distance from the earth to the moon using mirrors placed on the moon's surface by astronauts. Very powerful lasers can be used to cut and bore metals and are being developed as weapons against aircraft and tanks. Lasers are the heart of compact disc players (CDs) and of the computer systems in use at supermarket checkout counters. The world's most powerful laser (Fig. 28.18*a*) is being used in a research effort to

(a)

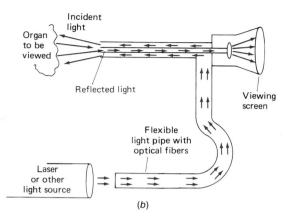

(b)

FIGURE 28.18 (a) The target chamber of NOVA, the world's most powerful laser, at the Lawrence Livermore National Laboratory in Livermore, California. This chamber is a 16-ft-diameter aluminum sphere inside of which 10 powerful laser beams converge simultaneously onto a small fuel pellet to produce fusion reactions. *(Courtesy of Lawrence Livermore National Laboratory.)* (b) A typical endoscope for seeing inside the human body (schematic).

produce controlled fusion reactions as a possible safe, economical, and inexhaustible energy source for future generations (Sec. 29.7).

Medical Applications   The dream of every medical doctor is to be able to see inside a patient's body and to perform necessary surgery without having to cut open the patient. This dream is now closer to reality because of the joining of lasers to fiber-optics technology.

For diagnostic purposes an *endoscope* (meaning "seeing within") like the one in Fig. 28.18b is routinely used. The endoscope consists of a bundle of optical fibers that carry light to the inside of the body and then transfer an image of the internal organs back outside the body for viewing, either directly or on a TV monitor. Endoscopes often use intense white-light sources, but for many purposes laser light in the visible region of the spectrum is preferable. Endoscopes can be inserted into the body through the mouth or rectum and moved along the esophagus, the gastrointestinal tract, the bronchus, or other internal passages to look for disease and blockages. Most endoscopes on the market are from 0.3 to 1.2 m in length and from 2.5 to 15 mm in diameter.

Pulsed ruby lasers have been used for many years to mend detached retinas, since the laser beam will pass through the transparent outer parts of the eye without damaging it and can then be focused on the defective area of the retina. A burst of photons generates sufficient heat to "weld" the retina to the choroid from which it has become detached.

Very high energy carbon dioxide (CO_2) lasers are now used to destroy cancerous tissue on the outside of the body or in other regions of the body accessible through natural body orifices. The 10,600-nm infrared light from a CO_2 laser is ideal for surgery because it is strongly absorbed by water, which makes up 80 to 90 percent of body tissue, including tumor tissue. Sufficient energy is delivered by the laser to heat the defective tissue to a temperature high enough to destroy it. At the same time the intense heat cauterizes the wound, prevents bleeding, and hinders the spread of the disease to normal tissues.

Fluorescence and Phosphorescence

These are other important processes involving the absorption and emission of light.

Fluorescence: The absorption of high-frequency photons by atoms or molecules, which then immediately radiate lower-frequency photons.

Figure 28.19 shows how fluorescence occurs. An atom in its ground state (1) absorbs a photon of energy $h\nu_1$ and makes a transition to state 3. The atom then returns by spontaneous emission to its ground state by two quick transitions, one from state 3 to state 2 and the second from state 2 to state 1. In the latter transition a photon of frequency ν_2 is emitted. If ν_1 is in the ultraviolet, then ν_2 may be in the visible region of the spectrum. Fluorescence therefore always leads to a lower-frequency photon than the one absorbed.

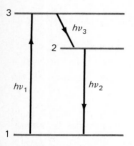

FIGURE 28.19 Fluorescence. An atom absorbs ultraviolet radiation at a frequency ν_1, and emits a visible photon of frequency ν_2.

Fluorescent light fixtures work by electrically exciting mercury atoms in an evacuated tube so that they emit ultraviolet radiation. This radiation is absorbed by a fluorescent material on the inner walls of the tube, which then emits the desired visible light. The fluorescent screen of a television set consists of a coating of chemicals that fluoresce under the impact of high-speed electrons.

Phosphorescence: Delayed fluorescence.

In phosphorescent materials an energy level like 2 in Fig. 28.19 is metastable and has a long lifetime. As a result, photons of frequency ν_2 are emitted for some time after the photon of frequency ν_1 is absorbed.

28.7 Quantum Numbers

The energy levels of the hydrogen atom are given by Eq. (28.9) as

$$E_n = \frac{-13.606 \text{ eV}}{n^2}$$

where n is called the *principal quantum number*. This quantum number n labels the discrete energy levels within the atom. All electrons that possess the same total energy E_n are said to be in the same energy *shell*. The lowest level, the ground state, corresponds to $n = 1$; the first excited state is $n = 2$; and so on. As n increases, the energy increases, since the numerical value of E_n in the above equation decreases. But the right side of the equation has a negative sign, and so the energy becomes progressively less negative as n increases. In other words, the energies of the excited states *increase with n*, as can be seen in Fig. 28.6.

Nature is always more complicated than our theories. It was soon found that the Bohr theory was not adequate even for the simplest of all atoms, hydrogen. When observed under high resolution, the lines in the hydrogen spectrum show a *fine structure* in which, instead of the single lines predicted by the Bohr theory, two or more lines are found very close together. The German physicist Arnold Sommerfeld (1868–1951) suggested that this might be because the electrons can move in elliptical orbits like those in which the planets move around the sun (a possibility mentioned by Bohr in his original paper). To a first approximation all the electrons in elliptic orbits corresponding to the same value of n would have the same energy, but the orbital angular momentum of the atom would vary with the shape (the *eccentricity*) of each ellipse. When the relativistic variation of mass with electron speed was taken into account (Sec. 26.7), Sommerfeld found that the different elliptical motions had slightly different energies and that these differences corresponded well to the observed fine structure in the hydrogen spectrum.

Sommerfeld was able to show that the angular momentum for these elliptical orbits could be specified by a second quantum number l, with the angular momentum given by $L = \sqrt{l(l + 1)}\hbar$. For any given n, the only possible values of l, called the *orbital angular momentum quantum number*, are

$$l = 0, 1, 2, 3, \ldots , n - 1$$

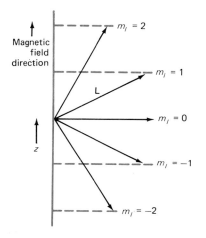

Still further complications arise. Since an electron has orbital angular momentum, we have a charged particle moving in a closed loop, which is the equivalent of a tiny bar magnet (see Sec. 19.7). In a magnetic field this bar magnet can line up with the applied field. According to quantum theory, however, there are only a limited number of quantized directions that the orbital angular momentum vector **L** can assume with respect to the magnetic field, as shown in Fig. 28.20 for $l = 2$. These directions are specified by a third quantum number, the *magnetic quantum number* m_l, that measures the component of the orbital angular momentum in the direction of the magnetic field, which we take to be the z direction. For this reason m_l can never be larger than l, and so the allowed values of m_l are $0, \pm 1, \pm 2, \pm 3, \ldots , \pm l$. The z component of the angular momentum is then $L_z = m_l \hbar$.

A final complication results from an observed doubling of spectral lines that could not be explained in terms of elliptic orbits and quantized orbital angular momentum. One possible explanation is that the electron itself is spinning about an axis through its center as it moves about the nucleus, in the same way that the earth spins on its axis as it rotates about the sun. Such a spinning, negatively charged electron has a spin angular momentum and a magnetic moment that could also line up either in the direction of an applied magnetic field or in the opposite direction to the field. These two possibilities are distinguished by a *spin quantum number* m_s that can have only one of the two values $+\frac{1}{2}$ (parallel to field) or $-\frac{1}{2}$ (antiparallel to field). It was found that this new quantum number could indeed account properly for the observed doubling of spectral lines.

In summary, then, instead of one quantum number n, we now have four quantum numbers n, l, m_l, and m_s that are needed to determine fully the energy state and the motion of the electron in the hydrogen atom. Table 28.2 summarizes

FIGURE 28.20 Quantized orientation of the orbital angular momentum **L** in a magnetic field (for $l = 2$).

TABLE 28.2   Quantum Numbers for a Single Electron in Hydrogen

Name of quantum number	Symbol	Physical significance	Allowed values
Principal quantum number	n	Determines energy shell of atom	$1, 2, 3, \ldots , \infty$
Orbital angular momentum quantum number	l	Determines angular momentum of electron in its orbit	$0, 1, 2, \ldots , n - 1$
Magnetic quantum number	m_l	Determines component of orbital angular momentum in direction of magnetic field	$0, \pm 1, +2, \ldots , \pm l$
Spin quantum number	m_s	Determines component of spin angular momentum of electron in direction of magnetic field	$+\frac{1}{2}, -\frac{1}{2}$

the significant facts about these four quantum numbers. Each individual combination of the quantum numbers n, l, m_l, and m_s specifies a particular energy state of an electron in an atom. The number of possible electronic states increases rapidly with n. Thus for $n = 1$ there are only two different states, but for $n = 2$ there are eight different states. (Try writing out for yourself the four quantum numbers specifying each of these states.) In general the number of states can be shown to be $2n^2$ for any principal quantum number n.

Example 28.5

(a) How many different energy states exist in the hydrogen atom when the electron is in the energy shell specified by principal quantum number $n = 3$? **(b)** Indicate how this number may be arrived at from a consideration of the possible values of the four quantum numbers n, l, m_l, and m_s.

SOLUTION

(a) As pointed out in the section above, the number of different energy states for principal quantum number n is $2n^2$. Here, with $n = 3$, we have

$$2(3)^2 = \boxed{18 \text{ states}}$$

(b) For $n = 3$, the possible values of l are 0, 1, and 2.

For $l = 0$: $\quad m_l = 0$

For $l = 1$: $\quad m_l = -1, 0, +1$

For $l = 2$: $\quad m_l = -2, -1, 0, +1, +2$

Adding the number of possible values of m_l we obtain $1 + 3 + 5 = 9$. Now for each of these values of m_l, the spin quantum number m_s can have either the value $+\frac{1}{2}$ or $-\frac{1}{2}$. Thus the total number of possible states is $2 \times 9 = 18$, which checks with the value obtained in part (a).

28.8 The Pauli Exclusion Principle and the Periodic Table

We now want to extend our discussion of the hydrogen atom to explain the structure of multiple-electron atoms. Such atoms contain orbiting electrons equal in number to the charge number Z of the nucleus, i.e., the number of elementary positive charges in the nucleus. For such atoms the general ideas about energy-level structure and the associated quantum numbers introduced above for H remain valid.

To explain, at least qualitatively, the structure of multiple-electron atoms, we will rely on two principles. The first is that, if left to itself, an atom will always arrange itself in the lowest energy state possible, just as a roller coaster car tries to reach the lowest point on its track. The atom's overall energy state depends on the individual, or one-electron, energy states (like those for H) in which each electron finds itself. The lowest energy state for the atom is called the *ground state*. This might lead us to believe that in the ground state all the electrons would crowd into the lowest possible one-electron energy states like the hydrogen ground state, with $n = 1$ and $l = 0$. That this does not happen is explained by a second principle first introduced in 1925 by the Austrian physicist Wolfgang Pauli (see Fig. 28.21) and called the *Pauli exclusion principle*.

FIGURE 28.21 Wolfgang Pauli (1900–1958). Pauli received the 1945 Nobel Prize in physics for his discovery of the exclusion principle. He was a brilliant theoretical physicist who first postulated the existence of the neutrino and made important contributions to statistical, nuclear, and elementary-particle physics. *(Photo by S. A. Goudsmit, courtesy of AIP Niels Bohr Library.)*

> *Pauli exclusion principle:* In one and the same atom no two electrons can have the same set of values for the four quantum numbers n, l, m_l, and m_s.

This means that no two electrons can be in the same energy state. As larger and larger atoms are built up by adding electrons, the added electrons must go into higher-energy states corresponding to larger values of n and, for a particular value of n, to larger values of l. All the electrons with the same value of n are said to be in the same *shell*. Those with the same values of n and l are said to be in the same *subshell*. When a shell or subshell contains the maximum number of electrons allowed, it is said to be *closed*. Any more electrons added after a shell or subshell is closed must therefore occupy higher-energy shells or subshells.

The Periodic Table

Let us apply these ideas to some simple atoms. In Table 28.3 we show the values of the four quantum numbers for the electrons in six atoms of the periodic table. We discuss only the ground states of these atoms.

Hydrogen ($Z = 1$) has its only electron in the $n = 1$ shell. Hence l and m_l must both be zero; m_s can be either $+\frac{1}{2}$ or $-\frac{1}{2}$, but this has no effect on the overall energy of the atom.

Helium ($Z = 2$) must have its second electron with a spin opposite to that of the first electron to satisfy the Pauli principle. We talk about the spins being *paired*. Since this completes the $n = 1$ shell, He is a closed-shell atom and is therefore very stable, does not enter into chemical reactions, and has a very high ionization potential (24.6 V).

Lithium ($Z = 3$) has a third electron outside the He closed shell. This electron must start a new ($n = 2$) shell. Such an electron has a large energy compared with the two $n = 1$ shell electrons and is far removed from the nucleus relative to them, as should be clear from our discussion of the energy levels of H. As a result, this electron is loosely bound, with an ionization potential of only 5.4 V. Lithium is therefore a very reactive element chemically, and is monovalent, because it is this outermost electron that enters into chemical reactions.

TABLE 28.3 Quantum Numbers for Electrons in Some Atoms

Element	Number of electrons	n	l	m_l	m_s	Electron configuration
		Quantum numbers				
H	1	1	0	0	$+\frac{1}{2}$	$(1s)$
He	2	1	0	0	$\pm\frac{1}{2}$	$(1s)^2$
		Closed shell				
Li	3	1	0	0	$\pm\frac{1}{2}$	$(1s)^2 2s$
		2	0	0	$+\frac{1}{2}$	
Be	4	1	0	0	$\pm\frac{1}{2}$	$(1s)^2(2s)^2$
		2	0	0	$\pm\frac{1}{2}$	
		Closed subshell				
C	6	1	0	0	$\pm\frac{1}{2}$	
		2	0	0	$\pm\frac{1}{2}$	$(1s)^2(2s)^2(2p)^2$
		2	1	$+1$	$\pm\frac{1}{2}$	
Na	11	1	0	0	$\pm\frac{1}{2}$	
		2	0	0	$\pm\frac{1}{2}$	
		2	1	$+1$	$\pm\frac{1}{2}$	
		2	1	0	$\pm\frac{1}{2}$	$(1s)^2(2s)^2(2p)^6(3s)$
		2	1	-1	$\pm\frac{1}{2}$	
		3	0	0	$+\frac{1}{2}$	

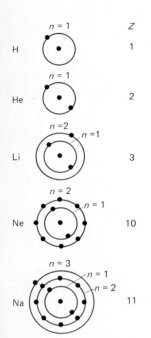

FIGURE 28.22 Schematic diagram of electron configurations for some common light atoms. (No attempt has been made to show the relative sizes of the atoms.)

After lithium the $n = 2$ shell fills up until at neon ($Z = 10$) the $n = 2$ shell has its full complement of $8\ (= 2n^2)$ electrons and the shell is closed. This makes Ne similar to He, since they both are closed-shell atoms. Since He and Ne have similar closed-shell configuration of electrons, they fall in the same column of the periodic table, along with the other inert gases like argon, krypton, and xenon.

Sodium ($Z = 11$) has one electron in the $n = 3$ shell outside two interior closed shells. Its electron configuration is therefore very similar to that of Li. It is reactive chemically, is monovalent, and has a low ionization potential (5.1 eV).

We could continue this process through the rest of the periodic table, but the principle on which this "building up" of the periodic table is based is the important factor here, not the details.

> *Building-up principle:* Atoms are built up by adding electrons one at a time to the lowest one-electron energy states permitted by the Pauli exclusion principle.

Diagrams of the electron configurations for some of the atoms discussed above are given in Fig. 28.22. The increased positive charge on the nucleus reduces the radius of the inner shells, but the Pauli principle requires that some of the electrons go into outer shells of greater radius. As a consequence, all atoms have roughly the same size.

The important quantum numbers in determining the energy shells and subshells of an atom are the principal quantum number n and the orbital angular momentum quantum number l. For historical reasons based on spectroscopy, values of l are specified by using the letters $s, p, d, f, \ldots$ to signify $l = 0, 1, 2, 3, \ldots$, respectively. The one electron in hydrogen that has $n = 1$ and $l = 0$ is called a $1s$ electron. Carbon, which has six electrons, has two $1s$ electrons, two $2s$ electrons, and two $2p$ electrons, and its electron configuration is written as $(1s)^2(2s)^2(2p)^2$. This symbolism is used to describe the electron configurations of the atoms in Table 28.3.

28.9 Successes and Failures of the Bohr Theory

One of the early successes of the Bohr theory was in explaining the spectra of hydrogen-like ions, i.e., ions which have only one electron but have nuclear charges greater than 1. Examples are He^+, Li^{2+}, and Be^{3+}. The spectra of such ions should be similar to the spectrum of H except that a factor of Z^2 is introduced into the Rydberg equation because the nuclear charge is no longer e, but Ze. The Rydberg equation (28.2) then becomes

$$\frac{1}{\lambda} = Z^2 R \left(\frac{1}{n_f^2} - \frac{1}{n_i^2} \right)$$ (28.14)

Good agreement was found for such one-electron ions between experimental results and those predicted by Bohr. This provided additional confirmation of the basic correctness of the Bohr theory.

The Characteristic X-Ray Spectrum

By experiment it is found that sharp lines occur in the x-ray region of the spectrum in both emission and absorption. As shown in Fig. 28.23, these lines are superimposed on the continuous x-ray spectrum. These lines are similar in some ways to those produced by the outer electrons of atoms, but there are also very significant differences between x-ray line spectra and those produced by optical transitions in the visible region of the spectrum. Among these differences are the following:

X-Ray Line Spectra

1 The lines are hardly at all dependent on the chemical state of the element.

2 The line spectrum is very similar from element to element, but there is a slight shift to higher frequencies as the atomic number increases. Also, new x-ray lines' appear as new energy shells are added.

3 The energy differences corresponding to the observed lines are thousands of electronvolts.

Optical Line Spectra

1 The lines are very dependent on the chemical state of the element.

2 The line spectrum differs greatly from element to element.

3 The energy differences corresponding to the observed lines are only a few electronvolts.

These experimental facts can be explained by assuming that x-ray line spectra are produced by electrons in shells deep inside the target atoms, just as the outer-shell electrons are responsible for optical line spectra. Such spectra vary with the nature of the target material (unlike continuous x-ray spectra) and are therefore called *characteristic x-ray spectra*.

Characteristic x-ray spectra can be well explained on the basis of the simple Bohr theory. Figure 28.24 shows the various electron shells corresponding to different principal quantum numbers n in a heavy atom. Shells with $n = 1, 2, 3, 4, \ldots$ are denoted as K, L, M, N, $\ldots$ shells, respectively. Suppose that, either by an electron collision process or by the absorption of an x-ray, one of the electrons in the K ($n = 1$) shell is removed from the atom. Then electrons can fall into the

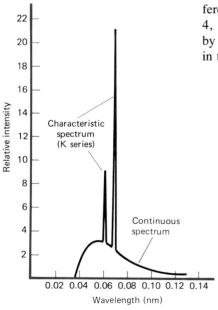

FIGURE 28.23 Characteristic lines in the x-ray spectrum of molybdenum (Mo).

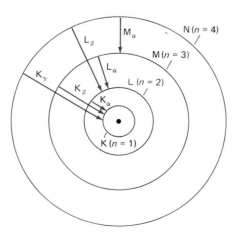

FIGURE 28.24 Bohr energy levels for a heavy atom. The lines connecting these levels indicate the transitions producing x-rays.

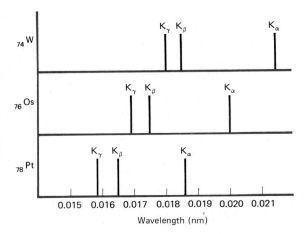

FIGURE 28.25 X-ray lines for three elements close together in the periodic table. Notice how the spectra are identical except that they shift to shorter wavelengths (higher frequencies) as the atomic number of the atom increases. The observed wavelengths can be obtained from the Bohr energy levels for each atom, as shown in Fig. 28.24.

K shell from the L, M, N shells, producing the lines marked K_α, K_β, K_γ in the diagram. Similarly, if an electron is missing from the L shell, electrons can fall in from the M and N shells, and the atom will radiate the L_α and L_β lines. Observed characteristic x-ray spectra agree well with this explanation. (See Fig. 28.25.)

Example 28.6

A lead target is bombarded by electrons of sufficient energy to knock a K-shell electron out of an atom. An L-shell electron then falls into the K shell to fill the original vacancy produced.

(a) What is the minimum energy required in the electron beam to remove the K-shell electron? **(b)** What is the approximate wavelength of the x-ray produced in the L → K transition?

SOLUTION

(a) The electrons in a lead atom ($_{82}$Pb) have energies given approximately by the Bohr theory, modified to allow for the fact that the nuclear charge is now not $Z = 1$ but $Z = 82$, as in Eq. (28.14). Therefore the energy states are given by an equation similar to (28.9):

$$E_n = \frac{-13.6Z^2}{n^2} \text{ eV}$$

The energy required to remove the K electron is then the difference between its energy in its final and initial states. For the K shell with $n = 1$, this is

$$E = E_f - E_i = 0 - E_1 = \frac{13.6Z^2}{n^2}$$

$$= \frac{13.6(82)^2}{1^2} = \boxed{9.14 \times 10^4 \text{ eV}}$$

(b) The wavelength of the emitted K line can be obtained directly from Eq. (28.13), which is as valid for x-ray line spectra as for spectra in the optical region:

$$\lambda = \frac{1.24 \times 10^{-6} \text{ m}}{\Delta E}$$

where ΔE is here the energy difference (in electronvolts) between the initial and final states. Therefore $\Delta E = E_2 - E_1 = -13.6Z^2(\frac{1}{4} - 1) = 6.86 \times 10^4$ eV, and so

$$\lambda = \frac{1.24 \times 10^{-6} \text{ m}}{6.86 \times 10^4} = \boxed{0.0181 \text{ nm}}$$

This is in reasonable agreement with the experimental value of 0.017 nm. The error is due to the screening effect of the other electrons on the Coulomb force experienced by the electron involved in the L → K transition.

Failures of the Bohr Theory

Although it is clear from the above that the Bohr theory was remarkably successful in explaining the structure and spectra of atoms qualitatively, it also had its share of failures:

1 The Bohr theory could not yield acceptable *quantitative* results for any but one-electron atoms because of the complexities introduced by the presence of more than one electron.

2   It could not explain the structure and spectra of molecules, although Bohr devoted a long section at the end of his 1913 paper to a vain attempt to do so.

3   It said nothing about spectral intensities, which were known to differ greatly from one spectral line to another, nor could it predict the polarization of the emitted radiation.

4   Even with regard to hydrogen it failed to account for the finer details of the spectrum.

Ad Hoc Nature of Bohr Theory

A more serious objection to the pioneering theory of Bohr is that it was an odd mixture of classical and quantum physics. It avoided contradictions simply by introducing ad hoc postulates to exclude them.

For example, Bohr's first postulate is in direct conflict with the well-verified classical result that all accelerated charges radiate. Bohr also gave no proof of his second postulate that the angular momentum of the H atom is quantized in units of $\hbar$. He introduced it because it led to discrete energy levels, and these energy levels produced a spectrum in agreement with experiment. A theory more fundamental and more rigorous than Bohr's was needed. To his great credit, this was recognized by Bohr, and he was delighted when such a theory was provided by de Broglie, Schrödinger, and Heisenberg in the years following 1923.

28.10   Wave Mechanics and Quantum Mechanics

The desired goal of a more fundamental quantum theory was achieved almost simultaneously along two very different but equivalent paths, that of wave mechanics and that of matrix, or quantum, mechanics.

Wave Mechanics

The serious study of wave mechanics began with de Broglie's insight in 1923 into the wave properties of particles (see Sec. 27.7). As we have seen, de Broglie predicted that particles should have both wave and particle properties interrelated by Planck's constant through the equation $\lambda = h/mv$. This led him to consider the motion of an electron in a Bohr atom in an attempt to understand why discrete energy states existed in the atom. The waves associated with discrete energy states could not be traveling waves, which constantly change with time, but must be *standing waves*, which constructively interfere with themselves and preserve their configuration as time goes on. De Broglie assumed, therefore, that the circumference of the circular orbit in which the electron moves in H must contain an integral number of de Broglie wavelengths, for only then is a standing wave pattern possible.

This is like forming a closed circle with a chain of equal-size links: there must be an integral number of links if the chain is to be closed, as in Fig. 28.26. So too a closed standing wave in a circle must contain an integral number of wavelengths so that $2\pi r = n\lambda$, with n an integer.

Some possible standing waves of this sort are shown in Fig. 28.27. These standing waves are specified by the value of n in the equation

$$2\pi r = n\lambda = \frac{nh}{mv}$$

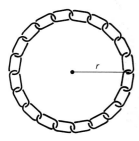

FIGURE 28.26   A circular chain consisting of links. If the chain is to be closed, there must be an integral number of links in the chain.

or $mvr = \dfrac{nh}{2\pi} = n\hbar$ **(28.15)**

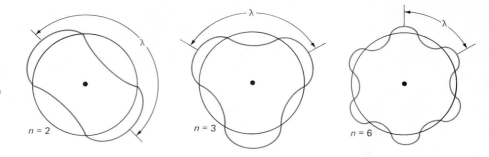

FIGURE 28.27 Possible standing wave patterns for an electron in a hydrogen atom. As n increases, the wavelength decreases in size.

FIGURE 28.28 Erwin Schrödinger (1887–1961), who received the 1933 Nobel Prize in physics for his development of wave mechanics. The equation he proposed in 1926, now called the Schrödinger equation, is the single most important equation in quantum mechanics. *(Photo by Pfaundler; courtesy of AIP Niels Bohr Library.)*

This expression for the angular momentum mvr derived by de Broglie is *exactly the same* as Eq. (28.3), which is an expression of Bohr's postulate that the angular momentum of the atom must be quantized.

Such standing waves would correspond to discrete energy states that are specified by the number of nodes in the standing wave. In this way de Broglie was able to explain both the quantized nature of the angular momentum and the discrete energy states needed to describe the hydrogen spectrum.

These ideas led de Broglie to suggest that what was needed was a new mechanics which would reduce to classical mechanics for ordinary macroscopic objects but would lead to observable wave properties for microscopic particles like electrons and neutrons. He saw in this an analogy with the field of optics, in which physical, or wave, optics reduces to ray optics when the wavelength of light is very small compared with the size of apertures or obstacles in the path of the light beam. As de Broglie wrote, what was needed was "a new mechanics with a wave character which would be with respect to the old mechanics what wave optics is with respect to geometrical optics."

This goal was achieved by the Austrian physicist Erwin Schrödinger (see Fig. 28.28) in a series of papers published in 1926. Schrödinger developed a single basic equation, now called the *Schrödinger wave equation*, which was similar to a classical wave equation of the kind that led Maxwell to his prediction of the speed of electromagnetic waves. The classical wave equation describes the variations in the electric field **E** as a function of time and position in space. The intensity of the wave at a particular time and place is then proportional to E^2. The Schrödinger equation, on the other hand, is the equation for something called the wave function ψ associated with an electron (or photon or other particle), and ψ^2 determines the *probability* of finding the electron at a particular point in space. Hence wave mechanics is a *probabilistic theory* that tells us, for example, where an electron in a hydrogen atom is most likely to be found, but can never tell us exactly where the electron is at any particular instant. For this reason Bohr's orbits must be abandoned as a true picture of what goes on inside an atom. Instead curves of the type shown in Fig. 28.29 indicate where the electron is likely to be found in the H atom in states with $n = 1, 2, 3$.

Schrödinger showed that, when applied to electrons in atoms, his equation had acceptable solutions only for certain discrete energy values, the so-called proper values, or *eigenvalues*, of the energy. Associated with each such energy was a *proper wave function*, or *eigenfunction* ψ. The solution of the Schrödinger equation led not only to the energy levels of the atom but also to an analytical expression for the wave function ψ. This analytical wave function could then be used to

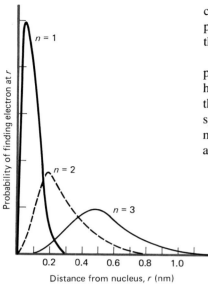

calculate all other important properties of the atom, such as the intensity and polarization of the radiation the atom emits and the change in this radiation when the atom is placed in an electric or magnetic field.

The solution of Schrödinger's equation thus led, in principle, to all the important physical and chemical properties of atoms and molecules, without any ad hoc postulates other than the Schrödinger equation itself. In practice, of course, the Schrödinger equation can be very difficult to solve, especially for many-particle systems like heavy atoms or molecules. Successful applications of this equation to many different fields of physics have demonstrated, however, its basic correctness and importance.

FIGURE 28.29 Probability of finding the electron in the H atom at various distances from the nucleus in the energy states with $n = 1$, 2, and 3, according to Schrödinger's wave mechanics.

Quantum Mechanics

In 1925 Werner Heisenberg (Fig. 27.28) began to publish his first papers on matrix, or quantum, mechanics. Heisenberg's theory is much more abstract than Schrödinger's and concentrates on the observed radiations from atoms and the knowledge they give us about energy states within the atoms. It omits completely any reference to the position or velocity of electrons in atoms, since these quantities cannot be observed and measured.

The Schrödinger and Heisenberg theories have now been blended together into what is called the *new quantum theory* (also called simply *quantum mechanics*) in contrast with the old quantum theory of Planck, Einstein, and Bohr. The many great successes of physics in the past half century are due, from a theoretical point of view, in great part to this new quantum theory developed by de Broglie, Schrödinger, and Heisenberg as a continuation of the ideas of Niels Bohr, and later extended to include the theory of relativity by P. A. M. Dirac (1902–1984).

Quantum mechanics produced an intellectual upheaval in both physics and philosophy. To many this revolutionary theory seemed to lessen the objectivity of the world by describing everything in terms of probabilities and by making the properties of matter depend on the act of observation. For this reason Einstein never accepted the usual interpretation of quantum mechanics (called the *Copenhagen interpretation* after Bohr's home city). The founder of relativity remained convinced to the end that quantum theory was incomplete and would some day be replaced by a more complete theory. That day has not yet come! During the last few years physicists and philosophers have again taken up this question of the meaning of quantum mechanics and the nature of reality with an ardor that rivals that of Einstein and Bohr in their heated debates on this subject more than 50 years ago.

All physicists agree that quantum mechanics works—and works extremely well—as a physical theory. There is, however, considerably less agreement about what the theory really means, and whether it is complete or only a stepping-stone to a better theory.

Example 28.7

For the energy state with principal quantum number $n = 3$ in hydrogen, find **(a)** the de Broglie wavelength λ of the electron in this orbit; **(b)** the value of the angular momentum for the electron in this orbit.

SOLUTION

(a) The radius of the Bohr orbit with $n = 3$ is

$$r_3 = (3)^2 r_1 = 9(0.0528) = 0.475 \text{ nm}$$

and so, from de Broglie's condition for standing waves,

$$\lambda = \frac{2\pi r_3}{n} = \frac{2\pi(0.475 \text{ nm})}{3} = \boxed{0.99 \text{ nm}}$$

There are therefore three complete wavelengths of this length in the circumference $2\pi r_3$.

(b) The angular momentum in this orbit is

$$n\hbar = 3\hbar = \boxed{3.2 \times 10^{-34} \text{ kg·m}^2/\text{s}}$$

Example 28.8

An excited H atom has an average lifetime in the $n = 4$ excited state of 10^{-8} s. **(a)** What is the frequency of the photon emitted when such an atom falls back to the $n = 2$ state of H? **(b)** What is the uncertainty in the frequency of this photon? **(c)** What is the width of the emitted line as a fraction of its central frequency?

SOLUTION

(a) The $n = 4$ to $n = 2$ transition in H can be seen in Fig. 28.6 to correspond to a wavelength of 486 nm. The corresponding frequency is therefore

$$\nu = \frac{c}{\lambda} = \frac{3.00 \times 10^8 \text{ m/s}}{486 \times 10^{-9} \text{ m}} = \boxed{6.17 \times 10^{14} \text{ Hz}}$$

(b) From the Heisenberg uncertainty principle (Sec. 27.10), $\Delta E \, \Delta t \geq h/4\pi$. We take the average lifetime as a measure of the incertainty Δt in the time at which the transition occurs, and so

$$\Delta \nu = \frac{\Delta E}{h} \geq \frac{1}{4\pi \, \Delta t} = \frac{1}{4\pi(10^{-8} \text{ s})} = \boxed{8.0 \times 10^6 \text{ Hz}}$$

Because of the uncertainty principle, therefore, the radiation from a group of excited H atoms does not have a precise frequency ν but has a range of frequencies between $\nu - \Delta \nu$ and $\nu + \Delta \nu$. This determines the *intrinsic width* of the spectral line.

(c) Since

$$\frac{\Delta \nu}{\nu} = \frac{8.0 \times 10^6}{6.2 \times 10^{14}} = \boxed{1.3 \times 10^{-8}}$$

the frequency spread is very small. In actual experiments larger contributions to the width of the spectral line are made by atomic collisions (*pressure broadening*) and the Doppler effect (*Doppler broadening*).

Summary: Important Definitions and Equations

Hydrogen spectrum

Rydberg equation

$$\frac{1}{\lambda} = R \left(\frac{1}{n_f^2} - \frac{1}{n_i^2} \right) \tag{28.2}$$

where $R = 1.0968 \times 10^7 \text{ m}^{-1}$.

For Lyman series (ultraviolet): $n_f = 1; n_i = 2, 3, 4, \ldots$
For Balmer series (visible): $n_f = 2; n_i = 3, 4, 5, \ldots$
For Paschen series (infrared): $n_f = 3; n_i = 4, 5, 6, \ldots$

Bohr theory of hydrogen spectrum

Rutherford-Bohr atom: A model of the atom consisting of a very small, positively charged nucleus, containing almost all the mass of the atom, surrounded by a number of moving electrons equal to the number of elementary positive charges in the nucleus.

Postulates of Bohr theory

1 Stationary states of discrete energies exist in the hydrogen atom in which the atom is stable and emits no radiation.
2 Angular momentum is quantized:

$$m\upsilon r = n\hbar \qquad \text{where } \hbar = h/2\pi \tag{28.3}$$

3 Bohr-Einstein frequency condition:

$$\nu = (E_i - E_f)/h. \tag{28.4}$$

Results of Bohr theory

Discrete energies:
$$E_n = -\frac{1}{n^2}\frac{mk_e^2 e^4}{2\hbar^2} \quad (28.8)$$

$$= \frac{-13.606 \text{ eV}}{n^2} \quad (28.9)$$

Discrete radii:
$$r_n = \frac{n^2\hbar^2}{mk_e e^2} = n^2 r_1 \quad (28.12)$$

where $r_1 = 0.0528$ nm.

Wavelengths:
$$\lambda = \frac{1.24 \times 10^{-6} \text{ m}}{\Delta E} \quad (28.13)$$

where ΔE is the energy difference $E_i - E_f$, in electron-volts, between the initial and final states.

Fluorescence: The absorption of high-frequency photons by atoms or molecules, which then immediately radiate lower-frequency photons.

Phosphorescence: Delayed fluorescence.

Franck-Hertz experiment: The excitation of atoms by inelastic electron collisions; a direct proof of the existence of discrete energy levels in atoms.

Excitation potential: The accelerating voltage required to give a colliding electron sufficient energy to raise atoms to an excited state. The corresponding energy is called the *excitation energy.*

Ionization potential: The accelerating voltage required to give a colliding electron sufficient energy to remove an electron completely from an atom. The corresponding energy is called the *ionization energy.*

Quantum numbers for a single electron

Quantum numbers	Allowed values
Principal quantum number n	1, 2, 3, . . .
Orbital angular momentum quantum number l	0, 1, 2, . . . , $n-1$
Magnetic quantum number m_l	0, ±1, ±2, ±3, . . . , $\pm l$
Spin quantum number m_s	$+\frac{1}{2}$, $-\frac{1}{2}$

Pauli exclusion principle: In one and the same atom no two electrons can have the same set of values for the four quantum numbers n, l, m_l, and m_s.

Building-up principle for periodic table of elements: Atoms are built up by adding electrons one at a time in the lowest one-electron energy states permitted by the Pauli exclusion principle.

Electron configuration of an atom: The assignment of electrons to the various energy shells and subshells by indicating the values of n and l for each electron. In so doing, values of 0, 1, 2, 3, . . . for l are indicated by the letters s, p, d, f, ...

X-ray line spectra: Discrete spectral lines with wavelengths in the nanometer region produced by transitions involving inner-shell electrons in atoms.

$$\frac{1}{\lambda} = Z^2 R\left(\frac{1}{n_f^2} - \frac{1}{n_i^2}\right) \quad (28.14)$$

Questions

1 Why do only gases and vapors emit line spectra, whereas solids emit a continuous spectrum?

2 Sunlight is passed through a long tube containing atomic hydrogen gas. It is found that both a strong Lyman series and a weak Balmer series appear in the absorption spectrum. What can be concluded about the temperature of the gas?

3 Why does the spectrum of hydrogen contain so many different lines and series if it is produced by only one electron?

4 Why would you expect from classical physics that an electron orbiting a nucleus in an atom would continuously lose energy by radiation?

5 Why would you expect to see more lines in the Balmer series of H in the spectrum of some celestial bodies than you can in vacuum discharges in the laboratory?

6 An astronomer wants to know if there is any iron in the atmosphere of a star. Can you suggest an experiment that might determine an answer to this question?

7 The lights above the meat counter in a supermarket often are "warm white" fluorescent lights that heighten the reddish color of the meat. Can you suggest how this is achieved?

8 The fluorescent spectrum of various atoms and molecules are often used like "fingerprints" to identify unknown compounds. Explain why this is possible.

9 Prove that $2n^2$ is the number of different possible sets of four quantum numbers corresponding to a principal quantum number n.

10 Indicate the electron configuration for chlorine ($_{17}$Cl) and show why Cl is similar to fluorine ($_9$F) in chemical and spectroscopic properties.

11 Explain on the basis of de Broglie's theory of matter waves why you would expect free electrons to have a continuum of energy values instead of discrete energy values.

12 Explain on the basis of energy considerations why an electron in an orbit close to the nucleus must move more rapidly than an electron in an orbit farther away from the nucleus.

13 Show how the Pauli exclusion principle helps explain the solidity and impenetrability of matter.

14 Observed spectral lines are said to be *Doppler broadened*, where the reference is to the Doppler effect discussed under sound. Can you suggest how this effect applied to the electromagnetic radiation emitted by a gas can explain the broadening of spectral lines?

Multiple-Choice and Simple Exercises

28.1 The ionization energy for the electron in a normal hydrogen atom is numerically equal to the energy of the atom in its:

(a) Ground state
(b) Lowest excited state
(c) Second lowest excited state
(d) Highest excited state
(e) None of the above

28.2 The wavelength of the first line in the Paschen series in hydrogen in the infrared is:

(a) 122 nm   (b) 656 nm   (c) 959 nm
(d) 380 nm   (e) 1875 nm

28.3 If we take as our zero of energy the energy of the state with the electron and proton separated and at rest, the energy of a H atom in a state characterized by the principal quantum number n equal to 10 is:

(a) $+2.80 \times 10^{-20}$ J   (b) -1.36 eV
(c) $+1.36$ eV   (d) $+0.136$ eV
(e) -2.18×10^{-20} J

28.4 The radius of the electron orbit in the energy state with $n = 100$ in the hydrogen atom is:

(a) 0.0528 nm   (b) 0.0428×10^{-9} m
(c) 528 nm   (d) 5.28 nm   (e) 5.28 m

28.5 What would the temperature of H gas have to be for the average kinetic energy of the gas atoms to correspond to the 10.2-eV energy difference between the ground state and the first excited state of hydrogen?

(a) 7.90×10^3 K   (b) 7.90×10^4 K
(c) 7.50×10^2 K   (d) 300 K
(e) None of the above

28.6 In a Franck-Hertz experiment on potassium (K) vapor, it is found that the current falls off rapidly at a voltage $V_1 = 1.62$ V. The following line would be expected to appear in the spectrum of K when that voltage is reached:

(a) 345 nm   (b) 464 nm   (c) 691 nm
(d) 767 nm   (e) 960 nm

28.7 A photon of wavelength 100 nm is incident on a hydrogen atom in its ground state, from which transitions to excited states correspond to wavelengths of 122, 103, 97.3, and 95.0 nm. The following is a correct statement of what happens:

(a) The photon is absorbed, and a photon of longer wavelength is emitted by the atom.
(b) The photon is absorbed, and the atom gains increased kinetic energy.
(c) The photon passes right through the atom and is not absorbed.
(d) The photon is absorbed, and the atom loses kinetic energy in the absorption process.
(e) The atom is stimulated to emit a second photon at exactly 100 nm.

28.8 In a subshell of an atom specified by $n = 4$, $l = 3$, the number of possible values of the magnetic quantum number m_l is:

(a) 1   (b) 3   (c) 6   (d) 7   (e) 4

28.9 Which of the following electron configurations is forbidden by the Pauli exclusion principle:

(a) $(1s)^2(2s)^2(2p)$ (b) $(1s)^2(2s)^2(2p)^5$
(c) $(1s)^2(2s)(2p)^6$ (d) $(1s)(2s)^2(2p)^5$
(e) $(1s)^2(2s)^3(2p)^4$

28.10 The atomic diameter of argon ($_{18}$A) is 0.382 nm and the diameter of calcium ($_{20}$Ca) is 0.393 nm. A likely value for the atomic diameter of potassium ($_{19}$K) is:

(a) 0.476 nm   (b) 0.275 nm   (c) 0.180 nm
(d) 4.76 nm   (e) 0.388 nm

28.11 X-ray line spectra are produced by:

(a) Valence electrons in atoms
(b) Compton scattering
(c) Complicated molecules formed when electrons strike the target in an x-ray tube
(d) Electrons in shells deep inside the atoms of the target material
(e) The rapid deceleration of the electrons in the beam striking the target

28.12 What is the series limit for the Paschen series in the infrared spectrum of hydrogen?

28.13 What is the magnitude of the Coulomb force required to keep an electron in the hydrogen atom moving in an orbit of radius 0.84 nm?

28.14 What is the speed of an electron moving in an orbit of radius 0.84 nm in a hydrogen atom?

28.15 A strong line in the spectrum of calcium (Ca) occurs when the atom makes a transition from an excited state 4.7 eV above the ground state to a lower excited state 1.9 eV above the ground state.

(a) What is the frequency of the light emitted in such a transition?
(b) What is the wavelength of the light?

28.16 It is found that monatomic hydrogen gas can be raised to its first excited state by absorbing 10.2-eV photons. The ionization potential of H is 13.6 V. How much energy would be required to remove the electron from the atom in its first excited state?

28.17 The ionization energy of the mercury atom from its ground state is 10.4 eV. At what wavelength would you expect the lines in the spectral series involving transitions from excited states to the ground state of Hg to merge into a continuum?

28.18 Why is the first ionization potential of $_4$Be (9.3 V) higher than that of its neighbors on either side, $_3$Li (5.4 V) and $_5$B (8.3 V), even though the $n = 2$ shell is not completely filled until $_{10}$Ne is reached?

28.19 (a) How many electrons are there in a filled subshell with orbital angular momentum quantum number equal to 4?
(b) Write out the quantum numbers l, m_l, and m_s for such a subshell.

28.20 Show numerically that the ratio of the frequencies of the K_α lines in tungsten ($_{74}$W) and platinum ($_{78}$Pt) is proportional to the squares of their atomic numbers (see Fig. 28.25).

28.21 What is the de Broglie wavelength of an electron in the orbit corresponding to the $n = 4$ stationary state in the H atom?

Problems

28.22 (*a*) Calculate the four longest wavelengths in the Lyman series of H and plot their positions on a linear horizontal scale.

(*b*) Indicate the series limit on the same plot.

(*c*) Are any of these spectral lines in the visible range?

28.23 (*a*) What is the radiation of longest wavelength that will ionize hydrogen atoms in their ground state?

(*b*) What wavelength radiation will ionize a hydrogen atom in its ground state and in addition give the electron 1.2 eV of kinetic energy after it leaves the atom?

28.24 The radius of the gold nucleus is about 7.0×10^{-15} m. What minimum energy must the alpha particle in a Rutherford scattering experiment have to come this close to the center of the nucleus?

28.25 Find the distance of closest approach for a 5.0-MeV proton making a head-on collision with a platinum nucleus (Z = 78) in a scattering target.

28.26 Calculate the fraction of an alpha particle's initial kinetic energy that is transferred to another particle initially at rest in a head-on elastic collision, if the other particle is (*a*) a gold nucleus, (*b*) a proton.

28.27 (*a*) Calculate the value of the orbital angular momentum of the electron in the four lowest energy states of the hydrogen atom, using the expression $L = I\omega$ for the angular momentum.

(*b*) Show that in each case this value is equal to $n\hbar$.

28.28 Calculate the values of (*a*) the energy, (*b*) the radius of the electron orbit, and (*c*) the speed of the electron in this orbit for the upper and lower energy states involved in the third line of the Paschen series.

28.29 Calculate (*a*) the energy, (*b*) the radius of the electron orbit, and (*c*) the speed of the electron in this orbit for the upper level involved in the fourth line of the Lyman series in hydrogen.

28.30 A negative mu meson, or muon, can be captured by a proton to form a so-called muonic atom, which is similar to a hydrogen atom in structure except that the muon rest mass is 207 times that of an electron.

(*a*) Calculate the radius of the first Bohr orbit for such a muonic atom.

(*b*) Calculate the energy of the ground state of such an atom.

(*c*) Find the wavelength of the first Balmer line for such an atom.

28.31 A hydrogen atom exists on the average for about 10^{-8} s in an excited state before making a transition back to its ground state.

(*a*) About how many revolutions will the electron make in the n = 2 state before the atom falls back to the ground state?

(*b*) What is the uncertainty in the energy of the atom in this excited state?

*28.32 Show that for large values of the principal quantum number *n*, the frequencies of revolution for an electron in adjacent energy levels of a hydrogen atom and the frequency of light radiated in a transition between these two levels approach the same value. (This is an example of *Bohr's correspondence principle*.)

*28.33 Show that for H the radius of the first Bohr orbit and the ground-state energy of the atom can be written in the following form:

$$r_1 = \frac{\hbar}{\alpha mc} = \frac{\lambda_0}{2\pi\alpha}$$

$$E_1 = \tfrac{1}{2}\alpha^2 mc^2$$

where $\lambda_0 = h/mc$ is the Compton wavelength of the electron (see Sec. 27.5) and $\alpha = k_e e^2/\hbar c$ is the fine-structure constant.

*28.34 (*a*) Prove that when allowance is made for the motion of both electron and nucleus about the common center of mass of a one-electron atom, the result for E_n, r_n, and v_n are the same as in the simple Bohr theory except that the mass of the electron *m* is replaced by the *reduced mass* $\mu = m/(1 + m/M)$, where *M* is the mass of the nucleus.

(*b*) Use the reduced mass to calculate accurate values of the Rydberg constant for H, He^+, and Li^{2+}.

(*c*) What is the percentage difference between the Rydberg constant for H and for He^+?

28.35 (*a*) Calculate the electric current produced by an electron moving in the first Bohr orbit.

(*b*) At the site of the proton in the hydrogen atom, what is the approximate value of the magnetic field produced by this orbiting electron?

28.36 Find the speed with which a hydrogen atom recoils when it emits the first line in the Balmer spectrum ($n_i = 3 \rightarrow n_f = 2$).

*28.37 A positronium atom is a hydrogenlike atom with a positron (a positive electron) replacing the proton as the nucleus of the atom.

(*a*) How does the ionization potential of positronium compare with that of the hydrogen atom?

(*b*) How does the wavelength of the $n_i = 4 \rightarrow n_f = 2$ transition in positronium compare with the wavelength of the same transition in hydrogen?

28.38 The energy separation between two states in the sodium atom is 3.36×10^{-19} J.

(*a*) Find the minimum voltage through which an electron must be accelerated to excite the Na atom from the lower to the upper state.

(*b*) Find the wavelength of the radiation which must be absorbed to excite the atom from the lower to the upper state.

(*c*) Will higher voltages than that found in (*a*) also be able to excite the Na atom from the lower to the upper state?

(*d*) Will shorter wavelengths than that found in (*b*) be able to do the same?

28.39 When ordinary table salt is put into a Bunsen burner flame, a characteristic yellow line of sodium at about 589 nm appears. This line arises when an electron returns from the first excited state to the ground state of sodium.

(*a*) What is the energy of the first excited state above the ground state in Na (in both joules and electronvolts)?

(*b*) What would the temperature of the vapor have to be to produce atoms with an average kinetic energy equal to the energy of this excited state?

(*c*) Since the average temperature of the molecules in the

Bunsen flame is only 2100 K, how do the Na atoms obtain sufficient energy to excite the 589-nm line?

28.40 Hydrogen gas is bombarded by an electron beam of just enough energy to excite the H_γ line ($n_i = 5 \rightarrow n_f = 2$) in the Balmer spectrum.

(a) What is the least energy required in the electron beam to do this?

(b) What other lines would you expect to see in the visible spectrum of H under these excitation conditions?

28.41 Mercury vapor is bombarded by electrons of energy 7.00 eV. What wavelengths would you expect to observe in the spectrum of Hg excited in this way? (See Fig. 28.12.)

28.42 How much energy is absorbed by an atom when mercury vapor is bombarded with (a) 5.52-eV electrons; (b) 5.52-eV photons? (See Fig. 28.12.)

28.43 The cesium atom has excited states at 1.38 and 2.30 eV above the ground state, and an ionization potential of 3.87 V. If cesium vapor in its ground state is bombarded with electrons accelerated through 3.90 V, what spectral lines might you expect to observe with a spectrograph?

28.44 In an electron-impact experiment it is found that radiation with a wavelength of 589 nm is emitted by sodium vapor when bombarded by 2.11-eV electrons. Compute the value of h/e from these data.

28.45 The ground state of potassium (K) has the following values of the four quantum numbers of the outermost electron: $n = 4$, $l = 0$, $m_l = 0$, $m_s = +\frac{1}{2}$.

(a) If this electron is excited to the energy state specified by the quantum numbers $n = 5$, $l = 1$, how many different sets of quantum numbers are possible in this state?

(b) What are the values of the four quantum numbers for each of these configurations?

28.46 In potassium the first exception to the filling in of the lowest possible one-electron energy levels occurs when the last electron goes into the $4s$ level while the $3d$ level is still vacant. Show that this makes the electron configuration of $_{19}K$ very similar to that of $_{11}Na$ and accounts for their similar physical and chemical properties.

28.47 If for some reason atoms could contain only electrons with principal quantum numbers n up to and including $n = 5$, and if all these five shells were completely filled, how many elements would there be in the periodic table?

28.48 What are the wavelengths of the following spectral lines?

(a) The line in the spectrum of He^+ which corresponds to the same transition as the one producing the first line in the Paschen series in H.

(b) The line in the spectrum of Li^{2+} which corresponds to the same transition as that producing the first line in the Balmer series in H.

(c) The line in the spectrum of Be^{3+} which corresponds to the same transition as that producing the first line in the Lyman series of H.

(d) Indicate in what regions of the spectrum these lines occur.

28.49 (a) How much energy is required to remove the electron from the ground state of He^+?

(b) What is the minimum frequency of an incident photon that can remove this electron from the ion?

28.50 At extremely high temperatures such as exist in the stars, collisions can strip heavy atoms of almost all their electrons. Suppose that potassium ($_{19}K$) vapor has all but one of its electrons stripped away.

(a) What is the wavelength of the first line of the Lyman series for such an ion?

(b) In what region of the spectrum does this wavelength fall?

28.51 An important quantity in the theory of atomic spectra is the fine-structure constant $\alpha = k_e e^2 / \hbar c$, where k_e is the Coulomb's law constant. Show (a) that α is a pure number, without dimensions; (b) that its numerical value is approximately 1/137.

*28.52 (a) Derive an approximate expression showing that the frequencies of the K_α lines in the x-ray spectra of the heavy elements are proportional to the squares of the atomic numbers of the elements.

(b) Calculate the constant of proportionality.

28.53 (a) Use Bohr theory to calculate an approximate value for the wavelength of the K_α line in tungsten ($_{74}W$). (See Fig. 28.24.) Compare your answer with the value indicated in Fig. 28.25.

(b) What must the minimum energy of the electrons bombarding the tungsten target be to produce this K_α line?

28.54 (a) Calculate, using Bohr theory, an approximate value for the wavelength of the K_α line in platinum ($_{78}Pt$), and compare your results with the value indicated in Fig. 28.25.

(b) If the platinum atom loses one electron from its K shell by the absorption of an x-ray photon that removes the electron entirely from the atom, what must be the maximum value of the wavelength of this incident photon?

28.55 (a) What is the de Broglie wavelength of an electron in the $n = 5$ quantum state in hydrogen?

(b) How many complete de Broglie wavelengths are there in the length of the electron's orbit?

(c) What is the value of the angular momentum in this orbit?

28.56 A man decides to lengthen his de Broglie wavelength by moving very slowly. His mass is 100 kg and he moves at a speed of 1 cm every 10 years.

(a) What is his de Broglie wavelength?

(b) How does this wavelength compare with the wavelength of an electron in the ground state of the hydrogen atom?

28.57 An atom has a lifetime of 2.0×10^{-8} s in a particular excited state; i.e., it remains in the excited state for 2.0×10^{-8} s, on the average, before falling back to its ground state. The excited state is 1.7 eV above the ground state.

(a) What is the uncertainty in the frequency emitted by the atom in the transition back to the ground state?

(b) What is the uncertainty in the wavelength of the emitted radiation?

*28.58 Estimate the uncertainty in the position of the electron in the hydrogen atom in its ground ($n = 1$) state.

28.59 A mercury atom emits a strong green line of wavelength 546 nm. Its average lifetime in the excited state involved in the transition producing this line is 0.80×10^{-9} s. Estimate the wavelength spread of the 546-nm line.

Additional Readings

Andrade, E. N. da C.: *Rutherford and the Nature of the Atom*, Doubleday Anchor, Garden City, N.Y., 1964. This brief, interesting life of Rutherford contains considerable material on Bohr's work in Manchester with Rutherford and on the development of Bohr's atomic theory.

——: "The Birth of the Nuclear Atom," *Scientific American*, vol. 195, no. 5, November 1956, pp. 93–104. A briefer account of the material contained in the preceding reference.

Boraiko, Allen A.: "Lasers—A Splendid Light," *National Geographic*, vol. 165, no. 3, March 1984, pp. 334–363. An up-to-date, well-illustrated article in the famous first issue with a hologram on the cover.

Cline, Barbara L.: *Men Who Made a New Physics*, University of Chicago Press, Chicago, 1988. A lively account of the development of quantum physics, with considerable insight into the personalities of the physicists involved in its development.

Deutsch, Thomas F.: "Medical Applications of Lasers," *Physics Today*, vol. 41, no. 10, October 1988, pp. 56–63. An account of the ever-increasing use of lasers for medical purposes.

Gamow, George: "The Exclusion Principle," *Scientific American*, vol. 201, no. 1, July 1959, pp. 74–86. A discussion of the impact of the Pauli principle on many different fields of physics.

Hänsch, Theodor W., Arthur L. Schawlow, and George W. Series: "The Structure of Atomic Hydrogen," *Scientific American*, vol. 240, no. 3, March 1979, pp. 94–110. An outstanding article by three physicists who have contributed much to our knowledge of spectra, indicating the importance and the complexity of the hydrogen spectrum.

Lasers and Light, Readings from *Scientific American*, Freeman, San Francisco, 1969. A collection of articles by some of the foremost research workers in laser optics.

Moore, Walter J.: *Schrödinger: Life and Thought*, Cambridge University Press, New York, 1989. The only complete biography of Schrödinger in English.

Physics Today, vol. 16, no. 10, October 1963, pp. 21–64. This memorial issue of a popular physics journal is devoted to Bohr and contains perceptive articles by physicists Felix Bloch, J. Rud Nielsen, Victor F. Weisskopf, and John A. Wheeler.

Rozental, S. (ed.): *Niels Bohr: His Life and Work as Seen by His Friends and Colleagues*, Interscience, New York, 1964. One of the best books on Bohr and his work.

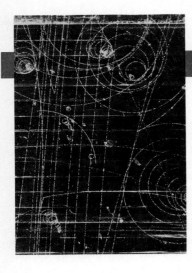

CHAPTER 29

Nuclear Physics

To everyone who, like myself, had the good fortune to visit the physical laboratories in Cambridge and Manchester about twenty years ago and work under the inspiration of the great leaders, it was an unforgettable experience to witness almost every day the disclosure of hitherto hidden features of nature. I remember, as if it were yesterday, the enthusiasm with which the new prospects for the whole of physical and chemical science, opened by the discovery of the atomic nucleus, were discussed in the spring of 1912 among the pupils of Rutherford.

Niels Bohr (1885–1962), in his 1930 Faraday Lecture

Once physicists understood atoms and molecules reasonably well, it was inevitable that they should next tackle the atomic *nucleus*. The nucleus is four orders of magnitude smaller than an atom, however, and unraveling its secrets is correspondingly more difficult. Nevertheless, great progress has been made in understanding nuclear structure and behavior, although many questions remain. The atomic nucleus has turned out to contain an enormous store of energy that is available for the use or the destruction of the world. This chapter emphasizes the basic physical principles needed to understand the nucleus and to weigh arguments concerning nuclear weapons and nuclear power.

29.1 The Structure of the Atomic Nucleus

The Bohr-Rutherford atom consists of light, negatively charged electrons in rapid motion about a dense, positively charged stationary nucleus. Since the atom is electrically neutral, there must be just as many positive charges in the nucleus as there are negative extranuclear electrons. This number is the charge number, or atomic number Z.

The early theories of the nucleus, which emanated from Rutherford's nuclear model of the atom, employed a proton-electron model. According to this view, the nucleus consisted of a number of protons equal to the atomic mass number A, and a number of electrons just sufficient to reduce the nuclear charge to Z. Each nucleus should therefore contain A protons and $A - Z$ electrons, leaving a net positive charge of Z on the nucleus.

The proton-electron model soon ran into severe difficulties. First of all, if an electron were confined inside the nucleus, the uncertainty principle predicts that it would have a kinetic energy above 10 MeV. Such an electron would be so energetic that it would quickly escape from the nucleus.

Second, nuclei have an associated spin angular momentum in the same way that atoms have an angular momentum produced by the orbital and spin motion of the electrons in the atom. Since the spin of each proton and each electron in the nucleus was known to be either $+\frac{1}{2}$ or $-\frac{1}{2}$, the nuclear spin would have to be the sum of these values. According to the proton-electron model a nitrogen nucleus ($^{14}_{7}N$) would have 14 protons and 7 electrons, and no matter how the half-integral spins of these 21 particles were added together, the sum would have to be a half-integer like $\frac{1}{2}$, $\frac{7}{2}$, or $\frac{21}{2}$. But it was well known from the spectrum of the N_2 molecule that the spin of the N nucleus was an integer (actually 1). Clearly something was wrong with the proton-electron model.

Discovery of the Neutron

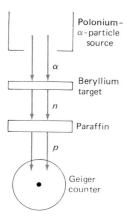

By 1930 evidence had accumulated that certain nuclei emitted penetrating particles which were not affected by electric and magnetic fields. These particles did not ionize atoms and molecules and so did not appear to be photons. In 1932 the English physicist James Chadwick (1891–1974) identified such particles by allowing the nuclear radiation emitted by beryllium when bombarded by alpha particles to strike a paraffin target, as in Fig. 29.1. Protons were knocked out of the paraffin and, because of their charge, were easily detected by a Geiger counter (discussed in Sec. 29.9). From energy and momentum measurements and the conservation principles, Chadwick was able to show that the protons in paraffin must have been hit by an uncharged particle of approximately the same mass as the proton. This particle had been knocked out of beryllium by an alpha particle. Here was the massive uncharged particle sought by physics for a decade. Because it was electrically neutral, Chadwick named it the *neutron*.

FIGURE 29.1 Chadwick's apparatus for detecting neutrons.

The Proton-Neutron Model of the Nucleus

Soon after Chadwick's discovery Heisenberg suggested a proton-neutron model of the nucleus, according to which the nucleus contains no electrons but only protons and neutrons. The charge on the nucleus is then given by the number of protons, which is equal to the atomic number (or charge number) Z, whereas the mass of the nucleus is the sum of the masses of the protons and neutrons. The mass number A then equals the total number of neutrons and protons in the nucleus. On this model the number of neutrons in the nucleus is $A - Z$. The protons and neutrons are commonly referred to collectively as *nucleons*, because they are found in the *nucleus*. The total number of nucleons in a $^{14}_{7}N$ nucleus would then be 14, and the overall spin would be integral in agreement with experiment.

The basic constituents of atoms, as known about 1930, are listed for convenience in Table 29.1. The symbols $_{-1}^{0}e$, $_{1}^{1}p$, and $_{0}^{1}n$ are used for the electron, proton,

TABLE 29.1   Properties of Elementary Particles in Atoms

Particle	Symbol	Charge (C)	Mass (kg)	Mass (u)
Electron	$_{-1}^{0}e$	$-1.602177 \times 10^{-19}$	9.10939×10^{-31}	5.48580×10^{-4}
Nucleons:				
Proton	$_{1}^{1}p$	$+1.602177 \times 10^{-19}$	1.672623×10^{-27}	1.007277
Neutron	$_{0}^{1}n$	0	1.674929×10^{-27}	1.008665

and neutron, respectively, where the superscript is the mass number of the particle and the subscript is its charge number. The masses are given in unified mass units u (See Sec. 9.3), as well as in kilograms.

Nuclear Size

In Sec. 28.2 we showed that Rutherford was able to arrive at a maximum size for atomic nuclei by studying the distribution of the alpha particles they scattered. He found that nuclei had diameters four orders of magnitude smaller than those of atoms. Since atomic diameters are about 0.1 nm, nuclei have diameters of about 10^{-5} nm, or 10^{-14} m. Hence, if an atom were expanded until it was 1 mi in diameter, the nucleus at the atom's center would be about the size of a grapefruit.

Experiments on the scattering of alpha particles, neutrons, and electrons by nuclei have shown that nuclei have an approximately spherical shape. As the mass number A of the nucleus increases (by the addition of protons and neutrons), the size of the nucleus increases so as to keep the nuclear density approximately constant. Since the nuclear density is $\rho = m/V \propto A/V$, we find that $V \propto A$, the mass number. But $V = \frac{4}{3}\pi r_0^3$, where r_0 is the nuclear radius, and so also $V \propto r_0^3$. We therefore conclude that $r_0 \propto A^{1/3}$. Scattering experiments have shown that the constant of proportionality is 1.2×10^{-15} m, so

$$r_0 = (1.2 \times 10^{-15} \text{ m})(A)^{1/3} \tag{29.1}$$

For a gold nucleus we find that

$$r_0 = (1.2 \times 10^{-15} \text{ m})(197)^{1/3} = 0.69 \times 10^{-14} \text{ m}$$

which is consistent with Rutherford's conclusions from his scattering experiments (Sec. 28.2). The volume of a nucleus is thus $V_0 = \frac{4}{3}\pi r_0^3$, or $(7.2 \times 10^{-45})A$ m^3, and is proportional to A and therefore to the number of nucleons in the nucleus.

Nucleons act like small marbles that are packed together as tightly as possible in the nucleus. The more nucleons there are, the larger the nucleus. This is *not* what happens with atoms, where the size of the atom varies *in a periodic fashion* with the number of electrons in the atom.

Example 29.1

Compare **(a)** the radius and **(b)** the volume of a uranium nucleus with that of a hydrogen nucleus.

SOLUTION

(a) Since the radius of a nucleus is given by Eq. (29.1), we have for the ratio of the radii of the two most common isotopes of uranium and hydrogen listed in Appendix 4,

$$\frac{r_0(^{238}_{92}\text{U})}{r_0(^1_1\text{H})} = \frac{(1.2 \times 10^{-15} \text{ m})(238)^{1/3}}{(1.2 \times 10^{-15} \text{ m})(1)^{1/3}} = (238)^{1/3} = 6.2$$

The uranium nucleus therefore has a radius about six times that of a hydrogen nucleus.

(b) The volume of the uranium nucleus is 238 times larger than the volume of the hydrogen nucleus, since the volume depends directly on the mass number A.

29.2 Nuclear Forces and Binding Energies

If nuclei are composed of only protons and neutrons, what holds them together? Why does the nucleus not fly apart, since the protons repel each other with very large Coulomb forces at the small distances separating them inside a nucleus? In the nucleus there must exist some new kind of attractive force between nucleons (now called the *strong nuclear force*), which is stronger than the electrostatic repulsion between protons. (This nuclear force is clearly not a gravitational force, since the gravitational force between two protons is about 35 orders of magnitude *weaker* than the electrostatic repulsion between them.) The nature and characteristics of this nuclear force are still not completely known, and many of the experiments being carried out in laboratories all over the world are directed toward a better understanding of this force.

We do know from experiment that the nuclear force is the same between any two nucleons, i.e., the same for *np*, *nn*, and *pp* interactions. In other words, it is *charge-independent*. We also know that this very strong force comes into play only when the nucleons are very close together, say, within about 10^{-15} m of each other, and that it drops rapidly to zero for larger distances, as in Fig. 29.2. It is thus a very short-range force compared with the long-range Coulomb force that falls off as $1/r^2$.

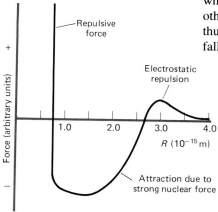

FIGURE 29.2   The strong nuclear force between two protons as a function of the distance R between the protons. This short-range force is attractive in the range of 1 to 2.5×10^{-15} m.

The short-range nature of the nuclear force causes this force to be important only for interactions between a nucleon and its nearest neighbors. In $^4_2$He each nucleon has three nearest neighbors compared to only two in $^3_2$He. For this reason a nucleon in $^4_2$He is more tightly bound than one in $^3_2$He.

However, in a heavy nucleus like $^{208}_{82}$Pb, most of the nucleons already have the maximum possible number of nearest neighbors. (We call this the *saturation* of the nuclear force.) Adding another proton, for example, does not increase much the contribution of the strong force to the stability of the nucleus; the added proton does interact repulsively with the other protons through the long-range Coulomb force, however, and this *reduces* the stability of the nucleus. For this reason light nuclei and heavy nuclei are less stable than those with intermediate masses.

Nuclear Binding Energy

Since we cannot express the nuclear force in the form of an equation, we must rely on experiment to determine the energy of nuclei. In doing this a most useful tool is Einstein's mass-energy relationship $E = m_0c^2$. It is much easier to determine the masses of atoms than it is to measure their energies directly. By determining the masses of the individual protons and neutrons making up a nucleus and then measuring the mass of the nucleus itself, the *mass defect* and the *nuclear binding energy* can be determined.

Mass defect Δm: The difference between the sum of the masses of the individual protons and neutrons making up a nucleus and the mass of the nucleus itself.

Nuclear binding energy: $E = \Delta mc^2$, where Δm is the mass defect. This is the energy required to tear a nucleus apart into its constituent protons and neutrons.

When protons and neutrons come together to form a stable nucleus, the total energy of the nucleus is less than the sum of the energies of the isolated nucleons by an amount equal to the *nuclear binding energy*. For nuclei, however, this energy is of the order of some megaelectronvolts (10^6 eV) rather than the few electronvolts required for molecular binding. From the point of view of Einstein's mass-energy relationship, this means that when the nucleus was formed, the excess mass (equivalent to the binding energy) was radiated away in the form of energy. Conversely, if we want to break up a nucleus into its individual protons and neutrons, and place all these particles at rest far apart from one another, we must supply an amount of energy numerically equal to the nuclear binding energy.

It is found experimentally (mostly from precise mass-spectrometer measurements) that the binding energy varies from nucleus to nucleus throughout the periodic table. To eliminate the effect of differing numbers of nucleons when comparing binding energies, we introduce the *binding energy per nucleon*, i.e., the nuclear binding energy for a particular nucleus divided by the number of nucleons in that nucleus. In Fig. 29.3 we plot the nuclear binding energy per nucleon as a function of the mass number A, that is, the number of nucleons in the nucleus. The greater the binding energy, the greater the energy that must be supplied to break up the nucleus into separated protons and neutrons, and so the more stable the nucleus.

The curve in Fig. 29.3 has some interesting features. It has a maximum in the middle of the periodic table, in the vicinity of the iron ($^{56}_{26}$Fe) nucleus. This $^{56}_{26}$Fe nucleus has the largest binding energy per nucleon, about 8.8 MeV, of any nucleus and is therefore unusually stable (this is one reason so much iron is found in the earth's crust). As we go to heavy nuclei like $^{235}_{92}$U, the binding energy changes slowly to about 7.6 MeV per nucleon, and at the low-mass end of the periodic table it changes more rapidly to values like 2.9 MeV per nucleon for tritium ($^3_1$H), indicating that both very heavy and very light nuclei are less stable than those in the middle of the periodic table, as discussed above.

The binding-energy curve is one of the most significant graphs in all physics. It accounts for the stability of the various elements and thus for the existence of

FIGURE 29.3 Graph of nuclear binding energies per nucleon against the mass number A. The higher the position of a nucleus on the graph, the more stable the nucleus.

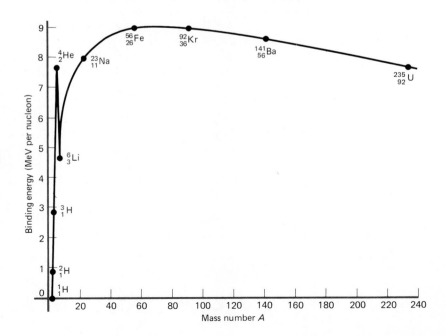

the varied forms of matter around us. It explains the fusion process that has powered the evolution of the universe and the fission process that has introduced a new energy source of tremendous value and tremendous danger.

Energy Units

Since very small mass differences between large masses can be converted into very large amounts of energy, it is necessary, in dealing with nuclear reactions, to retain more than two or three significant figures. We therefore need a conversion factor from mass to energy good to five significant figures.

The masses of the electron, proton, and neutron are given in unified mass units (Sec. 9.3) in Table 29.1. The conversion factor used in this table to convert masses from unified mass units to kilograms is given in Table B.2 as $1.00000 \text{ u} = 1.66054 \times 10^{-27} \text{ kg}$. We have, therefore, for the energy equivalent of 1.00000 u,

$$E = m_0 c^2 = (1.66054 \times 10^{-27} \text{ kg})(2.99792 \times 10^8 \text{ m/s})^2$$

$$= \frac{1.49241 \times 10^{-10} \text{ J}}{1.60218 \times 10^{-19} \text{ J/eV}} = 9.3149 \times 10^8 \text{ eV} = 931.49 \text{ MeV}$$

and so $\boxed{1.00000 \text{ u is equivalent to } 931.49 \text{ MeV}}$ **(29.2)**

Example 29.2

Calculate the Coulomb repulsive force between two protons separated by a distance of 10^{-15} m in a nucleus.

SOLUTION

From Coulomb's law we have

$$F_{el} = \frac{k_e q q'}{r^2} = \left(8.99 \times 10^9 \frac{\text{N·m}^2}{\text{C}^2}\right) \frac{(1.60 \times 10^{-19} \text{ C})^2}{(10^{-15} \text{ m})^2}$$

$$= \boxed{230 \text{ N}}$$

This is approximately the force exerted by the earth's gravitational field on a 20-kg mass, and, as anyone who has ever lifted a 20-kg (~45-lb) bucket of water knows, this is quite a substantial force.

Example 29.3

Calculate the binding energy per nucleon of the $^{23}_{11}$Na (sodium) nucleus and compare the result with that indicated on the binding-energy curve of Fig. 29.3. (The mass of the $^{23}_{11}$Na isotope is 22.98977 u, from the table in Appendix 4.)

SOLUTION

The $^{23}_{11}$Na isotope contains 11 protons and 12 neutrons. Hence the total mass of these nucleons is, from Table 29.1,

$11 \times (1.00728 \text{ u}) = 11.08008 \text{ u}$
$12 \times (1.00866 \text{ u}) = \underline{12.10392 \text{ u}}$

Total mass of nucleons = 23.18400 u

The mass of the ^{23}Na nucleus is the mass of the atom minus the mass of the 11 extranuclear electrons:

$22.98977 \text{ u} - 11(5.49 \times 10^{-4} \text{ u}) = 22.98373 \text{ u}$

The mass defect Δm is therefore

$\Delta m = 23.18400 \text{ u} - 22.98373 \text{ u} = 0.20027 \text{ u}$

and so

$$\text{Binding energy} = (0.20027 \text{ u})\left(\frac{931.49 \text{ MeV}}{1 \text{ u}}\right) = 186.55 \text{ MeV}$$

$$\text{Binding energy per nucleon} = \frac{186.55 \text{ MeV}}{23 \text{ nucleons}}$$

$$= \boxed{8.11 \text{ MeV/nucleon}}$$

This agrees well with the value indicated in Fig. 29.3.

29.3 Nuclear Reactions I: Natural Radioactivity

Most elements have nuclei that are stable and show no tendency to change spontaneously or "decay" into other kinds of nuclei. Some nuclei found in nature, however, especially those at the large-mass end of the periodic table, are unstable and decay into different nuclei by emitting highly energetic particles. Such nuclei are said to exhibit *natural radioactivity*, a name given to this phenomenon by the Polish-French physicist Marie Curie in 1898 (see Figs. 29.4 and 29.5).

The first evidence for natural radioactivity was found by the French physicist Antoine Henri Becquerel (1858–1908) in 1896, when he discovered that a chemical compound containing the element uranium was able to expose photographic film placed near it in a drawer, even though the film was tightly wrapped in black paper. Becquerel rightly guessed that some atoms in the uranium compound were emitting particles of sufficient energy to penetrate the wrapping paper and expose the photographic film. It was soon established that uranium atoms were the culprits. A few years later Marie Curie and her husband, Pierre Curie (1859–1906), were able to extract, using laborious chemical techniques, small quantities of the new elements polonium ($^{210}_{84}$Po) and radium ($^{226}_{88}$Ra) from tons of pitchblende and found that these elements were more highly radioactive than uranium itself. Polonium and radium are two typical naturally radioactive substances. At the present time some 30 natural elements have been found to possess radioactive isotopes, and over 50 naturally radioactive isotopes have been identified.

Radioactivity is the result of the disintegration or decay of an unstable nucleus. Certain nuclei are not sufficiently stable under the combined operation of the strongly attractive nuclear force and the strongly repulsive Coulomb force and spontaneously decay with the emission of nuclear particles of various kinds.

FIGURE 29.4 (left) Marie Sklodowska Curie (1867–1934) with her husband, Pierre Curie, in their laboratory in Paris. This husband and wife team discovered the elements radium and polonium and did much of the original work on radioactivity. Madame Curie received two Nobel Prizes, the 1903 prize in physics and the 1911 award in chemistry. *(Courtesy of AIP Niels Bohr Library.)*

FIGURE 29.5 (right) Marie Curie with her daughter, Irène Joliot-Curie (1897–1956). Marie Curie trained her daughter in scientific research so well that Irène and Irène's husband, Frédéric Joliot-Curie (1900–1958), received the 1935 Nobel Prize in chemistry for the discovery of artificial radioactivity. *(AIP Niels Bohr Library, William G. Meggers collection.)*

Types of Nuclear Emissions

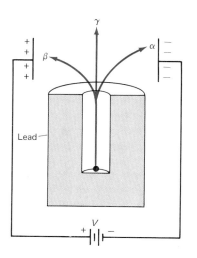

FIGURE 29.6 Particles emitted by radioactive nuclei, as separated by an electric field. Alpha and beta particles are attracted to charged plates. Gamma rays are not affected by charged plates.

What kinds of particles are emitted by naturally radioactive nuclei? It is easy to show by a simple experiment that they fall into three distinct classes. Suppose we put a piece of pitchblende (which contains uranium) at the bottom of a cylindrical cavity in a lead block, as in Fig. 29.6. At the top of the cylinder we set up two plates, with a voltage between them, so that one plate becomes positively charged and the other negatively charged. If we now move a particle detector across the top of the cylinder, we find that some emitted particles are positively charged and attracted to the negative plate, some are negatively charged and attracted to the positive plate, and some have no charge at all, for they are completely unaffected by the charged plates. In the early days of the study of radioactivity, when the nature of these particles was as yet unclear, Lord Rutherford differentiated them by the first three letters of the Greek alphabet as alpha (α), beta (β), and gamma (γ) rays.

We now know that the alpha rays are nothing but helium nuclei stripped of their electrons, that is $_2^4\mathrm{He}^{2+}$. These particles rapidly lose their kinetic energy by collisions; they travel on the average only an inch or so in air before they combine with free electrons and become normal helium atoms.

Beta rays are nothing but ordinary electrons, $_{-1}^{0}e$, which differ from the electrons in the exterior of atoms only in that they are produced in the decay of radioactive nuclei. Beta rays travel only a few feet in air before they are slowed down sufficiently by collisions to combine with positive ions and form neutral atoms.

Gamma rays are high-energy photons produced by transitions between nuclear energy levels in the same way that transitions between atomic energy levels produce visible light. These bundles of electromagnetic energy have no charge and no rest mass, are highly penetrating, and are the most difficult of all radioactive emissions against which to shield.

Half-Life

We do not know how long it will take any particular nucleus to decay, just as we do not know how long it will take for a particular atom to fall spontaneously to a lower-energy state and emit light. What we can determine experimentally, however, is how long it takes, *on the average*, for a large group of radioactive nuclei of the same isotope to decay to some other nuclear species. Suppose, for example, that we have 100 million $_{55}^{137}\mathrm{Cs}$ nuclei in a box. If we could go away and return 30 years later and again count the number of cesium nuclei present, we would find that we had only 50 million left. The other 50 million would have spontaneously transformed themselves into barium nuclei by the following beta-decay process*:

$$_{55}^{137}\mathrm{Cs} \rightarrow \ _{-1}^{0}e \ + \ _{56}^{137}\mathrm{Ba}$$

If this is the case, we say that the *half-life* of $_{55}^{137}\mathrm{Cs}$ is 30 years.

Half-life: The time required for half the nuclei in a sample of a particular nuclear species to decay radioactively.

The half-life is a measure of the stability of a particular nuclear species. Stable

---

*In beta-decay processes particles called neutrinos and antineutrinos are also produced. Since they carry neither charge nor measurable mass away from the reaction, for clarity we will postpone discussion of these particles until Chap. 31.

TABLE 29.2 Half-Lives of Some Important Radioactive Isotopes

Isotope	Half-life $T_{1/2}$
^3_1H	12.3 years
$^{14}_6\text{C}$	5730 years
$^{32}_{15}\text{P}$	14.3 days
$^{40}_{19}\text{K}$	10^9 years
$^{41}_{21}\text{Sc}$	0.6 s
$^{60}_{27}\text{Co}$	5.24 years
$^{82}_{34}\text{Se}$	1.1×10^{20} years
$^{85}_{36}\text{Kr}$	10.8 years
$^{88}_{36}\text{Kr}$	2.8 h
$^{90}_{38}\text{Sr}$	28 years
$^{129}_{53}\text{I}$	1.7×10^6 years
$^{131}_{53}\text{I}$	8 days
$^{137}_{55}\text{Cs}$	30 years
$^{222}_{86}\text{Rn}$	3.82 days
$^{226}_{88}\text{Ra}$	1600 years
$^{232}_{90}\text{Th}$	1.4×10^{10} years
$^{239}_{94}\text{Pu}$	2.4×10^4 years

nuclei have infinitely long half-lives. In one half-life the amount of the radioactive substances is cut in half; in two half-lives it is reduced by a factor of 4; in three half-lives by a factor of 8; and so on. This kind of decay is plotted in Fig. 29.7, where the symbol $T_{1/2}$ denotes half-life.

The half-lives for some important radioactive nuclei are given in Table 29.2. Note the great difference in the half-lives listed, ranging from a fraction of a second to 10^{20} years. Half-lives of artificially produced radioactive nuclei can be as small as 10^{-12} s. Despite these large differences, all radioactive materials decay in the same basic fashion, following the same exponential-decay law.

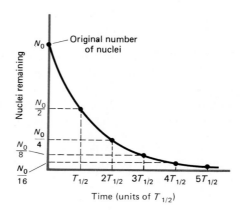

FIGURE 29.7 Exponential decay of a radioactive nucleus. In each half-life, half the nuclei present at the beginning of that half-life decay.

Exponential Decay

The decay curve shown in Fig. 29.7 is a typical exponential-decay curve,* with the property in this case that the height of the curve is reduced by a factor of 2 for each half-life along the horizontal axis. The equation of this curve is

$$N = N_0 e^{-\lambda t} \tag{29.3}$$

where N is the number of radioactive nuclei present at time t, N_0 is the number present at the beginning when $t = 0$, and λ is the *decay constant*, with units s^{-1}. The larger λ, the more rapidly the nuclei decay.

To obtain λ or t when we know N/N_0, we can transform Eq. (29.3) by taking natural logarithms (base e) of both sides of the equation and obtain

$$\ln N = \ln N_0 + \ln e^{-\lambda t} = \ln N_0 - \lambda t$$

or $\quad \lambda t = \ln N_0 - \ln N = \ln \dfrac{N_0}{N}$

Now, by the definition of the half-life ($T_{1/2}$), $N_0/N = 2$ when $t = T_{1/2}$, and so

$$T_{1/2} = \frac{\ln 2}{\lambda} = \frac{0.693}{\lambda} \quad \text{or} \quad \boxed{\lambda = \frac{0.693}{T_{1/2}}} \tag{29.4}$$

A reciprocal relationship therefore exists between the decay constant λ and the half-life $T_{1/2}$. If we know one, we can find the other. The larger λ, the shorter the half-life.

---

*See Appendix 1.E for the properties of exponential and logarithmic functions.

Nuclear Activity

The number of nuclei decaying in some short time interval Δt is clearly proportional both to the length of the time interval and to the number of nuclei present at the beginning of this time interval, so that

$$\Delta N \propto N\, \Delta t$$

The proportionality constant can be shown to be equal to the decay constant λ, and so we have $\Delta N = \lambda N\, \Delta t$, or

$$\frac{\Delta N}{\Delta t} = -\lambda N \tag{29.5}$$

where the minus sign indicates that the number N *decreases* in time. We thus have the following definition of *nuclear activity*:

Nuclear activity: The rate at which a particular nuclear species decays in time: $\Delta N/\Delta t = -\lambda N$, where λ is the decay constant and N is the number of radioactive nuclei present at the time at which the activity is calculated.

Since λ is a constant for any nuclear species and N decreases exponentially with time according to Eq. (29.3), we see that *the nuclear activity also decreases exponentially with time*.

Radiocarbon Dating

One especially useful isotope is $^{14}_{6}C$, which has a half-life of 5730 years and is used to date fossils, old manuscripts, works of art, and other objects made from once-living materials. The isotope $^{14}_{6}C$ is continuously produced in the air by cosmic rays and is breathed in by living creatures. Since $^{14}_{6}C$ makes up a tiny but constant fraction of all the carbon in the air (about 1.5×10^{-12}), all living creatures have the same fraction of $^{14}_{6}C$ in their bodies. Once death occurs, however, no more $^{14}_{6}C$ is added to the body, and the $^{14}_{6}C$ already present decays with a half-life of 5730 years according to the equation $^{14}_{6}C \rightarrow {}^{14}_{7}N + {}^{0}_{-1}e$. If the fraction of the radioactive carbon in a fossil is found to be half of that in a living creature today, the fossil is therefore presumed to be 5730 years old. This was the technique used to date the Dead Sea scrolls, which were found wrapped in linen. From an analysis of the $^{14}_{6}C$ content of the linen (made from fibers of flax—a living material), it was found that the linen was 2000 years old. It could therefore be concluded that the scrolls were approximately 2000 years old.

In the last 10 years high-energy particle accelerators (Sec. 31.2) have been used to separate the $^{14}_{6}C$ nuclei from the more abundant $^{12}_{6}C$ nuclei by bombardment with positive ions and deflection in a magnetic field. This method allows the use of samples 1000 to 10,000 times smaller than was previously possible with the C-14 dating technique.

Example 29.4

A sample of Kr gas contains 2.00×10^{20} atoms of $^{88}_{36}Kr$ at a time $t = 0$. **(a)** What is the decay constant for $^{88}_{36}Kr$? **(b)** At a time $t = 11.2$ h later, how many $^{88}_{36}Kr$ atoms remain? **(c)** What is the initial nuclear activity of the sample?

SOLUTION

It should be clear that not enough data are given to allow the problem to be solved. In any problem dealing with radioactive decay we need either the decay constant, the half-life, or some other data that will enable us to obtain one or other of these two quantities. Since neither is given, the presumption is that the necessary data must be obtained from an available table. In this case, Table 29.2 gives us the half-life of $^{88}_{36}Kr$ as 2.8 h, and so 11.2 h is just four half-lives.

(a) The decay constant is

$$\lambda = \frac{0.693}{T_{1/2}} = \frac{0.693}{(2.8 \text{ h})(3600 \text{ s/h})} = \boxed{6.88 \times 10^{-5} \text{ s}^{-1}}$$

From Eq. (29.5), $(1/N)(\Delta N/\Delta t) = -\lambda$, and so this means that the fraction of the $^{88}_{36}\text{Kr}$ atoms which decays each second is 6.88×10^{-5}.

(b) In four half-lives (11.2 h) the number of $^{88}_{36}\text{Kr}$ atoms decreases by a factor of $2^4 = 16$. Hence the number of atoms present after 11.2 h is

$$N = \frac{N_0}{16} = \frac{2.00 \times 10^{20} \text{ atoms}}{16} = \boxed{1.25 \times 10^{19} \text{ atoms}}$$

(c) The initial nuclear activity is, from Eq. (29.5), with $N = N_0$ when $t = 0$,

$$\left(\frac{\Delta N}{\Delta t}\right)_0 = -\lambda N_0 = -(6.88 \times 10^{-5} \text{ s}^{-1})(2.00 \times 10^{20} \text{ atoms})$$

$$= \boxed{-1.38 \times 10^{16} \text{ atoms per second}}$$

EXERCISE 1 What is the nuclear activity when $t = 11.2$ h?

Example 29.5

Cobalt-60 ($^{60}_{27}\text{Co}$) is frequently used as a radiation source by doctors. How long after a fresh sample of cobalt-60 is delivered to a hospital will the activity have decreased by a factor of 3 from its original value?

SOLUTION

The activity has been shown [Eqs. (29.5) and (29.3)] to be equal to

$$\frac{\Delta N}{\Delta t} = -\lambda N = -\lambda N_0 e^{-\lambda t}$$

The initial activity when $t = 0$ is, since $e^{-\lambda t} = 1$ and $N = N_0$ for $t = 0$,

$$\left(\frac{\Delta N}{\Delta t}\right)_0 = -\lambda N_0$$

When the activity has been reduced to one-third this original value, we must have

$$\frac{\Delta N}{\Delta t} = \frac{1}{3}\left(\frac{\Delta N}{\Delta t}\right)_0 = -\tfrac{1}{3}\lambda N_0 = -\lambda N_0 e^{-\lambda t}$$

and so from Eq. (29.4),

$$\tfrac{1}{3} = e^{-\lambda t} = e^{-(0.693/T_{1/2})t}$$

In this case we find, from Table 29.2, that $^{60}_{27}\text{Co}$ has a half-life $T_{1/2}$ of 5.24 years, and so

$$\tfrac{1}{3} = e^{-(0.693)/(5.24)t} = e^{-0.132t}$$

with t in years. Taking natural logarithms of both sides, we obtain

$$\ln \tfrac{1}{3} = -0.132t$$

$$-1.10 = -0.132t$$

$$t = \frac{1.10}{0.132} = \boxed{8.3 \text{ years}}$$

29.4 Rules Governing Nuclear Reactions

In the radioactive decay of nuclei all the basic conservation principles of physics must be satisfied. These include the conservation of electric charge, of mass, of energy, and of linear and angular momentum.

To guarantee the conservation of electric charge and of mass (neglecting for the present the conversion of mass into energy) in any nuclear reaction, the following rules must be satisfied in the equation describing the reaction:

1 The sum of the charge numbers on the left side of the equation must be equal to the sum of the charge numbers on the right side of the equation. This guarantees the conservation of charge.

2 The sum of the mass numbers on the left side of the equation must be equal to the sum of the mass numbers on the right side of the equation. This guarantees the conservation of the number of nucleons taking part in the reaction, or the conservation of *nucleon number*.

For example, the nuclear equation describing the alpha decay of a radium nucleus is

$$^{226}_{88}\text{Ra} \rightarrow ^{222}_{86}\text{Rn} + ^{4}_{2}\text{He}$$

where the radium is called the *parent nucleus* and the radon the *daughter nucleus*. It can be seen that both the mass numbers and the charge numbers on the two sides of this equation balance out, indicating that the reaction can occur without violating the conservation principles of physics.

The nuclear mass changes in such decays are very small compared to the nuclear masses involved in the reactions, and the energy arising from mass-energy conversion is carried off either as kinetic energy of the interacting particles or in the form of gamma rays or other massless particles. In all cases, however, spontaneous decay of a nucleus can only occur if the mass of the parent nucleus is greater than the sum of the masses of the daughter nucleus and the other particles produced, since some mass will always be converted into energy. If the parent has less mass than the products, then the decay cannot occur spontaneously without violating the mass-energy conservation principle.

Example 29.6

A polonium nucleus decays into lead in the reaction

$$^{210}_{84}\text{Po} \rightarrow ^{206}_{82}\text{Pb} + X$$

where X is an unknown particle. **(a)** What kind of particle is X? **(b)** The exact masses for the neutral atoms are given in Appendix 4. What is the kinetic energy of the products of this reaction?

SOLUTION

(a) From conservation of mass and conservation of charge, the particle must be $^{4}_{2}X$, which can be identified from Appendix 4 as $^{4}_{2}\text{He}$. The particle is therefore a helium nucleus, or *alpha particle*.

(b) We can consider the above equation as applying either to bare nuclei or to neutral atoms. Since there are the same number (84) of extranuclear electrons in the atoms on both sides of the above equation, the electron masses cancel out (as is usually the case), and we can use the masses of the neutral atoms instead of the masses of the bare nuclei in our calculation of the mass-energy conversion.

From Appendix 4 the total mass of the product nuclei is 209.97707 u, and the mass defect is Δm = 209.98288 u − 209.97707 u = 0.00581 u. This mass appears as kinetic energy of the daughter nucleus and the alpha particle, with most of it going to the alpha particle because of its smaller mass. The kinetic energy produced is

$$KE = (0.00581 \text{ u})\left(\frac{931.49 \text{ MeV}}{1 \text{ u}}\right) = \boxed{5.41 \text{ MeV}}$$

This then is the energy produced by the conversion of mass into energy in this particular nuclear reaction.

To balance both mass and energy in a nuclear equation like the one in this example, we should really write it as

$$^{210}_{84}\text{Po} \rightarrow ^{206}_{82}\text{Pb} + ^{4}_{2}\text{He} + 5.41 \text{ MeV}$$

This indicates that a small amount of mass has been converted into 5.41 MeV of energy in the reaction and that mass-energy is therefore conserved, in addition to charge and nucleon number.

29.5   Nuclear Reactions II: Artificial Radioactivity

It is possible to take extremely stable nuclei, such as $^{14}_{7}\text{N}$, the nuclei of the atoms forming nitrogen molecules in the air, and make them unstable by bombarding them with highly energetic particles like protons, deuterons, or alpha particles. These particles can force their way inside a stable nucleus and create a new, unstable nucleus, which then decays into yet another nucleus and other smaller particles. This process we call *artificial* (or *induced*) *radioactivity*.

Suppose we fire high-energy alpha particles at a nitrogen nucleus. One reaction that occurs is found by experiment to be

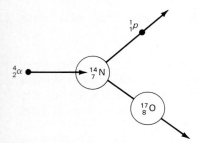

FIGURE 29.8 Alpha-particle bombardment of $^{14}_{7}\text{N}$, leading to the production of $^{17}_{8}\text{O}$.

$$^{4}_{2}\alpha + ^{14}_{7}\text{N} \rightarrow \{^{18}_{9}\text{F}\} \rightarrow ^{17}_{8}\text{O} + ^{1}_{1}p \tag{29.6}$$

In this reaction the alpha particle combines with the nitrogen nucleus to produce the highly unstable fluorine nucleus $^{18}_{9}\text{F}$, which then spontaneously decays into a rare isotope of oxygen and a proton, as shown in Fig. 29.8. The braces around the $^{18}_{9}\text{F}$ indicate that this "compound" nucleus cannot be observed because its lifetime is so short but, to make sense of the experimental data, must be presumed to have existed for a very small fraction of a second. Equation (29.6) is sometimes written in a shorthand notation as $^{14}_{7}\text{N}$ (α, p) $^{17}_{8}\text{O}$.

Another possibility is to use deuterons (nuclei of heavy hydrogen, $^{2}_{1}\text{H}$) as the bombarding particles. A possible reaction in this case is

$$^{2}_{1}d + ^{238}_{92}\text{U} \rightarrow \{^{240}_{93}\text{Np}\} \rightarrow ^{238}_{93}\text{Np} + 2^{1}_{0}n \tag{29.7}$$

Here we have converted uranium, the heaviest atom to occur naturally on earth, into the "transuranic" (i.e., beyond uranium in the periodic table) element neptunium by deuteron bombardment, as shown in Fig. 29.9. Again this can be written as $^{238}_{92}\text{U}$ (d, $2n$) $^{238}_{93}\text{Np}$.

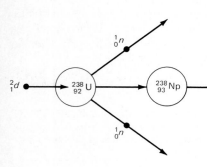

FIGURE 29.9 Deuteron bombardment of $^{238}_{92}\text{U}$, leading to the production of the transuranic element neptunium, $^{238}_{93}\text{Np}$.

Many other possibilities exist, but they all require bombardment with high-energy particles that have sufficient speeds to penetrate the Coulomb barrier of the bombarded nucleus and produce a new, unstable nucleus. To accomplish this goal, particle accelerators like cyclotrons and synchrotrons are needed.

Neutron Bombardment of Nuclei

There is another kind of bombarding particle of great importance both in the history of nuclear physics and in the practical use of nuclear energy today. This is the neutron, which has the great advantage that it is uncharged and so can get closer to the bombarded nucleus than can positively charged particles which are repelled electrostatically by the positively charged nucleus. On the other hand, the neutron has the compensating disadvantage that it cannot be accelerated by electric or magnetic fields, since it has no charge. Physicists desiring to use neutrons as nuclear "bullets," therefore, often use reactions like the one in Eq. (29.7) above, which has two energetic neutrons as end products.

In the early 1930s the Italian physicist Enrico Fermi (see Fig. 1.6 and accompanying biography) realized the importance of a careful study of the neutron bombardment of nuclei. He bombarded a series of elements from nitrogen to uranium with neutrons and studied the radioactive transformations produced. When he came to uranium, he expected to find a whole series of transuranic elements produced in reactions such as the following:

$$^{238}_{92}\text{U} + ^{1}_{0}n \rightarrow ^{239}_{92}\text{U} \tag{29.8}$$

The uranium nucleus would then be expected to decay spontaneously by successive beta emissions into isotopes of what we now call neptunium and plutonium by the following two processes*:

---

*Again note that we are for the present omitting the neutrinos produced in such beta-decay processes.

$$^{239}_{92}U \rightarrow \,^{239}_{93}Np + \,_{-1}^{\,0}e \qquad\qquad\qquad\qquad\qquad (29.9)$$

$$^{239}_{93}Np \rightarrow \,^{239}_{94}Pu + \,_{-1}^{\,0}e \qquad\qquad\qquad\qquad\qquad (29.10)$$

When Fermi carried out this experiment, however, he found neither neptunium nor plutonium but other nuclei that he could not identify. About this time he was forced to leave Italy with his wife because of the political situation under Mussolini and Hitler at the end of the 1930s.

A number of European physicists, among them Irène Joliot-Curie, daughter of Marie Curie (see Fig. 29.5), continued Fermi's work. The German chemists Otto Hahn and Fritz Strassman finally identified one of the nuclei produced when uranium was bombarded by neutrons as barium ($^{141}_{56}Ba$), an element very much lighter than uranium. The presence of barium in the reaction products was so unexpected, and so impossible to explain by the then-accepted theories of nuclear transformations by alpha, beta, and gamma emissions, that Hahn and Strassman were hesitant about publishing their findings, but did so anyway in 1939. Shortly thereafter the Austrian physicist Lise Meitner (see Fig. 29.10 and accompanying biography) and her nephew, Otto Frisch, interpreted Hahn and Strassman's experimental results as proof of a new kind of nuclear reaction they named *nuclear fission*.

Example 29.7

In the nuclear reaction described by Eq. (29.6), what is the *minimum* energy an alpha particle must have to make this reaction occur? Assume that the atomic masses involved are those given in Appendix 4.

SOLUTION

We write the equation for this reaction in the form $^4_2He +\,^{14}_7N \rightarrow \,^1_1H + \,^{17}_8O$, for then we can use a table of atomic masses, since the masses of the electrons on the two sides cancel out. The mass of the reacting atoms is $4.00260\ u + 14.00307\ u = 18.00567\ u$. The total mass of the products is the mass of the hydrogen atom, $1.00783\ u$, plus that of the oxygen atom, which is $16.99916\ u$, or $18.00699\ u$ in all.

The products therefore have a *greater* mass than the original atoms, and the reaction cannot take place unless the alpha particle brings at least enough energy to the collision to make the masses before and after the reaction the same. Otherwise mass-energy cannot be conserved in the collision.

The difference in the masses after and before is

$$18.00699\ u - 18.00567\ u = 0.00132\ u$$

The minimum kinetic energy needed by the alpha particle is, therefore,

$$KE = (0.00132\ u)\left(\frac{931.49\ MeV}{1\ u}\right) = 1.23\ MeV$$

Actually a kinetic energy somewhat greater than this will be needed, since, to conserve momentum, the hydrogen and oxygen atoms must carry away some kinetic energy from the reaction.

29.6 Nuclear Fission and Nuclear Fusion

Meitner and Frisch explained the presence of $^{141}_{56}Ba$ in the decay products of uranium bombarded by neutrons by assuming that the neutrons were absorbed by uranium nuclei. Niels Bohr and the American physicist John Wheeler (born 1911) later showed that the absorbing nuclei must not be the $^{238}_{92}U$ nuclei, which make up 99.3 percent of natural uranium, but rather $^{235}_{92}U$ nuclei, which make up only 0.72 percent of natural uranium. The result was the unstable isotope $^{236}_{92}U$, which then broke up, or "fissioned," into two smaller nuclei of about equal size. One possible reaction was the following:

$$^{235}_{92}U + \,^1_0n \rightarrow \{^{236}_{92}U\} \rightarrow \,^{141}_{56}Ba + \,^{92}_{36}Kr + 3\,^1_0n \qquad\qquad (29.11)$$

This would explain Hahn and Strassman's finding of barium among the product nuclei after the neutron bombardment of uranium. The fission products are highly

FIGURE 29.10 *(Courtesy of AIP Niels Bohr Library, Karl F. Herzfeld collection.)*

Lise Meitner (1878–1968)

Lise Meitner grew up in the Vienna of Emperor Franz Joseph and horse-drawn trolley cars. She was born there in 1878 into a well-to-do Jewish family and decided at an early age that she wanted to be a scientist like Madame Curie. In 1901 she entered the University of Vienna. There, in an environment where serious women students were considered odd, she was treated rudely by many of her fellow students. When she received her Ph.D. in physics in 1905, she was only the second woman in history to receive a Ph.D. from the University of Vienna.

In 1907 she went to Berlin to study under Max Planck, promising her devoted parents that she would be back in Vienna in 6 months at the most. She stayed 31 years! In Berlin Meitner met Otto Hahn, a professor of chemistry, and took a research position (without salary), helping Hahn with research on the chemistry of radioactive substances. At that time women were not allowed to work in the Chemical Institute, and she had to set up her laboratory in a carpenter's workshop outside the institute.

While continuing work with Hahn at the Kaiser Wilhelm Institute for Chemistry in Berlin-Dahlem, she served as assistant to Max Planck at the Institute for Theoretical Physics at the University of Berlin from 1912 to 1915, and in 1918 was appointed head of the physics department at the Kaiser Wilhelm Institute. During World War I she enlisted as a nurse in the Austrian army and served loyally throughout the war, as did her role model, Madame Curie, for the opposing side.

In 1922 Meitner delivered her inaugural lecture as a faculty member at the University of Berlin. The title of her talk was actually "Cosmic Physics," but it appeared in the newspapers as "Cosmetic Physics," an indication of the press's narrow-minded view of the scientific interests of women.

In the following years Hahn and Meitner did some very significant research on beta- and gamma-ray spectra. They discovered the new element protoactinium-91 and took up the work on the neutron bombardment of nuclei that Enrico Fermi had commenced in Rome. This work was suspended in 1938, however, since Meitner had to flee Germany when Hitler annexed Austria. She found asylum in Stockholm, Sweden, after spending a few months at Bohr's institute in Copenhagen.

At the end of 1938 Otto Hahn sent her a description of his experiments on the interaction of neutrons with uranium. He and Strassman had determined that one of the reaction products was clearly barium. Meitner was so excited about this that she showed Hahn's letter to her nephew, the physicist Otto Frisch, and suggested that they go for a long walk and sort out some of these ideas. During this walk all the pieces quickly fell into place and the idea of nuclear fission was born. Frisch then demonstrated in his laboratory the tremendous release of energy accompanying fission. A short paper by Meitner and Frisch in the British journal *Nature* in 1939 revealed the momentous idea of fission to the scientific world.

Although it was her basic insight which eventually led to the fission bomb dropped on Hiroshima, Meitner refused to work on the bomb and hoped that it would not work.

After the war, although now famous, she continued her laboratory research in Stockholm, interrupted only by trips to receive honorary degrees and other scientific accolades. In 1946 she spent a semester as a visiting professor at Catholic University in Washington, D.C. In 1960 she retired to Cambridge, England, to be near her nephew, Otto Frisch, but continued to travel, lecture, and attend concerts, for music was one of the great joys of her life. She died in 1968 at the age of 90.

A fitting epitaph for Lise Meitner might well be her own words, spoken in a lecture at Bryn Mawr College in 1959: "Life need not be easy, provided only that it is not empty."

radioactive and quickly decay into isotopes of other elements. For example, the half-life of $^{141}_{56}$Ba is 18 min, and of $^{92}_{36}$Kr only 2.4 s.

Physicists all over the world took up these ideas and soon showed in their laboratories that not only does the above reaction occur, but many reactions similar to the following occur:

$$^{235}_{92}\text{U} + {}^{1}_{0}n \rightarrow \{{}^{236}_{92}\text{U}\} \rightarrow {}^{137}_{53}\text{I} + {}^{97}_{39}\text{Y} + 2{}^{1}_{0}n \tag{29.12}$$

$$^{235}_{92}\text{U} + {}^{1}_{0}n \rightarrow \{{}^{236}_{92}\text{U}\} \rightarrow {}^{143}_{57}\text{La} + {}^{90}_{35}\text{Br} + 3{}^{1}_{0}n \tag{29.13}$$

All these reactions conserve charge and nucleon number, and all result in two isotopes each with roughly half the mass of the original uranium nucleus. All three of these reactions are examples of what is now called *nuclear fission*.

Nuclear fission: The breaking up of a heavy nucleus like uranium into two nuclei of intermediate masses, together with the release of two or more neutrons and large amounts of energy.

This definition points up two other important facts about fission. First, large amounts of energy (about 200 MeV) are released in each fission reaction, mostly in the form of the kinetic energies of the two heavy fission products and the neutrons. Second, for each neutron absorbed, at least two neutrons are produced in the fission process. These two facts immediately led Meitner, Frisch, and others to the startling conclusion that the fission of uranium might provide a tremendous new source of energy.

Chain Reactions

The key to releasing the tremendous store of energy in a $^{235}_{92}$U nucleus is a *chain reaction* of the kind shown in Fig. 29.11 for the fission of $^{235}_{92}$U into barium and krypton. A single neutron causes the fission of a $^{235}_{92}$U nucleus into $^{141}_{56}$Ba, $^{92}_{36}$Kr, and three neutrons. Even if one of these escapes from the reaction region, two are left to produce additional fission reactions. Each of these new fission processes produces enough neutrons to cause two additional fission processes, and so on. The energy thus released increases *exponentially*, since the number of fission reactions doubles at each stage of the chain. The result of this chain reaction is the

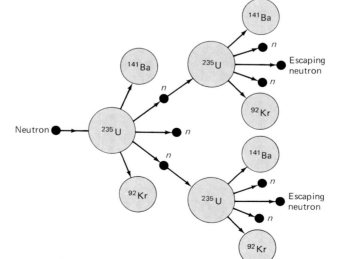

FIGURE 29.11 One possible nuclear chain reaction involving $^{235}_{92}$U.

release of a huge amount of energy in a very short time—the equivalent of a dynamite explosion of extraordinary destructive power.

The secret of a successful energy-releasing fission process is that for every neutron producing fission, at least one of the neutrons produced in that fission must be absorbed and produce fission at the next stage of the reaction.

Multiplication factor: The ratio of the number of neutrons producing fission at any stage of the reaction process to the number of neutrons producing fission in the immediately preceding stage.

When the multiplication factor becomes equal to unity, i.e., each fission produces one additional fission, the chain reaction is said to have gone *critical*. This is the case in power reactors in which we want a controlled release of heat. For nuclear bombs, on the other hand, the multiplication factor must be as large as possible to achieve the most complete energy release in the shortest possible time. For the chain reaction to achieve criticality, there is a minimum mass of fissionable material, called the *critical mass*, required. This minimum or critical mass varies with the way the fissionable material is distributed in space.

That such an abstract idea could lead to a practical source of energy was shown conclusively when the first nuclear reactor (which in this case was just a *pile* of uranium and graphite blocks) went critical at Stagg Field at the University of Chicago in 1942, with Enrico Fermi directing the operation. The fearful destructive power of nuclear fission was revealed to the world when the first uranium bomb was dropped on the city of Hiroshima in Japan in 1945.

Energy Considerations in Fission Processes

The secret of the energy released in a nuclear fission process lies in the binding-energy curve for nuclei. Suppose a $^{235}_{92}U$ nucleus fissions into barium and krypton nuclei, as described by Eq. (29.11). The barium and krypton nuclei have larger binding energies per nucleon and are therefore more stable than the uranium nucleus, as is clear from the binding-energy curve in Fig. 29.3. Hence the total energy of the barium and krypton nuclei and the neutrons produced in the reaction is less than the energy of the $^{236}_{92}U$ nucleus undergoing fission, and also less than the sum of the energies of the $^{235}_{92}U$ and the bombarding neutron that produced the $^{236}_{92}U$. The excess energy must, therefore, be released in the form of the kinetic energy of the particles produced in the reaction. This energy adds up to about 215 MeV per fission, as shown in Table 29.3, for the energy release averaged over all the different fission reactions that can occur in $^{235}_{92}U$. This corresponds to a mass of about 0.23 u, so that only about 0.1 percent of the mass of the $^{235}_{92}U$ nucleus is converted into energy.

TABLE 29.3   Energy Released in the Fission of $^{235}_{92}U$ by a Neutron

Form of energy released	Amount of energy released (MeV)
Kinetic energy of two fission fragments	168
Immediate gamma rays	7
Delayed gamma rays	3–12
Fission neutrons	5
Energy of decay products of fission fragments:	
Gamma rays	7
Beta rays	8
Neutrons	12
Total energy released	210–219 (Average = 215 MeV)

Although 215 MeV is only 3.4×10^{-11} J, this energy comes from a single nucleus, and so kilograms of $^{235}_{92}$U undergoing fission can easily produce enough energy for a nuclear reactor or a nuclear bomb.

Another way to look at the same process is to note that the mass of the fission products (including the three neutrons) is less than the mass of the $^{236}_{92}$U nucleus produced by the neutron bombardment. The mass defect (equivalent to the difference in binding energies) must therefore be converted into energy if mass-energy is to be conserved in the reaction.

Nuclear Fusion Reactions

Another source of nuclear energy comes from *fusion*, which is the reverse process of fission.

Nuclear fusion: A nuclear reaction in which two light nuclei are combined, or *fused*, to form a heavier nucleus, accompanied by the release of large amounts of energy.

From the binding-energy curve in Fig. 29.3 it is clear that energy can be released in such a fusion reaction, since the very light elements are less tightly bound than elements closer to the middle of the periodic table. Thus, when two light nuclei like a deuterium nucleus ($^2_1$H) and a tritium nucleus ($^3_1$H) fuse to produce a helium nucleus ($^4_2$He) and a neutron, it is found experimentally that 17.6 MeV of energy is released. This energy release is to be expected, since the helium nucleus is so much more stable than the deuterium and tritium nuclei, as is clear from the binding-energy curve.

Another way to look at this is to note that in the reaction

$$^2_1\text{H} + {}^3_1\text{H} \rightarrow {}^4_2\text{He} + {}^1_0 n \qquad (29.14)$$

the combined mass of the $^2_1$H and $^3_1$H is greater than the mass of the $^4_2$He plus the neutron, and to conserve mass-energy some mass must be converted into energy in the reaction. This is the source of the observed 17.6 MeV of energy released.

The problem with fusion processes is that they only occur at very high temperatures (10^7 to 10^8 K) such as exist in the sun or other stars, since the reacting nuclei must get very close together if fusion is to occur. Controlled fusion is, therefore, much more difficult to achieve than is controlled fission.

Example 29.8

From the atomic masses given in Appendix 4, find the energy released in the fusion reaction described by Eq. (29.14).

SOLUTION

In discussing the fusion reaction $^2_1\text{H} + {}^3_1\text{H} \rightarrow {}^4_2\text{He} + {}^1_0 n$, we use atomic masses, since the masses of the electrons cancel out. We have

$m(^2_1\text{H}) = 2.01410$ u $m(^4_2\text{He}) = 4.00260$ u

$m(^3_1\text{H}) = 3.01605$ u $m(^1_0 n) = 1.00866$ u

The mass defect is then

$m = (2.01410 + 3.01605)$ u $- (4.00260 + 1.00866)$ u

$\quad = 0.01889$ u

This is equivalent to an energy of

$$E = (0.01889 \text{ u})\left(\frac{931.49 \text{ MeV}}{1 \text{ u}}\right) = \boxed{17.60 \text{ MeV}}$$

This agrees well with the experimental value mentioned above. Most of this energy goes into the kinetic energy of the alpha particle and the neutron.

29.7 The Nucleus as a Practical Energy Source

Both fission and fusion have been used to make weapons of unparalleled destructive power. The use of these two processes as practical energy sources, however, has been marked by technical, economic, and political problems that are still far from being solved.

Fission Reactors

For a nuclear power reactor intended to produce electricity, a multiplication factor exactly equal to 1 is required, for then exactly one neutron from each fission event initiates another event, and energy is released at a steady rate. This energy is then converted into heat by a series of collisions of the fission products with the surrounding materials in the reactor, and this heat is used to produce steam and generate electricity using a conventional steam-electric turbine like that in Fig. 15.14. The fission process is therefore used basically to boil water, so the efficiency of a nuclear reactor is severely limited by the second law of thermodynamics.

The desired multiplication factor of unity is achieved by monitoring the kind, amount, and location of fissionable material in the reactor, by using a "moderator" to reduce the neutron velocities, and by manipulating control rods from outside the reactor.

Nuclear Fuel   There are only three isotopes that readily undergo fission when bombarded by neutrons. These are $^{233}_{92}U$, $^{235}_{92}U$, and $^{239}_{94}Pu$. These are called fissionable, or *fissile*, nuclei. Most operating reactors today use uranium in which the amount of $^{235}_{92}U$ has been increased to 3 percent of the total uranium present. The fuel, in the form of small uranium dioxide cylinders, is loaded into long, thin fuel rods in the core of the reactor. The core of a reactor may contain 20,000 or more fuel rods, in each of which nuclear fission processes are occurring and heat is being generated.

Moderator   The efficiency of neutrons in producing fission of $^{235}_{92}U$ nuclei depends on their speed. The more slowly the neutrons move, the longer the time they spend near a $^{235}_{92}U$ nucleus and the greater the chance they will be captured and produce fission. Very slow, or "thermal," neutrons are needed. In most fission reactors in the United States today the water coolant flowing in contact with the fuel rods also acts as the moderator to slow down the neutrons to thermal speeds and increase the probability of fission occurring. In Fermi's original nuclear "pile," graphite was used as the moderator*; graphite is also used in most British reactors.

Control Rods   A great concern about any nuclear fission reactor is that it will "run away," i.e., get out of control. When this happens, the multiplication factor exceeds unity, an exponential buildup of heat results, and the cooling system may not be able to handle the excess heat generated. There is no possibility of a nuclear

---

*Enrico Fermi and his coworkers in Rome first discovered the effectiveness of paraffin in slowing down neutrons on the morning of October 22, 1934. Fermi worked out the theory during his lunch hour, and returned to his laboratory in the afternoon with the confident prediction that water would work equally well, since, like paraffin, it contained a large number of hydrogen atoms. He and his colleagues immediately checked this out by immersing a neutron source and a silver target in the goldfish fountain in the private garden behind their laboratory. They found that the induced radioactivity of the silver was indeed increased. This was the same goldfish fountain in which the members of Fermi's group liked to sail tiny toy boats as a relaxation from their intense work in the laboratory.

explosion, since the concentration of $^{235}_{92}U$ present (3 percent) is insufficient for such an explosion. The fuel rods could melt from the heat, however, and burn through the reactor floor, leading to what is sometimes called the "China syndrome," i.e., the movement of hot fission products into the earth in the general direction of China. Functioning reactors are required to have backup cooling systems to handle such eventualities, but some of these have not worked too well when they were most needed.

To prevent runaways, control rods are used. These are rods of cadmium or boron, which absorb neutrons strongly and can be dropped into the reactor core to reduce the multiplication factor. If at any time during operation the reaction threatens to run out of control, the control rods fall into place automatically and shut down the reactor.

Pressurized-Water Reactor (PWR)

A PWR is illustrated in Fig. 29.12. In these reactors the water that comes in contact with the fuel rods is kept under very high pressure and never boils, even though it reaches a very high temperature. This high-temperature water is then used to convert unpressurized water in a conventional boiler to steam that is used to produce electricity. Such a reactor operates at about 315°C and is therefore more efficient than the older boiling-water reactors. In addition the water that comes in contact with the fuel rods does not flow through the turbine or condenser. This greatly reduces the possibility of leakage of radioactive materials from the reactor.

Most of the reactors in use in the United States today are PWRs, as are the very few additional reactors being constructed or designed at present. PWRs have also been used in nuclear submarines and nuclear aircraft carriers, where they have a truly remarkable record for safety, efficiency, and performance.

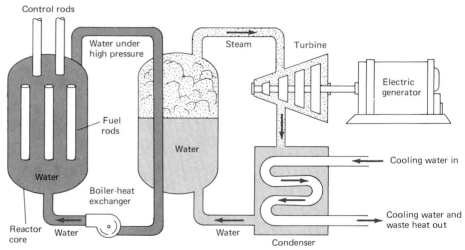

FIGURE 29.12 A pressurized-water reactor (PWR). The water in the loop on the left does not boil because it is under high pressure. This high-temperature water then converts water in the conventional boiler into steam that is used to produce electricity.

The Liquid-Metal Fast-Breeder Reactor (LMFBR)

One of the problems with conventional fission reactors is that it is unclear how long the world's supply of $^{235}_{92}U$ will last. (This lifetime would appear to be a few hundred years, but there are many uncertainties in reaching this number.) If we could find some way to use $^{238}_{92}U$ as fuel in a nuclear reactor, we could increase the lifetime of existing uranium fuel by a factor of about 100, since $^{238}_{92}U$ makes up 99.3 percent of all natural uranium. One way to use $^{238}_{92}U$ as a nuclear fuel is to convert it into a fissionable nucleus like $^{239}_{94}\text{Pu}$. This can be done by neutron bombardment of the uranium in a reactor, which produces $^{239}_{94}\text{Pu}$ by the following series of reactions:

$$^{238}_{92}\text{U} + {}^{1}_{0}n \rightarrow {}^{239}_{92}\text{U} \qquad {}^{239}_{92}\text{U} \rightarrow {}_{-1}^{0}e + {}^{239}_{93}\text{Np} \qquad {}^{239}_{93}\text{Np} \rightarrow {}_{-1}^{0}e + {}^{239}_{94}\text{Pu} \qquad \textbf{(29.15)}$$

In this way we can produce a fissionable nucleus, plutonium, from a nonfissile isotope of uranium. Such a process is called "breeding." The plutonium can then be used as a reactor fuel, since it can be fissioned by neutrons according to the following equation:

$$^{1}_{0}n + {}^{239}_{94}\text{Pu} \rightarrow \{{}^{240}_{94}\text{Pu}\} \rightarrow x + y + 2\,{}^{1}_{0}n + 200 \text{ MeV} \qquad \textbf{(29.16)}$$

Here x and y are fission products consisting of a variety of radioactive isotopes near the middle of the periodic table.

In breeder reactors the core is made up of $^{235}_{92}\text{U}$ or $^{239}_{94}\text{Pu}$, surrounded by a "blanket" of $^{238}_{92}\text{U}$, as shown in Fig. 29.13. Neutrons given off by the core fission processes both sustain these processes and convert some of the $^{238}_{92}\text{U}$ in the blanket into $^{239}_{94}\text{Pu}$. In principle such a breeder reactor can produce more fuel than it uses and thereby greatly extend our nuclear fuel supply.

Since fast neutrons are needed to produce the plutonium in a breeder reactor, no moderator is used, and a liquid metal like sodium is used as the coolant in place of water. The hot sodium then heats water to produce steam in a heat-exchange unit. The letters LMFBR refer to a breeder reactor (BR) using a liquid metal (LM) like sodium as the coolant and heat-transfer fluid, and fast neutrons (F) to convert $^{238}_{92}\text{U}$ to $^{239}_{94}\text{Pu}$.

Although the breeder-reactor program is "on hold" in the United States, the French government has invested large sums of money in breeder reactors as part of their extensive nuclear power program.

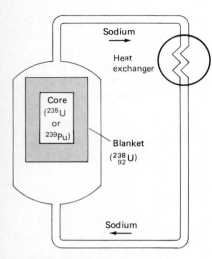

FIGURE 29.13 A breeder reactor. The $^{238}_{92}\text{U}$ in the blanket surrounding the core is gradually converted into fissionable $^{239}_{94}\text{Pu}$ by bombardment with neutrons produced in the reactor core.

Fusion Reactors

Whereas conventional nuclear fission reactors and the LMFBR are known to work, fusion reactors are still in process of development. It is known that fusion can lead to the release of large amounts of energy, as it has in the hydrogen bomb, or H-bomb.

The working idea behind a fusion reactor is to cause light nuclei to fuse in processes such as the following:

$$^{2}_{1}\text{H} + {}^{2}_{1}\text{H} \rightarrow {}^{3}_{2}\text{He} + {}^{1}_{0}n + 3.3 \text{ MeV} \qquad \textbf{(29.17)}$$

$$^{2}_{1}\text{H} + {}^{2}_{1}\text{H} \rightarrow {}^{3}_{1}\text{H} + {}^{1}_{1}\text{H} + 4.0 \text{ MeV} \qquad \textbf{(29.18)}$$

$$^{2}_{1}\text{H} + {}^{3}_{1}\text{H} \rightarrow {}^{4}_{2}\text{He} + {}^{1}_{0}n + 17.6 \text{ MeV} \qquad \textbf{(29.19)}$$

The first two of these reactions are called D-D reactions, and the third a D-T reaction, since D (deuterium) is the symbol for $^{2}_{1}\text{H}$, and T (tritium) the symbol for $^{3}_{1}\text{H}$. To obtain useful amounts of energy from these processes, however, large numbers of nuclei must be brought very close together at very high temperatures and held there long enough for fusion to occur. The temperatures required are about 4×10^7 °C for the process described by Eq. (29.19) and about 2×10^8 °C for those in Eqs. (29.17) and (29.18). (Note that these are very high temperatures indeed: the highest melting point of any element, that of tungsten, is only 3.4×10^3 °C.)

Much research is at present underway on techniques to achieve both the high temperatures and high densities needed for controlled fusion. These techniques include confining gas particles to small regions of space using strong magnetic fields and the compression of solid pellets containing fusionable isotopes by lasers or electron and ion beams (see Fig. 28.18a). Other more exotic techniques, like the use of mu-mesons to catalyze fusion reactions, are still in the discussion stage.

Even though considerable progress has been made on various aspects of the controlled fusion problem, it is still uncertain whether practical amounts of energy can be obtained at prices that will make fusion economically viable. Even if all the technical and engineering problems related to fusion should soon be solved, it is doubtful that we will have functioning fusion reactors before early in the twenty-first century.

Example 29.9

If a nuclear-electric power plant generates 1 GW of electric power, about how many kilograms of pure $^{235}_{92}U$ would be needed to produce electricity at this rate for a year, assuming that the power plant produces the electricity with an efficiency of about 33 percent?

SOLUTION

We have seen that the fission of one ^{235}U nucleus into barium and krypton releases about 215 MeV of energy. Hence 1 mol of pure $^{235}_{92}U$ will produce an amount of energy equal to

$$\mathscr{E} = (6.02 \times 10^{23})(215 \text{ MeV})\left(\frac{1.60 \times 10^{-13} \text{ J}}{\text{MeV}}\right)$$

$$= 2.07 \times 10^{13} \text{ J}$$

Now, the energy produced in 1 year by a 1-GW reactor operating at full capacity all year is

$$E' = \left(10^9 \frac{\text{J}}{\text{s}}\right)\left(\frac{3.16 \times 10^7 \text{ s}}{1 \text{ year}}\right)(1 \text{ year}) = 3.16 \times 10^{16} \text{ J}$$

Since the plant is only 33 percent efficient, the input energy to the steam-electric turbine must be

$$E = \frac{E'}{0.33} = \frac{3.16 \times 10^{16} \text{ J}}{0.33} = 9.49 \times 10^{16} \text{ J}$$

One mole of $^{235}_{92}U$ contains 0.235 kg of uranium. The mass of pure $^{235}_{92}U$ needed is, therefore,

$$M = \frac{E}{\mathscr{E}} = \frac{9.49 \times 10^{16} \text{ J}}{(2.07 \times 10^{13} \text{ J})/(0.235 \text{ kg})} = \boxed{1.08 \times 10^3 \text{ kg}}$$

Since 10^3 kg is often called a *metric ton*, a large 1-GW nuclear power requires about 1 metric ton of pure $^{235}_{92}U$ a year as fuel. Since $^{235}_{92}U$ is only about 0.7 percent of natural uranium, about 143 metric tons of natural uranium is needed.

29.8 Biological Effects of Nuclear Radiations

Nuclear radiations include the alpha, beta, and gamma rays emitted by nuclei, as well as protons, neutrons, and even x-rays, which have the same ionizing properties as nuclear gamma rays even though they do not come from the nucleus.

Effects of Nuclear Radiation on Matter

Charged particles like alpha and beta rays produce *ionization* in any material through which they pass, including living matter, as shown in Fig. 29.14. Since nuclear alpha and beta particles have energies in the MeV range, and less than 20 eV is needed to remove the outermost electron from most atoms, a single alpha or beta ray can produce thousands of ions.

Gamma rays and x-rays produce ionization by photoelectric and Compton scattering processes. Neutrons, on the other hand, produce artificial radioactive isotopes, which subsequently emit high-energy beta and gamma rays. These produce, in turn, additional ionization processes.

This ionizing property of nuclear radiation makes it harmful to living tissue. Ionization disrupts the normal chemical processes in a living cell and causes the cell to grow abnormally (as in cancerous tumors) or to die. In certain cases the

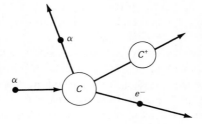

FIGURE 29.14 The ionization of a carbon atom when struck by an alpha particle.

radiation dose may cause *somatic* damage to an individual, that is, bodily damage observed either immediately or after a period of time in the form of cancers or leukemia. In other cases there is evidence that such radiation, especially when it is incident on the reproductive organs or on a human fetus, causes genetic *mutations*. These are changes in the DNA molecules (the molecules that carry genetic information) in an egg cell or sperm cell.

Units for Measuring Ionizing Radiation

There are a great variety of radiation units, but here we confine ourselves to the most important ones. These units refer to three quite different aspects of ionizing radiation:

1 Nuclear activity. As we have seen, *nuclear activity is a measure of the number of disintegrations that take place per second in a nuclear source* and is given by [see Eqs. (29.4) and (29.5)]

$$\text{Nuclear activity} = \frac{\Delta N}{\Delta t} = -\lambda N = -\frac{0.693}{T_{1/2}} N \qquad (29.20)$$

The SI unit for nuclear activity is the becquerel (Bq), where

$$1 \text{ Bq} = 1 \text{ nuclear disintegration per second} \qquad (29.21)$$

One gram of radium has 3.6×10^{10} nuclei decaying per second on the average, so the activity of a 1-g sample of radium is 3.6×10^{10} Bq.

There is an older unit for activity, the curie (Ci), which is equal to 3.7×10^{10} nuclear decays per second, and so

$$1 \text{ Ci} = 3.7 \times 10^{10} \text{ Bq} \qquad (29.22)$$

2 Absorbed dose. *Absorbed dose is a measure of the energy absorbed from a beam of radiation by a given mass of absorbing material.* It is expressed in joules per kilogram. The official SI unit for absorbed dose is the *gray* (Gy), where

$$1 \text{ Gy} = 1 \text{ J/kg} \qquad (29.23)$$

Older (and still frequently used) units for absorbed dose are the rad (*radiation absorbed dose*) and the roentgen, where

$$1 \text{ rad} = 10^{-2} \text{ Gy} = 10^{-2} \text{ J/kg} \qquad (29.24)$$
$$1 \text{ roentgen} = 0.87 \text{ rad} = 0.87 \times 10^{-2} \text{ Gy} = 0.87 \times 10^{-2} \text{ J/kg}$$

If, for example, a person were to stand 1 m away from a 1-Ci source of $^{60}_{27}\text{Co}$ for 1 h, the dose of gamma rays received at the front surface of the body would be 1.2×10^{-2} Gy, or 1.2 rad. Because gamma rays are rapidly attenuated by the body, the dose inside the body would be considerably less.

3   Biological effectiveness. *Biological effectiveness is a measure of the actual biological damage to be expected from the radiation incident on the body.* The SI unit is the *sievert* (Sv), where

$$\text{Absorbed dose in sieverts} = \text{absorbed dose in grays} \times \text{RBE} \qquad \textbf{(29.25)}$$

where RBE stands for *relative biological effectiveness* and is a measure of the relative biological damage produced by equal doses of different kinds of ionizing radition. A list of approximate RBEs for various nuclear particles is given in Table 29.4.

An older unit still frequently used is the *rem* (an acronym for *r*ad *e*quivalent *m*an), where

$$\text{Absorbed dose in rems} = \text{absorbed dose in rads} \times \text{RBE}$$

Since $1 \text{ rad} = 10^{-2} \text{ Gy}$, we have

$$1 \text{ rem} = 10^{-2} \text{ Sv} \qquad \textbf{(29.26)}$$

It is worth noting that a typical dental x-ray delivers about 1 mSv to the center of the patient's cheek.

TABLE 29.4   Relative Biological Effectiveness (RBE) for Nuclear Radiations

Type of ionizing radiation	RBE
Gamma rays (or x-rays)	1
Electrons (beta rays)	1–2
Protons	10
Alpha particles	10
Fast neutrons	10
Thermal neutrons	3
Heavy ions	20

Biological Effects of Radiation

There is no doubt that all forms of ionizing radiation are highly dangerous and should be avoided. The immediate somatic effects of ionizing radiation are better known because human beings have been exposed to very high levels of such radiation from the nuclear bombs dropped on Hiroshima and Nagasaki during World War II. Dosages of more than 7 Sv (700 rem) over the whole body are usually fatal; lesser whole-body doses above 0.1 Sv (10 rems) produce radiation sickness of a serious, but nonfatal, nature.

By *radiation sickness* we mean a collection of symptoms of excessive exposure to radiation, including nausea, vomiting, fatigue, loss of body hair, reddening of the skin, and a decrease in the white-blood-cell count.

Hard data on the delayed somatic effects (leukemia and cancers) and on genetic effects are more difficult to obtain. The consequences of exposure to radiation may be so long delayed that it is impossible to tell whether cancers and leukemia were caused by controllable ionizing radiations from nuclear bombs or reactors, by the natural background radiation to which we are all subjected, by medical or dental x-rays, or by the chemical pollutants which fill our air and which in some cases cause somatic and genetic effects similar to those caused by nuclear radiations.

Despite these ambiguities there is general agreement that exposure to strong ionizing radiation increases the likelihood of cancers, leukemia, and genetic defects. There is less agreement on the effects of accumulated low radiation doses over long periods of time, but a 1990 study by a Panel of the National Academy of Sciences* concluded that the risks of low-level radiation have been underestimated in past studies of this important question.

Table 29.5 lists the sources of ionizing radiation in the United States. Note that radon, which seeps into homes and other buildings from the ground, is now considered the single most prevalent source of ionizing radiation. It is also particularly hazardous because radon is a gas that is taken into the lungs when we breathe

---

Health Effects of Exposure to Low Levels of Ionizing Radiation (National Academy Press, Washington, D.C., 1990).

TABLE 29.5   Sources of Ionizing Radiation in the United States

Source	Percentage of Total Ionizing Radiation Produced	
Natural Sources:		Total: 82%
Radon	55%	
Emissions from inside human body	11%	
Cosmic rays from space	8%	
Emissions from radioactive minerals	8%	
Man-Made Sources:		Total: 18%
Medical and dental x-rays	11%	
Nuclear medicine (diagnosis and treatment)	4%	
Consumer products	2%	
All other	Less than 1%	
Nuclear fuel cycle	0.1%	

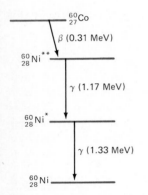

FIGURE 29.15 Decay of the radioactive nucleus $^{60}_{27}$Co to the ground state of $^{60}_{28}$Ni with the emission of two high-energy gamma rays. These gamma rays are frequently used for radiation therapy. The asterisks indicate excited nuclear states of $^{60}_{28}$Ni.

and can damage the lungs when it decays with the emission of alpha particles. Also notice that natural sources of radiation, and medical and dental x-rays, present a far greater danger than do well-functioning nuclear power plants.

Catastrophic accidents like the one which almost occurred at Three Mile Island in Pennsylvania in 1979 or the one which did occur at Chernobyl in the Soviet Union in 1986, however, can have grave consequences for the health of people living near the plant, and have cast a cloud over the future of nuclear power in the United States.

Ionizing radiation can cause cancer. But it can also be used to treat some cancers. Gamma rays from isotopes like $^{60}_{27}$Co, or x-rays produced by x-ray machines, can selectively destroy cancer cells that are fast-growing and therefore more susceptible to destruction by ionization. In so doing there is necessarily some damage to the surrounding healthy body cells, but this damage can be minimized by careful technique in administering the radiation.

Some of the more useful radioactive isotopes were listed in Table 29.2. Cobalt is particularly useful, since $^{60}_{27}$Co decays to an excited nuclear state of $^{60}_{28}$Ni, which then immediately decays to its ground state by emitting two gamma rays in succession, each with an energy above 1 MeV, as shown in Fig. 29.15. These gamma-ray energies are much larger than those produced by most x-ray machines, and $^{60}_{27}$Co is, therefore, a convenient and relatively inexpensive source of highly ionizing radiation.

Sometimes radioactive isotopes can be directly introduced into cancerous tissue to destroy it. A good example is the use of radioactive iodine $^{131}_{53}$I to treat thyroid cancer. Any iodine introduced into the body tends to be concentrated in the thyroid. Thus radioactive iodine injected into the blood stream is carried to the thyroid, where the ionizing radiation it emits destroys cancerous tissue.

Radiation therapy is most useful when the tumor or cancerous tissue is localized at one place in the body. If the cancer is spread throughout the body, radiation therapy is of little use, since in the process of killing the cancer cells the radiation will do excessive damage to healthy cells.

Example 29.10

A 1-mg sample of $^{60}_{27}$Co is being used for radiation therapy. What is the activity of such a sample in becquerels?

SOLUTION

To solve this problem we need first to find the number of atoms N of $^{60}_{27}Co$ in a 1-mg sample. Since 1 mol of ^{60}Co has a mass of 60 g and contains 6.02×10^{23} atoms, the number of atoms in 1 mg is

$$N = \left(\frac{10^{-3}\ g}{60\ g}\right)(6.02 \times 10^{23}\ atoms) = 1.0 \times 10^{19}\ atoms$$

The numerical value of the nuclear activity is, from Eq. (29.20),

$$\text{Nuclear activity} = \frac{\Delta N}{\Delta t} = -\frac{0.693N}{T_{1/2}}$$

This is the number of nuclei decaying per second. From Table

29.2 the half-life of ^{60}Co is 5.24 years. To obtain the activity in becquerels, the time must be expressed in seconds, since 1 Bq = 1 nuclear disintegration per second. We have, therefore, for the numerical value of the activity,

$$\text{Nuclear activity} = \frac{0.693(1.0 \times 10^{19}\ atoms)}{(5.24\ years)(3.16 \times 10^{7}\ s/year)}$$

$$= \frac{4.2 \times 10^{10}}{s} = \boxed{4.2 \times 10^{10}\ Bq}$$

EXERCISE 2   What is the activity of such a sample in curies?

Example 29.11

A radiologist decides to change the radiation treatment being given to a patient suffering from a cancerous tumor on the leg from x-rays to fast neutrons. If the original dosage was 10 Gy of x-rays, what must the neutron dosage be to produce the same effect on the tumor?

SOLUTION

Since

Absorbed dose in Sv = absorbed dose in Gy × RBE

in the first case the effective dose delivered to the tumor was

10 Gy × 1 = 10 Sv

To achieve the same effective dose with fast neutrons, which have an RBE of 10, we must have

$$\text{Absorbed dose in Gy} = \frac{\text{absorbed dose in Sv}}{\text{RBE}}$$

$$= \frac{10\ Sv}{10} = \boxed{1\ Gy}$$

Hence a 1-Gy absorbed dose of fast neutrons should be as effective in treating the cancer as a 10-Gy dose of x-rays.

29.9   Use of Radioactive Isotopes in Medicine and Technology

Radioactive isotopes enter into chemical reactions and form molecules in the same way as do nonradioactive stable isotopes of a given element. For this reason radioactive isotopes (or *radioisotopes*) can be used as *tracers* in chemical, biological, and industrial processes.

Tracer: A radioisotope that can be followed through the course of a chemical, biological, or other process by detecting the particles emitted in the radioactive decay of the isotope.

In addition to being used for therapeutic purposes, $^{131}_{53}I$ can be used in the diagnosis of thyroid abnormalities. By monitoring the rate at which $^{131}_{53}I$ accumulates in the thyroid, the functioning of this gland can be studied. Also by scanning the thyroid with a suitable detector, abnormalities in size or shape can be seen. Other popular tracer isotopes are $^{14}_{6}C$ and $^{3}_{1}H$ (tritium).

With such tracers it is possible to study how food is digested and in what parts of the body the molecules in the food concentrate. Thus the metabolism of iron in the human body has been studied by using $^{59}_{26}Fe$ as a tracer. Similar techniques can determine how the body synthesizes amino acids and other important compounds needed for human life.

Industry uses radioisotopes to study frictional wear between surfaces and moving parts by observing the transfer of radioisotopes from one surface to another. Makers of soap and detergents test the success of their products in removing dirt from clothes by adding radioactive tracers to the dirt and measuring the fraction of the tracers carried off in the soapy water.

There are countless other uses for radioactive tracers. Since most radioisotopes do not occur in nature and must be artificially produced, the use of tracers has increased greatly over the past 40 years because of the availability of nuclear reactors and high-energy particle accelerators to produce the needed radioactive isotopes.

Radiation Detectors

To monitor the passage of atoms or molecules containing tracer nuclei through the human body, and for nuclear physics research, sensitive detectors of the particles emitted by nuclei are needed. Three different kinds of detectors, described below, are available for detecting nuclear particles.

Gas-Filled Detectors   *All gas-filled detectors, such as ionization chambers, proportional counters,* and *Geiger counters,* consist of a cylindrical chamber with a thin wire along the axis of the cylinder, as in Fig. 29.16. The wire is kept at a high positive electric potential compared with the outer cylinder. The gases used (often argon) and the pressure maintained vary with the desired use of the counter, but all the counters work on the same basic principles: (1) Nuclear radiation passing into the chamber through the thin window ionizes some of the gas molecules. (2) The electric field pulls electrons to the central wire and positive ions to the outer cylinder, causing a current to flow in the external circuit. (3) The resulting current through the resistor R is measured by using a voltmeter to measure the voltage across R.

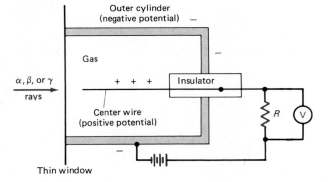

FIGURE 29.16   A gas-filled nuclear counter. Such counters are typically 10 inches long and 1 inch in diameter.

The main difference between ionization chambers, proportional counters, and Geiger counters is in the voltage applied across the electrodes. This results in a difference in the number of ions collected, as shown in Fig. 29.17 for incident alpha particles. For proportional counters the applied voltage is so high that collisions produce an avalanche effect, and a single incident particle may lead to 10^5 to 10^6 ion pairs. This makes it possible to count incident particles one by one. The size of the current pulse produced is approximately proportional to the energy of the ionizing particle, leading to the name *proportional counter.*

In a Geiger counter [named after the German physicist Hans Geiger (1882–1945)] the applied voltage is close to 1000 V. In this case any nuclear particle can initiate an avalanche of electrons and ions, and the current pulses are

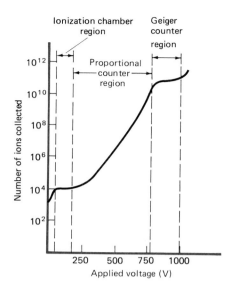

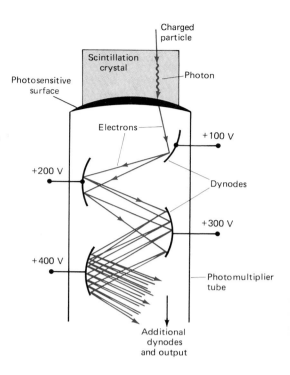

FIGURE 29.17 (left) Different voltage regions for the gas-filled counter of Fig. 29.16. Ionization chambers use voltages below about 250 V; Geiger counters use voltages close to 1000 V. The intermediate voltage region between these values is that used for proportional counters.

FIGURE 29.18 (right) A scintillation crystal with attached photomultiplier tube to amplify the output signal. At each stage the number of electrons produced increases greatly.

independent of the energy of the incident particle. It is therefore impossible to distinguish between various types of nuclear radiations with a Geiger counter unless counter windows of different thicknesses or external absorbers are used.

Scintillation Counters These *counters* use a phosphor or scintillator, which emits light when struck by high-energy particles. Scintillation counters generally use plastic scintillators for charged particles and sodium iodide (NaI) crystals for gamma rays. The light from the scintillator is then incident on a photocathode and releases an electron from this cathode by the photoelectric effect, as shown in Fig. 29.18. The photomultiplier tube multiplies the single electron into a measurable current pulse at its output. Amplifications of 10^9 are possible with photomultiplier tubes with 14 stages (each stage is called a *dynode*). Each incident particle can then be counted by a high-speed electric counter attached to the output of the photomultiplier tube.

Photographic Emulsions, Cloud Chambers, Bubble Chambers, and Spark Chambers Such devices make the paths of individual particles visible by causing observable changes in the gas or liquid through which the particles pass. These detectors are used more for physics research than for medical or industrial applications.

A charged particle passing through a thick *photographic emulsion* ionizes the atoms along its path. The resulting chemical changes show up the path of the particle when the film is developed, as shown in Fig. 29.19. *Cloud chambers* employ supercooled gases that produce droplets of condensed liquid along the path of a charged particle. *Bubble chambers* use a superheated liquid that evaporates and produces a vapor trail of bubbles along the path of the charged particle, as in

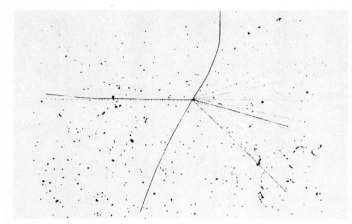

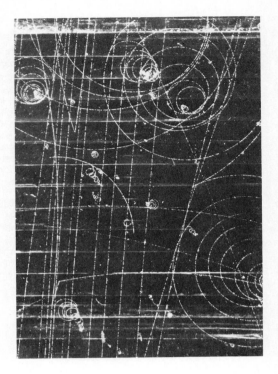

FIGURE 29.19 (above) Tracks of nuclear particles in a photographic emulsion. These tracks were produced by high-energy, heavy cosmic-ray particles. These particles struck the photographic emulsion, which had been carried to high altitudes by a balloon in a Brookhaven National Laboratory experiment. *(Courtesy of Brookhaven National Laboratory, Upton, N.Y.)*

FIGURE 29.20 (right) A typical bubble-chamber photograph showing the tracks of charged particles. The photograph was taken in the 80-in liquid-hydrogen bubble chamber at the Brookhaven National Laboratory, Upton, New York. *(Courtesy of Brookhaven National Laboratory.)*

Fig. 29.20. *Spark chambers* use high voltages across sets of closely spaced parallel plates to produce sparks when a charged particle passes through, making the path of the particle visible.

In cloud chambers, bubble chambers, and spark chambers, elaborate coincidence-counting systems outside the chamber trigger cameras to photograph the chamber only when the particles of interest are passing through. In this way permanent records of the particle's path are obtained.

Summary: Important Definitions and Equations

Structure of the nucleus

Nucleon: The generic name given to protons and neutrons, since they are found within the nucleus.

Radius of nucleus of mass number A:

$$r_0 = (1.2 \times 10^{-15} \text{ m})(A)^{1/3} \qquad (29.1)$$

Strong nuclear force: The short-range, charge-independent force that holds the nucleus together, despite the strong electrostatic repulsions between the protons in the nucleus.

Mass defect (Δm): The difference between the sum of the masses of the individual protons and neutrons making up a nucleus and the mass of the nucleus itself.

Nuclear binding energy: The energy required to tear a nucleus apart into its constituent isolated protons and neutrons; $E = \Delta m \, c^2$.

Isotopes: Two atoms having the same charge number Z but different mass numbers A.

Unified mass unit (u): A mass exactly equal to one-twelfth the mass of the carbon ($^{12}_{6}$C) atom.

1.00000 u is equivalent to 931.49 MeV (29.2)

Radioactivity

Natural radioactivity: The phenomenon peculiar to the nuclei of some atoms existing naturally on the earth in which they spontaneously emit alpha, beta, or gamma rays and change into other kinds of nuclei.

Alpha (α) particles = helium nuclei ($^4_2\text{He}^{2+}$)

Beta (β) particles = electrons

Gamma (γ) rays = high-energy photons

Half-life ($T_{1/2}$): The time required for half the nuclei in a sample of a particular nuclear species to decay radioactively.

Decay constant (λ): The constant in the exponential factor in the equation for the decay of a nuclear species:

$$N = N_0 e^{-\lambda t} \tag{29.3}$$

$$\lambda = \frac{0.693}{T_{1/2}} \tag{29.4}$$

Nuclear activity: The rate at which a particular nuclear species decays in time:

$$\frac{\Delta N}{\Delta t} = -\lambda N \tag{29.5}$$

where λ is the decay constant and N is the number of radioactive nuclei present at the time at which the activity is calculated.

Artificial (induced) radioactivity: The process in which energetic particles force their way inside a stable nucleus and create a new, unstable nucleus, which then decays into yet another nucleus and other smaller particles.

Special kinds of nuclear reactions

Nuclear fission: A nuclear reaction in which a heavy nucleus like uranium breaks up into two nuclei of intermediate masses, together with the release of two or more neutrons and large amounts of energy.
Fissile nuclei: Nuclei like $^{233}_{92}\text{U}$, $^{235}_{92}\text{U}$, and $^{239}_{94}\text{Pu}$ that undergo fission when bombarded by neutrons.
Chain reaction: The rapid buildup of energy that occurs in a nuclear bomb when the multiplication factor exceeds unity and the number of neutrons producing fission at successive stages of the reaction increases exponentially.

Nuclear reactor: A large boiler in which nuclear processes generate heat that is then converted into electricity by a steam-electric turbine.
Pressurized-water reactor: A nuclear reactor that uses ordinary water under high pressure as the moderator and cooling fluid. This water is then used to heat water in a separate unpressurized system to produce steam to drive an electric turbine.
Breeder reactor: A fission reactor that both produces electric power and "breeds" additional nuclear fuel.

Nuclear fusion: A nuclear reaction in which two light nuclei are combined, or "fused" together, to form a heavier nucleus, accompanied by the release of large amounts of energy.

Nuclear radiation

Ionizing radiation: The radiation emitted in the radioactive decay of nuclei, which ionizes atoms and molecules and damages living tissue.

Units for measuring ionizing radiation
Becquerel (Bq): A unit for measuring nuclear activity:

1 Bq = 1 nuclear disintegration per second (29.21)

1 curie (Ci) = 3.7×10^{10} Bq (29.22)

Gray (Gy): A unit for measuring the energy absorbed by a given mass of material:

1 Gy = 1 J/kg (29.23)

1 rad = 10^{-2} Gy = 10^{-2} J/kg (29.24)

Sievert (Sv): A unit for measuring the actual biological damage to be expected from radiation:

Absorbed dose in Sv =
absorbed dose in Gy × RBE (29.25)

1 rem = 10^{-2} Sv (29.26)

RBE (relative biological effectiveness): A measure of the relative biological damage produced by equal doses of different kinds of ionizing radiation.

Tracer: A radioisotope that can be followed through the course of a chemical, biological, or other process by detecting the particles emitted in the radioactive decay of the isotope.

Questions

1 (a) Why are the concentrations of most radioactive isotopes in nature so low?
(b) Suggest a process that might lead to a large concentration of a long-lived radioactive isotope in nature.
(c) Suggest a process that might lead to a large concentration of a short-lived radioactive isotope in nature.
2 When a chemical reaction like $2\text{H}_2 + \text{O}_2 \rightarrow 2\text{H}_2\text{O}$ occurs, is the mass of the products exactly equal to the mass of the reactants? If not, why cannot this difference in mass be used as an energy source?
3 Discuss the difference between nuclear reactions and chemical reactions with respect to:
(a) The behavior of atoms of a particular element during each kind of reaction.
(b) The behavior of different isotopes of the same element during each kind of reaction.

(c)   The size of the energy changes in each kind of reaction.

4   From the binding-energy curve (Fig. 29.3), would you expect fission or fusion to be the better energy-producing process?

5   (a)   Why are the names "atom bomb" and "atomic reactor" misnomers?
(b)   What are the more accurate names for these devices? Why?

6   What is the fundamental difference between a nuclear reactor and a nuclear bomb?

7   Why do moderators for nuclear reactors use relatively light atoms like $^1_1$H or $^2_1$H to slow down the neutrons rather than heavy atoms like gold or lead?

8   (a)   Would you expect some $^{239}_{94}$Pu to be produced by a conventional boiling-water reactor? Why?
(b)   What happens to this plutonium after it is produced?

9   (a)   What is the real source of energy in a LMFBR?
(b)   What is being consumed in the reaction?

(c)   Is there any violation of conservation of energy here?

10   (a)   What would be the advantages of using D–D fusion reactions [Eqs. (29.17) and (29.18)] in a fusion reactor rather than D–T reactions [Eq. (29.19)]?
(b)   What problems must be solved before the D–D reactor can be made to produce useful energy?

11   Why is it that some very long-lived radioactive nuclei like $^{40}_{19}$K ($T_{1/2} = 10^9$ years) present little danger to human life and health?

12   (a)   Would you expect any danger to the health of miners involved in the mining of uranium ore? Why?
(b)   How do you think this danger would compare with the dangers to which coal miners are subjected?

13   Explain in what respects gamma rays and x-rays are the same and in what respects they differ.

14   What would you expect to be the range of half-lives of the radioactive isotopes used for medical diagnosis? Give reasons for your answer.

Multiple-Choice and Simple Exercises

Note: To solve many of the problems in this chapter, you will need to refer to the periodic table (Table F.6, located inside the front cover), to the table of isotopes (in Appendix 4), and to the half-lives of isotopes given in Table 29.2.

29.1   The density of nuclei:
(a)   Varies greatly throughout the periodic table
(b)   Is much larger for uranium nuclei than for hydrogen nuclei
(c)   Is very different for isotopes of the same element
(d)   Is basically constant throughout the periodic table of elements
(e)   Is much smaller for uranium nuclei than for hydrogen nuclei

29.2   The mass of the $^{235}_{92}$U nucleus can be expected to be about:
(a)   92 u   (b)   235 u   (c)   235 g   (d)   92 kg
(e)   235 MeV

29.3   The most stable nucleus in the periodic table is
(a) $^{56}_{26}$Fe   (b) $^1_1$H   (c) $^4_2$He   (d) $^{20}_{10}$Ne
(e) $^{238}_{92}$U

29.4   A sample of the radioactive isotope $^{131}_{53}$I is received by a hospital for use in locating thyroid abnormalities, but is allowed to sit on the shelf for 80 days. The fraction of the originally present $^{131}_{53}$I remaining is about:
(a)   1/10   (b)   1/80   (c)   1/1000
(d) $1/10^6$ (e) $1/10^{10}$

29.5   In 20 half-lives the nuclear activity of a radioactive sample is reduced by about a factor of:
(a)   1/20   (b)   1/1000   (c) $1/20^2$
(d) $1/10^6$ (e) $1/e^{20}$

29.6   The nuclear activity of a sample containing 2.6×10^{18} atoms of the radioactive isotope $^{32}_{15}$P is:
(a) 1.5×10^{12} Ci   (b) 1.5×10^{12} Bq
(c) 1.3×10^{17} Gy   (d) 1.3×10^{17} Bq
(e) 1.3×10^{17} Ci

29.7   The energy released in the reaction $^{238}_{92}$U $+ ^1_0 n \rightarrow$ $^{239}_{92}$U would be about:
(a)   1.1 MeV   (b)   2.4 keV   (c)   4.8 MeV
(d)   5.2 keV   (e)   None of the above

29.8   In the nuclear equation, $^{237}_{93}$Np $\rightarrow ^4_2$He $+ ^A_Z X$, the unknown product nucleus must be:
(a) $^{233}_{91}$Pa   (b) $^{233}_{91}$U   (c) $^{233}_{91}$Pu   (d) $^{241}_{91}$Pa
(e) $^{235}_{89}$Ac

29.9   To produce the same biological damage to the human body as an x-ray source producing an absorbed dose of 10^{-1} Gy, an alpha-particle source would have to produce an absorbed dose of:
(a)   10 Gy   (b)   0.1 Gy   (c) 10^{-2} Gy
(d) 10^{-3} Gy   (e) 10^2 Gy

29.10   The actual biological damage to be expected from 1.6×10^{-3} Gy of protons absorbed by a person's body can be expressed as:
(a) 1.6×10^{-3} Sv   (b) 1.6×10^{-2} rem
(c) 1.6×10^{-2} Sv   (d) 1.6×10^{-4} Sv
(e) 1.6×10^{-3} rem

29.11   How many protons and neutrons are there in (a) $^{233}_{92}$U, (b) $^{238}_{92}$U, (c) $^{233}_{91}$Pa?

29.12   (a)   What is the approximate radius of the $^{235}_{92}$U nucleus?
(b)   What is its approximate volume?

29.13   By what percentage does the radius of the $^{233}_{92}$U nucleus differ from that of the $^{238}_{92}$U nucleus?

29.14   The strontium isotope $^{90}_{38}$Sr is one of the most dangerous of the radioactive isotopes produced in nuclear explosions. How long will it take for the amount of $^{90}_{38}$Sr produced in a nuclear bomb test to decay to 10^{-6} of its original value?

29.15   From the data given in Table 29.2, what are the decay constants of (a) $^{40}_{19}$K, (b) $^{137}_{55}$Cs?

29.16   What is the nuclear activity of a sample containing 3.0×10^{15} atoms of $^{88}_{36}$Kr?

29.17 In the radiocarbon dating of a parchment found in an Egyptian cave, it is found that the ratio of $^{14}_6$C to $^{12}_6$C is one-fourth that found in living specimens today. About how old is the parchment?

29.18 How much energy is released in the decay of a free neutron into a proton and an electron, if the equation for the reaction is $^1_0n \rightarrow\ ^1_1p\ +\ ^0_{-1}e$?

29.19 A person of mass 85 kg absorbs 6.0 J of energy from an x-ray machine. This energy is distributed over the person's entire body. What is the radiation dose absorbed by the person (a) in grays; (b) in rads?

29.20 Show that for thermal neutrons the biological effectiveness of an absorbed dose of 10 rad is 0.30 Sv.

Problems

29.21 Two very important elements in the semiconductor industry are silicon and germanium, with typical isotopes $^{28}_{14}$Si and $^{72}_{32}$Ge. How many electrons, protons, and neutrons does each of these isotopes contain?

29.22 (a) Find the density of the $^{206}_{82}$Pb nucleus.

(b) How much denser is the lead nucleus than ordinary water?

29.23 Calculate the atomic mass of chlorine as it is found in nature. Natural chlorine consists of 75.53% $^{35}_{17}$Cl and 24.47% $^{37}_{17}$Cl.

*__29.24__ (a) Show that, if an electron were confined to the region inside a nucleus of diameter 10^{-14} m, the uncertainty in its kinetic energy would be greater than 10 MeV.

(b) Calculate the potential energy of such an electron in the field of a proton if the average distance between the two is about 10^{-14} m.

(c) What can you conclude about the likelihood that there are electrons inside atomic nuclei?

29.25 Calculate the binding energy per nucleon for the $^{56}_{26}$Fe nucleus, and compare your result with that shown in the graph in Fig. 29.3.

29.26 Calculate the binding energy per nucleon for (a) the $^{16}_8$O nucleus and (b) the $^{15}_8$O nucleus, and compare the results.

29.27 (a) Calculate the percentage of its rest energy made up by the binding energy of a proton in a deuterium nucleus.

(b) Repeat this calculation for the binding energy of an electron in a hydrogen atom.

(c) Compare the two results.

29.28 Find the radius of the first Bohr orbit for the innermost electrons in the $_{92}$U atom, and compare its size with that of the $^{238}_{92}$U nucleus.

29.29 A sample of the frequently used radioactive isotope $^{137}_{55}$Cs is stored away in a protective lead vault for many years. If the half-life of $^{137}_{55}$Cs is 30 years, how long will it take for the activity of the sample to decrease (a) to one-fourth of its original value, (b) to one-fifth of its original value?

29.30 The half-life of $^{13}_7$N is 10 min.

(a) How many nitrogen atoms decay in 1.0 s in a 1.0-mg sample of this isotope?

(b) What is the activity of this sample in becquerels (Bq)?

29.31 How many half-lives are required for any radioactive sample to drop to 1 percent of its initial activity?

29.32 $^{238}_{92}$U is slightly radioactive, with a half-life of 4.5×10^9 years. Once $^{238}_{92}$U decays radioactively, a long chain of disintegrations occurs before the original ^{238}U nucleus is transformed into a final stable nucleus. If, in this process, 8

alpha particles and 6 beta particles are emitted, what is the final stable nucleus?

*__29.33__ It has been found that the oldest rocks containing uranium on earth contain about a 50-50 mixture of $^{238}_{92}$U and $^{206}_{82}$Pb. From the data given in Prob. 29.32, what is the approximate age of these rocks, if we assume that all the lead was produced by the radioactive decay of uranium?

29.34 What fraction of the $^{232}_{90}$Th atoms in existence at the beginning of the universe about 20 billion years ago still remain? Assume that the half-life of $^{232}_{90}$Th is 1.4×10^{10} years.

29.35 An animal bone found in a cave in southern France contained 200 g of carbon and had a $^{14}_6$C activity of 1600 decays per minute. If the $^{14}_6$C activity of a living organism is 16 decays per minute for each gram of carbon, how long ago did the animal die?

29.36 How many grams of $^{60}_{27}$Co are there in a sample whose activity is 2.0×10^7 Bq?

29.37 The silver isotope ^{108}Ag has a half-life of 2.4 min. Initially a sample contains 2.0×10^6 nuclei of $^{108}_{47}$Ag. How many radioactive nuclei remain after 1.2 min?

29.38 Twenty percent of a certain radioactive isotope decays in 6.0 h.

(a) What is the decay constant of the substance?

(b) What is the half-life?

*__29.39__ Use the abundance ratios and half-lives for $^{235}_{92}$U and $^{238}_{92}$U found in Appendix 4 to estimate the minimum age of the universe. Assume that when these two isotopes were formed they had the same natural abundance.

29.40 An air sample contains 1 million radioactive $^{137}_{55}$Cs nuclei. How many radioactive nuclei remain after 10 years?

29.41 A Geiger counter is immersed in a solution containing radiophosphorus, $^{32}_{15}$P, which has a half-life of 14.3 days. The counter records 1500 counts per minute. If the apparatus is left undisturbed for 60 days, what counting rate will be observed at the end of that time?

29.42 A piece of wood from a table found in an archaeological dig in Greece contains 100 g of carbon. The $^{14}_6$C counting rate from this sample is 240 counts per minute. How old is the sample?

29.43 Identify the particles resulting from these nuclear reactions:

(a) The beta decay of $^{137}_{55}$Cs (cesium).

(b) The alpha decay of $^{253}_{100}$Fm (fermium).

29.44 Since nuclear energies are in the MeV range, and gamma rays arise from transitions between two nuclear energy levels, what is roughly the wavelength of a gamma-ray photon?

29.45 The two isotopes $^{137}_{55}\text{Cs}$ and $^{90}_{38}\text{Sr}$ are produced by fission processes followed by beta decays. Determine x, y, x', and y' in the following formulas for these nuclear reactions (consider only particles having rest mass):

(a) $^{235}_{92}\text{U} + ^{1}_{0}n \rightarrow ^{143}_{57}\text{La} + ^{90}_{35}\text{Br} + x$

  $^{90}_{35}\text{Br} \rightarrow ^{90}_{38}\text{Sr} + y$

(b) $^{235}_{92}\text{U} + ^{1}_{0}n \rightarrow ^{137}_{53}\text{I} + ^{97}_{39}\text{Y} + x'$

  $^{137}_{53}\text{I} \rightarrow ^{137}_{55}\text{Cs} + y'$

29.46 Find the product nuclei in the following reactions:

(a) $^{12}_{6}\text{C}$ (d, n) (b) $^{130}_{52}\text{Te}$ $(d, 2n)$

(c) $^{59}_{27}\text{Co}$ (n, α) (d) $^{11}_{5}\text{B}$ (n, α)

29.47 (a) Show that $^{12}_{6}\text{C}$ is stable against decay into three alpha particles.

(b) Show that $^{8}_{4}\text{Be}$ is unstable against decay into two alpha particles.

29.48 A $^{238}_{92}\text{U}$ nucleus can spontaneously emit an alpha particle in the reaction

$$^{238}_{92}\text{U} \rightarrow ^{4}_{2}\text{He} + ^{234}_{90}\text{Th}$$

(a) If the uranium nucleus is originally at rest and the speed of the alpha particle is 1.5×10^7 m/s, what is the recoil velocity of the thorium nucleus?

(b) Do relativistic effects have to be taken into account here?

29.49 One way to enable Geiger counters to count slow neutrons is to fill the counter with boron trifluoride gas. Then the $^{10}_{5}\text{B}$ nucleus can capture the neutron and undergo the following reaction: $^{10}_{5}\text{B} + ^{1}_{0}n \rightarrow ^{4}_{2}\text{He} + ^{7}_{3}\text{Li}$. The alpha particles then produce ion pairs which can be counted by the Geiger counter. Find the energy of the product nuclei in this reaction.

29.50 Show that the fraction of the mass of the $^{235}_{92}\text{U}$ nucleus converted into energy in a fission reaction is only about 1/1000 of the original mass.

29.51 The following reaction occurs: $^{1}_{0}n + ^{10}_{5}\text{B} \rightarrow ^{7}_{3}\text{Li} + ^{4}_{2}\text{He}$. If the reactants are initially at rest, what is the speed of the alpha particle after the reaction?

29.52 In the fission of a uranium nucleus ($^{235}_{92}\text{U}$) by a neutron it is observed that one of the decay products is $^{143}_{57}\text{La}$. In this process three neutrons are also produced. What is the other intermediate-size nucleus produced in the fission process?

29.53 If a boiling-water reactor produces steam at 285°C, what is the maximum possible efficiency of a steam-electric turbine operating between this temperature and a water-exhaust temperature of 27°C?

29.54 The simplest kind of fusion reaction is the combination of a proton and a neutron to form a heavy hydrogen, or deuterium, nucleus ($^{2}_{1}\text{H}$). This reaction is described by the equation: $^{1}_{1}p + ^{1}_{0}n \rightarrow ^{2}_{1}\text{H} + $ energy. How much energy will be released in this fusion reaction?

29.55 (a) In 1.0 mol of natural uranium, how many atoms of $^{235}_{92}\text{U}$ and how many atoms of $^{238}_{92}\text{U}$ are found?

(b) How many grams of $^{235}_{92}\text{U}$ and $^{238}_{92}\text{U}$ are found in 1.0 mol of natural uranium?

29.56 The energy consumed in the United States each year is about 7.0×10^{19} J. How many kilograms of $^{235}_{92}\text{U}$ would be needed to release an amount of energy equal to this, if all the uranium nuclei are presumed to undergo fission according to Eq. (29.11)?

29.57 Hydrogen atoms are much more effective than lead atoms in slowing down neutrons to thermal velocities. Prove this by calculating the final speed of a neutron that originally had a speed of 2.0×10^6 m/s, after it makes a perfectly elastic head-on collision with (a) a $^{1}_{1}\text{H}$ nucleus that is initially at rest; (b) a $^{206}_{82}\text{Pb}$ nucleus initially at rest.

29.58 Write nuclear equations similar to those in Eq. (29.15) for the conversion of $^{232}_{90}\text{Th}$ into $^{233}_{92}\text{U}$ in a breeder reactor.

29.59 The complete D–D reaction chain in a fusion reactor would be a composite of the following reactions:

$$^{2}_{1}\text{H} + ^{2}_{1}\text{H} \rightarrow ^{3}_{2}\text{He} + ^{1}_{0}n + 3.3 \text{ MeV}$$

$$^{2}_{1}\text{H} + ^{2}_{1}\text{H} \rightarrow ^{3}_{1}\text{H} + ^{1}_{1}p + 4.0 \text{ MeV}$$

$$^{2}_{1}\text{H} + ^{3}_{1}\text{H} \rightarrow ^{4}_{2}\text{He} + ^{1}_{0}n + 17.6 \text{ MeV}$$

$$^{2}_{1}\text{H} + ^{3}_{2}\text{He} \rightarrow ^{4}_{2}\text{He} + ^{1}_{1}p + 18.3 \text{ MeV}$$

where the $^{3}_{2}\text{He}$ and $^{3}_{1}\text{H}$ required for the last two reactions are produced in the first two reactions.

(a) Add the above equations to obtain an equation for the net reaction occurring.

(b) Obtain the average energy produced in the change of each deuterium nucleus into a different nuclear species in the reaction.

29.60 In the complete D–D reaction chain in a fusion reactor, about 7.2 MeV of energy is produced from each deuterium nucleus. In the oceans the deuterium fraction in the hydrogen making up the ocean water is about 0.015 percent.

(a) Calculate the energy content of 1 km³ of ocean water.

(b) Compare your answer with the energy consumed in the United States in a year, which is about 7.0×10^{19} J.

29.61 A 1.0-μg sample of $^{131}_{53}\text{I}$ is being used in the treatment of thyroid cancer. What is the activity of such a sample in (a) becquerels, (b) curies?

29.62 A beam of fast neutrons passes through a man's body and deposits 0.052 J of energy in each kilogram of his body.

(a) Find the absorbed dose in grays.

(b) Find the biological effectiveness of this dose in sieverts.

(c) Express your answers to parts (a) and (b) in rads and rems.

29.63 A radiation dose of over 700 rem or 7.0 Sv is usually fatal to half the people exposed to it. This is equivalent to (a) how many grays of protons; (b) how many grays of gamma rays?

29.64 A whole-body dose of 7.0 Sv of gamma radiation causes death to half the people exposed to it. Calculate the energy (in joules) delivered to a 90-kg person by such a dose.

29.65 Scientific workers using gamma-ray sources are only permitted to experience dosages of about 5.5 millirad per hour.

(a) If a 1.0-Ci source of $^{60}_{27}\text{Co}$ would subject the workers to 1.2 rad during a 1-h exposure when they are 1.0 m from the source, what is the minimum safe distance for the workers?

(b) If a shield over the source reduces the gamma radiation to 5 percent of its original value, to what closer distance can the workers safely approach the source?

29.66 It is desired to inject enough iodine-131 into the bloodstream to produce a nuclear activity of 10^3 Bq, which can then be observed when the iodine collects in the thyroid. The $^{131}_{53}$I radioisotope sample is received directly from the reactor where it is produced, but is not used for 32 days. How many grams of iodine-131 must be injected into the bloodstream to produce the desired activity?

Additional Readings

Bethe, Hans A.: "The Necessity of Fission Power," *Scientific American*, vol. 234, no. 1, January 1976, pp. 21–31. A strong defense of nuclear energy by an eminent physicist and Nobel Laureate.

Cohen, Bernard L.: *The Heart of the Atom: The Structure of the Atomic Nucleus,* Doubleday Anchor, Garden City, N.Y., 1966. A brief elementary introduction to nuclear physics.

Crawford, Deborah: *Lise Meitner, Atomic Pioneer*, Crown, New York, 1969. A lively and very interesting account of a great scientist.

Curie, Eve: *Madame Curie*, Doubleday, Garden City, New York, 1937. The standard biography by Madame Curie's daughter.

Hecht, Selig: *Exploring the Atom*, Viking, New York, 1954. A simple and clear introduction to the essentials of atomic and nuclear physics.

Inglis, David R.: *Nuclear Energy—Its Physics and Its Social Challenge*, Addison-Wesley, Reading, Mass., 1973. A balanced discussion of nuclear energy, its promise and its problems.

Jaffe, C. Carl: "Medical Imaging," *American Scientist*, vol. 70, November–December 1982, pp. 576–585. A general discussion of a variety of techniques for obtaining information about the interior of the human body. Included are a number of nuclear techniques like radioisotope tracers, nuclear magnetic resonance (NMR), and positron emission tomography (PET).

Noz, Marilyn E., and Gerald O. Maguire, Jr.: *Radiation Protection in the Radiologic and Health Sciences*, Lea and Febiger, Philadelphia, 1979. A clear, careful presentation of the use of radiation in the health sciences.

Nuclear Energy (Readings from *Scientific American*, with introductions by Hans A. Bethe), W. R. Freeman and Co., San Francisco, 1983. A collection of articles on the use of nuclear fission as a power source.

O'Neill, Gerald K.: "The Spark Chamber," *Scientific American*, vol. 207, no. 2, August 1962, pp. 36–43. Includes discussions of the Geiger counter, the cloud chamber, the bubble chamber, and the spark chamber.

Rafelski, Johann, and Steven E. Jones: "Cold Nuclear Fusion," *Scientific American*, vol. 257, no. 1, July 1987, pp. 84–89. Describes an idea for producing nuclear fusion reactions at moderate temperatures, using muons as catalysts for the reaction.

Upton, Arthur C.: "The Biological Effects of Low-Level Ionizing Radiation," *Scientific American*, vol. 246, no. 2, February 1982, pp. 41–49. A discussion of the hazards of both background and human-produced low-level radiation. Upton was the chairman of the panel mentioned in Sec. 29.8.

Vandryes, Georges A.: "Superphenix: A Full-Scale Breeder Reactor," *Scientific American*, vol. 236, no. 3, March 1977, pp. 26–35. An account of France's progress in solving the country's energy problems by using breeder reactors.

Yost, Edna: *Women of Modern Science*, Dodd, Mead, New York, 1964. Includes brief biographies of Lise Meitner and of the well-known nuclear physicist C. S. Wu.

CHAPTER 30

Solid-State Physics

The seeds of great discoveries are constantly floating around us, but they only take root in minds well prepared to receive them.

Joseph Henry (1797–1878)

The properties of solid substances have always been of great interest to the intellectually curious. Our remote ancestors wondered why tiny bits of iron clung to a piece of lodestone. The architects and builders who planned the aqueducts and amphitheaters of ancient Rome probably debated the relative merits of stone, brick, and metal as building materials.

In a similar way, physicists have always been interested in solids. A new and most successful phase in the application of physics to solids came with the invention of the transistor in 1947 and the succession of revolutionary developments in electronics it has spawned. This development was based on a deeper understanding of the properties of solids at the atomic and molecular level, using the insights of quantum mechanics. This sophisticated understanding of solids is called *solid-state physics*, or *condensed-matter physics*.

This chapter could well have been placed immediately after Chap. 28 on the structure of atoms, since its content is the application of atomic theory to solids. Solid-state and elementary-particle physics, however, are both highly active, broad, and difficult fields of physics, in which theories are still being developed and sophisticated mathematical and experimental techniques applied to complicated problems. For this reason we provide in this and the next chapter only a qualitative survey of these important fields, with no worked examples or end-of-chapter problems.

These final two chapters emphasize ideas and an understanding of how these fields of physics have developed, what crucial problems remain, and what progress is being made in solving them. Such an approach may help the student finish the course with the realization that physics is not a closed subject, but one with many important and fascinating discoveries still to be made.

30.1 The Structure of Solids

A solid like iron differs greatly from a gas or liquid. First, the atoms or molecules in a solid are, on the average, much closer together than they are in a gas, with the separation between atoms being roughly the size of the electronic charge cloud surrounding the nucleus of each atom. For this reason the atoms interact with one another in somewhat the same fashion they do in molecules, and the whole solid may be considered one gigantic molecule. Second, atoms in a solid, unlike those in liquids and gases, do not move through the solid but remain in fixed positions.

There are two basic kinds of solids, *crystals* and *noncrystals*.

Crystal: A solid with a regular geometric shape in which the atoms making up the crystal arrange themselves in an ordered geometric pattern.

A crystal therefore forms a *lattice*, an orderly array of points in space at which atoms are located. Noncrystalline solids like glass and plastics, on the other hand, are amorphous and disordered.

In principle the structure of a crystal can be explained by applying quantum mechanics to the atoms or molecules making up the crystal and calculating the arrangement of atoms or molecules that minimizes the energy of the crystal. Once this structure is determined, it is possible to predict many of the mechanical, thermal, electrical, magnetic, and optical properties of the crystal. The high degree of order present in crystalline solids simplifies the calculation of their properties and makes them better understood than noncrystalline solids.

Types of Crystalline Solids

There are four basic types of crystalline solids, two of which have direct analogues in ionic- and covalent-bonded molecules.

Ionic Crystals   Ionic crystals are those in which alternate sites in the crystal lattice are occupied by ions of opposite charge. For example, in the salt (NaCl) crystal (and in all other alkali halide crystals like KCl, KI, RbCl) the Na atom loses an electron and becomes Na^+, while the Cl atom gains an electron and becomes Cl^-. These ions then arrange themselves in a cubic lattice, with the positive and negative ions occupying alternate positions in the lattice. In NaCl the positive and negative ions are separated by a lattice distance $d_0 = 0.282$ nm.

As shown in Fig. 30.1, each lattice plane (or face) of a NaCl crystal is made up of squares, with two Na^+ and two Cl^- ions at the corners of each square. An area of the plane that is two lattice distances ($2d_0 = 0.564$ nm) on a side therefore has four identical Na^+ ions at its corners and one Na^+ ion at the center of the square. For this reason this crystal structure is called a *face-centered cubic* lattice. The Cl^- ions form the same pattern in the lattice.

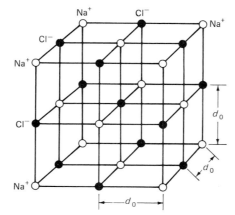

FIGURE 30.1   A face-centered cubic NaCl crystal. For NaCl, $d_0 = 0.282$ nm. (The dots are intended only to show the location of the ions in the lattice, not the size of the ions.)

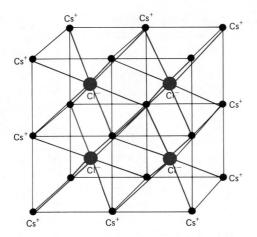

FIGURE 30.2 A body-centered cubic CsCl crystal. Notice how in this case the central atom in a cube with Cs$^+$ ions at the corners is a Cl$^-$ ion.

Calculations on the NaCl crystal show that the energy of the state in which a large number of ions are linked together to form a large lattice is lower than the state in which all the ions are separated by great distances. Since this energy decrease is greater than the energy expended in originally creating the ions by moving electrons from the Na to the Cl atoms, the NaCl crystal is stable.

All the alkali halides have similar face-centered cubic structures except for the cesium compounds. The Cs$^+$ ions are too large to fit into a face-centered cubic structure; CsCl and the other Cs halides therefore form the *body-centered cubic* lattice shown in Fig. 30.2.

Solid ionic crystals are electrical insulators, since they have no free charges to carry electric current. If, however, salt (NaCl) is poured into water, the ions separate when they go into solution and thus carry an electric current. Ionic salts in solution are therefore good conductors of electricity.

Covalent Crystals The bonding of neutral atoms to form covalent crystals is similar to the bonding of two oxygen atoms to form O$_2$. The electrons locate themselves between the nuclei in such a way that a net attractive force results between adjacent atoms in the crystal. This lowers the total energy of the crystal and leads to a stable crystal lattice.

The symmetry of a covalent crystal depends on the valence of the atoms making up the crystal and on the size of the atoms involved. Thus atoms in column IV of the periodic table, like carbon, silicon, and germanium, have valences of 4. They therefore usually form covalent crystals in which each atom has four nearest neighbors. Each atom shares one of its four outer electrons with a neighbor and thus forms four strong electron-pair bonds. This leads to the tetrahedral structure for these crystals shown in Fig. 30.3 for diamond (which is made up completely of carbon atoms), in which each carbon atom is bonded to four neighboring carbon atoms. Since all four valence electrons are localized in these four covalent bonds, there are no free charges to carry electric current, and covalent crystals are, like ionic crystals, poor conductors of electricity.

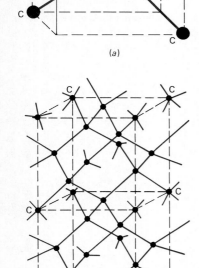

FIGURE 30.3 The crystal structure of diamond. (*a*) The tetrahedral bonding of one atom to four neighboring atoms in diamond. (*b*) The lattice structure of diamond, with each carbon atom bonded to four neighboring carbon atoms in a tetrahedral array. The incomplete bonds are directed toward more distant C atoms not shown in the figure.

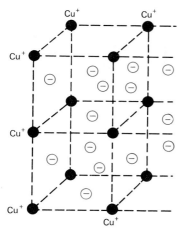

Metallic Crystals   Metallic crystals differ greatly from ionic and covalent crystals. Here the valence electrons are free to move throughout the whole crystal, while the positive ions remain fixed in position. One can view the electrons as a negatively charged fluid that "glues" together the positive ions in the crystal, as shown in Fig. 30.4 for copper. This results in a lowering of energy that makes the metallic crystal stable.

Since metallic crystals contain large numbers of free electrons, which can carry both charge and energy with them, metallic crystals are good conductors of both electricity and heat. A useful model that works well in describing this fluid of free electrons is called the *free-electron gas model*. It treats the metal as a "gas" of electrons free to move around in the periodic electric field provided by the stationary positive ions, as in Fig. 30.4.

FIGURE 30.4   A metallic crystal—copper, which has a face-centered cubic structure. The valence electrons of the copper atoms move freely through the crystal.

TABLE 30.1   Melting Points of Various Crystals

Type of crystal	Typical example	Melting point (K)
Ionic	NaCl	1074
Covalent	Diamond	4100
Metallic	Copper	1356
Molecular	O_2	55

Molecular Crystals   In molecular crystals the elementary units out of which the crystal is formed are not atoms but molecules. At low temperatures solid oxygen crystals can be formed. These consist of ordinary O_2 molecules loosely bound together by weak attractive forces (called *van der Waals forces*) to form a crystal. Van der Waals forces are electrical in nature and arise when one molecule with an electric dipole moment at some instant of time induces a dipole moment in another molecule. The two molecules therefore weakly attract each other. Because the bonds in molecular crystals are so weak, the melting point of O_2 is very low (55 K). Other molecular crystals include hydrogen, nitrogen, carbon dioxide, and most organic crystals. Table 30.1 shows the significant difference in the melting point of molecular crystals compared to the other crystal types discussed above.

30.2   Thermal Properties of Solids

TABLE 30.2   Thermal Conductivities K (at approximately 20°C)

Material	Thermal conductivity [W/(m·C°)]
Metallic crystals:	
Aluminum	238
Copper	380
Silver	417
Gold	290
Tungsten	169
Nonmetallic crystals:	
NaCl	6.7
CsI	1.3
KCl	6.7
MgO	33
SiO_2	7.1

In covalent and ionic crystals the principal mechanism for heat conduction is the vibration of the atoms in the crystal lattice. If the crystal is not in thermal equilibrium, the atoms in the hotter regions vibrate more strongly than those in the cooler regions, causing a transfer of energy as one atom passes on vibrational energy to its neighbors.

As Table 30.2 shows, the thermal conductivities of metallic crystals exceed those of ionic and covalent crystals by one or two orders of magnitude.* This is because of the contribution made to the thermal conductivity by the free electrons in the metals.

In 1853 the German physicists G. H. Wiedemann and R. Franz observed that a good thermal conductor is also a good electric conductor. This observation is incorporated into the following equation, referred to as the *Wiedemann-Franz law*:

$$\frac{K}{\sigma} = \frac{\pi^2}{3}\left(\frac{k}{e}\right)^2 T \tag{30.1}$$

where K is the thermal conductivity, σ the electric conductivity, k the Boltzmann

---

*The symbol K is used here for thermal conductivity to distinguish it from k, the Boltzmann constant, which also occurs in Eq. (30.1).

constant, e the charge on the electron, and T the absolute temperature. Interestingly, this equation predicts that the ratio of K to σ should be the same for all metals at the same temperature, independent of all other properties of the metal.

Molar Heat Capacities of Solids

Table 30.3 lists the *molar heat capacities* for some common solids. We see that many solids have a molar heat capacity close to 6 cal/(mol·C°). This rule for a time seemed to be so universal that it was called the *law* of Dulong and Petit. We now know that this "law" is an approximation valid only at high temperatures.

To calculate the molar heat capacity of a solid classically, we consider a crystalline solid consisting of N atoms bound together in a crystal by forces of the kind discussed in the preceding section. When any atom is displaced from its equilibrium position in the crystal lattice, it experiences a force that, to a first approximation, is proportional to the atom's displacement. The atom therefore executes simple harmonic motion, as do all its neighbors because they are coupled to the first atom by the interaction forces in the crystal. The simple harmonic motion of one atom is therefore passed on to the next atom, and an elastic wave propagates through the crystal like a sound wave moving through air.

The total energy content (which varies with temperature) of a crystal will then consist of the vibrational energy of the atoms in the crystal lattice. If the temperature changes, so too does the internal energy of the solid, and the change in internal energy per unit change in temperature is what we mean by the *heat capacity* of the solid.

The lattice heat capacity may be found by considering that the energy of the lattice is the sum of the energies of N simple harmonic oscillators. Since each atom in the lattice is free to move in three dimensions, we would expect it to have a kinetic energy of $\frac{3}{2}kT$, as would a gas atom at the same temperature (Sec. 9.8). As we saw in Sec. 11.2, the average kinetic energy and the average potential energy of a simple harmonic oscillator are equal, and hence the total energy should be $3kT$ per atom. Since there are N atoms, the total energy of the crystal is

$$E = N(3kT) = 3NkT \tag{30.2}$$

The classical heat capacity per mole (the *molar heat capacity*; see Sec. 14.9) is then

$$C_V = \frac{1}{n}\frac{\Delta E}{\Delta T} \tag{30.3}$$

where n is the number of moles. Now, from Eq. (30.2), we have

$$\Delta E = 3Nk\,\Delta T \qquad \text{or} \qquad \frac{\Delta E}{\Delta T} = 3Nk$$

and so

$$C_V = \frac{1}{n}(3\,Nk) = \frac{N}{n}(3k) = N_A(3k) = 3R \tag{30.4}$$

since Avogadro's number N_A is equal to N/n, the number of atoms in a mole. Also the gas constant per mole is

$$R = \left(8.31\,\frac{\text{J}}{\text{mol·C°}}\right)\left(\frac{1\text{ cal}}{4.18\text{ J}}\right) = 1.99\text{ cal/(mol·C°)}$$

and so $C_V = 3R = 5.97$ cal/(mol·C°) $\simeq 6$ cal/(mol·C°)

which is the "law" of Dulong and Petit.

TABLE 30.3 Molar Heat Capacities of Solids (20°C)

Substance	Molar heat capacity C_V [cal/(mol·C°)]
Aluminum	5.82
Copper	5.85
Silver	6.09
Tungsten	5.92
Lead	6.32
Carbon (diamond)	1.46
Silicon	4.73

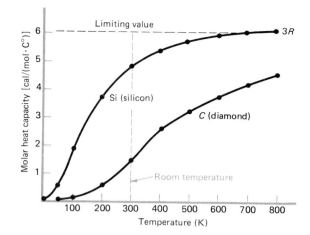

FIGURE 30.5 Variation of molar heat capacity with temperature for silicon and diamond.

Classical physics therefore predicts that the lattice molar heat capacity should be the same for all solids, in agreement with much of the data in Table 30.3. But classical physics cannot explain the low values of C_V for substances like diamond and Si at 20°C or the variation in the molar heat capacity with temperature shown in Fig. 30.5.

A successful theoretical treatment of the lattice molar heat capacity was given by Einstein in 1906. Peter Debye (1884–1966) extended Einstein's theory in 1912 and developed what is now usually referred to as the *Debye theory of specific heats*.

Debye Theory

The Debye theory is based on the same kind of consideration used by Max Planck in explaining blackbody radiation (Sec. 27.2). The atoms in a crystal can vibrate at a great variety of frequencies up to a cutoff frequency determined by the lattice spacing of the atoms and the sound velocity. The atomic oscillators have *quantized energies*, and cannot change their energies continuously, but only in quantum jumps of $nh\nu$, where n is an integer and ν is the frequency. At very low temperatures kT is very much less than $h\nu$ for most of the vibrational frequencies, and the atomic oscillators cannot absorb enough energy to make a quantum jump from a lower to a higher vibrational energy level. As a result, when the temperature changes, the energy of most atoms at low temperatures does not change with it, and so $\Delta E/\Delta T$ is not equal to $3Nk$ but to some smaller value. The molar heat capacity therefore starts from zero and gradually increases with temperature since, as kT increases, more and more oscillators can absorb enough energy to increase their vibrational energies by making the necessary quantum jumps. Finally the temperature becomes so high that all the atoms are able to absorb energy, $\Delta E/\Delta T$ becomes constant, and the molar heat capacity is equal to $3R$. This kind of behavior explains the experimental curves shown in Fig. 30.5.

This result was obtained in a rigorous, quantitative fashion by Debye, with results in excellent agreement with experiment. The molar heat capacity of solids thus provides another important confirmation of the validity of the quantum theory. The energy quanta associated with the vibrations of a crystal lattice are called *phonons* by analogy with the photons produced by electromagnetic vibrations.

30.3 The Band Theory of Solids

When an atom is placed in close proximity to other atoms in a crystal lattice, the interactions among the atoms' electrons change drastically the energy levels from those in the isolated atom. If there are N atoms in the crystal, then when the atoms are far apart, there are N electrons (one in each atom) occupying identical energy levels in the N atoms. When the atoms are arranged at their true interatomic distances in the crystal, however, each energy level in the crystal is broadened by the interactions into N very closely spaced levels. Since N is usually so large (of the order of Avogadro's number), the result is a continuous distribution of energies over a limited energy range. This range of allowed energies is called an *allowed energy band*. Electrons in solids can only possess energies in these allowed energy bands. The regions between these bands, or the energy gaps, represent energies that cannot be possessed by any electron in the crystal. Such regions are called *forbidden energy bands*, since quantum-mechanical calculations show that no electron can possess such energies. This is shown schematically in Fig. 30.6. Whereas in the Bohr atom we have discrete, allowed energy levels, with forbidden energy regions between them, in crystals we have energy bands with forbidden energy regions (or bands) between them.

The differences between electric conductors, insulators, and semiconductors are due basically to the arrangement of their energy bands and how these bands are filled by the available electrons. The conductivity of solid materials varies enormously from one substance to another. Thus the resistivity (the reciprocal of the conductivity) of a typical insulator such as quartz is about 10^{17} $\Omega \cdot$m, while that of most metals is on the order of 10^{-8} $\Omega \cdot$m, a ratio of 10^{25}.

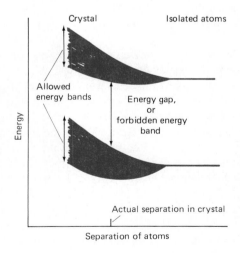

FIGURE 30.6 Allowed and forbidden energy bands in crystals. The interaction of the closely spaced atoms in the lattice broadens the atomic energy levels into energy bands as the atoms come closer together in the crystal.

Insulators

Suppose we have the situation shown in Fig. 30.7 for a particular solid crystal. The lower energy band, which is the band occupied by the outer electrons in the atoms and called the *valence band*, is completely filled with the number of electrons permitted by the Pauli exclusion principle for an array of N atoms. Above this band is another energy band, called the *conduction band*, which corresponds to an excited state of the system and which is almost empty. If the energy gap between the valence band and the conduction band is large, say, 4 to 5 eV, then the crystal is an *insulator*.

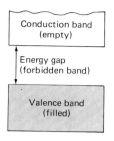

FIGURE 30.7 Energy-band structure for an insulator. (Energy increases vertically upward in this figure and in similar figures below).

Conductors

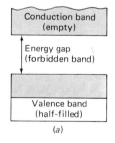

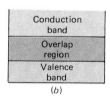

If an electric field is applied to such a crystal, electrons can only be accelerated by the field if they can increase their speed and hence their energy. But if all the electrons are already in filled energy bands, there are no empty higher energy states within reach that will enable the electrons to increase their energies. They therefore cannot be accelerated, and so no current flows. A current could flow only if the electrons were given enough energy to jump the energy gap from the valence to the conduction band, and usually not enough energy is available for this purpose.

An ionic crystal such as NaCl is an example of a good insulator. The energy bands of the crystal are derived from the energy levels of the isolated Na^+ and Cl^- ions. But both these ions have a closed-shell structure, so the valence band in NaCl is completely filled with electrons. Since there is a large energy gap between the valence band and the conduction band in the alkali halides, solid NaCl is an insulator.

In other cases, as in some metallic crystals, the valence band is not completely filled. Here electrons in the valence band can be raised to higher energy states in empty portions of the band by the application of an electric field. The field can therefore accelerate the electrons and produce an electric current.

The valence and conduction bands may also overlap. In this case excited states exist to which electrons can be raised by an electric field, and the material is again a conductor. Diagrams of these two kinds of band structure leading to electric conductivity are shown in Fig. 30.8.

The sodium crystal provides a good example. The sodium atom has an electron configuration $(1s)^2(2s)^2(2p)^6(3s)$. Since the $3s$ subshell can hold two electrons, this subshell is only half-filled. This situation carries over to the Na crystal, where the $3s$ energy band is also only half-filled. Electrons can therefore be promoted to higher energies within this $3s$ valence band and thus conduct current. Also, in Na the $3p$ conduction band overlaps the $3s$ band for the equilibrium atomic separation in the Na crystal. This allows electrons from the valence band to move into the conduction band and thus contribute to the current. Hence sodium is a good conductor. Magnesium, with a configuration $(1s)^2(2s)^2(2p)^6(3s)^2$, has its $3s$ band filled but is still expected to be a conductor because its $3p$ conduction band overlaps the $3s$ valence band. We know from experiment that magnesium is indeed a good conductor.

FIGURE 30.8 Two kinds of band structure leading to electric conductivity: (a) a crystal with a partially filled valence band; (b) a crystal in which the valence band and the conduction band overlap.

Semiconductors

A third possibility exists: There may be a gap between a filled valence band and an empty conduction band, but this gap may be very small, say, less than 1 eV, as in Fig. 30.9. In such cases we have a *semiconductor*. The energy gap is in this case small enough that an increase in temperature can provide sufficient thermal energy to excite electrons into the conduction band, where they can conduct current. Semiconductors therefore have electric conductivities larger than those of insulators but smaller than those of metallic conductors, as can be seen in Table 18.1.

The important semiconductors silicon and germanium are found in column IV of the periodic table. Carbon, which is also in column IV, has an electron configuration $(1s)^2(2s)^2(2p)^2$, and we might expect carbon to be a good conductor because

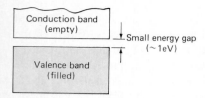

FIGURE 30.9 Band structure of a semiconductor. The valence band is filled, but the energy gap between the valence band and the conduction band is very small.

of the four unfilled $2p$ states in the atom. In diamond, however, carbon forms four covalent bonds with its four nearest neighbors, using its two $2s$ and two $2p$ electrons. This fills the valence band completely. Since there are no electrons left over, the conduction band remains empty. At the same lattice spacing (0.15 nm) in the diamond crystal, the energy gap is a large 7 eV, and so diamond is a very good insulator.

For silicon, which has two $3s$ and two $3p$ electrons, the band structure is the same as in carbon, but the binding is less tight, the lattice spacing being 0.23 nm, and the energy gap only about 1.1 eV. For germanium, with two $4s$ and two $4p$ electrons, the lattice spacing is 0.24 nm and the energy gap only 0.75 eV. Because of these small energy gaps in silicon and germanium, they are *semiconductors*. The resistivity of Ge is about 0.5 $\Omega \cdot$m and of Si about 200 $\Omega \cdot$m, but these values vary greatly with the purity of the material. In both cases the resistance falls with increasing temperature, since the increasing temperature creates both more electrons in the conduction band and more holes (see below) in the valence band.

Holes

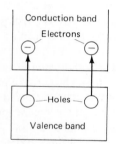

When an electron makes a transition from a normally full valence band to a normally empty conduction band in a crystal, it leaves behind a "hole" in the sea of electrons filling the valence band, as shown in Fig. 30.10. As a result an electron of lower energy in the valence band can increase its energy by moving up into this hole. This also corresponds to electric conduction, for the electron increases its energy by the move. In many ways the holes behave in a manner similar to positively charged particles, for the deficiency of one elementary negative charge in the valence band is the equivalent of a positive charge there. Thus both the electrons in the conduction band and the holes in the valence band help carry the electric current.

FIGURE 30.10   Production of electron-hole pairs.

30.4   Semiconductors

There are two basic kinds of semiconductors, intrinsic and extrinsic. *Intrinsic semiconductors* are materials, such as silicon or germanium, in which there are no impurities and therefore the concentrations of negative charge carriers (electrons) and positive charge carriers (holes) are the same. *Extrinsic* (or *impurity*) *semiconductors* are materials in which the concentration of charge carriers of one sign (positive or negative) is increased by introducing impurity atoms into the crystal. Extrinsic semiconductors can be further classified as *n*-type or *p*-type.

n-Type Extrinsic Semiconductors

The electrical conductivity of a silicon crystal can be increased if impurity atoms are artificially forced into the crystal lattice, a process called "doping." Suppose we dope pure silicon with phosphorus (P), which has an electron configuration $(1s)^2(2s)^2(2p)^6(3s)^2(3p)^3$; that is, it has one more $3p$ electron than silicon has. Only four of the five $3s$ and $3p$ electrons are needed for chemical bonding, as shown in Fig. 30.11, and one electron is therefore left over. Since such a crystal has an excess of electrons compared with the number in a pure Si crystal, a phosphorus-doped Si crystal is called an *n-type semiconductor* (where *n* means *negative*). Note that, although such a crystal has an excess of free electrons compared with pure Si, the total crystal is still electrically neutral; each P nucleus has one positive charge more than each of the Si nuclei, which compensates for the excess electrons.

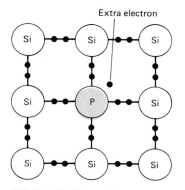

FIGURE 30.11 A silicon crystal doped with phosphorus.

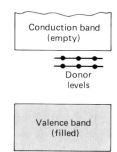

FIGURE 30.12 Donor levels in an *n*-type semiconductor.

These extra electrons occupy energy levels just slightly below the conduction band, since they are very loosely bound to their parent *P* atoms. These energy levels are called *donor levels* because the effect of the phosphorus impurity is to *donate* electrons to the conduction band without leaving holes in the valence band, as would happen in an intrinsic (pure) semiconductor. These donor levels are shown in Fig. 30.12. The conductivity of such a doped semiconductor can be controlled by the amount of impurity added. The addition of 1 part per million of phosphorus or arsenic can cause a significant change in the conductivity. Electrons in the donor levels can easily be raised to the conduction band by thermal or optical excitation, since the energy needed is only about 0.01 eV. These electrons then contribute to the current when an electric field is applied.

_p_-Type Extrinsic Semiconductors

Similarly, a boron-doped Si crystal has a deficiency of electrons compared with a normal Si crystal. The boron atom has an electron configuration $(1s)^2(2s)^2(2p)$ and so has one less valence electron than does silicon. When a boron atom replaces a Si atom in the crystal lattice, as in Fig. 30.13, electron deficiencies, or holes, are created in the crystal. Since a deficiency of negative charge is equivalent to an excess of positive charge, such a crystal is called a *p-type semiconductor* (with *p* signifying *positive*). Note that this does not mean that the Si crystal is charged, only that it has an excess of positive over negative charge carriers.

The effect of doping Si with boron is to create impurity *acceptor levels* just

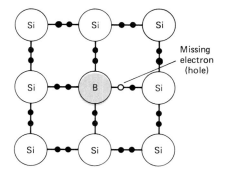

FIGURE 30.13 A silicon crystal doped with boron.

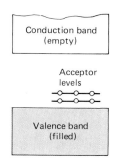

FIGURE 30.14 Acceptor levels in a *p*-type semiconductor.

above the filled valence band, as shown in Fig. 30.14. Electrons can then be excited thermally or optically from the filled valence band to these acceptor levels. This creates holes in the valence band that can be filled by other valence-band electrons moving to levels of higher energy in the band. This increase in energy of valence-band electrons when an electric field is applied leads to an electric current.

The number of extra electrons in an *n*-type semiconductor, or of holes in a *p*-type semiconductor, is usually much greater than the number of electron-hole pairs that can be created in a pure semiconductor crystal by temperature excitation of electrons from the valence band to the conduction band. Therefore in a doped semiconductor we have conduction due to both electrons and holes, with one dominating the other depending on the type of impurity added. If an electric field is applied, the current will consist of both *majority carriers* (electrons in *n*-type and holes in *p*-type semiconductors) and *minority carriers* (holes in *n*-type and electrons in *p*-type semiconductors).

These ideas can be checked by a Hall effect measurement of the type discussed in Sec. 19.3. Such a measurement verifies that the predominant charge carriers are truly negative in *n*-type and positive in *p*-type semiconductors. The measured Hall voltage also leads to a value for the number of charge carriers per unit volume and thus provides a convenient way of evaluating the success of any doping procedure in introducing *n*- and *p*-type carriers into a crystal.

30.5 Solid-State Devices

The movement of electrons and holes in semiconductors can be manipulated for practical purposes to rectify, amplify, and detect electric signals—tasks once handled exclusively by vacuum tubes. The main difference between a vacuum tube and a *solid-state device* like a transistor is that the flow of charge occurs not in a vacuum but through solid materials. Solid-state devices include *pn* junctions, diodes, transistors, and a large variety of more complicated devices.

pn Junctions

If a region of a semiconductor doped to be *p*-type is put in contact with a region of the same semiconductor doped to be *n*-type, we have a *pn junction*. The properties of such a junction are basic to the operation of all solid-state devices. A rough sketch of such a junction is shown in Fig. 30.15a. At the left side of the junction there is initially a surplus of holes to make the region *p*-type. At the right side of the junction there is a surplus of electrons, so the region is *n*-type. Both regions, however, are still electrically neutral, as shown, because of the charges on the impurity atoms in the crystal.

When these two regions are put in intimate contact, electrons from the *n*-type semiconductor flow across the junction and fill in holes in the *p*-type region, as

FIGURE 30.15 A *pn* junction. (a) Two pieces of semiconductor material—one *p*-type with an excess of holes and the other *n*-type with an excess of electrons—are placed in contact. The uncircled + and − signs indicate the charges on the impurity atoms that make the two sides of the junction electrically neutral. (b) After contact is made, electrons flow from *n*-type to *p*-type and an internal electric field directed from *n*-type to *p*-type results.

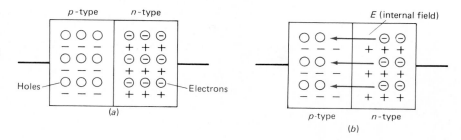

shown in Fig. 30.15*b*.* The *n*-type therefore becomes positively charged and the *p*-type negatively charged. This creates a situation similar to that in a charged parallel-plate capacitor, and an internal electric field *E* is set up from the *n*-type to the *p*-type regions, as shown in the figure. This voltage tends to prevent any further movement of charge across the junction, and all charge flow stops.

Junction Diodes

Suppose, now, that after all charge flow has ceased, we connect a battery across the junction. If the positive terminal of the battery is connected to the *n*-side and the negative terminal to the *p*-side, so as to increase the potential difference between the two regions (a condition called *reverse bias*), as shown in Fig. 30.16*a*, the electrons in the *n*-type are attracted to the positive pole of the battery and move away from the junction. Similarly, the holes in the *p*-type are attracted to the negative pole of the battery and also move away from the junction. Thus the junction area is emptied of both free electrons and holes, and so no current can flow through the junction. The *pn* junction acts as an insulator for an external voltage applied with reverse bias.

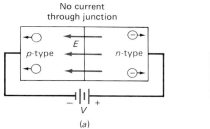 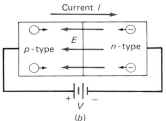

FIGURE 30.16 (*a*) a *pn* junction with reverse bias: The junction acts as an insulator. (*b*) A *pn* junction with forward bias: The junction acts as a conductor.

If, however, the polarity of the battery is reversed and the positive terminal is connected to the *p*-type side of the junction (a condition called *forward bias*), as shown in Fig. 30.16*b*, the electrons are pushed by the battery in the direction of the junction, and so are the holes. Electrons therefore combine with holes at the junction, while other electrons are driven by the battery through the external circuit out of the *p*-type and into the *n*-type material, creating new holes in the *p*-type region. As long as the battery is connected, this movement of electrons in one direction and holes in the other direction continues, and we have an electric current. The dependence of the current on the magnitude and direction of the applied voltage is shown in Fig. 30.17.

A *pn* junction therefore acts like a *rectifier* that allows current to flow in one direction but not in the other. If connected to a source of alternating current, the *pn* junction will rectify that current and convert it to pulsating direct current, as shown in Fig. 30.18. Such a *pn* junction has *two* essential parts and is therefore often called a junction *diode*, or *diode rectifier*. Each diode is represented by a symbol indicating the direction in which the diode will pass a conventional current.

FIGURE 30.17 Flow of current through a *pn* junction: The junction acts as a rectifier, which allows current to flow only in the direction of forward bias (note that the voltage scales differ for positive and negative voltages).

*Since the electron makes a transition from an energy state above the energy gap to a state below it, an amount of energy of the order of the gap energy is released in the process. This energy is released in the form of either phonons (thermal vibrations of the crystal) or photons. This photon emission is responsible for the light emitted by light-emitting diodes (LEDs) (Sec. 30.6).

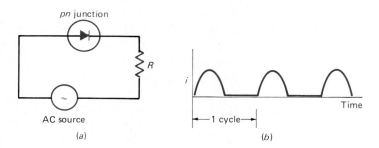

Junction Transistors

Junction transistors consist of a crystal of one type of semiconductor (*n*-type or *p*-type) sandwiched between two crystals of the opposite type to form transistors designated as either *pnp* or *npn* transistors. The three regions of the transistor are called the *emitter*, the *base*, and the *collector*, as shown in Fig. 30.19 for an *npn* transistor. In practice the emitter is much more heavily doped than are the base and the collector, and the base region between emitter and collector is very thin. The circuit symbols for *npn* and *pnp* transistors are given in Fig. 30.20, where the arrows indicate the direction of conventional current flow to or from the emitter.

Let us see how an *npn* transistor can be used as an amplifier. A constant voltage V_{CE} is maintained between collector and emitter by the battery of terminal voltage V_C in Fig. 30.21. Another constant voltage V_{BE} is applied between the base and the emitter by the battery V_B. Since the base is therefore maintained at a positive voltage with respect to the emitter, electrons flow from the emitter into the base. Because the base region is so thin, most of these electrons flow right across it into the collector, which is maintained at a positive voltage with respect to the emitter. A large electron current therefore flows from emitter to collector, which is equivalent to a conventional current I_C from collector to emitter. If a weak signal v_B is applied to the base, as shown in the diagram, it results in a large change in the collector current and therefore a large change in the voltage drop v_L across the output resistor R_L. The transistor therefore functions as an amplifier.

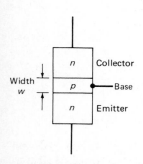

FIGURE 30.19 An *npn* transistor.

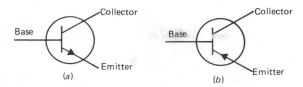

FIGURE 30.20 Circuit diagram symbols for (a) an *npn* transistor and (b) a *pnp* transistor. The arrows indicate the direction of the conventional current.

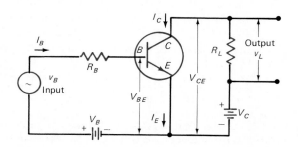

FIGURE 30.21 A transistor amplifier. A small input voltage v_B produces a large output voltage v_L.

FIGURE 30.22 Left to right, William Shockley (1910–1988), John Bardeen (born 1908), and Walter H. Brattain (born in 1902), who did the fundamental work on semiconductors that led to the development of the transistor late in 1947 at the Bell Telephone Laboratories in Murray Hill, New Jersey. This photograph was taken in 1948. For this work they received the 1956 Nobel Prize in physics. Bardeen was the first person to win two Nobel Prizes in physics, since he also shared in the 1972 prize for his theoretical work on superconductivity. *(Courtesy of AT&T Bell Laboratories.)*

A *pnp* transistor behaves in the same fashion except that holes move instead of electrons, and the voltages must be reversed to allow for this fact.

The junction transistor was first developed by Shockley, Bardeen, and Brattain at the Bell Telephone Laboratories in New Jersey in 1947 (see Fig. 30.22). Since this solid-state device "transferred" a current across a high-resistance material (a resistor), they called it a *transfer-resistor*, abbreviated to *transistor*.

30.6 Applications of Solid-State Devices in Modern Technology

Simple transistor voltage amplifiers are used a great deal in modern communication systems. Usually a number of amplifiers are connected in series so that the output of one amplifier is fed into another amplifier, resulting in large voltage gain and large output voltages. In this way the very small voltage produced by the motion of a needle over a record can lead to the large amounts of power needed to drive a system of stereo loudspeakers. This extra energy does not, of course, come from the transistors but rather from the dc power supplies applied to the transistors. The function of the transistor is to convert this energy input into the useful amplification of an input signal.

Because of their obvious advantages, transistors have almost completely replaced vacuum tubes for most purposes. First of all, transistors require no vacuum and no glass enclosures; they are therefore much sturdier than vacuum tubes. Second, they require no heated filament and so run at lower temperatures and need no "warmup" time. Third, they last longer, if properly treated, since there are no parts to wear out, in contrast to the filament in a vacuum tube. Most importantly, they take up much less space than do vacuum tubes.

Silicon Solar Cells

The *pn* junction (Fig. 30.15) is the basic element in a silicon solar cell. Suppose that light from the sun or some other source is incident on such a junction. If the photons in the light have energies above 1.1 eV, which corresponds to the energy gap between the valence and conduction bands in pure silicon, they raise electrons from the valence to the conduction band and thus create excess electrons in the

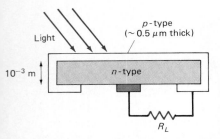

FIGURE 30.23 A photovoltaic (solar) cell. Light falling on the *p*-type region causes a current through the resistor R_L.

conduction band and excess holes in the valence band. Let us suppose that these electron-hole pairs are produced by light incident on the *p*-type region of the *pn* junction, as in Fig. 30.23. In this region the extra holes produced by the light are insignificant compared with the large number of holes already present in this region (about $10^{13}/cm^3$) due to the doping of the silicon. The extra electrons, however, are unusual in the *p*-type region, and even though some of them immediately combine with holes, others drift to the junction boundary. Now, as we saw in Sec. 30.5, there is an internal field E at the junction directed from the *n*-type to the *p*-type region. Such an electric field will therefore push electrons from the *p*-type side to the *n*-type side, and cause a buildup of extra negative charge in the *n*-type region and extra positive charge in the *p*-type region. The *n*-type region becomes like the negative terminal of a battery, and the *p*-type region like the positive terminal. The effect of the light on the *pn* junction is therefore to produce an electric voltage between the two sides of the junction. In practice this voltage is about 0.6 V. If an external circuit is connected between the two sides of the junction through a load resistor R_L, current continues to flow as long as light is incident on the junction.

A similar process occurs when light is incident on the *n*-type region (try to outline this process yourself).

The current that flows depends on the number of electron-hole pairs produced and hence on the intensity of the incident light. Since any imperfection in the body of the silicon crystal provides sites at which electrons and holes recombine before the electrons can flow through the external circuit, very pure silicon crystals are needed in Si solar cells.

Solar cells seem in many ways to be the ideal solution to our growing need for energy, particularly for electric energy, since they make use of energy that would otherwise be in great part wasted, and they do so without producing thermal, chemical, or radioactive pollution. The main problems remaining with solar cells are, first, the intermittent nature of the energy provided by the sun and our failure to develop large-capacity storage mechanisms for electric energy; and, second, the inefficiency and consequent high cost of solar cells, which make them at present noncompetitive with conventional sources of electricity except in unusual situations (see Fig. 30.24). This latter situation is gradually improving, and there is hope that solar cells will make significant contributions to our energy needs in the near future.

FITURE 30.24 A weather station powered by silicon solar cells. Remote locations like this one atop Mammoth Mountain in California are ideal for such uses of solar cells. The 12 solar-cell arrays shown here produce 5 W each, or 60 W in all. *(Courtesy of NASA.)*

Integrated Circuits

FIGURE 30.25 A silicon chip containing a full 32-bit microprocessor. This dime-sized chip contains 150,000 transistors and offers processing power comparable to today's minicomputers. *(Courtesy of AT&T Bell Laboratories.)*

When semiconductors replaced vacuum tubes in most electronic applications, the circuits were said to be "transistorized," which was another way of saying "miniaturized," since transistors occupy so little space compared with old-fashioned vacuum tubes. A further step in miniaturization has been the development of *integrated circuits* (ICs).

The resistance of a semiconductor material depends upon its impurity concentration. The capacitance of reverse-bias diodes can be changed by changing the voltage applied between the emitter and the base of a transistor. This makes it possible to combine an array of transistors, resistors, and capacitors into an *integrated circuit* on a single tiny "chip" of silicon. These integrated circuits can be mass-produced to accomplish almost any desired function. Integrated circuits containing large numbers of operating components can be assembled on a square silicon chip a few millimeters on a side (see Fig. 30.25). The drive toward smaller and smaller chips continues unabated.

Recently semiconductor companies have been building even more sophisticated and large-scale ICs. Such chips are very costly to develop, but once designed they can be stamped out by the millions, almost like cookies, at a very low cost per chip. This is the reason for the drastic lowering in the prices of hand calculators and digital wristwatches in recent years. ICs have improved the performance and reliability of a variety of electronic devices like computers and stereos, while at the same time lowering their prices.

Microprocessors

In 1969 a Japanese computer company asked the American firm Intel to develop a range of medium-size ICs for use in a series of calculators of gradually increasing complexity. Intel's engineers (in particular T. Hoff and F. Faggin) had the brilliant idea that it would be simpler in the long run to design a single general-purpose IC for all the calculators and to make this IC programmable, with the program unique to each kind of calculator and contained on another chip which in the final calculator would tell the IC what to do. Thus was born the *microprocessor*, an IC which can execute any sequence of coded instructions provided to it and which is extremely versatile. Thus the Pac-Man game in an arcade uses exactly the same microprocessor as a Timex, Heathkit, Radio Shack, or Xerox microcomputer. The so-called personal computer is nothing but a microprocessor chip joined to an array of memory ICs, a typewriter keyboard, and a video screen. All these developments are the results of spectacular improvements in semiconductor technology, and progress in this field continues. Personal computers augmented by very high speed integrated circuit (VHSIC) technology can now perform 200 million operations per second and have the power of the big mainframe computers that once filled large rooms (Fig. 30.26).

FIGURE 30.26 A memory chip that can store more than 4 million bits of data, the equivalent of 400 pages of double-spaced typewritten text. It operates very fast, requiring only 65 ns to access a bit of data stored in its memory. At this rate, it can read out the data in all its 4,194,304 memory cells in only 0.27 s. *(Courtesy of IBM Corporation.)*

Light-Emitting Diodes (LEDs)

LEDs consist of a *pn* junction on which has been impressed a large forward bias (Sec. 30.5). Electrons cross the junction, driven by the forward bias, to the *p* side, recombine with the holes there, and give up their energy in the form of radiation. Just as a photon is emitted when an electron in an atom falls from a higher energy state to a lower energy state, so too a photon can be emitted when an electron in the conduction band makes a transition to a hole in the valence band. By judicious choice of the semiconductor material and the impurities, radiation can be produced in the frequency range desired for a particular application.

LEDs are small, require little electric power, and never burn out. They are often used in displays on calculators, digital clocks, and other electronic instruments.

Extensions of these ideas have led to the solid-state laser, which produces coherent radiation of a fixed frequency. Such lasers are very small and are commonly used in compact disc (CD) players.

Solid-State Detectors

The development of semiconductors and ICs has led to great improvement in the accuracy and dependability of the instruments used in atomic, nuclear, and elementary-particle physics—detectors, pulse-height analyzers, microcomputers, etc. Of particular importance here is the semiconductor detector widely used in nuclear and elementary-particle physics.

The *semiconductor detector* (or *surface-barrier detector*) is simply a reverse-bias diode, or *pn* junction, in which a very thin region of *p*-type silicon is backed by a thicker region of *n*-type silicon. It is therefore very similar to the solar cell shown in Fig. 30.23. A voltage of about 100 V across a thickness of 0.1 mm produces a large electric field of 10^6 V/m in the *p*-type material, but no current flows because this applied voltage is opposed to the internal voltage appearing across the junction, as discussed in the preceding section. If, however, an ionizing particle produces electron-hole pairs in the *p*-type region, the electrons are pulled across the junction, flow through the external battery, and combine with the newly produced holes in the *p*-type region. We therefore have a sudden current pulse that can be used to count incident particles.

Such semiconductor detectors have the advantages of small size (about 2 cm in diameter and 1 to 2 mm thick); high efficiency, because the solid is much denser than the gas in gas-filled counters; fast action, with response times of a few nanoseconds; and better ability to distinguish between particles of different energies than is possible with gas-filled detectors. Silicon detectors are used to detect electrons and other charged particles; germanium detectors to which lithium ions have been added [so-called Ge(Li), or "jelly," detectors] are used to detect gamma rays and can provide an energy resolution better than 10 percent.

30.7 Superconductivity: Theory and Experiment

Superconductivity: A phenomenon in which the electric resistance of a material suddenly drops to zero at a temperature characteristic of the material.

This effect is illustrated for mercury, the first material in which it was observed, in Fig. 30.27. Mercury goes superconducting at a critical temperature (T_C) of about 4.2 K. It was the availability of liquid helium, which boils at 4.2 K, that made it possible to observe this unusual effect. A superconducting material below its critical temperature has no Joule heating (I^2R) losses because $R = 0$, and so the material can carry a current indefinitely without the need for a driving voltage.

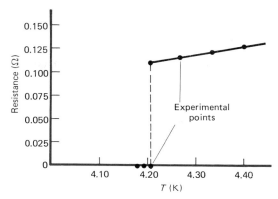

FIGURE 30.27 Superconductivity. The electric resistance of a mercury wire (below freezing point of mercury) falls suddenly to zero at about 4.20 K.

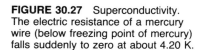

TABLE 30.4 Superconducting Temperatures Achieved before 1986

Metal	T_C (K)
Al (aluminum)	1.2
In (indium)	3.4
Sn (tin)	3.7
Pb (lead)	7.2
Nb_3Al	17.5
Nb_3Sn	17.9
Nb_3Ge	23.2

After the Dutch physicist H. Kamerlingh Onnes (1853–1926) discovered superconductivity in 1911, little progress was made until the 1930s in understanding superconductivity or in finding materials that became superconductors at higher temperatures than Hg. Since that time it has been found that some metal alloys have critical temperatures considerably higher than pure elements, the highest T_C reached by 1973 being 23.2 K for Nb_3Ge, as seen in Table 30.4. A fairly standard use of superconducting wire, in this case Nb_3Sn, to produce a magnetic field, is shown in Fig. 30.28.

This very complex phenomenon of superconductivity was explained in 1956 by the American physicists John Bardeen, Leon Cooper, and J. Robert Schrieffer (see Fig. 30.29) in a theory now called the *BCS* theory after its developers. They showed how, using quantum mechanics and our knowledge of the solid state, it is possible to explain the superconducting state as a large-scale quantum state in which the motions of all the electrons in the conduction band are locked together. In this state, at sufficiently low temperatures the electrons are coupled together in pairs. One electron interacts with the crystal lattice and "perturbs" it. The perturbed lattice then interacts with the second electron in the pair in such a way that there is an attraction between the two electrons which overcomes the Coulomb repulsion between them. This electron-electron interaction via the lattice produces an energy gap between the superconducting state, in which the electrons act collectively, and the normal (nonsuperconducting) state in which they act individually. The energy needed to excite the system to the higher, nonsuperconducting state and destroy the superconductivity is not available at very low temperatures, and the material therefore becomes a superconductor at such temperatures.

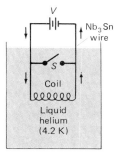

FIGURE 30.28 A superconducting coil made of Nb_3Sn wire, which has a superconducting temperature of 17.9 K. When immersed in a liquid helium bath at about 4 K, it can carry current indefinitely without any Joule heating losses, and produce a magnetic field.

FIGURE 30.29 John Bardeen (born 1908), Leon Cooper (born 1930), and J. Robert Schrieffer (born 1931), who developed the BCS theory of superconductivity at the University of Illinois in 1956. In 1972 they received the Nobel Prize in physics for this work. It is noteworthy that 1956 was the year Bardeen received his first Nobel Prize for his work on the transistor. *(Photo courtesy of AIP Niels Bohr Library.)*

High-Temperature Superconductivity

The BCS theory led some physicists to conclude that T_C could never exceed 30 K. Then in January 1986, K. A. Müller and J. G. Bednorz (Fig. 30.30) at the IBM Research Laboratory in Zurich, Switzerland, found that an oxide of barium, lanthanum, and copper became a superconductor at about 35 K. Such a discovery was completely unexpected, especially since this oxide was a ceramic and did not conduct electricity at room temperature. Many physicists had thought Müller and Bednorz were wasting their time working on such ceramics, but they were proved wrong when Müller and Bednorz received the Nobel Prize in physics for 1987—a tribute both to their physical insight and their psychological stamina.

Since 1986 many researchers around the world have reported seeing superconductivity above 90 K in a variety of other materials. The most famous ceramic discovered to be superconducting is yttrium-barium-copper oxide ($YBa_2Cu_3O_7$) shown by C. W. (Paul) Chu and his coworkers at the University of Houston to be superconducting at 95 K. The T_C's of the best-established high-T_C materials are plotted in Fig. 30.31. It is clear from the graph how rapidly progress has been made in this field in only a few years.

Some researchers claim to have seen superconductivity at temperatures as high as 240 K, but these claims are as yet unconfirmed. Before a material is accepted as a true high-T_C superconductor by physicists, four pieces of experimental evidence are needed: (1) The material must display zero resistance below its T_C; (2) it must expel any magnetic field from its interior below its critical temperature—a phenomenon called the *Meissner effect*; (3) the physical behavior of the material must be reproducible in time; (4) the physical behavior of the material must be reproducible when it is cycled above and below its T_C. In many claims for new high-T_C materials, all four of these criteria are not satisfied. There is such a tremendous upsurge of research activity in this field, however, that it is likely that materials with higher transition temperatures will be found in the years ahead.

These developments are of great fundamental and practical importance. Superconducting materials can be used to construct stronger, smaller, and cheaper

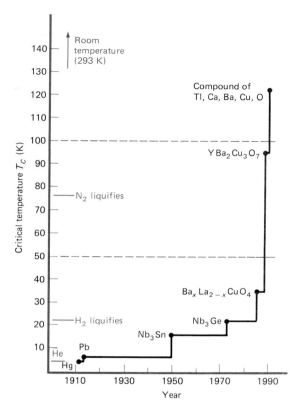

FIGURE 30.30   K. Alex Müller (left) and J. Georg Bednorz of the IBM Research Laboratory in Zurich, Switzerland. They received the 1987 Nobel Prize in physics for their discovery of a new class of high-temperature superconductors. *(Courtesy of IBM Corporation.)*

FIGURE 30.31   A plot of critical superconducting temperatures (T_C) as a function of the year in which the discovery was made. Notice the dramatic increases in T_C since the discovery by Müller and Bednorz of ceramic superconductors in 1986.

electromagnets, to eliminate power losses in underground transmission lines, to make electric cars and magnetically levitated trains more practical, and to reduce greatly the cost of nuclear magnetic resonance scanners for medical diagnosis. All these applications now require liquid-helium cooling to produce superconductivity in the wires used, and liquid helium is expensive and may some day be in short supply. Superconducting coils at 90 K can be cooled with liquid nitrogen, which costs no more than ordinary milk. And if room-temperature (293 K) superconductivity should be achieved, no coolant at all will be required. This is the goal of the thousands of physicists now working in this "hot" field of high-temperature superconductivity.

The BCS theory has been very successful in explaining the superconductivity of metals and alloys, but it is unclear that it will be able to explain superconductivity in ceramics. For one reason, the structure of these ceramics is still not fully understood, and until the location of all the atoms in the crystal lattice is clear, it is impossible to make much theoretical progress.

There are experimental and theoretical problems to be solved, and technological mountains to be climbed, before the great promise of high-temperature superconductivity will be realized. Most ceramic materials are very brittle and difficult to form into wire, coils, and other required shapes. It is also unclear whether these high-T_C superconductors can carry the sizable currents needed for some applications. Despite these problems, however, all indications are that high-temperature superconductors may some day be as important for the world's economy, health, and convenience as radio receivers, television sets, and electronic computers.

Josephson Effect

In 1962 the British physicist Brian Josephson (see Fig. 30.32) predicted an effect that was first observed the next year. If a very thin layer (about 2 nm) of an oxide (an insulator) is sandwiched between two superconductors and a voltage V is applied across the two superconductors, a current flows (see Fig. 30.33). This current has two components: a continuous dc component that is found to persist even after the voltage source is removed, and an ac component present when the voltage source is attached. Remarkably, the frequency of this ac component is always exactly $\nu = 2eV/h$, no matter what materials are used to construct the Josephson junction.

This phenomenon gives a new way of measuring the ratio e/h to high accuracy using the equation

$$\frac{e}{h} = \frac{\nu}{2V} \tag{30.5}$$

Since the frequency can be measured easily to 1 part in 10^{12}, the ratio of the charge on the electron to Planck's constant can be obtained to the same accuracy as that to which the voltage can be measured. This yields a value for e/h several orders of magnitude more precise than any other experimental method.

The reproducibility and precision of the results obtained with Josephson junctions are so great that the standard volt can now be defined in terms of a frequency, using Eq. (30.5). This eliminates the need for standard cells or other physical standards of voltage. As of January 1, 1990, the volt is defined as that voltage which, when applied across a Josephson junction, leads to an alternating current of frequency $\nu = 2e/h = 4.835979 \times 10^{14}$ Hz.

On a more applied level, Josephson junctions have led to the development of very sensitive detectors and of compact, fast, and efficient memory units for computers (see Fig. 30.34). SQUID (*superconducting quantum interference device*) detectors, based on the Josephson effect, are now used to measure poisonous levels of iron in the liver so that iron buildup can be treated before it does serious harm. SQUIDS are also being used to study electric currents in the human brain in an attempt to determine the causes of epilepsy.

FIGURE 30.32 Brian Josephson (born 1940). Making use of the BCS theory of superconductivity, Josephson predicted in 1962 that both a direct current and an alternating current would appear when a voltage is applied across a thin insulating layer separating two superconductors. For this work Josephson shared in the 1973 Nobel Prize in physics. *(Courtesy of AIP Meggers Gallery of Nobel Laureates.)*

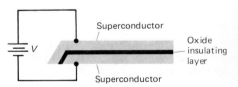

FIGURE 30.33 A Josephson junction. It consists of two superconducting layers separated by a very thin insulating layer.

FIGURE 30.34 A data processing chip containing 600 Josephson junctions to be used as fast switches in computers. The shirt button indicates the tiny size of the chip. *(Courtesy of AT&T Bell Laboratories.)*

Summary: Important Definitions and Equations

Crystal: A solid with a regular geometric shape in which the atoms making up the crystal arrange themselves in an ordered geometric pattern.

Lattice: An orderly array of points in space at which atoms are located.

Types of crystalline solids:
1 Ionic crystals: Those in which alternate sites in the lattice are occupied by ions of opposite charge.
2 Covalent crystals: Those in which all lattice sites are occupied by neutral atoms.
3 Metallic crystals: Those in which positive ions occupy all lattice sites, and the valence electrons move freely through the whole crystal.
4 Molecular crystal: Those in which all lattice sites are occupied by molecules, and in which the bonding between molecules is relatively weak.

Molar heat capacity of a solid

Law of Dulong and Petit (experimental): $C_V \simeq 6$ cal/(mol·C°)

Classical theory: $C_V = 3R \simeq 6$ cal/(mol·C°), independent of temperature

Debye theory: A quantum-mechanical theory which predicts that the molar heat capacity should be equal to zero for very low temperatures, and then increase gradually until, at very high temperatures, it approaches the classical value of 6 cal/(mol·C°).

Band theory of solids

Allowed energy bands: The values of the energy that are possible for an electron in a particular crystalline solid.

Forbidden energy bands: The gaps between the allowed energy bands; values of the energy that are not possible (or are forbidden) in a particular crystal.

Valence band: The energy band in which the valence electrons of the atoms in a solid are found.

Conduction band: An energy band in which the outermost electrons of the atoms are able to move freely through the crystal.

Insulator: A crystal in which the valence band is full and the conduction band is completely empty, and where there is a large energy gap between the valence and conduction bands.

Conductor: A crystal in which the valence band is not completely filled, or in which the valence and conduction bands overlap.

Semiconductor: A crystal in which the gap between the filled valence band and the empty conduction band is small, leading to properties intermediate between those of insulators and of conductors.

Hole: The absence of an electron from a previously filled band. Such a hole is equivalent in many ways to a positive charge.

Types of semiconductors

Intrinsic: Semiconductors in which there are no impurities and therefore the concentrations of negative and positive charge carriers are equal.

Extrinsic: Semiconductors in which the concentration of charge carriers of one sign has been increased by introducing impurity atoms into, or doping, the crystal.
n-Type: Semiconductors with an excess of electrons which occupy donor levels just below the conduction band and which can donate electrons to the conduction band.
p-Type: Semiconductors with an excess of holes. This leads to acceptor levels just above the filled valence band, to which electrons can be excited from the valence band.

Solid-state devices

pn Junction: A region in which a *p*-type semiconductor and an *n*-type semiconductor are in contact.

Junction diode: A *pn* junction that acts as a rectifier by allowing current to flow in only one direction through the junction.

Junction transistor: A crystal of one type of semiconductor material sandwiched between two crystals of the opposite type to form either a *pnp* or *npn* transistor.

Silicon solar cell: A *pn* junction that produces a voltage when visible light falls on it.

Integrated circuit (IC): A circuit that incorporates numerous solid-state components like diodes and transistors into one unit, often consisting of a single chip of silicon.

Microprocessor: A single, general-purpose IC that can be programmed to perform a variety of different electronic tasks.

Light-emitting diode: A forward-biased *pn* junction that emits radiation when electrons recombine with holes.

Semiconductor detector: A *pn* junction that produces a current pulse whenever an ionizing particle produces electron-hole pairs in the semiconductor.

Superconductivity: The property of certain metals, alloys, and metallic oxides by which their electric resistance falls to zero at a critical temperature (T_C) characteristic of the material.

Josephson effect: If a very thin layer of an insulating oxide is sandwiched between two superconductors and a constant voltage V is applied across the two superconductors, the current that flows has an ac component of frequency $\nu = 2\,eV/h$, regardless of what material is used in the Josephson junction.

Questions

1 Verify that Eq. (30.1) is dimensionally correct.

2 Compare the electric and thermal conductivities of copper with those of aluminum. Are the ratios of electric conductivity to thermal conductivity about the same in the two cases? Is this what you would expect? Why?

3 As the temperature increases, the lattice vibrations of a crystal increase and hence the collisions of the electrons with the positive ions in the lattice also increase. This would be expected to *increase* the resistivity, as it does in metals. Why, then, does the resistance of semiconductors like germanium and silicon *decrease* as the temperature increases?

4 In a lithium crystal, each atom has three electrons, each having three degrees of translational freedom. If the electrons in a lithium crystal are treated as a free-electron gas, what would be the expected molar heat capacity of the crystal on the basis of classical physics?

5 Why do all metals appear shiny when visible light falls on them?

6 How can an *n*-type semiconductor crystal have a surplus of free electrons if the crystal remains electrically neutral?

7 What elements are suitable to use as impurities in silicon to make it an *n*-type extrinsic semiconductor? To make it a *p*-type semiconductor?

8 Is it possible to use a junction diode as an amplifier? Why?

9 (*a*) Is the resistance of a *pn* junction the same for forward bias as for reverse bias?

(*b*) If not, in which direction is the resistance greater?

10 If the earth were covered with a huge number of solar cells to convert solar energy into electric energy, would this have any effect on the radiation balance between the earth and the sun?

11 The efficiency of silicon solar cells in converting light to electric energy is limited to about 20 percent. Can you provide some arguments to indicate what happens to the other 80 percent of the energy in the incident photons?

12 Good conductors like copper and gold are still not superconductors even at temperatures as low as 0.05 K. Such metals at normal temperatures have very weak interactions between the electrons and the lattice. This makes them good conductors. Can you explain, on the basis of the BCS theory, why such good conductors are poor superconductors?

Additional Readings

Chalmers, Bruce: "The Photovoltaic Generation of Electricity," *Scientific American*, vol. 235, no. 4, October 1976, pp. 34–43. A clear and well-illustrated account of the physics and technology of solar cells.

Geballe, Theodore H., and J. K. Hulm: "Superconductors in Electric Power Technology," *Scientific American*, vol. 243, no. 5, November 1980, pp. 138–172. A discussion of the many uses to which high-field, high-current superconductors can be put in the electric-power industry.

Hamakawa, Yoshihiro: "Photovoltaic Power," *Scientific American*, vol. 256, no. 4, April 1987, pp. 86–92. An optimistic report on the possibility of megawatt power plants based on solar cells within 15 years.

Hazen, Robert M.: *The Breakthrough: The Race for the Superconductor*. Summit Books, New York, 1988. A fascinating account of the race to produce and understand high-temperature superconductors, by a major actor in the drama.

Khurana, Anil: "Superconductivity Seen above the Boiling Point of Nitrogen," *Physics Today*, vol. 49, no. 4, April 1987, pp. 17–23. A factual account of recent stirring developments in the physics of superconductors.

Langenberg, Donald N., Douglas J. Scalapino, and Barry N. Taylor: "The Josephson Effects," *Scientific American*, vol. 214, no. 5, May 1966, pp. 30–39. A very clear discussion of the theory of the Josephson effect and of the experiments performed to verify this theory.

Maranto, Gino: "Superconductivity: Hype vs. Reality," *Discover*, August 1987, pp. 23–32. A well-illustrated article emphasizing the problems to be solved before high-temperature superconductors become a practical reality.

Pierce, John R.: *Quantum Electronics*, Doubleday Anchor, Garden City, N.Y., 1966. Contains a good elementary discussion of semiconductor devices.

Weber, Robert L.: *Pioneers of Science: Nobel Prize Winners in Physics*, The Institute of Physics, Bristol, England, 1980. This useful book contains brief biographies of many of the physicists whose work is described in this chapter, including Bardeen, Brattain, Shockley, Cooper, Schrieffer, and Josephson.

Wolsky, Alan M., Robert F. Giese, and Edward J. Daniels: "The New Superconductors: Prospects for Applications," *Scientific American*, vol. 260, no. 2, February 1989, pp. 61–69. An up-to-date account of the commercial prospects for high-temperature superconductors and of the problems involved in realizing these prospects.

In addition, the September 1977 issue of *Scientific American* (vol. 237, no. 3) contains 11 excellent articles on microelectronics.

CHAPTER 31

High-Energy and Elementary-Particle Physics

Can we ever hope to find the right way? Nay more, has this right way an existence outside our illusions? . . . I answer without hesitation that there is, in my opinion, a right way, and that we are capable of finding it. Our experience hitherto justifies us in believing that nature is the realisation of the simplest conceivable mathematical ideas. I am convinced that we can discover by means of purely mathematical constructions the concepts and the laws connecting them with each other, which furnish the key to the understanding of natural phenomena.

Albert Einstein (1879–1955)

Nature often turns out to be more complicated than physicists would like it to be. in the 1930s the accepted model of the atom was a marvelously simple one, consisting of only three "elementary" particles, the electron, the proton, and the neutron. We now know that both the proton and the neutron are not elementary at all but are made up of other particles, which may, in turn, be made up of additional particles not even dreamed of today. Over the last 30 years physicists have been uncovering a remarkable variety of particles that appear to make up the universe and have been attempting to introduce order into the seeming chaos which has resulted from their discovery. This is the field of *elementary-particle physics*, one of the most important frontiers of physics research today.

The goal of physicists working in elementary-particle physics is *understanding*, not applications. It is not clear that any practical applications will ever emerge from the research being done in this field, as did the laser from quantum and electromagnetic theory and the transistor from solid-state physics. But to rule out such practical applications at the present stage of our knowledge would be presumptuous. What is becoming clear is that a better understanding of elementary-particle physics will shed immense light on the origin and evolution of the universe, since they are united by the laws of physics, which are common to both.

In what follows we will give a broad overview of this complicated and developing field, without spending too much time on details that may change before this book is published. Our intention in what follows is to give the student a realization that many questions in physics remain to be answered.

31.1 The Four Forces of Physics

In previous chapters we discussed the four basic forces that govern all the interactions of physics. These are summarized in Table 31.1, together with their relative strengths (for two protons at a distance of 10^{-15} m) and their ranges. The two

TABLE 31.1 The Four Known Forces of Physics

Force	Present in	Relative strength	Range
Gravitational	All interactions	10^{-38}	Very great $\left(\dfrac{1}{r^2}\right)$
Weak nuclear	All weak interactions (e.g., beta decay)	10^{-13}	$\sim 10^{-18}$ m
Electromagnetic	All interactions involving two electrically charged particles	10^{-2}	Very great $\left(\dfrac{1}{r^2}\right)$
Strong nuclear	All strong interactions (e.g., nuclear binding)	1	$\sim 10^{-15}$ m

forces that most concern us here are the strong and weak nuclear forces. Both of these are very short-range forces, as is clear from experiment, but we are at present unable to write any equation describing the dependence of these forces on distance, since we do not understand them well enough to do so. For this reason elementary-particle physicists prefer to talk about "interactions" rather than the "forces" that cause these interactions. We follow this practice in the present chapter.

Strong interactions occur only within a class of particles called *hadrons* (to be defined later), of which the proton and the neutron are prime examples, and do not depend on the electric charge of the particles involved. *Weak interactions* occur within another class of particles called *leptons* (also to be defined later), of which the electron is the prime example. Weak interactions occur also between leptons and hadrons and between hadrons and hadrons, and they too are independent of electric charge.

The four basic forces of physics are necessary to explain the universe around us. The gravitational force holds the planets in their orbits about the sun and keeps us from flying off the earth into space. The electromagnetic force holds electrons in their atoms and binds atoms together into molecules. The strong force keeps protons and neutrons in stable nuclei, despite the electromagnetic repulsion between the protons. The weak force, in some ways the most mysterious of the four, governs the processes of radioactive beta decay of nuclei, is crucial to the energy-producing processes going on in the stars, and apparently played a very important role in the building up of the heavier elements from light nuclei when the universe began.

The Strong Nuclear Force

The strong nuclear force acts between the nucleons in atomic nuclei to hold them together. The approximate form of the force between two protons at short distances is shown in Fig. 31.1. It can be seen that only for distances R in the range from 1.0 to 2.5×10^{-15} m is this nuclear force sufficiently strong to counterbalance the repulsive electrostatic force between the two protons and produce a net attraction.

To learn more of this nuclear force, we need probes capable of "seeing" details in structure of size less than 10^{-15} m. This means particles with energies greater than about 100 MeV, for only such particles have de Broglie wavelengths as small as 10^{-15} m. This is the realm of *high-energy physics*. In recent years the availability of powerful particle accelerators to produce energies above 100 MeV has opened up this new energy domain and led to the discovery of many new particles. For this reason physicists use the terms *high-energy physics* and *elementary-particle physics* almost interchangeably to describe this exciting field.

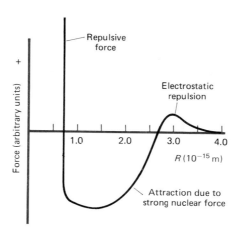

FIGURE 31.1 The force between two protons at short distances inside a nucleus. For distances greater than 3.0×10^{-15} m the electrostatic Coulomb repulsion dominates the interaction. For distances between about 1.0 and 2.5×10^{-15} m the strong nuclear force produces a strong attraction of the two protons for each other.

The Weak Nuclear Force and the Neutrino

The weak nuclear force is known to be responsible for the beta decay of radioactive nuclei. A typical beta-decay process is the one involved in radiocarbon dating:

$$^{14}_{6}C \rightarrow \, ^{14}_{7}N \, + \, ^{0}_{-1}e \tag{31.1}$$

Here both the charge and the number of nucleons are conserved, as required.

The fact that an electron is produced in this nuclear-decay process may seem to contradict our previous statement that there are no electrons in the nucleus. Electrons do not exist in nuclei any more than photons exist in atoms; both are produced just before they are observed. In beta decay a neutron in the nucleus is transformed into a proton and an electron inside the nucleus, and this electron is then immediately ejected from the nucleus as a beta ray. Free neutrons are known to decay in a similar fashion:

$$^{1}_{0}n \rightarrow \, ^{1}_{1}p \, + \, ^{0}_{-1}e \tag{31.2}$$

To determine the expected kinetic energy of the beta particle in Eq. (31.1), we use the masses of the neutral atoms. Since there are six extranuclear electrons in the carbon atom, after beta emission the N nucleus also has only six extranuclear electrons. A seventh electron is produced in the decay process as the emitted beta particle, however, and so the mass of $^{14}_{7}N$ (minus one electron) and $^{0}_{-1}e$ together is simply the mass of one neutral $^{14}_{7}N$ atom. The mass of $^{14}_{6}C$ is found experimentally to be 14.00324 u and that of $^{14}_{7}N$ to be 14.00307 u (see Appendix 4). Hence the mass defect is

$$\Delta m = 14.00324 \text{ u} - 14.00307 \text{ u} = 0.00017 \text{ u}$$

The energy equivalent of this, using 1 u = (931.5 MeV), is 158 keV. Almost all this goes into the kinetic energy of the electron, since the daughter nucleus recoils with a very low velocity because it has a large mass compared with that of the electron. We would therefore expect to detect a very large number of electrons all with the same kinetic energy, 158 keV, when a sample of $^{14}_{6}C$ decays.

In 1914 the British physicist James Chadwick (1891–1974), who later discovered the neutron, found that beta particles from $^{14}_{6}C$ did not have such well-defined energies. On the contrary their energy spectrum ranged continuously from practically zero to a maximum value close to 158 keV, as shown in Fig. 31.2. This seemed a direct violation of the principle of conservation of mass-energy in

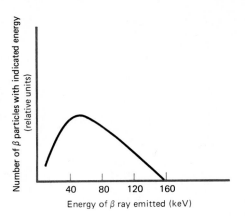

FIGURE 31.2 Observed energy spectrum of the beta particles emitted by $^{14}_{6}C$. The beta particles do not all have the same energy (158 keV, according to theoretical calculations) but a continuum of values ranging from near zero to 158 keV.

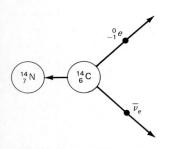

FIGURE 31.3 Beta decay of $^{14}_{6}C$. The antineutrino $\bar{\nu}_e$ carries both energy and momentum away from the reaction.

the beta-decay process, and at the time even Niels Bohr expressed doubts about the validity of the principle of conservation of energy in such processes.

Wolfgang Pauli could not accept such a violation of one of physics' most treasured laws, and in 1930 he postulated for beta decay the emission of another particle that has come to be called the *neutrino* (the name was given by Enrico Fermi and means ''the little neutral one''). According to Pauli, neutrinos are particles with no electric charge and zero (or very small) rest mass, but they are not photons. They are emitted paired to beta particles in such a way that the sum of the energies of the neutrino and the beta particle is always the same, for example, 158 keV for $^{14}_{6}C$. This reinstates the principle of conservation of mass-energy.

The requirements of the conservation laws for linear and angular momentum for beta decay are also met by the introduction of this elusive particle. In 1934 Enrico Fermi worked out a detailed theory of beta decay that agreed well with experimental results. Neutrinos are designated by the Greek letter *nu* (ν). We therefore write the equation for the decay of $^{14}_{6}C$ more properly as

$$^{14}_{6}C \rightarrow {}^{14}_{7}N + {}^{0}_{-1}e + \bar{\nu}_e \tag{31.3}$$

where the subscript on $\bar{\nu}_e$ means that this neutrino is emitted along with an electron and the bar indicates that it is actually an antineutrino (see Sec. 31.4). The complete reaction is thus as shown in Fig. 31.3.

Experiments on the decay of tritium ($^{3}_{1}H$) according to the equation

$$^{3}_{1}H \rightarrow {}^{3}_{2}He + {}^{0}_{-1}e + \bar{\nu}_e \tag{31.4}$$

indicate that the mass of the electron neutrino is no more than a few electronvolts, compared to the 0.5-MeV mass of the electron. Whether the electron neutrino mass is exactly zero is not as yet clear, but it seems certain that it is less than $10^{-4}m_e$.

Decay of the Free Neutron

Equation (31.2) for the decay of the free neutron must also be modified to include a neutrino and becomes

$$^{1}_{0}n \rightarrow {}^{1}_{1}p + {}^{0}_{-1}e + \bar{\nu}_e \tag{31.5}$$

The force that leads to the decay of the free neutron according to this equation is very weak. This can be seen from the fact that the average lifetime of a free neutron is about 17 min, which is extremely long on a nuclear scale, particularly when compared with lifetimes associated with strong nuclear reactions. For example, half-lives of artificially produced radioactive nuclei can be as small as 10^{-12} s. This is why the force involved in beta decay is called the *weak nuclear force*.

The reason the neutrino is so hard to detect is that it interacts with matter only through this weak nuclear force. As a result a lead shield used to absorb neutrinos effectively would have to be 2×10^{14} mi, or 34 light-years, in thickness! Also, billions of neutrinos pass through our bodies every second without leaving a trace.

Detection of the Neutrino

Because of its weak interaction with matter, it is very difficult to detect a neutrino, and it was in 1956, 26 years after Pauli postulated its existence, that the first neutrino was detected in the laboratory. This observation was made by the American physicists Clyde L. Cowan, Jr. (1919–1974) and Frederick Reines (born 1918), using the enormous flux of antineutrinos from a nuclear reactor in which neutron-rich nuclear fission products decay by emitting beta particles and antineutrinos. Cowan and Reines worked at the Savannah River reactor in South Carolina and detected the antineutrinos by their interaction with the protons in water molecules:

$$\bar{\nu}_e + {}^1_1 p \rightarrow {}^1_0 n + {}^{\ 0}_{+1} e \tag{31.6}$$

where ${}^{\ 0}_{+1} e$ (or e^+) is a *positron*, or positive electron, and $\bar{\nu}_e$ an antineutrino produced in a beta-decay process in the reactor.

Reines and Cowan used a gigantic tank containing layers of water mixed with cadmium chloride sandwiched between layers of a clear liquid scintillator, as in Fig. 31.4. In all they used some 10 tons of liquid scintillator and 500 photomultiplier tubes. Antineutrinos first were captured by the protons in water molecules, leading to the production of neutrons and positrons, in accordance with Eq. (31.6). The positrons were slowed down by collisions and brought almost to rest in about 10^{-9} s. These positrons then combined with electrons in the water, and a pair-annihilation process led to the production of two gamma rays. Since the two gamma-ray photons had to go off in opposite directions to conserve momentum, they passed through the layers of liquid scintillator on each side of the water and produced light flashes that could be seen by the photomultiplier tubes observing these layers. Then 10 μs or so later the neutron had slowed down sufficiently to

FIGURE 31.4 Schematic diagram of the apparatus used by Cowan and Reines to detect the neutrino. By observing the emitted gamma rays, they could argue back to the capture of an antineutrino by a proton in the water sandwiched between the two layers of liquid scintillator.

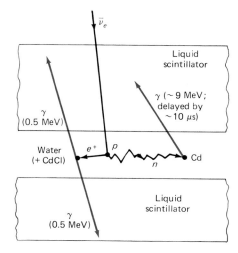

be captured by a cadmium atom in the water, leading to the emission of more gamma rays, which could again be detected by the scintillators and phototubes.

If the antineutrinos emerging from the reactor were actually taking part in the reaction indicated in Eq. (31.6), Reines and Cowan expected to find two simultaneous scintillation pulses in adjacent layers, followed by another scintillation pulse (or pulses) 10 μs or so later. Moreover, each of these three pulses would have a characteristic identifying energy which the photocells could measure—0.5 MeV for each of the annihilation photons and about 9 MeV for the neutron-produced photon (or photons). This is exactly what they observed, with one event being observed about every 20 min, as predicted by theory. This confirmed the weakness of the neutrino's interaction with other particles.

The pioneering work of Cowan and Reines in 1956 was the first convincing experimental evidence for the existence of the neutrino. More recently it has become possible to obtain huge numbers of neutrinos from the decay of particles produced in large accelerators and to study more easily their interactions with matter.

31.2 Particle Accelerators

The high-energy accelerators in use today are mainly of two types, linear accelerators, whose origins can be traced back to the Van de Graaff generator (Fig. 17.7), and circular accelerators, which are greatly improved versions of Lawrence's original cyclotron (Fig. 19.18).

Linear Accelerators

Electrons present unique problems as nuclear probes because of their small mass. If electrons are forced to move in a circular path in a particle accelerator, they soon reach such high speeds that they lose great amounts of energy by radiation, since they undergo continuous acceleration when moving in a circle. This imposes a serious limitation on the maximum energy that can be reached by electrons in a circular machine. As a result linear accelerators have until recently been the choice for accelerating electrons. The electron accelerator at the Stanford Linear Accelerator Center (SLAC) at Stanford University in Palo Alto, California (Fig. 31.5), accelerates electrons by subjecting them to voltage pulses as they move down a

FIGURE 31.5 Photo of the world's highest-energy linear electron accelerator at the Stanford Linear Accelerator Center (SLAC), Palo Alto, California. The straight section at the top of the photo is 2 mi long. (*Courtesy of SLAC.*)

straight, 2-mi-long evacuated tube. The SLAC accelerator can produce electrons of 47 GeV energy, and the energy available to produce new particles can be made double this by colliding the electron beam head-on with a 47-GeV positron beam in what is now called the Stanford Linear Collider (SLC). Note that 47 GeV is 90,000 times the rest energy of the electron and that at such energies the speed of the electron differs from that of light in vacuum by less than 1 part in 10^{10}.

Circular Accelerators: Synchrotrons

The particle accelerators that have achieved the highest energies are synchrotrons. A *synchrotron* keeps particles moving in a circular evacuated tube of fixed radius, with the confining magnetic field varying in time as the particles attain higher and higher speeds. This is accomplished by an automatic feedback system that monitors the motion of the circulating bunches of particles and adjusts the accelerating field to hold the beam radius constant as the magnetic field increases.

A diagram of a synchrotron is shown in Fig. 31.6. The evacuated tube, which is in the form of a circular ring, is divided into three different kinds of sections, one group of which provides the accelerating electric field, a second of which contains the magnets that bend the beam into a circular orbit, and a third of which uses different kinds of magnets to focus the particles into a narrow beam. The particles are introduced into the synchrotron by an injection system that often includes a smaller accelerator to give the beam an initial energy before it enters the synchrotron. The beam can be extracted when desired by turning on deflecting magnets to direct the beam out of the ring toward the target.

As we saw in the case of the cyclotron, the momentum of the accelerated particles is given by Eq. (19.6),

$$p = mv = Bqr$$

where B is the strength of the bending magnetic field and r is the radius of the orbit. Since in the synchrotron the particles finally move at speeds close to that of

FIGURE 31.6 Schematic diagram of the beam tube of a synchrotron, showing the three kinds of tube sections required to accelerate and focus the protons or other particles being used as probes.

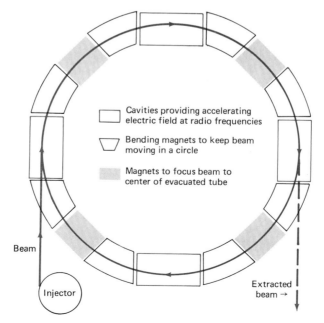

Cavities providing accelerating electric field at radio frequencies

Bending magnets to keep beam moving in a circle

Magnets to focus beam to center of evacuated tube

Beam

Injector

Extracted beam →

light, their total energy is given by Eq. (26.29), in which we can neglect E_0^2 compared to E^2 and obtain:

$$E = pc = Bqrc \tag{31.7}$$

The maximum energy attainable is, therefore, directly proportional to the radius of the fixed orbit in which the particles move and to the magnetic field. Physics laboratories around the world are engaged in a friendly rivalry to build the world's largest synchrotron in the hopes of unraveling more of nature's mysteries (see Sec. 31.9).

The world's largest synchrotron is now the Large Electron-Positron Collider (LEP) at the CERN (Centre Européen pour la Recherche Nucléaire) laboratory near Geneva, Switzerland. This 27-km-circumference accelerator is also by far the world's largest scientific instrument (Fig. 31.7). In head-on collisions of electrons with positrons, it has achieved energies of 100 GeV. Plans call for the LEP to reach energies of 200 GeV by 1992.

The largest synchrotron in the United States is the Tevatron Collider at Fermilab in Batavia, Illinois (Fig. 31.8). This 6.4-km-circumference machine has achieved energies of 1.8 TeV in head-on proton-antiproton collisions.

FIGURE 31.7 (left) Air view of the Large Electron-Positron Collider (LEP) at CERN, just outside Geneva, Switzerland, with the Jura mountains in the background. The large circle shows the extent of the LEP tunnel, 27 km in circumference, straddling the French-Swiss border. The smaller circle marked SPS is an older accelerator used to inject high-speed particles into the LEP. (*Courtesy of CERN.*)

FIGURE 31.8 (right) The proton synchrotron at Fermilab in Batavia, Illinois. The large circle is the accelerator that has a diameter of 2.0 km. (*Courtesy of Fermilab.*)

Types of Synchrotrons

Fixed-Target Machines A *fixed-target machine* (Fig. 31.9) fires the accelerated particles, usually protons, at a fixed target, such as liquid hydrogen, and the interactions of the protons with the nucleons in the target are studied. The problem with such an approach is that most of the beam energy goes into the kinetic energy of the particles emerging from the interaction, for momentum must be conserved. Since the energy E_u useful for creating new particles is the difference between the initial and final kinetic energies, too little energy remains for creating new particles of interest to researchers.

When the beam particles and the target particles have equal masses, as when protons bombard a liquid-hydrogen target, application of the principles of conservation of momentum and energy and of special relativity leads to the result that the energy E_u useful in producing new particles is related to the total energy of the bombarding particles E and to their rest masses m_0 by the equation

$$E_u = \sqrt{2m_0 c^2 E} \tag{31.8}$$

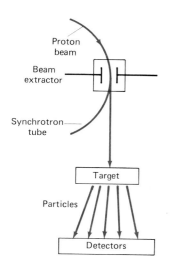

FIGURE 31.9 A fixed-target synchrotron. The exit beam from the synchrotron strikes a stationary target. In this case little energy is available to be converted into the mass of new particles.

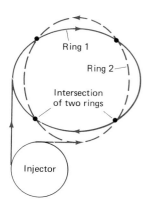

FIGURE 31.10 (above) A colliding-beam synchrotron. The diagram shows how two beams of protons or other particles traveling in opposite directions at close to the speed of light are made to collide at the intersection of two storage rings. In this case most of the energy is available for the creation of new particles.

FIGURE 31.11 (right) Photograph of one of the particle collision regions in the CERN proton accelerator. The two beams collide at the intersection of the two beam tubes at the center of the photo. (*Courtesy of CERN.*)

where E is the energy of the beam particles. For example, for a proton of energy 500 GeV from the Fermilab accelerator, the useful energy would be only

$$E_u = \sqrt{2m_0c^2E} = \sqrt{2(931 \text{ MeV})(500 \text{ GeV})} = 31 \text{ GeV}$$

and so only about 6 percent of the beam energy would go into creating new particles.

If it were possible to increase the beam energy to 1000 GeV, the useful energy would be only

$$E_u = \sqrt{2(931 \text{ MeV})(1000 \text{ GeV})} = \sqrt{2}(31 \text{ GeV}) = 44 \text{ GeV}$$

Hence the greatly increased size and cost of the 1000-GeV accelerator would increase the useful energy by only 13 GeV. To overcome this difficulty, colliding-beam machines are now used.

Colliding-Beam Machines In a *colliding-beam machine* (Fig. 31.10), two beams of high-speed particles traveling in opposite directions are collected in "storage rings" and then made to collide head-on. Since the total linear momentum in the head-on collision of two particles of equal mass is zero, very little of the energy goes into the kinetic energy of the scattered particles, and most of the energy is available to cause high-energy interactions that produce new particles. For example, two beams of 500-GeV protons would make available close to 1000 GeV for the creation of new particles.

In early 1983 in the CERN interacting storage rings (ISR) machine a 270-GeV proton beam was made to collide head-on with a 270-GeV antiproton beam, making available 540 GeV for the creation of new particles (Fig. 31.11). A 200-ton particle detector revealed the existence of heavy particles predicted by current theories of the elementary particles (see Sec. 31.8). The accelerator at SLAC has also been converted into a colliding-beam machine, the Stanford Linear Collider, and a similar conversion of the proton synchrotron at Fermilab has led to the Tevatron Collider.

31.3 Pions and Muons

Most high-energy collision processes of the kind discussed in the preceding section produce pi mesons, or (as they are usually called) *pions*, in addition to other particles. Pions have masses about one-seventh the mass of the proton, or approximately $275m_e$, and come in three varieties: positively charged, negatively charged, and neutral. All pions are unstable, with the charged pions having lifetimes of about 10^{-8} s and the neutral pions lifetimes of only 10^{-16} s. Pions are designated by the Greek letter *pi* (π).

The pion was predicted by the Japanese physicist H. Yukawa (see Fig. 31.12) in 1935. At that time Yukawa was even able to estimate its mass on the basis of the known range of the strong nuclear force, i.e., about 10^{-15} m. The pion was first seen experimentally in 1974 by the British physicist C. F. Powell (see Fig. 31.13) in photographic-emulsion tracks of cosmic-ray events, in which very high-energy particles from outer space, mainly protons, bombarded atoms in the earth's atmosphere and produced pions. Soon after Powell's discovery, for which he received the 1950 Nobel Prize, pions were also produced in high-energy particle accelerators.

Pion reactions that occur frequently when protons are bombarded by high-energy protons are the following:

$$p + p \rightarrow p + p + \pi^0 \qquad\qquad (31.9)$$

$$p + p \rightarrow p + n + \pi^+ \qquad\qquad (31.10)$$

Charge is always conserved in such interactions, as are momentum and mass-energy. The extra energy required to produce the pions, and to change a proton into a neutron, comes from the large kinetic energy of the bombarding proton.

The unstable pions have very short lives and decay into *muons* and neutrinos according to the reactions

$$\pi^+ \rightarrow \mu^+ + \nu \qquad \text{and} \qquad \pi^- \rightarrow \mu^- + \bar{\nu} \qquad\qquad (31.11)$$

A muon is lighter than a pion, and a negative muon is very similar to an electron

FIGURE 31.12 (left) Hideki Yukawa (1907–1981), the Japanese physicist who received the 1949 Nobel Prize in physics for his 1935 prediction of the pion. This photo was taken in 1950 in Yukawa's office at Columbia University in New York City. (*Photo by International News Photos; courtesy of Edward L. Bafford Photography Collection, University of Maryland Baltimore County Library.*)

FIGURE 31.13 (right) Cecil Powell (1903–1969), the British physicist who in 1947 discovered the pion, the particle predicted by Yukawa. (*Courtesy of AIP Meggers Gallery of Nobel Laureates.*)

in almost all respects except that a muon is unstable (with a lifetime of 2×10^{-6} s) and its mass is 207 times the electron mass. Muons are sometimes called "heavy electrons" for that reason. The muon was discovered in the cosmic radiation before the pion was discovered and was at first mistaken for the particle predicted by Yukawa. It was soon found, however, that the muon played no role in strong interactions and hence could not be the Yukawa particle.

31.4 Antiparticles and Neutrinos

In 1928 the British physicist P. A. M. Dirac (Fig. 31.14) fused special relativity and quantum theory into relativistic quantum mechanics. This theory predicted the existence of the *positron*, or antielectron, which is identical to an electron in every way except that it has a positive rather than a negative charge. The positron was first observed in 1932 in a cloud chamber exposed to cosmic rays by the American physicist Carl Anderson (see Fig. 31.15), who a few years later also discovered the muon.

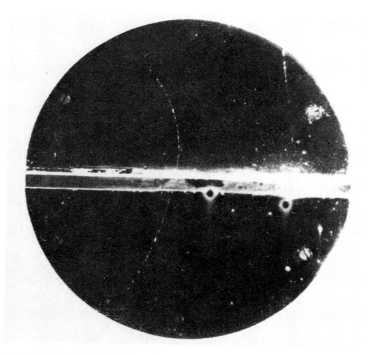

FIGURE 31.14 P. A. M. Dirac (1902–1984). He combined the theory of relativity with Schrödinger's wave mechanics to create relativistic wave mechanics. This led to his prediction of the existence of the positive electron (the positron), the negative proton (the antiproton), and other antiparticles. For these and other discoveries he shared with Schrödinger the 1933 Nobel Prize in physics. (*Photo courtesy of Florida State University.*)

FIGURE 31.15 A cloud-chamber photograph by Carl D. Anderson (born 1905), which led to his discovery of the positron, for which Anderson received the 1936 Nobel Prize in physics. There is a magnetic field directed into the plane of the paper. The positron comes from below and crosses a lead plate, where it is slowed down and therefore deflected in a tighter circle above the lead plate, since $r = mv/Be$. The direction of the force on the particle by the magnetic field proves that the positron is positively charged (*Photo by Carl D. Anderson: courtesy of AIP Niels Bohr Library.*)

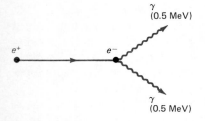

γ
(0.5 MeV)

e^+ e^-

γ
(0.5 MeV)

FIGURE 31.16 A pair-annihilation process. A positron, moving toward the right, collides with an electron, and all their mass is converted into energy in the form of two gamma rays, each with energy of about 0.5 MeV.

The positron is perfectly stable if left to itself, since there are no particles with lower mass than the positron except the neutrino and the photon, and these are uncharged. Hence a positron cannot decay into either of these without violating the principle of conservation of electric charge. If, however, a positron collides with an electron, they can "annihilate" each other, with their masses being entirely converted into energy in the form of two gamma rays with energies about 0.5 MeV each (Fig. 31.16). Similarly an electron-positron pair may be created from high-energy photons; if gamma rays with energies above 1 MeV pass through matter, electron-positron pairs are usually produced.

Positrons are not easily detected because they soon collide with the readily available free electrons and are annihilated. If positrons could be kept away from ordinary matter, however, they would last as long as electrons.

Other Antiparticles

Just as the electron has the positron as its antiparticle, so too all other particles like the proton, the neutron, and even the neutrino have their antiparticles, called the *antiproton*, the *antineutron*, and the *antineutrino*, respectively. These are usually designated by bars over the symbol for the particles, as $\bar{p}$, $\bar{n}$, and $\bar{\nu}$. For example, it is the antineutrino that arises from the beta decay of a neutron into a proton according to Eq. (31.5). The photon does not have an antiparticle, and is said to be its own antiparticle.

In 1955 Owen Chamberlain (born 1920) and Emilio Segrè (1905–1989) produced and detected antiprotons at the Bevatron accelerator in Berkeley, California. A short time later the antineutron was also found.

Atoms are possible that are made of nuclei containing antiprotons (negatively charged) and antineutrons, together with extranuclear positrons (positively charged). Such atoms make up what is called *antimatter*. As far as we can tell, there is no sizable amount of antimatter in the universe. If identical particles of matter and antimatter collide, they annihilate each other and the total mass of the two particles is converted into energy.

Different Kinds of Neutrinos

Neutrinos are always associated with weak interaction processes. In the beta decay of a nucleus either electrons or positrons are emitted. Accompanying the electrons are always antineutrinos, and accompanying the positrons are neutrinos. There are other neutrinos that accompany the muons produced in pion decays. These muon neutrinos are different from electron neutrinos, and again are of two varieties, as shown in Table 31.2. To complicate matters even further, tau particles (or *tauons*) have been discovered, and these have their own neutrinos.

TABLE 31.2 Types of Neutrinos

Kind of neutrino	Symbol	Associated with production of	Typical reaction
Electron neutrino	ν_e	Positrons	$p \rightarrow n + e^+ + \nu_e$
Electron antineutrino	$\bar{\nu}_e$	Electrons	$n \rightarrow p + e^- + \bar{\nu}_e$
Muon neutrino	ν_μ	μ^+	$\pi^+ \rightarrow \mu^+ + \nu_\mu$
Muon antineutrino	$\bar{\nu}_\mu$	μ^-	$\pi^- \rightarrow \mu^- + \bar{\nu}_\mu$
Tauon neutrino	ν_τ	τ^+	$\tau^+ \rightarrow e^+ + \nu_e + \bar{\nu}_\tau$
Tauon antineutrino	$\bar{\nu}_\tau$	τ^-	$\tau^- \rightarrow e^- + \bar{\nu}_e + \nu_\tau$

31.5 More New Particles

The tremendous energies available in cosmic rays and, since 1955, in particle accelerators have led to the discovery of large numbers of additional particles with exotic names like kaons, sigma particles, and eta mesons, most of which are highly unstable and have lifetimes of 10^{-8} s or less. Among these new particles are a group referred to as "strange particles" first detected in 1947 in cosmic rays. These particles participate in strong interactions but are *strange* in that they are always produced in pairs and have unexpected lifetimes.

All particles now known can be divided into three basic classes: the *photon*, which is unique since it is its own antiparticle; the *leptons*, which include the electron, the muon, and the neutrino; and the *hadrons*, which consist of a great variety of particles with masses equal to or greater than that of the proton.

Leptons: Particles like electrons, muons, and neutrinos, with spin $\frac{1}{2}$, that do not undergo strong interactions but do undergo weak, electromagnetic, and gravitational interactions.

Hadrons: Particles that undergo strong interactions in addition to the other three. Hadrons include the *mesons* and the *baryons*.

> *Mesons:* Hadrons with integral spin. Many, but not all, have masses between those of the electron and the proton.

> *Baryons:* Hadrons with half-integral spin like the proton, neutron, and many particles heavier than the proton and neutron.

Table 31.3 lists the more frequently observed hadrons. There are well over 100 hadrons in all, although many have very short lifetimes.

At one time nature seemed comparatively simple, with all the chemical elements made up of only three elementary particles. Now we have a bewildering array of new particles with a great variety of charges, masses, spins, and additional quantum numbers required to distinguish one from another. Where is the simplicity which physicists expect to find in nature? A ray of hope along these lines came in the early 1960s with independent proposals by the American physicists Murray Gell-Mann (see Fig. 31.17) and George Zweig that the many hadrons known to exist are actually made up of a small number of more fundamental particles that Gell-Mann called *quarks*. These quarks would, along with the photon and the leptons, be the true elementary particles.

FIGURE 31.17 Murray Gell-Mann (born 1929), an American theoretical physicist who is a world leader in elementary-particle physics. He received the 1969 Nobel Prize in physics for his many contributions to the theory of the elementary particles. (*Courtesy of AIP Meggers Gallery of Nobel Laureates.*)

TABLE 31.3   Properties of the Hadrons

Type of particle	Name of particle	Symbol	Antiparticle	Rest mass (kg)	Electric charge	Spin $\hbar$	Average lifetime
Mesons*	Pion (charged)	π^+	π^-	2.5×10^{-28}	± 1	0	2.6×10^{-8} s
	Pion (neutral)	π^0	Self	2.4×10^{-28}	0	0	0.8×10^{-16} s
Baryons†	Proton	p	$\bar{p}$	1.673×10^{-27}	± 1	$\frac{1}{2}$	More than 10^{32} years
	Neutron	n	$\bar{n}$	1.675×10^{-27}	0	$\frac{1}{2}$	1013 s

*Other mesons include the kaons and the eta meson.
†Other baryons include the lambda, sigma, xi, and omega particles.

31.6 Quarks

Quark: A fundamental particle of spin $\frac{1}{2}$ postulated to explain the structure of the hadrons.

Quarks are believed to have electric charges that are fractions (either one-third or two-thirds) of an elementary charge. Each quark also has its corresponding antiquark. The first quarks postulated were designated as u (up), with charge $+\frac{2}{3}$; d (down), with charge $-\frac{1}{3}$; and s (strange), with charge $-\frac{1}{3}$. The antiquarks to these three quarks were then indicated as $\bar{u}$, $\bar{d}$, and $\bar{s}$, with charges $-\frac{2}{3}$, $+\frac{1}{3}$, and $+\frac{1}{3}$, respectively.

In Gell-Mann's original scheme there were only these three distinct kinds of quark, but to explain the structure of all the hadrons more quarks than three are required, and the number is now up to six. Some quarks are called "charmed" and others "strange." Here *charmed* and *strange* are arbitrary terms for quantities which we cannot otherwise describe but which are needed to remove difficulties in the theoretical formulation of the structure of the elementary particles, in the same way that the spin quantum number was needed to remove difficulties in the building up of the elements in the periodic table.

According to the quark theory, the proton and the neutron are each made up of three quarks, the proton being *uud* and the neutron *udd*. These combinations yield the correct charges for these two baryons. In a similar fashion the π^+ would be $u\bar{d}$, to yield the correct charge of $+1$ for the π^+.

Similar procedures can be used to explain the structure of hadrons heavier than the proton. The simplification introduced into physics by this quark model is equivalent to the simplification introduced into chemistry by the idea of the chemical elements, for it reduces over 100 hadrons to combinations of a much smaller number of quarks, which are presumed to be elementary particles, or at least more elementary than any of the hadrons, including the proton and the neutron.

This quark scheme explains the hadrons quite well. There is strong evidence from the scattering of electrons and neutrinos by protons and neutrons that hadrons do possess a structure that is successfully explained by the quark model. This structure only becomes evident if the scattered particles have wavelengths small compared with the size of the proton. If, for example, the incident particle has a wavelength of about 10^{-15} m, which is about the proton size, the proton will appear to be a pointlike, featureless sphere, as in Fig. 31.18*a*. If, however, the incident particles have wavelengths an order of magnitude smaller, that is, 10^{-16} m, then the proton might appear as shown in Fig. 31.18*b*, where the three quarks making up the proton can be seen. The quarks themselves appear to be at most 10^{-17} m in diameter, so the proton, with a diameter of 10^{-15} m, again seems to consist mainly of empty space, just as an atom does when we consider the size of the nucleus compared with the overall size of the atom.

Research done at SLAC on deep inelastic scattering of fast electrons by nucleons first revealed particles much smaller than protons doing the scattering. Even stronger evidence for the presence of quarks in the proton comes from an experiment carried out on the older Super Proton Synchrotron (SPS) accelerator at CERN using a large bubble chamber as a detector. In the experiment 200-GeV neutrinos, which have wavelengths of slightly under 10^{-17} m, struck protons in the target and were scattered. The experimental results indicated an interaction between two pointlike particles, the one inside the proton being considerably smaller than the proton itself. This is presumed to be a quark. These experiments also indicate that

(a)

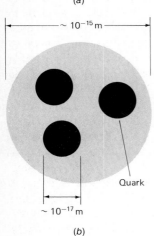

(b)

FIGURE 31.18 (a) The proton as seen by an incident particle of wavelength about 10^{-15} m. (b) The proton as seen by an incident particle of wavelength about 10^{-16} m.

the number of scattering centers inside the proton is close to 3, in good agreement with the quark theory.

The similarity of such experiments to the one on the scattering of alpha particles by nuclei (carried out many years earlier by Rutherford and his students) is worth noting.

Additional evidence for the presence of a substructure in the proton comes from the fact that most structured systems like atoms, molecules, and nuclei can exist in excited energy states, because the particles making up the systems can organize themselves in different ways that correspond to different energies. Scattering experiments reveal that the proton has excited states, just as does a mercury atom or a lithium nucleus, as shown in Fig. 31.19. It is therefore reasonable to assume that the proton consists of more than one particle, which can interact in different ways to produce different energy states of the system. These particles are presumed to be quarks.

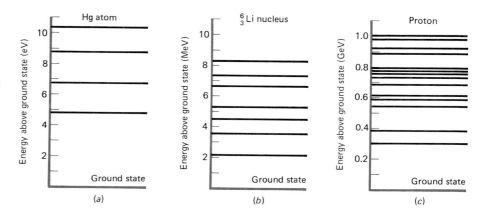

FIGURE 31.19 Excited states of (*a*) a mercury (Hg) atom, (*b*) a lithium (Li) nucleus, and (*c*) a proton. Notice how the energy scales change from eV to MeV to GeV as we go from atom to nucleus to proton.

Problems with the Quark Hypothesis

There were two main difficulties with the original quark hypothesis. First, no one had ever seen a free quark, and second, the quark hypothesis seemed to conflict with the Pauli exclusion principle as applied to particular hadrons.

The existence of free quarks is still an open question. If quarks existed for even a very short time outside a nucleus, they could easily be detected because of their fractional charge. Despite this fact no definitive observation of a free quark has ever been made. Theory explains this by predicting that the force between two quarks grows as they try to separate. This confines the quarks permanently within the hadron.

The conflict with the Pauli exclusion principle has been removed by assigning a new property called "color" to quarks, even though we are not sure what the word *color* refers to in this context, except that, like spin, it is a useful quantum number. In other words, elementary-particle theorists have saved the theory by introducing more complications into it. There is some concern that Gell-Mann's beautifully simple and elegant quark theory is becoming excessively complicated as it tries to accommodate the ever-growing collection of new particles and new data produced by high-energy experimental physics.

However, experiments carried out with colliding beams at SLAC and CERN in 1989 give some hope that order is finally emerging from the chaos of the

TABLE 31.4   Families of Elementary Particles (May 1990)

			Leptons				Quarks	
Family	Particle	Symbol	Charge (units of e) $= 1.6 \times 10^{-19}$ C)	Mass (units of m_e $= 9.1 \times 10^{-31}$ kg)	Particle	Symbol	Charge (units of e)	Mass† (units of m_e)
1	Electron	e^-	-1	1	Up	u	$+\frac{2}{3}$	6.1×10^2
	Electron neutrino	ν_e	0	<0.00004	Down	d	$-\frac{1}{3}$	6.1×10^2
2	Negative muon	μ^-	-1	207	Charmed	c	$+\frac{2}{3}$	2.9×10^3
	Muon neutrino	ν_μ	0	<0.49	Strange	s	$-\frac{1}{3}$	3.0×10^2
3	Negative tauon	τ^-	-1	3491	Top*	t	$+\frac{2}{3}$	$\geqslant 4.4 \times 10^4$
	Tauon neutrino	ν_τ	0	<68	Bottom	b	$-\frac{1}{3}$	9.8×10^3

*The top quark had still not been definitively observed by May 1990, but it is expected to turn up at higher accelerator energies.
†These quark masses are all rough estimates, since no physicist has ever carried out a measurement on a free quark.

elementary particles. There now appears to be no more than three "families" of ultimate elementary particles.

Everyday matter consists of four particles—two leptons and two quarks—belonging to the first family: the electron, the electron neutrino, and the up and down quarks. The second family can be found today only in atom smashers and the cosmic rays from space. This family consists of the muon, the muon neutrino, and the charmed and strange quarks. The particles in the third family—not discovered until the mid-1970s—are the negative tauon, the tauon neutrino, the bottom quark, and the yet-to-be-detected top quark. It is believed that the second and third families were much more prominent—and important—in the period immediately after the "big bang" than they are today.

Table 31.4 lists the ultimate elementary particles (except for the photon) that are believed to constitute all of matter. Each particle in the table has, of course, its associated antiparticle. The symmetry and conciseness of this table indicates that perhaps we are finally coming to an understanding of the ultimate elementary particles.

31.7   Conservation Principles for Elementary Particles

A conservation principle has the same significance for the elementary particles as for large-scale physical phenomena. It is a statement that some particular quantity of a system (charge, angular momentum, etc.) remains constant in time if the system is free from external interference. For the elementary particles there are the usual conservation principles that are absolute and must be satisfied for all reactions involving these particles. These include conservation of mass-energy, of linear momentum, of angular momentum, and of electric charge.

With the elementary particles there are other conservation principles introduced to describe what actually seems to happen in elementary-particle interactions, even though there is no fundamental understanding of why such principles should be valid. Physicists proceed on the assumption that any elementary-particle event that can happen will indeed be observed if we wait long enough and try hard enough to find it. If after all this effort the event is never seen, it is assumed that the event must be disallowed because it violates some new conservation principle, such as the principle of conservation of baryon number, which we discuss below. In this way empirical (i.e., based on experimental observations) and somewhat tentative conservation principles are introduced as ordering principles for sorting out the elementary particles and their interactions. In many cases it turns out that, although these principles provide good working guidelines, they may not be universally valid and may only apply to one class of interactions (e.g., the strong interactions) and not to another class of interactions (e.g., the weak interactions).

We discuss only two conservation principles of this latter, less general type—the conservation of baryon number and the conservation of parity. Other conservation principles frequently discussed in physics books and articles include the conservation of hypercharge, of strangeness, and of quark number.

Conservation of Baryon Number

The following reaction has never been observed, even though it could occur in such a way as to satisfy the four absolute conservation principles.

$$p + n \rightarrow p + p + \bar{p} \tag{31.12}$$

To account for the absence of this seemingly possible event, physicists have posited a new conservation principle, the *conservation of baryon number*:

In any elementary-particle event, the number of baryons must be the same before and after the event.

To allow for the presence of antiprotons, they are assigned baryon numbers of -1, whereas a proton has a baryon number of $+1$. Thus in Eq. (31.12) above, the baryon number on the left is 2, but on the right it is only 1. The principle of conservation of baryon number is therefore violated by such an interaction, and it does not occur.

Conservation of Parity

By *parity* we mean a fundamental property of a system that describes the observed behavior of the system when it is reflected in a mirror. The principle of conservation of parity is then stated as follows:

For a system in which parity is conserved, the mirror image of any physical process cannot be distinguished from the process itself.

Up until 1956 physicists assumed that parity was conserved in all nuclear and elementary-particle interactions. Then in 1956 the American physicists T. D. Lee and C. N. Yang (see Fig. 31.20) suggested that parity might only be conserved in strong interactions but not in weak interactions such as beta decay. In December 1956 a group of physicists at the then National Bureau of Standards, led by Professor C. S. Wu of Columbia University (Fig. 31.21), carried out an experiment to see if parity was indeed conserved in beta decay.

The experiment performed by Wu's group consisted of placing $^{60}_{27}$Co nuclei in a strong magnetic field in which the magnetic moment due to the spinning nucleus was lined up in the direction of the magnetic field. Suppose the nucleus emitted a beta ray (an electron) in the direction of the magnetic field, and hence in the direction of its nuclear spin vector. As seen in a mirror, the nucleus would appear

FIGURE 31.20 (above) C. N. Yang (left; born 1922) and T. D. Lee (born 1926). Lee came to the United States from China in 1946, a year after Yang had immigrated. They shared the 1957 Nobel Prize in physics and were the first scientists of Chinese birth to win the Nobel Prize. (*Photo by Alan Richards; courtesy of AIP Niels Bohr Library.*)

FIGURE 31.21 (right) C. S. Wu (born in China in 1912). Wu directed a group of physicists who showed experimentally that Lee and Yang were indeed right in their prediction that parity was not conserved in weak interactions. She has made several other important contributions to the study of beta decay. (*Photo courtesy of AIP Niels Bohr Library.*)

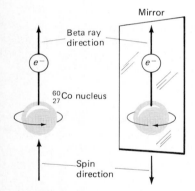

FIGURE 31.22 Conservation-of-parity experiment. On reflection in a mirror, the direction of the β particles is unchanged but the direction of the nuclear spin is reversed.

to be emitting the electron in the direction opposite to that of its spin, as shown in Fig. 31.22, since the reflection in the mirror would reverse the spin direction of the nucleus. The mirror image could only look the same as the original beta-decay process if the nucleus emitted as many electrons in the direction opposite to its spin as it did in the direction parallel to its spin, for the same distribution of electrons would then be observed when the spin was reversed on reflection in the mirror.

In most beta-decay experiments the emitting nuclei are oriented at random and hence it is impossible to tell if there is a preferred direction for the emerging electrons in beta decay. Wu and her colleagues overcame this problem by using a very strong magnetic field at a very low temperature (0.01 K) to align the cobalt nuclei. The experiment showed clearly that more β particles were emitted opposite to the spin direction of the nucleus than in the spin direction. Hence on reflection in a mirror more would be emitted in the spin direction than in the opposite direction. Thus the mirror image of the beta-decay process could be clearly distinguished from the process itself, and so parity would not be conserved.

Since beta decay is a weak interaction, the conclusion is that parity is not conserved in weak interactions. It appears that parity conservation remains perfectly valid, however, for strong interactions.

31.8 The Present State of Elementary-Particle Physics

The main problem faced by elementary-particle theory is that there are still too many unanswered questions about what goes on inside the nucleus. Considerable progress has been made in recent years in answering some of these questions by applying to high-energy interactions a theory called the *standard model*. The standard model consists of two parts, the *electroweak theory*, which applies to the

TABLE 31.5   Types of Interactions and Associated Quanta

Interaction	Particles involved	Quanta exchanged
Strong nuclear	Nucleons, quarks	Gluons
Electromagnetic	Leptons, nucleons	Photons
Weak nuclear	Leptons	Intermediate vector bosons ($W^{\pm}$, Z^0)
Gravitational	All particles	Gravitons

weak interactions (see Table 31.5), and *quantum chromodynamics* (QCD), which applies to the strong interactions.

Since both parts of the standard model use the concepts and insights of quantum electrodynamics, we digress briefly here to discuss that theory.

Quantum Electrodynamics (QED)

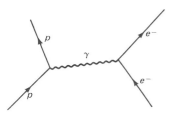

FIGURE 31.24   A Feynman graph of the interaction between a proton and an electron in a hydrogen atom. The exchange of photons between proton and electron produces the attractive force that holds the hydrogen atom together.

Electroweak Theory

Quantum electrodynamics is the relativistic quantum theory that describes the electromagnetic field and the electromagnetic interactions between charged particles. It was developed in the late 1940s by Richard Feynman (see Fig. 31.23 and accompanying biography), Julian Schwinger (born 1918), and Sin-Itiro Tomonaga (1906–1979) as an improvement on the earlier quantum field theory developed by Dirac, Heisenberg, and Pauli. QED has been remarkably accurate in its predictions (in some cases the agreement with experiment has been to within a few parts in 10^9), and is considered the most nearly perfect theory in contemporary physics.

QED explains the stability of a hydrogen atom in terms of the electromagnetic field that binds the proton and electron together. But associated with the electromagnetic field is a field quantum that in this case is the *photon*. The attractive force between the proton and the electron may therefore be considered as due to the constant interchange of photons between the proton and the electron. This interaction may be illustrated in a diagram called a *Feynman diagram* (after its originator) and shown in Fig. 31.24.

QED has influenced the style of theoretical physics from the late 1940s to the present. The electroweak theory, for which Steven Weinberg, Abdus Salam, and Sheldon Lee Glashow (see Fig. 31.25) shared the 1979 Nobel Prize in physics, tried to extend the ideas of QED to the weak interactions. This theory was able to combine the electromagnetic and weak interactions into a single "electroweak" interaction and thus show that these two forces are really different aspects of one basic force.

Just as the electromagnetic force between an electron and a proton can be understood as due to the exchange of a photon, so too the electroweak theory explains the weak nuclear force as involving the exchange of new particles called *intermediate vector bosons*. This is illustrated in the Feynman diagram in Fig. 31.26. The intermediate vector bosons act as the quanta of the field corresponding to the weak nuclear force, just as the photon acts as the quantum for the electromagnetic field.

The electroweak theory predicts that there should be three intermediate vector bosons, two charged bosons designated as W^+ and W^- (W means *weak*), with masses of about 81 GeV, or 86 proton masses,* and one uncharged boson (Z^0), with a mass of about 93 GeV, or 99 proton masses. Experiments at CERN in 1983,

---

*Note that this is about the mass of a krypton atom, with a nucleus $^{84}_{64}$Kr.

Richard P. Feynman (1918–1988)

FIGURE 31.23 (*Courtesy of AIP Niels Bohr Library.*)

When Richard Feynman died on February 15, 1988, the world lost one of its finest theoretical physicists and most inspiring teachers.

Richard Phillips Feynman was born on May 11, 1918, in Far Rockaway, New York, a beach community on the south shore of Long Island. He received his B.S. in physics from MIT in 1939 and his Ph.D. from Princeton in 1942, under the mentorship of Professor John A. Wheeler.

In 1942 Feynman was recruited to the research staff of the Manhattan Project at Los Alamos, New Mexico, where an all-out effort was underway to develop an atomic bomb before the Germans did. Feynman impressed the project director, J. Robert Oppenheimer, with his tremendous energy, original ideas, and ability to cut to the core of any physics problem, and was soon made a group leader in theoretical physics at Los Alamos.

In 1945 Feynman accepted a position as assistant professor of physics at Cornell University. Perhaps because of his 3 years dedicated to the war effort, he decided that in the future he was going to work on physics problems which interested him and pay no attention to their practicality or importance. One of his first ventures along these lines was his calculation of the motion of a plate thrown by a student across the Cornell dining hall!

While at Cornell, Feynman developed his own approach to quantum electrodynamics. He sought simple rules for the calculation of physically observable quantities, using what are now called *Feynman diagrams*, shorthand diagrams to guide the progress of the laborious calculations required in quantum electrodynamics. Feynman diagrams are still much used in elementary-particle and other fields of theoretical physics.

After 5 years at Cornell, Feynman had established a reputation as a brilliant theorist and in 1950 received an appointment as professor of physics at the California Institute of Technology. He was happy and successful at Cal Tech, and remained there for the rest of his life, receiving in 1959 the title of Richard Chase Tolman Professor of Theoretical Physics.

In the years 1953 to 1958, Feynman made important contributions to the theory of low-temperature phenomena, including the properties of liquid helium and superconductivity. But he is best remembered for his "fundamental work in quantum electrodynamics which involves profound consequences for elementary particle physics," in the words of his Nobel Prize citation in 1965.

Richard Feynman was a magnificent teacher; a showman in the classroom, funny, even a bit outrageous at times, he was also clear, profound, and eager to share his enthusiasm for physics with students. He related particularly well to young people and charmed them by his zest for life, impatience with pretension and hypocrisy, and boyish love of practical jokes, puzzles, and games. When he gave a guest lecture in a freshman course, it was never announced beforehand; otherwise the regularly enrolled students would not have been able to find seats.

Feynman's great legacy to teaching was *The Feynman Lectures on Physics*, a three-volume undergraduate physics text based on a course he taught from 1961 to 1963. This course may have been over the heads of many students in the class, but the textbook has conveyed to a generation of physics students and teachers Feynman's love for physics and brilliant insights into how it should be taught and learned.

There was a bit of the iconoclast in Feynman; he enjoyed being different: he played the bongo drums at parties and, while at Los Alamos, became an adept safecracker to show his reservations about the effectiveness of the laboratory's security measures. What people thought was not important to him; what was important was that he have the courage to follow his ideas wherever they led him.

In his last years, while very ill, he accepted an appointment to the commission investigating the 1986 *Challenger* spacecraft accident. His most memorable contribution to the commission's task was his demonstration (with a glass of ice water and a rubber O-ring) that the sealing ability of an O-ring was seriously impaired at low temperatures.

Richard Feynman was an honest, caring human being with a wonderful sense of humor, whose conviction never diminished that physics was meant to be fun. As he said in a 1955 address on "The Value of Science":

> Another value of science is the fun called intellectual enjoyment which some people get from reading and learning and thinking about it, and which others get from working at it.

(a) (b) (c)

FIGURE 31.25 Left to right, (a) Steven Weinberg (born 1933), (b) Abdus Salam (born in Pakistan in 1926), and (c) Sheldon Glashow (born 1932), who shared the 1979 Nobel Prize in physics for developing the electroweak theory. Weinberg and Glashow were classmates both at the Bronx High School of Science and at Cornell University. (*Courtesy of AIP Niels Bohr Library.*)

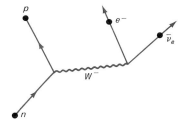

FIGURE 31.26 (left) A Feynman graph of a weak interaction in which a neutron decays into a proton, an electron, and an antineutrino. In this case the field quantum for the weak nuclear force is the intermediate vector boson W^-.

using the head-on collision of a 270-GeV proton beam with a 270-GeV antiproton beam, produced particles with almost exactly these masses. The electroweak theory appears, therefore, to be on very solid ground, and will probably outlast many other, less convincing parts of the standard model.

Quantum Chromodynamics (QCD)

This theory explains the strong forces between quarks as caused by the exchange of new particles called *gluons*, where a gluon is the quantum for the strong interactions, just as the photon is the quantum for the electromagnetic field. The strong force between nucleons is then explained as a spillover from the very strong forces between the quarks making up each nucleon, in somewhat the same fashion that attractive forces between atoms in molecules can be explained as spillovers from the electron-proton forces in each atom.

The name *quantum chromodynamics* has been chosen to indicate the correspondence between this quantum theory of the elementary particles and quantum electrodynamics. The word *chromo* means "color," indicating that the interactions are between quarks having a property called *color charge*, for want of a better name. "Color" seems an apt description in this case because there are similarities between the way particles with different values of color combine to form structures without color, just as the proper combination of colored lights leads to white (colorless) light.

The "standard model," which includes both the electroweak theory and quantum chromodynamics, has been remarkably successful in providing an accurate description of the fundamental particles and their interactions. Despite its many successes, few physicists regard the standard model as the ultimate theory of matter.

It is too complicated, includes too many arbitrary parameters, and thus leaves too much unexplained. Two major contributors to elementary-particle physics* have summed up the status of the standard model as follows: "Our present theory is incomplete, insufficient, and inelegant, though it may be long remembered as a significant turning point."

31.9 The Future of Elementary-Particle Physics

Theoretical

In recent years theoretical physicists have been working on grand unified theories (GUTs), of which there are many, in the hope of uniting the electroweak force with the strong nuclear force. If these grand unified theories should work out, the last step remaining would be to include the gravitational force in a theory embracing all four fundamental forces of nature. Whether this can be achieved is debatable, since the gravitational force is so weak compared with the other basic forces (see Table 31.1) that it is hard to find a role for it inside the nucleus. Albert Einstein devoted 35 years of his life to developing a unified theory of the gravitational and electromagnetic fields, and died before achieving his goal—an indication of the difficulty of the problem.

Three important predictions made by current GUTs are in process of being tested experimentally at the present time (May 1990).

1 The first prediction is that the proton may not be perfectly stable. Its lifetime may not be infinite, as for a perfectly stable particle, but only (!) about 10^{31} years. Therefore it is possible that the proton does infrequently decay into leptons and thus violate the conservation principle for the baryon number.

For example, the reaction $p \rightarrow e^{+} + \gamma$ has never been observed, even though it can satisfy the conservation principles for mass-energy, linear momentum, angular momentum, and electric charge. But, since this equation has one baryon on the left side and none on the right, it violates the conservation-of-baryon-number principle, and so the reaction $p \rightarrow e^{+} + \gamma$ should not occur. The new theories, on the other hand, predict that the proton does infrequently decay into leptons and photons, with a typical half-life of $T_{1/2} = 10^{31}$ years.

Eight different experiments are now being carried out to detect proton decay by physicists from India, France, Italy, the United States, Japan, and the Soviet Union in mines deep below the earth's surface. From 1000 tons of water about three proton decays a year may be expected if these theories are correct. These rare events must be sorted out from a variety of background events due to cosmic rays and other sources; the deep mines are used to provide shielding against these unwanted particles. Experimentalists are convinced that their equipment is sufficiently sensitive to observe proton decays if they do indeed occur. As of this date, however, experimental results indicate a lower limit for the mean life of the proton of 3.5×10^{32} years, a factor of 35 larger than the prediction of the simplest grand unified theory.

2 A second prediction of these unified theories is the existence of a magnetic monopole, that is, a separated magnetic north or south pole. Since separated posi-

---

*Glashow and Lederman, article listed at the end of this chapter.

tive and negative electric charges are common in nature, the symmetry between electricity and magnetism would seem to require separated north and south poles. But such have never been observed.

A useful modern technique for observing magnetic monopoles—if they exist—is to set up a large ring of superconducting material (see Sec. 30.7) and hope that a magnetic monopole in the cosmic radiation will induce a current in the ring when it passes through at high speeds.

On February 14, 1982, the current in such a superconducting ring at Stanford University in Palo Alto, California, jumped from zero to precisely the value to be expected from the transit of a single magnetic monopole. This was the first (and last) time that a careful experiment indicated the existence of a magnetic monopole. Most physicists consider it some kind of freak happening that they cannot explain.

So in this case, too, verification of the GUTs seems to be lacking.

3   A third prediction of the GUTs is that free neutrinos are not strictly massless. Such an idea has its attractions because it would help solve the problem of the "disappearing mass," which greatly disturbs astrophysicists. There simply is not enough mass in the observable universe to account for its observed large-scale behavior. Theoretical predictions indicate that there should be about as many neutrinos as photons in the universe and 10^9 to 10^{10} times more neutrinos than either protons or neutrons. As a result, even a small mass for the neutrino would contribute a substantial additional mass to the universe in the form of invisible "dark matter."

Again, all current experimental evidence indicates that the electron neutrino mass, if it is not zero, cannot be large enough to provide the dark matter needed. The experimental limits on the masses of the muon neutrino and the tauon neutrino are much higher (see Table 31.4), however, and might provide the missing mass. At the present time, then, the case is not closed, but there is no compelling experimental evidence to confirm this third prediction of the grand unified theories.

The Relationship between Elementary-Particle Physics and Astrophysics

In the light thrown by GUTs on this "disappearing mass" problem we see the close relationship developing at present between elementary-particle physics and astrophysics. The four basic forces of physics are related to the origin of the universe. It is thought that at the high temperatures which characterized the newborn universe, all four basic forces manifested themselves in exactly the same symmetrical form. As the universe cooled off, however, and the energy of the dense soup of elementary particles decreased, the four forces and their associated quanta became distinct, and the symmetry was "frozen out." Theoreticians refer to this as *symmetry breaking*. This connection indicates that learning more about the elementary particles will cast additional light on how the universe began. It will also intensify the drive toward unification that pervades physics today. If such a drive should reach pay dirt, we would have what some theoretical physicists have whimsically called "A Theory of Everything."

The close relationship of elementary-particle physics and astrophysics was also demonstrated by Supernova 1987A, the first exploding star visible to the naked eye in 384 years. (See the photograph on the first page of this chapter.) This supernova appeared in the heavens on February 23, 1987; but 3 hours before it was visible in the sky, two underground physics observatories (set up to detect proton decays) observed neutrinos from the supernova on their sensitive detectors. One of these was the IMB detector in a Morton Salt Company mine in Painesville, Ohio, 20 mi from Cleveland; the other was a modified version of the IMB in a zinc mine at Kamioka, Japan, at the other side of the earth. These observations provided clear evidence that the mass of neutrinos was close to zero, while at the same time helping astrophysicists understand more clearly what happens when a star explodes. We seem to have here the beginnings of a new field of neutrino astronomy, joining the already existing fields of optical, radio, microwave, infrared, ultraviolet, x-ray, and gamma-ray astronomy.

The Hubble Space Telescope (Sec. 25.5) is intended to provide information about the age, density, and rate of expansion of the universe. But it may also reveal evidence on the nature of the dark matter needed to explain the stable structure of the universe. If it is found, for instance, that dark matter consists of elementary particles like neutrinos, useful information may be obtained from the Space Telescope on the abundance, mass, and other properties of neutrinos.

Such connections between the incredibly small and the inconceivably large components of our universe indicate the promise of the rapidly developing interactions, both experimental and theoretical, between elementary-particle physics and astrophysics.

Experimental Work

The observation in early 1983 of the intermediate vector bosons in colliding beams at the CERN accelerator shows the power of modern, large-scale, high-energy physics. Over 1 billion collisions were automatically analyzed by counters and computers to turn up the small number of events needed to identify beyond any doubt the intermediate vector bosons.

Successes of this sort indicate that we may expect an increasingly detailed picture of the elementary particles as accelerators move to higher and higher energies and as methods of detection, counting, and data analysis improve.

The last 50 years have shown a very rapid increase in the energies achieved by particle accelerators and hence in their size and cost. Figure 31.27 shows this

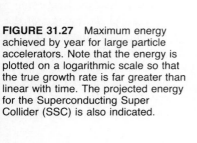

FIGURE 31.27 Maximum energy achieved by year for large particle accelerators. Note that the energy is plotted on a logarithmic scale so that the true growth rate is far greater than linear with time. The projected energy for the Superconducting Super Collider (SSC) is also indicated.

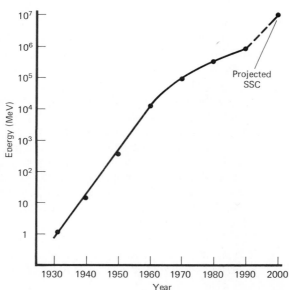

growth on a logarithmic scale. The growth rate is seen to have been much greater than linear with time, although it has slowed slightly in recent years.

The next quantum jump in the history of particle accelerators is expected to be the planned Superconducting Super Collider (SSC) to be built in Texas by 1994. This would have a beam energy of 20 TeV (Remember that 1 TeV $= 10^{12}$ eV!). The head-on collision of two proton beams would lead to an available energy of 40 TeV, which is expected to open up a whole new energy domain to physicists and answer many questions (and probably suggest some new ones). Such a powerful machine will permit study of the elementary particles at the energies they had at a time 10^{-16} s after the ''big bang'' and may therefore contribute greatly to our understanding of the universe's origin.

The SSC will occupy an underground tunnel 84 km in circumference at a location near Waxahachie, Texas, just south of Dallas. (Note that 84 km is half the distance from Philadelphia to New York City!) The cost is expected to be over $7.5 billion. In July 1983 the High Energy Physics Advisory Panel of the Department of Energy recommended that the highest priority be given to the SSC, and thus far the proposal has fared well in Washington. In fiscal year 1990 the U.S. Congress appropriated $225 million for the SSC, of which $135 million was intended for construction. But in these days of budget deficits, it is hard to predict whether the planned SSC will one day make the United States the world's undisputed leader in high-energy experimental physics.

Few countries can afford accelerators as large and costly as the SSC. It is likely that future accelerators will be limited to the United States and the consortium of European nations operating CERN. Some day, however, the construction of larger accelerators will become impossible for any single nation, and the result may well be one international center for research in high-energy physics.

The Future

Was Einstein's faith in the simplicity of the universe, expressed at the beginning of this chapter, justified? Only time will tell. Time is needed to decide whether elementary-particle physics is entering a period of triumphant clarification of the nature of the elementary particles or of growing confusion and disillusionment.

Physics has always progressed by asking the right questions. In elementary-particle physics we ask: Why are the masses of the electron and proton what they are? Why are there only three families of quarks and leptons? What are the truly elementary particles? Many extremely gifted physicists—both experimentalists and theoreticians—are striving earnestly to answer these questions. Most believe, with Einstein, that the answers will be uncovered when we succeed in peeling away the layers of matter and energy cloaking nature's simplicity.

It is remarkable that to understand the largest entity we know—our universe—we must discover the laws governing the smallest entities we know—the elementary particles. The continued quest by physicists for an understanding of the nature of the elementary particles and of the beginning of the universe has been well described by the great Greek philosopher Aristotle (384 B.C.–322 B.C.) in a quotation set in stone on the façade of the National Academy of Sciences building in Washington, D.C.:

> The search for truth is in one way hard and in another way easy, for it is evident that no one can master it fully or miss it wholly. But each adds a little to our knowledge of nature, and from all the facts assembled there arises a certain grandeur.

Summary: Important Definitions and Equations

The four forces of physics

Long-range forces
 Gravitational: Present in all interactions of any kind.
 Electromagnetic: Present in all interactions involving two electrically charged particles.

Short-range forces
 Strong nuclear: Present in all strong interactions (nuclear binding).
 Weak nuclear: Present in all weak interactions (beta decay).

Particle accelerators

Linear accelerators: Machines that accelerate charged particles (usually electrons) down a long, straight evacuated tube.

Synchrotrons (synchrocyclotrons): Machines that accelerate charged particles (usually protons) to high energies by using a magnetic field (B) to keep them moving in a circular orbit of fixed radius (r) as they acquire higher and higher speeds; $E = Bqrc$.

New particles

Hadrons
 Pions: Particles with masses about 275 times the electron mass.

Leptons
 Muons: Particles with masses about 207 times the electron mass.
 Positrons: The antiparticles of electrons; identical to an electron in every way except that their charge is positive.

Electron neutrinos: Particles of very small mass and no charge that are produced in beta decay. Of the particles emitted, neutrinos accompany the positrons and antineutrinos accompany the electrons.
Muon neutrinos: Particles of very small mass and no charge that accompany the emission of muons in pion decay. Neutrinos accompany the μ^+ and antineutrinos the μ^- produced.

Classes of elementary particles
1 Photon: A particle with zero rest mass; its own antiparticle.
2 Leptons: Particles with spin $\frac{1}{2}$ that undergo weak interactions but not strong interactions.
3 Hadrons: Particles that undergo strong interactions; mesons are hadrons with integral spin, and baryons are hadrons with half-integral spin.

Quarks: Elementary particles of spin $\frac{1}{2}$ postulated to explain the structure of the hadrons. Quarks have electric charges of either $+\frac{2}{3}$ or $-\frac{1}{3}$.

Standard model (for high-energy interactions)

Electroweak theory: A theory that combines the electromagnetic and weak interactions into a single "electroweak" interaction.

Quantum chromodynamics: A theory that explains the strong forces between quarks by the exchange of new particles called *gluons*.

Grand unified theories (GUTs): Attempts to unite the electroweak force with the strong nuclear force.

Questions

1 If the gravitational force is by far the weakest force known in nature, why has it been known longer than the electromagnetic or the strong and weak nuclear forces, and why do we experience it more in our daily lives than any of the other four forces?

2 List those interactions (from the four listed in Table 31.1) in which (*a*) the electron participates; (*b*) the proton participates; (*c*) the neutrino participates.

3 What would the temperature at the beginning of the universe have to have been to produce particles with random kinetic energies of 20 TeV? How does your result agree with the empirical rule often given that T (K) $= 10^{10}E$ (MeV)?

4 Why are linear accelerators better than synchrotrons for accelerating electrons?

5 Why is the head-on collision of two beams of 1-GeV protons more useful to high-energy physicists than the collision of a 2-GeV proton beam with a stationary target?

6 Explain how the introduction of the idea of the neutrino in beta decay can save the laws of conservation of energy and of angular momentum.

7 What particles do the following combinations of quarks correspond to: (*a*) *uud*, (*b*) *uus*, (*c*) *du*?

8 Indicate the combinations of quarks that produce (*a*) a neutron, (*b*) an antineutron.

9 Show that, if the photon is the quantum of the electromagnetic field, it must have zero mass.

10 Why would you expect a massless particle like the photon to be stable and, if left to itself, not to decay into other particles?

11 Comment on the following statement by the British novelist C. P. Snow: "For example, most physicists feel in their bones that the present bizarre assembly of nuclear particles, as grotesque as a stamp collection, can't possibly be, in the long run, the last word." (*The Physicists*, Little, Brown, New York, 1981, p. 182.)

(*a*) Do you think that Snow's statement is accurate?

(*b*) Do you find any convincing evidence from this chapter or from your other readings that physicists are making progress in bringing order out of the chaos of the elementary particles?

12 Do you think that relativity and quantum mechanics will remain valid as we get down to distances as small as 10^{-16} m inside the nucleus? Support your answer with evidence from the history of physics.

13 Do you think that the Superconducting Super Collider

(SSC) would be worth the investment of $7.5 billion needed to make it function properly? Justify your answer.

14 Estimate the maximum value of the magnetic field that the bending magnets in the SSC will have to provide.

Additional Readings

Feynman, Richard P.: *QED: The Strange Theory of Light and Matter*, Princeton University Press, Princeton, N.J., 1985. A wonderful series of lectures by an outstanding physicist, aimed at explaining quantum electrodynamics to intelligent nonscientists.

Glashow, Sheldon L., and Leon M. Lederman: "The SSC: A Machine for the Nineties," *Physics Today*, vol. 38, no. 3, March 1985, pp. 28–37. Contains a good summary of the present state of high-energy physics and a well-argued defense of the need for the SSC.

Hawking, Stephen W.: *A Brief History of Time*, Bantam Books, New York, 1988. A best-selling review by a prominent British physicist of attempts to create a unified theory of the universe.

Kernan, Anne: "The Discovery of the Intermediate Vector Bosons," *American Scientist*, vol. 74, no. 1, January–February 1986, pp. 21–28. A good account of one great success of the standard model.

Lederman, Leon M., and David N. Schramm: *From Quarks to the Cosmos: Tools of Discovery. Scientific American* Library, New York, 1989. In this exciting book a Nobel Laureate and a prominent cosmologist discuss the relationship between elementary-particle physics, astrophysics, and cosmology. The emphasis is on the gigantic research instruments in these fields.

Mann, Charles C., and Robert P. Crease: "Waiting for Decay," *Science 86*, vol. 7, March 1986, pp. 20–31. Describes the impressive apparatus being mounted to observe proton decay.

Morrison, Philip: "The Neutrino, *Scientific American*, vol. 194, no. 1, January 1956, pp. 58–68. An account of physicists' confidence in the existence of the neutrino long before it was discovered by Reines and Cowan in 1956.

Mulvey, J. H. (ed.): *The Nature of Matter*, Oxford University Press, New York, 1981. A series of outstanding popular lectures by leading elementary-particle physicists, including Nobel Laureates Salam and Gell-Mann.

Pagels, Heinz R.: *Perfect Symmetry: The Search for the Beginning of Time*, Simon and Schuster, New York, 1985. Much of this excellent book is devoted to the relationship between the elementary particles as we know them today and the beginning of the universe.

Sulak, Lawrence R.: "Waiting for the Proton to Decay," *American Scientist*, vol. 70, no. 6, 1982, pp. 616–625. An account of extremely difficult and costly experiments on the basis of which many modern theories of the elementary particles may stand or fall.

Trefil, James S.: *From Atoms to Quarks: An Introduction to the Strange World of Particle Physics*. Charles Scribner's Sons, New York, 1980. An accurate, popular book on the elementary particles.

Weinberg, Steven: *The First Three Minutes*, Basic Books, New York, 1977. A fascinating account of the connection between the first 3 min of the universe and the elementary particles that exist today.

Weisskopf, Victor F.: "The Three Spectroscopies," *Scientific American*, vol. 218, no. 5, May 1968, pp. 15–29. A stimulating discussion of the similarities that exist among the atomic, nuclear, and subnuclear domains of physics.

In addition, *Scientific American* contains a large number of articles by experts in the fields of high-energy and elementary-particle physics. Some of the best articles for the period 1953 to 1979 are included in *Particles and Fields* (readings from *Scientific American*, with an introduction by William J. Kaufmann III). Freeman, San Francisco, 1980. *Scientific American* articles since 1979 include Haim Harari, "The Structure of Quarks and Leptons," April 1983; Steven Weinberg, "The Decay of the Proton," June 1981; Robert R. Wilson, "The Next Generation of Particle Accelerators," January 1980; LoSecco, Reines, and Sinclair, "The Search for Proton Decay," June 1985; Jackson, Tigner, and Wojcicki, "The Superconducting Super Collider," March 1986; John R. Rees, "The Stanford Linear Collider," October 1989; and Stephen Myers and Emilio Picasso, "The LEP Collider," July 1990.

The February 1989 issue of *Physics Today* is dedicated to Richard Feynman and contains a number of interesting articles on various aspects of his career.

Review of Basic Mathematics

1.A Algebra

Simultaneous Linear Equations

Consider the linear equation $3x + 4y = 16$. There is no unique numerical solution to this equation. All we can do is solve for y in the form $y = (16 - 3x)/4$. Then for any value of x, we can find a corresponding value of y that will make our original equation an identity. If $x = 4$, we obtain $y = 1$; and if $x = 2$, $y = 2.5$.

If we have two independent linear equations for the same two variables x and y, however, a unique solution is possible. This can be found by solving these two equations simultaneously in either of two ways:

1   *Analytically.* We solve for y in terms of x in the first equation, as above. Then we substitute this value of y in the second equation. This leads to an equation in the one variable x, which equation can be solved for the unique value of x. Once x is determined, substituting this value in either equation will lead to the corresponding value of y.

2   *Graphically.* We plot y against x for the two equations on the same sheet of graph paper. The intersection of the two straight lines corresponds to the values of x and y that are simultaneous solutions to the two original equations.

Quadratic Equations

Any equation that may be reduced to the form $ax^2 + bx + c = 0$ is called a *quadratic equation.* Such an equation has the solution

$$x = \frac{-b \pm \sqrt{b^2 - 4ac}}{2a}$$

(A.1)

If $b^2 > 4ac$, the values of x obtained are real.

1.B Geometry

Length of arc of a circle that subtends an angle θ (in radians): $s = r\theta$

TABLE A.1   Areas and Volumes of Plane and Solid Objects

Plane object	Area	Circumference
Rectangle with sides a and b	ab	
Triangle with base b and altitude h	$\dfrac{bh}{2}$	
Circle of radius r	πr^2	$2\pi r$
Ellipse with semiaxes a and b	πab	$2\pi \left(\dfrac{a^2 + b^2}{2}\right)^{1/2}$

Solid object	Surface area	Volume	Cross-sectional area
Sphere of radius r	$4\pi r^2$	$\dfrac{4}{3}\pi r^3$	πr^2
Cylinder of radius r and height h	$2\pi r^2 + 2\pi rh$	$\pi r^2 h$	πr^2

Useful Theorems from Geometry

1 The sum of the angles in any triangle is 180°.

2 Two triangles are *similar* if any two angles in one triangle are equal to two angles in the other triangle.

3 The corresponding sides of similar triangles are proportional.

4 The two acute angles in a right triangle add to 90° and are therefore *complementary* angles.

5 Two angles are equal if any one of the following is true:
  a Their sides are parallel and in the same direction.
  b Their sides are mutually perpendicular.
  c Each of the two angles has the same complementary angle.
  d The two are vertical angles formed by the same two straight lines.

6 *Pythagorean theorem:* In a *right triangle* with sides a, b, and c, where c is the side opposite the right angle,

$$c^2 = a^2 + b^2 \tag{A.2}$$

1.C Trigonometry

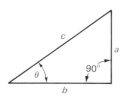

FIGURE A.1 A right triangle used to define the trigonometric functions.

In the right triangle shown in Fig. A.1, where c is the hypotenuse,

$$\text{sine } \theta = \sin \theta = \frac{\text{opposite side}}{\text{hypotenuse}} = \frac{a}{c}$$

$$\text{cosine } \theta = \cos \theta = \frac{\text{adjacent side}}{\text{hypotenuse}} = \frac{b}{c}$$

$$\text{tangent } \theta = \tan \theta = \frac{\text{opposite side}}{\text{adjacent side}} = \frac{a}{b}$$

Since, from Eq. (A.2), $c^2 = a^2 + b^2$, we have, from above,

$$c^2 = c^2 \sin^2 \theta + c^2 \cos^2 \theta$$

or $\quad \sin^2 \theta + \cos^2 \theta = 1 \qquad$ (for all angles) $\hfill$ (A.3)

For any angle θ,

$$\sin \theta = \cos (90° - \theta) = \sin (180° - \theta) \hfill \text{(A.4)}$$

$$\cos \theta = \sin (90° - \theta) = -\cos (180° - \theta) \hfill \text{(A.5)}$$

For any two angles α and β,

$$\sin (\alpha \pm \beta) = \sin \alpha \cos \beta \pm \cos \alpha \sin \beta \hfill \text{(A.6)}$$

$$\cos (\alpha \pm \beta) = \cos \alpha \cos \beta \mp \sin \alpha \sin \beta \hfill \text{(A.7)}$$

$$\cos \alpha + \cos \beta = 2 \cos \frac{\alpha + \beta}{2} \cos \frac{\alpha - \beta}{2} \hfill \text{(A.8)}$$

From the above equations, with $\alpha = \beta = \theta$, we find

$$\sin 2\theta = 2 \sin \theta \cos \theta \hfill \text{(A.9)}$$

$$\cos 2\theta = \cos^2 \theta - \sin^2 \theta = 2 \cos^2 \theta - 1 = 1 - 2 \sin^2 \theta \hfill \text{(A.10)}$$

For any triangle with sides a, b, and c opposite the angles α, β and γ, respectively,

Law of sines: $\quad \dfrac{a}{\sin \alpha} = \dfrac{b}{\sin \beta} = \dfrac{c}{\sin \gamma} \hfill \text{(A.11)}$

Law of cosines: $\quad c^2 = a^2 + b^2 - 2ab \cos \gamma \hfill \text{(A.12)}$

1.D Series Expansions

Binomial expansion:

$$(a + b)^n = a^n + na^{n-1}b + \frac{n(n - 1)}{2} a^{n-2}b^2 + \cdots \hfill \text{(A.13)}$$

Thus $\quad \left(\dfrac{a + b}{a}\right)^n = \left(1 + \dfrac{b}{a}\right)^n = 1 + \dfrac{nb}{a} + \dfrac{n(n - 1)}{2} \dfrac{b^2}{a^2} + \cdots$

If b/a is *small*, and we let $b/a = c$, then,

$$(1 + c)^n \simeq 1 + nc$$

$$(1 - c)^n \simeq 1 - nc$$

1.E Logarithms and Exponential Functions

By the *logarithm*, or *log*, of a number we mean the power to which some given base must be raised to yield that number. Ordinary or "common" logarithms use the number 10 as a base. Then, if $y = 10^x$, we have

$$\log_{10} y = \log y = x$$

since x is the power to which 10 must be raised to yield y. The subscript 10, included here to indicate clearly the base used, is usually omitted, since log is taken conventionally to mean "log to the base 10." For example, since $100 = 10^2$, we have $\log 100 = 2$, because 2 is the power to which the base 10 must be raised to yield the number 100.

There is another system of logarithms, called *natural*, or *Naperian*, logarithms, in which the base is not 10 but the exponential base $e = 2.718218$, which is the sum of the infinite series

$$e = 1 + \frac{1}{1} + \frac{1}{1 \cdot 2} + \frac{1}{1 \cdot 2 \cdot 3} + \frac{1}{1 \cdot 2 \cdot 3 \cdot 4} + \cdots$$

The natural logarithm is written as "ln" to distinguish it from a logarithm to the base 10. If $y = e^x$, then $\ln y = x$, for x is the power to which e must be raised to yield y. For example, suppose $y = e^2 = (2.718)^2 = 7.39.$* Then .

$$\ln y = 2 = \ln 7.39 \qquad \text{or} \qquad \ln 7.39 = 2$$

Since $\log 10 = 1$ and $\ln 10 = 2.3026$, we can write

$$\ln 10 = 2.3026 \log 10$$

This is true in general for any number N, and so $\ln N = 2.3026 \log N$.

Rules for Use of Logarithms (to Any Base)

1 Multiplication: If $N = AB$, then $\log N = \log (AB) = \log A + \log B$.

2 Division: If $N = \dfrac{A}{B}$, then $\log N = \log \dfrac{A}{B} = \log A - \log B$.

3 Exponentiation: If $N = A^n$, then $\log N = n \log A$.

For example,

$$\log 10^x = x \log 10 = x(1) = x \qquad\qquad \text{(A.14)}$$

$$\ln 10^x = 2.3026 \log 10^x = 2.3026x \qquad\qquad \text{(A.15)}$$

$$\ln e^x = x \ln e = x(1) = x \qquad\qquad \text{(A.16)}$$

$$\log e^x = \frac{1}{2.3026} \ln e^x = \frac{x}{2.3026} \qquad\qquad \text{(A.17)}$$

With hand and desk calculators so readily available, neither logarithms nor the engineering slide rule (which is based on logarithms) are used much any more to perform routine multiplications and divisions or to raise numbers to powers. Logarithms are still very useful in many fields of physics, however, where important physical quantities often increase or decrease *exponentially*, and in laboratory work where the functional dependence of one physical quantity on another is being investigated.

---

*It is useful to remember that $e^3 \simeq 20$. Thus, for example, $e^{18} = (e^3)^6 \simeq (20)^6 \simeq 6.4 \times 10^7$.

Functional Equations and Graphs

2.A Functional Equations

In mathematics courses you often come across expressions such as $y = x^2$, which states that "y is a function of x," in this case the square of x. This means that if x is equal to 3, y is equal to 9; if x is equal to 4, y is equal to 16; and so on. For example, the formula for the area of a square of side L is $A = L^2$. If the sides of the square are 4.0 cm in length, the area of the square is therefore $A = (4.0 \text{ cm})^2 = 16 \text{ cm}^2$. (Note that in physics not merely the numbers but also the units, in this case "cm," have to be squared, if the answer is to make good physical sense.)

The functional equation $y = x^2$ can be graphed by plotting x along the x axis over some reasonable range and then calculating and plotting the corresponding values of y along the y axis. The resulting graph often gives a more vivid picture of the dependence of y on x than does the simple functional equation $y = x^2$.

In some cases the functional dependence of y on x may assume a very simple form in which y varies either directly with x or inversely with x. These are called *direct proportions* and *inverse proportions*, respectively.

Direct Proportions

A good example of a direct proportion is the expression for the circumference of a circle, where $C = f(r)$ and r is the radius. In this case it is clear that C increases as r increases. If this dependence is linear, then we can write

$$C \propto r$$

which states that the circumference C is directly proportional to the radius r. This is not yet an equation, since it contains no equality sign; it is a *proportion*. If we graph C against r, we obtain a straight line, which can always be described by the equation

$$C = kr$$

By making use of the formula for the length of arc of a circle given in Appendix 1.B, $s = r\theta$, and noting that for a complete circle $\theta = 360°$, or 2π rad, we have $C = kr = 2\pi r$, which is the usual geometric expression for the circumference of a circle.

It is true in general that any direct proportion (such as $C \propto r$) can be converted into an equation by inserting the right constant of proportionality k (as in $C = kr = 2\pi r$). The actual value of k can be obtained as the slope of the straight line relating C to r. For any two points on the straight line designated by the values C_2, r_2 and C_1, r_1, the slope is given by $k = (C_2 - C_1)/(r_2 - r_1)$, as shown in

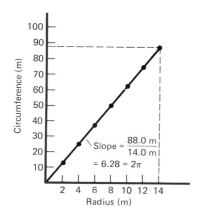

The value of k in an equation relating two physical quantities may vary in numerical value depending on the *units* used for the two quantities in the proportion. We will see many examples of this in our course.

FIGURE A.2 The circumference of a circle plotted against its radius. Since $C = 2\pi r$, the graph is a straight line of slope 2π.

Inverse Proportions

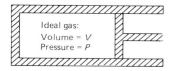

FIGURE A.3 A gas confined by a piston to a cylindrical container.

In other cases two physical quantities vary *inversely*; that is, y increases as x decreases, and vice versa. For example, suppose we take a gas enclosed in a cylindrical container and use a piston to change its volume V, as in Fig. A.3. We find that the volume of the gas is inversely proportional to its pressure, if the temperature remains constant. Expressed as a proportion, we have

$$V \propto \frac{1}{P} \quad \text{(for constant } T\text{)}$$

This proportion can again be converted into an equality by inserting the proper constant of proportionality so that

$$V = \frac{k}{P} \quad \text{or} \quad PV = k$$

In general, if $y \propto 1/x$, or $y = k/x$, then $xy = k =$ constant. If y is plotted against $1/x$, a straight line of slope k results. If y is plotted against x, the graph is a hyperbola, as in Fig. A.4. Whether a proportion is direct or inverse, this proportion can be changed to an equality by inserting the proper proportionality constant. From the data given in Fig. A.4, it can be seen that k is in this case equal to 10^5 N·m, since this is the product PV at any point on the graph.

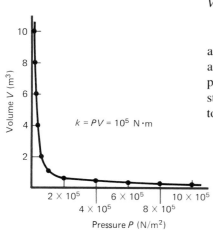

FIGURE A.4 A graph of the volume of a gas of fixed mass and at constant temperature against the pressure of the gas. Since $V \propto 1/P$, the graph is a hyperbola.

2.B Graphs

A graph is just a visual way of showing the functional dependence of one physical quantity on another, such as the speed of a car as a function of time or the volume of a gas as a function of its pressure.

Interpolation

Figure A.6 shows how the length of a piece of steel wire varies when it is supported in a clamp and weights are attached to its free end to stretch it, as in Fig. A.5.

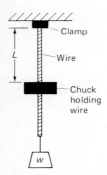

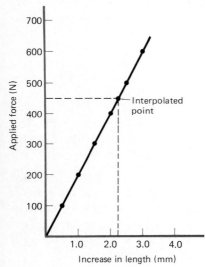

FIGURE A.5 An apparatus for stretching a steel wire. By adding additional weights *w* at the end of the wire, the length *L* of the wire can be increased.

For reasonable-size weights the increased length of the wire between the clamp and the chuck is found to be directly proportional to the applied force, and we can write

$$\Delta L \propto w \qquad \text{or} \qquad \Delta L = kw$$

where *w* is the weight attached. This is the equation of the straight line in the figure. The points marked on the straight line are the actual values of the applied force for which measurements of the length were made. If, now, we want to know what the length of the wire was for a weight of 450 N, we can *interpolate* between the measured points on the graph for 400 and 500 N and determine that the increased length of the wire was 2.25 mm when the applied force was 450 N.

We do this by finding the position where the line for a force of 450 N intercepts the graph, and then reading from that point down to the *x* axis to find the corresponding increase in length. We can be reasonably sure that the resulting length is accurate to within a small fraction of a millimeter, since we have data on both sides of the unknown point and the graph appears to be a straight line. Thus in this case interpolation appears to yield valid results. This is usually the case, unless we have a graph which is not a smooth curve but one on which the dependent variable fluctuates rapidly for small variations in the independent variable. In such cases interpolation can be a riskier proposition.

FIGURE A.6 (left) A graph of the applied force *w* = *mg* against the increase in length of the wire in Fig. A.5.

FIGURE A.7 (right) An invalid extrapolation of the increase in length of the wire in Fig. A.5. Experiment shows that the curve ceases to be a straight line for forces above about 700 N.

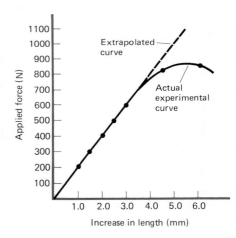

Extrapolation

Suppose, however, that we wanted the increased length of the wire for an applied force of 800 N. We could assume that the straight line continued beyond the last data point at 600 N and find the length of the wire from the graph by extending the straight line into this region, as in Fig. A.7. Such a process is called *extrapolation*. It is inherently risky and often leads to absurd results. For example, in the case of the wire under study, the application of a weight of 800 N would exceed the elastic limit of the wire and would begin to pull the molecules of the wire apart. As a result the graph would no longer be a straight line. Rather the curve would look something like the solid line in Fig. A.7. In this case extrapolation would lead to results in serious disagreement with experiment.

The fallacy involved in unreasonable extrapolations such as this is the assumption that the curve we are using will retain its shape outside the region where data points exist. Thus the last data point on the curve is at an applied force of 600 N. To assume that the curve will remain a straight line even at a force of 800 N is risky and in this case turns out to be simply wrong.

2.C Log-Log and Semilog Scales

Two of the most powerful tools for a physicist are a piece of graph paper and a sharp pencil. By plotting experimental data in graphical form, a physicist can uncover functional relationships between two physical quantities that could never be obtained merely by looking at tables of measured values for these two quantities. Graphical analysis is a way to "crack the code" beneath which nature hides its orderly behavior.

The simplest type of dependence of one physical quantity on another is the linear, or straight-line, relationship shown in Fig. A.6. If we plot measured values of the increased length of the wire against the applied force and draw a smooth curve through the points, we can easily tell whether the functional dependence is really linear over the range of the experiment. If the functional dependence is *not* linear, however, it is very difficult to find from an ordinary graph the exact functional dependence of one measured quantity on another. Only a straight-line dependence is easy to identify unambiguously.

For this reason *the key to graphical analysis is to reduce all graphs to straight lines.* There is unfortunately no completely general method for doing this, but we will consider a few particularly useful methods.

Power Laws

In physics, equations such as $T = 2\pi\sqrt{l/g}$ or $F = mv^2/R$ frequently occur. The first equation states that T varies directly with $\sqrt{l}$ and the second that F varies directly with v^2, all other quantities being held constant. Such equations, in which one physical quantity varies as a power of another quantity, are called *power-law equations* and can be represented in more general form as

$$y = kx^n \tag{A.18}$$

where n is the power which characterizes the functional dependence of y on x. Here n can be either positive or negative, integral or fractional, and k is a constant.

Suppose we do a laboratory experiment and measure a series of values of x and the corresponding values of y. The best way to determine the value of n in the power equation $y = kx^n$ is to take logarithms of both sides of this equation. We then have $\log y = \log k + n \log x$. If we plot $\log y$ against $\log x$, rather than y against x, we obtain a straight line of slope n, since the logarithmic equation is of the form $Y = nX + b$, where $Y = \log y$ and $X = \log x$, and $\log k = b$ is a constant. From the slope (n) of this straight line it is possible to find on what power of x the quantity y depends. In doing this either common (base 10) logarithms or natural (base e) logarithms may be used.

Log-Log Graph Paper

To make the technique for investigating power-law dependence easier, graph paper is available which is so constructed that, if we plot a number on either the horizontal or vertical scales, we are actually plotting the *logarithm* of the number. In this case, since the numbers being plotted on the two axes are really logarithms, they are pure numbers without units or dimensions. Hence the slope is obtained by merely taking the ratio of the measured distances along the two axes. We have for the slope, from the above equation,

$$n = \frac{\log y_2 - \log y_1}{\log x_2 - \log x_1}$$

where $\log y_2 - \log y_1$ is the measured distance between y_2 and y_1 and $\log x_2 - \log x_1$ is the measured distance between x_2 and x_1, when the values of x and y are plotted on log-log paper. The slope can therefore be read directly from the graph paper, as in Fig. A.8, where we plot the period (T) of the planets against the mean radius (R) of their orbits. The graph shows that $T \propto R^{3/2}$, or $T^2 \propto R^3$, which is one of Kepler's laws.

Log-log graph paper is usually described as two-cycle, three-cycle, four-cycle, etc., to describe the range of numbers that can be plotted. The logarithmic scales are so constructed that the range from 1 to 10 (one cycle) occupies the same space as from 10 to 100 (a second cycle) or from 100 to 1000 (a third cycle). Thus three-cycle log-log paper would have three regions of equal length along each axis and would allow us to plot numbers ranging from 1 to 1000. Since we want to plot our data on the largest possible scale to increase the accuracy of the graph, we should choose log-log paper with the proper number of cycles so that the data will fill up almost the whole sheet of graph paper rather than only a small portion of it. The graph paper used in Fig. A.8 is two-cycle log-log paper.

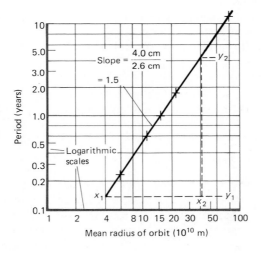

FIGURE A.8 A plot of the period of a planet against the radius of its orbit about the sun for the five planets nearest the sun. These data are plotted on two-cycle log-log graph paper and show that $T = kR^{1.5}$.

Exponential Relations

Many important physical quantities are related exponentially. A simple example is the decay of a sample of radioactive nuclei, for which the number of nuclei decreases in time according to the equation

$$N = N_0 e^{-\lambda t} \tag{A.19}$$

Here N is the number of nuclei remaining after time t, N_0 is the initial number of nuclei, and t is the time elapsed from the time when N was equal to N_0, that is, the time when t was equal to zero. The constant λ is called the *decay constant* of the radioactive sample and determines how rapidly the sample decreases in time. A graph of Eq. (A.19) is shown in Fig. 29.7.

An easy way to find the decay constant λ is to take natural logarithms of our exponential equation. We obtain

$$\ln N = \ln N_0 + \ln e^{-\lambda t} = \ln N_0 - \lambda t \tag{A.20}$$

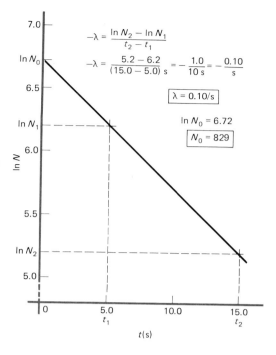

FIGURE A.9 A graph of the decay of a sample of radioactive nuclei, in which ln N is plotted against time. The slope of the straight line resulting is the decay constant λ, and the vertical intercept is ln N_0, where N_0 is the initial number of nuclei.

since ln $e^x = x$ from Eq. (A.16). Equation (A.20) represents a straight line, with intercept ln N_0 and slope $-\lambda$, when ln N is plotted against t.

For example, suppose we measure N at two times t_1 and t_2 and then plot ln N_1 and ln N_2 against t, as in Fig. A.9. Since in this case the decay is exponential, a straight line results. Its slope can be obtained from the graph and is

$$-\lambda = \frac{\ln N_2 - \ln N_1}{t_2 - t_1} = \frac{\ln (N_2/N_1)}{t_2 - t_1}$$

The slope of the straight line corresponding to Eq. (A.20) therefore leads directly to a value for the decay constant λ, as shown on the graph. The minus sign appears in front of λ because N is decreasing in time, and so ln $N_2 -$ ln N_1 is a negative number, which makes the decay constant λ positive. The units of λ must be s^{-1}, since λt appears as an exponent and all exponents are dimensionless.

Semilog Graph Paper

A special kind of graph paper is available to aid in the plotting of quantities related exponentially. It is called *semilog paper*, because one scale (usually the vertical) is ruled logarithmically and the other scale (the horizontal) in a normal linear fashion. Therefore, if we plot on the vertical scale the number of nuclei remaining, we are really plotting ln N. We therefore avoid the step of calculating the values of ln N from our measured values of N.

A more complete discussion of the use of simple graphical techniques in physics is to be found in Clifford Swartz, *Prelude to Physics* (Wiley, New York, 1983).

Solving Physics Problems

3.A Some General Hints on Problem Solving

Students in introductory physics courses often say that they understand the material but cannot do the assigned problems. Yet if they truly understood the material, they would be able to do the problems. This is not to say that physics problems are easy, but the more complete your understanding of the material, the easier the problems will be for you. In physics it is *understanding*, not memorization, that pays off. It is therefore a much better investment of time to study a chapter until you feel that you really understand it before tackling the assigned problems, rather than to try solving problems involving material which you have not yet mastered.

Solving physics problems is an art learned by practice. At first, the plethora of units and equations introduced in a course such as this one may make problems confusing, but the more problems you do, the simpler they will seem. For this reason do problems over and above those assigned in this course. A good source of problems, some fully worked out, some with only the answers given, is *Schaum's Outline of College Physics* (8th ed.) by Frederick J. Bueche (McGraw-Hill, New York, 1989).

The first step in tackling a set of problems is to read over the chaper in the text, your class notes, and anything else you may find helpful in understanding the material. Pay special attention to the examples worked out in the text, since the assigned problems will usually have some relationship to the examples. Once you feel that you *really understand* the chapter, the following steps may be useful in approaching problems. Note, however, that there are a great variety of physics problems and that no set procedure will work equally well in all cases.

1 First read the problem very carefully and be sure that you understand clearly the physical situation described in the statement of the problem. This is the most important step in arriving at a solution.

2 Form a simplified *model* of the physical situation that is capable of being treated quantitatively. Often drawing a diagram or picture of the situation helps. Textbook problems often force a simplified model of the situation by only giving you a limited number of known quantities to work with. Solving laboratory problems is more difficult, since you must decide for yourself what data you need.

3 Make a list of the quantities given and those to be determined. In some problems you may need additional data from tables in the text. In other cases some of the information given may not really be necessary for the problem. One key to understanding physics is to learn what physical quantities are important in a particular physical situation and what quantities are irrelevant.

4 Pick out one unknown quantity and try to recall how it is related physically to the given quantities by some basic law of physics or some general equation. The choice of a relevant equation to calculate the unknown should be based on an *understanding* of the physical situation, not on a hunt through a list of equations in the hope of finding one that works. If you are unable to determine a proper approach to the solution of the problem, remember that more than one step and more than one equation may be required. Looking at some worked examples in the text may help stimulate useful ideas.

5 If it does not appear possible to calculate the first unknown from the given quantities, go on to one of the other unknowns that may be easier to handle. Perhaps this other unknown must be calculated before the first unknown can be obtained. In some cases two equations may have to be solved simultaneously (Appendix 1; Sec. 1.A) to obtain two unknown quantities.

6 Reduce all formulas to the simplest possible analytical form before substituting numbers in them. When you have your final answer in the form of an equation, check to see that it is dimensionally correct (Sec. 1.6).

7 Substitute numbers in the equations, being sure that the units are consistent (all SI) and writing the units with every number. Solve for the numerical answer, canceling units throughout to give the correct units for the answer (Sec. 1.6).

8 In most numerical problems the data are given to only two or three significant figures, and answers should be to the same number of significant figures (Sec. 1.7).

9 Check the problem carefully to make sure you have made no mistakes in algebra or arithmetic (even the best physicists make mistakes at times). Are the units in the answer correct?

10 Look back at the original physical situation to make sure the answer is physically reasonable. (If you have a car going 3×10^5 m/s, it is not reasonable!)

11 Only at this stage should you look up the answer, if it is given in the back of the book. If your answer agrees to within a few percent with the listed answer, you can be fairly sure that your work is correct. In a text such as this, reasonably approximate answers to problems are perfectly acceptable. The method is the important thing, not great accuracy in the numerical result. Eventually you may build up enough confidence in your problem-solving ability to be convinced in certain cases that the answer given in the book is wrong. If so, discuss the matter with your instructor before handing in the assigned problem.

12 If the answer is not given in this book, try to check your work by doing the problem in a different way or working it backward to see that it checks out (see Example 2.10).

13 After you have worked out the problem on scrap paper and are convinced that it is right, transfer the essential work to good paper in as logical and clear a fashion as possible, leaving out all false starts. Indicate the calculations carried out, but not the actual arithmetic. The problem should be so clear that the corrector does not have to struggle to find out what you have done. Box the answers for the unknown, as is done in this text, or indicate clearly in some other way what your answers are.

The best way to learn how to do physics problems is to do lots of problems. An investment of time in doing extra problems at the beginning of this course will save you much time at the end of the semester, when time is more precious. The logical thought processes and techniques learned in solving physics problems will also prove useful to you in other college courses and in your future careers. For example, the thought process followed by a medical doctor in diagnosing what is wrong with a patient does not differ too much from the thought process you must follow in successfully solving a physics problem.

3.B Powers-of-10 Notation

Most of the physics problems in this book contain data given to only two, or at most three, significant figures, and answers are therefore expected only to two or three figures. If, for example, a car travels at 20 m/s for 10 min, the distance traveled is

$$\left(20 \ \frac{m}{s}\right)(10 \ min)\left(60 \ \frac{s}{min}\right) = 12,000 \ m$$

But it is incorrect to write the answer in this form, since it gives the answer to five figures, only two of which are really significant. For this reason it is much better to express this answer in powers-of-10 notation as 1.2×10^4 m.

For problems in this book, it is expected that calculations and answers will all be expressed in powers-of-10 notation. We now outline this useful tool for those students not already familiar with it.

If we raise 10 to various powers, we obtain the following results:

$$10^1 = 10 \qquad\qquad 10^{-1} = \frac{1}{10} \qquad = 0.1$$

$$10^2 = 100 \qquad\qquad 10^{-2} = \frac{1}{100} \qquad = 0.01$$

$$10^3 = 1000 \qquad\qquad 10^{-3} = \frac{1}{1000} \qquad = 0.001$$

$$10^4 = 10,000 \qquad\qquad 10^{-4} = \frac{1}{10,000} \qquad = 0.0001$$

$$10^5 = 100,000 \qquad\qquad 10^{-5} = \frac{1}{100,000} \qquad = 0.00001$$

$$10^6 = 1,000,000 \qquad\qquad 10^{-6} = \frac{1}{1,000,000} = 0.000001$$

This provides us with a shorthand method for writing both very large and very small numbers. Any number can be written as a decimal number between 1 and 10 followed by the appropriate power of 10. The power of 10 gives the number of places the decimal point must be moved to the right for positive powers and to the left for negative powers to convert to ordinary decimal form. Thus we may write large numbers as follows:

$$286,000 = 2.86 \times 100,000 = 2.86 \times 10^5$$

Note that we rewrite the number with the decimal point after the first digit, then multiply by the power of 10 corresponding to the number of digits to the right of the decimal point in the original number. Similarly, for very small numbers,

$$0.00312 = \frac{3.12}{1000} = \frac{3.12}{10^3} = 3.12 \times 10^{-3}$$

Note here that the negative power of 10 corresponds to the number of places the decimal point has to be moved to the left to convert back to ordinary decimal form.

This powers-of-10 notation is used throughout this book and is particularly useful in multiplying and dividing large numbers. In multiplying two numbers, we multiply the initial numbers and add the powers of 10, carefully watching the signs of the powers; in dividing two numbers, we divide the initial numbers and subtract the powers of 10. For example,

$$(2.86 \times 10^5)(3.12 \times 10^{-3}) = 8.92 \times 10^{5-3} = 8.92 \times 10^2$$

and

$$\frac{2.86 \times 10^5}{3.12 \times 10^{-3}} = 0.917 \times 10^{5-(-3)} = 0.917 \times 10^8 = 9.17 \times 10^7$$

In this way we can obtain a value for 10 raised to the zero power. Since $10^1 = 10$, $10^1/10^1 = 10/10$, or $10^{(1-1)} = 1$, and so $10^0 = 1$. As a matter of fact, it can be shown in the same way that any number raised to the zero power is equal to 1.

To add or subtract two numbers in powers-of-10 notation, the power of 10 must be made the same in both cases; then the numbers are added or subtracted as usual. Thus

$$(2.76 \times 10^{-5}) + (3.46 \times 10^{-6}) = (27.6 \times 10^{-6}) + (3.46 \times 10^{-6})$$
$$= (27.6 + 3.5) \times 10^{-6} = 3.11 \times 10^{-5}$$

This corresponds to the way we normally line up the decimal points in adding or subtracting decimal numbers. Note that we have rounded off 3.46 to 3.5, since the 6 loses its significance when added to the corresponding unknown digit in 27.6.

3.C Order-of-Magnitude Calculations

Often even two or three significant figures are excessive in the answer to a particular physics problem, for all we really need to know is the *order of magnitude* of the result. This is basically the result *to the nearest power of 10*. Thus we might say that the order of magnitude of an electron's speed is 10^5 m/s, meaning that the speed is probably somewhere between 5×10^4 and 5×10^5 m/s, but that we are not really interested in its exact value. For example, all we might want to know is whether the electron's speed is close enough to the speed of light (3×10^8 m/s) to make relativistic effects important. The knowledge that the speed is close to 10^5 m/s assures us that we can neglect such relativistic effects.

An order-of-magnitude calculation leads to a very approximate or, as physicists sometimes say, a "quick and dirty" answer to a physics problem. Two physicists can carry out separate order-of-magnitude calculations of the same quantity and, because they either made slightly different assumptions or used slightly different data or approximations, come up with results in disagreement by a factor of 2 or 3, or even 5, and we would still say that their results agreed "as to order of magnitude."

Order-of-magnitude calculations are very useful in solving physics problems. At the beginning of a problem we might want to know if we are on the right track by putting in rough numbers to see if they yield a reasonable value for the desired unknown quantity. At the end of the problem, if we are suspicious that we may have made a mistake in algebra or entered the wrong number into our hand calculator, an order-of-magnitude calculation may show us at what stage of the calculation we went wrong.

In the laboratory, physicists and physics students carry out order-of-magnitude calculations all the time to determine what range on an electric meter is needed for a particular measurement or to ascertain the error introduced into an experiment by a drop in power-line voltage from 120 to 115 V. Here the physicist is not looking for a highly precise result but for a "ballpark" figure to help decide the next step in the experiment. The physicist does not want to take the time to come up with an exact answer because at this stage of the research such an answer is not worth the effort involved. This is when order-of-magnitude calculations are quite helpful.

For example, suppose a swimming pool owner wants to know approximately how much water his pool holds. The pool is 20.4 m long and 6.3 m wide, and its depth varies linearly between 1.1 and 4.3 m. The average depth of the water is then 2.7 m, and an order-of-magnitude calculation of the volume of water in the pool leads to

$$V = (2 \times 10^1 \text{ m})(6 \text{ m})(3 \text{ m}) = 36 \times 10^1 \text{ m}^3 \simeq 4 \times 10^2 \text{ m}^3$$

This can be compared to the more exact value of

$$V = (20.4 \text{ m})(6.3 \text{ m})(2.7 \text{ m}) = 3.5 \times 10^2 \text{ m}^3$$

It can be seen that the order-of-magnitude calculation leads to a result differing by about 14 percent from the exact value. This is more than enough accuracy for an order-of-magnitude calculation.

In this example the time involved in the exact calculation is not much greater than the time required for the order-of-magnitude calculation. However, with more complicated physics problems order-of-magnitude calculations often take significantly less time than exact calculations and can thus be particularly useful.

Masses and Abundances of Some Important Isotopes

Atomic number Z	Element	Symbol	Mass number A*	Atomic mass† (u)	Percent abundance	Half-life (if radioactive)
0	Neutron	n	1	1.008665		10.6 min
1	Hydrogen	H	1	1.007825	99.985	
	Deuterium	D	2	2.014102	0.015	
	Tritium	T	3*	3.016049		12.33 years
2	Helium	He	3	3.016029	0.00014	
			4	4.002603	~100	
3	Lithium	Li	6	6.015123	7.5	
			7	7.016005	92.5	
4	Beryllium	Be	7*	7.016930		53.3 days
			8*	8.005305		6.7×10^{-17} s
			9	9.012183	100	
5	Boron	B	10	10.012938	19.8	
			11	11.009305	80.2	
6	Carbon	C	11*	11.011433		20.4 min
			12	12.000000	98.89	
			13	13.003355	1.11	
			14*	14.003242		5730 years
7	Nitrogen	N	13*	13.005739		9.96 min
			14	14.003074	99.63	
			15	15.000109	0.37	
8	Oxygen	O	15*	15.003065		122 s
			16	15.994915	99.76	
			18	17.999159	0.204	
9	Fluorine	F	19	18.998403	100	
10	Neon	Ne	20	19.992439	90.51	
			22	21.991384	9.22	
11	Sodium	Na	22*	21.994435		2.602 years
			23	22.989770	100	
			24*	23.990964		15.0 h
15	Phosphorus	P	31	30.973763	100	
			32*	31.973908		14.28 days
17	Chlorine	Cl	35	34.968853	75.77	
			37	36.965903	24.23	

Atomic number Z	Element	Symbol	Mass number A*	Atomic mass† (u)	Percent abundance	Half-life (if radioactive)
19	Potassium	K	39	38.963708	93.26	
			40*	39.964000		1.28×10^9 years
			41	40.961832	6.72	
26	Iron	Fe	56	55.934939	91.8	
27	Cobalt	Co	59	58.933198	100	
			60*	59.933820		5.24 years
36	Krypton	Kr	84	83.911506	57.0	
			89*	88.917563		3.2 min
38	Strontium	Sr	86	85.909273	9.8	
			88	87.905625	82.6	
			90*	89.907746		28.8 years
53	Iodine	I	127	126.904477	100	
			131*	130.906118		8.05 days
55	Cesium	Cs	133	132.90543	100	
			137*	136.90677		30 years
56	Barium	Ba	137	136.90582	11.2	
			138	137.90524	71.7	
			144*	143.922673		11.9 s
82	Lead	Pb	206	205.97447	24.1	
84	Polonium	Po	210*	209.98288		138.4 days
			214*	213.99519		164 μs
86	Radon	Rn	222*	222.017574		3.824 days
88	Radium	Ra	226*	226.025406		1.60×10^3 years
			228*	228.031069		5.70 years
90	Thorium	Th	228*	228.02873		1.91 years
			232*	232.038054	100	1.41×10^{10} years
92	Uranium	U	233*	233.039629		1.62×10^5 years
			234*	234.040904	0.0057	2.48×10^5 years
			235*	235.043925	0.715	7.13×10^8 years
			238*	238.050786	99.27	4.51×10^9 years
			239*	239.054291		23.5 min
93	Neptunium	Np	239*	239.052932		2.35 days
94	Plutonium	Pu	239*	239.052158		2.41×10^4 years

*An asterisk means that the isotope is radioactive.
†Masses include the electrons in the neutral atom. These atomic masses are in unified mass units u.

Answers to Odd-Numbered Exercises and Problems

CHAPTER 1

1.1 d
1.3 c
1.5 d
1.7 a
1.9 d
1.11 d
1.13 4.28 rad
1.15 1.693×10^2 m
1.17 (*a*) 9.2 m/s; (*b*) 21 mi/h
1.19 *x*-component: 6.09 cm; *y*-component: 49.6 cm
1.21 Moon: (*a*) 8.96×10^{-3} rad; (*b*) 0.513°
Sun: (*a*) 9.34×10^{-3} rad; (*b*) 0.535°
1.23 (*a*) $[L/T]$; (*b*) $[L/T^2]$; (*c*) $c_3:[L]$; $c_4: [1/T]$; (*d*) $[L]$
1.25 (*a*) 4.8 km; (*b*) 3.5 km at 31° S of E
1.27 (*a*) 13.6 cm at 16° below $+x$ axis; (*b*) 17.6 cm at 42° below $+x$ axis; (*c*) 3.9 cm at 106° above $+x$ axis; (*d*) 13.8 cm at 18° above $+x$ axis
1.29 (*a*) 324 m; (*b*) 19.1° N of E
1.31 (*a*) 8.77 cm; (*b*) No; a straight line is the shortest distance between two points; (*c*) Yes; any other path is longer.
1.33 $A_x = 10.6$ m; $A_y = 22.7$ m
1.35 (*a*) 6.40 cm at $+38.7°$; (*b*) 6.40 cm at $-38.7°$; (*c*) 1.81 cm at $-30.3°$; (*d*) 12.3 cm at $+46.6°$; (*e*) 8.5 cm at $+6.2°$
1.37 1.29×10^3 m^2
1.39 (*a*) $A_x = 8.60$ m; $A_y = 12.3$ m; (*b*) $A_x = -9.64$ m; $A_y = 11.5$ m; (*c*) $A_x = -6.34$ m; $A_y = -13.6$ m; (*d*) $A_x = 3.88$ m; $A_y = -14.5$ m
1.41 55.0 m in $+y$ direction
1.43 30.4 m at 99.5° clockwise from $+ x$ axis
1.45 $A'_x = 15.4$ m; $A'_y = -9.4$ m; $B'_x = 13.9$ m; $B'_y = 17.6$ m; $C'_x = 20.6$ m; $C'_y = 15.1$ m
1.47 35.5 m at 6.4° below $+x$ axis
1.49 (*a*) 17.7 cm; (*b*) 21.4 cm; (*c*) 120°
1.51 74 km at 22.5° above $+x$ axis

CHAPTER 2

2.1 c
2.3 b
2.5 e
2.7 e
2.9 b
2.11 (*a*) 5.43 m/s; (*b*) 184 s
2.13 6.0 s
2.15 (*a*) 25.5 m/s; (*b*) 33.1 m
2.17 19.0 m/s
2.19 (*a*) 1.56×10^4 m; (*b*) 18.6 m/s
2.21 0.86 m/s
2.23 (*a*) 80.0 m/s; (*b*) 325 m
2.25 3.46 m/s^2
2.27 61 m/s
2.29 5.5 s
2.31 25.0 m/s
2.33 $v_A = 1.3$ m/s; $v_B = 3.5$ m/s; $v_C = 0$; $v_D = -3.5$ m/s
2.35 (*a*) 34.0 m/s; (*b*) 6.28 km
2.37 (*a*) 7.50 m/s^2; (*b*) 202 m
2.39 0.34 s
2.43 (*a*) 62.5 m; (*b*) 35.0 m/s; (*c*) 7.14 s
2.45 27.0 m/s
2.47 (*a*) 247 m/s^2; (*b*) 25 g; (*c*) 400 m
2.49 (*a*) 3.17 s; (*b*) 31.7 m; (*c*) $x = 24.3$ m; $y = 20.4$ m
2.51 (*a*) 80.0 m; (*b*) 10.0 s
2.53 (*a*) -2.00 m/s^2; (*b*) 13.0 m
2.55 (*a*) 20.2 m/s; (*b*) 2.47 s
2.57 13.7 m/s
2.59 (*a*) 37.3 m; (*b*) 32.4 m; (*c*) 2.04 s
2.61 (*a*) They collide after 2.34 s; (*b*) 38.8 m; (*c*) Car: 8.15 m/s; Truck: 10.3 m/s

CHAPTER 3

3.1 d
3.3 d
3.5 e
3.7 e
3.9 b
3.11 100 N
3.13 1.50×10^3 N
3.15 5.6 m/s^2
3.17 30 N
3.19 (*a*) 4.14×10^2 N; (*b*) 8.88×10^2 N

3.21 5.59 m/s^2 at $63.4°$ above $+x$ axis
3.23 (a) 2.00 N in direction from m_1 to m_2; (b) 3.00 N in direction from m_2 to m_1; (c) They are not an action-reaction pair.
3.25 (a) 1.33 m/s^2; (b) 3.33 m/s
3.27 1.76×10^3 N down
3.29 (a) 3.1 m/s; (b) 0.50 m
3.31 1.20 m/s^2
3.33 6.82 kg
3.35 (a) 0.571 m/s^2; (b) 1.43×10^3 N
3.37 (b) 7.6 m/s^2; (c) 22 N
3.39 11 m/s
3.41 (a) 2.2×10^2 N (b) 18 m
3.43 (a) 8.5 kg; (b) $T_{BC} = 68$ N; $T_{AB} = 26$ N
3.45 23 m
3.47 (a) -2.5 m/s^2; (b) 3.0 m/s
3.49 (a) 1.1×10^{-5} N/(m/s); (b) 0.56 s
3.51 (a) 4.6 m/s^2 up the plane; (b) 3.7 m/s^2 up the plane
3.53 (a) 1.08×10^3 N; (b) 8.80×10^2 N; (c) 0
3.55 (a) 98.0 N; (b) 113 N; (c) 83.0 N
3.57 (a) 0.476 m/s^2; (b) 47.6 N
(c) Since the chain is not massless, an unbalanced force is required to accelerate it.
3.59 (a) 5.69×10^{-16} N;
(b) Yes; F_g is only 8.93×10^{-30} N
3.61 (a) 1.2 s; (b) 1.0 m

CHAPTER 4

4.1 d
4.3 b
4.5 e
4.7 a
4.9 a
4.11 0.55 s
4.13 (a) 0.33 s; (b) 3.0 rev/s; (c) 6π rad/s
4.15 $22°$
4.17 1.02×10^3 m/s
4.19 30.8 rad/s
4.21 (a) 983 N; (b) 371 N; (c) 2.69×10^3 N
4.23 0.065 m/s^2
4.25 $3.52°$ above the horizontal
4.27 (a) $x = 102$ m; $y = 116$ m; $v_x = 51$ m/s; $v_y = 48$ m/s; (b) $t = 6.9$ s; $y = 2.4 \times 10^2$ m;
(c) 7.1×10^2 m; (d) 85 m/s at $53°$ below the horizontal
4.29 80 m
4.31 (a) 42 m for both; (b) $t_{35°} = 2.5$ s; $t_{55°} = 3.5$ s
4.33 (a) No; an angle of $47°$ would be needed;
(b) 1.3×10^4 N; (c) 0.75
4.35 (b) 1.6 rad/s
4.37 $14°$
4.39 (a) $(5.9 \times 10^{-3}$ m)N, where m is your mass.
(b) $(5.9 \times 10^{-3}$ m)N
4.41 (a) 0.34%; (b) 0
4.43 (a) 2.21 m/s; (b) 0.196 N
4.45 7.0 m/s
4.47 (a) There must be a horizontal component to the

string's direction so that the centripetal force can keep the ball moving in a circle; (b) $42°$
4.49 (a) Their initial motion; conservation of angular momentum; (b) 4.11×10^{22} N; (c) 2.98×10^{27} kg
4.51 1.97×10^{30} kg
4.53 (a) 5.07×10^3 s; (b) 6.01×10^3 s
4.55 7.36×10^3 m/s
4.57 340 km
4.59 1.6×10^3 m/s

CHAPTER 5

5.1 c
5.3 d
5.5 c
5.7 d
5.9 $F_x = 23$ N; $F_y = 45$ N
5.11 392 N
5.13 (a) 5.7×10^2 N; (b) 1.13×10^3 N
5.15 57 kg
5.17 (a) 116 N; (b) 231 N
5.19 (b) $T_A = 124$ N; $T_B = 190$ N
5.21 77.7 N
5.23 980 N
5.25 $W_1 = 797$ N; $W_2 = 386$ N; $T_1 = 920$ N; $T_2 = 460$ N
5.27 $F_1 = 49$ N; $F_2 = 66$ N
5.31 $F_1 = 3.37 \times 10^5$ N; $F_2 = 3.24 \times 10^5$ N; both vertically upward
5.33 147 N; 343 N
5.35 (a) 781 N; (b) 1.58×10^3 N at $60.3°$ with respect to the ground
5.37 $d_1 = 2.15$ m; $d_2 = 1.85$ m, with 700-N girl closer to center
5.39 0.48
5.41 (a) 142 N; (b) 91.9 N at $5.4°$ above $+x$ axis
5.43 $\ell_1 = 0.824$ m; $\ell_2 = 0.593$ m
5.45 (a) 3.38×10^3 N;
(b) 5.87×10^3 N at $54.8°$ above ground
5.47 $\bar{x} = 0.404$ m; $\bar{y} = 0.433$ m
5.49 (a) 4.80×10^4 N; (b) 3.91 m from front wheels
5.51 1.00 cm from center of 18.0-cm disk along line joining the two centers
5.53 $18.4°$
5.55 (b) 0.28

CHAPTER 6

6.1 a
6.3 b
6.5 c
6.7 d
6.9 b
6.11 (a) 0; (b) 2.5×10^4 J; (c) 9.8×10^3 J
6.13 4.1×10^{-16} J
6.15 2.4×10^{10} J
6.17 0.63 J
6.19 3.9 hp

6.21 (a) 2.00×10^3 J; (b) 1.41×10^3 J;
(c) -2.00×10^3 J; (d) 0

6.23 (a) 1.0×10^3 J; (b) -4.9×10^2 J; (c) 5.1×10^2 J;
(d) $+5.1 \times 10^2$ J; (e) 10 m/s

6.25 (a) 5.0×10^2 J; (b) 2.9×10^2 J; (c) 2.1×10^2 J

6.27 11.0 m/s

6.29 2.50×10^4 N

6.31 (a) 0.625 J; (b) 0.625 J; (c) 0.625 J; (d) 0.500 m/s

6.33 (a) 1.96 J; (b) 1.96 J; (c) 1.40 m/s; (d) 4.90 m/s²;
(e) 1.84 J; (f) 0.123 J

6.35 (a) 8.85 m/s in horizontal direction;
(b) 10.8 m/s at 35.5° below horizontal

6.37 $wL(1 - \cos\theta)$

6.39 7.51 cm

6.41 (a) 1.54×10^4 J; (b) 736 J

6.43 (a) 1.40 m/s; (b) 1.96×10^{-2} J;
(c) PE = 9.80×10^{-3} J; KE = 9.80×10^{-3} J

6.45 0.39

6.47 (a) $KE_0 = 2.00 \times 10^4$ J = KE_f;
    $KE_{\text{top}} = 6.58 \times 10^3$ J
(b) $PE_0 = PE_f = 0$; $PE_{\text{top}} = 1.34 \times 10^4$ J
(c) 2.00×10^4 J at all points on the trajectory
(d) Total energy is the same, since the initial energy is
the same in the two cases and energy is conserved.

6.49 (a) 11; (b) 8.3; (c) 75%

6.51 2.34×10^4 W

6.53 3.1×10^7 W

6.55 (a) 3.71×10^2 J; (b) 92.8 W

6.57 2.1×10^7 J; 6.0 kWh

6.59 2.3×10^4 W = 31 hp

CHAPTER 7

7.1 e

7.3 a

7.5 c

7.7 d

7.9 a

7.11 20 kg·m/s

7.13 500 N

7.15 1.53 kg·m/s; 1.29 kg·m/s

7.17 1.0×10^6 N

7.19 0.33 m/s

7.21 (a) No change; (b) Momentum vector has been
deviated 60° from original direction.

7.23 F_{net} = 98.0 N downward; $\Delta p/\Delta t$ = 98.0 kg·m/s²
downward

7.25 (a) 15.8 kg·m/s; (b) 1.05×10^3 N

7.27 (a) 0.250 N; (b) 2.50×10^{-2} N·s

7.29 (a) Yes, because of momentum conservation;
(b) 2.9×10^2 s; (c) 0.086 m/s

7.31 3.1 m

7.33 (a) 1.50 m/s; (b) KE_2 = 2.25 J; KE_3 = 1.50 J

7.35 (b) 63.2 kg·m/s at 108.4° with respect to direction of
1.0-kg mass; (c) 4.26 kg

7.37 (a) v_{50} = 4.8 m/s; v_2 = 20.8 m/s; (b) 5.38 m/s

7.39 $m = m_p$

7.41 1.50×10^3 m/s

7.43 (a) 28% is lost; (b) 14

7.45 250 m/s

7.47 0.49 N

7.49 (a) 226 kg·m/s at 50.6° with respect to original
direction; (b) 513 J

7.51 (a) 68° with respect to car's direction;
(b) 18 m/s; (c) 3.9×10^4 N

7.53 2.48×10^5 m/s

7.55 (a) 0.77 m/s; (b) The earth recoils.

7.57 (a) 4.90×10^5 N; (b) 4.90×10^3 m/s;
(c) 3.10 m/s²

7.59 (a) Apply conservation of energy and momentum;
(b) Two balls will fly out at other end.

CHAPTER 8

8.1 d

8.3 d

8.5 a

8.7 b

8.9 e

8.11 25 rad/s

8.13 150 rad/s

8.15 1.33×10^2 hp

8.17 1.5×10^6 kg·m²/s

8.19 2.6 rev/s

8.21 (a) 0.500 rad/s²; (b) 2.25 m/s; (c) 7.50 rad/s;
(d) 225 rad

8.23 (a) 6.25×10^4 J; (b) 6.25×10^4 J;
(c) Yes; conservation of energy

8.25 7.28×10^{12} rad/s

8.27 (a) -9.42×10^{-2} N·m;
(b) 50.0 s after the switch is thrown to "off"

8.29 (a) 2.59 m/s²; (b) 0.721 N; (c) 25.9 m/s

8.31 (a) 0.251 kg·m²; (b) 0.40%

8.33 (a) 1.60 kg·m²; (b) 3.20 kg·m²/s

8.35 $2MR^2$

8.37 8.37 m/s

8.39 (a) v_{solid} = 3.97 m/s; v_{shell} = 3.44 m/s;
(b) Solid cylinder reaches bottom first; (c) 0.30 m

8.41 (a) 0.478 m; (b) No change; the height above ground
is the only important thing.

8.43 11.0 kg·m²

8.45 (a) 2.25 m/s²; (b) T_1 = 37.8 N; T_2 = 36.2 N;
(c) 4.50 rad/s² (d) Both are 22.1 J.

8.47 $\frac{2}{3}g$

8.49 5.42 rad/s

8.51 (a) 2.4 rev/s; (b) 28.4 J; (c) From work done by the
skater pulling arms in.

8.53 0.427 rad/s

8.55 (a) 1.20 m/s; (b) No; no torque acts; v changes due
to conservation of angular momentum.

8.57 (a) $\omega_1/3$; (b) No; $\frac{2}{3}$

CHAPTER 9

9.1 e

9.3 a

9.5 d

9.7   b
9.9   e
9.11   e
9.13   6.93
9.15   (a) 1.25×10^3 mol; (b) 0.851 mol; (c) 21.4 mol
9.17   0.017 m^3
9.19   73%
9.21 1.93×10^3 m/s
9.23   10.8
9.25   (a) 1.59×10^{26} u; (b) 0.264 kg
9.27   155 kg
9.29   (a) 1.6×10^{-4}; (b) Distance between molecules is large.
9.31   (a) 6.06×10^5 N/m²; (b) 6.52×10^5 N/m²
9.33 1.4×10^4 m^3
9.35   1.1 atm
9.37   34.2 lb/in²
9.39   0.90 kg
9.41 6.5×10^{-3} m^3
9.43   (a) 3.7×10^9 molecules; (b) 2.2×10^{-7} N/m²
9.45   2.6 atm
9.47   (a) 8.28×10^{-21} J; (b) 5.97×10^2 m/s
9.49   11.6 km
9.51 $KE_{trans} = 6.21 \times 10^{-21}$ J; $\Delta PE_{grav} = 3.25 \times 10^{-24}$ J
9.53   (a) 1.01×10^4 K; (b) 2.01×10^4 K
9.55   (a) 5.65×10^{-21} J for all three; (b) H_1: 1.85×10^3 m/s; H_2: 1.31×10^3 m/s; H_3: 1.07×10^3 m/s
9.57   The data given in Problem 9.58 can be used in the proof.
9.59   (a) 3.74×10^3 J; (b) 2.98×10^2 m/s; (c) Zero

CHAPTER 10

10.1   d
10.3   e
10.5   a
10.7   a
10.9   e
10.13 1.08×10^5 N/m²
10.15 4.5×10^4 N
10.17 4.00×10^3 kg/m³
10.19   0.43 cm³
10.21   (a) 2.97×10^5 N/m²; (b) 2.97×10^5 N/m²; (c) 1.99×10^5 N/m²
10.25   (a) 0.294 m; (b) The 100-cm² column is higher.
10.27   (a) 1.73×10^4 N/m²; 1.20×10^4 N/m²; (b) 1.18×10^5 N/m²; 1.13×10^5 N/m²; (c) 1.46×10^4 N/m²
10.29   (a) 2.00×10^3 kg/m³; (b) 2.00
10.31   (a) 6.7; (b) 6.7×10^3 kg/m³; (c) 30 cm³
10.33   11%
10.35 4.9×10^4 N
10.37   194% (i.e., the bubble has 2.94 times the volume it had at the bottom.)
10.39   1.00 m/s
10.41   20 m
10.43   0.80 m/s
10.45 5.3×10^3 N/m²
10.47   34 cm above bottom

10.49 8.0×10^{-4} m^3
10.53   0.073 N/m
10.55   2.2 cm
10.57   1.8 mm
10.59   8 mm

CHAPTER 11

11.1   e
11.3   c
11.5   b
11.7   b
11.9   c
11.11 9.0×10^{-3} J
11.13 $y = 0.020 \cos(5\pi t)$, with the amplitude in meters.
11.15   0.35 Hz
11.17   7.4 N
11.19   0.028 m
11.21   (a) 0.010 m; (b) 0.28 s; (c) 3.5 Hz; (d) $y = 0.010 \cos(7.0\pi t)$
11.23   Graphs are similar to those in Fig. 11.6.
11.25   1.6 Hz
11.27   1.3 Hz
11.29   (d) 0.020 J
11.31   (a) 1.4 Hz; (b) 0.125 m; (c) $y = 0.125 \cos(2.8\pi t)$
11.33   0.96 kg
11.35   2.7 m/s²
11.37   9.807 m/s²
11.41   1.7 kg·m/s
11.43   (a) 1.1 N; (b) 0.11 kg
11.45   (a) $y = 0.12 \cos(2\pi t)$; (b) $y = 0.12 \cos(4\pi/3 - 2\pi t)$; (c) 0.75 m/s
11.47   (a) 12 m/s; (b) 1.32 rad; (c) $y = 0.010 \cos[2\pi(x/0.10 - 120t) + 1.32]$ (d) 0.0025 m
11.49   (a) $2\pi/3$ or 120°; (b) 0.50 cm
11.51   (a) 224 m/s; (b) $f_1 = 112$ Hz; (c) $f_2 = 224$ Hz; $f_3 = 336$ Hz
11.53   156 Hz
11.55   264 Hz
11.57 $f_1 = Cm^{-1/2}L^{-1/2}F^{1/2}$, with $C = \frac{1}{2}$
11.59   (a) 10 cm; (b) 2.5 cm
11.61   (a) 10 Hz; (b) 20, 30, 40 Hz

CHAPTER 12

12.1   a
12.3   c
12.5   c
12.7   d
12.9   b
12.11   (a) 340 m/s; (b) 9.9×10^{-4} m/s²
12.13   182 Hz
12.15   434 and 446 Hz
12.17 1.0×10^{-4} W/m²
12.19   1.4
12.21   2.5 m/s; 4.1×10^3 m/s²
12.23 4.7×10^{10} N/m²

12.25 0.010%
12.27 8.6×10^7 N/m²
12.31 8.5 m in length
12.33 (a) 2.2×10^3 Hz; (b) 7.7 cm
12.35 170, 213, 340 Hz
12.37 (a) 1757 Hz; 1763 Hz; (b) 1757 Hz
12.39 95 Hz
12.41 (a) $2X_{\max}$; (b) 4 times greater; (c) 6.0 dB
12.43 3.2×10^{-5} W/m²
12.45 3.0 dB
12.47 20 W
12.49 (a) 455 Hz; (b) 450 Hz; (c) Moving source: 250 Hz; Moving observer: zero frequency; no sound heard.
12.51 1.54 m/s
12.53 (a) 6.8 cm; (b) 6.1 cm; (c) 5.6×10^3 Hz (d) 6.8 cm; (e) 5.5×10^3 Hz
12.55 33 or 37 m/s
12.57 4.8×10^2 m/s
12.59 No discrete frequency; a sonic boom

CHAPTER 13

13.1 c
13.3 b
13.5 b
13.7 a
13.9 b
13.11 $-460°F$; 80°F; 620°F
13.13 2192°F; 1473 K
13.15 0.13 m
13.17 (a) 4.4 kcal; (b) 11 kcal; (c) 20 kcal
13.19 138 J/(kg·K)
13.21 54 kcal
13.23 126 F°; 70 K
13.25 320°F (or 160°C)
13.29 36 cm³
13.31 125.02 cm
13.33 6.6×10^5 N
13.35 (a) 1.07×10^3 cm³; (b) 178 kcal
13.37 4.0×10^2 s
13.39 24°C
13.41 0.095 kcal/kg·C°
13.43 20°C
13.45 20.23°C
13.47 0.12 C°
13.49 0°C; the heat is not sufficient to melt all the ice.
13.51 (a) 0.080 kg; (b) 0.0125 kg
13.53 (a) 1.5×10^5 J; (b) 8.5×10^2 W

13.55

	H_2O	Hg	H_2
(a)	1.4	0.57	0.028
(b)	9.7	14	0.22

(All answers in units of kcal/mol)
The attractive forces between water molecules and between Hg atoms are about the same size and are much greater than those between H_2 molecules. This explains why at STP H_2O and Hg are both liquids, whereas H_2 is a gas.

13.57 3.4×10^4 m
13.59 (a) 1.5 kg/h; (b) 1.9%
13.61 $\overline{KE} = 7.7 \times 10^{-21}$ J; $Q_{vap} = 6.8 \times 10^{-20}$ J; $Q_{vap} = 8.8\overline{KE}$

CHAPTER 14

14.1 a
14.3 a
14.5 b
14.7 d
14.9 a
14.11 (a) 1.9×10^6 J; (b) 4.2×10^3 J; (c) 50 J
14.13 (a) 4.2×10^3 W; (b) thin
14.15 1.06×10^{-5} m
14.17 3.1×10^9 s (about 100 years)
14.19 (a) 150 J; (b) 0
14.21 (a) 1.0×10^4 J; (b) 11 m
14.23 2.2×10^2 W
14.25 (a) 2.5×10^3 C°/m; (b) 2.4×10^{-2} C°/m; (c) Heat flows much more slowly through the earth to its surface than through the glass window in Example 14.1.
14.27 17 s
14.29 7.7 W
14.33 2 mm Al, 1 mm brass
14.35 1.4 W/m²·C°
14.37 5.8×10^6 J
14.39 37°C
14.41 (a) About 6 hours; (b) The mass increases in proportion to the volume, not to the surface area.
14.43 (a) 6.6 h; (b) 11 h; (c) 17 h
14.47 7.4 kcal
14.49 (a) 390 cal; (b) 550 cal
14.51 3.84×10^3 cal
14.53 (a) $T_A = 750$ K; $T_B = 3750$ K; $T_C = 1500$ K; (b) $\Delta Q_{AB} = +600$ kcal; $\Delta Q_{BC} = -270$ kcal; $\Delta Q_{CD} = -240$ kcal; $\Delta Q_{DA} = +53$ kcal
14.55 604 m/s
14.57 (a) 1.87×10^4 J; (b) 0.75×10^4 J; (c) 2.62×10^4 J
14.59 (a) 1.8×10^4 J; (b) 1.8×10^4 J; (c) 1.9 km
14.61 1.7×10^6 J
14.63 536 cal
14.65 (a) 50 J; (b) 10 J
14.67 (a) 501 J (b) 501 J

CHAPTER 15

15.1 e
15.3 b
15.5 b
15.7 b
15.9 d
15.11 210 K
15.13 4.2×10^3 J/K
15.15 90%
15.17 0 K

15.19   16

15.21   (a) 1.52×10^{-3} m^3; (b) 227 K; (c) -37 J

15.23   (a) 296 K; (b) Yes; the condensation of the water releases heat.

15.25   (a) (All answers in units of 10^3 J)

	ΔU	ΔQ	ΔW
$1 \rightarrow 2$	3.74	3.74	0
$2 \rightarrow 3$	-1.81	0	$+1.81$
$3 \rightarrow 1$	-1.93	-3.22	-1.29
$1 \rightarrow 2 \rightarrow 3 \rightarrow 1$	0	0.52	0.52

(b) $P_2 = 2.0 \times 10^5$ Pa; $V_2 = 2.5 \times 10^{-2}$ m^3; $P_3 = 1.0 \times 10^5$ Pa; $V_3 = 3.8 \times 10^{-2}$ m^3

15.27   (a) 7.8 m^3; (c) Zero; $T_3 = T_1$; (d) 2.7×10^4 J; (e) 2.7×10^4 J

15.29   (a) 1; (b) 6 (out of 16 ways possible); (c) $S_1 = 0$; $S_2 = 1.8$ K

15.31 $+0.96$ kcal/K

15.33   Zero; $\Delta Q = 0$

15.35   (a) $\Delta S_1 = -7.1 \times 10^{-3}$ J/K; $\Delta S_2 = +8.0 \times 10^{-3}$ J/K; (b) $\Delta S = +0.9 \times 10^{-3}$ J/K

15.37   (a) 500 J; (b) on the gas; (c) -1.67 J/K

15.39   (a) $+4.2 \times 10^{-2}$ J/K; (b) 0; (c) -4.2×10^{-2} J/K

15.41   (a) 5.0°C; (b) -74.6 J/K; (c) $+75.9$ J/K; (d) $+1.3$ J/K; Entropy increases, indicating that process is irreversible.

15.43   (a) 59%; (b) $\Delta Q_H = 1.7 \times 10^{12}$ J; $\Delta Q_C = 0.7 \times 10^{12}$ J

15.45 1.2×10^2 m

15.47   (a) 67%; (b) 67%

15.49   626 K (353°C)

15.51   (a) $\Delta Q_H = 65 \Delta Q_C$; (b) $\Delta W = 64 \Delta Q_C$

15.55   (a) 2.15×10^3 J; (b) 0.15×10^3 J; (c) 14.6

15.57   (a) 5.6×10^4 J; (b) 2.2×10^3 J (c) 278; 11

CHAPTER 16

16.1   a

16.3   c

16.5   e

16.7   b

16.9   (a) 11 N toward the -10 μC charge; (b) 1.1×10^4 m/s^2

16.11   (a) 4.6×10^{-8} N toward the positively charged sphere; (b) 4.6×10^{-8} N in the opposite direction to that in part (a)

16.13   (a) 6.8×10^6 N/C away from the $+30$ μC charge; (b) 6.8×10^6 N/C toward the -30 μC charge

16.15   (a) 3.83×10^{10} m/s^2; (b) Acceleration of electron is 7.0×10^{13} m/s^2 in opposite direction; (c) 7×10^{12} times larger

16.17   (a) 1.5×10^3 N; (b) 79° above the line joining q_A and q_B

16.19 $F_e/F_g = 4.2 \times 10^{42}$

16.21   4.8 N along the extension of the diagonal away from the charge

16.23   (a) 5.6×10^2 N at 22° with the line joining the -8.0 μC and $+5.0$ μC charges, inside the triangle;

(b) 2.1×10^2 N at 38° from line joining the -8 μC and $+3$ μC charges, outside the triangle

16.25   (a) 8.2×10^{-8} N; (b) 2.2×10^6 m/s

16.27 3.4×10^{-8} C

16.29 -47 μC and -3.0 μC

16.31   (a) Zero; (b) 2.9×10^8 N/C from + charge to − charge

16.33   (a) 3.4 m outside the 2.0 μC charge; (b) Yes; 0.77 m from the 2.0 μC charge

16.35   (a) Zero; (b) 2.0×10^6 N/C directed along the diagonal toward the -5.0 μC charge

16.37   (a) Zero; (b) 2.2×10^4 N/C directed from the center of the hoop toward the missing piece

16.39   (a) 1.9×10^{-13} J; (b) 1.5×10^7 m/s

16.41 4.6×10^{-2} N/C

16.45   (a) Zero; (b) 3.4×10^7 N/C; (c) Zero

CHAPTER 17

17.1   d

17.3   a

17.5   c

17.7   a

17.9   b

17.11 9.6×10^{-19} J

17.13   38 V/m from + to −

17.15 8.0×10^{-9} C

17.17   12 μF

17.19   1.0 J

17.21   (a) -4.3×10^{-18} J; (b) $+27$ V

17.23   (a) -4.50×10^2 J; (b) $+4.50 \times 10^6$ V

17.25 1.57×10^{-19} C

17.27   (a) 7.7×10^{-19} J; (b) 9.6×10^{-21} J

17.29   (a) 8.00×10^{-17} J; (b) 500 eV; (c) 1.33×10^7 m/s

17.31 1.57×10^{-19} C

17.33   (a) 7.1×10^{-4} F; (b) 4.4×10^{15} electrons

17.37   (a) 8.85×10^{-11} F; (b) 5.3×10^{-8} C; (c) 2.0×10^5 V/m

17.39   *Before:* (a) 2.2×10^{-11} F; (b) 2.2×10^{-8} C; (c) 1000 V; *After:* (a) 9.9×10^{-11} F; (b) 2.2×10^{-8} C; (c) 2.2×10^2 V

17.43   0.21 μF

17.45   (a) 20.0 μF; (b) 2.25×10^{-2} C; 0.75×10^{-2} C; (c) The same 1500 V for each; (d) 16.9 J; 5.6 J

17.47   17 μC; 33 μC

17.49 1.2×10^{-5} C

17.51   8.7 J

CHAPTER 18

18.1   d

18.3   a

18.5   c

18.7   d

18.9   e

18.11 4.3×10^5 C

18.13   (a) 0.60 A; (b) 2.2×10^3 C

18.15 0.50 ohms
18.17 5.0 V
18.19 8.6×10^8 J
18.21 6.0×10^3 J
18.23 8.5×10^7 m/s
18.25 (a) 84 A; (b) 6.3×10^{-3} m³
18.27 19 min
18.29 6.4×10^{-3} ohms
18.31 468 ohms
18.33 Two separate coils in parallel: 600 W for one coil; 2,400 W for two in parallel.
18.35 (a) 0.15 A; (b) Decreasing; (c) Increasing
18.37 $I_1 = 1.05$ A; $I_2 = -0.42$ A; (c) $I_3 = 1.47$ A
18.39 (a) $I_1 = 0.92$ A; $I_2 = 0.97$ A; $I_3 = -0.05$ A; (b) 17 V
18.41 (a) $I_1 = 2.5$ A; $I_2 = -1.1$ A; (b) $I_3 = 1.4$ A; (c) $V_1 = V_2 = 7.1$ V
18.43 (a) 18 V; (b) C is at higher potential.
18.45 (a) 0.37 A; (b) $P = 1.8$ W
18.47 (a) 17 ohms; (b) 1.1 ohms
18.49 (a) 10 ohms; (b) Total current = 4.0 A; currents through resistors are, from left to right: 4.0 A, 2.4 A, 1.6 A, 0.8 A, 0.8 A.
18.51 16 ohms
18.53 36 μC
18.55 $107
18.57 (a) $R_{50} = 288$ ohms; $R_{100} = 144$ ohms; $R_{150} = 96$ ohms; (b) $1/R_{50} + 1/R_{100} = 1/R_{150}$
18.59 13 h
18.61 (a) 24 ohms; (b) 67 W
18.63 (a) 12.6 kWh; (b) 1.1×10^4 kcal

CHAPTER 19

19.1 e
19.3 a
19.5 c
19.7 e
19.9 c
19.11 (a) 0.40 N; (b) Out of the plane of paper
19.13 1.8×10^{-14} N
19.15 1.5×10^{-5} T
19.17 1.0×10^{-5} N/m; attractive force
19.19 0.031 T·m/A
19.21 0.053 N
19.23 0.20 T
19.25 (a) Spiral path in which the proton has a net speed of 2.5×10^6 m/s along the $+x$ axis and moves in clockwise circles of radius 4.5 cm around the direction of B. (b) Same as (a) except that circles are counterclockwise and of radius 2.5×10^{-3} cm. (c) No effect on the neutron.
19.27 10 cm
19.29 (a) 6.7×10^{-12} N; (b) 7.0×10^7 m/s; the magnetic field cannot change the speed of the electron, only its direction.
19.31 1.67×10^{11} C/kg
19.33 0.31 T along $-z$ axis
19.35 6.9×10^{-4} T

19.39 1.71×10^{11} C/kg
19.43 (a) 2.7×10^{-3} T; (b) 1.5×10^{-3} T out of page; (c) 3.0×10^{-3} T into page
19.45 (a) Zero; (b) 6.0×10^{-4} N/m to right
19.47 1.6×10^{-20} N away from wire
19.49 Zero
19.51 3.5 T
19.53 2.8×10^{-3} T
19.55 52 μT
19.57 5.9×10^{-7} N·m
19.59 (a) Add a 2.0×10^8 ohm resistor in series with galvanometer. (b) 2.0×10^8 ohms
19.61 (a) 1.0×10^8 ohms; (b) It draws so little current that it does not change the current being measured.

CHAPTER 20

20.1 b
20.3 c
20.5 c
20.7 b
20.9 e
20.11 e
20.13 c
20.15 0.10 Wb
20.17 4.5×10^{-2} H
20.19 2.5 J
20.21 20 s
20.23 40 V
20.25 0.30 V
20.27 (a) 8.5×10^{-3} V; (b) 3.6×10^{-6} J; (c) Into heat in the resistor
20.29 (a) 6.3×10^{-4} V, in same direction as original current; (b) 7.9×10^{-2} A
20.35 8.0 V
20.37 (a) 8.4×10^{-4} H; (b) 4.2×10^{-5} Wb
20.39 (a) 2.2×10^3; (b) 6.1×10^{-4} H
20.43 (a) -5.0×10^{-2} V; (b) In opposite direction (minus sign) to that of current; (c) 5.0×10^{-3} J
20.45 0.35 A
20.47 (a) 8.0×10^2; (b) 0.020 H
20.49 (a) 2.0×10^2 s; (b) 3.0×10^{-6} A; (c) 2.0×10^{-4} C
20.51 (a) 1.0×10^{-2} A; (b) 60 s
20.53 27 N·m
20.55 (a) 4.4×10^3 W; (b) 47 N·m
20.57 (a) 40 A; (b) 42 V
20.59 7.5 A

CHAPTER 21

21.1 e
21.3 a
21.5 c
21.7 a
21.9 d
21.11 b

21.13 (a) 1.6 A; (b) Zero; v and i are 90° out of phase
21.15 (a) 38 ohms; (b) 1.8×10^4 ohms;
(c) 1.8×10^4 ohms
21.17 1.13×10^3 Hz
21.19 4.0 A
21.21 (a) 0; (b) 0.71 A; (c) 0.50 A
21.23 (a) 0; (b) 1.5 A; (c) -1.5 A
21.25 (a) 1.28 mA; (b) 0; (c) -1.28 mA
21.27 (a) 1.02×10^{-2} C; (b) 13.2 A; (c) 205 Hz
21.29 (a) 3.0%; (b) 99.7%
21.31 (a) 159 Hz; (b) 2.00×10^{-3} J; (c) 2.00×10^{-3} J
21.33 43 ohms
21.35 (a) 71 ohms; (b) 37 μF
21.37 4.5 mA
21.39 (a) 2.0 A; (b) 71°; (c) 0.33; (d) 80 W
21.41 (a) 139 ohms; (b) 0.86 A;
(c) $V_R = 86$ V; $V_L = 540$ V; $V_C = 457$ V
21.43 (a) 6.0 ohms; (b) 62 mH
21.45 (a) 0.11 cm/V in both cases; (b) 2.3 A
21.47 503 Hz
21.49 (a) $V_L = 537$ V; $V_C = -537$ V;
(b) 120 V; (c) 120 V
21.51 (a) 130 Hz; (b) 1
21.53 (a) 13 ohms; (b) 0.20 H
21.55 (a) $N_2/N_1 = 1/22$; (b) $N_2/N_1 = 1/5.5$
21.57 (a) 50 turns; (b) 10.9 kW;
(c) $I_{\text{out}} = 83.3$ A; $I_{\text{in}} = 4.5$ A
21.59 (a) 4.0-ohm output; (b) 16-ohm output

CHAPTER 22

22.1 e
22.3 b
22.5 e
22.7 e
22.9 7.5×10^4 V/(m·s)
22.11 2.4×10^3 V/m
22.13 8.3×10^4 W/m²
22.15 2.90 m
22.17 2.3×10^{-4} s
22.19 (a) 1.8×10^{-10} F; (b) 2.0×10^{-9} A;
(c) 2.0×10^{-9} A; (d) 2.0×10^{-15} T;
(e) 2.0×10^{-15} T
22.21 1.5×10^{-2} A
22.23 (a) 8.3×10^{-12} T (b) 10^{-7} times as large as earth's magnetic field
22.25 (a) 8.65×10^{14} Hz; (b) 347 nm;
(c) 3.00×10^8 m/s
22.29 (a) 1.45×10^8 m/s; (b) 7.25×10^4 m
22.31 (a) 31 V/m; (b) 1.4×10^{-7} T
22.33 6.4 J
22.35 (a) $E_{\text{rms}} = 7.13 \times 10^2$ V/m; $B_{\text{rms}} = 2.38 \times 10^{-6}$ T; (b) 3.8×10^{26} W
22.37 (a) 3.3×10^{-9} T; (b) 2.7×10^{-3} W/m²;
(c) 7.5×10^4 W
22.39 (a) 4.74×10^{14} Hz; (b) 9.48×10^{14} Hz;
(c) 316 nm; ultraviolet
22.41 The husband, by about 0.12 s

22.43 2.3×10^{-6} H
22.45 2.2 to 3.3×10^{-12} F

CHAPTER 23

23.1 b
23.3 d
23.5 a
23.7 c
23.9 a
23.11 c
23.13 0.30 m
23.15 -3.0
23.17 (a) 0.18 m from lens on object side;
(b) Virtual, upright; $M = +0.091$
23.19 (a) -0.14 m; (b) -7.3 D
23.21 0.85 m
23.23 (a) 31.0°; (b) 2.29×10^8 m/s
23.25 Bends must make angle of incidence always greater than 53.5°.
23.27 (a) 58.0°; (b) 60.8°
23.29 1.40
23.31 (a) 52.2°; (b) 30.0°; (c) 52.2°; (d) 44.4°
23.33 (a) Real, inverted, 1.27 m from mirror;
(b) Real, inverted, 3.44 m from mirror; (c) Virtual, upright, 1.41 m behind mirror; (d) As object moves through focal point, image changes from real to virtual.
25.35 0.639 m
25.37 (a) Virtual, upright, 0.282 m behind mirror;
(b) $M = +1.38$.
23.39 (a) Virtual, upright, 0.375 m behind mirror; $M = 0.25$; (b) Virtual, upright, 0.300 m behind mirror; $M = 0.40$; (c) Virtual, upright, 0.207 m behind mirror; $M = 0.59$; (d) Image moves closer to mirror and becomes larger but is otherwise unchanged.
23.43 (a) Concave; (b) 9.45 cm;
(c) Head of hammer is down.
23.45 0.144 m
23.47 (b) 3.6 m from lens on opposite side from object; real and inverted; $M = -5.5$
23.49 2.16 m
23.51 (a) $f' \to \infty$; (b) $f' \to f$
23.53 70.1 cm to right of second lens; image real, inverted, and larger than object
23.55 (b) Virtual, upright image, 0.87 m from lens on object side; $M = 0.46$
23.57 At focal point of lens
23.59 -2.73 m
23.61 Image is real, upright, 29.5 cm from lens on same side as object, and 0.46 as large
23.63 (a) 66.4 cm from lens; (b) Letter T in sign is upside down.

CHAPTER 24

24.1 c
24.3 e
24.5 a

24.7   a
24.9   b
24.11 e
24.13 2.5×10^3 m
24.15 (*a*) 7.2 mm from center; (*b*) 14.4 mm from center
24.17 257 nm
24.19 36 cm from center
24.21 30 mW/cm^2
24.23 15
24.25 80 cm
24.27 0.76 cm
24.29 3.8 μm
24.31 4 orders
24.33 2.01×10^{-6} m
24.35 (*a*) 0.0017°; (*b*) 0.0034°; (*c*) No
24.37 202 nm
24.41 0.654 μm
24.43 1.18×10^{-5}/C°
24.45 (*a*) 127 nm; (*b*) 380 nm, 634 nm, etc.
24.47 112 nm
24.49 (*a*) 30; (*b*) 0.3; (*c*) 0.01; (*d*) 5×10^{-8};
          (*e*) Diffraction depends on ratio λ/w.
24.51 2.11 cm
24.53 (*a*) Central image is 7.8 mm wide and the dark fringes are 3.9 mm apart; (*b*) All distances in the fringe pattern are increased by a factor of 3.
24.55 (*a*) $\frac{3}{8}$; (*b*) 51°
24.57 36.9°
24.59 (*a*) 36.9°; (*b*) 49.9°; (*c*) 53.1°
24.61 226 m

CHAPTER 25

25.1   d
25.3   e
25.5   b
25.7   e
25.9   d
25.11 b
25.13 16 times brighter
25.15 6.3
25.17 8.6×10^4 m
25.19 6.32×10^4
25.21 1.7 s
25.23 60 cm
25.25 0.2 cm (from 5.5 to 5.7 cm from the film)
25.27 +2.0 D converging lens
25.29 (*a*) 33.3 cm; (*b*) 33.3 cm
25.31 −1.0 D top lens; +2.0 D bottom lens
25.33 5.0 times
25.35 167 times
25.39 (*a*) 23.8 cm; (*b*) 1.2 cm
25.41 13.3 times
25.43 0.050 mm
25.45 $f_{obj} = 94.1$ cm; $f_{eye} = 1.9$ cm
25.49 (*a*) 1.15; (*b*) 0.24 μm
25.51 1.5×10^{11} m
25.53 30 μm

25.55 (*a*) 2.01 mm; (*b*) 7,514
25.57 (*a*) The two wavelengths 589.0 and 589.6 nm cause destructive interference for certain path differences; (*b*) 0.29 mm

CHAPTER 26

26.1   b
26.3   c
26.5   b
26.7   b
26.9   c
26.11 b
26.13 1.1×10^{-14} L$_0$
26.15 (*a*) 4.1×10^{-8} s; (*b*) 0.44 L$_0$
26.17 9.5×10^{-31} kg
26.19 3.7×10^{-12} kg
26.21 3.28×10^{-13} J
26.29 2.3×10^4 y
26.31 About 1,000
26.33 5.8×10^2 μs
26.35 2.2×10^{-5} s
26.37 (*a*) Yes; (*b*) 1.8×10^8 m/s along $+x$ axis; (*c*) 400 m; (*d*) No such inertial frame exists; v would have to be greater than c.
26.39 (*a*) 1.1×10^{-14}; (*b*) Mass is 5 times greater.
26.41 (*a*) 11.4×10^{-31} kg; (*b*) 2.0×10^{-14} J
26.43 (*a*) 2.82×10^8 m/s; 2.69×10^{-30} kg; (*b*) 2.42×10^{-13} J
26.45 (*a*) $0.999999995c$; (*b*) 0.30 m
26.47 9.8×10^{-2} (i.e., 9.8%)
26.49 (*a*) 4.33×10^9 kg/s; (*b*) 1.5×10^{13} y
26.51 (*a*) 1.03×10^{-21} kg·m/s; (*b*) 2.90×10^8 m/s
26.53 (*a*) 1.6×10^{-13} J; (*b*) 7.6×10^{-22} kg·m/s
26.55 (*a*) 3.2×10^{-19} kg·m/s; (*b*) 2.8×10^{-11} J; (*c*) 0.54; (*d*) 1.19

CHAPTER 27

27.1   b
27.3   e
27.5   e
25.7   b
27.9   a
27.11 b
27.13 about 7.6×10^{11} photons
27.15 6.63×10^{-19} J (4.14 eV)
27.17 (*a*) 5.3×10^{-22} kg·m/s; (*b*) 2.4 nm
27.19 0.360 nm
27.21 about 1×10^{-34} J·s
27.23 (*a*) 1.99×10^{-19} J (1.24 eV); (*b*) 19.9×10^{-19} J (12.4 eV)
27.25 6.62×10^{-34} J·s
27.27 3.4 eV
27.29 cesium, rubidium, and potassium
27.31 (*a*) No electron is emitted; energy of photon is too low.

27.33 (*a*) 0.025 nm; (*b*) 1.2×10^{19} Hz
27.35 (*a*) $\nu_{max} = 2.41 \times 10^{14}$ V, with ν_{max} in Hz and V in volts. (*b*) 1.24×10^4 V
27.37 (*a*) 7.89×10^{-16} J; (*b*) 2.63×10^{-24} kg·m/s; (*c*) 0.256 nm; (*d*) 1.2×10^{-17} J
27.39 (*a*) 0.121 nm; (*b*) $-60°$ (i.e., 120° with respect to scattered x-ray photon)
27.41 1.3×10^{-15} J
27.43 55 MeV
27.45 (*a*) 1.2×10^3 m/s; (*b*) 0.64 m/s
27.47 0.012 nm
27.49 1.4×10^{-12} m
27.51 (*a*) 0.17 nm; (*b*) 62°
27.53 1.19×10^{-16} J
27.55 (*a*) 1.1×10^{-30} m/s; (*b*) 6.0×10^{22} y
27.57 (*a*) 32 m/s; (*b*) 5.8×10^4 m/s

CHAPTER 28

28.1 a
28.3 e
28.5 b
28.7 c
28.9 e
28.11 d
28.13 3.3×10^{-10} N
28.15 (*a*) 6.77×10^{14} Hz; (*b*) 443 nm
28.17 119 nm
28.19 (*a*) 18; (*b*) $l = 4$; $m_l = 4, 3, 2, 1, 0, -1, -2, -3, -4$; $m_s = \pm\frac{1}{2}$
28.21 1.33 nm
28.23 (*a*) 91.3 nm; (*b*) 83.9 nm
28.25 2.2×10^{-14} m
28.27 (*a*) 1.05×10^{-34} J·s; 2.10×10^{-34} J·s; 3.15×10^{-34} J·s; 4.20×10^{-34} J·s
28.29 (*a*) -0.544 eV; (*b*) 1.32 nm; (*c*) 4.37×10^5 m/s
28.31 (*a*) 8.3×10^6 rev; (*b*) 5.3×10^{-27} J (3.3×10^{-8} eV)
28.35 (*a*) 1.1×10^{-3} A; (*b*) 13 T
28.37 (*a*) 6.8 eV compared with 13.6 eV for H; (*b*) 972 nm compared with 486 nm for H.
28.39 (*a*) 2.1 eV (3.4×10^{-19} J); (*b*) 1.6×10^4 K; (*c*) Enough molecules have energies on the high side of the average to excite Na atoms by collisions.
28.41 255, 185, and 685 nm, of which only the latter is visible.
28.43 1348, 899, 790 nm (IR); 539, 498 nm (VIS); 320 nm (UV)

28.45 (*a*) 6; (*b*) $n, l, m_l, m_s = 5, 1, -1, \pm\frac{1}{2}$; 5, 1, 0, $\pm\frac{1}{2}$; 5, 1, +1, $\pm\frac{1}{2}$
28.47 110
28.49 (*a*) 54.4 eV; (*b*) 1.31×10^{16} Hz
28.53 (*a*) 0.022 nm; (*b*) 9.0×10^{-15} J (56 keV)
28.55 (*a*) 1.66 nm; (*b*) 5; (*c*) 5.3×10^{-34} kg·m²/s
28.57 (*a*) 4.0×10^6 Hz; (*b*) 7.1×10^{-6} nm
28.59 9.9×10^{-5} nm

CHAPTER 29

29.1 d
29.3 a
29.5 d
29.7 c
29.9 c
29.11 (*a*) $92p$, $141n$; (*b*) $92p$, $146n$; (*c*) $91p$, $142n$
29.13 0.71%
29.15 (*a*) 2.2×10^{-17}/s; (*b*) 7.3×10^{-10}/s
29.17 1.15×10^4 y
29.19 (*a*) 7.1×10^{-2} Gy; (*b*) 7.1 rad
29.21 Si: $14e$, $14p$, $14n$; Ge: $32e$, $32p$, $40n$
29.23 35.46 g
29.25 8.8 MeV per nucleon
29.27 (*a*) 0.24%; (*b*) 0.0027%; (*c*) Binding energy compared to rest energy is 100 times smaller for electron.
29.29 (*a*) 60 y; (*b*) 70 y
29.31 $6.6 \, T_{1/2}$
29.33 4.5×10^9 y
29.35 About 5.7×10^3 y
29.37 1.4×10^6
29.39 6×10^9 y
29.41 82 counts/min
29.43 (*a*)$^{137}_{56}$Ba + $_{-1}^{0}e$; (*b*) $^{249}_{98}$Cf + $^4_2$He
29.45 (*a*) $3n$, $3e$; (*b*) $2n$, $2e$
29.47 (*a*) $m(3\alpha) > m$ ($^{12}_6$C); (*b*) $m(^8_4$Be$) > m(2\alpha)$
29.49 2.8 MeV
29.51 9.2×10^6 m/s
29.53 46%
29.55 (*a*) 4.30×10^{21} atoms of $^{235}_{92}$U; 5.98×10^{23} atoms of $^{238}_{92}$U; (*b*) 1.7 g of $^{235}_{92}$U; 236 g of $^{238}_{92}$U
29.57 (*a*) 1.4×10^3 m/s; (*b*) 1.98×10^6 m/s
29.59 (*a*) 6^2_1H $\rightarrow 2^4_2$He + 2^1_1p + 2^1_0n + 43.2 MeV; (*b*) 7.2 MeV per nucleus
29.61 (*a*) 4.6×10^9 Bq; (*b*) 0.12 Ci
29.63 (*a*) 0.70 Gy; (*b*) 7.0 Gy
29.65 (*a*) 15 m; (*b*) 3.3 m

Captions and Credits for Chapter-Opening Photographs

(If a photograph is taken, in whole or in part, from another place in this book, only the number of the corresponding figure is given. If the photograph does not appear elsewhere in the book, its content and credits are given below.)

CHAP. 1: The world's first view of the earth as seen from the moon—August 23, 1966 (NASA). **CHAP. 2:** Junior Olympian swimmers moving in straight lines (Mimi Forsyth/Monkmeyer). **CHAP. 3:** A high-speed flash photograph of a tennis racket exerting a force on a tennis ball (Dr. Harold E. Edgerton). **CHAP. 4:** See Fig. 4.4. **CHAP. 5:** Bridges over the Chesapeake Bay (Maryland State Transportation Authority). **CHAP. 6:** Hoover Dam and Lake Mead in Arizona-Nevada (U.S. Department of Energy). **CHAP. 7:** Blast-off of Space Shuttle *Columbia* on April 12, 1981 (NASA). **CHAP. 8:** An old-fashioned windmill and a large wind turbine (U.S. Department of Energy). **CHAP. 9:** Scanning Tunneling Microscope photograph of a silicon crystal (Thomas J. Watson Research Center, IBM Corporation). **CHAP. 10:** See Fig. 10.25. **CHAP. 11:** A time-lapse photograph of a clock pendulum (Robert Mathena/Fundamental Photographs). **CHAP. 12:** See Fig. 12.23. **CHAP. 13:** An outdoor thermometer (Mimi Forsyth/Monkmeyer). **CHAP. 14:** Solar cell panels on NASA's Nimbus-5 Satellite (NASA). **CHAP. 15:** A state of very high entropy (David Powers/Stock, Boston). **CHAP. 16:** Nikola Tesla surrounded by man-made lightning in his laboratory (Burndy Library). **CHAP. 17:** An electrostatic machine for separating positive and negative charges, thus producing a potential difference between its two terminals (Burndy Library.) **CHAP. 18:** An experimental electric automobile running on batteries (Ai Research Manufacturing Co., courtesy of U.S. Department of Energy). **CHAP. 19:** See Fig. 19.5. **CHAP. 20:** Electric generating plant (Public Utility District of Grant County, Ephrata, Washington. **CHAP. 21:** Electric power lines (Jack Spratt/The Image Works). **CHAP. 22:** TV/Radio tower (Gerard Fritz/Monkmeyer). **CHAP. 23:** See Fig. 23.10. **CHAP. 24:** See Fig. 24.16. **CHAP. 25:** The 200-inch Hale Reflecting Telescope at Mount Palomar, California (Mount Palomar Observatories). **CHAP. 26:** Total Solar Eclipse of March 7, 1970, as taken at Miahuatlan, Mexico (NASA). **CHAP. 27:** See Fig. 27.27. **CHAP. 28:** Lasers in use in a physics laboratory (FPG International). **CHAP. 29:** See Fig. 29.20. **CHAP. 30:** A magnet floating above a superconductor (Thomas J. Watson Research Center, IBM Corporation). **CHAP. 31:** Photograph of Supernova of February 23, 1987 (Robert McNaught/Science Photo Library/Photo Researchers).

Index

Page numbers followed by *i*. indicate illustrations; page numbers followed by *t*. indicate tables of data; page numbers followed by *n*. indicate footnotes. F. and B. prefixes refer to front and back endpaper tables.

A

Aberrations:
  of lenses, 611–613
    chromatic, 612–613
    spherical, 612
  of mirrors, 597
Absolute electric potential, 434
Absolute pressure, 232–233
Absolute space, 680
Absolute (Kelvin) temperature scale, 231–232, 331, 396–397
Absolute zero, 231, 330, 397
  definition of, 330
  quest for, 330
Absorbed dose, 792
  unit of, 792
Absorption of photons by atoms, 746–747, 751
AC (*see* Alternating current entries)
Acceleration(s):
  adding, 39
  angular, 195
  average, 37–38, 44–45
  centripetal (*see* Centripetal acceleration)
  of charges as source of electromagnetic waves, 579–581
  constant, 38
  definition of, 37
  graphs of, 45, 50
  instantaneous, 38–39, 45
  tangential, 39
Acceleration due to gravity, 52–56
  value, 52, 93
  variation of, with position and altitude, 93
Acceleration vectors for motion in a plane, 37–39
Accelerators, particle:
  circular (synchrotrons), 833–835
  colliding-beam machines, 835
  energy achieved by, 850*i*.

fixed-target machines, 834–835
future of, 850–851
linear, 832–833
  SLAC, 832–833
  Van de Graaff, 437*i*., 832
Superconducting Super Collider (SSC), 851
Acceptor levels, 813–814
Accommodation of eye, 650–651
Accuracy of physical data, 14–15
Achromatic doublet, 612–613
Action, quantum of, 708, 742
Activity, nuclear (*see* Nuclear activity)
Adiabatic process, 368
  adiabats, 380–381
    equations for, 380–381
  in sound waves, 368
  throttling process, 368
  work done in, 381–383
Aether:
  failure to detect, 684
  luminiferous, 680
  and speed of light, 679–680
Air, 224–225
Air conditioners, 401
  coefficient of performance of, 401
Air resistance, 80–81
Airplane lift, 257
Airy disk, 660
Algebra, 854
Allowed energy bands, 810
Alpha particles, 738–741
Alphabet, Greek, F.4
Alternating current (ac) and dc power, 538, 558–559
Alternating current (ac) circuits, 538–559
  with capacitance only, 543–545
  energy transfer in, 546–547
  with inductance only, 541–542
  LC circuit, 545–548
  phase relations in, 544*t*.
  with resistance only, 538–539
  resonance in, 550–552
  RLC circuit, 547–553
    with applied emf, 548–553
    effect of resistance, 547
    phase relationships in, 548–549
  transformers in, 553–556

Alternating current (ac) generators, 527–528
Alternating current (ac) quantities:
  effective value, 539
  instantaneous value, 539
  peak value, 539
  rms value, 539–540
Ammeter, 502–503
Ampère, André-Marie, 479, 493
  explanation of magnetism by, 499
  force between current-carrying conductors, 493–494
Ampere (unit of current), 452
  definition of, 493–494
Ampère's circuital law, 495–497
Amplifier, transistor as, 816–817
Amplitude, 273
Amplitude modulation (AM), 582
Analytical methods for vectors, 21–27
Analyzer, 639
Anderson, Carl, 837
Angle measurement, 10–11
Angle of incidence, 589–590
Angle of reflection, 589–590
Angle of refraction, 591
Angles and hand calculators, 26–27
Angstrom unit, 619
Angular acceleration, 195
Angular magnification, 654–655
Angular momentum, 207–212
  conservation of, 210–212
  definition of, 207
  quantization of, 741–742
  vector nature of, 208–209
Angular quantities and linear quantities, 101, 195, 196*t*., 199
Angular size of image, 653
Angular velocity, 100–101, 195
  constant, motion with, 196–198
Antenna, 579
Antimatter, 838
Antineutrino, 831–832, 838
Antinodes, 291
Antiparticles, 837–838
  kinds of, 838
  positron, 837
Antireflection coatings for lenses and mirrors, 631

Arago, D. F., 637
Archimedes, 161, 250
Archimedes' principle, 250–252
    applied to solids, 251–252
    examples of, 252–253
    proof of, 250–251
Aristotle, 33, 80, 680, 851
Artificial radioactivity, 781–783
    by alpha-particle bombardment, 781–782
    by deuterons, 782
    by neutrons, 782–783
Asimov, Isaac, 479
Astigmatism in human eye, 652–653
Astronomical data on solar system, F.2t.
Astrophysics and elementary-particle
        physics, 849–850
Atmosphere, 233, 248
Atmospheric pressure, 246
Atomic clock, 8, 697
Atomic energy (see Nuclear fission; Nuclear
        fusion)
Atomic number, 221
Atomic reactors (see Reactors, nuclear,
        fission; fusion)
Atomic spectra (see Spectrum)
Atomicity of charge, 454–455
Atoms:
    absorption of photons by, 746–747, 751
    density of, 219
    diameter of, 219
    emission of photons by, 750–752
    particles in, 218–219, 771t.
    structure of, 218–222, 736–745
Atwood's machine, 74–75
Automobile tire pressure, 233
Avogadro, A., 223
Avogadro's number, 223–224

B

Back emf, 531
Balance, spring, 65, 94
    use of, to compare masses, 9
Ball at the end of a string, 105
Ballistic pendulum, 182–183
Balmer, J. J., 737
Balmer series, 737
    formula for, 737
Band theory of solids, 810–814
    conduction band, 810
    in conductors, 811
    energy bands, 810-814
        allowed, 810
        forbidden (energy gap), 810
    holes, 812
    in insulators, 810–811
    in semiconductors, 811–814
        silicon and germanium, 812
    valence band, 810

Bands, energy, 810
Banking of curves, 104–105
    banking angle, 104
Bardeen, John, 817i., 821, 822i.
    and BCS theory of superconductivity, 821
Barometers, 248
    mercury, 248
Baryon number, conservation of, 843
Baryons, 839
    properties of, 839t.
Base, 816
Basic mathematics, review of, 854–857
Batteries:
    combination of, 452–453
    dry cells, 452
    emf of, 451
    practical, 452–453
    storage, 452
BCS theory of superconductivity, 821
Beat frequency:
    definition of, 309
    use of, 309
Beats:
    definition of, 308
    frequency of, 309
    production of, 309i.
Becquerel, Antoine Henri, 675, 776
Becquerel (unit for nuclear activity), 792
Bednorz, J. Georg, 822, 823i.
Bel, 313
Bell, Alexander Graham, 313
Bellows camera, 648
Bennet, Abraham, 410
Bergstrand, E., and speed of light
        measurement, 576–577
Bernoulli, Daniel, 255
Bernoulli's equation, 255–258
    and airplane flight, 257
    application of, to hydrostatics, 256
    and baseball curves, 257
Beta decay, 829–832
    experiments on, 829–830
    Fermi's theory of, 830
    and the neutrino, 830
Beta rays:
    decay, 829–832
    emission of, by nuclei, 829
    nature of, 777
Bicycle wheel, motion of, 194, 197–198
Bifocal glasses, 652
"Big Bang" theory of origin of universe,
        1, 705
Bimetallic strip, 335–336
    and thermostats, 336
Binding energy:
    of nuclei, 773–775
    per nucleon, 774, 774i.
Binnig, Gerd, 725
Binomial expansion, 856
Biological effectiveness, 793

Biological effects of radiation, 791–795
Biological sciences and physics, 2
Birefringent crystals, 665n.
Blackbody, 461, 706
Blackbody radiation, 706–710
    equation for, Planck's, 708
    experiments on, 707
    Planck's quantum theory of, 707–710
    Rayleigh-Jeans theory of, 707
    Wien's theory of, 707
Blood, flow of, 255, 260–261
Blood pressure, 249–250
    diastolic, 249
    in the human circulatory system, 255
    systolic, 249
Body-centered cubic lattice, 806
Bohr, Niels, 726, 736, 748i., 763, 783,
        830
    biography of, 739
    and nuclear fission, 783
    and principle of complementarity,
        726–727
    and theory of hydrogen atom, 741–745
Bohr-Einstein frequency condition, 742
Bohr-Rutherford atom, 738–741
Bohr's theory of the hydrogen atom,
        741–745
    ad hoc nature of, 761
    disagreement of, with classical physics,
        742
    discrete energies, 743–744
    failures of, 760–761
    postulates of, 741
    prediction of value of Rydberg constant,
        743–744
    radii of electron orbits, 744–745
    successes of, 744, 758–760
Boiling point, 343–345
    effect of temperature on, 344–345
    of various substances, 344t.
Boltzmann, L., 235, 384
Boltzmann's constant, 235
Born, Max, 6i.
Boyle, R., 231
Boyle's law, 230–231
Bragg, William H., 721
Bragg, William L., 721
Bragg equation for x-rays, 722
Bragg planes, 721
    reflection of x-rays by, 721–722
Brahe, Tycho, 109
Braking radiation, 716
Brattain, Walter H., 817i.
Breaking point of a wire, 267
Breeder reactors (LMFBR), 789–790
Bremsstrahlung (braking radiation), 716
Brewster's angle, 640–641
Btu (British thermal unit), 339
    and kilocalorie, 339
Bubble chambers, 797–798

Building-up principle, 758
Bulk modulus, 265
Buoyancy, center of, 253
Butterfield, Herbert, 33

C

Caloric, 153
Calorie, 339
Cameras, 647–650
  bellows, 648
  f-number of, 648–649
  and human eye, 650
  pinhole, 647–648
  shutter speed of, 649–650
Capacitance, 437–440
  definition of, 438
  of parallel-plate capacitor, 438–439
  unit of, 438
Capacitive reactance, 543–545
  dependence of, on C and f, 544
Capacitors, 437–445
  in dc circuits, 468–469
  energy stored in, 444–445
  in parallel, 442–443
  parallel-plate, 438–439
  practical, 441
  and resistors in dc circuits, 468
  in series, 443
  tuning, 441
Cape Canaveral, 95, 407
Carbon atom, 221i.
Carbon dioxide, phase diagram of, 345
Carbon ($^{14}_{6}$C) nucleus, use of, in radiocarbon dating, 779
Carnot, Nicolas Léonard Sadi, 390
  biography of, 390
Carnot cycle, 390–393
  and second law of thermodynamics, 397
Carnot engine, 390–393
  efficiency of, 392
  practicality of, 393
Cassegrainian focus, 658, 659
Cathode-ray oscilloscope, 526–527
Cathode rays, 489
Cavendish, Henry, 91, 413
  measurement of gravitational constant by, 91–92
Cavendish balance, 91
Cell (see Batteries; Solar cells)
Celsius, Anders, 332
Celsius temperature scale, 332
Center:
  of buoyancy, 253
  of gravity, 130–131
  of mass, 132
Center-of-mass motion, 175
Centigrade temperature scale, 332
Centipoise, 259

Centrifuge, 105–107, 200
  effective g values of, 106
  physiological effects of, 107
  use of:
    in separating blood samples, 106
    in space program, 106i.
Centripetal acceleration, 101–103, 137
  equation for, 102
  and tangential acceleration, 103
Centripetal force, 103–107, 111
  equation for, 103
CERN (Centre Européen pour la Recherche Nucléaire), 834, 835, 840, 841
Cesium atomic clock, 8, 697
Chadwick, James, 771, 829
  and beta-decay experiments, 829
  and discovery of the neutron, 771
Chain reaction, 785–786
Chamberlain, Owen, 838
Change of phase, 342–345
Characteristic x-ray spectrum, 758–760
Charge, electric, 407–416
  atomicity of, 454–455
  conservation of, 409
  elementary, 413–414
    measurement of, by Millikan, 413–414
  induction of, 412
  interaction of, 409
  kinds of, 408–409
  on metal objects, 423
  positive and negative, 408–409
  on a spherical metal surface, 422–423
  storage of, 436–437
Charge number (see Atomic number)
Charge-to-mass ratio (q/m) for particles, 487
  determination for electron, 489–492
Charles, J., 231
Charles' law, 231–232
Chemical elements, 736
Chernobyl, 794
China syndrome, 789
Chip, silicon, 819i.
Chromatic aberration, 612–613
Circle of least confusion, 612
Circuit breakers, 557
  magnetic, 557
  thermal, 557
Circuital law (see Ampère's circuital law)
Circuits, electric:
  dc and ac, 469n.
  and Kirchhoff's rules, 460–469
  simple, 456–457
Circular motion:
  acceleration in, 101–103
  force in, 103–107
  uniform, 100–101
  in a vertical circle, 107–109
Circular obstacle, diffraction around, 636–637
Classical models, inadequacy of, 726–727

Clausius, Rudolf, 383n., 385
  and summary of thermodynamics, 388–389
Clocks:
  atomic, 8, 697
  pendulum, 148, 279
Cloud chambers, 797–798
Cobalt ($^{60}_{27}$Co) nuclei:
  gamma-rays produced by, 794
  and parity nonconservation, 843–844
Coefficient of linear expansion, 334
Coefficient of performance, 399–401
  of air conditioners, 401
  of heat pumps, 399–400
  of refrigerators, 401
Coefficient of volume expansion, 335
Coherence:
  and incoherence, 621
  in laser light, 621, 752
  in light sources, 621–622
Coils, circular, 496–497
  magnetic field at center of, 496
Coincidence-counting systems, 798
Collector, 816
Colliding-beam accelerators, 835
Collimator, 626
Collisions, 179–186
  in one dimension, 179–183
    elastic, 179–181
    inelastic, 181–183
  in two dimensions, 183–185
  of nuclear particles, 184–185, 835
Color sensitivity of eye, 653
Color temperatures of stars, 356t.
Communication channels:
  expansion of, 583–584
  use of visible light in, 584
Commutator, split-ring, 529
Compass needle, 492
Complementarity, Bohr's principle of, 726–727
Complete electromagnetic spectrum (see Electromagnetic spectrum)
Components of a vector, 21–22
Compound microscope, 655–657
  magnification of, 656
  resolution of, 660–661
Compressibility, 265
Compression, 121–122
Compressional waves, 285–286
  (See also Sound waves)
Compton, Arthur Holly, 717
Compton effect, 717–719
  apparatus for, 718
  Compton's experimental results on, 719i.
  Compton's explanation of, 717–718
  difference of, from photoelectric effect, 717n.
Compton wavelength of electron, 718
Concave mirrors, 596–601

Concrete, 121–122
Condensors (*see* Capacitors)
Conduction band, 810
Conduction of heat, 351–353
  in houses, 353–354
Conductivity, thermal (*see* Thermal
    conductivity)
Conductors, electric, 409–410
  band theory of, 810
  metals as, 410
Confinement of nuclei, 790–791
Conservation of energy:
  and first law of thermodynamics, 366–367
  general principle of, 160
  mechanical, 156–159
  in roller coaster, 157–158
Conservation of mass-energy in nuclear
    reactions, 781
Conservation of nucleon number in nuclear
    reactions, 780–781
Conservation of parity, 843–844
Conservation principles:
  angular momentum, 210–211
  in classical physics (*see* Conservation of
    energy; Linear momentum,
    conservation of)
  for elementary particles, 842–844
  energy, 160, 365–366, 530
  linear momentum, 174–177
  mass-energy, 781
Conservative forces, 149–152
Constancy of speed of light in special
    relativity, 684–685
Constant of proportionality, 858
Constants:
  numerical, F.3
  physical, B.2
Contact forces, 119–120
Continuity, equation of, 254–255
Continuum, 736, 746
Control rods, 788
Convection of heat, 354
Converging lenses, 604–607
Conversion of mass into energy (*see* Mass-
    energy conversion)
Conversion of units, 12
Conversion factors, B.1
Convex mirrors, 596, 602–603
Coolidge, William D., 715
Coolidge x-ray tube, 715–716
Cooling, Newton's law of, 358–359
Cooper, Leon, 822*i.*
  and BCS theory of superconductivity, 821
Coordinate system, 17
  right-handed and left-handed, 604
Copenhagen interpretation of quantum
    mechanics, 763
Cornea, 650–651
Corner-cube reflectors, 590
Coulomb, Charles-Augustin, 413

Coulomb (unit of charge), 413
Coulomb's law, 413–416
  compared with Newton's law of universal
    gravitation, 413–414
Coulomb's torsion balance, 413
Counters, gas-filled, 796–797
Covalent crystals, 806
Cowan, Clyde L., Jr., 831–832
Crick, Francis, 2, 722
Critical angle for total internal reflection,
    593
Critical mass, 786
Cross product, 126*n.*; 208*n.*; 485*n.*
Crystal, 805
Crystalline solids, 805–807
  covalent, 806
  ionic, 805–806
  metallic, 807
  molecular, 807
Curie, Marie, 776, 783
Curie, Pierre, 776
Curie (unit for nuclear activity), 792
Current, electric (*see* Electric current)
Current balance, 494
Current-carrying conductors:
  force on, in magnetic field, 481–482
  force between two, 493–494
Current loop, magnetic field of, 496
Curves:
  banking of, 104–105
    banking angle, 104
  baseball, 257
Cut-off (stopping) potential, 711
Cut-off wavelength for x-rays, 716
Cyclotron, 488–489
  principle of, 489

D

D-D and D-T reactions, 790
Damped vibrations, 280
Damping of ac circuit, 547
Dark matter in universe, 849
d'Arsonval, Jacques, 503
  and galvanometer, 503*i.*
  and ocean thermal energy conversion, 398
Data, physical, accuracy of, 14–15
Daughter nucleus, 781
Davisson, Clinton J., 723
DC (*see* Direct current entries)
Dead Sea Scrolls, 779
de Broglie, Louis Prince, 720–721, 723,
    761–763
de Broglie waves, 720–721, 761
  de Broglie wavelength, 721
  of electron, 721
  experimental verification, 723
  standing waves in atoms, 761
Debye, Peter, 809

Debye theory of specific heats, 809
Decay constant:
  and half-life, 778
  for radioactive decay, 777–778
Decibel, 312–313
Decibel scale, 312–313
Dees, 488–489
Defects of the eye, 651–653
Degrees of freedom, 372
Density:
  determination of, 250–252
  of a gas, 225
  linear, 285
  of a liquid, 252
  of liquids and gases, 244*t.*
  of solids, 263*t.*
Depth of field, 647, 649
Detection of radio waves, 581
  electric, 581
  magnetic, 581
Detection of sound:
  in human ear, 310–311
  microphones, 310
  transducers, 310
Detectors (*see* Radiation detectors)
Dewar, James, 359
Dewar (*see* Thermos bottles)
Diamagnetic materials, 498
Diamond, 594, 806
Dicke, Robert H., 699
Dielectric constants, 440
  table of, 440*t.*
Dielectric materials, 440–441
  molecular view of, 441
  speed of electromagnetic waves in, 572
Diesel engine, 395
Diffraction:
  of atoms and molecules, 726
  of electrons, 723–726
  of light, 632–638
    by circular obstacle, 636–637
    criterion for, 632–633
    definition of, 632
    and Huygens' principle, 633–634
    importance of, 642
    and interference, 637–638
    single-slit, 634–636
    by slits of varying width, 633
  of neutrons, 726
  of water waves, 633
  of x-rays, 721–722
Diffraction grating, 624–627
Diffraction pattern:
  of circular obstacle:
    experimental pattern, 637*i.*
    theory, 636–637
  of single slit:
    calculation of, 634–636
    experimental pattern, 636*i.*
Diffuse reflection, 590

Diffusion of gases, 236–237
Dimensional analysis, 13–14, 147, 174
Dimensionless constants, 13
Dimensions, 12–13
Diode rectifier, 815
Diopters, 610
Dip needle, 501
Dipole:
  electric, 419
    induced, 441
    permanent, 441
  magnetic, 496
Dipole antenna, 579
Dipole moment, magnetic, 496, 502
Dirac, P. A. M., 763, 837
  and the positron, 837
Direct current (dc) electricity, 450–472
Direct current (dc) generators, 529
Discrete energy states:
  in atoms, 745–750
  in nuclei, 841
Discrete spectrum, 736
Disorder, 383, 386
  and entropy, 383–384
  and probability, 384
Dispersion, 595
  by a prism, 595–596
Displacement in simple harmonic motion, 273
Displacement current (*see* Maxwell's displacement current)
Displacement vectors, 18–19, 41
Dissipative forces (*see* Nonconservative forces)
Distance and displacement, 18
Distribution of molecular velocities in gases, 237–238
Diverging lenses, 604, 607–608
DNA, double-helix theory of, 2
Dollond, John, 612
Domains, magnetic, 499–500
  growth of, 499
  rotation of, 499
  visibility of, 500*i.*
Donor levels, 813
Doping of semiconductors, 812
Doppler, Christian, 317
Doppler effect:
  in light, 319–320
  in sound, 317–320
    general equation, 319
    moving observer, 317–318
    moving source, 318–320
Dot product, 144, 147
Double-helix theory of DNA, 2
Double-slit interference, 622–624
Drift velocity, 457, 486
Dry cells, 452
Dulong and Petit, law of, 808–809
Dynamic equilibrium, 119, 134–136

Dynamics:
  definition of, 33
  equations of, 64–77
Dynode, 797

E

Ear, human, 310–311
  defects in, 311
  diagram of, 310*i.*
  frequency response of, 314
  hearing aids for, 311
  pressure on, 246–247
Earth satellites, 111–114, 323
  synchronous (stationary), 113–114
  weightlessness in, 112–113
Earth's magnetic field, 500–502
Eddy-current damping, 517
Eddy currents, 517
Edison, Thomas, 538, 558
Effective values:
  of ac current, 539–540
  of ac voltage, 539–540
Efficiency:
  of heat engines, 389
  of machines, 162
  of physical processes, 367
Eigenfunctions, 762
Eigenvalues, 762
Einstein, Albert, 2, 3*i.*, 675, 682*i.*, 748*i.*, 763, 809, 848, 851
  biography of, 681
  and photoelectric effect, 711–714
  and special theory of relativity, 684–698
  and unified field theory, 848
Elastic collisions, 179–181
Elastic forces as conservative forces, 152
Elastic limit, 267
Elastic modulus, 264–265
  of metals, 265*t.*
Elastic potential energy, 151–152
Elastic properties of solids, 264–267
  elastic moduli, 265*t.*
Electric chair, 538
Electric charge (*see* Charge, electric)
Electric circuit, simple, 456–457
Electric current, 451
  alternating (ac), 538–540
  conventional, 452
  definition of, 451
  direct (dc), 451–471
  direction of, 452
  displacement, 564–568
  in electrolytes, 453–455
  equivalent, 568
  in metals, 455–457
  rms value, 539–540
  unit of, 452
Electric dipole, 419, 441

Electric energy (*see* Energy, electric)
Electric field, 416–420
  of charged metal sphere, 422–423
  definition of, 417
  of electric dipole, 419
  and electric potential, 432
  energy stored in, 444–445
  and equipotential surfaces, 431–433
  lines of, 418–419
  of long, straight wire, 423–424
  between two charged plates, 419
  units of, 417
Electric flux, 421
Electric generators (dynamo), 527–530
  alternating current, 527–528
  direct current, 529
  emf produced by, 528
  and laws of electromagnetic induction, 532–533
  loading the, 529
  practical, 529
Electric meters, 502–503
Electric motors, 530–533
  ac and dc, 530–531
  back emf, 531–532
  and laws of electromagnetic induction, 532–533
Electric oscillations, comparison of, to mechanical oscillations, 547*t.*
Electric potential (*see* Potential, electric)
Electric power:
  in ac circuits, 549–550
  in dc circuits, 469–470
  and electric energy, 469
  home use of, 469–470
  practical aspects of, 556–559
  safety devices for, 557–558
  transmission of, 556
Electric shocks, 558–559
Electrical equivalent of heat, 470–471
  and conservation of energy, 471
  Joule's experiment, 471
  and laws of thermodynamics, 471
Electricity:
  alternating current, 538–559
  direct current, 450–472
  home usage of, 556–557
Electrocardiograph, 527
Electrodynamics, 493
Electrolytes, 453–455
  electric currents in, 453–454
  Faraday's laws of, 454
Electromagnetic induction, 510–519
  laws of, applied to motors and generators, 532–533
  mutual, 517–518
  self, 518–519
Electromagnetic spectrum:
  complete, 574–575
  table of wavelengths and frequencies, 575*t.*

Electromagnetic waves, 564–584
  energy in, 572–574
  experiments of Hertz, 571
  generation of, 579–581
    phase relations in, 580–581
  intensity of, 572–574
  nature of, 569
  properties of, 569–570
  relationship between electric and magnetic
    fields, 570
  speed of (*see* Speed of electromagnetic
    waves)
Electromagnets, 497–498
Electrometer, 411
Electron(s):
  charge on the, 218–219
  e/m for, 489–492
  mass of, 218–219
  in the nucleus, 770
  origin of name of, 492
Electron configurations, 758*i*.
Electron diffraction, 723–725
  comparison of, with x-ray diffraction,
    723*i*.
  equation for, 723
Electron impact experiments, 747–750
Electron microscope, 664, 724–725
  Nobel prizes for, 725
  scanning (SEM), 725
    SEM photo of silicon crystal, 725*i*.
  transmission (TEM), 724–725
    comparison of, with optical microscope,
      724*i*.
    resolution of, 725
    TEM photo of barium atoms, 725*i*.
Electron spin, 755
Electronvolt:
  definition of, 430
  unit of energy, 430
Electroscope:
  charging by induction, 412
  gold-leaf, 410–413
    use of, in electrostatics, 411–413, 436
Electrostatics, 407–445
  basic facts of, 407–409
  Benjamin Franklin's contributions to,
    408–409
  problems of, 415–416
Electroweak theory, 844–847
Elementary-particle physics, 827–851
  experimental work, 850–851
  future of, 848–851
  and high-energy physics, 828
  present state of, 844–848
  relationship of, to astrophysics, 849
Elementary particles, conservation principles
    for, 842–844
  conservation of baryon number, 843
  conservation of parity, 843–844
  families of (May, 1990), 842*t*.

Elements, periodic table of, F.6
Elevator:
  forces on a passenger in, 75–76
  and weightlessness, 76
Ellipse, 109
e/m for electron, 489–492
emf:
  induced, 511
  motional, 510–511
  source of, 450
  and terminal voltage, 464–465
  and voltage, 451
Emission of photons by atoms:
  forced (stimulated), 750–752
  spontaneous, 750–751
Emission spectrum of hydrogen, 736–738,
    741–745
Emissivity, 355
Emitter, 816
Endoscope, 594, 753–754, 753*i*.
Energy, 1, 145
  electric:
    in dc circuits, 469–470
    storage of:
      in capacitor, 444
      in electric field, 444–445
      in inductor, 519–520
      in magnetic field, 520–521
  internal, 365–366
  mechanical:
    conservation of, 156–159
    kinetic, 145–147
    potential:
      elastic, 151–152
      gravitational, 147–151
    (*See also* Conservation of energy;
      Work)
Energy bands, 810–814
Energy conservation, 160
  and first law of thermodynamics, 366–367
Energy density:
  in combined electric and magnetic fields,
    521
  in electric field, 444–445
  in magnetic field, 520–521
Energy gap in band theory of solids, 810
Energy levels, 745–750
  in hydrogen, 746–747
  in mercury, 749
  negative values of, 746
  origin of, 744
  proof of, 749
Energy shell, 754
Engines (*see* Heat engines)
Entropy, 383–388
  changes in, 385–387
    in irreversible processes, 385–386
    in reversible processes, 385
  definition of, 383–384
  and disorder, 383–384

  practical consequences of, 387–388
  statement of second law, 386
Equilibrium, 119–136
  conditions for, 122, 127
  dynamic, 119, 134–136
    rotational, 135–136
    translational, 134–135
  kinds of, 133–134
    neutral, 133
    stable, 133
    unstable, 133
  static, 119, 122–130
    rotational, 127–130
    translational, 122–124
Equipartition of energy, principle of, 372
Equipotential surfaces, 431–433
  and lines of electric field, 431–433,
    434*i*.
  in metals, 433
Equivalence principle in general relativity,
    699
Equivalent current, 568
Ether (*see* Aether)
Evaporation, 238
Excitation of atoms:
  by electron-impact, 747–749
  by heating, 747
  by photon absorption, 746–747
Excitation energy, 747
Excitation potential, 747
Excited states of nuclei, 841
Exclusion principle (*see* Pauli exclusion
    principle)
Expansion:
  linear, 334
  of liquids with temperature, 336
  of solids with temperature, 334–335
  volume, 335
Experiment, role of, in physics, 5
Experimental basis of quantum mechanics,
    705–730
Exploding shell, 177–178
Exponential curves, 522–525, 777–779,
    862–863
Exponential decay of radioactive nuclei,
    777–779
Exponential functions, 856–857
Extended work-energy theorem, 155
Extrapolation, 860–861
Extrinsic semiconductors, 812–814
Eye, human, 650–653
  accommodation of, 650–651
  and camera, 650
  color sensitivity of, 653
  cornea, 650–651
  defects of:
    astigmatism, 652–653
    farsightedness, 651–652
    nearsightedness, 651
  diagram of, 650*i*.

Eye, human, (cont.)
near point of, 650
retina, 650
Eyepiece lens, 655

F

f-number, 648–649
and brightness of image, 649
definition of, 649
and shutter speed, 649–650
Fabry, Charles, 668*n.*
Fabry-Perot interferometer, 668–669
interference fringes produced with, 669
(*See also front cover*)
Face-centered cubic crystals, 805
Face-centered cubic lattice, 805
Fahrenheit, Gabriel, 332
Fahrenheit temperature scale, 332
Falling objects, 52–56
energy analysis of, 156–159
Farad, 438
Faraday, Michael, 510–515, 564
biography of, 513
Faraday (amount of charge), 454
and Avogadro's number, 454
and charge on electron, 455
Faraday's induction law, 512–515, 527, 532–533
Faraday's laws of electrolysis, 454
Farsightedness, 651–652
Feedback mechanism, 336
Fermi, Enrico, 5, 7*i.*, 782
biography of, 7
and bombardment of nuclei, 782–783
and first nuclear pile, 786
and order-of-magnitude calculations, 7
and thermal neutrons, 788*n.*
and theory of beta-decay, 830
Fermi, Laura, 7
Fermi-Dirac statistics, 7
Fermilab Tevatron Collider, 834
Ferromagnetic materials, 499–500
Feynman, Richard P., 845
biography of, 846
Feynman diagrams, 845–847
Fiber optics, 594–595
use of:
in communications, 594–595
in medicine, 594, 753
Field, electric (*see* Electric field)
Field, magnetic (*see* Magnetic field)
Figure skaters, 210
Fine structure, 754
First Bohr radius, 744
First law of thermodynamics, 364–367
applications of, 368–370
and conservation of energy, 366–367
living systems and, 369–370

Fissile nuclei, 788
Fission, nuclear (*see* Nuclear fission)
FitzGerald, George Francis, 688
Fixed-target accelerator, 834–835
Fizeau, Armand, speed-of-light measurement, 575–576, 684
Flashlight, 456–457
Fluids, 245
hydrostatic pressure of, 245–249
in motion, 253–261
Fluorescence, 754
Fluorescent lights, 754
Flux:
electric, 421
magnetic, 512
Flywheels, 202, 210
Focal length:
of lens, 605
of spherical mirror, 596–597
Focus:
real, 597
virtual, 602
Focusing, 648
Foot-pound, 144
Forbidden energy bands, 810
Force:
and acceleration, 66
centripetal (*see* Centripetal force)
between current-carrying conductors, 493–494
definition of, 64
nature of, 64–65
net, 65
vector nature of, 65
Force constant of a spring, 151, 267
Force vectors, 72–73
Forced emission, 751–752
Forced vibrations:
resonant frequencies, 293
of a string, 292–293
Forces:
air resistance, 80–81
components of, 72–73
compressive, 121–122
conservative, 149–151
contact, 119–120
external, 150
frictional, 77–79
on an inclined plane, 81–84
internal, 150
kinds of, 64, 69, 827–832, 828*t.*
nonconservative, 151
nuclear (*see* Nuclear forces)
reaction, 119–120
resistive, 77–81
tensile, 121–122
Fossil-fuel power plants, 394
efficiency of, 394*i.*
Four forces of physics, 827–832
function of, 828

relationship of, to the origin of the universe, 849
relative strengths, 828*t.*
unification of, 845, 849
Frame of reference, 34, 676
inertial, 676
noninertial, 676
preferred, 678–679, 684
Franck, James, 747, 748*i.*
Franck-Hertz experiment, 747–750
apparatus for, 747–748
and photon emission, 749
results of, 748–749
significance of, 749
Franklin, Benjamin, 407–409
biography of, 408
Franklin, Rosalind, 722
Fraunhofer, Joseph von, 624
Fraunhofer lines, 747
Free-body diagrams, 72
Free-electron gas model, 807
Free fall, 52–56
acceleration in, 52
flash photo of, 53*i.*
Galileo's ideas on, 56–58
with initial velocity, 53–55
position, velocity and acceleration in, 53*i.*
Freezing point, effect of pressure on, 345
Frequencies used in radio and television, 583*t.*
Frequency, 8*n.*
in circular motion, 100
and energy, 706
fundamental and harmonics, 293
in simple harmonic motion, 276
ultrasonic, 322–323
unit of, 100
and wavelength of waves, 283, 574–575
Frequency bands, 583–584
Frequency modulation (FM), 582
Fresnel, Augustin, 637
Friction, 77–79
coefficients of, 78, 79*t.*
finding by experiment, 83
force of, 77–79
graph of forces, 78*i.*
on an inclined plane, 82–83
sliding (kinetic), 78
static, 78
Frisbee, motion of, 194
Frisch, Otto, 783–785
Froome, K. D., measurement of speed of microwaves, 577
Functional equations, 858–859
Fundamental frequency, 293
Fundamental quantities of mechanics, 6–10
Fuses, electric, 556–558
Fusion (change of state), 343
heat of, 343
Fusion, nuclear (*see* Nuclear fusion)

G

Galileo Galilei, 56–58, 57*i.*
  biography of, 57
  and motion on an inclined plane, 56, 58
Galvanometer, 502–502
  conversion of:
      to an ammeter, 503
      to a voltmeter, 503
  d'Arsonval, 503*i.*
Gamma rays, 575*t.*, 777
Gap, energy, 810
Gas constant:
  per mole, 233
  per molecule, 235
Gas-filled detectors, 796–797
Gas laws:
  Boyle's law, 230–231
  Charles' law, 231–232
  ideal gas law, 233–234
  pressure-temperature law, 232
Gaseous diffusion (*see* Diffusion of gases)
Gases:
  density and specific gravity, 244*t.*
  important physical quantities for,
      225–227
  kinetic energy and pressure of, 230
  kinetic theory of, 227–230, 235–237
  nature of, 224
  pressure of, 225
  structure of, 224–225
  types of, 224–225
Gauge pressure, 232–233
Gauss, Karl Friedrich, 421
Gauss (unit of magnetic field), 482
Gauss' law, 421–423, 438
Gaussian surface, 421, 438
  use of, in electrostatics, 422–424
Gay-Lussac, J., 232
Ge(Li) detectors, 820
Geiger, Hans, 740, 796
Geiger counters, 796–797
Gell-Mann, Murray, 839–841
General theory of relativity (*see* Relativity,
      general theory of)
Generators, electric (*see* Electric generators)
Genetic damage, 792–794
Geodimeter, 576–577
Geometrical optics, 588–613
  validity of, 588
Geometry:
  basic formulas from, 855
  useful theorems from, 855
Germanium, 222, 812
Germer, I. A., 723
Glashow, Sheldon Lee, 845–847, 848*n.*
Gluons, 845*t.*, 847
Goeppert-Mayer, Maria, 6*i.*
Grand unified theories (GUTs), 848–850
  predictions of, 848–849

Graphical methods:
  for rectilinear motion, 41–47
  for vectors, 19–21
Graphs, 859–863
  importance of straight line, 861
Grating, diffraction (*see* Diffraction grating)
Grating spectrometer, 626
Gravitation, Newton's law of universal (*see*
      Newton's law of universal
      gravitation)
Gravitational collapse, 211–212
Gravitational constant *(G)*, 91–92
  measurement of, 91–92
  units of, 92
  value of, 92
Gravitational force, 90–94
  conservative nature of, 149–150
Gravitational mass and inertial mass, 699
Gravity:
  center of, 130–131
      and center of mass, 132
      technique for finding, 132
  specific (*see* Specific gravity)
Gray (unit of absorbed dose), 792
Greek alphabet, F.4*t.*
Greenhouse effect, 361–362
Grimaldi, Francesco, 632, 635
Ground, electric, 412
Ground fault interrupter (GFI), 557–558
Ground state, 756
Grounding of appliances, 558–559
  3-pronged plugs, 558
Gyroscopes, 209–210

H

Hadrons, 828, 839
  definition of, 839
  properties of, 839*t.*
Hafele, J. C., 697–698
Hahn, Otto, 783
Hale reflecting telescope, 658, 659*i.*
Half-life of radioactive nuclei, 777–779,
      778*t.*
  and decay constant, 778
  and exponential decay, 778
Hall effect, 485–487
  and current carriers in metals, 485–486
  gaussmeter, 486
  measurements of, on semiconductors, 814
Hall voltage, 486
Hand calculators, 26–27
Harmonics, 293
  for vibrations of a string, 293
He$^+$ spectrum of, 758
  explanation of, by Bohr, 758
He–Ne gas laser, 751–752
  energy levels involved, 752
  schematic diagram of, 752*i.*

Hearing aids, 311
Heat, 337–345
  conduction of, 351–353
      in houses, 353–354
  convection of, 354
  definition of, 338
  electrical equivalent of, 471
  flow dependent on path, 363–364
  as a form of energy, 337–338
  mechanical equivalent of, 340–342
  radiation of, 355–362
      from sun, 360–361
  and temperature, 338
  and thermal energy, 337–338
  transfer of, 351–362
  and work, 362–364
Heat of fusion, 343
Heat of sublimation, 345
Heat of vaporization, 343
Heat capacities:
  molar, for ideal gas, 370–373
  specific, 338–340, 339*t.*
Heat conductivity (*see* Thermal conductivity)
Heat death, 388
Heat engines, 389–395
  Carnot, 390–393
  definition of, 389
  diesel, 395
  efficiency of, 389, 394*i.*
  internal-combustion, 395
  practical, 393–395
  steam, 393–394
      reciprocating, 393
  steam turbines, 394
Heat pumps, 398–399
  coefficient of performance of, 399–400
  practical, 400
Heisenberg, Werner, 727, 763, 771
  and proton-neutron nuclear model, 771
  and quantum mechanics, 763
  uncertainty principle, 727–730
Helium, 757–758
  liquid, 330
Helmholtz, Hermann von:
  analysis and synthesis of musical sounds
      by, 317
  biography of, 316
  and the ophthalmoscope, 665–666
  resonators, 317
Henry, Joseph, 512
Henry (unit of inductance), 518
Hero's engine, 72
Hertz, Gustav, 747
Hertz, Heinrich, 571, 680
  experiments with radio waves, 571
  and photoelectric effect, 711
Hertz (unit of frequency), 100
Hertzian dipole, 571, 579
High-energy physics, 827–851
  and elementary-particle physics, 828

High-temperature superconductivity, 822–823
  experimental evidence needed, 822
  practical applications of, 822–823
  practical problems of, 823
  progress, 822
  theoretical explanation of, 823
Holes:
  motion of, in electric fields, 812–814
  production of, in crystals, 812–814
Holography, 631–632
Home usage of electricity, 556–557
Hooke, Robert, 266
Hooke's law, 151, 266–267, 275
  and simple harmonic motion, 275
Horsepower, 163
Hubble Space Telescope, 659, 850
Human ear (*see* Ear, human)
Human eye (*see* Eye, human)
Huygens, Christiaan, 117–178, 620
  biography of, 178
Huygens' principle, 620–621
  and law of reflection, 621
Hydraulic devices, 247
  conservation of energy in, 247–248
Hydrogen atom:
  Balmer series, 737
  Bohr's theory of, 741–745
  discrete energies, 743–744
  emission spectrum of, 736–738
  energy-level diagram of, 744*i.*
  Rydberg equation for, 737
  spectral series, 737*t.*
  spectrum of 736–738, 741–745
  structure of, 738–741
Hydrogen-like atoms, spectrum of, 758
Hydrologic cycle, 148–149
Hydrostatic pressure, 245–246
Hydrostatics, 245–253
Hypothesis, 4

I

Ideal gas:
  definition of, 231
  kinetic theory of, 227–230
Ideal gas law, 233–235
  and kinetic theory, 235–237
Image distance, 597, 605
Image formation:
  by lenses, 605–608
    analytical method, 606–608
    graphical method, 605–606
  by mirrors, 598–601
    analytical method, 599–601
    graphical method, 598–599
Images:
  inverted, 599
  real, 598
  virtual, 598–599

Impedance, 548–549
Impedance matching, 554–556
Impulse, 173–174
  definition of, 173
  dimensions and units of, 173–174
  and impulse-momentum theorem, 173–174
  rotational, 207–208
Impulse-momentum theorem, 173–174
Inclined plane:
  dynamics of motion on an, 81–84
    with friction, 82–84
    without friction, 81–82
  kinematics of motion on an, 56–58
  as a simple machine, 160
Indeterminacy principle (*see* Uncertainty
    principle)
Index of refraction, 591
  of glasses, 596
  of important materials, 591*t.*
Induced charges on insulators, 412–413, 441
Induced radioactivity (*see* Artificial
    radioactivity)
Inductance:
  in dc circuits, 521–523
  mutual, 517–518
  self-, 519
  unit of, 518
Induction:
  of charges, 412–413
  electromagnetic (*see* Electromagnetic
    induction)
Inductive reactance, 541–543
  dependence of, on L and f, 542
Inductor, 519
  energy stored in, 519–520
Industrial revolution, 510
Inelastic collisions, 181–183
Inertia, 9, 66
  rotational (*see* Rotational inertia)
Inertial frame of reference, 70, 676
Inertial mass and gravitational mass, 699
Infrared, 575*t.*
Initial position, 48
Instantaneous acceleration, 38–39, 45
Instantaneous velocity, 36–37, 42–43
Insulators, electric, 410, 810–811
Integrated circuits (ICs), 819
Intensity:
  of electromagnetic waves, 572–574
  level, 312–313
  levels of common sounds, 313*t.*
  and loudness, 312–314
  of a sound wave, 311–315
    plane wave, 312
    spherical wave, 312
Interactions:
  and associated quanta, 845*i.*
  and forces, 828
    strong, 828
    weak, 829–832

Interference:
  of light, 622–632
    applications of, 631–632
    and diffraction, 637–638
    diffraction grating, 624–627
      grating equation, 626
    double-slit experiment, 622–624
      importance of, 642
    multislit, 624–627
    in thin films, 628–631
  of sound waves, 304–309
    of different frequencies, 308–309
    moving in opposite directions, 304–307
  of waves, 288–289
    constructive, 288–289
    destructive, 289
Interference fringes, produced by a double
    slit and a grating, 626
Interferometers, 9
  Fabry-Perot, 668–669
  Michelson, 666–668
Intermediate vector bosons, 845
  experimental evidence for, 845–847
Internal combustion engine, 395
Internal energy, 365–366
Internal resistance of a battery, 464–465
Interpolation, 859–860
Intrinsic semiconductors, 812
Invariance:
  definition of, 677
  in general relativity, 699
  of laws of mechanics under Galilean
    transformations, 677–678
  of Maxwell's equations, 679–680
  in special relativity, 684–687
Iodine ($^{131}_{53}$I) nucleus, use of, in thyroid
    treatment, 794
Ionic crystals, 805–806
Ionization chambers, 796
Ionization energy, 746
Ionization potential, 746
Ionizing radiation:
  effect of, on living matter, 791–794
  sources of, in U.S., 793*t.*
Ions, 221
Iron, soft and hard, 498*n.*
Iron ($^{56}_{26}$Fe) nucleus, 774
Irreversible processes, 379–381
  adiabatic, 380–381
  entropy changes in, 385–386
  and real physical processes, 380
Isobaric process, 369
Isochoric process, 369
Isothermal process, 368
  isotherm, 380
  work done in, 381–382, 383*n.*
Isotopes, 221–222
  abundances of, 869–870*t.*
  constituent parts of, 222*t.*
  masses of, 869–870*t.*

Isotopes, (cont.)
  meaning of name, 221
  radioactive (*see* Radioactive isotopes)
ISR (interacting storage rings) accelerator at
  CERN, 835

J

Jeans, Sir James, 707
Joliot-Curie, Frédéric, 776
Joliot-Curie, Irène, 776*i.*, 783
Josephson, Brian, 824*i.*
Josephson effect, 824
  frequency of ac current in, 824
  value of *e/h* from, 824
  as voltage standard, 824
Josephson junction, 824*i.*
Joule, James, 340–342, 362, 470
  biography of, 341
  paddle-wheel experiment by, 340–342
Joule (unit of work), 144
Joule heating, 470–471
Joule's equivalent, 341–342
  value of, 342
Junction diodes, 815
  as rectifiers, 815
Junction transistors, 816–817

K

Kamerlingh Onnes, H., 821
Keating, Richard E., 697–698
Kelvin, Lord (William Thomson), 231, 538,
  546, 675
Kelvin (unit of temperature), 231*n.*
Kelvin temperature, 231–232, 396–397
  significance of, 236, 337
Kepler, Johannes, 109
Kepler's laws, 109–111
  Newton's explanation of, 110
  Newton's proof of third law, 110–111
  proof of second law, 210–211
Kilocalorie, 339
Kilogram, 9
Kilogram-mole, 223*n.*
  (*See also* Mole)
Kilowatt, 163
Kilowatthour, 163–164, 470
Kinematics, 33
Kinetic energy, 145–147
  definition of, 146
  dimensions and units of, 146
  of ideal gas, 227–230
  rotational, 200–202
  in special relativity, 693
Kinetic theory of gases, 227–230, 235–237
  assumptions of, 228

and ideal gas law, 235–237
and pressure of gas, 228–230
Kirchhoff, Gustav:
  biography of, 461
  and blackbody radiation, 706
Kirchhoff's rules, 460–469
  applications of, 463–469
  junction rule, 460
  loop rule, 460–462
  sign conventions, 462–463
Klystron, 575
Kundt, A., 307
Kundt's tube, 307

L

Laminar flow, 253–254
Land, Edwin H., 638*n.*
Langevin, Paul, 720
Lasers, 751–754
  carbon dioxide, 753–754
  coherence of emitted light, 621, 752
  He–Ne gas laser, 751–752
  measurements of speed of light with,
    577–578
  medical applications of, 753–754
  nature of light output, 621, 752
  NOVA, 753*i.*
  pulsed ruby, 753
  required conditions, 751
  research with, 3*i.*
  solid-state, 820
  surgery, 346
  uses of, 584, 752–754
Lattice, crystal, 805
  body-centered cubic, 806
  face-centered cubic, 805
  tetrahedral, 806
Lawrence, Ernest O., 488*i.*
  and cyclotron, 488–489
Laws, physical, 4–5
LC circuit, 545–548
  analogy of, with SHM of a spring,
    545–547
  effect of resistance on, 547–548
  energy transfer in, 546–547
  natural frequency for, 545–546
Lederman, Leon M., 848*n.*
Lee, T. D., 843–844
Length, 8
  common units of, 8
  SI unit of, 8
Length contraction, 687–688
  experimental proof of, 696
Lens aberrations, 611–613
  chromatic aberration, 612–613
  spherical aberration, 612
Lens equation, 607
  examples of, 607–609

Lens power, 610
Lenses:
  combination of, 609–613
  converging, 604–607
    and concave mirrors, 605*i.*
  and diopters, 610
  diverging, 604, 607–608
    and convex mirrors, 608*i.*
  focal length, 605
  important equations and sign conventions
    for, 614*t.*
  magnification, 606
  ray diagrams, 604–608
  sign conventions, 605, 614*t.*
  thin, 604
  types of, 605*i.*
Lenz, Heinrich, 516
Lenz' law, 516–517, 531–533
Leonardo da Vinci, 77
LEP (Large Electron-Positron Collider) at
  CERN, 834
Leptons, 828, 839, 842*t.*
  definition of, 839
Lever arm, 124–125
Levers, 160–161
Leyden jar, 408, 437
Lifetime of proton, 848
Lift of airplanes, 257
Light:
  nature of, 619, 720
  particle properties of, 706–720
  place of, in complete electromagnetic
    spectrum, 619
  speed of, 569–572, 575–578, 578*t.*
  wave nature of, 619–641
Light-emitting diodes (LEDs), 815*n*, 820
Lightning, 407–408
Lightning rods, 408
Limiting speed in universe, 691
Line of action of a force, 124–125
Line spectrum, 736
Linear density, 285
Linear expansion, coefficient of, 334
Linear momentum, 171–188
  conservation of, 174–177
    and Newton's third law, 176
    recoil and rockets, 186–188
  definition of, 171
  effect of internal forces on, 176
  and impulse-momentum theorem, 173–174
  and Newton's second law, 172
  vector nature of, 171
Linearly polarized light, 638
Liquids, 243–262
  density of, 243, 244*t.*, 250–253
  evaporation of, 238
  flow of, 253–261
    laminar, 253
    turbulent, 254
  in motion, 253–261

Liquids, (cont.)
  Pascal's principle, 245–248
  pressure of, 245–250
  at rest, 245–253
  specific gravity of, 244*t.*
  surface tension of, 261–262
  viscosity of, 258–259
Living systems:
  and entropy, 387
  and first law of thermodynamics, 369–370
Loch Ness Monster (roller coaster), 157–158
Logarithms, 856–857
Log-log graph paper, 260–261, 861–862
Log-log scales, 861–862
Longitudinal waves, 285–286
  (*See also* Sound waves)
Loops, 291
Lorentz, Hendrik, 492, 687–688
Lorentz-FitzGerald contraction (*see* Length
    contraction)
Lorentz transformation equations, 686–687
  consequences of, 687–691
  and Galilean transformation equations, 686
  for a light pulse, 686–687
  for low speeds, 686–687
  and postulates of special relativity,
    686–687
  and time transformation, 686
Loudness:
  decibel scale, 312–313
  of sound, 312–314
  and sound intensity, 312
Low-temperature physics, 330
Ludington pumped-water storage plant,
    145–146
Lyman, Theodore, 737
Lyman series, 737

M

Mach number, 320–321
Machines, simple, 160–163
  efficiency of, 162
Magnesium fluoride, 631
Magnetic declination, 501–502
Magnetic dipole, 496
Magnetic domains, 499–500
Magnetic field, 479–481
  charged particles in a, 487–489
  of a current-carrying wire, 492–494
  definition of, 480
  of earth, 500–502
  energy stored in, 520–521
  force of:
    on a current-carrying conductor,
      481–483
    on a moving charged particle, 483–486
  lines of, 480–481
  units, 482

Magnetic flux, 512
Magnetic flux density, 513
Magnetic force, 482–486
  between current-carrying conductors,
    493–494
Magnetic inclination, 501–502
Magnetic monopoles, 480*n.*, 848–849
Magnetic poles:
  action of, on unmagnetized materials,
    480
  interaction of, 479–480
  kinds of, 478
  production of, 478
Magnetic properties of materials, 498–500
  diamagnetic, 498
  ferromagnetic, 499–500
  and magnetic domains, 499–500
  paramagnetic, 498–499
Magnetism, 479–503
  applications of, 502–503
  atomic explanation of, 499–500
  of earth, 500–502
  permanent, 500
Magnetron, 575
Magnets, 479–480
  electromagnets, 497–498
Magnification, 599, 606, 656, 658
Magnifying glass, 653–655
  angular magnification of, 654–655
Magnifying power:
  of simple magnifier, 654–655
  of telescope, 657–658
Majority carriers, 814
Malus, Étienne, 639
Malus' law, 639
Manometers, 248–649
  mercury, 249
Many-electron atoms, 756–758
  building-up principle and, 758
  Pauli exclusion principle and, 756–757
  quantum numbers for, 757*t.*
Marsden, E., 740
Mass:
  center of, 132–133
    and center of gravity, 132
  definition of, 9
  dependence of, on speed, 692, 698
    experimental, 698
    theoretical, 692
  and energy, 693–694
  of a gas, 225
  gravitational and inertial, 699
  and inertia, 9
  reduced, 744, 767
  relativistic, 691–693
  rest, 692
  standard kilogram, 9*i.*
  standards of, 9
  and weight, 93–94
Mass defect, 773–775

Mass-energy conservation:
  examples of, 694–695
  principle of, 693–694
Mass-energy conversion, 693–694
  experimental proof of, 698
  in nuclear fission, 786
  in nuclear fusion, 787
Mass-energy equivalence, 775
Mass number, 221
Mass spectrometer, 488
Mass unit, unified (*see* Unified mass unit)
Masses and abundances of important
    isotopes, 869–870
Matter, 1–2
  wave properties of, 720–726
Maxwell, James Clerk, 564, 680
  biography of, 567
Maxwell's displacement current, 564–568
Maxwell's distribution law, 238
Maxwell's equations, 564–568
  and electromagnetic waves, 569–570
  invariance under a Galilean
    transformation, 679–680
Measurements, physical, 14–15
  accuracy of, 14
  of angles, 10
Mechanical and electric oscillations, 547*t.*
Mechanical advantage:
  actual, 162
  ideal, 162
Mechanical energy (*see* Energy, mechanical)
Mechanical equivalent of heat, 340–342
Mechanics:
  definition of, 33
  fundamental quantities of, 6–10
Meggers, William F., 630*i.*
Meitner, Lise, 783–786
  biography of, 784
Melting point, 263, 343, 344*t.*
Mendeléev, D. I., 221
Mercury-in-glass thermometers, 331
Metallic crystals, 807
Metastable states, 751
Meter, SI definition of, 8, 578
Michelson, A. A., 628, 666, 675, 680, 682*i.*
  biography of, 667
  speed-of-light measurements by, 576
Michelson interferometer, 666–668
  fringes produced with, 668*i.*
  use of, in Michelson-Morley experiment,
    683–684
Michelson-Morley experiment, 680–684
  and length contraction, 696
  negative results of, 683–684
Micrometer caliper, 14, 15*i.*
Microphone, 310
Microprocessors, 819
Microscope:
  compound, 655–657
    magnification of, 656–657

Microscope, (cont.)
oil-immersion, 663–664
resolution of, 662–663
scanning electron (SEM), 725
specialized kinds of, 664–665
Microwave interferometer, 577
Microwaves, 575
Millikan, Robert A., 413, 414*i*., 682*i*., 713
experiments of, on photoelectric effect, 713
oil-drop experiment, 413–414
Minority carriers, 814
Mirror equation, 600, 614
sign conventions for, 600, 614*t*.
Mirrors, 596–604
concave, 596–601
convex, 596, 602–603
focal length of, 597
focal point of, 596–597
image formation by, 597–604
important equations and sign conventions, 614*t*.
magnification, 599
parabolic, 597
plane, 603–604
principal axis of, 596
radius of curvature of, 596
ray diagrams for, 598–599, 602–603
relationship of focal length to radius of curvature, 600
spherical, 596–603
vertex of, 596
Models, physical, 4, 726–727
Moderator, 180, 788
Modern physics, 675
Modulation of radio waves, 582
amplitude (AM), 582
frequency (FM), 582
Molar heat capacities:
for ideal gas, 370–373
at constant pressure, 370
at constant volume, 370
ratio of, 371–372
for selected gases, 371*t*.
of solids, 808*t*.
Mole, 223
Molecular crystals, 807
Molecular velocities, distribution of, in gases, 237–238
average, 238
most probable, 238
root-mean-square, 238
Molecules:
in gases, 228*n*.
structure of, 224–225
Moment of a force (*see* Torque)
Moment of inertia (*see* Rotational inertia)
Momentum:
angular, 207–212
and energy, 695–696

linear, 171–188
conservation of, 174–177
Newton's second law and, 172
Newton's third law and, 176
relativistic, 695
Morley, Edward W., 680
and Michelson-Morley experiment, 680–684
Morse code, 582
Motion:
Aristotle on, 33
on an inclined plane, 56–58, 81–84
in a plane, 34–39, 90–114
equations of motion for, 95
planetary, 109–111
rectilinear, 39–59
rotational, 100–101, 194–212
simple harmonic (*see* Simple harmonic motion)
uniformly accelerated (*see* Uniformly accelerated motion)
Motional emf, 510–511
Motors, electric (*see* Electric motors)
Mt. Palomar, 658–659
Müller, K. Alex, 822, 823*i*.
Multiplication factor in nuclear fission, 786
Muons, 836–837
and electrons, 837
lifetimes of, 696–697, 837
Musical instruments, 315–317
Musical sounds, analysis and synthesis of, 317
Mutations, 792–794
Mutual induction, 517–518

N

n-type semiconductor, 812–813
Nanometer, 619
National Academy of Sciences, 851
National Institute of Standards and Technology, 8, 9, 494, 577–578
National Physical Laboratory, 9, 577–578
Natural frequency of *LC* circuit, 546
Natural radioactivity, 776–780
exponential decay, 778–779
half-life, 777–778
Naval Observatory, U.S., 697
Near point of eye, 650
Nearsightedness, 651
Neon, 751–752
Neutrino, 777*n*.
and conservation principles, 143, 830
detection of, 831–832
mass of, 830–831, 849
postulated by Pauli, 830
types of, 838*t*.
Neutron:
decay of free, 829–831

discovery of, 771
mass of, 219
thermal, 788, 788*n*.
Neutron bombardment of nuclei, 782–785
Neutron diffraction, 726
Neutron star, 211
New quantum theory, 762–763
Newton, Sir Isaac, 67*i*, 109
biography of, 67
Newton (unit of force), 68
definition of, 68
Newtonian focus, 658
Newton's law of cooling, 358–359
Newton's law of universal gravitation, 90–92
importance of, 90
Newton's laws of motion, 65–77
first law, 65–66
for rotational motion, 202–204, 207
second law, 66–71
in component form, 69
and momentum, 172
relationship of, to first law, 70
steps in applying, 73–74
third law, 71–72, 415
action and reaction, 71
and conservation of momentum, 176
in electrostatics, 415
using, 72–77
Newton's rings, 629–630
experimental pattern, 630*i*.
Nicol prism, 639
Nitrogen molecule, vibration and rotation of, 337
Nodes, 291
Nonconservative forces, 151
Noninertial frame of reference, 676
Nonreflecting films, 631
multilayer, 631
Normal, 39*n*., 589
NOVA laser, 753*i*.
Nuclear accidents, 794
Nuclear activity, 779, 792
unit of, 792
Nuclear binding energy, 773–775
per nucleon, 774, 774*i*.
Nuclear emissions, 777
alpha particles, 777
beta rays, 777
gamma rays, 777
Nuclear energy (*see* Nuclear fission; Nuclear fusion)
Nuclear fission, 775, 783–787
chain reactions, 785–786
critical mass for, 786
definition of, 785
energy considerations in, 786–787
as practical energy source, 788–790
Nuclear forces, 772–775, 827–832
strong, 772–773
weak, 829–832

Nuclear fuel, 788
Nuclear fusion, 775, 787
  definition of, 787
  mass-energy conversion in, 787
  as practical energy source, 790–791
  temperatures needed for, 787
Nuclear physics, 770–798
Nuclear power plants, 394
  (*See also* Reactors, nuclear)
Nuclear radiations:
  biological effects of, 791–795
  effects of, on living matter, 791–794
  treatment of disease, 794
  units for measuring, 792–793
  useful isotopes, 778*t.*
Nuclear reactions:
  artificial radioactivity, 781–783
  conservation of charge, 780
  conservation of mass-energy, 781
  conservation of nucleon number, 780
  natural radioactivity, 776–780
  rules governing, 780–781
Nuclear reactors
      (*see* Reactors, nuclear)
Nuclear spin, 839
Nuclei, excited states of, 841
Nucleons, 219, 771
Nucleus, atomic:
  as practical energy source, 788–791
      fission, 788–790
      fusion, 790–791
  density of, 219, 772
  proton-electron model of, 770
      difficulties with, 770–771
  proton-neutron model of, 771–772
  size of, 740–741, 772
  structure of, 770–772
Numerical aperture, 663
Numerical constants, F.3

O

Object distance, 597, 605
Objective lens, 655
Ocean thermal energy conversion (OTEC),
    398
Ocular (*see* Eyepiece lens)
Oersted, Hans Christian, 479, 492–493
  experiment by, on the magnetic effects of
      a current, 492–493
Ohm, Georg Simon, 456
Ohm (unit of resistance), 456
Ohm's law, 455–456
Oil-drop experiment; 413–414, 455
Oil-immersion microscope, 663–664
Ophthalmoscope, 665–666
  direct, 665
  indirect, 665–666
Oppenheimer, J. Robert, 3*i.*, 681

Optical fibers, 584
  use of, in medicine, 753*i*, 754
Optical instruments, 647–669
  cameras, 647–650
  compound microscope, 655–657
  human eye, 650–653
  interferometers, 666–669
  magnifying glass, 653–655
  ophthalmoscope, 665–666
  specialized microscopes, 664–665
  telescopes, 657–660
Optical pumping, 751
Optical thickness, 628–629
Optics:
  ray (geometrical), 588–613
  wave (physical), 619–641
Order of interference, 669
Order-of-magnitude calculations, 867–868
Organ pipes, standing sound waves in,
    306–307
Oscillation, 272
  comparison of mechanical and electrical,
      547*t.*
Oscilloscope, 526–527
Overexposure, 649
Overlapping orders, method of, 627
Overtones, 293

P

p-type semiconductors, 813–814
Pair annihilation, 838
Pair creation, 838
Parabola, equation for, 96–97
Parabolic mirror, 597
Parallel-axis theorem, 206
Parallel circuits, 466
Paramagnetic materials, 498–499
Parent nucleus, 781
Parity:
  conservation of, 843–844
  experimental work, 843–844
Particle accelerators (*see* Accelerators,
    particle)
Particle properties of light, 706
Pascal, Blaise, 245
Pascal (unit of pressure), 226
Pascal's principle, 245–248
Paschen, Louis, 737
Paschen series, 737
Pauli, Wolfgang, 756, 830
Pauli exclusion principle, 756–758
  and periodic table, 757–758
Pendulum, simple, 277–280
  clocks, 148, 279
  forces on, 278
  period of, 279
  practical applications of, 279
  and simple harmonic motion, 278–279

Period, 6
  in circular motion, 100
  of simple harmonic motion, 276
Periodic motion, 272–280
  amplitude of, 273
  definition of, 272
  displacement in, 273
  energy transformations in, 273
  frequency of, 273
  period of, 272
  simple harmonic, 274–277
Periodic table of the elements, F.6; 221
  and Pauli exclusion principle, 756–758
Permeability:
  of free space, 494
  relative magnetic, 498
Permittivity:
  of dielectrics, 440
  of a vacuum, 439
Perot, Alfred, 668*n.*
Perpetual motion machines:
  of first kind, 367
  of second kind, 397
Phase (stage in a cycle):
  changes of, on reflection, 290
  definition of, 288
  differences, 288–289
Phase (state of matter):
  change of, 342–345
  diagrams of, 345
      for CO_2, 345*i.*
      for H_2O, 345*i.*
  transition, 263
Phase angle, 549
Phase-contrast microscope, 664–665
Phase relationships:
  in ac circuits, 538–552
      capacitance only, 543–545
      inductance only, 541–543
      resistance only, 538–541
      summary of, 544*t.*
  in electromagnetic waves, 580–581
Phonons, 809
Phosphorescence, 754
Photoelectric effect, 711–714
  apparatus for, 711
  comparison of theory and experiment for,
      712*t.*
  Einstein's explanation of, 713
  Einstein's photoelectric equation for, 713
  empirical equation for, 712
  experimental results for, 711–712
  Millikan's experiments for, 713
  threshold frequency, 712
  work function, 713
Photographic emulsions as particle detectors,
    797–798
Photomultiplier tube, 797*i.*
Photon, 706, 716, 839
  absorption of, by atoms, 746–747

Photon, (cont.)
  definition of, 706
  energy of, 706
  linear momentum of, 718
Photovoltaic cells (*see* Solar cell)
Physical constants, B.2
Physical data, accuracy of, 14–15
Physical laws, 4–5
Physical measurements, 14–15
Physical models, 4, 726–727
Physical theories, 4–5
  domains of validity, 705*i*.
Physicists:
  approach of, 2–4
  experimental, 5
  secret of success of, 4
  theoretical, 5
Physics:
  application of, 2
  and the biological sciences, 2
  definition of, 1
  and engineering, 2
  experiment in, 5
  incompleteness of, 4, 851
  and mathematics, 33
  method of, 4
  principles of, 1–2, 5
  progress of, 4
  as a science, 1
  scope of, 1–2
  theory in, 5
Physics problems, solving, 13–15,
    73–74, 122–124, 128–130,
    864–868
Physiological effects:
  of increased *g*, 107
  of weightlessness, 112–113
Piezoelectric crystals, 322
Pinhole camera, 647–648
  magnification by, 647
Pions, 836–837
  discovery of, by C. Powell, 836
  prediction of, by H. Yukawa, 836
Pitch of sound, 315–317
Pitchblende, 776
Planck, Max, 705–710, 763, 809
  biography of, 709
Planck's constant, 708
  numerical values, 710, 713
Planck's quantum theory of blackbody
    radiation, 707–710
Plane:
  inclined (*see* Inclined plane)
  motion in a, 34–39, 90–114
Plane mirrors, 603–604
Planetary motion, 109–111
Planets, data on, F.2
Plasma, 224*n*.
pn junctions, 814–815
Poincare, Henri, 687

Point charges:
  electric potential in vicinity of, 433–434
  problems involving, 415–416
Poise, 259
Poiseuille, J., 260
Poiseuille's law, 260–261
Poisson, Simeon, 637
Poisson's spot, 637
Polarization:
  of atoms, 412
  of dielectrics, 439–441
  of light, 638–641
    circularly polarized light, 638*n*.
    importance of, 642
    linearly polarized light, definition of,
      638
    by Nicol prisms, 639
    by Polaroid sheets, 638
    by reflection, 640–641
    unpolarized light, 638
Polarizer, 639
Polarizing microscope, 665
Polaroid, 638–641
Polonium, 776
Population inversion, 751–752
  produced by collisions, 751
  produced by optical pumping, 751
Positive and negative lenses, 611
Positron, 831, 837*i*., 838
Potential, electric:
  absolute, 434–435
  definition of, 429
  difference, 429
    and voltage, 430
  and the electric field, 431–432
  in vicinity of a point charge, 433–434
  (*See also* Voltage)
Potential energy:
  elastic, 151–152
  electric, 428–429
  gravitational, 147–151
    zero of, 148
Pound, 65
Powell, Cecil F., 836
Power, 163–165
  definition of, 163
  electric (*see* Electric power)
  of a lens, 610
  and velocity, 164–165
  units of, 163
Power factor, 549
Power laws, 861
Power losses in transmission lines, 553
Power plants, 393–394
  fossil-fuel, 393–394
    efficiency of, 394
  nuclear, 394
Powers-of-ten notation, 11, 866–867
Practical aspects of electric power, 556–559
Precession, 209

Pressure:
  absolute and gauge, 232–233
  atmospheric, 246
  in an automobile tire, 226, 233
  of blood (*see* Blood pressure)
  of a gas, 225
  in the human body, 246–247
  hydrostatic, 245–253
  of a liquid, 245–250
  molecular basis for gas, 228–230
  relationship of, to kinetic energy of a gas,
    230
  in a sound wave, 301
  units of, 233*t*.
Pressure units, conversion of, 233*t*.
Pressurized water reactor, 789
Principal quantum number, 754
Principle of equivalence, 699
Principle of invariance, 699
Principles of physics, 2, 5
Prism(s):
  dispersion by, 595–596
  reflection by, 593
  refraction by, 617
  totally reflecting, 593
Prism spectrometer, 596
Problems, useful techniques for solving
    physics, 13–15, 73–74, 864–868
Projectiles:
  independence of horizontal and vertical
    motion, 96
  motion of, 95–99
  range of, 98
  trajectory of, 95–98
Proper length, 688
Proper time, 689
Proportional counters, 796
Proportionality constant, 858
Proportions:
  direct, 858–859
  inverse, 859
Proton:
  charge of, 219
  excited states of, 841
  lifetime of, 848
  mass of, 219
  stability of, 848
Proton decay, search for, 848
Pulleys, 161–163
Pulse, 281
Pupin, Michael I., 715*i*.

Q

Q-factor of resonant circuit, 551
Quadratic equations, 854
Quality of sound, 315–317
Quanta, 706, 708
  associated with interactions, 845*i*.

Quantized energies, 706–710, 743–744
Quantum chromodynamics (QCD), 845–848
Quantum electrodynamics, 845, 846
Quantum mechanics, 221, 763
    domain of validity, 705
    experimental basis of, 705–726
    interpretation of, 763
Quantum nature of light, 705–726
Quantum numbers, 754–756
    for electrons in some atoms, 757t.
    in the hydrogen atom, 755t.
    magnetic, 755
    orbital angular momentum, 754–755
    and the periodic table, 756–758
    principal, 754
    spin, 755
Quantum theory:
    new, 763–764
    old, 706–761
Quarks, 414, 839–842
    charge of, 840
    definition of, 840
    evidence for, 840–841
    and hadron structure, 840
    kinds of, 840
    properties of, 842t.
    problems with quark hypothesis, 841–842
    search for, 841

R

Rabi, I. I., 748i.
Rad, 792
Radar, 378
    (See also Microwaves)
Radian, 10
    and the degree, 10
Radiation, emission and absorption of, 746–752
    (See also Blackbody radiation; Electromagnetic waves; Nuclear radiations; Radiation of heat; Radio waves; Solar radiation)
Radiation of heat, 355–362
    from human body, 357
    and radiation balance, 357–358
    from sun, 360–361
Radiation balance, 357–358
Radiation cooling, 361
Radiation detectors:
    gas-filled, 796–797
    other detectors, 797–798
    scintillation counters, 797
Radiation sickness, 793
Radiation therapy, 794
    Co-60 therapy, 794
Radio:
    frequency bands of, 583t.
    practical details of, 583–584

Radio waves, 579–584
    detection of, 581
    frequencies of, 583t.
    generation of, 579–581
    modulation of, 582
    nature of, 579–581
    phase relations in, 580–581
Radioactive isotopes:
    in medicine and technology, 795–796
    table of useful, 778t.
Radioactivity:
    artificial, 781–783
    natural, 776–780
Radiocarbon dating, 779–780
Radioisotopes, 794–796
Radium, 776
Radius of curvature of mirror, 596
Radius of nucleus, 740–741, 772
Radon, 794
Range:
    maximum, 98
    of a projectile, 98
Rarefaction, 299–301
Ratio of molar heat capacities, 371–372
Ray optics (see Geometrical optics)
Rayleigh, Lord, 660, 707
Rayleigh criterion, 661
Rays of light, 588–589
RC circuit, 524–526
    time constant for, 524–525
Reactance:
    capacitive, 543–545
        dependence of, on C and f, 544
    inductive, 541–543
        dependence of, on L and f, 542
Reaction, 71–72
Reaction forces, 119–120
Reactors, nuclear:
    fission:
        breeder, 789–790
        and collision processes, 180
        control rods, 788–789
        moderator, 788
        multiplication factor, 788
        nuclear fuel, 788–789
        pressurized water, 789
    fusion, 790–791
        confinement techniques for, 753i., 791
        temperatures required for, 790
Real image, 598
Recoil, 186
Rectangular components, 21–25
Rectangular resolution of vectors, 21–25
Rectilinear motion, 39–59
    equations for, 40
    graphical techniques for, 41–47
Red shift, 320
Reduced mass, 744, 767
Reference circle, 274–275
Reflection, phase changes on, 290, 304

Reflection of light:
    diffuse, 590
    laws of, 589
    polarization by, 640–641
    on a ray model, 589–590
    on a wave model, 621
Reflector (see Telescope, reflecting)
Refraction, index of, 591
    of different kinds of glass, 596i.
    of important materials, 591t.
Refraction of light:
    definition of, 591
    by a prism, 595–596
    on a ray model, 591–592
        and total internal reflection, 592–595
    Snell's law for refraction, 591–592
Refractor (see Telescopes, refracting)
Refrigerators, 401
    coefficient of performance of, 401
Reines, Frederick, 831–832
Relative biological effectiveness (RBE), 793
Relative magnetic permeability, 498, 499t.
Relativity:
    Galilean, 675–679
        equations for transformations, 677
        invariance of mechanics under such transformations, 677–678
        principle of, 678
    general theory of, 698–700
        experimental evidence for, 699–700
        postulates of, 699
    Newtonian (see Galilean, above)
    special theory of, 684–698
        experimental tests of, 696–698
        and length contraction, 687–689
        Lorentz transformation equations, 686–687
        mass, energy, and momentum in, 691–695
        and Michelson-Morley experiment, 685n.
        postulates of, 684–685
        principle of, 685–686
        and simultaneity, 690–691
        and time dilation, 689–690
rem, 793
Resistance, electric, 456–460
    factors determining, 458–459
    of human body, 558
        as diagnostic technique, 459
    internal, 464–465
    and resistivity, 458–459
    thermometers, 330, 459
    of a wire, 457
Resistance thermometers, 330, 459
Resistive forces, 77–81
    air resistance, 80–81
    friction, 77–79
    viscosity, 258–259

Resistivity, electric, 457, 458*t*.
  temperature coefficient of, 458*t*, 459
Resistors, 457
  examples of calculations for, 466–468
  in parallel, 466
  in series, 465–466
  shunt, 503
Resolution:
  of human eye, 662
  limitation on, 660, 663
  of microscopes, 662–663
    with oil-immersion, 663–664
  of telescopes, 660–661
    Rayleigh criterion, 660–661
Resolving power (*see* Resolution)
Resonance:
  definition of, 281
  examples of, 281
  in forced vibrations of a string, 292–293
  series (*see* Series resonance)
  for sound waves in organ pipes, 306–307
Rest energy, 693
Rest mass, 692
Retarding potential, 711
Retina, 650
Retroreflector, 590
Reversibility of light rays, principle of, 597
Reversible processes, 379–381
  adiabatic, 380–381
  entropy changes in, 385
  isothermal, 380
  work done in, 381–383
Revolution, 10
Reynold's number, 254, 257*n*.
Right-hand field rule, 492
Right-hand force rule, 482–487
  and cross product of vectors, 485*n*.
Rigid body, 195
  rotational motion of, 194–209
RL circuit, 521–523
  time constant of, 522–523
RLC series circuit:
  with applied ac voltage, 548–553
  phase relationships in, 548–549
  power factor of, 549
  resonance in, 550–552
  use of, in radio, 581
  voltages in, 549–550
rms currents and voltages (*see* Effective values)
Rockets, 187–188
  acceleration of, 187
  booster, 188
  thrust of, 187–188
Roentgen, Wilhelm, 675, 714
  discovery of x-rays by, 714
Roentgen (unit of radiation), 792
Rohrer, Heinrich, 725
Roller coaster, 157–158

Root-mean-square speeds, 236–237
  and kinetic energies, 237*t*.
Rotational equilibrium, 127–130
  condition for, 127
  of moving objects, 135–136
  solving problems involving, 128–130
Rotational impulse, 207–208
Rotational inertia, 201–202, 205–206
  and mass, 201
  significance of, 202
    for various solid objects, 205*t*.
Rotational kinetic energy, 201
  and work, 200–202
Rotational motion, 100–101, 194–212
  with constant angular acceleration, 198–200
  with constant angular velocity, 196–198
  Newton's second law for, 202–204, 207
  relationship of, to translational motion, 195–196
  of a rigid body, 194–209
Rotational speed, 195
Rumford, Count (*see* Thompson, Benjamin)
Ruska, Ernst, 725
Rutherford, Ernest, 219–220, 738, 739*i*., 777
  alpha-particle scattering experiments, 738–741
  biography of, 220
  model of a nucleus, 741
  scientific achievements of, 220
Rydberg, Johannes R., 737
Rydberg constant, 737
  calculated value for, 743
  experimental value for, 737
Rydberg equation, 737

S

Sailboat, 134–135
Salam, Abdus, 845–847
Satellites, earth, 111–114, 323
  synchronous (stationary), 113–114
  weightlessness in, 112–113
Sawtooth voltage, 526–527
Scalars, 16
  and vectors, 16–19
Scanning electron microscope (SEM), 725
Scanning tunneling microscope, 725*i*.
Schrieffer, J. Robert, 822*i*.
  and BCS theory of superconductivity, 821
Schrödinger, Erwin, 762–763
Schrödinger wave equation, 762
  eigenfunctions, 762
  eigenvalues, 762
  probabilistic nature of, 762
Schwinger, Julian, 845
Science and technology, 378
Scientific method, 4–5

Scintillation counters, 797
Scope of physics, 1–2
Second, definition of, 6–8
Second law of thermodynamics, 378–398
  applications of, 398–401
  and Carnot cycle, 397
  Clausius' statement, 388
  and entropy, 383–388
  entropy statement, 386
  Kelvin statement, 397
  perpetual-motion machine statement, 397
  statement of impossibility, 386, 388
Segrè, Emilio, 838
Self-inductance, 519
Semiconductor detectors (*see* Solid-state detectors)
Semiconductors, 811–814
  band structure of, 811–812
  extrinsic (impurity), 812–814
    n-type, 812–813
    p-type, 813–814
  intrinsic, 812
    germanium, 812
    silicon, 812
  majority carriers, 814
  minority carriers, 814
Semilog graph paper, 863
Semilog scales, 862–863
Series circuits, 465–466, 547–549
Seris expansions, 856
Series limit, 736
Series resonance, 550–552
  dangers of, 552–553
  definition of, 551
  frequency of, 551
  importance of, 552
  in radio and TV, 552
Shadows, 589
Shear modulus, 265
Shell, energy, 754, 757
Shock waves, 321–322
Shockley, William, 817*i*.
Short circuit, 557
Shutter speed, 649–650
SI (Système International) units, 11–12
  advantages of, 11
  complete list of, F.5
  prefixes and abbreviations used in , F.1
Sievert (unit of biological effectiveness), 793, 793*t*.
Sign conventions:
  for freely falling objects, 52–56
  for lenses and mirrors, 614*t*.
Significant figures, 14–16
  rules for handling, 14–15
Silicon, 222, 812
Silicon chip, 819*i*.
Silicon solar cells (*see* Solar cells)
Simple harmonic motion, 274–277
  acceleration in, 275–276

Simple harmonic motion, (cont.)
  definition of, 275
  equations for, 277
  graphs of, 277*i.*
  and Hooke's law, 275
  period of, 276
  reference circle for, 274–275
  velocity in, 276
Simple machines, 160–163
  efficiency of, 162
Simple magnifier (*see* Magnifying glass)
Simple pendulum (*see* Pendulum, simple)
Simultaneity, relativity of, 690–691
Sines, table of, for small angles, 279*t.*
Single-slit diffraction, 634–636
Size of nucleus, 740–741
Skylab 4 space mission, 113*i.*
SLAC (Stanford Linear Accelerator Center)
  accelerator, 832–833, 840
SLC (Stanford Linear Collider), 833, 841
Slope of a graph, 42–43
Smithsonian Institution, 512*n.*
Snell, W., 592
Snell's law, 591–592
Soap bubbles:
  colors of, 630
  interference in, 630
Sodium, 758
  band structure of, 811
Sodium chloride:
  crystal, 805–806
  in solution, 806
Solar cells, 817–818
  disadvantages of, 818
  uses of, 818
Solar energy, 360–361
Solar radiation, 360–361
Solar system, astronomical data on, F.2
Solenoid, 496–497
  magnetic field of, 496–497
Solid-state detectors, 820
  advantages of, 820
  Ge(Li) detector, 820
Solid-state devices, 814–817
  applications of, 816–820
  junction diodes, 815
  junction transistors, 816–817
  pn junctions, 814–815
Solid-state physics, 804–824
  definition of, 804
  impact of, on other fields, 817–824
Solids, 243, 263–267, 804–824
  band theory of (*see* Band theory of
    solids)
  density of, 263*t.*
  elastic properties of, 263–267
  molar heat capacities of, 807–809
  structure of, 263, 804–807
  thermal conductivities of, 807*t.*
  thermal properties of, 807–809

Solving physics problems, 13–15, 73–74,
  122–124, 128–130, 864–868
Somatic damage, 792–794
Sommerfeld, Arnold, 754
Sonar, 323
Sonic boom, 321
Sound spectrum of a piano, 315
Sound waves, 299–323
  definition of, 299–300
  detection of, 310–311
  displacement of particles in, 301
  Doppler effect for, 317–320
  equations for, 300–301
  intensity of, 311–315
  interference of, 308–309
  pitch of, 315–317
  pressure variations in, 301
  production of, 299–300
  propagation of, 301–302
  reflection of, 304
  speed of, 302–304, 303*t.*
    temperature dependence, 304
  standing, 304–307
Spark chambers, 797–798
Special theory of relativity (*see* Relativity,
  special theory of)
Specific gravity, 244
  of liquids and gases, 244*t.*
  of solids, 263
Specific heat capacity, 338–340
  definition of, 339
  for various substances, 339*t.*
Spectroscopic notation, 758
Spectrum:
  complete electromagnetic, 574–575
  of hydrogen atom, 736–738, 741–745
  line, 627
  produced by a grating, 626–627
  produced by a prism, 596*i.*, 627
Speed:
  average, 35, 238
  instantaneous, 36
  most probable, 238
  root-mean-square (rms), 236, 238
  and velocity, 35
Speed of electromagnetic waves, 569–578, 578*t.*
  and the aether, 679–680
  and definition of the meter, 8, 578
  in dielectric materials, 572
  experiments of Hertz, 571
  in free space, 570, 691
  measurement of, 571–572, 575–578, 578*t.*
  predicted by theory, 569–570
  standing wave measurements:
    Froome, 577
    laser methods, 577–578
  time-of-flight measurements:
    Bergstrand, 576–577
    Fizeau, 575–576
    Michelson, 576

Speed of sound, 303*t.*
  dependence of, on temperature, 304
  measurement of, 302–304
Speed of transverse waves on a string, 285
Spherical aberration:
  of lenses, 612
  of mirrors, 597
Spherical mirrors, 596–603
Spin:
  electron, 755
  nuclear, 839
Spontaneous emission, 750–751
Spring:
  horizontal, motion of, 273
  vertical, motion of, 273
Spring balance, 65, 94
SQUID detectors, 824
SSC (Superconducting Super Collider), 851
SST (supersonic transport) plane, 321
Standard model, 844
  successes of, 847
Standing sound waves, 304–307
  displacement nodes and antinodes,
    304–306
  equation for, 305
  in organ pipes, 304–306
  phase changes on reflection, 304
  pressure nodes and antinodes, 304–305
  resonant frequencies, 306–307
Standing waves:
  equation for, 291
  nodes and antinodes, 291
  production of, 291*i.*
  on a string, 290–292
Stars, color temperatures of, 356*t.*
State function, 365
State variables, 365
Static equilibrium, 119, 122–130
Stationary states, 741
Statistical mechanics, 328
Steam engines, 393–394
Steam turbines, 394
Stefan, Josef, 355
Stefan-Boltzmann constant, 356
Stefan's radiation law, 355–356, 707
Stethoscope, 306*n.*
Stimulated emission (*see* Forced emission)
Stoney, George, 492
Storage batteries, 452
Storage rings, 835
STP (standard temperature and pressure),
  225*n.*
Straight-line motion, 39–59
  (*See also* Motion)
Strain, 264–267
  change in length, 264
  change in shape, 264
  change in volume, 264
Strange particles, 839
Strassman, Fritz, 783

Streamline flow, 254
Stress, 263–265
String:
   forced vibrations of a, 292–293
   standing waves on a, 290–292
   waves on a, 281–284
Strong nuclear force, 773, 828–829
   charge-independence, 773
   dependence of, on distance, 773i.
   saturation of, 773
Structure of atoms, molecules, and gases,
   218–238, 736–763
Subatomic particles, 218–219
   properties of, 218t.
Sublimation, 345
   heat of, 345
Subshell, 757
Summation notation, 23
Sun, radiation from, 360–361
Superconducting coil, 821i.
Superconducting power lines, 823
Superconducting temperatures:
   of ceramic materials, 822–823
   of metals and alloys, 821t.
Superconductivity, 820–824
   in alloys, 821
   BCS theory of, 821
   discovery of, 821
   high-temperature (see High-temperature
      superconductivity)
   Josephson effect of, 824
   in metals, 821
   recent progress in field of, 823i.
Supernova 1987A, 849–850
Superposition principle, 287–289
   in electrostatics, 415
   in optics, 637–638
Supersonic speeds, 320–321
   and Mach number, 321
   and shock waves, 321
   and sonic booms, 321
Surface-barrier detectors (see Solid-state
      detectors)
Surface tension, 261–262
   temperature dependence, 262
Symmetry:
   breaking, 849
   in physics, 720
Synchrotrons, 833–835
   CERN Large Electron-Positron Collider, 834
   colliding-beam, 835
   Fermilab Tevatron Collider, 834
   fixed-target, 834–835
Système International (SI) units, 11–12

T

Tacoma Narrows Bridge, 281
Tangential acceleration, 39

Tangents, table of, for small angles, 279t.
Teflon, 79
Telephoto lenses, 609
Telescopes:
   Galileo and, 57
   Hubble Space Telescope, 659
   light-gathering power of, 657
   magnifying power of, 657–658
   new ideas about, 659
   reflecting, 658–659
   refracting, 657–658
   resolution of, 657, 660–662
Television, 526–527
   frequency bands of, 583–584
   practical details of, 583–584
Temperature, 226–227, 328–329
   and heat, 337–338
   intensive property, 227
   measuring, 226–227
   precise definition of, 328
   preliminary definition of, 226
   significance of, 236
Temperature expansion (see Expansion)
Temperature scales, 331–334
   absolute thermodynamic (Kelvin),
      231–232, 331, 396–397
   Celsius, 332
   centigrade, 332
   Fahrenheit, 332
   relations between, 332–333
Temperatures, important physical, 333i.
Tension, 121–122
Terminal velocity, 80–81
Terminal voltage, 464
   and emf of a battery, 464–465
Tesla, Nikola, 407i., 538
Tesla (unit of the magnetic field), 482
Tetrahedral lattice, 806
Tevatron Collider at Fermilab, 834
Theories in physics, 5
Thermal conductivity, 351–354
   coefficient of, 352
   in houses, 353
   values of, for some common materials,
      352t.
Thermal energy, 337–338
   definition of, 337
Thermal equilibrium, 226–227
Thermal neutrons, 788n.
Thermal radiation, 355
   laws governing, 355–356
   (See also Blackbody radiation)
Thermodynamic probability, 384
Thermodynamic state variables, 365
Thermodynamics, 328, 379
   and energy conservation, 365–367
   first law of, 364–367
      and Joule heating, 471
   second law of, 378–398
      and Joule heating, 471

   third law of, 330
   zeroth law of, 227
Thermography, 359–360
Thermometers, 227, 329–331
   bimetallic strip, 335
   gas, 329–330
      constant-volume, 329–330
   mercury-in-glass, 331
   resistance, 330, 459
Thermonuclear reactions (see Nuclear fusion)
Thermos bottles, 359
Thermostat, 335–336
Thin-film interference, 628–631
   air films, 628
   films other than air, 628–629
   Newton's rings, 629–630
   nonreflecting films, 631
   soap bubbles, 630
   wedge-shaped films, 629
Thin-lens equation, 607
Third law of thermodynamics, 330
Thompson, Benjamin, 154i.
   biography of, 154
Thomson, George P., 723
Thomson, J. J., 489, 698, 723
   biography of, 490
   determination of e/m for electron by,
      489–492
Thomson, William (see Kelvin, Lord)
Three Mile Island, 794
Threshold of hearing, 313
Threshold of pain, 313i., 314
Threshold frequency in photoelectric effect,
    712
Throttling process, 368, 369i.
Thrust of a rocket, 187–188
Thyratron sweep circuit, 526–527
Timbre (see Quality of sound)
Time:
   measurement of, 6
   in special relativity, 689–691
   unit of, 6–8
Time constant:
   for RC circuit, 524–525
   for RL circuit, 522–523
Time dilation, 689–690
   difference between stationary and moving
      clocks, 689
   experimental proofs of, 696–698
   lifetimes of muons, 696–697
   and space travel, 690
Time's arrow, 387
Tissue ablation, 346
Tomonaga, Sin-Itiro, 845
Torque, 124–127
   definition of, 125
   sign convention for, 125
   units of, 125
   vector nature of, 125–126, 126n.
Torr (unit of pressure), 248

Torricelli, Evangelista, 248
Torricelli's theorem, 258
Torsion balance, 413
Torsion modulus, 265
Total internal reflection, 592–595
    and fiber optics, 594–595
Tracers, 795
Trajectory, 95–98
Trampoline, 152
Transducer, 310
Transfer of heat, 351–362
    dependence of, on path, 363–364
Transformers, 515, 553–556
    definition of, 553
    equation for, 554
    step-down, 554
    step-up, 554
    turns ratio of, 554
    use of, 554–556
Transistors, 816–817
    discovery of, 817*i*.
    use of, as amplifiers, 816–817
    and vacuum tubes, 817
Translational equilibrium, 122–124
    condition for, 122
    of moving objects, 134–135
    solving problems on, 122–124
Transmission of electric power, 556
Transmission electron microscope (TEM),
    724–725
Transverse waves, 284–285
    examples of, 284
    speed of, on a string, 285
    (*See also* Electromagnetic waves)
Triangle method for adding vectors, 19
Trigonometry, 22, 855–856
Triple point, 345
Tritium (H-3) nucleus, 774
Turbines, steam, 394
    (*See also* Electric generators)
Turbulent flow, 254
Turns ratio, 554

U

Ultracentrifuges, 106–107
    effective *g* values, 107
Ultrasonic frequencies, 322–323
Ultrasonic imaging, 322*i*., 323
    in medicine, 323
Ultraviolet, 575*t*.
Ultraviolet catastrophe, 707–708
Ultraviolet microscope, 664
Uncertainty principle:
    for position and momentum, 727–728
    and principle of complementarity, 730
    for time and energy, 729–730
        classical expression of, 729–730
Underexposure, 649

Unification of forces in physics, 849
Unified mass unit, 223–224
    definition of, 223
Uniform circular motion, 100–101
Uniformly accelerated motion:
    under gravity, 52–56
    in a straight line, 47
        equations for, 48–49
        graphs of position, velocity, and
            acceleration for, 50*i*.
Units:
    British engineering, 11–12
    conversion of, 12
    and dimensions, 11–13
    SI (*see* SI units)
Universal gas constant, 233
Universal gravitation, Newton's law of,
    90–92
Uranium, 221–222
    U-235 nucleus, 783–787, 786*t*.

V

Valence band, 810
Van de Graaff generator, 437
Van der Waals forces, 807
Vapor pressure, 344–345
    and boiling point, 344–345
    definition of, 344
Vaporization, heat of, 343
Vectors, 16–27
    acceleration, 37–39
    addition of, 19–27, 415
        analytical methods for, 21–27
        graphical methods for, 19–21
        parallelogram method for, 20
        by rectangular resolution, 21–27
        tail-to-tip method for, 19–20
        triangle method for, 19
        trigonometric methods for,
            22–27
    definition of, 16
    displacement, 18–19
    force, 64–65
    position, 17
    rectangular components of, 21–22
    subtraction of, 20
    velocity, 34–37
Velocities, molecular (*see* Molecular
            velocities, distribution of, in gases)
Velocity, 34–37
    addition of, 36–37
    angular, 100–101, 195–198
    average, 34–35, 42
    constant, 35
    drift, 457
    instantaneous, 36–37, 42–43
    and speed, 35
    terminal, 80–81

versus time for three different
        accelerations, 50*i*.
    vectors for motion in a plane, 34–37
Vertex of mirror, 596
Vertical circle, motion in a, 107–109
Vibrations, 280–284
    damped, 280
    forced, 280–281, 292–293
        resonance for, 281, 292
    as source of waves, 281–284
Virtual focus, 602
Virtual image, 598–599
Viscosity, 258–259
    coefficient of, for liquids, 259, 259*t*.
    measurement of, 259
    temperature dependence of, 259
    units of, 259
Visible light, 619
Visual acuity, 661
Volt, 429
Voltage, 430–431
    and electric field, 432
    and potential difference, 430
    unit of, 429
Voltmeter, 502–503
Volume:
    dependence of, on temperature, 334–336
    of a gas, 225
    relationship of, to kinetic energy of a gas,
        230
Volume expansion, coefficient of, 335
von Laue, Max, 721
Vostok 6 (Russian earth satellite), 111–112
Voyager 2, 501

W

Water:
    density of, 336*i*.
    expansion of, with temperature, 336
    moderating effect of, on climate, 343
    phase diagram of, 345*i*.
    temperature dependence of phases,
        342*i*.
Water waves, diffraction of, 663*i*.
Watson, James D., 722
Watson-Crick theory of DNA, 2
Watt, Sir James, 163
Watt (unit of power), 163, 469
Wave equation, Schrödinger, 762
Wave fronts, 588–589
    definition of, 589
    for plane waves, 620
    for spherical waves, 620
Wave mechanics:
    de Broglie standing waves, 761
    probabilistic nature of, 762
    and quantized angular momentum, 762
    Schrödinger equation, 762

Wave motion:
  equation for, 286
  mathematical description of, 286–287
Wave-particle duality, 720*i.*, 726–727
  statement of, 726
Waveguide, 577*n.*
Wavelength, 282
  and frequency, 283
  of visible light, 619
Waves, 281–293
  definition of, 283
  electromagnetic (*see* Electromagnetic
    waves)
  energy propagation by, 283
  longitudinal, 284–286
  mathematical description of, 286–290
  relationship of wavelength and frequency
    in, 283*i.*
  sound, 299–323
  speed of, 283
  superposition of, 287–290
  transverse, 284–285
  wavelength of, 282
Weak nuclear force and the neutrino,
    829–832
Weber, Wilhelm, 513
Weber (unit of magnetic flux), 513
Weber-Fechner law, 312*n.*
Weight, 92–94
  and mass, 93–94
Weightlessness, 112–113
  physiological effects, 112–113
Weinberg, Steven, 845–847
Weisskopf, Victor F., 6*i.*
Westinghouse, George, 538
Wheel, 194, 198

Wheeler, John, 783
Wiedemann-Franz law, 807
Wien, Wilhelm, 356, 707–708
Wien displacement law, 356, 707
Wien's constant, 356
Wilkens, Maurice, 722
Work:
  conversion of:
    into elastic potential energy, 151–152
    into gravitational potential energy,
      147–151
    into kinetic energy, 145–147
    into thermal energy, 153–155
  definition of, 143
  dependence of, on path, 363
  dimensions of, 147
  done in a cyclic process, 363
  done by a gas on expanding, 362–363
  and heat, 362–364
  negative, 144–145
  and rotational kinetic energy, 200–202
  unit of, 143–144
Work-energy theorem, 155
  extended, 155
Work function of metals, 713*t.*
Wu, C. S., 843–844

X

X-ray microscope, 664
X-ray spectrometer, 722
X-rays, 714–717
  characteristic (line) spectrum, 758–759
    contrasted with optical line
      spectrum, 759

  source of, 759–760
continuous spectrum, 715–717
  experimental data, 715–716
  source of, 716–717
diffraction and interference of, 721–722
discovery of, by Roentgen, 714
properties of, 714
use of, in medicine, 715

Y

Yalow, Rosalyn, 2, 3*i.*
Yang, C. N., 843–844
Yeager, C., 320
Yerkes Observatory, 659
Young, Thomas, 622–624
  biography of, 625
Young's double-slit experiment, 622–624
  effect of single-slit diffraction on, 637
  theoretical and experimental patterns,
    623*i.*
  theory of, 622–624
Young's modulus, 265
  measurement of, 266–267
Yukawa, Hideki, 836

Z

Zeeman, Pieter, 687*i.*
Zernike, Fritz, 664
Zero, absolute (*see* Absolute zero)
Zeroth law of thermodynamics, 227
Zweig, George, 839

TABLE B.1 Useful Conversion Factors

Length
1 in = 0.0254 m; 1 m = 39.37 in
1 ft = 0.305 m; 1 m = 3.28 ft
1 mi = 1.61 km; 1 km = 0.621 mi
1 mi = 5280 ft = 1760 yd
1 Ångstrom (Å) = 10^{-10} m
1 light-year = 9.46×10^{15} m

Mass
1 u (unified mass unit) = 1.661×10^{-27} kg
1 kg = 6.022×10^{26} u
1 metric ton = 10^3 kg
(A mass of 1 kg corresponds to a weight of 2.21 lb; a
  weight of 1 lb corresponds to a mass of 0.454 kg)

Time
1 h = 60 min = 3600 s
1 day = 24 h = 1440 min = 8.64×10^4 s
1 year = 365.24 days = 3.156×10^7 s

Area
1 in^2 = 6.452×10^{-4} m^2; 1 m^2 = 1550 in^2
1 ft^2 = 9.29×10^{-2} m^2; 1 m^2 = 10.76 ft^2
1 km^2 = 10^6 m^2; 1 m^2 = 10^{-6} km^2

Volume
1 in^3 = 1.64×10^{-5} m^3; 1 m^3 = 6.10×10^4 in^3
1 ft^3 = 2.83×10^{-2} m^3; 1 m^3 = 35.3 ft^3
1 km^3 = 10^9 m^3; 1 m^3 = 10^{-9} km^3
1 liter = 10^3 cm^3 = 10^{-3} m^3
1 gallon = 3.786 liters = 3.786×10^{-3} m^3

Density
1 g/cm^3 = 10^3 kg/m^3; 1 kg/m^3 = 10^{-3} g/cm^3

Angular quantities
360° = 2π rad
1 rad = 57.30°; 1° = 1.745×10^{-2} rad
1 rev/min = 0.1047 rad/s

Velocity
1 ft/s = 0.305 m/s; 1 m/s = 3.28 ft/s
1 mi/h = 0.447 m/s; 1 m/s = 2.24 mi/h
1 mi/h = 1.61 km/h; 1 km/h = 0.621 mi/h
1 km/h = 0.278 m/s; 1 m/s = 3.60 km/h

Force
1 lb = 4.45 N; 1 N = 0.225 lb

Pressure
1 lb/in^2 = 6.90×10^3 N/m^2 (or Pa); 1 N/m^2 = 1.45×10^{-4} lb/in^2
1 lb/ft^2 = 47.9 N/m^2; 1 N/m^2 = 2.09×10^{-2} lb/ft^2
1 atm = 1.013×10^5 N/m^2; 1 N/m^2 = 9.87×10^{-6} atm
1 atm = 760 torr; 1 torr = 1.32×10^{-3} atm
1 torr = 133.3 N/m^2; 1 N/m^2 = 7.502×10^{-3} torr

Energy
1 ft·lb = 1.356 J; 1 J = 0.737 ft·lb
1 Btu = 1054 J; 1 J = 9.49×10^{-4} Btu
1 cal = 4.184 J; 1 J = 0.2390 cal
1 kcal = 4184 J; 1 J = 2.390×10^{-4} kcal
1 kWh = 3.60×10^6 J; 1 J = 2.78×10^{-7} kWh
1 eV = 1.602×10^{-19} J; 1 J = 6.242×10^{18} eV
1 eV/particle = 23.1 kcal/mol

Power
1 horsepower (hp) = 746 W; 1 W = 1.34×10^{-3} hp
1 ft·lb/s = 1.356 W; 1 W = 0.738 ft·lb/s
1 hp = 550 ft·lb/s; 1 ft·lb/s = 1.82×10^{-3} hp

Mass-energy
1 u is equivalent to 931.49 MeV
1 kg is equivalent to 5.6096×10^{29} MeV
1 m_e is equivalent to 0.51100 MeV
1 m_p is equivalent to 938.27 MeV
1 m_n is equivalent to 939.56 MeV

Magnetic field strength
1 T = 1 Wb/m^2 = 10^4 G; 1 G = 10^{-4} T